Lecture Notes in Computational Science and Engineering

Volume 145

This series contains monographs of lecture notes type, lecture course material, and high-quality proceedings on topics described by the term "computational science and engineering". This includes theoretical aspects of scientific computing such as mathematical modeling, optimization methods, discretization techniques, multiscale approaches, fast solution algorithms, parallelization, and visualization methods as well as the application of these approaches throughout the disciplines of biology, chemistry, physics, engineering, earth sciences, and economics.

Susanne C. Brenner • Eric Chung
Axel Klawonn • Felix Kwok • Jinchao Xu
Jun Zou
Editors

Domain Decomposition Methods in Science and Engineering XXVI

Springer

Editors

Susanne C. Brenner
Department of Mathematics and Center for Computation and Technology
Louisiana State University
Baton Rouge, LA, USA

Eric Chung
Department of Mathematics
The Chinese University of Hong Kong
Shatin, New Territories
Hong Kong SAR, China

Axel Klawonn
Department Mathematik/Informatik, Abteilung Mathematik
Universität zu Köln
Köln, Germany

Felix Kwok
Département de mathématiques et de statistique
Université Laval
Québec, QC, Canada

Jinchao Xu
Department of Mathematics
Pennsylvania State University
University Park, PA, USA

Jun Zou
Department of Mathematics
The Chinese University of Hong Kong
Shatin, New Territories
Hong Kong SAR, China

ISSN 1439-7358 ISSN 2197-7100 (electronic)
Lecture Notes in Computational Science and Engineering
ISBN 978-3-030-95024-8 ISBN 978-3-030-95025-5 (eBook)
https://doi.org/10.1007/978-3-030-95025-5

Mathematics Subject Classification (2020): 65F10, 65N30, 65N55

This Springer imprint is published by the registered company Springer Nature Switzerland AG
The registered company address is: Gewerbestrasse 11, 6330 Cham, Switzerland

Preface

This volume contains a selection of 84 papers submitted to the 26th International Conference on Domain Decomposition Methods, hosted by the Department of Mathematics at the Chinese University of Hong Kong, and held virtually on December 7–12, 2020.

Background of the Conference Series

With its first meeting in Paris in 1987, the International Conference on Domain Decomposition Methods has been held in 15 countries in Asia, Europe, and North America, and now for the first time in Hong Kong SAR, China. The conference is held at roughly 18-month intervals. A complete list of the 26 meetings appears below.

Domain decomposition is often seen as a form of divide-and-conquer for mathematical problems posed over a physical domain, reducing a large problem into a collection of smaller problems, each of which is much easier to solve computationally than the undecomposed problem, and most or all of which can be solved independently and concurrently, and then solved iteratively in a consistent way. Much of the theoretical interest in domain decomposition algorithms lies in ensuring that the number of iterations required to converge is very small. Domain decomposition algorithms can be tailored to the properties of the physical system, as reflected in the mathematical operators, to the number of processors available, and even to specific architectural parameters, such as cache size and the ratio of memory bandwidth to floating point processing rate, proving it to be an ideal paradigm for large-scale simulation on advanced architecture computers.

The principal technical content of the conference has always been mathematical, but its motivation has primarily been to make efficient use of distributed memory computers for complex applications arising in science and engineering. While research in domain decomposition methods is presented at numerous venues, the International Conference on Domain Decomposition Methods is the only regularly occurring international forum dedicated to interdisciplinary technical interactions between theoreticians and practitioners working in the development, analysis, software implementation, and application of domain decomposition methods.

As we approach the dawn of exascale computing, where we will command 10^{18} floating point operations per second, efficient and mathematically well-founded methods for the solution of large-scale systems will clearly become more and more important — as will their sound realization in the framework of modern HPC architectures. In fact, the massive parallelism, which makes exascale computing possible, requires the development of new solution methods, which are capable of efficiently exploiting this large number of cores as well as the connected hierarchies for memory access. Ongoing developments such as parallelization in time asynchronous iterative methods or nonlinear domain decomposition methods show that this massive parallelism not only demands new solution and discretization methods, but also fosters the development of new approaches.

Here is a list of the 26 conferences on Domain Decomposition:

1. Paris, France, January 7–9, 1987
2. Los Angeles, USA, January 14–16, 1988
3. Houston, USA, March 20–22, 1989
4. Moscow, USSR, May 21–25, 1990
5. Norfolk, USA, May 6–8, 1991
6. Como, Italy, June 15–19, 1992
7. University Park, Pennsylvania, USA, October 27–30, 1993
8. Beijing, China, May 16–19, 1995
9. Ullensvang, Norway, June 3–8, 1996
10. Boulder, USA, August 10–14, 1997
11. Greenwich, UK, July 20–24, 1998
12. Chiba, Japan, October 25–20, 1999
13. Lyon, France, October 9–12, 2000
14. Cocoyoc, Mexico, January 6–11, 2002
15. Berlin, Germany, July 21–25, 2003
16. New York, USA, January 12–15, 2005
17. St. Wolfgang–Strobl, Austria, July 3–7, 2006
18. Jerusalem, Israel, January 12–17, 2008
19. Zhangjiajie, China, August 17–22, 2009
20. San Diego, California, USA, February 7–11, 2011
21. Rennes, France, June 25–29, 2012
22. Lugano, Switzerland, September 16–20, 2013
23. Jeju Island, Korea, July 6–10, 2015
24. Spitsbergen, Svalbard, Norway, February 6–10, 2017
25. St. John's, Newfoundland, Canada, July 23–27, 2018
26. Hong Kong SAR (virtual format), China, December 7–12, 2020

International Scientific Committee on Domain Decomposition Methods

- Petter Bjørstad, University of Bergen, Norway
- Susanne Brenner, Louisiana State University, USA
- Xiao-Chuan Cai, CU Boulder, USA
- Martin J. Gander, University of Geneva, Switzerland
- Laurence Halpern, University Paris 13, France
- David Keyes, KAUST, Saudi Arabia
- Hyea Hyun Kim, Kyung Hee University, Korea
- Axel Klawonn, Universität zu Köln, Germany
- Ralf Kornhuber, Freie Universität Berlin, Germany
- Ulrich Langer, University of Linz, Austria
- Luca Pavarino, University of Pavia, Italy
- Olof Widlund, Courant Institute, USA
- Jinchao Xu, Penn State, USA
- Jun Zou, The Chinese University of Hong Kong, Hong Kong SAR, China

About the 26th Conference

The twenty-sixth International Conference on Domain Decomposition Methods had close to 250 participants from about 30 different countries. The conference contained 12 invited presentations selected by the International Scientific Committee, fostering both experienced and younger scientists, 22 minisymposia around specific topics and 6 contributed sessions. The present proceedings contain a selection of 84 papers grouped into three separate groups: 9 plenary papers, 60 minisymposium papers, and 15 contributed papers.

Sponsoring Organizations

- Department of Mathematics, The Chinese University of Hong Kong
- Department of Mathematics, Hong Kong Baptist University
- Faculty of Science, The Chinese University of Hong Kong
- United College, The Chinese University of Hong Kong
- The Hong Kong Mathematical Society

Local Organizing/Program Committee Members

- Jian-Feng Cai, Hong Kong University of Science and Technology
- Raymond Chan, City University of Hong Kong
- Zhiming Chen, Chinese Academy of Sciences
- Eric Chung, The Chinese University of Hong Kong
- Felix Kwok, Université Laval and Hong Kong Baptist University

- Lok Ming Lui, The Chinese University of Hong Kong
- Michael Ng, University of Hong Kong
- Jinchao Xu, Pennsylvania State University
- Tieyong Zeng, The Chinese University of Hong Kong
- Jun Zou, The Chinese University of Hong Kong

Plenary Presentations

- *Local Multiscale Model Reduction and Applications*, Eric Chung (The Chinese University of Hong Kong, Hong Kong SAR)
- *Robust Solvers for Time-Harmonic Wave Propagation Problems*, Victorita Dolean (Université Côte d'Azur, France and Strathclyde University, Scotland)
- *Improving Efficiency of Scalable TFETI/BETI Contact Solvers for Huge Problems*, Zdeněk Dostál (Technical University of Ostrava, Czech Republic)
- *An Efficient and High Order Accurate Direct Solution Technique for Variable Coefficient Elliptic Partial Differential Equations*, Adrianna Gillman (University of Colorado, Boulder, USA)
- *Fundamental Coarse Space Components for Schwarz Methods with Crosspoints*, Laurence Halpern (Université Paris 13, France)
- *Domain Decomposition Methods for Time Harmonic Wave Propagation Problems*, Patrick Joly (ENSTA ParisTech, France)
- *Multilevel Strategies for Non-Linear Problems and Machine Learning: On Non-Linear Preconditioning, Multilevel Optimization, and Multilevel Training*, Rolf Krause (University of Lugano, Switzerland)
- *Adaptive Space-Time Finite Element and Isogeometric Analysis*, Ulrich Langer (Johannes Kepler University Linz, Austria)
- *From Differential Equations to Deep Learning for Image Processing*, Carola-Bibiane Schönlieb (University of Cambridge, UK)
- *Nonoverlapping Domain Decomposition Methods for Saddle Point Problems*, Xuemin Tu (University of Kansas, USA)
- *Domain Decomposition for Modeling Two-Phase Flow in Porous Media*, Mary Wheeler (University of Texas at Austin, USA)
- *General Convection-Diffusion Problems: Robust Discretizations, Fast Solvers and Applications*, Shuonan Wu (Peking University, China)

Acknowledgments

The organizers would like to thank all the participants for their enthusiasm and carefully prepared contributions that made this meeting a very successful event. A warm thanks also to our sponsors that made the budget come together. We have all experienced a very unique meeting, which was held virtually.

Hong Kong, December 2021.

Susanne Brenner
Louisiana State University, USA

Eric Chung
The Chinese University of Hong Kong, Hong Kong SAR, China

Axel Klawonn
Universität zu Köln, Germany

Felix Kwok
Université Laval, Canada

Jinchao Xu
Pennsylvania State University, USA

Jun Zou
The Chinese University of Hong Kong, Hong Kong SAR, China

Contents

Part I Plenary Talks (PT)

Multiscale Model Reduction for a Class of Optimal Control Problems with Highly Oscillating Coefficients 3
Tak Shing Au Yeung and Eric Chung

Several Ways to Achieve Robustness When Solving Wave Propagation Problems 17
Niall Bootland, Victorita Dolean, Pierre Jolivet, Frédéric Nataf, Stéphane Operto, and Pierre-Henri Tournier

Scalable Hybrid TFETI-DP Methods for Large Boundary Variational Inequalities 29
Zdeněk Dostál, Tomáš Brzobohatý, David Horák, Jakub Kružík and Oldřich Vlach

Fundamental Coarse Space Components for Schwarz Methods with Crosspoints 41
François Cuvelier, Martin J. Gander, and Laurence Halpern

Nonoverlapping Domain Decomposition Methods for Time Harmonic Wave Problems 53
Xavier Claeys, Francis Collino, Patrick Joly, and Emile Parolin

Quantitative Analysis of Nonlinear Multifidelity Optimization for Inverse Electrophysiology 67
Fatemeh Chegini, Alena Kopaničáková, Martin Weiser, Rolf Krause

Adaptive Space-Time Finite Element and Isogeometric Analysis 79
Ulrich Langer

Nonoverlapping Domain Decomposition Methods for Saddle Point Problems 91
Jing Li and Xuemin Tu

Local Residual Minimization Smoothing for Improving Convergence Behavior of a Space-Time Domain Decomposition Method 103
Hanyu Li and Mary F. Wheeler

Part II Talks in Minisymposia

GenEO Coarse Spaces for Heterogeneous Indefinite Elliptic Problems . . . 117
Niall Bootland, Victorita Dolean, Ivan G. Graham, Chupeng Ma, and Robert Scheichl

Inexact Subdomain Solves Using Deflated GMRES for Helmholtz Problems 127
N. Bootland, V. Dwarka, P. Jolivet, V. Dolean, and C. Vuik

Non-Overlapping Domain Decomposition Methods with Cross-Points and Padé Approximants for the Helmholtz Equation 137
Yassine Boubendir and Tadanaga Takahashi

OSDS: A Sweeping Preconditioner for the Helmholtz Equation 145
Nacime Bouziani and Frédéric Nataf

Decomposition and Preconditioning of Deep Convolutional Neural Networks for Training Acceleration 153
Linyan Gu, Wei Zhang, Jia Liu, and Xiao-Chuan Cai

Numerical Calculation of the Portal Pressure Gradient of the Human Liver With a Domain Decomposition Method 161
Zeng Lin, Bokai Wu, Shanlin Qin, Xinhong Wang, Rongliang Chen, and Xiao-Chuan Cai

A Parallel Adaptive Finite Element Method for Modeling a Deformable Droplet Travelling in Air 169
Li Luo, Xiao-Chuan Cai, and David E. Keyes

On the Effect of Boundary Conditions on the Scalability of Schwarz Methods 177
Gabriele Ciaramella and Luca Mechelli

On the Asymptotic Optimality of Spectral Coarse Spaces 187
Gabriele Ciaramella and Tommaso Vanzan

Discrete Analysis of Schwarz Waveform Relaxation for a Simplified Air-Sea Coupling Problem with Nonlinear Transmission Conditions 197
S. Clement, F. Lemarié, and E. Blayo

A Posteriori Error Estimates in Maximum Norm for Interior Penalty Discontinuous Galerkin Approximation of the Obstacle Problem 205
B. Ayuso de Dios, T. Gudi, and K. Porwal

Spectral Equivalence Properties of Higher-Order Tensor Product Finite Elements 213
Clark R. Dohrmann

Optimizing Transmission Conditions for Multiple Subdomains in the Magnetotelluric Approximation of Maxwell's Equations 221
V. Dolean, M.J. Gander, and A. Kyriakis

Non-overlapping Spectral Additive Schwarz Methods for HDG and Multiscale Discretizations 229
Yi Yu, Maksymilian Dryja, and Marcus Sarkis

Robust BPX Solver for Cahn–Hilliard Equations 239
Siamak Faal, Adam Powell, and Marcus Sarkis

Natural Factor Based Solvers 247
Juan Galvis, Marcus Sarkis, and O. Andrés Cuervo

A Simple Finite Difference Discretization for Ventcell Transmission Conditions at Cross Points 257
Martin J. Gander and Laurence Halpern

Cycles in Newton–Raphson Preconditioned by Schwarz (ASPIN and Its Cousins) 265
Conor McCoid and Martin J. Gander

Should Multilevel Methods for Discontinuous Galerkin Discretizations Use Discontinuous Interpolation Operators? 273
Martin J. Gander and José Pablo Lucero Lorca

Domain Decomposition in Shallow Water Modelling of Dutch Lakes for Multiple Applications 281
Menno Genseberger, Asako Fujisaki, Christophe Thiange, Carlijn Eijsberg - Bak, Arnout Bijlsma, and Pascal Boderie

A Variational Interpretation of Restricted Additive Schwarz With Impedance Transmission Condition for the Helmholtz Problem 291
Shihua Gong, Martin J. Gander, Ivan G. Graham, and Euan A. Spence

Application of Multilevel BDDC to the Problem of Pressure in Simulations of Incompressible Flow 299
Martin Hanek and Jakub Šístek

Predicting the Geometric Location of Critical Edges in Adaptive GDSW Overlapping Domain Decomposition Methods Using Deep Learning 307
Alexander Heinlein, Axel Klawonn, Martin Lanser, and Janine Weber

Optimized Coupling Conditions for Discrete Fracture Matrix Models ... 317
Martin J. Gander, Julian Hennicker, and Roland Masson

Efficient Monolithic Solvers for Fluid-Structure Interaction Applied to Flapping Membranes 327
D. Jodlbauer, U. Langer, and T. Wick

Adaptive Nonlinear Elimination in Nonlinear FETI-DP Methods 337
Axel Klawonn, Martin Lanser, and Matthias Uran

Globalization of Nonlinear FETI–DP Methods 347
S. Köhler and O. Rheinbach

A Multilevel Active-Set Trust-Region (MASTR) Method for Bound Constrained Minimization 355
Alena Kopaničáková and Rolf Krause

A Multigrid Preconditioner for Jacobian-free Newton–Krylov Methods .. 365
Hardik Kothari, Alena Kopaničáková, and Rolf Krause

Overlapping DDFV Schwarz Algorithms on Non-Matching Grids 373
Martin J. Gander, Laurence Halpern, Florence Hubert, and Stella Krell

On the Nonlinear Dirichlet–Neumann Method and Preconditioner for Newton's Method 381
F. Chaouqui, M. J. Gander, P. M. Kumbhar, and T. Vanzan

A Nonlinear Optimized Schwarz Preconditioner for Elliptic Optimal Control Problems 391
Gabriele Ciaramella, Felix Kwok, and Georg Müller

SParse Approximate Inverse (SPAI) Based Transmission Conditions for Optimized Algebraic Schwarz Methods 399
Martin J. Gander, Lahcen Laayouni, and Daniel B. Szyld

A Parareal Architecture for Very Deep Convolutional Neural Networks .. 407
Chang-Ock Lee, Youngkyu Lee, and Jongho Park

Construction of 4D Simplex Space-Time Meshes for Local Bisection Schemes 417
David Lenz

Coefficient-Robust A Posteriori Error Estimation for H(curl)-elliptic Problems 425
Yuwen Li

Convergence of Parareal for a Vibrating String with Viscoelastic Damping 435
Martin J. Gander, Thibaut Lunet, and Aušra Pogoželskytė

Consistent and Asymptotic-Preserving Finite-Volume Robin Transmission Conditions for Singularly Perturbed Elliptic Equations . . . 443
Martin J. Gander, Stephan B. Lunowa, and Christian Rohde

Adaptive Schwarz Method for Crouzeix–Raviart Multiscale Problems in 2D 451
Leszek Marcinkowski, Talal Rahman, and Ali Khademi

An Overlapping Waveform Relaxation Preconditioner for Economic Optimal Control Problems With State Constraints 461
Gabriele Ciaramella and Luca Mechelli

Optimized Schwarz Methods With Data-Sparse Transmission Conditions 471
Martin J. Gander and Michal Outrata

Space-Time Finite Element Tearing and Interconnecting Domain Decomposition Methods 479
Douglas R.Q. Pacheco and Olaf Steinbach

Localized Reduced Basis Additive Schwarz Methods 487
Martin J. Gander and Stephan Rave

Micromechanics Simulations Coupling the deal.II Software Library With a Parallel FETI-DP Solver 495
S. Köhler, O. Rheinbach, and S. Sandfeld

A Three-Level Extension for Fast and Robust Overlapping Schwarz (FROSch) Preconditioners with Reduced Dimensional Coarse Space 505
Alexander Heinlein, Axel Klawonn, Oliver Rheinbach, and Friederike Röver

Space-Time Hexahedral Finite Element Methods for Parabolic Evolution Problems 515
Ulrich Langer and Andreas Schafelner

Towards a IETI-DP Solver on Non-Matching Multi-Patch Domains 523
Rainer Schneckenleitner and Stefan Takacs

The Parallel Full Approximation Scheme in Space and Time for a Parabolic Finite Element Problem 531
Oliver Sander, Ruth Schöbel, and Robert Speck

A New Coarse Space for a Space-Time Schwarz Waveform Relaxation Method 539
Martin J. Gander, Yao-Lin Jiang and Bo Song

On Space-Time Finite Element Domain Decomposition Methods for the Heat Equation 547
Olaf Steinbach and Philipp Gaulhofer

IETI-DP for Conforming Multi-Patch Isogeometric Analysis in Three Dimensions 555
Rainer Schneckenleitner and Stefan Takacs

Coupling of Navier–Stokes Equations and Their Hydrostatic Versions and Simulation of Riverbend Flow 563
Wenbin Dong, Hansong Tang, and Yingjie Liu

On the Links Between Observed and Theoretical Convergence Rates for Schwarz Waveform Relaxation Algorithm for the Time-Dependent Problems 571
Sophie Thery

Construction of Grid Operators for Multilevel Solvers: a Neural Network Approach 579
Claudio Tomasi and Rolf Krause

Coarse Corrections for Schwarz methods for Symmetric and Non-symmetric Problems 589
Martin J. Gander and Serge Van Criekingen

A Numerical Algorithm Based on Probing to Find Optimized Transmission Conditions 597
Martin J. Gander, Roland Masson, and Tommaso Vanzan

Additive Schwarz Preconditioners for C^0 Interior Penalty Methods for a State Constrained Elliptic Distributed Optimal Control Problem 607
Susanne C. Brenner, Li-Yeng Sung, and Kening Wang

Space-Time Finite Element Methods for Initial Temperature Reconstruction 617
Ulrich Langer, Olaf Steinbach, Fredi Tröltzsch, and Huidong Yang

Numerical Results for an Unconditionally Stable Space-Time Finite Element Method for the Wave Equation 625
Richard Löscher, Olaf Steinbach, and Marco Zank

Décomposition de Domaine et Problème de Helmholtz: Thirty Years After and Still Unique 633
Martin J. Gander and Hui Zhang

Part III Contributed Talks

Space-Time Parallel Methods for Evolutionary Reaction-Diffusion Problems 643
Andrés Arrarás, Francisco J. Gaspar, Laura Portero, and Carmen Rodrigo

Parallel Domain Decomposition Solvers for the Time Harmonic Maxwell Equations 653
Sven Beuchler, Sebastian Kinnewig, and Thomas Wick

Adaptive Finite Element Thin-Plate Spline With Different Data Distributions 661
Lishan Fang and Linda Stals

A Multirate Accelerated Schwarz Waveform Relaxation Method 671
Ronald D. Haynes and Khaled Mohammad

A Convergence Analysis of the Parallel Schwarz Solution of the Continuous Closest Point Method 679
Alireza Yazdani, Ronald D. Haynes, and Steven J. Ruuth

Dual-Primal Preconditioners for Newton–Krylov Solvers for the Cardiac Bidomain Model 689
Ngoc Mai Monica Huynh, Luca F. Pavarino, and Simone Scacchi

Domain Decomposition Algorithms for Physics-Informed Neural Networks 697
Hyea Hyun Kim and Hee Jun Yang

Numerical Study of an Additive Schwarz Preconditioner for a Thin Membrane Diffusion Problem 705
Piotr Krzyżanowski

Additive Schwarz Methods for Convex Optimization – Convergence Theory and Acceleration 715
Jongho Park

Non-local Impedance Operator for Non-overlapping DDM for the Helmholtz Equation 725
Francis Collino, Patrick Joly and Emile Parolin

Asynchronous Multi-Subdomain Methods With Overlap for a Class of Parabolic Problems 735
Mohamed Laaraj and Karim Rhofir

Toward a New Fully Algebraic Preconditioner for Symmetric Positive Definite Problems 745
Nicole Spillane

Aitken–Schwarz Heterogeneous Domain Decomposition for EMT-TS Simulation 753
H. Shourick, D. Tromeur-Dervout, and L. Chedot

Parareal Schwarz Waveform Relaxation Method for the Time-Periodic Parabolic Problem 763
Bo Song, Yao-Lin Jiang, and Kang-Li Xu

Acceleration of the Convergence of the Asynchronous RAS Method 773
D. Tromeur-Dervout

List of Contributors

Andrés Arrarás
Institute for Advanced Materials and Mathematics (INAMAT^2), Department of Statistics, Computer Science and Mathematics, Public University of Navarre (UPNA), 31006 Pamplona, Spain,
e-mail: andres.arraras@unavarra.es

Tak Shing Au Yeung
Department of Mathematics, The Chinese University of Hong Kong, Hong Kong SAR,
e-mail: iauyeung@math.cuhk.edu.hk

Sven Beuchler
Leibniz University Hannover, Institute of Applied Mathematics, Welfengarten 1, 30167 Hannover, Germany,
e-mail: beuchler@ifam.uni-hannover.de

Eric Blayo
Univ Grenoble Alpes, CNRS, Inria, Grenoble INP, LJK, Grenoble, France,
e-mail: eric.blayo@univ-grenoble-alpes.fr

Niall Bootland
University of Strathclyde, Department of Mathematics and Statistics, Glasgow, UK,
e-mail: niall.bootland@strath.ac.uk

Yassine Boubendir
New Jersey Institute of Technology, 323 Dr Martin Luther King Jr Blvd, Newark, NJ 07102,
e-mail: boubendi@njit.edu

Nacime Bouziani
Imperial College London, Department of Mathematics, London, SW7 2AZ, UK,
e-mail: n.bouziani18@imperial.ac.uk, nacime.bouziani@gmail.com

Susanne C. Brenner
Department of Mathematics and Center for Computation and Technology, Louisiana State University, Baton Rouge, LA 70803, USA,
e-mail: brenner@math.lsu.edu

Tomáš Brzobohatý
IT4Innovations - National Supercomputing Center, VSB-Technical University of Ostrava, Ostrava, Czech republic,
e-mail: tomas.brzobohaty@vsb.cz

Xiao-Chuan Cai
Department of Mathematics, University of Macau, Macau, China,
e-mail: xccai@um.edu.mo

Fayçal Chaouqui
Temple University, Philadelphia, USA,
e-mail: faycal.chaouqui@temple.edu,

Laurent Chedot
Supergrid-Institute, 14 rue Cyprien, 69200 Villeurbanne,
e-mail: laurent.chedot@supergrid-institute.com

Fatemeh Chegini
Zuse Institute Berlin, Germany, and Center for Computational Medicine in Cardiology, Università della Svizzera italiana, Switzerland,
e-mail: chegini@zib.de

Rongliang Chen
Shenzhen Institute of Advanced Technology, Chinese Academy of Sciences, Shenzhen, China,
e-mail: rl.chen@siat.ac.cn

Eric Chung
Department of Mathematics, The Chinese University of Hong Kong, Hong Kong SAR,
e-mail: tschung@math.cuhk.edu.hk

Gabriele Ciaramella
Politecnico di Milano,
e-mail: gabriele.ciaramella@polimi.it

Xavier Claeys
Sorbonne Université, Laboratoire Jacques-Louis Lions, and équipe Alpines, 75005 Paris, France,
e-mail: claeys@ann.jussieu.fr

Simon Clement
Univ Grenoble Alpes, CNRS, Inria, Grenoble INP, LJK, Grenoble, France,
e-mail: simon.clement@grenoble-inp.org

Francis Collino
POEMS, CNRS, INRIA, ENSTA Paris, Institut Polytechnique de Paris, 91120 Palaiseau, France,
e-mail: francis.collino@orange.fr

Serge Van Criekingen
CNRS/IDRIS and Maison de la Simulation,
e-mail: serge.van.criekingen@idris.fr

O. Andrés Cuervo
Departamento de Matemáticas, Universidad Nacional de Colombia and School of Engineering, Science and Technology, Universidad del Rosario, Bogotá, Colombia,
e-mail: omar.cuervo@urosario.edu.co

François Cuvelier
LAGA, Université Sorbonne Paris-Nord,
e-mail: cuvelier@math.univ-paris13.fr

Blanca Ayuso de Dios
Dipartimento di Matematica e Applicazioni, Università di Milano-Bicocca, Milan, Italy,
e-mail: blanca.ayuso@unimib.it

Clark R. Dohrmann
Sandia National Laboratories, Albuquerque, New Mexico, USA,
e-mail: crdohrm@sandia.gov

Victorita Dolean
University of Strathclyde, Dept. of Maths and Stats and University Côte d'Azur, CNRS, LJAD,
e-mail: work@victoritadolean.com

Wenbin Dong
Civil Engineering Department, City College, City University of New York, NY 10031, USA,
e-mail: wdong000@citymail.cuny.edu

Zdeněk Dostál
Department of Applied Mathematics and IT4Innovations - National Supercomputing Center, VSB-Technical University of Ostrava, 17. listopadu 15, Czech Republic,
e-mail: zdenek.dostal@vsb.cz

Maksymilian Dryja
Warsaw University,Banacha 2, 00-097 Warsaw, Poland,
e-mail: dryja@mimuw.edu.pl

Vandana Dwarka
Delft University of Technology, Delft Institute of Applied Mathematics, Delft, The Netherlands,
e-mail: v.n.s.r.dwarka@tudelft.nl

Siamak Faal
Worcester Polytechnic Institute, 100 Institute Rd, Worcester, USA,
e-mail: sghorbanifaal@wpi.edu

Lishan Fang
Mathematical Sciences Institute, Australian National University,
e-mail: Lishan.Fang@anu.edu.au

Juan Galvis
Departamento de Matemáticas, Universidad Nacional de Colombia, Bogotá, Colombia,
e-mail: jcgalvisa@unal.edu.co

Martin J. Gander
Université de Genève, Section de Mathématiques, Rue du Conseil-Général 7-9, CP 64, 1205 Genève, Suisse,
e-mail: martin.gander@unige.ch

Francisco J. Gaspar
Institute for Mathematics and Applications (IUMA), Department of Applied Mathematics, University of Zaragoza, 50009 Zaragoza, Spain,
e-mail: fjgaspar@unizar.es

Philipp Gaulhofer
Institut für Angewandte Mathematik, TU Graz, Steyrergasse 30, 8010 Graz, Austria,
e-mail: philipp.gaulhofer@student.tugraz.at

Menno Genseberger
Deltares, P.O. Box 177, 2600 MH Delft, The Netherlands,
e-mail: Menno.Genseberger@deltares.nl

Shihua Gong
Department of Mathematical Sciences, University of Bath, Bath BA2 7AY, UK.

Ivan G. Graham
Department of Mathematical Sciences, University of Bath, Bath BA2 7AY, UK,
e-mail: i.g.graham@bath.ac.uk

Linyan Gu
Shenzhen Institutes of Advanced Technology, Chinese Academy of Sciences, Shenzhen, 518055, China,
e-mail: ly.gu@siat.ac.cn

T. Gudi
Department of Mathematics, Indian Institute of Science, Bangalore - 560012,
e-mail: gudi@math.iisc.ac.in

Laurence Halpern
LAGA, Université Sorbonne Paris-Nord, 93430 Villetaneuse,
e-mail: halpern@math.univ-paris13.fr

Martin Hanek
Institute of Mathematics of the Czech Academy of Sciences, Žitná 25, Prague, Czech Republic and Czech Technical University in Prague, Technická 4, Prague, Czech Republic,
e-mail: martin.hanek@fs.cvut.cz

Ronald D. Haynes
Department of Mathematics & Statistics, Memorial University of Newfoundland, St. John's, Newfoundland, Canada, A1C 5S7,
e-mail: rhaynes@mun.ca

Alexander Heinlein
Delft University of Technology, Faculty of Electrical Engineering Mathematics & Computer Science, Delft Institute of Applied Mathematics, Mekelweg 4, 72628 CD Delft, Netherlands,
e-mail: a.heinlein@tudelft.nl

Julian Hennicker
Université du Luxembourg,
e-mail: julian.hennicker@uni.lu

David Horák
Department of Applied Mathematics, VSB-Technical University of Ostrava, Ostrava, Czech republic, and Institute of Geonics ASCR, Ostrava,
e-mail: david.horak@vsb.cz

Florence Hubert
Aix-Marseille Université, CNRS, Centrale Marseille, I2M UMR 7373, 39 rue F. Joliot Curie, 13453 Marseille, Cedex 13, FRANCE,
e-mail: florence.hubert@univ-amu.fr

Ngoc Mai Monica Huynh
Dipartimento di Matematica, UniversitÃ degli Studi di Pavia, via Ferrata 5 - 27100 Pavia, Italy,
e-mail: ngocmaimonica.huynh01@universitadipavia.it

Yao-Lin Jiang
School of Mathematics and Statistics, Xi'an Jiaotong University, Xi'an 710049, China,
e-mail: yljiang@mail.xjtu.edu.cn

Daniel Jodlbauer
Doctoral Program "Computational Mathematics" Johannes Kepler University Linz, Altenberger Str. 69, A-4040 Linz, Austria,
e-mail: daniel.jodlbauer@dk-compmath.jku.at

Pierre Jolivet
University of Toulouse, CNRS, IRIT, Toulouse, France,
e-mail: pierre.jolivet@enseeiht.fr

Patrick Joly
POEMS, CNRS, INRIA, ENSTA Paris, Institut Polytechnique de Paris, 91120 Palaiseau, France,
e-mail: patrick.joly@inria.fr

Stephan Köhler
Institut für Numerische Mathematik und Optimierung, Fakultät für Mathematik und Informatik, Technische Universität Bergakademie Freiberg, Akademiestr. 6, 09596 Freiberg,
e-mail: stephan.koehler@math.tu-freiberg.de

David E. Keyes
Extreme Computing Research Center, King Abdullah University of Science and Technology, Thuwal, Saudi Arabia,
e-mail: david.keyes@kaust.edu.sa

Ali Khademi
Faculty of Engineering and Science, Western Norway University of Applied Sciences, Inndalsveien 28, 5063 Bergen, Norway,
e-mail: Ali.Khademi@hvl.no(akhademi.math@gmail.com)

Hyea Hyun Kim
Department of Applied Mathematics and Institute of Natural Sciences, Kyung Hee University, Korea,
e-mail: hhkim@khu.ac.kr

Sebastian Kinnewig
Leibniz University Hannover, Institute of Applied Mathematics, Welfengarten 1, 30167 Hannover, Germany,
e-mail: kinnewig@ifam.uni-hannover.de

Axel Klawonn
Department of Mathematics and Computer Science, Division of Mathematics, University of Cologne, Weyertal 86-90, 50931 Cologne, Germany,
e-mail: axel.klawonn@uni-koeln.de

Alena Kopaničáková
Euler Institute, Università della Svizzera italiana, Switzerland,
e-mail: alena.kopanicakova@usi.ch

Hardik Kothari
Euler Institute, Università della Svizzera italiana, Switzerland,
e-mail: hardik.kothari@usi.ch

Rolf Krause
Center for Computational Medicine in Cardiology, Euler Institute, Università della Svizzera italiana, Switzerland,
e-mail: rolf.krause@usi.ch

Stella Krell
Université Côte d'Azur, Inria, CNRS, LJAD, France,
e-mail: stella.krell@univ-cotedazur.fr

Jakub Kružík
Department of Applied Mathematics, VSB-Technical University of Ostrava, Ostrava, Czech republic, and Institute of Geonics ASCR, Ostrava,
e-mail: jakub.kruzik@vsb.cz

Piotr Krzyżanowski
Faculty of Mathematics, Informatics and Mechanics, University of Warsaw, Poland,
e-mail: p.krzyzanowski@mimuw.edu.pl

Pratik M. Kumbhar
Karlsruher Institut für Technologie, Germany,
e-mail: pratik.kumbhar@kit.edu

Felix Kwok
Université Laval,
e-mail: felix.kwok@mat.ulaval.ca

Alexandros Kyriakis
University of Strathclyde,
e-mail: alexandros.kyriakis@strath.ac.uk

Mohamed Laaraj
Univ. Hassan II Mohammadia Casablanca ENSAM Avenue Nile 150, Casablanca, Morocco,
e-mail: mohamed.laaraj@gmail.com

Lahcen Laayouni
School of Science and Engineering, Al Akhawayn University, Avenue Hassan II, 53000 P.O. Box 1630, Ifrane, Morocco,
e-mail: L.Laayouni@aui.ma

Ulrich Langer
Johann Radon Institute for Computational and Applied Mathematics, Austrian Academy of Sciences, Altenbergerstr. 69, A-4040 Linz, Austria,
e-mail: ulrich.langer@ricam.oeaw.ac.at

Martin Lanser
Department of Mathematics and Computer Science, Division of Mathematics, University of Cologne, Weyertal 86-90, 50931 Cologne, Germany,
e-mail: martin.lanser@uni-koeln.de

Chang-Ock Lee
Department of Mathematical Sciences, KAIST, Daejeon 34141, Korea,
e-mail: colee@kaist.edu

Youngkyu Lee
Department of Mathematical Sciences, KAIST, Daejeon 34141, Korea,
e-mail: lyk92@kaist.ac.kr

Florian Lemarié
Univ Grenoble Alpes, Inria, CNRS, Grenoble INP, LJK, Grenoble, France,
e-mail: florian.lemarie@inria.fr

David Lenz
Argonne National Laboratory, Lemont, IL, USA,
e-mail: dlenz@anl.gov

Hanyu Li
Oden Institute for Computational Engineering and Sciences, 201 E 24th St, Austin, TX 78712,
e-mail: lihanyu234@utexas.edu

Jing Li
Department of Mathematical Sciences, Kent State University, Kent, OH 44242,
e-mail: li@math.kent.edu

Yuwen Li
Department of Mathematics, The Pennsylvania State University, University Park, PA 16802, USA,
e-mail: yuwenli925@gmail.com

Zeng Lin
Shenzhen Institute of Advanced Technology, Chinese Academy of Sciences, Shenzhen, China,
e-mail: zeng.lin@siat.ac.cn

Jia Liu
Shenzhen Institutes of Advanced Technology, Chinese Academy of Sciences, Shenzhen, 518055, China,
e-mail: jia.liu@siat.ac.cn

Yingjie Liu
School of Mathematics, Georgia Institute of Technology, Atlanta, GA 30332, USA,
e-mail: yingjie@math.gatech.edu

Richard Löscher
Fachbereich Mathematik, TU Darmstadt, Dolivostraße 15, 64293 Darmstadt, Germany,
e-mail: loescher@mathematik.tu-darmstadt.de

José Pablo Lucero Lorca
University of Colorado at Boulder, e-mail: pablo.lucero@colorado.edu

Thibaut Lunet
University of Geneva,
e-mail: thibaut.lunet@unige.ch

Stephan B. Lunowa
UHasselt – Hasselt University, Computational Mathematics, Agoralaan D, 3590 Diepenbeek, Belgium,
e-mail: stephan.lunowa@uhasselt.be

Li Luo
Faculty of Science and Technology, University of Macau, Macau, China,
e-mail: liluo@um.edu.mo

Georg Müller
Universität Konstanz,
e-mail: georg.mueller@uni-konstanz.de

Chupeng Ma
Heidelberg University, Inst. Applied Math.,
e-mail: chupeng.ma@uni-heidelberg.de

Leszek Marcinkowski
Faculty of Mathematics, University of Warsaw, Banacha 2, 02-097 Warszawa, Poland,
e-mail: Leszek.Marcinkowski@mimuw.edu.pl

Roland Masson
Université Côte d'Azur, CNRS, Inria, LJAD,
e-mail: roland.masson@univ-cotedazur.fr

Conor McCoid
University of Geneva,
e-mail: conor.mccoid@unige.ch

Luca Mechelli
Universität Konstanz,
e-mail: luca.mechelli@uni-konstanz.de

Khaled Mohammad
Department of Mathematics & Statistics, Memorial University of Newfoundland, St. John's, Newfoundland, Canada, A1C 5S7,
e-mail: km2605@mun.ca

Frédéric Nataf
Laboratory J.L. Lions, Sorbonne Université, Paris,
e-mail: frederic.nataf@sorbonne-universite.fr

Stéphane Operto
University Côte d'Azur, CNRS, Géoazur,
e-mail: stephane.operto@geoazur.unice.fr

Michal Outrata
University of Geneva,
e-mail: michal.outrata@unige.ch

Douglas R. Q. Pacheco
Institut für Angewandte Mathematik, TU Graz, Steyrergasse 30, 8010 Graz, Austria,
e-mail: pacheco@math.tugraz.at

Jongho Park
Natural Science Research Institute, KAIST, Daejeon 34141, Korea,
e-mail: jongho.park@kaist.ac.kr

Emile Parolin
POEMS, CNRS, INRIA, ENSTA Paris, Institut Polytechnique de Paris, 91120 Palaiseau, France,
e-mail: emile.parolin@inria.fr

Luca F. Pavarino
Dipartimento di Matematica, UniversitÃ degli Studi di Pavia, via Ferrata 5 - 27100 Pavia, Italy,
e-mail: luca.pavarino@unipv.it

Aušra Pogoželskytė
University of Geneva,
e-mail: ausra.pogozelskyte@unige.ch

Laura Portero
Institute for Advanced Materials and Mathematics (INAMAT^2), Department of Statistics, Computer Science and Mathematics, Public University of Navarre (UPNA), 31006 Pamplona, Spain,
e-mail: laura.portero@unavarra.es

K. Porwal
Department of Mathematics, Indian Institute of Technology Delhi - 110016,
e-mail: kamana@maths.iitd.ac.in

Adam Powell
Worcester Polytechnic Institute, 100 Institute Rd, Worcester, USA,
e-mail: acpowell@wpi.edu

Shanlin Qin
Shenzhen Institute of Advanced Technology, Chinese Academy of Sciences, Shenzhen, China,
e-mail: sl.qin@siat.ac.cn

Friederike Röver
Institut für Numerische Mathematik und Optimierung, Technische Universität Bergakademie Freiberg, Akademiestr. 6, 09599 Freiberg, Germany,
e-mail: friederike.roever@math.tu-freiberg.de

Talal Rahman
Faculty of Engineering and Science, Western Norway University of Applied Sciences, Inndalsveien 28, 5063 Bergen, Norway,
e-mail: Talal.Rahman@hvl.no

Stephan Rave
Mathematics Münster, University of Münster, Einsteinstrasse 62, 48149 Münster, Germany,
e-mail: Stephan.Rave@uni-muenster.de

Oliver Rheinbach
Institut für Numerische Mathematik und Optimierung, Technische Universität Bergakademie Freiberg, Akademiestr. 6, 09599 Freiberg, Germany,
e-mail: oliver.rheinbach@math.tu-freiberg.de

Karim Rhofir
Univ. Sultan Moulay Slimane LISERT-ENSA Bd. Beni Amir, BP. 77, Khouribga, Morocco,
e-mail: k.rhofir@usms.ma

Carmen Rodrigo
Institute for Mathematics and Applications (IUMA), Department of Applied Mathematics, University of Zaragoza, 50009 Zaragoza, Spain,
e-mail: carmenr@unizar.es

Christian Rohde
University of Stuttgart, Institute for Applied Analysis and Numerical Simulation, Pfaffenwaldring 57, 70569 Stuttgart, Germany,
e-mail: christian.rohde@mathematik.uni-stuttgart.de

Steven J. Ruuth
Simon Fraser University, 8888 University Dr., Burnaby, BC V5A 1S6,
e-mail: sruuth@sfu.ca

Oliver Sander
Faculty of Mathematics, TU Dresden, Germany,
e-mail: oliver.sander@tu-dresden.de

Stefan Sandfeld
Forschungszentrum Jülich, Institute for Advanced Simulation – Materials Data Science and Informatics (IAS-9), Wilhelm-Johnen-Straße, 52428 Jülich and RWTH Aachen University, Chair of Materials Data Science and Materials Informatics, Faculty 5, 52072 Aachen

Marcus Sarkis
Worcester Polytechnic Institute, 100 Institute Rd, Worcester, USA,
e-mail: msarkis@wpi.edu

Simone Scacchi
Dipartimento di Matematica, Università degli Studi di Milano, via Saldini 50 - 20133 Milano, Italy,
e-mail: simone.scacchi@unimi.it

Ruth Schöbel
Jülich Supercomputing Centre, Forschungszentrum Jülich GmbH, Germany,
e-mail: r.schoebel@fz-juelich.de

Andreas Schafelner
Doctoral Program "Computational Mathematics", Johannes Kepler University, Altenbergerstr. 69, 4040 Linz, Austria,
e-mail: andreas.schafelner@dk-compmath.jku.at

Robert Scheichl
Heidelberg University, Inst. Applied Math.,
e-mail: r.scheichl@uni-heidelberg.de

Rainer Schneckenleitner
Institute of Computational Mathematics, Johannes Kepler University Linz, Altenberger Straße 69, 4040 Linz, Austria,
e-mail: schneckenleitner@numa.uni-linz.ac.at

Héléna Shourick
Supergrid-Institute, 14 rue Cyprien, 69200 Villeurbanne,
e-mail: helena.shourick@supergrid-institute.com

Jakub Šístek
Institute of Mathematics of the Czech Academy of Sciences, Žitná 25, Prague, Czech Republic,
e-mail: sistek@math.cas.cz

Bo Song
School of Mathematics and Statistics, Northwestern Polytechnical University, Xi'an 710072, China,
e-mail: bosong@nwpu.edu.cn

Robert Speck
Jülich Supercomputing Centre, Forschungszentrum Jülich GmbH, Germany,
e-mail: r.speck@fz-juelich.de

Euan A. Spence
Department of Mathematical Sciences, University of Bath, Bath BA2 7AY, UK.

Nicole Spillane
CNRS, CMAP, Ecole Polytechnique, Institut Polytechnique de Paris, 91128 Palaiseau Cedex, France,
e-mail: nicole.spillane@cmap.polytechnique.fr

Linda Stals
Mathematical Sciences Institute, Australian National University,
e-mail: Linda.Stals@anu.edu.au

Olaf Steinbach
Institut für Angewandte Mathematik, TU Graz, Steyrergasse 30, 8010 Graz, Austria,
e-mail: o.steinbach@tugraz.at

Li-Yeng Sung
Department of Mathematics and Center for Computation and Technology, Louisiana State University, Baton Rouge, LA 70803, USA,
e-mail: sung@math.lsu.edu

Daniel B. Szyld
Department of Mathematics, Temple University (038-16), 1805 N. Broad Street, Philadelphia, Pennsylvania 19122-6094, USA,
e-mail: szyld@temple.edu

Stefan Takacs
RICAM, Austrian Academy of Sciences, Altenberger Straße 69, 4040 Linz, Austria,
e-mail: stefan.takacs@ricam.oeaw.ac.at

Tadanaga Takahashi
New Jersey Institute of Technology, 323 Dr Martin Luther King Jr Blvd, Newark, NJ 07102,
e-mail: tt73@njit.edu

Hansong Tang
Civil Engineering Department, City College, City University of New York, NY 10031, USA,
e-mail: htang@ccny.cuny.edu

Sophie Thery
Univ. Grenoble-Alpes, Inria, CNRS, Grenoble INP, LJK, Grenoble, France,
e-mail: sophie.thery@univ-grenoble-alpes.fr

Claudio Tomasi
Università della Svizzera Italiana,Via Buffi 13, CH-6904 Lugano,
e-mail: claudio.tomasi@usi.ch

Pierre-Henri Tournier
Sorbonne Université, CNRS, LJLL,
e-mail: tournier@ann.jussieu.fr

Fredi Tröltzsch
Institut für Mathematik, Technische Universität Berlin, Straße des 17. Juni 136, 10623 Berlin, Germany,
e-mail: troeltzsch@math.tu-berlin.de

Damien Tromeur-Dervout
University of Lyon, UMR5208 U.Lyon1-CNRS, Institut Camille Jordan,
e-mail: damien.tromeur-dervout@univ-lyon1.fr

Xuemin Tu
Department of Mathematics, University of Kansas, 1460 Jayhawk Blvd, Lawrence, KS 66045, USA,
e-mail: xuemin@ku.edu

Matthias Uran
Department of Mathematics and Computer Science, Division of Mathematics, University of Cologne, Weyertal 86-90, 50931 Cologne, Germany,
e-mail: m.uran@uni-koeln.de

Tommaso Vanzan
CSQI Chair, Institute de mathématiques, Ecole Polytechnique Fédérale de Lausanne,
e-mail: tommaso.vanzan@epfl.ch

Oldřich Vlach
Department of Applied Mathematics and IT4Innovations - National Supercomputing Center, VSB-Technical University of Ostrava, Czech Republic,
e-mail: oldrich.vlach@vsb.cz

Cornelis Vuik
Delft University of Technology, Delft Institute of Applied Mathematics, Delft, The Netherlands,
e-mail: c.vuik@tudelft.nl

Kening Wang
Department of Mathematics and Statistics, University of North Florida, Jacksonville, FL 32224, USA,
e-mail: kening.wang@unf.edu

Xinhong Wang
Department of Radiology, The Second Affiliated Hospital, School of Medicine, Zhejiang University, Hangzhou, China,
e-mail: 2611104@zju.edu.cn

Janine Weber
Department of Mathematics and Computer Science, University of Cologne, Weyertal 86-90, 50931 Köln, Germany,
e-mail: janine.weber@uni-koeln.de

Martin Weiser
Zuse Institute Berlin, Germany,
e-mail: weiser@zib.de

Mary F. Wheeler
Oden Institute for Computational Engineering and Sciences, 201 E 24th St, Austin, TX 78712,
e-mail: mfw@oden.utexas.edu

Thomas Wick
Institut für Angewandte Mathematik, Leibniz Universität Hannover, Welfengarten 1, 30167 Hannover, Germany,
e-mail: thomas.wick@ifam.uni-hannover.de

Bokai Wu
Shenzhen Institute of Advanced Technology, Chinese Academy of Sciences, Shenzhen, China,
e-mail: bk.wu@siat.ac.cn

Kang-Li Xu
School of Mathematics and Statistics, Xi'an Jiaotong University, Xi'an 710049, China,
e-mail: klxuqd@163.com

Hee Jun Yang
Department of Mathematics, Kyung Hee University, Korea,
e-mail: yhjj109@khu.ac.kr

Huidong Yang
Johann Radon Institute for Computational and Applied Mathematics, Altenberger Straße 69, 4040 Linz, Austria,
e-mail: huidong.yang@oeaw.ac.at

Alireza Yazdani
Simon Fraser University, 8888 University Dr., Burnaby, BC V5A 1S6,
e-mail: alirezay@sfu.ca

Yi Yu
Worcester Polytechnic Institute, 100 Institute Rd, Worcester, USA,
e-mail: yyu5@wpi.edu

Marco Zank
Fakultät für Mathematik, Universität Wien, Oskar–Morgenstern–Platz 1, 1090 Wien, Austria,
e-mail: marco.zank@univie.ac.at

Hui Zhang
Xi'an Jiaotong-Liverpool University, Department of Applied Mathematics & Laboratory for Intelligent Computing and Financial Technology (KSF-P-02, RDF-19-01-09), Suzhou 215123, China,
e-mail: mike.hui.zhang@hotmail.com

Wei Zhang
Shenzhen Institutes of Advanced Technology, Chinese Academy of Sciences, Shenzhen, 518055, China,
e-mail: wei.zhang@siat.ac.cn

Part I
Plenary Talks (PT)

Multiscale Model Reduction for a Class of Optimal Control Problems with Highly Oscillating Coefficients

Tak Shing Au Yeung and Eric Chung

1 Introduction

This paper is concerned with the discretization of a class of elliptic optimal control problems with highly heterogeneous coefficients

$$\inf J(u) = F(y, u) = \frac{1}{2}\|y - y_d\|^2_{L^2(\Omega)} + \frac{\nu}{2}\|u\|^2_{L^2(\Omega)} \tag{1}$$

subject to the state equations

$$-\mathrm{div}(\kappa(x)\nabla y) = u, \qquad \text{in } \Omega, \tag{2}$$

$$y = 0, \qquad \text{on } \Gamma, \tag{3}$$

and to the control constraints

$$a \le u(x) \le b \qquad \text{for a.e. } x \in \Omega, \tag{4}$$

where $\Omega \subset \mathbb{R}^2$ is a bounded polygonal Lipschitz domain and Γ is the boundary of Ω; $\kappa(x)$ is a high-contrast heterogeneous permeability field with $0 < \kappa_0 \le \kappa(x) \le \kappa_1$ and a, b are real numbers. In (1), we assume $y_d \in L^2(\Omega)$. Moreover, $\nu > 0$ is a fixed positive number. We denote the set of admissible controls by U_{ad}:

$$U_{ad} = \{u \in L^2(\Omega) : a \le u \le b, \text{ a.e. in } \Omega\}.$$

Tak Shing Au Yeung
Department of Mathematics, The Chinese University of Hong Kong, Hong Kong SAR,
e-mail: iauyeung@math.cuhk.edu.hk

Eric Chung
Department of Mathematics, The Chinese University of Hong Kong, Hong Kong SAR,
e-mail: tschung@math.cuhk.edu.hk

S. C. Brenner et al. (eds.), *Domain Decomposition Methods in Science and Engineering XXVI*, Lecture Notes in Computational Science and Engineering 145,
https://doi.org/10.1007/978-3-030-95025-5_1

In many practical situations, one may encounter heterogeneous media such as fractured media or porous media with high contrast channels. The coefficients for these problems usually contain scale disparity and high contrast regions. Solutions to the problems in these scenarios can contain multiple scales, and very fine computational meshes are typically needed in order to capture these scales. Because of this, some type of model reduction is crucial for these problems. These reduced models are usually constructed based on a coarse grid, whose size does not necessarily resolve any of the scales. In addition, the resulting solutions are required to be robust with respect to the scales and the contrasts of the media, which is the main challenge.

There are existing multiscale approaches, such as [1, 2, 4, 6, 7]. The method in this paper is based on the Constraint Energy Minimizing Generalized Multiscale Finite Element Method (CEM-GMsFEM) [3]. In general, the method has two computational stages, called the offline and the online stages. In the offline stage, some computations are performed once and the reduced model is obtained. In the online stage, the problem formulated using the reduced model is solved when the input arguments and source terms are provided. The key to the success of the method is that the reduced model is only computed once in the offline stage, and the model can be used repeatedly in the online stage for various choices of input parameters and sources. In the offline stage, we will construct some local multiscale basis functions. The construction begins with a local auxiliary space, which is defined for each coarse element. The local auxiliary space is determined using a local spectral problem, which is able to identify high contrast channelized networks and fractures, as well as some important modes of the solution. We will use the first few eigenfunctions corresponding to small eigenvalues as the local auxiliary functions. Next, for each auxiliary function on a target coarse element, we will define a corresponding target multiscale basis function. The multiscale basis function is obtained by minimizing an energy over an oversampling region, obtained by extending the target coarse element by a few coarse grid layers, subject to some orthogonality conditions. These orthogonality conditions require that the target multiscale basis function is orthogonal to all auxiliary functions except the one being selected. The resulting multiscale basis functions have several important properties. One of them is that these basis functions are localized, providing the foundation of computing numerically on local oversampling regions. Another property is that the resulting coarse model based on the Galerkin formulation is first order convergent with respect to the coarse mesh size in the natural energy norm. The error bound is independent of the heterogeneities and contrast of the medium parameter κ. Hence the reduced model is very robust.

2 Method description

In this section we will give the details of our multiscale method and state the main convergence results. First of all, we introduce the adjoint equation

$$-\text{div}(\kappa(x)\nabla p) = y - y_d, \qquad \text{in } \Omega, \tag{5}$$

$$p = 0, \qquad \text{on } \Gamma. \tag{6}$$

We call the solution y of (2)–(3) for a control u an associated state to u and denote it as $y(u)$. In the same way, we call the solution p of (5)–(6) corresponding to $y(u)$ an associated adjoint state to u and denote it as $p(u)$. We introduce the projection

$$\Pi_{[a,b]}(f(x)) = \max(a, \min(b, f(x))).$$

Then we can formulate the necessary and sufficient first order optimality condition for (1)–(4).

Lemma 1 *A necessary and sufficient condition for the optimality of a control $\bar{u}$ with corresponding state $\bar{y} = y(\bar{u})$ and adjoint state $\bar{p} = p(\bar{u})$, respectively, is that the equation*

$$\bar{u} = \Pi_{[a,b]}\left(-\frac{1}{\nu}\bar{p}\right) \tag{7}$$

holds, where the state and adjoint equations for control $\bar{u}$ is given by:

$$-\text{div}(\kappa(x)\nabla\bar{y}) = \bar{u}, \qquad \text{in } \Omega,$$

$$\bar{y} = 0, \qquad \text{on } \Gamma$$

and

$$-\text{div}(\kappa(x)\nabla\bar{p}) = \bar{y} - y_d, \qquad \text{in } \Omega,$$

$$\bar{p} = 0, \qquad \text{on } \Gamma.$$

Moreover, due to (7), we obtain $\bar{u} \in H^1(\Omega)$. See Theorem 2.28 in [8] for details.

We are now in a position to introduce the discretized problem. We apply a multiscale finite element-based approximation of the optimal control problem (1)–(4). First, the notions of fine and coarse grids are introduced. Let T_H be a conforming partition of Ω into finite elements. Here, H is the coarse-mesh size and this partition is called coarse grid. We let N_c be the number of vertices and N be the number of elements in the coarse mesh. We assume that each coarse element is partitioned into a connected union of fine-grid cells and this partition is called T_h. Note that T_h is a refinement of the coarse grid T_H with the mesh size h. It is assumed that the fine grid is sufficiently fine to resolve the solution.

Moreover, we set

$$\begin{aligned} U_H &= \{u \in L^\infty(\Omega) : u|_{T'} \text{ is constant on all } T' \in T_H\}, \\ U_H^{ad} &= U_H \cap U_{ad}, \\ V &= H_0^1(\Omega). \end{aligned}$$

For each $u_H \in U_H$, the solution $y(u_H)$ of (2)–(3) satisfies

$$a(y(u_H), v) = \int_\Omega u_H v \, \mathrm{d}x \qquad \forall v \in V, \tag{8}$$

where $a : V \times V \to \mathbb{R}$ is the bilinear form defined by $a(y, v) = \int_\Omega \kappa \nabla y \cdot \nabla v \, \mathrm{d}x$. We define the energy norm $\|y\|_a = a(y, y)^{\frac{1}{2}}$. Notice that our goal is to construct a numerical scheme that gives the cell average of the control on the coarse grid.

2.1 Multiscale basis functions

We will construct V_{ms}, which is the space spanned by all multiscale basis functions. Then the multiscale solution y_{ms} is defined as the solution of the following problem: find $y_{ms} \in V_{ms}$ such that

$$a(y_{ms}(u_H), v) = \int_\Omega u_H v \, \mathrm{d}x \qquad \forall v \in V_{ms}. \tag{9}$$

We will first construct our auxiliary multiscale basis functions, which will be constructed for each coarse cell K in the coarse grid. Let K_i be the i-th coarse cell and let $V(K_i)$ be the restriction of V on K_i, which is $H^1(K_i)$. Following the construction from [3], we need to solve a local spectral problem, which is to find a real number $\lambda_j^{(i)}$ and a function $\phi_j^{(i)} \in V(K_i)$ such that

$$a_i(\phi_j^{(i)}, w) = \lambda_j^{(i)} s_i(\phi_j^{(i)}, w), \qquad \forall w \in V(K_i), \tag{10}$$

where a_i is a symmetric non-negative definite bilinear operator and s_i is a symmetric positive definite bilinear operator defined on $V(K_i) \times V(K_i)$. We assume the normalization $s_i(\phi_j^{(i)}, \phi_j^{(i)}) = 1$. Notice that $\lambda_j^{(i)}$ depends on H. In the numerical implementation, we need a fine grid in order to compute $\phi_j^{(i)}$. Based on the analysis in [3], we can choose

$$a_i(v, w) = \int_{K_i} \kappa \nabla v \cdot \nabla w \, \mathrm{d}x, \;\; s_i(v, w) = \int_{K_i} \tilde{\kappa} v w \, \mathrm{d}x,$$

where $\tilde{\kappa} = \sum_{j=1}^{N_c} \kappa |\nabla \chi_j^{ms}|^2$ and $\{\chi_j^{ms}\}_{j=1}^{N_c}$ are the standard multiscale finite element (MsFEM) basis functions, or piecewise bilinear basis, which satisfy the partition of unity property. We note that $\tilde{\kappa}$ is positive and it is important in the estimate of localization of basis functions, see Lemma 3 of [3]. We let $\lambda_j^{(i)}$ be the eigenvalues of (10) arranged in ascending order. We will use the first l_i eigenfunctions to construct our local auxiliary multiscale space $V_{aux}^{(i)}$, where $V_{aux}^{(i)} = \text{span}\{\phi_j^{(i)} | j \leq l_i\}$. The precise choice of l_i is based on a given tolerance. In particular, we let $\Lambda = \min_{1\leq i \leq N} \lambda_{l_i+1}^{(i)}$. Then we can choose l_i so that Λ is less than a given tolerance, which can be chosen as $O(1)$. Such tolerances will be introduced in Theorem 1. The global auxiliary multiscale space V_{aux} is the sum of these local auxiliary multiscale spaces, namely $V_{aux} = \oplus_{i=1}^N V_{aux}^{(i)}$. This space is used to construct the target multiscale basis functions that are ϕ-orthogonal to the auxiliary space V_{aux}. The notion of ϕ-orthogonality will be defined next.

For the local auxiliary multiscale space $V_{aux}^{(i)}$, the bilinear form s_i in (10) defines an inner product with norm $\|v\|_{s(K_i)} = s_i(v,v)^{\frac{1}{2}}$. These local inner products and norms provide natural definitions of inner product and norm for the global auxiliary multiscale space V_{aux}, which are defined by

$$s(v,w) = \sum_{i=1}^{N} s_i(v,w), \quad \|v\|_s = s(v,v)^{\frac{1}{2}}, \quad \forall v \in V_{aux}.$$

We note that $s(v,w)$ and $\|v\|_s$ are also an inner product and norm for the space V. Using the above inner product, we can define the notion of ϕ-orthogonality in the space V. Given a function $\phi_j^{(i)} \in V_{aux}$, we say that a function $\psi \in V$ is $\phi_j^{(i)}$-orthogonal if

$$s(\psi, \phi_j^{(i)}) = 1, \quad s(\psi, \phi_{j'}^{(i')}) = 0, \ \text{if} j' \neq j \text{ or } i' \neq i.$$

We remark that the function $\phi_j^{(i)}$ has support K_i, and we assume that $\phi_j^{(i)}$ is zero outside K_i. Now, we let $\pi_i : L^2(K_i) \to V_{aux}^{(i)}$ be the projection with respect to the inner product $s_i(v,w)$. So, the operator π_i is given by

$$\pi_i(u) = \sum_{j=1}^{l_i} s_i(u, \phi_j^{(i)})\phi_j^{(i)}, \quad \forall u \in V.$$

In addition, we let $\pi : L^2(\Omega) \to V_{aux}$ be the projection with respect to the inner product $s(v,w)$. So, the operator π is given by

$$\pi(u) = \sum_{i=1}^{N}\sum_{j=1}^{l_i} s_i(u, \phi_j^{(i)})\phi_j^{(i)}, \quad \forall u \in V.$$

Note that $\pi = \sum_{i=1}^N \pi_i$.

We next present the construction of our multiscale basis functions. For each coarse element K_i, we define an oversampled domain $K_{i,m} \subset \Omega$ by enlarging K_i by m coarse grid layers, where $m \geq 1$ is an integer. An illustration of the fine grid, coarse grid, and oversampling domain are shown in Fig. 1. We emphasize that the

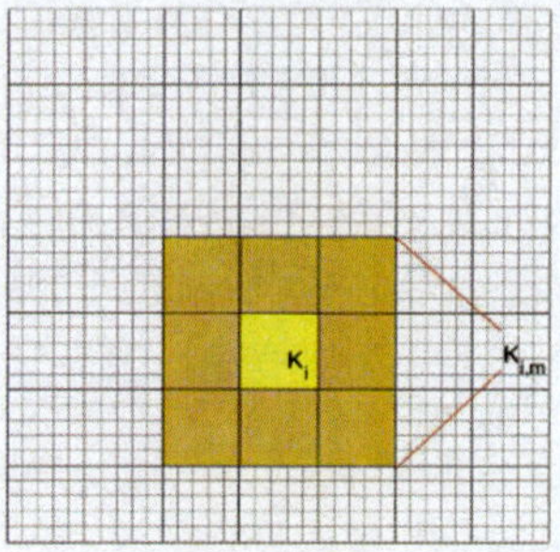

Fig. 1: Illustration of the coarse grid, fine grid and oversampling domain.

basis functions $\psi_{j,ms}^{(i)}$ are supported in the oversampling region $K_{i,m}$ with m being the number of oversampling layers. We will state in Theorem 1 the requirement on this integer m.

We next define the multiscale basis function $\psi_{j,ms}^{(i)} \in V_0(K_{i,m})$ by

$$\psi_{j,ms}^{(i)} = \text{argmin} \left\{ a(\psi, \psi) | \psi \in V_0(K_{i,m}), \quad \psi \text{ is } \phi_j^{(i)}\text{-orthogonal} \right\}, \tag{11}$$

where $V(K_{i,m})$ is the restriction of V in $K_{i,m}$ which is $H^1(K_{i,m})$, and $V_0(K_{i,m})$ is the subspace of $V(K_{i,m})$ with zero trace on $\partial K_{i,m}$, i.e. $V_0(K_{i,m}) = H_0^1(K_{i,m})$. Equivalently, we find $\psi_{j,ms}^{(i)} \in V_0(K_{i,m})$ and $\mu \in V_{aux}^{(i,m)}$ that satisfy the following

$$\begin{aligned} a(\psi_{j,ms}^{(i)}, v) + s(v, \mu) &= 0, \quad \forall v \in V_0(K_{i,m}), \\ s(\psi_{j,ms}^{(i)}, v) &= s(\phi_j^{(i)}, v), \quad \forall v \in V_{aux}^{(i,m)}. \end{aligned} \tag{12}$$

In the above, we define $V_{aux}^{(i,m)} = \oplus V_{aux}^{(j)}$ where the sum is over all $K_j \subset K_{i,m}$. Our multiscale finite element space V_{ms} is defined by

$$V_{ms} = \text{span} \left\{ \psi_{j,ms}^{(i)} | \ 1 \le j \le l_i, \ 1 \le i \le N \right\}.$$

Finally, we set

$$V_H = V_{ms} \subset H_0^1(\Omega).$$

This is the coarse space for the systems (2)–(3) and (5)–(6).

2.2 The proposed method

We will use the space U_H^{ad} for the approximation of the control u. For the state variables y and p, we will use the space V_H. For each $u_H \in U_H$, the approximate solution $y_H(u_H) \in V_H$ of (2)–(3) satisfies

$$a(y_H(u_H), v_H) = \int_\Omega u_H v_H \,\mathrm{d}x, \qquad \forall v_H \in V_H. \tag{13}$$

In other words, $y_H(u_H)$ is the approximated state associated with u_H. The finite-dimensional approximation of the optimal control problem is defined as: find $u_H \in U_H^{ad}$ such that it minimizes the following functional

$$J(u_H) = \frac{1}{2}\|y_H(u_H) - y_d\|^2_{L^2(\Omega)} + \frac{\nu}{2}\|u_H\|^2_{L^2(\Omega)}. \tag{14}$$

The adjoint equation is discretized in the same way: find $p_H(u_H) \in V_H$ such that

$$a(p_H(u_H), v_H) = \int_\Omega (y_H(u_H) - y_d)v_H \,\mathrm{d}x \qquad \forall v_H \in V_H. \tag{15}$$

From now on, we denote the optimal control of the discrete optimization problem as $\bar{u}_H$ and corresponding state and adjoint solutions as $\bar{y}_H$ and $\bar{p}_H$ respectively. That is, $\bar{y}_H = y_H(\bar{u}_H)$ and $\bar{p}_H = p_H(\bar{u}_H)$. We remark that the associated adjoint state $\bar{p}$ belongs to the space $H^1(\Omega)$. The optimal control $\bar{u}$ is obtained by the projection formula (7).

Let $\bar{u}$ be the solution of (1)–(4). We define a piecewise constant function by using the local mean value of $\bar{u}$:

$$w_H(x) = \frac{\int_{T_i} \bar{u}(x)\mathrm{d}x}{\int_{T_i} 1\mathrm{d}x} \qquad \text{if } x \in T_i \text{ where } T_i \in \mathcal{T}_H, \tag{16}$$

It is clear that $w_H \in U_H^{ad}$.

Now we are able to formulate our convergence result.

Theorem 1 *Let $\bar{u}_H$ be the solution of (14). Moreover, if the number of oversampling layers $m = O(\log(H^{-1}\kappa_0^{-1}\kappa_1))$ and χ_i are bilinear partition of unity, then we have the following error bound*

$$\|\bar{u}_H - \bar{u}\|_{L^2(\Omega)} + \|\bar{y}_H - \bar{y}\|_a + \|\bar{p}_H - \bar{p}\|_a \le CH\Lambda^{-\frac{1}{2}}\nu^{-1}. \tag{17}$$

where Λ is the minimal eigenvalue such that the corresponding eigenvector is not included in the auxiliary space, that is, $\Lambda = \min_{1\le i\le N} \lambda^{(i)}_{l_i+1}$. Moreover, the constant C is independent of the mesh size and the coefficient κ.

Note that the precise equation for m can be found at the end of Section 5 in [3].

2.3 Outline of error analysis

We will briefly outline the error analysis and a proof of Theorem 1. Using the results in [3], we obtain Lemmas 2–4.

Lemma 2 *Let $u \in L^2(\Omega)$. Moreover, if the number of oversampling layers $m = O(\log(H^{-1}\kappa_0^{-1}\kappa_1))$ and $\{\chi_i\}$ are bilinear partition of unity, then we have*

$$\|y(u) - y_H(u)\|_a \leq CH\Lambda^{-\frac{1}{2}}\|\kappa^{-\frac{1}{2}}u\|_{L^2(\Omega)}, \tag{18}$$

$$\begin{aligned}\|p(u) - p_H(u)\|_a &\leq CH\Lambda^{-\frac{1}{2}}(\|\kappa^{-\frac{1}{2}}y\|_{L^2(\Omega)} + \|\kappa^{-\frac{1}{2}}y_d\|_{L^2(\Omega)}),\\ &\leq CH\Lambda^{-\frac{1}{2}}\kappa_0^{-\frac{1}{2}}(\|u\|_{L^2(\Omega)} + \|y_d\|_{L^2(\Omega)}).\end{aligned} \tag{19}$$

Lemma 3 *Let w_H be the function defined by (16). In addition, suppose that the assumptions of Lemma 2 are fulfilled. Then we have*

$$\|y_H(\bar{u}) - y_H(w_H)\|_a \leq CH\kappa_0^{-\frac{1}{2}}\|\bar{u}\|_{H^1(\Omega)}, \tag{20}$$

$$\|p_H(\bar{u}) - p_H(w_H)\|_a \leq CH\kappa_0^{-\frac{1}{2}}\|\bar{u}\|_{H^1(\Omega)}, \tag{21}$$

$$\|\bar{p} - p_H(w_H)\|_a \leq CH\Lambda^{-\frac{1}{2}}\kappa_0^{-\frac{1}{2}}(\|\bar{u}\|_{H^1(\Omega)} + \|y_d\|_{L^2(\Omega)}). \tag{22}$$

Lemma 4 *The following variational inequalities are necessary and sufficient for the optimality of the unique solutions of (1)–(4) and (14):*

$$(\bar{p} + \nu\bar{u}, u - \bar{u})_{L^2(\Omega)} \geq 0 \qquad \forall u \in U_{ad}, \tag{23}$$

$$(p_H(\bar{u}_H) + \nu\bar{u}_H, \zeta_H - \bar{u}_H)_{L^2(\Omega)} \geq 0 \qquad \forall \zeta_H \in U_H^{ad}. \tag{24}$$

Now, we derive a variational inequality for the function w_H. We define a new function $\hat{p}$ by

$$\hat{p}(x) = \frac{\int_{T_i} \bar{p}(x)\,\mathrm{d}x}{\int_{T_i} 1\,\mathrm{d}x}, \qquad \text{where } x \in T_i \in T_H.$$

Then, using (23), we obtain

$$(\hat{p} + \nu w_H, \bar{u}_H - w_H)_{L^2(\Omega)} \geq 0. \tag{25}$$

Moreover, we can test inequality (24) with the function w_H and get

$$(p_H(\bar{u}_H) + \nu\bar{u}_H, w_H - \bar{u}_H)_{L^2(\Omega)} \geq 0. \tag{26}$$

Combining the results, we have

$$\nu\|w_H - \bar{u}_H\|^2_{L^2(\Omega)} \leq (\hat{p} - p_H(\bar{u}_H), \bar{u}_H - w_H)_{L^2(\Omega)}. \tag{27}$$

The right-hand side of (27) can be written as

$$(\hat{p} - p_H(\bar{u}_H), \bar{u}_H - w_H)_{L^2(\Omega)} = (p_H(w_H) - p_H(\bar{u}_H), \bar{u}_H - w_H)_{L^2(\Omega)} + (\bar{p} - p_H(w_H), \bar{u}_H - w_H)_{L^2(\Omega)} + (\hat{p} - \bar{p}, \bar{u}_H - w_H)_{L^2(\Omega)}. \tag{28}$$

Next we estimate these three terms. The first term on the right-hand side of (28) can be estimated as

$$\begin{aligned}&(p_H(w_H) - p_H(\bar{u}_H), \bar{u}_H - w_H)_{L^2(\Omega)}\\ &= (y_H(w_H) - y_H(\bar{u}_H), y_H(\bar{u}_H) - y_H(w_H))_{L^2(\Omega)}\\ &\leq 0.\end{aligned} \tag{29}$$

The second term on the right-hand side of (28) can be estimated using (22):

$$\begin{aligned}&(\bar{p} - p_H(w_H), u_H - w_H)_{L^2(\Omega)}\\ &\leq CH\Lambda^{-\frac{1}{2}}\kappa_0^{-\frac{1}{2}}(\|\bar{u}\|_{H^1(\Omega)} + \|y_d\|_{L^2(\Omega)}) \cdot \|w_H - u_H\|_{L^2(\Omega)}.\end{aligned} \tag{30}$$

The third term represents a formula for the numerical integration. Using that u_H and w_H are constant on each triangle T_i,

$$\begin{aligned}(\hat{p} - \bar{p}, \bar{u}_H - w_H)_{L^2(\Omega)} &= \sum_i \int_{T_i} ((\hat{p}(x) - \bar{p}(x))(\bar{u}_H(x) - w_H(x))\,\mathrm{d}x\\ &= \sum_i (\bar{u}_H(x) - w_H(x)) \int_{T_i} (\hat{p}(x) - \bar{p}(x))\,\mathrm{d}x\\ &= \sum_i (\bar{u}_H(x) - w_H(x))(\int_{T_i} \bar{p}(x)\,\mathrm{d}x - \int_{T_i} \bar{p}(x)\,\mathrm{d}x)\\ &= 0.\end{aligned} \tag{31}$$

Using (29)–(31) in (28), we get

$$\begin{aligned}&(\hat{p} - p_H(\bar{u}_H), u_H - w_H)_{L^2(\Omega)}\\ &\leq CH\Lambda^{-\frac{1}{2}}\kappa_0^{-\frac{1}{2}}(\|\bar{u}\|_{H^1(\Omega)} + \|y_d\|_{L^2(\Omega)}) \cdot \|w_H - \bar{u}_H\|_{L^2(\Omega)}.\end{aligned}$$

Note that, by the standard finite element interpolation theory, we have

$$\begin{aligned}\|\bar{u}_H - \bar{u}\|_{L^2(\Omega)} &\leq \|\bar{u}_H - w_H\|_{L^2(\Omega)} + \|\bar{u} - w_H\|_{L^2(\Omega)}\\ &\leq CH\Lambda^{-\frac{1}{2}}\kappa_0^{-\frac{1}{2}}\nu^{-1}(\|\bar{u}\|_{H^1(\Omega)} + \|y_d\|_{L^2(\Omega)}).\end{aligned}$$

By Lemma 2 and Lemma 3, we have

$$\begin{aligned}
&\|\bar{y}_H - \bar{y}\|_a \\
&\leq \|\bar{y}_H - y_H(w_H)\|_a + \|y_H(w_H) - y(w_H)\|_a + \|y(w_H) - \bar{y}\|_a \\
&\leq CH\Lambda^{-\frac{1}{2}}\kappa_0^{-\frac{1}{2}}\|\bar{u}\|_{H^1(\Omega)} + CH\Lambda^{-\frac{1}{2}}\|\kappa^{-\frac{1}{2}}w_H\|_{L^2(\Omega)} + CH\Lambda^{-\frac{1}{2}}\kappa_0^{-\frac{1}{2}}\|\bar{u}_H\|_{H^1(\Omega)} \\
&\leq CH\Lambda^{-\frac{1}{2}}.
\end{aligned}$$

Similarly, we have $\|\bar{p}_H - \bar{p}\|_a \leq CH\Lambda^{-\frac{1}{2}}$. This proves Theorem 1.

3 Numerical results

In this section, we will present some numerical tests to validate the convergence of the method. The optimization problems are solved numerically by a primal-dual active set strategy; see, for instance, [5]. The primal-dual active set strategy will be presented here. For this purpose we introduce the active and inactive sets for the solution and define

$$A_+^* = \{x \in \Omega : u^*(x) = b\}, \qquad A_-^* = \{x \in \Omega : u^*(x) = a\},$$
$$\text{and } I^* = \{x \in \Omega : a < u^*(x) < b\}.$$

Here and below, the set-theoretic definitions are understood in the almost everywhere sense. Given (u_{n-1}, λ_{n-1}), the active sets for the new iterate are chosen according to

$$A_n^+ = \left\{x \in \Omega : u_{n-1}(x) + \frac{\lambda_{n-1}(x)}{c} > b\right\}, \tag{32}$$

$$A_n^- = \left\{x \in \Omega : u_{n-1}(x) + \frac{\lambda_{n-1}(x)}{c} < a\right\}, \tag{33}$$

where $c > 0$. The update strategies for A_n^+ and A_n^- are the key ingredients of the proposed algorithm. The complete algorithm is specified in Algorithm 1.

In our simulations, we take the medium parameter κ shown in Fig. 2, and the contrast is 10^4. Note that, the state equation is given by

$$\begin{aligned}
-\mathrm{div}(\kappa\nabla y) &= u \quad &&\text{in } \Omega, \\
y &= 0 \quad &&\text{on } \Gamma
\end{aligned} \tag{34}$$

Define $u_f(x_1, x_2) = 2\pi x_1(1 - x_1)^2 \sin(\pi x_2)$. We construct the exact optimal control $\bar{u}$

$$\bar{u}(x_1, x_2) = \begin{cases} a & \text{if } u_f(x_1, x_2) < a, \\ u_f(x_1, x_2) & \text{if } u_f(x_1, x_2) \in [a, b], \\ b & \text{if } u_f(x_1, x_2) > b. \end{cases}$$

Algorithm 1 Primal-dual Active Set Strategy.

1: Initialization: Choose u_0 and λ_0, and set $n = 1$.
2: Determine the active sets according to (32)–(33), and set $I_n = \Omega \setminus (A_n^+ \cup A_n^-)$.
3: If $n \geq 2$, $A_n^+ = A_{n-1}^+$, $A_n^- = A_{n-1}^-$, and $I_n = I_{n-1}$, then STOP.
4: Else, find $(y_n, p_n) \in V_H \times V_H$ such that

$$\int_\Omega \kappa \nabla y_n \cdot \nabla v_H \,\mathrm{d}x = \int_\Omega u_n v_H \,\mathrm{d}x \qquad \forall v_H \in V_H$$

$$\int_\Omega \kappa \nabla p_n \cdot \nabla v_H \,\mathrm{d}x = \int_\Omega (y_n - y_d) v_H \,\mathrm{d}x \qquad \forall v_H \in V_H$$

where

$$u_n(x) = \begin{cases} b & \text{if } x \in A_n^+, \\ a & \text{if } x \in A_n^-, \\ -\frac{\int_T p_n \,\mathrm{d}x}{\nu \int_T 1 \,\mathrm{d}x} & \text{if } x \in I_n \cap T, \text{ where } T \in \mathcal{T}_H. \end{cases}$$

5: Set $\lambda_n = -p_n - \nu u_n$, update $n := n + 1$, and goto 2.

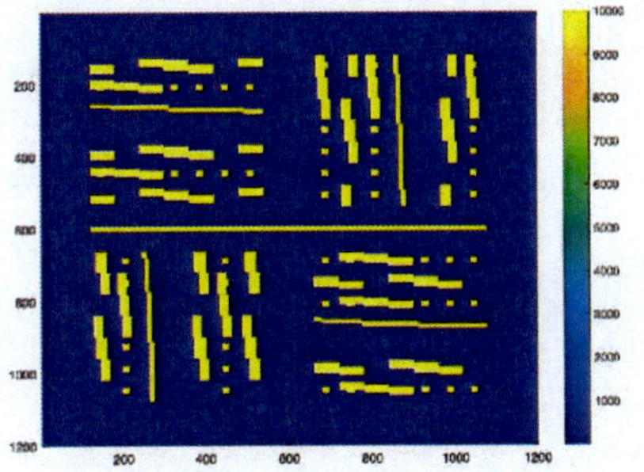

Fig. 2: The high contrast medium κ.

We also denote the optimal state $\bar{y}$ by solving (34). For the optimal adjoint state $\bar{p}$, we find

$$\bar{p}(x_1, x_2) = -2\pi\nu x_1(1 - x_1)^2 \sin(\pi x_2).$$

The desired state is given by

$$y_d(x_1, x_2) = \bar{y} + \operatorname{div}(\kappa \nabla \bar{p}).$$

It is easy to see that these functions fulfill the necessary and sufficient first order optimality conditions. Also, we take $c = 2$ and $\nu = 1$.

The solution $\bar{y}$ is calculated by the reference solution using a 200×200 fine mesh. We need 6 iterations to stop the primal-dual active set strategy. Also, if we solve the problem on the fine mesh, we need to solve problems with $2 \times 201 \times 201$ unknowns in one iteration but in our approach, we only need 9600 unknowns even for the finest case $H = \frac{1}{40}$. Fig. 3 and Fig. 4 show the numerical solutions $\bar{u}_H$ for $H = 0.05$ and

$H = 0.025$ respectively. Table 1 shows the relative L^2-norm error for $\bar{u} - \bar{u}_H$. The order of the L^2-error is about 1. Table 2 shows the same result with one more basis function per coarse element.

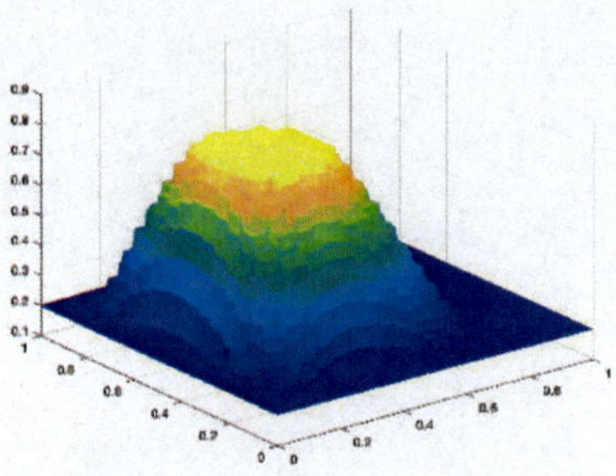

Fig. 3: $\bar{u}_H$ using $H = 0.05$.

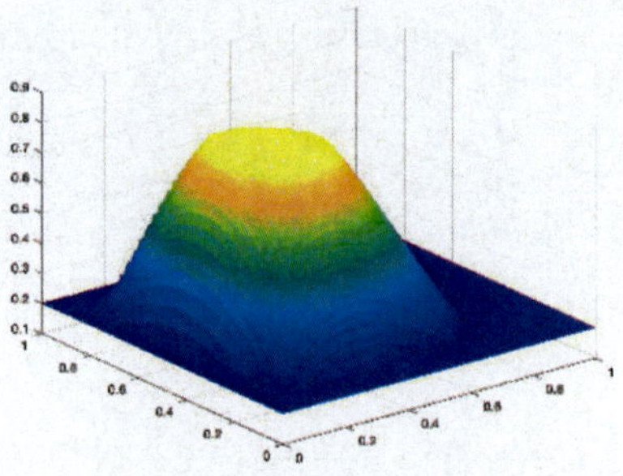

Fig. 4: $\bar{u}_H$ using $H = 0.025$.

Table 1: L^2 error with 3 basis functions per coarse element

size of basis	H	# oversample layers	$\frac{\|\bar{u}_H-\bar{u}\|_{L^2(\Omega)}}{\|\bar{u}\|_{L^2(\Omega)}}$
3	1/5	3	23.8716%
3	1/10	3	9.6973%
3	1/20	4	4.3582%
3	1/40	5	1.6717%

Table 2: L^2 error with 4 basis functions per coarse element

size of basis	H	# oversample layers	$\frac{\|\bar{u}_H-\bar{u}\|_{L^2(\Omega)}}{\|\bar{u}\|_{L^2(\Omega)}}$
4	1/5	3	21.9864%
4	1/10	3	9.6987%
4	1/20	4	4.3463%
4	1/40	5	1.6782%

Acknowledgements The research of Eric Chung is partially supported by the Hong Kong RGC General Research Fund (Project numbers 14304719 and 14302018) and the CUHK Faculty of Science Direct Grant 2020-21.

References

1. A. Abdulle, E. Weinan, B. Engquist and E. Vanden-Eijnden. The heterogeneous multiscale method. *Acta Numerica* **21**, 1–87 (2012).
2. E. Chung, Y. Efendiev and T.Y. Hou. Adaptive multiscale model reduction with generalized multiscale finite element methods. *Journal of Computational Physics* **320**, 69–95 (2016).
3. E.T. Chung, Y. Efendiev and W.T. Leung. Constraint energy minimizing generalized multiscale finite element method. *Computer Methods in Applied Mechanics and Engineering* **339**, 298–319 (2018).
4. T.J. Hughes, G.R. Feijoo, L. Mazzei and J.B. Quincy. The variational multiscale method – a paradigm for computational mechanics. *Computer methods in applied mechanics and engineering* **166**, 3–24 (1998).
5. K. Kunisch and A. Rösch. Primal-dual active set strategy for a general class of constrained optimal control problems. *SIAM Journal on Optimization* **13**, 321–334 (2002).
6. A. Malqvist and D. Peterseim. *Numerical homogenization by localized orthogonal decomposition.* Society for Industrial and Applied Mathematics (2020).
7. H. Owhadi. Multigrid with rough coefficients and multiresolution operator decomposition from hierarchical information games. *SIAM Review* **59**, 99–149 (2017).
8. F. Tröltzsch. *Optimal control of partial differential equations: theory, methods, and applications*. American Mathematical Soc. (2010).

Several Ways to Achieve Robustness When Solving Wave Propagation Problems

Niall Bootland, Victorita Dolean, Pierre Jolivet, Frédéric Nataf, Stéphane Operto, and Pierre-Henri Tournier

1 Motivation and challenges

Why do we need robust solution methods for wave propagation problems? Very often in applications, as for example in seismic inversion, we need to reconstruct the a priori unknown physical properties of an environment from given measurements. From a mathematical point of view, this means solving inverse problems by applying an optimisation algorithm to a misfit functional between the computation and the data. At each iteration of this algorithm we need to solve a boundary value problem involving the Helmholtz equation

$$-\Delta u - \frac{\omega^2}{c^2} u = f, \tag{1}$$

where $c = \sqrt{\rho c_P^2}$, ρ is the density of the medium and c_P is the speed of longitudinal waves. Here, ω is usually given as being the frequency of a localised source and we wish to reconstruct $c = \frac{1}{n}$ from the measured data (here, n is also called the refraction

Niall Bootland
University of Strathclyde, Dept. of Maths and Stats, e-mail: niall.bootland@strath.ac.uk

Victorita Dolean
University of Strathclyde, Dept. of Maths and Stats and University Côte d'Azur, CNRS, LJAD
e-mail: work@victoritadolean.com

Pierre Jolivet
University of Toulouse, CNRS, IRIT, e-mail: pierre.jolivet@enseeiht.fr

Frédéric Nataf
Sorbonne Université, CNRS, LJLL, e-mail: frederic.nataf@sorbonne-universite.fr

Stéphane Operto
University Côte d'Azur, CNRS, Géoazur, e-mail: stephane.operto@geoazur.unice.fr

Pierre-Henri Tournier
Sorbonne Université, CNRS, LJLL e-mail: tournier@ann.jussieu.fr

S. C. Brenner et al. (eds.), *Domain Decomposition Methods in Science and Engineering XXVI*, Lecture Notes in Computational Science and Engineering 145,
https://doi.org/10.1007/978-3-030-95025-5_2

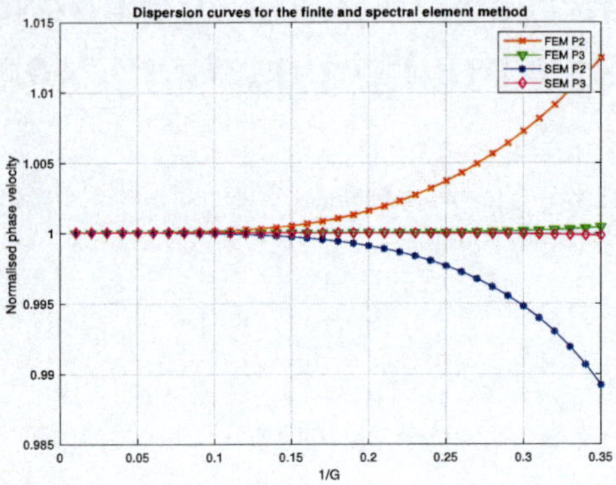

Fig. 1 Dispersion curves for finite elements and spectral elements of order 2 and 3: normalised phase velocity (the ratio between the numerical wave speed and the physical one) as a function of the reciprocal number of points per wavelength $\frac{1}{G} = \frac{\omega h}{2\pi}$ for different discretisations. Notice that the use of higher order elements minimises dispersion even for a low value of G.

index). The Helmholtz equation is also known as the reduced wave equation or time-harmonic wave equation. Solving this equation is mathematically difficult, especially for high wave number $k = \frac{\omega}{c}$, as the solution is oscillatory and becomes more so with increasing k. Note that the notion of a high frequency problem is to be understood relative to the size of the computational domain: meaning how many wavelengths are present in the latter. In geophysics, the typically large size of the domain, and therefore the presence of hundreds of wavelengths, makes the problem difficult.

1.1 Why the time-harmonic problem in mid and high frequency is hard

What happens if one wants to approximate this problem with a numerical method? A simple computation in the one-dimensional case shows that the numerical refraction index is different from the physical one and the error depends on the product between the spacing of the grid h and the frequency ω, in other words numerical waves travel at a different speed to physical waves and this is also reflected in the size of error. This is also called the pollution effect and was first highlighted in the seminal paper [3]. For quasi optimality in the finite element sense we require that $h^p \omega^{p+1}$ be bounded, where p is the order or the precision of the method, as shown in [21]. To summarise, the high-frequency solution u oscillates at a scale $\frac{1}{\omega}$, therefore the mesh size should be chosen as at least $h \sim \frac{1}{\omega}$, leading to a large number of degrees of freedom. The pollution effect requires $h \ll \frac{1}{\omega}$, namely $h \sim \omega^{-1-\frac{1}{p}}$, therefore in practice one needs an even larger number of degrees of freedom. Note that in order to get a bounded finite element error the constraint is weaker, being $h \sim \omega^{-1-\frac{1}{2p}}$, as shown in [14]. A trade-off should be found between the number of points per wavelength (ppwl) $G = \frac{\lambda}{h} = \frac{2\pi}{\omega h}$ and the polynomial degree p in order to minimise pollution, and this is usually the object of dispersion analysis [1]. This is illustrated in Figure 1, where we see that the best dispersion properties are achieved when we increase the order of the discretisation or we increase G.

Suppose now that we have discretised the equation following the previous rules. We end up with a huge linear system (for a typical application we should expect millions of unknowns) whose size increases with ω very quickly, especially with more spatial dimensions. The matrix is symmetric and non-Hermitian, which makes this system difficult to solve by standard iterative methods, as shown in the review paper by Ernst and Gander [15] or the most recent one by Gander and Zhang [16]. Our aim should be to find the solution in optimal time for large frequencies and our algorithms should not only have good parallel properties but they should also be robust with respect to heterogeneities.

It is well known that direct solvers, while being robust, have two main drawbacks: their high memory storage and poor parallel properties. On the other hand, iterative methods are not robust but very easy to parallelise. For this reason we consider hybrid methods, such as the naturally parallel compromise of domain decomposition methods, to obtain the best of both worlds. However, how large is truly large? In real applications, problems do not need to be over-resolved (for example, 4 ppwl are enough to perform Full Waveform Inversion with a finite-difference scheme that is specifically tuned to minimise numerical dispersion for this discretisation rule [2]) and time-harmonic Helmholtz equations with 50 million degrees of freedom were solved by a parallel direct method [20]. On the other side, when we consider much larger domains (for example via the use of a separate network of nodes rather than with cables) and that the number of nodes is limited, we must switch to iterative or hybrid methods of domain decomposition type. The methods we develop are not only motivated by the current trend in seismic imaging, meaning the development of sparse node devices (OBN) for data acquisition in the oil industry [4], but in the last decades, since the seminal work of Després [9], they have become the method of choice when solving the discretised Helmholtz equations.

2 What is the best coarse space for Helmholtz?

Consider the decomposition of the computational domain Ω into N overlapping subdomains Ω_j. The construction of these domains is explained later in Section 2.2 and illustrated in Figure 3. We usually solve the system $A\mathbf{u} = \mathbf{b}$ stemming from the finite element discretisation of (1) by a preconditioned GMRES method, e.g., in the form $M^{-1}A\mathbf{u} = M^{-1}\mathbf{b}$ with

$$M^{-1} = \sum_{j=1}^{N} R_j^T D_j B_j^{-1} R_j, \tag{2}$$

where $R_j : \Omega \to \Omega_j$ is the restriction operator, $R_j^T : \Omega_j \to \Omega$ the prolongation operator and D_j corresponds to the partition of unity, i.e., it is chosen such that $\sum_{j=1}^{N} R_j^T D_j R_j = I$. Note also that local matrices B_j are stiffness matrices of local Robin boundary problems

$$\begin{aligned} (-\Delta - k^2)(u_j) &= f && \text{in } \Omega_j, \\ \left(\frac{\partial}{\partial n_j} + ik\right)(u_j) &= 0 && \text{on } \partial\Omega_j \setminus \partial\Omega. \end{aligned}$$

We call (2) the one-level preconditioner, in particular it is the ORAS preconditioner.

Conventional wisdom in domain decomposition, backed by the definitions of strong and weak scaling, says that one-level preconditioners are not scalable (i.e., their behaviour deteriorates with the number of subdomains N). The crucial idea is to add a second level: that is, coarse information that is cheap to compute and immediately available to all subdomains/processors. Suppose that the coarse space is spanned by a matrix Z, then $E = Z^*AZ$ is the coarse matrix and $H = ZE^{-1}Z^*$ is the coarse space correction. This coarse space correction can be combined with the one-level preconditioner in an additive or hybrid manner via projectors P and Q ($P = Q = I$ for additive while $P = I - AH$, $Q = I - HA$ provides a hybrid variant)

$$M_2^{-1} = QM^{-1}P + H.$$

This coarse correction can be understood as a solution of a coarser problem on a geometrical grid with a larger spacing for example. For time-harmonic wave propagation problems, the size of the coarse grid is, however, constrained by the wave number. The theory of the grid CS (coarse space) has been introduced by Graham et al. [17] for a two-level approach to the Helmholtz problem using an equivalent problem with absorption; it has since been extended to the time-harmonic Maxwell equations [5]. This preconditioner is based on local Dirichlet boundary value problems within the one-level method. An extension to Robin transmission conditions was recently provided in [18].

The questions we would like to answer are the following: Is the grid coarse space the best choice for heterogeneous problems? Note also that the definition of the coarse space does not have to be geometrical, we can build more sophisticated coarse spaces based on solving eigenvalue problems. Can we further improve performance by extending the idea of spectral coarse spaces to Helmholtz problems? And, if yes, what kind of modes should be included in the coarse space?

2.1 Spectral coarse spaces for Helmholtz

There are already now a few spectral versions of two-level preconditioners and these are DtN, H-GenEO and Δ-GenEO. For the first two there is no theory available and, while a theory has been developed for the latter, this preconditioner works mainly for low frequency and mildly non-symmetric problems.

The idea of the DtN coarse space was first introduced in [22] for elliptic problems, further analysed in [13], and extended to the Helmholtz equation in [8]. Let $D \subset \Omega$ with internal boundary $\Gamma_D = \partial D \setminus \partial\Omega$ and $v_{\Gamma_D} : \Gamma_D \to \mathbb{C}$. Then the DtN operator is defined as $\mathrm{DtN}_D(v_{\Gamma_D}) = \frac{\partial v}{\partial n}|_{\Gamma_D}$, where $v\colon D \to \mathbb{C}$ is the Helmholtz extension of v_{Γ_D} (the solution to a local boundary value problem with Dirichlet value v_{Γ_D} on Γ_D). The DtN coarse space (introduced in [13]) is based on eigenvalue problems of the DtN operator local to each subdomain: find $(u_{\Gamma_j}, \lambda) \in V(\Gamma_j) \times \mathbb{C}$ such that

$$\mathrm{DtN}_{\Omega_j}(u_{\Gamma_j}) = \lambda u_{\Gamma_j}.$$

To provide the modes in the coarse space we use the Helmholtz extension v. We choose only eigenfunctions with λ such that $\mathrm{Re}(\lambda) < k_j$, where $k_j = \max_{x \in \Omega_j} k(x)$. Note that this criterion depends on the local heterogeneity in the problem and is purely heuristic (as explained in detail in [8]). In practice, finding the coarse space vectors amounts to solving local problems depending on Schur complements and mass matrices on the interfaces. By a local Helmholtz extension, we obtain vectors that, after multiplication by the partition of unity and extension by zero, form the matrix Z. We do this in each subdomain and combine to give the global coarse space.

The GenEO (Generalised Eigenproblems in the Overlap) coarse space was first developed in [24] for SPD problems with heterogeneous coefficients, where the heterogeneities do not align with the subdomain decomposition. More precisely, in each Ω_j we solve discrete eigenproblems with local Dirichlet matrices A_j weighted by the partition of unity on one side and the local Neumann matrix $\widetilde{A}_j$ on the other:

$$D_j A_j D_j u = \lambda \widetilde{A}_j u. \tag{3}$$

We then choose only eigenfunctions with eigenvalue λ such that $\lambda > \lambda_{\min}$. Note that if we try to replicate this exactly for Helmholtz, the method will fail. For this reason, we need to make some adaptations. The first idea is to use a nearby positive problem to build the coarse space and then use these modes for Helmholtz. This approach is called Δ-GenEO and it is amenable to theory. The second idea is more Helmholtz related in the sense that we only modify the right-hand side of the generalised eigenvalue problem (3) and thus the wave number k is included in the eigenproblem:

$$D_j L_j D_j u = \lambda \widetilde{A}_j u,$$

where L_j corresponds to the Laplacian part of the problem and $\widetilde{A}_j$ is the Neumann matrix for the Helmholtz operator. Eigenvectors associated to the eigenvalues with $\mathrm{Re}(\lambda) > \lambda_{\min}$ are now those put into the coarse space. We call this method H-GenEO.

2.2 Comparison of coarse spaces

We show how these three two-level methods (grid CS, DtN and H-GenEO) compare on the Marmousi[1] problem [25] (see also Figure 2), which is a 2D geophysical benchmark problem consisting of propagation of seismic waves in a heterogeneous medium from a point source situated towards the surface. This problem is high

[1] `https://reproducibility.org/RSF/book/data/marmousi/paper_html/node2.html`

Fig. 2: The real part of the solution to the Marmousi problem at 20 Hz. The domain is 9.2 km×3 km.

Fig. 3: Left to right: (a) the coarse mesh, (b) use of minimum overlap (refine the non-overlapping decomposition) and (c) use of coarse overlap (refine the coarse overlapping mesh directly).

frequency because of the large number of wavelengths in the domain. For more extensive results and comparative performance tests with these methods on other benchmark problems, see [6].

From the practical point of view, a coarse mesh is generated (from which we build the grid coarse space) and this coarse mesh is refined to give the fine mesh; see Figure 3. Alternatively, we can refine on the underlying non-overlapping decomposition and then take minimum overlap. For the discretisation by finite elements (here P2 Lagrange finite elements), we have used FreeFEM. For the domain decomposition and solver we use the FreeFEM library ffddm along with HPDDM and PETSc.[2]

Note that the grid CS is applied naturally to the configuration (c) (with coarse overlap), whereas for the one-level and spectral methods we can choose between minimum and coarse overlap. In the following tables we report the best results for each method in the most favourable configuration (overlap and number of modes for the spectral coarse spaces). These are the iteration counts of the GMRES method applied to the preconditioned problem with the two-level domain decomposition preconditioner in order to achieve a relative residual tolerance of 10^{-6}. We consider two scenarios: the *under-resolved* case with a discretisation of 5 points per wavelength (Table 1) and the *over-resolved* case with a discretisation of 10 points per wavelength (Table 2) and vary the frequency and the number of subdomains. Low resolution is motivated by applications where high precision is not needed, especially when solving inverse problems by FWI (Full Waveform Inversion). In this case, since the test cases are large, one needs to find a good trade-off between precision and the size of the system to be solved. We refer the reader to the references [11, 12], where a

[2] Software available at `FreeFem-sources/examples/ffddm` (ffddm) within FreeFEM, `https://github.com/hpddm/hpddm` (HPDDM), and `https://www.mcs.anl.gov/petsc` (PETSc).

Table 1: Results using the one-level and coarse grid methods for the Marmousi problem when using 5 points per wavelength, varying the frequency f and the number of subdomains N.

		One-level					Coarse grid					H-Geneo					DtN				
f	#dofs \ N	10	20	40	80	160	10	20	40	80	160	10	20	40	80	160	10	20	40	80	160
1	4×10^3	26	39	47	64	–	15	18	19	20	–	9	11	17	21	–	6	7	9	6	–
5	1×10^5	53	76	105	154	213	26	29	28	29	31	15	17	26	37	56	7	19	10	8	19
10	5×10^5	68	102	158	212	302	32	35	41	40	42	33	40	45	56	73	18	19	21	48	29
20	2×10^6	82	125	178	248	347	34	35	42	43	44	64	83	121	134	157	43	75	77	61	35

Table 2: Results using the one-level and coarse grid methods for the Marmousi problem when using 10 points per wavelength, varying the frequency f and the number of subdomains N.

		One-level					Coarse grid					H-Geneo					DtN				
f	#dofs \ N	10	20	40	80	160	10	20	40	80	160	10	20	40	80	160	10	20	40	80	160
1	2×10^4	30	43	63	97	–	16	18	19	21	–	7	8	8	13	–	4	7	5	6	–
5	5×10^5	58	87	126	175	246	29	29	34	34	36	10	9	10	10	12	10	11	12	17	24
10	2×10^6	78	124	172	251	346	35	41	43	46	45	20	16	14	13	13	19	23	25	25	24
20	8×10^6	92	142	198	272	389	39	47	48	49	49	45	40	34	25	19	35	46	48	56	59

more extensive numerical study was performed. We notice that in the first scenario the grid CS outperforms the spectral methods (with a slight advantage over the DtN method) whereas in the second scenario the H-GenEO method displays the best performance.

We conclude this comparison by noting that there is no clear advantage in one method over another, all depends on the frequency and precision desired. We have not sought an optimal implementation and the grid CS is the finest possible (which is in principle very expensive), in this sense the timings are not relevant, even if the cost per iteration might be different. In the case of multiple right-hand sides, spectral coarse spaces may have an advantage, although we have not studied this aspect here.

For large-scale geophysical example problems, we have explored extensively the performance of the grid coarse space in [10, 11]. A few conclusions are stated below:

- The use of higher order finite elements allow us to minimise dispersion with a minimum number of ppwl, as shown in Figure 1. A good compromise is the choice of P3 finite elements for which, with 5 ppwl on unstructured meshes, we note a reduction by a factor 2 in the number of degrees of freedom with respect to a finite difference discretisation on uniform meshes.
- Local solves in domain decomposition methods are usually done by direct methods such as Cholesky factorisation, which is part of the setup phase ahead of the application of the GMRES method. We can already improve performance by replacing the Cholesky method with incomplete Cholesky factorisation.
- Precision is also important in the parsimony of the computation and the use of single precision highly decreases both the setup and solution times.

3 Can we improve on the auxiliary subspace preconditioner?

Let us consider the positive (or time-discretised) Maxwell equations

$$\begin{aligned} \nabla \times (\mu_r^{-1} \nabla \times \mathbf{u}) + \alpha \varepsilon_r \mathbf{u} &= \mathbf{f} && \text{in } \Omega, \\ \mathbf{u} \times \mathbf{n} &= 0 && \text{on } \partial\Omega. \end{aligned}$$

Here $\mathbf{u}$ is the vector-valued electric field, $\mathbf{f}$ is the source term, $\alpha > 0$ is a constant (e.g., stemming from the time discretisation), and μ_r and ε_r are electromagnetic parameters which are uniformly bounded and strictly positive but which we allow to be heterogeneous. We suppose Ω is a polyhedral computational domain and $\mathbf{n}$ is the outward normal to $\partial\Omega$. After discretisation by Nédélec elements we obtain

$$A\mathbf{U} := (K + \alpha M)\mathbf{U} = \mathbf{b}, \tag{4}$$

where $K \in \mathbb{R}^{n \times n}$ represents the discretisation of the curl–curl operator $\nabla \times (\mu_r^{-1} \nabla \times)$ and $M \in \mathbb{R}^{n \times n}$ is the ε_r-weighted mass matrix computed in the edge element space. Note that matrix K has a huge kernel (all the gradients of H^1 functions are part of the kernel of the curl operator) so designing efficient preconditioners for this problem can be challenging.

There is a well-established preconditioner in the literature known as the (nodal) auxiliary space preconditioner (ASP) [19] which is based on a splitting of the space, here $\mathbf{H}(\mathbf{curl})$, by isolating the kernel. The auxiliary space then uses a nodal (Lagrangian) discretisation. The preconditioner is given as

$$\mathcal{M}_{ASP}^{-1} = \operatorname{diag}(A)^{-1} + P(\tilde{L} + \alpha \tilde{Q})^{-1} P^T + \alpha^{-1} C L^{-1} C^T,$$

where $\tilde{L}$ is the nodal discretisation of the μ_r^{-1}-weighted vector Laplacian operator, $\tilde{Q}$ is the nodal ε_r-weighted vector mass matrix, P is the matrix form of the nodal interpolation operator between the Nédélec space and nodal element space, and C is the "gradient matrix", which is exactly the null space matrix of A here.

The spectral condition number $\kappa_2(\mathcal{M}_{ASP}^{-1} A)$ of the preconditioned problem is independent of the mesh size but might depend on any heterogeneities present. The natural question is then whether we can improve upon this preconditioner in the case of heterogeneous Maxwell problems?

In order to do this, we use extensively the fictitious space lemma (FSL) of Nepomnyaschickh, which can be considered the Lax–Milgram theorem of domain decomposition [23].

Lemma 1 (Nepomnyaschickh, 1991) *Consider two Hilbert spaces H and H_D along with positive symmetric bilinear forms $a\colon H \times H \to \mathbb{R}$ and $b\colon H_D \times H_D \to \mathbb{R}$. The operators A and B are defined as follows*

- *$A\colon H \to H$ such that $(Au, v) = a(u, v)$ for all $u, v \in H$;*
- *$B\colon H_D \to H_D$ such that $(Bu_D, v_D)_D = b(u_D, v_D)$ for all $u_D, v_D \in H_D$.*

Suppose we have a linear surjective operator $\mathcal{R}\colon H_D \to H$ satisfying the properties

- *Continuity:* $\exists c_R > 0$ *such that* $\forall u_D \in H_D$ *we have*

$$a(\mathcal{R}u_D, \mathcal{R}u_D) \le c_R b(u_D, u_D).$$

- *Stable decomposition:* $\exists c_T > 0$ *such that* $\forall u \in H$ $\exists u_D \in H_D$ *with* $\mathcal{R}u_D = u$ *and*

$$c_T b(u_D, u_D) \le a(\mathcal{R}u_D, \mathcal{R}u_D) = a(u, u).$$

Consider the adjoint operator $\mathcal{R}^* : H \to H_D$ *given by* $(\mathcal{R}u_D, u) = (u_D, \mathcal{R}^* u)_D$ *for all* $u_D \in H_D$ *and* $u \in H$*. Then for all* $u \in H$ *we have the spectral estimate*

$$c_T a(u, u) \le a\left(\mathcal{R}B^{-1}\mathcal{R}^* A u,\ u\right) \le c_R a(u, u).$$

Thus, the eigenvalues of the preconditioned operator $\mathcal{R}B^{-1}\mathcal{R}^* A$ *are bounded from below by* c_T *and from above by* c_R.

In this lemma we have a few ingredients: two Hilbert spaces with the associated scalar products (that are linked by the surjective operator $\mathcal{R}$) and two symmetric positive bilinear forms. The first of each comes from our problem while the second is for the preconditioner. Under the assumptions of continuity and stable decomposition, the spectral estimate tells us that the spectral condition number of the preconditioned problem is bounded solely in terms of the constants c_R and c_T.

Discretised problems which are perturbations of a singular operator, such as the Maxwell problem in (4) when α is small, have a huge near-kernel $G \subset \mathbb{R}^n$ of A, given by the gradient of all $H^1(\Omega)$ functions for example. This near-kernel will be within a space $V_G \subset \mathbb{R}^n$, which is the vector space spanned by the sequence $(R_i^T D_i R_i G)_{1 \le i \le N}$ so that $G \subset V_G$. These spaces may not be equal due to the fact that not all the elements of $R_i^T D_i R_i G$ are in G, for example, corresponding to the degrees of freedom for which D_i is not locally constant. Nevertheless, since the D_i are related to a partition of unity, we guarantee the inclusion. The space V_G can now serve as a “free” coarse space. We denote the coarse space $V_0 := V_G$ and let $Z \in \mathbb{R}^{n_0 \times n}$ be a rectangular matrix whose columns are a basis of V_0. The coarse space matrix is then defined in the usual way by $E = Z^T A Z$.

We now need to define all the other ingredients in the FSL. The second Hilbert space is the product space of vectors stemming, for example, from the n_i degrees of freedom on the local subdomains Ω_i and the n_0 coarse space vectors

$$H_D := \mathbb{R}^{n_0} \times \prod_{i=1}^{N} \mathbb{R}^{n_i}.$$

The bilinear form b for the preconditioner is given by the sum of local bilinear forms b_i and the coarse space contribution

$$b(\mathcal{U}, \mathcal{V}) := (E\mathbf{U}_0, \mathbf{V}_0) + \sum_{i=1}^{N} b_i(\mathbf{U}_i, \mathbf{V}_i), \qquad b_i(\mathbf{U}_i, \mathbf{V}_i) := (R_i A R_i^T \mathbf{U}_i, \mathbf{V}_i),$$

for $\mathcal{U} = (\mathbf{U}_0, (\mathbf{U}_i)_{1\le i\le N}) \in H_D, \mathcal{V} = (\mathbf{V}_0, (\mathbf{V}_i)_{1\le i\le N}) \in H_D$. Finally, the surjective operator $\mathcal{R}_{AS}: H_D \longrightarrow H$ corresponding to the additive Schwarz method is given by

$$\mathcal{R}_{AS}(\mathcal{U}) := Z\mathbf{U}_0 + (I - P_0)\sum_{i=1}^{N} R_i^T \mathbf{U}_i,$$

where P_0 is the A-orthogonal projection on the coarse space V_0. By applying the FSL we obtain a spectral condition number estimate $\kappa(M_{AS}^{-1}A) \le C$, with a bound C that can be large due to heterogeneities in the problem.

How can we improve this preconditioner in this case in order to be robust? We simply build a GenEO space from local generalised eigenproblems in the orthogonal complement of the "free" coarse space: find $(\mathbf{V}_{jk}, \lambda_{jk}) \in \mathbb{R}^{n_j} \setminus \{0\} \times \mathbb{R}$ such that

$$(I - \xi_{0j}^T) D_j R_j A R_j^T D_j (I - \xi_{0j}) \mathbf{V}_{jk} = \lambda_{jk} \widetilde{A}_j \mathbf{V}_{jk},$$

where ξ_{0j} denotes the b_j-orthogonal projection from $\mathbb{R}^{n_j}$ on $G_j = R_j G$ and $\widetilde{A}_j$ is the local Neumann matrix for the problem. We define $V_{j,geneo}^{\tau} \subset \mathbb{R}^n$ to be the vector space spanned by the family of vectors $(R_j^T D_j (I - \xi_{0j}) \mathbf{V}_{jk})_{\lambda_{jk} > \tau}$ corresponding to eigenvalues larger than a chosen threshold parameter τ. Now, collecting over all subdomains j, we let V_{geneo}^{τ} be the span of all $(V_{j,geneo}^{\tau})_{1\le j\le N}$, which will lead to a new coarse space

$$V_0 := V_G + V_{geneo}^{\tau}.$$

Applying the FSL now yields a spectral condition number estimate of the resulting two-level Schwarz method which is independent of the heterogeneity in the problem.

Several other variants of this approach can be formulated, including with the use of inexact coarse solves in order to more efficiently handle the large coarse space; these theoretical advances can be found in our recent preprint [7].

4 General conclusions

In this short paper we have offered a brief overview of the main difficulties and some recent solution methods now available to solve Helmholtz equations in the mid and high frequency regimes, which occur in many applications and especially in geophysics. Although there is no established method as the go-to solver, we have proposed a number of different strategies based on two-level domain decomposition methods where the second level comes from the solution of local spectral problems. Indeed, spectral coarse spaces have shown excellent theoretically-proven results for symmetric positive definite problems and currently offer very promising directions to explore for the Helmholtz equation and other wave propagation problems.

The discretisation here is also intertwined with the solution method as solvability and accuracy are very important for wave propagation problems. However, problems in applications do not need to be over-resolved (for example, in full waveform inversion for a discretisation by a finite difference method minimising dispersion 4 ppwl are enough) as this can lead to increasingly large problems whose size is not fully justified by practical reasons. Further, multi-frontal direct solvers based on block low rank approximations have been developed in recent years and problems as large as 50 million unknowns can be tackled successfully by these methods. In this sense, domain decomposition solvers need to be designed with the idea to go beyond these limits while keeping the applicative context in mind.

Last but not least, while not of the same nature, positive Maxwell's equations present different challenges. Here, the auxiliary space preconditioner has successfully been applied to problems where the underlying operator has an infinite-dimensional kernel. By exploiting the idea of subspace decomposition together with spectral methods of GenEO type, a new generation of preconditioners, capable of tackling heterogeneous problems, has been introduced. Future work includes an extensive numerical exploration of such an approach on realistic example problems.

Wave propagation problems have been a key source of difficult problems, not just for domain decomposition but more widely in scientific computing. As large-scale computing infrastructure continues to evolve and practitioners become ever more ambitious, often driven by industrial challenges, robustness will remain a central theme when designing algorithms for the future. Our work here then contributes some of the most recent ideas towards achieving such desired robustness for domain decomposition methods applied to challenging applications in wave propagation.

Acknowledgements The first two authors gratefully acknowledge support from the EPSRC grant EP/S004017/1. The fifth author acknowledges support from the Wind project[3] funded by Shell, Total and Chevron.

References

1. M. Ainsworth and H.A. Wajid. Optimally blended spectral-finite element scheme for wave propagation and nonstandard reduced integration. *SIAM J. Numer. Anal.* **48**(1), 346–371 (2010).
2. P. Amestoy, R. Brossier, A. Buttari, J.-Y. L'Excellent, T. Mary, L. Métivier, A. Miniussi and S. Operto. Fast 3D frequency-domain full-waveform inversion with a parallel block low-rank multifrontal direct solver: Application to OBC data from the North Sea. *Geophysics* **81**(6), R363–R383 (2016).
3. I.M. Babuška and S.A. Sauter. Is the pollution effect of the FEM avoidable for the Helmholtz equation considering high wave numbers? *SIAM J. Numer. Anal.* **34**(6), 2392–2423 (1997).
4. J. Blanch, J. Jarvis, C. Hurren, Y. Liu and L. Hu. Designing an exploration scale OBN: Acquisition design for subsalt imaging and velocity determination. In *SEG Technical Program Expanded Abstracts 2019*, pp. 192–196, Society of Exploration Geophysicists (2019).

[3] https://www.geoazur.fr/WIND/

5. M. Bonazzoli, V. Dolean, I.G. Graham, E.A. Spence and P.-H. Tournier. Domain decomposition preconditioning for the high-frequency time-harmonic Maxwell equations with absorption. *Math. Comp.* **86**, 2089–2127 (2017).
6. N. Bootland, V. Dolean, P. Jolivet and P.-H. Tournier. A comparison of coarse spaces for Helmholtz problems in the high frequency regime. *Comput. Math. Appl.* 98, 239–253, doi:10.1016/j.camwa.2021.07.011 (2021).
7. N. Bootland, V. Dolean, F. Nataf and P.-H. Tournier. Two-level DDM preconditioners for positive Maxwell equations. *arXiv preprint* arXiv:2012.02388 (2020).
8. L. Conen, V. Dolean, R. Krause and F. Nataf. A coarse space for heterogeneous Helmholtz problems based on the Dirichlet-to-Neumann operator. *J. Comput. Appl. Math.* **271**, 83–99 (2014).
9. B. Després. Domain decomposition method for the Helmholtz problem: *C. R. Math. Acad. Sci. Paris. I Math.* **311**(6), 313–316 (1990).
10. V. Dolean, P. Jolivet, P.-H. Tournier and S. Operto. Iterative frequency-domain seismic wave solvers based on multi-level domain-decomposition preconditioners. *82nd EAGE Annual Conference & Exhibition 2020* (1), 1–5 (2020).
11. V. Dolean, P. Jolivet, P.-H. Tournier and S. Operto. Large-scale frequency-domain seismic wave modeling on h-adaptive tetrahedral meshes with iterative solver and multi-level domain-decomposition preconditioners. *SEG Technical Program Expanded Abstracts 2020*, 2683–2688 (2020).
12. V. Dolean, P. Jolivet, P.-H. Tournier, L. Combe, S. Operto and S. Riffo. Large-scale finite-difference and finite-element frequency-domain seismic wave modelling with multi-level domain-decomposition preconditioner. *arXiv preprint* arXiv:2103.14921 (2021).
13. V. Dolean, F. Nataf, R. Scheichl and N. Spillane. Analysis of a two-level Schwarz method with coarse spaces based on local Dirichlet–to–Neumann maps. *Comput. Methods Appl. Math.* **12**(4), 391–414 (2012).
14. Y. Du and H. Wu. Preasymptotic error analysis of higher order FEM and CIP-FEM for Helmholtz equation with high wave number. *SIAM J. Numer. Anal.* **53**(2), 782–804 (2015).
15. O.G. Ernst and M.J. Gander. Why it is difficult to solve Helmholtz problems with classical iterative methods. In: *Numerical Analysis of Multiscale Problems*, I. Graham, T. Hou, O. Lakkis, R. Scheichl (eds), LNCSE, vol 83, Springer, Berlin, Heidelberg (2012).
16. M.J. Gander and H. Zhang. A class of iterative solvers for the Helmholtz equation: Factorizations, sweeping preconditioners, source transfer, single layer potentials, polarized traces, and optimized Schwarz methods. *SIAM Rev.* **61**(1), 3–76 (2019).
17. I.G. Graham, E.A. Spence and E. Vainikko. Domain decomposition preconditioning for high-frequency Helmholtz problems with absorption. *Math. Comp.* **86**, 2089–2127 (2017).
18. I.G. Graham, E.A. Spence and J. Zou. Domain Decomposition with local impedance conditions for the Helmholtz equation with absorption. *SIAM J. Numer. Anal.* **58**(5), 2515–2543 (2020).
19. R. Hiptmair and J. Xu. Nodal auxiliary space preconditioning in $\mathbf{H}(\mathbf{curl})$ and $\mathbf{H}(\mathrm{div})$ spaces. *SIAM J. Numer. Anal.* **45**(6), 2483–2509 (2007).
20. T. Mary. *Block low-rank multifrontal solvers: complexity, performance and scalability*. PhD Thesis, Université de Toulouse (2017).
21. J.M. Melenk and S.A. Sauter. Wavenumber explicit convergence analysis for Galerkin discretizations of the Helmholtz equation. *SIAM J. Numer. Anal.* **49**(3), 1210–1243 (2011).
22. F. Nataf, H. Xiang, V. Dolean and N. Spillane. A coarse space construction based on local Dirichlet-to-Neumann maps. *SIAM J. Sci. Comput.* **33**(4), 1623–1642 (2011).
23. S.V. Nepomnyaschikh. Mesh theorems of traces, normalizations of function traces and their inversions. *Sov. J. Numer. Anal. Math. Modeling* **6**(3), 223–242 (1991).
24. N. Spillane, V. Dolean, P. Hauret, F. Nataf, C. Pechstein and R. Scheichl. Abstract robust coarse spaces for systems of PDEs via generalized eigenproblems in the overlaps. *Numer. Math.* **126**(4), 741–770 (2014).
25. R. Versteeg. The Marmousi experience: Velocity model determination on a synthetic complex data set. *The Leading Edge* **13**(9), 927–936, doi:10.1190/1.1437051 (1994).

Scalable Hybrid TFETI-DP Methods for Large Boundary Variational Inequalities

Zdeněk Dostál, Tomáš Brzobohatý, David Horák, Jakub Kružík and Oldřich Vlach

1 Introduction

Variants of the FETI (finite element tearing and interconnecting) methods introduced by Farhat and Roux [8] belong to the most powerful methods for the massively parallel solution of large discretized elliptic partial differential equations. The basic idea is to decompose the domain into subdomains connected by Lagrange multipliers and then eliminate the primal variables to get a small coarse problem and local problems that can be solved in parallel. If applied to variational inequalities, the procedure simultaneously transforms the general inequality constraints into bound constraints. This simple observation and development of specialized quadratic programming algorithms [2] with optimal convergence rate have been at the heart of the generalization of the classical scalability results to variational inequalities [4]. The algorithms have been applied to solve contact problems discretized by billions of nodal variables [7].

Z. Dostál
Department of Applied Mathematics and IT4Innovations - National Supercomputing Center, VSB-Technical University of Ostrava, 17. listopadu 15, Czech Republic,
e-mail. zdenek.dostal@vsb.cz

T. Brzobohatý
IT4Innovations - National Supercomputing Center, VSB-Technical University of Ostrava, Ostrava, Czech republic, e-mail: tomas.brzobohaty@vsb.cz

D. Horák
Department of Applied Mathematics, VSB-Technical University of Ostrava, Ostrava, Czech republic, and Institute of Geonics ASCR, Ostrava, e-mail: david.horak@vsb.cz

J. Kružík
Department of Applied Mathematics, VSB-Technical University of Ostrava, Ostrava, Czech republic, and Institute of Geonics ASCR, Ostrava, e-mail: jakub.kruzik@vsb.cz

O. Vlach
Department of Applied Mathematics and IT4Innovations - National Supercomputing Center, VSB-Technical University of Ostrava, Czech Republic, e-mail: oldrich.vlach@vsb.cz

S. C. Brenner et al. (eds.), *Domain Decomposition Methods in Science and Engineering XXVI*, Lecture Notes in Computational Science and Engineering 145,
https://doi.org/10.1007/978-3-030-95025-5_3

The bottleneck of the original FETI is caused by the coarse problem, which has a dimension which is proportional to the number of subdomains. The coarse problem is typically solved by a direct solver – its cost is negligible for a small number of subdomains. However, it starts to dominate when the number of subdomains is large, currently some tens of thousands of subdomains.

Here we introduce a model problem, the semi-coercive scalar variational inequality, describe its discretization and decomposition into subdomains and clusters, reduce the problem by duality to bound and equality constrained problems, give results on numerical scalability of the algorithms, and demonstrate their performance by numerical experiments. The analysis uses recently proved bounds on the spectrum of the Schur complements of the clusters interconnected by edge/face averages. The bounds for 2D and 3D scalar problems have been published in [5] and [3]; the development of the theory for elasticity is in progress. The results extend the scope of scalability of powerful massively parallel algorithms for the solution of variational inequalities [7] and show the unique efficiency of H-TFETI-DP coarse grid split between the primal and dual variables. We illustrate the analysis on a simple model problem but also include numerical experiments with 3D elastic contact problem with the clusters interconnected by average face rigid body motions.

Throughout the paper, we use the following notation. For any matrix $\mathbf{A} \in \mathbb{R}^{m\times n}$ and subsets $\mathcal{I} \subseteq \{1,\ldots,m\}$ and $\mathcal{J} \subseteq \{1,\ldots,n\}$, we denote by $\mathbf{A}_{\mathcal{I}\mathcal{J}}$ a submatrix of $\mathbf{A}$ with the rows $i \in \mathcal{I}$ and columns $j \in \mathcal{J}$. If $m = n$ and $\mathbf{A} = \mathbf{A}^T$, then $\lambda_i(\mathbf{A})$, $\lambda_{\min}(\mathbf{A})$, $\lambda_{\max}(\mathbf{A})$ denote the eigenvalues of $\mathbf{A}$,

$$\lambda_{\max}(\mathbf{A}) = \lambda_1(\mathbf{A}) \geq \lambda_2(\mathbf{A}) \geq \cdots \geq \lambda_n(\mathbf{A}) = \lambda_{\min}(\mathbf{A}).$$

The smallest nonzero eigenvalue of $\mathbf{A}$ is denoted by $\overline{\lambda}_{\min}(\mathbf{A})$. The Euclidean norm and zero vector are denoted by $\|\cdot\|$ and $\mathbf{o}$, respectively.

2 Model problem

For simplicity, we shall reduce our analysis to a simple model problem, but our reasoning is also valid for more general cases. Let $\Omega = \Omega^1 \cup \Omega^2$, where $\Omega^1 = (0,1)\times(0,1)$ and $\Omega^2 = (1,2)\times(0,1)$ denote square domains with the boundaries Γ^1, Γ^2; their parts Γ_u^i, Γ_f^i, Γ_c^i are formed by the sides of Ω^i, $i = 1, 2$, as in Fig. 1.

Let $H^1(\Omega^i)$, $i = 1,2$, denote the subspace of $L^2(\Omega^i)$ of elements with the first derivatives in $L^2(\Omega^i)$. Let

$$V^i = \left\{v^i \in H^1(\Omega^i) : v^i = 0 \quad \text{on} \quad \Gamma_u^i\right\}$$

denote closed subspaces of $H^1(\Omega^i)$, let $\mathcal{H} = H^1(\Omega^1)\times H^1(\Omega^2)$, and let

$$V = V^1 \times V^2 \qquad \text{and} \quad \mathcal{K} = \left\{(v^1, v^2) \in V : v^2 - v^1 \geq 0 \quad \text{on} \quad \Gamma_c\right\}$$

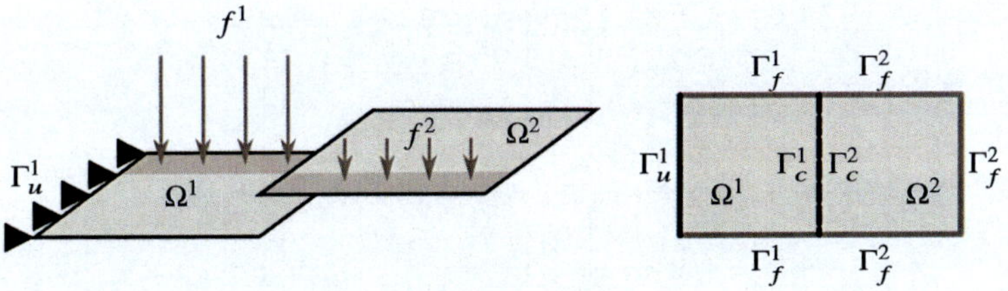

Fig. 1: Coercive model problem (left) and boundary conditions (right).

denote a closed subspace and a closed convex subset of $\mathcal{H}$, respectively. The relations on the boundaries are in terms of traces. On $\mathcal{H}$, we define a symmetric bilinear form

$$a(u,v) = \sum_{i=1}^{2} \int_{\Omega^i} \left(\frac{\partial u^i}{\partial x}\frac{\partial v^i}{\partial x} + \frac{\partial u^i}{\partial y}\frac{\partial v^i}{\partial y} \right) \mathrm{d}\Omega$$

and a linear form

$$\ell(v) = \sum_{i=1}^{2} \int_{\Omega^i} f^i v^i \mathrm{d}\Omega,$$

where $f^i \in L^2(\Omega^i)$, $i = 1, 2$, are nonzero and nonpositive. Thus we can define the problem of finding

$$\min \quad q(u) = \frac{1}{2}a(u,u) - \ell(u) \quad \text{subject to} \quad u \in \mathcal{K}. \tag{1}$$

The solution of the model problem can be interpreted as the displacement of two membranes under the traction f. The left edge of the right membrane cannot penetrate below the right edge of the left membrane.

3 Domain decomposition and discretization

To enable efficient application of domain decomposition methods, we optionally decompose each Ω^i into $p = 1/H_s \times 1/H_s, i = 1, 2,$ square subdomains. Misusing the notation a little, we assign to each subdomain of Ω^1 a unique number $i \in \{1, \ldots, p\}$ and to each subdomain of Ω^2 a unique number $i \in \{p+1, \ldots, s\}$, $s = 2p$, as in Fig. 2. We call H_s a *decomposition parameter*.

To get a variational formulation of the decomposed problem, let

$$V_D^i = \{v^i \in H^1(\Omega^i) : v^i = 0 \quad \text{on} \quad \Gamma_U \cap \Gamma^i\}, \quad i = 1, \ldots s,$$

denote the closed subspaces of $H^1(\Omega^i)$, and let

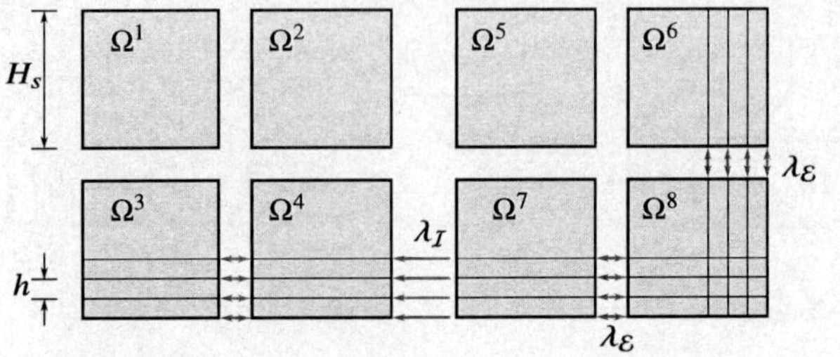

Fig. 2: Domain decomposition and discretization.

$$V_D = V_D^1 \times \cdots \times V_D^s,$$
$$\mathcal{K}_D^C = \left\{ v \in V_D : v^j - v^i \geq 0 \text{ on } \Gamma_C^1 \cap \Gamma_C^2, \quad i \leq p < j \right\},$$
$$\mathcal{K}_D = \left\{ v \in \mathcal{K}_D^C : v^i = v^j \text{ on } \Gamma^{ij} \right\}, \quad \Gamma^{ij} = \overline{\Gamma}^i \cap \overline{\Gamma}^j, \quad i, j \leq p \text{ or } \quad i, j > p.$$

On V_D, we define the scalar product

$$(u, v)_D = \sum_{i=1}^{s} \int_{\Omega^i} u^i v^i \mathrm{d}\Omega,$$

and the forms

$$a_D(u, v) = \sum_{i=1}^{s} \int_{\Omega^i} \left(\frac{\partial u^i}{\partial x_1} \frac{\partial v^i}{\partial x_1} + \frac{\partial u^i}{\partial x_2} \frac{\partial v^i}{\partial x_2} \right) \mathrm{d}\Omega \quad \text{and} \quad \ell_D(v) = (f, v)_D.$$

Using the above notation, it is a standard exercise [7, Sect. 10.2] to prove that (1) is equivalent to the problem of finding $u \in \mathcal{K}_D$ such that

$$q_D(u) \leq q_D(v), \quad q_D(v) = \frac{1}{2} a_D(v, v) - \ell_D(v), \quad v \in \mathcal{K}_D. \tag{2}$$

After introducing regular grids with the discretization parameter h in the subdomains Ω^i (see Fig. 2), and using Lagrangian finite elements for the discretization, we get the discretized version of problem (2) with auxiliary domain decomposition

$$\min \frac{1}{2} \mathbf{u}^T \mathbf{K} \mathbf{u} - \mathbf{f}^T \mathbf{u} \quad \text{s.t.} \quad \mathbf{B}_I \mathbf{u} \leq \mathbf{o} \quad \text{and} \quad \mathbf{B}_E \mathbf{u} = \mathbf{o}. \tag{3}$$

In (3), $\mathbf{K} \in \mathbb{R}^{n \times n}$ denotes a block diagonal SPS (symmetric positive semidefinite) stiffness matrix, the full rank matrices $\mathbf{B}_I$ and $\mathbf{B}_E$ describe the non-penetration and interconnecting conditions, respectively, and $\mathbf{f}$ represents the discretized linear form $\ell_D(u)$. We can write the stiffness matrix and the vectors in the block form

$$\mathbf{K} = \begin{bmatrix} \mathbf{K}_1 & \mathbf{O} & \dots & \mathbf{O} \\ \mathbf{O} & \mathbf{K}_2 & \dots & \mathbf{O} \\ \dots & \dots & \dots & \dots \\ \mathbf{O} & \mathbf{O} & \dots & \mathbf{K}_s \end{bmatrix}, \quad \mathbf{u} = \begin{bmatrix} \mathbf{u}_1 \\ \dots \\ \mathbf{u}_s \end{bmatrix}, \quad \mathbf{f} = \begin{bmatrix} \mathbf{f}_1 \\ \dots \\ \mathbf{f}_s \end{bmatrix}, \quad s = 2p.$$

After a suitable scaling of the rows of $\mathbf{B} = [\mathbf{B}_E^T, \mathbf{B}_I^T]^T$, we can achieve $\mathbf{B}\mathbf{B}^T = \mathbf{I}$.

4 TFETI problem

To reduce the problem to the subdomain boundaries using duality theory, let us introduce the Lagrangian associated with problem (3) by

$$L(\mathbf{u}, \lambda_I, \lambda_E) = \frac{1}{2}\mathbf{u}^T\mathbf{K}\mathbf{u} - \mathbf{f}^T\mathbf{u} + \lambda_I^T\mathbf{B}_I\mathbf{u} + \lambda_E^T\mathbf{B}_E\mathbf{u}, \tag{4}$$

where λ_I and λ_E are the Lagrange multipliers associated with the inequalities and equalities, respectively. Introducing the notation

$$\lambda = \begin{bmatrix} \lambda_I \\ \lambda_E \end{bmatrix} \quad \text{and} \quad \mathbf{B} = \begin{bmatrix} \mathbf{B}_I \\ \mathbf{B}_E \end{bmatrix},$$

we can observe that $\mathbf{B} \in \mathbb{R}^{m\times n}$ is a full rank matrix and write the Lagrangian as

$$L(\mathbf{u}, \lambda) = \frac{1}{2}\mathbf{u}^T\mathbf{K}\mathbf{u} - \mathbf{f}^T\mathbf{u} + \lambda^T\mathbf{B}\mathbf{u}.$$

The solution satisfies the KKT conditions, including

$$\mathbf{K}\mathbf{u} - \mathbf{f} + \mathbf{B}^T\lambda = \mathbf{o}. \tag{5}$$

Equation (5) has a solution if and only if $\mathbf{f} - \mathbf{B}^T\lambda \in \mathrm{Im}\mathbf{K}$, which can be expressed by means of a matrix $\mathbf{R}$ the columns of which span the null space of $\mathbf{K}$ as

$$\mathbf{R}^T(\mathbf{f} - \mathbf{B}^T\lambda) = \mathbf{o}. \tag{6}$$

The matrix $\mathbf{R}$ can be formed directly, and $\mathbf{R}^T\mathbf{B}^T$ is non-singular.

Now assume that λ satisfies (6), so that we can evaluate λ from (5) by means of any (left) generalized inverse matrix $\mathbf{K}^+$ which satisfies $\mathbf{K}\mathbf{K}^+\mathbf{K} = \mathbf{K}$. We can verify directly that if $\mathbf{u}$ solves (5), then there is a vector α such that

$$\mathbf{u} = \mathbf{K}^+(\mathbf{f} - \mathbf{B}^T\lambda) + \mathbf{R}\alpha. \tag{7}$$

After eliminating the primal variables $\mathbf{u}$, we can find λ by solving

$$\min\ \theta(\lambda) \quad \text{s.t.} \quad \lambda_I \geq \mathbf{o} \quad \text{and} \quad \mathbf{R}^T(\mathbf{f} - \mathbf{B}^T\lambda) = \mathbf{o}, \tag{8}$$

where

$$\theta(\lambda) = \frac{1}{2}\lambda^T \mathbf{BK}^+\mathbf{B}^T\lambda - \lambda^T\mathbf{BK}^+\mathbf{f}. \tag{9}$$

Once the solution $\widehat{\lambda}$ of (8) is known, $\widehat{\mathbf{u}}$ (3) can be evaluated by (7) and

$$\alpha = -(\mathbf{R}^T\widehat{\mathbf{B}}^T\widehat{\mathbf{B}}\mathbf{R})^{-1}\mathbf{R}^T\widetilde{\mathbf{B}}^T\widehat{\mathbf{B}}\mathbf{K}^+(\mathbf{f} - \mathbf{B}^T\widehat{\lambda}), \tag{10}$$

where $\widehat{\mathbf{B}} = [\widehat{\mathbf{B}}_I^T, \mathbf{B}_E^T]^T$, and the matrix $\widehat{\mathbf{B}}_I$ is formed by the rows $\mathbf{b}_i$ of $\mathbf{B}_I$ that correspond to the positive components of the solution $\widehat{\lambda}_I$ characterized by $\widehat{\lambda}_i > 0$. A more effective procedure avoiding manipulation with $\widehat{\mathbf{B}}$ can be found in [9].

To proceed further, let us define

$$\begin{aligned} \mathbf{F} = \mathbf{BK}^+\mathbf{B}^T = \widetilde{\mathbf{B}}\mathbf{S}^+\widetilde{\mathbf{B}}^T, \quad & \widetilde{\mathbf{d}} = \mathbf{BK}^+\mathbf{f}, \\ \widetilde{\mathbf{G}} = \mathbf{R}^T\mathbf{B}^T, \quad & \widetilde{\mathbf{e}} = \mathbf{R}^T\mathbf{f}, \end{aligned}$$

and let $\mathbf{T}$ denote a matrix that defines orthonormalization of the rows of $\widetilde{\mathbf{G}}$ so that the matrix $\mathbf{G} = \mathbf{T}\widetilde{\mathbf{G}}$ has orthonormal rows. After introducing $\mathbf{e} = \mathbf{T}\widetilde{\mathbf{e}}$, problem (8) reads

$$\min \quad \frac{1}{2}\lambda^T\mathbf{F}\lambda - \lambda^T\widetilde{\mathbf{d}} \quad \text{s.t.} \quad \lambda_I \geq \mathbf{o} \quad \text{and} \quad \mathbf{G}\lambda = \mathbf{e}. \tag{11}$$

After homogenization of the equality constraints and introducing orthogonal projectors, problem (11) turns into

$$\min \quad \overline{\theta}_\varrho(\lambda) \quad \text{s.t.} \quad \mathbf{G}\lambda = \mathbf{o} \quad \text{and} \quad \lambda_I \geq -\widetilde{\lambda}_I, \tag{12}$$

where ϱ is a positive constant, $\mathbf{G}\lambda_I = \mathbf{e}$, and

$$\overline{\theta}_\varrho(\lambda) = \frac{1}{2}\lambda^T\mathbf{H}_\varrho\lambda - \lambda^T\mathbf{Pd}, \quad \mathbf{H}_\varrho = \mathbf{PFP} + \varrho\mathbf{Q}, \quad \mathbf{Q} = \mathbf{G}^T\mathbf{G}, \quad \mathbf{P} = \mathbf{I} - \mathbf{Q}.$$

The matrices $\mathbf{P}$ and $\mathbf{Q}$ are the orthogonal projectors onto Ker$\mathbf{G}$ and Im$\mathbf{G}^T$, respectively. It has been proved (see, e.g., Brenner [1] or Pechstein [16]) that there are constants $0 < c < C$ that depend neither on h nor on H such that

$$c \leq \lambda_{\min}(\mathbf{H}_\varrho) \leq \max\{CH/h, \varrho\}.$$

5 Connecting subdomains into clusters

The bottleneck of classical FETI methods is the rank d of the projector $\mathbf{Q}$ which is equal to the defect of stiffness matrix $\mathbf{K}$, in our case $d = s$. To reduce the rank of $\mathbf{Q}$, we use the idea of Klawonn and Rheinbach [11] to interconnect some subdomains on the primal level into clusters so that the defect of the stiffness matrix of the cluster is equal to the defect of one of the subdomain stiffness matrices.

For example, to couple adjacent subdomains with common corners $\mathbf{x}, \mathbf{y} \in \overline{\Omega}^i \cap \overline{\Omega}^j$, we can transform the nodal variables associated with $\widetilde{\Omega}^q = \overline{\Omega}^i \times \overline{\Omega}^j$ by the expansion matrix $\mathbf{L}^q$ obtained by replacing two columns of the identity matrix associated with $\mathbf{x}, \mathbf{y}$ by one column obtained as a normalized sum of the columns associated with the displacements of nodes $\mathbf{x}$ and $\mathbf{y}$. Feasible variables $\mathbf{u}^q$ of the cluster are related to global variables $\widetilde{\mathbf{u}}^q$ by $\mathbf{u}^q = \mathbf{L}^q \widetilde{\mathbf{u}}^q$ and the stiffness matrix $\widetilde{\mathbf{K}}^q$ of such a cluster in global variables can be obtained by

$$\widetilde{\mathbf{K}}^q = (\mathbf{L}^q)^T \mathrm{diag}(\mathbf{K}^i, \mathbf{K}^j) \mathbf{L}^q.$$

Let us denote by $\mathbf{e}$ and $\overline{\mathbf{e}}$ the vectors with all components equal to 1 and $1/\|\mathbf{e}\|$, respectively. To describe the coupling by averages, we use the transformation of bases proposed by Klawonn and Widlund [12], see also Klawonn and Rheinbach [10] and Li and Widlund [14]. The basic idea is a rather trivial observation that if

$$[\mathbf{C}\, \overline{\mathbf{e}}] = [\mathbf{c}_1, \ldots, \mathbf{c}_{p-1}, \overline{\mathbf{e}}], \quad \overline{\mathbf{e}} = \frac{1}{\sqrt{p}} \mathbf{e},$$

denote an orthonormal basis of $\mathbb{R}^p$, then the last coordinate of a vector $\mathbf{x} \in \mathbb{R}^p$ in this basis is given by $x_p = \overline{\mathbf{e}}^T \mathbf{x}$. If we apply the transformation to the variables associated with the interiors of adjacent edges, we can join them by the expansion mapping $\mathbf{L}$ as above.

The procedure can be generalized to specify the feasible vectors of any cluster connected by the edge averages of adjacent edges. Using a proper numbering of variables by subdomains, in each subdomain setting first the variables that are not affected by the interconnecting, then the variables associated with the averages ordered by edges, we get the matrix $\mathbf{Z}$ with orthonormal columns the range of which represents the feasible displacements of the cluster,

$$\mathbf{Z} = \left[\, \mathbf{C}\, \mathbf{E} \,\right], \ \mathbf{C} = \mathrm{diag}(\mathbf{C}^1, \ldots \mathbf{C}^s), \ \mathbf{E} = 1/\sqrt{2} \begin{bmatrix} \cdot & & \cdot & \cdot \\ \cdot & \overline{\mathbf{e}}^{ij} & & \cdot \\ \cdot & & \cdot & \cdot \\ \cdot & \overline{\mathbf{e}}^{ji} & & \cdot \\ \cdot & & \cdot & \cdot \end{bmatrix}, \ (i, j) \in C. \tag{13}$$

In (13), C denotes a set of ordered couples of indices (i, j), $i < j$, s here denotes the number of subdomains in the cluster, and $\overline{\mathbf{e}}^{ij}$ denotes the basis vectors associated with the edge averages. Each couple $(i, j) \in C$ defines the connection of the adjacent edges of Ω^i and Ω^j by averages. The procedure is very similar to that described in the introduction of this section; the only difference is that we replace the expansion matrix $\mathbf{L}^q$ by the basis of feasible displacements of the cluster $\mathbf{Z}^q$. The feasible variables of the cluster are related to global variables $\widetilde{\mathbf{u}}^q$ by

$$\mathbf{u}^q = \mathbf{Z}^q \widetilde{\mathbf{u}}^q$$

and the Schur complement $\widetilde{\mathbf{S}}^q$ of such a cluster in global variables can be obtained by

$$\widetilde{\mathbf{S}}^q = (\mathbf{Z}^q)^T \mathrm{diag}(\mathbf{S}^i, \mathbf{S}^j, \ldots, \mathbf{S}^\ell)\mathbf{Z}^q.$$

Assuming that the set of all subdomains is decomposed into c clusters interconnected by the edge averages, we can use the global transformation matrix with orthonormal columns

$$\mathbf{Z} = \mathrm{diag}(\mathbf{Z}^1, \cdots, \mathbf{Z}^c)$$

to connect the groups of $m \times m$ subdomains into clusters to get the stiffness matrix

$$\widetilde{\mathbf{S}} = \mathbf{Z}^T \mathbf{S}\mathbf{Z} = \mathrm{diag}(\widetilde{\mathbf{S}}^1, \cdots \widetilde{\mathbf{S}}^c)$$

and the matrices

$$\widetilde{\mathbf{B}}, \quad \widetilde{\mathbf{R}} = \mathrm{diag}(\widetilde{\mathbf{e}}^1, \ldots, \widetilde{\mathbf{e}}^c), \quad \widetilde{\mathbf{G}} = \mathbf{T}\widetilde{\mathbf{R}}^T \widetilde{\mathbf{B}}^T,$$

where $\widetilde{\mathbf{B}}$ denotes a matrix that enforces interconnecting constraints that are not enhanced on the primal level and $\mathbf{T}$ denotes an orthogonalization matrix so that $\widetilde{\mathbf{G}}\widetilde{\mathbf{G}}^T = \mathbf{I}$. It is easy to achieve that

$$\widetilde{\mathbf{B}}\widetilde{\mathbf{B}}^T = \mathbf{I}. \tag{14}$$

Notice that $\widetilde{\mathbf{B}}$ enforces both constraints that connect subdomains into clusters and those connecting the clusters. Moreover, Ker$\widetilde{\mathbf{B}}$ =Ker$\mathbf{BZ}$, but $\mathbf{BZ}$ need not have orthonormal rows. Using the above transformation, we reduced problem (12) to

$$\min \quad \widetilde{\theta}_\varrho(\lambda) \quad \text{s.t.} \quad \widetilde{\mathbf{G}}\lambda = \mathbf{o} \quad \text{and} \quad \lambda_I \geq -\widetilde{\lambda}_I, \tag{15}$$

where ϱ is a positive constant and

$$\widetilde{\theta}_\varrho(\lambda) = \frac{1}{2}\lambda^T \widetilde{\mathbf{H}}_\varrho \lambda - \lambda^T \widetilde{\mathbf{P}}\widetilde{\mathbf{d}}, \tag{16}$$

$$\widetilde{\mathbf{H}}_\varrho = \widetilde{\mathbf{P}}\widetilde{\mathbf{F}}\widetilde{\mathbf{P}} + \varrho\widetilde{\mathbf{Q}}, \quad \widetilde{\mathbf{Q}} = \widetilde{\mathbf{G}}^T\widetilde{\mathbf{G}}, \quad \widetilde{\mathbf{P}} = \mathbf{I} - \widetilde{\mathbf{Q}} \quad \widetilde{\mathbf{F}} = \widetilde{\mathbf{B}}\widetilde{\mathbf{S}}^+\widetilde{\mathbf{B}}^T.$$

$\widetilde{\mathbf{P}}$ and $\widetilde{\mathbf{Q}}$ are the orthogonal projectors onto Ker $\widetilde{\mathbf{G}}$ and Im $\widetilde{\mathbf{G}}^T$, respectively.

Notice that the number of the rows of $\mathbf{G}$ is m^2 times larger than that of $\widetilde{\mathbf{G}}$, so that the cost of $(\mathbf{G}^T\mathbf{G})^{-1}$ is about m^4 times larger than that of $(\widetilde{\mathbf{G}}\widetilde{\mathbf{G}})^{-1}$.

6 Bounds on the spectrum of $\widetilde{\mathbf{H}}_\varrho$ and optimality

Using that Im $\widetilde{\mathbf{P}}$ and Im $\widetilde{\mathbf{Q}}$ are invariant subspaces of $\widetilde{\mathbf{H}}_\varrho$, it is easy to check that

$$\min\{\overline{\lambda}_{\min}(\widetilde{\mathbf{P}}\widetilde{\mathbf{F}}\widetilde{\mathbf{P}}), \varrho\} \leq \lambda_i(\widetilde{\mathbf{H}}_\varrho) \leq \max\{\|\widetilde{\mathbf{F}}\|, \varrho\}. \tag{17}$$

Applying standard arguments (see, e.g., [5, Lemma 3.1]), it is easy to reduce the problem of finding bounds on the spectrum of $\widetilde{\mathbf{H}}_\varrho$ to the problem of finding bounds on the spectrum of $\widetilde{\mathbf{S}}_i$. Some bounds were proved recently (see [5]):

Theorem 1 *For each integer $m > 1$, let $\widetilde{\mathbf{S}}$ denote the Schur complement of the cluster with the side-length H_c comprising $m \times m$ square subdomains of the side-length $H_s = H_c/m$. Let the subdomains be discretized by a regular grid with the step-length h and interconnected by the edge averages. Let $\overline{\lambda}_{\min}(\mathbf{S})$ denote the smallest nonzero eigenvalue of*

$$\mathbf{S} = \operatorname{diag}(\mathbf{S}^1, \ldots, \mathbf{S}^{m^2}),$$

where $\mathbf{S}^i$ denote the Schur complements of the subdomain stiffness matrices $\mathbf{K}^i$, $i = 1, \ldots, m^2$, with respect to the interior variables. Then

$$\|\mathbf{S}\| = \lambda_{\max}(\mathbf{S}) \geq \lambda_{\max}(\widetilde{\mathbf{S}}), \tag{18}$$

$$\overline{\lambda}_{\min}(\mathbf{S}) \geq \overline{\lambda}_{\min}(\widetilde{\mathbf{S}}) \geq \frac{2n_e}{n_s}\overline{\lambda}_{\min}(\mathbf{S}) \sin^2\left(\frac{\pi}{2m}\right) \approx \frac{1}{2}\overline{\lambda}_{\min}(\mathbf{S}) \left(\frac{\pi}{2m}\right)^2. \tag{19}$$

The spectrum of $\mathbf{S}$ can be bounded in terms of the decomposition and discretization parameters H_s and h, respectively – there are positive constants c, C such that

$$ch/H_s \leq \overline{\lambda}_{\min}(\mathbf{S}) \leq \|\mathbf{S}\| \leq C. \tag{20}$$

For the proof, see Pechstein [16, Lemma 1.59] or Brenner [1]. Since there are algorithms that can solve (15) with a rate of convergence that depends on the bounds on the spectrum of $\widetilde{\mathbf{H}}_\varrho$, we can formulate the following theorem.

Theorem 2 *Let $\varrho \approx \|\widetilde{\mathbf{F}}\|$ and let the parameters H_s, m, and h specify problem (15). Then there are constants $c, C > 0$ independent of H_s, m, h such that*

$$c \leq \lambda_{\min}(\widetilde{\mathbf{H}}_\varrho) \leq \|\widetilde{\mathbf{H}}_\varrho\| \leq CmH_s/h. \tag{21}$$

Moreover, there is a constant $M_{\max}$ such that if $C_1 > 2$ is an arbitrary constant and

$$mH_s/h \leq C_1,$$

then the SMALBE-M algorithm [7, Chap. 9] with the inner loop implemented by MPRGP [7, Chap. 8] can find an approximate solution of any problem (15) generated by the parameters H_s, m, h in at most $M_{\max}$ matrix–vector multiplications.

The proof is similar to the proof of optimality of TFETI for a variational inequality [7, Sect. 10.8] or contact problems [7, Sect. 11.10].

7 Numerical experiments

We carried out some numerical experiments to check the bounds and compare H-TFETI-DP with TFETI for both linear and non-linear problems. In all experiments, we use the relative precision stopping criterion with $\varepsilon = 10^{-4}$.

7.1 Comparing estimate and experiments

To compare estimates (19) with the real values, we have computed [5] the bounds on the extreme nonzero eigenvalues of the Schur complements of $m \times m$ clusters joined by edge averages using $m \in \{2, 4, 8, 16\}$, $H_c = 1$, $H_s = 1/m$, and $h = 1/64$. Some of the results are in Table 1. The results comply with those carried out by Klawonn and Rheinbach [11] and Lee [13].

Table 1: Regular condition number and extreme nonzero eigenvalues – edge averages.

m	2	4	8	16
H_s/h	32	16	8	4
$\overline{\lambda}_{\max}(\widetilde{\mathsf{S}})$	2.8235	2.8098	2.7638	2.6843
$\overline{\lambda}_{\min}(\widetilde{\mathsf{S}})$	0.0173	0.093	0.047	0,0022
$\overline{\lambda}_{\min}^{\mathrm{est}}(\widetilde{\mathsf{S}})$	0.0104	0.059	0.029	0.0012

7.2 Comparing linear unpreconditioned H-TFETI-DP and TFETI

We compared H-TFETI-DP with standard TFETI on the unit square Poisson benchmark discretized by Q1 finite elements on regular grid with parameters h and H_s, $H_s/h = 100$ [5]. We used the ESPRESO (ExaScale PaRallel FETI SOlver) package [15] developed at the Czech National Supercomputing Center in Ostrava. The domain was decomposed into $n_c \times n_c$ clusters, $n_c = 6, 18, 54$, each cluster comprising 15×15 square subdomains joined by edge averages. Notice that H-TFETI-DP outperforms TFETI due to the small coarse problem and cheap iterations.

Table 2: Billion variables Poisson - unpreconditioned H-TFETI-DP and TFETI, $m = 15$, see [5].

Clusters	Subdomains	Cores	Unknowns	H-TFETI-DP (iter/sec)	TFETI (iter/sec)
36	8,100	108	81,018,001	117/26.0	45/14.5
324	72,900	972	729,054,001	118/27.7	42/40.2
2,916	656,100	8.748	6,561,162,001	116/28.0	41/61.0

7.3 Model variational inequality and elastic body on rigid obstacle

We used the above procedure to get the discretized H-TFETI-DP QP problem (12) that we solved by a combination of the SMALBE-M (semimonotonic augmented Lagrangian) [7, Chap. 9] algorithm with the inner loop resolved by MPRGP (modified proportioning with reduced gradient projection) [7, Chap. 8]. We implemented both algorithms in the PETSc based package PERMON [6] developed at the Department of Applied Mathematics of the VSB-Technical University of Ostrava and the Institute of Geonics of the Czech Academy of Science.

Table 3: Semicoercive variational inequality, primal dimension 20,480,000, inequalities 3169.

$m=$	1	2	4	8
outer iter	52	25	16	12
matrix × vector	243	252	186	218
coarse problem dimension	2048	512	128	32

Our final benchmark is a clamped cube over a sinus-shape obstacle as in Fig. 3, loaded by own weight, decomposed into $4 \times 4 \times 4$ clusters, $H_s/h = 14$, using the ESPRESO [15] implementation of H-TFETI-DP for contact problems. We can see that TFETI needs a much smaller number of iterations, but H-TFETI-DP is still faster due to 64-times smaller coarse space and better exploitation of the node-core memory organization. In general, if we use $m \times m \times m$ clusters, the hybrid strategy reduces the dimension and the cost of the coarse problem by m^3 and m^6, respectively.

Table 4: Clamped elastic cube over a sinus-shaped obstacle, $m = 4$, $H_s/h = 14$.

Clusters	Subdomains	Cores	Unknowns (10^6)	H-TFETI-DP (iter/sec)	TFETI (iter/sec)
64	4,096	192	13	169/23.9	117/24.9
512	72,900	1536	99	208/30.2	152/115.1
1,000	656,100	3000	193	206/42.6	173/279.9

Acknowledgements This work was supported by the Ministry of Education, Youth and Sports of the Czech Republic through the e-INFRA CZ (ID:90140).

Fig. 3: Displacements of a clamped elastic cube over a sinus-shaped obstacle.

References

1. S.C. Brenner. The condition number of the Schur complement in domain decomposition. *Numerische Mathematik* **83**(2), 187–203 (1999). DOI:10.1007/s002110050446
2. Z. Dostál. *Optimal Quadratic Programming Algorithms: With Applications to Variational Inequalities*. 1st Edition. Springer, New York (2009).
3. Z. Dostál, T. Brzobohatý and O. Vlach. Schur complement spectral bounds for large hybrid FETI-DP clusters and huge three-dimensional scalar problems. *Journal of Numerical Mathematics*, to appear (2021). DOI:10.1515/jnma-2020-0048
4. Z. Dostál and D. Horák. Theoretically supported scalable FETI for numerical solution of variational inequalities. *SIAM Journal on Numerical Analysis* **45**(2), 500–513 (2007). DOI:10.2307/40232873
5. Z. Dostál, D. Horák, T. Brzobohatý and P. Vodstrčil. Bounds on the spectra of Schur complements of large H-TFETI-DP clusters for 2D Laplacian. *Numerical Linear Algebra with Applications* **28**(2), e2344 (2021). DOI:10.1002/nla.2344
6. Z. Dostál, D. Horák, V. Hapla, R. Sojka, M. Pecha and J. Kružík. PERMON - PETSc framework for numerical computations. `http://permon.vsb.cz`
7. Z. Dostál, T. Kozubek, M. Sadowská and V. Vondrák. *Scalable Algorithms for Contact Problems*. Springer, New York (2016).
8. C. Farhat and F.X. Roux. A method of Finite Element Tearing and Interconnecting and its parallel solution algorithm. *International Journal for Numerical Methods in Engineering* **32**(6), 1205–1227 (1991).
9. D. Horák, Z. Dostál and R. Sojka. On the Efficient Reconstruction of Displacements in FETI Methods for Contact Problems. *Advances in Electrical and Electronic Engineering* **15**(2) (2017). DOI:10.15598/aeee.v15i2.2322
10. A. Klawonn and O. Rheinbach. A Parallel Implementation of Dual-Primal FETI Methods for Three-Dimensional Linear Elasticity Using a Transformation of Basis. *SIAM Journal on Scientific Computing* **28**(5), 1886–1906 (2006).
11. A. Klawonn and O. Rheinbach. A hybrid approach to 3-level FETI. *Proceedings in Applied Mathematics and Mechanics* **8**(1), 10841–10843 (2009).
12. A. Klawonn and O. Widlund. Dual-primal FETI method for linear elasticity. *Communications on Pure and Applied Mathematics* **59**(11), 1523–1572 (2006).
13. J. Lee. Domain Decomposition Methods for Auxiliary Linear Problems of an Elliptic Variational Inequality. In: *Domain Decomposition Methods in Science and Engineering XX*, pp. 305–312, Lect Notes Comp Sci, Civil-Comp Press, Springer, Berlin-Heidelberg (2013).
14. J. Li and O.B. Widlund. FETI-DP, BDDC, and block Cholesky methods. *International Journal for Numerical Methods in Engineering* **66**(2), 250–271 (2005).
15. O. Meca, T. Brzobohatý, L. Říha, O. Vlach, M. Merta, T. Panoc and R. Vavřík. ESPRESO - Highly Parallel Framework for Engineering Applications. `http://numbox.it4i.cz`
16. C. Pechstein. *Finite and Boundary Element Tearing and Interconnecting Solvers for Multiscale Problems*. Springer, Heidelberg (2013).

Fundamental Coarse Space Components for Schwarz Methods with Crosspoints

François Cuvelier, Martin J. Gander, and Laurence Halpern

1 Introduction

Historically, coarse spaces for domain decomposition methods were based on a coarse grid, like in geometric multigrid methods, see e.g. [17, page 36]: "The subspace V_0 is usually related to a coarse problem, often built on a coarse mesh". More recently, a wealth of research has been devoted to designing new coarse spaces for high contrast problems: after the first steps in [3, 13], where volume eigenfunctions were used, see also [2], a coarse space using the eigenfunctions of the Dirichlet-to-Neumann maps on the boundary of each subdomain was developed in [1, 15]. This then led to the GenEO coarse space [16], and also motivated the ACMS based coarse space [11], all seminal for many further developments: for FETI, see for example [14], or for the GDSW coarse space, see [12]. A different idea for new coarse spaces is to first define an optimal coarse space, which makes the method a direct solver [5, 6], and then to approximate it, which led to the SHEM coarse space [7–10].

Our new idea here is to design a coarse space based on insights from the eigenmodes of the parallel Schwarz iteration operator that converge most slowly. We start with a numerical experiment for Laplace's equation on the unit square divided into 4×4 subdomains using the classical parallel Schwarz method of Lions with minimal overlap[1]. In Figure 1 we observe that the error in the iteration, after an initial transient phase, forms two typical distinct modes which converge most slowly: for the constant initial guess we see a continuous mode consisting of affine (harmonic)

François Cuvelier
LAGA, Université Sorbonne Paris-Nord, e-mail: cuvelier@math.univ-paris13.fr

Martin J. Gander
Section de Mathématiques, Université de Genève, e-mail: martin.gander@unige.ch

Laurence Halpern
LAGA, Université Sorbonne Paris-Nord, e-mail: halpern@math.univ-paris13.fr

[1] With minimal overlap (and only then) this is equivalent to Additive Schwarz without Conjugate Gradient acceleration.

S. C. Brenner et al. (eds.), *Domain Decomposition Methods in Science and Engineering XXVI*, Lecture Notes in Computational Science and Engineering 145,
https://doi.org/10.1007/978-3-030-95025-5_4

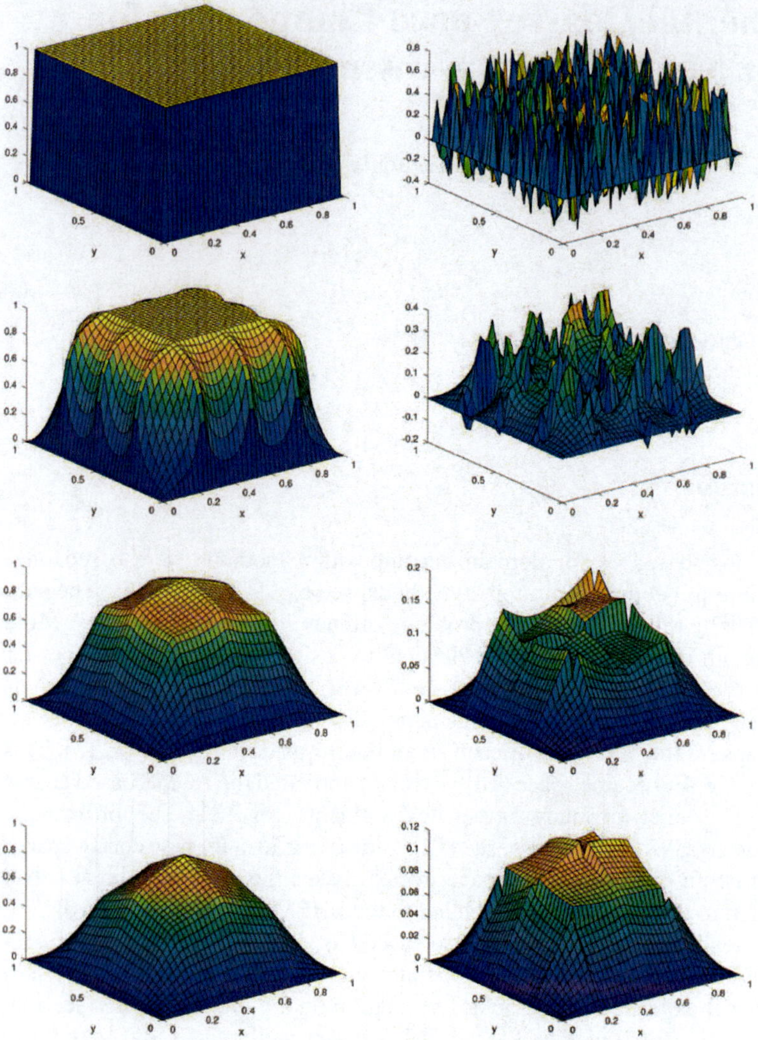

Fig. 1: Error for the parallel Schwarz method of Lions with 4×4 subdomains at iteration 0, 1, 10, 20 from top to bottom. Left: constant initial error, Right: random initial error.

functions in each subdomain, whereas for the random initial guess these functions seem to be discontinuous across subdomains. Our goal is to understand this behavior by studying the eigenmodes of the continuous parallel Schwarz iteration operator, and to deduce from this study a very effective new coarse space for Schwarz methods.

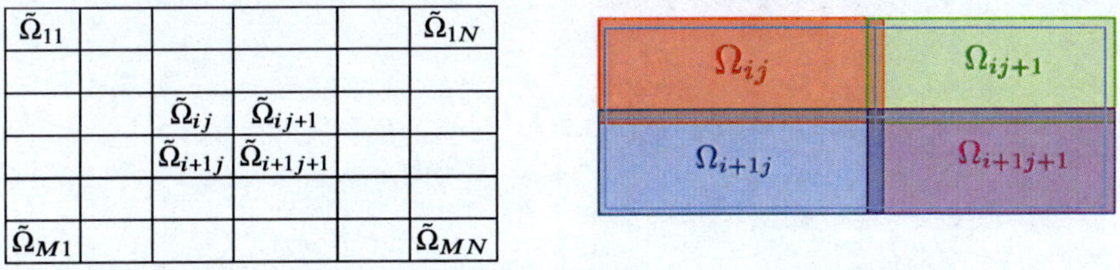

Fig. 2: Left: general decomposition of a rectangle into $M \times N$ subrectangles $\tilde{\Omega}_{ij}$. Right: adding $2L$ overlap to obtain the subdomains Ω_{ij}.

2 Modal analysis of the Schwarz iteration map

We consider a general decomposition of a rectangle into $M \times N$ smaller overlapping rectangles Ω_{ij}, as indicated in Figure 2. We denote by (x_j, y_i) the crosspoints of the nonoverlapping decomposition. The parallel Schwarz iteration from $\prod H^1(\Omega_{ij})$ into itself maps the old error iterate $u = \{u_{ij}\}$ which is harmonic in the subdomains, into a new error $v = \{v_{ij}\}$, also harmonic in the subdomains. We allow in our analysis the more general Robin transmission conditions, which on the vertical interfaces are

$$\begin{aligned} \partial_x v_{ij} + p v_{ij} &= \partial_x u_{ij+1} + p u_{ij+1}, \; x = x_j + L, \\ -\partial_x v_{ij+1} + p v_{ij+1} &= -\partial_x u_{ij} + p u_{ij}, \quad x = x_j - L, \end{aligned}$$

and similarly on the horizontal interfaces at $y_i \pm L$. For $p = +\infty$ and $L > 0$, our results will correspond to the classical parallel Schwarz method of Lions with Dirichlet transmission conditions. If $0 < p < +\infty$ and $L \geq 0$, our results will correspond to a possibly non-overlapping optimized parallel Schwarz method.

An eigenmode of the iteration map associated to an eigenvalue λ is defined by $v = \lambda u$, λ being the convergence factor of this mode. For simplicity, we study the case where all underlying nonoverlapping subdomains are squares of equal sides H. The error function u_{ij} in the subdomain Ω_{ij} is harmonic, and we use separation of variables,

$$u_{ij} = (a_{ij} \sin\zeta(x-x_{j-1}) + a'_{ij} \sin\zeta(x-x_j))(b_{ij} \sinh\zeta(y-y_{i-1}) + b'_{ij} \sinh\zeta(y-y_i))$$

for the oscillatory modes in x. Exchanging x and y gives the oscillatory modes in y. Affine modes are obtained by replacing $\sin\zeta(x - x_{j-1})$ by $(x - x_{j-1})$ for instance. By a lengthy, technical computation, we obtain

Theorem 1 (Eigenvalue-Frequency Relation) *Defining for each $\zeta \neq 0$ the quantities*

$$\begin{aligned} Z^- &:= \zeta \cos\zeta(H-L) - p \sin\zeta(H-L), & Z_h^- &:= \zeta \cosh\zeta(H-L) - p \sinh\zeta(H-L), \\ Z^+ &:= \zeta \cos\zeta(H+L) + p \sin\zeta(H+L), & Z_h^+ &:= \zeta \cosh\zeta(H+L) + p \sinh\zeta(H+L), \\ Z^0 &:= \zeta \cos\zeta L + p \sin\zeta L, & Z_h^0 &:= \zeta \cosh\zeta L + p \sinh\zeta L, \end{aligned}$$

the eigenvalue λ, the angular frequency ζ and the coefficients of the eigenmode are related by

$$
\begin{aligned}
\lambda(Z^+ + \delta_x^{(j)} Z^0) a_{ij} b_{ij} &= (Z^0 + \delta_x^{(j+1)} Z^-) a_{ij+1} b_{ij+1},\\
\lambda(Z^0 + \delta_x^{(j+1)} Z^+) a_{ij+1} b_{ij+1} &= (Z^- + \delta_x^{(j)} Z^0) a_{ij} b_{ij},\\
\lambda(Z_h^+ + \delta_y^{(i)} Z_h^0) a_{ij} b_{ij} &= (Z_h^0 + \delta_y^{(i+1)} Z_h^-) a_{i+1j} b_{i+1j},\\
\lambda(Z_h^0 + \delta_y^{(i+1)} Z_h^+) a_{i+1j} b_{i+1j} &= (Z_h^- + \delta_y^{(i)} Z_h^0) a_{ij} b_{ij},
\end{aligned}
\tag{1}
$$

where the numbers $\delta_x^{(j)} := \frac{a'_{ij}}{a_{ij}}$ *and* $\delta_y^{(i)} := \frac{b'_{ij}}{b_{ij}}$ *for* $j = 1, \ldots, N-1$ *and* $i = 1, \ldots, M-1$.

The dispersion relation (equation for the modes) is obtained from (1) by multiplying the equations pairwise, which leads to

Theorem 2 (Eigenvalues of the $M \times N$ Schwarz iteration map)

$$
\begin{aligned}
\lambda^2 &= \frac{Z^- + \delta_x^{(j)} Z^0}{Z^+ + \delta_x^{(j)} Z^0} \frac{Z^0 + \delta_x^{(j+1)} Z^-}{Z^0 + \delta_x^{(j+1)} Z^+}, \quad j = 1, \ldots, N-1,\\
\lambda^2 &= \frac{Z_h^- + \delta_y^{(i)} Z_h^0}{Z_h^+ + \delta_y^{(i)} Z_h^0} \frac{Z_h^0 + \delta_y^{(i+1)} Z_h^-}{Z_h^0 + \delta_y^{(i+1)} Z_h^+}, \quad i = 1, \ldots, M-1.
\end{aligned}
\tag{2}
$$

With Theorem 1 and Theorem 2, we thus have a complete characterization of the eigenmodes of the classical and optimized parallel Schwarz iteration map for decompositions of the form in Figure 2 for squares. The affine modes, some of which we observed in the numerical experiment in Figure 1, are obtained by letting ζ go to zero in (2), and we obtain by a direct calculation

Corollary 1 (Existence of affine Eigenmodes) *For $N \times N$ subdomains, there are $2(N-1)$ affine modes. There are no affine modes when $M \neq N$.*

For our initial experiment setting, $N = M = 4$, there are 6 affine eigenmodes, shown in Figure 3 for $p = 10^{15}$ to emulate classical parallel Schwarz, and overlap $L = 0.1$. We clearly recognize at the top left the slowest eigenmode we saw in the numerical experiment in Figure 1 on the left. We also see a corresponding discontinuous eigenmode just below on the left in Figure 3, responsible for the same slow convergence in our numerical experiment in Figure 1 on the right, since their eigenvalues are equal in modulus. It is therefore important for a good coarse space for Schwarz methods to contain both continuous and discontinuous harmonic functions per subdomain.

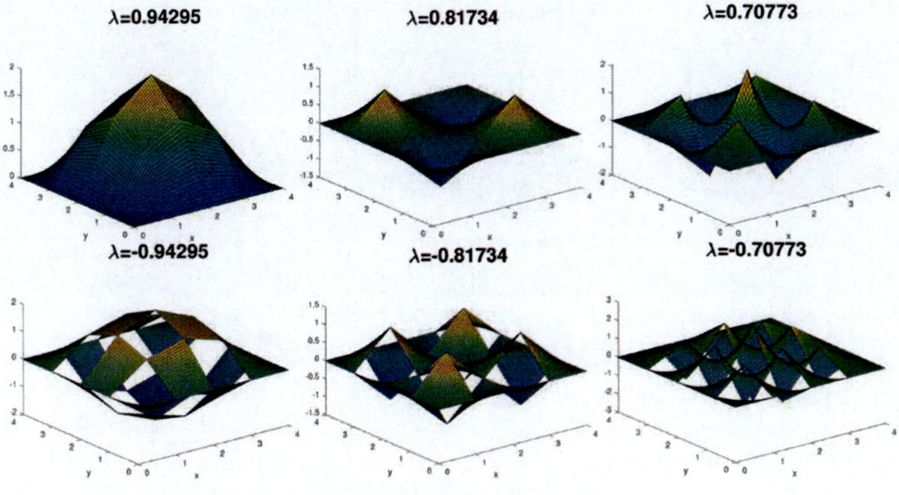

Fig. 3: Affine eigenmodes for 4×4 subdomains.

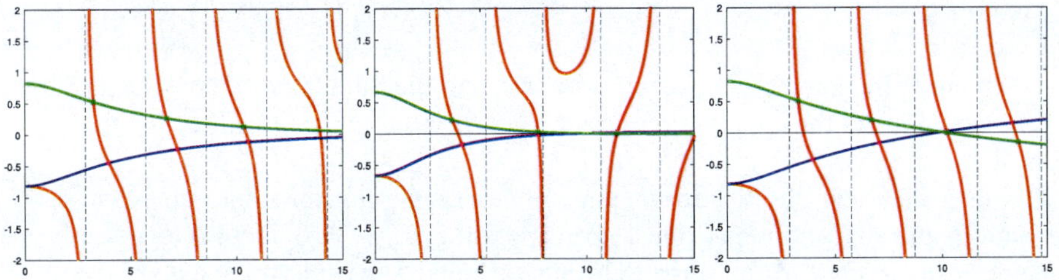

Fig. 4: Functions φ in red, φ_h in green and $-\varphi_h$ in blue. Left: classical parallel Schwarz method of Lions. Middle: overlapping optimized Schwarz. Right: Nonoverlapping optimized Schwarz.

3 The special case of 2 × 2 subdomains

For a 2×2 domain decomposition, the relation (2) between λ and ζ takes the simple form

$$\lambda^2 = (\varphi(\zeta))^2 = (\varphi_h(\zeta))^2, \quad \varphi(\zeta) = \frac{Z^-}{Z^+}(\zeta), \; \varphi_h(\zeta) = \frac{Z_h^-}{Z_h^+}(\zeta). \tag{3}$$

Then ζ is determined by either choosing the positive or negative sign,

$$\varphi(\zeta) = \varphi_h(\zeta), \qquad \varphi(\zeta) = -\varphi_h(\zeta). \tag{4}$$

Each of these equations has a sequence of solutions we denote by $\zeta_1^k(p, H, L)$ and $\zeta_2^k(p, H, L)$. We show in Figure 4 these functions of ζ, and the intersections thus represent solutions of (4). We chose subdomain length $H = 1$, and, if present, for the overlap $L = 0.1$ and the Robin parameter $p = 10$. The frequencies $\zeta_1^k(p, H, L)$ are at the intersection between the red and the green curve, while the frequencies

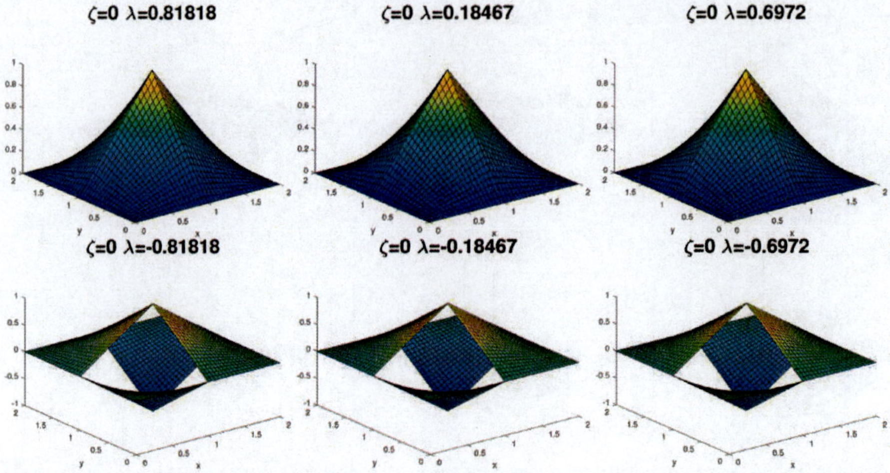

Fig. 5: Slowest, affine eigenmodes. Left: classical parallel Schwarz method of Lions. Middle: overlapping optimized Schwarz. Right: nonoverlapping optimized Schwarz.

$\zeta_2^k(p, H, L)$ are at the intersection between the red and the blue curve. The value of any of the functions at those points represents a corresponding eigenvalue λ of the Schwarz iteration map.

In Figure 5 we show the two affine eigenmodes, at the top the continuous and at the bottom the discontinuous ones, corresponding to $\zeta = 0$ in (4), together with the corresponding λ for all three Schwarz variants. These are the most slowly converging modes, and their corresponding eigenvalues in modulus show that the three different parallel Schwarz variants have very different convergence speeds: the classical parallel Schwarz method of Lions on the left converges most slowly, while optimized Schwarz with overlap in the middle is the fastest, followed by optimized Schwarz without overlap. The affine eigenmodes in Figure 5 look however very similar for all three Schwarz variants, an observation which is the basis for our new coarse space for Schwarz methods.

4 A new coarse space for parallel Schwarz methods

The affine eigenmodes in Figure 5 not only look very similar, they are asymptotically the same, and the following theorem shows that they are the basis to assemble such affine eigenfunctions for more general N.

Theorem 3 (Asymptotic Assembly Theorem) *When the overlap L is small, and/or the Robin parameter p is large, the affine modes for $N \times N$ subdomains are asymptotically special linear combinations of the two limiting affine functions Θ^c and Θ^d from the 2×2 decomposition modulo translations.*

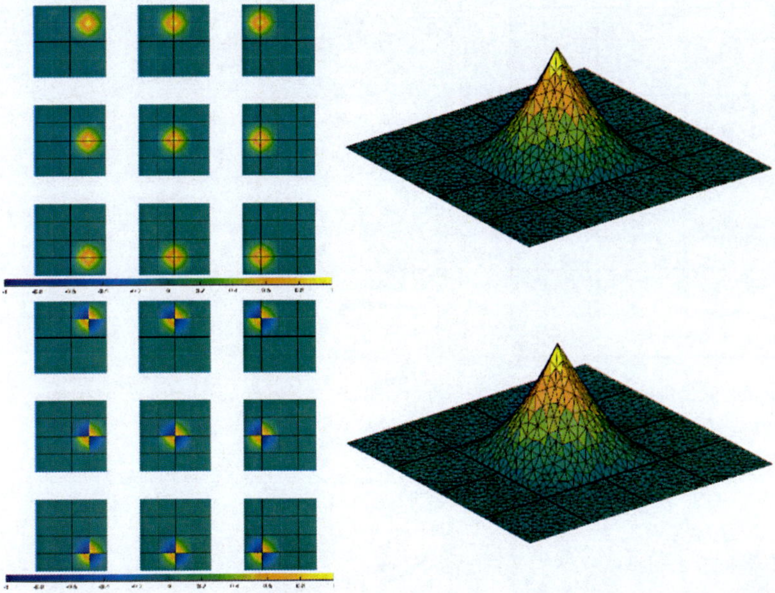

Fig. 6: Our new coarse space assembly for 4×4 subdomains. Top: continuous functions Θ^c_{ij}. Bottom: discontinuous functions Θ^d_{ij}.

For our initial 4×4 subdomain example, the precise asymptotic formulas, with respect to $X = \frac{H}{L}$ (classical parallel Schwarz method of Lions) or $X = pH$ (optimized Schwarz), are for the eigenvalues, with $\varepsilon = \pm 1$,

$$\lambda^{(1),\varepsilon} \sim \varepsilon \left(1 - \frac{2+\sqrt{2}}{X}\right), \quad \lambda^{(2),\varepsilon} \sim \varepsilon \left(1 - \frac{2}{X}\right), \quad \lambda^{(3),\varepsilon} \sim \varepsilon \left(1 - \frac{2-\sqrt{2}}{X}\right),$$

and for the corresponding six eigenfunctions shown in Figure 3 we have

$$\begin{aligned}
u^{(1),1} &\sim \Theta^c_{11} + \Theta^c_{13} + \Theta^c_{31} + \Theta^c_{33} + \sqrt{2}(\Theta^c_{12} + \Theta^c_{21} + \Theta^c_{23} + \Theta^c_{32}) + 2\Theta^d_{22},\\
u^{(1),-1} &\sim \Theta^d_{11} + \Theta^d_{13} + \Theta^d_{31} + \Theta^d_{33} - \sqrt{2}(\Theta^d_{12} + \Theta^d_{21} + \Theta^d_{23} + \Theta^d_{32}) + 2\Theta^d_{22},\\
u^{(2),1} &\sim \Theta^c_{11} - \Theta^c_{13} - \Theta^c_{31} + \Theta^c_{33},\\
u^{(2),-1} &\sim \Theta^d_{11} - \Theta^d_{13} - \Theta^d_{31} + \Theta^d_{33},\\
u^{(3),1} &\sim \Theta^c_{11} + \Theta^c_{13} + \Theta^c_{31} + \Theta^c_{33} - \sqrt{2}(\Theta^c_{12} + \Theta^c_{21} + \Theta^c_{23} + \Theta^c_{32}) + 2\Theta^c_{22},\\
u^{(3),-1} &\sim \Theta^d_{11} + \Theta^d_{13} + \Theta^d_{31} + \Theta^d_{33} + \sqrt{2}(\Theta^d_{12} + \Theta^d_{21} + \Theta^d_{23} + \Theta^d_{32}) + 2\Theta^d_{22}.
\end{aligned}$$

We therefore propose a new coarse space for Schwarz methods, based on assembling the continuous and discontinuous 'hat' functions Θ^c_{ij} and Θ^d_{ij} from the 2×2 subdomain decomposition, as illustrated for our example in Figure 6. We allow our new two-level Schwarz methods also to perform more than just $\nu = 1$ domain decompo-

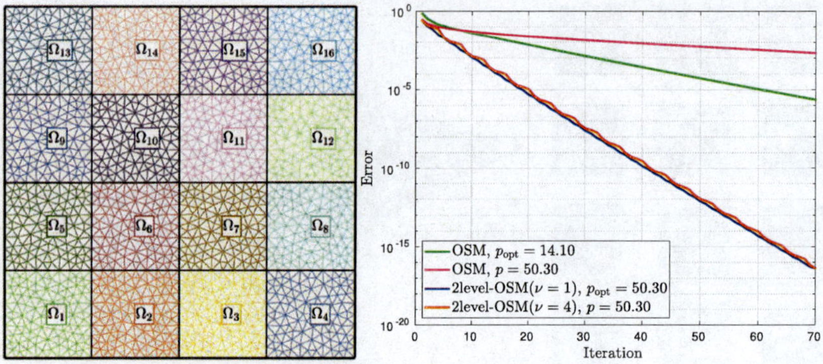

Fig. 7: Left: finite element setting for our 4×4 model problem. Right: convergence comparison of the one- and two-level optimized Schwarz methods.

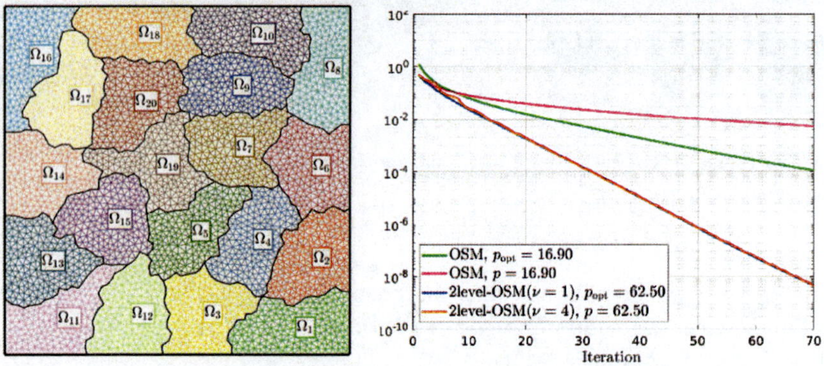

Fig. 8: Left: finite element example obtained from METIS. Right: convergence comparison of the one- and two-level optimized Schwarz methods.

sition iteration or smoothing step, since the new coarse space is so effective that it does not need to be used at every iteration, as we will see in the next section.

5 Numerical experiments

We start with a numerical experiment for our 4×4 example, running a nonoverlapping optimized Schwarz method (OSM) to solve Laplace's equation using now a finite element discretization, as indicated in Figure 7 on the left.

On the right, we show how the error decreases, both for the one-level OSM and 2-level-OSM with two different numbers of smoothing steps $\nu = 1, 4$. We see that it is sufficient to use a coarse correction with our new coarse space only every fourth Schwarz iteration with a two-level optimized parameter $p_{\text{opt}} = 50.3$, and this value

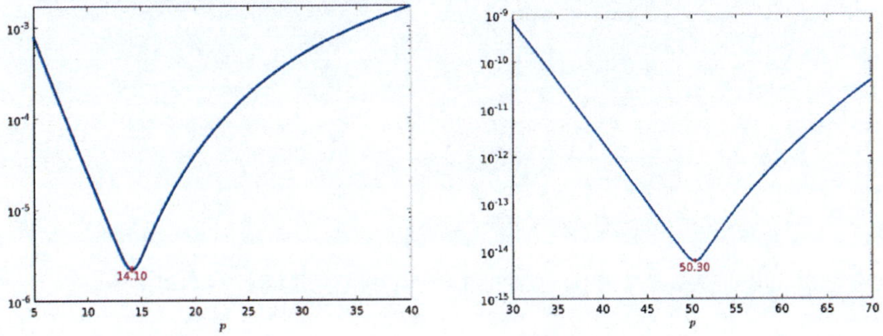

Fig. 9: Left: one-level OSM after 70 iterations. Right: two-level OSM with $\nu = 1$ after 60 iterations (since `macheps` has already been reached).

is very different from the one-level optimized parameter $p_{\text{opt}} = 14.1$, as one can see from the one-level convergence curves, see also Section 6.

We next show a numerical experiment for a more general decomposition obtained by METIS, shown in Figure 8. Here we constructed our new coarse space by generating harmonic functions in the subdomains from edge solutions of the Laplace–Beltrami operator, and we span both continuous and discontinuous 'hat' functions as in the rectangular decomposition. For cross points with an even number of incoming edges, we need again two functions, one continuous and one discontinuous, like in the rectangular case, and for cross points with an odd number of incoming edges, we need three functions, one continuous and two discontinuous, except when only three edges are incoming, for which case one continuous coarse function suffices! We see again a similar behavior in the convergence of the optimized new two-level Schwarz method, and a coarse correction every fourth iteration suffices with our new coarse space.

6 A note on the optimized Robin parameter

From the literature on optimized Schwarz methods, the optimized choice of the Robin parameter is known from two subdomain analysis [4], e.g. in the non-overlapping case $p^* \sim \frac{\pi}{\sqrt{Hh}}$. In the case with cross points, there are no results available so far. We first show a numerical experiment for our 4×4 original model problem from Figure 7 running the method for many values of the parameter p, and plotting the error as a function of p, see Figure 9. We clearly see that in both cases there is a best parameter p^*. This parameter is $p_1^* = 14.1$ for the one-level method, and $p_2^* = 50.3$ for the two-level method, also used in Figure 7.

In order to better understand this optimized choice, we return to the optimization of the convergence factor for 2×2 subdomains. Recall that the relevant frequencies ζ_j^k form a discrete set, defined in (4). We show in Figure 10 the convergence factor

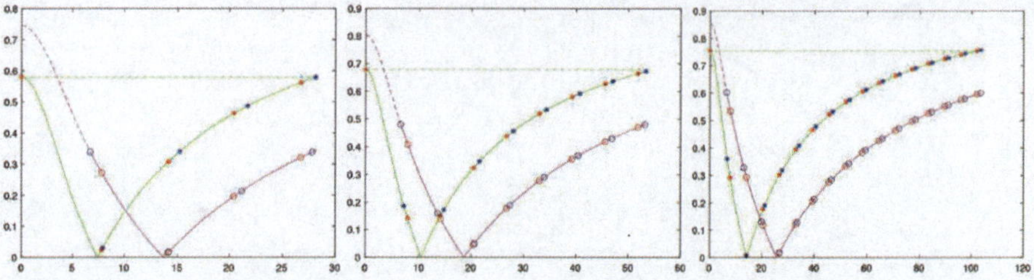

Fig. 10: Convergence factors $|\varphi_h(\zeta)|$ of our optimized Schwarz methods for $h = 1/2^4, 1/2^5, 1/2^6$ from left to right. Green-star: one level, magental-circle: two level method

$|\lambda| = \left|\frac{Z_h^-}{Z_h^+}\right|$ as a function of ζ, for our optimized Schwarz method, $H = 1/2$ and three different fine mesh parameters h. Best performance in optimized Schwarz methods is obtained by equioscillation in the convergence factor [4], in the non-overlapping case between the lowest and highest frequency (green curves in Figure 10). Since our new coarse space with affine modes removes the lowest frequency, the best parameter choice now only needs to equioscillate with the second lowest and the highest frequency (magenta curves in Figure 10), which explains why p^* for the two-level method with our new coarse space is larger than p^* for the one-level method. One can show that for given H and h, the highest frequency index is $k_0 = \frac{1}{2}\frac{H}{h} + 1$, and equating the values of $|\varphi_h(\zeta)|$ at $\zeta = 0$ and $\zeta = \zeta_2(k_0, p)$ gives the optimized parameter p_1^* for the one-level method. For the two-level method, equating the values of $|\varphi_h(\zeta)|$ at $\zeta = \zeta_2(1, p)$ and $\zeta = \zeta_2(k_0, p)$ yields the optimized value p_2^* for the two-level method. An asymptotic analysis gives

$$p_1^* \sim \sqrt{\frac{\pi}{2Hh}}, \quad p_2^* \sim \pi\sqrt{\frac{\coth \pi}{2Hh}}, \quad p^* \sim \pi\sqrt{\frac{\coth \pi}{Hh}}, \tag{5}$$

where p^* is the best parameter obtained for two subdomains. Naturally we can enrich our new coarse space with the next, non-affine modes that come in the Schwarz iteration spectrum, which we show in Figure 11 (the corresponding ones exchanging x and y are not shown), and then the optimized parameter would again increase further when an OSM is used. We see that these modes are very similar to the SHEM modes, but again they come in pairs, thus reducing the SHEM coarse space dimension by a factor of two. We see however that specific new modes appear, like the ones in the top row of Figure 11 which form a tip at the cross point, which were not in the spectral sine decomposition of the SHEM coarse space, an issue which merits further investigation.

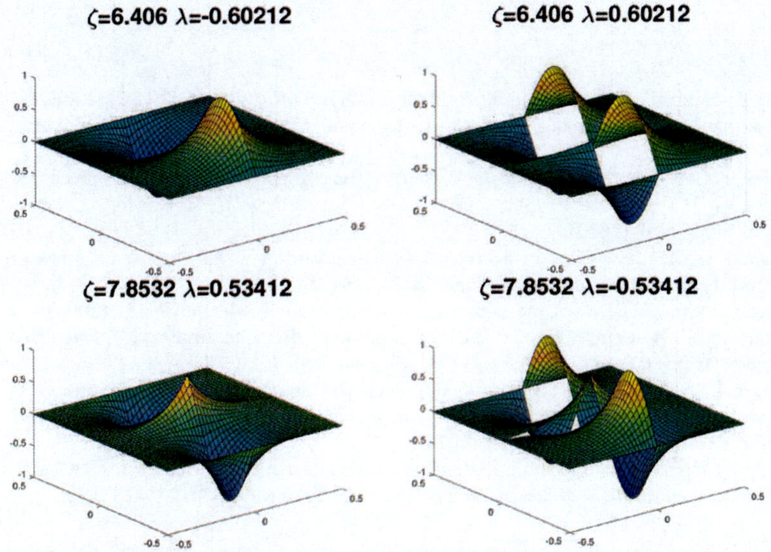

Fig. 11: First non-affine modes to enrich our new coarse space.

7 Conclusion

We designed a new coarse space for Schwarz methods, based on a spectral analysis of the parallel Schwarz iteration operator. Our new coarse space is assembled from continuous and discontinuous hat functions obtained from the eigenfunctions of local 2×2 subdomain decompositions. The new coarse space components are the same for the classical parallel Schwarz method of Lions, and overlapping and non-overlapping optimized Schwarz. We showed numerically that our new coarse space is also very effective on more general decompositions, like the ones obtained by METIS, and that using a coarse space modifies in an important way the optimized parameter in the Robin transmission conditions of the optimized Schwarz methods. Further enrichment is possible with known oscillatory enrichment functions, again from the analysis of local 2×2 subdomain decompositions.

Clearly our work is just a first step toward the construction of such new coarse spaces. Our approach can be used to detect good coarse space components for other types of partial differential equations, like problems with high contrast, advection diffusion problems, or also the much harder case of time harmonic wave propagation. This is possible also for situations where no general convergence theory for the associated Schwarz method is available, since it is based on a direct spectral study of the Schwarz iteration operator in a simplified setting.

References

1. V. Dolean, F. Nataf, R. Scheichl and N. Spillane. Analysis of a two-level Schwarz method with coarse spaces based on local Dirichlet-to-Neumann maps. *Comput. Methods Appl. Math.* **12**(4), 391–414 (2012).
2. Y. Efendiev, J. Galvis, R. Lazarov and J. Willems. Robust domain decomposition preconditioners for abstract symmetric positive definite bilinear forms. *ESAIM: Math. Model. Numer. Anal.* **46**(5), 1175–1199 (2012).
3. J. Galvis and Y. Efendiev. Domain decomposition preconditioners for multiscale flows in high-contrast media. *Multiscale Model. Simul.* **8**(4), 1461–1483 (2010).
4. M.J. Gander. Optimized Schwarz methods. *SIAM J. Numer. Anal.* **44**(2), 699–731 (2006).
5. M.J. Gander and L. Halpern. Méthode de décomposition de domaine. In: *Encyclopédie électronique pour les ingénieurs* (2012).
6. M.J. Gander, L. Halpern and K. Santugini Repiquet. Discontinuous coarse spaces for DD-methods with discontinuous iterates. In: *Domain Decomposition Methods in Science and Engineering XXI*, pp. 607–615, Springer (2014).
7. M.J. Gander, L. Halpern and K. Santugini Repiquet. A new coarse grid correction for RAS/AS. In: *Domain Decomposition Methods in Science and Engineering XXI*, pp. 275–283, Springer (2014).
8. M.J. Gander and A. Loneland. SHEM: An optimal coarse space for RAS and its multiscale approximation. In: *Domain Decomposition Methods in Science and Engineering XXIII*, pp. 281–288, Springer (2016).
9. M.J. Gander, A. Loneland and T. Rahman. Analysis of a new harmonically enriched multiscale coarse space for domain decomposition methods. *arXiv preprint* arXiv:1512.05285 (2015).
10. M.J. Gander and B. Song. Complete, optimal and optimized coarse spaces for additive Schwarz. In: *Domain Decomposition Methods in Science and Engineering XXIV*, pp. 301–309 (2019).
11. A. Heinlein, A. Klawonn, J. Knepper and O. Rheinbach. Multiscale coarse spaces for overlapping Schwarz methods based on the ACMS space in 2D. *ETNA* **48**, 156–182 (2018).
12. A. Heinlein, A. Klawonn, J. Knepper and O. Rheinbach. Adaptive GDSW coarse spaces for overlapping Schwarz methods in three dimensions. *SIAM J. Sci. Comput.* **41**(5), A3045–A3072 (2019).
13. C. Japhet. *Méthode de décomposition de domaine et conditions aux limites artificielles en mécanique des fluides : méthode optimisée d'ordre 2 (OO2).* PhD thesis, Université Paris 13 (1998). http://www.theses.fr/1998PA132044.
14. A. Klawonn, M. Kuhn and O. Rheinbach. Adaptive coarse spaces for FETI-DP in three dimensions. *SIAM J. Sci. Comput.* **38**(5), A2880–A2911 (2016).
15. F. Nataf, H. Xiang, V. Dolean and N. Spillane. A coarse space construction based on local Dirichlet-to-Neumann maps. *SIAM J. Sci. Comput.* **33**(4), 1623–1642 (2011).
16. N. Spillane, V. Dolean, P. Hauret, F. Nataf, C. Pechstein and R. Scheichl. Abstract robust coarse spaces for systems of PDEs via generalized eigenproblems in the overlaps. *Numer. Math.* **126**(4), 741–770 (2014).
17. A. Toselli and O. Widlund. *Domain Decomposition Methods – Algorithms and Theory*. Volume **34**, Springer Series in Computational Mathematics (2005).

Nonoverlapping Domain Decomposition Methods for Time Harmonic Wave Problems

Xavier Claeys, Francis Collino, Patrick Joly, and Emile Parolin

The domain decomposition method (DDM) was initially designed, in Schwarz's celebrated 1870 paper [24], as a theoretical tool for partial differential equations (PDEs), but since the advent of the computer and parallel computing techniques, it has become a major tool for the numerical solution of such PDEs, especially for large scale problems. Time harmonic wave problems offer a large spectrum of applications in various domains (acoustics, electromagnetics, geophysics, ...) and occupy a place of their own that shines, for instance, through the existence of a natural (possibly small) length scale for the solutions: the wavelength. Numerical DDMs were first invented for elliptic type equations (e.g. the Laplace equation), and even though the governing equations of wave problems (e.g. the Helmholtz equation) look similar, standard approaches do not work in general.

The objective of this work is to give a rapid, but hopefully pedagogical, survey of the research led mainly at INRIA (in the teams ONDES then POEMS and ALPINES) since 1990, on non-overlapping domain decomposition methods for time harmonic wave propagation problems, based on the notion of impedance transmission conditions. Our point of view, and we consider that this sets us apart from the rest of the wave DDM community, is theory-driven: we proposed and progressively developed a unified framework that guarantees the well-posedness and convergence of the related iterative algorithms in the **most general cases (geometry, variable coefficients, boundary conditions. . .**). This research was punctuated by four Phd theses.

Xavier Claeys
Sorbonne Université, Laboratoire Jacques-Louis Lions, and équipe Alpines, 75005 Paris, France, e-mail: claeys@ann.jussieu.fr

Francis Collino, Patrick Joly, Emile Parolin
POEMS, CNRS, INRIA, ENSTA Paris, 91120 Palaiseau, France,
e-mail: francis.collino@orange.fr; patrick.joly@inria.fr; emile.parolin@inria.fr

S. C. Brenner et al. (eds.), *Domain Decomposition Methods in Science and Engineering XXVI*, Lecture Notes in Computational Science and Engineering 145,
https://doi.org/10.1007/978-3-030-95025-5_5

- The PhD thesis of B. Després [10] (1991) is definitely a pioneering work which constitutes a decisive step. It is worth mentioning that P.L. Lions [18] (1988), [19] (1990) wrote his papers on the theory of DDMs for elliptic problems during the same period.
- With the PhD thesis of S. Ghanemi [15], at CERFACS in 1996, we developed our general theoretical framework, proposed using second order local transmission conditions and initiated non-local transmission conditions [7] (after [16, 21]).

  Then there was a big pause (about 15 years) in our activity, during which a huge literature was devoted to Optimized Schwarz Methods (OSMs) associated to local impedance operators (see also Section 2), up to the opportunity of a contract with CEA (French Nuclear Agency) which started the second phase of our activity.
- The PhD thesis of M. Lecouvez [17] (2015), in collaboration with CEA, permitted us to develop the notion of non-local transmission operators.
- The PhD thesis of E. Parolin [22] (2020) supported by the ANR Project NonlocalDD, which closes a chapter of the story with the notion of elliptic DtN operators, the treatment of Maxwell's equations and a solution to the cross points issue [3].

1 Elliptic equations versus the Helmholtz equation

In this section, we expose the general ideas, More formalism will be introduced in Section 2. To emphasize the difference of status between the two types of equations w.r.t. DDM, let us simply consider the equation with constant coefficients

$$- \Delta u + k^2 u = f, \quad \text{in } \mathbb{R}^d, \quad k \in \mathbb{C}, \quad \text{where} \tag{1}$$

- if $k \in \mathbb{R}^+$ then (1) is of a (strongly) elliptic nature
- if $k = i\omega,\ \omega \in \mathbb{R}^+$ (Helmholtz) then one models waves with frequency ω.

The distinction is important for DDMs: for instance, the classical overlapping Schwarz method converges (**linearly** in most cases) in the elliptic case but **does not converge** for the Helmholtz equation. In fact, in the elliptic case, the boundary value problems (BVPs) associated with (1) enjoy many nice properties including the H^1-coercivity of $a(u, v) = \int (\nabla u \cdot \nabla v + k^2\, u\, v)$, the associated bilinear form, and their solutions are often interpreted as the solutions of convex minimization problems. From this point of view, P.L. Lions gave a general proof of convergence of the Schwarz method by interpreting the error at each step of the algorithm as the result of successive orthogonal projections on two (with two subdomains) supplementary subspaces of H^1 [18]. These problems also benefit from the maximum principle, which also provides another way of proving the convergence of the Schwarz method.

However, if $k = i\omega$, then $a(u,v) = \int (\nabla u \cdot \nabla v - \omega^2 u\, v)$, the natural bilinear form for Helmholtz, is no longer coercive and there is no underlying variational principle for the corresponding BVPs. Also, there is no maximum principle: the (complex-valued) solutions naturally oscillate with the wavelength $\lambda = 2\pi/\omega$.

Fortunately, good news comes from the boundary: if u satisfies $-\Delta u - \omega^2 u = 0$ in a bounded domain Ω with boundary Γ and outgoing normal ν then (multiply the equation by $\overline{u}$, integrate over Ω, apply Green's formula and take the imaginary part)

$$\mathcal{I}m \int_\Gamma \partial_\nu u\, \overline{u} = 0, \quad \text{i.e.} \quad \mathcal{I}m \left\langle \partial_\nu u, u \right\rangle_\Gamma = 0, \tag{2}$$

with $\langle \cdot, \cdot \rangle_\Gamma$ the inner product in $L^2(\Gamma) \equiv L^2(\Gamma; \mathbb{C})$. This leads to the following isometry result, where $\| \cdot \|_\Gamma$ denotes the $L^2(\Gamma)$-norm

$$\| \partial_\nu u + i\omega\, u \|_\Gamma^2 = \| \partial_\nu u - i\omega\, u \|_\Gamma^2, \tag{3}$$

(simply note that the difference of the two sides of (3) is proportional to $\mathcal{I}m \langle \partial_\nu u, u \rangle_\Gamma$, which is 0 by (2)). One obtains many other isometry results by playing with identity (2): introducing a "boundary operator" Λ (which transforms a function defined on Γ into another function defined on Γ), assumed to be bijective (between appropriate spaces) with (formal) adjoint Λ^*, we note that

$$\mathcal{I}m \left\langle \partial_\nu u, u \right\rangle_\Gamma = 0 \Leftrightarrow \mathcal{I}m \left\langle \partial_\nu u, \Lambda^{-1}\Lambda u \right\rangle_\Gamma = 0 \Leftrightarrow \mathcal{I}m \left\langle (\Lambda^*)^{-1} \partial_\nu u, \Lambda u \right\rangle_\Gamma = 0,$$

from which we deduce the other isometry result

$$\| (\Lambda^*)^{-1} \partial_\nu u + i\omega\, \Lambda u \|_\Gamma^2 = \| (\Lambda^*)^{-1} \partial_\nu u - i\omega\, \Lambda u \|_\Gamma^2. \tag{4}$$

Introducing the positive definite self-adjoint boundary operator $T = \Lambda^* \Lambda$ (called an *impedance operator* in the sequel) and the associated norm

$$(\varphi, \psi) := \left\langle \psi, T^{-1} \varphi \right\rangle_\Gamma, \quad \|\varphi\|^2 := \left\langle \varphi, T^{-1} \varphi \right\rangle_\Gamma, \tag{5}$$

(4) rewrites as

$$\| \partial_\nu u + i\omega\, T u \|^2 = \| \partial_\nu u - i\omega\, T u \|^2. \tag{6}$$

This is one of the reasons which led us, in the context of iterative overlapping DDMs, denoting by $\{\Omega_j\}$ the subdomains (with outgoing normals ν_j), to propose

$$\partial_{\nu_j} u_j^n + i\omega\, T u_j^n = (rhs)_{n-1}, \quad u_j^n = u^n|_{\Omega_j}, \tag{7}$$

as a boundary condition in Ω_j, where $(rhs)_{n-1}$ is a quantity, depending on the previous iteration and the adjacent subdomain, providing the good continuity conditions at convergence (Section 2). An important consequence of the properties of T (symmetric positive definite) is that (7) is of an absorbing nature so that the **local problem** in Ω_j is automatically **well posed**. Moreover, as we shall see in Section 2, the **isometry result** (6) can be exploited to prove the **convergence** of the iterative algorithm.

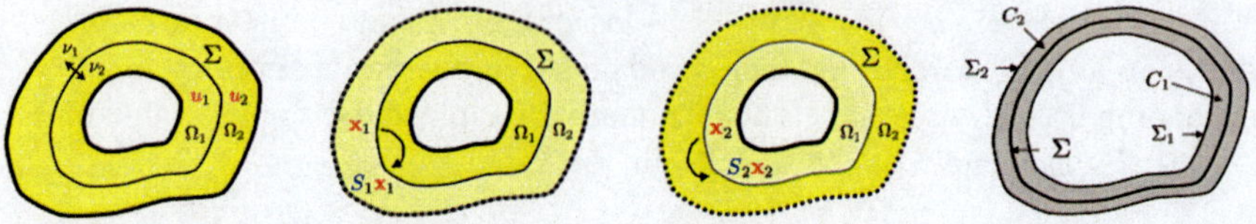

Fig. 1: The subdomains Ω_1 and Ω_2 (left). The scattering operators S_1 ans S_2 (center). The layers C_1 and C_2, cf, Section 5 (right)

2 Impedance-based transmission conditions and related DDM

Presentation of the method on a simple model. Let (BVP) consist in solving the Helmholtz equation in a $\Omega \subset \mathbb{R}^d$, bounded, with a perfectly reflecting inner boundary Γ_1 and absorbing outer boundary Γ_2:

$$(BVP) \quad -\Delta u - \omega^2 u = f, \text{ in } \Omega, \quad u = 0, \text{ on } \Gamma_1, \quad \partial_\nu u + i\,\omega\, u = 0, \text{ on } \Gamma_2.$$

Introducing an interface Σ that splits Ω into two subdomains Ω_1 (interior) and Ω_2 (exterior), see Figure 1 (left picture), (BVP) is equivalent to a transmission problem $(LP)+(TC)$ (local problem + transmission conditions) where, with obvious notation (in particular, ν_j is the unit normal vector to Σ, outgoing w.r.t. Ω_j)

$$(LP) \begin{cases} -\Delta u_j - \omega^2 u_j = f, & \text{in } \Omega_j, j = 1,2 \\ u_1 = 0, & \text{on } \Gamma_1, \\ \partial_\nu u_2 + i\,\omega\, u_2 = 0, & \text{on } \Gamma_2, \end{cases} \qquad (TC) \begin{cases} (\mathbf{n})\ u_1 = u_2, & \text{on } \Sigma, \\ (\mathbf{d})\ \partial_{\nu_1} u_1 + \partial_{\nu_2} u_2 = 0, & \text{on } \Sigma. \end{cases}$$

Given $s \in [0, 1/2]$, we introduce an impedance operator T with the property that

$$T \in \mathcal{L}\big(H^s(\Sigma), H^{-s}(\Sigma)\big) \text{ is a positive and self-adjoint isomorphism.} \qquad (8)$$

With this choice, the norm defined by (5) (with Γ replaced by Σ, and $\langle \cdot, \cdot \rangle_\Sigma$ understood as a duality bracket) is a **Hilbert space norm** in $H^{-s}(\Sigma)$.

Next, we rewrite (TC) in an equivalent way (thanks to the injectivity of T) by considering the two independent linear combinations $(TC)(\mathbf{n}) \pm i\omega\,(TC)(\mathbf{d})$, i.e.

$$\begin{cases} \partial_{\nu_1} u_1 + i\,\omega\, T u_1 = -\partial_{\nu_2} u_2 + i\,\omega\, T u_2, & (1) \\ \partial_{\nu_2} u_2 + i\,\omega\, T u_2 = -\partial_{\nu_1} u_1 + i\,\omega\, T u_1, & (2) \end{cases} \qquad (9)$$

where (9)-(j) is seen here as a boundary condition for u_j. The iterative DDM algorithm consists in applying a **fixed point** procedure (with relaxation) to (9). More precisely, we construct inductively two sequences $u_j^n \in H^1(\Omega_j)$, $j = 1, 2$, by imposing, at each step n, the local equations (LP) completed by the following boundary conditions on Σ (where $r \in\,]0, 1[$ is the relaxation parameter)

$$\begin{cases} \partial_{\nu_1} u_1^n + i\,\omega\, T u_1^n = r\,\big(-\,\partial_{\nu_2} u_2^{n-1} + i\,\omega\, T u_2^{n-1}\big) + (1-r)\,\big(\partial_{\nu_1} u_1^{n-1} + i\,\omega\, T u_1^{n-1}\big), \\ \partial_{\nu_2} u_2^n + i\,\omega\, T u_2^n = r\,\big(-\,\partial_{\nu_1} u_1^{n-1} + i\,\omega\, T u_1^{n-1}\big) + (1-r)\,\big(\partial_{\nu_2} u_2^{n-1} + i\,\omega\, T u_2^{n-1}\big). \end{cases} \tag{10}$$

The reader will notice that, by construction, the local problems in (u_1^n, u_2^n) are well posed, and can be solved in parallel.

A functional analytic observation. It is insightful to look at the quantities in (9) for the two extreme values for $s \in [0, 1/2]$. Given $u_j \in H^1(\Omega_j)$ with $\Delta u \in L^2(\Omega_j)$:

- if $s = 0$, for instance $T = I$, the identity: the combinations $\partial_\nu u_j \pm i\,\omega\, u_j$ are not well balanced since u naturally belongs to $H^{1/2}(\Sigma)$ while $\partial_\nu u$ only belongs to $H^{-1/2}(\Sigma)$,
- if $s = 1/2$: the presence of $T \in \mathcal{L}\big(H^{1/2}(\Sigma), H^{-1/2}(\Sigma)\big)$ re-equilibrates the combination as a sum of two terms in $H^{-1/2}(\Sigma)$.

In fact, a **misfit** is present as soon as $s \neq 1/2$ and one can thus anticipate that the best option should be $s = 1/2$. This will be confirmed by the analysis (Section 3).

A rapid guided tour of the bibliography. A lot of literature has been devoted to DDMs based on transmission written in impedance form.

- In the original work of B. Després [10] (or [11] for Maxwell), $T = \alpha\, I$, where α is a bounded strictly positive function, which fits (8) with $s = 0$.
- Since the mid 90s a huge literature has been devoted to "**local**" operators T as rational functions of the Laplace–Beltrami operator Δ_Σ [2, 13, 16, 21], with a great filiation with local absorbing conditions (Remark 1). These often do not satisfy (8) and a general theory (existence for local problems and convergence) is missing.
- In [8], we promote the use of non-local impedance operators T fitting (8) with $s = \frac{1}{2}$, in particular boundary integral operators issued from potential theory.

Some optimized Schwarz methods, for instance that of Boubendir–Antoine–Geuzaine, perform very well in practice (despite examples of failure, see [8], Section 8.2.3). However, they cannot lead to linear convergence (see [8], Thm 4.6).

Remark 1 : There is an ideal choice of transmission conditions with two (not one) operators, $\partial_{\nu_1} u_1 + i\omega\, T_1 u_1 = -\partial_{\nu_2} u_2 + i\omega\, T_1 u_2$ and $\partial_{\nu_2} u_2 + i\,\omega\, T_2 u_2 = -\partial_{\nu_1} u_1 + i\,\omega\, T_2 u_1$: take T_1 (resp. T_2) as the DtN operator, when it exists, associated to Ω_2 (resp. Ω_1) (see [8] Section 1.3.2 and [14]). Then Algorithm (10) with $r = 1$ converges in two iterations. In general, finding T_1 or T_2 is almost as difficult as the original problem. For two homogeneous half-spaces (plane interface), $T_1 = T_2$ with symbol $i\omega\sqrt{1 - |\xi|^2/\omega^2}$ (ξ is the space Fourier variable), whose rational approximations (Taylor, Padé, continued fraction expansions) give local operators, as for ABCs.

3 Convergence analysis

Interface formulation. For both the implementation and the analysis of our method, it is useful to reinterpret the problem and the algorithm on the interface Σ. To do so we introduce the interface auxiliary unknowns (where traces on Σ are implicitly considered), i.e. the outgoing traces x_j and incoming traces y_j:

$$x_j := \partial_{\nu_j} u_j + i\,\omega T\, u_j, \quad y_j := -\partial_{\nu_j} u_j + i\,\omega T\, u_j, \quad \text{in } H^{-s}(\Sigma). \tag{11}$$

Given x_1 and x_2, u_1 and u_2 can be seen as the solutions of the local problems

$$\begin{cases} -\Delta u_1 - \omega^2 u_1 = f, & \text{in } \Omega_1, \\ u_1 = 0, & \text{on } \Gamma_1, \\ \partial_{\nu_1} u_1 + i\,\omega T\, u_1 = x_1, & \text{on } \Sigma, \end{cases} \qquad \begin{cases} -\Delta u_2 - \omega^2 u_2 = f, & \text{in } \Omega_2, \\ \partial_{\nu} u_2 + i\,\omega u_2 = 0, & \text{on } \Gamma_2, \\ \partial_{\nu_2} u_2 + i\,\omega T\, u_2 = x_2, & \text{on } \Sigma, \end{cases} \tag{12}$$

and, exploiting the linearity of (12), the incoming traces y_j can be rewritten as

$$y_1 = S_1 x_1 + \widetilde{g}_1, \quad y_2 = S_2 x_2 + \widetilde{g}_2, \tag{13}$$

where, in an obvious manner, the source terms $\widetilde{g}_j$ are due to f (they are issued from (12) with $x_j = 0$) and the scattering operators S_j are constructed from the local problems (12) with $f = 0$. Next, the transmission conditions simply rewrite as

$$y_2 = x_1, \quad y_1 = x_2, \tag{14}$$

and the transmission problem (LP, TC) is equivalent to the system (13, 14) in $\mathbf{x} = (x_1, x_2)$ and $\mathbf{y} = (y_1, y_2)$: (13) takes account of local problems and (14) of transmission conditions. Eliminating $\mathbf{y}$ then leads to a problem in $\mathbf{x}$:

$$\text{Find } \mathbf{x} \in \mathbf{V} := H^{-s}(\Sigma) \times H^{-s}(\Sigma) \quad / \quad (\mathbf{I} - \mathbf{A})\,\mathbf{x} = \mathbf{g}, \quad \mathbf{g} = \mathbf{\Pi}\,\widetilde{\mathbf{g}}, \tag{15}$$

with the (T-dependent) **scattering** operator $\mathbf{S}$ and the **exchange** operators $\mathbf{\Pi}$:

$$\mathbf{S} := \begin{pmatrix} S_1 & 0 \\ 0 & S_2 \end{pmatrix}, \quad \mathbf{\Pi} := \begin{pmatrix} 0 & I \\ I & 0 \end{pmatrix}, \quad \text{thus} \quad \mathbf{I} - \mathbf{A} := \begin{pmatrix} I & -S_2 \\ -S_1 & I \end{pmatrix}. \tag{16}$$

Mathematical properties. In the following, we equip the Hilbert space $\mathbf{V}$ with the (T-dependent) norm naturally inherited from the H^{-s}-norm defined by (5), that we still denote by $\|\cdot\|$ for simplicity. From (8), it is clear that the operators $\mathbf{\Pi}$ and $\mathbf{S}$ are continuous in $\mathbf{V}$. Obviously, $\mathbf{\Pi}$ is an isometry while, from the identity (6) (applied in Ω_1 and Ω_2), we immediately infer that, for any $(x_1, x_2) \in \mathbf{V}$,

$$(a) \quad \|S_1 x_1\| = \|x_1\|, \quad (b) \quad \|S_2 x_2\| \le \|x_2\|, \tag{17}$$

where the inequality in (17)-(b) is due to the absorbing condition on Γ_2 for u_2 in (12). As a consequence, the operator $\mathbf{S}$, thus the operator $\mathbf{A}$, is **contractant** in $\mathbf{V}$. Concerning the **invertibility** of $\mathbf{I} - \mathbf{A}$, algebraic manipulations show that

$$\mathbf{z} = (\mathbf{I} - \mathbf{A})\,\mathbf{x} \quad \Leftrightarrow \quad x_j = \partial_{\nu_j} w_j + i\,\omega T w_j \text{ on } \Sigma,\ j = 1, 2, \tag{18}$$

where, denoting by ν the normal to Σ pointing towards Ω_2 and $[\cdot]_\Sigma$ the jump across Σ, $w \in H^1(\Omega_1 \cup \Omega_2)$ satisfies (H) $-\Delta w - \omega^2 w = f$ in $\Omega_1 \cup \Omega_2$, (BC) : $w = 0$ on Γ_1 and $\partial_\nu w + i\,\omega\, w = 0$ on Γ_2 and the "jump conditions", with $[\mathbf{z}] = z_1 - z_2$, $\{\mathbf{z}\} = \frac{1}{2}(z_1 + z_2)$:

$$[w]_\Sigma = \tfrac{1}{2i\omega} T^{-1}\,[\mathbf{z}], \qquad [\partial_n v]_\Sigma = \{\mathbf{z}\}. \tag{19}$$

The **injectivity** of $\mathbf{I} - \mathbf{A}$ is due to the **uniqueness** of a solution w of (H, BC, 19): this results from the uniqueness for the original problem. The **surjectivity** is related to the **existence** of v. Trace theorems require $\{\mathbf{z}\} \in H^{-1/2}(\Sigma)$, which holds since $s \le 1/2$, and $T^{-1}[\mathbf{z}] \in H^{1/2}(\Sigma)$. However, (8) only ensures $T^{-1}[\mathbf{z}] \in H^s(\Sigma)$: we recover the **misfit** mentioned in Section 1 unless $s = 1/2$:

Theorem 1 *The operator* $\mathbf{I} - \mathbf{A}$ *is injective in* $\mathbf{V}$ *and it is surjective if and only if* $s = 1/2$. *In this case, by Banach's theorem, there exists a* $\delta > 0$ *such that*

$$\forall\,\mathbf{x} \in \mathbf{V}, \quad \|(\mathbf{I} - \mathbf{A})\,\mathbf{x}\| \ge \delta\, \|\mathbf{x}\|, \quad \textit{(with } \delta \le 2 \textit{ because } \mathbf{A} \textit{ is contractant).} \tag{20}$$

Theorem 1 implies that, when $s = 1/2$, the interface problem (15) is a nice coercive problem in $\mathbf{V}$ (the lack of H^1-coercivity – emphasized in Section 1 – is hidden in the definition of $\mathbf{A}$). Indeed, from $\mathbf{A}\mathbf{x} = \mathbf{x} - (\mathbf{I} - \mathbf{A})\,\mathbf{x}$, we get (take the square norms) $\|\mathbf{A}\mathbf{x}\|^2 = \|\mathbf{x}\|^2 + \|(\mathbf{I} - \mathbf{A})\,\mathbf{x}\|^2 - 2\,\mathcal{R}e\,((\mathbf{I} - \mathbf{A})\,\mathbf{x}, \mathbf{x})$. Since $\|\mathbf{A}\mathbf{x}\|^2 \le \|\mathbf{x}\|^2$, we deduce

$$\forall\,\mathbf{x} \in \mathbf{V}, \quad \mathcal{R}e\,((\mathbf{I} - \mathbf{A})\,\mathbf{x}, \mathbf{x}) \ge (1/2)\, \|(\mathbf{I} - \mathbf{A})\,\mathbf{x}\|^2 \ge (\delta^2/2)\, \|\mathbf{x}\|^2. \tag{21}$$

Convergence. We go back to the iterative method (LP) + (10). If $\mathbf{x}^n := (x_1^n, x_2^n)$ with $x_j^n := \partial_{\nu_j} u_j^n + i\,\omega\, T\, u_j^n$, one easily sees that $\mathbf{x}^n$ satisfies the following Richardson algorithm (or relaxed Jacobi in reference to the block form (16) of $\mathbf{I} - \mathbf{A}$):

$$\mathbf{x}^n = (1 - r)\,\mathbf{x}^{n-1} + r\,\mathbf{A}\,\mathbf{x}^{n-1} + \mathbf{g}. \tag{22}$$

The error $\mathbf{e}^n = \mathbf{x}^n - \mathbf{x}$ satisfies $\mathbf{e}^n = (1 - r)\,\mathbf{e}^{n-1} + r\,\mathbf{A}\,\mathbf{e}^{n-1}$ $(*)$. From the identity $\|(1 - r)\,\mathbf{x} + r\,\mathbf{y}\|^2 = (1 - r)\,\|\mathbf{x}\|^2 + r\,\|\mathbf{y}\|^2 - r(1 - r)\,\|\mathbf{x} - \mathbf{y}\|^2$, we thus get

$$\begin{aligned} \|\mathbf{e}^n\|^2 &= (1 - r)\,\|\mathbf{e}^{n-1}\|^2 + r\,\|\mathbf{A}\mathbf{e}^{n-1}\|^2 - r(1 - r)\,\|(\mathbf{I} - \mathbf{A})\,\mathbf{e}^{n-1}\|^2 \\ &\le \|\mathbf{e}^{n-1}\|^2 - r(1 - r)\,\|(\mathbf{I} - \mathbf{A})\,\mathbf{e}^{n-1}\|^2, \quad \text{(contractivity of } \mathbf{A}). \end{aligned} \tag{23}$$

Thus $\|\mathbf{e}^n\|$ decreases and $\|(\mathbf{I} - \mathbf{A})\,\mathbf{e}^n\| \to 0$. By weak compactness in $\mathbf{V}$, at least for a subsequence, $\mathbf{e}^n \rightharpoonup \mathbf{e}$ (weakly) in $\mathbf{V}$. So $(\mathbf{I} - \mathbf{A})\,\mathbf{e} = 0$ thus (injectivity of $\mathbf{I} - \mathbf{A}$) $\mathbf{e} = 0$. This being true for any such subsequence, the whole sequence $\mathbf{e}^n$ converges and it is easy to infer that $(u_1^n, u_2^n) \to (u_1, u_2)$ in $L^2(\Omega_1) \times L^2(\Omega_2)$.

However, in the case $s = 1/2$, we have better since, using (20) again in (23)

$$\|\mathbf{e}^n\| \le \tau^n\, \|\mathbf{e}^0\|, \quad \tau := \sqrt{1 - r(1 - r)\,\delta^2} < 1, \tag{24}$$

i.e. the iterative algorithm converges linearly provided $s = 1/2$ and $0 < r < 1$.

GMRES algorithm. One can of course use more sophisticated algorithms than (22) to update the interface unknowns $\mathbf{x}^n$ (from which (u_1^n, u_2^n) are still reconstructed via the local problems (12)). This includes nonlinear algorithms such as GMRES [23], in which $\mathbf{x}^n$ is computed by minimizing $\mathbf{y} \mapsto \|(\mathbf{I}-\mathbf{A})\,\mathbf{y}-\mathbf{g}\|^2$, the square $\mathbf{V}$-norm of the residue, over the Krylov subspace generated by the n first iterates $\mathbf{x}^k, k \le n-1$ [9]. As a consequence, the corresponding error $\mathbf{e}^n$ is such that

$$\big\|(\mathbf{I}-\mathbf{A})\,\mathbf{e}^n\big\| = \min_{p\in\mathbb{P}_n} \big\|(\mathbf{I}-\mathbf{A})\,p(\mathbf{A})\,\mathbf{e}^0\big\|, \qquad \mathbb{P}_n = \{\text{ polynomials of degree} \le n\,\}.$$

Considering the polynomial $P(a) = (1-r+r\,a)^n$, which corresponds to Jacobi's algorithm (22), we deduce from Theorem 1, (24) and $\|\mathbf{I}-\mathbf{A}\| \le 2$ that, if $s = 1/2$,

$$\|\mathbf{e}^n\| \le (2/\delta)\,\big\|(\mathbf{I}-\mathbf{A})\,\mathbf{e}^n\big\| \le (2/\delta)\,\tau^n \quad \text{with } \tau \text{ as in (24)},$$

which means that the convergence rate of the GMRES algorithm is necessarily better than with (22). Numerical evidence show that it is strictly better and that it is worthwhile using GMRES despite the larger computational cost for each iteration.

4 Construction of appropriate impedance operators

According to what precedes, the question is to construct an impedance operator T satisfying (8) with $s = 1/2$, i.e. a positive self-adjoint pseudo-differential operator of order 1. A first mathematical fact is that such an operator **cannot be a local operator** in the sense of Section 2: this is clearly demonstrated in 2D circular geometries [8] with a Fourier modal expansion in the azimuthal variable θ . On the other hand, there exist many ways to construct good nonlocal operators. Let us describe some of them (see also [17], [8], [22]).

From Sobolev norms (A). The operator T is entirely defined by the scalar product (5), which is used for finite elements. A first choice is the following (if $\Omega \subset \mathbb{R}^3$):

$$\alpha \int_\Sigma \varphi\psi\,\mathrm{d}\sigma + \frac{\beta}{\omega}\iint_\Sigma \chi\Big(\frac{|x-y|}{L}\Big)\frac{\big(\varphi(x)-\varphi(y)\big)\big(\overline{\psi(x)}-\overline{\psi(y)}\big)}{|x-y|^3}\,\mathrm{d}\sigma_x\mathrm{d}\sigma_y \tag{25}$$

with $\alpha, \beta > 0$, $\chi(r) \ge 0$ a C^1 cut off function with support in $[0,1]$ and $\chi(r) = 1$ for $r < 1/2$, and $L > 0$. If $L = +\infty$, T is fully nonlocal and one recovers the usual Gagliardo–Niremberg norm in $H^{1/2}(\Sigma)$ if $\alpha = \beta = 1$. If not, T only couples points at a distance less than L apart and the (discretized) impedance condition is less costly.

From potential theory (B). An automatic way to build a good impedance operator is to take $T = \Lambda^*\Lambda$, with Λ an isomorphism from $H^{1/2}(\Sigma)$ in $L^2(\Sigma)$ provided by a Riesz-type potential: given $a, b > 0$, the associated bilinear form is given by

$$a \int_\Sigma \varphi\psi\,\mathrm{d}\sigma + \frac{b}{\sqrt{\omega}}\iint_\Sigma \chi\Big(\frac{|x-y|}{L}\Big)\frac{\mathrm{rot}_\Sigma\varphi(x)\cdot\mathrm{rot}_\Sigma\overline{\psi}(y)}{|x-y|^{1/2}}\,\mathrm{d}\sigma_x\mathrm{d}\sigma_y, \tag{26}$$

where rot_Σ denotes the usual tangential curl operator on Σ. Such operators are familiar to specialists in boundary integral equations, except the non-standard exponent $1/2$, which ensures that Λ is of order $1/2$. Contrary to **(A)**, Alternative **(B)** can be extended to Maxwell's equations [22]. In separable geometries, the convergence of (22) for **(A)** or **(B)** can be precisely quantified via a modal decomposition. This analysis also permits us to show that a good choice for L is $L \sim \lambda/2$ [8].

From local elliptic DtN operators (C). A more recently investigated option consists in building $T\varphi$ from the solution v^φ of an auxiliary elliptic problem posed in a layer $C_1 \cup C_2$ surrounding the interface Σ (Figure 1): given $B = I, \partial_\nu$ or $I + \omega^{-1}\,\partial_\nu$ (it can be shown [22] that the Robin operator $I + \omega^{-1}\,\partial_\nu$ is the best choice)

$$\begin{cases} T\varphi := \frac{1}{2}\left(\partial_{\nu_1} v_1^\varphi + \partial_{\nu_2} v_2^\varphi\right) \\ v_1^\varphi = v^\varphi|_{\Omega_1}, v_2^\varphi = v^\varphi|_{\Omega_2} \end{cases} \quad \text{where} \quad \begin{cases} -\Delta v + \omega^2\, v^\varphi = 0, & \text{in } C_1 \cup C_2, \\ v^\varphi = \varphi, & \text{on } \Sigma, \\ B\, v^\varphi = 0 & \text{on } \Sigma_j,\ j=1,2. \end{cases} \tag{27}$$

One advantage of such a DtN operator is that it is perfectly adapted to variable coefficients and other types of equations. Moreover it gives very good performances in practice. Let us consider the experiment of the scattering of a plane wave by a circular disk (see Figure 3): the interface is a circle of radius R and $\omega\, R = 9$. We use P_1 finite elements on a meshstep $h = 2\pi/(40\omega)$ and 0 as the initial guess. In Figure 2, we show the evolution of the relative $H^1(\Omega_1 \cup \Omega_2)$ norm of the error $u_h^n - u_h$, u_h being the solution of the undecomposed discrete problem, as a function of n for $T = I$ and T given by **(B)** or **(C)** with $C_1 = \Omega_1$ (red domain) $C_2 = \Omega_2$ (blue domain). This clearly shows the interest of non-local versus local and of strategy **(C)** versus strategy **(B)**. The picture on the right shows that, with non-local operators, the number of iterations needed to reach a given tolerance is independent of h (this can be proven, see [4], and reflects the linear convergence for the continuous problem) while, if $T = I$ (or more generally any local operator), it increases when one refines the mesh. In Figure 3, we show the spatial structure of the error after 80 iterations (be careful: the scales are different in the two pictures). With $T = I$, the error concentrates near the interface and highly oscillates (from one mesh point to the other) along the interface. This is representative of the incapacity of local operators to produce linear convergence at the continuous level and explained in circular geometry by the Fourier azimuthal analysis: the modal convergence rate τ_m for the mth mode in θ tends to 1 for large m. With the DtN operator, the error does not concentrate and oscillates, as explained again by the modal analysis, at the (quasi)-**resonant mode**: observe the $m = 9$ lobes $\Leftrightarrow \omega R = 9$.

5 The problem of cross points

Consider now a partition of Ω into $N \geq 2$ subdomains Ω_j, where, for simplicity, Ω_N is an exterior layer, with the possibility that more than 2 boundaries $\partial\Omega_j$ meet at a so-called **cross point**. Such points raise theoretical and practical questions for DDMs that deserve a special treatment [1, 12, 20]. Denoting by Σ_{ij} the interface $\partial\Omega_i \cup \partial\Omega_j$

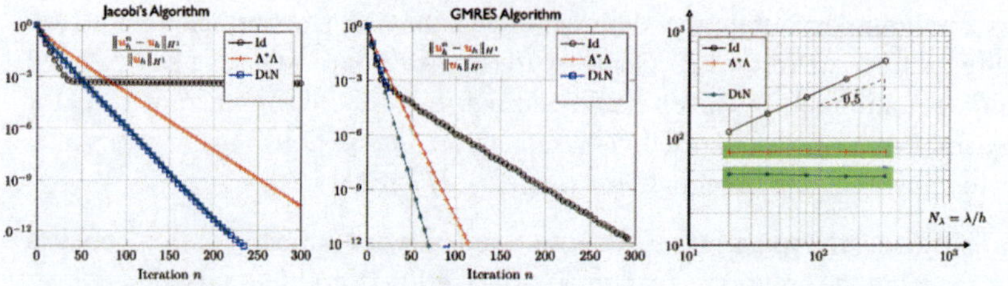

Fig. 2: Convergence histories (left and center). Iteration count versus mesh size (right)

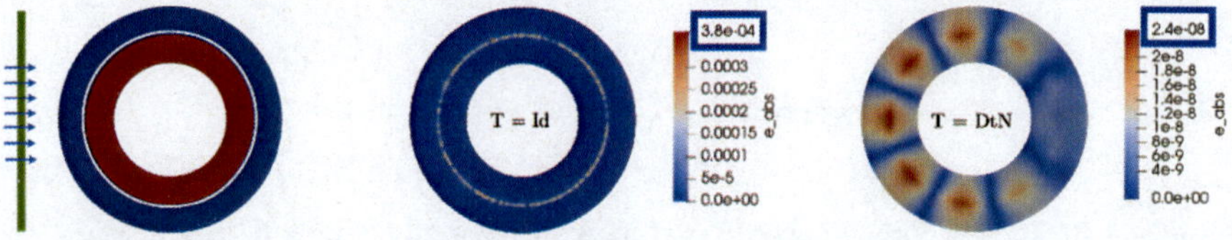

Fig. 3: Left: the experiment. Center, right: the errors after 80 iterations (the color bars differ!)

(possibly empty), the most naïve generalization of the transmission condition (9) consists in writing a transmission problem for $\{u_j\}$ with the transmission conditions

$$\begin{cases} \partial_{\nu_i} u_i + i\,\omega\, T_{ij}\, u_i = -\partial_{\nu_j} u_j + i\,\omega\, T_{ij}\, u_j, & \nu_i \text{ outgoing w.r.t. } \Omega_i, \\ \partial_{\nu_j} u_j + i\,\omega\, T_{ij}\, u_j = -\partial_{\nu_i} u_i + i\,\omega\, T_{ij}\, u_i, & \nu_j \text{ outgoing w.r.t. } \Omega_j, \end{cases} \tag{28}$$

where, aiming at achieving linear convergence, T_{ij} would be a **positive definite self-adjoint** operator from $H^{1/2}(\Sigma_{ij})$ in $H^{-1/2}(\Sigma_{ij})$. In this way, defining x_{ij} on Σ_{ij} similarly as (x_1, x_2) in (11) and $\mathbf{x}$ the collection of the $\{x_{ij}\}$, the transmission problem can be rewritten in an abstract form (15) with a natural generalization of the operator $\mathbf{A}$. The convergence of the DDM algorithm (22) is still guaranteed but the **linear convergence** faces the problem of the surjectivity of $\mathbf{I}-\mathbf{A}$, which relies on the existence of a solution to a generalized jump problem in Ω coupling the Helmholtz equation in each Ω_j with the inhomogeneous jump conditions:

$$\big[w\big]_{\Sigma_{ij}} = T^{-1}\Big(\frac{z_{ij} - z_{ji}}{2i\omega}\Big), \ \big[\partial_n w\big]_{\Sigma_{ij}} = z_{ij} - z_{ji}, \text{ given } (z_{ij}, z_{ji}) \in H^{-1/2}(\Sigma_{ij}). \tag{29}$$

Unfortunately, the inclusion of $\mathcal{T} := \{\gamma_J v := [v]_{\Sigma_{ij}} / v_i \in H^1(\Omega_i)\}$ in $\Pi\ H^{1/2}(\Sigma_{ij})$ is **strict**, with infinite codimension, if **cross points** exist [25]. This **defect of surjectivity** of the jump operator γ_J is an obstacle to the first condition in (29): we meet again a functional **misfit** as for the two domains case when $s < 1/2$ in (8).

In [3], a new paradigm was proposed, abandoning the interfaces Σ_{ij} to the profit of the boundaries $\Sigma_i = \partial\Omega_i$ $(i < N)$ and $\Sigma_N = \partial\Omega_N \setminus \partial\Omega$ and the skeleton $\mathbf{\Sigma} = \cup\,\Sigma_i$. This uses the concept of **multi-traces** developed for multi-domain boundary integral equations [5]: let $\Omega_{\mathbf{\Sigma}} := \Omega \setminus \mathbf{\Sigma}$ and (γ_D, γ_N) be the two **surjective** (multi)-trace operators

$$\begin{cases} u \in H^1(\Omega_\Sigma) & \mapsto \gamma_D u = \{u_i|_{\Sigma_i}\} \quad \in \mathcal{M}_D(\Sigma) := \Pi\, H^{\frac{1}{2}}(\Sigma_i), \\ \mathbf{v} \in H(\mathrm{div}, \Omega_\Sigma) & \mapsto \gamma_N \mathbf{v} = \{\mathbf{v}_i \cdot \nu_i|_{\Sigma_i}\} \in \mathcal{M}_N(\Sigma) := \Pi\, H^{-\frac{1}{2}}(\Sigma_i). \end{cases} \tag{30}$$

Note that $\mathcal{M}_N(\Sigma)$ is the dual space of $\mathcal{M}_D(\Sigma)$ and we shall denote by $\langle\cdot,\cdot\rangle_\Sigma$ the natural duality bracket that extends the $L^2(\Sigma)$ inner product. As $H^1(\Omega) \subset H^1(\Omega_\Sigma)$ and $H(\mathrm{div}\,\Omega) \subset H(\mathrm{div}\,\Omega_\Sigma)$, we can define

$$\mathcal{S}_D(\Sigma) := \gamma_D\big[H^1(\Omega)\big] \subset \mathcal{M}_D(\Sigma), \quad \mathcal{S}_N(\Sigma) := \gamma_N\big[H(\mathrm{div}, \Omega)\big] \subset \mathcal{M}_N(\Sigma).$$

The idea is to reformulate the classical Dirichlet and Neumann transmission conditions for $u = \{u_i\} \in H^1(\Omega_\Sigma)$, namely $[u]_{\Sigma_{ij}} = 0$ and $[\partial_\nu u]_{\Sigma_{ij}} = 0$, in a non-standard form expressed in terms of the traces $\gamma_D u$ and $\gamma_N(\nabla u)$ that writes as

$$-\Delta u_i - \omega^2 u_i = f, \quad \textbf{(D)} \quad \gamma_D u \in \mathcal{S}_D(\Sigma), \quad \textbf{(N)} \quad \gamma_N(\nabla u) \in \mathcal{S}_N(\Sigma).$$

To recover the framework of Section 3, we first express **(D)** and **(N)** in an impedance form. To do so, we introduce **positive self-adjoint** impedance operators associated to the Σ_i's (and no longer the Σ_{ij}'s), $T_i \in \mathcal{L}(H^{1/2}(\Sigma_i), H^{-1/2}(\Sigma_i))$, where each T_i is an isomorphism, so that, if $\mathbf{T} = \mathrm{diag}\, T_i \in \mathcal{L}(\mathcal{M}_D(\Sigma), \mathcal{M}_N(\Sigma))$,

$$(\varphi, \psi) := \langle \varphi, \mathbf{T}^{-1} \psi \rangle_\Sigma \quad \text{is an Hilbert inner product in } \mathbf{V} := \mathcal{M}_N(\Sigma). \tag{31}$$

Mimicking (11), we set (S) : $\mathbf{x} := \gamma_N \nabla u + i\,\omega\, \mathbf{T}\gamma_D u$ and $\mathbf{y} := -\gamma_N \nabla u + i\,\omega\,\mathbf{T}\gamma_D u$, the skeleton unknowns in $\mathbf{V}$. Let $\mathbf{S} = \mathrm{diag}\, S_i \in \mathcal{L}(\mathcal{M}_N(\Sigma))$, where each S_i is defined as in (12) (in Ω_i and T_i instead of T). Each S_i is isometric for the T_i-norm – (5) for $T = T_i$ – except $\mathcal{S}_N$, which is contractant. The Helmholtz equations in Ω_i rewrite as (13), namely $\mathbf{y} = \mathbf{S}\mathbf{x} + \widetilde{\mathbf{g}}$. It then remains to account for **(D)** and **(N)**. This relies on a key result of [5] characterizing $\mathcal{S}_D(\Sigma)$ and $\mathcal{S}_N(\Sigma)$ as "orthogonal" to each other:

Lemma 1 [5] *Let* $\varphi \in \mathcal{M}_D(\Sigma)$ *and* $\psi \in \mathcal{M}_N(\Sigma))$. *Then*

$$(i) \quad \varphi \in \mathcal{S}_D(\Sigma) \iff \langle \psi_N, \varphi \rangle_\Sigma = 0, \quad \forall\, \psi_N \in \mathcal{S}_N(\Sigma),$$
$$(ii) \quad \psi \in \mathcal{S}_N(\Sigma) \iff \langle \psi, \varphi_D \rangle_\Sigma = 0, \quad \forall\, \varphi_D \in \mathcal{S}_D(\Sigma).$$

This lemma is a direct consequence of Green's identity, in which the left-hand side vanishes if $u \in H^1(\mathbb{R}^d)$ or $\mathbf{v} \in H(\mathrm{div}, \mathbb{R}^d)$ (below $\mathbb{R}^d_\Sigma = \mathbb{R}^d \setminus \Sigma$):

$$\forall\, (u, \mathbf{v}) \in H^1(\mathbb{R}^d_\Sigma) \times H(\mathrm{div}, \mathbb{R}^d_\Sigma), \quad \sum_i \int_{\Omega_i} (\nabla u_i \cdot \mathbf{v}_i + u_i\, \mathrm{div}\, \mathbf{v}_i) = \langle \gamma_N \mathbf{v}, \gamma_D u \rangle_\Sigma.$$

Theorem 2 [3] *Let* $\mathbf{P}_N$ *be the orthogonal projector (in* $\mathcal{M}_N(\Sigma)$ *equipped with* (31)*) on* $\mathcal{S}_N(\Sigma)$. *The transmission conditions* **(D)** *and* **(N)** *are satisfied if and only if the unknowns* $\mathbf{x}$ *and* $\mathbf{y}$ *are related by* $\mathbf{y} = \mathbf{\Pi}\, \mathbf{x}$, *where* $\mathbf{\Pi} = \mathbf{I} - 2\,\mathbf{P}_N$.

Proof Let $\varphi := \gamma_D u$ and $\psi := \gamma_N u$. By (S), **(N)** is equivalent to $\mathbf{y} - \mathbf{x} \in \mathcal{S}_N(\Sigma)$ while **(D)** is equivalent to $\mathbf{T}^{-1}(\mathbf{x} + \mathbf{y}) \in \mathcal{S}_D(\Sigma)$, that is to say, by Lemma 1 and (31), to $(\mathbf{y} + \mathbf{x}, \psi_N) = 0$, $\forall\, \psi_N \in \mathcal{S}_N(\Sigma)$. Thus, writing $\mathbf{y} + \mathbf{x} = (\mathbf{y} - \mathbf{x}) + 2\mathbf{x}$, this gives

$$\big((\mathbf{y} - \mathbf{x}) + 2\mathbf{x}, \psi_N\big) = 0, \quad \forall\, \psi_N \in \mathcal{S}_N(\Sigma). \tag{32}$$

Since $\mathbf{y} - \mathbf{x} \in \mathcal{S}_N(\mathbf{\Sigma})$, this is nothing but $\mathbf{y} - \mathbf{x} = \mathbf{P}_N(-2\mathbf{x})$. □

Proceeding as in Section 3 to eliminate $\mathbf{y}$, the problem in $\mathbf{x}$ rewrites as in (15), with $\mathbf{V} := \mathcal{M}_N(\mathbf{\Sigma})$ and $\mathbf{A} = \mathbf{\Pi}\,\mathbf{S}$, the exchange operator (16) being replaced by $\mathbf{\Pi} = \mathbf{I} - 2\,\mathbf{P}_N$. The reader will notice that, as the exchange operator, $\mathbf{\Pi}$ is **isometric** and **involutive**. As a consequence, $\mathbf{A}$ is contractant. The invertibility of $\mathbf{I} - \mathbf{A}$ is linked to a generalized jump problem across the skeleton (instead of (29)), a solution of which is ensured by the **surjectivity** of γ_D and γ_N (30): the misfit due to the defect of surjectivity of the operator γ_J in the interface approach has been eliminated. The conditions for linear convergence of (22) are thus satisfied.

It is worth mentioning that the evaluation of $\mathbf{\Pi x}$ amounts to solving the (coercive and $\mathbf{T}$ dependent) variational problem (32) on $\mathbf{\Sigma}$ for $\mathbf{y} - \mathbf{x}$. Even though each T_i is local to Σ_i, being posed in $\mathcal{S}_N(\mathbf{\Sigma})$, the problem is non-local over $\mathbf{\Sigma}$. Thus, $\mathbf{\Pi x}$ couples all Σ_i's: rather than an exchange across interfaces, it is a **communication** operator (but without a cross point, a "natural" choice for T_i gives back the exchange). Working in $\mathbf{V} = \mathcal{M}_N(\mathbf{\Sigma})$ means that the Neumann condition (**N**) is handled in a **strong sense** while the Dirichlet condition (**D**) is handled **weakly** via (32). The (dual) opposite choice is possible, see [6]. In our case, the space discretization of the problem uses a finite element space $V_h(\Omega)$ for $H(\mathrm{div}, \Omega)$ and a natural candidate for an approximation space of $\mathcal{S}_N(\mathbf{\Sigma})$ is $\mathcal{S}_N^h(\mathbf{\Sigma}) := \gamma_N[V_h(\Omega)]$. In Figure 4, we demonstrate that the developments of this section are not only a question of mathematical beauty. On the model problem of Section 4 and a partition of Ω into 10 subdomains with one cross point, we compare Després's condition (a), non-local interface operators T_{ij} (b) and finally the **multi-trace** method (c) showing the error after 10 iterations. In case (b), we see that the non-local interface operators solve most of the problems with $T = I$ but produce an important error (the big peak) concentrated around the cross point, an error which is eliminated with the **multi-trace** strategy!

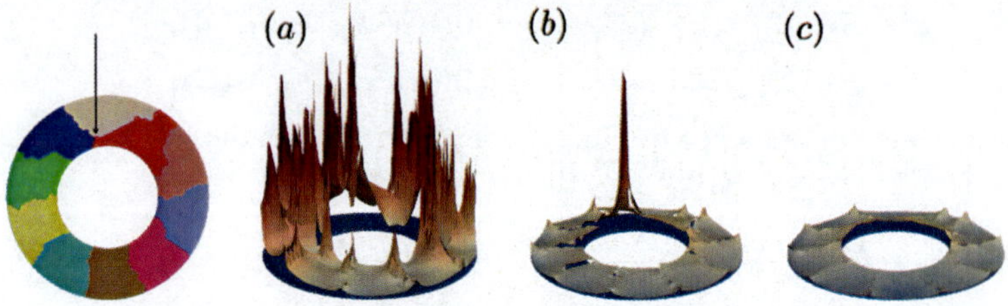

Fig. 4: Left: the 10 subdomains with one cross point (the arrow). Right: the errors after 10 iterations

Acknowledgements This work was supported by the research grant ANR-15-CE23-0017-01.

References

1. A. Bendali and Y. Boubendir. Non-overlapping Domain Decomposition Method for a Nodal Finite Element Method. *Numerische Mathematik* **103**(4), 515–537 Jun (2006).
2. Y. Boubendir, X. Antoine and C. Geuzaine. A Quasi-Optimal Non-Overlapping Domain Dec. Algorithm for the Helmholtz Equation. *J. Comp. Phys.* **213**(2), 262–280 (2012).
3. X. Claeys. Non-local variant of the Optimised Schwarz Method for arbitrary non-overlapping subdomain partitions. *ESAIM: M2AN* **55**(2), 429–448 (2021).
4. X. Claeys, F. Collino, P. Joly and E. Parolin. A discrete domain decomposition method for acoustics with uniform exponential rate of convergence using non-local impedance operators. In: *DDMs in Science and Engineering XXV*, Springer International Publishing (2020).
5. X. Claeys and R. Hiptmair. Multi-trace boundary integral formulation for acoustic scattering by composite structures. *Comm. Pure Appl. Math.* **66**(8), 1163–1201 (2013).
6. X. Claeys and E. Parolin. Robust treatment of cross points in Optimized Schwarz Methods. *arXiv preprint* arXiv.2003.06657 (2020).
7. F. Collino, S. Ghanemi and P. Joly. Domain decomposition method for harmonic wave propagation: a general presentation. *CMAME* **184**(24), 171–211 (2000).
8. F. Collino, P. Joly and M. Lecouvez. Exponentially convergent non overlapping domain decomposition methods for the Helmholtz equation. *ESAIM: M2AN* **54**(3), 775–810 (2020).
9. M. Crouzeix and A. Greenbaum. Spectral Sets: Numerical Range and Beyond. *SIAM J. on Matrix Anal. and Appl.* **40**(3), 1087–1101 (2019).
10. B. Després. *Méthodes de décomposition de domaine pour la propagation d'ondes en régime harmonique*. PhD thesis, Université Paris IX Dauphine (1991).
11. B. Després, P. Joly, and J.E. Roberts. A domain decomposition method for the harmonic Maxwell equations. In: *Iterative Methods in Linear Algebra (Brussels, 1991)*, pp. 245–252 (1992).
12. B. Després, A. Nicolopoulos and B. Thierry. Corners and stable optimized domain decomposition methods for the Helmholtz problem. *preprint HAL hal-02612368* (2020).
13. M. Gander, F. Magoulès and F. Nataf. Optimized Schwarz Methods without Overlap for the Helmholtz Equation. *SIAM J. on Sci. Comp.* **24**(1), 38–60 (2002).
14. M.J. Gander and H. Zhang. A Class of Iterative Solvers for the Helmholtz Equation: Factorizations, Sweeping Preconditioners, Source Transfer, Single Layer Potentials, Polarized Traces, and Optimized Schwarz Methods. *SIAM Rev.* **61**(1), 3–76 (2019).
15. S. Ghanemi. *Méthodes de DDM avec conditions de transmission non locales pour des problèmes de propagation d'ondes*. PhD thesis, Université Paris IX Dauphine (1996).
16. T. Hagstrom, R.P. Tewarson and A. Jazcilevich. Numerical experiments on a domain decomposition algorithm for nonlinear elliptic boundary value problems. *Applied Mathematics Letters* **1**(3), 299–302 (1988).
17. M. Lecouvez. *Méthodes itératives de DDM sans recouvrement avec convergence géométrique pour l'équation de Helmholtz*. PhD thesis, École polytechnique (2015).
18. P.-L. Lions. On the Schwarz alternating method. I. In: *First international symposium on DDMs for PDEs*, volume 1, pp. 1–42 (1988).
19. P.-L. Lions. On the Schwarz alternating method. III: a variant for nonoverlapping subdomains. In: *Third international symposium on DDMs for PDEs*, volume 6, pp. 202–223 (1990).
20. A. Modave, C. Geuzaine and X. Antoine. Corner treatments for high-order local absorbing boundary conditions in high-frequency acoustic scattering. *J. Comp. Phys.* **401**, 109029 (2020).
21. F. Nataf, F. Rogier and E. De Sturler. Optimal Interface Conditions for Domain Decomposition Methods. *Tech. Report, CMAP Ecole Polytechnique* (1994).
22. E. Parolin. *Non-overlapping domain decomposition methods with non-local transmission operators for harmonic wave propagation problems*. PhD thesis, Inst. Polytech. de Paris (2020).
23. Y. Saad. *Iterative Methods for Sparse Linear Systems*. SIAM (2003).

24. H. Schwarz. *Ueber einen Grenzübergang durch alternirendes Verfahren.* Zürcher u. Furrer (1870).
25. T. Von Petersdorff. Boundary integral equations for mixed Dirichlet, Neumann and transmission problems. *M2AS* (1989).

Quantitative Analysis of Nonlinear Multifidelity Optimization for Inverse Electrophysiology

Fatemeh Chegini, Alena Kopaničáková, Martin Weiser, Rolf Krause

1 Introduction

Reliable cardiac excitation predictions depend not only on accurate geometric and physiological models, usually formulated as PDEs, and our ability to solve those faithfully, but also on the model's correct parameterization. One critical parameter is the tissue conductivity. Its correct identification from measurement data can provide valuable information about the location and size of scars, which would be beneficial for diagnosis and treatment of several heart diseases [13].

One approach to parameter identification in electrocardiography is minimizing the mismatch between simulated and measured voltages on the heart's inner surface [28]. This inverse problem can be formulated as a PDE constrained optimization problem and has been addressed, for example, by using BFGS for a reduced problem formulation [32]. A related general framework for multilevel parameter optimization can be found in, e.g., [23].

Solving this optimization problem is, however, a major computational challenge, since the forward models describing the electrical excitation of the heart exhibit very different temporal and spatial scales and therefore require the use of fine meshes and short time steps. Together with a considerable number of optimization iterations, the

Fatemeh Chegini
Zuse Institute Berlin, Germany, and Center for Computational Medicine in Cardiology, Università della Svizzera italiana, Switzerland, e-mail: chegini@zib.de

Alena Kopaničáková
Euler Institute, Università della Svizzera italiana, Switzerland,
e-mail: alena.kopanicakova@usi.ch

Martin Weiser
Zuse Institute Berlin, Germany, e-mail: weiser@zib.de

Rolf Krause
Center for Computational Medicine in Cardiology, Euler Institute, Università della Svizzera italiana, Switzerland, e-mail: rolf.krause@usi.ch

S. C. Brenner et al. (eds.), *Domain Decomposition Methods in Science and Engineering XXVI*, Lecture Notes in Computational Science and Engineering 145,
https://doi.org/10.1007/978-3-030-95025-5_6

resulting computational complexity is a major hurdle for widespread practical application. Consequently, several attempts have been made to reduce the computational effort, including model reduction by proper orthogonal decomposition and empirical interpolation [33], Gaussian process surrogate models [11], and topological derivative formulations [2].

A nonlinear multilevel approach based on heterogeneous model hierarchies has recently been proposed by the authors [6]. In the present study, we analyze the performance benefits and relative merits of different hierarchies quantitatively, and obtain insights concerning the behavior of nonlinear multilevel approaches applied to the inverse problem at hand.

The remainder of the paper is organized as follows. Mathematical models for cardiac electrophysiology are briefly recalled in Section 2, while Section 3 formalizes the inverse problem under consideration. In Section 4, the recursive multilevel trust-region (RMTR) method and the model hierarchies are described. Section 5 contains the numerical results for single-level trust-region, RMTR with multigrid, and RMTR with heterogeneous model hierarchies, using limited memory BFGS.

2 Electrophysiological models

Excitation of cardiac tissue occupying the domain $\Omega \subset \mathbb{R}^d$ in terms of the transmembrane voltage v between intracellular and extracellular domain is usually described by the bidomain model or its monodomain and eikonal simplifications [9]. For simplicity, we will consider only monodomain and eikonal models here.

The monodomain system consists of a nonlinear parabolic reaction-diffusion equation for the transmembrane voltage $v : \Omega \to \mathbb{R}$ and a system of ordinary differential equations (ODEs) describing the dynamics of the ion channels, which regulate the transmembrane current, in terms of gating variables $w : \Omega \to \mathbb{R}$:

$$\begin{aligned}
\operatorname{div}(\sigma \nabla v) &= \chi(C_m \dot{v} + I_{\text{ion}}(v, w)) && \text{in } \Omega \times [0, T] \\
\dot{w} &= f(v, w) && \text{in } \Omega \times [0, T] \\
\boldsymbol{n}^T \sigma \nabla v &= 0 && \text{on } \partial\Omega \times [0, T] \\
v|_{t=0} &= v_0 && \text{in } \Omega \\
w|_{t=0} &= w_0 && \text{in } \Omega.
\end{aligned} \tag{1}$$

Here, $\boldsymbol{n}$ is the unit outer normal vector to Ω, σ a symmetric positive definite conductivity tensor, χ the membrane surface area per unit volume, and C_m the membrane capacity per unit area. I_{ion} denotes the transmembrane current density and f the gating dynamics, both defined by an electrophysiological membrane model (2).

Many different membrane models have been developed [17, 19]. Here, we use the modified Fitzhugh–Nagumo (FHN) model by [31],

$$I_{\text{ion}}(v,w) = \eta_0 v \left(1 - \frac{v}{v_{\text{th}}}\right)\left(1 - \frac{v}{v_{\text{pk}}}\right) + \eta_1 v w$$
$$f(v,w) = \eta_2 \left(\frac{v}{v_{\text{pk}}} - \eta_3 w\right), \tag{2}$$

with positive coefficients $\eta_0, \eta_1, \eta_2, \eta_3, v_{th}, v_{pk}$. In particular, peak and threshold potential are given by $v_{\text{pk}} > v_{\text{th}}$, respectively.

Eikonal models derived from bidomain or monodomain models [7, 8, 29] consider only the activation time $u(x)$ of the tissue at a particular spatial position x, and recover the transmembrane voltage by the travelling wave ansatz

$$v(x,t) = v_m(t - u(x)), \tag{3}$$

which depends on some fixed activating front shape v_m of usually hyperbolic tangent or sigmoid structure. This ansatz results in a nonlinear elliptic equation for the activation time,

$$c_0\sqrt{\nabla u \cdot \sigma \nabla u} - \nabla \cdot (\sigma \nabla u) = \tau_m \qquad \text{on } \Omega, \tag{4}$$

where c_0 and τ_m are parameters used for fitting the eikonal model to mono- or bidomain models.

As the eikonal equation is stationary and activation times are significantly smoother than the transmembrane voltage, eikonal solutions can be obtained much faster and on coarser grids than monodomain solutions. Nevertheless, they are a rather good approximation of the more involved models in many cases.

3 Inverse problem of Conductivity Identification

Here we turn to the prototypical inverse problem of estimating a scalar conductivity $\sigma \in H^1(\Omega)$ from N_σ voltages $\hat{v}_i$ given at disjoint open surface patches $\Gamma_i \subset \partial\Omega$ by minimizing the mismatch between simulated voltages $v|_{\Gamma_i}$ and measurements. Writing $\Gamma = \bigcup_i \Gamma_i$, the resulting optimization problem reads

$$\begin{aligned} \min_{v,\sigma} \hat{J}(v,\sigma) &= \frac{1}{2}\|v - \hat{v}\|^2_{L^2(\Gamma \times [0,T])} + R(\sigma,\beta) \\ \text{subject to} \quad C(v,\sigma) &= \mathbf{0} \\ \sigma &\in \mathcal{F} = \{s \in L^2(\Omega) \mid \sigma_{\min} \leq s \leq \sigma_{\max}\}. \end{aligned} \tag{5}$$

$C(v,\sigma)$ is the monodomain model (1), and $\sigma_{\min}$ and $\sigma_{\max}$ are the lower and upper bounds of the conductivity. As the problem is ill-posed, a regularization term R is added [30] in order to reduce high-frequent solution components amplified by measurement noise. Here, we choose $R(\sigma,\beta) = \frac{1}{2}\|\beta_1(\sigma - \overline{\sigma})\|^2_{L^2(\Omega)} + \frac{1}{2}\|\beta_2 \nabla \sigma\|^2_{L^2(\Omega)}$, where $\overline{\sigma}$ is an a priori reference conductivity. The regularization parameters β_i can be determined, e.g., by the L-curve method [5] or Morozov's discrepancy principle. For a more detailed discussion of modeling aspects we refer to [6].

Reduced problem

In order to avoid a large 4D discretization of the space-time problem resulting from the first order necessary optimality conditions, we resort to the reduced problem by eliminating the transmembrane voltage v explicitly as $v(\sigma)$ satisfying $C(v(\sigma),\sigma) = 0$, and obtain

$$\begin{aligned} \min_{\sigma\in H^1(\Omega)} \quad & J(\sigma) = \hat{J}(v(\sigma), s) \\ \text{subject to} \quad & \sigma \in \mathcal{F}. \end{aligned} \tag{6}$$

This bound-constrained problem can then be solved by gradient type algorithms such as steepest descent or quasi-Newton methods. The gradient of the reduced objective J with respect to σ can be obtained efficiently by solving the adjoint equation

$$\begin{aligned} -\chi C_m \dot{\lambda} &= \operatorname{div}(\sigma\nabla\lambda) - \chi I_{\text{ion},v}(v,w)\lambda - f_v(v,w)\eta \\ -\dot{\eta} &= \chi I_{\text{ion},w}(v,w)\lambda + f_w(v,w)\eta \end{aligned} \tag{7}$$

with terminal and boundary conditions

$$\begin{aligned} \lambda(T) = 0, \quad \eta(T) &= 0 \\ \boldsymbol{n}^T\sigma\nabla\lambda &= 0 && \text{on } (\partial\Omega\backslash\Gamma)\times[0,T] \\ \boldsymbol{n}^T\sigma\nabla\lambda &= \hat{v} - v && \text{on } \Gamma\times[0,T] \end{aligned}$$

backwards in time and then computing

$$\nabla J = \int_0^T \nabla\lambda^T \nabla v \,\mathrm{d}t + \nabla R. \tag{8}$$

Since the state v, η enters as data into the adjoint equation, the whole 4D trajectory still needs to be stored. This can be done efficiently by error-controlled lossy data compression [14]. When using the eikonal equation for describing cardiac excitation, the reduced gradient ∇J can be computed analogously. Conveniently, the adjoint equation is then again a single and much simpler stationary equation.

Discretization

For the spatial discretization of the conductivity σ, the transmembrane voltage v, the gating variables w, and the adjoint states λ and η, and the activation time u we employ standard linear finite elements on a simplicial grid covering the domain Ω. The time integration is done by a common equidistant implicit-explicit Euler scheme with operator splitting for both the monodomain problem (1) and adjoint equation (7).

Denoting by $\boldsymbol{x} \in \mathbb{R}^N$ the coefficient vector of the conductivity σ, we obtain thus, with a slight abuse of notation, the discretized version of (6) as

$$\begin{aligned} &\min_{\boldsymbol{x} \in \mathbb{R}^N} \ J(\boldsymbol{x}) \\ &\text{subject to} \quad \boldsymbol{x} \in \mathcal{F}. \end{aligned} \tag{9}$$

Due to the use of Lagrangian finite elements, the continuous feasible set $\mathcal{F}$ for σ translates into component-wise bounds on $\boldsymbol{x}$, such that (9) is again a bound-constrained problem.

4 Multilevel Quasi-Newton Trust-region Method

In this section, we discuss how to minimize (9) using a multilevel solution strategy, namely the recursive multilevel trust-region (RMTR) method [15]. The RMTR method combines the global convergence properties of the trust-region method with the efficiency of multilevel methods. In this work, we consider three different approaches for obtaining the multilevel hierarchy: i) multi-resolution, ii) multi-model, and iii) combined (multi-resolution and multi-model) approach.

Quasi-Newton trust-region method

A trust-region method (TR) is an iterative method which generates a sequence $\{\boldsymbol{x}_i\}$ of iterates converging to a first-order critical point [10]. At each iteration i, the TR method approximates the objective function J by a quadratic model

$$m_i(\boldsymbol{x}_i + \mathbf{p}) = J(\boldsymbol{x}_i) + J'(\boldsymbol{x}_i)\mathbf{p} + \frac{1}{2}\mathbf{p}^T\mathbf{H}_i\mathbf{p}$$

around the current iterate $\boldsymbol{x}_i$. For the Hessian approximation $\mathbf{H}_i$ we employ a memory-efficient quasi-Newton approach known as the L-BFGS(m) [3, 4], where only the m most recent gradients are taken into account in order to update the Hessian $\mathbf{H}_i \approx J(\boldsymbol{x}_i)''$ recursively using a rank-two update formula [24]. For $m \ll n$, significantly less storage is needed compared to the dense Hessian approximations used in [32].

Being based on a Taylor-like approximation, the model m_i is considered to be an adequate representation of the objective J only in a certain region, called the trust-region. The trust-region $\mathcal{B}_i := \{\boldsymbol{x}_i + \mathbf{p} \in \mathbb{R}^n \mid \|\mathbf{p}\| \leq \Delta_i\}$ is defined around the current iterate, with a size prescribed by the trust-region radius $\Delta_i > 0$ and a shape defined by the choice of norm. Here, we employ the maximum norm $\|\cdot\|_\infty$, which simplifies the step computation in bound-constrained problems compared to the Euclidean norm. The trial step $\mathbf{p}_i$ is determined by solving the constrained minimization problem

$$\min_{\mathbf{p}_i \in \mathbb{R}^n} \; m_i(\boldsymbol{x}_i + \mathbf{p}_i) \quad \text{subject to} \quad \boldsymbol{x}_i + \mathbf{p}_i \in \mathcal{F}, \qquad \|\mathbf{p}_i\|_\infty \le \Delta_i. \tag{10}$$

The first constraint in (10) ensures the feasibility of the iterates throughout the solution process, while the second constraint restricts the size of the trial step $\mathbf{p}_i$. Both constraints are defined component-wise, such that (10) is a bound-constrained problem with easily computable bounds.

To ensure global convergence, it is sufficient to solve the trust-region subproblems (10) approximately, such that an approximate solution $\mathbf{p}_i$ of (10) satisfies the so-called sufficient decrease condition (SDC), see [10]. An obtained step $\mathbf{p}_i$ is accepted, if the actual decrease in the objective, $J(\boldsymbol{x}_i) - J(\boldsymbol{x}_i + \mathbf{p}_i)$, agrees sufficiently well with the predicted decrease $m_i(\boldsymbol{x}_i) - m_i(\boldsymbol{x}_i + \mathbf{p}_i)$. This is quantified in terms of the trust-region ratio

$$\rho_i = \frac{J(\boldsymbol{x}_i) - J(\boldsymbol{x}_i + \mathbf{p}_i)}{m_i(\boldsymbol{x}_i) - m_i(\boldsymbol{x}_i + \mathbf{p}_i)}. \tag{11}$$

If ρ_i is close to unity, there is a good agreement between the objective J and the model m_i and it is therefore safe to accept the step $\mathbf{p}_i$. More precisely, the step $\mathbf{p}_i$ is accepted, only if $\rho_i > \eta_1$, where $0 < \eta_1 < 1$. In addition, the trust-region radius has to be adjusted accordingly.

Remark 1 It is important to update the approximation $\mathbf{H}_i$ even if the trial step $\mathbf{p}_i$ is rejected, since the rejection might indicate that the current $\mathbf{H}_i$ is not an adequate approximation of the true Hessian $J''(\boldsymbol{x}_i)$.

Remark 2 Using the L-BFGS method, the implementation of the trust-region algorithm can be realized in a matrix-free way. The operations involving $\mathbf{H}_i$, or its inverse $(\mathbf{H}_i)^{-1}$, can be implemented using the approach proposed in [25] and the two-loop recursion algorithm developed in [27], respectively.

Recursive multilevel trust-region method

The computational cost of the trust-region method is dominated by evaluating the objective in (11) and the reduced gradient $J'(\boldsymbol{x}_i)$ via (8), which incurs the solution of at least two parabolic equations per accepted trial step. Reducing the computational effort requires a decrease of the number of steps, which in turn is only possible if the quadratic models m_i of the objective are replaced or complemented by models J_i^- that approximate J on larger trust-regions, but are nevertheless significantly cheaper to minimize than the original objective. Approximate models that we consider here are (i) the monodomain equation on coarser grids and (ii) an eikonal model.

These models can be defined on the same discretization, i.e. approximation space, for σ of size n, or on a coarser one of size $n^- < n$, in which case a transfer between the original problem and the model is necessary. This affects the transfer of the current iterate x_i (projection) and the gradient ∇J (restriction) to the model J_i^-, and the transfer of the model's minimizer back to the original problem (prolongation). Note that, if the models J_i^- are formulated on the same discretization, all these transfers are trivial. Otherwise, we define both the prolongation $\mathbf{I} : \mathbb{R}^{n^-} \to \mathbb{R}^n$ and the projection $\mathbf{P} : \mathbb{R}^n \to \mathbb{R}^{n^-}$ as a pseudo-L^2-projection, as proposed in [18] and successfully applied in [21]. We assemble these transfer operators using the library MOONoLith [22]. As usual, the restriction is defined as the adjoint of the prolongation, i.e. $\mathbf{R} = \mathbf{I}^T$.

Naturally, we intend J_i^- to approximate J well. Therefore, we enforce first-order consistency between both models i.e., the gradients of both models shall coincide locally as far as possible. As is common for nonlinear multilevel schemes, the model functions J_i^- can be defined in terms of some computationally cheaper/coarse approximation j^- of the objective J by means of the additive approach [26] as

$$J_i^-(x^-) = j^-(x^-) + (x^- - \mathbf{P}x_i)^T \underline{(\mathbf{R}\nabla J(x_i) - \nabla j^-(\mathbf{P}x_i))}. \tag{12}$$

Alternatively, we can exploit a multiplicative approach [1, 20] and define the models

$$J_i^-(x^-) = \beta(x_i, x^-) j^-(x^-) \tag{13}$$

with

$$\beta(x, x^-) = \frac{J(x)}{j^-(\mathbf{P}x)} + (x^- - \mathbf{P}x)^T \underline{\left(\frac{1}{j^-(\mathbf{P}x)} \mathbf{R}\nabla J(x) - \frac{J(x)}{(j^-(\mathbf{P}x))^2} \nabla j^-(\mathbf{P}x) \right)}.$$

Both approaches employ a so-called coupling term (underlined), which takes into account the difference between restricted original gradient $\mathbf{R}\nabla J(x_i)$ and initial coarse gradient $\nabla j^-(\mathbf{P}x_i)$. The use of this coupling term guarantees that the first-order behavior of J and J^- is locally coherent in the neighborhood of x_i and $\mathbf{P}x_i$, respectively [26].

At each iteration i, the trial step $\mathbf{p}_i \in \mathbb{R}^{n^-}$ is obtained either by approximately solving the quadratic trust-region subproblem (10) or the coarse subproblem

$$\min_{\mathbf{p} \in \mathbb{R}^{n^-}} J_i^-(\mathbf{P}x_i + \mathbf{p}), \quad \text{subject to} \quad x_i + \mathbf{I}\mathbf{p} \in \mathcal{F}, \qquad \|\mathbf{I}\mathbf{p}_i\|_\infty \le \Delta_i. \tag{14}$$

As is common for trust-region methods, it is not necessary to solve the problem (14) exactly. Indeed, it is sufficient that an approximate minimizer $\mathbf{p}$ of (14) satisfies the SDC condition. Here, we solve the nonlinear problem (14) iteratively by employing a few steps of the trust-region method. This gives rise to a recursive multilevel trust-region (RMTR) scheme [16]. A line search-based alternative would be the multilevel model correction (MMC) method [23].

Potentially, we can utilize a hierarchy of multiple coarse models $\{j^l\}_{l=1}^L$, where $L > 1$, which gives rise to a truly multilevel method. In this work, we obtain models $\{j^l\}_{l=1}^L$ by exploring the following alternatives:

1. **Multi-resolution:** We uniformly coarsen finite element grids (by factor of 2) in order to discretize the monodomain equation entering the reduced objective J. Consequently, the coarse-level models are computationally cheaper to optimize. *Note that a certain mesh resolution is required to reasonably resolve the monodomain model, such that mesh coarsening is limited.*
2. **Multi-model:** The eikonal model is used instead of monodomain on coarser grids. This model is significantly cheaper and a better global approximation model for the monodomain model compared to the standard quadratic model.
3. **Combined:** Combinations of multi-resolution and multi-model variants are also possible. For instance, one can obtain a hierarchy of models $\{j^l\}_{l=1}^{l=L}$ by first coarsening the spatial-resolution and then changing the model complexity.

Finally, we highlight the fact that the overall efficiency of the multilevel algorithm is determined by how many times the respective coarse and fine level models are minimized. For instance, in the multi-resolution approach, it is crucial to alternate between both models, such that the components of the error associated with a given level are effectively eliminated.

5 Numerical results

Here, we focus on numerical results for different algorithmic configurations on a simple 2D geometry. For the numerical tests, different synthetic transmembrane voltage data $\hat{v}$ have been created by simulations on a finer mesh. For illustration, we also present some reconstruction results for scar tissue on a 3D ventricular geometry. For a more detailed discussion of reconstruction quality we refer to [6].

5.1 Patient-specific geometry

As an example from clinical practice, we use the ventricular geometry of a patient with a nontransmural scar located on the left endocardium. In Fig. 1, scar tissue is shown in blue, sparse endocardial measurement locations by yellow spheres, and the reconstructed conductivity color-coded on the right. The reconstruction quality depends on the quantity and location of available cardiac mapping data. Due to the stability of excitation propagation, reliable results can in general only be expected in the vicinity of measurement locations. In this case, the small number of measured data on the left endocardium is not enough to reconstruct the scar shape accurately.

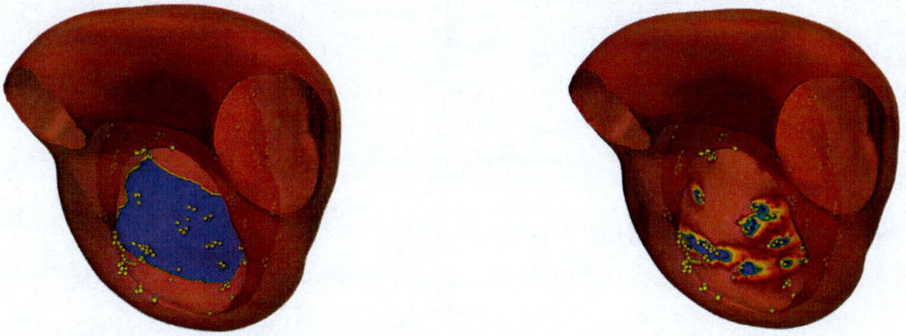

Fig. 1: *Left:* Target conductivity with marked measured data. *Right:* Solution with marked measured data.

5.2 Convergence study

In this section, we compare the convergence behavior of single-level trust-region methods with several RMTR variants on a simpler idealized 2D cross-section of a left ventricle. We use L-BFGS(m) with $m = 1$ or $m = 8$ secant pairs, and a termination criterion $\|\mathcal{P}(\boldsymbol{x} - \nabla J(\boldsymbol{x})) - \boldsymbol{x}\| < 10^{-4}$ based on the projected gradient expressed in terms of an orthogonal projection $\mathcal{P}$ onto the feasible set $\mathcal{F}$. The arising quadratic trust-region subproblems (10) are solved using the MPRGP method [12]. The RMTR method is configured with additive coarse level models (12) for multi-resolution variants, while multiplicative coarse level models (13) are employed for multi-model variants. Solution strategies are implemented as part of the open-source library UTOPIA [34], while the implementation of inverse problems, including monodomain and eikonal models, is part of our framework HEART. All simulations have been run using 10 nodes (XC50, 12 cores) of the Piz Daint supercomputer (CSCS, Switzerland).

To provide a more robust insight, four different sets of simulated measurement data $\hat{v}$ have been used: generated with monodomain on a finer mesh, with additional Gaussian noise, with slightly changed membrane area χ per volume, and generated with eikonal on a finer grid. We provide averaged iteration counts and run times in Tab. 1 for single level trust-region with different L-BFGS memory size, and for RMTR with two or three levels of monodomain on coarser grids or eikonal models.

The convergence results suggest that both monodomain multigrid and heterogeneous monodomain-eikonal multilevel methods lead to a significant reduction of iteration count by a factor between 3 and 6. For the used grid resolution, monodomain multigrid is more effective by a factor of 1.5 to 2 in reducing iteration counts. Since the coarse level subproblems are more expensive to solve, the heterogeneous multilevel approach is almost as efficient. We can also observe a slight convergence rate deterioration of the heterogeneous approach in the asymptotic phase, probably due to a less accurate Hessian approximation of the eikonal model. The three-level

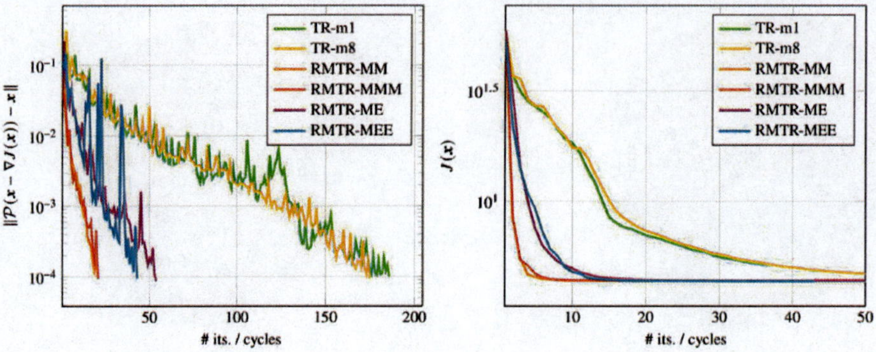

Fig. 2: The convergence history in terms of the projected gradient (left) and the objective function J. The measurement data were generated using the monodomain model on a finer mesh.

multigrid approach appears to be less effective than the two-level method, probably because the monodomain model deteriorates quickly for coarser grids.

	models	meshes	m	# its/cycles	time (minutes)
TR	mono (TR-m1)	$\mathcal{T}^3$	1	179 ± 29	149 ± 35
	mono (TR-m8)	$\mathcal{T}^3$	8	148 ± 19	127 ± 44
RMTR	mono-mono (RMTR-MM)	$\mathcal{T}^3, \mathcal{T}^2$	8	26 ± 5*	80 ± 49
	mono-mono-mono (RMTR-MMM)	$\mathcal{T}^3, \mathcal{T}^2, \mathcal{T}^1$	8	31 ± 11*	88 ± 56
	mono-eiko (RMTR-ME)	$\mathcal{T}^3, \mathcal{T}^3$	8	51 ± 6	97 ± 35
	mono-eiko-eiko (RMTR-MEE)	$\mathcal{T}^3, \mathcal{T}^3, \mathcal{T}^2$	8	47 ± 8	85 ± 32

Table 1: The average computational cost required by trust-region and RMTR method. The results are obtained by averaging over four datasets. The symbol * indicates that for one dataset the termination criterion was not satisfied within 500 cycles.

6 Conclusion

Identifying tissue conductivities using monodomain models from surface measurements is computationally expensive and calls for acceleration. Multilevel methods can be effective in two ways: First, the classical multigrid method based on a Galerkin projection of the Hessian improves the convergence rate of steepest descent or similar smoothers, which suffer from ill-conditioning. Second, nonlinear multilevel methods aim at improving the objective reduction also in the pre-asymptotic phase, where the progress of first or second order methods is limited due to high nonlinearity.

The numerical results suggest that the RMTR method used here is effective in both regimes and leads to a clear reduction of iterations. Due to the overhead of the subproblems, the reduction of run time is not as large, but still significant.

Acknowledgements Funding by EU and BMBF via the JU Euro-HPC project MICROCARD (grant agreement No 955495), Swiss National Science Foundation project ML^2 (197041), and CCMC (Fidinam, Horten), is gratefully acknowledged.

References

1. N.M. Alexandrov and R.M. Lewis. An overview of first-order model management for engineering optimization. *Optimization and Engineering* **2**(4), 413–430 (2001).
2. E. Beretta, C. Cavaterra and L. Ratti. On the determination of ischemic regions in the monodomain model of cardiac electrophysiology from boundary measurements. *Nonlinearity* **33**(11), 5659–5685 (2020).
3. R. Byrd, P. Lu, J. Nocedal and C. Zhu. A limited memory algorithm for bound constrained optimization. *SIAM Journal on Scientific Computing* **16**(5), 1190–1208 (1995).
4. R. Byrd, J. Nocedal and R. Schnabel. Representations of quasi-Newton matrices and their use in limited memory methods. *Mathematical Programming* **63**(1-3), 129–156 (1994).
5. D. Calvetti, B. Lewis and L. Reichel. GMRES, L-curves, and discrete ill-posed problems. *BIT Numerical Mathematics* **42**(1), 44–65 (2002).
6. F. Chegini, A. Kopaničáková, R. Krause and M. Weiser. Efficient identification of scars using heterogeneous model hierarchies. *EP Europace* **23**, i113–i122 (2021).
7. P. Colli Franzone, L. Guerri and S. Rovida. Wavefront propagation in an activation model of the anisotropic cardiac tissue: asymptotic analysis and numerical simulations. *Journal of mathematical biology* **28**(2), 121–176 (1990).
8. P. Colli Franzone, L. Guerri and B. Taccardi. Modeling ventricular excitation: axial and orthotropic anisotropy effects on wavefronts and potentials. *Mathematical biosciences* **188** (1-2), 191–205 (2004).
9. P. Colli Franzone, L. Pavarino and S. Scacchi. *Mathematical cardiac electrophysiology*. Springer (2014).
10. A. Conn, N. Gould and P. Toint. *Trust region methods*. SIAM (2000).
11. J. Dhamala, H.J. Arevalo, J. Sapp, M. Horacek, K.C. Wu, N.A. Trayanova and L. Wang. Spatially adaptive multi-scale optimization for local parameter estimation in cardiac electrophysiology. *IEEE Transactions on Medical Imaging* **36**(9), 1966–1978 (2017).
12. Z. Dostál. *Optimal quadratic programming algorithms: with applications to variational inequalities*, vol. 23. Springer Science & Business Media (2009).
13. J. Fernández-Armenta, A. Berruezo, L. Mont, M. Sitges, D. Andreu, E. Silva, J. Ortiz-Pérez, J. Tolosana, T. de Caralt, R. Perea, N. Calvo, E. Trucco, R. Borrás, M. Matas and J. Brugada. Use of myocardial scar characterization to predict ventricular arrhythmia in cardiac resynchronization therapy. *EP Europace* **14**(11), 1578–1586 (2012).
14. S. Götschel, N. Chamakuri, K. Kunisch and M. Weiser. Lossy compression in optimal control of cardiac defibrillation. *J. Sci. Comp.* **60**(1), 35–59 (2014).
15. S. Gratton, M. Mouffe, P. Toint and M. Weber-Mendonca. A recursive trust-region method for bound-constrained nonlinear optimization. *IMA Journal of Numerical Analysis* **28**(4), 827–861 (2008).
16. S. Gratton, A. Sartenaer and P. Toint. Recursive trust-region methods for multiscale nonlinear optimization. *SIAM Journal on Optimization* **19**(1), 414–444 (2008).
17. J.L. Greenstein and R.L. Winslow. An integrative model of the cardiac ventricular myocyte incorporating local control of Ca2+ release. *Biophysical journal* **83**(6), 2918–2945 (2002).

18. C. Groß and R. Krause. A recursive trust-region method for non-convex constrained minimization. In: *Domain Decomposition Methods in Science and Engineering XVIII*, pp. 137–144, Springer (2009).
19. A.L. Hodgkin and A.F. Huxley. A quantitative description of membrane current and its appl. to conduction and excitation in nerve. *The Journal of physiology* **117**(4), 500–544 (1952).
20. A. Kopaničáková. *Multilevel minimization in trust-region framework: algorithmic and software developments*. Ph.D. thesis, Università della Svizzera italiana (2020).
21. A. Kopaničáková and R. Krause. A recursive multilevel trust region method with application to fully monolithic phase-field models of brittle fracture. *Computer Methods in Applied Mechanics and Engineering* **360**, 112720 (2020).
22. R. Krause and P. Zulian. A parallel approach to the variational transfer of discrete fields between arbitrarily distributed unstructured finite element meshes. *SIAM Journal on Scientific Computing* **38**(3), C307–C333 (2016).
23. J. Li and J. Zou. A multilevel model correction method for parameter identification *Inverse Problems* **23**(5), 1759–1786 (2007).
24. D. Liu and J. Nocedal. On the limited memory BFGS method for large scale optimization. *Mathematical programming* **45**(1-3), 503–528 (1989).
25. D. Mahidhara and L. Lasdon. An SQP algorithm for large sparse nonlinear programs. Austin, MSIS Department–School of Business Administration, University of Texas (1991).
26. S. Nash. A multigrid approach to discretized optimization problems. *Optimization Methods and Software* **14**(1-2), 99–116 (2000). DOI 10.1080/10556780008805795
27. J. Nocedal. Updating quasi-Newton matrices with limited storage. *Mathematics of computation* **35**(151), 773–782 (1980).
28. A. Pullan, L. Cheng, M. Nash, A. Ghodrati, R. MacLeod and D. Brooks. The inverse problem of electrocardiography. In: *Comprehensive Electrocardiology*, pp. 299–344, Springer (2010).
29. A. Pullan, K. Tomlinson and P. Hunter. A finite element method for an eikonal equation model of myocardial excitation wavefront propagation. *SIAM J. Appl. Math.* **63**(1), 324–350 (2002).
30. A. Tikhonov. On the solution of ill-posed problems and the method of regularization. In: *Doklady Akademii Nauk*, vol. 151, pp. 501–504, Russian Academy of Sciences (1963).
31. A. Xu and M. Guevara. Two forms of spiral-wave reentry in an ionic model of ischemic ventricular myocardium. *Chaos: An Interdisciplinary Journal of Nonlinear Science* **8**(1), 157–174 (1998).
32. H. Yang and A. Veneziani. Estimation of cardiac conductivities in ventricular tissue by a variational approach. *Inverse Problems* **31**(11), (2015).
33. H. Yang and A. Veneziani. Efficient estimation of cardiac conductivities via POD-DEIM model order reduction. *Applied Numerical Mathematics* **115**, 180–199 (2016).
34. P. Zulian, A. Kopaničáková, M.C.G. Nestola, A. Fink, N. Fadel, A. Rigazzi, V. Magri, T. Schneider, E. Botter, J. Mankau and R. Krause. Utopia: A C++ embedded domain specific language for scientific computing. Git repository. `https://bitbucket.org/zulianp/utopia` (2016).

Adaptive Space-Time Finite Element and Isogeometric Analysis

Ulrich Langer

1 Introduction

The traditional approaches to the numerical solution of initial-boundary value problems (IBVP) for parabolic or hyperbolic Partial Differential Equations (PDEs) are based on the separation of the discretization in time and space leading to time-stepping methods; see, e.g., [20]. This separation of time and space discretizations comes along with some disadvantages with respect to parallelization and adaptivity. To overcome these disadvantages, we consider completely unstructured finite element (fe) or isogeometric (B-spline or NURBS) discretizations of the space-time cylinder and the corresponding stable space-time variational formulations of the IBVP under consideration. Unstructured space-time discretizations considerably facilitate the parallelization and the simultaneous space-time adaptivity. Moving spatial domains or interfaces can easily be treated since they are fixed in the space-time cylinder. Beside initial-boundary value problems for parabolic PDEs, we will also consider optimal control problems constrained by linear or non-linear parabolic PDEs. Here unstructured space-time methods are especially suited since the reduced optimality system couples two parabolic equations for the state and adjoint state that are forward and backward in time, respectively. In contrast to time stepping methods, one has to solve one big linear or non-linear system of algebraic equations. Thus, the memory requirement is an issue. In this connection, adaptivity, parallelization, and matrix-free implementations are very important techniques to overcome this bottleneck. Fast parallel solvers like domain decomposition and multigrid solvers are the most important ingredients of efficient space-time methods.
This paper is partially based on joint works with Svetlana Kyas (Matculevich) and Sergey Repin on adaptive space-time IGA based on functional a posteriori error estimators [10, 11], Martin Neumüller and Andreas Schafelner on adaptive space-time

Ulrich Langer
Institute for Computational Mathematics, Johannes Kepler University Linz, Altenbergerstr. 69, A-4040 Linz, Austria, e-mail: ulanger@numa.uni-linz.ac.at

S. C. Brenner et al. (eds.), *Domain Decomposition Methods in Science and Engineering XXVI*, Lecture Notes in Computational Science and Engineering 145,
https://doi.org/10.1007/978-3-030-95025-5_7

FEM [13, 14], and Olaf Steinbach, Fredi Tröltzsch and Huidong Yang on space-time FEM for optimal control problems [15, 16].

2 Space-time variational formulations

Let us consider the parabolic IBVP, find u such that

$$\partial_t u - \mathrm{div}_x(\alpha \nabla_x u) = f + \mathrm{div}_x(\mathbf{f}) \text{ in } Q, \; u = 0 \text{ on } \Sigma, \; u = u_0 := 0 \text{ on } \Sigma_0, \tag{1}$$

as a typical model problem, where $Q = \Omega \times (0,T)$, $\Sigma = \partial\Omega \times (0,T)$, $\Sigma_0 = \Omega \times \{0\}$, $\Omega \subset \mathbb{R}^d$, $d = 1, 2, 3$, denotes the spatial domain that is assumed to be bounded and Lipschitz, $T > 0$ is the terminal time, $f \in L_2(Q)$ and $\mathbf{f} \in L_2(Q)^d$ are given sources, and $\alpha \in L_\infty(Q)$ is a given uniformly bounded and positive coefficient (matrix) that may discontinuously depend on the spatial variable $x = (x_1, \ldots, x_d)$ and the time variable t (non-autonomous case). The standard variational formulation of the IBVP (1) in Bochner spaces reads as follows [17]: Find $u \in U_0 := \{v \in U := \{w \in V := L_2(0,T; H_0^1(\Omega)) : \partial_t w \in V^* := L_2(0,T; H^{-1}(\Omega))\} : v = 0 \text{ on } \Sigma_0\}$ such that

$$a(u,v) = \ell(v) \quad \forall v \in V, \tag{2}$$

where the bilinear form $a(\cdot,\cdot)$ and the linear form $\ell(\cdot)$ are defined by the identities

$$a(u,v) := \int_Q [\partial_t u(x,t) v(x,t) + \alpha(x,t) \nabla_x u(x,t) \cdot \nabla_x v(x,t)] \, dQ \text{ and}$$

$$\ell(v) := \int_Q [f(x,t) v(x,t) - \mathbf{f}(x,t) \cdot \nabla_x v(x,t)] \, dQ, \text{ respectively.}$$

We note that $U = W(0,T)$ is continuously embedded into $C([0,T], L_2(\Omega))$; see [17]. Alternative space-time variational formulations of the IBVP (1) in anisotropic Sobolev spaces on Q are discussed in [9]. The textbook proof of existence and uniqueness of a weak solution is based on Galerkin's method and a priori estimates; see, e.g., [17] and [9]. Alternatively one can use the Banach–Nečas–Babuška (BNB) theorem (see, e.g., [3, Theorem 2.6]) that provides sufficient and necessary conditions for the well-posedness of variational problems like (2). Indeed, Steinbach proved in [19] for $\alpha = 1$ that the bilinear form $a(\cdot,\cdot)$ fulfills the following three conditions:

(BNB1) boundedness: $|a(u,v)| \le \sqrt{2}\, \|u\|_U \|v\|_V$, $\forall u \in U_0, v \in V$,
(BNB2) inf-sup condition: $\inf_{u \in U_0 \setminus \{0\}} \sup_{v \in V \setminus \{0\}} \frac{a(u,v)}{\|u\|_U \|v\|_V} \ge 1/(2\sqrt{2})$,
(BNB3) injectivity of A^*: For every $v \in V \setminus \{0\}$, there exists $u \in U_0$: $a(u,v) \ne 0$,

which are sufficient and necessary for the well-posedness of (2), in other words, the operator $A : U_0 \to V^*$, defined by $a(\cdot,\cdot)$, is an isomorphism. Moreover, $\|u\|_{U_0} \le 2\sqrt{2}\, \|\ell\|_{V^*}$. The norms in the spaces U_0, U, and V are defined as follows:

$$\|u\|_{U_0}^2 = \|u\|_U^2 := \|u\|_V^2 + \|\partial_t u\|_{V^*}^2 = \|\nabla_x u\|_{L_2(Q)}^2 + \|\nabla_x w_u\|_{L_2(Q)}^2,$$

where $w_u \in V$ such that $\int_Q \nabla_x w_u \cdot \nabla_x v \, dQ = \langle \partial_t u, v \rangle_Q$ for all $v \in V$. Here, $\langle \cdot, \cdot \rangle_Q := \langle \cdot, \cdot \rangle_{V^* \times V}$ denotes the duality product on $V^* \times V$.

In the following two sections, *maximal parabolic regularity* plays an important role when deriving locally stabilized isogeometric and finite element schemes. Let us assume that $\mathbf{f} = \mathbf{0}$ and that the coefficient $\alpha = \alpha(x,t)$ fulfills additional conditions (see, e.g., [2]) such that the solution $u \in U_0$ of (2) belongs to the space

$$H_0^{L,1}(Q) = \{v \in V : \partial_t v,\ L_x v := \mathrm{div}_x(\alpha \nabla_x u) \in L_2(Q)\}.$$

Hence, the PDE $\partial_t u - L_x u = f$ holds in $L_2(Q)$. The maximal parabolic regularity even remains true for inhomogeneous initial data $u_0 \in H_0^1(\Omega)$. We also refer the reader to the classical textbook [9], where the case $\alpha = 1$ was considered.

3 Space-time isogeometric analysis

Let us assume that $\mathbf{f} = \mathbf{0}$ and that α fulfills conditions such that maximal parabolic regularity holds, i.e. the parabolic PDE (1) can be treated in $L_2(Q)$. The time variable t can be considered as just another variable, say, x_{d+1}, and the term $\partial_t u$ can be viewed as convection in the direction x_{d+1}. Thus, we can multiply the parabolic PDE (1) by a time-upwind test function $v_h + \lambda \partial_t v_h$ in order to derive stable discrete schemes, where v_h is a test function from some finite-dimensional test space V_{0h}, and $\lambda \geq 0$ is an appropriately chosen scaling parameter. This choice of test functions is motivated by the famous SUPG method, introduced by Hughes and Brooks for constructing stable fe schemes for stationary convection-diffusion problems [4], and which was later used by Johnson and Saranen [7] for transient problems; see also [6] for the related Galerkin Least-Squares finite element methods. Instead of fe spaces V_{0h}, we can also use IGA (B-splines, NURBS) spaces that have some advantages over the more classical fe spaces; see [5] where IGA was introduced. In particular, in the single patch case, one can easily construct IGA spaces $V_{0h} \subset C^{k-1}(\overline{Q})$ of $(k-1)$-times continuously differentiable B-splines of underlaying polynomial degree k. These B-splines of highest smoothness have asymptotically the best approximation properties per degree of freedom. In [12], we used such IGA spaces to derive stable space-time IGA schemes provided that $\lambda = \theta h$ with a fixed constant $\theta > 0$, where h denotes the mesh-size.

In order to construct stable adaptive space-time IGA schemes, we replaced the global scaling parameter λ by a local scaling function $\lambda(x,t)$ that is changing on the mesh according to the local mesh sizes [10, 11]. Let us describe the construction of these locally stabilized space-time IGA more precisely. In IGA, we use the same basis functions for describing both the geometry and IGA spaces V_{0h}. Thus, we assume that the physical computational domain $Q = \Phi(\widehat{Q})$ is the image of the parameter domain $\widehat{Q} := (0,1)^{d+1}$ using the geometrical mapping $\Phi(\xi) = \sum_{i \in \mathcal{I}} \widehat{B}_{i,k}(\xi) \mathbf{P}_i$, where $\{\mathbf{P}_i\}_{i \in \mathcal{I}} \subset \mathbb{R}^{d+1}$ are the control points, and $\widehat{B}_{i,k}, i \in \mathcal{I}$, are the multivariate B-Splines or NURBS. Now we can define the finite-dimensional space

$$V_{0h} = \{v_h \in V_h : v_h = 0 \text{ on } \overline{\Sigma} \cup \overline{\Sigma}_0\} = \text{span}\{\varphi_i : i \in \mathcal{I}_0\} \tag{3}$$

by means of the same basis functions, i.e.,

$$V_h = \mathcal{S}_h^k = \mathcal{S}_{k-1,h}^k = \text{span}\{\varphi_i = \hat{\varphi}_i \circ \Phi^{-1} : i \in \mathcal{I}\},$$

where $\hat{\varphi}_i(\xi) = \widehat{B}_{i,k}(\xi)$, $i \in \mathcal{I}$. We now test the PDE $\partial_t u - L_x u = f$ restricted to a mesh element K from the set of all mesh elements $\mathcal{K}_h = \{K = \Phi(\widehat{K})\}$, into which Q is decomposed, by $v_h + \lambda_K \partial_t v_h$, yielding

$$\big(\partial_t u - L_x u, v_h + \lambda_K\, \partial_t v_h\big)_{L_2(K)} = (f, v_h + \lambda_K\, \partial_t v_h)_{L_2(K)} \quad \forall v_h \in V_{0h}.$$

Summing up over all $K \in \mathcal{K}_h$, and integrating by parts, we get the variational consistency identity

$$a_h(u, v_h) = \ell_h(v_h) \quad \forall v_h \in V_{0h}, \tag{4}$$

where the bilinear form and the linear form are defined by the identities

$$\begin{aligned} a_h(u, v_h) =& (\partial_t u, v_h)_{L_2(Q)} + (\alpha \nabla_x u, \nabla_x v_h)_{L_2(Q)} \\ &+ \sum_{K \in \mathcal{K}_h} \lambda_K \Big((\partial_t u, \partial_t v_h)_{L_2(K)} - (L_x u, \partial_t v_h)_{L_2(K)} \Big) \end{aligned} \tag{5}$$

and

$$\ell_h(v_h) := (f, v_h)_{L_2(Q)} + \sum_{K \in \mathcal{K}_h} \lambda_K\, (f, \partial_t v_h)_{L_2(K)},$$

respectively. Now, the corresponding consistent IGA scheme reads as follows: Find $u_h \in V_{0h}$ such that

$$a_h(u_h, v_h) = \ell_h(v_h) \quad \forall v_h \in V_{0h}. \tag{6}$$

The following three properties are fundamental for the derivation of error estimates:

1. Galerkin orthogonality: $a_h(u - u_h, v_h) = 0 \quad \forall\, v_h \in V_{0h}$,
2. V_{0h}-coercivity: $a_h(v_h, v_h) \geq \mu_c \|v_h\|_h^2 \quad \forall\, v_h \in V_{0h}$,
3. Extended boundedness: $|a_h(u, v_h)| \leq \mu_b \|u\|_{h,*}\, \|v_h\|_h \quad \forall\, u \in V_{0h,*},\ v_h \in V_{0h}$,

provided that $\lambda_K = \theta_K h_K$ with $\theta_K = c_K^{-2} \overline{\alpha}_K^{-1} h_K$, where $h_K = \text{diam}(K)$ denotes the local mesh-size, $\overline{\alpha}_K$ is an upper bound of α on K, and c_K is the computable constant (upper bound) in the local inverse inequality $\|\text{div}_x \nabla_x v_h\|_{L_2(K)} \leq c_K h_K^{-1} \|\nabla_x v_h\|_{L_2(K)}$. Then we get $\mu_c = 1/2$. The boundedness constant μ_b can also be computed; see [10, 11]. The norms $\|\cdot\|_h$ and $\|\cdot\|_{h,*}$ are defined as follows:

$$\|v\|_h^2 := \sum_{K \in \mathcal{K}_h} \Big[\|\alpha^{1/2} \nabla_x v\|_{L_2(K)}^2 + \lambda_K \|\partial_t v\|_{L_2(K)}^2 \Big] + \frac{1}{2} \|v\|_{L_2(\Sigma_T)}^2, \tag{7}$$

$$\|v\|_{h,*}^2 := \|v\|_h^2 + \sum_{K \in \mathcal{K}_h} \big[\lambda_K^{-1} \|v\|_{L_2(K)}^2 + \lambda_K \|\text{div}_x(\alpha \nabla_x v)\|_{L_2(K)}^2 \big]. \tag{8}$$

We mention that both norms are not only well defined on the IGA space V_{0h} but also on the extended space $V_{0h,*} = V_{0h} + H_0^{L,1}(Q)$ to which the solution u belongs in the maximal parabolic regularity setting considered here. The Galerkin orthogonality directly follows from subtracting (6) from (4). The proof of the other two properties is also elementary; see [10, 11].

From the V_{0h}-coercivity of the bilinear form $a_h(\cdot,\cdot)$, we conclude that the solution u_h of the IGA scheme (6) is unique, and, therefore, it exists. In other words, the corresponding linear system of IGA equations

$$K_h \underline{u}_h = \underline{f}_h \tag{9}$$

has a unique solution $\underline{u}_h = (u_i)_{i=1}^{N_h} \in \mathbb{R}^{N_h=|\mathcal{I}_0|}$. The coefficients (control points) u_i then uniquely define the solution $u_h = \sum_{i=1}^{N_h} u_i \varphi_i$ of the IGA scheme (6). The system matrix K_h is non-symmetric, but positive definite due to the V_{0h}-coercivity.

The following best-approximation estimate directly follows from properties 1–3 given above:

Theorem 1 *Let $u \in U_0 \cap H_0^{L,1}(Q)$ be the solution of the IBVP (2), and $u_h \in V_{0h}$ the solution of space-time IGA schemes (6). Then the best-approximation estimate*

$$\|u - u_h\|_h \leq \inf_{v_h \in V_{0h}} \left(\|u - v_h\|_h + \tfrac{\mu_b}{\mu_c} \|u - v_h\|_{h,*} \right) \tag{10}$$

holds.

The best-approximation estimate (10) finally yields convergence rate estimates in terms of h respectively the local mesh-sizes h_K, $K \in \mathcal{K}_h$, provided that u has some additional regularity; see [10, 11].

In practical applications, the use of adaptive IGA schemes is more attractive than uniform mesh refinement. In order to drive adaptivity, we need local error indicators, a marking strategy, and the possibility to refine the mesh locally. In IGA, which starts from a tensor-product setting, local mesh refinement is more involved than in the FEM. However, nowadays, several refinement techniques are available; see [10] and the references given therein. Local error indicators $\eta_K(u_h)$, $K \in \mathcal{K}_h$, should be derived from a posteriori error estimators. We here consider functional error estimators that provide an error bound for any conform approximation v to the solution u of (2). Of course, we are interested in the case $v = u_h \in V_{0h}$. We get the following functional error estimator for a special choice of parameters from [18]:

$$|\!|\!| u - u_h |\!|\!|^2 \leq \overline{\mathfrak{M}}^2(\beta, u_h, \mathbf{y}) := \sum_{K \in \mathcal{K}_h} \eta_K^2(\beta, u_h, \mathbf{y}), \tag{11}$$

where the norm is defined by $|\!|\!| w |\!|\!|^2 := \|\sqrt{\alpha}\nabla_x w\|^2_{L_2(Q)} + \|w\|^2_{L_2(\Sigma_T)}$, β is a fixed positive scaling parameter (function [18]), and $\mathbf{y} \in H(\mathrm{div}_x, Q)$ is a suitable flux reconstruction. The local error indicator $\eta_K^2(\beta, u_h, \mathbf{y}) := \eta^2_{K,\mathrm{flux}}(\beta, u_h, \mathbf{y}) + \eta^2_{K,\mathrm{pde}}(\beta, u_h, \mathbf{y})$ consists of the parts

$$\eta^2_{K,\mathrm{flux}}(\beta, u_h, \mathbf{y}) := \int_K (1 + \beta)|\mathbf{y} - \alpha\nabla_x u_h|^2 \mathrm{d}K \quad \text{and} \tag{12}$$

$$\eta_{K,\text{pde}}^2(\beta, u_h, \mathbf{y}) := c_{F\Omega}^2 \int_K \left(\frac{1+\beta}{\beta} |f - \partial_t u_h + \text{div}_x \mathbf{y}|^2 \right) \mathrm{d}K \tag{13}$$

evaluating the errors in the flux and in the residual of the PDE, where $c_{F\Omega}$ denotes the constant in the inequality $\|v\|_{L_2(Q)} \leq c_{F\Omega} \|\sqrt{\alpha}\nabla_x v\|_{L_2(Q)}$ for all $v \in V$. For $\alpha = 1$, $c_{F\Omega}$ is nothing but the Friedrichs constant in $H_0^1(\Omega)$. In contrast to the FEM (see Sect. 4), the IGA flux $\alpha\nabla_x u_h$ belongs to $H(\text{div}_x, Q)$ provided that α is sufficiently smooth, and $V_{0h} \subset C^1(\overline{Q})$, which is ensured for $k \geq 2$. Then we can choose $\mathbf{y} = \alpha\nabla_x u_h$ yielding $\eta_{K,\text{flux}}(\beta, u_h, \mathbf{y}) = 0$ and, therefore, $\eta_K(\beta, u_h, \mathbf{y}) = \eta_{K,\text{pde}}(\beta, u_h, \mathbf{y})$. A more sophisticated flux reconstruction was proposed by Kleiss and Tomar for elliptic boundary value problems in [8]. Following this idea, we also propose to reconstruct the flux $\mathbf{y}$ from the minimization of the majorant $\overline{\mathfrak{M}}^2(\beta, u_h, \mathbf{y})$ in an IGA space $(S_{l-1,H}^l)^d$ on a coarser mesh with some mesh-size $H \geq h$ and with smoother splines of the underlying degree $l \geq k$. In [10, 11], we present and discuss the results of many numerical experiments showing the efficiency of this technique for constructing adaptive space-time IGA methods using different marking strategies. Here we only show an example from [1] with the manufactured solution $u(x,t) = x^{5/2}(1-x)t^{3/4}$ of (1) with $Q = (0,1) \times (0,2)$, $\alpha = 1$, and $\mathbf{f} = 0$. The uniform mesh refinement yields $O(h^{3/4})$ in the $\|\cdot\|_h$ norm for $k = 2$, whereas the adaptive version with THB-splines recovers the full rate $O(h^2)$, where $h = N_h^{-2}$ and $k = 2$; see Fig. 1.

4 Space-time finite element analysis

We can construct locally stabilized space-time finite element schemes in the same way as in the IGA case, replacing the IGA space (3) by the finite element space

$$V_{0h} = \{v_h \in C(\overline{Q}) : v_h(x_K(\cdot)) \in \mathbb{P}_k(\hat{K}),\ \forall K \in \mathcal{K}_h,\ v_h = 0 \text{ on } \overline{\Sigma} \cap \overline{\Sigma}_0\}, \tag{14}$$

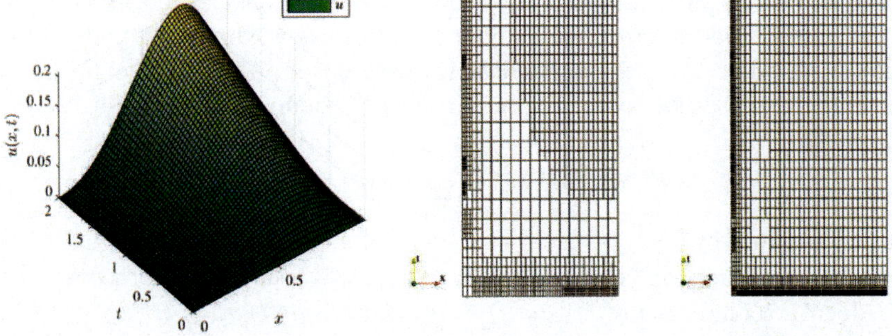

Fig. 1: Solution $u(x,t)$ (right), mesh after 6 (middle) and 8 (right) adaptive refinement levels.

where $\mathcal{K}_h$ is a shape regular decomposition of the space-time cylinder Q into simplicial elements, i.e., $\overline{Q} = \bigcup_{K\in\mathcal{K}_h} \overline{K}$, and $K \cap K' = \emptyset$ for all K and K' from $\mathcal{K}_h$ with $K \neq K'$ (see, e.g., [3] for details), $x_K(\cdot)$ denotes the map from the reference element $\hat{K}$ (unit simplex) to the finite element $K \in \mathcal{K}_h$, and $\mathbb{P}_k(\hat{K})$ is the space of polynomials of degree k on the reference element $\hat{K}$. For the space-time finite element solution $u_h \in V_{0h}$ of (6), we can derive the same best-approximation estimate as given in Theorem 1, from which we get convergence rate estimates under additional regularity assumptions; see [13, Theorem 13.3]. The case of special distributional sources $\mathbf{f}$, the divergence of which exists in $L_2(Q_i)$ on subdomains Q_i of a non-overlapping domain decomposition of the space-time cylinder $\overline{Q} = \bigcup_{i=1}^m \overline{Q}_i$, and the case of low-regularity solutions are investigated in [14]. In [13] and [14], we also present numerical results for different benchmark examples exhibiting different features in space and time. We compare uniform and adaptive refinement. In the finite element case, the corresponding system (9) of algebraic equations is always solved by a parallel AMG preconditioned GMRES. We use *BoomerAMG*, provided by the linear solver library hypre[1], to realize the AMG preconditioner. The adaptive version can be based on different local error indicators; see [13, 14]. Below we show an example where we compare uniform refinement with the adaptive refinement that is based on Repin's first functional error estimate (11). It was already mentioned in Sect. 3 that, in the FEM, we cannot take $\mathbf{y} = \alpha\nabla_x u_h$ because the finite element flux does not belong to $H(\mathrm{div}_x, Q)$. Therefore, we first recover an appropriate flux $\mathbf{y}_h = R_h(\alpha\nabla_x u_h) \in (V_h)^d \subset H(\mathrm{div}_x, Q)$ by nodal averaging à la Zienkiewicz and Zhu (ZZ). One can use this $\mathbf{y}_h$ as $\mathbf{y}$, or one can improve this $\mathbf{y}_h$ by performing some CG minimization steps on the majorant $\overline{\mathfrak{M}}^2(\beta, u_h, \mathbf{y})$ in $(V_h)^d$ with the initial guess $\mathbf{y}_h$. Finally, one minimizes with respect to β. We mention that the local ZZ-indicator is nothing but $\eta_{K,\mathrm{flux}}(0, u_h, R_h(\alpha\nabla_x u_h))$.

Let us now consider the parabolic NIST Benchmark *Moving Circular Wave Front*[2] for testing our adaptive locally stabilized space-time fe method. We again consider the parabolic IBVP (1) with the following data: $d = 2$, $Q = (0, 10) \times (-5, -5) \times (0, T) \subset \mathbb{R}^3$, $T = 10$, $\alpha = 1$, $\mathbf{f} = \mathbf{0}$, and the manufactured exact solution

$$u(x,t) = (x_1 - 0)(x_1 - 10)(x_2 + 5)(x_2 - 5)\tan^{-1}(t)\left(\frac{\pi}{2} - \tan^{-1}(\zeta(r - t))\right)/C$$

with $r = \sqrt{(x_1 - x_{1c})^2 + (x_2 - x_{2c})^2}$, where the parameters (x_{1c}, x_{2c}) and ζ describe the center and the steepness of the circular wave front, respectively. We choose $(x_{1c}, x_{2c}) = (0, 0)$ and $\zeta = 20$ (mild wave front). The scaling parameter C is equal to 10000. The space-time adaptivity is driven by the local error indicators $\eta_{K,\mathrm{flux}}(\beta, u_h, \mathbf{y}_h)$ using Dörfler's marking. Fig. 2 shows the adaptive meshes after a cut through the space-time cylinder Q at $t = 0, 2.5, 5, 7.5$, and 10. In Fig. 3, we compare the convergence history for uniform and adaptive refinements for the polynomial degrees $k = 1, 2, 3$. In the adaptive case, we use Dörfler's marking with the bulk parameter 0.25. The solution has steep gradients in the neighborhood of

[1] https://computing.llnl.gov/projects/hypre

[2] https://math.nist.gov/cgi-bin/amr-display-problem.cgi

the wave front that is perfectly captured by the adaptive procedure. This adaptive procedure quickly leads to the optimal rates $O(h^k)$, and dramatically reduces the error in the $\|\cdot\|_h$ norm, where $h = (N_h)^{-1/(d+1)} = N_h^{-1/3}$ in the adaptive case. Fig. 4 shows the corresponding efficiency indices $I_{\text{eff}} = \eta_{\text{flux}}(0, u_h, \mathbf{y}_h)/\|u - u_h\|_h$, where $\eta^2_{\text{flux}}(\beta, u_h, \mathbf{y}_h) = \sum_{K\in\mathcal{K}_h} \eta^2_{K,\text{flux}}(\beta, u_h, \mathbf{y}_h)$.

5 Space-time optimal control

The optimal control of evolution equations turns out to be interesting from both a mathematical and a practical point of view. Indeed, there are many important applications in the technical, natural, and life sciences. Let us first consider the following space-time tracking optimal control problem: For a given target function $u_d \in L_2(Q)$ (desired state) and for some appropriately chosen regularization (cost) parameter $\varrho > 0$, find the state $u \in U_0$ and the control $z \in Z$ minimizing the cost functional

$$J(u, z) = \frac{1}{2}\int_Q |u - u_d|^2 \, dQ + \frac{\varrho}{2} R(z) \tag{15}$$

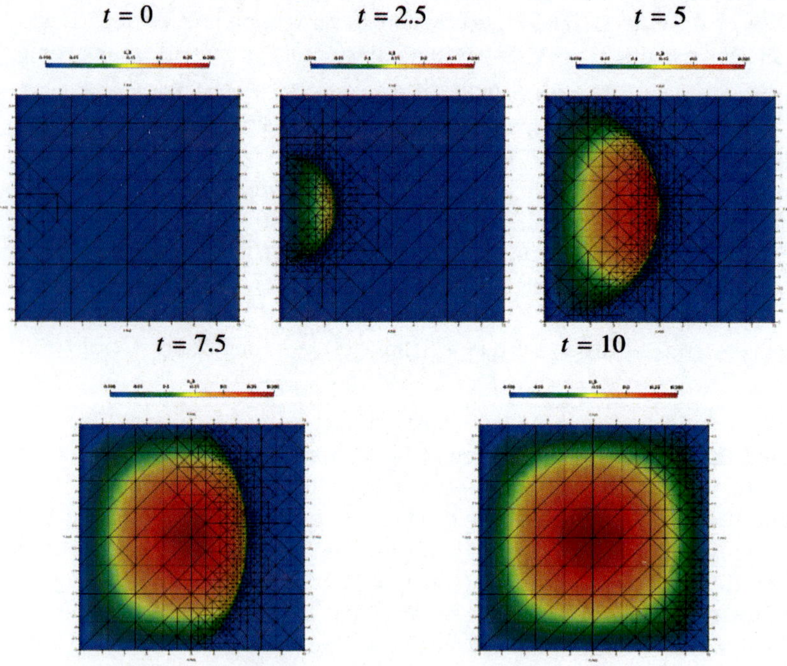

Fig. 2: Adaptive space-time meshes at the cuts $t = 0, 2.5, 5, 7.5$, and 10 through $Q \subset \mathbb{R}^3$.

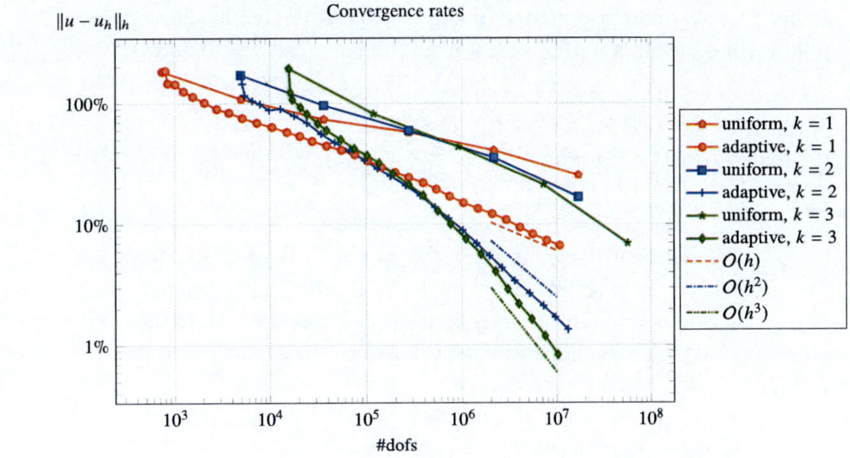

Fig. 3: Comparison of uniform and adaptive refinements for $k = 1, 2, 3$.

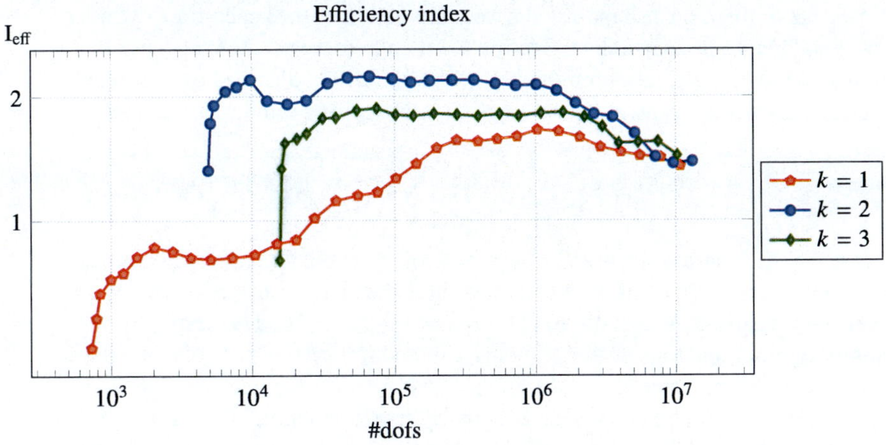

Fig. 4: Efficiency indices I_{eff} for Dörfler's marking with bulk parameter 0.25

subject to the linear parabolic IBVP (1) respectively its variational formulation (2). The regularization term $R(z)$ is usually chosen as the $L_2(Q)$-norm $\|z\|^2_{L_2(Q)}$, and, thus, $Z = L_2(Q)$, whereas the control z acts as the right-hand side f in (1) respectively (2), and $\mathbf{f} = \mathbf{0}$. Since the state equation (2) has a unique solution $u \in U_0$, one can reason the existence of a unique control $z \in Z$ minimizing the quadratic cost functional $J(S(z), z)$, where S is the solution operator mapping $z \in Z$ to the unique solution $u \in U_0$ of (2); see, e.g., [17] and [21]. On the other side, the solution of the quadratic optimization problem $\min_{z \in Z} J(S(z), z)$ is equivalent to the solution of the first-order optimality system. After eliminating the control z from the optimality system by means of the gradient equation $p + \varrho z = 0$, we arrive at the reduced optimality system: Find the state $u \in U_0$ and the adjoint state $p \in P_T$ such that

$$\begin{aligned} \varrho \int_Q \Big[\partial_t u\, v + \alpha \nabla_x u \cdot \nabla_x v\Big] dQ + \int_Q p\, v\, dQ &= 0, \\ - \int_Q u\, q\, dQ + \int_Q \Big[-\partial_t p\, q + \alpha \nabla_x p \cdot \nabla_x q\Big] dQ &= -\int_Q u_d\, q\, dQ, \end{aligned} \tag{16}$$

holds for all $v, q \in V$, where $P_T := \{p \in W(0,T) : p = 0 \text{ on } \Sigma_T\}$. Now the well-posedness of (16) can again be proved by means of the BNB theorem verifying the corresponding conditions (BNB1) – (BNB3); see [16, Theorem 3.3]. In the same paper, we analyze the finite element Galerkin discretization of the reduced optimality system: Find $(u_h, p_h) \in U_{0h} \times P_{Th}$ such that

$$B(u_h, p_h; v_h, q_h) = -(u_d, q_h)_{L_2(Q)} \quad \forall (v_h, q_h) \in V_{0h} \times V_{Th}, \tag{17}$$

where the bilinear form $B(\cdot,\cdot)$ results from adding the left-hand sides of (16). The finite element subspace spaces $U_{0h} = V_{0h} = S_h^k(Q) \cap U_0$ and $P_{Th} = V_{Th} = S_h^k(Q) \cap P_T$ are defined on a shape-regular decomposition of the space-time cylinder Q in simplicial elements as usual; cf. Section 4. Of course, we can here also use IGA instead of FEM as discretization method; cf. Section 3. In [16], we show a *discrete inf-sup condition* which leads to a best-approximation error estimate of the form

$$\sqrt{\varrho \|u - u_h\|_V^2 + \|p - p_h\|_V^2} \le c \inf_{(v_h, q_h) \in U_{0h} \times P_{Th}} \sqrt{\|u - v_h\|_{U_0}^2 + \|p - q_h\|_{P_T}^2} \tag{18}$$

for the case $\alpha = 1$, where $c = 1 + 2\sqrt{2} c_B(\varrho)$ and $c_B(\varrho)$ is the boundedness constant of the bilinear form $B(\cdot,\cdot)$. If u and p have additional regularity, we easily get convergence rate estimates, e.g., $O(h)$ if $u, p \in H^2(Q)$; see [16, Theorem 3.5].

In some applications, one wants to restrict the action of the control z in space and time. Thus, in the case of partial control, we have to replace the right-hand side $f = z$ by $f = \chi_{Q_c} z$, where χ_{Q_c} is the characteristic function of the space-time control domain $Q_c \subset Q$. Then we can again derive the reduced optimality system, and solve it by means of the space-time finite element method. Let us consider a concrete example. In this example, we consider the spatial domain $\Omega = (0,1)^2$ and the terminal time $T = 1$. Therefore, we have $Q = (0,1)^3$. The control subdomain is given as $Q_c = (0.25, 0.75)^2 \times (0,T)$. A smooth target $u_d = \sin(\pi x)\sin(\pi y)\sin(\pi t)$ is used, and the regularization (cost) parameter $\varrho = 10^{-5}$. Fig. 5 presents the state u_h and the control z_h for partial (up) and full (down) distributed controls. We use continuous, piecewise linear finite element approximations on a quasi-uniform decomposition of Q into tetrahedral elements.

Finally, we mention that, in [15], we introduce and investigate the space-time energy regularization $R(z) = \|z\|^2_{L_2(0,T;H^{-1}(\Omega))}$, and compare it to the $L_2(Q)$ and the sparse regularization. Furthermore, the space-time approach can easily be generalized to other observations like terminal time observation, the control via boundary conditions, the control via initial conditions (inverse heat conduction problem), and, last but not least, the control of non-linear parabolic IBVP with box constraints imposed on the control [16].

Acknowledgements The author would like to thank his coworkers mentioned in the *Introduction* for their collaboration on finite element and isogeometric space-time methods. Furthermore, this research was supported by the Austrian Science Fund (FWF) through the projects NFN S117-03 and DK W1214-04. This support is gratefully acknowledged.

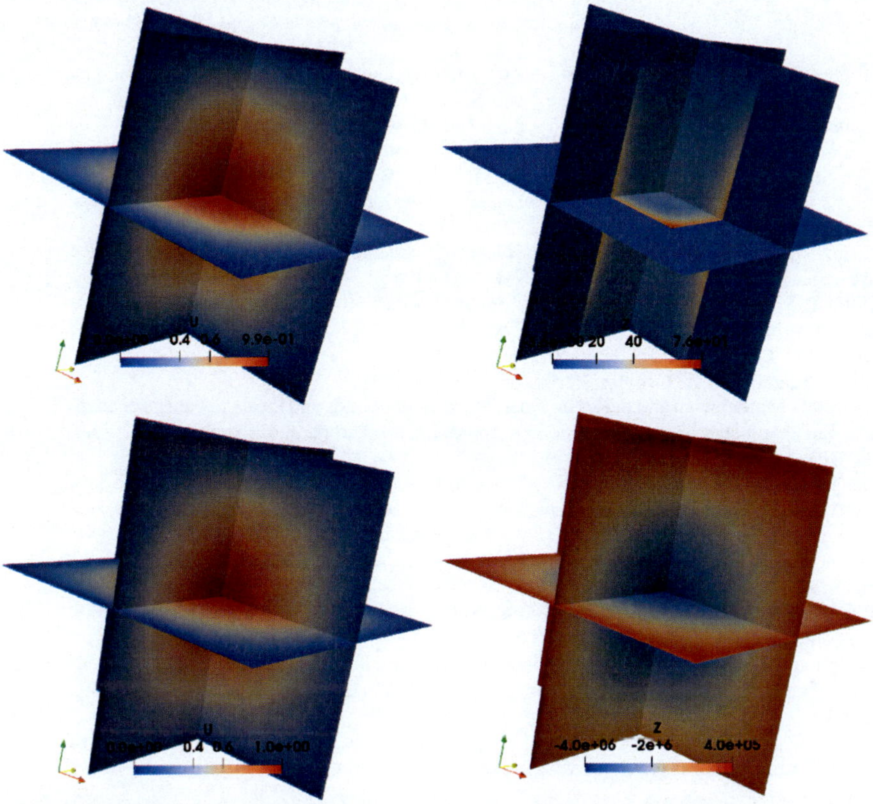

Fig. 5: The state u (left) control z (right) for partial control (up) and full control (down).

References

1. D. Devaud and C. Schwab. Space–time hp-approximation of parabolic equations. *Calcolo* **55**(3), 1–23 (2018).
2. D. Dier. Non-autonomous maximal regularity for forms of bounded variation. *J. Math. Anal. Appl.* **425**, 33–54 (2015).
3. A. Ern and J.-L. Guermond. *Theory and Practice of Finite Elements*. Springer, NY (2004).
4. T. Hughes and A. Brooks. A multidimensional upwind scheme with no crosswind diffusion. In: *Finite Element Methods for Convection Dominated Flows (New York, 1979)*, T. Hughes, Ed., vol. 34 of AMD, ASME (1979).
5. T. Hughes, J. Cottrell and Y. Bazilevs. Isogeometric analysis: CAD, finite elements, NURBS, exact geometry and mesh refinement. *Comput. Methods Appl. Mech. Engrg.* **194**, 4135–4195 (2005).
6. T. Hughes, L. Franca and G. Hulbert. A new finite element formulation for computational fluid dynamics: VIII. The Galerkin/least-squares method for advection-diffusive equations. *Comput. Methods Appl. Mech. Engrg.* **73**, 173–189 (1989).
7. C. Johnson and J. Saranen. Streamline diffusion methods for the incompressible Euler and Navier–Stokes equations. *Math. Comp.* **47**(175), 1–18 (1986).
8. S. Kleiss and S. Tomar. Guaranteed and sharp a posteriori error estimates in isogeometric analysis. *Comput. Math. Appl.* **70**(3), 167–190 (2015).
9. O. Ladyžhenskaya. *The boundary value problems of mathematical physics*. Vol. 49 of Applied Mathematical Sciences, Springer-Verlag, New York (1985).
10. U. Langer, S. Matculevich and S. Repin. Adaptive space–time Isogeometric Analysis for parabolic evolution problems. In: *Space–Time Methods: Applications to Partial Differential Equations*, U. Langer and O. Steinbach, Eds., vol. 25 of Radon Series on Computational and Applied Mathematics, de Gruyter, Berlin, pp. 155–200 (2019).
11. U. Langer, S. Matculevich and S. Repin. Guaranteed error bounds and local indicators for adaptive solvers using stabilised space-time IgA approximations to parabolic problems. *Comput. Math. with Appl.* **78**, 2641–2671 (2019).
12. U. Langer, ,S. Moore and M. Neumüller. Space-time isogeometric analysis of parabolic evolution equations. *Comput. Methods Appl. Mech. Engrg.* **306**, 342–363 (2016).
13. U. Langer, M. Neumüller and A. Schafelner. Space-time finite element methods for parabolic evolution problems with variable coefficients. In: *Advanced Finite Element Methods with Applications – Selected Papers from the 30th Chemnitz Finite Element Symposium 2017*, T. Apel, U. Langer, A. Meyer and O. Steinbach, Eds., vol. 128 of LNCSE. Springer, Berlin, Heidelberg, New York, ch. 13, pp. 247–275 (2019).
14. U. Langer and A. Schafelner. Adaptive space-time finite element methods for nonautonomous parabolic problems with distributional sources. *Comput. Methods Appl. Math.* **20**(4), 677–693 (2020).
15. U. Langer, O. Steinbach, F. Tröltzsch and H. Yang. Space-time finite element discretization of parabolic optimal control problems with energy regularization. *SIAM J. Numer. Anal.* **59**(2), 660–674 (2021).
16. U. Langer, O. Steinbach, F. Tröltzsch and H. Yang. Unstructured space-time finite element methods for optimal control of parabolic equation. *SIAM J. Sci. Comput.* **43**(2), A744–A771 (2021).
17. J. Lions. *Optimal control of systems governed by partial differential equations*. Grundlehren der mathematischen Wissenschaften Vol. 170, Springer, Berlin (1971).
18. S. Repin. *A posteriori estimates for partial differential equations*. Vol. 4 of Radon Series on Computational and Applied Mathematics, de Gruyter, Berlin (2008).
19. O. Steinbach. Space-time finite element methods for parabolic problems. *Comput. Methods Appl. Math.* **15**(4), 551–566 (2015).
20. V. Thomée. *Galerkin finite element methods for parabolic problems*. Second ed. Vol. 25 of Springer Series in Computational Mathematics, Springer-Verlag, Berlin (2006).
21. F. Tröltzsch. *Optimal control of partial differential equations: Theory, methods and applications*. American Mathematical Society, Providence, Rhode Island (2010).

Nonoverlapping Domain Decomposition Methods for Saddle Point Problems

Jing Li and Xuemin Tu

1 Introduction

Domain decomposition methods have been applied extensively for the saddle point problems arising from mixed finite element discretizations. Overlapping methods have been studied by many researchers such as [1, 3, 4, 6, 7, 15]. Some of these algorithms can be applied for both continuous and discontinuous pressure discretizations, however, the convergence analyses are available only for the methods with discontinuous pressure, to the best of our knowledge.

Most nonoverlapping domain decomposition methods are based on the benign subspace idea which was successfully used by [21] for the Stokes problem, followed by [10–12, 14, 16, 18, 22, 24, 26] for different nonoverlapping domain decomposition algorithms and different saddle point problems. In this approach, the original saddle point problems can be reduced to positive definite problems in the benign subspace with subdomain interface velocity and constant subdomain pressure variables. Therefore a conjugate gradient method (CG) can be used to accelerate the convergence. Most of the above-mentioned applications and analyses require discontinuous pressures to be used in the discretization. Several domain decomposition algorithms allow the use of continuous pressures such as [2, 13, 23], but the convergence rate analyses of those approaches are not available. [17, 27, 28] have proposed and analyzed a FETI-DP algorithm for solving the incompressible Stokes equation, which allowed the use of both discontinuous and continuous pressures in the discretization. There, the Lagrange multipliers are introduced to enforce the continuity of the velocity variables across the subdomain interface. Recently, this FETI-DP algorithm has

Jing Li
Department of Mathematical Sciences, Kent State University, Kent, OH 44242,
e-mail: li@math.kent.edu

Xuemin Tu
Department of Mathematics, University of Kansas, 1460 Jayhawk Blvd, Lawrence, KS 66045,
U.S.A. e-mail: xuemin@ku.edu

S. C. Brenner et al. (eds.), *Domain Decomposition Methods in Science and Engineering XXVI*, Lecture Notes in Computational Science and Engineering 145,
https://doi.org/10.1007/978-3-030-95025-5_8

been applied to almost incompressible elasticity with isogeometric discretization by [32].

In this paper, we show for both BDDC and FETI-DP algorithms how the original saddle point problems can be reduced to positive definite problems using either primal or dual variable approaches, outline their analyses, and make the connections between these two approaches.

The rest of this paper is organized as follows. The saddle problems are described in Section 2. In Section 3, the domain decomposition is introduced and the original system is reduced to Schur complements or a system of the Lagrange multiples and pressure. The positive definite formulations are discussed in Section 4 and the condition number estimates are outlined in Section 5. Finally, we summarize some differences and connections of these two methods in Section 6.

2 Problem setting

We consider the following saddle point problem: find $\mathbf{u}_h \in \mathbf{W}$ and $p_h \in Q$, such that,

$$\begin{cases} a(\mathbf{u}_h, \mathbf{v}_h) + b(\mathbf{v}_h, p_h) = (\mathbf{f}_h, \mathbf{v}_h), \ \forall\, \mathbf{v}_h \in \mathbf{W}, \\ b(\mathbf{u}_h, q_h) \qquad\qquad\quad\;\; = (g_h, q_h), \ \forall\, q_h \in Q, \end{cases} \tag{1}$$

where $\mathbf{W}$ and Q are finite element spaces. The continuous bilinear forms $a(\mathbf{u}_h, \mathbf{v}_h)$ and $b(\mathbf{u}_h, q_h)$ can come from the variational formulation of the Stokes equation or the Darcy problem. We call $\mathbf{u}_h$ velocity variables and p_h pressure variables, respectively.

The system (1) can be written as

$$\begin{bmatrix} A & B^T \\ B & 0 \end{bmatrix} \begin{bmatrix} \mathbf{u} \\ p \end{bmatrix} = \begin{bmatrix} \mathbf{f} \\ g \end{bmatrix}. \tag{2}$$

Here A is symmetric positive definite but B is rank deficient. Ker (B^T), the kernel of B^T, includes all constant pressures in Q. Im (B), the range of B, includes all vectors in Q with zero average. We note that Im (B) is orthogonal to Ker (B^T). Under the assumption that $g \in$ Im (B), i.e., g has zero average, the solution of (2) is uniquely determined if the pressure is restricted to the quotient space $Q/\text{Ker}\,(B^T)$.

We assume that $\mathbf{W}$ and Q are inf-sup stable: there exists a positive constant β, independent of h, such that

$$\sup_{\mathbf{w} \in \mathbf{W}} \frac{\langle q, B\mathbf{w} \rangle^2}{\langle \mathbf{w}, A\mathbf{w} \rangle} \geq \beta^2 \langle q, Zq \rangle, \qquad \forall q \in Q/\text{Ker}\,(B^T), \tag{3}$$

where Z is the so-called mass matrix on Q, i.e., $\|q\|_{L^2}^2 = \langle q, Zq \rangle$, $\forall q \in Q$.

3 Domain decomposition

We decompose the domain Ω into N nonoverlapping polygonal/polyhedral subdomains Ω_i, $i = 1, 2, ..., N$. We assume that each subdomain is a union of a bounded number of elements, with typical diameter H. The subdomain interface nodes $\Gamma = (\cup \partial\Omega_i)\backslash\partial\Omega$ includes the subdomain faces, which are open sets and shared by two subdomains, the subdomain edges, which are open sets and shared by more than two subdomains; and the subdomain vertices, which are end points of edges.

Denote the subdomain interior velocity spaces by $\mathbf{W}_I^{(i)}$ and subdomain interior pressure spaces by $Q_I^{(i)}$, respectively. The subdomain boundary velocity space is denoted by $\mathbf{W}_\Gamma$, which is shared by neighboring subdomains, while Q_Γ contains the subdomain boundary pressure degrees of freedom shared by neighboring subdomains. Let

$$\mathbf{W}_I = \bigoplus_{i=1}^{N} \mathbf{W}_I^{(i)}, \quad Q_I = \bigoplus_{i=1}^{N} Q_I^{(i)}.$$

We decompose the velocity and pressure finite element spaces $\mathbf{W}$ and Q into the subdomain interior and interface subspaces,

$$\mathbf{W} = \mathbf{W}_I \bigoplus \mathbf{W}_\Gamma, \quad Q = Q_I \bigoplus Q_\Gamma,$$

respectively, and write (2) as

$$\begin{bmatrix} A_{II} & B_{II}^T & A_{I\Gamma} & B_{\Gamma I}^T \\ B_{II} & 0 & B_{I\Gamma} & 0 \\ A_{I\Gamma}^T & B_{I\Gamma}^T & A_{\Gamma\Gamma} & B_{\Gamma\Gamma}^T \\ B_{\Gamma I} & 0 & B_{\Gamma\Gamma} & 0 \end{bmatrix} \begin{bmatrix} \mathbf{u}_I \\ p_I \\ \mathbf{u}_\Gamma \\ p_\Gamma \end{bmatrix} = \begin{bmatrix} \mathbf{f}_I \\ g_I \\ \mathbf{f}_\Gamma \\ g_\Gamma \end{bmatrix}, \tag{4}$$

which can be assembled from the subdomain problems, defined as below

$$\begin{bmatrix} A_{II}^{(i)} & B_{II}^{(i)^T} & A_{I\Gamma}^{(i)} & B_{\Gamma I}^{(i)^T} \\ B_{II}^{(i)} & 0 & B_{I\Gamma}^{(i)} & 0 \\ A_{I\Gamma}^{(i)^T} & B_{I\Gamma}^{(i)^T} & A_{\Gamma\Gamma}^{(i)} & B_{\Gamma\Gamma}^{(i)^T} \\ B_{\Gamma I}^{(i)} & 0 & B_{\Gamma\Gamma}^{(i)} & 0 \end{bmatrix} \begin{bmatrix} \mathbf{u}_I^{(i)} \\ p_I^{(i)} \\ \mathbf{u}_\Gamma^{(i)} \\ p_\Gamma^{(i)} \end{bmatrix} = \begin{bmatrix} \mathbf{f}_I^{(i)} \\ g_I^{(i)} \\ \mathbf{f}_\Gamma^{(i)} \\ g_\Gamma^{(i)} \end{bmatrix}. \tag{5}$$

We note that the blocks corresponding to $\mathbf{u}_I$ and p_I in (4) can be arranged in subdomain wise. As long as p_Γ contains at least one pressure variables from each subdomain, we can eliminate $\mathbf{u}_I$ and p_I by solving independent subdomain problems and obtain the following global Schur complement system

$$\begin{bmatrix} S_\Gamma & T_{\Gamma\Gamma}^T \\ T_{\Gamma\Gamma} & -C_{\Gamma\Gamma} \end{bmatrix} \begin{bmatrix} \mathbf{u}_\Gamma \\ p_\Gamma \end{bmatrix} = \begin{bmatrix} \mathbf{f}_s \\ g_s \end{bmatrix}, \tag{6}$$

where

$$S_\Gamma = A_{\Gamma\Gamma} - [A_{\Gamma I}\ B_{I\Gamma}^T] \begin{bmatrix} A_{II} & B_{II}^T \\ B_{II} & 0 \end{bmatrix}^{-1} \begin{bmatrix} A_{I\Gamma} \\ B_{I\Gamma} \end{bmatrix}, \tag{7}$$

$$C_{\Gamma\Gamma} = [B_{\Gamma I}\ 0] \begin{bmatrix} A_{II} & B_{II}^T \\ B_{II} & 0 \end{bmatrix}^{-1} \begin{bmatrix} B_{\Gamma I}^T \\ 0 \end{bmatrix}, \tag{8}$$

$$T_{\Gamma\Gamma} = B_{\Gamma\Gamma} - [B_{\Gamma I}\ 0] \begin{bmatrix} A_{II} & B_{II}^T \\ B_{II} & 0 \end{bmatrix}^{-1} \begin{bmatrix} A_{I\Gamma} \\ B_{I\Gamma} \end{bmatrix}, \tag{9}$$

and

$$\begin{bmatrix} \mathbf{f}_s \\ g_s \end{bmatrix} = \begin{bmatrix} \mathbf{f}_\Gamma \\ g_\Gamma \end{bmatrix} - \begin{bmatrix} A_{I\Gamma}^T & B_{I\Gamma}^T \\ B_{\Gamma I} & 0 \end{bmatrix} \begin{bmatrix} A_{II} & B_{II}^T \\ B_{II} & 0 \end{bmatrix}^{-1} \begin{bmatrix} \mathbf{f}_I \\ g_I \end{bmatrix}.$$

We note that S_Γ can be assembled from the local subdomain Schur complements $S_\Gamma^{(i)}$ defined from (5) as:

$$\begin{bmatrix} A_{II}^{(i)} & B_{II}^{(i)^T} & A_{I\Gamma}^{(i)} \\ B_{II}^{(i)} & 0 & B_{I\Gamma}^{(i)} \\ A_{I\Gamma}^{(i)^T} & B_{I\Gamma}^{(i)^T} & A_{\Gamma\Gamma}^{(i)} \end{bmatrix} \begin{bmatrix} \mathbf{u}_I^{(i)} \\ p_I^{(i)} \\ \mathbf{u}_\Gamma^{(i)} \end{bmatrix} = \begin{bmatrix} 0 \\ 0 \\ S_\Gamma^{(i)} \mathbf{u}_\Gamma^{(i)} \end{bmatrix}. \tag{10}$$

We call (6) the primal approach. To formulate the preconditioners of (6) and describe the domain decomposition algorithms using the dual approach, we introduce a partially sub-assembled interface velocity space

$$\widetilde{\mathbf{W}}_\Gamma = \mathbf{W}_\Pi \bigoplus \mathbf{W}_\Delta = \mathbf{W}_\Pi \bigoplus \left(\bigoplus_{i=1}^{N} \mathbf{W}_\Delta^{(i)} \right).$$

Here, $\mathbf{W}_\Pi$ is the continuous coarse level velocity space, whose elements are shared by neighboring subdomains. The complimentary space $\mathbf{W}_\Delta$ is the direct sum of subdomain remaining interface velocity spaces $\mathbf{W}_\Delta^{(i)}$, whose elements vanish at the primal degrees of freedom. In general the functions $\mathbf{w}_\Delta$ in $\mathbf{W}_\Delta$ are not continuous across the subdomain interface Γ and we need to introduce Lagrange multipliers to enforce their continuity. We construct a boolean matrix J_Δ such that $J_\Delta \mathbf{w}_\Delta = 0$ implies the continuity of $\mathbf{w}_\Delta$ cross subdomain interface, see [8, 9] for details. We choose J_Δ to have full row rank and denote the range of J_Δ applied on $\mathbf{W}_\Delta$ by Λ.

The original fully assembled linear system (2) is equivalent to: find $(\mathbf{u}_I, p_I, \mathbf{u}_\Delta, \mathbf{u}_\Pi, p_\Gamma, \lambda) \in \mathbf{W}_I \bigoplus Q_I \bigoplus \mathbf{W}_\Delta \bigoplus \mathbf{W}_\Pi \bigoplus Q_\Gamma \bigoplus \Lambda$, such that

$$\begin{bmatrix} A_{II} & B_{II}^T & A_{I\Delta} & A_{I\Pi} & B_{\Gamma I}^T & 0 \\ B_{II} & 0 & B_{I\Delta} & B_{I\Pi} & 0 & 0 \\ A_{\Delta I} & B_{I\Delta}^T & A_{\Delta\Delta} & A_{\Delta\Pi} & B_{\Gamma\Delta}^T & J_\Delta^T \\ A_{\Pi I} & B_{I\Pi}^T & A_{\Pi\Delta} & A_{\Pi\Pi} & B_{\Gamma\Pi}^T & 0 \\ B_{\Gamma I} & 0 & B_{\Gamma\Delta} & B_{\Gamma\Pi} & 0 & 0 \\ 0 & 0 & J_\Delta & 0 & 0 & 0 \end{bmatrix} \begin{bmatrix} \mathbf{u}_I \\ p_I \\ \mathbf{u}_\Delta \\ \mathbf{u}_\Pi \\ p_\Gamma \\ \lambda \end{bmatrix} = \begin{bmatrix} \mathbf{f}_I \\ g_I \\ \mathbf{f}_\Delta \\ \mathbf{f}_\Pi \\ g_\Gamma \\ 0 \end{bmatrix}, \tag{11}$$

which can be reduced to

$$G \begin{bmatrix} p_\Gamma \\ \lambda \end{bmatrix} = g_g, \tag{12}$$

where

$$G = B_C \widetilde{A}^{-1} B_C^T, \quad g_g = B_C \widetilde{A}^{-1} f - \begin{bmatrix} g_\Gamma \\ 0 \end{bmatrix}, \tag{13}$$

$$\widetilde{A} = \begin{bmatrix} A_{II} & B_{II}^T & A_{I\Delta} & A_{I\Pi} \\ B_{II} & 0 & B_{I\Delta} & B_{I\Pi} \\ A_{\Delta I} & B_{I\Delta}^T & A_{\Delta\Delta} & A_{\Delta\Pi} \\ A_{\Pi I} & B_{I\Pi}^T & A_{\Pi\Delta} & A_{\Pi\Pi} \end{bmatrix}, \quad B_C = \begin{bmatrix} B_{\Gamma I} & 0 & B_{\Gamma\Delta} & B_{\Gamma\Pi} \\ 0 & 0 & J_\Delta & 0 \end{bmatrix}, \quad f = \begin{bmatrix} \mathbf{f}_I \\ 0 \\ \mathbf{f}_\Delta \\ \mathbf{f}_\Pi \end{bmatrix}. \tag{14}$$

Since (12) is a system related to the Lagrange multipliers λ, we call it the dual approach.

4 Positive definite formulations

We have reduced the original saddle point problem to two systems: the primal system (6) and the dual system (12). Even though none of them is positive definite, they can be reduced to positive definite problems in certain special subspaces.

4.1 The primal system (6)

For a general pressure space Q, it is not easy to formulate the Schur complement system (6) as a positive definition system. However, when Q is a discontinuous finite element space, one can decompose Q properly and make (6) positive definite in a special subspace.

When p is discontinuous, subdomains do not share any pressure degrees of freedom on the subdomain boundary. We can take Q_Γ as the subspace of Q with constant values $p_0^{(i)}$, the average of the pressure in the subdomain Ω_i, satisfying $\sum_{i=1}^N p_0^{(i)} m(\Omega_i) = 0$, where $m(\Omega_i)$ is the measure of the subdomain Ω_i. The elements of $Q_I^{(i)}$ are the restrictions of the pressure variables to Ω_i, which satisfy $\int_{\Omega_i} p_I^{(i)} = 0$. Since p_Γ is a constant pressure on each subdomain, $B_{\Gamma I} = 0$. Using this fact in (8) and (9), we have $C_{\Gamma\Gamma} = 0$ and $T_{\Gamma\Gamma} = B_{\Gamma\Gamma}$ and therefore the system (6) can be simplified as

$$\begin{bmatrix} S_\Gamma & B_{\Gamma\Gamma}^T \\ B_{\Gamma\Gamma} & 0 \end{bmatrix} \begin{bmatrix} \mathbf{u}_\Gamma \\ p_\Gamma \end{bmatrix} = \begin{bmatrix} \mathbf{f}_s \\ g_s \end{bmatrix}. \tag{15}$$

For the applications with $g_s \neq 0$, one can find a special $\mathbf{u}_\Gamma^*$ such that $B_{\Gamma\Gamma}\left(\mathbf{u}_\Gamma - \mathbf{u}_\Gamma^*\right) = 0$, see [25, Section 4.8] for details. From now on we assume $g_s = 0$.

The system matrix of (6) is positive definite in the space with $B_{\Gamma\Gamma}\mathbf{u}_\Gamma = 0$. Since p_Γ contains pressure variables which are constant in each subdomain, to make $B_{\Gamma\Gamma}\mathbf{u}_\Gamma = 0$, we only need require $\int_{\partial\Omega_i} \mathbf{u}_\Gamma^{(i)} \cdot \mathbf{n} = 0$, where $\mathbf{n}$ is the normal direction to $\partial\Omega_i$.

We still need to construct a preconditioner to solve (6). Let $\overline{R}_\Gamma^{(i)}$ map $\widetilde{\mathbf{W}}_\Gamma$ to $\mathbf{W}_\Delta^{(i)} \bigoplus \mathbf{W}_\Pi^{(i)}$ and $\overline{R}_\Gamma$ is a direct sum of $\overline{R}_\Gamma^{(i)}$. We can define

$$\widetilde{S}_\Gamma = \overline{R}_\Gamma^T \mathrm{diag}\left(S_\Gamma^{(1)}, \cdots, S_\Gamma^{(N)}\right) \overline{R}_\Gamma.$$

$\widetilde{B}_{\Gamma\Gamma}$ is defined on $\widetilde{\mathbf{W}}_\Gamma$ and is assembled from $B_{\Gamma\Gamma}^{(i)}$ given in (5). The BDDC preconditioned system of (6) can be written as

$$M_B^{-1} S \begin{bmatrix} \mathbf{u}_\Gamma \\ p_\Gamma \end{bmatrix} = M_B^{-1} \begin{bmatrix} \mathbf{f}_s \\ g_s \end{bmatrix}, \tag{16}$$

where $M_B^{-1} = \begin{bmatrix} \widetilde{R}_{D,\Gamma} & 0 \\ 0 & I \end{bmatrix}^T \begin{bmatrix} \widetilde{S}_\Gamma & \widetilde{B}_{\Gamma\Gamma}^T \\ \widetilde{B}_{\Gamma\Gamma} & 0 \end{bmatrix}^{-1} \begin{bmatrix} \widetilde{R}_{D,\Gamma} & 0 \\ 0 & I \end{bmatrix}$, $S = \begin{bmatrix} S_\Gamma & B_{\Gamma\Gamma}^T \\ B_{\Gamma\Gamma} & 0 \end{bmatrix}$, $\widetilde{R}_\Gamma$ maps $\mathbf{W}_\Gamma$ to $\widetilde{\mathbf{W}}_\Gamma$ and $\widetilde{R}_{D,\Gamma}$ is scaled operator obtained from $\widetilde{R}_\Gamma$ with the scaling D. The matrix D should provide a partition of unity:

$$\widetilde{R}_{D\Gamma}^T \widetilde{R}_\Gamma = \widetilde{R}_\Gamma^T \widetilde{R}_{D\Gamma} = I.$$

See [5, 18, 19] for more details about the construction of the BDDC preconditioners. See [20, 31, 33, 34] for different scaling options.

We define two subspaces of $\mathbf{W}_\Gamma$ and $\widetilde{\mathbf{W}}_\Gamma$, respectively, as

$$\mathbf{W}_{\Gamma,B} = \{\mathbf{u}_\Gamma \in \mathbf{W}_\Gamma \mid B_{\Gamma\Gamma}\mathbf{u}_\Gamma = 0\}, \quad \widetilde{\mathbf{W}}_{\Gamma,B} = \{\mathbf{u}_\Gamma \in \widetilde{\mathbf{W}}_\Gamma \mid \widetilde{B}_{\Gamma\Gamma}\mathbf{u}_\Gamma = 0\}.$$

They are called benign subspaces.

It is easy to see that the BDDC preconditioned system (16) is positive definite in the benign subspace $\mathbf{W}_{\Gamma,B}$. In order to use the conjugate gradient method (CG) to solve (16), we need to ensure all CG iterates are in $\mathbf{W}_{\Gamma,B}$ for any initial guess in $\mathbf{W}_{\Gamma,B}$.

We can choose a proper $\mathbf{W}_\Pi$ such that

$$\int_{\partial\Omega_i} \mathbf{w}_\Delta^{(i)} \cdot \mathbf{n} = 0 \tag{17}$$

is satisfied for all $\mathbf{w}_\Delta^{(i)} \in \mathbf{W}_\Delta^{(i)}$. By [18, Lemma 6.2], all CG iterates will stay in $\mathbf{W}_{\Gamma,B}$ if the initial initial guess lies in $\mathbf{W}_{\Gamma,B}$.

The choice of $\mathbf{W}_\Pi$ to satisfy (17) depends on the original problem (1) and the finite element spaces $\mathbf{W}$, namely the discretization methods. See [18, Section 7] for incompressible Stokes problems; [29] for Stokes with the weak Galerkin discretization and [30] for the hybridizable discontinuous Galerkin discretizations; and [24, 34] for the Darcy problem.

4.2 The dual system (12)

Similar to (6), (12) can be positive definite in a special subspace.

If $\widetilde{A}$, defined in (14), is nonsingular, by Sylvester's law of inertia, we know that G is symmetric positive semi-definite. Let 1_v denote the constant vector 1 which has the same dimension as v and let $J_{\Delta,D}$ be obtained by scaling J_Δ with the scaling matrix D. The null space of G is given by

$$\left(1_{p_\Gamma}, -J_{\Delta,D}\left[B_{I\Delta}^T \;\; B_{\Gamma\Delta}^T\right]\begin{bmatrix}1_{p_I}\\1_{p_\Gamma}\end{bmatrix}\right).$$

Let $X = Q_\Gamma \bigoplus \Lambda$ and $\mathrm{Im}\,(G)$ be the range space of G, which is a subspace of X. $\mathrm{Im}\,(G)$ is orthogonal to the null space of G and can be written as

$$\mathrm{Im}\,(G) = \left\{\begin{bmatrix}g_{p_\Gamma}\\g_\lambda\end{bmatrix} \in X : g_{p_\Gamma}^T 1_{p_\Gamma} - g_\lambda^T\left(J_{\Delta,D}\left[B_{I\Delta}^T \;\; B_{\Gamma\Delta}^T\right]\begin{bmatrix}1_{p_I}\\1_{p_\Gamma}\end{bmatrix}\right) = 0\right\}. \tag{18}$$

The restriction of G to its range space $\mathrm{Im}\,(G)$ is positive definite. By [27], we know g_g, defined in (13), belongs to $\mathrm{Im}\,(G)$. All CG iterates will be in $\mathrm{Im}\,(G)$ if the CG method is used to solve (12) with zero initial guess.

Block preconditioners, proposed in [17, 27, 28], are used to solve (12). The preconditioned system can be written as

$$M_F^{-1}G\begin{bmatrix}p_\Gamma\\\lambda\end{bmatrix} = M_F^{-1}g_g, \quad M_F^{-1} = \begin{bmatrix}M_p^{-1} & \\ & M_\lambda^{-1}\end{bmatrix}. \tag{19}$$

$M_p^{-1} = \frac{1}{h^n} I_{p_\Gamma}$ for the Stokes problem and M_λ^{-1} can be either lumped or Dirichlet preconditioners for λ. [32] defines M_p^{-1} to be a BDDC preconditioner for isogeometric discretization for almost incompressible elasticity and deluxe scaling is used. All these additional techniques ensure the algorithms are robust in the presence of discontinuous material parameters, which is not considered for the algorithms in [17, 27, 28] for the Stokes problem. In [34], deluxe scaling and local generalized eigenvalue problems are also used to further enhance the performance of algorithms for (16). However, some special designs of these techniques are needed to make sure these additional primal variables lie in the benign subspace.

We note that for (12), we do not require the pressure to be discontinuous for the positive definite formulation. Moreover, we do not need to choose a proper primal space $\mathbf{W}_\Pi$ to ensure the CG iterates in the subspace. The choices of $\mathbf{W}_\Pi$ for (12) only ensure the nice bound for the condition number of the preconditioned operator. This fact makes the algorithms much simpler, especially for three-dimensional problems.

However, we do need to define a subspace $\widetilde{V}_0$ for the convergence analysis, which plays a similar role as the benign subspaces. Let $\widetilde{V} = \mathbf{W}_I \bigoplus Q_I \bigoplus \widetilde{\mathbf{W}}_\Gamma$ and define its subspace

$$\widetilde{V}_0 = \left\{ v = (\mathbf{w}_I,\ p_I,\ \mathbf{w}_\Delta,\ \mathbf{w}_\Pi) \in \widetilde{V} \;\middle|\; B_{II}\mathbf{w}_I + B_{I\Delta}\mathbf{w}_\Delta + B_{I\Pi}\mathbf{w}_\Pi = 0 \right\}. \tag{20}$$

For any $v \in \widetilde{V}_0$, $\langle \cdot, \cdot \rangle_{\widetilde{A}}$ defines a semi-inner product on $\widetilde{V}_0$, see [28] for details.

5 Condition number estimates

Since both (16) and (19) are symmetric positive definite in the special subspaces, we can use the CG methods to solve them. For the convergence analysis of the CG methods, we only need to bound the maximum and minimum eigenvalues of the preconditioned operators. Here we only outline the analyses, see, for example, [18] and [28] for details.

We first define two useful operators E_D and P_D. Unlike the E_D and P_D defined for elliptic problems in [19], our E_D and P_D are defined on different subspaces. The matrix S in (16) is defined with S_Γ and $B_{\Gamma\Gamma}$, which is for the variables $\mathbf{u}_\Gamma$ and p_Γ. The matrix G in (19) is defined with $\widetilde{A}$, which is for the variables $\mathbf{u}_I$, p_I, and $\mathbf{u}_\Gamma$.

E_D is an averaging operator, defined by

$$E_D = \widetilde{R}\widetilde{R}_D^T = \begin{bmatrix} \widetilde{R}_\Gamma & \\ & I \end{bmatrix} \begin{bmatrix} \widetilde{R}_{D,\Gamma}^T & \\ & I \end{bmatrix}.$$

It maps $\widetilde{\mathbf{W}}_\Gamma \times Q_\Gamma$ to itself and computes a weighted average for the velocity across the subdomain interface Γ, and then distributes the average back to the original degrees of freedom on the interfaces while keeping the pressure variables unchanged.

Similarly, P_D is a jump operator which maps $\widetilde{V}$ to itself. Here we only define the jump operator related to solving a Dirichlet problem on each subdomain. For any given $v = (\mathbf{w}_I,\ p_I,\ \mathbf{w}_\Delta,\ \mathbf{w}_\Pi) \in \widetilde{V}$, $P_D v = (\mathbf{u}_I,\ 0,\ \mathbf{u}_\Delta,\ 0) \in \widetilde{V}$, where each $\mathbf{u}_I^{(i)}$ is the harmonic extension, with given subdomain boundary velocity $\mathbf{u}_\Delta^{(i)} = J_{\Delta,D}^{(i)^T} J_\Delta \mathbf{w}_\Delta$ and $\mathbf{u}_\Pi^{(i)} = 0$. Here $J_{\Delta,D}^{(i)^T}$ represents restriction of $J_{\Delta,D}^T$ on subdomain Ω_i and is a map from Λ to $\mathbf{W}_\Delta^{(i)}$.

We assume that the interface averaging operator E_D and the jump operator P_D satisfy the following bounds:

$$|E_D\mathbf{w}|_{\widetilde{S}}^2 \le C_{ED}(H,h)|\mathbf{w}|_{\widetilde{S}}^2, \quad \forall\, \mathbf{w} = (\mathbf{u}_\Gamma,\ q_0) \in \widetilde{\mathbf{W}}_{\Gamma,B} \times Q_\Gamma, \tag{21}$$

and

$$|P_D v|_{\widetilde{A}}^2 \le C_{PD}(H,h)|v|_{\widetilde{A}}^2, \quad \forall\, v \in \widetilde{V}_0, \tag{22}$$

where $C_{ED}(H, h)$ and $C_{PD}(H, h)$ are positive constants dependent on the subdomain size H and mesh size h.

Theorem 1 *For any* $\mathbf{w} = (\mathbf{u}_\Gamma, p_\Gamma) \in \mathbf{W}_{\Gamma,B} \times Q_\Gamma$,

$$\langle \mathbf{w}, \mathbf{w} \rangle_S \leq \left\langle \mathbf{w}, M_B^{-1} S\mathbf{w} \right\rangle_S \leq C_{ED} \langle \mathbf{w}, \mathbf{w} \rangle_S ,$$

where $C_{ED}(H, h)$ *is the bound of the average operator, given in (21).*

Theorem 2 *For any* x *in the range of* $M_F^{-1} G$,

$$c(\beta) \langle M_F x, x \rangle \leq \langle G x, x \rangle \leq (C C_{PD}(H, h)) \langle M_F x, x \rangle ,$$

where $c(\beta)$ *is a function of the inf-sup constant* β*, defined in (3),* C *is a positive constant, and* C_{PD} *is the bound of the jump operator, given in (22).*

6 Connections and differences

One of the big advantages of using (19) is that the formulation can be applied to both continuous and discontinuous pressure discretizations. The algorithms can be applied to the problems discretized with the widely used Taylor–Hood finite elements and isogeometric discretizations. Moreover, since the formulation does not put any constraints on the velocity variable $\mathbf{u}$ for its positive definite formulation, we can relax the divergence-free constraints defined in (17), which can be quite complicated to enforce, see [18, Section 7]. The coarse problem resulting from (19) can be positive definite, which can be the same as those for simple elliptic problems.

Both (16) and (19) can be applied to discontinuous pressures. When Q is discontinuous, there are two choices of p_Γ in (19), as discussed in detail in [27]. When p_Γ is taken as an empty set, (19) become a system for the Lagrange multiplier λ only. If the Stokes extension is used in the jump operator P_D instead of the harmonic extension, the divergence free condition will be required and it has been proved in [18, Theorem 8.1] that both (16) and (19) have the same nonzero eigenvalues with the possible exception of 1. However, the Stokes extension and divergence free condition are not necessary for (19). Harmonic extension will make the algorithms more efficient.

From the analysis point of view, if (16) can be applied, the minimal eigenvalue of the preconditioned operator is always 1, as stated in Theorem 1. One only needs to estimate the bound C_{ED} of the average operator E_D, defined in (21). For the analysis of (19), one needs to estimate the bound C_{PD} of the jump operator P_D, defined in (22), which is similar to the estimate of E_D. Moreover, the lower bound in Theorem 2 has to be established, which is not as easy as for (16).

There are many discretizations with discontinuous pressure spaces such as the family of discontinuous Galerkin methods. (16) has been applied in [29, 30] for some of these discretizations, where the primal constraints, required by the bound of E_D, also ensure the divergence free conditions, which makes the algorithms simpler than those with standard finite element discretizations, especially in three dimensions.

The difficulty for those applications is to estimate the bound for the average operator E_D, where properties of the discretizations have to be explored carefully.

Acknowledgements This work was supported in part by National Science Foundation Contract No. DMS-1723066.

References

1. G.R. Barrenechea, M. Bosy and V. Dolean. Numerical assessment of two-level domain decomposition preconditioners for incompressible Stokes and elasticity equations. *Electron. Trans. Numer. Anal.* **49**, 41–63 (2018).
2. H. Benhassine and A. Bendali. A non-overlapping domain decomposition method for continuous-pressure mixed finite element approximations of the Stokes problem. ESAIM *Math. Model. Numer. Anal.* **45**(4), 675–696 (2011).
3. M. Cai and L.F. Pavarino. Hybrid and multiplicative overlapping Schwarz algorithms with standard coarse spaces for mixed linear elasticity and Stokes problems. *Commun. Comput. Phys.* **20**(4), 989–1015 (2016).
4. M. Cai, L.F. Pavarino, and O.B. Widlund. Overlapping Schwarz methods with a standard coarse space for almost incompressible linear elasticity. *SIAM J. Sci. Comput.* **37**(2), A811–A830 (2015).
5. C. Dohrmann. A preconditioner for substructuring based on constrained energy minimization. *SIAM J. Sci. Comput.* **25**(1), 246–258 (2003).
6. C.R. Dohrmann and O.B. Widlund. An overlapping Schwarz algorithm for almost incompressible elasticity. *SIAM J. Numer. Anal.* **47**(4), 2897–2923 (2009).
7. C.R. Dohrmann and O.B. Widlund. Hybrid domain decomposition algorithms for compressible and almost incompressible elasticity. *Internat. J. Numer. Methods Engrg.* **82**(2), 157–183 (2010).
8. C. Farhat, M. Lesoinne, P. LeTallec, K. Pierson and D. Rixen. FETI-DP: a dual-primal unified FETI method. I. A faster alternative to the two-level FETI method. *Internat. J. Numer. Methods Engrg.* **50**(7), 1523–1544 (2001).
9. C. Farhat, M. Lesoinne and K. Pierson. Ascalable dual-primal domain decomposition method. *Numer. Linear Algebra Appl.* **7**(7–8), 687–714 (2000).
10. P. Goldfeld, L. Pavarino and O. Widlund. Balancing Neumann-Neumann preconditioners for mixed approximations of heterogeneous problems in linear elasticity. *Numer. Math.* **95**(2), 283–324 (2003).
11. H.H. Kim and C. Lee. A Neumann-Dirichlet preconditioner for a FETI-DP formulation of the two-dimensional Stokes problem with mortar methods. *SIAM J. Sci. Comput.* **28**(3), 1133–1152 (electronic) (2006).
12. H.H. Kim and C. Lee. A FETI-DP formulation for the three-dimensional Stokes problem without primal pressure unknowns. *SIAM J. Sci. Comput.* **32**(6), 3301–3322 (2010).
13. H.H. Kim and C. Lee. A two-level nonoverlapping Schwarz algorithm for the Stokes problem: numerical study. *Comput. Methods Appl. Mech. Engrg.* **223/224**, 153–160 (2012).
14. H.H. Kim, C. Lee and E. Park. A FETI-DP formulation for the Stokes problem without primal pressure components. *SIAM J. Numer. Anal.* **47**(6), 4142–4162 (2010).
15. A. Klawonn and L. Pavarino. Overlapping Schwarz methods for mixed linear elasticity and Stokes problems. *Comput. Methods Appl. Mech. Engrg.* **165**, 233–245 (1998).
16. J. Li. A Dual-Primal FETI method for incompressible Stokes equations. *Numer. Math.* **102**, 257–275 (2005).
17. J. Li and X. Tu. A nonoverlapping domain decomposition method for incompressible Stokes equations with continuous pressures. *SIAM J. Numer. Anal.* **51**(2), 1235–1253 (2013).

18. J. Li and O. Widlund. BDDC algorithms for incompressible Stokes equations. *SIAM J. Numer. Anal.* **44**(6), 2432–2455 (2006).
19. J. Li and O. Widlund. FETI–DP, BDDC, and block Cholesky methods. *Internat. J. Numer. Methods Engrg.* **66**, 250–271 (2006).
20. D. Oh, O. Widlund, S. Zampini and C. Dohrmann. BDDC algorithms with deluxe scaling and adaptive selection of primal constraints for Raviart–Thomes vector fields. *Math. Comp.* **87**, 659–692 (2018).
21. L. Pavarino and O. Widlund. Balancing Neumann-Neumann methods for incompressible Stokes equations. *Comm. Pure Appl. Math.* **55**(3), 302–335 (2002).
22. L. Pavarino, O. Widlund and S. Zampini. BDDC preconditioners for spectral element discretizations of almost incompressible elasticity in three dimensions. *SIAM J. Sci. Comput.* **32**(6), 3604–3626 (2010).
23. J. Sístek, B. Sousedík, P. Burda, J. Mandel and J. Novotný. Application of the parallel BDDC preconditioner to the Stokes flow. *Comput. & Fluids* **46**, 429–435 (2011).
24. X. Tu. A BDDC algorithm for a mixed formulation of flows in porous media. *Electron. Trans. Numer. Anal.* **20**, 164–179 (2005).
25. X. Tu. *BDDC Domain Decomposition Algorithms: Methods with Three Levels and for Flow in Porous Media.* PhD thesis, Courant Institute, New York University, January (2006).
26. X. Tu. A BDDC algorithm for flow in porous media with a hybrid finite element discretization. *Electron. Trans. Numer. Anal.* **26**, 146–160 (2007).
27. X. Tu and J. Li. A unified dual-primal finite element tearing and interconnecting approach for incompressible Stokes equations. *Internat. J. Numer. Methods Engrg.* **94**(2), 128–149 (2013).
28. X. Tu and J. Li. A FETI-DP type domain decomposition algorithm for three-dimensional incompressible Stokes equations. *SIAM J. Numer. Anal.* **53**(2), 720–742 (2015).
29. X. Tu and B. Wang. A BDDC algorithm for the Stokes problem with weak Galerkin discretizations. *Comput. Math. Appl.* **76**(2), 377–392 (2018).
30. X. Tu, B. Wang and J. Zhang. Analysis of BDDC algorithms for Stokes problems with hybridizable discontinuous Galerkin discretizations. *Electron. Trans. Numer. Anal.* **52**, 553–570 (2020).
31. O.B. Widlund. BDDC domain decomposition algorithms. In: *75 years of mathematics of computation*, volume 754 of Contemp. Math., pp. 261–281, Amer. Math. Soc., Providence, RI (2020).
32. O.B. Widlund, S. Zampini, S. Scacchi and L.F. Pavarino. Block FETI–DP/BDDC preconditioners for mixed isogeometric discretizations of three-dimensional almost incompressible elasticity. *Math. Comp.* Accepted. (2021).
33. O.B. Widlund and C.R. Dohrmann. BDDC deluxe domain decomposition. In: *Domain decomposition methods in science and engineering XXII*, volume 104 of Lect. Notes Comput. Sci. Eng., pp. 93–103, Springer, Cham (2016).
34. S. Zampini and X. Tu. Addaptive multilevel BDDC deluxe algorithms for flow in porous media. *SIAM J. Sci. Comput.* **39**(4), A1389–A1415 (2017).

Local Residual Minimization Smoothing for Improving Convergence Behavior of a Space-Time Domain Decomposition Method

Hanyu Li and Mary F. Wheeler

1 Introduction

Complex multiphase flow and reactive transport in subsurface porous media is mathematically modeled by systems of nonlinear equations. Due to significant nonlinearity, solving such systems with Newton's method requires small time steps for stable numerical convergence, resulting in significant computational load. Our space-time domain decomposition method addresses this difficulty by allowing different time scales for different spatial subdomains of the system, thus distributing computing resources according to load requirements.

Many space-time domain decomposition approaches have been proposed in the past. To mention a few works, such as [1, 8, 9], space-time finite elements were introduced for elastodynamics with discontinuous Galerkin (DG) in time. The space-time method has also been applied to other systems such as reaction-diffusion problems, with different time discretization schemes [3, 10–12]. Regarding flow in porous media, [7] focused on linear single phase flow and transport problems where flow is naturally decoupled from advection-diffusion transport. In [17] a space-time approach for nonlinear coupled multiphase flow and transport problems on a static grid using an enhanced velocity method is formulated, an MFE variant [2, 18, 20], where the continuity of fluxes at non-matching space-time interfaces was strongly enforced.

Although space-time domain decomposition methods can provide tremendous computational speedup, initiating such systems properly has always been a challenge, especially for nonlinear problems with rough coefficients. The main issue is, for

Hanyu Li
Oden Institute for Computational Engineering and Sciences, 201 E 24th St, Austin, TX 78712, e-mail: lihanyu234@utexas.edu

Mary F. Wheeler
Oden Institute for Computational Engineering and Sciences, 201 E 24th St, Austin, TX 78712 e-mail: mfw@oden.utexas.edu

S. C. Brenner et al. (eds.), *Domain Decomposition Methods in Science and Engineering XXVI*, Lecture Notes in Computational Science and Engineering 145,
https://doi.org/10.1007/978-3-030-95025-5_9

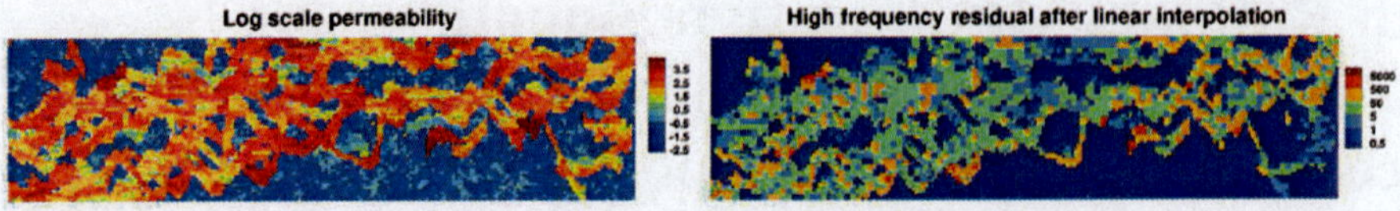

Fig. 1: High frequency residual after linear interpolation of a multigrid method for rough coefficient cases.

subdomains with local time steps being solved in parallel, initiating all the local time steps with the solution at the previous space-time slab, which is similar to the procedure in traditional time-stepping schemes, frequently leads to non-convergence. In [13] a geometric multigrid type of approach was adopted, which starts solving each space-time slab with the coarsest resolution and sequentially refines the mesh in certain subdomains to the finest resolution. After each refinement, the unknowns on the finer mesh are generated by linear interpolation (prolongation) of the solution on the coarse mesh and the Newton iteration continues. The sequential refinement provides an initial guess of the unknowns close enough to the true solution to prevent convergence failure. However, like all multigrid methods, the linear interpolation causes high frequency residuals to appear sporadically throughout the entire domain, especially for problems with rough coefficients. An example is shown in Fig. 1. Here a flow in subsurface porous media problem with channelized permeability as coefficients is presented. The discontinuity of the permeability is clearly observed at the channel boundary. On the right-hand side, we demonstrate the initial residual of a typical space-time slab after the prolongation step. The high frequency residuals colored in red appear sporadically throughout the system, typically on the channel boundary. As Newton's method continues, the first few iterations focus on reducing such high frequency residuals while not much effort is devoted to the rest of the system. Therefore, it is critical to remove these high frequency residuals before continuing the Newton iteration, to harness the full potential of space-time domain decomposition.

In this paper, we introduce a local residual minimization algorithm to improve Newtonian convergence behavior of a space-time geometric multigrid method. In Section 2, we present the model problem followed by the smoothing algorithm in Section 3. Results from numerical experiments using the proposed algorithm are discussed in Section 4. A summary of our findings follows in Section 5.

2 Flow model problem

We use miscible multiphase flow in porous media as the model problem due to its extensive nonlinearity imposing significant numerical convergence challenges. Since miscible flow involves multiple components coexisting in a single fluid phase, we need to write the governing equation of the model in component form. A simplified version of such a model is the black-oil model widely accepted in the petroleum engineering industry, which we use to demonstrate our numerical results.

To start off, assuming no dispersion, the conservation equation of a component existing in a given fluid phase is stated as follow

$$\frac{\partial}{\partial t}(\varphi \rho_\alpha \xi_{c\alpha} s_\alpha) + \nabla \cdot (\xi_{c\alpha} \boldsymbol{u}_\alpha) = q_{c\alpha} + r_{c\alpha} \,. \tag{1}$$

Here, φ is the porosity, ρ_α, s_α and $\boldsymbol{u}_\alpha$ are the density, saturation and velocity of the fluid phase, $\xi_{c\alpha}$ is the fraction of component c included in phase α, in either mass or molar basis, $q_{c\alpha}$ is the source/sink and $r_{c\alpha}$ is the increase/decrease rate of component c in phase α due to phase changes. The rate of phase change and mass/molar fraction of component c obeys the following constraints.

$$\sum_\alpha r_{c\alpha} = 0 \,, \tag{2}$$

$$\sum_c \xi_{c\alpha} = 1, \; \xi_{c\alpha} \geq 0 \,. \tag{3}$$

We sum Eqn. (1) over the total number of phases (N_p) to acquire the component mass conservation equation as

$$\frac{\partial}{\partial t}\left(\varphi \sum_\alpha \rho_\alpha \xi_{c\alpha} s_\alpha\right) + \nabla \cdot \left(\sum_\alpha \xi_{c\alpha} \boldsymbol{u}_\alpha\right) = \sum_\alpha q_{c\alpha} \,. \tag{4}$$

To simplify the notation, let us define the component concentration and flux in a given phase by $n_{c\alpha} = \rho_\alpha \xi_{c\alpha} s_\alpha$ and $\boldsymbol{u}_{c\alpha} = \xi_{c\alpha} \boldsymbol{u}_\alpha$, and the total component concentration by $N_c = \sum_\alpha n_{c\alpha}$. Then we can rewrite Eqn. (4) as

$$\frac{\partial}{\partial t}\left(\varphi \sum_\alpha n_{c\alpha}\right) + \nabla \cdot \left(\sum_\alpha \boldsymbol{u}_{c\alpha}\right) = \sum_\alpha q_{c\alpha} \,. \tag{5}$$

The boundary and initial conditions are

$$\boldsymbol{u}_\alpha \cdot \boldsymbol{\nu} = 0 \quad \text{on } \partial\Omega \times J \,, \tag{6}$$

$$\begin{cases} p_\alpha = p_\alpha^0 \\ N_c = N_c^0 \end{cases} \quad \text{at } \Omega \times \{t = 0\} \,, \tag{7}$$

where $J = (0, T]$ is the time domain of interest and Ω is the spatial domain. The phase velocity is given by Darcy's law as

$$\boldsymbol{u}_\alpha = -K \rho_\alpha \frac{k_{r\alpha}}{\mu_\alpha} (\nabla p_\alpha - \rho_\alpha \boldsymbol{g}) \,, \tag{8}$$

in which K is the absolute permeability while $k_{r\alpha}$, μ_α and p_α are the relative permeability, viscosity and pressure for the given fluid phase. The relative permeability and capillary pressure are functions of phase saturations.

In complex compositional simulations, the saturations are estimated by concentrations after finding the equilibrium hydrocarbon component distribution. This procedure is called flash vaporization calculation and readers can refer to [16] for details. In this paper, we avoid such complexity and use the black-oil model as the simplified compositional model to further introduce our concept. The black-oil model allows a maximum number of three phases in the system, namely oleic, aqueous and gaseous. The components contained within are water, a hydrocarbon oil which mainly consists of heavy non-volatile molecules and a hydrocarbon gas which mostly includes light volatile molecules. Consequently the hydrocarbon gas can exist as either free gas or dissolved gas in the oleic phase. This results in the following relations on the component fractions:

$$\begin{aligned} &\xi_{1o} + \xi_{3o} = 1, \quad \xi_{2o} = 0 \\ &\xi_{1w} = 0, \quad \xi_{2w} = 1, \quad \xi_{3w} = 0 \\ &\xi_{1g} = 0, \quad \xi_{2g} = 0, \quad \xi_{3g} = 1. \end{aligned} \tag{9}$$

ξ_{3o} is commonly referred to as the solution gas-oil ratio and is usually a function of pressure, but it remains constant after the oleic phase reaches the bubble point pressure. We remark that if the hydrocarbon oil component also contains some medium weight molecules and thus is able to vaporize into the gaseous phase, then we obtain the volatile oil model. The dissolved gas causes the oleic phase to swell, thus decreasing its density. Since the hydrocarbon oil component itself is slightly compressible, such a swelling effect can be described by the following equation:

$$\rho_o = \rho_{o,std} \cdot (e^{-c_o p_o} + \beta \xi_{3o})^{-1} . \tag{10}$$

The aqueous and gaseous phase, which contain only water and the hydrocarbon gas component, are slightly compressible and fully compressible, respectively. Therefore, the two phase densities are given as follow:

$$\rho_w = \rho_{w,std} \cdot e^{c_w p_w} , \tag{11}$$

$$\rho_g = \rho_{g,std} \cdot c_g p_g . \tag{12}$$

Then saturations can be related to concentrations by

$$\begin{cases} s_o = \dfrac{N_1}{\rho_o(1-\xi_{3o})} \\ s_w = \dfrac{N_2}{\rho_w} \\ s_g = \dfrac{1}{\rho_g}\left(N_3 - N_1 \dfrac{\xi_{3o}}{1-\xi_{3o}}\right) \end{cases} , \tag{13}$$

with the following constraint

$$\sum_{\alpha} s_\alpha = 1,\; s_\alpha \geq 0\,. \tag{14}$$

Let $\boldsymbol{V} = H(\text{div};\Omega)$, $W = L^2(\Omega)$ with $\boldsymbol{V}_h$ and W_h the finite-dimensional subspaces. Let $J_n = (t_n, t_{n+1}]$ be the nth partition of the time domain of interest. Then for each space-time slab $J_n \times \Omega$, we define the velocity and pressure/saturation spaces as, for any element $E = F_E \times T_E$,

$$\boldsymbol{V}_h^n = \left\{ \boldsymbol{v} \in L^2\Big(J_n; H(\text{div};\Omega)\Big) : \boldsymbol{v}(\cdot,\boldsymbol{x})\Big|_{F_E} \in \boldsymbol{V}_h,\; \boldsymbol{v}(t,\cdot)\Big|_{T_E} = \sum_{a=1}^{l} \boldsymbol{v}_a t^a \;\&\; \boldsymbol{v}_a \in \boldsymbol{V}_h \right\},$$

$$W_h^n = \left\{ w \in L^2\Big(J_n; L^2(\Omega)\Big) : w(\cdot,\boldsymbol{x})\Big|_{F_E} \in W_h,\; w(t,\cdot)\Big|_{T_E} = \sum_{a=1}^{l} w_a t^a \;\&\; w_a \in W_h \right\}.$$

Functions in $\boldsymbol{V}_h^n$ and W_h^n along the time dimension are represented by polynomials with degrees up to l. We formulate the space-time enhanced velocity variational formulation as: find $\boldsymbol{u}_{\alpha,h}^n, \tilde{\boldsymbol{u}}_{\alpha,h}^n \in \boldsymbol{V}_h^n$ and $p_{\alpha,h}^n, s_{\alpha,h}^n, \xi_{c\alpha,h}^n \in W_h^n$ such that

$$\int_{J_n}\int_{\Omega} \partial_t\Big(\varphi \sum_{\alpha} n_{c\alpha,h,\tau}^n\Big) w + \int_{J_n}\int_{\Omega}\Big(\nabla \cdot \sum_{\alpha} \boldsymbol{u}_{up,c\alpha,h}^n\Big) w \tag{15}$$
$$= \int_{J_n}\int_{\Omega}\Big(\sum_{\alpha} q_{c\alpha}\Big) w \quad \forall w \in W_h^n\,,$$

$$\int_{J_n}\int_{\Omega} K^{-1}\tilde{\boldsymbol{u}}_{\alpha,h}^n \cdot \boldsymbol{v} = \int_{J_n}\int_{\Omega} p_{\alpha,h}^n \nabla \cdot \boldsymbol{v} \quad \forall \boldsymbol{v} \in \boldsymbol{V}_h^n\,, \tag{16}$$

$$\int_{J_n}\int_{\Omega} \boldsymbol{u}_{\alpha,h}^n \cdot \boldsymbol{v} = \int_{J_n}\int_{\Omega} \lambda_\alpha \tilde{\boldsymbol{u}}_{\alpha,h}^n \cdot \boldsymbol{v} \quad \forall \boldsymbol{v} \in \boldsymbol{V}_h^n\,. \tag{17}$$

The phase mobility ratio in Eqn. (17) is defined as

$$\lambda_\alpha = \rho_\alpha \frac{k_{r\alpha}}{\mu_\alpha}\,, \tag{18}$$

and $\boldsymbol{u}_{up,\alpha,h}^n, \boldsymbol{u}_{up,c\alpha,h}^n$ are the upwind velocities calculated by

$$\int_{J_n}\int_{\Omega} \boldsymbol{u}_{up,\alpha,h}^n \cdot \boldsymbol{v} = \int_{J_n}\int_{\Omega} \lambda_\alpha^* \tilde{\boldsymbol{u}}_{\alpha,h}^n \cdot \boldsymbol{v} \quad \forall \boldsymbol{v} \in \boldsymbol{V}_h^n\,, \tag{19}$$

$$\int_{J_n}\int_{\Omega} \boldsymbol{u}_{up,c\alpha,h}^n \cdot \boldsymbol{v} = \int_{J_n}\int_{\Omega} \xi_{c\alpha}^* \boldsymbol{u}_{up,\alpha,h}^n \cdot \boldsymbol{v} \quad \forall \boldsymbol{v} \in \boldsymbol{V}_h^n\,. \tag{20}$$

The additional auxiliary phase fluxes $\tilde{\boldsymbol{u}}_{\alpha,h}^n$ are used to avoid inverting zero phase relative permeability [15]. Calculation of the upwind properties $(\lambda_\alpha^*, \xi_{c\alpha}^*)$ is done by using saturations and component fractions from the grid cell on the upwind direction

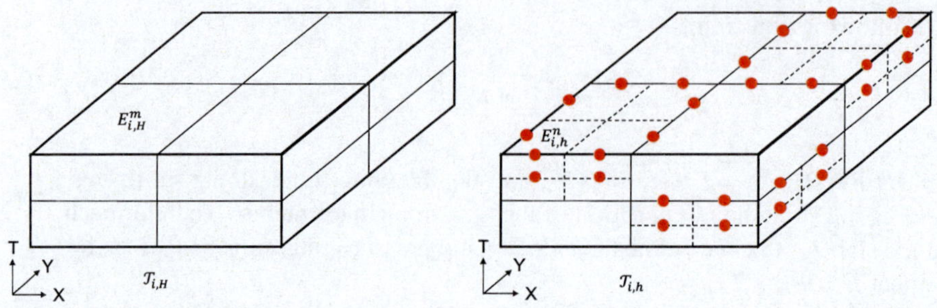

Fig. 2: Coarse and refined partition of subdomain $I_i \times \Omega_i$ with boundary interpolation nodes (red circles).

of the pressure gradient. We choose pressure and saturations as primary unknowns to solve. In case of phase disappearance, the solution gas-oil ratio (ξ_{3o}) needs to replace gaseous phase saturation (s_g) as the new primary unknown and vice versa.

3 Local residual minimization

Previous work regarding residual smoothing mainly involved linear problems with rough coefficients. Such pre-processing serves as a preconditioner for iterative linear solvers, such as the conjugate gradient method, and reduces their iteration counts significantly. In [19], an energy minimization method was introduced, which solves for a coarse basis function that minimizes the energy functional on the fine grid. However, direct application of such approach on nonlinear transport is problematic since no energy functional can be constructed due to the degenerate coefficients. Therefore we propose the local residual minimization approach.

Consider $J_n \times \Omega$ as a union of some non-overlapping subdomains $\{I_i \times \Omega_i\}$, namely $J_n \times \Omega = \cup_i (I_i \times \Omega_i)$, where $I_i = (\tau_i, \tau_{i+1}]$ is a sub-interval of $J_n = (t_n, t_{n+1}]$ and Ω_i is a subdomain of Ω. Now let $\mathcal{T}_{i,H}$ be a coarse rectangular partition of $I_i \times \Omega_i$, $E^m_{i,H} = T^m_i \times F^m_i$ be a space-time element in such partition with $T^m_i = (\tau^m_{i,0}, \tau^m_{i,1}]$. Consider $\mathcal{T}_{i,H}$ to be partially refined resulting in a finer rectangular partition $\mathcal{T}_{i,h}$ with elements $E^n_{i,h}$. We define the linear interpolation of a piecewise constant function (pressure and saturation) in space-time slab as f_ζ. We then construct the local problem as follows:

$$\int_{E^m_{i,H}} \sum_\alpha \left(\partial_t \left(\varphi n^n_{c\alpha,h,\tau} \right) + \nabla \cdot \boldsymbol{u}^n_{up,c\alpha,h} - q_{c\alpha} \right) w = 0 \qquad (21)$$

$$\forall E^m_{i,H} = \cup_{E^n_{i,h} \subsetneq E^m_{i,H}} E^n_{i,h} \,,$$

subject to

$$\begin{cases} p_\alpha = p_{\alpha,\zeta} \\ s_\alpha = s_{\alpha,\zeta} \end{cases} \quad \text{on } \partial E^m_{i,H} \,. \qquad (22)$$

Fig. 2 demonstrates the two partitions and the boundary interpolation nodes necessary for solving the local system. The boundary nodes appear on the top of the time level due to the discontinuous Galerkin of order zero discretization scheme.

If the interpolated pressure and saturations on the boundary are exact, then Eqn. (21) is well-posed and provides a unique solution that matches the global solution on the local subdomain. Unfortunately, providing exact boundary saturations by linear interpolation of the coarse solution is hardly achievable in nonlinear transport. The main reason being the transport and advection-diffusion process are closely coupled, making the variations of the pressure field across different grid resolutions have a strong influence on transport flow equation. Therefore, the local problem tends to be ill-posed. In response, we rewrite Eqn. (21) into a minimization problem as follow:

$$\min_{p_{\alpha,h}, s_{\alpha,h}} \left\{ \left\| \int_{E^m_{i,H}} \sum_{\alpha} \left(\partial_t \left(\varphi n^n_{c\alpha,h,\tau} \right) + \nabla \cdot \boldsymbol{u}^n_{up,c\alpha,h} - q_{c\alpha} \right) w \right\|_\infty \right\} \tag{23}$$
$$\forall E^m_{i,H} = \cup_{E^n_{i,h} \subsetneq E^m_{i,H}} E^n_{i,h} \, .$$

Since the goal of the minimization is only to remove the high frequency residuals (smoothing), to prevent over-working the problem, the algorithm is stopped once the average background residual is reached, instead of the absolute minimum.

Like solving the global problem, we use Newton's method to reduce the high frequency local residual. However, with a reduced problem size providing fewer constraints on the system, a "soft" Jacobian is likely to produce a solution outside the acceptable range (e.g. saturations must be in $[0, 1]$). Therefore, we apply the line search algorithm introduced in [14] to prevent divergence during the Newton iteration. Line search scales back the update when the Jacobian appears to be too "soft", by setting the update direction orthogonal to the post update residual.

4 Numerical results

We apply the SPE10 dataset [4] to conduct our numerical experiments. The fluid data are listed in Table. 1. The solution gas-oil ratio is estimated by

$$\xi_{3o} = \begin{cases} 1 - e^{n_{rs} p_o} & \text{if } p_o < p_b \\ 1 - e^{n_{rs} p_b} & \text{if } p_o \geq p_b \end{cases} , \tag{24}$$

with n_{rs} and p_b being the exponent and bubble point pressure respectively. We take $\beta = 2$ for Eqn. (10) to calculate oil density. For nonlinear transport, we use the Brooks–Corey model illustrated in Fig. 3 for both relative permeability and capillary pressure, which is described by Eqn. (25) and (26):

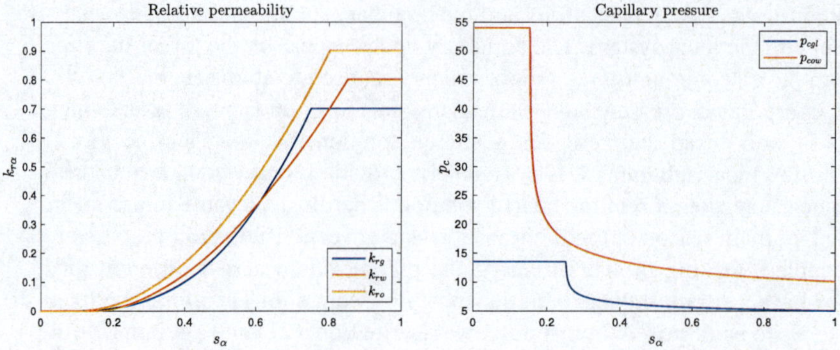

Fig. 3: Relative permeability (left) and capillary pressure (right) curve for the numerical experiment.

Table 1: Fluid data for the numerical experiment.

Parameter	Value	Unit
Gas compressibility (c_g)	5.0×10^{-2}	psi^{-1}
Oil compressibility (c_o)	1.0×10^{-4}	psi^{-1}
Water compressibility (c_w)	3.0×10^{-6}	psi^{-1}
Gas viscosity (μ_g)	0.03	cp
Oil viscosity (μ_o)	3.0	cp
Water viscosity (μ_w)	1.0	cp
Gas standard density ($\rho_{g,std}$)	0.1	lb/ft^3
Oil standard density ($\rho_{o,std}$)	53	lb/ft^3
Water standard density ($\rho_{w,std}$)	64	lb/ft^3
Solution gas-oil ratio exponent (n_{rs})	-1.5×10^{-4}	
Bubble point pressure (p_b)	3000.0	psi

$$\begin{cases} k_{rg} = k_{rg}^0 \left(\dfrac{s_g - s_{gr}}{1 - s_{gr} - s_{or} - s_{wr}} \right)^{n_g} \\ k_{ro} = k_{ro}^0 \left(\dfrac{s_o - s_{or}}{1 - s_{gr} - s_{or} - s_{wr}} \right)^{n_o} \\ k_{rw} = k_{rw}^0 \left(\dfrac{s_w - s_{wr}}{1 - s_{gr} - s_{or} - s_{wr}} \right)^{n_w} \end{cases}, \tag{25}$$

$$\begin{cases} p_{cgl} = p_{en,cgl} \left(\dfrac{1 - s_{or} - s_{wr}}{s_o + s_w - s_{or} - s_{wr}} \right)^{l_{cgl}} \\ p_{cow} = p_{en,cow} \left(\dfrac{1 - s_{wr}}{s_w - s_{wr}} \right)^{l_{cow}} \end{cases}. \tag{26}$$

The endpoint values for relative permeability are $k_{rg}^0 = 0.7$, $k_{ro}^0 = 0.9$, $k_{rw}^0 = 0.8$, $s_{gr} = 0.05$, $s_{or} = 0.1$, $s_{wr} = 0.15$ and the exponents are $n_g = 2.5$, $n_o = 2.0$, $n_w = 1.8$. There are capillary pressures on both water-oil and gas-liquid interfaces. The entry pressures are $p_{en,cow} = 10$ psi, $p_{en,cgl} = 5$ psi and the exponents are

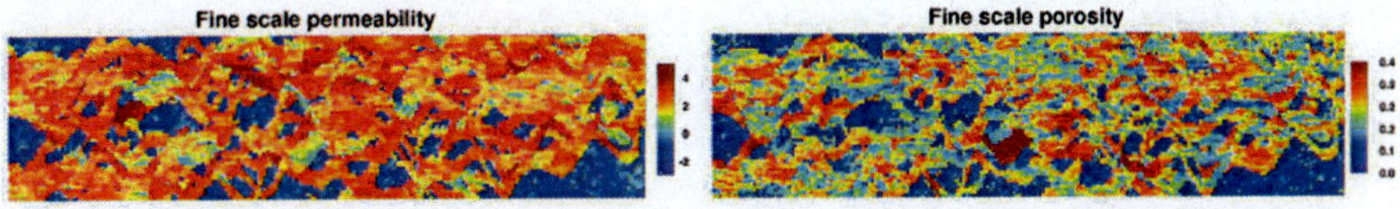

Fig. 4: Channelized fine scale permeability (left) and porosity (right) distribution.

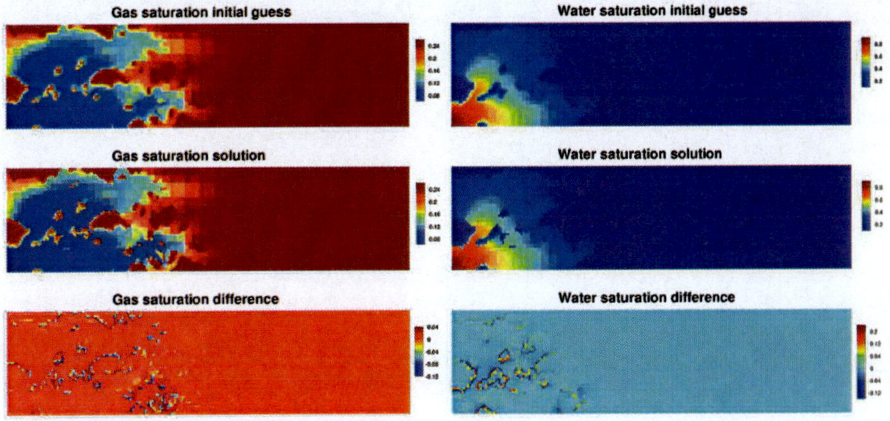

Fig. 5: Difference between saturation initial guess and solution without local residual minimization.

$l_{cow} = 0.25, l_{cgl} = 0.15$. The reservoir size is 56ft×216ft×1ft. We place a water rate specified injection well at the bottom left corner and a pressure specified production well at the upper right corner. The water injection rate is 1 ft^3/day and production pressure is 2000 psi. Furthermore, the initial pressure, gas saturation and water saturation are set to be 2000 psi, 0.25 and 0.15 respectively.

The experiment uses the bottom layer of the SPE10 dataset as petrophysical property input. The fine scale data are shown in Fig. 4 with clear discontinuity at the channel boundary. We use the algorithm described in [13] to solve the system. The number of refinement levels is set to three in both space and time and the refinement ratio is set to 2 uniformly. A numerical homogenization algorithm introduced in [5] and [6] is used to compute coarse resolution data.

The main cause of high frequency residual is the inaccurate initial estimate of saturations due to the discontinuous nature of the solution. A sample snapshot during the numerical experiment is shown in Fig. 5. Here, the initial guess of the saturations is compared to their respective final solutions and the difference between the two is calculated. We observe major discrepancies along the channel boundary since the saturation solution is discontinuous in such regions while the linear interpolation of the coarse solution provides a continuous transition. Now we apply the local residual minimization algorithm as a pre-processing step and compare the saturation difference to the same quantity without smoothing. The result is illustrated in Fig. 6. We observe that the discrepancies along the channel boundary have been reduced significantly. Most parts of the system show no sign of inconsistency between the

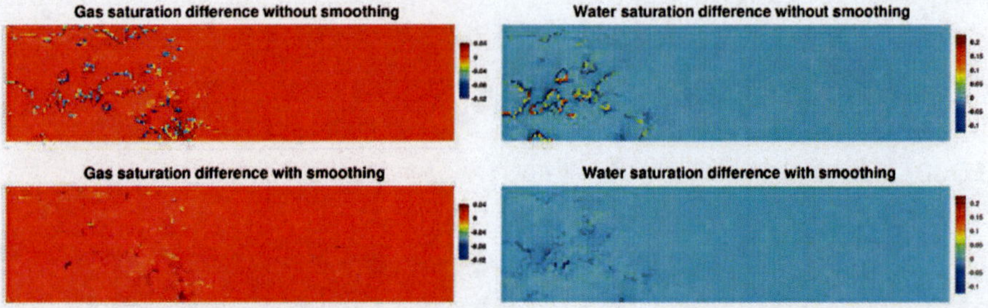

Fig. 6: Difference between saturation initial guess and solution with and without local residual minimization.

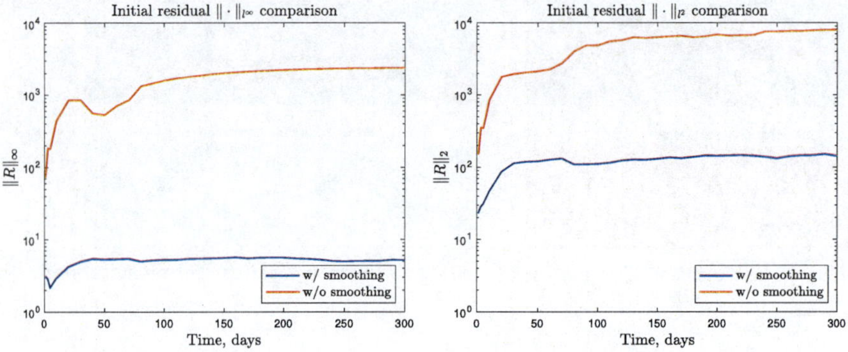

Fig. 7: l_∞ and l_2 norm of initial residual with and without local residual minimization.

initial guess and the solution. Some mismatch still exists, typically in regions with complex channel structure. Minimizing local residuals in these regions is unstable since a clear flow direction cannot be determined when only a small subdomain of the system is provided. Applying an oversampling technique could improve the stability of the minimization process.

We also quantify the reduction in initial residual when the Newton iteration continues after grid refinement and the result is demonstrated in Fig. 7. We observe that by applying the minimization algorithm as a smoothing pre-process, the initial residual has been reduced by approximately two orders of magnitude. As a result, the global system enters the quadratic convergence region sooner during Newton iteration and therefore improves convergence behavior. The number of iterations required to achieve convergence is reduced by roughly 40%.

5 Conclusions

In this paper, a new space-time compositional model is considered. We introduce the local residual minimization algorithm as a pre-processing step for geometric multigrid type methods to remove high frequency residuals after grid refinement. The minimization is approached by solving the same global physical system using Newton's method in the local subdomain with boundary conditions set to linear interpolation of the coarse scale solution before mesh refinement. The iteration is terminated once the residual in the subdomain reaches global background residual instead of the absolute minimum, to prevent over-working the ill-posed local problem. Results from a numerical experiment using a black-oil model is presented. We observe that after residual smoothing, the difference between saturation initial guess and solution is diminished significantly. The initial residual norm has been reduced by approximately 2 orders of magnitude. This improvement helps Newton's method to enter the quadratic convergence region and therefore cuts the number of iterations required to achieve nonlinear convergence by 40%. The algorithm performance in regions with a complex coefficient structure is sub-optimal, which can be improved by applying over-sampling techniques.

References

1. R. Abedi, B. Petracovici and R.B. Haber. A space–time discontinuous Galerkin method for linearized elastodynamics with element-wise momentum balance. *Computer Methods in Applied Mechanics and Engineering* **195**, 3247–3273, May (2006).
2. Y. Amanbek, G. Singh, G. Pencheva and M.F. Wheeler. Error indicators for incompressible Darcy flow problems using enhanced velocity mixed finite element method. *Computer Methods in Applied Mechanics and Engineering* **363**, 112884, May (2020).
3. M. Bause, F.A. Radu and U. Köcher. Space-time finite element approximation of the Biot poroelasticity system with iterative coupling. *Computer Methods in Applied Mechanics and Engineering* **320**, 745–768, June (2017).
4. M.A. Christie and M.J. Blunt. Tenth spe comparative solution project: A comparison of upscaling techniques. *SPE Reservoir Evaluation and Engineering* **4**(04), 308–317, Aug (2001).
5. E. Chung, Y. Efendiev and T.Y. Hou. Adaptive multiscale model reduction with generalized multiscale finite element methods. *Journal of Computational Physics* **320**, 69–95, Sept (2016).
6. Y. Efendiev and T.Y. Hou. *Multiscale finite element methods: Theory and applications*, volume 4. Springer Science and Business Media (2009).
7. T. Hoang, C. Japhet, M. Kern and J.E. Roberts. Space-time domain decomposition for advection–diffusion problems in mixed formulations. *Mathematics and Computers in Simulation* **137**, 366–389, July (2017).
8. T.J.R. Hughes and G.M. Hulbert. Space-time finite element methods for elastodynamics: Formulations and error estimates. *Computer Methods in Applied Mechanics and Engineering* **66**(3), 339–363, Feb (1988).
9. G.M. Hulbert and T.J.R. Hughes. Space-time finite element methods for second-order hyperbolic equations. *Computer Methods in Applied Mechanics and Engineering* **84**(3), 327–348, Dec (1990).
10. U. Köcher and M. Bause. Variational space–time methods for the wave equation. *Journal of Scientific Computing* **61**(2), 424–453, Nov (2014).

11. D. Krause and R. Krause. Enabling local time stepping in the parallel implicit solution of reaction–diffusion equations via space-time finite elements on shallow tree meshes. *Applied Mathematics and Computation* **277**, 164–179, Mar (2016).
12. U. Langer, S.E. Moore and M. Neumüller. Space–time isogeometric analysis of parabolic evolution problems. *Computer Methods in Applied Mechanics and Engineering* **306**, 342–363, July (2016).
13. H. Li, W.T. Leung and M.F. Wheeler. Sequential local mesh refinement solver with separate temporal and spatial adaptivity for non-linear two-phase flow problems. *Journal of Computational Physics* **403**, 109074, Feb (2020).
14. H. Matthies and G. Strang. The solution of nonlinear finite element equations. *International Journal for Numerical Methods in Engineering* **14**(11), 1613–1626, Dec (1979).
15. M. Peszyńska, M.F. Wheeler and I. Yotov. Mortar upscaling for multiphase flow in porous media. *Computational Geosciences* **6**(1), 73–100, Mar (2006).
16. G. Singh and M.F. Wheeler. Compositional flow modeling using a multi-point flux mixed finite element method. *Computational Geosciences* **20**, 421–435, Oct (2015).
17. G. Singh and M.F. Wheeler. A space-time domain decomposition approach using enhanced velocity mixed finite element method. *Journal of Computational Physics* **374**, 893–911, Dec (2018).
18. S.G. Thomas and M.F. Wheeler. Enhanced velocity mixed finite element methods for modeling coupled flow and transport on non-matching multiblock grids. *Computational Geosciences* **15**(4), 605–625, Sept (2011).
19. W.L. Wan, T.F. Chan and B. Smith. An energy-minimizing interpolation for robust multigrid methods. *SIAM Journal of Scientific Computing* **21**(4), 1632–1649, Dec (1998).
20. J.A. Wheeler, M.F. Wheeler and I. Yotov. Enhanced velocity mixed finite element methods for flow in multiblock domains. *Computational Geosciences* **6**, 315–332, Jan (2002).

Part II
Talks in Minisymposia

GenEO Coarse Spaces for Heterogeneous Indefinite Elliptic Problems

Niall Bootland, Victorita Dolean, Ivan G. Graham, Chupeng Ma, and Robert Scheichl

1 Introduction and motivations

For domain decomposition preconditioners, the use of a coarse correction as a second level is usually required to provide scalability (in the weak sense), such that the iteration count is independent of the number of subdomains, for subdomains of fixed dimension. In addition, it is desirable to guarantee robustness with respect to strong variations in the physical parameters. Achieving scalability and robustness usually relies on sophisticated tools such as spectral coarse spaces [4, 5]. In particular, we can highlight the GenEO coarse space [9], which has been successfully analysed and applied to highly heterogeneous positive definite elliptic problems. This coarse space relies on the solution of local eigenvalue problems on subdomains and the theory in the SPD case is based on the fact that local eigenfunctions form an orthonormal basis with respect to the energy scalar product induced by the bilinear form.

Our motivation here is to gain a better insight into the good performance of spectral coarse spaces even for highly indefinite high-frequency Helmholtz problems with absorbing boundary conditions, as observed in [3] (for the Dirichlet-to-Neumann coarse space) and more recently in [2] for coarse spaces of GenEO type. While a rigorous analysis for Helmholtz problems still lies beyond reach (see also [6] for the

Niall Bootland
University of Strathclyde, Dept. of Maths and Stats, e-mail: niall.bootland@strath.ac.uk

Victorita Dolean
University of Strathclyde, Dept. of Maths and Stats and University Côte d'Azur, CNRS, LJAD
e-mail: work@victoritadolean.com

Ivan G. Graham
University of Bath, Dept. Math. Sci., e-mail: i.g.graham@bath.ac.uk

Chupeng Ma
Heidelberg University, Inst. Applied Math., e-mail: chupeng.ma@uni-heidelberg.de

Robert Scheichl
Heidelberg University, Inst. Applied Math., e-mail: r.scheichl@uni-heidelberg.de

S. C. Brenner et al. (eds.), *Domain Decomposition Methods in Science and Engineering XXVI*, Lecture Notes in Computational Science and Engineering 145,
https://doi.org/10.1007/978-3-030-95025-5_10

challenges), we present here numerical results, showing the benefits of GenEO-type coarse spaces for the heterogeneous symmetric indefinite elliptic problem

$$-\nabla \cdot (A(\boldsymbol{x})\nabla u) - \kappa u = f \ \text{ in } \Omega, \qquad \text{subject to} \qquad u = 0 \ \text{ on } \partial\Omega, \tag{1}$$

in a bounded domain Ω with homogeneous Dirichlet boundary conditions on $\partial\Omega$, thus extending the results of [9] to this case. The coefficient function A in (1) is a symmetric positive-definite matrix-valued function on $\Omega \rightarrow \mathbb{R}^{d\times d}$ (where d is the space dimension) with highly varying but bounded values ($a_{\min}|\xi|^2 \leq A(\boldsymbol{x})\xi \cdot \xi \leq a_{\max}|\xi|^2, \boldsymbol{x} \in \Omega, \xi \in \mathbb{R}^d$) and κ is an $L^\infty(\Omega)$ function which can have positive or negative values. We assume throughout that problem (1) is well-posed and that there is a unique weak solution $u \in H_0^1(\Omega)$, for all $f \in L^2(\Omega)$.

We propose two types of spectral coarse spaces, one built from local spectra of the whole indefinite operator on the left-hand side of (1), and the other built using only the second-order operator in (1). For the latter, the analysis in [1] will apply, while the better performance of the former for large $\|\kappa\|_\infty$ provides some insight into the good performance of the $\mathcal{H}$-GenEO method introduced in [2] for high-frequency Helmholtz problems, even though it is not amenable to the theory in [1].

The problem (1) involves a Helmholtz-type operator (although this term would normally be associated with the case when κ has a positive sign and (1) would normally be equipped with an absorbing boundary condition rather than the Dirichlet condition here). In the special case $A = I, \kappa = k^2$ with k constant, the assumption of well-posedness of the problem is equivalent to the requirement that k^2 does not coincide with any of the Dirichlet eigenvalues of the operator $-\Delta$ in the domain Ω. In this case, for large k^2, the solution of (1) will be rich in modes corresponding to eigenvalues near k^2 and thus will have oscillatory behaviour, increasing as k increases. The Helmholtz problem with $A = I$ and $\kappa = \omega^2 n$ (with ω real and n a function), together with an absorbing far-field boundary condition appears regularly in geophysical applications; here n is the refractive index or 'squared slowness' of waves and ω is the angular frequency.

To solve discretisations of (1), we consider an additive Schwarz (AS) method with a GenEO-like coarse space and study the performance of this solver methodology for some heterogeneous test cases. GenEO coarse spaces have been shown theoretically and practically to be very effective for heterogeneous positive definite problems. Here, our main focus is to investigate how this approach performs in the indefinite case (1). We now review the underlying numerical methods that are used.

2 Discretisation and domain decomposition solver

We suppose that the domain Ω is a bounded Lipschitz polygon/polyhedron in 2D/3D. To discretise the problem we use the Lagrange finite element method of degree p on a conforming simplicial mesh T^h of Ω. Denote the finite element space by $V^h \subset H_0^1(\Omega)$. The finite element solution $u_h \in V^h$ satisfies the weak formulation

$b(u_h, v_h) = F(v_h)$, for all $v_h \in V^h$, where

$$b(u,v) = \int_\Omega (A(\boldsymbol{x})\nabla u \cdot \nabla v - \kappa uv)\,\mathrm{d}\boldsymbol{x} \qquad \text{and} \qquad F(v) = \int_\Omega f v\,\mathrm{d}\boldsymbol{x}. \tag{2}$$

Using the standard nodal basis for V^h we can represent the solution u_h through its basis coefficients $\boldsymbol{u}$ and reduce the problem to solving the symmetric linear system

$$B\boldsymbol{u} = \boldsymbol{f} \tag{3}$$

where B comes from the bilinear form $b(\cdot,\cdot)$ and $\boldsymbol{f}$ from the linear functional $F(\cdot)$. Note that B is symmetric but generally indefinite. For sufficiently small fine-mesh diameter h, problem (3) has a unique solution $\boldsymbol{u}$; see [8]. To solve (3), we utilise a two-level domain decomposition preconditioner within a Krylov method.

Consider an overlapping partition $\{\Omega_j\}_{1\le j\le N}$ of Ω, where each Ω_j is assumed to have diameter H_j and H denotes the maximal diameter of the subdomains. For each j we define $\widetilde{V}_j = \{v|_{\Omega_j} : v \in V^h\}$, $V_j = \{v \in \widetilde{V}_j : \mathrm{supp}(v) \subset \Omega_j\}$, and for $u, v \in \widetilde{V}_j$

$$b_j(u,v) := \int_{\Omega_j} (A(\boldsymbol{x})\nabla u \cdot \nabla v - \kappa uv)\,\mathrm{d}\boldsymbol{x} \quad \text{and} \quad a_j(u,v) := \int_{\Omega_j} A(\boldsymbol{x})\nabla u \cdot \nabla v \mathrm{d}\boldsymbol{x}.$$

Let $\mathcal{R}_j^T : V_j \to V^h$, $1 \le j \le N$, denote the zero-extension operator, let R_j^T denote its matrix representation with respect to the nodal basis and set $R_j = (R_j^T)^T$. The classical one-level additive Schwarz preconditioner is

$$M_{\mathrm{AS}}^{-1} = \sum_{j=1}^{N} R_j^T B_j^{-1} R_j, \qquad \text{where} \qquad B_j = R_j B R_j^T. \tag{4}$$

It is well-known that one-level additive Schwarz methods are not scalable with respect to the number of subdomains in general, since information is exchanged only between neighbouring subdomains. Thus, we introduce the two-level additive Schwarz method with GenEO coarse space first proposed in [9]. To this end, for $1 \le j \le N$, let $\{\varphi_1^j, \ldots, \varphi_{\widetilde{n}_j}^j\}$ be a nodal basis of $\widetilde{V}_j$, where $\widetilde{n}_j = \dim(\widetilde{V}_j)$.

Definition 1 (Partition of unity) Let $\mathrm{dof}(\Omega_j)$ denote the internal degrees of freedom (nodes) on subdomain Ω_j. For any degree of freedom i, let μ_i denote the number of subdomains Ω_j for which i is an internal degree of freedom, i.e., $\mu_i := \#\{j : 1 \le j \le N,\ i \in \mathrm{dof}(\Omega_j)\}$. Then, for $1 \le j \le N$, the *local partition of unity operator* $\Xi_j : \widetilde{V}_j \to V_j$ is defined by

$$\Xi_j(v) := \sum_{i\in\mathrm{dof}(\Omega_j)} \frac{1}{\mu_i} v_i \varphi_i^j, \qquad \text{for all} \qquad v = \sum_{i=1}^{\widetilde{n}_j} v_i \varphi_i^j \in \widetilde{V}_j. \tag{5}$$

The operators Ξ_j form a partition of unity, i.e., $\sum_{j=1}^N R_j^T \Xi_j(v|_{\Omega_j}) = v$, $\forall v \in V^h$ [9].

For each j, we define the following generalised eigenvalue problems:

$$\text{find } p \in \widetilde{V}_j \backslash \{0\},\ \lambda \in \mathbb{R}: \quad a_j(p, v) = \lambda\, a_j(\Xi_j(p), \Xi_j(v)), \quad \text{for all } v \in \widetilde{V}_j, \tag{6}$$

$$\text{find } q \in \widetilde{V}_j \backslash \{0\},\ \lambda \in \mathbb{R}: \quad b_j(q, v) = \lambda\, a_j(\Xi_j(q), \Xi_j(v)), \quad \text{for all } v \in \widetilde{V}_j, \tag{7}$$

where Ξ_j is the local partition of unity operator from Definition 1.

Definition 2 (Δ-GenEO and $\mathcal{H}$-GenEO coarse spaces) For each j, $1 \leq j \leq N$, let $(p_l^j)_{l=1}^{m_j}$ and $(q_l^j)_{l=1}^{m_j}$ be the eigenfunctions of the eigenproblems (6) and (7) corresponding to the m_j smallest eigenvalues, respectively. Then we define the Δ-*GenEO* and $\mathcal{H}$-*GenEO* coarse spaces, respectively, by

$$V_\Delta^0 := \operatorname{span}\{\mathcal{R}_j^T \Xi_j(p_l^j) : l = 1, \ldots, m_j;\ j = 1, \ldots, N\} \text{ and} \tag{8}$$

$$V_{\mathcal{H}}^0 := \operatorname{span}\{\mathcal{R}_j^T \Xi_j(q_l^j) : l = 1, \ldots, m_j;\ j = 1, \ldots, N\}. \tag{9}$$

Note that here and subsequently, the subscript Δ refers to the GenEO coarse space (8) based on (6), the eigenproblem with respect to the 'Laplace-like' operator induced by the bilinear form a_j, while the subscript $\mathcal{H}$ refers to the $\mathcal{H}$-GenEO coarse space (9) based on (7), with the 'Helmholtz-like' operator appearing in b_j.

Since $V_\Delta^0, V_{\mathcal{H}}^0 \subset V^h$, we can introduce the natural embeddings $\mathcal{R}_{0,\Delta}^T : V_\Delta^0 \to V^h$ and $\mathcal{R}_{0,\mathcal{H}}^T : V_{\mathcal{H}}^0 \to V^h$, with matrix representations $R_{0,\Delta}^T$ and $R_{0,\mathcal{H}}^T$, respectively, and set $R_{0,\Delta} = (R_{0,\Delta}^T)^T$ and $R_{0,\mathcal{H}} = (R_{0,\mathcal{H}}^T)^T$ to obtain the following two-level extensions of the one-level additive Schwarz method (4):

$$M_{\mathrm{AS},\Delta}^{-1} = M_{AS}^{-1} + R_{0,\Delta}^T B_{0,\Delta}^{-1} R_{0,\Delta} \quad \text{and} \quad M_{\mathrm{AS},\mathcal{H}}^{-1} = M_{AS}^{-1} + R_{0,\mathcal{H}}^T B_{0,\mathcal{H}}^{-1} R_{0,\mathcal{H}}, \tag{10}$$

where $B_{0,\Delta} := R_{0,\Delta} B R_{0,\Delta}^T$ and $B_{0,\mathcal{H}} := R_{0,\mathcal{H}} B R_{0,\mathcal{H}}^T$.

3 Theoretical results

The theoretical properties of the preconditioner $M_{\mathrm{AS},\Delta}^{-1}$ are studied in the forthcoming paper [1]. There, the PDE studied is a generalisation of (1), which also allows the inclusion of a non-self-adjoint first order convection term. The important parameters in the preconditioner are the coarse mesh diameter H and the 'eigenvalue tolerance'

$$\Theta := \max_{1 \leq j \leq N} \left(\lambda_{m_j+1}^j\right)^{-1},$$

where $\{\lambda_m^j : m = 1, 2, \ldots\}$ are the eigenvalues of the generalised eigenproblem (6), given in non-decreasing order. We now highlight a special case of the results in [1].

Theorem 1 *Let the fine-mesh diameter h be sufficiently small. Then there exist thresholds $H_0 > 0$ and $\Theta_0 > 0$ such that, for all $H \le H_0$ and $\Theta \le \Theta_0$: the matrices B_j and $B_{0,\Delta}$ appearing in (4) and (10) are non-singular. Moreover, if problem (3) is solved by GMRES with left preconditioner $M_{\mathrm{AS},\Delta}^{-1}$ and residual minimisation in the energy norm $\|u\|_a := \left(\int_\Omega \nabla u \cdot A \nabla u\right)^{1/2}$, then there exists a constant $c \in (0, 1)$, which depends on H_0 and Θ_0 but is independent of all other parameters, such that we have the robust GMRES convergence estimate*

$$\|r_\ell\|_a^2 \le \left(1 - c^2\right)^\ell \|r_0\|_a^2, \tag{11}$$

for $\ell = 0, 1, \ldots$, where r_ℓ denotes the residual after ℓ iterations of GMRES.

In fact, the paper [1] will investigate in detail how the thresholds H_0 and Θ_0 depend on the heterogeneity and indefiniteness of (1). For example, if the problem is scaled so that $a_{\min} = 1$, then as $\|\kappa\|_\infty$ grows, H_0 and Θ_0 have to decrease to maintain the convergence rate of GMRES:

$$H_0 \lesssim \|\kappa\|_\infty^{-1} \qquad \text{and} \qquad \Theta_0 \lesssim C_{\mathrm{stab}}^{-2} \|\kappa\|_\infty^{-4}, \tag{12}$$

where $C_{\mathrm{stab}} = C_{\mathrm{stab}}(A, \kappa)$ denotes the stability constant for problem (1), i.e., the solution u satisfies $\|u\|_{H^1(\Omega)} \le C_{\mathrm{stab}} \|f\|_{L^2(\Omega)}$ for all $f \in L^2(\Omega)$ and the hidden constants are independent of h, H, $a_{\max}$ and κ. Thus, as $\|\kappa\|_\infty$ gets smaller, the indefiniteness diminishes and the requirements on H_0 and Θ_0 are relaxed.

4 Numerical results

We give results for a more efficient variant of the preconditioner described in §2. Instead of (10), we here use the *restricted additive Schwarz* (RAS) method, with the GenEO coarse space incorporated using a deflation approach, yielding:

$$M^{-1} = M_{\mathrm{RAS}}^{-1}(I - BQ_0) + Q_0, \qquad \text{where} \quad M_{\mathrm{RAS}}^{-1} = \sum_{j=1}^{N} R_j^T D_j B_j^{-1} R_j. \tag{13}$$

Here, D_j is the matrix form of the partition of unity operator Ξ_j. Moreover, we have $Q_0 = R_0^T B_0^{-1} R_0$ with $B_0 = R_0 B R_0^T$ and either $R_0 = R_{0,\Delta}$ or $R_0 = R_{0,\mathcal{H}}$, depending on whether we use Δ-GenEO or $\mathcal{H}$-GenEO. We include all eigenfunctions p_l^j or q_l^j in V_Δ^0 or $V_\mathcal{H}^0$ corresponding to eigenvalues $\lambda_l^j < \lambda_{\max}$, for Δ-GenEO or $\mathcal{H}$-GenEO, respectively. In $\mathcal{H}$-GenEO this includes all eigenfunctions corresponding to negative eigenvalues. Unless otherwise stated, the eigenvalue threshold is $\lambda_{\max} = \frac{1}{2}$.

As a model problem, we consider (1) on the unit square $\Omega = (0, 1)^2$, take κ constant, and define A to model various layered media, as depicted in Fig. 1. The right-hand side f is taken to be a point source at the centre $(\frac{1}{2}, \frac{1}{2})$. To discretise, we use a uniform square grid with n_{glob} points in each direction and triangulate along

(a) Alternating layers (b) Diagonal layers (c) Inclusions ($N = 9$)

Fig. 1: Piecewise constant profiles $a(\boldsymbol{x})$, where $A(\boldsymbol{x}) = a(\boldsymbol{x})I$. For the darkest shade $a(\mathbf{x}) = 1$ while for the lightest shade $a(\mathbf{x}) = a_{\max}$. Profiles (a) and (b) are fixed while in (c) the interfaces of a are the same per subdomain, although the value of a depends on height (case $N = 3^2$ is shown).

one set of diagonals to form P1 elements. We further use a uniform decomposition into N square subdomains and throughout use minimal overlap (non-overlapping subdomains are extended by adding only the fine-mesh elements which touch them).

Our computations are performed using FreeFem (`http://freefem.org/`), in particular using the `ffddm` framework. We use preconditioned GMRES with residual minimisation in the Euclidean norm and a relative residual tolerance of 10^{-6}. We have assumed $a_{\min} = 1$; otherwise a rescaling will ensure this. The indefiniteness is controlled by κ, taken here to be a positive constant. Although estimate (11) describes GMRES implemented in the energy inner product, we use here the standard Euclidean implementation and prove in [1, §4] that (for quasi-uniform meshes) the latter algorithm requires at most $O(\log(a_{\max}/h))$ more iterations than the former to achieve the same residual reduction. Experiments for Helmholtz problems in [7] showed that the two approaches performed almost identically.

In Table 1 we provide GMRES iteration counts for Δ-GenEO and $\mathcal{H}$-GenEO as N varies in two cases: in case (i), on the left, we use profile (a) and increase $a_{\max}$ while in case (ii), on the right, we use profile (c) and increase $n_{\text{glob}} = h^{-1}$. In (i) we see clear robustness to increasing the contrast parameter $a_{\max}$. In (ii) we observe robustness to decreasing h, with markedly better performance for $\mathcal{H}$-GenEO. In (ii), the coefficient $a(\boldsymbol{x})$ (and hence the problem itself) becomes more complicated as N increases since the geometry of the coefficient remains identical in each subdomain.

In Table 2 we illustrate the effect of increasing κ, giving iteration counts and (in brackets) coarse space sizes. Here we see the substantial advantage of $\mathcal{H}$-GenEO over Δ-GenEO: much better iteration counts are obtained, yet the coarse space size increases only modestly. As N increases, although the dimension of the coarse space grows, the number of eigenfunctions per subdomain decreases. For very large κ, neither method is fully robust while, for small κ, both methods perform similarly. This leads to the interesting question of whether robustness to κ can be gained by taking more eigenfunctions in the coarse space. Table 3 gives results for a sequence of increasing values of κ for the diagonal layers problem, in which we simultaneously increase $\lambda_{\max}$, indicating (apparent almost) robustness with respect to κ.

These observations align with the fact that eigenfunctions appear qualitatively similar for Δ-GenEO and $\mathcal{H}$-GenEO when κ is small. As seen in Fig. 2, once κ increases the $\mathcal{H}$-GenEO eigenfunctions change: the type of eigenfunctions produced

Table 1: GMRES iteration counts with $\lambda_{\max} = \frac{1}{2}$. Left-hand table: Alternating layers problem, varying $a_{\max} > 1$ and N, with fixed $\kappa = 100$ and $n_{\text{glob}} = 400$. Right-hand table: Inclusions problem, varying n_{glob} and N, with fixed $\kappa = 1000$ and $a_{\max} = 50$.

	Δ-GenEO				$\mathcal{H}$-GenEO			
$a_{\max} \setminus N$	16	36	64	100	16	36	64	100
10	10	9	9	10	9	9	9	9
50	9	9	9	9	9	9	9	9
200	9	9	9	9	9	9	9	9
1000	9	9	9	9	9	9	9	9

	Δ-GenEO				$\mathcal{H}$-GenEO			
$n_{\text{glob}} \setminus N$	16	36	64	100	16	36	64	100
200	24	16	26	22	10	10	10	11
400	23	14	19	18	8	9	9	8
600	21	14	18	19	8	9	8	10
800	22	14	19	20	9	9	9	10

Table 2: GMRES iteration counts and (in brackets) coarse space dimension for the diagonal layers problem with $\lambda_{\max} = \frac{1}{2}$, varying κ and N, with fixed $a_{\max} = 5$ and $n_{\text{glob}} = 600$.

	Δ-GenEO				$\mathcal{H}$-GenEO			
$\kappa \setminus N$	16	36	64	100	16	36	64	100
10	9 (627)	9 (1050)	9 (1468)	9 (1804)	9 (627)	9 (1050)	9 (1468)	9 (1804)
100	10 (627)	9 (1050)	9 (1468)	9 (1804)	9 (627)	9 (1052)	9 (1473)	9 (1814)
1000	36 (627)	43 (1050)	35 (1468)	28 (1804)	13 (674)	11 (1083)	9 (1520)	10 (1877)
10000	215 (627)	339 (1050)	437 (1468)	506 (1804)	27 (1256)	33 (1651)	40 (2139)	18 (2549)

Table 3: GMRES iteration counts and (in brackets) coarse space dimension for $\mathcal{H}$-GenEO for the diagonal layers problem, varying N, with fixed $n_{\text{glob}} = 600$ and $a_{\max} = 5$ and increasing eigenvalue threshold $\lambda_{\max}$ as κ increases, aiming to control iteration counts as κ increases.

		$\mathcal{H}$–GenEO			
$\lambda_{\max}$	$\kappa \setminus N$	16	36	64	100
0.1	10	23 (108)	23 (199)	25 (214)	23 (324)
0.1	100	23 (111)	24 (201)	28 (223)	27 (324)
0.2	1000	19 (265)	27 (418)	20 (574)	20 (684)
0.6	10000	24 (1430)	25 (2129)	28 (2680)	15 (3252)

by Δ-GenEO remain, albeit perturbed, but now we have further eigenfunctions which include more oscillatory behaviour in the interior of the subdomain; such features are not found with Δ-GenEO where higher oscillations only appear near the boundary.

5 Conclusions

In this work we have summarised how the forthcoming analysis in [1] can be applied to a GenEO-type coarse space for heterogeneous indefinite elliptic problems. We provide numerical evidence supporting these results and a comparison with a more

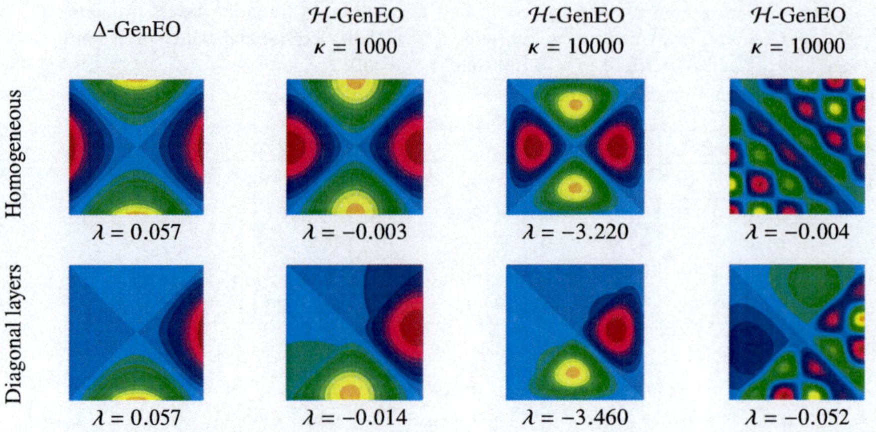

Fig. 2: Example eigenfunctions on the central subdomain when $N = 25$ and $n_{\text{glob}} = 800$: In the first three columns, we plot qualitatively similar eigenfunctions, computed (left-to-right) by Δ-GenEO, $\mathcal{H}$-GenEO when $\kappa = 1000$, and $\mathcal{H}$-GenEO when $\kappa = 10000$. This illustrates how eigenfunctions of (7) are affected by the indefiniteness in b_j, relative to the size of κ. In addition, as κ increases the $\mathcal{H}$-GenEO eigenproblem enriches the coarse space with "wave-like" eigenfunctions that are not seen for Δ-GenEO; one of the many examples when $\kappa = 10000$ is plotted in the final column. While the top row explores the homogeneous case, the bottom row demonstrates the effect of heterogeneity in $a(\boldsymbol{x})$ for the diagonal layers problem: For $N = 25$, $a(\boldsymbol{x}) = a_{\max} = 10$ in the upper-left triangle ($x_2 > x_1$) and $a(\boldsymbol{x}) = a_{\min} = 1$ in the lower-right triangle ($x_2 < x_1$) of the central subdomain. Note that variation in the eigenfunctions is mainly confined to the low coefficient region.

effective GenEO-type method for highly indefinite problems but for which no theory is presently available. For mildly indefinite problems these two approaches perform similarly, providing the first theoretical insight towards explaining the good behaviour of the $\mathcal{H}$-GenEO method for challenging heterogeneous Helmholtz problems.

References

1. N. Bootland, V. Dolean, I.G. Graham, C. Ma and R. Scheichl. Overlapping Schwarz methods with GenEO coarse spaces for indefinite and non-self-adjoint problems. *arXiv preprint* arXiv:2110.13537 (2021).
2. N. Bootland, V. Dolean, P. Jolivet and P.H. Tournier. A comparison of coarse spaces for Helmholtz problems in the high frequency regime. *Comput. Math. Appl.* **98**, 239–253 (2021).
3. L. Conen, V. Dolean, R. Krause and F. Nataf. A coarse space for heterogeneous Helmholtz problems based on the Dirichlet-to-Neumann operator. *J. Comput. Appl. Math.* **271**, 83–99 (2014).
4. V. Dolean, P. Jolivet and F. Nataf. *An Introduction to Domain Decomposition Methods: Algorithms, Theory, and Parallel Implementation*. SIAM, Philadelphia, PA (2015).
5. J. Galvis and Y. Efendiev. Domain decomposition preconditioners for multiscale flows in high contrast media. *Multiscale Model. Sim.* **8**(4), 1461–1483 (2010).

6. M. Gander and O. Ernst. Why it is difficult to solve Helmholtz problems with classical iterative methods? In: I.G. Graham, T.Y. Hou, O. Lakkis, R. Scheichl (eds.) *Numerical Analysis of Multiscale Problems*, pp. 325–363. Springer, Berlin, Heidelberg (2011).
7. I.G. Graham, E.A. Spence and E. Vainikko. Domain decomposition preconditioning for high-frequency helmholtz problems with absorption. *Math. Comp.* **86**, 2089–2127 (2017).
8. A.H. Schatz and J.P. Wang. Some new error estimates for Ritz–Galerkin methods with minimal regularity assumptions. *Math. Comput.* **213**, 19–27 (1996).
9. N. Spillane, V. Dolean, P. Hauret, F. Nataf, C. Pechstein and R. Scheichl. Abstract robust coarse spaces for systems of PDEs via generalized eigenproblems in the overlaps. *Numer. Math.* **126**(4), 741–770 (2014).

Inexact Subdomain Solves Using Deflated GMRES for Helmholtz Problems

N. Bootland, V. Dwarka, P. Jolivet, V. Dolean, and C. Vuik

1 Introduction

In recent years, domain decomposition based preconditioners have become popular tools to solve the Helmholtz equation. Notorious for causing a variety of convergence issues, the Helmholtz equation remains a challenging PDE to solve numerically. Even for simple model problems, the resulting linear system after discretisation becomes indefinite and tailored iterative solvers are required to obtain the numerical solution efficiently. At the same time, the mesh must be kept fine enough in order to prevent numerical dispersion 'polluting' the solution [4]. This leads to very large linear systems, further amplifying the need to develop economical solver methodologies.

Domain decomposition (DD) techniques combined with Krylov solvers provide a way to deal with these large systems [6]. While the use of two-level deflation and DD techniques have been explored before, their expedience has primarily been measured in terms of providing a way to add a coarse space to obtain a two-level DD

Niall Bootland
University of Strathclyde, Department of Mathematics and Statistics, Glasgow, UK
e-mail: niall.bootland@strath.ac.uk

Vandana Dwarka
Delft University of Technology, Delft Institute of Applied Mathematics, Delft, The Netherlands
e-mail: v.n.s.r.dwarka@tudelft.nl

Pierre Jolivet
University of Toulouse, CNRS, IRIT, Toulouse, France
e-mail: pierre.jolivet@enseeiht.fr

Victorita Dolean
University of Strathclyde, Department of Mathematics and Statistics, Glasgow, UK
Université Côte d'Azur, CNRS, Laboratoire J.A. Dieudonné, Nice, France
e-mail: work@victoritadolean.com

Cornelis Vuik
Delft University of Technology, Delft Institute of Applied Mathematics, Delft, The Netherlands
e-mail: c.vuik@tudelft.nl

S. C. Brenner et al. (eds.), *Domain Decomposition Methods in Science and Engineering XXVI*, Lecture Notes in Computational Science and Engineering 145,
https://doi.org/10.1007/978-3-030-95025-5_11

preconditioner [1, 2]. Without a coarse space, DD methods typically do not scale with the number of subdomains. Moreover, subdomain sizes need to be relatively small in order to optimally use local memory and direct solution methods on subdomains.

In this work we focus on the subdomain solves. Instead of using a direct solution method, we solve the local systems using GMRES preconditioned by a two-level deflation approach. As a result, similar to the inclusion of a coarse space on the fine-level, we obtain a two-level method on each subdomain as well. The inexact solve on the subdomains will allow for larger subdomains by reducing computing and memory requirements, especially in 3D. In order to allow for inexact subdomain solves we require a flexible wrapper for the outer iteration—the application of the DD preconditioner—and so we will use FGMRES. Our local subdomain solves will then employ a preconditioned GMRES iteration. It is well known that the cost of GMRES increases with each iteration. Thus, in order to mitigate the number of iterations at the subdomain level, we use a two-level deflation preconditioner [5, 8].

The techniques proposed here will feature as the topic of future research on large-scale 3D applications using pollution-free meshes. In this work we introduce the key ideas and begin an initial exploration by considering a simple 2D model problem.

2 Model problem, discretisation and preconditioning strategies

Our model problem consists of the Helmholtz equation posed on the unit square:

$$-\Delta u - k^2 u = f \qquad \text{in } \Omega = (0,1)^2, \tag{1a}$$

$$u = 0 \qquad \text{on } \partial\Omega. \tag{1b}$$

Here, the parameter k denotes the wave number. Problem (1) is well-posed so long as k^2 is not a Dirichlet eigenvalue of the corresponding Laplace problem. Solving the problem with Dirichlet conditions provides a more robust test for the solver, as there is no shift keeping the spectrum away from the origin [3, 8]. In this work we will assume the problem, and any sub-problems, are well-posed. To discretise (1) we use piecewise linear (P1) finite elements on a uniform grid with mesh spacing given by $h = 2\pi k^{-1} n_{\text{ppwl}}^{-1}$, where n_{ppwl} is the number of (grid) points per wavelength (hereinafter referred to as "ppwl"). To test the solver performance we initially ensure $n_{\text{ppwl}} \approx 10$. We then double this to approximately 20 ppwl to obtain more accurate numerical solutions. Letting $V^h \subset H_0^1(\Omega)$ denote the space of piecewise linear functions on our finite element mesh T^h of Ω, our discrete solution $u_h \in V^h$ satisfies the weak formulation $a(u_h, v_h) = F(v_h)\ \forall\, v \in V^h$, where

$$a(u,v) = \int_\Omega \left(\nabla u \cdot \nabla v - k^2 uv\right) \mathrm{d}\mathbf{x} \qquad \text{and} \qquad F(v) = \int_\Omega f v \,\mathrm{d}\mathbf{x}. \tag{2}$$

With the standard basis for V^h, we can write the weak formulation as finding the solution to the linear system $A\mathbf{u} = \mathbf{f}$. We now consider how to solve such systems.

The global matrix A is preconditioned by a one-level domain decomposition method. To construct the decomposition, we define an overlapping partition $\{\Omega_j\}_{j=1}^N$ of Ω together with a restriction operator R_j to move from the global level to the subdomain level. We only consider Cartesian (rectangular) subdomains. Using this decomposition, the restricted additive Schwarz (RAS) preconditioner is defined by

$$M_{RAS}^{-1} = \sum_{j=1}^{N} R_j^T D_j A_j^{-1} R_j, \tag{3}$$

where D_j are diagonal matrices representing a partition of unity ($\sum_{j=1}^N R_j^T D_j R_j = I$) and $A_j = R_j A R_j^T$ are the local Dirichlet matrices. Each subdomain solve requires the solution of a local auxiliary linear system, which we denote by $A_j \tilde{\mathbf{u}}_j = \tilde{\mathbf{f}}_j$ as a general case. We solve these systems using a two-level deflation approach, that is, deflation is used to accelerate the convergence of GMRES by removing the near-zero eigenvalues. For normal matrices it has been shown that convergence can be directly related to the behaviour of these near-zero eigenvalues [7].

The two-level deflation preconditioner is defined as a projection operator P_j which leads to solving $P_j A_j \tilde{\mathbf{u}}_j = P_j \tilde{\mathbf{f}}_j$, where

$$P_j = I - A_j Q_j \qquad \text{with} \qquad Q_j = Z_j E_j^{-1} Z_j^T \qquad \text{and} \qquad E_j = Z_j^T A_j Z_j. \tag{4}$$

The rectangular matrix Z_j in this particular setting is called the deflation matrix and its columns span the deflation space. The choice of Z_j strongly dictates the overall convergence behaviour. Here, we use quadratic rational Bézier curves, as they have been shown to provide satisfactory convergence [5]. Consequently, if we let $\tilde{u}_j^i$ represent the i-th degree of freedom (DOF) on subdomain Ω_j, then in 1D Z_j maps these nodal approximations onto their coarse-grid counterpart as follows

$$\left[Z_j \tilde{u}_j\right]^i = \frac{1}{8}\left(\tilde{u}_j^{2i-2} + 4\,\tilde{u}_j^{2i-1} + 6\,\tilde{u}_j^{2i} + 4\,\tilde{u}_j^{2i+1} + \tilde{u}_j^{2i+2}\right). \tag{5}$$

As such, Z_j can be constructed using the following 1D stencil $\frac{1}{8}\left[1\ 4\ 6\ 4\ 1\right]$. The dimension of Z_j will then be $n_j \times \frac{n_j}{2}$, where n_j is the size of the local 1D system.

In 2D on Cartesian grids this can be naturally extended by using the Kronecker product. For a rectangular subdomain, letting Z_j^x denote the 1D deflation matrix in the x-direction and Z_j^y that in the y-direction, the 2D deflation matrix is given by

$$Z_j = Z_j^y \otimes Z_j^x, \tag{6}$$

assuming lexicographic ordering running through x coordinates first.

3 Numerical results

We now provide numerical results for our model problem on the unit square. We take the right-hand side f to be given by a point-source at the centre of the domain. Unless stated otherwise, 10 ppwl are used to construct the mesh and we let n_{glob} be the number of DOFs along each edge of the square. For the domain decomposition, we use a uniform decomposition into N (square) subdomains. Overlap is added by appending one layer of mesh elements in a Cartesian manner (note that this means subdomains touching only one edge of Ω are rectangular rather than square).

For the outer solve we use preconditioned FGMRES with the one-level RAS preconditioner (3). The tolerance for the relative residual has been set at 10^{-6}. For the inner solve on the subdomain level we use preconditioned GMRES with the two-level deflation preconditioner (4) instead of a direct solver. Note that subdomains systems are decoupled and so can be solved locally in parallel. We will vary the inner tolerance for the relative residual between 10^{-10} and 10^{-2} in order to assess an appropriate level of accuracy needed when solving the subdomain problems. The solver is equipped to deal with both symmetric (Dirichlet) and non-symmetric systems (Sommerfeld), as we use the RAS preconditioner together with GMRES, which do not require symmetry.

All matrices are constructed using FreeFem (`http://freefem.org/`) while the solvers are then implemented using PETSc (`http://www.mcs.anl.gov/petsc/`). Computations are carried out on a laptop with an i7-10850H processor having 6 cores (12 threads).

3.1 Direct subdomain solves

We start by constructing a benchmark where we use a direct solution method for the subdomain solves, namely via an LU-decomposition. Table 1 shows that the number of iterations does not scale as the number of subdomains N increases, in agreement with the literature. The number of iterations also rapidly increases with the wave number k. The inclusion of a coarse space to improve both k- and N-scalability on top of inexact subdomain solves will be explored in future research. An interesting observation is that increasing n_{ppwl} leads to a higher iteration count. The opposite effect has been observed when using a two-level deflation preconditioner [5, 8]. There, a finer mesh leads to a smaller number of iterations as the mapping of the eigenvectors from the fine- and coarse-grid becomes more accurate.

Table 1: FGMRES iteration counts using the one-level RAS preconditioner with direct subdomain solves.

10 ppwl

		N			
k	n_{glob}	4	9	16	25
20	30	20	27	48	45
40	60	31	60	85	101
80	120	64	133	191	216
160	240	159	262	365	495

20 ppwl

		N			
k	n_{glob}	4	9	16	25
20	60	20	40	42	59
40	120	37	66	89	115
80	240	76	131	189	255
160	480	130	289	398	520

3.2 Inexact subdomain solves

The direct subdomain solves remain feasible for medium-size problems. Once we move to high-frequency 3D problems the subdomain systems become larger and the direct solver will start to become inefficient and consume more computing power and memory. In order to assess the feasibility of inexact solves, we will use the benchmarks from Section 3.1 and compare with our iterative method. The aim of these experiments is twofold. First, we want to examine the scalability with respect to the number of subdomains once we substitute the direct solution method. Secondly, we want to observe what level of accuracy is needed at the subdomain level such that the outer number of iterations remains within a satisfactory range.

High-tolerance: 10^{-10}

We start with a tolerance of 10^{-10} with results given in Table 2. This case is the closest to the use of a direct solver (see Table 1). Comparing, we observe that the results are almost identical when using 4 subdomains. Once we increase the number of subdomains N, the number of FGMRES iterations increases for both 10 and 20 ppwl. However, the increase is more noticeable when using 10 ppwl. For example, when $k = 160$ and $N = 9$ a direct solver on the subdomains leads to 262 iterations while the inexact approach converges in 301 outer iterations. However, when we double n_{ppwl} to 20, we go from 289 to 292 outer iterations. In all cases, as expected, the number of outer iterations increases with the wave number k. A similar yet slower increase in iteration counts is observed for the average number of inner iterations required by the deflated GMRES approach on the subdomains. As mentioned previously, the deflation preconditioner becomes more efficient on finer meshes. This can also be observed in our results: while the number of outer FGMRES iterations (using RAS) increases when moving from 10 ppwl to 20 ppwl, the number of inner GMRES iterations (using two-level deflation) decreases as the local subdomain systems become larger. Additionally, in this case, the number of inner iterations appears to be scaling better with the wave number k.

Table 2: FGMRES iteration counts using the one-level RAS preconditioner with subdomain problems solved inexactly to a relative tolerance of 10^{-10} using GMRES with a two-level deflation preconditioner. In parentheses we display the average number of GMRES iterations per subdomain solve.

10 ppwl

		N			
k	n_{glob}	4	9	16	25
20	30	20 (23)	27 (25)	56 (24)	46 (21)
40	60	32 (35)	62 (29)	86 (26)	101 (25)
80	120	65 (45)	137 (32)	192 (32)	221 (29)
160	240	160 (63)	301 (58)	373 (36)	518 (33)

20 ppwl

		N			
k	n_{glob}	4	9	16	25
20	60	20 (30)	43 (26)	42 (25)	59 (23)
40	120	37 (31)	66 (32)	93 (27)	112 (27)
80	240	75 (36)	132 (30)	191 (30)	268 (27)
160	480	131 (53)	292 (47)	407 (31)	530 (28)

Medium-tolerance: 10^{-5}

In Table 3 we report the results when lowering the inner tolerance to 10^{-5}. We compare with the results reported in Table 2. A general observation is that as k increases, so does the number of outer FGMRES iterations. Naturally, lowering the inner tolerance ensures we require less iterations to converge on the subdomains.

For the largest wave number reported and 9 subdomains, we needed 262 outer iterations when using a direct solver. For the inexact approach with tolerance 10^{-5} the number of outer iterations increases to 301, which is the same as when using a tolerance of 10^{-10}. However, for the finer mesh with 20 ppwl the number of outer iterations goes up from 292 to 308. At the same time, the number of inner iterations reduces accordingly: from 58 to 40 for 10 ppwl and from 47 to 31 for 20 ppwl.

If we increase the number of subdomains from 9 to 25, the outer number of FGMRES iterations increases more rapidly. If we use 20 ppwl, the direct local solves lead to 520 outer iterations. This goes up to 555 when we use the iterative approach and a tolerance of 10^{-5}. Note that the extra outer iterations compared to a tolerance of 10^{-10} is surmountable as relaxing the tolerance by 5 orders of magnitude leads to an increase of 25 iterations (from 530 to 555). Moreover, if we compare the number of inner iterations, a finer mesh works better with two-level deflation preconditioned GMRES on the subdomains, since we now need 15 iterations on average.

Similarly, on the finer mesh the number of inner iterations scales better with increasing wave number k by adding more subdomains. Contrary to the results for $k < 160$, moving from 10 ppwl to 20 ppwl with 25 subdomains leads to less outer iterations. Thus, for larger wave numbers, using a finer mesh with more subdomains leads to a smaller number of outer and inner iterations. This effect is not observed with respect to the direct solves on the subdomains and/or the use of the tolerance 10^{-10} (see Table 1 and Table 2): here as we go from 10 ppwl to 20 ppwl, the number of outer iterations always increases.

Table 3: FGMRES iteration counts using the one-level RAS preconditioner with subdomain problems solved inexactly to a relative tolerance of 10^{-5} using GMRES with a two-level deflation preconditioner. In parentheses we display the average number of GMRES iterations per subdomain solve.

10 ppwl

k	n_{glob}	N = 4	9	16	25
20	30	20 (12)	27 (14)	65 (14)	46 (12)
40	60	34 (20)	76 (15)	95 (14)	122 (14)
80	120	72 (26)	154 (17)	210 (19)	262 (17)
160	240	175 (44)	301 (40)	398 (19)	572 (20)

20 ppwl

k	n_{glob}	N = 4	9	16	25
20	60	20 (15)	50 (13)	43 (13)	59 (12)
40	120	37 (16)	76 (18)	104 (14)	118 (14)
80	240	86 (19)	148 (16)	218 (16)	314 (14)
160	480	144 (34)	308 (31)	431 (16)	555 (15)

Low-tolerance: 10^{-2}

Finally, we reduce the inner tolerance to just 10^{-2} and report results in Table 4. While the overall observations follow a similar trend to the previous case, the number of inner iterations are reduced drastically. This comes at the expense of a higher number of outer iterations as the wave number and number of subdomains increase.

The most noticeable result is again for the highest wave number, $k = 160$. If we use 20 ppwl, the direct local solves lead to 520 outer iterations. This goes up to 584 when we use the iterative approach with a tolerance of 10^{-2}. The extra outer iterations compared to a tolerance of 10^{-10} is again surmountable as relaxing the tolerance by 8 orders of magnitude leads to an increase of 54 outer iterations (from 530 to 584). Meanwhile, the average inner iterations goes from 28 (for 10^{-10}) to 15 (for 10^{-5}), and finally to 6 iterations when using a tolerance of 10^{-2}.

Analogous to the case where we set the tolerance to 10^{-5}, we again observe that, as an exception to the rule that increasing n_{ppwl} leads to more outer iterations, the number of outer iterations actually decreases when using 20 ppwl instead of 10 ppwl. This effect is only observed for the iterative approach on the subdomains in combination with a sufficiently low tolerance, here 10^{-5} or 10^{-2}.

An important take-away message here is that the outer iteration of the one-level RAS preconditioned FGMRES method is able to reach convergence even when the subdomain systems are solved only to a relatively low level of accuracy.

Table 4: FGMRES iteration counts using the one-level RAS preconditioner with subdomain problems solved inexactly to a relative tolerance of 10^{-2} using GMRES with a two-level deflation preconditioner. In parentheses we display the average number of GMRES iterations per subdomain solve.

10 ppwl

k	n_{glob}	N = 4	9	16	25
20	30	20 (5)	27 (6)	73 (7)	50 (5)
40	60	42 (8)	87 (7)	109 (7)	126 (6)
80	120	84 (13)	172 (8)	241 (10)	298 (8)
160	240	211 (30)	332 (22)	451 (9)	1007 (8)

20 ppwl

k	n_{glob}	N = 4	9	16	25
20	60	21 (7)	59 (5)	49 (5)	72 (5)
40	120	44 (7)	84 (8)	124 (6)	134 (6)
80	240	94 (9)	154 (7)	229 (8)	333 (6)
160	480	154 (20)	327 (19)	450 (7)	584 (6)

To provide some perspective on these results, we repeat the 10 ppwl experiment but now use GMRES preconditioned by ILU(0) as the subdomain solution method (with tolerance 10^{-2}). The results in Table 5 show that the average number of inner iterations drastically increases. Further, for $k = 160$ simulation run times were noticeably increased. Note that, for $k = 160$ and $N = 25$, using the two-level deflation preconditioner with 20 ppwl leads to both a lower inner and outer iteration count.

Table 5: FGMRES iteration counts using the one-level RAS preconditioner with subdomain problems solved inexactly to a relative tolerance of 10^{-2} using GMRES with an ILU(0) preconditioner. In parentheses we display the average number of GMRES iterations per subdomain solve. Here we use 10 ppwl.

		N			
k	n_{glob}	4	9	16	25
20	30	28 (21)	42 (12)	74 (10)	53 (7)
40	60	44 (64)	80 (36)	110 (25)	131 (17)
80	120	89 (250)	177 (121)	239 (83)	303 (55)
160	240	221 (983)	344 (502)	475 (260)	658 (199)

4 Conclusions

In this work we examined the utility of the one-level RAS preconditioner together with FGMRES to solve the 2D homogeneous Helmholtz equation when using an inexact solution method for the subdomain solves. Our results support the notion that the direct solve can be substituted by an efficient iterative solver. By using two-level deflation as a local preconditioner, we are able to keep the number of inner iterations on the subdomains low and scalable with respect to the wave number k.

The next step would be to include a coarse space and experiment with a two-level RAS preconditioner combined with inexact solves on the subdomains. Adding a coarse space would reduce the impact on the number of outer iterations when substituting direct solves for inexact solves on the subdomains. The trade-off between a higher number of outer iterations and a fast and memory efficient local subdomain solve needs to be analysed in large-scale applications to determine the break-even point in terms of wall-clock time, identifying where the iterative approach can be beneficial. Especially in high-frequency 3D applications, the inclusion of the coarse space can become a bottleneck. To reduce the outer iteration count, we either need to solve with a large coarse space or on larger subdomains. Both options can be costly when using a direct method and so an inexact solver is likely more suitable. Further, in the iterative approach the inner Krylov solvers for the subdomain problems may also benefit from the use of recycling techniques, which could further reduce the number of inner iterations and increase efficiency.

References

1. N. Bootland and V. Dolean. On the Dirichlet-to-Neumann coarse space for solving the Helmholtz problem using domain decomposition. In: N*umerical Mathematics and Advanced Applications ENUMATH 2019*, Lect. Notes Comput. Sci. Eng., vol. 139, pp. 175–184. Springer, Cham (2021).
2. N. Bootland, V. Dolean, P. Jolivet and P.H. Tournier. A comparison of coarse spaces for Helmholtz problems in the high frequency regime. *Comput. Math. Appl.* **98**, 239–253 (2021).
3. N. Bootland, V. Dolean, A. Kyriakis and J.Pestana. Analysis of parallel Schwarz algorithms for time-harmonic problems using block Toeplitz matrices. *Electron. Trans. Numer. Anal.* **55**, 112–141 (2022).
4. A. Deraemaeker, I. Babuška and P. Bouillard. Dispersion and pollution of the FEM solution for the Helmholtz equation in one, two and three dimensions. *Internat. J. Numer. Methods Engrg.* **46**(4), 471–499 (1999).
5. V. Dwarka and C. Vuik. Scalable convergence using two-level deflation preconditioning for the Helmholtz equation. *SIAM J. Sci. Comput.* **42**(2), A901–A928 (2020).
6. M.J. Gander and H. Zhang. A class of iterative solvers for the Helmholtz equation: factorizations, sweeping preconditioners, source transfer, single layer potentials, polarized traces, and optimized Schwarz methods. *SIAM Rev.* **61**(1), 3–76 (2019).
7. G. Meurant and J. Duintjer Tebbens. The role eigenvalues play in forming GMRES residual norms with non-normal matrices. *Numer. Algorithms* **68**(1), 143–165 (2015).
8. A.H. Sheikh, D. Lahaye, L. Garcia Ramos, R. Nabben and C. Vuik. Accelerating the shifted Laplace preconditioner for the Helmholtz equation by multilevel deflation. *J. Comput. Phys.* **322**, 473–490 (2016).

Non-Overlapping Domain Decomposition Methods with Cross-Points and Padé Approximants for the Helmholtz Equation

Yassine Boubendir and Tadanaga Takahashi

1 Introduction

We present a new non-overlapping domain decomposition method (NDDM) based on the square-root transmission conditions and the utilization of an appropriate technique dealing with the so-called cross-points problem in the context of a nodal finite element method (FEM). The square-root operator is localized using the Padé Approximants technique. In addition, we use a Krylov solver to accelerate the iterative procedure. Several numerical results are displayed to validate this new algorithm.

2 Model problem

Consider an obstacle S with a smooth boundary condition $\Gamma = \partial S$. We are solving for the scattered field solution u of the Helmholtz equation equipped with the Sommerfeld radiation condition

$$\begin{cases} \Delta u + k^2 u = 0 \text{ in } \mathbb{R}^2 \backslash \mathcal{S} \\ \partial_{\boldsymbol{n}_S} u = f \text{ on } \Gamma := \partial \mathcal{S} \\ \lim\limits_{|\boldsymbol{x}| \to \infty} |\boldsymbol{x}|^{1/2} \left(\nabla u \cdot \dfrac{\boldsymbol{x}}{|\boldsymbol{x}|} - iku \right) = 0, \end{cases} \tag{1}$$

Yassine Boubendir
New Jersey Institute of Technology, 323 Dr Martin Luther King Jr Blvd, Newark, NJ 07102,
e-mail: boubendi@njit.edu

Tadanaga Takahashi
New Jersey Institute of Technology, 323 Dr Martin Luther King Jr Blvd, Newark, NJ 07102,
e-mail: tt73@njit.edu

S. C. Brenner et al. (eds.), *Domain Decomposition Methods in Science and Engineering XXVI*, Lecture Notes in Computational Science and Engineering 145,
https://doi.org/10.1007/978-3-030-95025-5_12

where $\boldsymbol{n}_S$ indicates the outward unit normal to Γ and f is a given plane wave function $f = -\partial_{\boldsymbol{n}_S} e^{-ik\boldsymbol{d}\cdot\boldsymbol{x}}$, with $\boldsymbol{x} = (x_1, x_2) \in \mathbb{R}^2$ and $\mathrm{i} = \sqrt{-1}$. The incidence angle $\boldsymbol{d}$ is normalized on the unit sphere $|\boldsymbol{d}| = 1$ and k denotes the wavenumber.

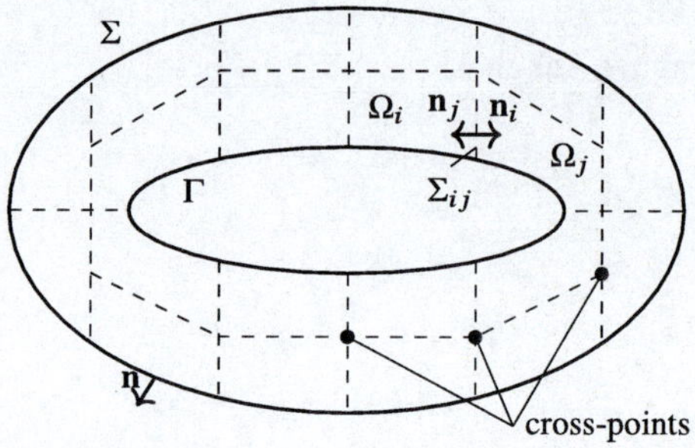

Fig. 1: Sketch of a non-overlapping domain decomposition of the domain Ω.

To solve problem (1), we truncate the original computational domain using an artificial interface Σ on which an absorbing boundary condition is posed, see Fig. 1. Therefore, problem (1) is reduced to the following system

$$\begin{cases} \Delta u + k^2 u = 0 \text{ in } \Omega \\ \partial_{\boldsymbol{n}} u = f \text{ on } \Gamma \\ \partial_{\boldsymbol{n}} u - iku = 0 \text{ on } \Sigma, \end{cases} \tag{2}$$

where $\boldsymbol{n}$ represents the normal derivative pointing outward from Ω.

3 Non-overlapping domain decomposition algorithm

The first step of this method consists of splitting the domain Ω into N_{dom} disjoint subdomains Ω_i, $i = 1, \ldots, N_{\text{dom}}$ such that:

- $\overline{\Omega} = \bigcup_{i=1}^{N_{\text{dom}}} \overline{\Omega}_i, \quad i = 1, \ldots, N_{\text{dom}}$
- $\Omega_i \cap \Omega_j = \emptyset, \quad \forall i \neq j, \quad i, j = 1, \ldots, N_{\text{dom}}$
- $\partial\Omega_i \cap \partial\Omega_j = \overline{\Sigma}_{ij} = \overline{\Sigma}_{ji}, \quad i, j = 1, \ldots, N_{\text{dom}}.$

We define $u_i, \Gamma_i, \Sigma_i, f_i$ to be their respective original definitions but restricted to $\overline{\Omega_i}$. Let $\mathbf{n}_i$ be the outward unit normal to $\partial\Omega_i$ and let Λ_i be the set of all indices of subdomains adjacent to Ω_i. Following Després' NDDM framework [5], we solve at each step $n + 1$ and for each subdomain $i = 1, \ldots, N_{\text{dom}}$, the local problem:

$$\begin{cases} \Delta u_i^{(n+1)} + k^2 u_i^{(n+1)} = 0 & x \in \Omega_i \\ \partial_{\boldsymbol{n}_i} u_i^{(n+1)} = f_i & x \in \Gamma_i \\ \partial_{\boldsymbol{n}_i} u_i^{(n+1)} - iku_i^{(n+1)} = 0 & x \in \Sigma_i \\ \partial_{\boldsymbol{n}_i} u_i^{(n+1)} + \mathcal{B}u_i^{(n+1)} = g_{ij}^{(n)} & x \in \Sigma_{ij} : j \in \Lambda_i, \end{cases} \tag{3}$$

where $g_{ij}^{(n)}$ represent the transmitting quantities along the common interfaces defined by

$$g_{ij}^{(n)} = -\partial_{\boldsymbol{n}_j} u_j^{(n)} + \mathcal{B}u_j^{(n)} = 2\mathcal{B}u_j{}^{(n)} - g_{ji}^{(n-1)}. \tag{4}$$

Several methods have been proposed in the past regarding the choice of the operator $\mathcal{B}$ in order to improve the convergence of the Després NDDM [2–4, 6, 8]. In this paper, we are interested in the following transmission operator

$$\mathcal{B}^{\mathrm{sq},\varepsilon} u = -ik\sqrt{1 + \mathrm{div}_{\mathcal{S}}\left(\frac{1}{k_\varepsilon^2}\nabla_{\mathcal{S}}\right)}u, \tag{5}$$

where $\mathrm{div}_{\mathcal{S}}$ and $\nabla_{\mathcal{S}}$ represent the surface divergence and surface gradient of a surface $\mathcal{S}$, respectively, ε is a parameter which may depends on $\mathcal{S}$, and $k_\varepsilon = k + \mathrm{i}\varepsilon$ is its corresponding complexified wavenumber. Operator (5) is non-local but it can be localized using Padé approximants. The approximate square-root transmission operator of order N_{p} has the form

$$\mathcal{B}^{N_{\mathrm{p}},\alpha,\varepsilon} u_i = -ik\left(C_0 u_i + \sum_{\ell=1}^{N_{\mathrm{p}}} A_\ell \,\mathrm{div}_{\mathcal{S}}\left(\frac{1}{k_\varepsilon^2}\nabla_{\mathcal{S}}\varphi_{i,\ell}\right)\right), \tag{6}$$

where the auxiliary unknowns $\varphi_{i,\ell}$ for $\ell = 1, \ldots, N_{\mathrm{p}}$ satisfy

$$\left(1 + B_\ell \,\mathrm{div}_{\mathcal{S}}\left(\frac{1}{k_\varepsilon^2}\nabla_{\mathcal{S}}\right)\right)\varphi_{i,\ell} = u_i. \tag{7}$$

The C_0, A_ℓ, B_ℓ are complex Padé coefficients depending on a branch cut rotation parameter α. We refer to [2] for the details of this operator.

4 Nodal FEM-NDDM and the cross-points problem

The algorithm developed in [4] is based on the modification of the Padé transmission conditions introduced in [2]. The main goal of these modified conditions resides in reducing the cost of local problems because of the resolution of a series of equations on the artificial interfaces related to the auxiliary functions $\varphi_{i,\ell}$ (7) that are coupled to each local problem. In addition, this modification [4] leads to transmission conditions where the transmitting operator $\mathcal{B}$ is a scalar. In this case, it is possible to use the approach dealing with the so-called *cross-points problem* developed for nodal FEM-NDDM [3]. The main idea of this approach consists of preserving the finite element equations at the level of these points, i.e., of taking a common value for the degree of freedom located on the nodes at the junction of several subdomains. The novelty of the method presented here consists of effectively extending this technique to the original Padé algorithm [2], i.e., in the case where the transmitting operator $\mathcal{B}$ is given by (6). Cross-points are corner nodes shared by multiple domains. The remainder of this section describes the steps of the nodal FEM-NDDM in [1, 3] adapted to (6)–(7).

Let $\mathcal{T}^h$ and X^h be, respectively, a global, non-degenerate triangular mesh of Ω and its associated $\mathbb{P}_1$-continuous finite element space. The discrete formulation of problem (2) is defined as follows

$$a_\Omega\left(u^h, v^h\right) = Lv^h, \qquad u^h \in X^h, \quad \forall v^h \in X^h \tag{8}$$

where

$$\begin{aligned} a_\Omega\left(u^h, v^h\right) &:= \int_\Omega \left(\nabla u^h \cdot \nabla v^h - k^2 u^h v^h\right) \mathrm{d}\Omega - \mathrm{i}k \int_\Sigma u^h v^h \,\mathrm{d}\Sigma \\ Lv^h &:= \int_\Gamma f v^h \,\mathrm{d}\Sigma \end{aligned} \tag{9}$$

and u^h is the FEM solution. Consider now the discrete solution to the local problem (3). Let $\mathcal{T}_i^h$ and X_i^h be, respectively, a non-degenerate triangular mesh of Ω_i and its associated $\mathbb{P}_1$-continuous finite element space which conforms to the global mesh. The discrete form of (3) with the Padé square root operator is

$$u_i^h \in X_i^h, \quad \forall v_i^h \in X_i^h$$
$$\begin{cases} a_i(u_i^h, v_i^h) + \displaystyle\sum_{\ell=1}^{N_\mathrm{p}} p_{i,\ell}(\varphi_{i,\ell}^h, v_i^h) = L_i(v_i^h) \\ q_i(u_i^h, v_i^h) + r_{i,\ell}(\varphi_{i,\ell}^h, v_i^h) = 0 \quad \forall \ell = 1, \ldots, N_\mathrm{p}, \end{cases} \tag{10}$$

where

$$a_i(u_i, v_i) = \int_{\Omega_i} (\nabla u_i \cdot \nabla v_i - k^2 u_i v_i)\mathrm{d}x - \mathrm{i}k \int_{\Sigma_i} u_i\, v_i \mathrm{d}s - \mathrm{i}kC_0 \sum_{j\in\Lambda_i} \int_{\Sigma_{ij}} u_i v_i \mathrm{d}s \tag{11}$$

$$p_{i,\ell}(\varphi_{i,\ell}, v_i) = \sum_{j\in\Lambda_i} A_\ell \frac{\mathrm{i}k}{k_\varepsilon^2} \int_{\Sigma_{ij}} \nabla_{\Sigma_{ij}} \varphi_{i,\ell} \cdot \nabla_{\Sigma_{ij}} v_i \,\mathrm{d}s \tag{12}$$

$$q_i(u_i, v_i) = \sum_{j\in\Lambda_i} \int_{\Sigma_{ij}} u_i v_i \,\mathrm{d}s \tag{13}$$

$$r_{i,\ell}(\varphi_{i,\ell}, v_i) = -\sum_{j\in\Lambda_i} \int_{\Sigma_{ij}} \varphi_{i,\ell} v_i \,\mathrm{d}s + \frac{B_\ell}{k_\varepsilon^2} \int_{\Sigma_{ij}} \nabla_{\Sigma_{ij}} \varphi_{i,\ell} \cdot \nabla_{\Sigma_{ij}} v_i \mathrm{d}s \tag{14}$$

$$L_i(v_i) = \int_{\Gamma_i} f_i\, v_i\, \mathrm{d}s + \sum_{j\in\Lambda_i} \int_{\Sigma_{ij}} g_{ij} v_i\, \mathrm{d}s, \tag{15}$$

with $\varphi_{i,\ell} = 0$ on $\partial\Sigma_{ij}$. The method proposed here consists in relating the discrete original problem with the discrete local problems. We start by classifying the nodes of our mesh as one of the following:

- Independent: these nodes are interior to Ω_i, Σ_i, and Γ_i.
- Shared: these nodes are interior to Σ_{ij}.
- Cross-points: these are points where any of the curves Γ_i, Σ_{ij}, or Σ_i meet.

Any discrete local test function $v_i^h \in X_i^h$ can be decomposed as follows:

$$v_i^h = v_{i\mathrm{I}}^h + \sum_{j\in\Lambda_i} v_{ij}^h + v_{\mathrm{c}}^h, \tag{16}$$

where $v_{i\mathrm{I}}^h$ is supported only on independent nodes, v_{ij}^h is supported only on shared nodes, and v_{c}^h is supported only on cross-points. Let us introduce the broken space X_{B}^h defined as the span of the function

$$v^h = \sum_{i=1}^{N_{dom}} \left(v_{i\mathrm{I}}^h + \sum_{j\in\Lambda_i} v_{ij}^h \right) + v_{\mathrm{c}}^h, \tag{17}$$

see [1, 3] for more details. We also define the series of functions $(\varphi_{i,1}^h, \ldots, \varphi_{i,N_{\mathrm{p}}}^h)$ in the space $(\Phi_i^h)^{N_{\mathrm{p}}}$, where Φ_i^h represents a subspace of X_i^h supported only on the shared nodes. Finally, using conditions (4), the decompositions (16) and (17), and following a similar derivation to the algorithm described in [1], we can see that problem (8) is reduced to solving the following system

$$\left.\begin{array}{c} a_i\left(u_{i\mathrm{I}}^h + \sum_{j\in\Lambda_i} u_{ij}^h + u_{\mathrm{c}}^h, v_{i\mathrm{I}}^h\right) = L_i(v_{i\mathrm{I}}^h), \quad \forall v_{i\mathrm{I}}^h \in X_i^h \\ a_i\left(u_{i\mathrm{I}}^h + \sum_{j\in\Lambda_i} u_{ij}^h + u_{\mathrm{c}}^h, v_{ij}^h\right) + \sum_{\ell=1}^{N_{\mathrm{p}}} p_{i,\ell}(\varphi_{i,\ell}^h, v_{ij}^h) = L_i(v_{ij}^h), \\ q_i(u_{ij}^h, v_{ij}^h) + r_{i,\ell}(\varphi_{i,\ell}^h, v_{ij}^h) = 0, \quad \forall \ell = 1, \ldots, N_{\mathrm{p}}, \\ \forall v_{ij}^h \in X_i^h, \quad j \in \Lambda_i \end{array}\right\}, i = 1, \ldots, N_{\mathrm{dom}}$$

$$\sum_{i=1}^{N} a_i\left(u_{i\mathrm{I}}^h + \sum_{j\in\Lambda_i} u_{ij}^h + u_{\mathrm{c}}^h, v_{\mathrm{c}}^h \right) = \sum_{i=1}^{N} L_i(v_{\mathrm{c}}^h) \quad \forall v_{\mathrm{c}}^h \in X_c^h, \tag{18}$$

and this system can be put in matrix form (with N subdomains) as follows

$$
\begin{bmatrix}
A_{11} & P_1 & & & & A_{1c} \\
Q_1 & R_1 & & & & 0 \\
& & \ddots & & & \vdots \\
& & & A_{NN} & P_N & A_{Nc} \\
& & & Q_N & R_N & 0 \\
A_{c1} & 0 & \dots & A_{cN} & 0 & A_{cc}
\end{bmatrix}
\begin{bmatrix} \boldsymbol{u_1} \\ \boldsymbol{\varphi_1} \\ \vdots \\ \boldsymbol{u_N} \\ \boldsymbol{\varphi_N} \\ \boldsymbol{u_c} \end{bmatrix}
=
\begin{bmatrix} \boldsymbol{g_1} \\ \boldsymbol{0} \\ \vdots \\ \boldsymbol{g_N} \\ \boldsymbol{0} \\ \boldsymbol{g_c} \end{bmatrix}
\tag{19}
$$

with the quantities $\boldsymbol{g}_i$ being computed from equation (15), where g_{ij} is updated on Σ_{ij} at each step $(n+1)$ as follows

$$
g_{ij}^{(n+1)} = -g_{ji}^{(n)} - 2ik\left(C_0 u_i^{(n+1)} + \sum_{\ell=1}^{N_p} A_\ell \operatorname{div}_{\Sigma_{ij}} \left(\frac{1}{k_\varepsilon^2} \nabla_{\Sigma_{ij}} \varphi_{i,\ell} \right) \right), \tag{20}
$$

where $\varphi_{i,\ell}$ are the auxiliary functions corresponding to the solution $u_i^{(n+1)}$. The system (19) is similar to the one numbered (21) in [3] obtained in the case where $\mathcal{B}$ is a scalar. We use the same procedure (based on a Schur complement) described in [3] to deal with (19). For that, one needs to change the notation in this description so that each block A_{ii} is composed by the sub-blocks A_{ii}, P_i, Q_i, and R_i.

5 Numerical results

This section is devoted to validating the proposed method through numerical simulations. We generate a triangular mesh controlled by 2 parameters: wavelength λ and points per wavelength n_λ. We use both a Krylov subspace solver (Orthodir) and successive approximation (Jacobi) methods to solve the iteration operator [7]. For its performance, we choose to compare our new method (19) with the evanescent mode damping algorithm (EMDA) [3] which consists of choosing $\mathcal{B} = -ik(1+i\mathcal{X})$. We evaluate the performance based on the number of iterations until *convergence* (initial residue decreases by a factor of 10^{-6}). For the Padé parameters, we chose [2]

Fig. 2: Geometries used for the numerical simulation: pie (left) and layered (right) configurations.

$\alpha = \pi/4$ and $\varepsilon = 0.6k^{1/3}$. For the EMDA, we chose $\mathcal{X} = 1/2$. These simulations are performed on the two geometries shown in Fig. 2.

In the pie configuration (left of Fig. 2), the annulus is split radially into N_{dom} equal subdomains. The inner radius is fixed $r_1 = 1$ and the outer radius varies with frequency $r_2 = r_1 + 2\pi/k$. In the first experiment, we fixed $N_{\mathrm{dom}} = 5$, $k = \pi$, and varied $n_\lambda = 12, 16, 20, 24$ and the Padé orders $N_{\mathrm{p}} = 1, 2, 4, 8$. Table 1 shows the obtained results. In the second experiment, we determine how the wavenumber affects the convergence rate. Table 2 displays the results for fixed $N_{\mathrm{dom}} = 5$, $n_\lambda = 16$ and varied wavenumber $k = \pi, 2\pi, 3\pi, 4\pi$. In the third experiment, we varied the number of subdomains $N_{\mathrm{dom}} = 2, 4, 6, 12$ with fixed $k = \pi$ and $n_\lambda = 16$. These results are summarized in Table 3.

Table 1: Number of iterations until convergence with respect to points per wavelength n_λ.

	Jacobi					Orthodir				
n_λ	EMDA	Padé1	Padé2	Padé4	Padé8	EMDA	Padé1	Padé2	Padé4	Padé8
12	105	33	25	25	25	28	18	15	14	14
16	136	44	28	24	24	32	20	17	14	14
20	159	52	34	23	23	35	22	18	15	15
24	192	63	42	23	23	38	24	20	16	15

Table 2: Number of iterations until convergence with respect to wavelength k.

	Jacobi					Orthodir				
k	EMDA	Padé1	Padé2	Padé4	Padé8	EMDA	Padé1	Padé2	Padé4	Padé8
π	136	44	28	24	24	32	20	17	14	14
2π	124	40	25	22	22	29	19	16	14	14
3π	127	41	26	24	24	30	20	16	15	15
4π	120	39	25	24	24	29	19	16	15	15

Table 3: Number of iterations until convergence with respect to number of subdomains N_{dom}.

	Jacobi					Orthodir				
N_{dom}	EMDA	Padé1	Padé2	Padé4	Padé8	EMDA	Padé1	Padé2	Padé4	Padé8
2	136	44	28	17	17	20	14	13	12	11
4	137	44	27	20	20	29	18	15	13	13
6	139	44	28	25	25	32	20	17	15	15
12	139	67	54	50	49	39	28	24	23	23

Table 4: Results for the cross-point configuration experiments.

	Jacobi			Orthodir		
n_λ	EMDA	Padé2	Padé4	EMDA	Padé2	Padé4
12	110	60	59	36	24	24
16	123	70	68	40	25	25
20	144	76	73	43	27	26
24	165	81	79	46	29	27

	Jacobi			Orthodir		
k	EMDA	Padé2	Padé4	EMDA	Padé2	Padé4
π	123	70	68	40	25	25
2π	115	44	44	42	31	31
3π	110	53	53	45	37	37
4π	165	56	56	46	42	42

The layered configuration (right of Fig. 2) is a six-domain annulus in which the radii $r_1 = 1, r_2 = 3$ and N_{dom} are fixed. For this geometry, we first varied $n_\lambda = 12, 16, 20, 24$ with fixed $k = \pi$. We then tested several wavenumbers $k = \pi, 2\pi, 3\pi, 4\pi$ with fixed $n_\lambda = 16$. All the obtained results are listed in Table 4.

These tests demonstrate the effectiveness of the combination of the square root transmission operator, localized using Padé approximants, with the treatment of the cross-points. We observe stability and consistency in terms of convergence, in particular when the Krylov solver Orthodir is applied.

Acknowledgements Y. Boubendir's work is supported by the NSF through Grants DMS- 1720014 and DMS-2011843.

References

1. A. Bendali and Y. Boubendir. Non-overlapping domain decomposition method for a nodal finite element method. *Numerische Mathematik* **103**(4), 515–537, June (2006).
2. Y. Boubendir, X. Antoine and C. Geuzaine. A quasi-optimal non-overlapping domain decomposition algorithm for the Helmholtz equation. *J. Comput. Phys.* **231**(2), 262–280, January (2012).
3. Y. Boubendir, A. Bendali and M.B. Fares. Coupling of a non-overlapping domain decomposition method for a nodal finite element method with a boundary element method. *Int. J. Numer. Methods in Engrg.* **73**(11), 1624–1650, March (2008).
4. Y. Boubendir and D. Midura. Non-overlapping domain decomposition algorithm based on modified transmission conditions for the Helmholtz equation. *Computers & Mathematics with Applications* **75**(6), 1900–1911 (2018).
5. B. Desprès. *Méthodes De Décomposition de Domaine pour les Problemes De Propagation D'ondes en Régime Harmonique*. PhD thesis, University of Paris (1991).
6. M.J. Gander, F. Magoulès and F. Nataf. Optimized Schwarz methods without overlap for the Helmholtz equation. *SIAM J. Sci. Comput.* **24**(1), 38–60 (2002).
7. Y. Saad. *Iterative Methods for Sparse Linear Systems*. 2nd edition. SIAM (2003).
8. C.C. Stolk. A rapidly converging domain decomposition method for the Helmholtz equation. *J. Comput. Phys.* **241**, 240–252 (2013).

OSDS: A Sweeping Preconditioner for the Helmholtz Equation

Nacime Bouziani and Frédéric Nataf

Domain decomposition algorithms have become popular in solving the Helmholtz equation since Després' seminal paper [3]. Although it is known that the presence of overlaps helps to speed up the convergence for domain decomposition methods, nonoverlapping based methods are often used to avoid dealing with the construction of the normal derivative of the solution. For decompositions into vertical strips, a sweeping algorithm was first proposed and analyzed in [6] for convection-diffusion operators. Recently, sweeping methods have gained interest due to their capability to achieve nearly-linear asymptotic complexity, see e.g. the double sweep preconditioner of Vion and Geuzaine for non-overlapping decomposition with high order interface conditions [8, 9], the PML-based sweep method of Stolk [2], and the polarized traces method of Zepeda-Núñez and Demanet [10].

We consider a decomposition of the domain into layers where the local subproblems are equipped with interface conditions, also called absorbing boundary conditions (ABCs). In practice, the exact ABCs (which are also the optimal interface conditions, see [7]) procedure is tedious to implement and computationally expensive. As a consequence, the boundary conditions at the interfaces produce spurious reflected waves that significantly increase the number of iterations to converge, in particular for heterogeneous media and high frequency regimes.

We propose to precondition the discrete Helmholtz system by an overlapping splitting double-sweep algorithm that allows for overlapping subdomains and prevents spurious interface reflections from hindering the convergence. Using overlapping subdomains allows us to leverage its beneficial effect on the damping of high frequency modes of the error, while splitting prevents its adversary effect on the convergence of propagative modes. This is highly beneficial since in the non-overlapping

Nacime Bouziani
Imperial College London, Department of Mathematics, London, SW7 2AZ, UK,
e-mail: n.bouziani18@imperial.ac.uk, nacime.bouziani@gmail.com

Frédéric Nataf
Laboratory J.L. Lions, Sorbonne Université, Paris,
e-mail: frederic.nataf@sorbonne-universite.fr

S. C. Brenner et al. (eds.), *Domain Decomposition Methods in Science and Engineering XXVI*, Lecture Notes in Computational Science and Engineering 145,
https://doi.org/10.1007/978-3-030-95025-5_13

approach [8, 9], the quality of the ABCs is nearly the only way of impacting the convergence of the algorithm, and when dealing with more complex problems such as Maxwell equations high order ABCs are harder to handle.

1 Statement of the problem and some algorithms

We consider the Helmholtz equation in a bounded domain $\Omega \subset \mathbb{R}^2$ with frequency ω, velocity c and wavenumber k defined by $k^2 = \omega^2/c^2$:

$$\begin{aligned} &\left(-k^2 - \Delta\right) u = f \text{ in } \Omega \\ &+ \text{appropriate boundary conditions on } \partial\Omega\,. \end{aligned} \tag{1}$$

We consider a layered decomposition of Ω into N slices $(\Omega_i)_{1 \le i \le N}$, with or without overlap, see Figure 1. More precisely, for each $1 \le i \le N$, $\Omega \setminus \Omega_i$ is written as the disjoint union of two open subsets $\Omega_{i,l}$ and $\Omega_{i,r}$, where $\Omega_{i,l}$ is on the left of Ω_i and $\Omega_{i,r}$ on its right. The boundary $\partial\Omega_i \setminus \partial\Omega$ is written as the disjoint union of $\Gamma_{i,l}$ and $\Gamma_{i,r}$ where $\Gamma_{i,l}$ is on the left of Ω_i and $\Gamma_{i,r}$ is on its right ($\Omega_{1,l} = \emptyset$ and $\Omega_{N,r} = \emptyset$) (see Figure 2). The outward normal from Ω_i on $\Gamma_{i,l}$ (resp. $\Gamma_{i,r}$) is denoted by $\boldsymbol{n}_{i,l}$ (resp. $\boldsymbol{n}_{i,r}$). The problem (1) can be solved iteratively using a domain decomposi-

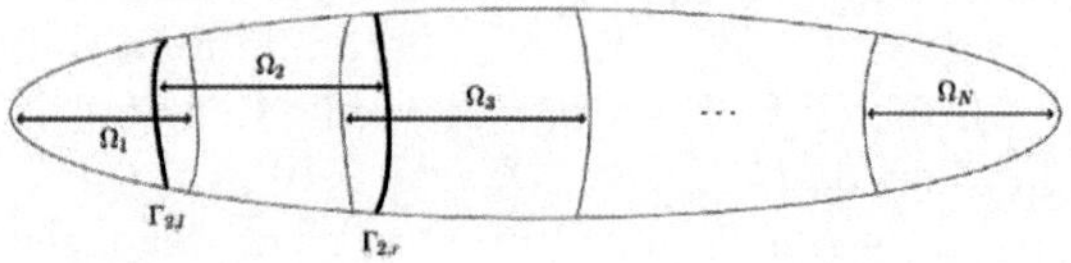

Fig. 1: Decomposition into vertical strips.

tion method where we solve locally on each subdomain Ω_i the equation (1) with appropriate boundary conditions on the physical boundaries and interfaces [3]. The method reads:

Solve in parallel:

$$\begin{cases} \left(-k^2 - \Delta\right) u_i^{n+1} = f \ \text{ in } \Omega_i, \ 1 \le i \le N \\ \quad \mathcal{B}_{i,l}\left(u_i^{n+1}\right) = \mathcal{B}_{i,l}\left(u_{i-1}^{n}\right) \ \text{on } \Gamma_{i,l}, \ 2 \le i \le N \\ \quad \mathcal{B}_{i,r}\left(u_i^{n+1}\right) = \mathcal{B}_{i,r}\left(u_{i+1}^{n}\right) \ \text{on } \Gamma_{i,r}, \ 1 \le i \le N-1 \\ \qquad\qquad + \text{ appropriate boundary conditions on } \partial\Omega \cap \partial\Omega_i\,, \end{cases} \tag{2}$$

where $\mathcal{B}_{i,l}$ and $\mathcal{B}_{i,r}$ are the interface conditions. For the sake of simplicity, we consider first-order ABCs as interface conditions:

$$\begin{cases} \mathcal{B}_{i,l} = \partial_{\boldsymbol{n}_{i,l}} + Ik \\ \mathcal{B}_{i,r} = \partial_{\boldsymbol{n}_{i,r}} + Ik \end{cases} \tag{3}$$

where $I^2 = -1$ and $\boldsymbol{n}_{i,r}$ (resp. $\boldsymbol{n}_{i,l}$) is the outward normal to domain Ω_i on $\Gamma_{i,r}$ (resp. $\Gamma_{i,l}$). It is known that higher-order ABCs lead to significant improvement of the convergence speed, see e.g. [1, 4].

A more efficient variant of algorithm 2 was introduced in [6]. It consists in double sweeps over the subdomains:

Left to right sweep:

$$\begin{cases} \left(-k^2 - \Delta\right) u_i^{n+1/2} = f \text{ in } \Omega_i,\ 1 \le i \le N \\ \mathcal{B}_{i,l}\left(u_i^{n+1/2}\right) = \mathcal{B}_{i,l}\left(u_{i-1}^{n+1/2}\right) \text{ on } \Gamma_{i,l},\ 2 \le i \le N \\ \mathcal{B}_{i,r}\left(u_i^{n+1/2}\right) = \mathcal{B}_{i,r}\left(u_{i+1}^{n}\right) \text{ on } \Gamma_{i,r},\ 1 \le i \le N-1 \\ \quad + \text{ appropriate boundary conditions on } \partial\Omega \cap \partial\Omega_i\,. \end{cases} \tag{4}$$

Right to left sweep:

$$\begin{cases} \left(-k^2 - \Delta\right) u_i^{n+1} = f \text{ in } \Omega_i,\ 1 \le i \le N \\ \mathcal{B}_{i,l}\left(u_i^{n+1}\right) = \mathcal{B}_{i,l}\left(u_{i-1}^{n+1/2}\right) \text{ on } \Gamma_{i,l},\ 2 \le i \le N \\ \mathcal{B}_{i,r}\left(u_i^{n+1}\right) = \mathcal{B}_{i,r}\left(u_{i+1}^{n+1}\right) \text{ on } \Gamma_{i,r},\ 1 \le i \le N-1 \\ \quad + \text{ appropriate boundary conditions on } \partial\Omega \cap \partial\Omega_i\,. \end{cases} \tag{5}$$

2 Overlapping splitting double sweep

In this section, we define a variant of algorithm (4)–(5) which has a superior convergence. Numerical results will show that it benefits better from the overlap and has better parallelism. This algorithm is written in terms of the substructured problem that we define first.

2.1 Substructuring

Substructuring algorithm (2), the iterative method can be reformulated considering only surfacic unknowns on the interfaces:

$$\begin{cases} h_{i,l}^n := \mathcal{B}_{i,l}\left(u_i^n\right), & \text{on } \Gamma_{i,l} \ \text{ for } 2 \le i \le N, \\ h_{i,r}^n := \mathcal{B}_{i,r}\left(u_i^n\right), & \text{on } \Gamma_{i,r} \ \text{ for } 1 \le i \le N-1\,. \end{cases} \tag{6}$$

Considering the global vector $h^n := (h^n_{2,l}, \dots, h^n_{N,l}, h^n_{N-1,r}, \dots, h^n_{1,r})^T$, containing first the local unknowns $(h^n_{i,l})_{2\leq i\leq N}$ and then in reverse order $(h^n_{i,r})_{1\leq i\leq N-1}$, we can reformulate the parallel Schwarz method (2) as a Jacobi algorithm on h^n: $h^{n+1} := \mathcal{T}(h^n) + G$, where the iteration operator $\mathcal{T}$ can be written in the form of an operator-valued matrix and G refers to the contribution of the right-hand side f, see [7]. Therefore, we look for a vector h such that

$$(\mathrm{Id} - \mathcal{T})\,(h) = G\,. \tag{7}$$

In order to define the operator $\mathcal{T}$ more precisely, we introduce for each subdomain an operator S_i which takes three arguments, two surfacic functions h_l and h_r and a volume function f:

$$S_i(h_{i,l}, h_{i,r}, f) := v, \tag{8}$$

where $v : \Omega_i \mapsto \mathbb{C}$ satisfies:

$$\begin{cases} \left(-k^2 - \Delta\right) v = f \ \text{ in } \Omega_i \\ \quad \mathcal{B}_{i,l}(v) = h_{i,l} \text{ on } \Gamma_{i,l} \quad (2 \leq i \leq N) \\ \quad \mathcal{B}_{i,r}(v) = h_{i,r} \text{ on } \Gamma_{i,r} \quad (1 \leq i \leq N-1) \\ \qquad\qquad + \text{ appropriate boundary conditions on } \partial\Omega \cap \partial\Omega_i\,, \end{cases} \tag{9}$$

for $1 < i < N$ (see Figure 2). For $i = 1$, the definition of S_1 is similar except that it takes only the two arguments $(h_{1,r}, f)$ since domain Ω_1 has no left interface and similarly operator S_N takes only the two arguments $(h_{N,l}, f)$ since domain Ω_N has no right interface. *As of now, for the sake of simplicity and by abuse of notation, $S_1(h_{1,l},\ h_{1,r},\ f)$ (resp. $S_N(h_{N,l},\ h_{N,r},\ f)$) will refer to $S_1(h_{1,r}, f)$ (resp. $S_N(h_{N,l},\ f)$).*

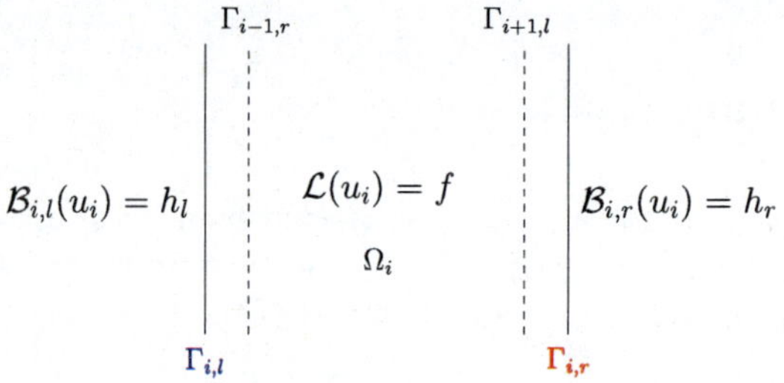

Fig. 2: Local problem on the subdomain Ω_i.

Next, we introduce the surfacic right-hand side G by

$$\begin{aligned} G_{i+1,l} &:= \mathcal{B}_{i+1,l}(S_i(0,0,\ f)),\ \ 1 \leq i \leq N-1\,, \\ G_{i-1,r} &:= \mathcal{B}_{i-1,r}(S_i(0,0,\ f)),\ \ 2 \leq i \leq N\,, \end{aligned} \tag{10}$$

and the substructured operator $\mathcal{T}$ by:

$$\begin{aligned}\mathcal{T}(h)_{i+1,l} &:= \mathcal{B}_{i+1,l}(S_i(h_{i,l},\ h_{i,r},0)),\ \ 1 \le i \le N-1\,,\\ \mathcal{T}(h)_{i-1,r} &:= \mathcal{B}_{i-1,r}(S_i(h_{i,l},\ h_{i,r},0)),\ \ 2 \le i \le N\,.\end{aligned} \tag{11}$$

We can now write the substructured form of the *double sweep algorithm* as:

Forward sweep

$$\begin{aligned}h^{n+1/2}_{i+1,l} &:= \mathcal{B}_{i+1,l}(S_i(h^{n+1/2}_{i,l},\ h^{n}_{i,r}, f))\,,\\ h^{n+1/2}_{i-1,r} &:= \mathcal{B}_{i-1,r}(S_i(h^{n+1/2}_{i,l},\ h^{n}_{i,r}, f))\,,\end{aligned} \tag{12}$$

followed by a Backward sweep

$$\begin{aligned}h^{n+1}_{i+1,l} &:= \mathcal{B}_{i+1,l}(S_i(h^{n+1/2}_{i,l},\ h^{n+1}_{i,r}, f))\,,\\ h^{n+1}_{i-1,r} &:= \mathcal{B}_{i-1,r}(S_i(h^{n+1/2}_{i,l},\ h^{n+1}_{i,r}, f))\,.\end{aligned} \tag{13}$$

As for the Jacobi method, by introducing an operator $\mathcal{T}_{DS}$, this algorithm can be written in a compact form $h^{n+1} = h^n + (I - \mathcal{T}_{DS})^{-1}(G - (I - \mathcal{T})(h^n))$, see [6].

2.2 Overlapping Splitting Double Sweep preconditioner (OSDS)

We explain now the rationale behind the overlapping splitting double sweep preconditioner that we define in this section. Note first that by linearity of the operators $(S_i)_{1\le i\le N}$, the contribution of each subdomain can be split into two contributions, one for each of its two interfaces:

$$\begin{aligned}\mathcal{T}(h)_{i+1,l} &= \mathcal{B}_{i+1,l}(S_i(h_{i,l},\ 0,\ 0))\ +\ \mathcal{B}_{i+1,l}(S_i(0,\ h_{i,r},\ 0)),\ \ 1 \le i \le N-1\,,\\ \mathcal{T}(h)_{i-1,r} &= \mathcal{B}_{i-1,r}(S_i(0,\ h_{i,r},\ 0))\ +\ \mathcal{B}_{i-1,r}(S_i(h_{i,l},\ 0,\ 0)),\ \ 2 \le i \le N\,.\end{aligned} \tag{14}$$

Had we used exact absorbing (a.k.a transparent or non-reflecting) boundary conditions (EABC) $\mathcal{B}^{EABC}$ instead of the zero-th order ones (3) in equations (8)–(9), two terms in (14) would vanish, namely:

$$\begin{aligned}\mathcal{B}^{EABC}_{i+1,l}(S^{EABC}_i(0,\ h_{i,r},0)) &= 0,\ \ 1 \le i \le N-1\,,\\ \mathcal{B}^{EABC}_{i-1,r}(S^{EABC}_i(h_{i,l},0,0)) &= 0,\ \ 2 \le i \le N\,.\end{aligned} \tag{15}$$

The corresponding operator in (14) would thus only contain one term,

$$\begin{aligned}\mathcal{T}^{EABC}(h)_{i+1,l} &= \mathcal{B}^{EABC}_{i+1,l}(S^{EABC}_i(h_{i,l},0,0)),\ \ 1 \le i \le N-1\,,\\ \mathcal{T}^{EABC}(h)_{i-1,r} &= \mathcal{B}^{EABC}_{i-1,r}(S^{EABC}_i(0,\ h_{i,r},0)),\ \ 2 \le i \le N\,.\end{aligned} \tag{16}$$

Then thanks to our numbering of h, the operator-valued matrix $\mathcal{T}^{EABC}$ is a 2×2 block diagonal matrix where each block is subdiagonal. As a consequence, for some vector G, computing $(I - \mathcal{T}^{EABC})^{-1}\,G$ can be performed by two parallel forward substitutions, which amounts to a single double sweep over the subdomains.

In practice, the absorbing boundary conditions are non-exact, therefore we have

$$\begin{aligned} \mathcal{B}_{i+1,l}(S_i(0,\ h_{i,r},0)) &\neq 0, \quad 1 \leq i \leq N-1\,, \\ \mathcal{B}_{i-1,r}(S_i(h_{i,l},0,0)) &\neq 0, \quad 2 \leq i \leq N\,, \end{aligned} \tag{17}$$

and we lose the block diagonal structure of $\mathcal{T}$. This led us to define a new operator $\mathcal{T}_{OSDS}$

$$\begin{aligned} \mathcal{T}_{OSDS}(h)_{i+1,l} &:= \mathcal{B}_{i+1,l}(S_i(h_{i,l},0,0)), \quad 1 \leq i \leq N-1\,, \\ \mathcal{T}_{OSDS}(h)_{i-1,r} &:= \mathcal{B}_{i-1,r}(S_i(0,\ h_{i,r},0)), \quad 2 \leq i \leq N\,, \end{aligned} \tag{18}$$

which by definition has the same structure as $\mathcal{T}^{EABC}$. We propose to use this newly defined operator to build a preconditioner for (7). The right-preconditioned solves reads: Find the solution $\tilde{h}$ to

$$(\mathrm{Id} - \mathcal{T})\,(\mathrm{Id} - \mathcal{T}_{OSDS})^{-1}(\tilde{h}) = G\,, \tag{19}$$

followed by $h := (\mathrm{Id} - \mathcal{T}_{OSDS})^{-1}(\tilde{h})$.

More intuitively, the key idea is to cancel out the reverse contribution at the interfaces that should not happen for the exact ABC case in order to prevent spurious interface reflections from hindering the convergence. In fact, these boundary conditions at the interfaces produce spurious reflected waves that significantly increase the number of iterations to converge, in particular for heterogeneous media and high frequency regimes. Note that for a non-overlapping domain decomposition, the OSDS algorithm is similar to the double sweep method of [8, 9]. Our approach addresses the case of overlapping subdomains that benefits the convergence rate.

3 Numerical results

In this section, we present numerical results when solving the substructured equation (7) with the GMRES algorithm right preconditioned by Id (Jacobi method), $(\mathrm{Id} - \mathcal{T}_{DS})^{-1}$ (Double sweep algorithm) and $(\mathrm{Id} - \mathcal{T}_{OSDS})^{-1}$ (Overlapping Splitting Double sweep algorithm). Note that the Jacobi method requires N subdomain solves per iteration instead of $2\,N$ for the sweeping methods. The Helmholtz equation is discretized with a P1 finite element using FreeFem++ [5]. Note that we use a careful variational discretisation of the normal derivative ensuring that the solution obtained converges to the solution of the problem without domain decompositions.

3.1 Wedge test

We consider the classical test case of the wedge, see e.g. [8], a rectangular domain $[0, 600]\times[0, 1000]$ with three different velocities in regions separated by non-parallel boundaries (Fig. 3 left). Starting from the top, the velocities are $c = 2000$, $c = 1500$ and $c = 3000$. Sommerfeld conditions are imposed on the bottom, right and left

Fig. 3: Heterogeneous media: Wedge (**Left**: Velocity model, **Right**: Solution (real) for $\omega = 160\pi$).

boundaries. The abrupt variations of the wavenumber produce internal reflections in different directions. A typical solution is shown in Figure 3 right.

Iteration counts are given in Table 1. The OSDS method is clearly superior to the Jacobi and DS methods. When increasing the number of subdomains, the ratio in favor of the OSDS method compared to the DS method increases up to reaching a value of nearly 4 for a domain decomposition into 40 vertical strips. Interestingly, we see that for a low tolerance on the residual (TOL=10^{-3}), the OSDS iteration counts are almost independent of the number of subdomains.

N	$\omega = 40\pi$			$\omega = 60\pi$		
	Jacobi	DS	OSDS	Jacobi	DS	OSDS
5	28 (17)	19 (11)	13 (6)	28 (15)	18 (10)	12 (5)
10	55 (31)	31 (16)	14 (7)	56 (30)	31 (15)	14 (6)
20	110 (55)	58 (29)	18 (8)	111 (53)	57 (28)	18 (7)
40	203 (88)	103 (47)	27 (9)	206 (85)	111 (55)	30 (9)

Table 1: Wedge, $\omega = 40\pi$ and 60π, $\delta = 16h$, TOL=$10^{-6}(10^{-3})$, nppwl = 24 , P1.

3.2 Influence of the overlap

We have also tested the effect of the width of the overlap on the convergence. We considered two test cases: the homogeneous waveguide and the wedge (see Table 2) that is defined in more detail in § 3.1. We observe that for the waveguide solved by the Overlapping Splitting Double Sweep method, the iteration count decreases significantly with increasing overlap. This monotonic decrease in the iteration count contrasts with the behaviour of the other two methods. We see that for the Jacobi and double sweep (DS) methods, the overlap has very little effect. For the Jacobi method it improves slightly the iteration counts whereas for the DS method, it might deteriorate the iteration count. For the wedge test case, all methods benefit monotonically from

δ	Homogeneous waveguide ($\omega = 20$)			Wedge ($\omega = 40\pi$)		
	Jacobi	DS	OSDS	Jacobi	DS	OSDS
2	159	69	27	259	127	97
4	165	74	23	245	117	83
8	160	76	20	221	105	69
16	143	73	18	202	91	53

Table 2: Influence of the overlap, δ varies, TOL=10^{-6}, nppwl = 24 , P1.

the size of the overlap but once again the reduction in the iteration count is more pronounced for the Overlapping Splitting Double Sweep method where the iteration count is reduced by a factor 1.83 when the overlap is increased from $2h$ to $16h$.

4 Conclusion

We have introduced an overlapping splitting double sweep algorithm which yields improved convergence for various problems. Many aspects deserve further investigations: higher-order ABC instead of the zero-th order one considered here and the introduction of a pipelining technique that can be applied to multiple right-hand sides problems to improve parallelism and achieve significant speed-ups, see [9].

References

1. X. Antoine, Y. Boubendir and C. Geuzaine. A quasi-optimal non-overlapping domain decomposition algorithm for the Helmholtz equation. *Journal of Computational Physics* **231**(2), 262–280 (2012).
2. C.C. Stolk. A rapidly converging domain decomposition method for the Helmholtz equation. *Journal of Computational Physics* **241**, 240–252, May (2013).
3. B. Després. Domain decomposition method and the helmholtz problem. In: *Mathematical and numerical aspects of wave propagation phenomena (Strasbourg, 1991)*, pp. 44–52, SIAM, Philadelphia, PA (1991).
4. M.J. Gander, F. Magoulès and F. Nataf. Optimized Schwarz methods without overlap for the Helmholtz equation. *SIAM J. Sci. Comput.* **24**(1), 38–60 (2002).
5. F. Hecht. New development in Freefem++. *J. Numer. Math.* **20**(3-4), 251–265 (2012).
6. F. Nataf and F. Nier. Convergence rate of some domain decomposition methods for overlapping and nonoverlapping subdomains. *Numerische Mathematik* **75**(3), 357–77 (1997).
7. F. Nataf, F. Rogier and E. de Sturler. Optimal interface conditions for domain decomposition methods. Technical Report 301, CMAP (Ecole Polytechnique) (1994).
8. A. Vion and C. Geuzaine. Double sweep preconditioner for optimized Schwarz methods applied to the Helmholtz problem. *Journal of Computational Physics* **266**, 171–190 (2014).
9. A. Vion and C. Geuzaine. Parallel Double Sweep Preconditioner for the Optimized Schwarz Algorithm Applied to High Frequency Helmholtz and Maxwell Equations. In: *Domain Decomposition Methods in Science and Engineering XXII*, Springer (2016).
10. L. Zepeda-Núñez and L. Demanet. The method of polarized traces for the 2D Helmholtz equation. *Journal of Computational Physics* **308**, 347–388, March (2016).

Decomposition and Preconditioning of Deep Convolutional Neural Networks for Training Acceleration

Linyan Gu, Wei Zhang, Jia Liu, and Xiao-Chuan Cai

1 Introduction

Deep convolutional neural networks (DCNNs) [7] have brought significant improvements to the field of computer vision for a wide range of problems [3, 7]. Larger models and larger datasets have led to breakthroughs in accuracy; however, it results in much longer training time and memory intensity, which negatively impact the development of CNNs [1, 9]. There are two ways to parallelize training: model parallelism and data parallelism [1]. Model parallelism partitions the network into pieces, and different processors train different pieces. In model parallelism, frequent communication between different processors is needed since the calculation of the next layer usually requires the outputs of the previous layer. In data parallelism, the dataset is partitioned into parts stored in each processor which has a local copy of the network with its parameters. However, scaling the training to a large number of processors means an increase in the batch size, which results in poor generalization. New training methods are developed to avoid this problem [1, 9].

In this paper, we propose a method to parallelize the training of DCNNs by decomposing and preconditioning them motivated by the idea of domain decomposition methods [8]. Domain decomposition methods are a family of highly parallel methods for solving partial differential equations on large scale computers, which is based on the divide and conquer philosophy for solving a problem defined on a global domain by iteratively solving subproblems defined on smaller subdomains [8]. The advantages of domain decomposition methods consist of the straightforward

Linyan Gu, Wei Zhang, Jia Liu
Shenzhen Institutes of Advanced Technology, Chinese Academy of Sciences, Shenzhen, 518055, China, e-mail: ly.gu@siat.ac.cn, wei.zhang@siat.ac.cn, jia.liu@siat.ac.cn

Xiao-Chuan Cai
Faculty of Science and Technology, University of Macau, Avenida da Universidade, Taipa, Macao, China, e-mail: xccai@um.edu.mo

S. C. Brenner et al. (eds.), *Domain Decomposition Methods in Science and Engineering XXVI*, Lecture Notes in Computational Science and Engineering 145,
https://doi.org/10.1007/978-3-030-95025-5_14

applicability for parallel computing [4] and the localized treatment for the specificity of subdomain problems [8].

First, motivated by the domain decomposition methods, a DCNN (also called a global network) is decomposed into sub-networks by partitioning the width of the network while keeping the depth constant. All the sub-networks are individually trained, in parallel without any interprocessor communication, with the corresponding decomposition of the input samples. Then, following the idea of nonlinear preconditioning of Newton's method [5], which replaces the standard initial guess by an initial guess that satisfies some of the constraints locally, we propose a sub-network transfer learning strategy in which the weights of the trained sub-networks are composed to initialize the global network, which is then further trained. There are some differences between the proposed preconditioning of DCNNs and the standard nonlinear domain decomposition preconditioners. For example, we use the stochastic gradient descent (SGD) method instead of Newton's method. Furthermore, the nonlinear preconditioner of DCNNs (i.e., compositing the sub-networks to initialize the global network) is applied only once, while the nonlinear preconditioner is applied in every iteration (or some iterations) in the nonlinear preconditioning of Newton's method.

The rest of this paper is organized as follows. Section 2 proposes a new method to parallelize the training of DCNNs by decomposing and preconditioning them. Section 3 provides some experiments, followed by our conclusions in Section 4. Additionally, we have submitted some parts of this work to a special issue of *Electronic Transactions on Numerical Analysis* [2], where more details, additional theoretical discussions and more experimental results are included.

2 Proposed approaches

In this section, we propose and study a new method to parallelize the training of DCNNs by decomposing and preconditioning DCNNs. We consider a DCNN for a classification consisting of some convolutional layers and some fully connected (FC) layers, followed by a classification module which is usually a softmax layer.

Notation. Denote an L-layer DCNN as $F(\boldsymbol{x};\Theta)$ with input $\boldsymbol{x}$ and set of parameters Θ. The output of each layer is called a feature map and is a 3D tensor, where the third dimension of the tensor is the number of independent maps, and the first and the second are the height and the width, respectively. The kernel of the l-th layer is a 4D tensor and can be denoted by $\boldsymbol{w}^l \in \mathbb{R}^{t_1^l \times t_2^l \times c_{\text{in}}^l \times c_{\text{out}}^l}$, where c_{in}^l and c_{out}^l are the number of input and output channels, respectively, and t_1^l, t_2^l are the kernel widths. An FC layer can be regarded as a special case of a convolutional layer.

Assume $\boldsymbol{x} \in \mathbb{R}^{H\times W\times D}$ is a 3D tensor with element $x_{i,j,k}$ where $(i,j,k) \in \Omega$, $\Omega = [H]\times[W]\times[D]$ with $[H] = \{1,\cdots,H\}$. Given a Cartesian product $\tilde{\Omega} \subset \Omega$, we call $\tilde{\boldsymbol{x}} = \mathcal{D}_{\tilde{\Omega}}(\boldsymbol{x})$ a subdomain of $\boldsymbol{x}$ with element $\tilde{x}_{\tilde{i},\tilde{j},\tilde{k}} = x_{i,j,k}$ for all $(i,j,k) \in \tilde{\Omega}$, where the elements of $\mathcal{D}_{\tilde{\Omega}}(\boldsymbol{x})$ remain order-preserving (cf. [2]). Given a set of Cartesian products $\{\Omega_k\}_{k=1}^K$ satisfying

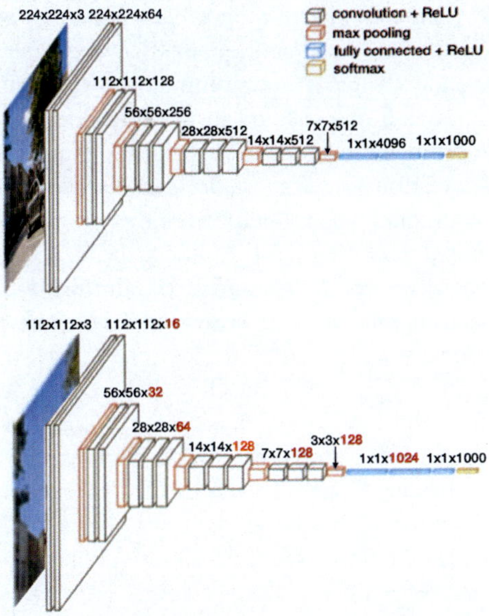

Fig. 1: Illustration of the decomposition of a DCNN. The global network and input is uniformly decomposed into 4 partitions. The architecture of VGG16 (top) and one of sub-networks (bottom); cf. [2].

$$\Omega_i \cap \Omega_j = \emptyset, \bigcup_{k=1}^{K} \Omega_k = \Omega,$$

we call $\{\tilde{x}_k = \mathcal{D}_{\Omega_k}(\boldsymbol{x}) \mid k \in [K]\}$ a complete decomposition of $\boldsymbol{x}$. Besides, given a subdomain $\tilde{\Omega}$ of the input, the activation field of $\tilde{\Omega}$ in the l-th layer, which is denoted by $\mathcal{G}_{F,l}(\tilde{\Omega})$, represents the largest subdomain of the output of this layer that only responds to $\tilde{\Omega}$; see [2] for more formal details.

2.1 Decomposing a DCNN into sub-networks

We consider a global network for a classification task with samples $X = \{\boldsymbol{x}_i\}_i$ and their corresponding labels. Given a set of Cartesian products $\{\Omega_k\}_{k=1}^{K}$, the samples are decomposed into K subdomains denoted by $X_k = \{\mathcal{D}_{\Omega_k}(\boldsymbol{x}_i)\}_i$ for $k \in [K]$. For a natural RGB image (i.e., $D = 3$), we decompose $\boldsymbol{x}_i$ in the first and second dimensions but not the third dimension. Correspondingly, a global network is decomposed into K sub-networks by partitioning the **width** (i.e., along the channel dimension) of the network while keeping the depth constant; see more formal details in [2]. Then, the samples of each subdomain and their ground truth are used to train the corresponding sub-network. The training of sub-networks can be performed completely independently on parallel computers. Compared with existing distributed training methods for DCNNs, the parallel training of sub-networks is more closely related to model parallelism than to data parallelism. However, the sub-networks are trained completely independently and the communication between different sub-networks in the proposed approach only occurs in the initialization of the global network, which is different from the current model parallelism that suffers from excessive inter-GPU communication since the model part trained by one processor usually requires output from a model part trained by another processor.

Fig. 1 shows the decomposition of VGG16 [7], where the VGG16 and the input samples are uniformly decomposed into $K = 4$ partitions; uniformly means that the decomposition of inputs is a complete decomposition, and the channel number and the input are partitioned uniformly.

The number of floating point operations (FLOPs) [6] is used to estimate the computational complexity of a network, in which each multiplication or addition is counted as one FLOP. Assume that the global network and the inputs are uniformly decomposed into $K = n^2$ partitions. Generally, the FLOPs of each sub-network is approximately $1/K^3$ of that of the global network, since the convolutional layers contain the vast majority of computations; see [2] for more details.

2.2 Preconditioning of DCNNs

We propose an algorithm for composing the trained sub-networks to initialize the global network, which we call the sub-network transfer learning strategy, and then the global network is further trained. In a reverse process of the decomposition, the weights of the sub-networks are composed along the channel dimension but with additional connections between the sub-networks initialized to zero. Taking VGG16 as an example, Fig. 2 shows how to compose 4 sub-networks into one global network. More formally, denoting the weights in l-th layer of the global network and the k-th sub-networks by $\boldsymbol{w}^l$ and $\{\boldsymbol{w}_k^l\}_K$, respectively, $\boldsymbol{w}^l$ is initialized as follows (cf. [2]):

- For the first layer,

$$\begin{aligned} \mathcal{D}_{\Omega_k'}(\boldsymbol{w}^1) &= \boldsymbol{w}_k^1, \text{ for } k \in [K], \\ \Omega_k' &= [t_1^1] \times [t_2^1] \times [c_{\text{in}}^1] \times \{1 + \sum\nolimits_{i=1}^{k-1} c_{i,\text{out}}^1 : \sum\nolimits_{i=1}^{k} c_{i,\text{out}}^1\}. \end{aligned} \tag{1}$$

- For the first FC layer,

$$\begin{aligned} &\mathcal{D}_{\Omega_k'}(\boldsymbol{w}^l) = \boldsymbol{w}_k^l, \text{ for } k \in [K], \mathcal{D}_{\Omega''}(\boldsymbol{w}^l) = \boldsymbol{0}, \\ &\Omega_k' = (\mathcal{G}_{F,l-1}(\Omega_k) \cap ([t_1^l] \times [t_2^l] \times \{1 + \sum_{i=1}^{k-1} c_{i,\text{in}}^l : \sum_{i=1}^{k} c_{i,\text{in}}^l\})) \times \{1 + \sum_{i=1}^{k-1} c_{i,\text{out}}^l : \sum_{i=1}^{k} c_{i,\text{out}}^l\}, \\ &\Omega'' = ([t_1^l] \times [t_2^l] \times [c_{\text{in}}^l] \times [c_{\text{out}}^l]) \setminus \cup_{k=1}^K \Omega_k'. \end{aligned} \tag{2}$$

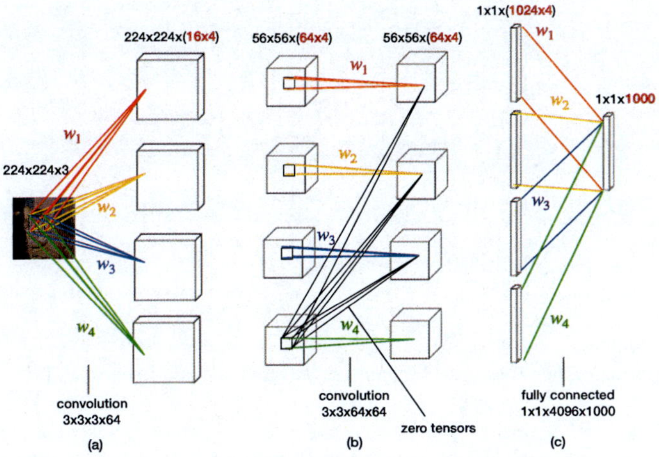

Fig. 2: Illustration of preconditioning the global network by composing 4 sub-networks. $\boldsymbol{w}_k$ denotes the weights of the k-th sub-network (using the same notation for different layers for simplicity). (a) The first convolutional layer; (b) One of the intermediate convolutional layers. Note that some of the connections with zero weights are omitted for simplicity; (c) The last FC layer; cf. [2].

- For the last FC layer,

$$\mathcal{D}_{\Omega'_k}(\boldsymbol{w}^L) = \boldsymbol{w}_k^L, \text{ for } k \in [K],$$
$$\Omega'_k = [t_1^L] \times [t_2^L] \times \{1 + \sum_{i=1}^{k-1} c_{i,\text{in}}^L : \sum_{i=1}^{k} c_{i,\text{in}}^L\} \times [C]. \tag{3}$$

- For other convolutional layers and other FC layers,

$$\mathcal{D}_{\Omega'_k}(\boldsymbol{w}^l) = \boldsymbol{w}_k^l, \text{ for } k \in [K], \mathcal{D}_{\Omega''}(\boldsymbol{w}^l) = \boldsymbol{0},$$
$$\Omega'_k = [t_1^l] \times [t_2^l] \times \{1 + \sum_{i=1}^{k-1} c_{i,\text{in}}^l : \sum_{i=1}^{k} c_{i,\text{in}}^l\} \times \{1 + \sum_{i=1}^{k-1} c_{i,\text{out}}^l : \sum_{i=1}^{k} c_{i,\text{out}}^l\},$$
$$\Omega'' = ([t_1^l] \times [t_2^l] \times [c_{\text{in}}^l] \times [c_{\text{out}}^l]) \setminus \cup_{k=1}^{K} \Omega'_k. \tag{4}$$

3 Experiments

In this section, some experiments on image classification tasks are carried out to evaluate the proposed approach by observing the training time and the classification accuracy. The experiments are carried out using the TensorFlow library on a workstation with 4 NVIDIA Tesla V100 32G GPUs. We compare the performances between two training strategies: 1) to train the global network with the parameters randomly initialized (referred to as "GNet-R"), 2) to train the sub-networks (referred to as "SNets") in parallel and then further train the global network initialized by the sub-networks transfer learning method (referred to as "GNet-T"). The sub-networks

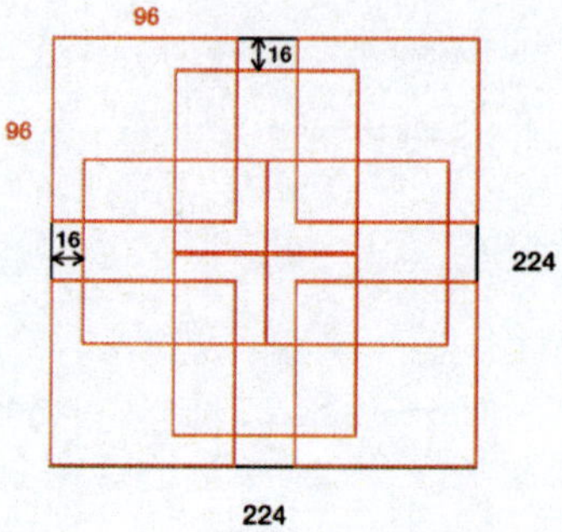

Fig. 3: Illustration of the way of partitioning each image into 8 sub-images; cf. [2].

Table 1: The FLOPs and the number of parameters of the global network ("GNet") and one sub-network ("SNet", and "4", "8" and "16" mean 4, 8 and 16 partitions, respectively); the training time of the global networks and 4 (or 8, 16) sub-networks for 10 epochs. "M" means 10^6.

	GNet	SNet-4	SNet-8	SNet-16
# param	17.22M	1.08M	0.27M	0.07M
FLOPs	5015.56M	154.57M	19.69M	4.05M
training time	289 s	137 s	101 s	82 s

and the global networks are trained using the same computing resources. The global networks are trained using the data-parallel strategy. The sub-networks are trained in parallel by a multiprocessing strategy, with GPUs uniformly assigned to sub-networks. In addition, the numbers of training iterations in the two strategies are the same. For the strategy 1), the GNet-R is trained for 200 epochs. For the strategy 2), the SNets are trained for 100 epochs and then the GNet-T is trained for 100 epochs.

The experiments are carried out on the dataset which contains 4323 images of flowers in 5 categories[1]. The dataset is split into training, validation and testing sets in the ratio of about 70:15:15, and the images are all resized to 224×224. We use a residual network of 18 layers in [3]. Additionally, the network and the input images are decomposed into 4, 8, and 16 partitions. For 4 (or 16) partitions, the input images are cropped 24 pixels on the boundaries, which are then decomposed into 4 (or 16) sub-images of size 140×140 (or 70×70) by decomposing into 2 (or 4) partitions in both the width and height dimensions and then applying overlap between each pair of neighbouring subdomains; for 8 partitions, the decomposition is illustrated in Fig. 3.

Table 1 shows the FLOPs and the number of parameters of the global network and one sub-network, and the training times of the two training strategies, which indicates that 1) the number of parameters of the sub-network is approximately $1/K^2$ of that of the corresponding global network, 2) for 4 partitions, the computation of the sub-network is approximately $1/2^5$ of the corresponding global network; for

[1] https://www.kaggle.com/alxmamaev/flowers-recognition.

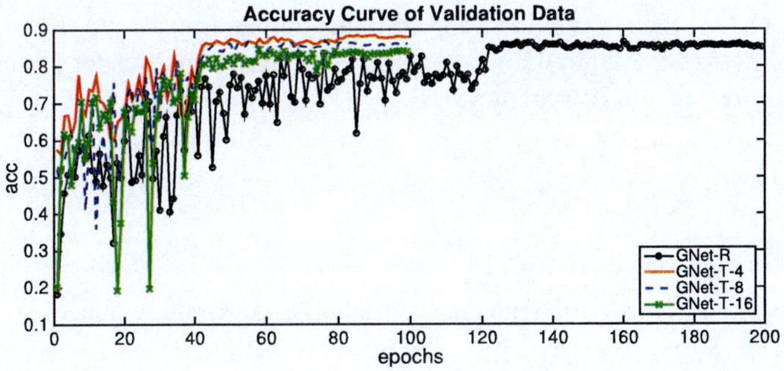

Fig. 4: Classification accuracy curves for validation data during training with varying partition numbers. Comparisons between the two training strategies (i.e., "GNet-R" and "GNet-T").

Table 2: The comparison of the classification accuracy of the testing data between the global networks of the two training strategies (i.e., "GNet-T" and "GNet-R") with varying partition numbers. "Initialized" means that the global networks are initialized by the sub-network transfer learning strategy (i.e., "GNet-T") or randomly initialized (i.e., "GNet-R") without being further trained, and "Trained" means that the global networks are trained.

Initialized (%)				Trained (%)			
GNet-R	GNet-T-4	GNet-T-8	GNet-T-16	GNet-R	GNet-T-4	GNet-T-8	GNet-T-16
17.39	77.73	57.45	49.46	82.80	83.56	82.49	79.26

8 and 16, this ratio decreases to $1/2^8$ and $1/2^{10}$, and 3) for the same number of iterations, the training time of K sub-networks is less than $1/2$ of that of the global network; thus, the sub-network transfer learning strategy saves more than $1/4$ of the training time.

Fig. 4 and Table 2 show the comparisons of the classification accuracy between the two training strategies, which shows that 1) in general, as the number of partitions increases, the initialization seem to be worse and the accuracy of GNet-T after further training decreases, and 2) after further training, the sub-network transfer learning strategy shows almost no loss of accuracy, except for the case of 16 partitions. These results indicate that a decomposition into too many partitions may reduce the quality of the initialization and also perform poorly after further training.

4 Conclusion

In this paper, inspired by the idea of domain decomposition methods and nonlinear preconditioning, we propose and study a new method of decomposing and preconditioning DCNNs for the purpose of parallelizing their training. The global network is firstly decomposed into sub-networks that are trained independently without any interprocessor communication, which are then recomposed to initialize the global network via the transfer learning strategy. The experimental results show that the proposed approach can indeed provide good initialization and accelerate the training of the global network. Additionally, after further training, the transfer learning strategy shows almost no loss of accuracy.

References

1. P. Goyal, P. Dollár, R. Girshick, P. Noordhuis, L. Wesolowski, A. Kyrola, A. Tulloch, Y. Jia, and K. He. Accurate, large minibatch SGD: Training imagenet in 1 hour. *arXiv preprint* arXiv:1706.02677 (2017).
2. L. Gu, W. Zhang, J. Liu and X.-C. Cai. Decomposition and composition of deep convolutional neural networks and training acceleration via sub-network transfer learning. Submitted to *ETNA* (2020).
3. K. He, X. Zhang, S. Ren and J. Sun. Deep residual learning for image recognition. In: *CVPR*, pp. 770–778 (2016).
4. Z.-J. Liao, R. Chen, Z. Yan and X.-C. Cai. A parallel implicit domain decomposition algorithm for the large eddy simulation of incompressible turbulent flows on 3d unstructured meshes. *Int. J. Numer. Methods Fluids* **89**(9), 343–361 (2019).
5. L. Luo, W. Shiu, R. Chen and X.-C. Cai. Anonlinear elimination preconditioned inexact Newton method for blood flow problems in human artery with stenosis. *J. Comput. Phys.* **399**, 108926 (2019).
6. P. Molchanov, S. Tyree, T. Karras, T. Aila and J. Kautz. Pruning convolutional neural networks for resource efficient inference. In: *ICLR* (2017).
7. K. Simonyan and A. Zisserman. Very deep convolutional networks for large-scale image recognition. *arXiv preprint* arXiv:1409.1556 (2014).
8. A. Toselli and O.B. Widlund. *Domain Decomposition Methods – Algorithms and Theory.* Springer (2005).
9. Y. You, Z. Zhang, C.-J. Hsieh, J. Demmel and K. Keutzer. Imagenet training in minutes. In: *ICPP*, pp. 1–10 (2018).

Numerical Calculation of the Portal Pressure Gradient of the Human Liver With a Domain Decomposition Method

Zeng Lin, Bokai Wu, Shanlin Qin, Xinhong Wang, Rongliang Chen, and Xiao-Chuan Cai

1 Introduction

Portal hypertension (PH) refers to the abnormal increase of the portal venous pressure, which is a common chronic liver disease with clinical consequences of cirrhosis, such as hepatic encephalopathy, variceal hemorrhage and ascites [7, 11]. Fig. 1 shows the portal vein and hepatic vein extracted from CT images, from the figure we also see the single inlet and multiple outlets structure of the portal vein and the multiple inlets and single outlet characteristics of the hepatic vein. The portal pressure gradient (PPG) is defined as the difference in the pressure between the inlet of the portal vein and the outlet of the inferior vena cava. PH refers to the situation that PPG is greater than 5 mmHg [4]. When the value of PPG is higher than 10 mmHg, the PH is

Zeng Lin
Shenzhen Institute of Advanced Technology, Chinese Academy of Sciences, Shenzhen, China, e-mail: zeng.lin@siat.ac.cn

Bokai Wu
Shenzhen Institute of Advanced Technology, Chinese Academy of Sciences, Shenzhen, China, e-mail: bk.wu@siat.ac.cn

Shanlin Qin
Shenzhen Institute of Advanced Technology, Chinese Academy of Sciences, Shenzhen, China, e-mail: sl.qin@siat.ac.cn

Xinhong Wang
Department of Radiology, The Second Affiliated Hospital, School of Medicine, Zhejiang University, Hangzhou, China, e-mail: 2611104@zju.edu.cn

Rongliang Chen
Shenzhen Institute of Advanced Technology, Chinese Academy of Sciences, Shenzhen, China, e-mail: rl.chen@siat.ac.cn

*Corresponding Author: Xiao-Chuan Cai
Department of Mathematics, University of Macau, Macau, China, e-mail: xccai@um.edu.mo

S. C. Brenner et al. (eds.), *Domain Decomposition Methods in Science and Engineering XXVI*, Lecture Notes in Computational Science and Engineering 145,
https://doi.org/10.1007/978-3-030-95025-5_15

called the clinically significant portal hypertension. If PPG is higher than 12 mmHg, variceal hemorrhage may occur [11].

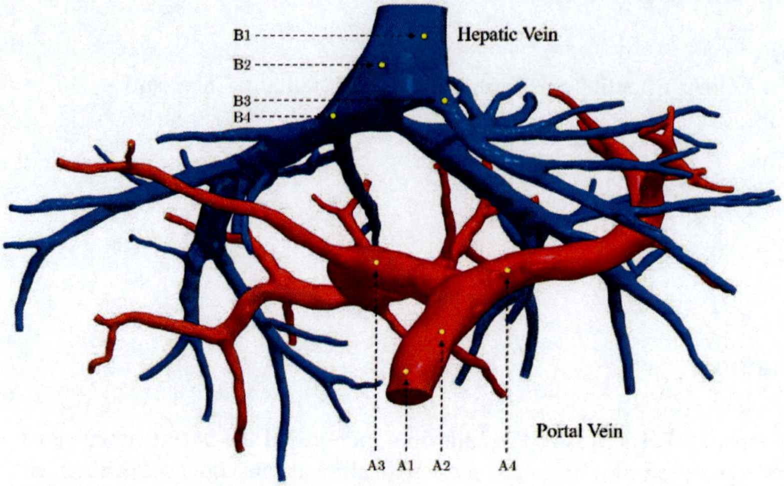

Fig. 1: The segmented portal vein and hepatic vein.

In clinical applications, the common approach to measure the PPG is the transjugular route, which requires the insertion of a radiopaque catheter into the right hepatic vein via the jugular vein under fluoroscopic guidance. The method is invasive and sometimes impractical for routine clinical practice. Recently, a technology based on computational fluid dynamics (CFD) [5, 6, 14] has been introduced as an alternative approach to measure the pressure difference non-invasively. With CFD, several desired pathological values, such as pressure, velocity and wall shear stress (WSS) can be easily computed.

In this work, we model the blood flow by a system of Navier–Stokes equations which is discretized by a fully implicit finite element method on a fully unstructured mesh, and solved by an efficient and highly parallel domain decomposition method [9]. With this method, a simulation of a full 3D patient-specific hepatic flow can be realized in a few hours. The numerical experiments are carried out on a cluster of computers with near 2000 processor cores and the parallel efficiency is higher than 60%. The computed PPG values are within the normal range of published data.

2 Numerical method

The blood flows in the hepatic vessels are described by the unsteady incompressible Navier–Stokes equations:

$$\begin{cases} \rho\dfrac{\partial \boldsymbol{u}}{\partial t} + \rho(\boldsymbol{u}\cdot\nabla)\boldsymbol{u} - \nabla\cdot\boldsymbol{\sigma} = \boldsymbol{f} \;\; in \;\; \Omega\times(0,T], \\ \nabla\cdot\boldsymbol{u} = 0 \;\; in \;\; \Omega\times(0,T]. \end{cases} \tag{1}$$

Here $\boldsymbol{u}$ denotes the velocity vector, ρ the blood density, $\boldsymbol{f}$ the external force and $\boldsymbol{\sigma}$ is the Cauchy stress tensor defined as:

$$\boldsymbol{\sigma} = -p\boldsymbol{I} + 2\mu\boldsymbol{\varepsilon}(\boldsymbol{u}), \tag{2}$$

where p is the pressure, $\boldsymbol{I}$ is the identity tensor, μ is the dynamic viscosity and $\boldsymbol{\varepsilon}$ is the deformation tensor defined as $\boldsymbol{\varepsilon}(\boldsymbol{u}) = 1/2(\nabla\boldsymbol{u} + \nabla\boldsymbol{u}^T)$.

The initial condition is imposed by a given function. The velocity boundary conditions are imposed for the inlets of the portal vein. No slip condition is applied on the wall. The resistance boundary conditions are used for the outlets of the portal vein and hepatic vein [8].

The weak form of (1) reads: Find $\boldsymbol{u} \in V$ and $p \in P$ such that $\forall \boldsymbol{v} \in V_0$ and $\forall q \in P$,

$$\mathbf{B}(\{\boldsymbol{u}, p\}, \{\boldsymbol{v}, q\}) = 0, \tag{3}$$

where

$$\begin{aligned} \mathbf{B}(\{\boldsymbol{u}, p\}, \{\boldsymbol{v}, q\}) = \; & \rho\int_\Omega \frac{\partial \boldsymbol{u}}{\partial t}\cdot\boldsymbol{v}\mathrm{d}\Omega + \rho\int_\Omega (\boldsymbol{u}\cdot\nabla)\boldsymbol{u}\cdot\boldsymbol{v}\mathrm{d}\Omega \\ - \; & \int_\Omega p(\nabla\cdot\boldsymbol{v})\mathrm{d}\Omega + 2\mu\int_\Omega \boldsymbol{\varepsilon}(\boldsymbol{u}) : \boldsymbol{\varepsilon}(\boldsymbol{v})\mathrm{d}\Omega \\ + \; & \int_\Omega (\nabla\cdot\boldsymbol{u})q\mathrm{d}\Omega + \int_{\Gamma_O} (\boldsymbol{\sigma}\boldsymbol{n})\cdot\boldsymbol{v}\mathrm{d}\Gamma - \rho\int_\Omega \boldsymbol{f}\cdot\boldsymbol{v}\mathrm{d}\Omega. \end{aligned} \tag{4}$$

Here Γ_O is the outlet boundary and $\boldsymbol{n}$ is the outward normal vector of the outlet. The functional spaces V, V_0 and P are defined in detail in [5].

The computational domain Ω is covered with a fully unstructured tetrahedral mesh on which we introduce $P_1 - P_1$ finite element function spaces. As the $P_1 - P_1$ pair doesn't satisfy the Ladyzhenskaya–Babuska–Brezzi (LBB) [2] condition, some stabilization terms are added in the weak form (4) when applied to finite element functions. More details about the stabilization parameters can be found in [2]. Then (4) can be rewritten as a time-dependent nonlinear algebraic system

$$\frac{\mathrm{d}\mathcal{X}(t)}{\mathrm{d}t} = \mathcal{N}(\mathcal{X}), \tag{5}$$

where $\mathcal{X}(t)$ is the vector of the nodal values of the velocity $\boldsymbol{u}$ and pressure p, $\mathcal{N}(\cdot)$ is the nonlinear function representing the spatial discretization of (4). (5) can be further discretized by the fully implicit backward Euler method in time

$$\frac{\mathcal{X}^n - \mathcal{X}^{n-1}}{\Delta t} = \mathcal{N}(\mathcal{X}^n), \tag{6}$$

where $\mathcal{X}^n$ is the value of $\mathcal{X}(t)$ at the n-th time step and Δt is the time step size.

For simplicity, (6) can be rearranged into a nonlinear system

$$\mathcal{F}^n(\mathcal{X}^n) = 0 \tag{7}$$

to be solved at each time step.

In this work, the nonlinear system (7) will be solved by the Newton–Krylov–Schwarz algorithm [13]. The algorithm includes three components, an inexact Newton [3] as the nonlinear solver, a preconditioned Krylov subspace method (GMRES) [12] as the linear solver at each Newton step, and an overlapping Schwarz method [1] as the preconditioner. More details about the algorithm are available in [8].

3 Numerical experiments

In this section, we present some numerical experiments for blood flows in the portal vein and hepatic vein, and also the parallel performance of the algorithm with respect to the number of processor cores. Lastly, the PPG values will be calculated based on the simulation of blood flows in a patient-specific portal vein and hepatic vein.

In all the numerical experiments, $\rho = 1.05\text{g/cm}^3$ and $\mu = 0.038\text{cm}^2/\text{s}$ [10] are used to characterize the properties of the hepatic blood. The algorithm is implemented with the Portable Extensible Toolkit for Scientific computation (PETSc) library. In the experiments, the relative stopping condition for Newton is set to be 1.0×10^{-6} and the relative stopping condition for GMRES is 1.0×10^{-3}. Incomplete LU (ILU) is used to solve the subdomain problems in the additive Schwarz preconditioner. "ILU(l)" represents ILU with l level of fill-ins, "np" means the number of processor cores, "Newton" stands for the average number of Newton iterations per time step, "GMRES" denotes the average number of GMRES iterations per Newton step, "Time" is the average wall clock time in seconds spent per time step, "Memory" indicates the memory consumption in megabyte per processor core per time step, "Speedup" denotes the speedup ratio and "Efficiency" means the parallel efficiency.

A sample finite element mesh for the portal vein and hepatic vein is shown in Fig. 2. The portal vein has 1 inlet and 25 outlets and the hepatic vein has 47 inlets and 1 outlet. The clinically measured flow velocity [10] is used for the inflow boundary condition and the total resistance is chosen such that the computed pressures are within the ranges of typical adult patients.

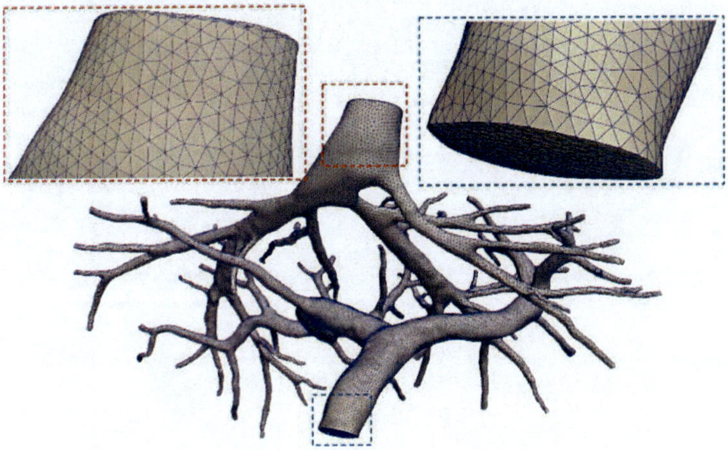

Fig. 2: A sample finite element mesh for the portal vein and hepatic vein.

Table 1: Parallel performance using different numbers of processor cores.

np	Newton	GMRES	Memory (MB)	Time (s)	Speedup	Efficiency
240	3.10	404.52	450.89	160.39	1	100%
480	3.10	452.68	250.55	93.97	1.71	86%
960	3.10	457.98	143.06	51.93	3.09	77%
1920	3.10	462.02	78.84	30.14	5.32	67%

A parallel scalability study. The parallel scalability is investigated on a cluster of computers, and each compute node of the computer has two Intel Xeon processors and 64GB of shared memory. The performance of the algorithm in terms of the number of Newton iterations per time step, the number of GMRES iterations per Newton step, the total memory per processor core per time step, the total compute time per time step, the speedup ratio and the parallel efficiency are presented in Table 1. A mesh with 9.96×10^6 elements is utilized for the numerical tests, where the largest size of the elements is 0.85mm, the smallest is 0.09mm and the average is 0.26mm. The time step size is set as $\Delta t = 1.00 \times 10^{-3}$s, the subdomain solver is ILU(1) and the overlapping size is 2. The scalability about the linear and nonlinear algebraic solvers are clearly observed, wherein the number of Newton iterations and GMRES iterations change only slightly as the number of processor cores increase, especially for the Newton iterations. It can be seen that when the number of processor cores increases from 240 to 1920, the compute time reduces to 30.14s and the parallel efficiency reduces to 67%, which is quite good considering the fact that the geometry of the problem is rather complicated.

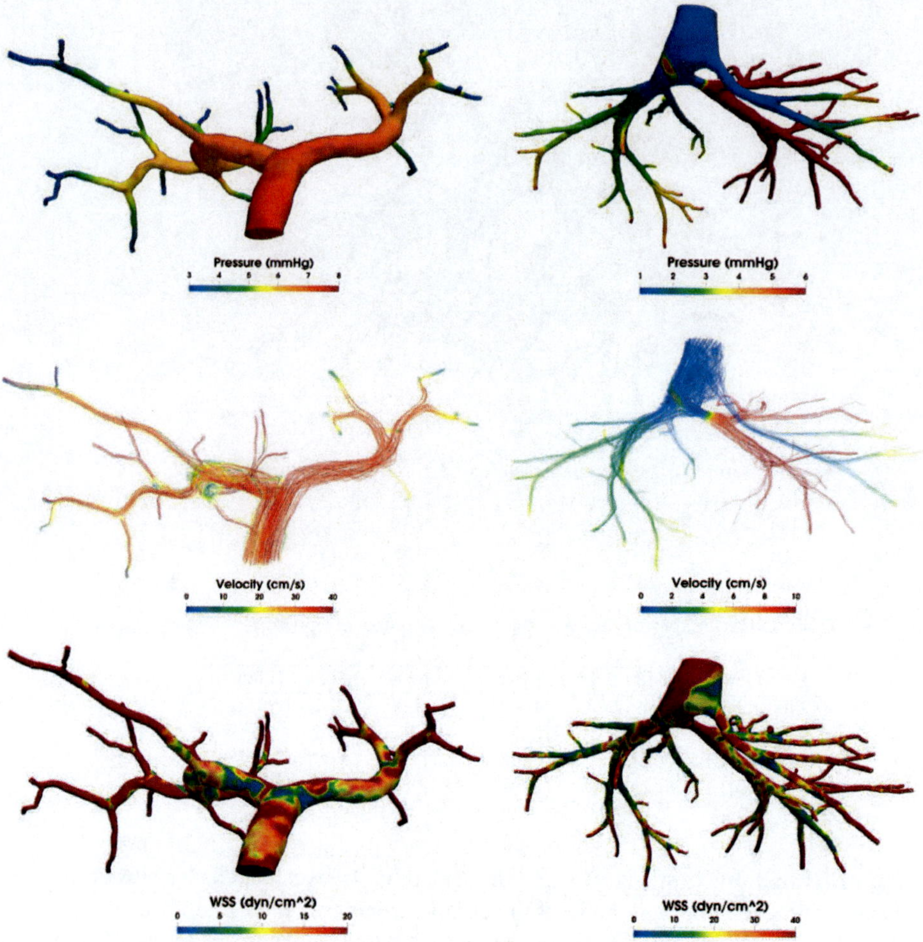

Fig. 3: The pressure, velocity and WSS distribution of the computed flow in the portal vein and hepatic vein at $t = 0.5$s.

The portal pressure gradient. Next, we present a numerical calculation of PPG. Firstly, the pressure, velocity and WSS distributions of the blood flow in the portal vein and hepatic vein at $t = 0.5$s are plotted in Fig. 3. Then we pick several pairs of points (A1,B1), (A2,B2), (A3,B3) and (A4,B4) as marked in Fig. 1 to compute the difference in the pressure between the portal vein and the hepatic vein, i.e., the PPG, for three cardiac cycles. The portal vein pressure at points A1, A2, A3 and A4 are drawn in the top-left sub-figure of Fig. 4. Meanwhile, the hepatic vein pressure at points B1, B2, B3 and B4 are illustrated in the top-right sub-figure of Fig. 4. Then their PPG values of the pairs (A1,B1), (A2,B2), (A3,B3) and (A4,B4) are plotted in the bottom-left sub-figure of Fig. 4. Finally, the time-averaged PPG (TAPPG) values are presented in the bottom-right sub-figure of Fig. 4. It is clear that all four approximations are within the normal ranges as indicated in [8].

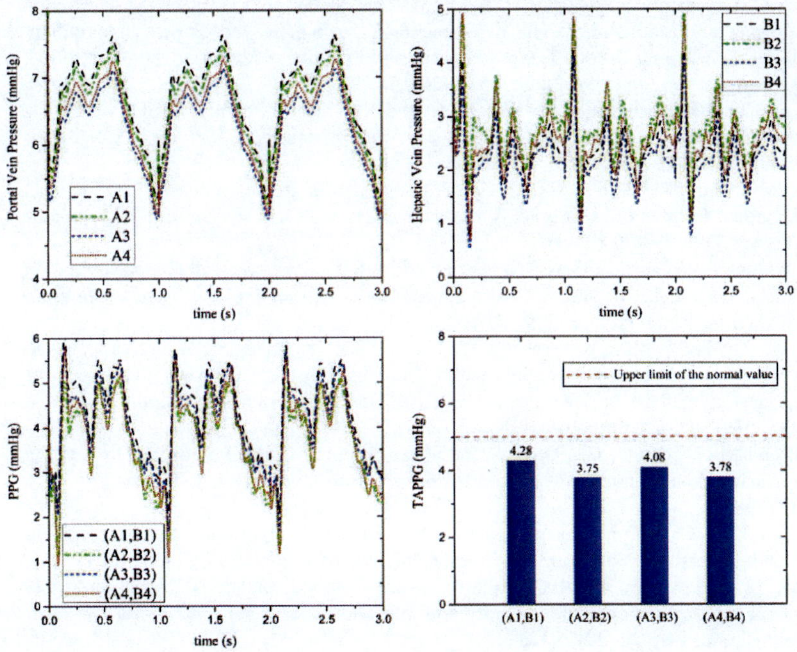

Fig. 4: The computed portal vein pressure, hepatic vein pressure, PPG and TAPPG values for three cardiac cycles.

References

1. X. Cai and M. Sarkis. A restricted additive schwarz preconditioner for general sparse linear systems. *SIAM Journal on Scientific Computing* **21**(2), 792–797 (1999).
2. R. Chen, Y. Wu, Z. Yan, Y. Zhao and X.-C. Cai. A parallel domain decomposition method for 3D unsteady incompressible flows at high Reynolds number. *Journal of Scientific Computing* **58**(2), 275–289 (2014).
3. S.C. Eisenstat and H.F. Walker. Choosing the forcing terms in an inexact Newton method. *SIAM Journal on Scientific Computing* **17**(1), 16–32 (1996).
4. J.Y. Huang, J.B. Samarasena, T. Tsujino, J. Lee, K.-Q. Hu, C.E. McLaren, W.-P. Chen and K.J. Chang. EUS-guided portal pressure gradient measurement with a simple novel device: a human pilot study. *Gastrointestinal endoscopy* **85**(5), 996–1001 (2017).
5. F. Kong, V. Kheyfets, E. Finol and X.-C. Cai. An efficient parallel simulation of unsteady blood flows in patient-specific pulmonary artery. *International journal for numerical methods in biomedical engineering* **34**(4), e2952 (2018).

6. F. Kong, V. Kheyfets, E. Finol and X.-C. Cai. Simulation of unsteady blood flows in a patient-specific compliant pulmonary artery with a highly parallel monolithically coupled fluid-structure interaction algorithm. *International Journal for Numerical Methods in Biomedical Engineering* **35**(7), e3208 (2019).
7. J.-Y. Lin, C.-H. Zhang, L. Zheng, H.-J. Li, Y.-M. Zhu, X. Fan, F. Li, Y. Xia, M.-Z. Huang, S.-H. Yang et al. Establishment and assessment of the hepatic venous pressure gradient using biofluid mechanics (hvpgbfm): protocol for a prospective, randomised, non-controlled, multicentre study. *BMJ open* **9**(12), e028518 (2019).
8. Z. Lin, R. Chen, B. Gao, S. Qin, B. Wu, J. Liu and X.-C. Cai. A highly parallel simulation of patient-specific hepatic flows. *International Journal for Numerical Methods in Biomedical Engineering*, e3451 (2021).
9. L. Luo, W.-S. Shiu, R. Chen and X.-C. Cai. A nonlinear elimination preconditioned inexact Newton method for blood flow problems in human artery with stenosis. *Journal of Computational Physics* **399**, 108926 (2019).
10. R. Ma, P. Hunter, W. Cousins, H. Ho, A. Bartlett and S. Safaei. Anatomically based simulation of hepatic perfusion in the human liver. *International journal for numerical methods in biomedical engineering* **35**(9), e3229 (2019).
11. X. Qi, W. An, F. Liu, R. Qi, L. Wang, Y. Liu, C. Liu, Y. Xiang, J. Hui, Z. Liu, et al. Virtual hepatic venous pressure gradient with CT angiography (CHESS 1601): a prospective multicenter study for the noninvasive diagnosis of portal hypertension. *Radiology* **290**(2), 370–377 (2019).
12. Y. Saad and M.H. Schultz. GMRES: a generalized minimal residual algorithm for solving nonsymmetric linear systems. *SIAM Journal on Scientific and Statistical Computing* **7**(3), 856–869 (1986).
13. Y. Wu and X.-C. Cai. A parallel two-level method for simulating blood flows in branching arteries with the resistive boundary condition. *Computers & fluids* **45**(1), 92–102 (2011).
14. N. Xiao, J.D. Humphrey and F.C. Alberto. Multi-scale computational model of three-dimensional hemodynamics within a deformable full-body arterial network. *Journal of Computational Physics* **244**, 22–40 (2013).

A Parallel Adaptive Finite Element Method for Modeling a Deformable Droplet Travelling in Air

Li Luo, Xiao-Chuan Cai, and David E. Keyes

1 Introduction

Violent respiratory events such as coughing and sneezing can contribute to the transmission of infectious diseases from host to host. The dynamics of droplet transfer between individuals and the range of contamination are extremely complex and remain unclear [3]. Studying the fluid dynamics of pathogen-laden droplets is critically important to controlling the covid19 pandemic.

Fluid dynamics studies of violent ejections are presented in [1, 4]. These studies focus on analytical modeling of the puff evolution or the transport of inertial spherical droplets to understand the available quantitative relationships. CFD simulation has been conducted to investigate the dispersion of airborne particles by using a Lagrangian-based model for particle motion [10] or by coupling the Navier–Stokes equations with an additional transport equation for a scalar concentration field [9]. Although the transport of particles in a crowd are detailed in these studies, the travelling process of an individual droplet and its dynamics subject to the combined effect of size, gravitational settling, surface tension, and aerodynamic forces are not given particular attention. To address these issues, incorporating multi-phase flow physics in the modeling is necessary [3].

In this work, we study the process of a deformable droplet travelling over a long distance based on two-phase flow simulation, with focus on the two-way coupling between the droplet dynamics and the ambient airflow through advection and surface tension, in order to provide some numerical understanding of the transmission

Li Luo
Faculty of Science and Technology, University of Macau, Macau, China, e-mail: liluo@um.edu.mo

Xiao-Chuan Cai
Faculty of Science and Technology, University of Macau, Macau, China, e-mail: xccai@um.edu.mo

David E. Keyes
Extreme Computing Research Center, King Abdullah University of Science and Technology, Thuwal, Saudi Arabia, e-mail: david.keyes@kaust.edu.sa

S. C. Brenner et al. (eds.), *Domain Decomposition Methods in Science and Engineering XXVI*, Lecture Notes in Computational Science and Engineering 145,
https://doi.org/10.1007/978-3-030-95025-5_16

of covid19. A phase-field model consisting of the coupled Cahn–Hilliard–Navier–Stokes equations with appropriate boundary conditions is used to describe the two-phase flow. Due to the vast difference between the size of the droplets and the long trajectories they travel (over 1000 times), the problem is computationally very expensive and rarely addressed by previous studies of phase-field methods. To tackle this issue, we develop an efficient adaptive finite element method based on a posterior error estimate to refine elements near the interface, while using coarse elements elsewhere to save computation. In the numerical experiments, we are mainly concerned with: 1) the influence of the droplet size on its shape dynamics and travelling path; 2) the influence of the droplet motion on the surrounding airflow; and 3) the lift and drag forces acting on the droplet through the trajectory.

2 A mathematical model based on the Cahn–Hilliard–Navier–Stokes equations

In a bounded domain $\Omega \subset \mathbb{R}^d$ $(d = 2, 3)$, the system of two immiscible incompressible fluids can be described by the coupled Cahn–Hilliard–Navier–Stokes equations:

$$\frac{\partial \varphi}{\partial t} + \mathbf{u} \cdot \nabla \varphi = L_d \Delta \mu, \qquad \mu = -\epsilon \Delta \varphi - \frac{\varphi}{\epsilon} + \frac{\varphi^3}{\epsilon}, \tag{1}$$

$$Re\rho \left(\frac{\partial \mathbf{u}}{\partial t} + (\mathbf{u} \cdot \nabla)\mathbf{u} \right) = \nabla \cdot \boldsymbol{\sigma} - \frac{Re\rho}{Fr^2} \mathbf{e}_g, \qquad \nabla \cdot \mathbf{u} = 0. \tag{2}$$

Here, a phase field variable φ is introduced to describe the transition between the two homogeneous equilibrium phases $\varphi_\pm = \pm 1$. μ is the chemical potential, ϵ is the ratio between the interface thickness and the characteristic length. $\boldsymbol{\sigma} = -p\mathbf{I} + \eta D(\mathbf{u}) - B\epsilon(\nabla\varphi \otimes \nabla\varphi)$ is the total stress tensor, where p is the pressure, $\mathbf{u}$ is the fluid velocity field, and $D(\mathbf{u}) = \nabla\mathbf{u} + (\nabla\mathbf{u})^T$ is the rate of strain tensor. The term $\epsilon(\nabla\varphi \otimes \nabla\varphi)$ represents the capillary force. The mass density ρ and the dynamic viscosity η are interpolation functions of φ between fluid 1 and fluid 2, i.e. $\rho = \frac{1+\varphi}{2} + \lambda_\rho \frac{1-\varphi}{2}$, $\eta = \frac{1+\varphi}{2} + \lambda_\eta \frac{1-\varphi}{2}$, where $\lambda_\rho = \rho_2/\rho_1$ is the ratio of density between the two fluids and $\lambda_\eta = \eta_2/\eta_1$ is the ratio of viscosity. $\mathbf{e}_g$ is the unit gravitational vector and Fr is the Froude number. L_d is the phenomenological mobility coefficient, Re is the Reynolds number, and B measures the strength of the capillary force compared to the Newtonian fluid stress.

We assume $\partial\Omega = \Gamma_i \cup \Gamma_o \cup \Gamma_w$, where Γ_i denotes the inflow boundary, Γ_o denotes the outflow boundary, and Γ_w denotes the solid surface. Given functions φ_i and $\mathbf{u}_i$, the boundary conditions on Γ_i are stated as

$$\varphi = \varphi_i, \quad \mu = 0, \quad \mathbf{u} = \mathbf{u}_i, \qquad \text{on } \Gamma_i. \tag{3}$$

On Γ_o, we consider the following outflow boundary conditions [5],

$$\partial_n \varphi = 0, \quad \partial_n \mu = 0, \qquad \text{on } \Gamma_o, \tag{4}$$

$$-(p + BF(\varphi))\mathbf{n} + \eta \mathbf{n} \cdot D(\mathbf{u}) - \frac{Re\rho}{2} |\mathbf{u}|^2 \chi(\mathbf{u} \cdot \mathbf{n})\mathbf{n} = \mathbf{0}, \qquad \text{on } \Gamma_o, \tag{5}$$

where $F(\varphi) = \frac{\epsilon}{2}|\nabla\varphi|^2 + \frac{1}{4\epsilon}(\varphi^2 - 1)^2$ is the free energy of the two-phase system. $\chi(\mathbf{u}\cdot\mathbf{n}) = \frac{1}{2}\left(1 - \tanh\frac{\mathbf{u}\cdot\mathbf{n}}{U\delta}\right)$ is a smoothed step function, where U is a characteristic velocity scale (here $U = 1$), and $\delta > 0$ is a non-dimensional constant that is sufficiently small. As $\delta \to 0$, χ takes a unit value in regions where $\mathbf{u}\cdot\mathbf{n} < 0$ and vanishes elsewhere.

On Γ_w, we consider the generalized Navier boundary conditions [6]:

$$\frac{\partial\varphi}{\partial t} + \mathbf{u}_\tau \cdot \nabla_\tau \varphi = -V_s L(\varphi), \quad \partial_n \mu = 0, \quad \mathbf{u}\cdot\mathbf{n} = 0, \qquad \text{on } \Gamma_w, \tag{6}$$

$$\Big((L_s l_s)^{-1}\mathbf{u}_\tau - BL(\varphi)\nabla_\tau\varphi/\eta + \mathbf{n}\cdot D(\mathbf{u})\Big)\times\mathbf{n} = \mathbf{0}, \qquad \text{on } \Gamma_w, \tag{7}$$

where $\mathbf{n}$ is the unit outward normal vector and τ is the unit tangential vector of the boundary. $\mathbf{u}_\tau = \mathbf{u} - (\mathbf{n}\cdot\mathbf{u})\mathbf{n}$, $\nabla_\tau = \nabla - (\mathbf{n}\cdot\nabla)\mathbf{n}$. V_s is a phenomenological parameter, $L(\varphi) = \epsilon\partial_n\varphi + \partial\gamma_{wf}(\varphi)/\partial\varphi$, and $\gamma_{wf}(\varphi) = -\frac{\sqrt{2}}{3}\cos\theta_s \sin(\frac{\pi}{2}\varphi)$, where θ_s is the static contact angle. L_s is the slip length of liquid, $l_s = \frac{1+\varphi}{2} + \lambda_{l_s}\frac{1-\varphi}{2}$, and $\lambda_{l_s} = l_2/l_1$.

3 A parallel, semi-implicit solution algorithm based on an adaptive finite element discretization, and an overlapping Schwarz preconditioned GMRES

We apply a second-order semi-implicit time discretization scheme to decouple φ, $\mathbf{u}$, and p at each time step [6]. Specifically, we apply a convex-splitting of the free energy functional and treat the nonlinear term explicitly so that the resulting matrix has constant coefficients. In addition, we consider a pressure-stabilized method to decouple the Navier–Stokes equations into a convection-diffusion equation for velocity and a Poisson equation for pressure. Then, the pressure equation results in a constant matrix and can be solved efficiently.

The resulting decoupled systems are discretized by a finite element method on unstructured meshes. We use P1-P1 finite element spaces for the Cahn–Hilliard equation and P2-P1 for the Navier–Stokes equations. Let T_h be a triangulation of Ω with h be the mesh size of an element T. We denote by φ_h^n, μ_h^n, $\mathbf{u}_h^n$, p_h^n the finite element interpolations of φ, μ, $\mathbf{u}$, p at the nth time step, respectively. In this work, we use the adaptive mesh refinement (AMR) method to accurately capture the phase field variable φ within the thin interface between the two phases. The AMR procedure is performed in an iterative manner. At each adaptive step, we introduce a physics-informed approach to refine the elements repeatedly if they are fully inside the interface region (i.e., $-0.9 \le \varphi_h^{n+1}|_T \le 0.9$) and their sizes are considered to be large (i.e., $\max_{e\in\partial T}|e| > \bar{e}$, where $\bar{e}$ is a given scale). Meanwhile, we merge adjacent elements if they are divided from the same "parent" and their error indicator Θ_T is sufficiently small, i.e.,

$$\Theta_T < \gamma_c \max_{T\in T_h}\{\Theta_T\}, \quad \text{where} \quad \Theta_T = \left(\sum_{e\in\partial T}\int_e \frac{|e|}{24}\left[\frac{\nabla\varphi_h^{n+1}\cdot\mathbf{n}_e}{2}\right]^2 de\right)^{\frac{1}{2}}.$$

Here, Θ_T is the gradient jump of φ_h^{n+1} on the interface of adjacent elements [7]. $[\cdot]$ denotes the jump on the element boundary, $\mathbf{n}_e$ is the unit outward normal vector on e, and γ_c is a given parameter. The iteration of refinement and coarsening is stopped when the maximum error indicator $\max_{T\in T_h}\{\Theta_T\} < tol$, where tol is a prescribed tolerance. Combining the above techniques, we present the overall numerical algorithm as follows:

Algorithm 1 A decoupled solution algorithm based on an adaptive finite element method

Set initial values $\varphi_h^0 (= \varphi_h^{-1})$, $\mathbf{u}_h^0 (= \mathbf{u}_h^{-1})$, p_h^0, and $t = 0$.
Loop in time for $n = 0, \cdots$

1 Solve the Cahn–Hilliard system to update φ_h^{n+1} and μ_h^{n+1}.
2 Loop in AMR for $k = 0, \cdots$
   (a) Compute Θ_T for all $T \in T_h$, if $\max_{T\in T_h}\{\Theta_T\} < tol$, go to step 3.
   (b) Refine the elements repeatedly if $\varphi_h^{n+1}|_T \in [-0.9, 0.9]$ and $\max_{e\in\partial T}|e| > \bar{e}$.
   (c) Merge the adjacent elements if each of them yields $\Theta_T < \gamma_c \max_{T\in T_h}\{\Theta_T\}$.
   (d) Update φ_h^{n-1}, φ_h^n, $\mathbf{u}_h^{n-1}$, $\mathbf{u}_h^n$, and p_h^n on the new mesh.
   (e) Solve the Cahn–Hilliard system to update φ_h^{n+1} and μ_h^{n+1}.
3 Compute ρ_h^{n+1}, η_h^{n+1}, $l_{s\,h}^{\,n+1}$ using φ_h^{n+1}.
4 Solve the velocity system to update $\mathbf{u}_h^{n+1}$.
5 Solve the pressure system to update p_h^{n+1}.

End time loop

For the purpose of efficiency, we perform the AMR method every n_{skip} time steps, and terminate the AMR loop in step 2 when $\max_{T\in T_h}\{\Theta_T\}$ does not decrease any more. Because the matrices arising from the discretization of the Cahn–Hilliard equation and the pressure equation involve only constant coefficients, they need to be rebuilt only when refinement or coarsening occurs. The decoupled solution algorithm requires to solve three linear systems at each time step. We employ a restricted additive Schwarz preconditioned GMRES method to solve the Cahn–Hilliard system and the velocity system. For the pressure Poisson equation, we use an aggregation-based algebraic multigrid preconditioned GMRES method. As far as we know, no existing combination of the above algorithms has been presented for the concerned problem.

4 Numerical experiments

The proposed algorithm is implemented using libMesh [8] for the generation of finite element stiffness matrices, and PETSc [2] for the preconditioned Krylov subspace solvers. The overall algorithm is implemented on a parallel computer with distributed memory.

In this section, we present 2D numerical experiments for a droplet travelling in a scenario when two people begin to talk face to face at $t = 0$, and an airflow is expelled horizontally from one's mouth (Γ_i: $x = 10$, $y \in [62.536, 64]$). The airflow has a parabolic profile with initial speed $V = 5$ m/s. The computational domain is $[0, 35] \times [50, 80]$ and the unit is 2.5 cm, as shown in Fig. 1. A nonuniform triangular mesh is generated such that the mesh is finer between the two people. The initial mesh has 56,568 elements and 28,285 vertices. The densities for the droplet ($\varphi = -1$) and air ($\varphi = 1$) are 10^3 kg/m^3 and 1.2 kg/m^3, the viscosities for the droplet and air are 10^{-3} Pa·s and 1.8×10^{-5} Pa·s. The interfacial tension is 0.072 N/m. The gravity constant is 9.8 m/s^2. By taking V as the characteristic velocity and the opening size of mouth 2.5 cm as the characteristic length, we obtain the following dimensionless numbers: $\lambda_\rho = 830$, $\lambda_\eta = 55$, $Re = 8333.25$, $\mathcal{B} = 707.2$, and $Fr = 10.1$. The thickness of the interface is $\epsilon = 0.002$. The static contact angle is taken as 90°. Other physical parameters are taken as in [6]. For the numerical parameters, we choose $\delta = 0.05$, $\bar{e} = 0.002, \gamma_c = 0.01, tol = 0.01, n_{\text{skip}} = 15$, and the time step size $\Delta t = 5\times10^{-4}$. For the inflow boundary condition, we consider a droplet that is ejected from Γ_i along with the airflow at $t_s = 0.05$ ms, and its initial size is determined by the ejection time δt_e, that is, $\varphi = -1$ if $x = 10, y \in [63.238, 63.298]$ and $t \in [t_s, t_s + \delta t_e]$ whereas $\varphi = 1$ on Γ_i. We consider three cases of ejection time: a. $\delta t_e = 1$ ms, b. $\delta t_e = 0.5$ ms, and c. $\delta t_e = 0.1$ ms.

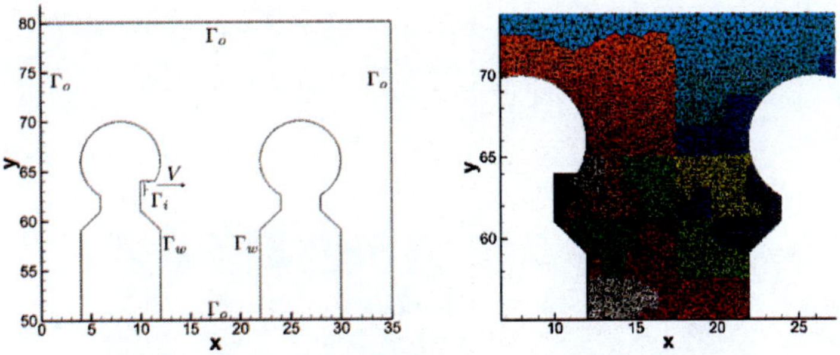

Fig. 1: (left) Computational domain and (right) a sample partition of the computational domain into 16 subdomains for the Schwarz preconditioning.

Fig. 2 shows the streamlines colored by velocity magnitude at 1.25 ms and 75 ms for case b. At the early stage, the parabolic velocity profile leads to a natural expansion of airflow. As the flow evolves, it is angled down due to the gravitational pull and generates two primary vortices, one on either side.

From Fig. 3 we see that the droplets travel ballistically subject to inertia and gravity. They overshoot the airflow stream and can reach the recipients' mucosa directly or settle on surfaces to be later picked up by the recipients. While all droplets evolve to a circle shape with the effect of surface tension, the large droplets undergo a more obvious topological change than the smaller droplets. The bottom row of the figure shows the effectiveness of the AMR method in tracking the moving

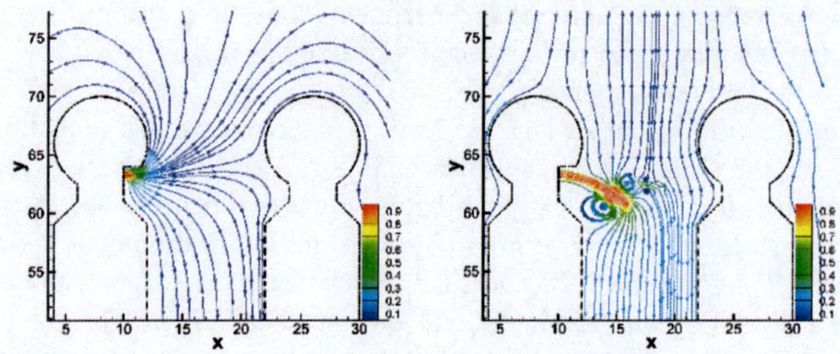

Fig. 2: Streamlines colored by velocity magnitude at (left) 1.25 ms and (right) 75 ms for case b.

interface. For these cases, usually 3 or 4 adaptive iterations are needed for each application of AMR.

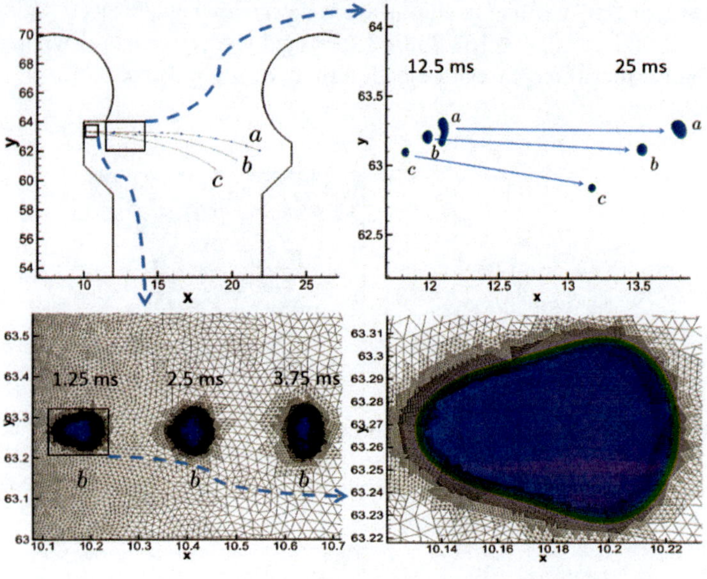

Fig. 3: (top left) Trajectory of the droplets, (top right) snapshot of droplets at 12.5 ms and 25 ms, (bottom left) adaptive mesh refinement for case b at 1.25 ms, 2.5 ms, and 3.75 ms, and (bottom right) enlarged view of the mesh at 1.25 ms for case b.

In the presented two-way coupling model, the airflow is affected by the motion of the droplet due to the viscosity contrast and surface tension, especially when the droplet is large. This is evidenced by the streamlines near the droplet a in Fig. 4 (left), one can observe a vortex street generated behind the droplet. In contrast, a smaller droplet does not influence the airflow much as shown in Fig. 4 (right).

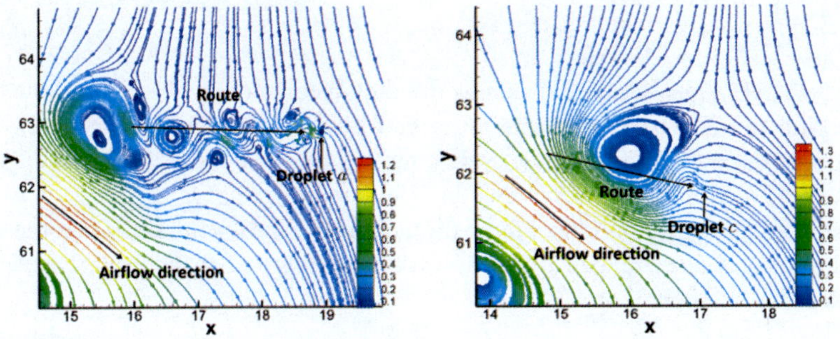

Fig. 4: Streamlines colored by velocity magnitude at 75 ms for (left) case a and (right) case c.

In Fig. 5, we show the time histories of the lift coefficient $C_l = \frac{2F_l}{\rho_1 U^2 A}$ and the drag coefficient $C_d = \frac{2F_d}{\rho_1 U^2 A}$ which evaluate the combined effect of surface tension and aerodynamic forces acting on the droplets. Here $\rho_1 = 1$, $U = 1$ are dimensionless constants. F_l, F_d, and A can be computed using the integral transformation with the surface delta function $d = \frac{1-\varphi}{2}$:

$$F_\alpha = -\frac{1}{Re}\int_\Omega \sigma \cdot \nabla d \cdot \mathbf{e}_\alpha \mathrm{d}\Omega, \quad \text{and} \quad A = -\int_\Omega \nabla d \cdot \mathbf{n}_\varphi \mathrm{d}\Omega,$$

where $\alpha = l, d$, $\mathbf{e}_d = \mathbf{i}$, $\mathbf{e}_l = \mathbf{j}$, and $\mathbf{n}_\varphi = \frac{\nabla\varphi}{|\nabla\varphi|}$. The forces exerted on the droplets exhibit a oscillatory nature similar to the case of flow around a stationary circular cylinder, but with more irregular patterns here because of the shape dynamics of the droplets and the instability of the high Reynolds flows. The magnitude of the oscillation generally decreases as the size of the droplet becomes smaller.

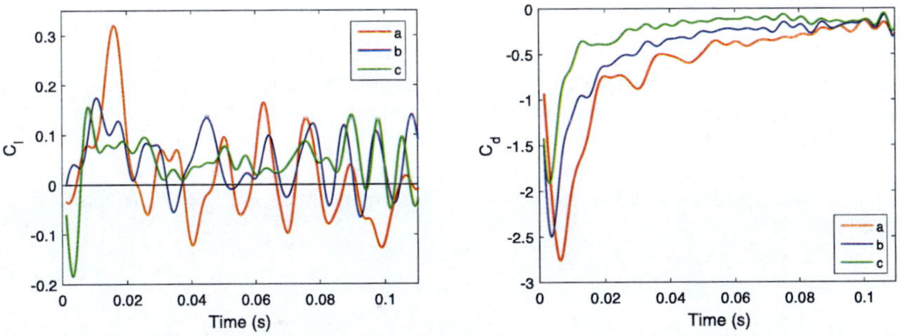

Fig. 5: Time histories of (left) lift coefficient C_l and (right) drag coefficient C_d.

5 Conclusions

We present a parallel adaptive finite element method for the modeling of a deformable droplet travelling in air. The problem is described by the Cahn–Hilliard–Navier–Stokes equations that account for the two-way coupling between the airflow and the droplet through advection and surface tension. The parallelization is realized via a Schwarz type overlapping domain decomposition method. Our results show that the size of the droplet has a significant impact on its travelling path, shape dynamics, and the ambient airflow behavior.

References

1. S. Balachandar, S. Zaleski, A. Soldati, G. Ahmadi and L. Bourouiba. Host-to-host airborne transmission as a multiphase flow problem for science-based social distance guidelines. *Int. J. Multiph. Flow* **132**, 103439 (2020).
2. S. Balay, S. Abhyankar, M.F. Adams, J. Brown, P. Brune and K. Buschelman et al. PETSc Users Manual. Technical Report ANL-95/11-Revision 3.15, Argonne National Laboratory (2021).
3. L. Bourouiba. The fluid dynamics of disease transmission. *Annu. Rev. Fluid Mech.* **53**, 473–508 (2021).
4. C.P. Cummins, O.J. Ajayi, F.V. Mehendale, R. Gabl and I.M. Viola. The dispersion of spherical droplets in source-sink flows and their relevance to the COVID-19 pandemic. *Phys. Fluids* **32**, 083302 (2020).
5. S. Dong. An outflow boundary condition and algorithm for incompressible two-phase flows with phase field approach. *J. Comput. Phys.* **266**, 47–73 (2014).
6. M. Gao and X.-P. Wang. An efficient scheme for a phase field model for the moving contact line problem with variable density and viscosity. *J. Comput. Phys.* **272**, 704–718 (2014).
7. D.W. Kelly, J.P. Gago, O.C. Zienkiewicz and I. Babuska. A posteriori error analysis and adaptive processes in the finite element method: part I error analysis. *Int. J. Numer. Methods Eng.* **19**, 1593–1619 (1983).
8. B.S. Kirk, J.W. Peterson, R.H. Stogner and G.F. Carey. libMesh: A C++ library for parallel adaptive mesh refinement/coarsening simulations. *Eng. Comput.* **22**, 237–254 (2006).
9. V. Vuorinena, M. Aarniob and M. Alavah et al. Modelling aerosol transport and virus exposure with numerical simulations in relation to SARS-CoV-2 transmission by inhalation indoors. *Saf. Sci.* **130**, 104866 (2020).
10. M.P. Wan, G.N.S. To, C.Y.H. Chao, L. Fang and A. Melikov. Modeling the fate of expiratory aerosols and the associated infection risk in an aircraft cabin environment. *Aerosol Sci. Technol.* **43**, 322–343 (2009).

On the Effect of Boundary Conditions on the Scalability of Schwarz Methods

Gabriele Ciaramella and Luca Mechelli

1 Introduction

This work is concerned with convergence and weak scalability[1] analysis of the one-level parallel Schwarz method (PSM) and optimized Schwarz method (OSM) for the solution of the problem

$$\begin{aligned} -\Delta u = f \text{ in } \Omega, \quad & u(a_1, y) = u(b_N, y) = 0 \;\; y \in (0,1), \\ \mathcal{B}_b(u)(x) = \mathcal{B}_t(u)(x) = 0 & \;\; x \in (a_1, b_N), \end{aligned} \tag{1}$$

where Ω is the domain depicted in Fig. 1, and $\mathcal{B}_b$ and $\mathcal{B}_t$ are Dirichlet, Neumann or Robin operators:

$$\begin{aligned} \text{Dirichlet: } & \mathcal{B}_b(u)(x) = u(0,x), & \mathcal{B}_t(u)(x) &= u(1,x), \\ \text{Neumann: } & \mathcal{B}_b(u)(x) = \partial_y u(0,x), & \mathcal{B}_t(u)(x) &= \partial_y u(1,x), \\ \text{Robin: } & \mathcal{B}_b(u)(x) = qu(0,x) - \partial_y u(0,x), & \mathcal{B}_t(u)(x) &= qu(1,x) + \partial_y u(1,x). \end{aligned}$$

Here, $q > 0$ and the subscripts 'b' and 't' stand for 'bottom' and 'top'. As shown in Fig. 1, the domain Ω is the union of subdomains Ω_j, $j = 1, \dots, N$, defined as $\Omega_j := (a_j, b_j) \times (0,1)$, where $a_1 = 0$, $a_j = L + a_{j-1}$ for $j = 2, \dots, N+1$ and $b_j = a_{j+1} + 2\delta$ for $j = 0, \dots, N$. Hence, the length of each subdomain is $L + 2\delta$ and the length of the overlap is 2δ with $\delta \in (0, L/2)$.

Gabriele Ciaramella
Politecnico di Milano e-mail: gabriele.ciaramella@polimi.it

Luca Mechelli
Universität Konstanz e-mail: luca.mechelli@uni-konstanz.de

[1] Here, weak scalability is understood in the sense that the contraction factor does not deteriorate as the number N of subdomains increases and, hence, the number of iterations, needed to reach a given tolerance, is uniformly bounded in N; see, e.g., [3].

S. C. Brenner et al. (eds.), *Domain Decomposition Methods in Science and Engineering XXVI*, Lecture Notes in Computational Science and Engineering 145,
https://doi.org/10.1007/978-3-030-95025-5_17

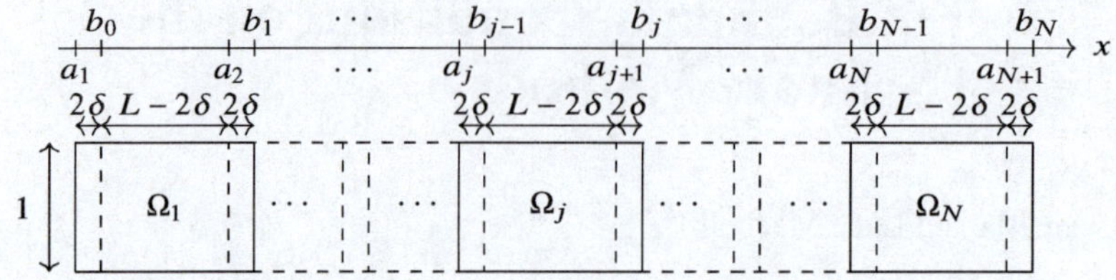

Fig. 1: Two-dimensional chain of N rectangular fixed-sized subdomains.

It is well known that one-level Schwarz methods are not weakly scalable, if the number of subdomains increases and the whole domain Ω is fixed. However, the recent work [2], published in the field of implicit solvation models used in computational chemistry, has drawn attention to the opposite case in which the number of subdomains increases, but their size remains unchanged, and, as a result, the size of the whole domain Ω increases. In this setting, weak scalability of the PSM and OSM for (1) with Dirichlet boundary conditions is studied in [3, 4]. Scalability results for the PSM in the case of more general geometries of the (sub)domains are presented in [5–7]. In these works, only external Dirichlet conditions are discussed and, in such a case, weak scalability is shown; see also [11] for a scalability analysis of the classical (alternating) Schwarz method. A short remark about the non-scalability in case of external Neumann conditions is given in [3]. Similar results have been recently presented in [1] for time-harmonic problems. Moreover, very similar results to the ones of [3] are obtained a few years later in [9]. The goal of this work is to study the effect of different (possibly mixed) external boundary conditions on convergence and scalability of the PSM and OSM. In particular, we will show that only in the case of (both) external Neumann conditions at the top and the bottom of Ω, the PSM and OSM are not scalable. External Dirichlet conditions lead to the fastest convergence, while external Robin conditions lead to a convergence that depends heavily on the parameter q.

The one-level PSM and OSM for the solution of (1) are

$$\begin{aligned} -\Delta u_j^n &= f_j \text{ in } \Omega_j, \\ \mathcal{B}_b(u_j^n)(x) &= \mathcal{B}_t(u_j^n)(x) = 0 \;\; x \in (a_1, b_N), \\ \mathcal{T}_\ell(u_j^n)(a_j) &= \mathcal{T}_\ell(u_{j-1}^{n-1})(a_j), \quad \mathcal{T}_r(u_j^n)(b_j) = \mathcal{T}_r(u_{j+1}^{n-1})(b_j), \end{aligned} \tag{2}$$

for $j = 1, \dots, N$, where $\mathcal{T}_\ell$ and $\mathcal{T}_r$ are Dirichlet trace operators,

$$\mathcal{T}_\ell(u_j^n)(a_j) = u_j^n(a_j, y) \text{ and } \mathcal{T}_r(u_j^n)(b_j) = u_j^n(b_j, y), \tag{3}$$

for the PSM, and Robin trace operators,

$$\mathcal{T}_\ell(u_j^n)(a_j) = pu_j^n(a_j,y) - \partial_x u_j^n(a_j,y) \text{ and } \mathcal{T}_r(u_j^n)(b_j) = pu_j^n(b_j,y) + \partial_x u_j^n(b_j,y), \tag{4}$$

with $p > 0$ for the OSM. The subscript 'ℓ' and 'r' stand for 'left' and 'right'. For $j = 1$ the condition at a_1 must be replaced by $u_1^n(a_1, y) = 0$ and for $j = N$ the condition at b_N must be replaced by $u_N^n(b_N, y) = 0$. In this paper, 'external conditions' and 'transmission conditions' will always refer to the conditions obtained by the pairs $(\mathcal{B}_b, \mathcal{B}_u)$ and $(\mathcal{T}_\ell, \mathcal{T}_r)$, respectively. Note that the Robin parameter p of the OSM can be chosen independently of the Robin parameter q used for the operators $\mathcal{B}_b$ and $\mathcal{B}_t$. We analyze convergence of the PSM and OSM by a Fourier analysis in Section 3. For this purpose, we use the solutions of eigenproblems of the 1D Laplace operators with mixed boundary conditions. These are studied in Section 2. Finally, results of numerical experiments are presented in Section 4.

2 Laplace eigenpairs for mixed external conditions

Consider the 1D eigenvalue problem

$$\varphi''(y) = -\lambda\varphi(y), \text{ for } y \in (0, 1), \quad \mathcal{B}_b(\varphi)(0) = \mathcal{B}_t(\varphi)(1) = 0, \tag{5}$$

and six pairs of boundary operators $(\mathcal{B}_b, \mathcal{B}_t)$:

(DD) $\mathcal{B}_b(\varphi)(0) = \varphi(0)$, $\quad \mathcal{B}_t(\varphi)(1) = \varphi(1)$,
(DR) $\mathcal{B}_b(\varphi)(0) = \varphi(0)$, $\quad \mathcal{B}_t(\varphi)(1) = q\varphi(1) + \varphi'(1)$,
(DN) $\mathcal{B}_b(\varphi)(0) = \varphi(0)$, $\quad \mathcal{B}_t(\varphi)(1) = \varphi'(1)$,
(RR) $\mathcal{B}_b(\varphi)(0) = q\varphi(0) - \varphi'(0)$, $\quad \mathcal{B}_t(\varphi)(1) = q\varphi(1) + \varphi'(1)$,
(NR) $\mathcal{B}_b(\varphi)(0) = \varphi'(0)$, $\quad \mathcal{B}_t(\varphi)(1) = q\varphi(1) + \varphi'(1)$,
(NN) $\mathcal{B}_b(\varphi)(0) = \varphi'(0)$, $\quad \mathcal{B}_t(\varphi)(1) = \varphi'(1)$,

where $q > 0$ and 'D', 'R' and 'N' stand for 'Dirichlet', 'Robin' and 'Neumann'. For all these six cases the eigenvalue problem (5) is solved by orthonormal (in $L^2(0, 1)$) Fourier basis functions.

Theorem 1 (Eigenpairs of the Laplace operator) *Let $q > 0$. The eigenproblems (5) with the above external conditions are solved by the non-trivial eigenpairs (φ_k, λ_k) given by*

(DD) $\varphi_k(y) = \sqrt{2}\sin(\pi k y)$, $\lambda_k = \pi^2 k^2$, $k = 1, 2, \ldots$

(DR) $\varphi_k(y) = \sqrt{\frac{4\mu_k}{2\mu_k - \sin(2\mu_k)}}\sin(\mu_k y)$, $\lambda_k = \mu_k^2$, $k = 1, 2, \ldots$, *where* $\mu_k \in (k\pi - \pi/2, k\pi)$, $k = 1, 2, \ldots$, *are roots of* $\widehat{d}(x) := q\sin(x) + x\cos(x)$. *Moreover,* $\lim_{q\to 0}\mu_1(q) = \pi/2$ *and* $\lim_{q\to\infty}\mu_1(q) = \pi$.

(DN) $\varphi_k(y) = \sqrt{2}\sin(\frac{2k+1}{2}\pi y)$, $\lambda_k = \frac{(2k+1)^2}{4}\pi^2$, $k = 0, 1, 2, \ldots$

(RR) $\varphi_k(y) = \sqrt{\frac{4\tau_k}{(\tau_k^2 - q^2)\sin(2\tau_k) + 4q\tau_k\sin(\tau_k)^2 + 2\tau_k^3 + 2q^2\tau_k}}\left(q\sin(\tau_k y) + \tau_k\cos(\tau_k y)\right)$, $\lambda_k = \tau_k^2$, $k = 1, 2, \ldots$, *where* $\tau_k \in (0, \pi)$, $k = 1, 2, \ldots$, *are roots of* $\widetilde{d}(x) := 2qx\cos(x) + (q^2 - x^2)\sin(x)$. *Moreover,* $\lim_{q\to 0}\tau_1(q) = 0$ *and* $\lim_{q\to\infty}\tau_1(q) = \pi$.

(NR) $\varphi_k(y) = \sqrt{\frac{4\nu_k}{2\nu_k+\sin(2\nu_k)}}\cos(\nu_k y)$, $\lambda_k = \nu_k^2$, $k = 1, 2, \ldots$, *where* $\nu_k \in ((k-1)\pi, (k-\frac{1}{2})\pi)$, $k = 1, 2, \ldots$, *are roots of* $d(x) := x\sin(x) - q\cos(x)$. *Moreover,* $\lim_{q\to 0}\nu_1(q) = 0$ *and* $\lim_{q\to\infty}\nu_1(q) = \pi/2$.

(NN) $\varphi_k(y) = \sqrt{2}\cos(\pi k y)$, $\lambda_k = \pi^2 k^2$, $k = 0, 1, 2, \ldots$

Proof If we multiply (5) by φ, integrate over $[0, 1]$, and integrate by parts, we get $\lambda \int_0^1 |\varphi(y)|^2 \mathrm{d}y = \int_0^1 |\varphi'(y)|^2 \mathrm{d}y - \varphi'(1)\varphi(1) + \varphi'(0)\varphi(0)$. Using any of the above external conditions (and that $q > 0$, for the Robin conditions) one gets $\lambda \geq 0$. We refer to, e.g., [10, Section 4.1] for similar discussions. Now, all the cases can be proved by using the ansatz $\varphi(y) = A\cos(\sqrt{\lambda}y) + B\sin(\sqrt{\lambda}y)$, which clearly satisfies (5), and computing, e.g., A and λ in such a way that $\varphi(y)$ satisfies the two external conditions and B as a normalization factor. □

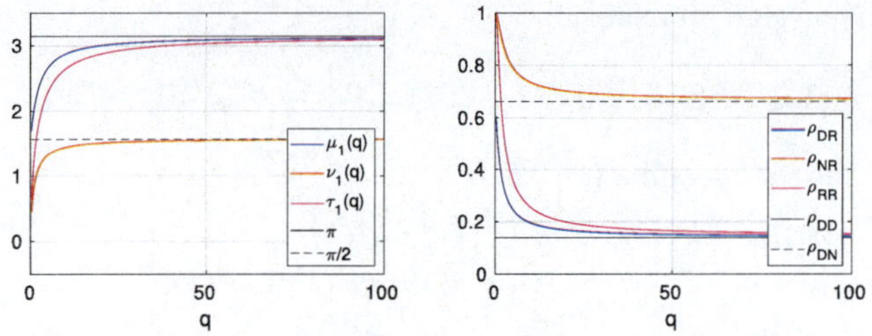

Fig. 2: Left: Maps $q \mapsto \mu_1(q)$, $q \mapsto \nu_1(q)$ and $q \mapsto \tau_1(q)$. Right: ρ_{DR}, ρ_{NR}, ρ_{DD} and ρ_{DN} as functions of q and for $\delta = 0.1$ and $L = 1.0$.

The coefficients ν_1, μ_1 and τ_1 as functions of q are shown in Fig. 2 (left), where we can observe that $\nu_1(q) < \frac{\pi}{2} < \mu_1(q) < \pi$ and $0 < \tau_1(q) < \pi$, and that the maps $q \mapsto \nu_1(q)$, $q \mapsto \mu_1(q)$ and $q \mapsto \tau_1(q)$ increase monotonically and approach, respectively, $\frac{\pi}{2}$ and π as $q \to \infty$. Hence, by taking the limit $q \to 0$, one can pass from the conditions (DR), (RR) and (NR) to (DN), (NN) and (NN), respectively. Similarly, by taking the limit $q \to \infty$, the conditions (DR), (RR) and (NR) become (DD), (DD) and (DN), respectively.

3 Convergence and scalability

Consider the Schwarz method (2) and any pair $(\mathcal{B}_b, \mathcal{B}_t)$ of operators as in Section 2. The Fourier expansions of $u_j^n(x, y)$, $j = 1, \ldots, N$, are

$$u_j^n(x, y) = \sum_k \widehat{u}_j^n(x, \lambda_k)\varphi_k(y), \tag{6}$$

where the sum is over $k = 1, 2, \dots$ for (DD), (DR), (RR) and (NR), and over $k = 0, 1, 2, \dots$ for (DN) and (NN). The functions φ_k depend on the external boundary conditions and are the ones obtained in Theorem 1. The Fourier coefficients $\widehat{u}_j^n(x, \lambda_k)$ satisfy[2]

$$\begin{aligned} -\partial_{xx}\widehat{u}_j^n(x,\lambda_k) + \lambda_k \widehat{u}_j^n(x,\lambda_k) &= \widehat{f}_j(x,\lambda_k) \text{ in } (a_j, b_j), \\ \mathcal{T}_\ell(\widehat{u}_j^n(\cdot,\lambda_k))(a_j) &= \mathcal{T}_\ell(\widehat{u}_{j-1}^{n-1}(\cdot,\lambda_k))(a_j), \\ \mathcal{T}_r(\widehat{u}_j^n(\cdot,\lambda_k))(b_j) &= \mathcal{T}_r(\widehat{u}_{j+1}^{n-1}(\cdot,\lambda_k))(b_j), \end{aligned} \tag{7}$$

for $j = 1, \dots, N$. For $j = 1$, the condition at a_1 must be replaced by $u_1^n(a_1) = 0$ and for $j = N$ the condition at b_N must be replaced by $u_N^n(b_N) = 0$. If the operators $\mathcal{T}_\ell$ and $\mathcal{T}_r$ correspond to Dirichlet conditions (see (3)), then (7) is a PSM. If they correspond to Robin conditions (see (4)), then (7) is an OSM. The convergence of the iteration (7) is analyzed in Theorem 2.

Theorem 2 (Convergence of Schwarz methods in Fourier space) *The contraction factors of the Schwarz methods*[3] *(7) are bounded by*

$$\rho(\lambda_k, \delta) = \frac{e^{2\lambda_k \delta} + e^{\lambda_k L}}{e^{2\lambda_k \delta + \lambda_k L} + 1}. \tag{8}$$

Moreover, it holds that $\rho(\lambda_k, \delta) \in [0, 1]$ with $\rho(0, \delta) = 1$ (independently of N), and that $\lambda \mapsto \rho(\lambda, \delta)$ is strictly monotonically decreasing.

Proof The Dirichlet case follows from [4, Lemma 2 and Theorem 3]. See also [3, Lemma 2 and Theorem 1]. We focus here on the Robin case. From Theorem 3 in [3] and the corresponding proof we have that the contraction factor of the OSM is bounded by $\max\{\varphi(\lambda, \delta, p), |\zeta(\lambda, \delta, p)|\}$ where

$$\varphi(\lambda,\delta,p) := \frac{(\lambda+p)^2 e^{2\delta\lambda} - (\lambda-p)^2 e^{-2\delta\lambda} + (\lambda+p)|\lambda-p|(e^{\lambda L} - e^{-\lambda L})}{(\lambda+p)^2 e^{\lambda L + 2\lambda\delta} - (\lambda-p)^2 e^{-\lambda L - 2\lambda\delta}} \geq 0,$$

$$\zeta(\lambda,\delta,p) := \frac{(\lambda+p)e^{-\lambda L} + (\lambda-p)e^{\lambda L}}{(\lambda+p)e^{\lambda(L+2\delta)} + (\lambda-p)e^{-\lambda(L+2\delta)}},$$

with $\varphi(\lambda,\delta,p) \leq \varphi(\lambda,\delta,0) = \lim_{\widetilde{p}\to\infty} \varphi(\lambda,\delta,\widetilde{p}) = \frac{e^{2\delta\lambda} - e^{-2\delta\lambda} + e^{\lambda L} - e^{-\lambda L}}{e^{\lambda L + 2\delta\lambda} - e^{-\lambda L - 2\delta\lambda}}$ for all $\lambda \geq 0$ and $\delta > 0$. If we compute the derivative of $\lambda \mapsto \varphi(\lambda, \delta, 0)$ we get

$$\partial_\lambda \varphi(\lambda,\delta,0) = -\frac{L(e^{4\delta\lambda + L\lambda} - e^{L\lambda}) + 2\delta(e^{2\delta\lambda + 2L\lambda} - e^{2\delta\lambda})}{(e^{2\delta\lambda + L\lambda} + 1)^2},$$

which is negative for any $\lambda \geq 0$ and $\delta > 0$. Thus, $\lambda \mapsto \varphi(\lambda, \delta, 0)$ is strictly monotonically decreasing. Let us now study the function $\zeta(\lambda, \delta, p)$. Direct calculations

[2] Notice that the procedure to obtain (7) is standard. We refer to, e.g., [10] for more details and examples.

[3] The contraction factor for (7) (corresponding to the k-th Fourier component) is the spectral radius of the Schwarz iteration matrix; see [3, 4].

reveal that $\partial_p \zeta(\lambda, \delta, p) = -\frac{2\lambda e^{2\delta\lambda}(e^{4\lambda(\delta+L)}-1)}{((\lambda+p)e^{4\delta\lambda+2L\lambda}+\lambda-p)^2}$, which is negative for any $\lambda \geq 0$ and $\delta > 0$, and $\zeta(\lambda, \delta, 0) = \frac{(e^{2L\lambda}+1)e^{2\delta\lambda}}{e^{4\delta\lambda+2L\lambda}+1} > 0$ and $\lim_{p\to\infty} \zeta(\lambda, \delta, p) = -\frac{(e^{2L\lambda}-1)e^{2\delta\lambda}}{e^{4\delta\lambda+2L\lambda}-1} < 0$ for any $\lambda \geq 0$ and $\delta > 0$. These observations imply that $p \mapsto \zeta(\lambda, \delta, p)$ is strictly monotonically decreasing and attains its maximum at $p = 0$. Finally, a direct comparison shows that $\varphi(\lambda, \delta, 0) \geq \zeta(\lambda, \delta, 0) \geq \lim_{p\to\infty} |\zeta(\lambda, \delta, p)|$ and the result follows, because $\varphi(\lambda, \delta, 0) = \frac{e^{2\delta\lambda}-e^{-2\delta\lambda}+e^{\lambda L}-e^{-\lambda L}}{e^{\lambda L+2\delta\lambda}-e^{-\lambda L-2\delta\lambda}} = \frac{e^{2\lambda\delta}+e^{\lambda L}}{e^{2\lambda\delta+\lambda L}+1}$. □

Theorem 2 gives the same bound (8) for the convergence factors of the PSM and OSM. This fact is not surprising. First, it is well known that the OSM converges faster than the PSM for $\delta > 0$. Hence, a convergence bound for the PSM is a valid bound also for the OSM. Second, in the above proof the convergence bound for the OSM is obtained for $p \to \infty$, which corresponds to passing from Robin transmission conditions to Dirichlet transmission conditions. The bound (8) is based on the ones obtained in [3, 4]. These are quite sharp for large values of N; see, e.g., [3, Fig. 4 and Fig. 5].

We can now prove our main convergence result, which allows us to study convergence and scalability of the PSM and OSM for all the external conditions considered in Section 2.

Theorem 3 (Convergence of PSM and OSM) *The contraction factors (in the L^2 norm) of the PSM and OSM for the solution to* (1) *are bounded by*

$$\begin{aligned} &\text{(DD) } \rho_{\text{DD}}(\delta) := \rho(\pi^2, \delta), && \text{(DR) } \rho_{\text{DR}}(\delta, q) := \rho(\mu_1(q)^2, \delta), \\ &\text{(DN) } \rho_{\text{DN}}(\delta) := \rho(\pi^2/4, \delta), && \text{(RR) } \rho_{\text{RR}}(\delta, q) := \rho(\tau_1(q)^2, \delta), \\ &\text{(NR) } \rho_{\text{NR}}(\delta, q) := \rho(\nu_1(q)^2, \delta), && \text{(NN) } \rho_{\text{NN}}(\delta) := \rho(0, \delta) = 1, \end{aligned}$$

where $q \in (0, \infty)$ and $\rho(\lambda, \delta)$ is defined in Theorem 2. Moreover, for any $\delta > 0$ we have that

$$\rho_{\text{DD}}(\delta) < \rho_{\text{DR}}(\delta, q) < \rho_{\text{DN}}(\delta) < \rho_{\text{NR}}(\delta, q) < \rho_{\text{NN}}(\delta) = 1, \tag{9}$$

$$\rho_{\text{DD}}(\delta) < \rho_{\text{RR}}(\delta, q) < \rho_{\text{NN}}(\delta) = 1. \tag{10}$$

Proof According to Theorem 2, the bounds of the Fourier contraction factor $\rho(\lambda, \delta)$ is monotonically decreasing in λ. Therefore, an upper bound for the convergence factor of the PSM and OSM (in the L^2 norm) can be obtained by taking the maximum over the admissible Fourier frequencies λ_k and invoking Parseval's identity (see, e.g., [4]). Recalling Theorem 1, these maxima are attained at $\lambda_1 = \pi^2$ for (DD), $\lambda_1 = \mu_1^2$ for (DR), $\lambda_0 = \pi^2/4$ for (DN), $\lambda_1 = \tau_1^2$ for (RR), $\lambda_1 = \nu_1^2$ for (NR), and $\lambda_0 = 0$ for (NN). The inequalities (9) and (10) follow from the monotonicity $\lambda \mapsto \rho(\lambda, \delta)$ and the fact that $\nu_1(q) < \frac{\pi}{2} < \mu_1(q) < \pi$ and $\tau_1(q) \in (0, \pi)$. □

bottom \ top	Dirichlet	Robin	Neumann
Dirichlet	yes	yes	yes
Robin	yes	yes	yes
Neumann	yes	yes	no

bottom \ top	Dirichlet	Robin	Neumann
Dirichlet	-	yes	-
Robin	yes	no	no
Neumann	-	no	-

Table 1: Left: Scalability of the PSM and OSM for different external conditions (for a fixed and finite $q > 0$) applied at the top and at the bottom of the domain. Right: Robustness of the PSM and OSM with respect to $q \in [0, \infty]$.

The inequalities (9) and (10) imply that the contraction factor is bounded, independently of N, by a constant strictly smaller than 1 for all the cases except (NN). In the case (NN), the first Fourier frequency is $\lambda_0 = 0$. Hence, the coefficients $\widehat{u}_j^n(x, \lambda_0)$ are generated by the 1D Schwarz method

$$
\begin{aligned}
-\partial_{xx}\widehat{u}_j^n(x,\lambda_0) &= \widehat{f}_j(x,\lambda_0) \text{ in } (a_j, b_j),\\
\mathcal{T}_\ell(\widehat{u}_j^n(\cdot,\lambda_0))(a_j) &= \mathcal{T}_\ell(\widehat{u}_{j-1}^{n-1}(\cdot,\lambda_0))(a_j),\\
\mathcal{T}_r(\widehat{u}_j^n(\cdot,\lambda_0))(b_j) &= \mathcal{T}_r(\widehat{u}_{j+1}^{n-1}(\cdot,\lambda_0))(b_j),
\end{aligned}
\tag{11}
$$

which is known to be not scalable; see, e.g., [3, 8]. The scalability of the PSM and OSM for different external conditions applied at the top and at the bottom of the domain is summarized in Table 1. Inequalities (9) and (10) lead to another interesting observation. The contraction factors are clearly influenced by the external boundary conditions. Dirichlet conditions lead to faster convergence than Robin conditions, which in turn lead to faster convergence than Neumann conditions. For example, if one external condition is of the Dirichlet type, then the PSM and OSM converge faster if the other condition is of the Dirichlet type, slower if it is of Robin type and even slower if it is of Neumann type. The case (RR) is slightly different, because the corresponding convergence of the PSM and OSM depends heavily on the Robin parameter q. The behavior of the bounds $\rho_{\text{RR}}(\delta, q)$, $\rho_{\text{DR}}(\delta, q)$ and $\rho_{\text{NR}}(\delta, q)$ with respect to q is depicted in Fig. 2 (right), which shows the bounds discussed in Theorem 3 as functions of q (recall that $\rho_{\text{NN}} = 1$). Here, we can observe that the inequalities (9) and (10) are satisfied and that

- As q increases the Dirichlet part of the Robin external condition dominates. In addition, the bounds ρ_{RR} and ρ_{DR} decrease and approach ρ_{DD} as $q \to \infty$. Similarly, ρ_{NR} decreases and approaches ρ_{DN}.
- As q decreases the Neumann part of the Robin external condition dominates. In addition, the bounds ρ_{NR} and ρ_{RR} decrease and approach $\rho_{\text{NN}} = 1$ as $q \to 0$. Similarly, ρ_{DR} increases and approaches ρ_{DN}.

These observations lead to Tab. 1 (right), where we summarize the robustness of the PSM and OSM with respect to the parameter q. The methods are robust with respect to q only if one of the two external boundary conditions is of Dirichlet type. This is due to the fact that Robin conditions become Neumann conditions for $q \to 0$.

N	DD	DR(10)	DN	RR(10)	NR(10)	NN	RR(0.1)
3	12 - 9	13 - 10	27 - 19	14 - 10	26 - 19	77 - 54	65 - 45
4	13 - 9	14 - 10	29 - 21	15 - 11	29 - 21	130 - 90	95 - 66
5	13 - 9	14 - 10	31 - 22	15 - 11	31 - 22	194 - 134	124 - 86
10	13 - 10	14 - 10	33 - 24	15 - 11	34 - 24	>401 - >401	227 - 155
30	13 - 10	14 - 10	34 - 24	15 - 11	35 - 24	>401 - >401	311 - 210
50	13 - 10	14 - 10	34 - 24	15 - 11	35 - 24	>401 - >401	319 - 216

Table 2: Number of iterations of the PSM (left) and OSM (right) needed to reduce the norm of the error below a tolerance of 10^{-6} for increasing number N of fixed-sized subdomains. The maximum number of allowed iterations is 401. This limit is only reached in the (NN) case, for which the PSM and OSM are not scalable.

4 Numerical experiments

In this section, we test the scalability of the PSM and OSM by numerical simulations. For this purpose, we run the PSM and OSM for all the external boundary conditions discussed in this paper and measure the number of iterations required to reach a tolerance on the error of 10^{-6}. To guarantee that the initial errors contain all frequencies, the methods are initialized with random initial guesses. In all cases, each subdomain is discretized with a uniform grid of size 90 interior points in direction x and 50 interior points in direction y. The mesh size is $h = \frac{L}{51}$, with $L = 1$, and the overlap parameter is $\delta = 10h$. For the OSM the Robin parameter is $p = 10$. The Robin parameter q of the external Robin conditions is $q = 10$, and the (RR) case is also tested with $q = 0.1$. The results of our experiments are shown in Tab. 2 and confirm the theoretical results discussed in the previous sections.

Acknowledgements G. Ciaramella is a member of INdAM GNCS.

References

1. N. Bootland, V. Dolean, A. Kyriakis and J. Pestana. Analysis of parallel Schwarz algorithms for time-harmonic problems using block Toeplitz matrices. *Electron. Trans. Numer. Anal.* **55**, 112–141 (2022).
2. E. Cancès, Y. Maday and B. Stamm. Domain decomposition for implicit solvation models. *J. Chem. Phys.* **139**, 054111 (2013).
3. F. Chaouqui, G. Ciaramella, M.J. Gander and T. Vanzan. On the scalability of classical one-level domain-decomposition methods. *Vietnam J. Math.* **46**(4), 1053–1088 (2018).
4. G. Ciaramella and M.J. Gander. Analysis of the parallel Schwarz method for growing chains of fixed-sized subdomains: Part I. *SIAM J. Numer. Anal.* **55**(3), 1330–1356 (2017).
5. G. Ciaramella and M.J. Gander. Analysis of the parallel Schwarz method for growing chains of fixed-sized subdomains: Part II. *SIAM J. Numer. Anal.* **56**(3), 1498–1524 (2018).
6. G. Ciaramella and M.J. Gander. Analysis of the parallel Schwarz method for growing chains of fixed-sized subdomains: Part III. *Electron. Trans. Numer. Anal.* **49**, 210–243 (2018).

7. G. Ciaramella, M. Hassan and B. Stamm. On the scalability of the parallel Schwarz method in one-dimension. In: *Domain Decomposition Methods in Science and Engineering XXV*, pp. 151–158, Springer International Publishing (2020).
8. G. Ciaramella, M. Hassan and B. Stamm. On the scalability of the Schwarz method. *SMAIJCM* **6**, 33–68 (2020).
9. M. El Haddad, J.C. Garay, F. Magoulès and D.B. Szyld. Synchronous and asynchronous optimized Schwarz methods for one-way subdivision of bounded domains. *Numer. Linear Algebra Appl.* **27**(2), e2279 (2020).
10. P.J. Olver. *Introduction to Partial Differential Equations*. Undergraduate Texts in Mathematics, Springer International Publishing (2013).
11. A. Reusken and B. Stamm. Analysis of the Schwarz domain decomposition method for the conductor-like screening continuum model. *SIAM J. Num. Anal.* **59**(2), 769–796 (2021).

On the Asymptotic Optimality of Spectral Coarse Spaces

Gabriele Ciaramella and Tommaso Vanzan

1 Introduction

The goal of this work is to study the asymptotic optimality of spectral coarse spaces for two-level iterative methods. In particular, we consider a linear system $A\mathbf{u} = \mathbf{f}$, where $A \in \mathbb{R}^{n\times n}$ and $\mathbf{f} \in \mathbb{R}^n$, and a two-level method that, given an iterate $\mathbf{u}^k$, computes the new vector $\mathbf{u}^{k+1}$ as

$$\mathbf{u}^{k+1/2} = G\mathbf{u}^k + M^{-1}\mathbf{f}, \qquad \text{(smoothing step)} \qquad (1)$$

$$\mathbf{u}^{k+1} = \mathbf{u}^{k+1/2} + PA_c^{-1}R(\mathbf{f} - A\mathbf{u}^{k+1/2}). \qquad \text{(coarse correction)} \qquad (2)$$

The smoothing step (1) is based on the splitting $A = M - N$, where M is the preconditioner, and $G = M^{-1}N$ the iteration matrix. The correction step (2) is characterized by prolongation and restriction matrices $P \in \mathbb{R}^{n\times m}$ and $R = P^\top$, and a coarse matrix $A_c = RAP$. The columns of P are linearly independent vectors spanning the coarse space $V_c := \text{span}\{\mathbf{p}_1, \dots, \mathbf{p}_m\}$. The convergence of the one-level iteration (1) is characterized by the eigenvalues of G, λ_j, $j = 1, \dots, n$ (sorted in descending order by magnitude). The convergence of the two-level iteration (1)–(2) depends on the spectrum of the iteration matrix T, obtained by substituting (1) into (2) and rearranging terms:

$$T = [I - P(RAP)^{-1}RA]G. \qquad (3)$$

The goal of this short paper is to answer, though partially, the fundamental question: **given an integer m, what is the coarse space of dimension m which minimizes the spectral radius $\rho(T)$?** Since step (2) aims at correcting the error components

Gabriele Ciaramella
Politecnico di Milano, e-mail: gabriele.ciaramella@polimi.it

Tommaso Vanzan
CSQI Chair, EPFL Lausanne, e-mail: tommaso.vanzan@epfl.ch

S. C. Brenner et al. (eds.), *Domain Decomposition Methods in Science and Engineering XXVI*, Lecture Notes in Computational Science and Engineering 145,
https://doi.org/10.1007/978-3-030-95025-5_18

that the smoothing step (1) is not able to reduce (or eliminate), it is intuitive to think that an optimal coarse space V_c is obtained by defining $\mathbf{p}_j$ as the eigenvectors of G corresponding to the m largest (in modulus) eigenvalues. We call such a V_c a *spectral coarse space*. Following the idea of correcting the 'badly converging' modes of G, several papers proposed new, and in some sense optimal, coarse spaces. In the context of domain decomposition methods, we refer, e.g., to [1–3], where efficient coarse spaces have been designed for parallel, restricted additive and additive Schwarz methods. In the context of multigrid methods, it is worth mentioning the work [5], where the interpolation weights are optimized using an approach based on deep-neural networks. Fundamental results are presented in [7]: for a symmetric A, it is proved that the coarse space of size m that minimizes the energy norm of T, namely $\|T\|_A$, is the span of the m eigenvectors of $\overline{M}A$ corresponding to the m lowest eigenvalues. Here, $\overline{M} := M^{-1}+M^{-\top}-M^{-\top}AM^{-1}$ is symmetric and assumed positive definite. If M is symmetric, a direct calculation gives $\overline{M}A = 2M^{-1}A - (M^{-1}A)^2$. Using that $M^{-1}A = I - G$, one can show that the m eigenvectors associated to the lowest m eigenvalues of $\overline{M}A$ correspond to the m largest modes of G. Hence, the optimal coarse space proposed in [7] is a spectral coarse space. The sharp result of [7] provides a concrete optimal choice of V_c minimizing $\|T\|_A$. This is generally an upper bound for the asymptotic convergence factor $\rho(T)$. As we will see in Section 2, choosing the spectral coarse space, one gets $\rho(T) = |\lambda_{m+1}|$. The goal of this work is to show that this is not necessarily the optimal asymptotic convergence factor. In Section 2, we perform a detailed optimality analysis for the case $m = 1$. The asymptotic optimality of coarse spaces for $m \geq 1$ is studied numerically in Section 3. Interestingly, we will see that by optimizing $\rho(T)$ one constructs coarse spaces that lead to preconditioned matrices with better condition numbers.

2 A perturbation approach

Let G be diagonalizable with eigenpairs $(\lambda_j, \mathbf{v}_j)$, $j = 1, \dots, n$. Suppose that $\mathbf{v}_j$ are also eigenvectors of A: $A\mathbf{v}_j = \widetilde{\lambda}_j \mathbf{v}_j$. Concrete examples where these hypotheses are fulfilled are given in Section 3. Assume that $\operatorname{rank} P = m$ ($\dim V_c = m$). For any eigenvector $\mathbf{v}_j$, we can write the vector $T\mathbf{v}_j$ as

$$T\mathbf{v}_j = \sum_{\ell=1}^{n} \widetilde{t}_{j,\ell} \mathbf{v}_\ell, \; j = 1, \dots, n. \tag{4}$$

If we denote by $\widetilde{T} \in \mathbb{R}^{n \times n}$ the matrix of entries $\widetilde{t}_{j,\ell}$, and define $V := [\mathbf{v}_1, \dots, \mathbf{v}_n]$, then (4) becomes $TV = V\widetilde{T}^\top$. Since G is diagonalizable, V is invertible, and thus T and $\widetilde{T}^\top$ are similar. Hence, T and $\widetilde{T}$ have the same spectrum. We can now prove the following lemma.

Lemma 1 (Characterization of $\widetilde{T}$) *Given an index $\widetilde{m} \geq m$, assume that $V_c :=$ span $\{\mathbf{p}_1, \ldots, \mathbf{p}_m\}$ satisfies*

$$V_c \subseteq \text{span}\,\{\mathbf{v}_j\}_{j=1}^{\widetilde{m}} \text{ and } V_c \cap \{\mathbf{v}_j\}_{j=\widetilde{m}+1}^{n} = \{0\}. \tag{5}$$

Then, it holds that

$$\widetilde{T} = \begin{bmatrix} \widetilde{T}_{\widetilde{m}} & 0 \\ X & \Lambda_{\widetilde{m}} \end{bmatrix}, \qquad \begin{array}{l} \Lambda_{\widetilde{m}} = \text{diag}\,(\lambda_{\widetilde{m}+1}, \ldots, \lambda_n), \\ \widetilde{T}_{\widetilde{m}} \in \mathbb{R}^{\widetilde{m}\times\widetilde{m}}, X \in \mathbb{R}^{(n-\widetilde{m})\times\widetilde{m}}. \end{array} \tag{6}$$

Proof The hypothesis (5) guarantees that span $\{\mathbf{v}_j\}_{j=1}^{\widetilde{m}}$ is invariant under the action of T. Hence, $T\mathbf{v}_j \in \text{span}\,\{\mathbf{v}_j\}_{j=1}^{\widetilde{m}}$ for $j = 1, \ldots, \widetilde{m}$, and, using (4), one gets that $\widetilde{t}_{j,\ell} = 0$ for $j = 1, \ldots, \widetilde{m}$ and $\ell = \widetilde{m}+1, \ldots, n$. Now, consider any $j > \widetilde{m}$. A direct calculation using (4) reveals that $T\mathbf{v}_j = G\mathbf{v}_j - P(RAP)^{-1}RAG\mathbf{v}_j = \lambda_j\mathbf{v}_j - \sum_{\ell=1}^{\widetilde{m}} x_{j-\widetilde{m},\ell}\mathbf{v}_\ell$, where $x_{i,k}$ are the elements of $X \in \mathbb{R}^{(n-\widetilde{m})\times\widetilde{m}}$. Hence, the structure (6) follows. □

Notice that, if (5) holds, then Lemma 1 allows us to study the properties of T using the matrix $\widetilde{T}$ and its structure (6), and hence $\widetilde{T}_{\widetilde{m}}$.

Let us now turn to the questions posed in Section 1. Assume that $\mathbf{p}_j = \mathbf{v}_j$, $j = 1, \ldots, m$, namely $V_c = \text{span}\,\{\mathbf{v}_j\}_{j=1}^{m}$. In this case, (5) holds with $\widetilde{m} = m$, and a simple argument[1] leads to $\widetilde{T}_{\widetilde{m}} = 0$, $\widetilde{T} = \begin{bmatrix} 0 & 0 \\ X & \Lambda_{\widetilde{m}} \end{bmatrix}$. The spectrum of $\widetilde{T}$ is $\{0, \lambda_{m+1}, \ldots, \lambda_n\}$. This means that $V_c \subset \text{kern}\,T$ and $\rho(T) = |\lambda_{m+1}|$. Let us now perturb the coarse space V_c using the eigenvector $\mathbf{v}_{m+1}$, that is $V_c(\varepsilon) := \text{span}\,\{\mathbf{v}_j + \varepsilon\,\mathbf{v}_{m+1}\}_{j=1}^{m}$. Clearly, $\dim V_c(\varepsilon) = m$ for any $\varepsilon \in \mathbb{R}$. In this case, (5) holds with $\widetilde{m} = m + 1$ and $\widetilde{T}$ becomes

$$\widetilde{T}(\varepsilon) = \begin{bmatrix} \widetilde{T}_{\widetilde{m}}(\varepsilon) & 0 \\ X(\varepsilon) & \Lambda_{\widetilde{m}} \end{bmatrix}, \tag{7}$$

where we make explicit the dependence on ε. Notice that $\varepsilon = 0$ clearly leads to $\widetilde{T}_{\widetilde{m}}(0) = \text{diag}\,(0, \ldots, 0, \lambda_{m+1}) \in \mathbb{R}^{\widetilde{m}\times\widetilde{m}}$, and we are back to the unperturbed case with $\widetilde{T}(0) = \widetilde{T}$ having spectrum $\{0, \lambda_{m+1}, \ldots, \lambda_n\}$. Now, notice that $\min_{\varepsilon\in\mathbb{R}} \rho(\widetilde{T}(\varepsilon)) \leq \rho(\widetilde{T}(0)) = |\lambda_{m+1}|$. Thus, it is natural to ask the question: is this inequality strict? Can one find an $\widetilde{\varepsilon} \neq 0$ such that $\rho(\widetilde{T}(\widetilde{\varepsilon})) = \min_{\varepsilon\in\mathbb{R}} \rho(\widetilde{T}(\varepsilon)) < \rho(\widetilde{T}(0))$ holds? If the answer is positive, then we can conclude that choosing the coarse vectors equal to the dominating eigenvectors of G is not an optimal choice. The next key result shows that, in the case $m = 1$, the answer is positive.

[1] Let $\mathbf{v}_j$ be an eigenvector of A with $j \in \{1, \ldots, m\}$. Denote by $\mathbf{e}_j \in \mathbb{R}^n$ the jth canonical vector. Since $P\mathbf{e}_j = \mathbf{v}_j$, $RAP\mathbf{e}_j = RA\mathbf{v}_j$. This is equivalent to $\mathbf{e}_j = (RAP)^{-1}RA\mathbf{v}_j$, which gives $T\mathbf{v}_j = \lambda_j(\mathbf{v}_j - P(RAP)^{-1}RA\mathbf{v}_j) = \lambda_j(\mathbf{v}_j - P\mathbf{e}_j) = 0$.

Theorem 1 (Perturbation of V_c) *Let $(\mathbf{v}_1, \lambda_1)$, $(\mathbf{v}_2, \lambda_2)$ and $(\mathbf{v}_3, \lambda_3)$ be three real eigenpairs of G, $G\mathbf{v}_j = \lambda_j \mathbf{v}_j$ such that with $0 < |\lambda_3| < |\lambda_2| \le |\lambda_1|$ and $\|\mathbf{v}_j\|_2 = 1$, $j = 1, 2$. Denote by $\widetilde{\lambda}_j \in \mathbb{R}$ the eigenvalues of A corresponding to $\mathbf{v}_j$, and assume that $\widetilde{\lambda}_1\widetilde{\lambda}_2 > 0$. Define $V_c := \operatorname{span}\{\mathbf{v}_1 + \varepsilon\mathbf{v}_2\}$ with $\varepsilon \in \mathbb{R}$, and $\gamma := \mathbf{v}_1^\top \mathbf{v}_2 \in [-1, 1]$. Then*

(A) *The spectral radius of $\widetilde{T}(\varepsilon)$ is $\rho(\widetilde{T}(\varepsilon)) = \max\{|\lambda(\varepsilon, \gamma)|, |\lambda_3|\}$, where*

$$\lambda(\varepsilon, \gamma) = \frac{\lambda_1\widetilde{\lambda}_2\varepsilon^2 + \gamma(\lambda_1\widetilde{\lambda}_2 + \lambda_2\widetilde{\lambda}_1)\varepsilon + \lambda_2\widetilde{\lambda}_1}{\widetilde{\lambda}_2\varepsilon^2 + \gamma(\widetilde{\lambda}_1 + \widetilde{\lambda}_2)\varepsilon + \widetilde{\lambda}_1}. \tag{8}$$

(B) *Let $\gamma = 0$. If $\lambda_1 > \lambda_2 > 0$ or $0 > \lambda_2 > \lambda_1$, then $\min_{\varepsilon \in \mathbb{R}} \rho(\widetilde{T}(\varepsilon)) = \rho(\widetilde{T}(0))$.*

(C) *Let $\gamma = 0$, If $\lambda_2 > 0 > \lambda_1$ or $\lambda_1 > 0 > \lambda_2$, then there exists an $\widetilde{\varepsilon} \neq 0$ such that $\rho(\widetilde{T}(\widetilde{\varepsilon})) = |\lambda_3| = \min_{\varepsilon \in \mathbb{R}} \rho(\widetilde{T}(\varepsilon)) < \rho(\widetilde{T}(0))$.*

(D) *Let $\gamma \neq 0$. If $\lambda_1 > \lambda_2 > 0$ or $0 > \lambda_2 > \lambda_1$, then there exists an $\widetilde{\varepsilon} \neq 0$ such that $|\lambda(\widetilde{\varepsilon}, \gamma)| < |\lambda_2|$ and hence $\rho(\widetilde{T}(\widetilde{\varepsilon})) = \max\{|\lambda(\widetilde{\varepsilon}, \gamma)|, |\lambda_3|\} < \rho(\widetilde{T}(0))$.*

(E) *Let $\gamma \neq 0$. If $\lambda_2 > 0 > \lambda_1$ or $\lambda_1 > 0 > \lambda_2$, then there exists an $\widetilde{\varepsilon} \neq 0$ such that $\rho(\widetilde{T}(\widetilde{\varepsilon})) = |\lambda_3| = \min_{\varepsilon \in \mathbb{R}} \rho(\widetilde{T}(\varepsilon)) < \rho(\widetilde{T}(0))$.*

Proof Since $m = 1$, a direct calculation allows us to compute the matrix

$$\widetilde{T}_{\widetilde{m}}(\varepsilon) = \begin{bmatrix} \lambda_1 - \frac{\lambda_1\widetilde{\lambda}_1(1+\varepsilon\gamma)}{g} & -\varepsilon\frac{\lambda_1\widetilde{\lambda}_1(1+\varepsilon\gamma)}{g} \\ -\frac{\lambda_2\widetilde{\lambda}_2(\varepsilon+\gamma)}{g} & \lambda_2 - \frac{(\varepsilon\lambda_2\widetilde{\lambda}_2)(\varepsilon+\gamma)}{g} \end{bmatrix},$$

where $g = \widetilde{\lambda}_1 + \varepsilon\gamma[\widetilde{\lambda}_1 + \widetilde{\lambda}_2] + \varepsilon^2\widetilde{\lambda}_2$. The spectrum of this matrix is $\{0, \lambda(\varepsilon, \gamma)\}$, with $\lambda(\varepsilon, \gamma)$ given in (8). Hence, point (A) follows recalling (7).

To prove points (B), (C), (D) and (E) we use some properties of the map $\varepsilon \mapsto \lambda(\varepsilon, \gamma)$. First, we notice that

$$\lambda(0, \gamma) = \lambda_2, \lim_{\varepsilon \to \pm\infty} \lambda(\varepsilon, \gamma) = \lambda_1, \lambda(\varepsilon, \gamma) = \lambda(-\varepsilon, -\gamma). \tag{9}$$

Second, the derivative of $\lambda(\varepsilon, \gamma)$ with respect to ε is

$$\frac{\mathrm{d}\lambda(\varepsilon, \gamma)}{\mathrm{d}\varepsilon} = \frac{(\lambda_1 - \lambda_2)\widetilde{\lambda}_1\widetilde{\lambda}_2(\varepsilon^2 + 2\varepsilon/\gamma + 1)\gamma}{(\widetilde{\lambda}_2\varepsilon^2 + \gamma(\widetilde{\lambda}_1 + \widetilde{\lambda}_2)\varepsilon + \widetilde{\lambda}_1)^2}. \tag{10}$$

Because of $\lambda(\varepsilon, \gamma) = \lambda(-\varepsilon, -\gamma)$ in (9), we can assume without loss of generality that $\gamma \ge 0$.

Let us now consider the case $\gamma = 0$. In this case, the derivative (10) becomes $\frac{\mathrm{d}\lambda(\varepsilon,0)}{\mathrm{d}\varepsilon} = \frac{(\lambda_1-\lambda_2)\widetilde{\lambda}_1\widetilde{\lambda}_2 2\varepsilon}{(\widetilde{\lambda}_2\varepsilon^2+\widetilde{\lambda}_1^2)^2}$. Moreover, since $\lambda(\varepsilon, 0) = \lambda(-\varepsilon, 0)$ we can assume that $\varepsilon \ge 0$.

Case (B). If $\lambda_1 > \lambda_2 > 0$, then $\frac{d\lambda(\varepsilon,0)}{d\varepsilon} > 0$ for all $\varepsilon > 0$. Hence, $\varepsilon \mapsto \lambda(\varepsilon,0)$ is monotonically increasing, $\lambda(\varepsilon,0) \geq 0$ for all $\varepsilon > 0$ and, thus, the minimum of $\varepsilon \mapsto |\lambda(\varepsilon,0)|$ is attained at $\varepsilon = 0$ with $|\lambda(0,0)| = |\lambda_2| > |\lambda_3|$, and the result follows. Analogously, if $0 > \lambda_2 > \lambda_1$, then $\frac{d\lambda(\varepsilon,0)}{d\varepsilon} < 0$ for all $\varepsilon > 0$. Hence, $\varepsilon \mapsto \lambda(\varepsilon,0)$ is monotonically decreasing, $\lambda(\varepsilon,0) < 0$ for all $\varepsilon > 0$ and the minimum of $\varepsilon \mapsto |\lambda(\varepsilon,0)|$ is attained at $\varepsilon = 0$.

Case (C). If $\lambda_1 > 0 > \lambda_2$, then $\frac{d\lambda(\varepsilon,0)}{d\varepsilon} > 0$ for all $\varepsilon > 0$. Hence, $\varepsilon \mapsto \lambda(\varepsilon,0)$ is monotonically increasing and such that $\lambda(0,0) = \lambda_2 < 0$ and $\lim_{\varepsilon\to\infty} \lambda(\varepsilon,0) = \lambda_1 > 0$. Thus, the continuity of the map $\varepsilon \mapsto \lambda(\varepsilon,0)$ guarantees the existence of an $\widetilde{\varepsilon} > 0$ such that $\lambda(\widetilde{\varepsilon},0) = 0$. Analogously, if $\lambda_2 > 0 > \lambda_1$, then $\frac{d\lambda(\varepsilon,0)}{d\varepsilon} < 0$ for all $\varepsilon > 0$ and the result follows by the continuity of $\varepsilon \mapsto \lambda(\varepsilon,0)$.

Let us now consider the case $\gamma > 0$. The sign of $\frac{d\lambda(\varepsilon,\gamma)}{d\varepsilon}$ is affected by the term $f(\varepsilon) := \varepsilon^2 + 2\varepsilon/\gamma + 1$, which appears at the numerator of (10). The function $f(\varepsilon)$ is strictly convex, attains its minimum at $\varepsilon = -\frac{1}{\gamma}$, and is negative in $(\bar{\varepsilon}_1, \bar{\varepsilon}_2)$ and positive in $(-\infty, \bar{\varepsilon}_1) \cup (\bar{\varepsilon}_2, \infty)$, with $\bar{\varepsilon}_1, \bar{\varepsilon}_2 = -\frac{1\mp\sqrt{1-\gamma^2}}{\gamma}$.

Case (D). If $\lambda_1 > \lambda_2 > 0$, then $\frac{d\lambda(\varepsilon,\gamma)}{d\varepsilon} > 0$ for all $\varepsilon > \bar{\varepsilon}_2$. Hence, $\frac{d\lambda(0,\gamma)}{d\varepsilon} > 0$, which means that there exists an $\widetilde{\varepsilon} < 0$ such that $|\lambda(\widetilde{\varepsilon},\gamma)| < |\lambda(0,\gamma)| = |\lambda_2|$. The case $0 > \lambda_2 > \lambda_1$ follows analogously.

Case (E). If $\lambda_1 > 0 > \lambda_2$, then $\frac{d\lambda(\varepsilon,\gamma)}{d\varepsilon} > 0$ for all $\varepsilon > 0$. Hence, by the continuity of $\varepsilon \mapsto \lambda(\varepsilon,\gamma)$ (for $\varepsilon \geq 0$) there exists an $\widetilde{\varepsilon} > 0$ such that $\lambda(\widetilde{\varepsilon},\gamma) = 0$. The case $\lambda_2 > 0 > \lambda_1$ follows analogously. □

Theorem 1 and its proof say that, if the two eigenvalues λ_1 and λ_2 have opposite signs (but they could be equal in modulus), then it is always possible to find an $\varepsilon \neq 0$ such that the coarse space $V_c := \text{span}\{\mathbf{v}_1 + \varepsilon\mathbf{v}_2\}$ leads to a faster method than $V_c := \text{span}\{\mathbf{v}_1\}$, even though both are one-dimensional subspaces. In addition, if $\lambda_3 \neq 0$ the former leads to a two-level operator T with a larger kernel than the one corresponding to the latter. The situation is completely different if λ_1 and λ_2 have the same sign. In this case, the orthogonality parameter γ is crucial. If $\mathbf{v}_1$ and $\mathbf{v}_2$ are orthogonal ($\gamma = 0$), then one cannot improve the effect of $V_c := \text{span}\{\mathbf{v}_1\}$ by a simple perturbation using $\mathbf{v}_2$. However, if $\mathbf{v}_1$ and $\mathbf{v}_2$ are not orthogonal ($\gamma \neq 0$), then one can still find an $\varepsilon \neq 0$ such that $\rho(\widetilde{T}(\varepsilon)) < \rho(\widetilde{T}(0))$.

Notice that, if $|\lambda_3| = |\lambda_2|$, Theorem 1 shows that one cannot obtain a $\rho(T)$ smaller than $|\lambda_2|$ using a one-dimensional perturbation. However, if one optimizes the entire coarse space V_c (keeping m fixed), then one can find coarse spaces leading to better contraction factor of the two-level iteration, even though $|\lambda_3| = |\lambda_2|$. This is shown in the next section.

3 Optimizing the coarse-space functions

Consider the elliptic problem

$$-\Delta u + c\,(\partial_x u + \partial_y u) = f \text{ in } \Omega = (0,1)^2, \quad u = 0 \text{ on } \partial\Omega. \tag{11}$$

Using a uniform grid of size h, the standard second-order finite-difference scheme for the Laplace operator and the central difference approximation for the advection terms, problem (11) becomes $A\mathbf{u} = \mathbf{f}$, where A has constant and positive diagonal entries, $D = \text{diag}(A) = 4/h^2 I$. A simple calculation shows that, if $c \geq 0$ satisfies $c \leq 2/h$, then the eigenvalues of A are real. The eigenvectors of A are orthogonal if $c = 0$ and non-orthogonal if $c > 0$.

One of the most used smoothers for (11) is the damped Jacobi method: $\mathbf{u}^{k+1} = \mathbf{u}^k + \omega D^{-1}(\mathbf{f} - A\mathbf{u}^k)$, where $\omega \in (0,1]$ is a damping parameter. The corresponding iteration matrix is $G = I - \omega D^{-1} A$. Since $D = 4/h^2 I$, the matrices A and G have the same eigenvectors. For $c = 0$, it is possible to show that, if $\omega = 1$ (classical Jacobi iteration), then the nonzero eigenvalues of G have positive and negative signs, while if $\omega = 1/2$, the eigenvalues of G are all positive. Hence, the chosen model problem allows us to work in the theoretical framework of Section 2.

To validate numerically Theorem 1, we set $h = 1/10$ and consider $V_c := \{\mathbf{v}_1 + \varepsilon \mathbf{v}_2\}$. Figure 1 shows the dependence of $\rho(T(\varepsilon))$ and $|\lambda(\varepsilon,\gamma)|$ on ε and γ. On the top left panel, we set $c = 0$ and $\omega = 1/2$ so that the hypotheses of point (B) of Theorem 1 are satisfied, since $\gamma = 0$ and $\lambda_1 \geq \lambda_2 > 0$. As point (B) predicts, we observe that $\min_{\varepsilon \in \mathbb{R}} \rho(T(\varepsilon))$ is attained at $\varepsilon = 0$, i.e. $\min_{\varepsilon \in \mathbb{R}} \rho(T(\varepsilon)) = \rho(T(0)) = \lambda_2$. Hence, adding a perturbation does not improve the coarse space made only by $\mathbf{v}_1$. Next, we consider point (C), by setting $c = 0$ and $\omega = 1$. Through a direct computation we get $\lambda_1 = -0.95$, $\lambda_2 = -\lambda_1$ and $\lambda_3 = 0.90$. The top-right panel shows, on the one hand, that for several values of ε, $\rho(T(\varepsilon)) = \lambda_3 < \lambda_2$, that is with a one-dimensional perturbed coarse space, we obtain the same contraction factor we would have with the two-dimensional spectral coarse space $V_c = \text{span}\,\{\mathbf{v}_1, \mathbf{v}_2\}$. On the other hand, we observe that there are two values of ε such that $\lambda(\varepsilon,\gamma) = 0$, which (recalling (4) and (6)) implies that T is nilpotent over the span$\{\mathbf{v}_1, \mathbf{v}_2\}$. To study point (D), we set $c = 10$, $\omega = 1/2$, which lead to $\lambda_1 = 0.92$, $\lambda_2 = \lambda_3 = 0.90$. The left-bottom panel confirms there exists an $\varepsilon^* < 0$ such that $|\lambda(\varepsilon^*,\gamma)| \leq \lambda_2$, which implies $\rho(T(\varepsilon^*)) \leq \lambda_2$. Finally, we set $c = 10$ and $\omega = 1$. Point (E) is confirmed by the right-bottom panel, which shows that $|\lambda(\varepsilon,\gamma)| < |\lambda_2|$, and thus $\min_\varepsilon \rho(T(\varepsilon)) = |\lambda_3|$, for some values of ε.

We have shown both theoretically and numerically that the spectral coarse space is not necessarily the one-dimensional coarse space minimizing $\rho(T)$. Now, we wish to go beyond this one-dimensional analysis and optimize the entire coarse space V_c keeping its dimension m fixed. This is equivalent to optimizing the prolongation operator P whose columns span V_c. Thus, we consider the optimization problem

$$\min_{P \in \mathbb{R}^{n \times m}} \rho(T(P)). \tag{12}$$

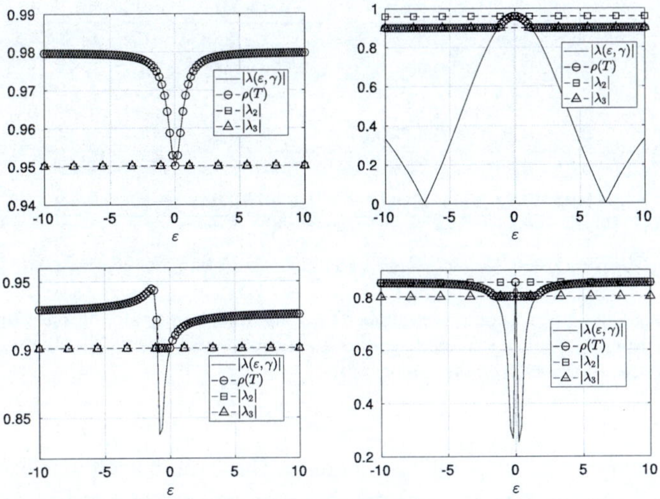

Fig. 1: Behavior of $|\lambda(\varepsilon, \gamma)|$ and $\rho(T(\varepsilon))$ as functions of ε for different c and γ. Top left panel: $c = 0$, $\omega = 1/2$; top right panel: $c = 0$, $\omega = 1$; bottom left panel: $c = 10$, $\omega = 1/2$; bottom right panel: $c = 10$, $\omega = 1$.

To solve (12) approximately, we follow the approach proposed by [5]. Due to the Gelfand formula $\rho(T) = \lim_{k\to\infty} \sqrt[k]{\|T^k\|_F}$, we replace (12) with the simpler optimization problem $\min_P \|T(P)^k\|_F^2$ for some positive k. Here, $\|\cdot\|_F$ is the Frobenius norm. We then consider the unbiased stochastic estimator [4]

$$\|T^k\|_F^2 = \text{trace}\left((T^k)^\top T^k\right) = \mathbb{E}_{\mathbf{z}}\left[\mathbf{z}^\top (T^k)^\top T^k \mathbf{z}\right] = \mathbb{E}_{\mathbf{z}}\left[\|T^k \mathbf{z}\|_2^2\right],$$

where $\mathbf{z} \in \mathbb{R}^n$ is a random vector with Rademacher distribution, i.e. $\mathbb{P}(\mathbf{z}_i = \pm 1) = 1/2$. Finally, we rely on a sample average approach, replacing the unbiased stochastic estimator with its empirical mean such that (12) is approximated by

$$\min_{P\in\mathbb{R}^{n\times m}} \frac{1}{N}\sum_{i=1}^{N} \|T(P)^k \mathbf{z}_i\|_F^2, \tag{13}$$

where $\mathbf{z}_i$ are a set of independent, Rademacher distributed, random vectors. The action of T onto the vectors $\mathbf{z}_i$ can be interpreted as the feed-forward process of a neural net, where each layer represents one specific step of the two-level method, that is the smoothing step, the residual computation, the coarse correction and the prolongation/restriction operations. In our setting, the weights of most layers are fixed and given, and the optimization is performed only on the weights of the layer representing the prolongation step. The restriction layer is constrained to have as weights the transpose of the weights of the prolongation layer. The cost of constructing coarse spaces using deep neural networks can be very high, and not

	c	ω	$m = 1$	$m = 5$	$m = 10$	$m = 15$
$\rho(T)$	0	1/2	0.95 - 0.95	0.90 - 0.90	0.82 - 0.83	0.76 - 0.78
	0	1	0.95 - 0.90	0.90 - 0.80	0.80 - 0.65	0.74 - 0.53
	10	1/2	0.90 - 0.90	0.85 - 0.82	0.79 - 0.74	0.73 - 0.68
	10	1	0.85 - 0.80	0.80 - 0.67	0.71 - 0.55	0.66 - 0.37
$\|\|T\|\|_A$	0	1/2	0.95 - 0.95	0.90 - 0.90	0.82 - 0.84	0.76 - 0.77
	0	1	0.95 - 0.95	0.90 - 0.94	0.80 - 0.88	0.74 - 0.88
κ_2	0	1	46.91 - 29.45	18.48 - 14.40	9.37 - 8.22	6.69 - 8.53
	10	1	27.25 - 23.98	22.44 - 12.36	17.34 - 11.35	13.06 - 9.71

Table 1: Values of $\rho(T)$, $\|T\|_A$ and condition number κ_2 of the matrix A preconditioned by the two-level method for different c and ω and using either a spectral coarse space (left number), or the coarse space obtained solving (13) (right number).

practical if the problem needs to be solved only once. However, our interest here is in theoretical aspects, and deep neural networks are used only to show the existence of coarse spaces (asymptotically) better than the spectral ones.

We solve (13) for $k = 10$ and $N = n$ using Tensorflow [6] and its stochastic gradient descent algorithm with learning parameter 0.1. The weights of the prolongation layer are initialized with an uniform distribution. Table 1 reports both $\rho(T(P))$ and $\|T(P)\|_A$ using a spectral coarse space and the coarse space obtained solving (13). We can clearly see that there exist coarse spaces, hence matrices P, corresponding to values of the asymptotic convergence factor $\rho(T(P))$ much smaller than the ones obtained by spectral coarse spaces. Hence, Table 1 confirms that a spectral coarse space of dimension m is not necessarily a (global) minimizer for $\min_{P \in \mathbb{R}^{n \times m}} \rho(T(P))$. This can be observed not only in the case $c = 0$, for which the result of [7, Theorem 5.5] states that (recall that M is symmetric) the spectral coarse space minimizes $\|T(P)\|_A$, but also for $c > 0$, which corresponds to a nonsymmetric A. Interestingly, the coarse spaces obtained by our numerical optimizations lead to preconditioned matrices with better condition numbers, as shown in the last row of Table 1, where the condition number κ_2 of the matrix A preconditioned by the two-level method (and different coarse spaces) is reported.

Acknowledgements G. Ciaramella is a member of INdAM GNCS.

References

1. M.J. Gander, L. Halpern and K. Repiquet. A new coarse grid correction for RAS/AS. In: *Domain Decomposition Methods in Science and Engineering XXI*, pp. 275–283, Springer (2014).
2. M.J. Gander, L. Halpern and K. Santugini-Repiquet. On optimal coarse spaces for domain decomposition and their approximation. In: *Domain Decomposition Methods in Science and Engineering XXIV*, pp. 271–280, Springer (2018).

3. M.J. Gander and B. Song. Complete, optimal and optimized coarse spaces for additive Schwarz. In: *Domain Decomposition Methods in Science and Engineering XXIV*, pp. 301–309, Springer (2019).
4. M.F. Hutchinson. A stochastic estimator of the trace of the influence matrix for Laplacian smoothing splines. *Commun. Stat.-Simul. C.* **18**(3), 1059–1076 (1989).
5. A. Katrutsa, T. Daulbaev and I. Oseledets. Deep multigrid: learning prolongation and restriction matrices. *arXiv preprint* arXiv:1711.03825 (2017).
6. TensorFlow: Large-scale machine learning on heterogeneous systems (2015). Software available from `tensorflow.org`.
7. J. Xu and L. Zikatanov. Algebraic multigrid methods. *Acta Numer.* **26**, 591–721 (2017).

Discrete Analysis of Schwarz Waveform Relaxation for a Simplified Air-Sea Coupling Problem with Nonlinear Transmission Conditions

S. Clement, F. Lemarié, and E. Blayo

1 Introduction

Schwarz-like domain decomposition methods are very popular in mathematics, the computational sciences and engineering, notably for the implementation of coupling strategies. Such an iterative method has recently been applied in a state-of-the-art Earth System Model (ESM) to evaluate the consequences of inaccuracies in the usual ad hoc ocean-atmosphere coupling algorithms used in realistic models [2]. For such a complex application it is challenging to have an a priori knowledge of the convergence properties of the Schwarz method. Indeed coupled problems arising in ESMs often exhibit sharp turbulent boundary layers whose parameterizations lead to peculiar transmission conditions. The objective in this paper is to study a model problem representative of the coupling between the ocean and the atmosphere, including discretization and so-called bulk interface conditions which are analogous to a quadratic friction law. Such a model is introduced in Sec. 2 and its discretization, as done in state-of-the-art ESMs, is described in Sec. 3. In the semi-discrete case in space we conduct in Sec. 4 a convergence analysis of the model problem first with a linear friction and then with a quadratic friction linearized around equilibrium solutions. Finally, in Sec. 5, numerical experiments in the linear and nonlinear case are performed to illustrate the relevance of our analysis.

S. Clement
Univ Grenoble Alpes, CNRS, Inria, Grenoble INP, LJK, Grenoble, France,
e-mail: simon.clement@grenoble-inp.org

F. Lemarié
Univ Grenoble Alpes, Inria, CNRS, Grenoble INP, LJK, Grenoble, France,
e-mail: florian.lemarie@inria.fr

E. Blayo
Univ Grenoble Alpes, CNRS, Inria, Grenoble INP, LJK, Grenoble, France,
e-mail: eric.blayo@univ-grenoble-alpes.fr

S. C. Brenner et al. (eds.), *Domain Decomposition Methods in Science and Engineering XXVI*, Lecture Notes in Computational Science and Engineering 145,
https://doi.org/10.1007/978-3-030-95025-5_19

2 Model problem for ocean-atmosphere coupling

We focus on the dynamical part of the oceanic and atmospheric primitive equations and neglect the horizontal variations of the velocity field, which leads to a model problem depending on the vertical direction only. This assumption, commonly made to study turbulent mixing in the boundary layers near the air-sea interface, is justified because of the large disparity between the vertical and the horizontal spatial scales in these layers. We consider the following diffusion problem accounting for Earth's rotation (f is the Coriolis frequency and $\mathbf{k}$ a vertical unit vector):

$$\begin{cases} \partial_t \mathbf{u} + f\mathbf{k} \times \mathbf{u} - \partial_z \left(\nu(z,t)\partial_z \mathbf{u}\right) = \mathbf{g}, & \text{in } \Omega \times (0,T), \\ \mathbf{u}(z,0) = \mathbf{u}_0(z), & \forall z \text{ in } \Omega, \\ \mathbf{u}(H_o,t) = \mathbf{u}_o^\infty(t),\ \mathbf{u}(H_a,t) = \mathbf{u}_a^\infty(t), & t \in (0,T), \end{cases}$$

with $\mathbf{u} = (u,v)$ the horizontal velocity vector, $\nu(z,t) > 0$ the turbulent viscosity and $\Omega = (H_o, H_a)$ a bounded open subset of $\mathbb{R}$ containing the air-sea interface $\Gamma = \{z = 0\}$. In the ocean and the atmosphere, which are turbulent fluids, the velocity field varies considerably in the few meters close to the interface (in a region called the *surface layer*). The cost of an explicit representation of the surface layer in numerical simulations being unaffordable, this region is numerically accounted for using wall laws, a.k.a. log laws (e.g. [4]). This approach, traditionally used to deal with solid walls, is also used in the ocean-atmosphere context, with additional complexity arising from the stratification effects [5]. In this context wall laws are referred to as *surface layer* parameterizations. The role of such parameterizations is to provide $\nu\partial_z\mathbf{u}$ on the upper and lower interfaces of the surface layer as a function of the difference of fluid velocities. Thus the coupling problem of interest should be understood as a domain decomposition with three non-overlapping subdomains. For the sake of convenience the velocity vector $\mathbf{u} = (u,v)$ is rewritten as a complex variable $U = u + iv$. Then the model problem reads

$$\begin{aligned} \partial_t U_j + ifU_j - \partial_z \left(\nu_j(z,t)\partial_z U_j\right) &= g_j, \qquad (j = o,a) && \text{in } \Omega_j \times (0,T) \\ U_j(H_j,t) &= U_j^\infty(t), && t \in (0,T), \\ U_j(z,0) &= U_0(z), && \forall z \text{ in } \Omega_j, \\ \rho_o\nu_o\partial_z U_o(\delta_o,t) = \rho_a\nu_a\partial_z U_a(\delta_a,t) &= \mathcal{F}_{\text{sl}}(U_a(\delta_a,t) - U_o(\delta_o,t)), && t \in (0,T), \end{aligned} \tag{1}$$

where $\Omega_o = (H_o, \delta_o)$, $\Omega_a = (\delta_a, H_a)$, and $\mathcal{F}_{\text{sl}}$ is a parameterization function for the surface layer extending over $\Omega_{\text{sl}} = (\delta_o, \delta_a)$. A typical formulation for $\mathcal{F}_{\text{sl}}$ is

$$\mathcal{F}_{\text{sl}}(U_a(\delta_a,t) - U_o(\delta_o,t)) = \rho_a C_D |U_a(\delta_a,t) - U_o(\delta_o,t)|(U_a(\delta_a,t) - U_o(\delta_o,t)),$$

which corresponds to a quadratic friction law with C_D a drag coefficient (assumed constant in the present study). Geostrophic winds and currents are used in this study as source terms and boundary conditions. Geostrophic equilibrium is the stationary state for which the Coriolis force compensates for the effects of gravity. It corresponds to the large scale dynamics of ocean and atmosphere, and leads to reasonable values of the solution U.

The well-posedness of (1) has been studied in [6], where it is proved that its stationary version admits a unique solution for realistic values of the parameters. The study of the nonstationary case is much more challenging: numerical experiments tend to confirm this well-posedness, but with no theoretical proof.

3 Discretized coupled problem

3.1 Implementation of the surface layer

As described in Sec. 2, the full domain Ω is split into three parts: Ω_o in the ocean, Ω_a in the atmosphere and Ω_{sl} a thin domain containing the interface (see Fig. 1). The role of Ω_{sl} is to provide $\rho_j \nu_j \partial_z U_j$ at $z = \delta_j$ $(j = o, a)$ as a function of fluid velocities at the same locations. However, in state-of-the-art climate models, the discretization is based on an approximate form of the coupled problem (1). For practical reasons, the computational domains are $\widetilde{\Omega}_o = (H_o, 0) = \Omega_o \bigcup (\delta_o, 0)$ and $\widetilde{\Omega}_a = (0, H_a) = (0, \delta_a) \bigcup \Omega_a$, and the locations of the lower and upper boundaries of the surface layer ($z = \delta_j$) are assimilated to the centers of the first grid cells (i.e. $\delta_o = -h_{\text{o}}/2$ and $\delta_a = h_{\text{a}}/2$ with h_{o} and h_{a} the thicknesses of the first grid cell in each subdomain), where the values of the velocity closest to the interface are available. Typical resolutions in the models are $\delta_a = h_{\text{a}}/2 = 10$ m and $\delta_o = -h_{\text{o}}/2 = -1$ m. At a discrete level, the transmission condition in (1) is replaced by

$$\rho_{\text{o}} \nu_{\text{o}} \partial_z U_{\text{o}}(0, t) = \rho_{\text{a}} \nu_{\text{a}} \partial_z U_{\text{a}}(0, t) = \rho_{\text{a}} \alpha \left(U_{\text{a}} \left(\frac{h_{\text{a}}}{2}, t \right) - U_{\text{o}} \left(-\frac{h_{\text{o}}}{2}, t \right) \right), \tag{2}$$

where $\alpha = C_D \left| U_{\text{a}} \left(\frac{h_{\text{a}}}{2}, t \right) - U_{\text{o}} \left(-\frac{h_{\text{o}}}{2}, t \right) \right|$ for the *nonlinear* case. In the following, for the analysis in Sec. 4, we consider a *linear* friction where α is assumed constant and a quadratic friction *linearized* around equilibrium solutions.

3.2 Schwarz Waveform Relaxation

As discussed for example in [2], current ocean-atmosphere coupling methods can actually be seen as a single iteration of a Schwarz Waveform Relaxation (SWR) algorithm. SWR applied to the coupling problem presented in Sec. 2 with the transmission conditions (2) and constant viscosity in each subdomain reads:

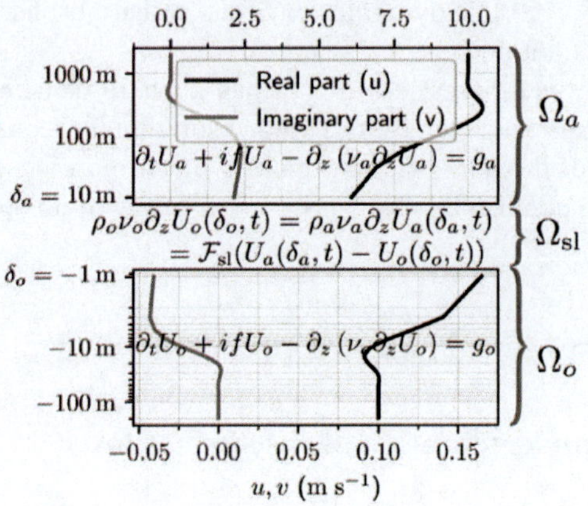

Fig. 1: Discrete representation of the three domains $\Omega_a, \Omega_{sl}, \Omega_o$ together with a typical stationary state. Note the different scales for (u, v) in the ocean and in the atmosphere.

$$(\partial_t + if)U_j^k - \nu_j \partial_z \varphi_j^k = g_j, \qquad \text{in } \widetilde{\Omega}_j \times (0, T) \tag{3a}$$

$$U_j^k(z, 0) = U_0(z), \qquad \forall z \in \widetilde{\Omega}_j \tag{3b}$$

$$U_j^k(H_j, t) = U_j^\infty, \qquad t \in [0, T] \tag{3c}$$

$$\nu_a \varphi_a^k(0, t) = \alpha^{k-1}\left(U_a^{k-1+\theta}\left(\frac{h_a}{2}, t\right) - U_o^{k-1}\left(-\frac{h_o}{2}, t\right)\right), \quad t \in [0, T] \tag{3d}$$

$$\rho_o \nu_o \varphi_o^k(0, t) = \rho_a \nu_a \varphi_a^k(0, t), \qquad t \in [0, T] \tag{3e}$$

where $j = \mathrm{a, o}$, $\varphi_j = \partial_z U_j$, and $U_a^{k-1+\theta} = \theta U_a^k + (1 - \theta) U_a^{k-1}$ with θ a relaxation parameter (interpolation for $0 \le \theta \le 1$ or extrapolation for $\theta > 1$). At each iteration, (3e) ensures that the kinetic energy is conserved at the machine precision in the coupled system which is a major constraint for climate models. In (3d), the presence of the parameter θ makes it resemble a Dirichlet–Neumann Waveform Relaxation algorithm. Indeed, if (3d) is replaced by $U_a^k = \theta U_o^{k-1} + (1 - \theta) U_a^{k-1}$ the DNWR algorithm is retrieved, as examined in the continuous case in [1] and in the discrete case in [3]. However (3d) involves both φ_a^k and $U_a^{k-1+\theta}$: the θ parameter thus appears here within (close to Robin) condition ($\nu_a \varphi_a(0) - \alpha \theta U_a(h_a/2) = \ldots$), i.e. the relaxation is not performed directly on the converging variable, which leads to convergence properties different from the DNWR case, as shown in Sec. 4.

In the following, centered finite difference schemes in space are used with constant space steps h_j. The derivatives are $\varphi_j(z, t) = \frac{U_j(z+h_j/2,t) - U_j(z-h_j/2,t)}{h_j}$ and the semi-discrete version of (3a) in the homogeneous case is

$$(\partial_t + if)U_j(z, t) = \nu_j \frac{\varphi_j(z + h_j/2, t) - \varphi_j(z - h_j/2, t)}{h_j}. \tag{4}$$

4 Convergence analysis

In this section we conduct a convergence analysis of the SWR algorithm (3) first with α a constant and then in a more complicated case where the problem is linearized around its equilibrium solutions. In the following we systematically make the assumption that the space domain is of infinite size (i.e. $H_j \to \infty$) for the sake of simplicity.

Linear friction case (α = const) We assume in this paragraph that $\alpha = \alpha_c$ with α_c a constant independent of U_j and we study the system satisfied by the errors (i.e. $g_j, U_0, U^\infty = 0$). The Fourier transform in time of the finite difference scheme (4) yields $\widehat{U}_{\text{a}}(h_{\text{a}}/2) = \nu_{\text{a}} \frac{\widehat{\varphi}_{\text{a}}(h_{\text{a}}) - \widehat{\varphi}_{\text{a}}(0)}{i(\omega+f)h_{\text{a}}}$ with $\omega \in \mathbb{R}$ the frequency variable. After simple algebra, the transmission condition (3d) in Fourier space expressed in terms of the $\widehat{\varphi}_j$ is

$$\begin{aligned}\left(\frac{\chi_a \nu_a}{h_a} + \theta\alpha_c\right)\widehat{\varphi}_{\text{a}}^k(0) - \theta\alpha_c\widehat{\varphi}_{\text{a}}^k(h_a) =& (1-\theta)\alpha_c(\widehat{\varphi}_{\text{a}}^{k-1}(h_{\text{a}}) - \widehat{\varphi}_{\text{a}}^{k-1}(0)) \\ & - \alpha_c \frac{h_a \nu_o}{h_o \nu_a}(\widehat{\varphi}_{\text{o}}^{k-1}(0) - \widehat{\varphi}_{\text{o}}^{k-1}(-h_{\text{o}}))\end{aligned} \tag{5}$$

with $\chi_j = \frac{i(\omega+f)h_j^2}{\nu_j}$. A discrete analysis of the finite difference scheme (4) in the frequency domain (e.g. [7]) leads to $\widehat{\varphi}_{\text{o}}^k(-mh_{\text{o}}) = A_k(\lambda_{\text{o}}+1)^m$ and $\widehat{\varphi}_{\text{a}}^k(mh_{\text{a}}) = B_k(\lambda_{\text{a}}+1)^m$ with $\lambda_j = \frac{1}{2}\left(\chi_j - \sqrt{\chi_j}\sqrt{\chi_j+4}\right)$ and m the space index. The convergence factor of SWR is then the rate at which A_k or B_k tends to 0. Combining (5) with the Fourier transform in time of (3e), we get the evolution of B_k which eventually leads to the following convergence factor:

$$\xi = \left|\frac{B_k}{B_{k-1}}\right| = \left|\frac{(1-\theta) + \epsilon\frac{h_{\text{a}}\lambda_{\text{o}}}{h_{\text{o}}\lambda_{\text{a}}}}{\frac{\nu_{\text{a}}\chi_{\text{a}}}{\alpha_c h_{\text{a}}\lambda_{\text{a}}} - \theta}\right|, \tag{6}$$

where $\epsilon = \frac{\rho_{\text{a}}}{\rho_{\text{o}}} \approx 10^{-3}$ in the ocean-atmosphere context. Note that the convergence factor (6) differs significantly from the semi-discrete convergence factor $\xi_{\text{DNWR}} = |1 - \theta_{\text{DNWR}}(1 - \epsilon h_{\text{a}}\lambda_{\text{o}}/(\lambda_{\text{a}}h_{\text{o}}))|$ of the DNWR algorithm. Moreover, it can be found that

$$\lim_{(\omega+f)\to 0} \xi = \frac{1}{\theta}\left|1 - \theta + \epsilon\sqrt{\frac{\nu_{\text{a}}}{\nu_{\text{o}}}}\right| = \xi_0, \qquad \lim_{(\omega+f)\to\infty} \xi = 0.$$

As $\omega + f \to 0$ the asymptotic value ξ_0 depends on θ: it is $+\infty$ for $\theta = 0$ (i.e. a fast divergence), and $\xi_0 = \epsilon\sqrt{\frac{\nu_{\text{a}}}{\nu_{\text{o}}}}$ for $\theta = 1$. When $\omega \to \infty$, the convergence factor tends to zero (i.e. the convergence is fast for high frequencies). Whatever ω, it can be shown that the value ξ_0 is an upper bound of the convergence factor when $\theta \leq 1$ if $\sqrt{\frac{\nu_{\text{o}}}{\nu_{\text{a}}}} \leq \frac{h_{\text{o}}}{h_{\text{a}}}$, the latter condition being easily satisfied. Since we have $\epsilon \approx 10^{-3}$, the convergence is fast for $\theta = 1$ whereas ϵ does not play any role for $\theta = 0$. The optimal parameter θ_{opt} for low frequencies is $1 + \epsilon\sqrt{\frac{\nu_{\text{a}}}{\nu_{\text{o}}}}$, which is very close to 1.

Linearized quadratic friction case The analysis of the nonlinear quadratic friction case (i.e. with $\alpha = C_D \mid U_\mathrm{a}(h_\mathrm{a}/2, t) - U_\mathrm{o}(-h_\mathrm{o}/2, t) \mid$) cannot be pursued through a Fourier transform. We thus consider the linearization of the problem around a stationary state U_j^e, φ_j^e satisfying (1): assuming that $U_j^k(\pm h_j/2, t)$ is in a neighborhood of $U^e(\pm h_j/2)$, the modulus in α is non-zero and we can differentiate α. Differences with the stationary state are denoted $\delta\varphi_j^k = \varphi_j^k(0,t) - \varphi_j^e(0)$ and $\delta U_j^k = U_j^k(\pm h_j/2, t) - U_j^e(\pm h_j/2)$. After some algebra, the linearized transmission operator reads

$$\nu_\mathrm{a}\delta\varphi_\mathrm{a}^k = \alpha^e \left(\left(\frac{3}{2} - \theta \right) \delta U_\mathrm{a}^{k-1} + \theta\, \delta U_\mathrm{a}^k - \frac{3}{2} \delta U_\mathrm{o}^{k-1} + \frac{1}{2} \frac{U_\mathrm{a}^e - U_\mathrm{o}^e}{\overline{U_\mathrm{a}^e - U_\mathrm{o}^e}} \overline{\delta U_\mathrm{a}^{k-1} - \delta U_\mathrm{o}^{k-1}} \right) \tag{7}$$

with $\alpha^e = C_D \left| U_\mathrm{a}^e(h_\mathrm{a}/2) - U_\mathrm{o}^e(-h_\mathrm{o}/2) \right|$. Following the derivation in the previous paragraph, we find that the convergence factor ξ^q in the linearized quadratic friction case differs from one iteration to another (it is indeed a function of $\overline{\frac{B_{k-1}(-\omega)}{B_{k-1}(\omega)}}$). However, for $(\omega + f) \to 0$ the term $\frac{1}{2}\frac{U_\mathrm{a}^e - U_\mathrm{o}^e}{\overline{U_\mathrm{a}^e - U_\mathrm{o}^e}} \overline{\delta U_\mathrm{a}^{k-1} - \delta U_\mathrm{o}^{k-1}}$ vanishes, therefore the asymptotic convergence rate ξ_0^q is independent of the iterate:

$$\lim_{(\omega+f)\to 0} \xi^q = \frac{1}{\theta} \left| \frac{3}{2} - \theta + \frac{3}{2}\epsilon \sqrt{\frac{\nu_\mathrm{a}}{\nu_\mathrm{o}}} \right| = \xi_0^q, \qquad \lim_{(\omega+f)\to\infty} \xi^q = 0.$$

The convergence is fast for high frequencies, as in the linear friction case. However the optimal parameter for $(\omega + f) \to 0$ is here $\theta_\mathrm{opt}^q = \frac{3}{2} + \frac{3}{2}\epsilon\sqrt{\frac{\nu_\mathrm{a}}{\nu_\mathrm{o}}}$. It is different from the optimal parameter θ_opt obtained with linear friction: for typical values of the ocean-atmosphere coupling problem, θ_opt^q is close to $\frac{3}{2}$. The asymptotic value ξ_0^q is not an upper bound of the convergence factor but it is a good choice for θ_opt^q.

5 Numerical experiments

The aim of this section is to illustrate the influence of the parameter θ, in the linear and quadratic friction cases. The stationary state U_j^e is used to compute $\alpha_c = \alpha^e = C_D|U_\mathrm{a}^e(\frac{h_\mathrm{a}}{2}) - U_\mathrm{o}^e(\frac{h_\mathrm{o}}{2})|$ in the linear case. Parameters of the problem are taken as realistic: $C_D = 1.2\times 10^{-3}$, the space steps are $\frac{h_\mathrm{a}}{2} = 10$ m, $\frac{h_\mathrm{o}}{2} = 1$ m, the time step is 60 s, the size of the time window T is 1 day ($1440\Delta t$) and the computational domains sizes are $H_o = H_a = 2000$ m (100 and 1000 nodes respectively in Ω_a and Ω_o). The Coriolis parameter is $f = 10^{-4}$ s^{-1} and the diffusivities are $\nu_\mathrm{a} =$ 1 m^2 s^{-1}, $\nu_\mathrm{o} = 3\times 10^{-3}$ m^2 s^{-1}. U_j^∞ are set to constant values of 10 m s^{-1} in the atmosphere and 0.1 m s^{-1} in the ocean, while the forcing terms $g_j = ifU_j^\infty$ and the initial condition $U_0(z) = U_j^e(z)$. SWR is initialized at the interface with a white noise around the interface value of the initial condition. Figure 2 shows the evolution of the error for two choices of θ. The theoretical convergence according to ξ_0 is also displayed: $\sup_\omega \xi$ is an upper bound of the L^2 convergence factor [6] and ξ_0 is an approximation of $\sup_\omega \xi$. Both ξ_0 and ξ_0^q are close to the convergence rate, with the exception of ξ_0^q, which predicts much faster convergence than observed when $\theta = 1.5$. This shows that the maximum of the convergence factor is not reached when $(\omega + f) \longrightarrow 0$ in this case. Figure 2 confirms the results of Sec. 4: when considering $\alpha = \alpha_c$ constant, the fastest convergence is achieved when θ is close to 1, similarly to the DNWR algorithm. However this does not translate into the nonlinear case, which converges faster with $\theta = 1.5$. Figure 3 shows that the convergence behavior with the linearized transmission condition is similar to the nonlinear case. As expected, the convergence is faster for $\theta = 1.5$ than for $\theta = 1$. We observed that those results are robust to changes in the values of the parameters in the range of interest. Linearized transmission conditions are hence relevant to the theoretical study of the convergence properties of our nonlinear problem.

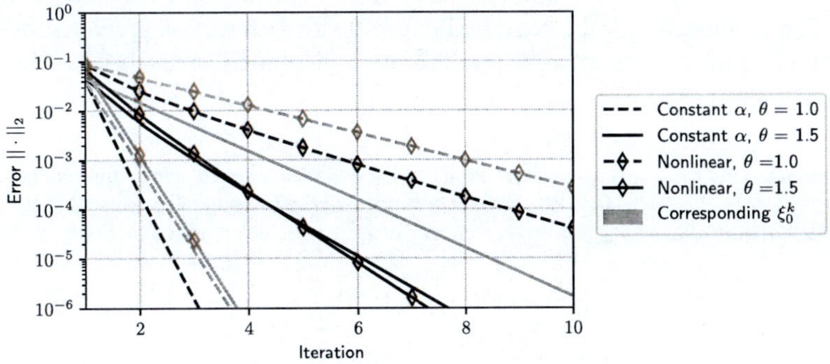

Fig. 2: Evolution of the L^2 norm of the errors. Black lines represent the observed convergence; grey lines are the estimated convergence with slopes ξ_0 for linear cases and ξ_0^q for quadratic cases.

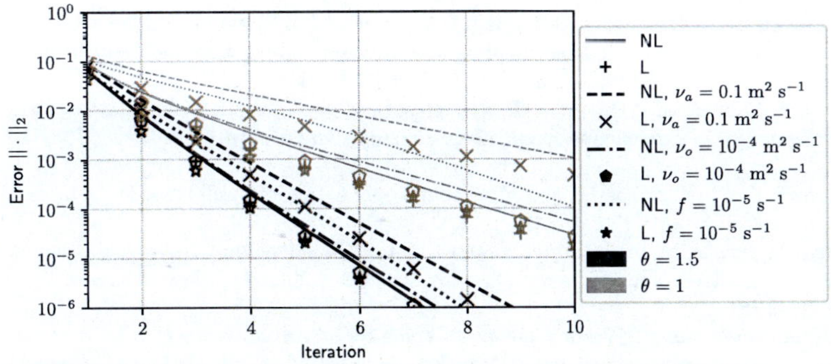

Fig. 3: Evolution of the L^2 norm of the errors with linearized (L) and nonlinear (NL) transmission conditions. The legend indicates the changes in the parameters for each case.

6 Conclusion

In this paper, we studied an SWR algorithm applied to a simplified ocean-atmosphere problem. This problem considers nonlinear transmission conditions arising from wall laws representative of the ones used in Earth-System Models and analogous to a quadratic friction law. We motivated the fact that the convergence analysis of such problems can only be done at a semi-discrete level in space due to the particular practical implementation of continuous interface conditions in actual climate models. Then we analytically studied the convergence properties in a case with linear friction and in a case with linearized quadratic friction. We formulated the problem with a relaxation parameter θ in the transmission conditions and systematically assessed its impact on the convergence speed. For the two cases of interest, the convergence factors are derived and the asymptotic limits for small values of the frequency $\omega + f$ are given. This asymptotic limit allowed us to choose appropriate values for the parameter θ to guarantee fast convergence of the algorithm. The behavior of the algorithm for linear friction and linearized quadratic friction turns out to be different, which leads to different "optimal" values of θ. Numerical experiments in the nonlinear case showed that the observed convergence behaves as predicted by the linearized quadratic friction case, whose thorough theoretical analysis is left for future work.

Acknowledgements This work was supported by the French national research agency through the ANR project COCOA (grant ANR-16-CE01-0007). Part of this study was carried out within the project PROTEVS under the auspices of French Ministry of Defense/DGA, and led by Shom.

References

1. M.J. Gander, F. Kwok and B. Mandal. Dirichlet-Neumann and Neumann-Neumann waveform relaxation algorithms for parabolic problems. *Electron. Trans. Numer. Anal.* **45**, 424–456 (2016).
2. O. Marti, S. Nguyen, P. Braconnot, S. Valcke, F. Lemarié and E. Blayo. A Schwarz iterative method to evaluate ocean–atmosphere coupling schemes: implementation and diagnostics in IPSL-CM6-SW-VLR. *Geosci. Model Dev.* **14**, 2959–2975 (2021).
3. P. Meisrimel, A. Monge and P. Birken. A time adaptive multirate Dirichlet-Neumann waveform relaxation method for heterogeneous coupled heat equations. *arXiv preprint* arXiv:2007.00410 (2020).
4. B. Mohammadi, O. Pironneau and F. Valentin. Rough boundaries and wall laws. *Int. J. Numer. Methods Fluids* **27**(1-4), 169–177 (1998).
5. C. Pelletier, F. Lemarié, E. Blayo, M.-N. Bouin and J.-L. Redelsperger. Two-sided turbulent surface-layer parameterizations for computing air–sea fluxes. *Quart. J. Roy. Meteorol. Soc.* **47**(736), 1726–1751 (2021).
6. S. Thery. *Étude numérique des algorithmes de couplage océan-atmosphère avec prise en compte des paramétrisations physiques de couches limites*. Phd thesis, Université Grenoble Alpes (2021). https://tel.archives-ouvertes.fr/tel-03164786.
7. S.-L. Wu and M. Al-Khaleel. Optimized waveform relaxation methods for RC circuits: Discrete case. *Esaim Math. Model. Numer. Anal.* **51**(1), 209–222 (2017).

A Posteriori Error Estimates in Maximum Norm for Interior Penalty Discontinuous Galerkin Approximation of the Obstacle Problem

B. Ayuso de Dios, T. Gudi, and K. Porwal

1 Introduction

The adaptive finite element method (AFEM) is an effective numerical tool for solving linear and nonlinear PDEs. A proper local refinement plays a key role in AFEM and relies on proper a posteriori error estimators. In this contribution, we introduce a pointwise a posteriori error estimator for the symmetric interior penalty discontinuous Galerkin (SIPG) approximation of the elliptic obstacle problem. The elliptic obstacle problem is a prototype of the elliptic variational inequalities of the first kind. This problem exhibits the free boundary and appears in various processes in engineering and the physical sciences such as elasto-plasticity, dam problems and mathematical finance [3]. A posteriori error analysis in the maximum norm for conforming approximation of obstacle problems is given in the seminal works [6, 7]. For discontinuous Galerkin (DG) approximation, the a posteriori error analysis in the energy norm is contained in [4]. In the maximum norm, to the best of our knowledge, the results in [1] are the first in this direction. Here, due to space limitation, we state the reliability result and focus on its numerical verification and validation. Details on the analysis as well as further discussion can be found in [1].

Blanca Ayuso de Dios
Dipartimento di Matematica e Applicazioni, Università di Milano-Bicocca, Milan, Italy,
e-mail: blanca.ayuso@unimib.it

T. Gudi
Department of Mathematics, Indian Institute of Science, Bangalore - 560012,
e-mail: gudi@math.iisc.ac.in

K. Porwal
Department of Mathematics, Indian Institute of Technology, Delhi - 110016,
e-mail: kamana@maths.iitd.ac.in

S. C. Brenner et al. (eds.), *Domain Decomposition Methods in Science and Engineering XXVI*, Lecture Notes in Computational Science and Engineering 145,
https://doi.org/10.1007/978-3-030-95025-5_20

2 The elliptic obstacle problem

Let $\Omega \subset \mathbb{R}^d$, $d = 2, 3$, be a bounded, polygonal ($d = 2$) or polyhedral ($d = 3$) domain with boundary $\partial\Omega$. Let $f \in L^\infty(\Omega)$ and the obstacle $\chi \in H^1(\Omega) \cap C^0(\overline{\Omega})$ be such that $\chi \le 0$ on $\partial\Omega$. The variational formulation of the obstacle problem then reads: find $u \in \mathcal{K}$ such that

$$\int_\Omega \nabla u \cdot \nabla(u - v)\, dx \le (f, u - v) \quad \forall v \in \mathcal{K} := \{v \in H_0^1(\Omega) : v \ge \chi \ \text{ a.e. in}\, \Omega\}\,, \quad (1)$$

where $(\cdot,\cdot)$ refers to the $L^2(\Omega)$ inner-product and $\mathcal{K}$ is the so-called set of admissible displacements which is a non-empty, closed and convex set. In (1), the solution u could be regarded as the equilibrium position of an elastic membrane subject to the load f whose boundary is held fixed ($u \in H_0^1(\Omega)$) and which is constrained to lie above the given obstacle χ. Such constraint results in non-linearity inherent to the PDE. The contact and non-contact sets of the exact solution u are defined as

$$\mathbb{C} := \{x \in \Omega : u(x) = \chi(x)\}^o, \qquad \mathbb{N} := \{x \in \Omega : u(x) > \chi(x)\}.$$

The continuous Lagrange multiplier $\sigma(u) \in H^{-1}(\Omega)$ is defined by

$$\langle \sigma(u), v\rangle = (f, v) - (\nabla u, \nabla v), \quad \forall v \in H_0^1(\Omega), \quad (2)$$

where $\langle \cdot,\cdot\rangle$ denotes the duality pairing of $H^{-1}(\Omega)$ and $H_0^1(\Omega)$. From (2) and (1), it follows that

$$\langle \sigma(u), v - u\rangle \le 0, \quad \forall v \in \mathcal{K}.$$

In particular, $\sigma(u) = 0$ on the non-contact set $\mathbb{N}$. The classical theory of Stampacchia [3, Chapter 1, page 4] guarantees the existence and uniqueness of the solution. Notice, however that the solution operator is not only non-linear and non-differentiable, but it is strikingly not one-to-one (observe that any variation in f within the contact set might or might not result in a variation in the solution u).

3 The Symmetric Interior Penalty method

Basic Notation and Finite Element spaces

Let $\mathcal{T}_h$ be a shape-regular family of partitions of Ω into triangles or tetrahedra T and let h_T denote the diameter of each $T \in \mathcal{T}_h$ and set $h_{min} = \min\{h_T : T \in \mathcal{T}_h\}$. We denote by $\mathcal{E}_h^o$ and $\mathcal{E}_h^\partial$ the sets of all interior and boundary edges/faces, respectively, and we set $\mathcal{E}_h = \mathcal{E}_h^o \cup \mathcal{E}_h^\partial$. The *average* and *jump* trace operators are defined in the usual way: let T^+ and T^- be two neighbouring elements, and $\mathbf{n}^+$, $\mathbf{n}^-$ be their outward normal unit vectors, respectively ($\mathbf{n}^\pm = \mathbf{n}_{T^\pm}$) and let $\zeta^\pm$ be the restriction of ζ to $T^\pm$. We set:

$$2\{\zeta\} = (\zeta^+ + \zeta^-), \quad [[\,\zeta\,]] = \zeta^+\mathbf{n}^+ + \zeta^-\mathbf{n}^- \qquad \text{on } w \in \mathcal{E}_h^o,$$

and on $e \in \mathcal{E}_h^\partial$ we set $[[\,\zeta\,]] = \zeta\mathbf{n}$. We will also use the notations

$$(u,w)_{\mathcal{T}_h} = \sum_{T\in\mathcal{T}_h}\int_T uw dx, \qquad \langle u,w\rangle_{\mathcal{E}_h} = \sum_{e\in\mathcal{E}_h}\int_e uw ds \quad \forall u,w \in V.$$

Let $\mathbb{P}^1(T)$ be the space of linear polynomials on T and $\mathcal{V}_T$ denotes the set of vertices of the simplex T. We denote by V_h and V_h^{conf} the discontinuous and conforming finite element spaces defined respectively, by

$$V_h = \left\{v \in L^2(\Omega) \;:\; v_{|_T} \in \mathbb{P}^1(T)\ \forall T \in \mathcal{T}_h\right\}, \quad V_h^{conf} = V_h \cap H_0^1(\Omega)\,. \tag{3}$$

Let $\chi_h \in V_h^{conf}$ be the nodal Lagrange linear interpolant of χ. We define the discrete analogue of $\mathcal{K}$ by

$$\mathcal{K}_h := \{v_h \in V_h : v_h|_T(p) \geq \chi_h(p),\ \ \forall p \in \mathcal{V}_T,\ \ \forall T \in \mathcal{T}_h\} \neq \emptyset,$$

which is a nonempty, closed and convex subset of V_h. Note that, $\mathcal{K}_h \not\subseteq \mathcal{K}$.

• *SIPG method:* The method reads: find $u_h \in \mathcal{K}_h$ such that

$$\mathcal{A}_h(u_h, u_h - v_h) \leq (f, u_h - v_h) \qquad \forall v_h \in \mathcal{K}_h, \tag{4}$$

where the SIPG bilinear form $\mathcal{A}_h(\cdot,\cdot)$ is defined as:

$$\mathcal{A}_h(u,w) = (\nabla u, \nabla w)_{\mathcal{T}_h} - \langle\{\nabla u\}, [[\,w\,]]\rangle_{\mathcal{E}_h} - \langle[[\,u\,]], \{\nabla w\}\rangle_{\mathcal{E}_h} + \langle S_e[[\,u\,]], [[\,w\,]]\rangle_{\mathcal{E}_h}\,, \tag{5}$$

with $S_e = \alpha_e h_e^{-1}$, $\alpha_e \geq \alpha^* > 0$, $\forall\, e \in \mathcal{E}_h$ and h_e the length of the edge/face e. Following [4, 8], we define the discrete Lagrange multiplier $\sigma_h \in V_h$:

$$\langle\sigma_h, v_h\rangle_h := (f, v_h) - \mathcal{A}_h(u_h, v_h) \quad \forall v_h \in V_h, \tag{6}$$

where $\langle\cdot,\cdot\rangle_h$ is given by

$$\langle w_h, v_h\rangle_h := \sum_{T\in\mathcal{T}_h}\int_T \mathcal{I}_h(w_h|_T v_h|_T)\,dx = \sum_{T\in\mathcal{T}_h}\frac{|T|}{d+1}\sum_{p\in\mathcal{V}_T} w_h(p)v_h(p),$$

with $\mathcal{I}_h$ denoting the nodal Lagrange linear interpolation operator. The use of the $\langle\cdot,\cdot\rangle_h$ inner product in the definition (6) of σ_h allows for localizing σ_h at the vertices of the partition, which facilitates the implementation. The discrete contact and non-contact sets relative to u_h are defined by:

$$\mathbb{C}_h := \{T \in \mathcal{T}_h : u_h(p) = \chi_h(p)\ \forall\, p \in \mathcal{V}_T\}, \quad \mathbb{N}_h := \{T \in \mathcal{T}_h : u_h(p) > \chi_h(p)\ \forall\, p \in \mathcal{V}_T\},$$

and the free boundary set is given by $\mathbb{M}_h = \mathcal{T}_h \setminus (\mathbb{C}_h \cup \mathbb{N}_h)$. Using (6) and the discrete problem (4), we obtain that $\langle\sigma_h, v_h - u_h\rangle_h \leq 0\ \forall\ v_h \in \mathcal{K}_h$, from which it can be further deduced that $\sigma_h(p) = 0$ on vertices p of $T \subset \mathbb{N}_h$.

4 Reliable a posteriori error estimates in the maximum norm

We now define the error estimators that enter in the full error estimator η_h:

$$\eta_1 = \max_{T\in\mathcal{T}_h} \|h_T^2(f-\sigma_h)\|_{L^\infty(T)}, \qquad \eta_2 = \max_{T\in\mathbb{C}_h\cup\mathbb{M}_h} \|h_T^2\nabla\sigma_h\|_{L^d(T)},$$
$$\eta_3 = \max_{e\in\mathcal{E}_h^o} \|h_e[[\,\nabla u_h\,]]\|_{L^\infty(e)}, \qquad \eta_4 = \|[[\,u_h\,]]\|_{L^\infty(\mathcal{E}_h)},$$
$$\eta_5 = \|(\chi-u_h)^+\|_{L^\infty(\Omega)}, \qquad \eta_6 = \|(u_h-\chi)^+\|_{L^\infty(\{\sigma_h<0\})}\,.$$

The full a posteriori error estimator η_h is then defined as:

$$\eta_h = |\log h_{min}|\Big(\eta_1+\eta_2+\eta_3+\eta_4\Big)+\eta_5+\eta_6.$$

Theorem 1 *Let $u\in\mathcal{K}$ and $u_h\in\mathcal{K}_h$ be the solution of (1) and (4), respectively. Then,*

$$\|u-u_h\|_{L^\infty(\Omega)} \lesssim \eta_h.$$

The proof of the theorem is technical and we refer to [1] for the details as well as the results regarding local efficiency of the estimator.

5 Numerical results

To solve (4), we use the iterative *primal dual active set method* [5]. We briefly describe the algorithm in the present setting.
Primal dual active set method: Let $\lambda_h\in V_h$ be defined by setting for every $T\in\mathcal{T}_h,\ p\in\mathcal{V}_T$ $\lambda_h(p) := \frac{|T|}{d+1}\sigma_h(p)$. Then equation (6) can be rewritten as

$$\mathcal{A}_h(u_h,v_h)+\sum_{T\in\mathcal{T}_h}\sum_{p\in\mathcal{V}_T}\lambda_h(p)v_h(p) = (f,v_h) \quad \forall v_h\in V_h. \tag{7}$$

The so-called complementarity conditions are then given by: $\forall p\in\mathcal{V}_T,\ T\in\mathcal{T}_h$

$$\lambda_h(p)\le 0, \quad u_h(p)\ge\chi_h(p) \quad \text{and} \quad \sum_{T\in\mathcal{T}_h}\sum_{p\in\mathcal{V}_T}\lambda_h(p)(u_h(p)-\chi_h(p))=0\,. \tag{8}$$

After choosing a Lagrangian linear basis for V_h in (3), with $N=\dim(V_h)$, we denote by $\mathbb{A}\in\mathbb{R}^{N\times N}$ and $\boldsymbol{F}\in\mathbb{R}^N$ the matrix and vector representation of $\mathcal{A}_h(\cdot,\cdot)$ in (5) and the right-hand side in (4), respectively. Similarly, $\boldsymbol{U},\boldsymbol{\chi},\boldsymbol{\Lambda}\in\mathbb{R}^N$ denote respectively the vector representations of u_h, χ_h and λ_h. The algebraic formulation of (7)–(8) reads:

$$\mathbb{A}\boldsymbol{U}+\mathbb{I}\Lambda=\boldsymbol{F}, \quad (\Lambda,\boldsymbol{U}-\boldsymbol{\chi})=0, \quad \Lambda\le 0, \quad \boldsymbol{U}\ge\boldsymbol{\chi}, \tag{9}$$

where $\mathbb{I} \in \mathbb{R}^{N\times N}$ is the identity matrix and $(\cdot,\cdot)$ the standard $\mathbb{R}^N$-scalar product. By defining

$$C(U,\Lambda) := \Lambda - \min(0, \Lambda + (\boldsymbol{U} - \boldsymbol{\chi})), \tag{10}$$

the complementarity conditions in (9) reduce to $C(U,\Lambda) = 0$. Indeed, from the definition (10), notice that if $\Lambda + (\boldsymbol{U} - \boldsymbol{\chi}) < 0$ then $C(U,\Lambda) = (\boldsymbol{\chi} - \boldsymbol{U})$ and so, $C(U,\Lambda) = 0$ implies $\boldsymbol{U} = \boldsymbol{\chi}$, which together with $\Lambda + (\boldsymbol{U} - \boldsymbol{\chi}) < 0$ gives $\Lambda < 0$. Similarly, $\Lambda + (\boldsymbol{U} - \boldsymbol{\chi}) > 0$ would imply $C(U,\Lambda) = \Lambda$. In this case the complementarity condition $C(U,\Lambda) = 0$ gives $\Lambda = 0$ which, together with $\Lambda + (\boldsymbol{U} - \boldsymbol{\chi}) > 0$, yields $\boldsymbol{U} > \boldsymbol{\chi}$. Hence, the solution of (9) is reduced to solving the system

$$\mathbb{A}\,\boldsymbol{U} + \mathbb{I}\,\Lambda = \boldsymbol{F}, \qquad C(U,\Lambda) = 0. \tag{11}$$

The primal-dual active set algorithm solves (11) iteratively:

(i) Set $k = 0$, Initialise $\boldsymbol{U}^{(0)}$, $\Lambda^{(0)}$.
(ii) Find the sets of vertices $\mathbb{AC}^{(k)}$ and $\mathbb{D}^{(k)}$ defined as

$$\mathbb{AC}^{(k)} = \{1 \le j \le N : \Lambda_j^{(k)} + (U_j^{(k)} - \chi_j^{(k)}) < 0\} \quad \text{indices in active set,}$$
$$\mathbb{D}^{(k)} = \{1 \le j \le N : \Lambda_j^{(k)} + (U_j^{(k)} - \chi_j^{(k)}) \ge 0\} \quad \text{indices not in active set.}$$

(iii) Solve for $(\boldsymbol{U}^{(k+1)}, \Lambda^{(k+1)})$ from the following system:

$$\mathbb{A}\,\boldsymbol{U}^{(k+1)} + \mathbb{I}\,\Lambda^{(k+1)} = \boldsymbol{F}, \quad U_j^{(k+1)} = \chi_j \;\forall j \in \mathbb{AC}^{(k)}, \qquad \Lambda_j^{(k+1)} = 0 \;\forall j \in \mathbb{D}^{(k)}.$$

(iv) Set $k = k + 1$. Go to Step (ii) and compute $\mathbb{AC}^{(k)}$ and its complementary $\mathbb{D}^{(k)}$.

The iteration is stopped when $\mathbb{AC}^{(k)} = \mathbb{AC}^{(k+1)}$. The set $\mathbb{AC}^{(k)}$ contains the indices for the vertices in the discrete contact set $\mathbb{C}_h$; $\mathbb{D}^{(k)}$ contains the indices of the remaining nodes.

5.1 Numerical experiments

We present now some test examples to illustrate the performance of an a posteriori error estimator. For the adaptive refinement, we use the paradigm

SOLVE ⟶ ESTIMATE ⟶ MARK ⟶ REFINE

In the step SOLVE, we compute u_h using the primal-dual active set algorithm as described before. Thereafter, we compute the error estimator η_h on each element $T \in \mathcal{T}_h$ and use the maximum marking strategy with parameter $\theta = 0.4$. Finally, the mesh is refined using the newest vertex bisection algorithm. In all examples, we set $\alpha_e = 25$ and $\mathbb{AC}$ refers to the discrete active set (depicted in yellow).

Example 1: Madonna's obstacle: (scaled version of [7, example 3.1])). Let $\Omega = (0,1)^2$, $f = 0$ and $r^2 = (x - 1/2)^2 + (y - 1/2)^2$, $x, y \in \Omega$

$$\chi = 1 - 4r, \qquad u = \begin{cases} 1 - 4r, & r < 1/4 \\ -(\log(r) + 2\log(2)), & r \geq 1/4. \end{cases}$$

In Figure 1a we report the error and the estimator η_h. This graphic indicates a rate $(1/\mathrm{DOF})$ with respect to degrees of freedom (DOF). The single estimators η_i, $i = 1, ..., 6$ are plotted in Figure 1b. Both graphics confirm the reliability of the estimator. In Figure 2 are depicted the efficiency indices (leftmost subfigure), the adaptive mesh refinement at level 20 (center) and the discrete contact set $\mathbb{C}_h$ (rightmost figure). Note that the solution is singular in the contact set $\mathbb{C}$ due to the singularity of the obstacle therein, which leads to more refinement in $\mathbb{C}_h$. Also as expected, we observe more refinement near the free boundary.

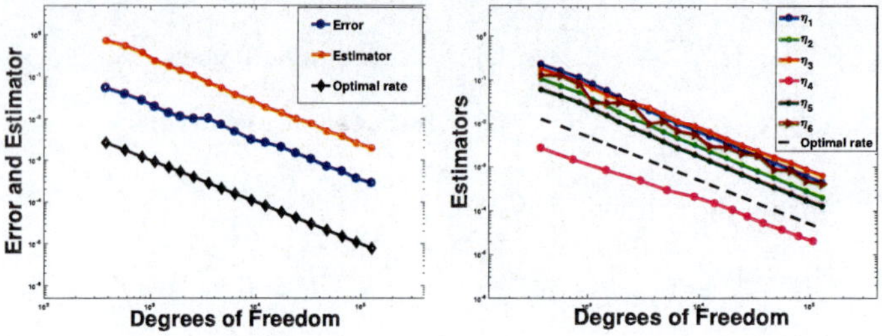

(a) Error and Estimator

(b) Estimator contributions

Fig. 1: Example 1: Error, Estimator and estimator contributions

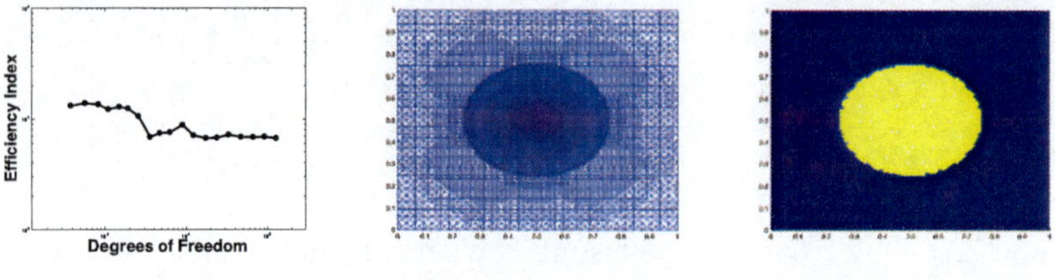

Fig. 2: Example 1: Efficiency index, adaptive mesh and $\mathbb{AC}$

Example 2: non-convex domain [2]:

$$\Omega = (-2,2)^2 \setminus [0,2) \times (-2,0], \qquad \chi = 0,$$

$$u = r^{2/3}\sin(2\theta/3)\gamma_1(r), \qquad r^2 = x^2 + y^2\,, \qquad \tilde{r} = 2(r - 1/4)$$

$$f = -\,r^{2/3}\sin(2\theta/3)\Big(\frac{\gamma_1'(r)}{r} + \gamma_1''(r)\Big) - \frac{4}{3}r^{-1/3}\sin(2\theta/3)\gamma_1'(r) - \gamma_2(r)$$

$$\gamma_1(r) = \begin{cases} 1, & \tilde{r} < 0 \\ -6\tilde{r}^5 + 15\tilde{r}^4 - 10\tilde{r}^3 + 1, & 0 \leq \tilde{r} < 1 \\ 0, & \tilde{r} \geq 1, \end{cases} \qquad \gamma_2(r) = \begin{cases} 0, & r \leq \frac{5}{4} \\ 1, & \text{otherwise.} \end{cases}$$

In Figure 3(a) we compare the estimator η_h with the one in the energy norm. From this graphic, it is evident that the error and the estimator converge with rate $1/\text{DOF}$ in L^∞ and $1/\sqrt{\text{DOF}}$ in energy norm. The convergence behaviour of the single estimators η_i, $i = 1, ..., 6$, is given in Figure 3(b). Note that η_5 is zero since $\chi = \chi_h = 0$ in this example. Figure 4 confirms the efficiency of the estimator η_h. In Figure 4 are also given the adaptive mesh refinement at the level 24 and $\mathbb{C}_h$. We observe that the estimator captures well the singular behavior of the solution. The mesh refinement near the free boundary is higher due to the large jump in gradients.

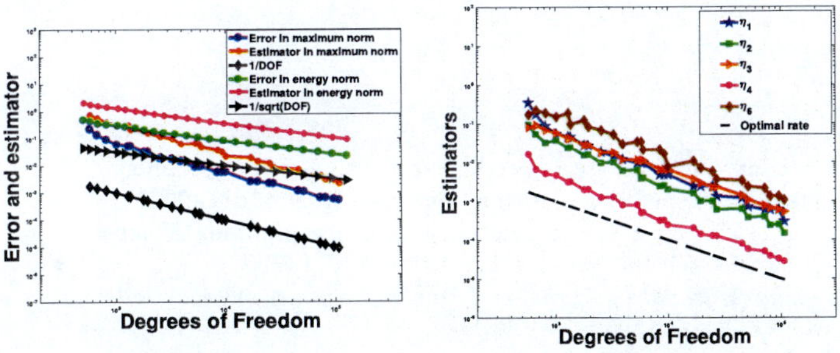

(a) Error and Estimator

(b) Estimators in maximum norm

Fig. 3: Example 2: Error and estimator in energy norm and maximum norm and single estimators in maximum norm.

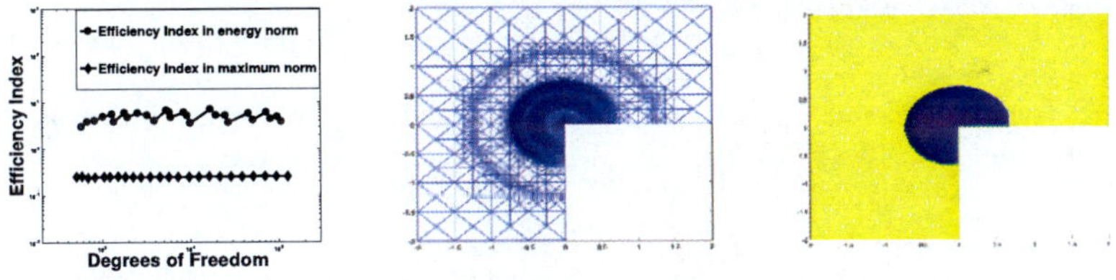

Fig. 4: Example 2: Efficiency index in energy norm and maximum norm, adaptive mesh and $\mathbb{AC}$.

Example 3: taken from [6] (Lipschitz obstacle):

$$\Omega = \{(x, y) \in \mathbb{R}^2 : |x| + |y| < 1\}, \quad f = -5, \qquad \chi = \text{dist}(x, \partial\Omega) - 1/5.$$

In Figure 5 we have reported the estimator η_h and the single estimators η_i, $i = 1, ..., 6$, ($\eta_5 = 0$ since χ is piecewise linear) (leftmost) together with the adaptive mesh at refinement level 7 (center) and $\mathbb{C}_h$ (rightmost). It can be observed that the estimator converges with the optimal rate. The obstacle function is in the shape of a pyramid and the continuous Lagrange multiplier has support along the edges of the obstacle which justifies the refinement along the edges of the obstacle in the contact region.

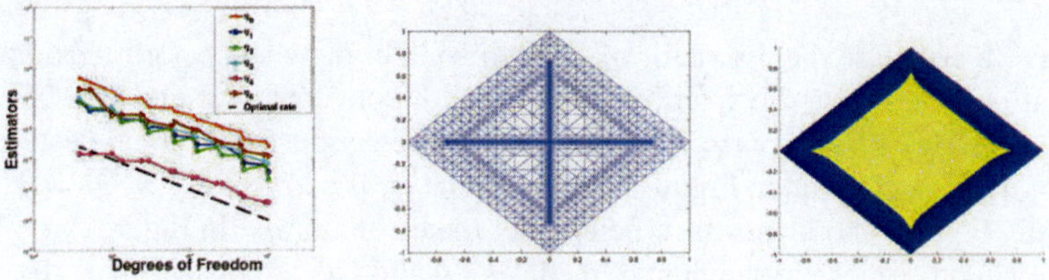

Fig. 5: Example 3: Estimators, Adaptive Mesh and $\mathbb{AC}$.

References

1. B. Ayuso de Dios, T. Gudi and K. Porwal. Pointwise a posteriori error analysis of a discontinuous Galerkin method for the elliptic obstacle problem. *arXiv preprint* arXiv:2108.11611 (2021).
2. D. Braess, C. Carstensen and R.H.W. Hoppe. Convergence analysis of a conforming adaptive finite element method for an obstacle problem. *Numer. Math.* **107**, 455–471 (2007).
3. R. Glowinski. *Numerical Methods for Nonlinear Variational Problems*. Springer-Verlag, Berlin (2008).
4. T. Gudi and K. Porwal. A posteriori error control of discontinuous Galerkin methods for elliptic obstacle problems. *Math. Comp.* **83**, 579–602 (2014).
5. M. Hintermüller, K. Ito and K. Kunish. The primal-dual active set strategy as a semismooth Newton method. *SIAM J. Optim.* 13, 865–888 (2003).
6. R.H. Nochetto, K.G. Siebert and A. Veeser. Pointwise a posteriori error control for elliptic obstacle problems. *Numer. Math.* **95**, 163–195 (2003).
7. R.H. Nochetto, K.G. Siebert and A. Veeser. Fully localized a posteriori error estimators and barrier sets for contact problems. *SIAM J. Numer. Anal.* **42**(5), 2118–2135 (2005).
8. A. Veeser. Efficient and reliable a posteriori error estimators for elliptic obstacle problems. *SIAM J. Numer. Anal.* **39**, 146–167 (2001).

Spectral Equivalence Properties of Higher-Order Tensor Product Finite Elements

Clark R. Dohrmann

1 Introduction

The focus of this study is on spectral equivalence results for higher-order tensor product finite elements in the $H(\text{curl})$, $H(\text{div})$, and L^2 function spaces. For certain choices of the higher-order shape functions, the resulting mass and stiffness matrices are spectrally equivalent to those for an assembly of lowest-order edge-, face- or interior-based elements on the associated Gauss–Lobatto–Legendre (GLL) mesh. This equivalence will help enable the development of efficient domain decomposition or multigrid preconditioners. Specifically, preconditioners for the equivalent lowest-order linear system can be used for the higher-order problem and avoid the demands of assembling a higher-order coefficient matrix.

Using assemblies of lowest-order (linear) elements for efficient preconditioning of higher-order discretizations in the function space H^1 is not new. We refer the interested reader to Section 7.1 of [10] or the introduction of [2] for a discussion of the pioneering work by Orszag [9], Deville and Mund [3, 4], Canuto [1] and others. We are, however, not aware of similar approaches for problems using higher-order edge- (Nédélec), face- (Raviart–Thomas) or interior-based elements. We note for the case of nodal elements that the degrees of freedom (DOFs) for a higher order element and its equivalent assembly of lowest-order elements are nodal values in both cases. This natural one-to-one correspondence of DOFs can be realized for edge-, face- and interior-based elements by using shape functions (bases) associated with integrals and introduced by Gerritsma [6].

For edge-based elements, the DOFs for the shape functions are associated with integrals of tangential components of a vector field along each edge of the associated GLL mesh (see Figure 1 left). Similarly, DOFs for face-based elements correspond to integrals of the normal component of a vector field over individual faces of the GLL mesh (see Figure 1 right). For completeness, we also present shape functions

Clark R. Dohrmann
Sandia National Laboratories, Albuquerque, New Mexico, USA, e-mail: crdohrm@sandia.gov

S. C. Brenner et al. (eds.), *Domain Decomposition Methods in Science and Engineering XXVI*, Lecture Notes in Computational Science and Engineering 145,
https://doi.org/10.1007/978-3-030-95025-5_21

and equivalence results for related interior-based elements. For these elements, the DOFs correspond to integrals of a scalar function over individual elements of the GLL mesh. We note in all three cases that the shape functions can be expressed simply in terms of one-dimensional interpolatory nodal functions at the GLL points along with a one-dimensional function which enables the correspondence between DOFs of the higher- and lowest-order elements.

The paper is organized as follows. Shape functions for edge-, face-, and interior-based elements are described in §2. This is followed in §3 by a presentation of spectral equivalence results between higher-order elements and their lowest-order counterparts. Numerical results are presented in §4 which confirm these results. A more comprehensive report [5] can be consulted for complete proofs and applications of the spectral equivalence results to preconditioning.

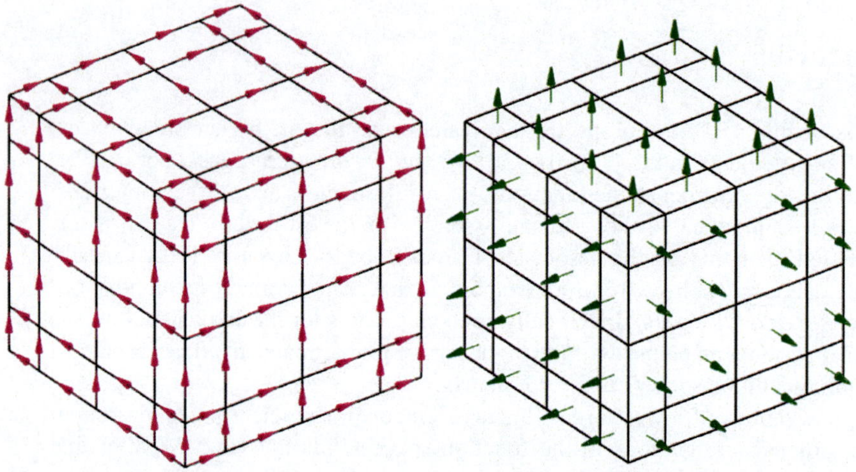

Fig. 1: Edge (left) and face (right) locations on three faces of a cube for a higher-order element of degree $p = 4$. Also shown is the corresponding assembly of p^3 lowest-order elements on the associated Gauss–Lobatto–Legendre (GLL) mesh.

2 Shape functions

Following the notation in [8], let $Q_{i,j,k}$ denote the space of polynomials in reference element coordinates $(\eta_1, \eta_2, \eta_3) \in [-1, 1]$ for which the maximum degree is i in η_1, j in η_2 and k in η_3.

As is commonly done for nodal elements, one-dimensional GLL shape functions $\varphi_0, \ldots, \varphi_p$ are used to construct higher-order shape functions in three dimensions. See Figure 2 (left) for the case of degree $p = 4$. Notice that these functions are simply interpolatory (Lagrange) shape functions at the GLL points $x_0 = -1$, $x_p = 1$, and $x_{i-1} < x_i$ for $i = 1, \ldots, p$. We remark that the internal GLL points $x_1, \ldots, x_{p-1}$ are the roots of L'_{p-1}, where L_p is the Legendre polynomial of degree p.

Shape functions for edge-, face-, and interior-based elements based on the work of Gerritsma [6] are described next. Although different from the shape functions in [8], they span the same polynomial spaces and are conforming between elements.

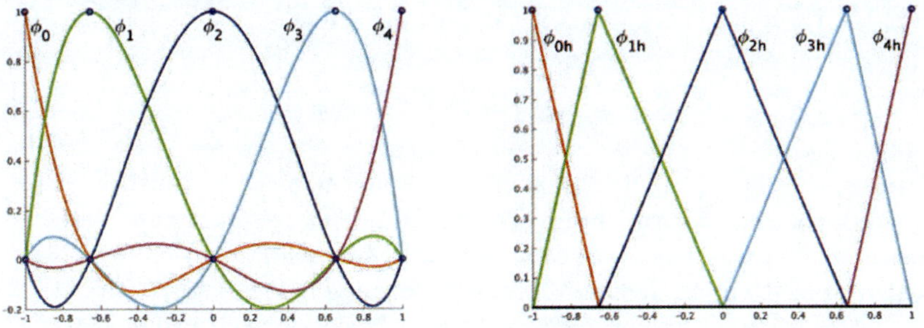

Fig. 2: One-dimensional higher-order (left) and linear (right) shape functions associated with GLL points for degree $p = 4$.

2.1 Edge shape functions

The vector field for an edge-based finite element of degree p can be expressed in terms of the reference element coordinates as

$$\boldsymbol{u}_p^{\mathrm{e}} = u_{1p}^{\mathrm{e}} \boldsymbol{b}_1 + u_{2p}^{\mathrm{e}} \boldsymbol{b}_2 + u_{3p}^{\mathrm{e}} \boldsymbol{b}_3,$$

where $u_{1p}^{\mathrm{e}} \in Q_{p-1,p,p}$, $u_{2p}^{\mathrm{e}} \in Q_{p,p-1,p}$, $u_{3p}^{\mathrm{e}} \in Q_{p,p,p-1}$, and $\boldsymbol{b}_1, \boldsymbol{b}_2, \boldsymbol{b}_3$ are unit vectors associated with the element coordinates (see e.g. [8]).

Our present focus is on edges aligned with the $\boldsymbol{b}_1$ direction; similar constructions of shape functions hold for edges aligned with the other two directions. For each $i \in \{0, \ldots, p-1\}$ define

$$\psi_i(\eta_1) = \sum_{m=0}^{p} a_{im} \varphi_m(\eta_1), \qquad a_{im} = \begin{cases} 0 & m \leq i \\ 1 & m > i. \end{cases}$$

Since $\psi_i(x_{m+1}) - \psi_i(x_m)$ is 1 for $m = i$ and 0 for $m \neq i$, it follows that

$$\int_{x_m}^{x_{m+1}} \psi_i' \, \mathrm{d}x = \delta_{im}, \tag{1}$$

where δ_{im} is the Kronecker delta function. The edge functions $\psi_0', \ldots, \psi_{p-1}'$ and their application to tensor product finite elements are discussed in [6].

Let $\mathcal{E}_{ijk}$ denote the edge with $\eta_1 \in (x_i, x_{i+1})$, $\eta_2 = x_j$ and $\eta_3 = x_k$. The shape function associated with this edge is given by

$$\boldsymbol{\varphi}^{\mathsf{e}}_{ijk}(\eta_1, \eta_2, \eta_3) = \psi_i'(\eta_1)\varphi_j(\eta_2)\varphi_k(\eta_3)\boldsymbol{b}_1. \tag{2}$$

Notice that $\boldsymbol{\varphi}^{\mathsf{e}}_{ijk} \cdot \boldsymbol{b}_2 = \boldsymbol{\varphi}^{\mathsf{e}}_{ijk} \cdot \boldsymbol{b}_3 = 0$. Thus, the tangential component of $\boldsymbol{\varphi}^{\mathsf{e}}_{ijk}$ vanishes along all edges not in the $\boldsymbol{b}_1$ direction. Consider the integral $a^{\mathsf{e}}_{lmn} := \int_{\mathcal{E}_{lmn}} \boldsymbol{\varphi}^{\mathsf{e}}_{ijk} \cdot \boldsymbol{b}_1 \, \mathrm{d}x$. Since $\varphi_j(x_m) = \delta_{jm}$ and $\varphi_k(x_n) = \delta_{kn}$, we find using (1) that

$$a^{\mathsf{e}}_{lmn} = \delta_{jm}\delta_{kn} \int_{x_l}^{x_{l+1}} \psi_i'(\eta_1) \, \mathrm{d}x = \delta_{il}\delta_{jm}\delta_{kn}.$$

In other words, the integral of the tangential component of $\boldsymbol{\varphi}^{\mathsf{e}}_{ijk}$ vanishes over all edges except for $\mathcal{E}_{ijk}$, for which this integral is 1. This feature ensures linear independence of the shape functions. Moreover, arguments similar to those in [8] can be used to show the finite element space is conforming in the space $H(\text{curl}; \hat{\Omega})$, where $\hat{\Omega} := (-1, 1)^3$. Using the curl-conserving transformation described in §3.9 of [7], the finite elements are also conforming in the space $H(\text{curl}; \Omega)$, where Ω is the domain of the higher-order finite element mesh.

2.2 Face shape functions

The vector field for a face-based finite element of degree p can be expressed in terms of the element coordinates as

$$\boldsymbol{u}^{\mathsf{f}}_p = u^{\mathsf{f}}_{1p}\boldsymbol{b}_1 + u^{\mathsf{f}}_{2p}\boldsymbol{b}_2 + u^{\mathsf{f}}_{3p}\boldsymbol{b}_3,$$

where $u^{\mathsf{f}}_{1p} \in \mathcal{Q}_{p,p-1,p-1}$, $u^{\mathsf{f}}_{2p} \in \mathcal{Q}_{p-1,p,p-1}$, and $u^{\mathsf{f}}_{3p} \in \mathcal{Q}_{p-1,p-1,p}$ (again, see e.g. [8]).

Our present focus is on faces aligned with the $\boldsymbol{b}_3$ direction; similar constructions of shape functions hold for faces aligned with the other two directions. Let $\mathcal{F}_{ijk}$ denote the face with $\eta_1 \in (x_i, x_{i+1})$, $\eta_2 \in (x_j, x_{j+1})$, and $\eta_3 = x_k$. The shape function associated with this face is given by

$$\boldsymbol{\varphi}^{\mathsf{f}}_{ijk}(\eta_1, \eta_2, \eta_3) = \psi_i'(\eta_1)\psi_j'(\eta_2)\varphi_k(\eta_3)\boldsymbol{b}_3. \tag{3}$$

Notice that $\boldsymbol{\varphi}^{\mathsf{f}}_{ijk} \cdot \boldsymbol{b}_1 = \boldsymbol{\varphi}^{\mathsf{f}}_{ijk} \cdot \boldsymbol{b}_2 = 0$. Thus, the normal component of $\boldsymbol{\varphi}^{\mathsf{f}}_{ijk}$ vanishes over all faces with normals not in the $\boldsymbol{b}_3$ direction. Next, consider the area integral $a^{\mathsf{f}}_{lmn} := \int_{\mathcal{F}_{lmn}} \boldsymbol{\varphi}^{\mathsf{f}}_{ijk} \cdot \boldsymbol{b}_3 \, \mathrm{d}x$. Since $\varphi_k(x_n) = \delta_{kn}$, we find using (1) that

$$\begin{aligned} a^{\mathsf{f}}_{lmn} &= \int_{\mathcal{F}_{lmk}} \delta_{kn} \psi'_i(\eta_1) \varphi'_j(\eta_2)\, \mathrm{d}x \\ &= \delta_{kn} \int_{x_l}^{x_{l+1}} \psi'_i(\eta_1)\, \mathrm{d}\eta_1 \int_{x_m}^{x_{m+1}} \psi'_j(\eta_2)\, \mathrm{d}\eta_2 = \delta_{il}\delta_{jm}\delta_{kn}. \end{aligned}$$

In other words, the integral of the normal component of $\boldsymbol{\varphi}^{\mathsf{f}}_{ijk}$ vanishes over all faces except for $\mathcal{F}_{ijk}$, for which this integral is 1. Again, this ensures linear independence of the shape functions, and arguments similar to those in [8] can be used to show the finite element space is conforming in $H(\mathrm{div}; \hat{\Omega})$. Using the divergence-conserving transformation described in §3.9 of [7], the finite elements are also conforming in the space $H(\mathrm{div}; \Omega)$, where Ω is the domain of the higher-order finite element mesh.

2.3 Interior shape functions

The scalar field of an interior-based element is approximated by functions $u^{\mathsf{v}}_p \in \mathcal{Q}_{p-1,p-1,p-1}$. Let V_{ijk} denote the cell with $\eta_1 \in (x_i, x_{i+1})$, $\eta_2 \in (x_j, x_{j+1})$, and $\eta_3 \in (x_k, x_{k+1})$. The shape function associated with this cell is given by

$$\varphi^{\mathsf{v}}_{ijk}(\eta_1, \eta_2, \eta_3) = \psi'_i(\eta_1)\psi'_j(\eta_2)\psi'_k(\eta_3). \tag{4}$$

Consider the volume integrals $a^{\mathsf{v}}_{lmn} := \int_{V_{lmn}} \varphi^{\mathsf{v}}_{ijk}\, \mathrm{d}x$. We find using (1) that

$$a^{\mathsf{v}}_{lmn} = \int_{x_l}^{x_{l+1}} \psi'_i(\eta_1) \int_{x_m}^{x_{m+1}} \psi'_j(\eta_2) \int_{x_n}^{x_{n+1}} \psi'_k(\eta_3)\, \mathrm{d}x = \delta_{il}\delta_{jm}\delta_{kn}.$$

In other words, the integral of $\varphi^{\mathsf{v}}_{ijk}$ vanishes over all regions except for V_{ijk}, for which this integral is 1. This ensures linear independence of the shape functions. Further, a polynomial function $u^{\mathsf{v}}_p \in \mathcal{Q}_{p-1,p-1,p-1}$ can be expressed in terms of the shape functions as

$$u^{\mathsf{v}}_p = \sum_{i,j,k=0}^{p-1} c^{\mathsf{v}}_{ijk} \psi'_i(\eta_1)\psi'_j(\eta_2)\psi'_k(\eta_3), \qquad c^{\mathsf{v}}_{ijk}(u^{\mathsf{v}}_p) = \int_{V_{ijk}} u^{\mathsf{v}}_p\, \mathrm{d}x.$$

Remark 1 Starting with the edge shape function $\boldsymbol{\varphi}^{\mathsf{e}}_{ijk}$ in (2), notice that the face shape function $\boldsymbol{\varphi}^{\mathsf{f}}_{ijk}$ in (3) is obtained simply by replacing $\varphi_j(\eta_2)\boldsymbol{b}_1$ with $\psi'_j \boldsymbol{b}_3$. Likewise, $\varphi^{\mathsf{v}}_{ijk}$ in (4) is obtained from $\boldsymbol{\varphi}^{\mathsf{f}}_{ijk}$ simply by replacing $\varphi_k(\eta_3)\boldsymbol{b}_3$ with $\psi'_k(\eta_3)$.

2.4 Lowest-order shape functions

The lowest-order counterparts of the one-dimensional higher-order shape functions $\varphi_0, \ldots, \varphi_p$ are piecewise linear and are denoted by $\varphi_{0h}, \ldots \varphi_{ph}$ (see Figure 2 (right) for the case of $p = 4$). Analogous to the higher-order edge, face, and interior shape functions, we may define the lowest-order counterparts of (2), (3) and (4) as

$$\boldsymbol{\varphi}^{\mathsf{e}}_{ijkh}(\eta_1, \eta_2, \eta_3) = \psi'_{ih}(\eta_1)\varphi_{jh}(\eta_2)\varphi_{kh}(\eta_3)\boldsymbol{b}_1, \tag{5}$$

$$\boldsymbol{\varphi}^{\mathsf{f}}_{ijkh}(\eta_1, \eta_2, \eta_3) = \psi'_{ih}(\eta_1)\psi'_{jh}(\eta_2)\varphi_{kh}(\eta_3)\boldsymbol{b}_3, \tag{6}$$

$$\varphi^{\mathsf{v}}_{ijkh}(\eta_1, \eta_2, \eta_3) = \psi'_{ih}(\eta_1)\psi'_{jh}(\eta_2)\psi'_{kh}(\eta_3), \tag{7}$$

where ψ_{ih} is defined analogously to ψ_i as

$$\psi_{ih}(\eta_1) = \sum_{m=0}^{p} a_{im}\varphi_{mh}(\eta_1).$$

By construction, the lowest-order edge, face, and interior shape functions in (5)–(7) have similar interpolatory properties to their higher-order counterparts. For example, the integrated tangential component of $\boldsymbol{\varphi}^{\mathsf{e}}_{ijkh}$ is 1 along edge $\mathcal{E}_{ijk}$ and vanishes along all other edges of the GLL mesh just like the higher-order shape function $\boldsymbol{\varphi}^{\mathsf{e}}_{ijk}$.

3 Spectral equivalence results

In this section, we summarize the spectral equivalence of mass and stiffness matrices of higher-order edge, face and interior-based elements with their assembled lowest-order counterparts on the GLL mesh. By spectral equivalence we mean that constants in the estimates are independent of the polynomial degree. In three dimensions, the constants for the equivalence are independent of element aspect ratios for mass matrices, while stiffness matrices have a weak dependence for edge-based elements but no dependence for face-based elements. More details, including proofs of the results, can be found in [5]. We use the notational convention $f \simeq g$ to mean that there exist positive constants c and C, independent of polynomial degree, such that $cg \leq f \leq Cg$ for non-negative scalars f and g.

3.1 Mass matrix equivalence

We follow closely in [5] the development given on pages 16 and 17 of [1] to show spectral equivalence of mass matrices. Based on these results, spectral equivalence for stiffness matrices is shown to follow.

Lemma 1 *Let $\boldsymbol{u}_h^{\mathsf{k}}$ denote the lowest-order interpolant of the higher-order vector function $\boldsymbol{u}_p^{\mathsf{k}}$, where $\mathsf{k} \in \{\mathsf{e}, \mathsf{f}\}$. Similarly, let u_h^{v} denote the lowest-order interpolant of the higher-order scalar function u_p^{v}. It holds that*

$$\|\boldsymbol{u}_h^{\mathsf{e}}\|_{L^2(\hat{\Omega})} \simeq \|\boldsymbol{u}_p^{\mathsf{e}}\|_{L^2(\hat{\Omega})}, \tag{8}$$

$$\|\boldsymbol{u}_h^{\mathsf{f}}\|_{L^2(\hat{\Omega})} \simeq \|\boldsymbol{u}_p^{\mathsf{f}}\|_{L^2(\hat{\Omega})}, \tag{9}$$

$$\|u_h^{\mathsf{v}}\|_{L^2(\hat{\Omega})} \simeq \|u_p^{\mathsf{v}}\|_{L^2(\hat{\Omega})}. \tag{10}$$

3.2 Stiffness matrix equivalence

The stiffness matrix for a higher-order edge-based element is associated with the curl semi-norm of $\boldsymbol{u}_p^{\mathsf{e}}$, which we denote by $|\nabla \times \boldsymbol{u}_p^{\mathsf{e}}|_{L^2(\hat{\Omega})}$. Similarly, the stiffness matrix for a higher-order face-based element is associated with the divergence semi-norm of $\boldsymbol{u}_p^{\mathsf{f}}$, which we denote by $|\nabla \cdot \boldsymbol{u}_p^{\mathsf{f}}|_{L^2(\hat{\Omega})}$.

Lemma 2 *Let $\boldsymbol{u}_h^{\mathsf{k}}$ denote the lowest-order interpolant of $\boldsymbol{u}_p^{\mathsf{k}}$, where $\mathsf{k} \in \{\mathsf{e}, \mathsf{f}\}$. It holds that*

$$|\nabla \times \boldsymbol{u}_h^{\mathsf{e}}|_{L^2(\hat{\Omega})} \simeq |\nabla \times \boldsymbol{u}_p^{\mathsf{e}}|_{L^2(\hat{\Omega})}, \tag{11}$$

$$|\nabla \cdot \boldsymbol{u}_h^{\mathsf{f}}|_{L^2(\hat{\Omega})} \simeq |\nabla \cdot \boldsymbol{u}_p^{\mathsf{f}}|_{L^2(\hat{\Omega})}. \tag{12}$$

4 Numerical results

Numerical support for the estimates in (8)–(12) is provided in this section. For each of these estimates, we consider a generalized eigenvalue problem of the form $B_p x = \lambda B_h x$, where B_p and B_h are the higher- and lowest-order element mass or stiffness matrices corresponding to the estimate. Notice that B_p and B_h are singular for (11) and (12), with null spaces corresponding to gradients of node-based finite element functions and curls of edge-based finite element functions, respectively. For these two cases, we confirmed that the null spaces for B_p and B_h are identical. Further, the generalized eigenvalue problem was solved in a space orthogonal to the null space.

The smallest and largest eigenvalues corresponding to (8)–(10) are shown in Figure 3 (left) for elements in three dimensions. For completeness, results are also shown for node-based elements in the space H^1. Notice in all cases that the smallest and largest eigenvalues are bounded by those for node-based elements. This provides numerical support for (8)–(10) based on node-based spectral equivalence results in [1]. Similar results are shown in Figure 3 (right) which correspond to (11)–(12).

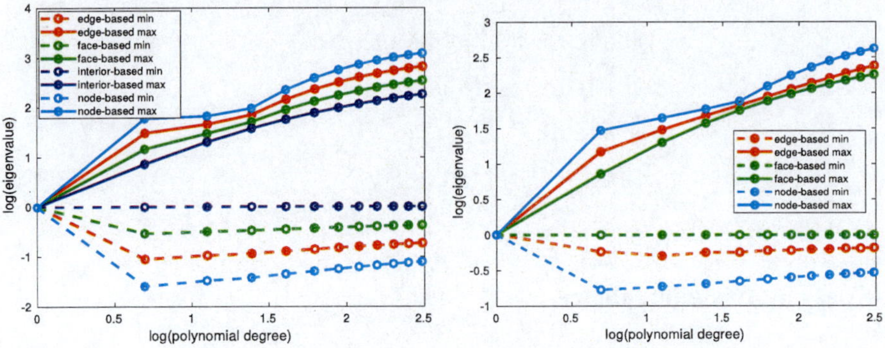

Fig. 3: Generalized eigenvalues associated with mass (left) and stiffness (right) matrices in three dimensions.

Acknowledgements Sandia National Laboratories is a multimission laboratory managed and operated by National Technology and Engineering Solutions of Sandia, LLC, a wholly owned subsidiary of Honeywell International Inc., for the U.S. Department of Energy's National Nuclear Security Administration under contract DE-NA0003525.

References

1. C. Canuto. Stabilization of spectral methods by finite element bubble functions. *Comput. Methods Appl. Mech. Engrg.* **116**, 13—26 (1994).
2. C. Canuto, P. Gervasio and A. Quarteroni. Finite-element preconditioning of G-NI spectral methods. *SIAM J. Sci. Comput.* **31**(6), 4422–4451 (2010).
3. M.O. Deville and E.H. Mund. Chebyshev pseudospectral solution of second-order elliptic equations with finite element preconditioning. *J. Comput. Phys.* **60**(3), 517–533 (1985).
4. M.O. Deville and E.H. Mund. Finite-element preconditioning for pseudospectral solutions of elliptic problems. *SIAM J. Sci. Stat. Comput.* **11**(2), 311–342 (1990).
5. C.R. Dohrmann. Spectral equivalence properties of higher-order tensor product finite elements and applications to preconditioning. Technical Report SAND2021-0323, Sandia National Laboratories (2021).
6. M. Gerritsma. Edge functions for spectral element methods. In: *Spectral and High Order Methods for Partial Differential Equations*, pp. 199–207, Springer (2011).
7. P. Monk. *Finite Element Methods for Maxwell's Equations*. Oxford University Press, Oxford (2003).
8. J.-C. Nédélec. Mixed finite elements in $\mathbb{R}^3$. *Numer. Math.* **35**, 315–341 (1980).
9. S.A. Orszag. Spectral methods for problems in complex geometry. *J. Comput. Phys.* **37**(1), 70–92 (1980).
10. A. Toselli and O. Widlund. *Domain Decomposition Methods – Algorithms and Theory*. Volume 34 of Springer Series in Computational Mathematics, Springer-Verlag, Berlin, Heidelberg, New York (2005).

Optimizing Transmission Conditions for Multiple Subdomains in the Magnetotelluric Approximation of Maxwell's Equations

V. Dolean, M.J. Gander, and A. Kyriakis

1 Optimized Schwarz for the Magnetotelluric Approximation

Wave propagation phenomena are ubiquitous in science and engineering. In Geophysics, the magnetotelluric approximation of Maxwell's equations is an important tool to extract information about the spatial variation of electrical conductivity in the Earth's subsurface. This approximation results in a complex diffusion equation [4],

$$\Delta u - (\sigma - i\varepsilon)u = f, \quad \text{in a domain } \Omega, \tag{1}$$

where f is the source function, and σ and ε are strictly positive constants.[1]

To study Optimized Schwarz Methods (OSMs) for (1), we use a rectangular domain Ω given by the union of rectangular subdomains $\Omega_j := (a_j, b_j) \times (0, \hat{L})$, $j = 1, 2, \ldots, J$, where $a_j = (j-1)L - \frac{\delta}{2}$ and $b_j = jL + \frac{\delta}{2}$, and δ is the overlap, like in [2]. Our OSM computes for iteration index $n = 1, 2, \ldots$

$$\begin{aligned} \Delta u_j^n - (\sigma - i\varepsilon)u_j^n &= f && \text{in } \Omega_j, \\ -\partial_x u_j^n + p_j^- u_j^n &= -\partial_x u_{j-1}^{n-1} + p_j^- u_{j-1}^{n-1} && \text{at } x = a_j, \\ \partial_x u_j^n + p_j^+ u_j^n &= \partial_x u_{j+1}^{n-1} + p_j^+ u_{j+1}^{n-1} && \text{at } x = b_j, \end{aligned} \tag{2}$$

where p_j^- and p_j^+ are strictly positive parameters in the so-called 2-sided OSM, see e.g. [6], and we have at the top and bottom homogeneous Dirichlet boundary

Alexandros Kyriakis
University of Strathclyde, e-mail: alexandros.kyriakis@strath.ac.uk

Victorita Dolean
University of Strathclyde and University Côte d'Azur e-mail: work@victoritadolean.com

Martin J. Gander
Université de Genève e-mail: martin.gander@unige.ch

[1] In the magnetotelluric approximation we have $\sigma = 0$, but we consider the slightly more general case here. Note also that the zeroth order term in (1) is much more benign than the zeroth order term of opposite sign in the Helmholtz equation, see e.g. [5].

S. C. Brenner et al. (eds.), *Domain Decomposition Methods in Science and Engineering XXVI*, Lecture Notes in Computational Science and Engineering 145,
https://doi.org/10.1007/978-3-030-95025-5_22

conditions, and on the left and right homogeneous Robin boundary conditions, i.e. we put for simplicity of notation $u_0^{n-1} = u_{J+1}^{n-1} = 0$ in (2). Note that the parameters p_j^-, p_j^+ are real and not complex (as one would expect in the case of a complex problem) for the sake of simplicity in our analysis. The Robin parameters are fixed at the domain boundaries $x = a_1$ and $x = b_J$ to $p_1^- = p_a$ and $p_J^- = p_b$. As p_a, p_b tend to infinity, this is equivalent to imposing Dirichlet conditions. By linearity, it suffices to study the homogeneous equations, $f = 0$, and analyze convergence to zero of the OSM (2). Expanding the homogeneous iterates in a Fourier series $u_j^n(x, y) = \sum_{m=1}^{\infty} v_j^n(x, \tilde{k}) \sin(\tilde{k}y)$ where $\tilde{k} = \frac{m\pi}{\tilde{L}}$ to satisfy the homogeneous Dirichlet boundary conditions at the top and bottom, we obtain for the Fourier coefficients the equations

$$
\begin{aligned}
\partial_{xx} v_j^n - (\tilde{k}^2 + \sigma - i\varepsilon) v_j^n &= 0 && x \in (a_j, b_j),\\
-\partial_x v_j^n + p_j^- v_j^n &= -\partial_x v_{j-1}^{n-1} + p_j^- v_{j-1}^{n-1} && \text{at } x = a_j,\\
\partial_x v_j^n + p_j^+ v_j^n &= \partial_x v_{j+1}^{n-1} + p_j^+ v_{j+1}^{n-1} && \text{at } x = b_j.
\end{aligned}
\tag{3}
$$

The general solution of the differential equation is $v_j^n(x, \tilde{k}) = \tilde{c}_j e^{-\lambda(\tilde{k})x} + \tilde{d}_j e^{\lambda(\tilde{k})x}$, where $\lambda = \lambda(\tilde{k}) = \sqrt{\tilde{k}^2 + \sigma - i\varepsilon}$. We next define the Robin traces, $\mathcal{R}_-^{n-1}(a_j, \tilde{k}) := -\partial_x v_{j-1}^{n-1}(a_j, \tilde{k}) + p_j^- v_{j-1}^{n-1}(a_j, \tilde{k})$ and $\mathcal{R}_+^{n-1}(b_j, \tilde{k}) := \partial_x v_{j+1}^{n-1}(b_j, \tilde{k}) + p_j^+ v_{j+1}^{n-1}(b_j, \tilde{k})$. Inserting the solution into the transmission conditions in (3), a linear system arises where the unknowns are $\tilde{c}_j$ and $\tilde{d}_j$, whose solution is

$$
\begin{aligned}
\tilde{c}_j &= \frac{1}{D_j}\left(e^{\lambda b_j}(p_j^+ + \lambda)\mathcal{R}_-^{n-1}(a_j, \tilde{k}) - e^{\lambda a_j}(p_j^- - \lambda)\mathcal{R}_+^{n-1}(b_j, \tilde{k})\right),\\
\tilde{d}_j &= \frac{1}{D_j}\left(-e^{-\lambda b_j}(p_j^+ - \lambda)\mathcal{R}_-^{n-1}(a_j, \tilde{k}) + e^{-\lambda a_j}(p_j^- + \lambda)\mathcal{R}_+^{n-1}(b_j, \tilde{k})\right),
\end{aligned}
$$

where $D_j := (\lambda + p_j^+)(\lambda + p_j^-)e^{\lambda(L+\delta)} - (\lambda - p_j^+)(\lambda - p_j^-)e^{-\lambda(L+\delta)}$. We thus arrive for the Robin traces in the OSM at the iteration formula

$$
\begin{aligned}
\mathcal{R}_-^n(a_j, \tilde{k}) &= \alpha_j^- \mathcal{R}_-^{n-1}(a_{j-1}, \tilde{k}) + \beta_j^- \mathcal{R}_+^{n-1}(b_{j-1}, \tilde{k}),\ j = 2, \ldots, J,\\
\mathcal{R}_+^n(b_j, \tilde{k}) &= \beta_j^+ \mathcal{R}_-^{n-1}(a_{j+1}, \tilde{k}) + \alpha_j^+ \mathcal{R}_+^{n-1}(b_{j+1}, \tilde{k}),\ j = 1, \ldots, J-1,
\end{aligned}
$$

where

$$
\begin{aligned}
\alpha_j^- &:= \frac{(\lambda + p_{j-1}^+)(\lambda + p_j^-)e^{\lambda\delta} - (\lambda - p_{j-1}^+)(\lambda - p_j^-)e^{-\lambda\delta}}{(\lambda+p_{j-1}^+)(\lambda+p_{j-1}^-)e^{\lambda(L+\delta)} - (\lambda-p_{j-1}^+)(\lambda-p_{j-1}^-)e^{-\lambda(L+\delta)}},\ j = 2, \ldots, J,\\
\alpha_j^+ &:= \frac{(\lambda + p_{j+1}^-)(\lambda + p_j^+)e^{\lambda\delta} - (\lambda - p_{j+1}^-)(\lambda - p_j^+)e^{-\lambda\delta}}{(\lambda+p_{j+1}^+)(\lambda+p_{j+1}^-)e^{\lambda(L+\delta)} - (\lambda-p_{j+1}^+)(\lambda-p_{j+1}^-)e^{-\lambda(L+\delta)}},\ j = 1, \ldots, J-1,
\end{aligned}
$$

$$
\beta_j^- := \frac{(\lambda + p_j^-)(\lambda - p_{j-1}^-)e^{-\lambda L} - (\lambda - p_j^-)(\lambda + p_{j-1}^-)e^{\lambda L}}{(\lambda+p_{j-1}^+)(\lambda+p_{j-1}^-)e^{\lambda(L+\delta)} - (\lambda-p_{j-1}^+)(\lambda-p_{j-1}^-)e^{-\lambda(L+\delta)}},\ j = 2, \ldots, J,
$$

$$\beta_j^+ := \frac{(\lambda + p_j^+)(\lambda - p_{j+1}^+)e^{-\lambda L} - (\lambda - p_j^+)(\lambda + p_{j+1}^+)e^{\lambda L}}{(\lambda+p_{j+1}^+)(\lambda+p_{j+1}^-)e^{\lambda(L+\delta)} - (\lambda-p_{j+1}^+)(\lambda-p_{j+1}^-)e^{-\lambda(L+\delta)}}, \; j = 1, \ldots, J-1.$$

Defining the matrices

$$T_j^1 := \begin{bmatrix} \alpha_j^- & \beta_j^- \\ 0 & 0 \end{bmatrix}, \; j = 2, .., J \quad \text{and} \quad T_j^2 := \begin{bmatrix} 0 & 0 \\ \beta_j^+ & \alpha_j^+ \end{bmatrix}, \; j = 1, .., J-1,$$

we can write the OSM in substructured form (keeping the first and last rows and columns to make the block structure appear), namely

$$\underbrace{\begin{bmatrix} 0 \\ \mathcal{R}_+^n(b_1, \tilde{k}) \\ \mathcal{R}_-^n(a_2, \tilde{k}) \\ \mathcal{R}_+^n(b_2, \tilde{k}) \\ \vdots \\ \mathcal{R}_-^n(a_j, \tilde{k}) \\ \mathcal{R}_+^n(b_j, \tilde{k}) \\ \vdots \\ \mathcal{R}_-^n(a_{N-1}, \tilde{k}) \\ \mathcal{R}_+^n(b_{N-1}, \tilde{k}) \\ \mathcal{R}_-^n(a_N, \tilde{k}) \\ 0 \end{bmatrix}}_{\mathcal{R}^n} = \underbrace{\begin{bmatrix} & T_1^2 & & & & \\ T_2^1 & & T_2^2 & & & \\ & \ddots & & \ddots & & \\ & & T_j^1 & & T_j^2 & \\ & & & \ddots & & \ddots \\ & & & T_{N-1}^1 & & T_{N-1}^2 \\ & & & & T_N^1 & \end{bmatrix}}_{T} \underbrace{\begin{bmatrix} 0 \\ \mathcal{R}_+^{n-1}(b_1, \tilde{k}) \\ \mathcal{R}_-^{n-1}(a_2, \tilde{k}) \\ \mathcal{R}_+^{n-1}(b_2, \tilde{k}) \\ \vdots \\ \mathcal{R}_-^{n-1}(a_j, \tilde{k}) \\ \mathcal{R}_+^{n-1}(b_j, \tilde{k}) \\ \vdots \\ \mathcal{R}_-^{n-1}(a_{N-1}, \tilde{k}) \\ \mathcal{R}_+^{n-1}(b_{N-1}, \tilde{k}) \\ \mathcal{R}_-^{n-1}(a_N, \tilde{k}) \\ 0 \end{bmatrix}}_{\mathcal{R}^{n-1}}. \tag{4}$$

If the parameters $p_j^\pm$ are constant over all the interfaces, and we eliminate the first and the last row and column of T, T becomes a block Toeplitz matrix. The best choice of the parameters minimizes the spectral radius $\rho(T)$ over a numerically relevant range of frequencies $K := [\tilde{k}_{\min}, \tilde{k}_{\max}]$ with $\tilde{k}_{\min} := \frac{\pi}{\hat{L}}$ (or 0 for simplicity) and $\tilde{k}_{\max} := \frac{M\pi}{\hat{L}}$, $M \sim \frac{1}{h}$, where h is the mesh size, and is thus solution of the min-max problem $\min_{p_j^\pm} \max_{\tilde{k} \in K} |\rho(T(\tilde{k}, p_j^\pm))|$.

The traditional approach to obtain optimized transmission conditions for optimized Schwarz methods is to optimize performance for a simple two subdomain model problem, and then to use the result also in the case of many subdomains. We want to study here if this approach is justified, by directly optimizing the performance for two and more subdomains, and then comparing the results. We obtain our results from insight by numerical optimisation for small overlap, in order to find asymptotic formulas for the convergence factor and the parameters involved. The constants in the asymptotic results are then obtained by rigorous analytical computations of asymptotic series. We thus do not obtain existence and uniqueness results, but our asymptotically optimized convergence factors equioscillate as one would expect. For Robin conditions with complex parameters for two subdomains, existence and uniqueness results can be found in B. Delourme and L. Halpern [3].

2 Optimization for 2, 3, 4, 5 and 6 subdomains

For two subdomains, the general substructured iteration matrix becomes

$$T = \begin{bmatrix} 0 & \beta_1^+ \\ \beta_2^- & 0 \end{bmatrix}.$$

The eigenvalues of this matrix are $\pm\sqrt{\beta_1^+\beta_2^-}$ and thus the square of the convergence factor is $\rho^2 = |\beta_1^+\beta_2^-|$.

Theorem 1 (Two Subdomain Optimization) *Let $s:=\sqrt{\sigma - i\varepsilon}$, where the complex square root is taken with a positive real part, and let C be the real constant*

$$C:=\Re\frac{s((p_b+s)(p_a+s)-(s-p_b)(s-p_a)\mathrm{e}^{-4sL})}{((s-p_a)\mathrm{e}^{-2sL}+s+p_a)((s-p_b)\mathrm{e}^{-2sL}+s+p_b)}, \tag{5}$$

where p_a and p_b are the Robin parameters at the outer boundaries. Then for two subdomains with $p_1^+ = p_2^- =: p$ and $\tilde{k}_{\min} = 0$, the asymptotically optimized parameter p for small overlap δ and associated convergence factor are

$$p = 2^{-1/3}C^{2/3}\delta^{-1/3}, \quad \rho = 1 - 2\cdot 2^{1/3}C^{1/3}\delta^{1/3} + O(\delta^{2/3}). \tag{6}$$

If $p_1^+ \neq p_2^-$ and $\tilde{k}_{min} = 0$, the asymptotically optimized parameters for small overlap δ and associated convergence factor are

$$p_1^+ = 2^{-2/5}C^{2/5}\delta^{-3/5},\ p_2^- = 2^{-4/5}C^{4/5}\delta^{-1/5},\ \rho = 1-2\cdot 2^{-1/5}C^{1/5}\delta^{1/5}+O(\delta^{2/5}). \tag{7}$$

Proof From numerical experiments, we obtain that the solution of the min-max problem equioscillates, $\rho(0) = \rho(\tilde{k}^*)$, where $\tilde{k}^*$ is an interior maximum point, and asymptotically $p = C_p\delta^{-1/3}$, $\rho = 1 - C_R\delta^{1/3} + O(\delta^{2/3})$, and $\tilde{k}^* = C_k\delta^{-2/3}$. By expanding for δ small, and setting the leading term in the derivative $\frac{\partial\rho}{\partial\tilde{k}}(\tilde{k}^*)$ to zero, we get $C_p = \frac{C_k^2}{2}$. Expanding the maximum leads to $\rho(\tilde{k}^*) = \rho(C_k\delta^{-2/3}) = 1-2C_k\delta^{1/3}+O(\delta^{2/3})$, therefore $C_R = 2C_k$. Finally the solution of the equioscillation equation $\rho(0) = \rho(\tilde{k}^*)$ determines uniquely $C_k = 2^{1/3}C^{1/3}$.

In the case with two parameters, we have two equioscillations, $\rho(0) = \rho(\tilde{k}_1^*) = \rho(\tilde{k}_2^*)$, where $\tilde{k}_j^*$ are two interior local maxima, and asymptotically $p_1 = C_{p1}\delta^{-3/5}$, $p_1 = C_{p1}\delta^{-1/5}$, $\rho = 1 - C_R\delta^{1/5} + O(\delta^{2/5})$, $\tilde{k}_1^* = C_{k1}\delta^{-2/5}$ and $\tilde{k}_2^* = C_{k2}\delta^{-4/5}$. By expanding for δ small, and setting the leading terms in the derivatives $\frac{\partial\rho}{\partial\tilde{k}}(\tilde{k}_{1,2}^*)$ to zero, and we get $C_{p1} = C_{k2}^2$, $C_{p2} = \frac{C_{k1}^2}{C_{k2}^2}$. Expanding the maxima leads to $\rho(\tilde{k}_1^*) = \rho(C_k\delta^{-2/5}) = 1 - 2\frac{C_{k1}}{C_{k2}^2}\delta^{1/5} + O(\delta^{2/5})$ and $\rho(\tilde{k}_2^*) = \rho(C_k\delta^{-4/5}) = 1 - 2C_{k_2}\delta^{1/5} + O(\delta^{2/5})$ and equating $\rho(\tilde{k}_1^*) = \rho(\tilde{k}_2^*)$ we get $C_{k1} = C_{k2}^3$ and $C_R = 2C_{k2}$. Finally equating $\rho(0) = \rho(\tilde{k}_2^*)$ asymptotically determines uniquely $C_{k2} = 2^{-1/5}C^{1/5}$ and then $C_{k1} = C_{k2}^3$ and $C_{p1} = C_{k2}^2$, $C_{p2} = C_{k2}^4$.

Corollary 1 (Two Subdomains with Dirichlet outer boundary conditions) *The case of Dirichlet outer boundary conditions can be obtained by letting p_a and p_b go to infinity, which simplifies (5) to*

$$C = \Re \frac{s(1 + e^{2sL})}{(e^{2sL} - 1)} \tag{8}$$

and the asymptotic results in Theorem 1 simplify accordingly.

For three subdomains, the general substructured iteration matrix becomes

$$T = \begin{bmatrix} 0 & \beta_1^+ & \alpha_1^+ & 0 \\ \beta_2^- & 0 & 0 & 0 \\ 0 & 0 & 0 & \beta_2^+ \\ 0 & \alpha_3^- & \beta_3^- & 0 \end{bmatrix},$$

and we obtain for the first time an optimization result for three subdomains:

Theorem 2 (Three Subdomain Optimization) *For three subdomains with equal parameters $p_1^+ = p_2^- = p_2^+ = p_3^- = p$, the asymptotically optimized parameter p for small overlap δ and associated convergence factor are*

$$p = 2^{-1/3}C^{2/3}\delta^{-1/3}, \quad \rho = 1 - 2 \cdot 2^{1/3}C^{1/3}\delta^{1/3} + O(\delta^{2/3}), \tag{9}$$

where C is a real constant that can be obtained in closed form. If the parameters are different, their asymptotically optimized values for small overlap δ are such that

$$p_1^+, p_2^+, p_2^-, p_3^- \in \{2^{-2/5}C^{2/5}\delta^{-3/5}, 2^{-4/5}C^{4/5}\delta^{-1/5}\}, \; p_1^+ \neq p_2^-, \; p_2^+ \neq p_3^-, \tag{10}$$

and the associated convergence factor is

$$\rho = 1 - 2 \cdot 2^{-1/5}C^{1/5}\delta^{1/5} + O(\delta^{2/5}). \tag{11}$$

Proof The characteristic polynomial of the iteration matrix is

$$G(\mu) = \mu^4 - (\beta_2^-\beta_1^+ + \beta_3^-\beta_2^+)\mu^2 - \alpha_3^-\beta_2^-\alpha_1^+\beta_2^+ + \beta_3^-\beta_2^+\beta_2^-\beta_1^+.$$

This biquadratic equation has the roots $\mu_1 = \pm\sqrt{\frac{m_1+\sqrt{m_2}}{2}}$, $\mu_2 = \pm\sqrt{\frac{m_1-\sqrt{m_2}}{2}}$ where

$$m_1 = \beta_2^-\beta_1^+ + \beta_3^-\beta_2^+, \; m_2 = 4\alpha_3^-\beta_2^-\alpha_1^+\beta_2^+ + (\beta_2^-\beta_1^+ - \beta_3^-\beta_2^+)^2.$$

Therefore $\rho(T) = \max\{|\mu_1|, |\mu_2|\}$. Following the same reasoning as in the proof of Theorem 1, we observe that the solution equioscillates, and minimizing the maximum asymptotically for δ small then leads to the desired result, for more details, see [7].□

Notice that the optimized parameters and the relation between them is the same as in the two-subdomain case, the only difference is the equation whose solution gives the exact value of the constant C. The only difference between a two subdomain optimization and a three subdomain optimization is therefore the constant.

Table 1: Asymptotic results for four subdomains: $\sigma = \varepsilon = 1, L = 1, p_a = p_b = 1$

	Many parameters							One parameter	
δ	ρ	p_1^+	p_2^-	p_2^+	p_3^-	p_3^+	p_4^-	ρ	p
$1/10^2$	0.5206	13.1269	1.2705	10.1871	0.7748	16.5975	2.1327	0.6202	2.8396
$1/10^3$	0.6708	37.9717	1.4208	42.9379	1.6005	68.1923	2.4896	0.8022	6.0657
$1/10^4$	0.7789	152.9323	2.3266	152.0873	3.1841	161.0389	2.4919	0.9029	13.0412
$1/10^5$	0.8510	651.7536	4.1945	645.0605	4.1519	649.8928	4.1828	0.9537	28.0834

Table 2: Asymptotic results for five subdomains : $\sigma = \varepsilon = 1, L = 1, p_a = p_b = 1$

	Many parameters									One parameter	
δ	ρ	p_1^+	p_2^-	p_2^+	p_3^-	p_3^+	p_4^-	p_4^+	p_5^-	ρ	p
$1/10^2$	0.5273	8.5648	1.4619	9.1763	0.8030	9.1398	0.8426	15.5121	2.2499	0.6290	2.6747
$1/10^3$	0.7333	24.6097	0.9209	23.4189	0.4499	37.2200	0.8433	34.8142	0.9181	0.8072	5.7261
$1/10^4$	0.7769	156.0648	2.4223	156.0502	2.4221	161.2036	2.5009	166.3478	2.5941	0.9055	12.3166
$1/10^5$	0.8547	704.4063	4.3378	611.3217	3.7296	611.3217	3.7296	690.8837	4.2116	0.9550	26.5260

Table 3: Asymptotic results for six subdomains: $\sigma = \varepsilon = 1, L = 1, p_a = p_b = 1$

δ	ρ	p_1^+	p_2^-	p_2^+	p_3^-	p_3^+	p_4^-	p_4^+	p_5^-	p_5^+	p_6^-
$1/10^2$	0.5460	10.5283	1.4526	7.7653	1.2124	8.2834	0.6573	7.6445	1.3410	8.0029	0.9586
$1/10^3$	0.7011	30.3314	0.9049	30.3452	1.1096	30.3010	0.9363	30.3458	0.8901	30.1139	1.1307
$1/10^4$	0.7837	145.7147	2.1126	146.4533	2.1231	145.7147	2.1126	149.1802	2.1743	146.7200	2.1909
$1/10^5$	0.8553	660.5326	3.9932	611.9401	3.7012	606.1453	3.6661	606.1144	3.6659	606.0914	3.8534

Corollary 2 (Three subdomains with Dirichlet outer boundary conditions) *When Dirichlet boundary conditions are used at the end of the computational domain, we obtain for the constant*

$$C = \Re \frac{s(e^{2sL} - e^{sL} + 1)}{e^{2sL} - 1}, \tag{12}$$

which is different from the two subdomain constant in (8).

For four subdomains, we show in Table 1 the numerically optimized parameter values when the overlap δ becomes small. We observe that again the optimized parameters behave like in Theorem 1 and Theorem 2 when the overlap δ becomes small. It is in principle possible to continue the asymptotic analysis from two and three subdomains, but this is beyond the scope of the present paper. Continuing the numerical optimization for five and six subdomains, we get the results in Table 2 and Table 3, which show again the same asymptotic behavior. We therefore conjecture the following two results for an arbitrary fixed number of subdomains:

1. When all parameters are equal to p, then the asymptotically optimized parameter p for small overlap δ and the associated convergence factor have the same form as for two-subdomains (6) in Theorem 1, only the constant is different.
2. If all parameters are allowed to be different, the optimized parameters behave for small overlap δ like

$$p_j^+, p_{j+1}^- \in \{2^{-2/5}C^{2/5}\delta^{-3/5}, 2^{-4/5}C^{4/5}\delta^{-1/5}\} \text{ and } p_j^+ \neq p_{j+1}^- \,\forall j = 1.., J-1,$$

as we have seen in the three subdomain case in Theorem 2, and we have again the same asymptotic convergence factor as for two and three subdomains, only the constant is different.

3 Optimization for many subdomains

In order to obtain a theoretical result for many subdomains, we use the technique of limiting spectra [1] to derive a bound on the spectral radius which we can then minimize. The technique of limiting spectra allows us to get an estimate of the spectral radius when the matrix size goes to infinity. To do so, we must however assume that the outer Robin boundary conditions use the same optimized parameter as at the interfaces, in order to have the Toeplitz structure needed for the limiting spectrum approach.

Theorem 3 (Many Subdomain Optimization) *With all Robin parameters equal, $p_j^- = p_j^+ = p$, the convergence factor of the OSM satisfies the bound*

$$\rho = \lim_{N\to+\infty} \rho(T_{2d}^{OS}) \leq \max\Big\{|\alpha - \beta|, |\alpha + \beta|\Big\} < 1,$$

where $\alpha = \frac{(\lambda+p)^2 e^{\lambda\delta} - (\lambda-p)^2 e^{-\lambda\delta}}{(\lambda+p)^2 e^{\lambda(L+\delta)} - (\lambda-p)^2 e^{-\lambda(L+\delta)}}$, $\beta = \frac{(\lambda-p)(\lambda+p)(e^{-\lambda L} - e^{\lambda L})}{(\lambda+p)(\lambda+p)e^{\lambda(L+\delta)} - (\lambda-p)(\lambda-p)e^{-\lambda(L+\delta)}}$. The asymptotically optimized parameter and associated convergence factor are

$$p = 2^{-1/3}C^{2/3}\delta^{-1/3}, \quad \rho = 1 - 2 \cdot 2^{1/3}C^{1/3}\delta^{1/3} + O(\delta^{2/3}) \tag{13}$$

with the constant $C := \Re\frac{s(1-e^{-sL})}{1+e^{-sL}}$. If we allow two-sided Robin parameters, $p_j^- = p^-$ and $p_j^+ = p^+$, the OSM convergence factor satisfies the bound

$$\rho = \lim_{N\to+\infty} \rho(T_{2d}^{OS}) \leq \max\Big\{\Big|\alpha - \sqrt{\beta_-\beta_+}\Big|, \Big|\alpha + \sqrt{\beta_-\beta_+}\Big|\Big\} < 1,$$

where $\alpha = \frac{(\lambda+p^+)(\lambda+p^-)e^{\lambda\delta} - (\lambda-p^+)(\lambda-p^-)e^{-\lambda\delta}}{D}$, $\beta^\pm = \frac{(\lambda^2-(p^\mp)^2)(e^{-\lambda L} - e^{\lambda L})}{D}$, with $D = (\lambda+p^+)(\lambda+p^-)e^{\lambda(L+\delta)} - (\lambda-p^+)(\lambda-p^-)e^{-\lambda(L+\delta)}$. The asymptotically optimized parameter choice $p^- \neq p^+$ and the associated convergence factor are

$$p^-, p^+ \in \Big\{C^{2/5}\delta^{-3/5}, C^{4/5}\delta^{-1/5}\Big\}, \quad \rho = 1 - 2C^{1/5}\delta^{1/5} + O(\delta^{2/5}),$$

with the same constant $C := \Re \frac{s(1-e^{-sL})}{1+e^{-sL}}$ *as for one parameter.*

Proof As in the case of two and three subdomains, we observe equioscillation by numerical optimization, and asymptotically that $p = C_p \delta^{-1/3}$, $\rho = 1 - C_R \delta^{1/3} + O(\delta^{2/3})$ and the convergence factor has a local maximum at the point $\tilde{k}^* = C_k \delta^{-2/3}$. By expanding for small δ, the derivative $\frac{\partial \rho}{\partial k}(\tilde{k}^*)$ needs to have a vanishing leading order term, which leads to $C_p = \frac{C_k^2}{2}$. Expanding the convergence factor at the maximum point $\tilde{k}^*$ gives $\rho(\tilde{k}^*) = \rho(C_k \delta^{-2/3}) = 1 - 2C_k \delta^{1/3} + O(\delta^{2/3})$, and hence $C_R = 2C_k$. Equating now $\rho(0) = \rho(\tilde{k}^*)$ determines uniquely C_k and then $C_p = \sqrt{C_k/2}$ giving (13). By following the same lines as for two and three subdomains, we also get the asymptotic result in the case of two different parameters. □

We can therefore safely conclude that for the magnetotelluric approximation of Maxwell's equations, which contains the important Laplace and screened Laplace equation as special cases, it is sufficient to optimize transmission conditions for a simple two subdomain decomposition in order to obtain good transmission conditions also for the case of many subdomains, a result which was previously unknown.

References

1. N. Bootland, V. Dolean, A. Kyriakis and J. Pestana. Analysis of parallel Schwarz algorithms for time-harmonic problems using block Toeplitz matrices. *Electronic Transactions on Numerical Analysis* **55**, 112–141 (2022).
2. F. Chaouqui, G. Ciaramella, M.J. Gander and T. Vanzan. On the scalability of classical one-level domain-decomposition methods. *Vietnam J. Math.* **46**(4), 1053–1088 (2018).
3. B. Delourme and L. Halpern. A complex homographic best approximation problem. Application to optimized Robin-Schwarz algorithms, and optimal control problems. *SIAM Journal on Numerical Analysis* **59**(3), 1769–1810 (2021).
4. F. Donzelli, M.J. Gander and R.D. Haynes. A Schwarz method for the magnetotelluric approximation of Maxwell's equations. In: *Domain Decomposition Methods in Science and Engineering XXV*, pp. 417–424, Springer (2020).
5. O.G. Ernst and M.J. Gander. Why it is difficult to solve Helmholtz problems with classical iterative methods? In: *Numerical analysis of multiscale problems*, Lect. Notes Comput. Sci. Eng., vol. 83, pp. 325–363, Springer, Heidelberg (2012).
6. M.J. Gander, L. Halpern and F. Magoulès. An optimized Schwarz method with two-sided Robin transmission conditions for the Helmholtz equation. *Internat. J. Numer. Methods Fluids* **55**(2), 163–175 (2007).
7. A. Kyriakis. *Scalable domain decomposition methods for time-harmonic wave propagation problems*. Ph.D. thesis, University of Strathclyde (2021).

Non-overlapping Spectral Additive Schwarz Methods for HDG and Multiscale Discretizations

Yi Yu, Maksymilian Dryja, and Marcus Sarkis

1 Introduction

In this paper, we design and state some theoretical results for the exact and inexact versions of Non-overlapping Spectral Additive Schwarz Methods (NOSAS) in the framework of Hybridizable Discontinuous Galerkin (HDG) discretizations and multiscale discretizations for the following elliptic problem:

$$\begin{aligned} \rho(x)^{-1}\mathbf{q} + \nabla u &= 0 && \text{in } \Omega, \\ \nabla \cdot \mathbf{q} &= f && \text{in } \Omega, \\ u &= 0 && \text{on } \partial\Omega, \end{aligned} \tag{1}$$

where $\rho(x) \in L^\infty(\Omega)$, $\rho(x) \geq \rho_0 > 0$, $f \in L^2(\Omega)$ and Ω is a polyhedral domain in $\mathbb{R}^d (d \geq 2)$. The problem (1) has a unique solution $(\mathbf{q}, u) \in \mathbf{H}(\text{div}, \Omega) \times H_0^1(\Omega)$, where $\mathbf{H}(\text{div}, \Omega) := \{\mathbf{q} \in L^2(\Omega)^d,\ \text{div}\, \mathbf{q} \in L^2(\Omega)\}$.

We begin by describing the HDG discretization. Consider a partitioning of the domain Ω into a conforming mesh $\mathcal{T}_h$ with elements K. We assume that the partition $\mathcal{T}_h$ is shape regular and quasi-uniform of size $O(h)$. A face of K is denoted by F and $\mathcal{E}_h$ denotes the set of all faces of $\mathcal{T}_h$ excluding the ones on $\partial\Omega$. The HDG yields a scalar approximation u_h to u, a vector approximation $\mathbf{q}_h$ to $\mathbf{q}$, and a scalar approximation λ_h to the trace of u on element faces, in the spaces of $\mathbf{Q}_h = \{\mathbf{p} \in L^2(\mathcal{T}_h)^d : \mathbf{p}|_K \in \mathbf{P}_k(K), \forall K \in \mathcal{T}_h\}$, $W_h = \{w \in L^2(\mathcal{T}_h) : w|_K \in P_k(K), \forall K \in \mathcal{T}_h\}$ and $M_h = \{\mu \in L^2(\mathcal{E}_h) : \mu|_F \in P_k(F), \forall F \in \mathcal{E}_h\}$, respectively. Here $\mathbf{P}_k(K) = P_k(K)^d$ and $P_k(K)$ is the space of polynomials of order at most k on K.

Yi Yu
Worcester Polytechnic Institute, 100 Institute Rd, Worcester, USA, e-mail: yyu5@wpi.edu

Maksymilian Dryja
Warsaw University, Banacha 2, 00-097 Warsaw, Poland, e-mail: dryja@mimuw.edu.pl

Marcus Sarkis
Worcester Polytechnic Institute, 100 Institute Rd, Worcester, USA, e-mail: msarkis@wpi.edu.

S. C. Brenner et al. (eds.), *Domain Decomposition Methods in Science and Engineering XXVI*, Lecture Notes in Computational Science and Engineering 145,
https://doi.org/10.1007/978-3-030-95025-5_23

To cope with the heterogeneous coefficients for each element, we define the numerical flux $\hat{\mathbf{q}}_h$, which is a double-valued vector function on mesh interfaces as follow:

$$\hat{\mathbf{q}}_h \cdot \mathbf{n} = \mathbf{q}_h \cdot \mathbf{n} + \tau_K \rho_K (u_h - \lambda_h) \qquad \text{on } \mathcal{E}_h. \tag{2}$$

Here $\tau := \tau_K \rho_K$ is called the stabilizer, and ρ_K is a constant which approximates $\rho(x)$ in element K, the nonnegative constant function τ_K defined on $\mathcal{E}_h$ can be either a single or a double valued function on the element interfaces and τ_K above denotes the τ-value on the ∂K. The novelty of writing $\tau = \tau_K \rho_K$ is that we will introduce an equivalent norm of $a(\cdot, \cdot)$ independently of the coefficients. With the definitions of the numerical flux $\hat{\mathbf{q}}_h$, the HDG discretization of problem (1) can be written as: find $(\mathbf{q}_h, u_h, \lambda_h) \in \mathbf{Q}_h \times W_h \times M_h$ such that, for all $(\mathbf{p}, w, \mu) \in \mathbf{Q}_h \times W_h \times M_h$:

$$(u_h, \nabla \cdot \mathbf{p})_{\mathcal{T}_h} - (\rho(x)^{-1} \mathbf{q}_h, \mathbf{p})_{\mathcal{T}_h} - \langle \lambda_h, \mathbf{p} \cdot \mathbf{n} \rangle_{\partial \mathcal{T}_h} = 0, \tag{3a}$$
$$-(\mathbf{q}_h, \nabla w)_{\mathcal{T}_h} + \langle \hat{\mathbf{q}}_h \cdot \mathbf{n}, w \rangle_{\partial \mathcal{T}_h} = (f, w)_{\mathcal{T}_h}, \tag{3b}$$
$$\langle \hat{\mathbf{q}}_h \cdot \mathbf{n}, \mu \rangle_{\partial \mathcal{T}_h \backslash \partial \Omega} = 0. \tag{3c}$$

It is proved in [1] that the system (3) is uniquely solvable and can be reduced to the matrix form of the following problem: find $\lambda_h \in M_h$ such that

$$a(\lambda_h, \mu) = b(\mu), \quad \forall \mu \in M_h. \tag{4}$$

Here

$$a(\eta, \mu) = \sum_{K \in \mathcal{T}_h} a_K(\eta, \mu) = \sum_{K \in \mathcal{T}_h} (\rho_K^{-1} Q\eta, Q\mu)_K + \langle \tau_K \rho_K (U\eta - \eta), (U\mu - \mu) \rangle_{\partial K}$$

and $b(\mu) = \sum_{K \in \mathcal{T}_h} b_K(\mu) = \sum_{K \in \mathcal{T}_h} (f, U\mu)_K$, where $Qv \in \mathbf{Q}_h$ and $Uv \in W_h$ are the unique solution of the local element problem (3) with $\lambda_h = v$ and right-hand side $f = 0$. We note that once we get λ_h, the solution of (3) can be completed by computing $\mathbf{q}_h$ and u_h in each element separately. Note that the bilinear form $a(\cdot, \cdot)$ is positive definite. Let us define the norm $|||\cdot|||_{\rho,h}$ as follows:

$$|||\lambda|||_{\rho,h} = \left(\sum_{K \in \mathcal{T}_h} \frac{\rho_K}{h} ||\lambda - m_K(\lambda)||^2_{L^2(\partial K)} \right)^{1/2}, \tag{5}$$

where $m_K(\lambda) = \frac{1}{|\partial K|} \int_{\partial K} \lambda ds$. The next theorem shows the norm $|||\cdot|||_{\rho,h}$ is equivalent to the energy norm $a(\cdot, \cdot)$. For the proof, see [1].

Theorem 1 *For all $\lambda \in M_h$, there are positive constants C_1, C_2, independent of h and ρ_K, such that*

$$C_1 |||\lambda|||^2_{\rho,h} \leq a(\lambda, \lambda) \leq C_2 \gamma \, |||\lambda|||^2_{\rho,h},$$

where $\gamma = 1 + \max_{K \in \mathcal{T}_h} \tau_K^* h$, and τ_K^* denotes the second largest value of τ_K on ∂K.

NOSAS were first introduced for Continuous Galerkin (CG) discretizations in [7, 8] as domain decomposition preconditioners designed for elliptic problems with highly heterogeneous coefficients. NOSAS are non-overlapping Schwarz preconditioners where the subdomain interactions are via the coarse problem. The coarse problem involves local and global interactions. The global component is introduced to guarantee the robustness of the preconditioners for any coefficients $\rho(x)$ and number of subdomains. The proposed global problem is built from generalized eigenfunctions on the subdomains. The size of the global problem is equal to the total number of those eigenfunctions and is only related to the number of islands or channels with high-contrast coefficients that touch the boundary of the subdomains, see [9]. Additionally, the inexact version of NOSAS has good parallelization properties. The main goal of this paper is to design and show results of NOSAS for HDG and multiscale discretizations. We note that other kinds of domain decomposition preconditioners for HDG were introduced in [1, 6].

2 Domain decomposition setting

We decompose Ω into N non-overlapping polygonal subdomains Ω_i of size $O(H)$. The local space V_i $(1 \le i \le N)$ is the restriction of M_h to Ω_i, vanishing on $\partial\Omega_i$, and the coarse space V_0 is the restriction of M_h to the interface of all subdomains. Then M_h admits the following direct sum decomposition:

$$M_h = R_0^T V_0 \oplus R_1^T V_1 \oplus \cdots \oplus R_N^T V_N.$$

The local extrapolation operators $R_i^T : V_i \longrightarrow M_h$ $(1 \le i \le N)$ is the extension by zero outside of Ω_i. The coarse extrapolation operator $R_0^T : V_0 \rightarrow M_h$ is the core of NOSAS. We will define it and state some theoretical results in Section 3.

For $1 \le i \le N$, denote by A_i the matrix corresponding to the exact local bilinear form:

$$a_i(u, v) = v^T A_i\, u = a(R_i^T u, R_i^T v) \qquad u, v \in V_i.$$

For $i = 0$, we first consider the matrix A_0 corresponding to the exact bilinear form:

$$a_0(u, v) = v^T A_0\, u = a(R_0^T u, R_0^T v) \qquad u, v \in V_0.$$

We will also consider the inexact bilinear form $\hat{a}_0(\cdot,\cdot)$ later in this paper. Then the non-overlapping Schwarz preconditioners have the following forms:

$$T_A = B^{-1} A, \qquad B^{-1} = R_0^T A_0^{-1} R_0 + \sum_{i=1}^{N} R_i^T A_i^{-1} R_i.$$

We note that if we had chosen R_0^T as the a-discrete harmonic extension, then the above preconditioners would become a direct solver and it would be too expensive to solve the coarse problem. The core of NOSAS is to use a low-rank a-discrete harmonic extension R_0^T, which makes the coarse problem inexpensive to solve, and which guarantees good condition numbers.

3 NOSAS with exact and inexact solvers

The linear system $A\lambda_h = b$ corresponding to (4) can be assembled by the Neumann matrix $A^{(i)}$ and $b^{(i)}$ in each subdomain Ω_i. We decompose $A^{(i)}$ into block matrices $(A_{\Gamma\Gamma}^{(i)}\ A_{\Gamma I}^{(i)}\ ;\ A_{I\Gamma}^{(i)}\ A_{II}^{(i)})$ and $b^{(i)}$ into $(b_\Gamma^{(i)}\ ;\ b_I^{(i)})$, where the subscripts Γ, I respectively denote the parts associated with the interface and interior of subdomains. The Schur complements of A and $A^{(i)}$ are denoted by S and $S^{(i)}$, respectively.

For the NOSAS exact solver, solve the generalized eigenvalue problem in each subdomain ($i = 1, \cdots, N$) separately:

$$S^{(i)}\xi_j^{(i)} := (A_{\Gamma\Gamma}^{(i)} - A_{\Gamma I}^{(i)}(A_{II}^{(i)})^{-1}A_{I\Gamma}^{(i)})\xi_j^{(i)} = \Lambda_j^{(i)}A_{\Gamma\Gamma}^{(i)}\xi_j^{(i)} \quad (1 \le j \le n_i), \tag{6}$$

where n_i is the degrees of freedom on $\Gamma_i := \Gamma \cap \partial\Omega_i$. Note that the eigenvalues lie in $[0, 1]$ for the above generalized eigenvalue problem. We fix a threshold $\delta < 1$ and pick the smallest k_i eigenvalues $\le \delta$ and corresponding eigenvectors to construct the eigenfunctions space $Q^{(i)}$ and harmonic extension $P^{(i)}$ as follows:

$$Q^{(i)} = [\xi_1^{(i)}, \xi_2^{(i)}, \cdots, \xi_{k_i}^{(i)}] \qquad \text{and} \qquad P^{(i)} = -(A_{II}^{(i)})^{-1}A_{I\Gamma}^{(i)}Q^{(i)}.$$

We also define $D^{(i)} = \text{diagonal}(1 - \Lambda_1^{(i)}, 1 - \Lambda_2^{(i)}, \cdots, 1 - \Lambda_{k_i}^{(i)}) = I - \Lambda^{(i)}$.

For $u_0 \in V_0$, we define the global extension $R_0^T : V_0 \to M_h$ as:

$$R_0^T u_0 = \begin{bmatrix} u_0 \\ -\sum_{i=1}^N P^{(i)}(P^{(i)^T}A_{II}^{(i)}P^{(i)})^{-1}P^{(i)^T}A_{I\Gamma}^{(i)}u_0^{(i)} \end{bmatrix}$$
$$= \begin{bmatrix} u_0 \\ \sum_{i=1}^N P^{(i)}(Q^{(i)^T}A_{\Gamma\Gamma}^{(i)}Q^{(i)})^{-1}Q^{(i)^T}A_{\Gamma\Gamma}^{(i)}u_0^{(i)} \end{bmatrix},$$

where $u_0^{(i)}$ is the restriction of u_0 on Γ_i. Below $v_0^{(i)}$ denotes the restriction of v_0 to Γ_i. Next $\forall u_0, v_0 \in V_0$, we define the exact coarse bilinear form as:

$$\begin{aligned} a_0(u_0, v_0) &= a(R_0^T u_0, R_0^T v_0) \\ &= \sum_{i=1}^N v_0^{(i)^T}(A_{\Gamma\Gamma}^{(i)} - A_{\Gamma I}^{(i)}P^{(i)}(P^{(i)^T}A_{II}^{(i)}P^{(i)})^{-1}P^{(i)^T}A_{I\Gamma}^{(i)})u_0^{(i)} \\ &= \sum_{i=1}^N a_0^{(i)}(u_0^{(i)}, v_0^{(i)}) \end{aligned}$$

$$= \sum_{i=1}^{N} v_0^{(i)^T} (A_{\Gamma\Gamma}^{(i)} - A_{\Gamma\Gamma}^{(i)} Q^{(i)} D^{(i)} (Q^{(i)^T} A_{\Gamma\Gamma}^{(i)} Q^{(i)})^{-1} Q^{(i)^T} A_{\Gamma\Gamma}^{(i)}) u_0^{(i)}.$$

Above, $a_0(\cdot,\cdot)$ is the global bilinear form and $a_0^{(i)}(\cdot,\cdot)$ is the bilinear form on Γ_i locally. The next lemma shows that $a_0^{(i)}(\cdot,\cdot)$ is equivalent to the Schur complement $S^{(i)}$ in the span of $Q^{(i)}$, and an extension by zero for the orthogonal complement subspace.

Lemma 1 [9] *Let* $\Pi_S^{(i)} u_0^{(i)}$ *be the projection of* $u_0^{(i)}$ *onto Span*$\{Q^{(i)}\}$. *That is,* $\Pi_S^{(i)} u_0^{(i)} := Q^{(i)} (Q^{(i)^T} A_{\Gamma\Gamma}^{(i)} Q^{(i)})^{-1} Q^{(i)^T} A_{\Gamma\Gamma}^{(i)} u_0^{(i)}$. *Then:*

$$a_0^{(i)}(u_0^{(i)}, v_0^{(i)}) = (\Pi_S^{(i)} v_0^{(i)})^T S^{(i)} (\Pi_S^{(i)} u_0^{(i)}) + (v_0^{(i)} - \Pi_S^{(i)} v_0^{(i)})^T A_{\Gamma\Gamma}^{(i)} (u_0^{(i)} - \Pi_S^{(i)} u_0^{(i)}).$$

Lemma 2 [9] *Let* $u_0 \in V_0$ *then*

$$a_0(u_0, u_0) = \sum_{i=1}^{N} a_0^{(i)}(u_0^{(i)}, u_0^{(i)}) \le \sum_{i=1}^{N} \frac{1}{\delta} u_0^{(i)^T} S^{(i)} u_0^{(i)} = \frac{1}{\delta} u_0^T S u_0.$$

Using Lemma 1 and Lemma 2 and the classical Schwarz Theory [5] we have:

Theorem 2 [9] *For any* $u \in M_h$, *the following holds:*

$$\left(2 + \frac{3}{\delta}\right)^{-1} a(u,u) \le a(T_A u, u) \le 2a(u,u) \Longrightarrow k(T_A) \le 2\left(2 + \frac{3}{\delta}\right).$$

In the implementation of NOSAS, the complexity of the coarse problem involves computing $A_{\Gamma\Gamma}^{-1}$ and this complexity can be reduced if we replace $A_{\Gamma\Gamma}$ by its diagonal $\hat{A}_{\Gamma\Gamma}$. This version is called the inexact NOSAS with $\tilde{T}_A$ as the preconditioner. The generalized eigenvalue problem is now given by:

$$S^{(i)} \hat{\xi}_i^{(i)} := (A_{\Gamma\Gamma}^{(i)} - A_{\Gamma I}^{(i)} (A_{II}^{(i)})^{-1} A_{I\Gamma}^{(i)}) \hat{\xi}_j^{(i)} = \hat{\Lambda}_j^{(i)} \hat{A}_{\Gamma\Gamma}^{(i)} \hat{\xi}_j^{(i)}.$$

And for $u_0, v_0 \in V_0(\Omega)$, we define the inexact coarse solver as:

$$\hat{a}_0(u_0, v_0) = \sum_{i=1}^{N} v_0^{(i)^T} (\hat{A}_{\Gamma\Gamma}^{(i)} - \hat{A}_{\Gamma\Gamma}^{(i)} \hat{Q}^{(i)} \hat{D}^{(i)} (\hat{Q}^{(i)^T} \hat{A}_{\Gamma\Gamma}^{(i)} \hat{Q}^{(i)})^{-1} \hat{Q}^{(i)^T} \hat{A}_{\Gamma\Gamma}^{(i)}) u_0^{(i)},$$

where $\hat{Q}^{(i)}$ are the generalized eigenvectors and

$$\hat{D}^{(i)} = \text{diagonal}(1 - \hat{\Lambda}_1^{(i)}, \cdots, 1 - \hat{\Lambda}_{k_i}^{(i)}).$$

Then, we obtain the following condition number estimate:

Theorem 3 [9] *For any $u \in M_h$, the following holds:*

$$\left(2+\frac{5}{\delta}\right)^{-1} a(u,u) \leq a(\tilde{T}_A u,u) \leq 3a(u,u) \Longrightarrow k(\tilde{T}_A) \leq 3\left(2+\frac{5}{\delta}\right).$$

4 Multiscale discretization methods

The idea of multiscale methods [2, 3] is to use $\lambda_{ms} \in V_{\text{off}}$ to approximate the exact solution λ_h from $a(\lambda_h,\mu) = b(\mu), \forall \mu \in M_h$, where V_{off} is the space of multiscale basis functions. The following procedures show how we construct V_{off}. The first step, a snapshot space $V_{\text{snapshots}}$ is constructed from the solutions of local problems. In our NOSAS methods, we construct the snapshot space by $V_{\text{snapshots}} = \mathcal{H}^T V_0$, where $\mathcal{H}^T$ is the a-discrete harmonic extension. Notice that the dimension of $V_{\text{snapshots}}$ can be extremely large. The next step is to construct V_{off} from $V_{\text{snapshots}}$, which can be used to generate an efficient and accurate approximation to the multiscale solution. We choose the offline space $V_{\text{off}} = R_0^T V_0$, where R_0^T is the global extension for NOSAS. We use the following outline of the Generalized Multiscale Finite Element Method (GMsFEM) to show the coarse space of NOSAS is a multiscale discretization.

Offline stages:

1. Mesh partitioning to obtain the subdomains.
2. Construct $V_{\text{snapshots}}$ that will be used to compute an offline space.
3. Construct a low-dimensional offline space V_{off} by performing dimension reduction in the space of local snapshots. This is done by choosing a threshold δ and then computing multiscale basis functions.
4. Build the coarse and local matrices and factorize.

Online stages:

1. Given f, solve the local problems inside each subdomain and update the residual on the interfaces.
2. Solve a coarse problem. Add coarse and local solutions.

In the offline stages, we construct the $V_{\text{off}} = R_0^T V_0$ by NOSAS and also compute the factorization of matrices A_0 and A_i $(1 \leq i \leq N)$. In the online stage, we first solve N local problems in parallel:

$$A_i \lambda_i = R_i b = b_i \qquad 1 \leq i \leq N.$$

Then, we use the local solutions to form and solve the following coarse problem:

$$A_0 \lambda_0 = R_0 (b - A \sum_{i=1}^{N} R_i^T \lambda_i).$$

Finally, $\lambda_{ms} = R_0^T \lambda_0 + \sum_{i=1}^N R_i^T \lambda_i$ is obtained.

We note that we obtain similar numerical results if we use $A_0 \lambda_0 = R_0 b$ as the coarse problem. Additionally, we can also develop a multiscale technique to reduce the dimension of the local problems, see [4].

Theorem 4 *For NOSAS methods with* $\lambda_{ms} = R_0^T \lambda_0 + \sum_{i=1}^N R_i^T \lambda_i$,

$$a(\lambda_h - \lambda_{ms}, \lambda_h - \lambda_{ms}) \leq (1 - \delta)\, a(\mathcal{H}^T \lambda_\Gamma, \mathcal{H}^T \lambda_\Gamma),$$

where $\mathcal{H}^T$ *is the a-discrete harmonic extension and* λ_Γ *the restriction of* λ_h *on* Γ.

The proof follows from Lemma 1. This bound is not sharp since we can see via numerical experiments for heterogeneous coefficients that a small δ can give small relative errors.

5 Numerical experiments

We first show the results of problem (1) for a square domain with side length 1, $f \equiv 1$ and with highly heterogeneous coefficients in the following mesh (see Figure 1). We choose $k = 0$ for HDG spaces $\mathbf{Q}_h$, W_h, and M_h. We divide the square domain into $H \times H$ congruent square subdomains, and we fix the number of stripes in each subdomain. That means we always have two horizontal stripes and two vertical stripes in each subdomain. The coefficients are $\rho(x) = 1$ in the green stripes and $\rho(x) = 10^6$ in the white regions. We do not consider $\rho(x) = 10^6$ in stripes and $\rho(x) = 1$ elsewhere, because it is robust without the generalized eigenfunctions. In Table 2 we show numerical results for HDG with different values of $\tau = \tau_K \rho_K$ in this mesh using NOSAS with exact solvers. By choosing $\delta = \frac{1}{4}\frac{h}{H}$ the NOSAS will not deteriorate due to the small eigenvalues related to the jumps of coefficients, and the condition number is $O(H/h)$. We also note that we see little difference in numerical results when we use inexact diagonal solvers or exact solvers (we do not include the results here.) In Table 1, we show that the size of the global part for the coarse problem of NOSAS is proportional to the number of subdomains and does not depend on H/h.

Finally, Table 3 shows results for multiscale in HDG. We use a similar mesh in Figure 1. However, the number of stripes is fixed in the whole domain. That means we always have eight horizontal stripes and eight vertical stripes with width h in the whole domain. The coefficients are $\rho(x) = 1$ in the stripes and $\rho(x) = 10^6$ elsewhere. We fix h and $\tau = \rho_K$, and choose different δ to show the relative error of $\lambda_h - \lambda_{ms}$ and the number of coarse basis functions per subdomain. Since we use the exact local solver in each subdomain, we expect that the relative error of $\lambda_h - \lambda_{ms}$ will increase if we decrease H because the error arises from approximating the exact local solution in each subdomain.

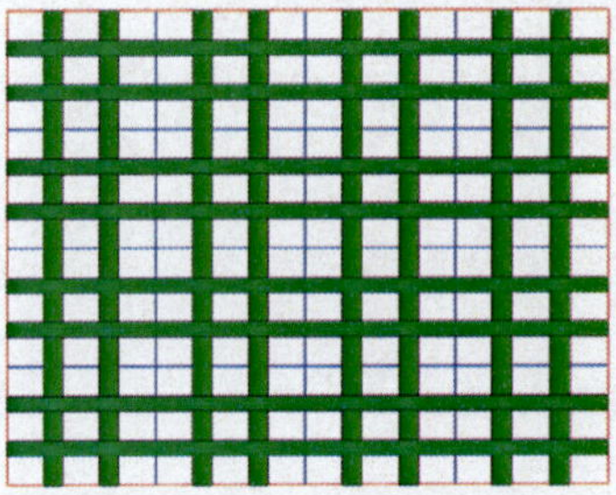

Fig. 1: Coefficients $\rho(x) = 1$ in green stripes and $\rho(x) = 10^6$ in white regions.

	$H = 1/2$	$H = 1/4$	$H = 1/8$	$H = 1/16$
$H/h = 8$	12	84	420	1860
$H/h = 16$	12	84	420	1860

Table 1: The number of eigenfunctions N_E for NOSAS with the exact solver. The size of the global problem is $N_E \times N_E$.

$\tau = \rho_K$	$H = \frac{1}{2}$	$H = \frac{1}{4}$	$H = \frac{1}{8}$	$H = \frac{1}{16}$
$\frac{H}{h} = 8$	14 (10.2596)	17 (10.2686)	17 (10.2732)	19 (10.2756)
$\frac{H}{h} = 16$	20 (20.8039)	25 (20.8062)	26 (20.8074)	29 (20.8080)

(a) $\tau = \tau_K \rho_K$ with $\tau_K = 1$.

$\tau = \frac{2}{\rho_{K_1}^{-1} + \rho_{K_2}^{-1}}$	$H = \frac{1}{2}$	$H = \frac{1}{4}$	$H = \frac{1}{8}$	$H = \frac{1}{16}$
$\frac{H}{h} = 8$	14 (10.2516)	17 (10.2646)	18 (10.2713)	18 (10.2746)
$\frac{H}{h} = 16$	20 (20.7990)	24 (20.8038)	27 (20.8062)	27 (20.8074)

(b)τ is the harmonic mean of ρ_K in adjacent elements.

$\tau = \frac{\rho_{K_1} + \rho_{K_2}}{2}$	$H = \frac{1}{2}$	$H = \frac{1}{4}$	$H = \frac{1}{8}$	$H = \frac{1}{16}$
$\frac{H}{h} = 8$	14 (10.2562)	17 (10.2669)	17 (10.2724)	17 (10.2752)
$\frac{H}{h} = 16$	19 (20.8019)	25 (20.8052)	26 (20.8069)	26 (20.8078)

(c)τ is the arithmetic mean of ρ_K in adjacent elements.

Table 2: NOSAS with exact solver for HDG with different choices of τ. The number of iterations of the PCG required to reduced the residual by 10^{-6} and the condition number (in parentheses).

	$H = \frac{1}{2}$	$H = \frac{1}{4}$	$H = \frac{1}{8}$	$H = \frac{1}{16}$
$\delta = 10^{-5}$	7 ($3.21e^{-5}$)	5.25 ($7.17e^{-5}$)	3.06 ($1.55e^{-4}$)	1.89 ($7.54e^{-4}$)
$\delta = 1/2$	24.25 ($3.21e^{-5}$)	18.56 ($7.15e^{-5}$)	11.26 ($1.54e^{-4}$)	5.05 ($7.50e^{-4}$)
$\delta = 3/4$	63 ($4.10e^{-18}$)	45.75 ($1.40e^{-14}$)	24.93 ($1.14e^{-13}$)	11.48 ($8.03e^{-13}$)

Table 3: NOSAS as a multiscale methods with fix $h = 1/64$. The number of global basis functions per subdomain and the relative energy error $\lambda_h - \lambda_{ms}$ with respect to λ_h (in parentheses).

References

1. B. Cockburn, O. Dubois, J. Gopalakrishnan and S. Tan. Multigrid for an HDG method. *IMA Journal of Numerical Analysis* **34**(4), 1386–1425 (2014).
2. Y. Efendiev, J. Galvis and T.Y. Hou. Generalized multiscale finite element methods (GMsFEM). *Journal of Computational Physics* **251**, 116–135 (2013).
3. S. Fu, E. Chung and G. Li. Edge multiscale methods for elliptic problems with heterogeneous coefficients. *Journal of Computational Physics* **396**, 228–242 (2019).
4. A.L. Madureira and M. Sarkis. Hybrid localized spectral decomposition for multiscale problems. *SIAM J. Numer. Anal.* **59**(2), 829–863 (2021).
5. A. Toselli and O. Widlund. *Domain decomposition methods – algorithms and theory*. Volume 34 of Springer Series in Computational Mathematics, Springer Science & Business Media (2006).
6. X. Tu and B. Wang. A BDDC algorithm for second-order elliptic problems with hybridizable discontinuous Galerkin discretizations. *Electronic Transactions on Numerical Analysis* **45**, 354–370 (2016).
7. Y. Yu, M. Dryja and M. Sarkis. Non-overlapping spectral additive Schwarz methods. In: *International Conference on Domain Decomposition Methods Proceedings of DD25*, pp. 375–382, Springer (2018).
8. Y. Yu, M. Dryja and M. Sarkis. From additive average Schwarz methods to non-overlapping spectral additive Schwarz methods. *SIAM Journal on Numerical Analysis* **59**(5), 2608–2638 (2021).
9. Y. Yu, M. Dryja and M. Sarkis. Non-overlapping spectral additive Schwarz methods for HDG discretizations. In preparation (2021).

Robust BPX Solver for Cahn–Hilliard Equations

Siamak Faal, Adam Powell, and Marcus Sarkis

1 Introduction

Since their introduction in the late 1950s, the Cahn–Hilliard equations have played an important role in understanding phase transition phenomena that is observed in materials. In particular, Cahn–Hilliard equations describe the process of phase separation in which a mixture of two materials separate or fuse to form pure material domains. The main purpose of these proceedings is to develop robust solvers for a well-known unconditionally stable time-stepping discretization.

Let $\Omega \subset \mathbb{R}^d$, $d \leq 3$, be a polygon or polyhedron with boundary denoted by $\partial\Omega$. We focus on Cahn–Hilliard equations [4] with initial and boundary conditions defined as

$$u_t = \Delta w, \qquad x \in \Omega,\ t > 0, \tag{1a}$$

$$w = \Psi'(u) - \gamma \Delta u, \qquad x \in \Omega,\ t > 0, \tag{1b}$$

$$u(x,0) = u_0(x), \qquad x \in \Omega, \tag{1c}$$

$$\nabla u \cdot n = \nabla w \cdot n = 0, \qquad x \in \partial\Omega,\ t > 0, \tag{1d}$$

where u is the order parameter, such as concentration in a binary compound, with $u = \pm 1$ indicating pure states, $\Psi(u)$ is the potential function and $\gamma > 0$ is related to interfacial width between the two phases. n denotes the normal vector to $\partial\Omega$. In this article we consider the logarithmic nonlinear potential defined as

Siamak Faal
Worcester Polytechnic Institute, 100 Institute Rd, Worcester, USA,
e-mail: sghorbanifaal@wpi.edu

Adam Powell
Worcester Polytechnic Institute, 100 Institute Rd, Worcester, USA, e-mail: acpowell@wpi.edu

Marcus Sarkis
Worcester Polytechnic Institute, 100 Institute Rd, Worcester, USA, e-mail: msarkis@wpi.edu

S. C. Brenner et al. (eds.), *Domain Decomposition Methods in Science and Engineering XXVI*, Lecture Notes in Computational Science and Engineering 145,
https://doi.org/10.1007/978-3-030-95025-5_24

$$\Psi(u) := \frac{\theta}{2}\big[(1+u)\ln(1+u) + (1-u)\ln(1-u)\big] + \frac{\theta_c}{2}(1-u^2), \quad 0 < \theta < \theta_c.$$

Following the splitting scheme presented in [5], we decompose Ψ into two functions $\Psi = \Psi_i + \Psi_e$ with Ψ_i convex and Ψ_e concave, and $\psi := \Psi' = \psi_i + \psi_e$, where

$$\Psi_i(u) := \frac{\theta}{2}\big[(1+u)\ln(1+u) + (1-u)\ln(1-u)\big], \qquad \psi_i(u) = \frac{\theta}{2}\ln\left(\frac{1+u}{1-u}\right),$$
$$\Psi_e(u) := \frac{\theta_c}{2}(1-u^2), \qquad \psi_e(u) = -\theta_c u.$$

We use the notation $i \sim$ and $e \sim$ to indicate implicit and explicit treatment of Ψ_i and Ψ_e in the time-stepping discretization, respectively, which leads to an unconditionally stable time discretization.

Let $T > 0$ denote a finite time, then the weak formulation of (1) demands finding $(u, w) \in L^\infty(0,T; H^1(\Omega)) \times L^2(0,T; H^1(\Omega))$ such that $u(x,0) = u_0(x)$ and

$$\langle u_t, v\rangle + (\nabla w, \nabla v) = 0, \qquad \forall v \in H^1(\Omega), \tag{2a}$$
$$(w, z) - (\psi, z) - \gamma(\nabla u, \nabla z) = 0, \qquad \forall z \in H^1(\Omega), \tag{2b}$$

where $\langle\cdot,\cdot\rangle$ denotes the duality pairing between $H^1(\Omega)$ and $H^{-1}(\Omega)$; and $(\cdot,\cdot)$ denotes the inner product in $L^2(\Omega)$.

The existence and uniqueness of this system for a class of Ψ has been shown in [8]. This system also has two interesting properties: Conservation of Mass and dissipative Free Energy $\mathcal{E}(u)$. Indeed, substituting $v \equiv 1$ in (2a) leads to $\int_\Omega u(t)\,\mathrm{d}x = \int_\Omega u(0)\,\mathrm{d}x$ and substituting $v = w$ and $z = u_t$ in (2) gives

$$\frac{\mathrm{d}}{\mathrm{d}t}\mathcal{E}(u(t)) = -\int_\Omega \gamma\,|\nabla w(t)|^2\,\mathrm{d}x \quad \text{where} \quad \mathcal{E}(u) := \int_\Omega \frac{1}{2}|\nabla u|^2 + \Psi(u)\,\mathrm{d}x.$$

2 Numerical approximation

In order to formulate a finite element approximation of problem (2), we focus on a polygon domain Ω, and construct a quasi-uniform family of triangulation $\mathbf{T}_h$ such that $\Omega = \bigcup_{\kappa\in\mathbf{T}_h}\kappa$ and $h := \max_{\kappa\in\mathbf{T}_h} h_\kappa$, where $h_\kappa := \mathrm{diam}(\kappa)$. We define the finite element space $S_h := \mathrm{span}(\{\varphi_r \in C(\Omega) : \varphi_r(x_k) = \delta_{rk}, \text{ for } r,k = 1,\dots,N_h\})$, where φ_r are the standard piecewise and continuous affine nodal basis functions and $\{x_r\}$ denotes the set of vertices. Let $\{t_n = n\tau : n = 0,1,\dots,N, \text{ and } N\cdot\tau = T\}$ denote a discretization of the time interval $[0,T]$ and τ the time-stepping size, and consider the first-order approximation of u_t in time as $u_t(t_n) \approx [u(t_n) - u(t_{n-1})]/(t_n - t_{n-1})$. Our goal is to find $(\hat{u}^n, \hat{w}^n) \in S_h \times S_h$, where $\hat{u}^n := \hat{u}(t_n)$ and $\hat{w}^n := \hat{w}(t_n)$, such that for all $0 < n \le N$,

$$(\hat{u}^n - \hat{u}^{n-1}, \hat{v})_h + \tau(\nabla \hat{w}^n, \nabla \hat{v}) = 0, \qquad \forall \hat{v} \in S_h \tag{3a}$$
$$(\hat{w}^n, \hat{z})_h - (\psi_i(\hat{u}^n) + \psi_e(\hat{u}^{n-1}), \hat{z})_h - \gamma(\nabla \hat{u}^n, \nabla \hat{z}) = 0, \qquad \forall \hat{z} \in S_h. \tag{3b}$$

where $(\cdot, \cdot)_h$ is the lumped discrete inner product defined as

$$(\hat{v}, \hat{z})_h := \int_\Omega I_h(\hat{v}\,\hat{z})\,\mathrm{dx} = \sum_{r=1}^{N_h} m_r\,\hat{v}(x_r)\,\hat{z}(x_r), \quad \forall \hat{v}, \hat{z} \in S_h, \tag{4}$$

and I_h is the nodewise linear interpolation given by

$$I_h : C(\bar{\Omega}) \to S_h, \quad I_h v(x_r) = v(x_r) \text{ for } r = 1, \ldots, N_h, \tag{5}$$
$$m_r = (\varphi_r, \varphi_r)_h. \tag{6}$$

The proof of the existence of the discrete problem (3) follows from the convexity of the Ψ_i and lumped mass matrix approach considered in the discretization. The proof of the unconditionally time-stepping stability is similar to [5].

Theorem 1 *There exists a unique solution $(\hat{u}^n, \hat{w}^n) \in S_h \times S_h$ to the finite element problem (3).*

Proof The proof is a simple modification of the proof given in [5]. Let $(\hat{u}_1^n, \hat{w}_1^n)$ and $(\hat{u}_2^n, \hat{w}_2^n)$ be two solutions of (3). By subtracting one solution from the other we get

$$(\hat{y}_u^n, \hat{v}) + \tau(\nabla \hat{y}_w^n, \nabla \hat{v}) = 0, \qquad \forall \hat{v} \in S_h, \tag{7a}$$
$$(\hat{y}_w^n, \hat{z}) - (\psi_i(\hat{u}_2^n) - \psi_i(\hat{u}_1^n), \hat{z}) - \gamma(\nabla \hat{y}_u^n, \nabla \hat{z}) = 0, \qquad \forall \hat{z} \in S_h, \tag{7b}$$

where $\hat{y}_u^n := \hat{u}_2^n - \hat{u}_1^n$ and $\hat{y}_w^n := \hat{w}_2^n - \hat{w}_1^n$. Substituting $\hat{v}$ and $\hat{z}$ into (7) with $\hat{y}_w^n$ and $\hat{y}_u^n$, respectively, leads to

$$\tau|\hat{y}_w^n|_1^2 + \gamma|\hat{y}_u^n|_1^2 = (\psi_i(\hat{u}_1^n) - \psi_i(\hat{u}_2^n)_h, \hat{y}_u^n),$$

where $|\cdot|_1$ denotes the seminorm of $H^1(\Omega)$. Applying the mean value theorem to ψ_i gives

$$(\psi_i(s_1) - \psi_i(s_2))(s_2 - s_1) = -\psi_i'(c)(s_2 - s_1)^2,$$

for every $s_1, s_2 \in (-1, 1)$ and c between s_1 and s_2. Based on the convexity of ψ_i we get $0 \le \psi_i'(s)$ for $s \in (-1, 1)$, which implies

$$(\psi_i(\hat{u}_1^n) - \psi_i(\hat{u}_2^n), \hat{y}_u^n)_h \le 0,$$

and accordingly $\tau\,|\hat{y}_w^n|_1^2 + \gamma\,|\hat{y}_u^n|_1^2 \le 0$. Moreover, since $(\hat{y}_u^n, 1)_h = 0$, the Poincaré inequality implies that $\|\hat{y}_u^n\| = 0$. To show the uniqueness of $\hat{w}^n$ it suffices to set $\hat{z}$ equal to $\hat{y}_w^n$ in (7b) and get $\|\hat{y}_w^n\| = 0$. This concludes the proof of uniqueness. □

3 Newton's method

Our objective is to solve the nonlinear system (3) using Newton's method. Since the nonlinearity is associated with the potential function Ψ, for every Newton iteration j and time step n, we set

$$\hat{u}_j^n = \hat{u}_{j-1}^n - \delta\hat{u}_j^n. \tag{8}$$

Substituting $(\hat{u}^n, \hat{w}^n)$ with $(\hat{u}_j^n, \hat{w}_j^n)$ in (3) and using the linearization

$$\psi_i(\hat{u}_j^n) \approx \psi_i(\hat{u}_{j-1}^n) - \psi_i'(\hat{u}_{j-1}^n)\,\delta\hat{u}_j^n, \tag{9}$$

the system of equation in (3) leads to

$$(\delta\hat{u}_j^n, \hat{v})_h - \tau\,(\nabla\hat{w}_j^n, \nabla\hat{v}) = \phi_1(\hat{v}), \qquad \forall\,\hat{v} \in S_h, \tag{10a}$$

$$\left(\psi_i'(\hat{u}_{j-1}^n)\,\delta\hat{u}_j^n, \hat{z}\right)_h + \gamma\,(\nabla\delta\hat{u}_j^n, \nabla\hat{z}) + (\hat{w}_j^n, \hat{z})_h = \phi_2(\hat{z}), \qquad \forall\,\hat{z} \in S_h, \tag{10b}$$

where

$$\phi_1(\hat{v}) := (\hat{u}_{j-1}^n - \hat{u}^{n-1}, \hat{v})_h, \tag{11a}$$

$$\phi_2(\hat{z}) := \left(\psi_i(\hat{u}_{j-1}^n) + \psi_e(\hat{u}^{n-1}), \hat{z}\right)_h + \gamma\,(\nabla\hat{u}_{j-1}^n, \nabla\hat{z}). \tag{11b}$$

Let $\mathbf{u}_j^n$, $\mathbf{w}_j^n$ and $\boldsymbol{\delta}\mathbf{u}_j^n$ in $\mathbb{R}^{N_h}$ denote the vectors composed of the values of $\hat{u}_j^n$, $\hat{w}_j^n$ and $\delta\hat{u}_j^n$ evaluated at every vertex x_r, that is $[\mathbf{u}_j^n]_r = \hat{u}_j^n(x_r)$, $[\mathbf{w}_j^n]_r = \hat{w}_j^n(x_r)$ and $[\boldsymbol{\delta}\mathbf{u}_j^n]_r = \delta\hat{u}_j^n(x_r)$. In addition, let $[M]_{r,k} := (\varphi_r, \varphi_k)_h$, $[K]_{r,k} := (\nabla\varphi_r, \nabla\varphi_k)$ and

$$[J_{j-1}]_{r,k} := \Big(\psi_i'\big(\sum_{l=1}^{N_h}[\mathbf{u}_{j-1}^n]_l\varphi_l\big)\varphi_r, \varphi_k\Big)_h, \qquad \tilde{K} := K,$$

$$[\mathbf{q}_{j-1}]_r := \Big(\psi_i\big(\sum_{l=1}^{N_h}[\mathbf{u}_{j-1}^n]_l\varphi_l\big), \varphi_r\Big)_h, \qquad [\mathbf{p}]_r := \Big(\psi_e\big(\sum_{l=1}^{N_h}[\mathbf{u}^{n-1}]_l\varphi_l\big), \varphi_r\Big)_h.$$

Then, the discrete representation of (10) yields

$$M\,\boldsymbol{\delta}\mathbf{u}_j^n - \tau\,K\,\mathbf{w}_j^n = \mathbf{f}_{j-1} \tag{12a}$$

$$M^T\,\mathbf{w}_j^n + (J_{j-1} + \gamma\,\tilde{K})\,\boldsymbol{\delta}\mathbf{u}_j^n = \mathbf{g}_{j-1}, \tag{12b}$$

$$\mathbf{u}_j^n = \mathbf{u}_{j-1}^n - \boldsymbol{\delta}\mathbf{u}_j^n, \tag{12c}$$

where $\mathbf{f}_{j-1} = M(\mathbf{u}_{j-1}^n - \mathbf{u}^{n-1})$ and $\mathbf{g}_{j-1} = \mathbf{q}_{j-1} + \mathbf{p} + \gamma\,\tilde{K}\,\mathbf{u}_{j-1}^n$.

Let $\Omega = (0,1)^2$ and $\mathbf{T}_h$ be a uniform triangulation with 45-degree triangles where $h = 1/32$, $\tau = 0.01$, $\gamma = h^2$ with a random initial condition $\mathbf{u}^0$. Fig. 1 illustrates the numerical solution at six time instances. The Newton's method is stopped when $\|\boldsymbol{\delta}\mathbf{u}^n\|_{\ell_\infty} \leq 10^{-10}$. The color map that varies from blue (dark) to yellow (light) depicts values of $\mathbf{u}^n(x_r)$ close to -1^+ and $+1^-$, respectively.

4 Preconditioned iterative solver

We consider solving (12) simultaneously to find $\mathbf{w}_j^n$ and $\delta\mathbf{u}_j^n$ as a solution to

$$A\,\mathbf{x} = \begin{bmatrix} -\tau K & M \\ M^T & J_{j-1} + \gamma \tilde{K} \end{bmatrix} \begin{bmatrix} \mathbf{w}_j^n \\ \delta\mathbf{u}_j^n \end{bmatrix} = \begin{bmatrix} \mathbf{f}_{j-1} \\ \mathbf{g}_{j-1} \end{bmatrix} = \mathbf{b}. \tag{13}$$

Since the associated matrix is symmetric, we can use the MinRes algorithm to find $(\mathbf{w}_j^n, \delta\mathbf{u}_j^n)$ at each Newton iteration. We now propose a preconditioner based on the Schur complement of the first block of (13) given by

$$B_1 = \begin{bmatrix} -M(J_{j-1} + \gamma\tilde{K})^{-1}M^T - \tau K & 0 \\ 0 & J_{j-1} + \gamma\tilde{K} \end{bmatrix}. \tag{14}$$

In what follows, we investigate the performance of the solver based on the proposed

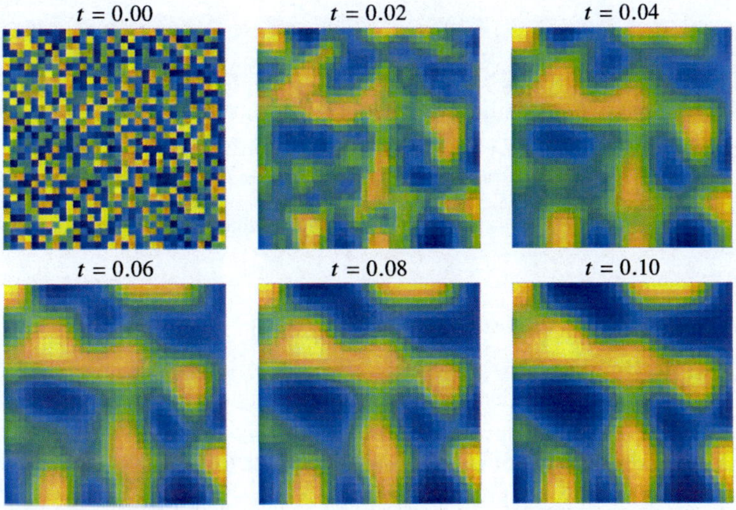

Fig. 1: numerical solution of (12) at $t = 0, 0.02, 0.04, 0.06, 0.08,$ and 0.1 for $\Omega = (0,1)^2$, $h = 1/32$, $\tau = 0.01$, $\gamma = h^2$ and a random initial condition $\mathbf{u}^0$.

$P_1\,\mathbf{x} = B_1^{-1}\,A\,\mathbf{x} = B_1^{-1}\,\mathbf{b}$. Note that, since the lumped inner product is utilized, M and J_{j-1} are diagonal matrices. See also [3] for other classes of preconditioners.

Table 1 demonstrates the behavior of the preconditioned system P_1. As listed, the preconditioned system requires six Newton iterations for the first, and four iterations for the second time step. The table includes the number of MinRes iterations that reduce the preconditioned residue by a factor of 10^{-6}. As time goes on, the solution becomes more regular (see Fig. 1) and demands fewer Newton iterations. Note that A is symmetric and B_1 is symmetric positive definite. We note that, without utilizing the preconditioner, the solver may not converge or the convergence is much slower. In Table 2 we consider $P_2\,\mathbf{x} = B_2^{-1}\,A\,\mathbf{x} = B_2^{-1}\,\mathbf{b}$ where B_2^{-1} is a BPX [2] type

n	j	$\Vert\delta\mathbf{u}_j^n\Vert_\infty$	relres	gmres itrs
1	1	9.342e-01	6.272e-07	25
1	2	4.144e-01	6.748e-07	27
1	3	1.548e-02	8.128e-07	26
1	4	3.020e-05	8.109e-07	26
1	5	1.133e-10	8.109e-07	26
1	6	2.112e-16	8.109e-07	26
2	1	2.651e-01	7.713e-07	27
2	2	4.494e-03	6.102e-07	28
2	3	3.427e-06	6.140e-07	28
2	4	2.031e-12	6.140e-07	28

Table 1: Results of solving (13) using preconditioned system P_1 with lumped matrices M and J_{j-1} for $h = 1/32$, $\tau = 0.01$, $\gamma = h^2$ and a random initial condition $\mathbf{u}^0$.

preconditioner. The BPX method we consider here follows the strategy in [6, 7] that deals with rational functions of Sobolev norms; this work was motivated by the BPX introduced in [1] that deals with sums of Sobolev norms of the same sign. The multilevel preconditioner is defined as $B_2^{-1} = \text{diag}\{D_1, D_2\}$ with

$$D_1 = -\sum_k R_k^T\left[\left(\text{diag}_k\{R_k M^{-1}(J_{j-1} + \gamma\tilde{K})M^{-1}R_k^T\}\right)^{-1} + \text{diag}_k\{R_k \tau\, K R_k^T\}\right]^{-1} R_k,$$

$$D_2 = \sum_k R_k^T\left(\text{diag}_k\{R_k(J_{j-1} + \gamma\tilde{K})R_k^T\}\right)^{-1} R_k,$$

where D_1 and D_2 are diagonal matrices, R_k^T is the prolongation operator from k-level to fine level and $\text{diag}_k\{C_k\}$ is the diagonal of the matrix C_k defined on the k-level.

As listed in Table 2, the number of iterations using the inexact BPX preconditioner (B_2^{-1}) is comparable with the exact preconditioner B_1^{-1}. See also [3] where the Schur complement of the second block is considered. In Table 3, we test the behavior of preconditioned systems P_1 and P_2 as we vary τ for fixed $\gamma = h^2 = (1/16)^2$. As listed, the number of MinRes iterations for P_1 and P_2 is smaller for larger time-stepping size τ. This is not surprising, since, as we increase τ, the Schur complement $M\,(J_{j-1} + \gamma\,\tilde{K})^{-1}\,M^T + \tau\,K$ becomes more positive definite. The result of changing $\gamma = h^2$ for fixed $\tau = 0.01$ is listed in Table 4. As depicted in the table, the number of MinRes iterations of both P_1 and P_2 is very robust with respect to mesh size h. We note that the condition $\gamma = O(h^2)$ is a reasonable choice since the size of the transition layer of u from -1 to 1 is $O(\sqrt{\gamma})$. Note that we use a random initial condition $\mathbf{u}^0$ and utilize the lumped matrices[1] when defining M and J_{j-1}.

[1] This is a consequence of using the inner product $(\cdot,\ \cdot)_h$ as defined in (4).

n	j	$\|\delta \mathbf{u}_j^n\|_\infty$	relres	gmres itrs
1	1	9.342e-01	7.115e-07	50
1	2	4.144e-01	7.393e-07	58
1	3	1.548e-02	7.555e-07	57
1	4	3.020e-05	7.575e-07	51
1	5	1.132e-10	7.575e-07	57
1	6	2.202e-16	7.575e-07	57
2	1	2.651e-01	7.712e-07	59
2	2	4.494e-03	8.443e-07	59
2	3	3.427e-06	8.430e-07	59
2	4	2.025e-12	8.430e-07	59

Table 2: Results of solving (13) using preconditioned system P_2 with lumped matrices M and J_{j-1} for $h = 1/32$, $\tau = 0.01$, $\gamma = h^2$ and a random initial condition $\mathbf{u}^0$.

τ	n	Exact P_1		Additive multigrid P_2	
		Newton stp	solver itr.	Newton stp	solver itr.
0.1	1	6	18	6	32
	2	4	19	4	34
	3	4	19	4	34
	4	5	19	5	34
0.01	1	5	24	5	51
	2	4	26	4	53
	3	4	26	4	53
	4	4	26	4	53
0.001	1	5	26	5	87
	2	4	27	4	90
	3	4	27	4	90
	4	4	27	4	90

Table 3: Results of varying τ for fixed $\gamma = h^2 = (1/16)^2$. The number of solver iterations (solver itr.) reported in the table is the average over the Newton iterations at every time step n. Preconditioned systems P_1 and P_2 are constructed using lumped matrices M and J_{j-1} and the initial condition for all is a fixed random $\mathbf{u}^0$.

5 Conclusions

Our main contribution is the development of a robust multilevel PBX-type preconditioner for a finite element approximation of the Cahn–Hilliard problem. The numerical results provided demonstrate the robustness of the preconditioned systems with respect to mesh size and time steps.

h	n	Exact P_1		Additive multigrid P_2	
		Newton stp	solver itr.	Newton stp	solver itr.
1/16	1	5	24	5	51
	2	4	26	4	53
	3	4	26	4	53
	4	4	26	4	53
1/32	1	6	26	6	56
	2	4	28	4	59
	3	4	29	4	60
	4	4	28	4	60
1/64	1	6	27	6	58
	2	4	29	4	61
	3	5	30	5	62
	4	5	30	5	62

Table 4: Results of varying $\gamma = h^2$ for fixed $\tau = 0.01$. The number of solver iterations (solver itr.) reported in the table is the average over the Newton iterations at every time step n. Preconditioned systems P_1 and P_2 are constructed using lumped matrices M and J_{j-1} and the initial condition for all is a fixed random $\mathbf{u}^0$.

References

1. J.H. Bramble, J.E. Pasciak and P.S. Vassilevski. Computational scales of Sobolev norms with application to preconditioning. *Math. Comp.* **69**(230), 463–480 (2000).
2. J.H. Bramble, J.E. Pasciak and J. Xu. Parallel multilevel preconditioners. *Math. Comp.* **55**(191), 1–22 (1990).
3. S.C. Brenner, A.E. Diegel and L.-Y. Sung. A robust solver for a mixed finite element method for the Cahn–Hilliard equation. *Journal of Scientific Computing* **77**(2), 1234–1249 (2018).
4. J.W. Cahn and J.E. Hilliard. Free energy of a nonuniform system. I. Interfacial free energy. *The Journal of Chemical Physics* **28**(2), 258–267 (1958).
5. M.I.M. Copetti and C.M. Elliott. Numerical analysis of the Cahn–Hilliard equation with a logarithmic free energy. *Numer. Math.* **63**(1), 39–65 (1992).
6. E. Gonçalves and M. Sarkis. Analysis of robust parameter-free multilevel methods for Neumann boundary control problems. *Comput. Methods Appl. Math.* **13**(2), 207–235 (2013).
7. E. Gonçalves and M. Sarkis. Robust parameter-free multilevel methods for Neumann boundary control problems. In: *Domain Decomposition Methods in Science and Engineering XX*, volume 91 of Lect. Notes Comput. Sci. Eng., pp. 111–118, Springer, Heidelberg (2013).
8. A. Miranville and S. Zelik. Exponential attractors for the Cahn–Hilliard equation with dynamic boundary conditions. *Math. Methods Appl. Sci.* **28**(6), 709–745 (1992).

Natural Factor Based Solvers

Juan Galvis, Marcus Sarkis, and O. Andrés Cuervo

1 Summary

We consider parametric families of partial differential equations (PDEs) where the parameter κ modifies only the (1,1) block of a saddle point matrix product of a discretization. The main goal is to develop an algorithm that removes some of the dependence of iterative solvers on the parameter κ. The algorithm we propose requires only one matrix factorization which does not depend on κ, and therefore allows it to be reused to quickly solve a large number of discrete PDEs for different κ and forcing terms. The design of the proposed algorithm is motivated by previous work on natural factor formulations of the stiffness matrices and their stable numerical solvers. As an application, in two dimensions, we consider an iterative preconditioned solver based on the null space of the Crouzeix–Raviart discrete gradient represented as the discrete curl of P_1 conforming finite element functions. For the numerical examples, we consider the case of a random coefficient pressure equation where the permeability is modeled by an stochastic process. We note that in contrast to recycling Krylov subspace techniques, the proposed algorithm does not require fixed forcing terms.

Juan Galvis
Departamento de Matemáticas, Universidad Nacional de Colombia, Bogotá, Colombia,
e-mail: jcgalvisa@unal.edu.co

Marcus Sarkis
Department of Mathematical Sciences, Worcester Polytechnic Institute, Worcester USA,
e-mail: msarkis@wpi.edu

Andrés Cuervo and Juan Galvis
School of Engineering, Science and Technology, Universidad del Rosario, Bogotá, Colombia,
e-mail: omar.cuervo@urosario.edu.co

S. C. Brenner et al. (eds.), *Domain Decomposition Methods in Science and Engineering XXVI*, Lecture Notes in Computational Science and Engineering 145,
https://doi.org/10.1007/978-3-030-95025-5_25

2 Introduction

The general form of a saddle point system of linear equations we consider is

$$\begin{bmatrix} D(\kappa)^{-1} & G \\ G^T & 0 \end{bmatrix} \begin{bmatrix} q \\ u \end{bmatrix} = \begin{bmatrix} r \\ b \end{bmatrix}, \tag{1}$$

where the matrix D is symmetric positive definite. This form is standard in the formulation of mixed finite elements. What is not very well known, as pointed out by Argyris and Brønlund [1], is that classical conforming and nonconforming finite element methods (FEMs) can also be written in the form (1) with $r = 0$; see Section 3 for the case of the Crouzeix–Raviart FEM and Section 4 for the P_1 conforming FEM. We show that the stiffness matrix associated to the Crouzeix–Raviart FEM element discretization for the PDE (2) with isotropic coefficients $\kappa(x)$ has a natural factor of the form $A_{CR} = G_{CR}^T D(\kappa) G_{CR}$, where G_{CR} is the discrete gradient (not affected by the parameter κ) and $D(\kappa)$ is a diagonal matrix with entries depending of the integration of κ in each element, hence, it is easy to update the natural factor if κ is modified. Due to the superior numerical stability with respect to roundoff errors when operating with G^T, $D(\kappa)$ and G rather than the assembled stiffness matrix, several works [2–5] have been dedicated to solving the saddle point problem (1) or associated SVD and diagonalization. In Sections 5 and 6 we review some aspects of these works. The methods start by representing q on the range of the matrix $[G\ Z]$ where Z is such that $Q = [G\ Z]$ is a square invertible matrix; two common choices of Z are $Z^T G = 0$ and $Z^T D^{-1} G = 0$. These works generate very stable algorithms for ill-conditioned κ, however, they do not remove the dependence of the factorizations on D, hence, they do not fit our goal of reusing the same factorization for different values of κ. In Section 7 we propose our method. We first use discrete Hodge Laplacian ideas to choose $Z = C_L$ as the curl of P_1 conforming piecewise linear basis functions, hence $G_{CR}^T C_L = 0$. Then we consider the coupled system

$$(\operatorname{grad} u_{CR} + \operatorname{curl} w_{P_1}, \kappa\, (\operatorname{grad} v_{CR} + \operatorname{curl} v_{P_1}))_{L^2(\Omega)}$$

as a preconditioner for the uncoupled system

$$(\operatorname{grad} u_{CR}, \kappa \operatorname{grad} v_{CR})_\kappa + (\operatorname{curl} w_{P_1}, \kappa \operatorname{curl} v_{P_1})_{L^2(\Omega)}.$$

3 Crouzeix–Raviart nonconforming finite elements

Consider the heterogeneous diffusion equation

$$\begin{cases} -\partial_1(\kappa(x)\partial_1 u(x)) - \partial_2(\kappa(x)\partial_2 u(x)) = f(x), & x \in \Omega, \\ u(x) = 0, & x \in \partial\Omega, \end{cases} \tag{2}$$

where $\Omega \subseteq \mathbb{R}^2$ and $\kappa : \Omega \to \mathbb{R}^+$, $f : \Omega \to \mathbb{R}$ are given.

In particular, in the target application $\kappa(x)$ is a random field that describes the permeability and allows the modeling of lack of data and uncertainties of the problem (e.g., subsurface flow). The forcing term f may also be a random field. In general, in many practical situations we must solve (2) for a large family of coefficients κ and forcing terms f. See Section 8.

Let us introduce a triangulation $\mathcal{T}^h$ of Ω. Discretize (2) by the Crouzeix–Raviart (CR) non-conforming finite element space. Define the CR space $\widetilde{V}^{CR}$ as the space of all piecewise linear functions with respect to $\mathcal{T}^h$ that are continuous at the midpoints of interior edges. The degrees of freedom are located at the midpoints of the edges of $\mathcal{T}^h$. Let $V_{CR} \subseteq \widetilde{V}^{CR}$ be the subspace of functions in $\widetilde{V}^{CR}$ with zero value at the midpoint of boundary edges. The approximation $u_{CR} \in V_{CR}$ of the solution of (2) is the solution of

$$\sum_{T\in\mathcal{T}^h}\int_T \kappa(x)(\partial_1 u_{CR}(x)\partial_1 v(x) + \partial_2 u_{CR}(x)\partial_2 v(x))\mathrm{d}x = \int_\Omega f(x)v(x)\mathrm{d}x,$$

for all $v \in V^{CR}$. The linear system of the CR approximation is given by

$$A_{CR}u_{CR} = b_{CR}, \tag{3}$$

where $A_{CR} = \left[a_{ei}^{CR}\right]_{e,i=1}^{N_e}$ and $b_{CR} = [b_e]_{e=1}^{N_e}$. Here, N_e denotes the number of interior edges of $\mathcal{T}^h$, $b_e = \int_\Omega f(x)\varphi_e^{CR}(x)\mathrm{d}x$ and

$$a_{ei}^{CR} = \sum_{T\in\mathcal{T}^h}\int_T \kappa(x)\Big(\partial_1\varphi_e^{CR}(x)\partial_1\varphi_i^{CR}(x) + \partial_2\varphi_e^{CR}(x)\partial_2\varphi_i^{CR}(x)\Big)\mathrm{d}x.$$

Let x_T denote the barycenter of a triangle $T \in \mathcal{T}^h$. Piecewise gradients of functions in V^{CR} are piecewise constant vector functions and then

$$a_{ei}^{CR} = \sum_{T\in\mathcal{T}^h}\kappa_T|T|\partial_1\varphi_e(x_T)\partial_1\varphi_i(x_T) + \sum_{T\in\mathcal{T}^h}\kappa_T|T|\partial_2\varphi_e(x_T)\partial_2\varphi_i(x_T), \tag{4}$$

where κ_T is the average value of $\kappa(x)$ in T. Therefore, we can write (see [1])

$$A_{CR} = G_{CR}^T D G_{CR} = G_{CR,1}^T D_1 G_{CR,1} + G_{CR,2}^T D_2 G_{CR,2},$$

where $G_{CR,l} = \left[g_{e,T}^{CR,l}\right]_{N_T\times N_e} = \left[\sqrt{|T|}\partial_l\varphi_e^{CR}(x_T)\right]_{N_T\times N_e}$, and N_T denotes the number of triangles in $\mathcal{T}^h$ and $l = 1, 2$. Furthermore, write,

$$D_l = \mathrm{diag}(\kappa_T)_{T\in\mathcal{T}^h},\quad D = \mathrm{diag}(D_1, D_2) \text{ and } G_{CR} = \begin{bmatrix} G_{CR,1} \\ G_{CR,2}\end{bmatrix}_{2N_T\times N_e}. \tag{5}$$

We can write the matrix formulation as

$$G_{CR}^T D G_{CR} u_{CR} = b_{CR}. \tag{6}$$

We see that problem (6) is the Schur complement of the saddle point problem

$$\begin{bmatrix} D^{-1} & G_{CR} \\ G_{CR}^T & 0 \end{bmatrix} \begin{bmatrix} q \\ u_{CR} \end{bmatrix} = \begin{bmatrix} 0 \\ -b_{CR} \end{bmatrix}. \tag{7}$$

4 Conforming finite elements P_1

Let $\widetilde{V}^L = P_1(\mathcal{T}^h) = \{v : \Omega \to \mathbb{R} |\ v|_T \text{ is linear for all } T \in \mathcal{T}^h\} \cap C^0(D)$. The space $\widetilde{V}^L$ has a base $\{\varphi_i^L\}_{i=1}^{\widetilde{N}_v}$, where $\widetilde{N}_v$ is the number of vertices and φ_i^L is the function that takes value 1 at the i-th node and 0 at the other nodes. Also define $V^L = \widetilde{V}^L \cap H_0^1(\Omega)$ and N_v the number of interior vertices.

The approximation u_L of the solution of (2) is: find $u_L \in V^L$ such that

$$\int_\Omega \kappa(x)(\partial_1 u_L(x)\partial_1 v(x) + \partial_2 u_L(x)\partial_2 v(x))\mathrm{d}x = \int_\Omega f(x)v(x)\mathrm{d}x,$$

for all $v \in V^L$, with matrix form $A_L u_L = b_L$, where, $A_L = [a_{ij}^L]_{i,j=1}^{N_v}$ and $b_L = [b_i^L]_{i=1}^{N_v}$ with

$$b_i^L = \int_D f(x)\varphi_i^L(x)\mathrm{d}x \quad \text{and}$$
$$a_{ij}^L = \int_\Omega \kappa(x)\Big(\partial_1\varphi_i^L(x)\partial_1\varphi_j^L(x) + \partial_2\varphi_i^L(x)\partial_2\varphi_j^L(x)\Big)\mathrm{d}x.$$

As before, we have (see [1])

$$A_L = G_L^T D G_L = G_{L,1}^T D_1 G_{L,1} + G_{L,2}^T D_2 G_{L,2},$$

where $G_{L,l} = \left[g_{e,v}^{L,l}\right]_{N_T \times N_v} = \left[\sqrt{|T|}\partial_l \varphi_v^L(x_T)\right]_{N_T \times N_v}$ and $G_L = \begin{bmatrix} G_{L,1} \\ G_{L,2} \end{bmatrix}_{2N_T \times N_v}$.
We can write the matrix formulation as

$$G_L^T D G_L u_L = b_L, \tag{8}$$

and the corresponding saddle point problem is $\begin{bmatrix} D^{-1} & G_L \\ G_L^T & 0 \end{bmatrix} \begin{bmatrix} q \\ u_L \end{bmatrix} = \begin{bmatrix} 0 \\ -b_L \end{bmatrix}$.

5 The null space method

A method for solving the saddle point problem (1) is called the null space method, see [3]. We split (1) into two equations, $D^{-1}q + Gu = r$ and $G^T q = b$. The null space method consists in finding Z that represents the null space of G^T, $G^T Z = 0$, and such that $[G \;\; Z]$ is a non-singular square matrix. Therefore, we can change variables to potentials χ and ψ such that

$$q = [G \;\; Z] \begin{bmatrix} \chi \\ \psi \end{bmatrix} = G\chi + Z\psi. \tag{9}$$

From (9) and $G^T Z = 0$ we have $G^T q = G^T G\chi$ and from $G^T q = b$ we have $b = G^T G\chi$, giving $\chi = (G^T G)^{-1} b$, which can be pre-computed. On the other hand, from $D^{-1}q + Gu = r$ and (9) we have that $D^{-1}G\chi + D^{-1}Z\psi + Gu = r$, which gives $Z^T D^{-1} Z\psi = Z^T r - Z^T D^{-1} G\chi$ and if we let $c = Z^T r - Z^T D^{-1} G\chi$, we can write the system as

$$Z^T D^{-1} Z\psi = c. \tag{10}$$

This is the null space system. It is similar to the Schur complement of (1). See (8).

6 Range null-space hybrid

Now we combine the first equation of (1) and (9). We have $D^{-1}(G\chi + Z\psi) + Gu = r$, which gives $Z\psi + DGu = Dr - G\chi$, allowing us to write the system as ([4, 5])

$$[DG \;\; Z] \begin{bmatrix} u \\ \psi \end{bmatrix} = Dr - G\chi. \tag{11}$$

We note that the matrix $[DG \;\; Z]$ is a square matrix. This system is called a range space scaled system. The related matrix $[G \;\; D^{-1}Z]$ is called a null space scaled matrix. The algorithm is called "hybrid" because it uses both the range-space and the null-space. See [4, 5].

Alternatively, we can write $Z^T(D^{-1}q + Gu) = Z^T r$, which gives $Z^T D^{-1} q = Z^T r$ and together with (9) gives the system

$$\begin{bmatrix} G^T \\ Z^T D^{-1} \end{bmatrix} q = \begin{bmatrix} b \\ Z^T r \end{bmatrix}. \tag{12}$$

Note that matrices (11) and (12) have a dependence on D, however, for numerical stability purposes it is very efficient since the matrix is based on discrete gradient times D rather than the assembled second-order derivatives with D.

7 An auxiliary problem and 2 × 2 systems

Recall that for a scalar w, $\overrightarrow{\text{curl}}\, w = (\partial_2 w, -\partial_1 w)$ and for a vector $\vec{q} = (q_1, q_2)$, $\text{curl}\, \vec{q} = \partial_1 q_2 - \partial_2 q_1$. Consider now the elliptic equation

$$\begin{cases} -\text{curl}\,(\kappa(x)\overrightarrow{\text{curl}}\, w(x)) = g(x), & x \in \Omega, \\ \kappa(x)\text{curl}\, w(x) \cdot \boldsymbol{\tau} = 0, & x \in \partial\Omega, \end{cases}$$

where τ is the tangential vector on the boundary of Ω. Note that we have $\text{curl}\,(\kappa(x)\overrightarrow{\text{curl}}\, w(x)) = -\partial_1(\kappa(x)\partial_1 w(x)) - \partial_2(\kappa(x)\partial_2 w(x))$ and $\kappa(x)\overrightarrow{\text{curl}}\, w(x) \cdot \boldsymbol{\tau} = -n_2\kappa(x)\partial_1 w(x) - n_1\kappa(x)\partial_2 w(x) = -\kappa(x)\nabla w(x) \cdot \boldsymbol{\eta}$, where $\boldsymbol{\eta}$ is the normal vector. We approximate this problem by conforming elements. Let $\widetilde{V}^L = P_1(\mathcal{T}^h) = \{v : \Omega \to \mathbb{R}$ such that $v|_T$ is linear for all $T \in \mathcal{T}^h\} \cap C^0(\Omega)$. The approximation of the problem above is: Find $\widetilde{w}_L \in \widetilde{V}^L$ such that

$$\int_\Omega \kappa(x)\overrightarrow{\text{curl}}\, \widetilde{w}_L(x) \cdot \overrightarrow{\text{curl}}\, v(x)\text{d}x = \int_\Omega g(x)v(x)\text{d}x \qquad \text{for all } v \in \widetilde{V}^L,$$

with additional requirement that $\int_\Omega \widetilde{w}_L(x)\text{d}x = 0$. The matrix form is

$$\widetilde{A}_L \widetilde{w}_L = \widetilde{b}_L,$$

where $\widetilde{A}_L = [a_{ij}^L]_{\widetilde{N}_v \times \widetilde{N}_v}$ and $\widetilde{b}_L = [b_i^L]_{\widetilde{N}_v \times 1}$ with entries defined by $a_{ij}^L = \int_\Omega \kappa(x)\overrightarrow{\text{curl}}\, \varphi_i^L(x) \cdot \overrightarrow{\text{curl}}\, \varphi_j^L(x)\text{d}x$ and $b_i^L = \int_\Omega g(x)\varphi_i^L(x)\text{d}x$. Here $\widetilde{N}_v$ is the number of vertices in $\mathcal{T}^h$. As before, we have

$$\widetilde{A}_L = \widetilde{C}_L^T D \widetilde{C}_L = \widetilde{G}_{L,2}^T D_1 \widetilde{G}_{L,2} + (-\widetilde{G}_{L,1})^T D_2 (-\widetilde{G}_{L,1}),$$

where $\widetilde{G}_{L,l} = \left[g_{e,v}^{L,l}\right]_{N_T \times \widetilde{N}_v} = \left[\sqrt{|T|}\partial_l \varphi_v^L(x_T)\right]$ and $\widetilde{C}_L = \begin{bmatrix} \widetilde{G}_{L,2} \\ -\widetilde{G}_{L,1} \end{bmatrix}_{2N_T \times \widetilde{N}_v}$. Note that $(u_{CR}, \widetilde{w}_L)$ satisfy the 2×2 system $\begin{bmatrix} A_{CR} & 0 \\ 0 & \widetilde{A}_L \end{bmatrix} \begin{bmatrix} u_{CR} \\ \widetilde{w}_L \end{bmatrix} = \begin{bmatrix} b_{CR} \\ \widetilde{b}_L \end{bmatrix}$.

Let

$$\widehat{A} = \begin{bmatrix} A_{CR} & 0 \\ 0 & \widetilde{A}_L \end{bmatrix}, \quad \widehat{u} = \begin{bmatrix} u_{CR} \\ \widetilde{w}_L \end{bmatrix} \text{ and } \widehat{b} = \begin{bmatrix} b_{CR} \\ \widetilde{b}_L \end{bmatrix} \tag{13}$$

and introduce the matrices $H = [G_{CR} \ \ \widetilde{C}_L]$ and

$$M = H^T D H = \begin{bmatrix} A_{CR} & G_{CR}^T D \widetilde{C}_L \\ \widetilde{C}_L^T D G_{CR} & \widetilde{A}_L \end{bmatrix}. \tag{14}$$

The preconditioned system is given by

$$M^{-1}\widehat{A}\widehat{u} = M^{-1}\widehat{b}. \tag{15}$$

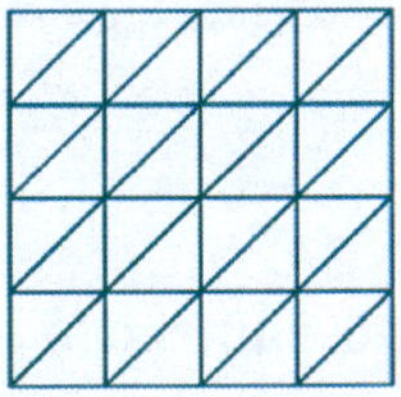

Fig. 1: Triangulation of $D = [0, 1]^2$.

For any planar triangulation (with triangular elements) of a simply connected domain we have $2N_T = N_e + \widetilde{N}_v - 1$ (where N_e is the number of interior edges and $\widetilde{N}_v$ is the number of vertices). See Figure 1 for the particular case of $\Omega = [0, 1]^2$ and $\mathcal{T}^h$ constructed by dividing Ω into n^2 squares and further dividing each square into two triangles by adding an edge from the left-bottom vertex to right-top one. The following lemma shows that no extra computation is required to obtain a basis of null spaces. Also, recall that $G^T G$ is the stiffness matrix of the Laplace operator.

Lemma 1 *We have:*

(a) $H = [G_{CR}\ \ \widetilde{C}_L]$ *is a square matrix of size* $2N_T \times 2N_T$.
(b) $G_{CR}^T \widetilde{C}_L = 0$.
(c) Because of (b), H *is non-singular and* $\widetilde{C}_L$ *spans the kernel of* G_{CR}^T. *Also* G_{CR} *spans the kernel of* $\widetilde{C}_L^T$.
(d) $M = H^T D H$ *is the product of three square matrices. Therefore the solution of* $M\widehat{v} = \widehat{r}$ *can be computed as* $\widehat{v} = H^{-1} D^{-1} H^{-T} \widehat{r}$.

Proof We prove (b). Let e be an interior edge and v a vertex of $\mathcal{T}^h$. Then

$$\begin{aligned}
(G_{CR}^T \widetilde{C}_L)_{e,v} &= \sum_{T\in\mathcal{T}^h} g_{e,T}^{CR,1} g_{v,T}^{L,2} - \sum_{T\in\mathcal{T}^h} g_{e,T}^{CR,2} g_{v,T}^{L,1} \\
&= \sum_{T\in\mathcal{T}^h} |T| \left[\partial_1\varphi_e^{CR}(x_T)\partial_2\varphi_v^L(x_T) - \partial_2\varphi_e^{CR}(x_T)\partial_1\varphi_v^L(x_T)\right] \\
&= \sum_{T\in\mathcal{T}^h} \int_T \nabla\varphi_e^{CR}(x)\cdot \overrightarrow{\mathrm{curl}}\,\varphi_v^L(x)\,\mathrm{d}x \\
&= \sum_{T\in\mathcal{T}^h} \int_{\partial T} \varphi_e^{CR}(x)\,\overrightarrow{\mathrm{curl}}\,\varphi_v^L(x)\cdot\boldsymbol{\eta}\,\mathrm{d}x = 0.
\end{aligned}$$
□

We have the following condition number bound.

Theorem 1 *Let* $\kappa_{\min} \leq \kappa(x) \leq \kappa_{\max}$ *and* $\eta = \kappa_{\max}/\kappa_{\min}$ *the contrast. Then* $cond\,(H^{-1}D^{-1}H^{-T}A) \leq 2\eta - 1$.

Proof Let $s = u_{CR}^T A_{CR} u_{CR} + \widetilde{w}_L^T \widetilde{A}_L \widetilde{w}_L$, using Lemma 1 (b), the result follows from $2|u_{CR}^T G_{CR}^T D\widetilde{C}_L\widetilde{w}_L| = 2|u_{CR}^T G_{CR}^T (D - D(k_{\min}))\widetilde{C}_L\widetilde{w}_L| \leq (1 - 1/\eta)s$. □

8 PCG for the block system and numerical experiments

We propose to solve $\widehat{A}\widehat{u} = \widehat{b}$ with $\widehat{A}$ and $\widehat{b}$ defined in (13) with $\widetilde{b}_L = 0$ using PCG with preconditioner M in (14). See (15). Recall that we use the construction in Section 7. For the numerical test we compute an LU or QR factorization for H and apply $M^{-1} = H^{-1}D^{-1}H^{-T}$. Note that M^{-1} depends on the coefficient κ only through the matrix $D = D(\kappa)$. See (5).

	Condition	**Iterations**	**Contrast**
Mean	1.79	7.32	5.65
Variance	0.23	1.46	23.91

Table 1: Condition number, number of iterations and coefficient contrast in the CG method for the Monte Carlo computation of $\overline{u}(x)$ for (2). The log-coefficient c is given by a truncated KL expansion with $K = 15$ terms with covariance function shown in (16). We use $N = 40$ elements in each direction and $R = 1000$ realizations.

Numerical tests for exponential covariance function. For problem (2) we consider a coefficient κ of the form $\kappa(x, \omega) = \mathrm{e}^{c(x,\omega)}$, where the stochastic process c is defined by the Karhunen–Loève expansion with associated covariance function

$$\mathbf{c}(x, x') = \exp\left(-\frac{1}{2} \parallel x - x' \parallel^2\right). \tag{16}$$

We approximate the expected value $\overline{u}(x)$ of the solution (2), through a Monte Carlo method with R realizations. In Table 1 we show the mean and variance of the condition number of the preconditioned system, the number of iterations and the contrast $\max_x \kappa(x, \omega)/\min_x \kappa(x, \omega)$ during the Monte Carlo solve. The small variance in the condition number indicates low dependence of the method on the parameter κ.

Matérn class of covariance functions. Now, the coefficient κ is defined with the Matérn class of covariance functions

$$\mathbf{c}_{\text{Matern}}(x, x') = \frac{2^{1-\nu}}{\Gamma(\nu)}\left(\frac{\sqrt{2\nu}\|x - x'\|}{l}\right)^{\nu} K_\nu\left(\frac{\sqrt{2\nu}\|x - x'\|}{l}\right) \tag{17}$$

with (probabilistic) parameters $\nu, l > 0$ and K_ν is the modified Bessel function of the second kind. With this function in the KL expansion, we obtain the results in Table 2. In Table 2 we show the dependence of the condition number, the number of iterations and the coefficient contrast. We note that the small variance of the number of iterations and the value of the condition number indicate that the iteration do not depend much on the parameter $\kappa(x, \omega)$. Additional experiments and results are the object of current research and will be presented elsewhere.

	Condition	**Iterations**	**Contrast**
Mean	3.07	11.1	11.18
Variance	0.73	1.3	67.28

Table 2: Condition number, number of iterations and contrast of coefficient κ in the CG method in the Monte Carlo computation of the solution $\overline{u}(x)$ of (2). The log-coefficient c is given as a truncated KL expansion with $K = 30$ terms constructed from the covariance function shown in (17) with $\nu = 0.5$ and $l = 1$. We use $N = 20$ elements in each direction and $R = 1000$ realizations.

Acknowledgements The authors are grateful to Professor Zlatko Drmac from the University of Zagreb for introducing M. Sarkis to the natural factor formulation of the stiffness matrices in finite element computations during our discussions in Rio de Janeiro. J. Galvis thanks the European Union's Horizon 2020 research and innovation programme for their partial support under the Marie Sklodowska-Curie grant agreement No 777778 (MATHROCKS).

References

1. J.H. Argyris and O.E. Brønlund. The natural factor formulation of the stiffness for the matrix displacement method. *Computer Methods in Applied Mechanics and Engineering* **5**(1), 97–119 (1975).
2. Z. Drmač. Numerical methods for accurate computation of the eigenvalues of hermitian matrices and the singular values of general matrices. *SeMA Journal* **78**(1), 53–92 (2021).
3. T. Rees and J. Scott. A comparative study of null-space factorizations for sparse symmetric saddle point systems. *Numerical Linear Algebra with Applications* **25**(1), e2103 (2018).
4. S.A. Vavasis. Stable numerical algorithms for equilibrium systems. *SIAM Journal on Matrix Analysis and Applications* **15**(4), 1108–1131 (1994).
5. S.A. Vavasis. Stable finite elements for problems with wild coefficients. *SIAM journal on numerical analysis* **33**(3), 890–916 (1996).

A Simple Finite Difference Discretization for Ventcell Transmission Conditions at Cross Points

Martin J. Gander and Laurence Halpern

1 Introduction

Our main focus here is on cross points in non-overlapping domain decomposition methods, but our techniques can also be applied to cross points in overlapping domain decomposition methods, which can be an issue as indicated already by P.L. Lions in his seminal paper [17], see Figure 1. The Additive Schwarz method

ON THE SCHWARZ ALTERNATING METHOD. I 19

As soon as $m \geq 3$, the situation becomes more interesting. And even if, as we will see in section II, each sequence u_n^i converges in $\mathcal{O}_i$ to u, this method does not have always a variational interpretation in terms of iterated projections. A related difficulty is that, using the sequences $(u_n^1)_n$, $(u_n^2)_n$, ..., $(u_n^m)_n$ it is not always possible to define a single-valued function defined on the whole domain $\mathcal{O}$ in a continuous way. In fact, the necessary and sufficient condition for these two difficulties not to happen is that

$$(36)\qquad \begin{cases} \text{For all distinct } i,j,k \in \{1,\dots,m\}, \text{ if } \mathcal{O}_i \cap \mathcal{O}_j \neq \emptyset,\ \mathcal{O}_i \cap \mathcal{O}_k \neq \emptyset \\ \text{then } \qquad \mathcal{O}_j \cap \mathcal{O}_k = \emptyset \end{cases}$$

Fig. 1: Lions' comment from the first international conference on domain decomposition methods in Paris in 1987 on the difficulty of cross point situations for the parallel overlapping Schwarz method (m is the number of subdomains, O_i a subdomain, and n the iteration index).

Martin J. Gander
University of Geneva, e-mail: martin.gander@unige.ch

Laurence Halpern
Université Paris 13, e-mail: halpern@math.univ-paris13.fr

S. C. Brenner et al. (eds.), *Domain Decomposition Methods in Science and Engineering XXVI*, Lecture Notes in Computational Science and Engineering 145,
https://doi.org/10.1007/978-3-030-95025-5_26

[7] for example leaves the treatment of the divergent modes around cross points[1] to Krylov acceleration, which leads to the coloring constant in the condition number estimate. A partition of unity can however be used to make the method convergent, as in Restricted Additive Schwarz, see [10, 11] for more information.

For non-overlapping domain decomposition methods, Dean and Glowinski proposed in 1993 [4] already a cross point treatment with specific Lagrange multipliers for wave equations, and FETI-DP treats cross points by imposing continuity there [8, 20], see also [2] for the Helmholtz case. At the continuous level, the seminal energy estimates of Lions in [18] and Després in [5] showed that Robin transmission conditions do not pose any problem at cross points, but when discretized, standard energy estimates do not work anymore [14], and one needs to use methods like auxiliary variables or complete communication to treat cross points [15], see also [19]. In an algebraic setting the optimized Robin parameter can also require a different weight at cross points [13]. Less was known historically for higher order transmission conditions containing tangential derivatives in the presence of cross points. For an early approach at the continuous level, see [22][2]. More recently, cross points have become a focus of attention in the domain decomposition community: in [21] a new approach at the cross points based on a corner treatment developed for absorbing boundary conditions is proposed for higher order transmission conditions for lattice type partitions; in [6] a new technique with quasi-continuity relations is proposed for polygonal domains; and in [3] cross points are treated with a non-local problem in the context of a multi-trace formulation and non-local transmission conditions, an approach related to the algebraic non-local approach in [12] which leads to a direct solver without approximation, independent of the number of subdomains and type of PDE solved.

Often however in the above references, several difficulties are mixed: the domain decomposition method is for high frequency wave propagation instead of simple Laplace problems, or non-local transmission conditions instead of local ones are used, which can make the cross point difficulties which exist already for Laplace problems appear less clearly.

2 Optimized Schwarz with Ventcell transmission conditions

We consider an optimized Schwarz method (OSM) with Ventcell transmission conditions [9] for the Laplace problem and the decomposition of a square domain Ω into four square subdomains Ω_ℓ, as shown in Figure 2,

$$\begin{aligned} \Delta u_\ell^n &= f && \text{in } \Omega_\ell, \\ (\partial_{n_\ell} + p - q\partial_\tau^2)u_\ell^n &= (\partial_{n_\ell} + p - q\partial_\tau^2)u_l^{n-1} && \text{on } \Gamma_{\ell,l}, \end{aligned} \tag{1}$$

[1] For an illustration of these modes, see [10, Figure 3.2].

[2] Note that the term 'additive' in this reference does not refer to the additive Schwarz method!

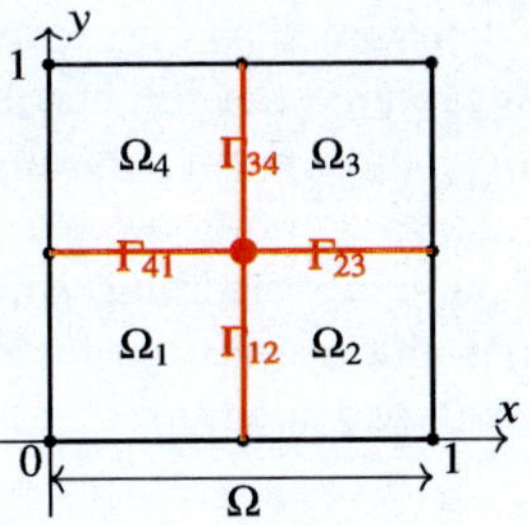

Fig. 2: Model problem domain and decomposition.

where n is the iteration index, $\ell, l \in \{1,2,3,4\}$ are the subdomain indices, ∂_{n_ℓ} is the normal and ∂_τ the tangential derivative, and p, q are the Ventcell transmission parameters (or Robin if $q = 0$) that can be optimized for best performance of the OSM [9]. A standard second order five point finite difference discretization, omitting the subdomain and iteration indices to avoid cluttering the notation, leads for generic grid point indices i in x and j in y to

$$\Delta u \approx \frac{u_{i+1,j} + u_{i-1,j} - 4u_{i,j} + u_{i,j+1} + u_{i,j-1}}{h^2}$$

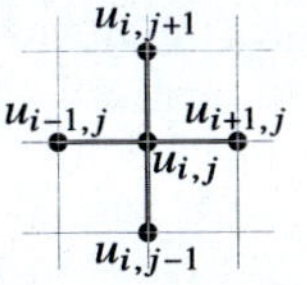

$$\partial_n u \approx \frac{u_{i+1,j} - u_{i-1,j}}{2h}$$

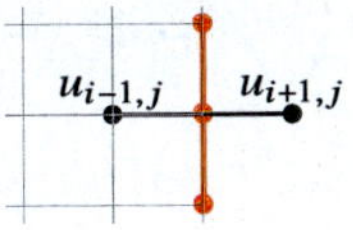

$$\partial_\tau^2 u \approx \frac{u_{i,j+1} - 2u_{i,j} + u_{i,j-1}}{h^2}$$

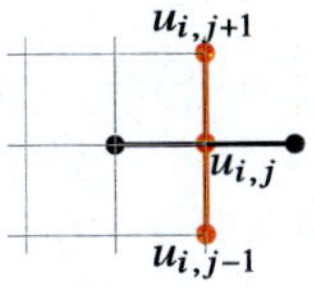

where we indicated the vertical interface in red. A problem is that for the normal derivative approximation, one point lies outside of the domain, here $u_{i+1,j}$ on the right. The value of this so-called ghost point is however also involved in the interior five point finite difference stencil when evaluated at the interface,

$$\frac{u_{i+1,j} + u_{i-1,j} - 4u_{i,j} + u_{i,j+1} + u_{i,j-1}}{h^2} = f_{i,j}$$

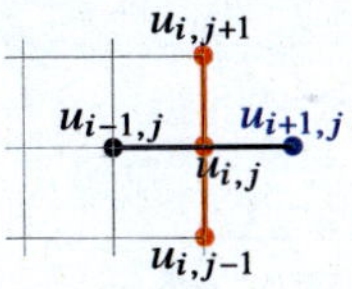

$$\frac{u_{i+1,j} - u_{i-1,j}}{2h} = g_j$$

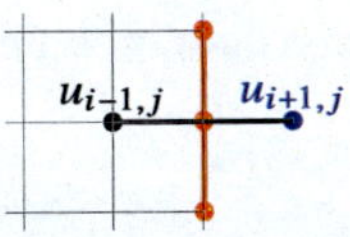

Hence the ghost point value $u_{i+1,j}$ is determined by the approximation of the boundary condition $\partial_n u = g$ imposed in a centered fashion at the vertical red interface, and the scheme is complete. The same approach can naturally be used for a centered approximation of the more general Ventcell condition $(\partial_n + p + q\partial_\tau^2)u = g$.

Now at the 90° cross point in Figure 2, something special happens with this discretization: we have for example for subdomain Ω_1 the interior five point Laplacian at the cross point (i, j) with the two Ventcell conditions

$$\frac{u_{i+1,j} + u_{i-1,j} - 4u_{i,j} + u_{i,j+1} + u_{i,j-1}}{h^2} = f_{i,j}$$

$$\frac{u_{i+1,j}-u_{i-1,j}}{2h} + pu_{i,j} - q\frac{u_{i,j+1}-2u_{i,j}+u_{i,j-1}}{h^2} = g_j$$

$$\frac{u_{i,j+1}-u_{i,j-1}}{2h} + pu_{i,j} - q\frac{u_{i+1,j}-2u_{i,j}+u_{i-1,j}}{h^2} = \tilde{g}_i$$

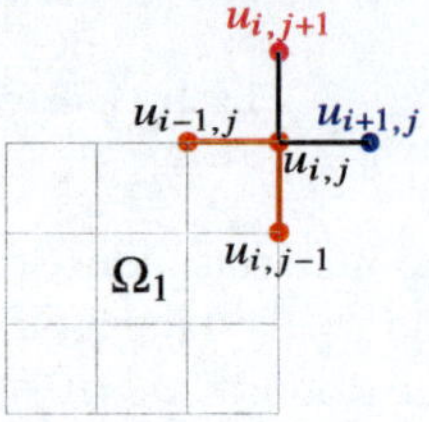

and thus the scheme is complete for the subdomain solve: we have the two equations from the two discretized boundary conditions for the two ghost points in color.

Once the subdomain solution is known, the ghost point values are known as well, and one can easily extract the values to be transmitted to the neighboring subdomains, again in the form of the centered discretized Ventcell conditions,

For Ω_2: $\quad \frac{u_{i-1,j}-u_{i+1,j}}{2h} + pu_{i,j} - q\frac{u_{i,j+1}-2u_{i,j}+u_{i,j-1}}{h^2}$

For Ω_4: $\quad \frac{u_{i,j-1}-u_{i,j+1}}{2h} + pu_{i,j} - q\frac{u_{i+1,j}-2u_{i,j}+u_{i-1,j}}{h^2}$

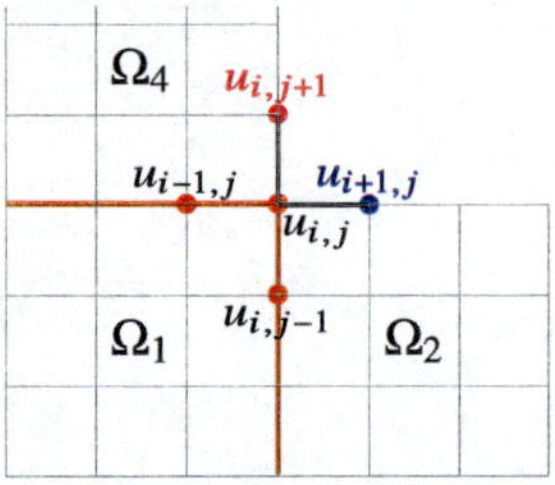

The complete discrete OSM algorithm is thus for example for subdomain Ω_1 given by solving at iteration n for $i = 1 \dots I$, $j = 1 \dots J$ for $u^n_{1,i,j}$ the discrete equations

$$\frac{u^n_{1,i+1,j} + u^n_{1,i-1,j} - 4u^n_{1,i,j} + u^n_{1,i,j+1} + u^n_{1,i,j-1}}{h^2} = f_{i,j},$$

with the transmission condition for $j = 1 \dots J$ on the right,

$$\frac{u^n_{1,i+1,j} - u^n_{1,i-1,j}}{2h} + pu^n_{1,i,j} - q\frac{u^n_{1,i,j+1} - 2u^n_{1,i,j} + u^n_{1,i,j-1}}{h^2}$$

$$= \frac{u^{n-1}_{2,i+1,j} - u^{n-1}_{2,i-1,j}}{2h} + pu^{n-1}_{2,i,j} - q\frac{u^{n-1}_{2,i,j+1} - 2u^{n-1}_{2,i,j} + u^{n-1}_{2,i,j-1}}{h^2},$$

and the transmission condition for $i = 1 \dots I$ at the top,

$$\frac{u^n_{1,i,j+1}-u^n_{1,i,j-1}}{2h}+pu^n_{1,i,j}-q\frac{u^n_{1,i+1,j}-2u^n_{1,i,j}+u^n_{1,i-1,j}}{h^2}$$
$$=\frac{u^{n-1}_{4,i,j+1}-u^{n-1}_{4,i,j-1}}{2h}+pu^{n-1}_{4,i,j}-q\frac{u^{n-1}_{4,i+1,j}-2u^{n-1}_{4,i,j}+u^{n-1}_{4,i-1,j}}{h^2},$$

and analogously for the other three subdomains. We thus have a very simple finite difference scheme for the OSM with Ventcell transmission conditions, which takes advantage of the rectangular structure of the Laplace operator at such rectangular cross points.

3 Numerical experiments

We show in Figure 3 the error in the first few iterates of the OSM with Ventcell transmission conditions in the left column, and for comparison in the right column the case of optimized Robin transmission conditions, i.e. $q = 0$. We used as mesh parameter $h = 1/16$, and solve the error equations directly, starting with a random initial guess; for the importance of this, see [10, Section 5.1]. We observe that the OSM is converging nicely also at the cross point, both for Robin and Ventcell transmission conditions, and convergence is much faster for the Ventcell transmission conditions. This appears even more clearly in the convergence plot shown in Figure 4. We see that the OSM with optimized Ventcell transmission conditions converges almost four times faster than with optimized Robin transmission conditions, at the same cost per iteration, and Krylov acceleration with GMRES only leads to little further improvement for the decay of the error in the iterations, especially for the Ventcell transmission condition (as already seen in [10, Fig 5.1] for the two subdomain case).

4 Conclusions

We presented a simple finite difference discretization of optimized Schwarz methods with Ventcell transmission conditions for the Laplace problem in the presence of cross points in the decomposition. The discretization takes advantage of the rectangular structure of the Laplace operator and only works for rectangular cross points as in Figure 2. For more general cross point situations, auxiliary variables or complete communication can be used [15], but only in the simpler case of Robin conditions.

For a rectangular cross point, our technique can also be used in a variational formulation: multiplying by a test function v and integrating by parts on Ω_1, we get using the Ventcell condition

$$\int_{\Omega_1}\nabla u_1\nabla v+p\int_{\Gamma_{12}\cup\Gamma_{14}}u_1 v-q\int_{\Gamma_{12}\cup\Gamma_{14}}\partial^2_\tau u_1 v,$$

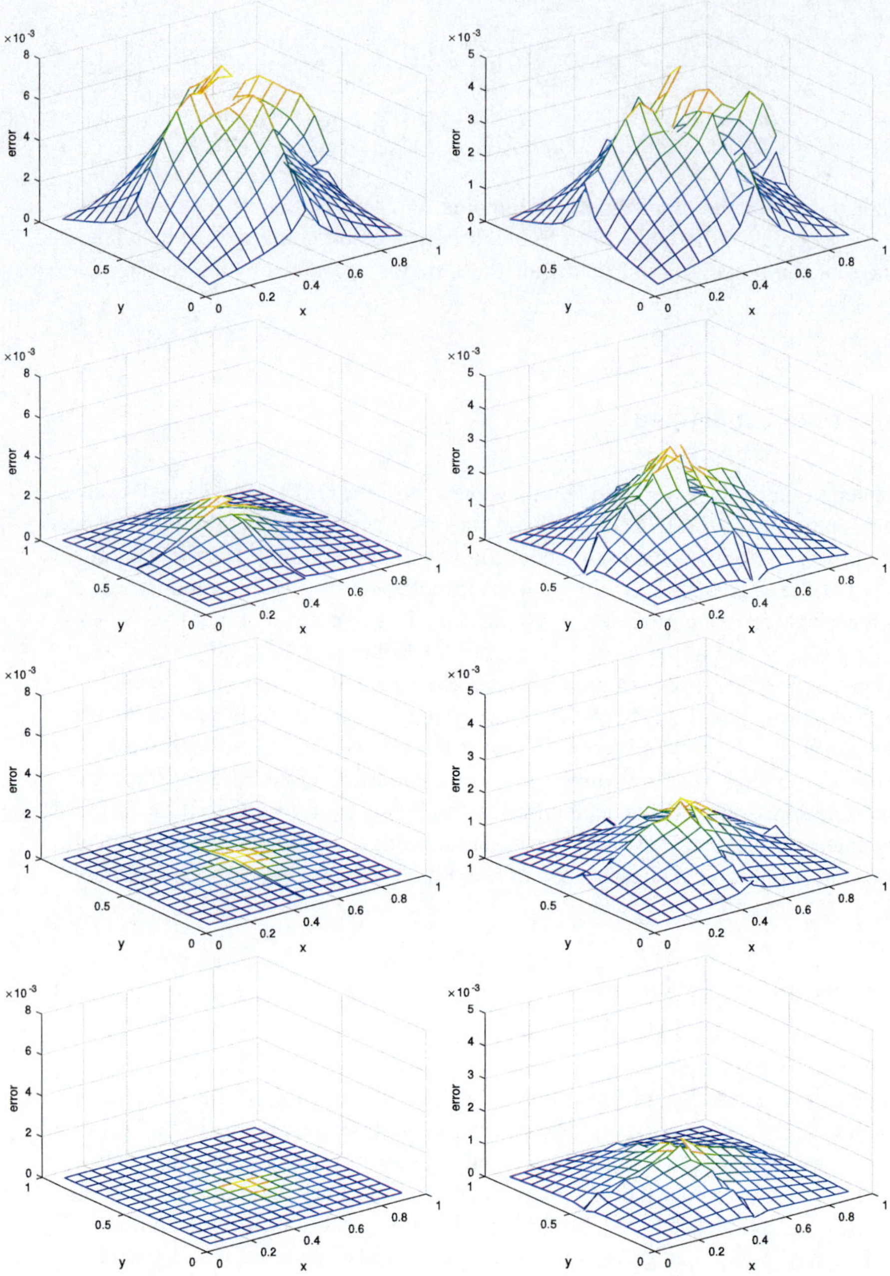

Fig. 3: Iterates of OSM with Ventcell transmission conditions on the left, and with Robin transmission conditions on the right.

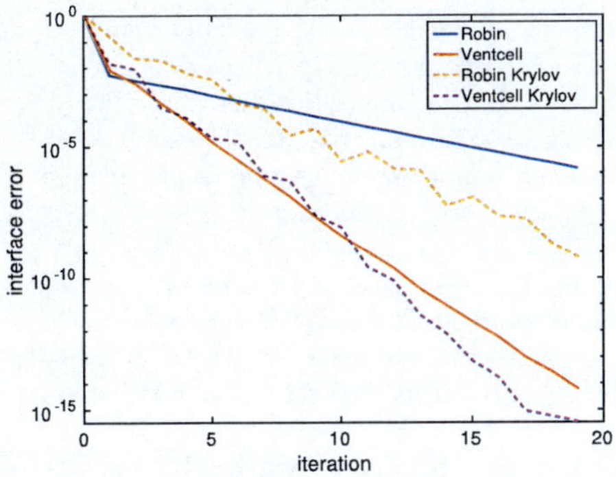

Fig. 4: Comparison of the decay of the error in the OSM as function of the iteration index n for optimized Robin and Ventcell transmission conditions.

and the last term gives, when integrating by parts, and using the fact that the test function v vanishes on the outer boundary due to the Dirichlet condition there,

$$\int_{\Gamma_{12}\cup\Gamma_{14}} \partial_\tau^2 u_1 v = \partial_y u_1(\tfrac{1}{2},\tfrac{1}{2})v(\tfrac{1}{2},\tfrac{1}{2}) - \int_{\Gamma_{12}} \partial_y u_1 \partial_y v + \partial_x u_1(\tfrac{1}{2},\tfrac{1}{2})v(\tfrac{1}{2},\tfrac{1}{2}) - \int_{\Gamma_{14}} \partial_x u_1 \partial_x v.$$

Due to the rectangular nature of the cross point, the two remaining terms there are well defined using the equation and the Ventcell condition at the cross point, as in the finite difference discretization earlier,

$$\partial_x^2 u_1 + \partial_y^2 u_1 = f, \quad (\partial_x + p - q\partial_y^2)u_1 = g, \quad (\partial_y + p - q\partial_x^2)u_1 = \tilde{g},$$

since solving the Ventcell conditions for the second-order derivative terms and inserting into the equation evaluated at the cross point leads to

$$\partial_x u_1(\tfrac{1}{2},\tfrac{1}{2}) + \partial_y u_1(\tfrac{1}{2},\tfrac{1}{2}) = -2pu_1(\tfrac{1}{2},\tfrac{1}{2}) + qf(\tfrac{1}{2},\tfrac{1}{2}) + g(\tfrac{1}{2},\tfrac{1}{2}) + \tilde{g}(\tfrac{1}{2},\tfrac{1}{2}),$$

and the variational formulation is complete (for the time-dependent case see [1], and for well posedness in the Helmholtz case [16]). Analogously this can be done for the other subdomains, and also the data for the next iteration can be extracted in this way, which leads to a natural finite element discretization for Ventcell transmission conditions at cross points. *see* aso [21] for a similar approach.

References

1. A. Bamberger, P. Joly and J.E. Roberts. Second-order absorbing boundary conditions for the wave equation: a solution for the corner problem. *SIAM J. Num. Anal.* **27**(2), 323–352 (1990).
2. Y. Boubendir and A. Bendali. Dealing with cross-points in a non-overlapping domain decomposition solution of the Helmholtz equation. In: *Mathematical and Numerical Aspects of Wave Propagation WAVES 2003*, pp. 319–324, Springer (2003).

3. X. Claeys and E. Parolin. Robust treatment of cross points in optimized Schwarz methods. *arXiv preprint* arXiv:2003.06657 (2020).
4. E.J. Dean and R. Glowinski. A domain decomposition method for the wave equation. In: *Les Grands Systemes des Sciences et de la Technologie*, Masson Paris (1993).
5. B. Després. *Méthodes de décomposition de domaine pour la propagation d'ondes en régime harmonique. Le théorème de Borg pour l'équation de Hill vectorielle.* PhD thesis, Paris 9 (1991).
6. B. Després, A. Nicolopoulos and B. Thierry. Corners and stable optimized domain decomposition methods for the Helmholtz problem. *hal-02612368* (2020).
7. M. Dryja and O. Widlund. An additive variant of the Schwarz alternating method for the case of many subregions. Technical report, UltracomputerResearch Laboratory, Courant Institute (1987).
8. Ch. Farhat, M. Lesoinne, P. LeTallec, K. Pierson and D. Rixen. FETI-DP: a dual–primal unified FETI method – part I: A faster alternative to the two-level FETI method. *International journal for numerical methods in engineering* **50**(7), 1523–1544 (2001).
9. M.J. Gander. Optimized Schwarz methods. *SIAM Journal on Numerical Analysis* **44**(2), 699–731 (2006).
10. M.J. Gander. Schwarz methods over the course of time. *Electronic transactions on numerical analysis* **31**, 228–255 (2008).
11. M.J. Gander. Does the partition of unity influence the convergence of Schwarz methods? In: *International Conference on Domain Decomposition Methods*, pp. 3–15, Springer (2018).
12. M.J. Gander and F. Kwok. Optimal interface conditions for an arbitrary decomposition into subdomains. In: *Domain Decomposition Methods in Science and Engineering XIX*, pp. 101–108, Springer (2011).
13. M.J. Gander and F. Kwok. Best Robin parameters for optimized Schwarz methods at cross points. *SIAM Journal on Scientific Computing* **34**(4), A1849–A1879 (2012).
14. M.J. Gander and F. Kwok. On the applicability of Lions' energy estimates in the analysis of discrete optimized Schwarz methods with cross points. In: *Domain Decomposition Methods in Science and Engineering XX*, pp. 475–483, Springer (2013).
15. M.J. Gander and K. Santugini-Repiquet. Cross-points in domain decomposition methods with a finite element discretization. *ETNA* **45**, 219–240 (2016).
16. P. Joly, S. Lohrengel and O. Vacus. Un résultat d'existence et d'unicité pour l'équation de Helmholtz avec conditions aux limites absorbantes d'ordre 2. *Comptes Rendus de l'Académie des Sciences-Series I – Mathematics* **329**(3), 193–198 (1999).
17. P.-L. Lions. On the Schwarz alternating method. I. In: *First international symposium on domain decomposition methods for partial differential equations*, volume 1, p. 42, Paris, France (1988).
18. P.-L. Lions. On the Schwarz alternating method. III: a variant for nonoverlapping subdomains. In: *Third international symposium on domain decomposition methods for partial differential equations*, volume 6, pp. 202–223, SIAM Philadelphia, PA (1990).
19. S. Loisel. Condition number estimates for the nonoverlapping optimized Schwarz method and the 2-Lagrange multiplier method for general domains and cross points. *SIAM Journal on Numerical Analysis* **51**(6), 3062–3083 (2013).
20. J. Mandel and R. Tezaur. On the convergence of a dual-primal substructuring method. *Numerische Mathematik* **88**(3), 543–558 (2001).
21. A. Modave, A. Royer, X. Antoine and C. Geuzaine. A non-overlapping domain decomposition method with high-order transmission conditions and cross-point treatment for Helmholtz problems. *Computer Methods in Applied Mechanics and Engineering* **368**, 113162 (2020).
22. F. Nataf. A Schwarz additive method with high order interface conditions and nonoverlapping subdomains. *ESAIM: Mathematical Modelling and Numerical Analysis-Modélisation Mathématique et Analyse Numérique* **32**(1), 107–116 (1998).

Cycles in Newton–Raphson Preconditioned by Schwarz (ASPIN and Its Cousins)

Conor McCoid and Martin J. Gander

1 Introduction

ASPIN [3], RASPEN [5], and MSPIN [8] rely on various Schwarz methods to precondition either Newton–Raphson or inexact Newton. While *a priori* convergence criteria have been found for the underlying Schwarz methods, so far none exist for their combination with Newton–Raphson.

Like in the linear case when combining a Krylov method and a Schwarz method, there is an equivalence between preconditioning Newton–Raphson with a Schwarz method and accelerating that same Schwarz method with Newton–Raphson [6]: A domain is first subdivided into subdomains, the problem solved on each subdomain, and the resulting formulation iterated through Krylov and Newton–Raphson, respectively.

We examine cycling behaviour in alternating Schwarz in one dimension that has been accelerated by applying Newton–Raphson. We begin by presenting the algorithm for alternating Schwarz and how it is accelerated by Newton–Raphson. Suppose we seek to solve the boundary value problem

$$F(x,u,u',u'')=0, \quad x\in[a,b], \quad u(a)=A, \quad u(b)=B$$

for some function $F(x,u,v,w)$. Then an iteration of alternating Schwarz with subdomains (a,β) and (α,b), $\alpha<\beta$, is comprised of the following three steps:

$$\begin{aligned}
&(1) \quad F(x,u_1,u_1',u_1'')=0, && u_1(a)=A, && u_1(\beta)=\gamma_n,\\
&(2) \quad F(x,u_2,u_2',u_2'')=0, && u_2(\alpha)=u_1(\alpha), && u_2(b)=B,\\
&(3) \quad \gamma_{n+1}=u_2(\beta)=G(\gamma_n).
\end{aligned}$$

Conor McCoid
University of Geneva, e-mail: conor.mccoid@unige.ch

Martin J. Gander
University of Geneva, e-mail: martin.gander@unige.ch

S. C. Brenner et al. (eds.), *Domain Decomposition Methods in Science and Engineering XXVI*, Lecture Notes in Computational Science and Engineering 145,
https://doi.org/10.1007/978-3-030-95025-5_27

The function $G(\gamma)$ thus represents one iteration of alternating Schwarz in substructured form. The process is repeated until convergence, i.e.

$$(G \circ G \circ \cdots \circ G)(\gamma) = G^n(\gamma) \approx G^{n+1}(\gamma) = (G \circ G^n)(\gamma).$$

This is naturally a fixed point iteration applied to the function $G(\gamma)$.

To accelerate the method one applies Newton–Raphson to the function $f(\gamma) = G(\gamma) - \gamma$, which has a root at the fixed point. If the fixed point is unique, this is the only root of $f(\gamma)$. To apply Newton–Raphson, one needs to know the value of $G'(\gamma)$, which may be found by adding two new steps, (1') and (2'), to alternating Schwarz:

$$\begin{aligned}
&(1) && F(x, u_1, u_1', u_1'') = 0, && u_1(a) = A, && u_1(\beta) = \gamma_n,\\
&(1') && J(u_1)\cdot(v_1, v_1', v_1'') = 0, && v_1(a) = 0, && v_1(\beta) = 1,\\
&(2) && F(x, u_2, u_2', u_2'') = 0, && u_2(\alpha) = u_1(\alpha), && u_2(b) = B,\\
&(2') && J(u_2)\cdot(v_2, v_2', v_2'') = 0, && v_2(\alpha) = 1, && v_2(b) = 0,\\
&(3) && \gamma_{n+1} = \gamma_n - \frac{u_2(\beta) - \gamma_n}{v_1(\alpha)v_2(\beta) - 1} = && \gamma_n - \frac{G(\gamma_n) - \gamma_n}{G'(\gamma_n) - 1},
\end{aligned}$$

where $v_i(x) = \partial u_i(x)/\partial\gamma$ and $J(u_i)$ is the Jacobian of $F(x, u_i, u_i', u_i'')$.

2 Convergence of generic fixed point iterations and Newton–Raphson

A generic fixed point iteration $x_{n+1} = g(x_n)$ converges when $|g(x_n) - x^*| < |x_n - x^*|$, where x^* is the fixed point. This occurs when $g(x)$ lies between x and $2x^* - x$. The convergence or divergence of the fixed point iteration is monotonic if $\text{sign}(g(x) - x^*) = \text{sign}(x - x^*)$ and oscillatory otherwise. This creates four lines, $y = x$, $y = 2x^* - x$, $y = x^*$ and $x = x^*$, that divide the plane into octants. The four pairs of opposite octants form four regions with distinct behaviour of the fixed point iteration, see left of Figure 1 or Figure 5.7 from [7]:

1, $g(x) < x < x^*$ or $g(x) > x > x^*$: monotonic divergence;
2, $x < g(x) < x^*$ or $x > g(x) > x^*$: monotonic convergence;
3, $x < x^* < g(x) < 2x^* - x$ or $x > x^* > g(x) > 2x^* - x$: convergent oscillations;
4, $x < x^* < 2x^* - x < g(x)$ or $x > x^* > 2x^* - x > g(x)$: divergent oscillations.

If the function $g(x)$ intersects the line $y = x$ at a point other than x^* then there are additional fixed points that the method can converge towards. If it intersects the line $y = 2x^* - x$ then a stable cycle can form. A fixed point iteration is therefore only guaranteed to converge if $g(x)$ lies entirely between the lines $y = x$ and $y = 2x^* - x$, i.e. within regions 2 and 3.

Newton–Raphson can make use of this analysis by considering it as a fixed point iteration:

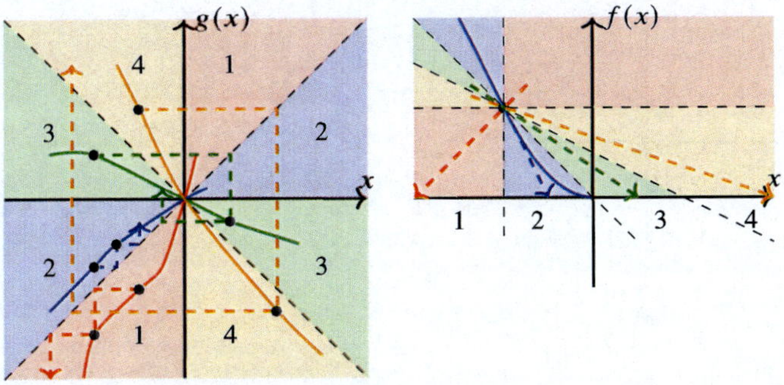

Fig. 1: Left: Behaviour of the fixed point iteration $x_{n+1} = g(x_n)$, where the origin is the fixed point, $g(0) = 0$. **Right:** Regions of Newton–Raphson, $x_{n+1} = x_n - f(x_n)/f'(x_n)$, where the origin is the root, $f(0) = 0$. The tangent line to $f(x)$ can be traced from $(x, f(x))$ towards the line $y = 0$. Where it lands on this line indicates which fixed point iteration behaviour occurs.

$$x_{n+1} = x_n - \frac{f(x_n)}{f'(x_n)} = g_f(x_n).$$

The borders between the regions no longer depend solely on the value of $f(x)$ but also $f'(x)$. The right of Figure 1 shows which type of behaviour Newton–Raphson will have based on where the tangent line points.

As stated, if $g_f(x)$ intersects the line $y = x$ there are additional fixed points, and if it intersects $y = 2x^* - x$ there may be stable cycles. For guaranteed convergence $g_f(x)$ must lie between these lines. Intersections of $g_f(x)$ with $y = x$ occur only if $f(x) = 0$ and $f(x)$ has additional roots or $f'(x) = \infty$. Both circumstances are assumed not to occur. Intersections of $g_f(x)$ with $y = 2x^* - x$ may be represented as a first order ODE:

$$f_C'(x) = -\frac{f_C(x)}{2(x^* - x)}, \quad f_C(x^*) = 0.$$

The solution to this ODE is $f_C(x) = C\sqrt{|x - x^*|}$ where $C \in \mathbb{R}$. If a function $f(x)$ with root x^* is tangential to $f_C(x)$ for any value of C then $g_f(x)$ intersects the line $y = 2x^* - x$. The left of Figure 2 shows the functions $f_C(x)$.

A function $f(x)$ that is monotonic with respect to this geometry has guaranteed convergence under Newton–Raphson. That is, if $f(x)$ is nowhere tangential to $f_C(x)$ in a given domain containing x^* for any value of C then $g_f(x)$ converges to the root for any initial guess in that domain. Since $f_C'(x^*) = \infty$ and $f(x^*) = 0$ there is always a region around the root x^* where $f(x)$ crosses all of these lines monotonically. This conforms with the theory on Newton–Raphson.

The corresponding geometry for a fixed point function accelerated by Newton–Raphson is skewed such that the line $y = 0$ is aligned to $y = x$, as seen in the right of Figure 2. The lines of this figure are the functions $g_C(x) = f_C(x) + x$. A function $g(x)$ must be monotonic in this geometry or Newton–Raphson applied to $g(x) - x$ may exhibit cycling behaviour.

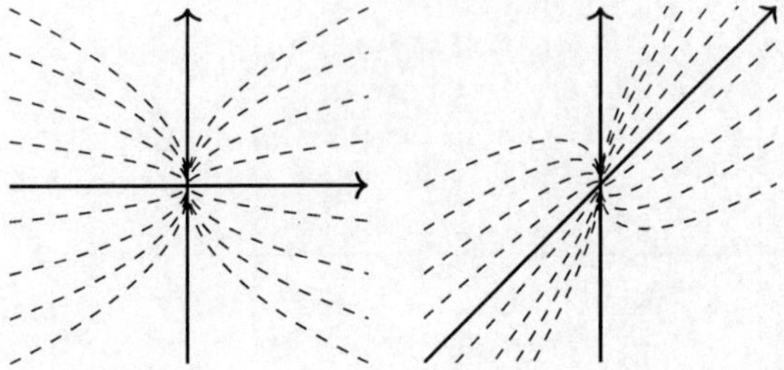

Fig. 2: Left: Solutions $f_C(x)$ such that $g_f(x)$ intersects $y = 2x^* - x$ for all x. **Right:** Functions $g_C(x) = f_C(x) + x$ such that $g_f(x)$ for $f(x) = g_C(x) - x$ intersects $y = 2x^* - x$ for all x.

Table 1: Conditions for convergent behaviour of Newton–Raphson applied to $g(x) - x$.

$g(x)$ lies in	Necessary condition	Sufficient condition
1	$g'(x) > 1$	
2	$g'(x) < 1$	$g'(x) < 1/2$
3	$g'(x) < 1/2$	$g'(x) < 0$
4	$g'(x) < 0$	

If it is known in which fixed point region of the left of Figure 1 $g(x)$ lies then one can find necessary and, in some cases, sufficient conditions for Newton–Raphson to have convergent behaviour based on the slopes of the lines $g_C(x)$. For example, in region 2 the maximum of $g'_C(x)$ is 1. If $g(x)$ lies in region 2 then its slope must therefore be less than 1 everywhere or there will be a point where $g(x)$ runs tangent to $g_C(x)$ for some C. Moreover, the minimum of $g'_C(x)$ is 1/2. If $g(x)$ has a slope less than 1/2 then it cannot run tangent to $g_C(x)$ for any C. The list of these conditions is summarized in Table 1.

3 The fixed point iteration of alternating Schwarz

We now seek to apply this theory to alternating Schwarz. As stated earlier, we consider alternating Schwarz as a function $G(\gamma)$, taking as input the value of $u_1(\beta)$ and as output the value of $u_2(\beta)$. Under reasonable conditions we can prove a number of useful properties of $G(\gamma)$ without prior knowledge of the fixed point γ^*.

Theorem 1 *If the problem $F(x, u, u', u'') = 0$ for $x \in \Omega$, $u(x) = h(x)$ for $x \in \partial\Omega$ has a unique solution on $\Omega = [a, \alpha]$ and $\Omega = [\beta, b]$ and the continuations of these solutions are also unique, then the function $G(\gamma)$ is strictly monotonic.*

Proof It suffices to show that $G(\gamma_1) = G(\gamma_2)$ implies $\gamma_1 = \gamma_2$. Let u_1^j solve the problem on $[a, \beta]$ with $u_1^j(\beta) = \gamma_j$. Likewise, u_2^j solves the problem on $[\alpha, b]$ with $u_2^j(\alpha) = u_1^j(\alpha)$. Suppose $u_2^1(\beta) = u_2^2(\beta)$. Then both u_2^1 and u_2^2 solve the same problem on $[\beta, b]$. By assumption, this must mean $u_2^1 = u_2^2$ and $u_1^1(\alpha) = u_1^2(\alpha)$. By a similar argument, this implies u_1^1 and u_1^2 solve the same problem on $[a, \alpha]$. Again by assumption $u_1^1 = u_1^2$ and $\gamma_1 = \gamma_2$. □

We can even prove that $G(\gamma)$ is restricted to region 2 with additional properties. As an example, we reprove a result from Lui [9].

Theorem 2 (Theorem 2 from [9]) *Consider the equation* $u''(x) + f(x, u, u') = 0$ *for* $x \in (a, b)$, $u(a) = u(b) = 0$ *under the assumptions that*

- $f \in C^1([a, b] \times \mathbb{R} \times \mathbb{R})$,
- $\frac{\partial f(x,v,v')}{\partial u} \leq 0$ *for all* $x \in [a, b]$ *and* $v \in H_0^1([a, b])$,
- $|f(x, v, v')| \leq C(1 + |v'|^\eta)$ *for all* $x \in [a, b]$ *and* $v \in H_0^1([a, b])$ *and some* $C > 0$, $0 < \eta < 1$.

The problem is solved using alternating Schwarz with two subdomains and Dirichlet transmission conditions. Then $G(\gamma)$ *for this problem lies within region 2.*

Proof It suffices to prove that the problem is well posed and $0 < G'(\gamma) < 1$ for all $\gamma \in \mathbb{R}$. The well-posedness of the problem is guaranteed by Proposition 2 from [9]. As Lui points out, this also means the problem is well posed on any subdomain. Using Theorem 1 this gives monotonicity of $G(\gamma)$. Moreover, if $u(x) = 0$ for any $x \in (a, b)$ then the problem would be well posed on the domains $[a, x]$ and $[x, b]$. As such, $u(x)$ has the same sign as γ and $G'(\gamma) > 0$.

Consider the problem in g_1:

$$g_1''(x) + \frac{\partial f}{\partial u} g_1 + \frac{\partial f}{\partial u'} g_1' = 0, \quad x \in [a, \beta], \quad g_1(a) = 0, \quad g_1(\beta) = 1.$$

From the second assumption on f the operator on g_1 satisfies a maximum principle (see, for example, [9]). Therefore, $g_1(x) < 1$ for all $x \in (a, \beta)$. By the same reasoning, $g_2(x) < g_1(\alpha) < 1$ for all $x \in (\alpha, b)$ and $G'(\gamma) < 1$. Incidentally, the same maximum principle applies for the operator on $-g_1$ and $-g_2$, and so $G'(\gamma) > 0$ as we had before. □

This provides guaranteed convergence of alternating Schwarz. However, it does not guarantee the convergence when one accelerates it through Newton–Raphson. Using Table 1 we know that such convergence is assured if $G'(\gamma) < 1/2$ for all γ, but this is not true in all cases and cannot be determined *a priori*.

Take as an example the following second order nonlinear differential equation

$$u''(x) - \sin(au(x)) = 0, \quad x \in (-1, 1), \tag{1}$$

with homogeneous Dirichlet boundary conditions. The problem is well posed and admits only the trivial solution $u(x) = 0$. It is easy to see that this equation satisfies the conditions of Theorem 2. Therefore, the alternating Schwarz fixed point iteration, $G(\gamma)$, lies within region 2 and is guaranteed to converge to the fixed point. Sadly, its Newton–Raphson acceleration will not do so for all initial conditions. Take $a = 3.6$ with an overlap of 0.4 and symmetric regions. The results of the Newton–Raphson

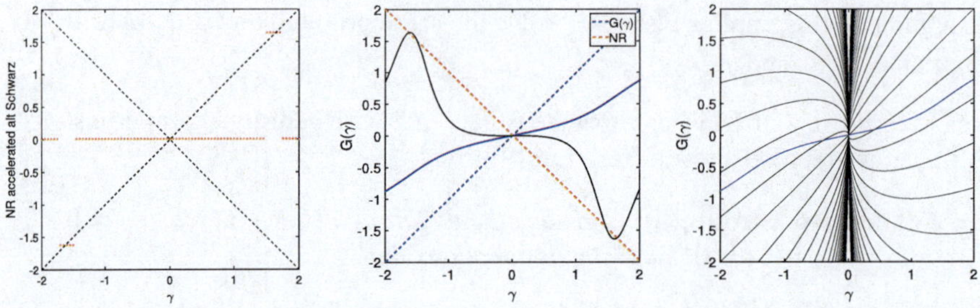

Fig. 3: Left: Results of Newton–Raphson accelerated alternating Schwarz as a function of initial condition in solving equation (1). The value of a is 3.6 and the subdomains are $\Omega_1 = (-1, 0.2)$ and $\Omega_2 = (-0.2, 1)$. **Middle:** $G(\gamma)$ and its Newton–Raphson acceleration. **Right:** $G(\gamma)$ plotted with the geometry of the right of Figure 2.

acceleration are found in Figure 3 (left). While for most initial values of γ the method converges to the correct solution $u = 0$ there are two small intervals where the method enters a stable cycle.

The function $G(\gamma)$ can be plotted numerically, along with its Newton–Raphson acceleration, see Figure 3 (middle), which shows that $G(\gamma)$ does indeed lie within region 2 as predicted by Theorem 2. However, $G(\gamma)$ runs tangential to one of the lines $g_C(\gamma)$, see Figure 3 (right), and so its Newton–Raphson acceleration crosses into region 4. Due to symmetry, there is a 2-cycle at each crossing. Depending on the slope of the acceleration as it crosses into region 4 this cycle may be stable.

Where stable cycles exist so too must there be period doubling bifurcation. Changing the value of the parameter a we find that the 2-cycle found in Figure 3 (left) becomes two 2-cycles, then two 4-cycles, and so on until it devolves into chaos, see

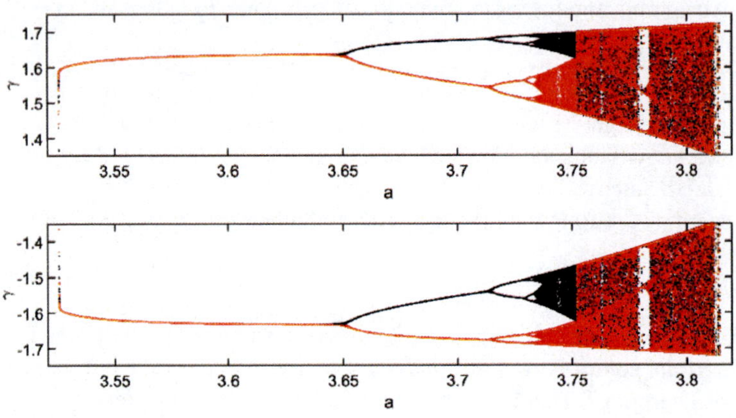

Fig. 4: Period doubling bifurcation in the example caused by Newton–Raphson acceleration.

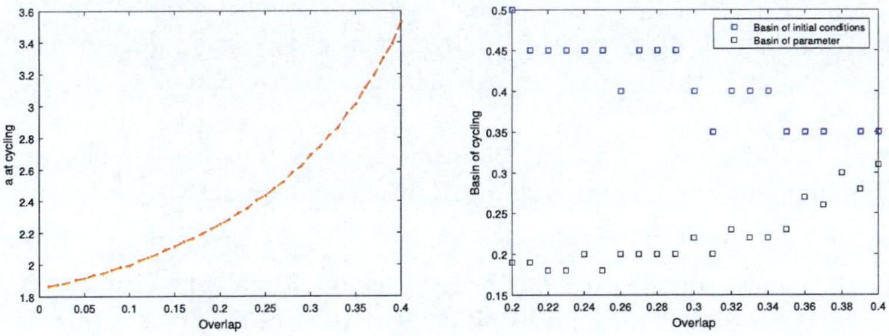

Fig. 5: Left: value of a at which bifurcation starts. **Right:** width of basin of cycling in γ and a.

Figure 4. With enough chaos the cycles are no longer stable and the acceleration exits into a convergent region.

While a change in the parameter a is the most obvious way to alter the dynamics, one can also change the size of the overlap. This has a direct effect on the basin of cycling in the spaces of both initial condition γ and the parameter a. Figure 5 (left) shows a nonlinear relationship between the first value of a at which cycling is observed and the size of the overlap. As the overlap grows the parameter a must be larger and larger for cycling to occur. Figure 5 (right) indicates that the interval of initial conditions that result in cycling shrinks as the overlap grows. Meanwhile, the length of the bifurcation diagram increases, meaning there are more values of a with stable cycling.

4 Accelerated alternating Schwarz with guaranteed convergence

Given Theorem 2 and the conditions of Table 1 one can construct a series of tests to see if the Newton–Raphson acceleration is suitable for a given iteration. We present one further useful trick to strengthen convergence, a correction to Newton–Raphson due to Davidenko and Branin [1, 2, 4]. We replace step (3) in the algorithm with

$$(3*) \quad \tilde{\gamma}_n = \gamma_n - \frac{G(\gamma_n) - \gamma_n}{|G'(\gamma_n) - 1|}.$$

For $G(\gamma)$ within region 2 the Newton–Raphson acceleration will now always march in the direction of the fixed point. It may still overshoot and cycle but the direction will always be correct.

For a problem satisfying the conditions of Theorem 2 or similar that guarantees that $G(\gamma)$ lies in region 2 the algorithm proceeds as follows:

1. Select some $\gamma_0 \in \mathbb{R}$. Set $n = 0$.
2. Calculate $G(\gamma_n)$ and $G'(\gamma_n)$. If $G'(\gamma_n) = 1$ then set $\gamma_{n+1} = G(\gamma_n)$, increment n and return to step 2. If this is not true, proceed to step 3.

3. Perform step (3*), which is the Newton–Raphson acceleration using the Davidenko–Branin trick. If $|G'(\gamma_n) - 1| \geq 1/2$ then set $\gamma_{n+1} = \tilde{\gamma}_n$, increment n and return to step 2. If this is not true, calculate $\hat{\gamma}_n$, the average of γ_n and $\tilde{\gamma}_n$, and proceed to step 4.
4. Calculate $G(\hat{\gamma}_n)$. If $G(\hat{\gamma}_n) - \hat{\gamma}_n$ has the same sign as $G(\gamma_n) - \gamma_n$ then set $\gamma_{n+1} = \tilde{\gamma}_n$, increment n and return to step 2. If this is not true, set $\gamma_{n+1} = G(\gamma_n)$, increment n and return to step 2.

Each of steps 2, 3 and 4 contain a test of whether Newton–Raphson will converge. In step 2, Newton–Raphson will not converge if the derivative of $G(\gamma) - 1$ is zero. In step 3, convergence is guaranteed if $G'(\gamma) \leq 1/2$ based on Table 1. The Davidenko–Branin trick strengthens this and also guarantees convergence if $G'(\gamma) \geq 3/2$.

In step 4 we test the point halfway between the starting value γ_n and the Newton–Raphson acceleration $\tilde{\gamma}_n$, denoted $\hat{\gamma}_n$. Since $G(\gamma)$ is in region 2 if $G(\gamma) > \gamma$ then $\gamma < \gamma^*$ and vice versa. Therefore, we can easily determine whether $\hat{\gamma}_n$ is on the same side of the fixed point as γ_n. If it is, then the fixed point γ^* lies on the same side of $\hat{\gamma}_n$ as $\tilde{\gamma}_n$, and so $\tilde{\gamma}_n$ is closer to γ^* than γ_n. If it is not, then γ^* lies between γ_n and $\hat{\gamma}_n$. Since $\tilde{\gamma}_n$ is on the other side of $\hat{\gamma}_n$ it is further from γ^* than γ_n and we have divergence. In such a case, the fixed point iteration should be used.

Note that while $G(\gamma)$ represents alternating Schwarz in this context, it may be exchanged for any fixed point iteration, in particular any Schwarz method. All that is required for the algorithm to function is for $G(\gamma)$ to be within region 2. For Schwarz methods, this would necessitate a theorem similar to Theorem 2.

References

1. F.H. Branin, Jr. Widely convergent method for finding multiple solutions of simultaneous nonlinear equations. *IBM J. Res. Develop.* **16**, 504–522 (1972).
2. R.P. Brent. On the Davidenko–Branin method for solving simultaneous nonlinear equations. *IBM J. Res. Develop.* **16**, 434–436 (1972).
3. X-C. Cai and D.E. Keyes. Nonlinearly preconditioned inexact Newton algorithms. *SIAM J. Sci. Comput.* **24**(1), 183–200 (2002).
4. D.F. Davidenko. On a new method of numerical solution of systems of nonlinear equations. *Doklady Akad. Nauk SSSR (N.S.)* **88**, 601–602 (1953).
5. V. Dolean, M.J. Gander, W. Kheriji, F. Kwok and R. Masson. Nonlinear preconditioning: how to use a nonlinear Schwarz method to precondition Newton's method. *SIAM J. Sci. Comput.* **38**(6), A3357–A3380 (2016).
6. M.J. Gander. On the origins of linear and non-linear preconditioning. In: *Domain Decomposition Methods in Science and Engineering XXIII*, volume 116 of Lect. Notes Comput. Sci. Eng., pp. 153–161, Springer, Cham (2017).
7. W. Gander, M.J. Gander and F. Kwok. *Scientific Computing – An introduction using Maple and MATLAB*. Volume 11 of Texts in Computational Science and Engineering. Springer, Cham (2014).
8. L. Liu and D.E. Keyes. Field-split preconditioned inexact Newton algorithms. *SIAM J. Sci. Comput.* **37**(3), A1388–A1409 (2015).
9. S.H. Lui. On Schwarz alternating methods for nonlinear elliptic PDEs. *SIAM J. Sci. Comput.* **21**(4), 1506–1523 (1999/2000).

Should Multilevel Methods for Discontinuous Galerkin Discretizations Use Discontinuous Interpolation Operators?

Martin J. Gander and José Pablo Lucero Lorca

1 Discontinuous interpolation for a model problem

Interpolation operators are very important for the construction of a multigrid method. Since multigrid's inception by Fedorenko [7], interpolation was identified as key, deserving an entire appendix in Brandt's seminal work [5]: *'[...] even a small and smooth residual function may produce large high-frequency residuals, and significant amount of computational work will be required to smooth them out.'*

For discontinuous Galerkin (DG) discretizations [2], the problem of choosing an interpolation becomes particularly interesting. A good interpolation operator will not produce undesirable high frequency components in the residual. In an inherited (Galerkin) coarse operator, the choice of restriction and prolongation operators defines the coarse space itself, and then convergence of multigrid algorithms with classical restriction and interpolation operators for DG discretizations of elliptic problems cannot be independent of the number of levels [1]. In 1D, the reason for this was identified in [9, §4.3]): the DG penalization is doubled at each coarsening, causing the coarse problem to become successively stiffer.

A simple classical interpolation operator is linear interpolation: in 1D one takes the average from two adjacent points in the coarser grid and sets the two DG degrees of freedom at the midpoint belonging to the fine mesh to this same value, therefore imposing continuity at that point and discontinuity at coarse grid points. But why should continuity be imposed on the DG interpolated solution on the fine mesh? Can solver performance be improved with a discontinuous interpolation operator?

Convergence of two-level methods for DG discretizations has been analyzed for continuous interpolation operators using classical analysis, see [3, 8] and references therein, and also Fourier analysis [9–12]. We use Fourier analysis here to investigate

Martin J. Gander
University of Geneva, e-mail: martin.gander@unige.ch

José Pablo Lucero Lorca
University of Colorado at Boulder, e-mail: pablo.lucero@colorado.edu

S. C. Brenner et al. (eds.), *Domain Decomposition Methods in Science and Engineering XXVI*, Lecture Notes in Computational Science and Engineering 145,
https://doi.org/10.1007/978-3-030-95025-5_28

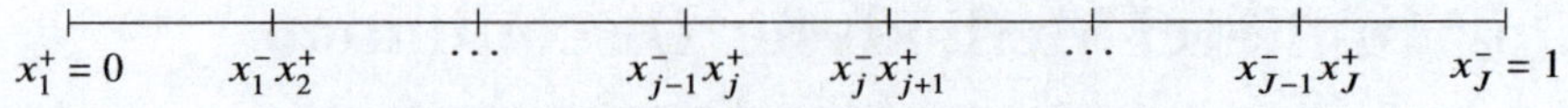

Fig. 1: Mesh for our DG discretization of the Poisson equation.

the influence of a discontinuous interpolation operator on the two level solver performance. We consider a symmetric interior penalty discontinuous Galerkin (SIPG) finite element discretization of the Poisson equation as in [2],

$$a_h(u,v) := \int_{\mathbb{T}} \nabla u \cdot \nabla v \mathrm{d}x + \int_{\mathbb{F}} \left([[u]] \left\{ \frac{\partial v}{\partial n} \right\} + \left\{ \frac{\partial u}{\partial n} \right\} [[v]] \right) \mathrm{d}s + \int_{\mathbb{F}} \delta \, [[u]] \, [[v]] \, \mathrm{d}s, \quad (1)$$

on a 1D mesh as shown in Fig. 1. The resulting linear system is (for details see [9])

$$A\boldsymbol{u} = \frac{1}{h^2} \begin{pmatrix} \ddots & \ddots & -\frac{1}{2} & & & \\ \ddots & \delta_0 & 1-\delta_0 & -\frac{1}{2} & & \\ -\frac{1}{2} & 1-\delta_0 & \delta_0 & & -\frac{1}{2} & \\ & -\frac{1}{2} & & \delta_0 & 1-\delta_0 & -\frac{1}{2} \\ & & -\frac{1}{2} & 1-\delta_0 & \delta_0 & \ddots \\ & & & -\frac{1}{2} & \ddots & \ddots \end{pmatrix} \begin{pmatrix} \vdots \\ u_{j-1}^+ \\ u_j^- \\ u_j^+ \\ u_{j+1}^- \\ \vdots \end{pmatrix} = \begin{pmatrix} \vdots \\ f_{j-1}^+ \\ f_j^- \\ f_j^+ \\ f_{j+1}^- \\ \vdots \end{pmatrix} =: \boldsymbol{f}, \quad (2)$$

where the top and bottom blocks will be determined by the boundary conditions, h is the mesh size, $\delta_0 \in \mathbb{R}$ is the DG penalization parameter, $\boldsymbol{f} = (\ldots, f_{j-1}^+, f_j^-, f_j^+, f_{j+1}^-, \ldots) \in \mathbb{R}^{2J}$ is the source vector, analogous to the solution $\boldsymbol{u}$.

The two-level preconditioner M^{-1} we study consists of a cell-wise nonoverlapping Schwarz (a *cell* block-Jacobi) smoother D_c^{-1}, since the discretization leads to a block matrix (see [6, 8])[1], and a new discontinuous interpolation operator P with discontinuity parameter c, i.e.

$$D_c^{-1}\boldsymbol{u} := h^2 \begin{pmatrix} \ddots & & & \\ & \delta_0 & & \\ & & \delta_0 & \\ & & & \ddots \end{pmatrix}^{-1} \begin{pmatrix} \vdots \\ u_j^+ \\ u_j^- \\ \vdots \end{pmatrix}, \quad P := \begin{pmatrix} 1 & & \\ c & 1-c & \\ 1-c & c & \\ & 1 & \ddots \\ & & \ddots \end{pmatrix}, \quad (3)$$

where $c = \frac{1}{2}$ gives a continuous interpolation on the nodes not present in the coarse mesh, and discontinuous elsewhere. The restriction operator is $R := \frac{1}{2}P^{\top}$, and we use $A_0 := RAP$. The action of our two-level preconditioner M^{-1}, with one presmoothing step and a relaxation parameter α, acting on a residual $\boldsymbol{g}$, is given by

[1] In 1D this is simply a Jacobi smoother, which is not the case in higher dimensions.

1. compute $\boldsymbol{x} := \alpha D_c^{-1}\boldsymbol{g}$,
2. compute $\boldsymbol{y} := \boldsymbol{x} + PA_0^{-1}R(\boldsymbol{g} - A\boldsymbol{x})$,
3. obtain $M^{-1}\boldsymbol{g} = \boldsymbol{y}$.

2 Study of optimal parameters by Local Fourier Analysis

In [9] we described in detail, for classical interpolation, how Local Fourier Analysis (LFA) can be used to block diagonalize all the matrices involved in the definition of M^{-1} by using unitary transformations. The same approach still works with our new discontinuous interpolation operator, and we thus use the same definitions and notation for the block-diagonalization matrices Q, Q_l, Q_r, Q_0, Q_{l0} and Q_{r0} from [9], working directly with matrices instead of stencils in order to make the important LFA more accessible to our linear algebra community. We extract a submatrix $\widetilde{A}$ containing the degrees of freedom of two adjacent cells from the SIPG operator defined in (2),

$$\widetilde{A} = \frac{1}{h^2}\begin{pmatrix} -\frac{1}{2} & 1-\delta_0 & \delta_0 & 0 & -\frac{1}{2} & & & \\ & -\frac{1}{2} & 0 & \delta_0 & 1-\delta_0 & -\frac{1}{2} & & \\ & & -\frac{1}{2} & 1-\delta_0 & \delta_0 & 0 & -\frac{1}{2} & \\ & & & -\frac{1}{2} & 0 & \delta_0 & 1-\delta_0 & -\frac{1}{2} \end{pmatrix},$$

which we can block-diagonalize, $\widehat{A} = Q_l\widetilde{A}Q_r$, to obtain

$$\widehat{A} = \frac{1}{h^2}\begin{pmatrix} \delta_0 + \cos(2\pi(k-J/2)h) & 1-\delta_0 & & \\ 1-\delta_0 & \delta_0 + \cos(2\pi(k-J/2)h) & & \\ & & \delta_0 - \cos(2\pi kh) & 1-\delta_0 \\ & & 1-\delta_0 & \delta_0 - \cos(2\pi kh) \end{pmatrix}.$$

The same mechanism can be applied to the smoother, $\widehat{D}_c = Q_l\widetilde{D}_cQ_r = \frac{\delta_0}{h^2}I$, where I is the 4×4 identity matrix, and also to the restriction and prolongation operators, $\widehat{R} = \frac{1}{2}Q_{l0}\widetilde{R}Q_r$ with

$$\widehat{R} = \frac{1}{\sqrt{2}}\begin{pmatrix} 1+(c-1)e^{\frac{2i\pi k}{J}} & -ce^{\frac{2i\pi k}{J}} & (-1)^j\left(1-(c-1)e^{\frac{2i\pi k}{J}}\right) & (-1)^j ce^{\frac{2i\pi k}{J}} \\ (-1)^j ce^{-\frac{2i\pi k}{J}} & (-1)^j\left(1+(c-1)e^{-\frac{2i\pi k}{J}}\right) & ce^{-\frac{2i\pi k}{J}} & 1-(c-1)e^{-\frac{2i\pi k}{J}} \end{pmatrix},$$

and $P = 2R^\top$, $\widehat{P} = Q_l\widetilde{P}Q_{r0} = 2\widehat{R}^*$. Finally, for the coarse operator, we obtain $Q_0^*A_0Q_0 = Q_0^*RAPQ_0 = Q_0^*RQQ^*AQQ^*PQ_0$, and thus $\widehat{A}_0 = \widehat{R}\widehat{A}\widehat{P}$ with

$$\widehat{A}_0 =$$

$$\frac{1}{H^2}\begin{pmatrix} \frac{1}{2}\left(c\left(4(c-1)\delta_0-2c+3\right)+(c-1)\cos\left(\frac{4\pi k}{J}\right)+2\delta_0-1\right) & \frac{1}{2}(-1)^j\left(-(2c-1)\left(c\left(2\delta_0-1\right)-\delta_0+1\right)\mathrm{e}^{\frac{4i\pi k}{J}}-c-\delta_0+1\right) \\ \frac{1}{2}(-1)^j\left(-(2c-1)\left(c\left(2\delta_0-1\right)-\delta_0+1\right)\mathrm{e}^{-\frac{4i\pi k}{J}}-c-\delta_0+1\right) & \frac{1}{2}\left(c\left(4(c-1)\delta_0-2c+3\right)+(c-1)\cos\left(\frac{4\pi k}{J}\right)+2\delta_0-1\right) \end{pmatrix},$$

where $H = 2h$. We notice that the coarse operator is different for j even and j odd; however, the matrices obtained for both cases are similar, with similarity matrix $(-1)^j I$ where I is the identity matrix, and therefore have the same spectrum. In what follows we assume j to be even, without loss of generality. This means that we will be studying a node that is present in both the coarse and fine meshes.

The error reduction capabilities of our two level preconditioner M^{-1} are given by the spectrum of the stationary iteration operator

$$E = (I - PA_0^{-1}RA)(I - \alpha D_c^{-1}A),$$

and as in [9], the 4-by-4 block Fourier-transformed operator from LFA,

$$\widehat{E}(k) = (I - \widehat{P}(k)\widehat{A}_0^{-1}(k)\widehat{R}(k)\widehat{A}(k))(I - \alpha\widehat{D}_c^{-1}(k)\widehat{A}(k)),$$

has the same spectrum. Thus, we focus on studying the spectral radius $\rho(\widehat{E}(k))$ in order to find the optimal choices for the relaxation parameter α, the penalty parameter δ_0 and the discontinuity parameter c. The non-zero eigenvalues of $\widehat{E}(k)$ are of the form $\lambda_\pm := c_1 \pm \sqrt{\frac{c_2}{c_3}}$, with

$$c_1 = \begin{cases} \Big\{ -\alpha\left(3c^2\delta_0\left(4\delta_0-3\right)+c\left(-12\delta_0^2+9\delta_0+1\right)+4\delta_0^2-2\delta_0-1\right) \\ +\delta_0\left(c^2\left(8\delta_0^2-4\delta_0-1\right)+c\left(-8\delta_0^2+4\delta_0+2\right)+2\delta_0^2-1\right) \\ +(1-c)\left(\alpha+\alpha c\left(\delta_0-2\right)+(c-1)\delta_0\right)\cos\left(\frac{4\pi k}{J}\right)\Big\} \Big/ \\ \left(2\delta_0^2-1+\delta_0 c^2\left(8\delta_0^2-4\delta_0-1\right)+\delta_0 c\left(-8\delta_0^2+4\delta_0+2\right)-\delta_0(c-1)^2\cos\left(\frac{4\pi k}{J}\right)\right), \end{cases}$$

$$c_2 = \begin{cases} 2\alpha^2\Big(16(c-1)^2c^2\delta_0^4-2(c-1)^2\left(4c^2+c+2\right)\delta_0-8(c-1)c(3(c-1)c-1)\delta_0^3 \\ +\left(c(17c+8)(c-1)^2+2\right)\delta_0^2+2(c-1)^2((c-1)c+1)\Big) \\ +4\alpha^2\left(4(c-1)c\delta_0^2-3(c-1)c\delta_0+c+\delta_0-1\right)\left(c\left(3(c-1)\delta_0-2c+3\right)+\delta_0-1\right)\cos\left(\frac{4\pi k}{J}\right) \\ +2\alpha^2(c-1)^2c\left(c\left(\left(\delta_0-4\right)\delta_0+2\right)+2\left(\delta_0-1\right)\right)\cos^2\left(\frac{4\pi k}{J}\right), \end{cases}$$

$$c_3 = \begin{cases} \delta_0^2\left(4c(c-1)\delta_0-2(1-2c)^2\delta_0^2+(c-1)^2\right)^2 \\ +2\delta_0^2\left(-2\left(2c^2-3c+1\right)^2\delta_0^2+4c(c-1)^3\delta_0+(c-1)^4\right)\cos\left(\frac{4\pi k}{J}\right) \\ +(c-1)^4\delta_0^2\cos^2\left(\frac{4\pi k}{J}\right). \end{cases}$$

A first approach to optimize would be to minimize the spectral radius for all frequency parameters k, but if we can find a combination of the parameters (α, δ_0, c) such that the eigenvalues of the error operator do not depend on the frequency parameter k, then the spectrum of the iteration operator, and therefore the preconditioned system becomes perfectly clustered , i.e. only a few eigenvalues repeat many times, regardless of the size of the problem. The solver then becomes mesh independent, and the preconditioner very attractive for a Krylov method that will converge in a finite number of steps.

For these equations not to depend on k, they must be independent of $\cos\left(\frac{4\pi k}{J}\right)$, and to achieve this, we impose three conditions on the coefficients accompanying the cosine, and we deduce a combination of the parameters (α, δ_0, c) which we verify *a posteriori* fall into the allowed range of values for each parameter. Our conditions are:

1. Set the coefficient accompanying the cosine in the numerator of c_1 to zero.
2. Since the denominator of c_1 also contains the cosine, set the rest of the numerator of c_1 to zero in order to get rid of c_1 entirely. Note that this requirement immediately implies an equioscillating spectrum (i.e. the maximum and the minimum have equal absolute value), which often is characterizing the solution minimizing the spectral radius, see e.g. [9].
3. c_2 and c_3 are second order polynomials in the cosine variable, if we want the quotient to be non-zero and independent of the cosine, we need the polynomials to simplify and for that, they must differ only by a multiplying factor independent of the cosine. We then equate the quotient of the quadratic terms with the quotient of the linear terms and verify *a posteriori* that c_2/c_3 becomes indeed independent of the cosine.

These three conditions lead to the nonlinear system of equations

$$\begin{cases} \alpha + \alpha c\,(\delta_0 - 2) + (c-1)\,\delta_0 = 0, \\[2ex] \alpha\left(3c^2\delta_0\,(4\delta_0 - 3) + c\left(-12\delta_0^2 + 9\delta_0 + 1\right) + 4\delta_0^2 - 2\delta_0 - 1\right) \\ = \delta_0\left(c^2\left(8\delta_0^2 - 4\delta_0 - 1\right) + c\left(-8\delta_0^2 + 4\delta_0 + 2\right) + 2\delta_0^2 - 1\right), \\[2ex] \dfrac{2\alpha^2(c-1)^2 c\,(c\,((\delta_0 - 4)\,\delta_0 + 2) + 2\,(\delta_0 - 1))}{(c-1)^4\,\delta_0^2} \\ = \dfrac{4\alpha^2\,(4(c-1)c\,\delta_0^2 - 3(c-1)c\,\delta_0 + c + \delta_0 - 1)\,(c\,(3(c-1)\,\delta_0 - 2c + 3) + \delta_0 - 1)}{2\delta_0^2\left(-2\,(2c^2 - 3c + 1)^2\,\delta_0^2 + 4c(c-1)^3\,\delta_0 + (c-1)^4\right)}. \end{cases}$$

This system of equations can be solved either numerically or symbolically. After a significant effort, the following values solve our nonlinear system:

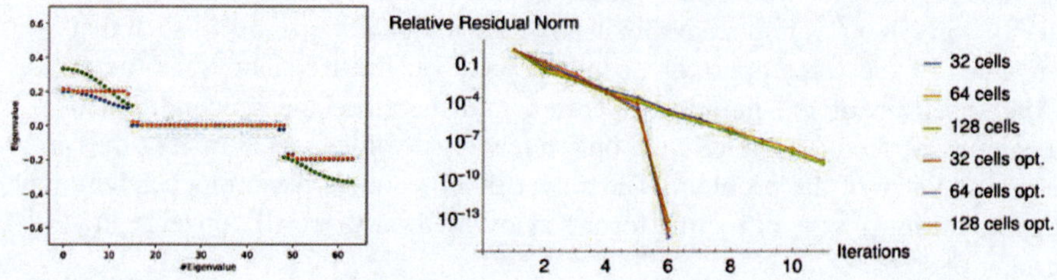

Fig. 2: Solving $-\Delta u = 1$ in 1D with Dirichlet boundary conditions. Left: eigenvalues of the error operator E, for a 32-cell mesh. Top curve at 0: optimizing α for $\delta_0 = 2$ (classical choice). Bottom curve at 0: optimizing α and δ_0. Middle curve at 0: optimizing α, δ_0 and c. Right: GMRES iterations for classical interpolation $c = 0.5$, with $\delta_0 = 2$ and $\alpha = 8/9$, and for the optimized clustering choice, leading to finite step convergence.

$$
\begin{aligned}
c &= \text{Root of } 3 - 8\tilde{c} + 8\tilde{c}^2 - 8\tilde{c}^3 + 4\tilde{c}^4 \text{ such that } \tilde{c} \in \mathbb{R} \text{ and } 0 < \tilde{c} < 1,\\
\delta_0 &= \text{Root of } -1 - 4\tilde{\delta_0} + 24\tilde{\delta_0}^2 - 32\tilde{\delta_0}^3 + 12\tilde{\delta_0}^4 \text{ such that } \tilde{\delta_0} \in \mathbb{R} \text{ and } 1 < \tilde{\delta_0}, \text{ and}\\
\alpha &= \text{Root of } -1 - 40\tilde{\alpha} + 214\tilde{\alpha}^2 - 352\tilde{\alpha}^3 + 183\tilde{\alpha}^4 \text{ such that } \tilde{\alpha} \in \mathbb{R} \text{ and } 0 < \tilde{\alpha} < 1.
\end{aligned}
$$

The corresponding numerical values are approximately

$$c \approx 0.564604, \quad \delta_0 \approx 1.516980, \quad \alpha \approx 0.908154,$$

and we see that indeed the interpolation should be discontinuous! We have found a combination of parameters that perfectly clusters the eigenvalues of the iteration operator of our two level method, and therefore also the spectrum of the preconditioned operator. Such clustering is not very often possible in preconditioners, a few exceptions are the HSS preconditioner in [4], and some block preconditioners, see e.g. [13]. Furthermore, the spectrum is equioscillating, which often characterizes the solution minimizing the spectral radius of the iteration operator.

3 Numerical results

We show in Fig. 2 on the left the eigenvalues of the iteration operator for a 32-cell mesh in 1D with Dirichlet boundary conditions, for continuous interpolation and $\delta_0 = 2$ optimizing only α, optimizing both α and δ_0, and the optimized clustering choice. We clearly see the clustering of the eigenvalues, including some extra clusters due to the Dirichlet boundary conditions. We also note that the spectrum is nearly equioscillating due to condition (1) and (2), which delivers visibly an optimal choice in the sense of minimizing the spectral radius of the error operator. With periodic boundary conditions, the spectral radius for the optimal choice of α, δ_0 and c is

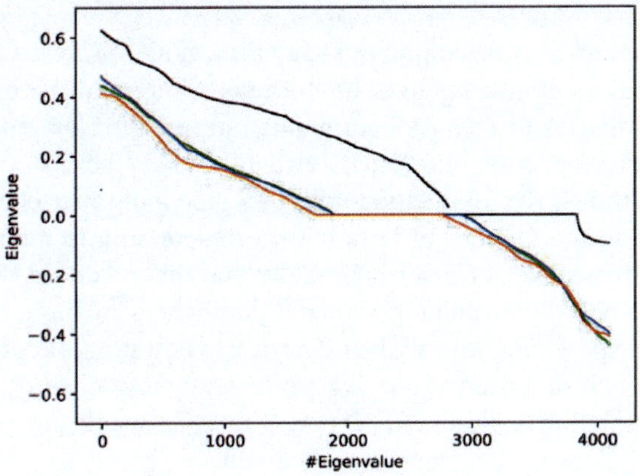

Fig. 3: Spectrum of the iteration operator for a 32-by-32 square 2D mesh. First curve at 0 from the top: optimizing α for $\delta_0 = 2$ (classical choice) in 1D. Third curve at 0 from the top: optimizing α and δ_0 in 1D. Second curve at 0 from the top: optimizing α, δ_0 and c in 1D. Fourth curve at 0 from the top: numerically optimizing α, δ_0 and c in 2D.

0.19732, while only optimizing α and δ_0 it is 0.2. The eigenvalues due to the Dirichlet boundary conditions are slightly larger than 0.2, but tests with periodic boundary conditions confirm that then these larger eigenvalues are not present. Refining the mesh conserves the shape of the spectrum shown in Fig. 2 on the left, but with more eigenvalues in each cluster, except for the clusters related to the Dirichlet boundary conditions. Note also that since the error operator is equioscillating around zero, the spectrum of the preconditioned system is equioscillating around one, and since the spectral radius is less than one, the preconditioned system has a positive spectrum and is thus invertible.

In Fig. 2 on the right we show the GMRES iterations needed to reduce the residuals by 10^{-8} for different parameter choices and the clustering choice, for different mesh refinements. We observe that the GMRES solver becomes exact after six iterations for the clustering choice.

We next perform tests in two dimensions using an interpolation operator with a stencil that is simply a tensor product of the 1D stencil $\begin{pmatrix} 1 & 0 \\ c & 1-c \\ 1-c & c \\ 0 & 1 \end{pmatrix} \otimes \begin{pmatrix} 1 & 0 \\ c & 1-c \\ 1-c & c \\ 0 & 1 \end{pmatrix}$, where $\otimes$ stands for the Kronecker product. This is very common in DG methods where even the cell block-Jacobi matrix can be expressed as a Kronecker sum for fast inversion. We show in Fig. 3 the spectrum for different optimizations in two dimensions. We observe that the clustering is not present, however as shown in detail in [9] for classical interpolation, the optimal choice from the 1D analysis is also here very close to the numerically calculated optimum in 2D.

4 Conclusion

We showed for a one-dimensional discontinuous Galerkin model problem that the optimization of a two grid method leads to a discontinuous interpolation operator, and its performance is superior to using a continuous interpolation operator. The discontinuous interpolation operator allowed us also to cluster the spectrum for our model problem, and thus a Krylov method with this preconditioner becomes a direct solver, converging in the number of iterations corresponding to the number of clusters in exact arithmetic. We showed numerically that this is indeed the case, and that when using the one-dimensional optimized parameters in higher spatial dimensions, we still get a spectrum close to the numerically best possible one, even though the spectrum is not clustered any more. We are currently investigating if there exist discontinuous interpolation operators in 2D that cluster the spectrum, and what their influence is on the Galerkin coarse operator obtained.

References

1. P.F. Antonietti, M. Sarti and M. Verani. Multigrid algorithms for hp-discontinuous Galerkin discretizations of elliptic problems. *SIAM J. on Numer. Anal.* **53**(1), 598–618 (2015).
2. D.N. Arnold. An interior penalty finite element method with discontinuous elements. *SIAM J. Numer. Anal.* **19**(4), 742–760 (1982).
3. B. Ayuso de Dios and L. Zikatanov. Uniformly convergent iterative methods for discontinuous Galerkin discretizations. *J. Sci. Comput.* **40**, 4–36 (2009).
4. M. Benzi, M.J. Gander and G.H. Golub. Optimization of the hermitian and skew-hermitian splitting iteration for saddle-point problems. *BIT Numerical Mathematics* **43**(5), 881–900 (2003).
5. A. Brandt. Multi-level adaptive solutions to boundary-value problems. *Mathematics of Computation* **31**(138), 333–390 (1977).
6. M. Dryja and P. Krzyżanowski. A massively parallel nonoverlapping additive Schwarz method for discontinuous Galerkin discretization of elliptic problems. *Numerische Mathematik* **132**(2), 347–367, February (2016).
7. R.P. Fedorenko. The speed of convergence of one iterative process. *Zh. Vychisl. Mat. Mat. Fiz.* **4**, 559–564 (1964).
8. X. Feng and O. Karakashian. Two-level non-overlapping Schwarz methods for a discontinuous Galerkin method. *SIAM J. Numer. Anal.* **39**(4), 1343–1365 (2001).
9. M.J. Gander and J.P.L. Lorca. Optimization of two-level methods for DG discretizations of reaction-diffusion equations. *arXiv preprint* arxiv.2004.14100 (2020).
10. P. Hemker. Fourier analysis of grid functions, prolongations and restrictions. Report NW 98, CWI, Amsterdam (1980).
11. P. Hemker, W. Hoffmann and M. van Raalte. Two-level fourier analysis of a multigrid approach for discontinuous Galerkin discretization. *SIAM J. on Sci. Comp.* **3**(25), 1018–1041 (2003).
12. P.W. Hemker, W. Hoffmann and M.H. van Raalte. Fourier two-level analysis for discontinuous Galerkin discretization with linear elements. *Numerical Linear Algebra with Applications* **5-6**(11), 473–491 (2004).
13. D. Silvester and A. Wathen. Fast iterative solution of stabilised Stokes systems Part II: Using general block preconditioners. *SIAM J. on Num. Anal.* **31**(5), 1352–1367 (1994).

Domain Decomposition in Shallow Water Modelling of Dutch Lakes for Multiple Applications

Menno Genseberger, Asako Fujisaki, Christophe Thiange, Carlijn Eijsberg - Bak, Arnout Bijlsma, and Pascal Boderie

1 Introduction

1.1 Area of interest

Lake IJssel, Lake Marken, and Veluwerandmeren originated from the construction of dams and land reclamation of an inland sea in the Netherlands (see Fig. 4). For Lake Marken and Veluwerandmeren the dynamic behavior is mainly governed by wind driven waves and flow of water. For Lake IJssel the discharge of River IJssel (in the south) and the flushing of water towards the Wadden Sea (in the north) play a role. Proper computational modelling of the dynamics of waves and flow of water is a challenge. This is of importance for different societal aspects of these lakes: in safety assessments of the primary water defences, operational forecasting of flooding [8], and water quality and ecological studies.

1.2 Previous approaches

Previously, for modelling the hydrodynamic flow in lakes, two shallow water solvers were used: Delft3D-FLOW and WAQUA [15]. Delft3D-FLOW is the depth averaged (2DH) and three-dimensional (3D) shallow water solver in the modelling suite Delft3D [7]. Delft3D is open source and used worldwide. WAQUA is the 2DH shallow water solver in the modelling suite SIMONA. SIMONA is maintained for Dutch public works and only applied to the Dutch main waters (coastal area, rivers, and lakes). The computational kernels of Delft3D-FLOW and WAQUA are quite similar:

Menno Genseberger
Deltares, P.O. Box 177, 2600 MH Delft, The Netherlands,
e-mail: Menno.Genseberger@deltares.nl

S. C. Brenner et al. (eds.), *Domain Decomposition Methods in Science and Engineering XXVI*, Lecture Notes in Computational Science and Engineering 145,
https://doi.org/10.1007/978-3-030-95025-5_29

both use the same ADI (Alternating Direction Implicit) time integration method on a staggered curvilinear computational grid.

In the first instance, for Delft3D-FLOW the focus was on modelling flexibility and for WAQUA on good parallel performance. Parallel implementation of WAQUA was developed [14, 18] based on domain decomposition with an overlap of one subdomain. In the same period, non-overlapping domain decomposition with optimized coupling/absorbing boundary conditions was considered for Delft3D-FLOW [5, 16, 17]. Ideas of the latter were adapted for incorporation in WAQUA to enable more modelling flexibility and further improvement of the parallel performance, see [1]. For application of this approach to operational forecasting of flooding on Lake IJssel and Lake Marken, see [4]. The WAQUA shallow water solver is suitable for safety assessments of the primary water defences and operational forecasting of flooding. Water quality and ecological studies require the more advanced modelling flexibility of Delft3D-FLOW (see for instance [3] for a typical application in Lake Marken). However, Delft3D-FLOW does not have such a good parallel behavior like WAQUA. This is a bottleneck for applications that require highly detailed modelling.

1.3 New approach

Currently there is a transition from Delft3D-FLOW and WAQUA/TRIWAQ to the shallow water solver for unstructured computational grids in the Delft3D FM (Flexible Mesh) suite [6, 10]. To enable the use of unstructured computational grids, the computational kernel of Delft3D FM is different from Delft3D-FLOW and WAQUA/TRIWAQ.

Delft3D FM solves the shallow-water equations with the spatial discretisation being achieved by a staggered finite volume method on an unstructured mesh of cells of varying complexity (triangles to hexagons). The discretised shallow water equations for the water levels are solved implicitly in time, with momentum advection treated explicitly. The velocities and fluxes are then obtained by back substitution. After linearisation of the temporal discretisation, the resulting systems are solved with a semi-implicit method. This involves a linear system which is currently solved by a minimum degree algorithm to reduce system size and a Conjugate Gradient iterative solver with block Jacobi preconditioner and ILU(0) factorization on the blocks as implemented in PETSc [13]. (Note that for this new simulation software for shallow water we are still working on major improvements, Delft3D FM is open source to enable collaboration world-wide.)

Because of these differences, a novel approach is required for model development and model application with Delft3D FM [6, 10]. In this paper we illustrate this for the new shallow water models of Lake IJssel, Lake Marken, and Veluwerandmeren. The aim is an integrated approach in which the Delft3D FM models can be used as a basis for the different societal aspects of these lakes. For that purpose we want to take advantage of the enhanced modelling possibilities of an unstructured computational grid. Therefore our focus is on the computational grid and we developed a strategy

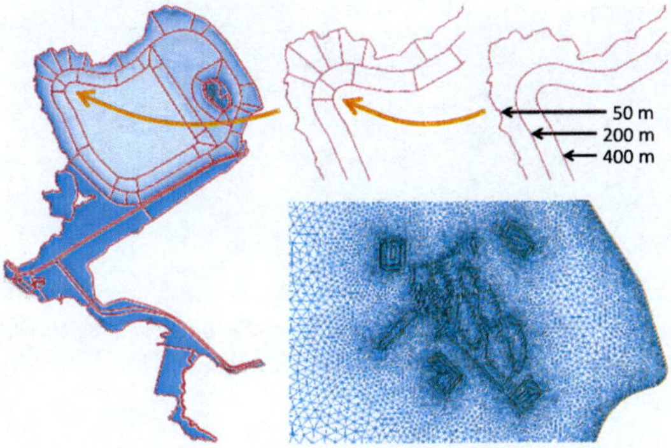

Fig. 1: Top: an illustration of the generation of the boundary fitted computational grid for Lake Marken. With 50 m triangular grid cells near the coast and 400 m triangular grid cells in the middle of the lake by using polygons. Bottom right: a local update of the computational grid near Marker Wadden.

to generate this, as outlined in Section 2. For the different applications on the lakes, Section 3 illustrates how domain decomposition in Delft3D FM enables parallel computing for practical use. We end with some concluding remarks in Section 4.

2 Computational grids

Lake IJssel, Lake Marken, and Veluwerandmeren are quite shallow lakes with local depth variations (due to navigation channels, pits for sand mining or that remained from the old inland sea, land reclamation for housing or nature). For accurate modelling of the flow of water, these local depth variations and structures like dams and sluices should be incorporated properly. Also projection of topography on the computational grid needs special care. The required accuracy can differ in the application to different societal aspects. However, our aim is an integrated approach in which the model can serve as a basis for different model applications.

The key idea is to have enough grid resolution with 50 m triangular cells near the dikes (important for dike safety assessments and operational forecasting) and a coarser grid resolution with 400 m triangular cells where possible in order to save computational time (important both for operational forecasting and water quality studies). For this we used polygons to force the required local resolutions. See the top of Fig. 1 for an illustration for Lake Marken of the strategy to generate the computational grid. From an initial pilot we learned that we can make a Delft3D FM boundary fitted triangular grid that has similar accuracy as a uniform boundary fitted grid with 50 m triangular cells but with more than three times less grid cells and, as a consequence, computational times that are more than three times lower. With the grid generation strategy, in addition to properly fitting the grid to the boundary,

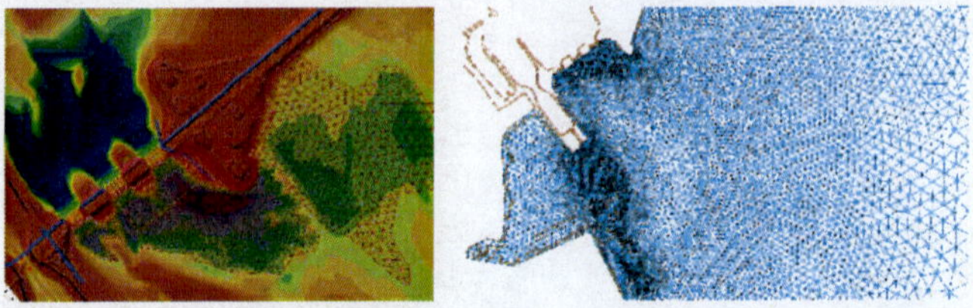

Fig. 2: Example near the discharge sluices of Den Oever (in the north west part) of taking into account for the grid generation the erosion pits near the Afsluitdijk (left) and the resulting highly detailed parts of the grid (right).

we were able to incorporate important details of new land reclamation projects, like IJburg for housing and Marker Wadden for nature in Lake Marken. See Fig. 4 for the resulting computational grids.

Furthermore, the strategy enables the local adaptation of the computational grid later on. For Lake Marken this is important as several infrastructural projects are still running or being started in the near future. On the bottom right of Fig. 1 an example of such an adaptation for Lake Marken is shown. It shows the local update of the computational grid near Marker Wadden by following the local structures for the most recent outline of the islands (which are being built between 2016 and 2019) and pits around the islands (topography obtained from recent surveys with high resolution multibeam depth samples). This update is important for modelling (in combination with in situ measurements and remote sensing images) the effect of Marker Wadden on sediments in Lake Marken in water quality and ecological studies. For Lake IJssel the computational grid was locally adapted near the Afsluitdijk in the north for salt intrusion via the locks from the Wadden Sea. This resulted in highly detailed parts of the grid covering deep pits and navigation channels to better represent steep gradients in the topography, see the example in Fig. 2. Fig. 3 shows model results for Lake IJssel for the 2018 drought. Then, very little fresh water entered the lake from River IJssel. Therefore, it was not possible to flush the more saline water that accumulated in the deep pits to the Wadden Sea. As a consequence chloride spread all over the lake and chloride concentrations exceeded average norms at drinking water intakes in the mid west of the lake. To monitor this situation during the 2018 drought many measurement campaigns were performed on Lake IJssel. These measurements were used to validate the model, they are plotted as bullets on the model results in Fig. 3.

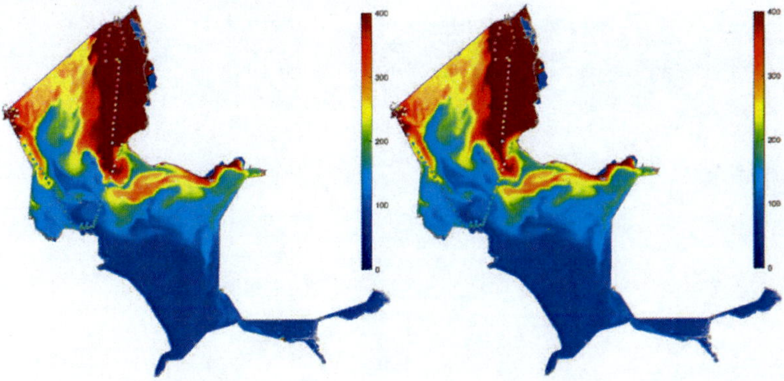

Fig. 3: Results of new model and measurements (bullets) during 2018 drought, chloride [mg/l] near bottom (left) and near surface (right).

3 Domain decomposition for parallel computing in Delft3D FM

The new shallow water models with Delft3D FM of Lake IJssel, Lake Marken, and Veluwerandmeren incorporate important details by local grid refinement.

As a consequence, compared with the previous shallow water solvers, for Lake IJssel and Lake Marken the corresponding horizontal computational grids are about a factor 4 larger. The WAQUA model of Lake IJssel contains 111 763 horizontal grid elements, whereas the Delft3D FM model 369 683. For Lake Marken these numbers are 109 793 and 33 5141, respectively. The WAQUA model of Veluwerandmeren contains 64 616 horizontal grid elements, the Delft3D FM model 91 291. To be able to use these models for the different applications in practice, we want to apply parallel computing with Delft3D FM.

Current parallelisation of Delft3D FM is via domain decomposition with METIS [11] to distribute the computational work. At the interfaces between subdomains, halo regions are defined using degree 4 neighbours for a proper representation of discretised stencils (see [6, 10] for more details) at the interfaces and communication between subdomains via MPI [9]. On the right in Fig. 4 an example is shown of such a decomposition for the shallow water model of Lake Marken.

The shallow water models of Lake Marken and Lake IJssel were two of the real life test cases to study the current parallelization of Delft3D FM in two PRACE projects. These projects investigated possible improvements, amongst others strategies for automatic partitioning into subdomains. For more details we refer to [2, 12]. In the present paper we show results for the current parallelisation in the standard version of Delft3D FM.

To investigate the parallel performance of the new Delft3D FM shallow water models we run tests both in depth-averaged mode (2DH) and in three-dimensional mode (3D). For Lake IJssel in 3D the model was run with both hydrodynamics and salinity for a part of the drought period in 2018 with 5 boundary fitted layers in

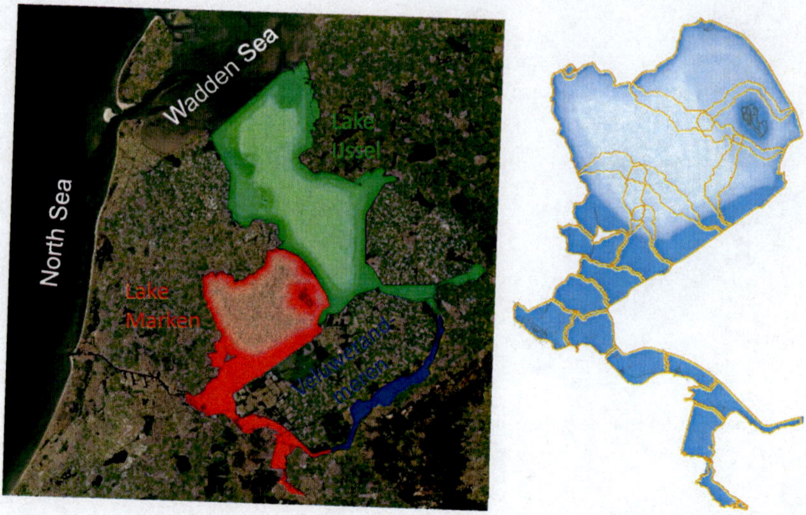

Fig. 4: On the left the area of interest with the computational grids of the lakes projected on a satellite image (by Copernicus Sentinel-2 from ESA at June 30th 2018, `https://scihub.copernicus.eu/dhus`). On the right an example for Lake Marken of automatic partitioning by domain decomposition with METIS into 16 subdomains. The overlap/halo regions are highlighted by orange lines (light gray in black and white print). Note that gridcells are 400 m in the middle of the lake and 50 m near the borders of the lake.

the vertical. For Lake Marken in 3D the model was run with hydrodynamics for the second half year of 2011 with 7 boundary fitted layers in the vertical. In 2DH the Lake IJssel and Lake Marken models were run with hydrodynamics for a storm in January 2007. For Lake Veluwerandmeren the model was run with hydrodynamics for a storm in December 2013 both in 2DH and 3D with 5 boundary fitted layers in the vertical. Tests were run on the Cartesius supercomputer with 2 Intel Xeon E5-2697A v4 processors and 32 cores per node and InfiniBand and Intel MPI between the nodes (Bull B720 bullx system, SURF, the Netherlands).

Fig. 5 shows the speedup compared to computations on 1 node. The Veluwerandmeren model is relatively small. In 2DH parallel scaling stops after about 4 nodes (with 128 cores), for 3D there is more computational work per horizontal grid point and computational times can still be lowered by incorporating more nodes/cores. The Lake IJssel and Lake Marken models have comparable horizontal grid sizes and speedup also shows similar behavior. In 2DH parallel scaling stops at about 16 nodes (with 512 cores), for 3D parallel scaling continues even beyond 16 nodes. This last observation is important for application in 3D for real life problems in these lakes with salinity, nutrients, sediments, and algae. The current numerical implementation of Delft3D FM uses a time integration method with automatic time stepping. The time step (that is used for the whole model domain) is determined with a local CFL criterion for which small grid cells may result in relatively small time steps. The computational grids for Lake IJssel and Lake Marken contain highly detailed parts which may lead to such small time steps. To finish the required simulation periods, which may be typically a year for applications in water quality and ecology, this accumulates to a lot of time steps to be taken. But as the scaling in 3D is still good, computational times can be further lowered by incorporating more nodes/cores. Note that, because of complexity due to different processes modeled and the automatic time stepping approach, it is hard to present generic and characteristic numbers that relate problem size and computational performance. For a model that only involves the shallow water equations, about 40 % to 60 % of the wall clock time is due to the solver part, see [12] for more details. This contribution is much lower when incorporating additional processes like salt intrusion for Lake IJssel. Profiling results for this and other models (amongst others for the North Sea with transport of nutrients and algal blooms) are currently being analyzed in a running PRACE project.

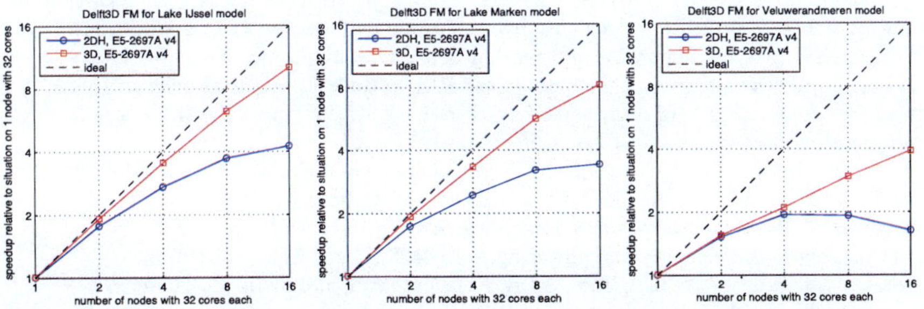

Fig. 5: Speed up of Delft3D FM shallow water models for Lake IJssel (left), Lake Marken (middle), and Veluwerandmeren (right) compared to computations on 1 node on the Cartesius supercomputer of SURF.

4 Conclusions and outlook

In this paper we illustrated the development of new shallow water models of Lake IJssel, Lake Marken, and Veluwerandmeren with Delft3D FM. The aim is an integrated approach in which the models can be used as a basis for the different societal aspects of these lakes: in safety assessments of the primary water defences, operational forecasting of flooding, and water quality and ecological studies. For that purpose domain decomposition in the current numerical implementation of Delft3D FM enables parallel computing for practical use. However, the time integration method used with automatic time stepping may become a bottleneck in the near future for these models due to the highly detailed parts in the computational grids. Therefore, a next step would be to make a more implicit time integration method available in Delft3D FM. That may also require more advanced non-overlapping domain decomposition techniques with optimized coupling/absorbing boundary conditions, as applied before in the previous shallow water solvers for these lakes.

Acknowledgements This paper presents results from projects financed by the Dutch Ministry of Infrastructure and the Environment. We acknowledge PRACE for awarding us access to resource Cartesius based in The Netherlands at SURF. The technical support of Maxime Mogé from SURF, The Netherlands and Andrew Emerson from CINECA, Italy is gratefully acknowledged.

References

1. M. Borsboom, M. Genseberger, B. van ´t Hof and E. Spee. Domain decomposition in shallow-water modelling for practical flow applications. In: J. Erhel et al. (eds) *Domain Decomposition Methods in Science and Engineering XXI*, Springer, Berlin (2014).
2. M. Genseberger, M. Mogé, M.J. Russcher and A. Emerson. Towards scalable Delft3D Flexible Mesh on PRACE infrastructure for real life hydrodynamic and water quality applications. Poster presented at 26th International Conference on Domain Decomposition Methods (2020).
3. M. Genseberger, R. Noordhuis, C.X.O. Thiange and P.M.A. Boderie. Practical measures for improving the ecological state of lake Marken using in-depth system knowledge. In: *Lakes & Reservoirs: Research & Management* **21**(1), 56–64 (2016).
4. M. Genseberger, E. Spee and L. Voort. Domain Decomposition in Shallow Lake Modelling for Operational Forecasting of Flooding. In: T. Dickopf, M. Gander, L. Halpern, R. Krause, L. Pavarino (eds) *Domain Decomposition Methods in Science and Engineering XXII*, Lecture Notes in Computational Science and Engineering, vol 104, Springer, Cham (2016).
5. E.D. De Goede, J. Groeneweg, K.H. Tan, M.J.A. Borsboom and G.S. Stelling. A domain decomposition method for the three-dimensional shallow water equations. In: *Simulation Practice and Theory* **3**, 307–325 (1995).
6. Delft3D FM Suite website, `https://www.deltares.nl/en/software/delft3d-flexible-mesh-suite`
7. Delft3D open source website, `https://oss.deltares.nl/web/delft3d/home`
8. M. Genseberger, A. Smale and H. Hartholt. Real-time forecasting of flood levels, wind driven waves, wave runup, and overtopping at dikes around Dutch lakes. In: *Proceedings 2nd European Conference on FLOODrisk Management,* pp. 1519–1525, Taylor & Francis Group (2013).
9. W. Gropp, S. Huss-Ledermann, A. Lumsdaine, E. Lusk, B. Nitzberg, W. Saphir and M. Snir. *MPI: The Complete Reference.* Vol. 2. MIT Press (1998).
10. H.W.J. Kernkamp, A. van Dam, G.S. Stelling and E.D. de Goede. Efficient scheme for the shallow water equations on unstructured grids with application to the Continental Shelf. In: *Ocean Dynamics* **61**(8), 1175–1188 (2011).
11. `http://glaros.dtc.umn.edu/gkhome/views/metis`
12. M. Mogé, M.J. Russcher, A. Emerson and M. Genseberger. Scalable Delft3D Flexible Mesh for Efficient Modelling of Shallow Water and Transport Processes. *PRACE White Paper* 284 (2019). `https://prace-ri.eu/wp-content/uploads/WP284.pdf`
13. `https://www.mcs.anl.gov/petsc`
14. M.R.T. Roest. *Partitioning for parallel finite difference computations in coastal water simulation.* Ph.D. thesis, Delft University of Technology, The Netherlands (1997).
15. SIMONA. WAQUA/TRIWAQ - two- and three-dimensional shallow water flow model, Technical documentation, SIMONA report number 99-01, Rijkswaterstaat, latest online version 3.17 from November 2016 at `http://simona.deltares.nl/release/doc/techdoc/waquapublic/sim1999-01.pdf`
16. K.H. Tan. *Local coupling in domain decomposition.* Ph.D. thesis, Utrecht University, The Netherlands (1995).

17. K.H. Tan and M.J.A. Borsboom. On generalized Schwarz coupling applied to advection-dominated problems. In: *Proc. 7th Int. Conf. on Domain Decomposition*, AMS (1994).
18. E.A.H. Vollebregt. *Parallel software development techniques for shallow water model*. Ph.D. thesis, Delft University of Technology, The Netherlands (1997).

A Variational Interpretation of Restricted Additive Schwarz With Impedance Transmission Condition for the Helmholtz Problem

Shihua Gong, Martin J. Gander, Ivan G. Graham, and Euan A. Spence

1 The Helmholtz problem

Motivated by the large range of applications, there is currently great interest in designing and analysing preconditioners for finite element discretisations of the Helmholtz equation

$$-(\Delta + k^2)u = f \quad \text{on} \quad \Omega, \tag{1}$$

on a d-dimensional domain Ω ($d = 2, 3$), with k the (assumed constant, but possibly large) angular frequency. While the methods presented easily apply to quite general scattering problems and geometries, we restrict attention here to the interior impedance problem, where Ω is bounded, and the boundary condition is

$$\left(\frac{\partial}{\partial n} - \mathrm{i}k\right) u = g \quad \text{on} \quad \partial\Omega, \tag{2}$$

where $\partial u / \partial n$ is the outward-pointing normal derivative of u on Ω.

The weak form of problem (1), (2) is to seek $u \in H^1(\Omega)$ such that

$$a(u, v) = F(v) := \int_\Omega f\bar{v} \,\mathrm{d}x + \int_{\partial\Omega} g\bar{v} \,\mathrm{d}s, \tag{3}$$

$$\text{where} \qquad a(u, v) := \int_\Omega (\nabla u.\nabla\bar{v} - k^2 u\bar{v}) - \mathrm{i}k \int_{\partial\Omega} u\bar{v}, \quad \text{for} \quad u, v \in H^1(\Omega).$$

Shihua Gong, Ivan G. Graham and Euan A. Spence
Department of Mathematical Sciences, University of Bath, Bath BA2 7AY, UK.

Martin J. Gander
Department of Mathematics, University of Geneva, Switzerland.

S. C. Brenner et al. (eds.), *Domain Decomposition Methods in Science and Engineering XXVI*, Lecture Notes in Computational Science and Engineering 145,
https://doi.org/10.1007/978-3-030-95025-5_30

2 Parallel iterative Schwarz method

To solve (1), (2), we shall consider domain decomposition methods, based on a set of Lipschitz polyhedral subdomains $\{\Omega_\ell\}_{\ell=1}^N$, forming an overlapping cover of Ω and equipped with a partition of unity: $\{\chi_\ell\}_{\ell=1}^N$, such that

$$\left.\begin{array}{ll}\text{for each} & \ell: \ \operatorname{supp}\chi_\ell \subset \overline{\Omega_\ell}, \quad 0 \le \chi_\ell(\boldsymbol{x}) \le 1 \text{ when } \boldsymbol{x} \in \overline{\Omega_\ell}, \\ \text{and} & \sum_\ell \chi_\ell(\boldsymbol{x}) = 1 \ \text{ for all } \boldsymbol{x} \in \overline{\Omega}.\end{array}\right\} \tag{4}$$

Then, the parallel Schwarz method for (1), (2) with Robin (impedance) transmission conditions is: given u^n defined on Ω, we solve the local problems:

$$-(\Delta + k^2)u_\ell^{n+1} = f \qquad \text{in } \Omega_\ell\,, \tag{5}$$

$$\left(\frac{\partial}{\partial n_\ell} - ik\right) u_\ell^{n+1} = \left(\frac{\partial}{\partial n_\ell} - ik\right) u^n \qquad \text{on } \partial\Omega_\ell \backslash \partial\Omega\,, \tag{6}$$

$$\left(\frac{\partial}{\partial n_\ell} - ik\right) u_\ell^{n+1} = g \qquad \text{on } \partial\Omega_\ell \cap \partial\Omega. \tag{7}$$

Then the next iterate is the weighted sum of the local solutions

$$u^{n+1} := \sum_\ell \chi_\ell u_\ell^{n+1}. \tag{8}$$

Information is shared between neighbouring subdomains at each iteration via (8).

In [6], we analyse the iteration (5)–(8) in the function space

$$U(\Omega) := \left\{v \in H^1(\Omega) : \Delta v \in L^2(\Omega),\ \partial v/\partial n \in L^2(\partial\Omega)\right\},$$

and its local analogues $U(\Omega_\ell)$. Using the fact that any function $v \in U(\Omega_\ell)$ has impedance trace $(\partial/\partial n - ik)v \in L^2(\Gamma)$ on any Lipschitz curve $\Gamma \subset \Omega_\ell$, we prove in [6] that (5)–(8) is well-defined in the space $U(\Omega)$. Moreover, introducing $e_\ell^n = u|_{\Omega_\ell} - u_\ell^n$, and letting $\mathbf{e}^n = (e_1^n, \ldots, e_N^n)$, we prove in [6] that $\mathbf{e}^{n+1} = \mathcal{T}\mathbf{e}^n$, where under certain geometric assumptions, $\mathcal{T}$ has the 'power contraction' property

$$\|\mathcal{T}^N\| \ll 1, \tag{9}$$

with respect to the product norm on $\prod_\ell U_0(\Omega_\ell)$, where $U_0(\Omega_\ell)$ is the subspace of functions $v \in U(\Omega_\ell)$, for which $\Delta v + k^2 v = 0$ on Ω_ℓ. Analogously to [1], the norm of v is the L^2 norm of its impedance data on $\partial\Omega_\ell$. See the remarks ras §5, especially (24), for a more precise explanation of (9).

The aim of this note is to show that a natural finite element analogue of (5)–(8) corresponds to a preconditioned Richardson-type iterative method for the finite element approximation of (1), (2), where the preconditioner is a Helmholtz-orientated version of the popular Restricted Additive Schwarz method. This preconditioner is given several different names in the literature – WRAS-H (Weighted RAS for Helmholtz)

[9], ORAS (Optimized Restricted Additive Schwarz) [2, 5, 10], IMPRAS1 (RAS with impedance boundary condition) [7]. However it has not previously been directly connected via a variational argument to the iterative method (5)–(8) in the Helmholtz case, although there are algebraic discussions (e.g., [3], [2, §2.3.2]). We also demonstrate numerically in §5, that the finite element analogue of (5)–(8) inherits the property (9) proved at the continuous level in [6].

Method (5)–(8) is an example of methods studied more generally in the Optimized Schwarz literature (e.g., [4, 10]), where Robin (or more sophisticated) transmission conditions are constructed with the aim of optimizing convergence rates. Although the transmission condition (6) above can be justified directly as a first order absorbing condition for the local Helmholtz problem (5) (without considering optimization), this method is still often called 'Optimized Restricted Additive Schwarz' (or 'ORAS') and we shall continue this naming convention here. ORAS is arguably the most successful one-level parallel method for Helmholtz problems. It can be applied on very general geometries, does not depend on parameters, and can even be robust to increasing k [5]. More generally it can be combined with coarse spaces to improve its robustness properties.

3 Variational formulation of RAS with impedance transmission condition (ORAS)

Here we formulate a finite element approximation of (1), (2) and show that it coincides with ORAS. We introduce a nodal finite element space $\mathcal{V}^h \subset H^1(\Omega)$ consisting of continuous piecewise polynomials of total degree $\leq p$ on a conforming mesh $\mathcal{T}^h$. Functions in $\mathcal{V}^h$ are uniquely determined by their values at nodes in $\overline{\Omega}$, denoted $\{x_j : j \in \mathcal{I}\}$, for some index set $\mathcal{I}$. The local space on $\overline{\Omega_\ell}$ is $\mathcal{V}_\ell^h := \{v_h|_{\overline{\Omega_\ell}} : v_h \in \mathcal{V}^h\}$ with corresponding nodes denoted $\{x_j : j \in \mathcal{I}_\ell\}$, for some $\mathcal{I}_\ell \subset \mathcal{I}$.

Using the sesquilinear form a and right-hand side F appearing in (3), we can define the discrete operators $\mathcal{A}_h, F_h : \mathcal{V}^h \mapsto (\mathcal{V}^h)'$ by

$$(\mathcal{A}_h u_h)(v_h) := a(u_h, v_h) \quad \text{and} \quad F_h(v_h) = F(v_h), \quad \text{for all} \quad u_h, v_h \in \mathcal{V}_h. \tag{10}$$

Analogously, on each subdomain Ω_ℓ, we define $\mathcal{A}_{h,\ell} : \mathcal{V}_\ell^h \to (\mathcal{V}_\ell^h)'$ by $(\mathcal{A}_{h,\ell} u_{h,\ell})(v_{h,\ell}) := a_\ell(u_{h,\ell}, v_{h,\ell})$. We also need prolongations $\mathcal{R}_{h,\ell}^\top, \widetilde{\mathcal{R}}_{h,\ell}^\top : \mathcal{V}_\ell^h \to \mathcal{V}^h$ defined for all $v_{h,\ell} \in \mathcal{V}_\ell^h$ by

$$(\mathcal{R}_{h,\ell}^\top v_{h,\ell})(x_j) = \begin{cases} v_{h,\ell}(x_j) & j \in \mathcal{I}_\ell, \\ 0 & \text{otherwise}, \end{cases} \quad \text{and} \quad \widetilde{\mathcal{R}}_{h,\ell}^\top v_{h,\ell} = \mathcal{R}_{h,\ell}^\top (\chi_\ell v_{h,\ell}). \tag{11}$$

Note the subtlety in (11): The extension $\mathcal{R}_{h,\ell}^\top v_{h,\ell}$ is defined *nodewise*: It coincides with $v_{h,\ell}$ at nodes in $\overline{\Omega_\ell}$ and vanishes at nodes in $\Omega \backslash \overline{\Omega_\ell}$. Thus $\mathcal{R}_{h,\ell}^\top v_{h,\ell} \in$

$\mathcal{V}^h \subset H^1(\Omega)$. This is an H^1– conforming finite element approximation of the zero extension of $v_{h,\ell}$ to all of Ω. (The zero extension is not in $H^1(\Omega)$ in general.) We define the restriction operator $\mathcal{R}_{h,\ell} : \mathcal{V}_h' \to \mathcal{V}_{h,\ell}'$ by duality, i.e., for all $F_h \in \mathcal{V}_h'$,

$$(\mathcal{R}_{h,\ell} F_h)(v_{h,\ell}) := F_h(\mathcal{R}_{h,\ell}^\top v_{h,\ell}), \quad v_{h,\ell} \in \mathcal{V}_\ell^h.$$

Then the ORAS preconditioner is the operator $\mathcal{B}_h^{-1} : \mathcal{V}_h' \to \mathcal{V}_h$ defined by

$$\mathcal{B}_h^{-1} := \sum_\ell \widetilde{\mathcal{R}}_{h,\ell}^\top \mathcal{A}_{h,\ell}^{-1} \mathcal{R}_{h,\ell}. \tag{12}$$

This preconditioner can also be written in terms of operators $Q_{h,\ell} : \mathcal{V}^h \to \mathcal{V}_\ell^h$ defined for all $u_h \in \mathcal{V}^h$ by

$$a_\ell(Q_{h,\ell} u_h, v_{h,\ell}) = a(u_h, \mathcal{R}_{h,\ell}^\top v_{h,\ell}), \quad \text{for all} \quad v_{h,\ell} \in \mathcal{V}_\ell^h, \tag{13}$$

where $\mathcal{R}_{h,\ell}^\top$ is defined in (11), and then $\mathcal{B}_h^{-1} = \sum_\ell \widetilde{\mathcal{R}}_{h,\ell}^\top Q_{h,\ell}$. The corresponding preconditioned Richardson iterative method can be written as

$$u_h^{n+1} = u_h^n + \mathcal{B}_h^{-1}(F_h - \mathcal{A}_h u_h^n). \tag{14}$$

The matrix realisation of (14) is given in §5.

4 Connecting the parallel iterative method with ORAS

In this section, we show that a natural finite element approximation of (5)–(8) yields (14). First, to write (5)–(8) in a residual correction form, we introduce the "corrections" $\delta_\ell^n := u_\ell^{n+1} - u^n|_{\overline{\Omega_\ell}}$. With this definition we have

$$-(\Delta + k^2)\delta_\ell^n = f + (\Delta + k^2)u^n \quad \text{in } \Omega_\ell, \tag{15}$$

$$\left(\frac{\partial}{\partial n_\ell} - \mathrm{i}k\right)\delta_\ell^n = 0 \quad \text{on } \partial\Omega_\ell \backslash \partial\Omega, \tag{16}$$

$$\left(\frac{\partial}{\partial n_\ell} - \mathrm{i}k\right)\delta_\ell^n = g - \left(\frac{\partial}{\partial n_\ell} - \mathrm{i}k\right)u^n \quad \text{on } \partial\Omega_\ell \cap \partial\Omega, \tag{17}$$

$$\text{and then} \qquad u^{n+1} = u^n + \sum_\ell \chi_\ell \delta_\ell^n. \tag{18}$$

Note, there is more subtlety here: Because of (8), $u^n|_{\overline{\Omega_\ell}}$ is *not* the same as u_ℓ^n. The theory in [6] can be used to show that (15)–(18) is still well-posed in $U(\Omega)$. Multiplying (15) by $v_\ell \in H^1(\Omega_\ell)$, integrating by parts and using (16), (17), δ_ℓ^n satisfies, for $v_\ell \in H^1(\Omega_\ell)$,

$$a_\ell(\delta_\ell^n, v_\ell) = \int_{\Omega_\ell} f\,\overline{v_\ell} + \int_{\partial\Omega_\ell\cap\partial\Omega} g\,\overline{v_\ell} + \int_{\Omega_\ell} (\Delta + k^2)u^n\,\overline{v_\ell} - \int_{\partial\Omega_\ell\cap\partial\Omega} \left(\frac{\partial}{\partial n_\ell} - \mathrm{i}k\right) u^n\,\overline{v_\ell}. \tag{19}$$

To implement the finite element discretization of this, we will need to handle the case when u^n on the right-hand side is replaced by a given iterate $u_h^n \in \mathcal{V}^h$ and when the test function $v_\ell \in H^1(\Omega_\ell)$ is replaced by $v_{h,\ell} \in \mathcal{V}_\ell^h$. The third term on the right-hand side of (19) then requires integration by parts to make sense. Using the nodewise extension $\mathcal{R}_{h,\ell}^\top$ we replace the third and fourth terms in (19) by

$$\int_\Omega (\Delta + k^2)u^n\,\overline{\mathcal{R}_{h,\ell}^\top v_{h,\ell}} - \int_{\partial\Omega} \left(\frac{\partial}{\partial n} - \mathrm{i}k\right) u^n\,\overline{\mathcal{R}_{h,\ell}^\top v_{h,\ell}} = -a(u^n, \mathcal{R}_{h,\ell}^\top v_{h,\ell}), \tag{20}$$

where the right-hand side is obtained from the left via integration by parts over Ω. This leads to the FEM analogue of (15)–(18): Suppose $u_h^n \in \mathcal{V}^h$ is given. Then

$$u_h^{n+1} := u_h^n + \sum_\ell \widetilde{\mathcal{R}}_{h,\ell}^\top \delta_{h,\ell}^n, \tag{21}$$

where (using (19), (20) and (10)), $a_\ell(\delta_{h,\ell}^n, v_{h,\ell}) = \mathcal{R}_{h,\ell} F_h(v_{h,\ell}) - a(u^n, \mathcal{R}_{h,\ell}^\top v_{h,\ell})$. Thus,

$$\delta_{h,\ell}^n = \mathcal{A}_{h,\ell}^{-1} \mathcal{R}_{h,\ell} (F_h - \mathcal{A}_h u_h^n).$$

Combining this with (21), we obtain exactly (14).

5 Numerical results

Denoting the nodal bases for $\mathcal{V}^h$ and $\mathcal{V}_\ell^h$ by $\{\varphi_j\}$ and $\{\varphi_{\ell,j}\}$ respectively, we introduce stiffness matrices $\mathsf{A}_{i,j} := a(\varphi_j, \varphi_i)$ and $(\mathsf{A}_\ell)_{i,j} := a_\ell(\varphi_{\ell,j}, \varphi_{\ell,i})$, and the load vector $f_i := F_h(\varphi_i)$. Then we can write (14) as

$$\mathbf{u}^{n+1} = \mathbf{u}^n + \mathsf{B}^{-1}(\mathbf{f} - \mathsf{A}\mathbf{u}^n). \tag{22}$$

Here $\mathbf{u}^n$ is the coefficient vector of u_h^n with respect to the nodal basis of $\mathcal{V}^h$, and

$$\mathsf{B}^{-1} = \sum_\ell \tilde{\mathsf{R}}_\ell^\top \mathsf{A}_\ell^{-1} \mathsf{R}_\ell,$$

where $(\mathsf{R}_\ell^\top)_{p,q} := (\mathcal{R}_\ell^\top \varphi_{\ell,q})(x_p)$, $(\tilde{\mathsf{R}}_\ell^\top)_{p,q} := (\widetilde{\mathcal{R}}_\ell^\top \varphi_{\ell,q})(x_{\ell,p})$, and $\mathsf{R}_\ell = (\mathsf{R}_\ell^\top)^\top$.

In this section, (motivated by (9)), we numerically investigate the contractive property of the ORAS iteration (22). Letting $\mathbf{u}$ be the solution of $\mathsf{A}\mathbf{u} = \mathbf{f}$, we can combine with (22) to obtain the error propagation equation

$$\mathbf{u}^{n+1} - \mathbf{u} = \mathsf{E}(\mathbf{u}^n - \mathbf{u}), \quad \text{where} \quad \mathsf{E} = \mathsf{I} - \mathsf{B}^{-1}A.$$

Since $\sum_\ell \tilde{\mathsf{R}}_\ell^\top \mathsf{R}_\ell = \mathsf{I}$, we can write

$$\mathsf{E} = \sum_\ell \tilde{\mathsf{R}}_\ell^\top (\mathsf{R}_\ell - \mathsf{A}_\ell^{-1}\mathsf{R}_\ell \mathsf{A}) = \tilde{\mathbf{R}}^\top(\mathbf{R} - \mathbf{Q}),$$

where $\tilde{\mathbf{R}}^\top$ is the row vector of matrices: $\tilde{\mathbf{R}}^\top = (\tilde{\mathsf{R}}_1^\top, \tilde{\mathsf{R}}_2^\top, \cdots, \tilde{\mathsf{R}}_N^\top)$, and $\mathbf{R} = (\mathsf{R}_1; \mathsf{R}_2; \cdots; \mathsf{R}_N)$, and $\mathbf{Q} = (\mathsf{A}_1^{-1}\mathsf{R}_1\mathsf{A}; \mathsf{A}_2^{-1}\mathsf{R}_2\mathsf{A}, \cdots, \mathsf{A}_N^{-1}\mathsf{R}_N\mathsf{A})$ are column vectors. Then it is easily seen that $\mathsf{E}\tilde{\mathbf{R}}^\top = \tilde{\mathbf{R}}^\top\mathbf{T}$, where $\mathbf{T} := (\mathbf{R} - \mathbf{Q})\tilde{\mathbf{R}}^\top$. Moreover, since $\tilde{\mathbf{R}}^\top\mathbf{R} = \mathsf{I}$, we have $\mathbf{T}\mathbf{R}\tilde{\mathbf{R}}^\top = \mathbf{T}$, and so it follows that

$$\mathsf{E}^s = \tilde{\mathbf{R}}^\top\mathbf{T}^s\mathbf{R} \quad \text{for any} \quad s \geq 1. \tag{23}$$

As explained in [6, §5.1], $\mathbf{T}$ is a discrete version of the operator $\mathcal{T}$ appearing in (9) above. In [6], we study fixed point iterations with matrix $\mathbf{T}$ and use these to illustrate various properties of the fixed point operator $\mathcal{T}$ in the product norm described above. In this paper we consider only the norms of E^s. By (23), if $\mathbf{T}^s$ is sufficiently contractive, then E^s will also be contractive.

To compute the norm of E^s, we introduce the vector norm: $\|\mathbf{u}\|_{1,k}^2 = \mathbf{u}^*\mathsf{D}_k\mathbf{u}$, for $\mathbf{u} \in \mathbb{C}^M$, where $M = \dim(\mathcal{V}^h)$ and, for all nodes x_p, x_q of $\mathcal{V}^h$, $(\mathsf{D}_k)_{p,q} = \int_\Omega \nabla\varphi_p \cdot \nabla\varphi_q + k^2\varphi_p\varphi_q \,\mathrm{d}x,$. This is the matrix induced by the usual k-weighted H^1 inner product on $\mathcal{V}^h$. We shall compute

$$\|\mathsf{E}^s\| := \max_{0 \neq \mathbf{v} \in \mathbb{C}^M} \frac{\|\mathsf{E}^s\mathbf{v}\|_{1,k}}{\|\mathbf{v}\|_{1,k}}, \quad \text{for integers} \quad s \geq 1,$$

which is equal to the square root of the largest eigenvalue of the matrix $\mathsf{D}_k^{-1}(\mathsf{E}^*)^s\mathsf{D}_k\mathsf{E}^s$. This is computed using the SLEPc facility within the package FreeFEM++ [8]. In the following numerical experiments, done on rectangular domains, we use conforming Lagrange elements of degree 2, on uniform meshes with mesh size decreasing with $h \sim k^{-5/4}$ as k increases, sufficient for avoiding the pollution effect.

We consider two different examples of domain decomposition. First we consider a long rectangle of size $(0, \frac{2}{3}N) \times (0, 1)$, partitioned into N non-overlapping strips of equal width $2/3$. We then extend each subdomain by adding neighbouring elements whose distance from the boundary is $\leq 1/6$. This gives an overlapping cover, with each subdomain a unit square, except for the subdomains at the ends, which are rectangles with aspect ratio $6/5$. For this example, a rigorous estimate ensuring (9) is proved in [6]. The result implies that

$$\|\mathcal{T}^N\| \leq C(N-1)\rho + O(\rho^2). \tag{24}$$

Here, ρ is the maximum of the L^2 norms of the 'impedance maps' which describe the exchange of impedance data between boundaries of overlapping subdomains within a single iteration. The constant C is independent of N, but the hidden constant may

N	2			4			8			16		
k	$\|\|E\|\|$	$\|\|E^s\|\|$	$\|\|E^{s+1}\|\|$	$\|\|E\|\|$	$\|\|E^{s-1}\|\|$	$\|\|E^s\|\|$	$\|\|E\|\|$	$\|\|E^{s-1}\|\|$	$\|\|E^s\|\|$	$\|\|E\|\|$	$\|\|E^{s-1}\|\|$	$\|\|E^s\|\|$
20	5.6	0.52	0.05	5.8	5.24	0.18	5.8	4.5	0.11	5.9	3.4	0.17
40	9.0	1.0	0.094	9.1	8.5	0.46	9.1	8.1	0.34	9.1	7.6	0.36
80	14.3	1.9	0.17	14.3	13.1	0.78	14.3	13.0	0.61	14.3	12.6	0.66

Table 1: Strip partition of $(0, \frac{2}{3}N) \times (0, 1)$: Norms of powers of E ($s = N$)

depend on N. Thus for small enough ρ, $\mathcal{T}^N$ is a contraction. Conditions ensuring this are explored in [6].

In Table 1 we observe the rapid drop in the norm of $||\mathsf{E}^s||$ compared with $||\mathsf{E}^{s-1}||$ (with $s = N$). Moreover E^N is a contraction when $N = 4, 8, 16$. When $N = 2$ we do not have E^2 contracting, but E^3 certainly is. Although $||E^N||$ is increasing (apparently linearly) with k, $||E^s||$ decreases rapidly for $s > N$, when k is fixed. Note that $||\mathsf{E}||$ can be quite large, and is growing as k increases: thus the error of the iterative method may grow initially before converging to zero. Also, although the right-hand side of (24) grows linearly in N for fixed ρ, the norm of E^N does not exhibit substantial growth. Thus we conclude that (24) may be pessimistic in its N-dependence. In fact sharper estimates are proved and explored computationally in [6]. An interesting open question is to find a lower bound for s as a function of N and k which ensures contractivity.

In [6] it is shown that the computation of ρ, or related more detailed quantities can be done by solving eigenvalue problems on subdomains. This, combined with estimates like (24) could be seen as an *a priori* condition for convergence, rather like convergence predictions via condition number estimates. These always give a sufficient condition for good performance (which is often not sharp).

In the next experiment the domain Ω is the unit square, divided into $N \times N$ equal square subdomains in a "checkerboard" domain decomposition. Each subdomain is extended by adding neighbouring elements a distance $\leq 1/4$ of the width of the non-overlapping subdomains, thus yielding an overlapping domain decomposition with "generous" overlap. In Table 2 we tabulate $||\mathsf{E}^{s-1}||$ and $||\mathsf{E}^s||$, for $s = N^2$

$N \times N$	2×2		4×4		6×6		8×8		
k	$\|\|E^{s-1}\|\|$	$\|\|E^s\|\|$	$\|\|E^{s-1}\|\|$	$\|\|E^s\|\|$	$\|\|E^{s-1}\|\|$	$\|\|E^s\|\|$	$\|\|E^{s-1}\|\|$	$\|\|E^s\|\|$	GMRES
20	4.0e-1	8.8e-2	2.3e-3	1.2e-3	38	41	1.2e6	1.4e6	34
40	7.2e-1	1.6e-1	4.4e-2	2.8e-3	1.5e-3	1.0e-3	6.4e-5	5.3e-5	28
80	1.0	2.4e-1	1.5e-2	9.8e-3	3.9e-4	2.8e-4	1.9e-6	9.2e-7	26
160	1.8	5.0e-1	1.1e-2	6.3e-3	7.3e-4	5.3e-4	9.2e-5	7.5e-5	24

Table 2: Checkerboard partition of the unit square: Norms of powers of E ($s = N^2$),

(i.e., the total number of subdomains). Here we do not see such a difference between these two quantities, but we do observe very strong contractivity for E^s, except in the case of k small and N large. In the latter case the problem is not very indefinite: and GMRES iteration counts are modest even though the norm of E^s is large (we give

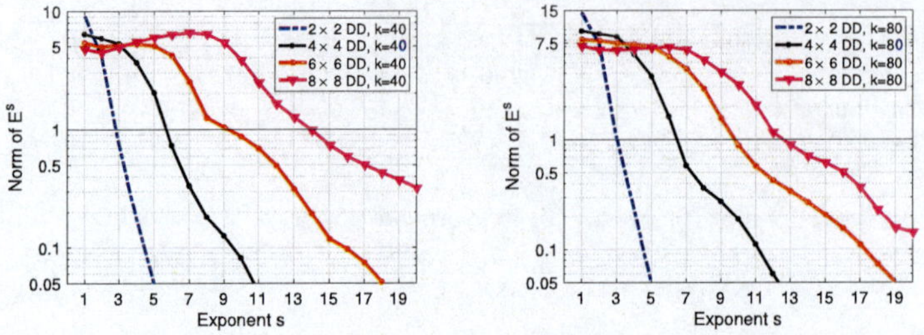

Fig. 1: Norm of the power of the error propagation matrix (left: $k = 40$, right: $k = 80$)

these for the case $N = 8$ in the column headed GMRES). In most of the experiments in the checkerboard case, E^s is contracting when s is much smaller that N^2. In Figure 1, we plot $\|\mathsf{E}^s\|$ against s and observe that $\|\mathsf{E}^s\| < 1$ for exponents $s \ll N^2$.

Acknowledgements SG thanks the Section de Mathématiques, University of Geneva for their hospitality during his visit in early 2020. We gratefully acknowledge support from the UK Engineering and Physical Sciences Research Council Grants EP/R005591/1 (EAS) and EP/S003975/1 (SG, IGG, and EAS).

References

1. J-D. Benamou and B. Després. A domain decomposition method for the Helmholtz equation and related optimal control problems. *J. Comp. Phys.*, **136**(1), 68–82 (1997).
2. V. Dolean, P. Jolivet and F. Nataf. *An introduction to domain decomposition methods: algorithms, theory, and parallel implementation.* SIAM (2015).
3. E. Efstathiou and M.J. Gander. Why restricted additive Schwarz converges faster than additive Schwarz. *BIT Numerical Mathematics* **43**, 945–959 (2003).
4. M.J. Gander. Optimized Schwarz methods. *SIAM J. Numer. Anal.* **44**(2), 699–731 (2006).
5. S. Gong, I.G. Graham and E.A. Spence. Domain decomposition preconditioners for high-order discretisations of the heterogeneous Helmholtz equation. *IMA J. Numer. Anal.* https://doi.org/10.1093/imanum/draa080, (2020).
6. S. Gong, M.J. Gander, I.G. Graham, D. Lafontaine and E.A. Spence. Convergence of overlapping domain decomposition methods for the Helmholtz equation. *arXiv preprint* arXiv:2106.05218 (2021).
7. I.G. Graham, E.A. Spence and E. Vainikko. Recent results on domain decomposition preconditioning for the high-frequency Helmholtz equation using absorption. In: Domenico Lahaye, Jok Tang, and Kees Vuik, editors, *Modern solvers for Helmholtz problems*, Birkhäuser (2017).
8. F. Hecht. Freefem++ manual (version 3.58-1) (2019).
9. J-H. Kimn and M. Sarkis. Restricted overlapping balancing domain decomposition methods and restricted coarse problems for the Helmholtz problem. *Comput. Method Appl. Mech. Engrg.* **196**(8), 1507–1514 (2007).
10. A. St-Cyr, M.J. Gander and S.J. Thomas. Optimized multiplicative, additive, and restricted additive Schwarz preconditioning. *SIAM J. Sci. Comput.* **29**(6), 2402–2425 (2007).

Application of Multilevel BDDC to the Problem of Pressure in Simulations of Incompressible Flow

Martin Hanek and Jakub Šístek

1 Introduction

We deal with the numerical solution of problems of incompressible flows and investigate the applicability of the Balancing Domain Decomposition by Constraints (BDDC) by [2] for solving the arising linear systems. In [5], we extended the multilevel version of BDDC ([7, 12]) to nonsymmetric problems arising from steady problems described by the Navier–Stokes equations. In the present contribution, we are interested in solving unsteady problems, for which we employ the pressure-correction operator-splitting scheme (see e.g. the overview paper by [4]). It presents a very efficient approach for solving the problem by transforming the coupled Navier–Stokes equations into a sequence of a scalar convection-diffusion problem for each velocity component, a Poisson problem for the pressure corrector, and an L_2-projection problem in each time step.

In [10], we studied efficient solution techniques for the arising systems based on Krylov subspace methods with one-level domain decomposition (DD) preconditioners from the PETSc library. A conclusion of the study was that while these relatively simple preconditioners work well for the nonsymmetric problems for velocities and the L_2-projection problem, the known dependence of one-level DD methods on the number of subdomains made the pressure Poisson problem increasingly difficult for a solution with growing problem size, eventually becoming a bottleneck of the simulations.

Martin Hanek
Institute of Mathematics of the Czech Academy of Sciences, Žitná 25, Prague, Czech Republic
Czech Technical University in Prague, Technická 4, Prague, Czech Republic,
e-mail: martin.hanek@fs.cvut.cz

Jakub Šístek
Institute of Mathematics of the Czech Academy of Sciences, Žitná 25, Prague, Czech Republic,
e-mail: sistek@math.cas.cz

S. C. Brenner et al. (eds.), *Domain Decomposition Methods in Science and Engineering XXVI*, Lecture Notes in Computational Science and Engineering 145,
https://doi.org/10.1007/978-3-030-95025-5_31

In this paper, we want to investigate the applicability of several variants of BDDC for the problem of the pressure corrector. As long as the mesh is not changed, the matrix and the preconditioner are set up just once for all time steps. This makes it interesting to use variants of BDDC with more expensive setup, saving the number of iterations in each time step, such as the BDDC method with the adaptive selection of constraints by [8], and its combination with the multilevel extension [11] implemented in our BDDCML library.

Another strategy worth investigating for sequences of algebraic problems is a recycling of the Krylov subspace across time steps, proposed e.g. by [3]. It has been shown in the literature that if the differences of the successive right-hand sides are not large, after expanding the new right-hand side in the pre-existing Krylov basis, one may require only very few or even no additional iterations for convergence to the full accuracy. Hence, it is another aim of this paper to investigate the benefits of the approach by [3] to the present problem.

2 The pressure-correction method

We consider a domain $\Omega \subset \mathbb{R}^3$ with its boundary Γ consisting of three disjoint parts Γ_S, Γ_∞, and Γ_O, $\Gamma = \Gamma_S \cup \Gamma_\infty \cup \Gamma_O$. Part Γ_S is the interface between fluid and the rigid body, Γ_∞ is the inflow free-stream boundary, and Γ_O is the outflow boundary. The flow is governed by the Navier–Stokes equations of an incompressible viscous fluid,

$$\begin{aligned} \frac{\partial \boldsymbol{u}}{\partial t} + (\boldsymbol{u} \cdot \nabla)\boldsymbol{u} - \nu\Delta\boldsymbol{u} + \nabla p = \boldsymbol{0} \quad &\text{in } \Omega, \\ \nabla \cdot \boldsymbol{u} = 0 \quad &\text{in } \Omega, \end{aligned} \tag{1}$$

where $\boldsymbol{u}$ is the velocity vector of the fluid, t denotes time, ν is the kinematic viscosity of the fluid and p is the kinematic pressure. System (1) is complemented by the following initial and boundary conditions: $\boldsymbol{u}(t = 0, \boldsymbol{x}) = \boldsymbol{0}$ in Ω, $\boldsymbol{u}(t, \boldsymbol{x}) = \boldsymbol{u}_\infty$ on Γ_∞, $\boldsymbol{u}(t, \boldsymbol{x}) = \boldsymbol{0}$ on Γ_S, and $-\nu(\nabla\boldsymbol{u})\boldsymbol{n} + p\boldsymbol{n} = \boldsymbol{0}$ on Γ_O, with $\boldsymbol{n}$ being the unit outer normal vector of Γ.

System (1) can be efficiently solved with a pressure-correction method. In particular, we use the incremental pressure-correction method in the *rotational form* discussed by [4]. Details of our implementation can be found in [10].

In this approach, we first define the pressure increment (corrector) $\psi^{n+1} = p^{n+1} - p^n + \nu\nabla \cdot \boldsymbol{u}^{n+1}$. In order to compute the velocity and pressure fields $(\boldsymbol{u}^{n+1}, p^{n+1})$ at time t^{n+1}, three subproblems are subsequently solved.

1. The velocity field $\boldsymbol{u}^{n+1}$ is obtained by solving the convection-diffusion problem for each component of velocity

$$\frac{1}{\Delta t}\boldsymbol{u}^{n+1} + (\boldsymbol{u}^n \cdot \nabla)\boldsymbol{u}^{n+1} - \nu\Delta\boldsymbol{u}^{n+1} = \frac{1}{\Delta t}\boldsymbol{u}^n - \nabla(p^n + \psi^n) \quad \text{in } \Omega \tag{2}$$

for $\boldsymbol{u}^{n+1} = \boldsymbol{u}_\infty$ on Γ_∞, $\boldsymbol{u}^{n+1} = \boldsymbol{0}$ on Γ_S, and $\nu(\nabla\boldsymbol{u}^{n+1})\boldsymbol{n} = p^n\boldsymbol{n}$ on Γ_O.

2. Next, the pressure corrector ψ^{n+1} is obtained by solving the Poisson problem

$$-\Delta\psi^{n+1} = -\frac{1}{\Delta t}\nabla \cdot \boldsymbol{u}^{n+1} \quad \text{in } \Omega \tag{3}$$

for $\frac{\partial \psi^{n+1}}{\partial n} = 0$ on $\Gamma_\infty \cup \Gamma_S$ and $\psi^{n+1} = 0$ on Γ_O.
3. Finally, the pressure field p^{n+1} is updated with

$$p^{n+1} = p^n + \psi^{n+1} - \nu \nabla \cdot \boldsymbol{u}^{n+1}. \tag{4}$$

Problems (2), (3), and (4) are solved by the finite element method (FEM) using Taylor–Hood $Q_2 - Q_1$ hexahedral elements. In the resulting finite element mesh, there are $n_{\boldsymbol{u}}$ nodes with velocity unknowns and n_p nodes with pressure unknowns, with the ratio $n_{\boldsymbol{u}}/n_p$ being approximately 8.

For solving the algebraic problems arising from (2) and (4), we use the methods identified as optimal by [10]. In particular, the Generalized Minimal Residual method (GMRES) is used for solving problem (2), and the Conjugate Gradient (CG) method is used for problem (4). Block Jacobi preconditioner using ILU(0) on subdomains is used for both problems.

The main focus of this study is a scalable solution of the Poisson problem for the pressure corrector (3). We apply different one-level domain decomposition preconditioners from the `PETSc`[1] library and compare them with several settings of the BDDC method from the `BDDCML`[2] library. Each preconditioner is combined with the CG method.

Problem (3) translates to an algebraic system with a discrete Laplacian matrix of size $n_p \times n_p$ which is symmetric and positive definite for $\Gamma_O \neq \emptyset$, i.e. a nonempty part with 'do-nothing' boundary condition. This is a well-studied case from the point of view of DD methods, which are very suitable solvers for this task.

For a fixed mesh, only the right-hand side of (3) differs in the sequence for the subsequent time steps. Hence, this problem offers large room for reusing information across all time steps. For example, one may afford a preconditioner with a more expensive setup if this leads to a lower number of iterations as long as each iteration does not get much more expensive. This is our motivation for experimenting also with the adaptive selection of constraints for BDDC.

3 Numerical results

We evaluate the strategies for solving (3) in the case of a flow past a sphere at Reynolds number 300. In our simulations, we consider two sizes of the problem mesh with the same geometry (see Fig. 1). The sphere diameter is 1 m, and the solution domain is a cylinder with radius 6 m and length 25 m. The centre of the

---

[1] `https://www.mcs.anl.gov/petsc` (version 3.10.4)

[2] `https://users.math.cas.cz/~sistek/software/bddcml.html` (version 2.6)

sphere lies on the cylinder axis and 5 m from the front wall of the cylinder. The far-field velocity of the fluid is $\boldsymbol{u}_\infty = (1,0,0)^\mathrm{T}$ m s^{-1}, and the kinematic viscosity is $\nu = 0.00333$ m^{2}s^{-1}, so that the Reynolds number defined as $\mathrm{Re} = \frac{|\boldsymbol{u}_\infty|\, d}{\nu}$ is equal to 300. The external boundary of the cylinder is considered as Γ_∞ except the rear face, which represents Γ_O with 'do-nothing' boundary condition. A zero Dirichlet boundary condition is prescribed on the surface of the sphere Γ_S.

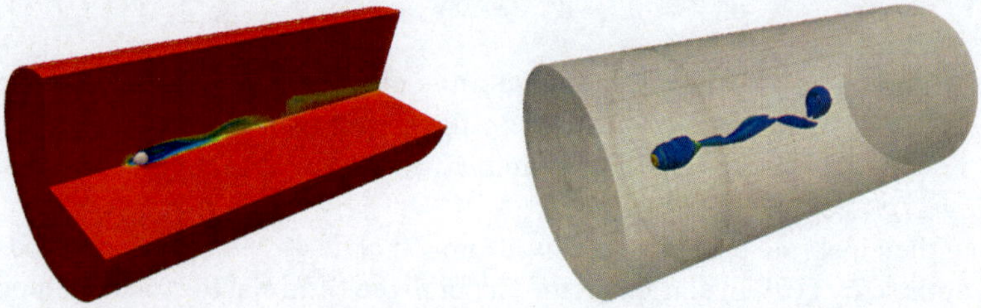

Fig. 1: Computational domain with meshing corresponding to Mesh 1, Reynolds number 300. Velocity magnitude (left) and vortical structures illustrated by isosurfaces of the average corotation ([6]) coloured by the magnitude of vorticity (right).

The computational Mesh 1 consists of 1.8 million unknowns for velocity in each component and 225 thousand for pressure, and it corresponds to the mesh used in [9]. Mesh 2 is obtained by doubling the number of elements in each direction. Hence, it has approx. 15 million unknowns for each velocity component and 1.9 million for pressure. The meshes were created in the `Gmsh`[3] generator and divided into 16 and 128 subdomains, respectively, by the `METIS`[4] graph partitioner to maintain approximately the same size of subdomain problems (approx. 15 thousand unknowns per subdomain). The problems on Mesh 1 and Mesh 2 were solved using 16 and 128 CPU cores of the *Salomon* supercomputer at the IT4Innovations National Supercomputing Center in Ostrava, Czech Republic. The computational nodes of Salomon are equipped with two 12-core Intel Xeon E5-2680v3 2.5 GHz processors and 128 GB RAM.

This kind of simulation is usually performed for thousands of time steps. We formally employ the non-dimensional time $t' = \frac{t|\boldsymbol{u}_\infty|}{d}$, although for our setting of $|\boldsymbol{u}_\infty| = 1$ and $d = 1$, the values are the same as for the physical time t.

In particular, the simulation of 200 s on Mesh 1 performed with time-step size $\Delta t = 0.05$ results in 4000 time steps, while the simulation on Mesh 2 with time-step size $\Delta t = 0.025$ results in 8000 time steps. The different values of the time steps are motivated by an approximate preservation of the Courant number $\frac{|\boldsymbol{u}_\infty|\,\Delta t}{h}$ coupling the resolution in time and space.

Since our aim is to test the behaviour for different preconditioners, we compute only 30 time steps, and we report and compare the numbers of linear iterations and

[3] `https://gmsh.info/`

[4] `http://glaros.dtc.umn.edu/gkhome/metis/metis/overview`

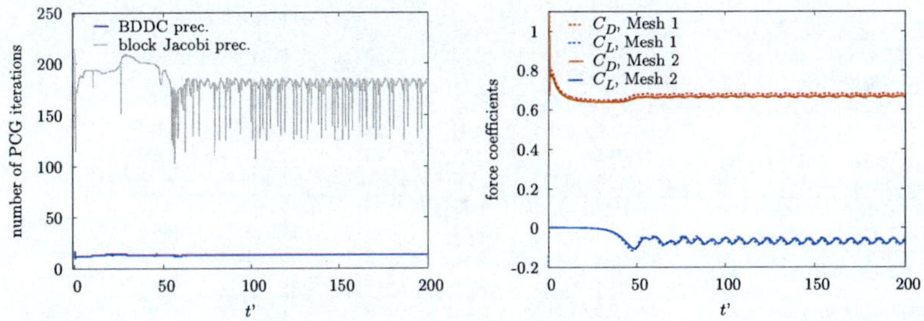

Fig. 2: Number of linear iterations in each time step (left), and values of drag (C_D) and lift (C_L) force coefficients during the whole simulation (right). Results for Mesh 1 are by [9] and for Mesh 2 from current simulations.

times for all time steps excluding the first time step. As you can see in the left part of Fig. 2, the number of iterations stays almost constant during the whole simulation for the BDDC method and within the same range for the block Jacobi method. This justifies using the first 30 iterations for our comparisons. We also compare the drag and lift force coefficients acting on the sphere with the results from [9] in Fig. 2. We have got a good agreement for the two resolutions.

In the first iteration, the setup of the preconditioner and the factorization of interior blocks of subdomain matrices are included. These operations are performed just once for all time steps. Hence, the number of linear iterations and the time of the whole solve (setup and iterations) is reported separately.

method	#its. min–max(avg.)	t./step [s]	#its. step 1	t. step 1 [s]	est. sim. [s]
block Jacobi + ILU(0)	63-169(108.3)	**0.16**	167	**0.25**	**640**
block Jacobi + ILU(1)	46-131(82.7)	0.25	130	0.37	1000
block Jacobi + ILU(2)	44-119(76.0)	0.46	118	0.61	1840
ASM–1 + ILU(0)	102-216(170.2)	0.36	209	0.43	1440
ASM–1 + ILU(1)	81-146(120.5)	0.48	140	0.54	1920
ASM–2 + ILU(0)	103-237(184.4)	0.48	230	0.57	1920
ASM–2 + ILU(1)	63-156(122.4)	0.60	148	0.67	2400
3-l. ad. BDDC + diag.	8-10(**9.6**)	0.32	**14**	75.74	1355
3-l. BDDC + diag.	10-12(11.5)	0.34	21	1.51	1361

Table 1: Mesh 1: Comparison of the number of linear iterations (minimum–maximum(average) across all time steps), average time for solving one time step, values for the first step, and estimated time of all time steps computed as the average time per step × 4000 + time for step 1. Here 'diag.' means scaling by diagonal entries of subdomain matrices, '2-l.' and '3-l.' stand for 2-level and 3-level variants of the BDDC method, respectively, and 'ad.' denotes the adaptive version of BDDC.

The results of our simulations are summarized in Tables 1 and 2. The tested preconditioners include block Jacobi and Additive Schwarz methods (ASM) from PETSc. For ASM, we compare one and two layers of overlap (ASM–1 and ASM–2). On subdomains, the incomplete LU factorization with different levels of allowed fill-in (ILU(0), ILU(1), and ILU(2)) is considered. As for the BDDC options, we use

method	#its. min–max(avg.)	t./step [s]	#its. step 1	t. step 1 [s]	est. sim. [s]
block Jacobi + ILU(0)	461-623(498.6)	1.15	611	1.41	9201
block Jacobi + ILU(1)	115-260(185.9)	0.71	258	**0.98**	5580
block Jacobi + ILU(2)	107-238(175.1)	1.08	236	1.39	8641
ASM–1 + ILU(0)	401-569(433.9)	1.60	557	2.05	12802
ASM–1 + ILU(1)	216-309(241.6)	1.36	300	1.64	10881
ASM–2 + ILU(0)	385-575(428.3)	1.95	550	2.43	15602
ASM–2 + ILU(1)	218-301(240.5)	1.79	294	2.14	14322
3-l. ad. BDDC + diag., $r = 0$	14-19(15.6)	0.62	**19**	157.77	5118
3-l. ad. BDDC + diag., $r = 50$	11-13(12.3)	**0.44**	**19**	132.88	**3652**
3-l. ad. BDDC + diag., $r = 100$	11-13(11.9)	0.54	**19**	159.48	4479
3-l. ad. BDDC + diag., $r = 200$	7-12(**9.9**)	0.51	**19**	160.92	4241
3-l. BDDC + diag., $r = 0$	29-42(32.8)	1.18	42	2.86	9443
3-l. BDDC + diag., $r = 50$	17-23(20.4)	0.89	40	3.14	7123
3-l. BDDC + diag., $r = 100$	17-20(18.7)	0.75	40	2.83	6003
3-l. BDDC + diag., $r = 200$	14-19(15.6)	0.81	40	2.92	6483
3-l. BDDC + arith., $r = 0$	14-19(17.1)	0.59	30	4.87	4724
3-l. BDDC + arith., $r = 50$	14-18(17)	0.55	29	4.07	4404
3-l. BDDC + arith., $r = 100$	14-17(15.9)	0.55	30	5.00	4405
3-l. BDDC + arith., $r = 200$	14-17(15.7)	0.55	30	5.04	4405
3-l. BDDC + diag., $r = 50$	17-25(22.6)	0.71	40	5.31	5685
2-l. ad. BDDC + diag., $r = 0$	14-19(15.2)	0.81	**19**	157.72	6638
2-l. ad. BDDC + diag., $r = 50$	11-13(12.5)	0.68	**19**	158.83	5599
2-l. BDDC + arith., $r = 50$	13-18(16.5)	0.80	27	4.27	6404
2-l. BDDC + diag., $r = 50$	17-13(20.6)	1.01	38	4.80	8085

Table 2: Mesh 2: Comparison of the number of linear iterations (minimum–maximum(average) across all time steps), average time for solving one time step, values for the first step, and estimated time of all time steps computed as the average time per step × 8000 + time for step 1. Here 'diag.' means scaling by diagonal entries of subdomain matrices, 'arith.' means scaling by arithmetic averaging, '2-l.' and '3-l.' stand for 2-level and 3-level variants of the BDDC method, respectively, and 'ad.' denotes the adaptive version of BDDC. Parameter r represents the maximum number of the stored Krylov basis vectors in recycling the Krylov subspaces.

several settings of the `BDDCML` library. Namely, we consider the 2- and 3-level BDDC methods, potentially with the adaptive selection of constraints for the coarse problem as in [11]. Two sequential instances of the MUMPS sparse direct solver (version 5.1.2, [1]) are used for each subdomain, namely a Cholesky LL^T decomposition of the block of unknowns interior to the subdomain, and an LDL^T factorization of the saddle-point problems of BDDC (see [2] for details). In addition, a distributed memory instance of MUMPS is used for the final coarse problem.

The following coarse spaces are considered in the BDDC method. For the non-adaptive version, values at corners and arithmetic averages on each subdomain edge and face are taken as the continuous coarse degrees of freedom. In the adaptive case, a maximum of ten adaptive constraints is also considered on the faces.

We also compare results for two types of interface scaling, the standard one based on arithmetic averages (*arith*) and the one based on diagonal entries of the subdomain matrices (*diag*). In our computations, only the diagonal scaling is compatible with the adaptive BDDC method, while the arithmetic scaling gives better results for the non-adaptive version.

We also test several values of the number of stored Krylov basis vectors r in the approach to recycling the Krylov subspace by [3]. However, we observe little difference among the values of $r = 0$, $r = 50$, $r = 100$, and $r = 200$. We have chosen $r = 50$ for the other simulations with the BDDC method, which is the default for

BDDCML. We have observed a larger improvement for reusing the solution from one time step as the starting approximation for the subsequent problem. This effect can be observed from the difference between the number of iterations in the first time step and their average number. The iterations are terminated when the relative norm of the residual gets below 10^{-6}.

An estimated cost of solving the pressure problem for all time steps (*est. sim.*) is also included in Tables 1 and 2. It is obtained as the time for the first step added to the average time per other steps multiplied by the number of time steps.

We can see that for the smaller problem, the most efficient method is the block Jacobi preconditioner with ILU(0) on subdomains, followed by the same preconditioner with ILU(1). The two configurations of the BDDC preconditioner are less efficient than these options.

However, for the larger problem, the most efficient method becomes BDDC with adaptive constraints, and also the non-adaptive 3-level BDDC method is more efficient than the one-level DD preconditioners, out of which the block Jacobi with ILU(1) requires the least time.

4 Conclusions

We have applied several variants of the BDDC method and one-level DD methods to the Poisson problem for the pressure corrector within a solution of an unsteady problem of incompressible flow with two different meshes.

We have seen that while for a smaller problem, a simple one-level DD method (block Jacobi) provides the fastest solution, the adaptive BDDC method becomes advantageous for larger problems divided into more subdomains. Although the setup of the preconditioner is significantly more expensive, its price gets outweighed by the lower number of CG iterations required in each time step. In addition, recycling the Krylov subspace basis is also slightly beneficial for a reasonable size of the stored basis (50 vectors in our experiments).

The results are encouraging, and we can expect that for even larger problems divided into more subdomains, the adaptive-multilevel BDDC method will be even more beneficial. Confirming this expectation will be a subject of a future study as well as other selection strategies for a suitable recycling basis.

Acknowledgements This research was supported by the Czech Science Foundation through grant 20-01074S, by the Czech Academy of Sciences through RVO:67985840, and by the Czech Technical University in Prague through the student project *SGS19/154/OHK2/3T/12*. Computational time on the Salomon supercomputer has been provided thanks to the support of The Ministry of Education, Youth and Sports from the Large Infrastructures for Research, Experimental Development and Innovations project "IT4Innovations National Supercomputing Center – LM2015070".

References

1. P.R. Amestoy, A. Buttari, J.-Y. L'Excellent and T. Mary. Performance and Scalability of the Block Low-Rank Multifrontal Factorization on Multicore Architectures. *ACM Trans. Math. Softw.* **45**(1), 1–23 (2019).
2. C.R. Dohrmann. A preconditioner for substructuring based on constrained energy minimization. *SIAM J. Sci. Comput.* **25**(1), 246–258 (2003).
3. C. Farhat, L. Crivelli and F.X. Roux. Extending substructure based iterative solvers to multiple load and repeated analyses. *Comput. Methods Appl. Mech. Engrg.* **117**, 195–209 (1994).
4. J.L. Guermond, P. Minev and J. Shen. An overview of projection methods for incompressible flow. *Comput. Methods Appl. Mech. Engrg.* **195**, 6011–6045 (2006).
5. M. Hanek, J. Šístek and P. Burda. Multilevel BDDC for incompressible Navier–Stokes equations. *SIAM J. Sci. Comput.* **42**(6), C359–C383 (2020).
6. V. Kolář, J. Šístek, F. Cirak and P. Moses. Average corotation of line segments near a point and vortex identification. *AIAA J.* **51**(11), 2678–2694 (2013).
7. J. Mandel, B. Sousedík and C.R. Dohrmann. Multispace and multilevel BDDC. *Computing* **83**(2-3), 55–85 (2008).
8. J. Mandel and B. Sousedík. Adaptive selection of face coarse degrees of freedom in the BDDC and the FETI-DP iterative substructuring methods. *Comput. Methods Appl. Mech. Engrg.* **196**(8), 1389–1399 (2007).
9. J. Šístek. A parallel finite element solver for unsteady incompressible Navier–Stokes equations. In: D. Šimurda and T. Bodnár, editors, *Proceedings of Topical Problems of Fluid Mechanics 2015*, pp. 193–198, Institute of Thermomechanics AS CR (2015).
10. J. Šístek and F. Cirak. Parallel iterative solution of the incompressible Navier–Stokes equations with application to rotating wings. *Comput. Fluids* **122**, 165–183 (2015).
11. B. Sousedík, J. Šístek and J. Mandel. Adaptive-Multilevel BDDC and its parallel implementation. *Computing* **95**(12), 1087–1119 (2013).
12. X. Tu. Three-level BDDC in three dimensions. *SIAM J. Sci. Comput.* **29**(4), 1759–1780 (2007).

Predicting the Geometric Location of Critical Edges in Adaptive GDSW Overlapping Domain Decomposition Methods Using Deep Learning

Alexander Heinlein, Axel Klawonn, Martin Lanser, and Janine Weber

1 Introduction

For complex model problems with coefficient or material distributions with large jumps along or across the domain decomposition interface, the convergence rate of classic domain decomposition methods for scalar elliptic problems usually deteriorates. In particular, the classic condition number bounds [1, 12] will depend on the contrast of the coefficient function. As a remedy, different adaptive coarse spaces, e.g. [4, 13], have been developed which are obtained by solving certain generalized eigenvalue problems on local parts of the interface, i.e., edges and/or faces. A selection of the resulting eigenmodes, based on a user-defined tolerance, is then used to enrich the coarse space and retain a robust convergence behavior. However, the setup and the solution of the eigenvalue problems usually take up a significant amount of time in a parallel computation, and for many realistic coefficient distributions, a relatively high number of the eigenvalue problems is unnecessary since they do not result in any additional coarse basis functions. Unfortunately, it is not known a priori which eigenvalue problems are unnecessary and thus can be omitted.

In order to reduce the number of eigenvalue problems, we have proposed to train a neural network to make an automatic decision which of the eigenvalue

Alexander Heinlein
Delft University of Technology, Faculty of Electrical Engineering Mathematics & Computer Science, Delft Institute of Applied Mathematics, Mekelweg 4, 72628 CD Delft, Netherlands,
e-mail: a.heinlein@tudelft.nl

Axel Klawonn, Martin Lanser
Department of Mathematics and Computer Science, University of Cologne, Weyertal 86-90, 50931 Köln, Germany, e-mail: {axel.klawonn,martin.lanser}@uni-koeln.de,
url: http://www.numerik.uni-koeln.de; Center for Data and Simulation Science, University of Cologne, url: http://www.cds.uni-koeln.de

Janine Weber
Department of Mathematics and Computer Science, University of Cologne, Weyertal 86-90, 50931 Köln, Germany, e-mail: janine.weber@uni-koeln.de

S. C. Brenner et al. (eds.), *Domain Decomposition Methods in Science and Engineering XXVI*, Lecture Notes in Computational Science and Engineering 145,
https://doi.org/10.1007/978-3-030-95025-5_32

problems can be omitted in a preprocessing step. In [5, 7, 10], we have applied this approach to a certain adaptive FETI-DP (Finite Element Tearing and Interconnecting - Dual Primal) method [13] for elliptic model problems in two dimensions and investigated the effect of different training data sets and different sizes of input data for the neural network. In [8], we have additionally extended our approach to three-dimensional model problems for the corresponding adaptive FETI-DP method in three dimensions [11]. In [9], for the first time, we additionally applied our proposed machine learning framework to an overlapping domain decomposition method, i.e., the adaptive GDSW (Generalized Dryja–Smith–Widlund) method [3]. The purpose of [9] was to provide a general overview of methods combining machine learning with domain decomposition methods, and thus, we have solely presented some preliminary results for adaptive GDSW. Here, we extend the results shown in [9] by providing numerical experiments for additional test problems. Furthermore, we take a closer look at the choice of the ML threshold which is used for the classification between critical edges, for which the eigenvalue problem is necessary, and edges where the eigenvalue problem can be omitted. The specific choice of the threshold is now, for the first time, motivated by the corresponding receiver operating characteristic (ROC) curve and the precision-recall graph (please refer to [14, Sec. 5] for a definition of a precision-recall graph and a ROC curve).

We focus on a stationary diffusion problem in two dimensions and the adaptive GDSW method [3]. The diffusion coefficient function is defined on the basis of different subsections of a microsection of a dual-phase steel material.

2 Model problem and adaptive GDSW

As a model problem, we consider a stationary diffusion problem in two dimensions with various heterogeneous coefficient functions $\rho : \Omega := [0,1] \times [0,1] \rightarrow \mathbb{R}$, i.e., the weak formulation of

$$\begin{aligned} -\operatorname{div}(\rho \nabla u) &= 1 \text{ in } \Omega \\ u &= 0 \text{ on } \partial\Omega. \end{aligned} \tag{1}$$

In this paper, we apply the proposed machine learning-based strategy to an adaptive GDSW method. We decompose the domain Ω into $N \in \mathbb{N}$ nonoverlapping subdomains $\Omega_i, i = 1, \ldots, N$, such that $\overline{\Omega} = \bigcup_{i=1}^{N} \overline{\Omega}_i$. Next, we introduce overlapping subdomains Ω_i', $i = 1, ..., N$, which can be obtained from Ω_i, $i = 1, ..., N$ by recursively adding k layers of finite elements. In the numerical experiments presented in this paper, we always choose an overlap of width $\delta = h$; this corresponds to choosing $k = 1$. Due to space limitations, we do not describe the standard GDSW preconditioner in detail; see, e.g., [1] for a detailed description.

As discussed in [4], the condition number bound for the standard GDSW preconditioner generally depends on the contrast of the coefficient function for completely arbitrary coefficient distributions. As a remedy, additional coarse basis functions resulting from the eigenmodes of local generalized eigenvalue problems are employed to compute an adaptive coarse space which is robust and yields a coefficient contrast-independent condition number bound. In two dimensions, each of these eigenvalue problems is associated with a single edge and its two neighboring subdomains. Thus, the main idea for the adaptive GDSW (AGDSW) coarse space [3] is to build edge basis functions based on local generalized eigenvalue problems. In particular, the coarse basis functions are defined as discrete harmonic extensions of certain corresponding edge eigenmodes. The specific eigenmodes which are necessary to retain a robust convergence behavior are chosen depending on a user-defined tolerance $tol_{\mathcal{E}} \geq 0$, which has to be chosen in relation to the spectrum of the preconditioned system. For a detailed description of the specific local edge eigenvalue problems and the computation of the discrete harmonic extensions, we refer to [3]. In particular, in the AGDSW approach, all eigenmodes with eigenvalues lower or equal to $tol_{\mathcal{E}}$ are chosen to build the adaptive coarse space. Since the left-hand side of the edge eigenvalue problem is singular (cf. [3, Sec. 5]), for each edge, we always obtain one eigenvalue equal to zero. It corresponds to the null space of the Neumann matrix of (1), which consists of the constant functions. The corresponding coarse basis function is also part of the standard GDSW coarse space, and we denote it as the *first coarse basis function* in this paper. Let us note that the first coarse basis function is always necessary for the scalability of the approach, even for the case of a constant coefficient function. However, since it corresponds to the constant function on the edge, it is known a priori and can be computed without actually solving the eigenvalue problem. This is different to the ML-FETI-DP method since, for adaptive FETI-DP [13], the eigenvalue equal to zero does not occur in the eigenvalue problem as it is already captured by the primal vertex constraints; see also Section 3 for more details.

As for most adaptive domain decomposition methods, for AGDSW, it is generally not known a priori on which edges additional coarse basis functions are necessary in order to obtain robustness. In general, building the adaptive coarse space, i.e, the setup and the solution of the eigenvalue problems as well as the computation of the discrete harmonic extensions, can make up the larger part of the time to solution in a parallel implementation. Since the computation of the adaptive GDSW coarse space is – similarly to the adaptive FETI-DP methods – based on local eigenvalue problems associated with edges, we can apply the same machine learning strategy introduced in [5, 7] to predict the location of necessary eigenvalue problems.

3 Machine learning for adaptive GDSW

Our approach is to train a neural network to make an automatic decision whether it is necessary to solve a local eigenvalue problem for a specific edge to retain a robust AGDSW algorithm. We denote this approach, which is inspired by the ML-FETI-DP approach introduced in [5, 7], as ML-AGDSW. In particular, we use a dense feedforward neural network, or more precisely, a multilayer perceptron [2, 14] to make this decision. Since each eigenvalue problem for AGDSW is associated with a single edge and both neighboring subdomains, we use samples of the coefficient

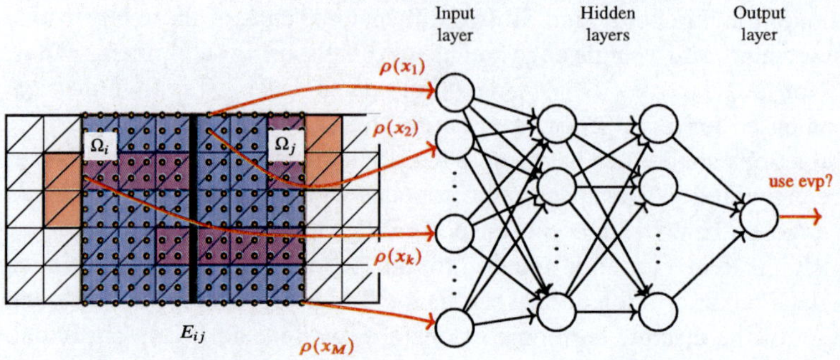

Fig. 1: Sampling of the coefficient function; white color corresponds to a low coefficient and red color to a high coefficient. In this representation, the samples are used as input data for a neural network with two hidden layers. Only sampling points from slabs around the edge are chosen. Taken from [10, Fig. 1].

function within the two adjacent subdomains as input data for the neural network; cf. Fig. 1. In particular, we apply a sampling approach which is independent of the finite element mesh, using a fixed number of sampling points for all mesh resolutions; this is reasonable as long as we can resolve all geometric features of the coefficient function. For more details on the computation of the sampling grid and its generalization to more general subdomain geometries than square subdomains; see [5].

As output for the neural network, we save the classification whether an adaptive basis function has to be computed for the specific edge or not. As already mentioned, in AGDSW, the first coarse basis function is always necessary but can be computed without actually solving the eigenvalue problem. Hence, an eigenvalue problem will only be marked as necessary in our approach if more than one coarse basis function corresponds to an eigenvalue lower than the chosen tolerance $tol_{\mathcal{E}}$. Therefore, for ML-AGDSW, all critical edges, where more than the single constant constraint is necessary, are classified as class 1. All other edges are classified as class 0. Let us note that this is different to the definition of class 1 for ML-FETI-DP introduced in [5, 7], where the eigenvalue 0 corresponding to the constant functions does not occur in the eigenvalue problem.

For the numerical results presented in this paper, we train the neural network on two regular subdomains sharing a straight edge and different types of coefficient functions. Using the same techniques as in [7, 9], we have generated a training and validation data set of 4 500 randomized coefficient distributions. In particular, the coefficient distributions are not completely random but we impose some sort of structure on the coefficients; see also [7] for a detailed discussion. For the first part of this training set, we randomly generate the coefficient for each pixel, consisting of two triangular finite elements, independently and only control the ratio of high and low coefficient values. Here, we use 30%, 20%, 10%, and 5% of high coefficient values. For the second part, we also control the distribution of the coefficients to a certain degree by randomly generating either horizontal or vertical stripes of a maximum

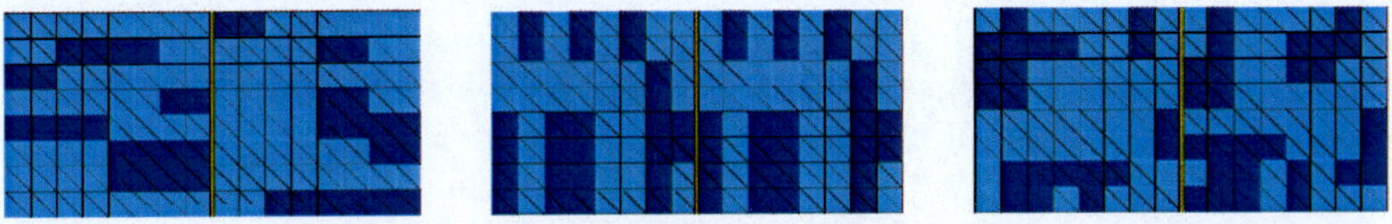

Fig. 2: Examples of three different randomly distributed coefficient functions obtained by using the same randomly generated coefficient for a horizontal (**left**) or vertical (**middle**) stripe of a maximum length of four finite element pixels, as well as by pairwise superimposing (**right**).

length of four or eight pixels, respectively; see Fig. 2. Additionally, we generate new coefficient distributions by superimposing pairs of coefficient distributions with horizontal and vertical stripes. We denote the resulting training data set by *R1'*. Let us note that the generation of the randomized coefficient distributions as training data for AGDSW is in complete analogy to our randomized training data for the ML-FETI-DP approach in [7]. However, we explicitly built a separate set of labels for the training and validation data for AGDSW since the classification of critical edges can be different for adaptive GDSW and adaptive FETI-DP.

To generate the output data for the neural network, we solve the eigenvalue problem as described in [4] for each edge in the aforementioned training and validation data. For all our training and validation data, we use a tolerance of $tol_{\mathcal{E}} = 0.01$ to generate the output for each edge.

Note that, for ML-FETI-DP, we additionally considered the extension to three classes, where we distinguished between zero, one, or more than one constraints. For the edges which require only one constraint, we used frugal constraints [6] instead of solving the eigenvalue problem; see [5] for more details. Consequently, the eigenvalue problem only had to be solved for edges with more than one constraint. However, this approach does not easily extend to AGDSW since we always obtain at least one a priori known coarse basis function on each edge; as mentioned earlier, we always obtain a constant eigenfunction corresponding to eigenvalue 0.

4 Numerical results

In this section, we apply our machine learning approach to AGDSW. We will present numerical results both for the training and validation data as well as for a specific test problem and compare the resulting condition number estimates and iteration counts with those obtained using both standard and adaptive GDSW; we use pcg with a relative residual reduction of $1e-8$. For the numerical experiments, we consider a discretization of the model problem Eq. (1) by piecewise linear finite elements.

First, we present results for the complete set of training data R1' using cross-validation and a fixed ratio of 20% validation data in Table 1. We observe that choosing the ML threshold as $\tau = 0.5$ to distinguish between class 0 and 1, i.e., assuming an equal distribution among the two classes, results in an accuracy which is comparable to the corresponding ML-FETI-DP approach; see [7, 9]. Besides the accuracy values for the training data in Table 1, we also provide the ROC curve and a precision-recall plot in Fig. 3. Both curves provide an evidence whether we obtain a reliable machine learning model [14, Sec. 5]; see also Section 1. As mentioned in Section 3, our aim is to identify all critical edges where an adaptive coarse basis function is necessary for robustness. For the remainder of this paper, we will refer to these critical edges as 'positive' or 'positive edges' and to edges where the eigenvalue problem is unnecessary as 'negative' or 'negative edges'. Thus, only false negative edges are critical for the convergence of ML-AGDSW, whereas false positive edges correspond to some unnecessary eigenvalue problems. Solving the eigenvalue problems on false positive edges increases the computational effort of our algorithm but does not negatively affect its convergence behavior; note that the additionally computed eigenfunctions will not enter the coarse space since the tolerance criterion will not be satisfied. When considering the precision-recall plot in Fig. 3 (right) we observe that using the ML threshold $\tau = 0.4$ compared to $\tau = 0.45$ results in a higher recall for the validation data while preserving nearly the same precision value. This is caused by a decrease in the number of false negative edges compared to $\tau = 0.45$. Moreover, the precision for both training and validation data strongly decreases when using ML thresholds smaller than 0.4. Since our predominant aim is to avoid false negative edges while still preserving a sufficient accuracy of the classification, using the ML threshold $\tau = 0.4$ seems to work best for our purpose. Besides, for $\tau = 0.4$ the ROC curve for the validation data in Fig. 3 (left) is close to the respective curve for the training data which suggests that we obtain a model with good generalization properties. We will thus use $\tau = 0.4$ for the classification of our test problems and also provide comparative results for $\tau = 0.5$.

As a test problem for our trained neural network, we use 10 different randomly chosen subsections of a microsection of a dual-phase steel as shown in Fig. 4 (right). In all presented computations, we consider $\rho = 1e6$ in the black part of the microsection and $\rho = 1$ elsewhere. We use a regular decomposition of the domain $\Omega := [0, 1] \times [0, 1]$ into 8×8 square subdomains with an overlap of $\delta = h$, a subdomain size of $H/h = 56$, and a tolerance of $tol_{\mathcal{E}} = 0.01$. For the test data, we only solve the local eigenvalue problem on edges which are classified as class 1 by the neural network. For all edges classified as class 0, we do not solve the eigenvalue problem and only enforce the constant constraint on the respective edge. When considering the results for one specific microsection in Table 2 as well as the average values for all 10 different subsections in Table 3, we observe that, in both cases, we are able to obtain no false negative edges for the classification using the ML threshold $\tau = 0.4$. Analogously to the training and validation data in Table 1, using the lower threshold $\tau = 0.4$ compared to $\tau = 0.5$ decreases the false negative rate of the predictions and thus increases the robustness of our algorithm. In particular, in Table 3, we obtain zero false negative edges for all 10 different microsection subsections when using $\tau = 0.4$. On the other hand, on average, we only solve 5.2 unnecessary eigenvalue problems. This implies that our framework is robust for different heterogeneous coefficient distributions and can successfully be applied to AGDSW.

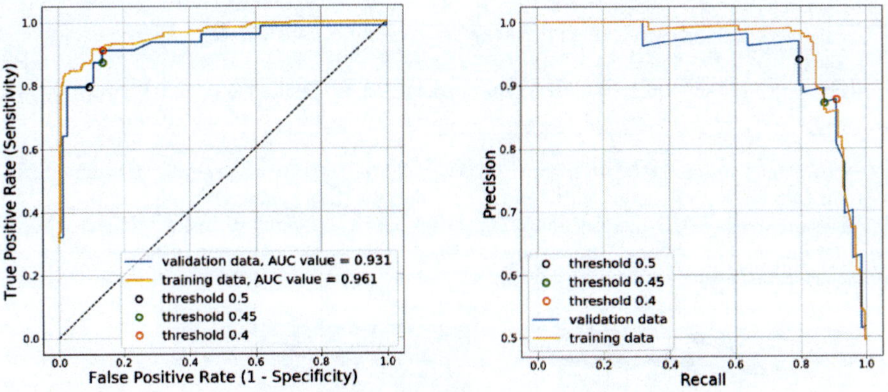

Fig. 3: ROC curve (left) and precision-recall plot (right) for the ML-AGDSW method. We define precision as *true positives* divided by (*true positives* + *false positives*), and recall as *true positives* divided by (*true positives* + *false negatives*). The thresholds used in Section 4 are indicated as circles.

training configuration	threshold	fp	fn	acc
R1', full sampling	0.4	11.5%	2.7%	85.8%
	0.5	6.7%	7.1%	86.2%

Table 1: Results on the complete training data set for the GDSW method and **stationary diffusion**; the numbers are averages over all training configurations. See Table 2 for the column labeling.

Model Problem	Algorithm	τ	cond	it	evp	fp	fn	acc
Microsection Problem	standard GDSW	-	3.66e06	500	0	-	-	-
	adaptive GDSW	-	162.60	95	112	-	-	-
	ML-AGDSW	0.5	9.64e4	98	25	2	2	0.95
	ML-AGDSW	0.4	163.21	95	29	6	0	0.95

Table 2: Comparison of standard GDSW, adaptive GDSW, and ML-AGDSW for a **regular domain decomposition** with 8×8 subdomains and $H/h = 56$ for the **two-class model**, with $tol_{\mathcal{E}} = 0.01$. We show the ML threshold (τ), the condition number (**cond**), the number of CG iterations (**it**), the number of solved eigenvalue problems (**evp**), the number of false positives (**fp**), the number of false negatives (**fn**), and the accuracy in the classification (**acc**). We define the accuracy (acc) as the number of true positives and true negatives divided by the total number of edges.

References

1. C.R. Dohrmann, A. Klawonn and O.B. Widlund. Domain decomposition for less regular subdomains: overlapping Schwarz in two dimensions. *SIAM J. Numer. Anal.* **46**(4), 2153–2168 (2008).
2. I. Goodfellow, Y. Bengio and A. Courville. *Deep learning, volume 1*. MIT press Cambridge (2016).
3. A. Heinlein, A. Klawonn, J. Knepper and O. Rheinbach. An adaptive GDSW coarse space for two-level overlapping Schwarz methods in two dimensions. In: *Domain Decomp. Meth. Sci. Eng. XXIV*, volume 125 of LNCSE, pp. 373–382, Springer, Cham (2018).

Alg.	τ	cond	it	evp	fp	fn	acc
standard	-	4.7e06 (5.11e06)	511.2 (518)	0	-	-	-
adaptive	-	178.6 (181.4)	87.2 (98)	112.0 (112)	-	-	-
ML-AGDSW	0.5	7.8e04 (9.2e04)	92.2 (102)	26.4 (29)	1.6 (2)	1.8 (3)	0.96 (0.95)
	0.4	178.7 (181.4)	87.3 (98)	33.4 (36)	**5.2 (8)**	**0 (0)**	0.95 (0.94)

Table 3: Comparison of standard GDSW, adaptive GDSW, and ML-AGDSW for a **regular domain decomposition** with 8 × 8 subdomains and $H/h = 56$ for the **two-class model**, with $tol_{\mathcal{E}} = 0.01$, for 10 different subsections of the microsection in Fig. 4 (right) for a **stationary diffusion** problem. See Table 2 for the column labeling. The numbers in brackets show the maximum or minimum values for the respective average values.

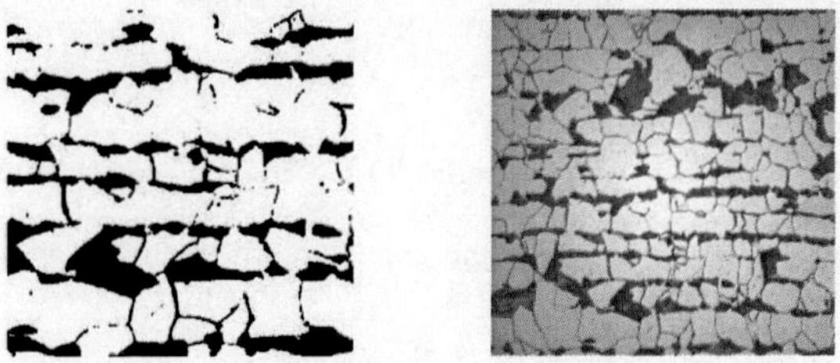

Fig. 4: Left: Subsection of a microsection of a dual-phase steel obtained from the image on the right. We consider $\rho = 1e6$ in the black part and $\rho = 1$ elsewhere. **Right:** Complete microsection of a dual-phase steel. Right image: Courtesy of Jörg Schröder, University of Duisburg-Essen, Germany, orginating from a cooperation with ThyssenKruppSteel.

4. A. Heinlein, A. Klawonn, J. Knepper and O. Rheinbach. Adaptive GDSW Coarse Spaces for Overlapping Schwarz Methods in Three Dimensions. *SIAM J. Sci. Comput.* **41**(5), A3045–A3072 (2019).
5. A. Heinlein, A. Klawonn, M. Lanser and J. Weber. Machine Learning in Adaptive Domain Decomposition Methods – Predicting the Geometric Location of Constraints. *SIAM J. Sci. Comput.* **41**(6), A3887–A3912 (2019).
6. A. Heinlein, A. Klawonn, M. Lanser and J. Weber. A Frugal FETI-DP and BDDC Coarse Space for Heterogeneous Problems. *Electr. Trans. Numer. Anal.* **53**, 562–591 (2020).
7. A. Heinlein, A. Klawonn, M. Lanser and J. Weber. Machine Learning in Adaptive FETI-DP – A Comparison of Smart and Random Training Data. In: *Domain Decomposition Methods in Science and Engineering XXV*, volume 138 of LNCSE, pp. 218–226, Springer, Cham (2020).
8. A. Heinlein, A. Klawonn, M. Lanser and J. Weber. Combining machine learning and adaptive coarse spaces – a hybrid approach for robust FETI-DP methods in three dimensions. *SIAM Journal on Scientific Computing* **43**(5), S816–S838 (2021).
9. A. Heinlein, A. Klawonn, M. Lanser and J. Weber. Combining machine learning and domain decomposition methods for the solution of partial differential equations – a review. *GAMMMitt.* **44**(1), e202100001–28 (2021).
10. A. Heinlein, A. Klawonn, M. Lanser and J. Weber. Machine Learning in Adaptive FETI-DP – Reducing the Effort in Sampling. In: *Numerical Mathematics and Advanced Applications ENUMATH 2019*, volume 139 of LNCSE, pp. 218–226, Springer, Cham (2021).
11. A. Klawonn, M. Kühn and O. Rheinbach. Adaptive coarse spaces for FETI-DP in three dimensions. *SIAM J. Sci. Comput.* **38**(5), A2880–A2911 (2016).
12. A. Klawonn and O.B. Widlund. Dual-primal FETI methods for linear elasticity. *Comm. Pure Appl. Math.* **59**(11), 1523–1572 (2006).

13. J. Mandel and B. Sousedík. Adaptive selection of face coarse degrees of freedom in the BDDC and the FETI-DP iterative substructuring methods. *Comput. Methods Appl. Mech. Engrg.* **196**(8), 1389–1399 (2007).
14. A. Müller and S. Guido. *Introduction to Machine Learning with Python: A Guide for Data Scientists*. O'Reilly Media (2016).

Optimized Coupling Conditions for Discrete Fracture Matrix Models

Martin J. Gander, Julian Hennicker, and Roland Masson

1 Introduction

In [7, 8], we derived and studied an asymptotic model for Darcy flow in fractured porous media, when the fracture aperture δ approaches zero. We showed that our new, general models coincide in special cases with common models from the literature, such as, for instance [1, 2, 10, 11]. Our general modeling approach leads to coupling conditions, which are suitable for small fracture apertures and for a resolution of low frequencies k. It also permits several adaptations, one of which we explore here, namely new coupling conditions with extended range of validity, obtained by replacing the parameters in the asymptotic coupling conditions by new parameters, which we then optimize w.r.t. the error for a given range of frequency components $k \in [k_{\min}, k_{\max}]$ present in the numerical solution to be computed. Our results are based on the explicit formula from [8] for the error for the solution of the asymptotic model in Fourier space, which we adapt to generalized parameters. In order to obtain explicit formulas for the optimized parameters, we make some simplifying assumptions, and then solve the resulting optimization problem analytically using asymptotic techniques for small fracture apertures. Our approach could also be adapted to more general situations, and we could have chosen to use expansions for $\delta \to \delta_0$ or $k \to k_\infty$, with δ_0 or k_∞ a fixed constant, for example. In this sense, we want to outline conceptually a technique to improve the model accuracy for the model in [7], which can be adapted by the reader to the situation at hand. An ad hoc generalisation to fracture networks would be to apply the matrix-fracture coupling conditions, as derived in our manuscript, to each of the fracture segments and to

Martin J. Gander
Université de Genève, e-mail: martin.gander@unige.ch

Julian Hennicker
Université du Luxembourg, e-mail: julian.hennicker@uni.lu

Roland Masson
Université Côte d'Azur, CNRS, Inria, LJAD, e-mail: roland.masson@univ-cotedazur.fr

S. C. Brenner et al. (eds.), *Domain Decomposition Methods in Science and Engineering XXVI*, Lecture Notes in Computational Science and Engineering 145,
https://doi.org/10.1007/978-3-030-95025-5_33

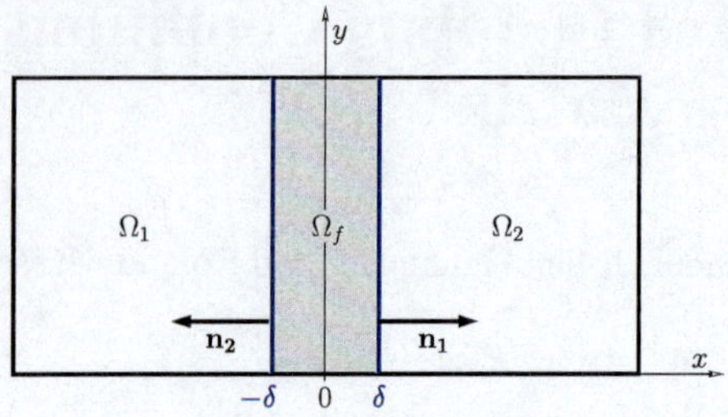

Fig. 1: Geometry of the domain under consideration.

impose pressure continuity and flux conservation at the fracture intersections (see e.g. [9], or [3] for an alternative formulation in the case of highly contrasted fracture permeabilities). A rigorous treatment of cross points in domain decomposition is a topic of substantial interest for current research (cf. [6], and references therein), and its application to fracture intersections is a project for future work.

2 Model problem

In the domains illustrated in Fig. 1, we consider the system of PDEs

$$-\mathrm{div}\mathbf{q}_j + \frac{\mathbf{b}_j}{2} \cdot \nabla u_j + \left(\eta_j - \mathrm{div}\frac{\mathbf{b}_j}{2}\right) u_j = h_j \quad \text{in} \quad \Omega_j,\ j = 1, 2, f, \tag{1}$$

$$\mathbf{q}_j = \left(\mathbf{A}_j \nabla - \frac{\mathbf{b}_j}{2}\right) u_j \quad \text{in} \quad \Omega_j,\ j = 1, 2, f, \tag{2}$$

connected at $x = \pm\delta$ with the coupling conditions

$$u_j = u_f \qquad \text{on} \quad \partial\Omega_j \cap \partial\Omega_f,\ j = 1, 2, \tag{3}$$

$$\mathbf{q}_j \cdot \mathbf{n}_j = \mathbf{q}_f \cdot \mathbf{n}_j \qquad \text{on} \quad \partial\Omega_j \cap \partial\Omega_f,\ j = 1, 2. \tag{4}$$

The model coefficients are $\eta_j : \Omega_j \to \mathbb{R}_{\geq 0}$, $\mathbf{b}_j : \Omega_j \to \mathbb{R}^2$, such that $\eta_j - \mathrm{div}\mathbf{b}_j \geq 0$, and the coercive matrices $\mathbf{A}_j : \Omega_j \to \mathbb{R}^{2\times 2}$. The model unknowns are $\mathbf{q}_j$ and u_j. For this problem, we can eliminate the fracture unknowns in Fourier space, as described in [8]: applying a Fourier transform in the direction tangential to the fracture, the fracture Fourier coefficients have to satisfy specific ODEs which can be solved using two of the four coupling conditions at the interfaces. Then, the fracture solution is substituted into the remaining two coupling conditions. The resulting equations at $x = \pm\delta$ for the coupling between the matrix domains, when the fracture has been eliminated, are

$$\hat{\mathbf{q}}_2 \cdot \mathbf{n}_2 + \hat{\mathbf{q}}_1 \cdot \mathbf{n}_1 = -a_{11}\sqrt{\frac{a_{22}}{a_{11}}k^2} \tanh\left(\delta\sqrt{\frac{a_{22}}{a_{11}}k^2}\right)(\hat{u}_1 + \hat{u}_2), \tag{5}$$

$$\hat{\mathbf{q}}_2 \cdot \mathbf{n}_2 - \hat{\mathbf{q}}_1 \cdot \mathbf{n}_1 = \frac{a_{11}\sqrt{\frac{a_{22}}{a_{11}}k^2}}{\tanh\left(\delta\sqrt{\frac{a_{22}}{a_{11}}k^2}\right)}(\hat{u}_2 - \hat{u}_1), \tag{6}$$

under the simplifying assumption that $h_f \equiv 0$, $\mathbf{b}_f = 0$, $\eta_f = 0$, and $\mathbf{A}_f$ being diagonal.

Asymptotic coupling for small δ. We recall first the asymptotic coupling conditions for small δ presented in [8]. For $\delta \to 0$, we can expand

$$\tanh\left(\delta\sqrt{\frac{a_{22}}{a_{11}}k^2}\right) = \delta\sqrt{\frac{a_{22}}{a_{11}}k^2} - \delta^3\frac{1}{3}\sqrt{\frac{a_{22}}{a_{11}}k^2}^{\,3} + O(\delta^5). \tag{7}$$

Truncation after the next-to-leading-order term yields at $x = \pm\delta$ the reduced order coupling conditions

$$\hat{\mathbf{q}}_1^{\text{red}} \cdot \mathbf{n}_1 + \hat{\mathbf{q}}_2^{\text{red}} \cdot \mathbf{n}_2 = -\delta a_{22}k^2(\hat{u}_1^{\text{red}} + \hat{u}_2^{\text{red}}), \tag{8}$$

$$\hat{\mathbf{q}}_1^{\text{red}} \cdot \mathbf{n}_1 - \hat{\mathbf{q}}_2^{\text{red}} \cdot \mathbf{n}_1 = \frac{a_{11}}{\delta}(\hat{u}_2^{\text{red}} - \hat{u}_1^{\text{red}}). \tag{9}$$

3 Generalized coupling conditions and their optimization

The coupling conditions (8) and (9) are by construction most suitable for small values of δ, and also for small values of k, due to a symmetry between δ and k. In practical numerical computations, the solution sought has however a certain range of frequencies, $k \in [k_{\min}, k_{\max}]$, not only low ones. To treat such a wider range of frequencies, we use now a common technique from Optimized Schwarz methods in domain decomposition [4, 5], which consists in keeping the structure of the reduced order coupling conditions, and introducing new parameters as d.o.f. for a subsequent optimization. In our case, the coupling conditions are of Robin type, and we replace the occurring parameters in (8) and (9), δa_{22} and $\frac{a_{11}}{\delta}$, by newly introduced parameters p and q, which gives the optimizable reduced coupling conditions

$$\hat{\mathbf{q}}_1^{\text{red}} \cdot \mathbf{n}_1 + \hat{\mathbf{q}}_2^{\text{red}} \cdot \mathbf{n}_2 = -pk^2(\hat{u}_1^{\text{red}} + \hat{u}_2^{\text{red}}),$$

$$\hat{\mathbf{q}}_1^{\text{red}} \cdot \mathbf{n}_1 - \hat{\mathbf{q}}_2^{\text{red}} \cdot \mathbf{n}_1 = q(\hat{u}_2^{\text{red}} - \hat{u}_1^{\text{red}}).$$

In [8], the error at the interfaces of the δ-asymptotic reduced order solution in Fourier space was derived, see the result after eq. (7.5) therein. The errors for our generalized reduced order model can analogously be obtained, and we get for $j = 1, 2$

$$\begin{aligned} \hat{e}_j &:= \hat{u}_j - \hat{u}_j^{\text{red}} \\ &= \rho(k,p)(\hat{u}_2 + \hat{u}_1) + (-1)^{j+1}\tau(k,q)(\hat{u}_2 - \hat{u}_1), \end{aligned} \tag{10}$$

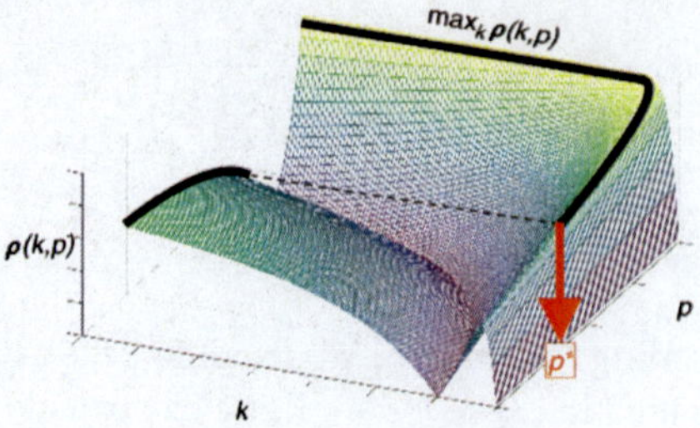

Fig. 2: Illustration of how to find a solution to (14) and (15).

where

$$\rho(k,p) = -\frac{1}{2}\frac{\sqrt{a_{11}a_{22}k^2}\tanh(\delta\sqrt{\frac{a_{22}}{a_{11}}k^2}) - pk^2}{\sqrt{k^2}+pk^2}, \tag{11}$$

$$\tau(k,q) = \frac{1}{2}\frac{\frac{\sqrt{a_{11}a_{22}k^2}}{\tanh(\delta\sqrt{\frac{a_{22}}{a_{11}}k^2})} - q}{\sqrt{k^2}+q}. \tag{12}$$

In order to minimize the error for a range of frequencies in a simulation, we need to solve

$$\min_{p,q} \max_{k\in(k_{\min},k_{\max})} |\hat{e}_j(k,p,q)|, \tag{13}$$

for small $\delta \ll k_{\max}^{-1}$. Since $(\hat{u}_2 + \hat{u}_1)$ and $(\hat{u}_2 - \hat{u}_1)$ are linearly independent, our objective functions to be minimized are $|\rho|$ and $|\tau|$. The following lemma will be applied without proof.

Lemma 1 *The solution* (k^*, p^*, q^*) *to*

$$\partial_k \rho(k^*, p^*) = 0 \tag{14}$$

$$|\rho(k_{\max}, p^*)| = |\rho(k^*, p^*)| \tag{15}$$

$$|\tau(k_{\max}, q^*)| = |\tau(k_{\min}, q^*)| \tag{16}$$

solves the relevant min-max problem (13).

We will first solve for the equation (16), and then for the independent problem (14) and (15), cf. Fig. 2. Since we are interested in the case of fracture apertures, which are not resolved by the mesh, i.e. $\delta \ll k_{\max}^{-1}$, we will solve the problem asymptotically in δ, for the leading and next-to-leading order terms of the expansions.

First, using the asymptotic expansion (7) in (12) yields

$$\tau(k,q) = \frac{1}{2}\frac{\frac{a_{11}}{\delta} + \frac{a_{22}\delta k^2}{3} - q}{|k|+q} + O(\delta^3). \tag{17}$$

Inserting this into (16) implies

$$\frac{\frac{a_{11}}{\delta} + \frac{a_{22}\delta k_{\min}^2}{3} - q^*}{|k_{\min}| + q^*} + \frac{\frac{a_{11}}{\delta} + \frac{a_{22}\delta k_{\max}^2}{3} - q^*}{|k_{\max}| + q^*} = O(\delta^3). \quad (18)$$

Hence,

$$\begin{aligned} q^* &= \frac{a_{11}}{2\delta} + \frac{a_{22}\delta k_{\max}^2 + k_{\min}^2}{12} - \frac{k_{\max} + k_{\min}}{4} \\ &+ \Big[(\frac{a_{11}}{2\delta})^2 + \frac{a_{11}(k_{\min} + k_{\max})}{4\delta} + \Big(\frac{k_{\min} + k_{\max}}{4}\Big)^2 + \frac{a_{11}a_{22}(k_{\min}^2 + k_{\max}^2)}{12} \\ &+ \frac{a_{22}\delta}{12}\Big(\frac{1}{2}(k_{\min} + k_{\max})^3 - (k_{\min}^3 + k_{\max}^3)\Big) + \Big(\frac{a_{22}\delta}{12}(k_{\min} + k_{\max})\Big)^2\Big]^{\frac{1}{2}}. \end{aligned} \quad (19)$$

We can now derive an asymptotic formula for the optimized error in the jump of u across the fracture by substituting the optimized parameter q^* into τ, at $k = k_{max}$ or equivalently at $k = k_{\min}$, and obtain

$$\begin{aligned} &\min_q \max_{k\in(k_{\min},k_{\max})} |\tau(k,q)| \\ &= |\tau(k_{\max}, q^*)| = |\tau(k_{\min}, q^*)| = \frac{a_{22}(k_{\max}^2 - k_{\min}^2)}{12a_{11}}\delta^2 + O(\delta^3). \end{aligned} \quad (20)$$

This result can further be compared to the corresponding error of the original model,

$$\max_{k\in(k_{\min},k_{\max})} |\tau(k, \frac{a_{11}}{d})| = \frac{a_{22}k_{\max}^2}{6a_{11}}\delta^2 + O(d^3). \quad (21)$$

We observe that the asymptotic constant in (20) is approximately half the value of the asymptotic constant in (21). For solving for (14) and (15), we can proceed analogously: first, we use the expansion (7) in (11), and obtain

$$\rho(k,p) = -\frac{1}{2}\frac{a_{22}\delta - \frac{a_{22}^2\delta^3k^2}{3a_{11}} - p}{(p + \frac{1}{|k|})} + O(\delta^4). \quad (22)$$

Substituting (22) into (14) and (15) implies

$$-\frac{-3a_{11}\,(a_{22}d - p^*) + 2a_{22}^2d^3k^{*2}\,(k^*p^* + 1) + a_{22}^2d^3k^{*2}}{3a_{11}\,(k^*p^* + 1)^2} = O(\delta^4), \quad (23)$$

$$\frac{a_{22}\delta - \frac{a_{22}^2\delta^3k^{*2}}{3a_{11}} - p^*}{(p^* + \frac{1}{|k^*|})} + \frac{a_{22}\delta - \frac{a_{22}^2\delta^3k_{\max}^2}{3a_{11}} - p^*}{(p^* + \frac{1}{|k_{\max}|})} = O(\delta^4). \quad (24)$$

Solving (23) and (24), we obtain the optimized parameters

$$k^* = \frac{k_{\max}}{2} + O(\delta^4) \quad \text{and} \quad p^* = a_{22}\delta - \frac{a_{22}^2 \delta^3 k_{\max}^2}{4a_{11}} + O(\delta^4). \tag{25}$$

Finally, we obtain an asymptotic formula for the optimized error in the averaged traces of u at the interface, by substituting the optimized parameters into ρ,

$$\min_p \max_{k \in (k_{\min}, k_{\max})} |\rho(k,p)| = |\rho(k_{\max}, p^*)| = |\rho(k^*, q^*)| = \frac{a_{22}^2 k_{\max}^3}{24 a_{11}} \delta^3 + O(\delta^4). \tag{26}$$

We can again compare this to the error of the original model,

$$\max_{k \in (k_{\min}, k_{\max})} |\rho(k, a_{22}\delta)| = \frac{a_{22}^2 k_{\max}^3}{6 a_{11}} \delta^3 + O(\delta^4), \tag{27}$$

and observe that the asymptotic constant in (26) is a fourth of the value of the asymptotic constant in (27).

4 Numerical results

We will now illustrate our results numerically and compare the theoretical error of the optimized problem with parameters p^* and q^*, for which we have the expressions (25) and (19), with the theoretical error of the asymptotic model (8), (9) from [7, 8], which employs the parameters

$$q^{\text{red}} = \frac{a_{11}}{\delta} \quad \text{and} \quad p^{\text{red}} = a_{22}\delta.$$

These parameters have been calculated analytically, for small fracture apertures. On the other hand, we can solve the problem (13) numerically for any given data, and thus obtain general optimized parameters, which will serve as reference parameters, and which we will denote by p^{opt} and q^{opt}. We will also show plots of the corresponding errors

$$\max_{k \in (k_{\min}, k_{\max})} |\tau(k,q)| \quad \text{and} \quad \max_{k \in (k_{\min}, k_{\max})} |\rho(k,p)|,$$

for $q \in \{q^{\text{opt}}, q^*, q^{\text{red}}\}$ and $p \in \{p^{\text{opt}}, p^*, p^{\text{red}}\}$. When interpreting the results, the reader is referred to (10). Please also note that the jump $\hat{u}_2 - \hat{u}_1$ is of order δ, as shown in [8]. We present three different cases: homogeneous isotropic fractures, fracture barriers, and fracture conduits. The fracture apertures are from 10^{-2} to 10^{-5} and the frequency range is set to $[k_{\min} = 0, k_{\max} = \pi]$, on an infinite domain.

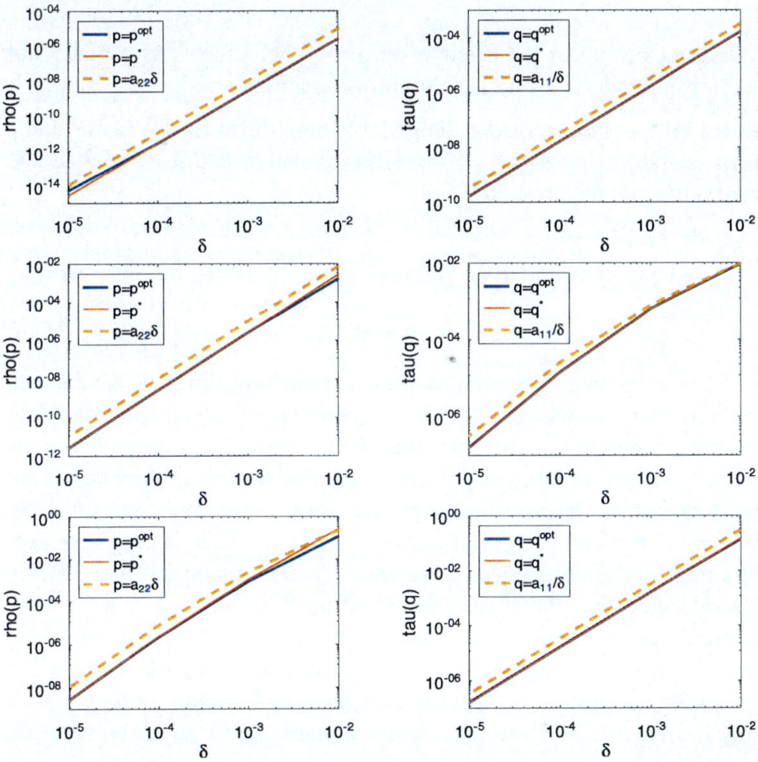

Fig. 3: Isotropic fracture, fracture barrier and fracture conduit (from top to bottom). Exact errors for the asymptotic, asymptotic optimized, and numerically optimized parameters.

Homogeneous isotropic fracture. This is a fracture with the same properties as the bulk domain, i.e. $a_{11} = a_{22} = 1$. The plots in Fig. 3 show the theoretical errors of the reduced order solutions, and their convergence to the reference solution, with $\delta \to 0$. We observe that the error of the asymptotic optimized model is in very good agreement with the error of the numerically optimized model for all δ. The slight difference in ρ for $\delta = 10^{-5}$ is due to round-off error, as we have reached machine precision. The error plots also reveal an advantage of the optimized models over the asymptotic model from [7]. The gain in accuracy can be analytically quantified by the ratios of asymptotic constants in (20) and (21) for τ, and in (26) and (27) for ρ.

Fracture barrier. Let us consider anisotropic diffusion coefficients in the fracture: a very low normal diffusion $a_{11} = 10^{-3}$ and a homogeneous tangential diffusion $a_{22} = 1$. Similar to the isotropic test case, we observe from the plots in Fig. 3 an advantage of the optimized models over the asymptotic model from [7], which can be quantified by looking at the asymptotic coefficients in (20) and (21) for τ, and in (26) and (27) for ρ. We observe that the error of the asymptotic optimized model is in very good agreement with the error of the numerically optimized model for all

δ, except for $\delta = 10^{-2}$, where there is a small difference. This is due to the strong heterogeneity and anisotropy of the fracture diffusion coefficients, which have not been accounted for in the derivation of the optimized parameters.

Fracture conduit. Let us now consider a high tangential diffusion $a_{22} = 10^3$ and a homogeneous normal diffusion $a_{11} = 1$. The results shown in Fig. 3 are comparable to the results from the previous test case.

5 Conclusion

We presented a new way to generalize the coupling conditions from [7, 8] for discrete fracture matrix models to a wider range of frequencies arising in the numerical solution. To do so, we conserved the structure of the original coupling conditions obtained for small fracture apertures, but optimized the occurring parameters for a given range of numerical frequencies, with the error as the objective function. This led to the new optimized parameters given in (19) and (25), which minimize the error committed by the reduced order model. We also quantified the error by comparing the asymptotic coefficients in the equations (20) and (21) for the error in the pressure jump across the fracture, and in (26) and (27) for the error in the averaged pressure across the fracture. This comparison shows that the error using the optimized coupling conditions is two to four times smaller than for the original ones. We finally illustrated the theoretical results numerically for several test cases.

References

1. P. Angot, F. Boyer and F. Hubert. Asymptotic and numerical modelling of flows in fractured porous media. *ESAIM: Mathematical Modelling and Numerical Analysis* **43**(2), 239–275 (2009).
2. E. Flauraud, F. Nataf, I. Faille and R. Masson. Domain decomposition for an asymptotic geological fault modeling. *Comptes Rendus Mécanique* **331**(12), 849–855, Dec (2003).
3. L. Formaggia, A. Fumagalli, A. Scotti and P. Ruffo. A reduced model for Darcy's problem in networks of fractures. *ESAIM: Mathematical Modelling and Numerical Analysis* **48**(4), 1089–1116 (2014).
4. M.J. Gander. Optimized Schwarz methods. *SIAM J. Numer. Anal.* **44**(2), 699–731 (2006).
5. M.J. Gander. Schwarz methods over the course of time. *ETNA* **31**, 228–255 (2008).
6. M.J. Gander and L. Halpern. A simple finite difference discretization for Ventcell transmission conditions at cross points. In: *Domain Decomposition Methods in Science and Engineering XXVI*, LNCSE, Springer-Verlag (2022).
7. M.J. Gander, J. Hennicker and R. Masson. Asymptotic analysis for the coupling between subdomains in discrete fracture matrix models. In: *Domain Decomposition Methods in Science and Engineering XXV*, LNCSE. Springer-Verlag (2019).
8. M.J. Gander, J. Hennicker and R. Masson. Modeling and analysis of the coupling in discrete fracture matrix models. *SIAM J. Numer. Anal.* **59**(1), 195–218 (2021).
9. J. Hennicker. *Hybrid dimensional modeling of multi-phase Darcy flows in fractured porous media*. PhD thesis, Université Côte d'Azur (2017). https://tel.archives-ouvertes.fr/tel-01614307

10. V. Martin, J. Jaffré and J.E. Roberts. Modeling fractures and barriers as interfaces for flow in porous media. *SIAM Journal on Scientific Computing* **26**(5), 1667–1691 (2005).
11. E. Sánchez-Palencia. Problèmes de perturbations liés aux phénomènes de conduction à travers des couches minces de grande résistivité. *J. Math. Pures et Appl.* **53**(9), 251–269 (1974).

Efficient Monolithic Solvers for Fluid-Structure Interaction Applied to Flapping Membranes

D. Jodlbauer, U. Langer, and T. Wick

1 Introduction

This work is devoted to the efficient solution of variational-monolithic fluid-structure interaction (FSI) initial-boundary value problems. Solvers for such monolithic systems were developed, e.g., in [2, 3, 5, 7, 9, 11–13, 15]. Due to the interface coupling conditions, the development of robust scalable parallel solvers remains a challenging task, and to the best of our knowledge only semi-cost optimal parallel approaches could be derived [4, 9]. The main purpose of this work consists in further numerical studies of the solver, developed in [9], for a benchmark problem that is motivated by hemodynamic applications. Specifically, we consider channel flow with elastic membranes and elastic solid walls. This situation is challenging because of the thin elastic flaps and was the motivation for fluid-structure interaction models such as immersed methods [6, 14]. However, we use arbitrary Lagrangian–Eulerian coordinates (see e.g., [8]), because of the high accuracy of the coupling conditions as the interface is tracked. For a careful evaluation of the performance of our physics-based block FSI preconditioner from [9], we use sparse direct solvers for the mesh, solid, and fluid subproblems. These sparse direct solvers should be replaced by iterative solvers in the case of large-scale problems with a high number of degrees of freedom. Therein, the flow part with well-known saddle-point structure becomes very critical, which was not yet the case for our solver applied to the FSI benchmarks in [9, 11]. The performance of our block FSI preconditioner and overall linear GMRES solver

D. Jodlbauer
Doctoral Program "Computational Mathematics" Johannes Kepler University Linz, Altenberger Str. 69, A-4040 Linz, Austria, e-mail: daniel.jodlbauer@dk-compmath.jku.at

U. Langer
Johann Radon Institute for Computational and Applied Mathematics, Austrian Academy of Sciences, Altenbergerstr. 69, A-4040 Linz, Austria, e-mail: ulrich.langer@ricam.oeaw.ac.at

T. Wick
Institut für Angewandte Mathematik, Leibniz Universität Hannover, Welfengarten 1, 30167 Hannover, Germany, e-mail: thomas.wick@ifam.uni-hannover.de

S. C. Brenner et al. (eds.), *Domain Decomposition Methods in Science and Engineering XXVI*, Lecture Notes in Computational Science and Engineering 145,
https://doi.org/10.1007/978-3-030-95025-5_34

is evaluated in terms of iteration numbers as well as memory storage. Moreover, iteration numbers of the nonlinear Newton solver are monitored. Finally, a computational convergence analysis for flap tip displacements, drag and lift for different spatial mesh levels is conducted.

2 FSI model

Let the function spaces $\widehat{X}$ (including extensions of non-homogeneous Dirichlet conditions) and $\hat{X}^0$ (homogeneous Dirichlet conditions) be given. Our variational-monolithic arbitrary Lagrangian–Eulerian FSI model from [17] (see also [9]) reads in the space-time formulation as follows: Find a global vector-valued velocity $\hat{v}$, global vector-valued displacements $\hat{u} = \hat{u}_s + \hat{u}_f$, and a scalar-valued fluid pressure $\hat{p}_f$, i.e., $\hat{U} := (\hat{v}, \hat{u}, \hat{p}_f) \in \widehat{X}$ such that the fluid/solid momentum equation

$$\int_I \Big((\hat{J}\hat{\rho}_f \partial_t \hat{v}, \hat{\psi}^v)_{\hat{\Omega}_f} + (\hat{\rho}_f \hat{J}(\hat{F}^{-1}(\hat{v} - \hat{w}) \cdot \hat{\nabla})\hat{v}), \hat{\psi}^v)_{\hat{\Omega}_f} + (\hat{J}\hat{\sigma}_f \hat{F}^{-T}, \hat{\nabla}\hat{\psi}^v)_{\hat{\Omega}_f}$$
$$+ \langle \hat{\rho}_f \nu_f \hat{J}(\hat{F}^{-T}\hat{\nabla}\hat{v}^T \hat{n}_f)\hat{F}^{-T}, \hat{\psi}^v \rangle_{\hat{\Gamma}_{\text{out}}} + (\hat{\rho}_s \partial_t \hat{v}, \hat{\psi}^v)_{\hat{\Omega}_s} + (\hat{F}\hat{\Sigma}, \hat{\nabla}\hat{\psi}^v)_{\hat{\Omega}_s} \Big)\, \mathrm{d}t$$
$$+ (\hat{J}(\hat{v}(0) - \hat{v}_0), \hat{\psi}^v(0))_{\hat{\Omega}_f} + (\hat{v}(0) - \hat{v}_0, \hat{\psi}^v(0))_{\hat{\Omega}_s} = 0,$$

the 2nd solid eq. $$\int_I \Big(\hat{\rho}_s (\partial_t \hat{u}_s - \hat{v}|_{\hat{\Omega}_s}, \hat{\psi}^u_s)_{\hat{\Omega}_s} \Big)\, \mathrm{d}t + (\hat{u}_s(0) - \hat{u}_{s,0}, \hat{\psi}^u_s(0)) = 0,$$

the mass conservation $$\int_I \Big((\widehat{\mathrm{div}}\, (\hat{J}\hat{F}^{-1}\hat{v}), \hat{\psi}^p_f)_{\hat{\Omega}_f} \Big)\, \mathrm{d}t = 0,$$

and the mesh motion $$\int_I (\hat{\sigma}_{\text{mesh}}, \hat{\nabla}\hat{\psi}^u_f)_{\hat{\Omega}_f}\, \mathrm{d}t = 0,$$

hold for all $\hat{\Psi} = (\hat{\psi}^v, \hat{\psi}^u, \hat{\psi}^p_f) \in \hat{X}^0$, with $\hat{\psi}^u = \hat{\psi}^u_f + \hat{\psi}^u_s$. Furthermore,

$$\begin{aligned}
\hat{F} &= \hat{I} + \hat{\nabla}\hat{u}, \\
\hat{J} &= \det(\hat{F}), \\
\hat{\sigma}_f &= -\hat{p}_f \hat{I} + \hat{\rho}_f \nu_f (\hat{\nabla}\hat{v}_f \hat{F}^{-1} + \hat{F}^{-T}\hat{\nabla}\hat{v}_f), \\
\hat{\Sigma} &= 2\mu_s \hat{E} + \lambda_s \mathrm{tr}(\hat{E})\hat{I}, \\
\hat{E} &= 0.5(\hat{F}^T\hat{F} - \hat{I}), \\
\hat{\sigma}_{\text{mesh}} &= \alpha_u \hat{\nabla}\hat{u}_f,
\end{aligned}$$

$\hat{\rho}_s, \hat{\rho}_f$ are the densities,

ν_f is the kinematic viscosity, and

μ_s, λ_s are the Lamé parameters [9].

In compact form, the above problem reads: Find $\hat{U} \in \widehat{X}$ such that $\hat{A}(\hat{U})(\hat{\Psi}) = 0 \quad \forall \hat{\Psi} \in \hat{X}^0$, where the FSI equations are combined in the semi-linear form $\hat{A}(\hat{U})(\hat{\Psi})$.

3 Numerical solution and physics-based preconditioners

3.1 Newton linearization

The previous FSI model is discretized in time by an A stable implicit finite difference scheme and in space by Galerkin finite elements on quadrilaterals. The temporal and spatial discretization parameters are denoted by k and h, respectively. At time step t_n, we need to solve for U_h^{n+1} at t_{n+1}, for which we utilize Newton's method. At each Newton step (index j), we have to solve a linear variational problem of the form

$$\underbrace{A'(U_h^{n,j})(\delta U_h, \Psi_h)}_{=A\delta U} = \underbrace{-A(U_h^{n,j})(\Psi_h)}_{=B} \ \forall \Psi_h \in \hat{X}_h^0 \subset \hat{X}^0,$$

$$U_h^{n,j+1} = U_h^{n,j} + \lambda \delta U_h, \quad \lambda \in (0, 1],$$

until $|B^j| \leq 10^{-6}|B^0|$. The linesearch parameter is $\lambda = 1$ in our simulations. Thus, we finally obtain the linear system of finite element equations

$$A\delta U = B$$

for determining the Newton correction δU. We note that the finite element functions and operators are identified with the corresponding matrix and vector representations via the finite element isomorphism. Up to 10^5 unknowns in $2d$, respectively, 10^4 unknowns in 3D, sparse direct solvers work still fine in the context of FSI problems. However, for large-scale problems with considerable more unknowns, we should use preconditioned iterative solvers in order to reduce the memory demand and the computational costs in terms of arithmetical operations required.

3.2 Block structure of linear systems

Since the FSI problem is non-symmetric, a GMRES scheme (generalized minimal residual) is a classical choice for the overall solution of the linear system arising at each Newton iteration. In order to reduce the number of GMRES iterations, one needs a suitable preconditioner P for the system matrix A. In [9], we have constructed a (left) preconditioner P such that

$$P^{-1}A\delta U = P^{-1}B$$

with $P^{-1} \approx A^{-1}$ in the sense that $P^{-1}A$ is close to the identity matrix I. We refer the reader to [16] for GMRES convergence results.

Observing the previous FSI model, we have three unknowns when global continuity of the displacements $\hat{u}_f$ and $\hat{u}_s$ and $\hat{v}_f$ and $\hat{v}_s$ is realized, which is due to the variational-monolithic coupling scheme. Consequently, $\hat{u}, \hat{v}, \hat{p}$ are obtained from three principal problems: (m) mesh motion, (f) fluid, (s) solid. This results in the following 3×3 block system:

$$A := \begin{bmatrix} \mathcal{M} & C_{ms} & 0 \\ C_{sm} & \mathcal{S} & C_{sf} \\ C_{fm} & C_{fs} & \mathcal{F} \end{bmatrix}.$$

A brief analysis yields that the principal problems appear on the diagonal. The coupling terms C_{**} are on the off-diagonals. In [7], details on the influence of these were studied on the overall solver behavior. Aiming for cost-optimal parallel schemes, the interface coupling terms play however a crucial rule [9].

3.3 Physics-based preconditioner

We now concentrate on the construction of the preconditioner P^{-1}, which is based on a simplified LDU block factorization

$$A \approx \begin{bmatrix} I & 0 & 0 \\ 0 & I & 0 \\ C_{fm}\mathcal{M}^{-1} & \tilde{C}_{fs}\mathcal{S}^{-1} & I \end{bmatrix} \begin{bmatrix} \mathcal{M} & 0 & 0 \\ 0 & \mathcal{S} & 0 \\ 0 & 0 & \mathcal{F} \end{bmatrix} \begin{bmatrix} I & \mathcal{M}^{-1}C_{ms} & 0 \\ 0 & I & \mathcal{S}^{-1}C_{sf} \\ 0 & 0 & I \end{bmatrix} = LDU = P,$$

where we neglect the coupling term C_{sm}. We have (see [11][Section 6.4.3]) $\tilde{C}_{fs} = C_{fs} - C_{fm}\mathcal{M}^{-1}C_{ms}$. Having such a decomposition, it is easy to compute the action of the inverse. We note that, in Krylov subspace methods, we only need the action of P^{-1} on the residual r.

From linear algebra we know that $P^{-1}r = U^{-1}D^{-1}L^{-1}r$ with $P = LDU$ from above. Consecutively solving with L, D and U yields the following result:

Algorithm Evaluation of $P^{-1}r$ (matrix-vector multiplications):

1. Solve $x_m = \mathcal{M}^{-1}r_m$
2. Solve $x_s = \mathcal{S}^{-1}r_s$
3. Solve $x_f = \mathcal{F}^{-1}(r_f - C_{fm}x_m - C_{fs}x_s)$
4. Update $x_s = x_s - \mathcal{S}^{-1}C_{sf}x_f$
5. Update $x_m = x_m - \mathcal{M}^{-1}C_{ms}x_s$ □

It remains to discuss the solutions of the subproblems with the system matrices $\mathcal{M}$, $\mathcal{S}$ and $\mathcal{F}$. In our $2d$ numerical example presented in Sect. 4, we use the sparse direct solver MUMPS[1] that solves these smaller subproblems very efficiently. However, if

[1] http://mumps.enseeiht.fr/

the subproblems are larger, we should replace the direct solvers for $\mathcal{M}^{-1}$, $\mathcal{S}^{-1}$ and $\mathcal{F}^{-1}$ by preconditioned iterative solvers $\widetilde{\mathcal{M}}^{-1}$, $\widetilde{\mathcal{S}}^{-1}$ and $\widetilde{\mathcal{F}}^{-1}$; see [9, 11], where we used AMG-based solvers for the subproblems. The implementation is based on the open-source finite element package deal.II [1].

4 Flapping membranes with elastic solid walls

This example was originally inspired by [6], later extended by ourselves, and the current configuration was recently used in [18] for optimal control with fluid-structure interaction. The geometry is shown in Figure 1 (left). It consists of the fluid domain $\widehat{\Omega}_{\text{Fluid}} := (0,8) \times (0.0, 1.61) \setminus \widehat{\Omega}_{\text{Flaps}}$ with inscribed flaps $\widehat{\Omega}_{\text{Flaps}} := (1.9788, 2.0) \times ((0,0.7) \cup (0.91, 1.61))$. It is further surrounded by elastic arteries $\widehat{\Omega}_{\text{Artery}} := (0,8) \times ((-0.1, 0.0) \cup (1.61, 1.71))$ on the top and bottom of $\widehat{\Omega}$.

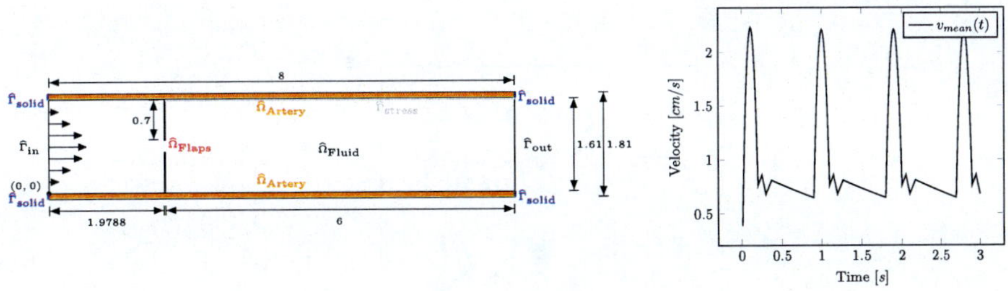

Fig. 1: Geometry with inflow profile (left) and mean inflow velocity (right).

On the inflow boundary, $\widehat{\Gamma}_{\text{in}} := \{0\} \times (0, 1.61)$, we prescribe a parabolic inflow profile $\hat{v}(0, y, t) := 6(1.61)^{-2} y(1.61 - y) v_{\text{mean}}(t)$ for $t \in I := [0, 3.6]$, where $v_{\text{mean}}(t)$ is given by the profile in Figure 1 (right). At the outflow boundary the do-nothing outflow condition $\widehat{\Gamma}_{\text{out}}$ is prescribed for $\hat{v}$ and $\hat{p}$. The elastic walls are fixed at the left and right, i.e., on $\widehat{\Gamma}_{\text{solid, left}} := \{0\} \times ((-0.1, 0.0) \cup (1.61, 1.71))$ and $\widehat{\Gamma}_{\text{solid, right}} := \{8\} \times ((-0.1, 0.0) \cup (1.61, 1.71))$, we prescribe $\hat{u} = 0$ and $\hat{v} = 0$.

The computations are performed on the time interval $I = (0, 3.6\text{s})$. The fluid parameters are given by the kinematic viscosity $\nu_f = 10^{-1}\text{cm}^2\,\text{s}^{-1}$, and density $\hat{\rho}_f = 10^2\text{g cm}^{-3}$. In the solid domains $\widehat{\Omega}_{\text{Flaps}}$ and $\widehat{\Omega}_{\text{Artery}}$, we use a Poisson ratio $\nu = 0.4$, and density $\rho_s = 10^2\text{g cm}^{-3}$. The Lamé parameters are given by $\mu_s^{\text{flaps}} = 2.0 \cdot 10^7\text{g cm}^{-1}\,\text{s}^{-2}$ in $\widehat{\Omega}_{\text{Flaps}}$, and $\mu_s^{\text{walls}} = 1.0 \cdot 10^9\text{g cm}^{-1}\,\text{s}^{-2}$ in $\widehat{\Omega}_{\text{Artery}}$.

We are interested in evaluating the number of GMRES iterations per linear solve in each Newton step to achieve a reduction of 10^{-4}. Moreover, we monitor the number of nonlinear iterations, the position of the tip $(2, 0.91)$ of the upper elastic flap, and the drag and lift $(F_D, F_L) = \int_{\widehat{\Gamma}_{\text{stress}}} \hat{J}\left(-\hat{p}I + \hat{\rho}_f \hat{\nu}_f (\hat{\nabla}\hat{v}\hat{F}^{-1} + \hat{F}^{-T}\nabla v^T)\right)\hat{F}^{-T}\hat{n}\,\mathrm{d}\hat{s}$ with $\widehat{\Gamma}_{\text{stress}} := (2, 8) \times \{1.61\}$.

Figure 2 shows that, during the whole simulation, we require an almost constant number of 4 to 6 Newton iterations. Similarly, the average number of linear GMRES iterations stays between 8 and 11 during refinement, although a slight increase can be observed on the finer grids. The computational aspects of certain parts of our simulation are summarized in Table 1. Our proposed iterative solver achieves similar performance as the direct solver on the coarsest grid. On the finest grid, with about 2 million dofs, the iterative variant is already about a factor of 2.3-times faster. Furthermore, the memory footprint of the iterative variant is roughly halved compared to the sparse direct solver; see Table 2. We note that for $2d$ problems, sparse direct solvers are hard to beat in terms of performance. For larger problems, we can split the application of the direct solver to the respective subproblems. This reduces the amount of memory and flops required to compute the factorization.

The resulting drag and lift values are visualized in Figure 3, the elongation of the tip is plotted in Figure 4. All these functional evaluations show surprisingly good agreement throughout the various levels of refinement. Only small differences are visible at the tips. As expected due to the symmetry of the configuration, evaluating the displacement, drag, or lift in the lower or upper part does not make a difference.

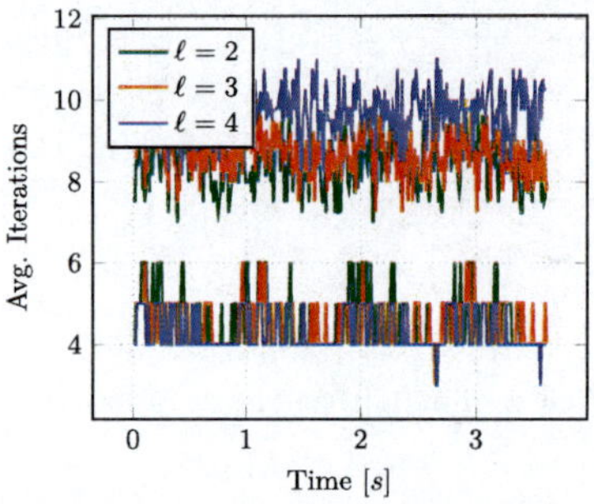

Fig. 2: Number of GMRES (higher values) and Newton iterations (lower values).

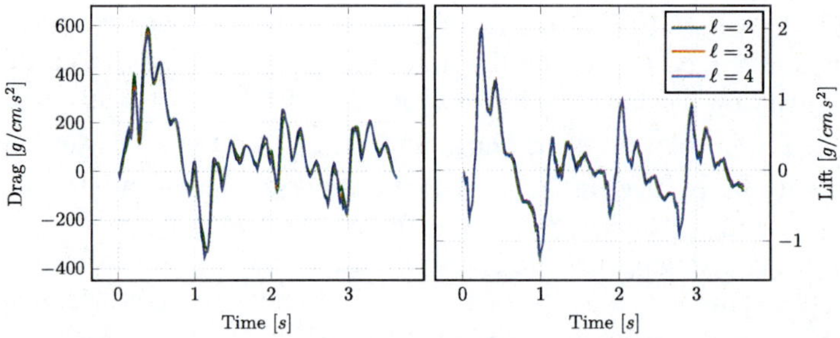

Fig. 3: Drag (left) and lift (right) evaluated at the artery behind the top flap, i.e., $(2.0, 8.0) \times \{1.61\}$.

ℓ	DoFs	Assemble [s]	Factorization [s]	Application [s]	Total [s]
2	$8.7 \cdot 10^3$	$1.6 \cdot 10^{-1}$	$3.2 \cdot 10^{-1}$	$6.7 \cdot 10^{-3}$	$3.3 \cdot 10^{-1}$
3	$3.4 \cdot 10^4$	$6.2 \cdot 10^{-1}$	$1.7 \cdot 10^0$	$2.8 \cdot 10^{-2}$	$1.7 \cdot 10^0$
4	$1.3 \cdot 10^5$	$2.5 \cdot 10^0$	$9.5 \cdot 10^0$	$1.2 \cdot 10^{-1}$	$9.6 \cdot 10^0$
5	$5.4 \cdot 10^5$	$1.0 \cdot 10^1$	$5.7 \cdot 10^1$	$5.4 \cdot 10^{-1}$	$5.7 \cdot 10^1$
6	$2.1 \cdot 10^6$	$4.1 \cdot 10^1$	$4.2 \cdot 10^2$	$2.3 \cdot 10^0$	$4.2 \cdot 10^2$

ℓ	DoFs	Assemble [s]	Factorization [s]	Application [s]	Total [s]
2	$8.7 \cdot 10^3$	$1.6 \cdot 10^{-1}$	$1.1 \cdot 10^{-1}$	$2.0 \cdot 10^{-1}$	$3.1 \cdot 10^{-1}$
3	$3.4 \cdot 10^4$	$6.3 \cdot 10^{-1}$	$6.7 \cdot 10^{-1}$	$5.7 \cdot 10^{-1}$	$1.2 \cdot 10^0$
4	$1.3 \cdot 10^5$	$2.5 \cdot 10^0$	$3.6 \cdot 10^0$	$2.7 \cdot 10^0$	$6.3 \cdot 10^0$
5	$5.4 \cdot 10^5$	$1.0 \cdot 10^1$	$2.0 \cdot 10^1$	$1.3 \cdot 10^1$	$3.3 \cdot 10^1$
6	$2.1 \cdot 10^6$	$4.1 \cdot 10^1$	$1.1 \cdot 10^2$	$6.9 \cdot 10^1$	$1.8 \cdot 10^2$

Table 1: Timings of a direct solver for the full FSI system (top) and our preconditioner with direct solvers for the fluid, solid, and mesh subproblems. Average time for the assembly, factorization, application of the preconditioner, and the total time for the a single linear system are given.

ℓ	DoFs	Matrix [B]	Fluid [B]	Mesh [B]	Solid [B]	P^{-1} [B]	Full [B]
2	$8.7 \cdot 10^3$	$6.3 \cdot 10^6$	$4.0 \cdot 10^6$	$4.0 \cdot 10^6$	$3.0 \cdot 10^6$	$1.1 \cdot 10^7$	$2.1 \cdot 10^7$
3	$3.4 \cdot 10^4$	$2.5 \cdot 10^7$	$1.8 \cdot 10^7$	$1.6 \cdot 10^7$	$1.4 \cdot 10^7$	$4.8 \cdot 10^7$	$9.3 \cdot 10^7$
4	$1.3 \cdot 10^5$	$1.0 \cdot 10^8$	$8.2 \cdot 10^7$	$7.6 \cdot 10^7$	$6.5 \cdot 10^7$	$2.2 \cdot 10^8$	$4.3 \cdot 10^8$

Table 2: Memory requirements using a direct solver (Full) for the whole system compared to our preconditioner P^{-1}, which uses direct solvers for fluid, solid and mesh.

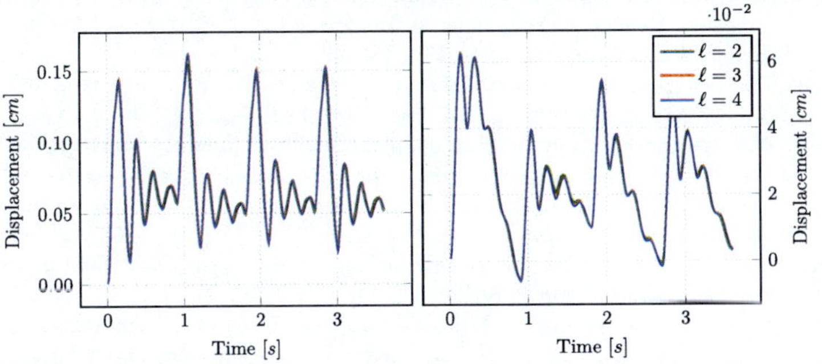

Fig. 4: Displacement of the top flap at (2.0, 0.91) in x-direction (left) and y-direction (right).

5 Conclusions and outlook

We presented a preconditioner based on a block-LDU-decomposition of the linear systems for a challenging 2*d* FSI problem. For a small number of degrees of freedoms, a sparse direct solver for the full problem is hard to beat. Nonetheless, the reduction of the sparse direct solver to the separate subproblems already leads to an improvement of a factor 2 in terms of memory requirements. For large systems, the storage cost and computational complexity of sparse direct solvers becomes a prohibitive barrier. Replacing the solvers for the fluid, solid and mesh problems by iterative or matrix-free techniques may solve this issue. Implementing matrix-free solvers for FSI is a very challenging task, mainly caused by the difficulty of treating the fluid subproblem. In [10], we have applied the matrix-free technique successfully to fracture propagation.

Acknowledgements This work has been supported by the Austrian Science Fund (FWF) grant P29181 'Goal-Oriented Error Control for Phase-Field Fracture Coupled to Multiphysics Problems', and by the Doctoral Program W1214-03 at the Johannes Kepler University Linz.

References

1. D. Arndt, W. Bangerth, D. Davydov, T. Heister, L. Heltai, M. Kronbichler, M. Maier, J.-P. Pelteret, B. Turcksin and D. Wells. The deal.II finite element library: Design, features, and insights. *Comput. Math. with Appl.* **81**, 407–422 (2021).
2. D. Balzani, S. Deparis, S. Fausten, D. Forti, A. Heinlein, A. Klawonn, A. Quarteroni, O. Rheinbach and J. Schröder. Numerical modeling of fluid–structure interaction in arteries with anisotropic polyconvex hyperelastic and anisotropic viscoelastic material models at finite strains. *Int. J. Numer. Methods Biomed. Eng.* **32**(10), e02756 (2016).
3. A.T. Barker and X.-C. Cai. Scalable parallel methods for monolithic coupling in fluid-structure interaction with application to blood flow modeling. *J. Comp. Phys.* **229**(3), 642–659 (2010).
4. P. Crosetto, S. Deparis, G. Fourestey and A. Quarteroni. Parallel algorithms for fluid-structure interaction problems in haemodynamics. *SIAM J. Sci. Comput.* **33**(4), 1598–1622 (2011).
5. M.W. Gee, U. Küttler and W.A. Wall. Truly monolithic algebraic multigrid for fluid–structure interaction. *Comput. Methods Appl. Mech. Engrg.* **85**(8), 987–1016 (2011).
6. A.J. Gil, A.A. Carreno, J. Bonet and O. Hassan. The immersed structural potential method for haemodynamic applications. *J. Comput. Phys.* **229**, 8613–8641 (2010).
7. M. Heil. An efficient solver for the fully coupled solution of large-displacement fluid-structure interaction problems. *Comput. Methods Appl. Mech. Engrg.* **193**, 1–23 (2004).
8. T.J.R. Hughes, W.K. Liu and T. Zimmermann. Lagrangian–Eulerian finite element formulation for incompressible viscous flows. *Comput. Methods Appl. Mech. Engrg.* **29**, 329–349 (1981).
9. D. Jodlbauer, U. Langer and T. Wick. Parallel block-preconditioned monolithic solvers for fluid-structure interaction problems. *Int. J. Numer. Methods Eng.* **117**(6), 623–643 (2019).
10. D. Jodlbauer, U. Langer and T. Wick. Matrix-free multigrid solvers for phase-field fracture problems. *Comput. Methods Appl. Mech. Engrg.* 372:113431 (2020).
11. D. Jodlbauer and T. Wick. A monolithic FSI solver applied to the FSI 1,2,3 benchmarks. In: S. Frei, B. Holm, T. Richter, T. Wick and H. Yang, editors, *Fluid-Structure Interaction: Modeling, Adaptive Discretization and Solvers*, volume 20 of Radon Series on Computational and Applied Mathematics, pp. 193–234, de Gruyter, Berlin (2017).
12. U. Langer and H. Yang. Robust and efficient monolithic fluid-structure-interaction solvers. *Int. J. Numer. Methods Eng.* **108**(4), 303–325 (2016).
13. U. Langer and H. Yang. Recent development of robust monolithic fluid-structure interaction solvers. In: S. Frei, B. Holm, T. Richter, T. Wick and H. Yang, editors, *Fluid-Structure Interaction: Modeling, Adaptive Discretization and Solvers*, volume 20 of Radon Series on Computational and Applied Mathematics, pp. 169–191, de Gruyter, Berlin (2017).
14. C.S. Peskin. Flow patterns around heart valves: A numerical method. *J. Comp. Phys.* **10**(2), 252–271 (1972).
15. T. Richter. A monolithic geometric multigrid solver for fluid-structure interactions in ALE formulation. *Int. J. Numer. Methods Eng.* **104**(5), 372–390 (2015).
16. Y. Saad. *Iterative Methods for Sparse Linear Systems*. SIAM, Philadelphia (2003).

17. T. Wick. Fluid-structure interactions using different mesh motion techniques. *Comput. Struct.* **89**(13-14), 1456–1467 (2011).
18. T. Wick and W. Wollner. Optimization with nonstationary, nonlinear monolithic fluid-structure interaction. *Int. J. Numer. Methods Eng.* **122**(19), 5430–5449 (2021).

Adaptive Nonlinear Elimination in Nonlinear FETI-DP Methods

Axel Klawonn, Martin Lanser, and Matthias Uran

1 Introduction

In recent years, we have formulated a unified framework that covers all nonlinear FETI-DP as well as nonlinear BDDC methods; see [3]. Both belong to the class of non-overlapping domain decomposition methods and can be used for the solution of discrete nonlinear problems of the form $A(\bar{u}) = 0$. For example, such systems arise from the discretization of nonlinear partial differential equations. In contrast to the traditional Newton–Krylov-DD approach (see [3]), where we first linearize the problem and then decompose it into subdomains, the order of operations is turned around in nonlinear domain decomposition methods. A nonlinear elimination of a subset of finite element unknowns before linearization allows us to interpret nonlinear FETI-DP methods as nonlinear right-preconditioned Krylov methods; see [3]. Although the unified framework covers arbitrary choices of elimination sets, only a few different types of elimination sets have been considered so far. All of them are based on the classification in interior, dual, and primal variables, which is a natural thing to do in FETI-DP methods but obviously not problem-dependent. In order to design a nonlinear FETI-DP method that fits optimally to an arbitrary problem, it is necessary to use problem-dependent or adaptive elimination sets. In this article, we describe how to use the residual of the nonlinear FETI-DP saddle point system to choose the elimination set. First studies were performed under our guidance as part of a master's thesis [6] and can also be found in [7]. The idea of

Axel Klawonn, Martin Lanser, Matthias Uran
Department of Mathematics and Computer Science, Division of Mathematics, University of Cologne, Weyertal 86-90, 50931 Cologne, Germany,
e-mail: axel.klawonn@uni-koeln.de, martin.lanser@uni-koeln.de, m.uran@uni-koeln.de, url: https://www.numerik.uni-koeln.de

and
Center for Data and Simulation Science, University of Cologne, Germany,
url: https://www.cds.uni-koeln.de

S. C. Brenner et al. (eds.), *Domain Decomposition Methods in Science and Engineering XXVI*, Lecture Notes in Computational Science and Engineering 145,
https://doi.org/10.1007/978-3-030-95025-5_35

using the residual to determine an elimination set is adapted from Cai and Gong in [1], where they introduced the idea in the context of inexact Newton methods.

2 Nonlinear FETI-DP

Before we describe the process of determining problem-dependent elimination sets, let us first recall the most relevant ideas of nonlinear domain decomposition methods and of the unified framework of nonlinear FETI-DP methods to introduce a suitable notation. For a detailed description, we also refer to [2, 3] and the references therein.

Throughout this paper, we assume that we have a computational domain $\Omega \subset \mathbb{R}^d$, $d = 2, 3$, which is divided into N non-overlapping subdomains Ω_i, i.e., $\Omega = \bigcup_{i=1}^N \Omega_i$. Each subdomain is the union of finite elements and the associated finite element spaces are denoted by $W^{(i)}$. We denote the product space of all finite element spaces as $W = W^{(1)} \times \cdots \times W^{(N)}$. In FETI-DP methods, we partition all variables into interior (I), dual (Δ), and primal (Π) variables, where only continuity in the primal variables is prescribed and continuity in the dual variables is enforced by Lagrange multipliers λ iteratively. Therefore, we further introduce a subspace $\widetilde{W} \subset W$ of all finite element functions from W that are continuous in the primal variables. A simple choice of primal variables are subdomain vertices. For completeness, we also introduce the subspace $\widehat{W} \subset W$, which contains all finite element functions that are continuous across the complete interface and it holds that $\widehat{W} \subset \widetilde{W} \subset W$.

As was shown in [2], finding the solution of the fully assembled finite element problem is equivalent to solving the nonlinear FETI-DP saddle point system

$$A(\tilde{u}, \lambda) = \begin{bmatrix} \widetilde{K}(\tilde{u}) + B^T \lambda - \tilde{f} \\ B\tilde{u} \end{bmatrix} = \begin{bmatrix} 0 \\ 0 \end{bmatrix}, \tilde{u},\ \tilde{f},\ \widetilde{K}(\tilde{u}) \in \widetilde{W}. \tag{1}$$

This system is the basis for all nonlinear FETI-DP methods. Here, the linear constraints $B\tilde{u} = 0$ together with Lagrange multipliers $\lambda \in V := \text{range}(B)$ enforce continuity in all dual variables.

As introduced in [3, 4], we use a nonlinear right-preconditioner $M(\tilde{u}, \lambda)$ that is nonlinear in $\tilde{u}$ and linear in λ; see [3, 4] for some desirable properties of M. Instead of $A(\tilde{u}, \lambda) = 0$, we now solve $A(M(\tilde{u}, \lambda)) = 0$ with a Newton–Krylov method.

Following [3], the application of a nonlinear right-preconditioner can be interpreted as a (partial) nonlinear elimination process (see also [5]), where different choices of M lead to different elimination sets. With this interpretation, it is natural to divide the overall set of variables into two different subsets E and L, where E contains all the variables that should be nonlinearly eliminated by the preconditioner M, and L contains the remaining variables in which it will be linearized.

After an appropriate rearrangement, we can represent all quantities in Eq. (1) according to the variable split into the subsets E and L. For example, we obtain $\tilde{f} = \begin{bmatrix} \tilde{f}_E^T\ \tilde{f}_L^T \end{bmatrix}^T$ and $B = \begin{bmatrix} B_E\ B_L \end{bmatrix}$. Thus, we can write the nonlinear saddle point system (Eq. (1)) as

$$A(\tilde{u}_E, \tilde{u}_L, \lambda) = \begin{bmatrix} \widetilde{K}_E(\tilde{u}_E, \tilde{u}_L) + B_E^T \lambda - \tilde{f}_E \\ \widetilde{K}_L(\tilde{u}_E, \tilde{u}_L) + B_L^T \lambda - \tilde{f}_L \\ B_E \tilde{u}_E + B_L \tilde{u}_L \end{bmatrix} = \begin{bmatrix} 0 \\ 0 \\ 0 \end{bmatrix}.$$

With the application of the nonlinear right-preconditioner, we now aim to eliminate all variables $\tilde{u}_E$, which correspond to the subset E. Thus, our preconditioner is implicitly defined by solving the nonlinear equation

$$\widetilde{K}_E(M_{\tilde{u}_E}(\tilde{u}_L, \lambda), \tilde{u}_L) + B_E^T \lambda - \tilde{f}_E = 0, \qquad (2)$$

where we have $M(\tilde{u}_E, \tilde{u}_L, \lambda) := (M_{\tilde{u}_E}(\tilde{u}_L, \lambda), \tilde{u}_L, \lambda)$, since, by construction, M is linear in $\tilde{u}_L$ and λ. After we have computed M by solving Eq. (2) with Newton's method, we obtain the nonlinear Schur complement system

$$S_L(\tilde{u}_L, \lambda) := \begin{bmatrix} \widetilde{K}_L(M_{\tilde{u}_E}(\tilde{u}_L, \lambda), \tilde{u}_L) + B_L^T \lambda - \tilde{f}_L \\ B_E M_{\tilde{u}_E}(\tilde{u}_L, \lambda) + B_L \tilde{u}_L \end{bmatrix} = \begin{bmatrix} 0 \\ 0 \end{bmatrix}.$$

This can be solved with the traditional Newton–Krylov-FETI-DP approach ([2]); see [3]. Putting this all together, in each of these (outer) Newton iterations, M has to be recomputed, resulting in two nested Newton loops. In the case of an indefinite tangent matrix, this approach might require further investigation since there are currently no provable convergence statements for FETI-DP for indefinite linear problems. At least, one should use GMRES instead of the CG method as the Krylov subspace method.

3 A problem-dependent choice of the elimination set

In [3], we have considered four different variants of Nonlinear-FETI-DP which are denoted as NL-i, $i = 1, \ldots, 4$. In all these methods, the elimination set is chosen a priori with respect to the sets I, Δ, and Π. We have $E_{\text{NL-1}} = \emptyset$, $E_{\text{NL-2}} = \left[I \, \Delta \, \Pi\right]$, $E_{\text{NL-3}} = \left[I \, \Delta\right]$, and $E_{\text{NL-4}} = \left[I\right]$. Let us note that the elimination sets are thus chosen statically and that NL-4 has the smallest set among these variants. In our earlier experiments, these methods often improved the nonlinear convergence behavior compared to the traditional Newton–Krylov-FETI-DP approach; see [3]. Furthermore, NL-3 and NL-4 show a high potential in reducing the computing time for large problems since the nonlinear elimination can be carried out completely independently for each subdomain without the need for communication and synchronization. For further information, we refer to [3] and the references therein.

However, we have also considered a model problem in [3] for which the performance of NL-4 is worse than the traditional NK-FETI-DP approach. This demonstrates that the choice of a good elimination set is essential for the performance of nonlinear FETI-DP methods. At the same time, it also suggests that there are problems for which the other NL-FETI-DP variants might perform poorly. Accordingly, we should incorporate information about the problem into the choice of the elimination set in order to construct a nonlinear FETI-DP method that is tailored to the specific problem in the best possible way.

In this paper, we introduce a Nonlinear-FETI-DP method with problem-dependent or adaptive elimination sets, which are determined with respect to the residual of the nonlinear saddle point system Eq. (1). This strategy is inspired by an article by Gong and Cai [1], where a similar approach was presented in the context of a nonlinear elimination preconditioned inexact Newton method. The underlying idea is that the elimination set contains all variables corresponding to large absolute values in the nonlinear residual. First studies for the use in nonlinear FETI-DP methods are also presented in [6, 7].

Let us first specify the residual that we shall consider. As usual, we are interested in finding the solution (u^*, λ^*) of $A(u^*, \lambda^*) = 0$ with $(u^*, \lambda^*) = M(\tilde{u}, \lambda)$. We are especially interested in the first component u^*, since the Lagrange multipliers are only introduced to guarantee continuity of the final solution across the interface. Therefore, we do not consider the complete residual of the nonlinear saddle point system but only the part belonging to $\tilde{u}$. Let us assume that we have finished the k-th outer iteration, i.e., we have computed $\lambda^{(k)} = \lambda^{(k-1)} - \delta\lambda^{(k-1)}$ and $\tilde{u}^{(k)} = g^{(k-1)} - \delta\tilde{u}^{(k-1)}$, where $g^{(k-1)} := \left[M_{\tilde{u}_E}\left(\tilde{u}_L^{(k-1)}, \lambda^{(k-1)}\right), \tilde{u}_L^{(k-1)} \right]$ is the vector after eliminating $\tilde{u}_E$ and $\delta\lambda^{(k-1)}, \delta\tilde{u}^{(k-1)}$ are the corresponding Newton updates. Thus, $g^{(k-1)}$ includes the solution of the inner Newton method in the k-th outer loop. Then, the elimination set for iteration $k+1$ is built with respect to the residual

$$A\left(\tilde{u}^{(k)}, \lambda^{(k)}\right)\Big|_{\tilde{u}} = \widetilde{K}\left(\tilde{u}^{(k)}\right) + B^T\lambda^{(k)} - \tilde{f};$$

cf. the first line of Eq. (1). As the tilde indicates, all quantities are only assembled in the primal variables and might have different values in a physical point belonging to more than one subdomain. To obtain a single value for each global degree of freedom, we make use of the dual assembly operator $R_\Delta^T : \widetilde{W} \rightarrow \widehat{W}$, which yields the residual

$$\overline{r}^{(k)} := R_\Delta^T \cdot A\left(\tilde{u}^{(k)}, \lambda^{(k)}\right)\Big|_{\tilde{u}}.$$

From $R_\Delta^T B^T \lambda^{(k)} = 0$, we obtain

$$\overline{r}^{(k)} = R_\Delta^T \widetilde{K}\left(\tilde{u}^{(k)}\right) - R_\Delta^T \tilde{f} = R_\Delta^T R_\Pi^T K\left(R_\Pi \tilde{u}^{(k)}\right) - R_\Delta^T R_\Pi^T f,$$

where $R_\Pi^T : W \rightarrow \widetilde{W}$ is the assembly operator in the primal variables; see, e.g. [2]. From the last line of [3, Eq. 17], we obtain

$$Bg^{(k-1)} - B_E \delta\tilde{u}_E^{(k-1)} - B_L \delta\tilde{u}_L^{(k-1)} = B\left(g^{(k-1)} - \delta\tilde{u}^{(k-1)}\right) = 0,$$

which automatically implies that $\tilde{u}^{(k)}$ is continuous across the interface. Thus, the residual is identical to the fully assembled residual $R_\Delta^T R_\Pi^T K\left(R_\Pi R_\Delta \tilde{u}^{(k)}\right) - R_\Delta^T R_\Pi^T f =$

$R^T K\left(R\bar{u}^{(k)}\right) - R^T f$ as long as we use a step length equal to 1 in the outer Newton iteration, which we assume throughout this article for simplicity.

Next, we describe the process how to assign variables to the elimination set used for the outer iteration $k+1$. Similar to [7], we introduce the following notation. We assume that we have n finite element nodes with l degrees of freedom each and introduce the two index sets $\mathcal{N} := \{1, \ldots, n\}$ and $\mathcal{D} := \{1, \ldots, m\}$, where the overall number of degrees of freedom belonging to $\bar{u}$ or $\bar{r}^{(k)}$ computes as $m = n \cdot l$. Since we have l degrees of freedom for each finite element node, the residual vector $\bar{r}^{(k)}$ decomposes into n subvectors $\bar{r}^{(k)}_{(i)} \in \mathbb{R}^l, i \in \mathcal{N}$, where the entries $\bar{r}^{(k)}_{(i)_j}$, $j = 1, \ldots, l$, belong to the corresponding degrees of freedom of finite element node i. Analogously to [1], the idea is to assign those degrees of freedom to the elimination set $E^{(k+1)}$ which correspond to a finite element node i with at least one degree of freedom with a high absolute residual value, i.e., $\|\bar{r}^{(k)}_{(i)}\|_\infty \geq \rho_{\text{res}} \cdot \|\bar{r}^{(k)}\|_\infty$, where $\rho_{\text{res}} \in (0, 1]$ is a tolerance specified by the user. Let us note that thus all degrees of freedom belonging to the same physical node are either all assigned to $E^{(k+1)}$ or not. Hence, the size of the elimination set increases with a decreasing tolerance. Consequently, the index set of degrees of freedom that belong to the elimination set reads as

$$\mathcal{D}_E^{(k+1)} := \left\{ i_1, \ldots i_l \in \mathcal{D} \mid i \in \mathcal{N}, \|\bar{r}^{(k)}_{(i)}\|_\infty \geq \rho_{\text{res}} \cdot \|\bar{r}^{(k)}\|_\infty \right\}.$$

For the final elimination set $E^{(k+1)}$, we introduce a $\delta_{\text{res}} \in \mathbb{R}$ and extend the index set $\mathcal{D}_E^{(k+1)}$ with the indices of degrees of freedom belonging to finite element nodes with a distance of at most δ_{res} to any finite element node whose degrees of freedom have been assigned to $\mathcal{D}_E^{(k+1)}$; see c) and d) in Fig. 1. Denoting the coordinates of finite element node i by v_i, the final elimination set reads as

$$E^{(k+1)} := \mathcal{D}_E^{(k+1)} \bigcup \left\{ i_1, \ldots, i_l \in \mathcal{D} \;\middle|\; \begin{array}{c} i \in \mathcal{N},\ \|\bar{r}^{(k)}_{(i)}\|_\infty < \rho_{\text{res}} \cdot \|\bar{r}^{(k)}\|_\infty, \\ \exists\, s \in \mathcal{N},\ \|\bar{r}^{(k)}_{(s)}\|_\infty \geq \rho_{\text{res}} \cdot \|\bar{r}^{(k)}\|_\infty \ :\ \text{dist}(v_i, v_s) \leq \delta_{\text{res}} \end{array} \right\}.$$

Following [1], this δ_{res} is introduced to avoid sharp jumps in the residual function. With this strategy, we are able to construct a new elimination set $E^{(k)}$ in each outer Newton iteration. However, if the problem at hand is completely unknown and the initial value is somehow random, it might be disadvantageous to choose an elimination set based on the initial residual. In such cases, we recommend choosing $E^{(1)} = \emptyset$ in the first iteration before switching to the elimination strategy.

4 Numerical Results

In this section, we present numerical results for a first problem-dependent nonlinear FETI-DP variant. Since the elimination set is build with respect to the nonlinear residual, we refer to this method as the Nonlinear-FETI-DP-Res method or, shorter, NL-Res. Within this section, we discuss different variants of NL-Res specified by different choices of ρ_{res} and δ_{res}. Moreover, for a single $(\rho_{\text{res}}, \delta_{\text{res}})$ pair, we compare the numerical results to those of NL-i, $i = 1, \ldots, 4$. We do not compare to the traditional NK-FETI-DP approach since the NL-1 method without the computation of an initial value (see [2, 3]) is closely related to it. To distinguish between different variants of NL-Res in our tables and figures, we introduce the notation NL-R($\rho_{\text{res}}, \eta_{\text{res}}$), with $\eta_{\text{res}} \cdot h = \delta_{\text{res}}$ and h the diameter of a finite element.

The results shown in this section have all been computed using our sequential MATLAB implementation. If we exceed 80 inner Newton iterations within a single elimination process or if more than 40 outer Newton iterations are required, the simulation is terminated and considered as diverged. Inner and outer Newton iterations reach convergence if $\|\widetilde{K}_E(M_{\tilde{u}_E}(\tilde{u}_L, \lambda), \tilde{u}_L) + B_E^T \lambda - \tilde{f}_E\|_{L_2} \leq 1e-12$ (see Eq. (2)) and $\|A(\tilde{u}, \lambda)\|_{L_2} \leq 1e-12$, respectively. Here, we consider two-dimensional scalar model problems of the form

$$
\begin{aligned}
-\alpha \Delta_4 u - \beta \Delta_2 u &= 1 \text{ in } \Omega, \\
u &= 0 \text{ on } \partial\Omega,
\end{aligned}
$$

where $\alpha, \beta : \Omega \to \mathbb{R}$ and $\Delta_p u$ is the p-Laplace operator with $p = 2, 4$. For model problems from nonlinear elasticity (2D) with and without contact, we refer to [6, 7].

As a computational domain Ω, we always consider the unit square and a decomposition into equally sized square subdomains of diameter H. Each subdomain is discretized by equally sized piecewise linear finite elements (P1) of diameter h. As primal variables, we exclusively use subdomain vertices which is the most simple choice. Analogously to [4], we have to measure the parallel potential of our non-

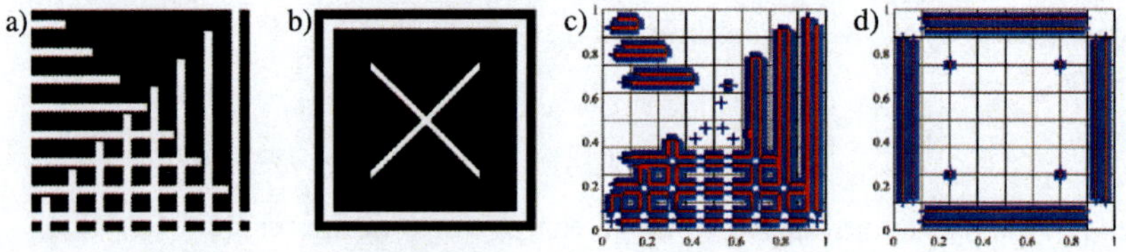

Fig. 1: a) and **b):** Different types of coefficient distributions. We have $\alpha = 1, \beta = 0$ in the white areas and $\alpha = 0, \beta = 1$ in the remaining (black) part. All channels as well as the cross have a width of $H/3$. **c)** and **d):** First elimination sets for NL-Res(0.1,3) for coefficients presented in a) and b). Red points belong to $\mathcal{D}_E^{(1)}$ and blue points are added due to a distance along main axes not larger than $3h$ to a red point; see Section 3.

linear FETI-DP methods by considering different metrics and indicators due to our sequential MATLAB implementation. However, we have to look at slightly different indicators compared to [4], since the structure of the elimination set of NL-Res is flexible and not known a priori. As before, we measure the need for global communication by counting the number of Krylov iterations, which are denoted as "# Krylov Its.". In addition to that, we also count inner ("# Inner Its.") and outer ("# Outer Its.") Newton iterations. Note that each outer Newton iteration requires a factorization of the FETI-DP coarse problem, which is also true for each iteration in the elimination

Table 1: Simulation results of different variants of the NL-Res approach as well as the NL-1 method without computation of an initial value and the best nonlinear FETI-DP method with a constant non-empty elimination set which is NL-4 in this case. For the distribution of the coefficients; see Fig. 1 a). Computational domain $\Omega = [0, 1]^2$; 8×8 square subdomains; P1 elements; $H/h = 16$.

	NL-1 no Init.	NL-4	NL-R (0.8,0)	NL-R (0.8,3)	NL-R (0.8,5)	NL-R (0.5,0)	NL-R (0.5,3)	NL-R (0.5,5)	NL- R (0.1,0)	NL-R (0.1,3)	NL-R (0.1,5)	NL-R (0.01,0)	NL-R (0.01,3)	NL-R (0.01,5)
Inner Its.	-	37	55	62	61	56	68	72	122	no	71	63	no	**37**
Outer Its.	15	8	14	13	13	13	13	12	18	conv.	11	12	conv.	**6**
Krylov Its.	307	155	287	270	267	268	256	240	355	inner	220	207	inner	**112**
Avg. Size E [%]	0.00	89.44	0.04	0.33	0.51	0.17	1.41	2.10	1.48	loop	9.46	9.43	loop	23.75

process of NL-2. In contrast to this, in NL-3 and NL-4 no coarse components are eliminated and thus a coarse factorization is only necessary in the outer loop. This property offers a higher potential for parallelization and we therefore precisely distinguished in [3] between the number of necessary coarse and local factorizations to measure the performance of the different nonlinear FETI-DP methods. In NL-Res, the elimination set is chosen problem-dependent and can contain arbitrary parts of the coarse problem and arbitrary parts of the local subdomains. Simply counting local and coarse factorizations is thus not sufficient anymore. Here, to measure the cost of the inner Newton iteration, we introduce the average size of the elimination set as an additional indicator, which allows us to evaluate the efficiency of our nonlinear FETI-DP variants. A single iteration of the elimination process is expected to be cheaper for a small elimination set. Accordingly, the most efficient nonlinear FETI-DP method has minimal inner and outer iteration numbers and, at the same time, the smallest average size of the elimination set.

First studies regarding the NL-Res approach have been carried out for the p-Laplace problem in [6, 7]. For relatively simple distributions of nonlinearity, parameters ρ_{res} and δ_{res} have been found in [7] such that the NL-Res variant yields quite similar iteration numbers compared to the best NL-FETI-DP-i method, $i = 2, 3, 4$, but using a significantly smaller average size of the elimination set E for each outer Newton iteration. Additionally, for most tested pairs of ρ_{res} and δ_{res} the NL-Res method was at least robust and converged in an acceptable number of iterations. However, in preliminary considerations of more complex distributions of nonlinearity, we already observed a significant influence of the choice of parameters on the convergence behavior of NL-Res, which complicates the right choice. The focus of this article is to discuss this observation in detail. Therefore, we consider two very complex distributions of nonlinearity; see a) and b) in Fig. 1. For both problems, we obtain similar results; see Table 1 as well as Fig. 2. It turns out that NL-4 is the best variant of the more traditional nonlinear FETI-DP methods. Compared to NL-1 without the computation of an initial value, the number of outer Newton iterations is reduced by a factor of 2 to 3 for the largest problem sizes for both model problems.

The performance of the various NL-Res methods can be summarized as follows: if the combination of ρ_{res} and δ_{res} leads to extremely small elimination sets, the performance of NL-Res is quite similar to that of NL-1 without the computation of an initial value. However, with the right choice of parameters, we also find variants of NL-Res that give iteration numbers at least as good as NL-4. In that case NL-Res is superior due to the much smaller average size of E. Let us remark that NL-Res(0.01,5) seems to be a good choice for both problems. However, as already mentioned, finding the right parameters is difficult. This is demonstrated by the results presented in Table 1, where a small change in δ_{res} turns the best NL-Res methods (NL-Res(0.01,5) and NL-Res(0.1,5)) into non-convergent variants (NL-Res(0.01,3) and NL-Res(0.1,3)). This hints that the elimination set cannot be chosen completely arbitrarily and especially the optimal selection of parameters has to be further analyzed. In ongoing research, we want to develop a heuristic that leads to a more effective choice of elimination sets. This will also include adjustments of the parameters during runtime in the case of poor or no convergence. To summarize,

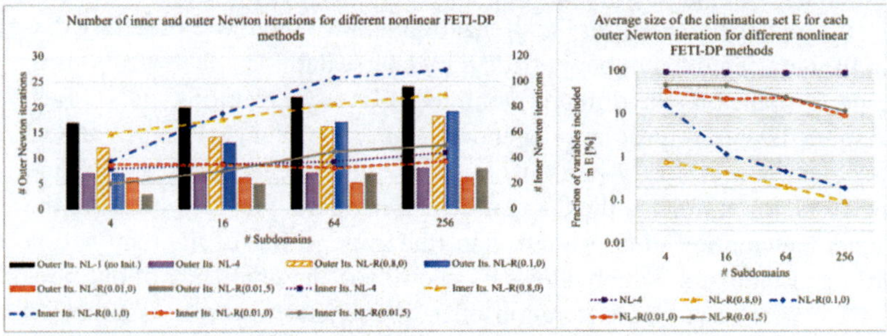

Fig. 2: Simulation results of different nonlinear FETI-DP methods including different variants of NL-Res with problem-dependent choices of the elimination set for the p-Laplace equation with a coefficient distribution as presented in Fig. 1 b); square subdomains; P1 finite elements; $H/h = 16$.

choosing the right parameters is crucial for the performance of NL-Res methods, but with the right parameters, NL-Res yields similar iteration numbers compared to the best of the more traditional NL-FETI-DP-i methods, $i = 2, 3, 4$. The advantage is a variable elimination set, which is formed depending on the problem. This results in a significantly smaller average size of the elimination set and thus less computational effort in the inner loops.

References

1. S. Gong and X.C. Cai. A nonlinear elimination preconditioned inexact Newton method for heterogeneous hyperelasticity. *SIAM J. Sci. Comput.* **41**(5), S390–S408 (2019).
2. A. Klawonn, M. Lanser and O. Rheinbach. Nonlinear FETI-DP and BDDC Methods. *SIAM Journal on Scientific Computing* **36**(2), A737–A765 (2014).
3. A. Klawonn, M. Lanser, O. Rheinbach and M. Uran. Nonlinear FETI-DP and BDDC methods: a unified framework and parallel results. *SIAM J. Sci. Comput.* **39**(6), C417–C451 (2017).

4. A. Klawonn, M. Lanser, O. Rheinbach and M. Uran. On the Accuracy of the Inner Newton Iteration in Nonlinear Domain Decomposition. In: P.E. Bjørstad, S.C. Brenner, L. Halpern, H.H. Kim, R. Kornhuber, T. Rahman, O.B. Widlund (eds.) *Domain Decomposition Methods in Science and Engineering XXIV*, pp. 435–443, Springer International Publishing, Cham (2018).
5. P.J. Lanzkron, D.J. Rose and J.T. Wilkes. An Analysis of Approximate Nonlinear Elimination. *SIAM Journal on Scientific Computing* **17**(2), 538–559 (1996).
6. F. Piechulla. *Residuenbasierte Eliminationsstrategien für nichtlineare FETI-DP Gebietszerlegungsverfahren*. Master's thesis, Universität zu Köln (08/2020).
7. M. Uran. *High-Performance Computing Two-Scale Finite Element Simulations of a Contact Problem Using Computational Homogenization – Virtual Forming Limit Curves for Dual-Phase Steel*. Ph.D. thesis, Universität zu Köln (2020).

Globalization of Nonlinear FETI–DP Methods

S. Köhler and O. Rheinbach

1 Introduction

Nonlinear FETI-DP (Finite Element Tearing and Interconnection - Dual Primal) methods [10] are nonlinear generalizations of linear FETI-DP domain decomposition methods [5, 16]. Nonlinear FETI-DP domain decomposition methods have shown their robustness and scalability, e.g., for linear and nonlinear structural mechanics problems [11], where results for up to 786 432 cores were presented. Related nonlinear domain decomposition methods (DDMs) are nonlinear BDDC methods [10] (derived from linear Balancing Domain Decomposition by Constraints [4]), nonlinear FETI-1 methods [15] and the ASPIN approach (Additive Schwarz Preconditioned Inexact Newton) method [2, 8, 9].

The idea of nonlinear FETI-DP methods is to decompose the global problem $\widehat{K}(\hat{u}) = \hat{f}$ into local nonlinear problems $K_i(u_i) = f_i$, $i = 1, \ldots, N$, defined on nonoverlapping subdomains $\Omega_i = 1, \ldots, N$, and to enforce continuity on the interface as $\Gamma := \cup_i^N \partial\Omega_i \cap \partial\Omega$ using subassembly of primal variables and Lagrange multipliers λ.

Nonlinear FETI-DP methods make use of nonlinear elimination, where different methods result from different elimination sets. In [12], four different types of static elimination sets were introduced, referred to as Nonlinear-FETI-DP-x (*NL*-x), where $x = 1$ (no elimination is applied), $x = 2$ (primal, dual and inner variables are eliminated), $x = 3$ (dual and inner variables are eliminated) and $x = 4$ (only the inner variables are eliminated). Other choices of elimination sets include automatic strategies to determine the elimination set [7, 18].

Stephan Köhler and Oliver Rheinbach
Institut für Numerische Mathematik und Optimierung, Fakultät für Mathematik und Informatik, Technische Universität Bergakademie Freiberg, Akademiestr. 6, 09596 Freiberg
e-mail: oliver.rheinbach@math.tu-freiberg.de,
stephan.koehler@math.tu-freiberg.de,
url: http://www.mathe.tu-freiberg.de/nmo/mitarbeiter/oliver-rheinbach

S. C. Brenner et al. (eds.), *Domain Decomposition Methods in Science and Engineering XXVI*, Lecture Notes in Computational Science and Engineering 145,
https://doi.org/10.1007/978-3-030-95025-5_36

If a tangent is available, nonlinear problems are typically solved by Newton's method or related methods such as quasi-Newton, inexact Newton or Newton-like methods [14, 17]. However, without globalization Newton's method may fail to converge.

Common globalization methods are trust-region methods or line search methods. In this paper, we study line search methods for the globalization of nonlinear FETI-DP methods for nonlinear structural mechanics problems. We use an exact differentiable penalty function [1] related to the augmented Lagrange approach, but we can use the Hessian of the standard Lagrange function for a Newton-like descent method.

2 Nonlinear FETI-DP

Nonlinear FETI-DP methods are methods to solve the nonlinear saddle point problem

$$\begin{aligned} \widetilde{K}(\tilde{u}) + B^T\lambda &= \tilde{f}, \\ B\tilde{u} &= 0, \end{aligned} \tag{1}$$

which directly corresponds to the linear FETI-DP saddle point problem [16]. Here, B is the FETI-DP jump operator (as in linear FETI-DP methods), and λ is the vector of corresponding Lagrange multipliers. The nonlinear operator $\widetilde{K}(\tilde{u}) := R_{\Pi}^T K(R_{\Pi}\tilde{u})$ is obtained from finite element subassembly of the block operator $K(u) = [K_1(u_1), \ldots, K_N(u_N)]^T$ in the primal variables using the operator R_{Π}^T as in linear FETI-DP methods [16]. Here, this coupling provides a nonlinear coarse problem for the method. Thus, $\widetilde{K}$ represents a nonlinear coarse approximation of the original problem.

Next, we perform the nonlinear elimination: we split the first row in (1) according to disjoint index sets E, L (eliminate or linearize) and solve in a first step

$$\widetilde{K}_E(\tilde{u}_E, \tilde{u}_L) - \tilde{f}_E + B_E^T\lambda = 0, \tag{2}$$

for $\tilde{u}_E$, given $\tilde{u}_L$ and λ. Then, we can insert $\tilde{u}_E$ into the remaining equations and solve by linearization in $\tilde{u}_L$ and λ, and using the implicit function theorem. Let us recall, that for *NL*-1 we have $E = \emptyset$. For *NL*-2 the elimination set E contains all variables and $L = \emptyset$, for *NL*-3 we eliminate the inner and dual variables [16], and for *NL*-4 we eliminate only the inner variables. Automatic strategies to determine the elimination set E can also be considered but are not discussed here. Note that the local nonlinear elimination uses an exact Newton method in the sense that we perform a Newton iteration using a direct sparse solver for the Newton equation. This can be afforded since this is an operation local to a subdomain. For *NL*-2 the elimination also involves the (small) coarse space.

3 Exact differentiable penalty method with nonlinear elimination

For $\nabla \tilde{J}(\tilde{u}) = \widetilde{K}(\tilde{u}) - \tilde{f}$ the equations in (1) are the first order optimality conditions for the minimization of the energy $\min \tilde{J}(\tilde{u})$ subject to the continuity constraint $B\tilde{u} = 0$, where $\tilde{J}(\tilde{u}) := J(R_\Pi \tilde{u})$ is obtained from the global energy $J(u) = \sum_{i=1}^N J^{(i)}(u_i)$.

Exact penalty method in nonlinear FETI-DP To be consistent with the vast literature in optimization, we will now use the notation

$$\min_{x \in \mathbb{R}^n} J(x) \qquad \text{subject to (s.t.)} \qquad c_i(x) = 0, \quad i = 1, \ldots, p, \tag{3}$$

where $J, c_i \in C^3(\mathbb{R}^n), i = 1, \ldots, p$. In the FETI-DP context, x is $\tilde{u}$, and $c(x) = 0$ are the continuity constraints $B\tilde{u} = 0$.

Penalty methods replace the original constrained minimization problem by a sequence of unconstrained minimization problems, where a penalty term, which measures the constraint violation, is added to the objective function. In [3] the exact differentiable penalty function

$$P(x, \lambda; \mu, M) = \mathcal{L}(x, \lambda) + \frac{\mu}{2} \|c(x)\|^2 + \frac{1}{2} \|M(x) \nabla_x \mathcal{L}(x, \lambda)\|^2 \tag{4}$$

was introduced, where $\mathcal{L}$ is the Lagrange function, $\mu > 0$ and $M : \mathbb{R}^n \to \mathbb{R}^{m \times n}$, $p \leq m \leq n$. This penalty function is exact in the sense that for each local solution $\hat{x}$ of the original constrained minimization problem and the related Lagrange multipliers $\hat{\lambda}$, a finite penalty parameter $\overline{\mu}$ exists such that for $\mu > \overline{\mu}$ the point $\hat{x}$ is the first component of a local minimum $(\hat{x}, \hat{\lambda})$ of the penalty function $P(x, \lambda)$. In this sense, $\mu \to \infty$ is not needed. The function $P(\cdot, \cdot; \mu, M)$ is closely related to augmented Lagrange methods, but there are some differences. The most import advantage, compared to standard augmented Lagrange, especially in the nonlinear FETI-DP context, is the fact that we can use the standard Lagrange–Newton equation

$$\begin{bmatrix} \nabla^2_{xx} \mathcal{L}(x, \lambda) & \nabla^2_{x\lambda} \mathcal{L}(x, \lambda) \\ \nabla^2_{\lambda x} \mathcal{L}(x, \lambda) & O \end{bmatrix} \begin{bmatrix} \delta x \\ \delta \lambda \end{bmatrix} = - \begin{bmatrix} \nabla_x \mathcal{L}(x, \lambda) \\ \nabla_\lambda \mathcal{L}(x, \lambda) \end{bmatrix} \tag{5}$$

see, e.g. [1], to compute a Newton-like search direction. Therefore, we do not need to modify the Hessian of $\mathcal{L}$, as in the standard augmented Lagrange method.

A detailed analysis of P can be found in [1, Chapter 4.3], including a proof of the exactness of P on X^*, where $X^* := \{x \in \mathbb{R}^n \mid \nabla c(x) \text{ has rank } p\}$, under the assumptions that $M \in C^1(X^*)$ and $M \nabla c$ is invertible on X^*.

We see that (5) is not affected by the penalty parameter μ. Indeed, μ only affects the acceptance criterion for this direction. Let us remark that in our context a good choice for M is $M(x) = \nabla c(x)^T$. Note that we assume that ∇c has full rank.

Let us remark that the standard method for the update of the penalty parameter in [1] needs to compute $(\nabla c(x)^T M(x)^T)^{-1} c(x)$. In our context, this is computationally expensive. Instead, we consider an update strategy inspired by augmented Lagrange methods [6]: Set $\mu_{k+1} = \varepsilon_{\text{update}} \, \mu_k$ whenever $\|c(x^{(k)})\| \geq \rho \, \|c(x^{(k+1)})\|$

for $\varepsilon_{\text{update}} > 1$ and $\rho \in (0, 1)$. The drawback is that we cannot guarantee any more that μ is increased only a finite number of times, which holds for the method suggested in [1].

A standard convergence result (every limit point of a Newton-like algorithm is a stationary point for P) can be obtained under standard assumptions, see e.g. [1], quite similar to Assumption 3.1, which we use later on.

We recall that by nonlinear elimination of x_E, we refer to solving

$$\nabla_{x_E}\mathcal{L}(x_E, x_L, \lambda) = \nabla_{x_E}J(x_E, x_L) + \nabla_{x_E}c(x_E, x_L)\lambda = 0 \tag{6}$$

for x_E, given x_L and λ, which defines the implicit function $g_E(x_L, \lambda)$. For simplicity, we now write ∇_E instead of ∇_{x_E} and ∇_L instead of ∇_{x_L}. We allow $E = \emptyset$ or $L = \emptyset$, then the related matrices or vectors are empty.

Combination with Nonlinear Elimination For the combination of $P(\cdot, \cdot; \mu, M)$ with nonlinear elimination, we replace x_E by the elimination g_E and define the functions $\mathfrak{L}(x_L, \lambda) := \mathcal{L}(g_E(x_L, \lambda), x_L, \lambda)$, $C(x_L, \lambda) := c(g_E(x_L, \lambda), x_L)$, and the penalty function

$$\begin{aligned}
&\mathcal{P}(x_L, \lambda; \mu, \mathcal{M}) \\
&= \mathfrak{L}(x_L, \lambda) + \frac{\mu}{2}\|C(x_L, \lambda)\|^2 + \frac{1}{2}\|\mathcal{M}(x_L, \lambda)\nabla_L \mathfrak{L}(x_L, \lambda)\|^2,
\end{aligned} \tag{7}$$

where $\mu > 0$ and $\mathcal{M} : \mathbb{R}^{n_L} \times \mathbb{R}^p \to \mathbb{R}^{p \times n_L}$. According to the considerations above, we define $\mathcal{M}$ as $\mathcal{M}(x_L, \lambda) := \nabla_L c\big|^T_{(g_E(x_L,\lambda), x_L)}$. By $\nabla_L c\big|_{(g_E(x_L,\lambda), x_L)}$ we mean the evaluation of $\nabla_L c$ at the point $(g_E(x_L, \lambda), x_L)$. By our assumptions on c it follows that $\mathcal{M} \in C^1(X_L^* \times \Lambda^*)$, where $X_L^* \times \Lambda^* := \{(x_L, \lambda) \mid \nabla c\big|_{(g_E(x_L,\lambda), x_L)}$ has rank $p\}$.

The special choice of $\mathcal{M}$ has the advantage of being consistent with the case $E = \emptyset$, $L = \mathbb{R}^n$. In this situation, we have $\mathcal{P}(\cdot, \cdot; \mu, \mathcal{M}) = P(\cdot, \cdot; \mu, M)$. The drawback is that for general selections of E, L we cannot guarantee that $\mathcal{M}$ has full rank. In the context of four nonlinear FETI-DP *NL*-1, 2, 3, 4 methods this means that only for *NL*-4 (*NL*-1) the matrix $\mathcal{M}$ has full rank. In *NL*-3 the matrix $\mathcal{M}$ has only zero entries and is empty in *NL*-2.

We cannot expect that all theoretical properties of P are transferred to $\mathcal{P}$. However, the exactness remains valid as well as some other properties.

Theorem 1 [13] *If $(x_E^*, x_L^*, \lambda^*)$ is a KKT point of (3) and $(x_L^*, \lambda^*) \in X_L^* \times \Lambda^*$, then (x_L^*, λ^*) is a stationary point of $\mathcal{P}(\cdot, \cdot; \mu, \mathcal{M})$ and*

$$\mathcal{P}(x_L^*, \lambda^*; \mu, \mathcal{M}) = \mathcal{J}(x_L^*, \lambda^*) = J(g_E(x_E^*, \lambda^*), x_L^*) = J(x_E^*, x_L^*).$$

Furthermore, if $\nabla^2_{xx}\mathcal{L}\big|_{(x_E^*, x_L^*, \lambda^*)}$ is positive definite on $\ker(\nabla c\big|^T_{(x_E^*, x_L^*)})$, then there exists a $\overline{\mu} > 0$ such that (x_L^*, λ^*) is a local minimum of $\mathcal{P}(\cdot, \cdot; \mu, \mathcal{M})$ for all $\mu > \overline{\mu}$.

Since $\mathcal{P}$ is an exact penalty function, we consider the unconstrained minimization problem $\min_{x_L, \lambda} \mathcal{P}(x_L, \lambda; \mu, \mathcal{M})$ to solve (3).

Init: $(x_L^{(0)}, \lambda^{(0)}) \in \mathbb{R}^{n_L} \times \in \mathbb{R}^{p}, \beta, \eta_1, \rho \in (0,1)$, $\varepsilon_{\text{update}} > 1$, $\varepsilon_{\text{tol}}, \mu_0, \eta_2, \eta_3, p > 0$.

for $k = 0, 1, \ldots$ until convergence **do**

1. If $\| \nabla \mathcal{P}^{(k)} \|_\infty \leq \varepsilon_{\text{tol}}$, *STOP*.
2. (a) Compute $\nabla \mathfrak{L}^{(k)}$ and $\nabla^2 \mathfrak{L}^{(k)}$.

   (b) Solve $\begin{bmatrix} \nabla^2_{LL} \mathfrak{L}^{(k)} & \nabla^2_{L\lambda} \mathfrak{L}^{(k)} \\ \nabla^2_{\lambda L} \mathfrak{L}^{(k)} & \nabla^2_{\lambda\lambda} \mathfrak{L}^{(k)} \end{bmatrix} \begin{bmatrix} \delta x_L^{(k)} \\ \delta \lambda^{(k)} \end{bmatrix} = - \begin{bmatrix} \nabla_L \mathfrak{L}^{(k)} \\ \nabla_\lambda \mathfrak{L}^{(k)} \end{bmatrix}$.

   (c) Set $d^{(k)} := \begin{bmatrix} \delta x_L^{(k)} \\ \delta \lambda^{(k)} \end{bmatrix}$.

   if $\nabla \mathcal{P}^{(k)T} d^{(k)} \leq - \min\{\eta_1, \eta_2 \|d^{(k)}\|^p\} \|d^{(k)}\|^2$. **then**

   Set $\begin{bmatrix} \delta x_L^{(k)} \\ \delta \lambda^{(k)} \end{bmatrix} := - \begin{bmatrix} \nabla_L \mathcal{P}^{(k)} \\ \nabla_\lambda \mathcal{P}^{(k)} \end{bmatrix}$.

   end
3. Compute the largest number $\alpha_k \in \{\beta^l \mid l = 0, 1, 2, \ldots\}$ such that the Armijo rule

$$\mathcal{P}(x_L^{(k)} + \alpha_k \delta x_L^{(k)}, \lambda^{(k)} + \alpha_k \delta \lambda^{(k)}; \mu_k, \mathcal{M}) - \mathcal{P}(x_L^{(k)}, \lambda^{(k)}; \mu_k, \mathcal{M})$$
$$\leq \eta_3 \, \alpha_k \left(\nabla_L \mathcal{P}^{(k)T} \delta x_L^{(k)} + \nabla_\lambda \mathcal{P}^{(k)T} \delta \lambda^{(k)} \right)$$

   holds.
4. Set $x_L^{(k+1)} = x_L^{(k)} + \alpha_k \delta x_L^{(k)}$ and $\lambda^{(k+1)} = \lambda^{(k)} + \alpha_k \delta \lambda^{(k)}$.
5. **if** $\| C(x_L^{(k+1)}, \lambda^{(k+1)}) \| \geq \rho \| C(x_L^{(k)}, \lambda^{(k)}) \|$ **then**

   Set $\mu_{k+1} = \varepsilon_{\text{update}} \mu_k$.

   else

   Set $\mu_{k+1} = \mu_k$.

   end

end

Fig. 1: Newton-like algorithm for the computation of stationary points of $\mathcal{P}$.

The same arguments, which show that (5) is a Newton-like direction for P, imply that

$$\begin{bmatrix} \delta x_L \\ \delta \lambda \end{bmatrix} = - \begin{bmatrix} \nabla^2_{LL} \mathfrak{L}(x_L, \lambda) & \nabla^2_{L\lambda} \mathfrak{L}(x_L, \lambda) \\ \nabla^2_{\lambda L} \mathfrak{L}(x_L, \lambda) & \nabla^2_{\lambda\lambda} \mathfrak{L}(x_L, \lambda) \end{bmatrix}^{-1} \begin{bmatrix} \nabla_L \mathfrak{L}(x_L, \lambda) \\ \nabla_\lambda \mathfrak{L}(x_L, \lambda) \end{bmatrix} \tag{8}$$

is a Newton-like direction for $\mathcal{P}(\cdot, \cdot; \mu, \mathcal{M})$ at (x_L, λ). Let us remark that the solution of (8) is equivalent to the solution of the standard Lagrange–Newton equation at the point $(g_E(x_L, \lambda)), x_L, \lambda)$.

We outline a Newton-like minimization algorithm for $\mathcal{P}$ in Figure 1, where we define

$$\nabla \mathfrak{L}^{(k)} := \nabla \mathfrak{L}(x_L^{(k)}, \lambda^{(k)}),$$
$$\nabla^2 \mathfrak{L}^{(k)} := \nabla^2 \mathfrak{L}(x_L^{(k)}, \lambda^{(k)}),$$
$$\nabla \mathcal{P}^{(k)} := \nabla \mathcal{P}(x_L^{(k)}, \lambda^{(k)}; \mu_k, \mathcal{M})$$

and the blocks $\nabla^2_{LL} \mathfrak{L}^{(k)}$, etc.

For the main convergence result of the algorithm presented in Figure 1 we need the following assumptions:

Assumption 3.1 *The sequence* $\left((x_L^{(k)}, \lambda^{(k)})\right)_k$ *generated by the Algorithm in Figure 1 is contained in a convex set* $\Omega_L \times \Lambda$ *and the following properties hold:*

(a) The nonlinear elimination $g_E(x_L, \lambda)$ *exists for all* $(x_L, \lambda) \in \Omega_L \times \Lambda$.
(b) The functions J *and* c_i, $i = 1, \ldots, p$ *and their first, second and third derivatives are bounded on* $g_E(\Omega_L \times \Lambda) \times \Omega_L$.
(c) The sequence $(\mu_k)_k$ *is bounded.*

The boundedness assumption 3.1(b) is needed to ensure that 2.(c) in the algorithm of Figure 1 is a generalized angle condition. Furthermore, we need 3.1(c) to prove the main convergence result.

Theorem 2 [13] *Let Assumption 3.1 be fulfilled. Then every limit point of the sequence* $((x_L^{(k)}, \lambda^{(k)}))_k$ *generated by the algorithm presented in Figure 1 is a stationary point of* $\mathcal{P}$.

4 Numerical results

We consider a neo-Hookean benchmark problem using stiff or almost incompressible inclusions embedded in each subdomain. The strain energy density function for the compressible part is given by $J(x) = \frac{\mu}{2}(\mathrm{tr}(F(x)^T F(x)) - 2) - \mu \log(\psi(x)) + \frac{\lambda}{2}(\log(\psi(x)))^2$, where $\psi(x) = \det(F(x))$, $F(x) = \nabla\varphi(x)$, $\varphi(x) = x + u(x)$, $u(x)$ denotes the displacement and μ and λ are the Lamé constants. The nearly incompressible part is given by $J(x) = \frac{\mu}{2}(\mathrm{tr}(\frac{1}{\psi(x)} F(x)^T F(x)) - 2) + \frac{\kappa}{2}(\psi(x) - 1)^2$, where $\kappa = \frac{\lambda(1+\mu)}{3\mu}$, see, e.g. [18]. As material parameters, we use $E = 210$ and $\nu = 0.3$ for the matrix material, $E = 210\,000$ and $\nu = 0.3$ for the stiff inclusions, and, finally, $E = 210$ and $\nu = 0.499$ for the (mildly) almost incompressible inclusions. For the discretization, we use $P2$ elements, which are not stable for the incompressible case.

As Krylov methods, we use GMRES or CG: During the factorizations, it is detected whether $D\widetilde{K}$ is positive definite; in this case, we use CG, otherwise GMRES is used. In Table 1 we see that Newton's method, without globalization, will not converge in the case without inclusions for the body force $(0, -20)^T$, and in the cases with inclusions even for the smaller body force $(0, -10)^T$. In Table 2 we see that, using the algorithm in Figure 1 using the four different nonlinear FETI-DP methods *NL*-1, *NL*-2, *NL*-3, and *NL*-4, we have convergence even for the higher body force $(0, -60)^T$. The cases $(0, -10)^T$ and $(0, -20)^T$ converge as well, but are not presented here. The failure of *NL*-1, 2 to converge despite globalization is due to the fact that we reached the stopping criterion, $\frac{\max\{\|x^{(k+1)} - x^{(k)}\|_\infty, \|\lambda^{(k+1)} - \lambda^{(k)}\|_\infty\}}{\max\{\|x^{(k)}\|_\infty, \|\lambda^{(k)}\|_\infty\}} < 10^{-8}$. This indicates that no sufficient progress is reached, and we abort the simulation since we are limited to machine precision. This example also illustrates that nonlinear elimination can help to achieve convergence.

Fig. 2: Model problem with 4×4 subdomains.
Left: Start configuration; *blue*: matrix material, *red*: inclusions.
Right: Deformed state

Table 1: Nonlinear FETI-DP-1 or NL–1; $H/h \approx 8$, see Fig. 2; Newton's method without globalization; the number of Newton iterations is shown; no conv.: $\|\nabla \mathcal{L}^{(k)}\|_\infty \geq 1e5\, \|\nabla \mathcal{L}^{(0)}\|_\infty$

No Globalization					
		body force			
		f=(0, −10)^T	f=(0, −20)^T	f=(0, −10)^T	f=(0, −10)^T
d.o.f.	#Sub.	no incl.		comp. incl. (E = 210 000)	incomp. incl. (ν = 0.499)
16 642	16	5	**no conv.**	**no conv.**	**no conv.**
25 922	25	5	**no conv.**	**no conv.**	**no conv.**
37 250	36	5	**no conv.**	**no conv.**	**no conv.**
50 626	49	5	**no conv.**	**no conv.**	**no conv.**
66 050	64	5	**no conv.**	**no conv.**	**no conv.**

Table 2: Nonlinear-FETI-DP-1,2,3,4 or NL–1, 2, 3, 4; body force $f = (0, -60)^T$; $H/h \approx 8$; globalized Newton-like method; number of Newton iterations is shown; stopping criterion: $\|\nabla \mathcal{L}^{(k)}\|_\infty \leq 1e-6\, \|\nabla \mathcal{L}^{(0)}\|_\infty$. Globalization using the algorithm in Figure 1 is used

Using Globalization; see Figure 1													
		body force f = (0, −60)^T											
		no incl.				comp. incl. (E = 210 000)				incomp. incl. (ν = 0.499)			
d.o.f.	#Sub.	NL-1	NL-2	NL-3	NL-4	NL-1	NL-2	NL-3	NL-4	NL-1	NL-2	NL-3	NL-4
16 642	16	8	5	5	7	18	6	5	13	-	5	8	19
25 922	25	8	5	6	8	16	6	6	9	-	5	8	21
37 250	36	8	7	5	8	16	7	5	11	-	6	8	21
50 626	49	8	6	6	8	16	7	5	15	-	6	9	22
66 050	64	8	-	7	8	16	7	5	23	-	9	9	23

References

1. D.P. Bertsekas. *Constrained optimization and Lagrange multiplier methods*. Computer Science and Applied Mathematics. Academic Press, Inc. [Harcourt Brace Jovanovich, Publishers], New York-London (1982).
2. X.C. Cai and D.E. Keyes. Nonlinearly Preconditioned Inexact Newton Algorithms. *SIAM J. Sci. Comput.* **24**(1), 183–200 (2002). DOI:10.1137/S106482750037620X
3. G. Di Pillo and L. Grippo. A new class of augmented Lagrangians in nonlinear programming. *SIAM J. Control Optim.* **17**(5), 618–628 (1979). DOI:10.1137/0317044
4. C.R. Dohrmann. A preconditioner for substructuring based on constrained energy minimization. *SIAM J. Sci. Comput.* **25**(1), 246–258 (2003).

5. C. Farhat, M. Lesoinne and K. Pierson. A Scalable Dual-Primal Domain Decomposition Method. *Numer. Linear Algebra Appl.* **7**(7–8), 687–714 (2000). DOI:10.1002/1099-1506(200010/12)7: 7/8%3C687::AID-NLA219%3E3.0.CO;2-S
6. C. Geiger and C. Kanzow. Theorie und Numerik restringierter Optimierungsaufgaben. Springer (2002). DOI:10.1007/1978-3-642-56004-0
7. S. Gong and X.C. Cai. A Nonlinear Elimination Preconditioned Inexact Newton Method for Heterogeneous Hyperelasticity. *SIAM J. Sci. Comput.* **41**(5), S390–S408 (2019). DOI:10.1137/18M1194936
8. C. Groß. *A Unifying Theory for Nonlinear Additively and Multiplicatively Preconditioned Globalization Strategies*. Universitäts-und Landesbibliothek Bonn (2009).
9. C. Groß and R. Krause. On the globalization of aspin employing trust-region control strategies – convergence analysis and numerical examples. Tech. Rep. 2011-03, Institute of Computational Science, Universita della Svizzera italiana (2011).
10. A. Klawonn, M. Lanser and O. Rheinbach. Nonlinear FETI–DP and BDDC methods. *SIAM J. Sci. Comput.* **36**(2), A737–A765 (2014). DOI:10.1137/130920563
11. A. Klawonn, M. Lanser and O. Rheinbach. FE2TI: Computational Scale Bridging for Dual-Phase Steels (2015). In: *PARCO*, pp. 797–806 (2015). DOI:10.3233/978-1-61499-621-7-797
12. A. Klawonn, M. Lanser, O. Rheinbach and M. Uran. Nonlinear FETI–DP and BDDC methods: a unified framework and parallel results. *SIAM J. Sci. Comput.* **39**(6), C417–C451 (2017). DOI:10.1137/16M1102495
13. S. Köhler and O. Rheinbach. Globalization of Nonlinear FETI–DP Methods. In Prep.
14. J. Nocedal and S. Wright. *Numerical Optimization*. Springer Science & Business Media (2006). DOI:10.1007/978-0-387-40065-5
15. J. Pebrel, C. Rey and P. Gosselet. A Nonlinear Dual-Domain Decomposition Method: Application to Structural Problems with Damage. *Int. J. Multiscale Comput. Eng.* **6**(3), 251–262 (2008). DOI:10.1615/IntJMultCompEng.v6.i3.50
16. A. Toselli and O. Widlund. *Domain decomposition methods – algorithms and theory*. Springer Series in Computational Mathematics, vol. 34, Springer, Berlin (2005).
17. M. Ulbrich and S. Ulbrich. Nichtlineare Optimierung. Birkhäuser Basel (2012). DOI:10.1007/978-3-0346-0654-7
18. M. Uran. *High-Performance Computing Two-Scale Finite Element Simulations of a Contact Problem Using Computational Homogenization*. Ph.D. thesis, Universität zu Köln (2020).

A Multilevel Active-Set Trust-Region (MASTR) Method for Bound Constrained Minimization

Alena Kopaničáková and Rolf Krause

1 Introduction

We consider a minimization problem of the following type:

$$\begin{aligned} \min_{\mathbf{x}\in\mathbb{R}^n} \quad & f(\mathbf{x}) \\ \text{subject to} \quad & \mathbf{x} \in \mathcal{F}, \end{aligned} \tag{P}$$

where $f : \mathbb{R}^n \to \mathbb{R}$ is a possibly non-convex, but twice continuously differentiable objective function. The feasible set $\mathcal{F} := \{\mathbf{x} \in \mathbb{R}^n | \, \mathbf{l} \leq \mathbf{x} \leq \mathbf{u}\}$ is defined in terms of the pointwise lower bound $\mathbf{l} \in \mathbb{R}^n$ and the upper bound $\mathbf{u} \in \mathbb{R}^n$. We assume that the function f arises from the finite element (FE) discretization of a partial differential equation (PDE). Here, $n \in \mathbb{N}$ denotes the dimension of the finite element space and it is typically very large. Problems of this type arise commonly in many scientific applications, for example in fracture mechanics [11].

Multilevel methods are known to be optimal solution strategies for systems arising from the discretization of, usually elliptic, PDEs, as their convergence rate is often independent of the problem size and the number of required arithmetic operations grows proportionally with the number of unknowns. These methods were originally designed for unconstrained PDEs [2]. Their extension to constrained settings is not trivial as the coarse levels are often not capable of resolving the finest-level constraints sufficiently well, especially if the constraints are oscillatory [13]. The initial attempts to incorporate the constraints into the multilevel framework were associated with solving linear complementarity problems, see for instance [1, 5, 8, 14]. The devised methods employed various constraint projection rules for constructing the coarse-level variable bounds, such that coarse-level corrections are admissible at the

Alena Kopaničáková
Università della Svizzera italiana, Switzerland, e-mail: alena.kopanicakova@usi.ch

Rolf Krause
Università della Svizzera italiana, Switzerland, e-mail: rolf.krause@usi.ch

S. C. Brenner et al. (eds.), *Domain Decomposition Methods in Science and Engineering XXVI*, Lecture Notes in Computational Science and Engineering 145,
https://doi.org/10.1007/978-3-030-95025-5_37

finest level. Unfortunately, these projection rules tend to be overly restrictive. As a consequence, the resulting multilevel methods converge significantly more slowly than the standard linear multigrid method.. In order to enhance the convergence speed, Kornhuber proposed an active-set multigrid method [12]. The method utilizes a truncated basis approach and recovers the convergence rate of the unconstrained multigrid, once the exact active-set is detected [9, 12].

In the field of nonlinear optimization, very few existing nonlinear multilevel algorithms can be readily employed. For instance, Vallejos proposed a gradient projection based multilevel method [15]. Two multilevel line-search methods, designed for convex optimization problems, are proposed in [10]. These methods utilize constraint projection rules developed in [8] and a variant of the active-set strategy from [12]. In the context of non-convex optimization problems, Youett et al. proposed a filter trust-region algorithm [16], which employs an active-set multigrid method [12] for the solution of arising linearized problems. Furthermore, Gratton et al. proposed a variant of the recursive multilevel trust-region (RMTR) method [7] by utilizing the constraint projection rules from [5]. To our knowledge, this is currently the only inherently nonlinear multilevel method, which provides global convergence guarantees for non-convex bound constrained optimization problems.

In the present work, we propose to enhance the convergence speed of the RMTR method introduced in [7]. More precisely, we present an active-set variant, called the Multilevel Active-Set Trust-Region (MASTR) method. The MASTR method employs an active-set strategy, which determines which components of the fine-level solution vector are active. These components are then held fixed and cannot be altered by the coarser levels. With this aim, we have to construct coarse-level models such that their minimization yields corrections, which fulfill this requirement. Here, we employ coarse-level models of the Galerkin type together with the truncated basis method [12]. In contrast to [10], the practical implementation of the proposed coarse-level models does not require any unconventional modifications to existing FE software packages. As our numerical results will demonstrate, employing the active-set approach leads to significant speedup of the RMTR method.

2 The recursive multilevel trust-region (RMTR) method

In this work, we minimize (P) using a novel variant of the RMTR method [7]. RMTR combines the global convergence properties of the trust-region (TR) method with the efficiency of multilevel methods. By design, the RMTR method employs a hierarchy of L levels. Each level l is associated with a mesh $\mathcal{T}^l$, which encapsulates the computational domain $\Omega \in \mathbb{R}^d$, where $d \in \mathbb{N}$. The mesh $\mathcal{T}^l$ is used to construct the first-order finite-element space V^l, spanned by the basis functions $\{N_k^l\}_{k \in \mathcal{N}^l}$, where $\mathcal{N}^l$ denotes the set of nodes of the mesh $\mathcal{T}^l$. The support of a given basis function N_k^l is defined as $\omega_k^l = \overline{\{x \in \Omega \mid N_k^l(x) \neq 0\}}$.

The transfer of data between subsequent levels of the multilevel hierarchy is carried out using three transfer operators, namely prolongation $\mathbf{I}_l^{l+1} : \mathbb{R}^{n^l} \rightarrow \mathbb{R}^{n^{l+1}}$, restriction $\mathbf{R}_{l+1}^{l} := (\mathbf{I}_{l}^{l+1})^T$ and projection $\mathbf{P}_{l+1}^{l} : \mathbb{R}^{n^{l+1}} \to \mathbb{R}^{n^l}$. In this work, all transfer operators are assembled using L^2-projection.

Algorithm

On each level l, the RMTR method approximates (P) by means of some level-dependent objective function $h^l : \mathbb{R}^{n^l} \to \mathbb{R}$ and feasible set $\mathcal{F}^l := \{\mathbf{x}^l \in \mathbb{R}^{n^l} \mid \mathbf{l}^l \leq \mathbf{x}^l \leq \mathbf{u}^l\}$. The function h^l is approximately minimized in order to obtain a coarse-level correction. This correction is then interpolated to the subsequent finer level, $l+1$, where it is used to improve the current iterate.

More precisely, the algorithm starts on the finest level, $l = L$, with an initial iterate $\mathbf{x}_0^L$ and passes through all levels until the coarsest level, $l = 1$, is reached. On each level l, the algorithm performs μ_1 pre-smoothing steps to improve the current iterate $\mathbf{x}_0^l$. The smoothing is performed using the TR method [3]. Thus, at each TR iteration i, the search direction $\mathbf{s}_i^l$ is obtained by approximately solving the following minimization problem:

$$
\begin{aligned}
\min_{\mathbf{s}_i^l \in \mathbb{R}^{n^l}} \ & m_i^l(\mathbf{x}_i^l + \mathbf{s}_i^l) := h^l(\mathbf{x}_i^l) + \langle \nabla h^l(\mathbf{x}_i^l), \mathbf{s}_i^l \rangle + \frac{1}{2}\langle \mathbf{s}_i^l, \nabla^2 h^l(\mathbf{x}_i^l)\, \mathbf{s}_i^l \rangle, \\
& \text{such that } \ \mathbf{x}_i^l + \mathbf{s}_i^l \in \mathcal{F}^l, \\
& \|\mathbf{s}_i^l\|_\infty \leq \Delta_i^l,
\end{aligned} \tag{1}
$$

where m_i is the second-order Taylor approximation of h^l. The symbol $\Delta_i^l > 0$ denotes a TR radius, which controls the size of the correction $\mathbf{s}_i^l$. In contrast to line-search methods, the correction $\mathbf{s}_i^l$ is used only if $\rho_i^l > \eta_1$, where $\rho_i^l = \frac{h^l(\mathbf{x}_i^l) - h^l(\mathbf{x}_i^l + \mathbf{s}_i^l)}{m^l(\mathbf{x}_i^l) - m^l(\mathbf{x}_i^l + \mathbf{s}_i^l)}$ and $\eta_1 > 0$. Otherwise, $\mathbf{s}_i^l$ is disposed of and the size of the TR radius is reduced. The result of the pre-smoothing, the iterate $\mathbf{x}_{\mu_1}^l$, is then used to initialize the solution vector on the subsequent coarser level, i.e., $\mathbf{x}_0^{l-1} := \mathbf{P}_l^{l-1}\mathbf{x}_{\mu_1}^l$.

Once the coarsest level is reached, we apply μ^1 steps of the TR method to obtain the updated iterate $\mathbf{x}_{\mu^1}^1$. The algorithm then returns to the finest level. To this end, the correction obtained on level l, i.e., $\mathbf{x}_{\mu^l}^l - \mathbf{x}_0^l$, is transferred to level $l+1$, by means of the prolongation operator, thus $\mathbf{s}_{\mu_1+1}^{l+1} := \mathbf{I}_l^{l+1}(\mathbf{x}_{\mu^l}^l - \mathbf{x}_0^l)$. Here, the symbol μ^l denotes a sum of all iterations taken on a given level l. However, the quality of the prolongated coarse-level correction $\mathbf{s}_{\mu_1+1}^{l+1} := \mathbf{I}_l^{l+1}(\mathbf{x}_{\mu^l}^l - \mathbf{x}_0^l)$ has to be assessed before it is accepted on level $l+1$. For this reason, we define a multilevel TR ratio as $\rho_{\mu_1+1}^{l+1} := \frac{h^{l+1}(\mathbf{x}_{\mu_1}^{l+1}) - h^{l+1}(\mathbf{x}_{\mu_1}^{l+1} + \mathbf{s}_{\mu_1+1}^{l+1})}{h^l(\mathbf{x}_0^l) - h^l(\mathbf{x}_{\mu^l}^l)}$. The correction $\mathbf{s}_{\mu_1+1}^{l+1}$ is accepted if $\rho_{\mu_1+1}^{l+1} > \eta_1$. If $\rho_{\mu_1+1}^{l+1} \leq \eta_1$, the correction $\mathbf{s}_{\mu_1+1}^{l+1}$ is rejected. Additionally, the TR radius has to be updated accordingly. At the end, the RMTR algorithm performs μ_2 post-smoothing steps at a given level l. This process is repeated on every level until the finest level is reached.

Construction of level-dependent objective functions and feasible sets

In this section, we discuss how to construct the objective function h^l and feasible set $\mathcal{F}^l$. For the finest level, $l = L$, we assume that $h^L := f$ and $\mathcal{F}^L := \mathcal{F}$. In coherence with the RMTR convergence theory [7], the coarse-level functions $\{h^l\}_{l=1}^{L-1}$ have to be constructed such that they are at least twice continuously differentiable, and at least first-order consistent with the function h^{l+1}. Here, we create a level-dependent objective function h^l as follows:

$$h^l(\mathbf{x}^l + \mathbf{s}^l) := \langle \mathbf{R}_{l+1}^l \nabla h_{\mu_1}^{l+1}, \mathbf{s}^l \rangle + \frac{1}{2} \langle \mathbf{s}^l, (\mathbf{R}_{l+1}^l \nabla^2 h_{\mu_1}^{l+1} \mathbf{I}_l^{l+1}) \mathbf{s}^l \rangle, \tag{2}$$

where $\mathbf{s}^l = \mathbf{x}^l - \mathbf{x}_0^l$, and $\mathbf{R}_{l+1}^l \nabla h_{\mu_1}^{l+1}$, $\mathbf{R}_{l+1}^l \nabla^2 h_{\mu_1}^{l+1} \mathbf{I}_l^{l+1}$ represent the restricted gradient and the Hessian from level $l+1$, evaluated after μ_1 pre-smoothing steps, respectively. As we will see in Section 3, employing coarse-level models of this particular type allows for straightforward incorporation of the active set strategy within the multilevel settings.

For all levels $l < L$, the level-dependent feasible set $\mathcal{F}^l$ is created by intersecting the set $\mathcal{L}^l$ with the set $\mathcal{S}^l$, thus as $\mathcal{F}^l := \mathcal{L}^l \cap \mathcal{S}^l$. The purpose of the set $\mathcal{S}^l := \{\mathbf{x}^l \in \mathbb{R}^{n^l} \mid \mathbf{tl}^l \leq \mathbf{x}^l \leq \mathbf{tu}^l\}$ is to ensure that the size of the prolongated coarse-level correction remains bounded by the TR radius $\Delta_{\mu_1}^l$, i.e., $\|\mathbf{I}_l^{l+1} \mathbf{s}^l\|_\infty \leq \Delta_{\mu_1}^l$. To this end, we construct $\mathcal{S}^l$ by employing the projection rules especially designed for TR bounds in [7].

The function of the set $\mathcal{L}^l := \{\mathbf{x}^l \in \mathbb{R}^{n^l} \mid \mathbf{vl}^l \leq \mathbf{x}^l \leq \mathbf{vu}^l\}$ is to guarantee that the prolongated coarse-level correction produces a feasible trial point, i.e., $\mathbf{x}_{\mu_1}^{l+1} + \mathbf{I}_l^{l+1} \mathbf{s}^l \in \mathcal{F}^{l+1}$. Following [5, 7], we can construct $\mathbf{vl}^l$, $\mathbf{vu}^l$ in a component-wise manner as

$$\begin{aligned} (\mathbf{vl}^l)_k &:= (\mathbf{x}_0^l)_k + \max_{j \in \mathcal{N}^{l+1} \cap \mathring{\omega}_k^l} [(\mathbf{vl}^{l+1} - \mathbf{x}_{\mu_1}^{l+1})_j], \\ (\mathbf{vu}^l)_k &:= (\mathbf{x}_0^l)_k + \min_{j \in \mathcal{N}^{l+1} \cap \mathring{\omega}_k^l} [(\mathbf{vu}^{l+1} - \mathbf{x}_{\mu_1}^{l+1})_j], \end{aligned} \tag{3}$$

where $(\cdot)_k$ denotes the k-th component of a given vector. Note that the support ω_k^l of the basis function N_k^l (associated with k-th node of the mesh $\mathcal{T}^l$) determines which components of the variable bounds $\mathbf{vl}^{l+1}$ and $\mathbf{vu}^{l+1}$ have to be taken into account while constructing $\mathbf{vl}^l$, $\mathbf{vu}^l$.

3 The multilevel active-set trust-region (MASTR) method

In this section, we present how to incorporate the active-set strategy into the RMTR framework. The devised algorithm has a form of the standard V-cycle. The key idea behind the proposed MASTR method is to identify an active-set

$$\mathcal{A}^l(\mathbf{x}^l_{\mu_1}) := \{k \in \{1,\dots,n^l\} \mid (\mathbf{vl}^l)_k = (\mathbf{x}^l_{\mu_1})_k \text{ or } (\mathbf{vu}^l)_k = (\mathbf{x}^l_{\mu_1})_k\}, \tag{4}$$

before descending to the coarser level. Here, the vectors $\mathbf{vl}^l, \mathbf{vu}^l$ denote lower and upper bounds that define the set $\mathcal{L}^l$ and are obtained using formula (3). The components of the solution vector $\mathbf{x}^l_{\mu_1}$ that are active are then held fixed and cannot be altered by the coarser levels. To this end, the level-dependent objective functions $\{h^a\}_{a=1}^{l-1}$ and feasible sets $\{\mathcal{L}^a\}_{a=1}^{l-1}$ have to be constructed such that the minimization process on a given level yields coarse-level corrections, which fulfil this requirement. Following [9, 12], we construct $\{h^a\}_{a=1}^{l-1}$ and $\{\mathcal{L}^a\}_{a=1}^{l-1}$ using a truncated basis method.

Construction of truncated FE spaces

The truncated basis method [12] constructs truncated FE spaces $\{\widetilde{\mathcal{X}}^l\}_{l=1}^{L-1}$, spanned by truncated basis functions $\{\widetilde{N}^l_k\}_{k\in\mathcal{N}^l}$, by exploiting the fact that the basis functions on level l can be written as linear combination of basis functions at level $l+1$, i.e., $N^l_k = \sum_{p=1}^{n^{l+1}} (\mathbf{I}_l^{l+1})_{pk} N_p^{l+1}$. In contrast to the classical multilevel approaches with canonical Galerkin restriction, the actual shape and support of the truncated basis functions depend on the current fine-level iterate. In particular, the value of the truncated basis functions is set to zero at all active nodes of the finer levels, i.e., their support vanishes at the active nodes. More precisely, we can construct truncated basis functions in a recursive manner as

$$\widetilde{N}^l_k = \sum_{p=1}^{n^{l+1}} (\widetilde{\mathbf{I}}_l^{l+1})_{pk} \widetilde{N}_p^{l+1}, \tag{5}$$

where $\widetilde{\mathbf{I}}_l^{l+1}$ is truncated prolongation operator defined by

$$(\widetilde{\mathbf{I}}_l^{l+1})_{pk} = \begin{cases} 0, & \text{if } p \in \mathcal{A}^{l+1}(\mathbf{x}^l_{\mu_1}), \\ (\mathbf{I}_l^{l+1})_{pk}, & \text{otherwise.} \end{cases} \tag{6}$$

The operator $\widetilde{\mathbf{I}}_l^{l+1}$ is obtained from the prolongation operator $\mathbf{I}_l^{l+1}$ by setting the p-th row of $\mathbf{I}_l^{l+1}$ to zero, for all $p \in \mathcal{A}^{l+1}(\mathbf{x}^l_{\mu_1})$. The application of $\widetilde{\mathbf{I}}_l^{l+1}$ in (5) removes contributions of basis functions associated with active nodes on level $l+1$, defined by the set $\mathcal{A}^{l+1}(\mathbf{x}^l_{\mu_1})$.

Remark 1 For the level $L-1$, the recursive formula (5) employs $\{N^L_k\}_{k\in\mathcal{N}^L}$, instead of $\{\tilde{N}^L_k\}_{k\in\mathcal{N}^L}$.

Construction of level-dependent objective functions and feasible sets

Using truncated FE spaces $\{\widetilde{\mathcal{X}}^l\}_{l=1}^{L-1}$, we can now construct level-dependent objective functions $\{h^l\}_{l=1}^{L-1}$ and feasible sets $\{\mathcal{F}^l\}_{l=1}^{L-1}$. In particular, for a given level $l < L$, the level-dependent objective function $h^l : \mathbb{R}^{n^l} \to \mathbb{R}$ is created as follows:

$$h^l(\mathbf{x}^l + \mathbf{s}^l) := \langle (\widetilde{\mathbf{I}}_l^{l+1})^T \nabla h_{\mu_1}^{l+1}, \mathbf{s}^l \rangle + \frac{1}{2} \langle \mathbf{s}^l, (\widetilde{\mathbf{I}}_l^{l+1})^T \nabla^2 h_{\mu_1}^{l+1} \widetilde{\mathbf{I}}_l^{l+1}) \mathbf{s}^l \rangle, \tag{7}$$

where we used the truncated transfer operator $\widetilde{\mathbf{I}}_l^{l+1}$ to restrict the gradient $\nabla h_{\mu_1}^{l+1}$ and Hessian $\nabla^2 h_{\mu_1}^{l+1}$ from level $l+1$ to level l. The application of $\widetilde{\mathbf{I}}_l^{l+1}$ in (7) removes the components of the fine-level gradient/Hessian associated with the active-set $\mathcal{A}^{l+1}(\mathbf{x}_{\mu_1}^l)$. Please note that the formulation (7) does not require an explicit representation of $\{\widetilde{\mathcal{X}}^l\}_{l=1}^{L-1}$.

The construction of each level-dependent feasible set $\mathcal{L}^l$ can be performed using projection rules defined by (3). However, formulas (3) are now determined by the support of the truncated basis functions, spanning $\widetilde{\mathcal{X}}^l$. Since the support of basis functions spanning $\widetilde{\mathcal{X}}^l$ is different from the support of the basis functions spanning $\mathcal{X}^l$, fewer components of the fine-level variable bounds are taken into account by (3). This yields less restrictive coarse-level constraints and allows for larger coarse grid corrections.

4 Numerical results

We study the performance of the proposed MASTR method using three numerical examples. Examples are defined on the domain $\Omega := [0,1]^2$ with boundary $\Gamma = \partial\Omega$, decomposed into three parts: $\Gamma_l = \{0\} \times [0,1]$, $\Gamma_r = \{1\} \times [0,1]$, and $\Gamma_f = [0,1] \times \{0,1\}$. The discretization is performed using a uniform mesh and $\mathbb{Q}_1$ Lagrange finite elements.

Ex.1. MEMBRANE: Let us solve the following minimization problem [4]:

$$\begin{aligned} &\min_{u \in \mathcal{X}} \ f(u) := \frac{1}{2} \int_\Omega \|\nabla u(x)\|^2 \, \mathrm{d}x + \int_\Omega u(x) \, \mathrm{d}x, \\ &\text{subject to } \ \mathrm{lb}(x) \leq u, \quad \text{on } \Gamma_r, \end{aligned} \tag{8}$$

where $\mathcal{X} := \{u \in H^1(\Omega) \mid u = 0 \text{ on } \Gamma_l\}$. The lower bound lb is defined on the right part of the boundary, Γ_r, by the upper part of the circle with the radius, $r = 1$, and the center, $C = (1; -0.5; -1.3)$. Thus, the lower bound is defined as

$$\mathrm{lb}(x) = \begin{cases} (-2.6 + \sqrt{2.6^2 - 4((x_2 - 0.5)^2 - 1.0 + 1.3^2)})/2, & \text{if } x = 1, \\ -\infty, & \text{otherwise,} \end{cases}$$

where the symbols x_1, x_2 denote spatial coordinates.

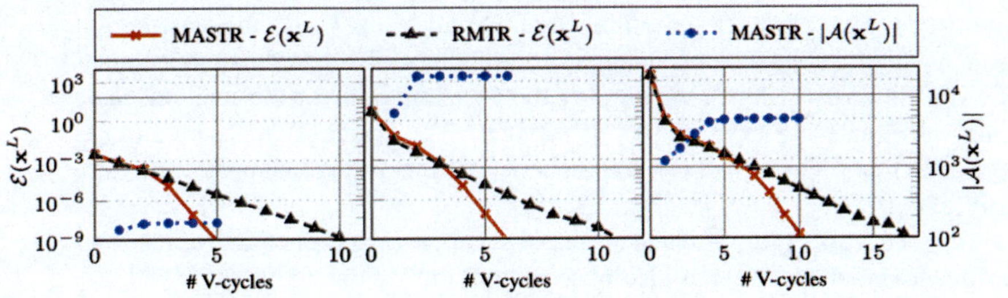

Fig. 1: The convergence of the MASTR method (red color) and RMTR without an active-set strategy (black color). The blue color illustrates the size of an active set detected by the MASTR algorithm. *Left:* MEMBRANE. *Middle:* IGNITION. *Right:* MOREBV.

Ex.2. IGNITION: Following [2, 10], we minimize following optimization problem:

$$\min_{u \in \mathcal{X}} \; f(u) := \frac{1}{2}\int_\Omega \|\nabla u(x)\|^2 - (ue^u - e^u)\,dx - \int_\Omega f(x)u\,dx, \qquad (9)$$
$$\text{subject to } \; \text{lb}(x) \le u \le \text{ub}(x), \quad \text{a.e. in } \Omega.$$

The variable bounds and right-hand side are defined as

$$\text{lb}(x) = -8(x_1 - 7/16)^2 - 8(x_2 - 7/16)^2 + 0.2, \quad \text{ub}(x) = 0.5,$$
$$f(x) = (9\pi^2 + e^{(x_1^2 - x_1^3)\sin(3\pi x_2)}(x_1^2 - x_1^3) + 6x_1 - 2)\sin(3\pi x_1),$$

where $\mathcal{X} := \{u \in H^1(\Omega) \mid u = 0 \text{ on } \Gamma\}$, and $f \in L^2(\Omega)$.

Ex.3. MOREBV: We consider the following non-convex minimization problem [6]:

$$\min_{u \in \mathcal{X}} \; f(u) := \int_\Omega \|\Delta u(x) - 0.5(u(x) + \langle e, x\rangle + 1)^3\|_2^2\,dx, \qquad (10)$$
$$\text{subject to } \; \text{lb}(x) \le u, \quad \text{a.e. in } \Omega,$$

where $\mathcal{X} := \{u \in H^1(\Omega) \mid u = 0 \text{ on } \Gamma\}$ and e denotes a unit vector. The lower bound is defined as $\text{lb}(x) = \sin(5\pi x_1)\sin(\pi x_2)\sin(\pi(1 - x_1))\sin(\pi(1 - x_2))$, where x_1, x_2 denote spatial coordinates.

Convergence study

We compare the convergence behavior of the proposed MASTR method with the standard RMTR method (without the active-set strategy). Both methods are implemented as part of the open-source library UTOPIA [17].

For all experiments, we consider the RMTR method configured with 6-levels. The multilevel hierarchy is obtained by uniformly refining the coarsest level mesh, consisting of 10×10 elements. This gives rise to 289×289 elements on the finest level, which corresponds to 83,521 dofs. The performed study considers an RMTR setup with one trust-region pre/post-smoothing step. On all levels $l > 1$, the trust-region subproblems (1) are solved using one iteration of successive coordinate minimization [7]. On the coarsest level, we employ the Semismooth-Newton method. The algorithms terminate if $\mathcal{E}(\mathbf{x}^L) < 10^{-9}$ is satisfied. The criticality measure $\mathcal{E}(\mathbf{x})$ is defined as $\mathcal{E}(\mathbf{x}) := \|\mathcal{P}(\mathbf{x} - \nabla f(\mathbf{x})) - \mathbf{x}\|$, where $\mathcal{P}$ is the orthogonal projection onto the feasible set $\mathcal{F}$.

As we can see from Figure 1, using an active-set approach is beneficial, as it allows for significant speed up. We can also observe that during the active-set identification phase (the first few V-cycles), both approaches are comparable. However, once the exact active-set is detected, MASTR accelerates and converges faster than standard RMTR.

Funding This work was funded by the Swiss National Science Foundation (SNF) under the project ML2 (grant no. 197041), Platform for Advanced and Scientific Computing (PASC) under the project ExaTrain and by the MATH+ (distinguished scholar R. Krause).

References

1. A. Brandt and C.W. Cryer. Multigrid algorithms for the solution of linear complementarity problems arising from free boundary problems. *SIAM journal on scientific and statistical computing* **4**(4), 655–684 (1983).
2. W.L.Briggs, S.F. McCormick, et al. A multigrid tutorial. SIAM (2000).
3. A.R. Conn, N.I.M. Gould and P.L. Toint. *Trust Region Methods*. MOS-SIAM Series on Optimization. SIAM (2000). DOI 10.1137/1.9780898719857
4. M. Domorádová and Z. Dostál. Projector preconditioning for partially bound-constrained quadratic optimization. *Numerical Linear Algebra with Applications* **14**(10), 791–806 (2007).
5. E. Gelman and J. Mandel. On multilevel iterative methods for optimization problems. *Mathematical Programming* **48**(1-3), 1–17 (1990).
6. S. Gratton, M. Mouffe, A. Sartenaer, P.L. Toint and D. Tomanos. Numerical experience with a recursive trust-region method for multilevel nonlinear bound-constrained optimization. *Optimization Methods and Software* **25**(3), 359–386 (2010). DOI 10.1080/10556780903239295
7. S. Gratton, M. Mouffe, P. Toint and M. Weber-Mendonça. A recursive ℓ_∞-trust-region method for bound-constrained nonlinear optimization. *IMA Journal of Numerical Analysis* **28**(4), 827–861 (2008).
8. W. Hackbusch and H.D. Mittelmann. On multi-grid methods for variational inequalities. *Numerische Mathematik* **42**(1), 65–76 (1983).
9. R.H. Hoppe and R. Kornhuber. Adaptive multilevel methods for obstacle problems. *SIAM journal on numerical analysis* **31**(2), 301–323 (1994).
10. M. Kočvara and S. Mohammed. A first-order multigrid method for bound-constrained convex optimization. *Optimization Methods and Software* **31**(3), 622–644 (2016).
11. A. Kopaničáková and R. Krause. A recursive multilevel trust region method with application to fully monolithic phase-field models of brittle fracture. *Computer Methods in Applied Mechanics and Engineering* **360**, 112720 (2020).
12. R. Kornhuber. Monotone multigrid methods for elliptic variational inequalities I. *Numerische Mathematik* **69**(2), 167–184 (1994).
13. R. Kornhuber and R. Krause. Adaptive multigrid methods for Signorini's problem in linear elasticity. *Computing and Visualization in Science* **4**(1), 9–20 (2001).
14. J. Mandel. A multilevel iterative method for symmetric, positive definite linear complementarity problems. *Applied mathematics and optimization* **11**(1), 77–95 (1984).

15. M. Vallejos. MGOPT with gradient projection method for solving bilinear elliptic optimal control problems. *Computing* **87**(1-2), 21–33 (2010).
16. J. Youett, O. Sander and R. Kornhuber. A globally convergent filter-trust-region method for large deformation contact problems. *SIAM Journal on Scientific Computing* **41**(1), B114–B138 (2019).
17. P. Zulian, A. Kopaničáková, M.C.G. Nestola, A. Fink, N. Fadel, A. Rigazzi, V. Magri, T. Schneider, E. Botter, J. Mankau and R. Krause. Utopia: A C++ embedded domain specific language for scientific computing. Git repository. `https://bitbucket.org/zulianp/utopia` (2016).

A Multigrid Preconditioner for Jacobian-free Newton–Krylov Methods

Hardik Kothari, Alena Kopaničáková, and Rolf Krause

1 Introduction

The numerical solution of partial differential equations (PDEs) is often carried out using discretization techniques, such as the finite element method (FEM), and typically requires the solution of a nonlinear system of equations. These nonlinear systems are often solved using some variant of the Newton method, which utilizes a sequence of iterates generated by solving a linear system of equations. However, for problems such as inverse problems, optimal control problems, or higher-order coupled PDEs, it can be computationally expensive, or even impossible to assemble a Jacobian matrix.

The Jacobian-free Newton–Krylov (JFNK) methods exploit the finite difference method to evaluate the action of a Jacobian on a vector, without requiring the knowledge of the analytical form of the Jacobian, and still retain the local quadratic convergence of the Newton method. Even though JFNK methods are quite effective, the convergence properties of the Krylov subspace methods deteriorate with increasing problem size. Hence, it is desirable to reduce the overall computational cost by accelerating the convergence of the Krylov methods. To this end, many preconditioning strategies have been proposed in the literature, see e.g., [5]. We aim to employ multigrid (MG) as a preconditioner to accelerate the convergence of the Krylov subspace methods. Unfortunately, it is not straightforward to incorporate the MG method into the JFNK framework, as the standard implementations of the MG

Hardik Kothari
Euler Institute, Università della Svizzera italiana, Switzerland, e-mail: hardik.kothari@usi.ch

Alena Kopaničáková
Euler Institute, Università della Svizzera italiana, Switzerland,
e-mail: alena.kopanicakova@usi.ch

Rolf Krause
Euler Institute, Università della Svizzera italiana, Switzerland, e-mail: rolf.krause@usi.ch

S. C. Brenner et al. (eds.), *Domain Decomposition Methods in Science and Engineering XXVI*, Lecture Notes in Computational Science and Engineering 145,
https://doi.org/10.1007/978-3-030-95025-5_38

method require either a matrix representation of the Jacobian or an analytical form of the Jacobian.

In this work, we propose a matrix-free geometric multigrid preconditioner for the Krylov methods used within the JFNK framework. The proposed method exploits the finite difference technique to evaluate the action of a Jacobian on a vector on all levels of the multilevel hierarchy and does not require explicit knowledge of the Jacobian. Additionally, we employ polynomial smoothers which can be naturally extended to a matrix-free framework. Compared to other matrix-free MG preconditioners proposed in the literature, e.g., [1, 2, 6, 7], our method does not require the knowledge of the analytical form of the Jacobian, and no additional modifications are required in the assembly routine to compute the action of a Jacobian on a vector.

Jacobian-free Newton–Krylov methods: The Newton method is the most frequently used iterative scheme for solving nonlinear problems. Newton's method is designed to find a root $\boldsymbol{x}^* \in \mathbb{R}^n$ of some nonlinear equation $F(\boldsymbol{x}^*) = 0$. The iteration process has the following form:

$$\boldsymbol{x}^{(k+1)} = \boldsymbol{x}^{(k)} + \alpha\delta\boldsymbol{x}^{(k)}, \quad \text{for } k = 0, 1, 2, \dots,$$

where $\alpha > 0$ denotes a line-search parameter and $\delta\boldsymbol{x}^{(k)}$ denotes a Newton direction. The correction $\delta\boldsymbol{x}^{(k)}$ is obtained by solving the following linear system of equations: $J(\boldsymbol{x}^{(k)})\delta\boldsymbol{x}^{(k)} = -F(\boldsymbol{x}^{(k)})$, where $J(\boldsymbol{x}^{(k)}) = \nabla F(\boldsymbol{x}^{(k)})$. In the context of this work, we assume that the F is obtained as a gradient of some energy functional Ψ, i.e., $F(\boldsymbol{x}^{(k)}) \equiv \nabla\Psi(\boldsymbol{x}^{(k)})$. In this way, the Jacobian J will be a symmetric matrix, which in turn allows us to use a multigrid preconditioner. In the JFNK methods [5], the solution process is performed without explicit knowledge of the Jacobian J. Instead, the application of a Jacobian to a vector is approximated using the finite difference scheme, given as $J(\boldsymbol{x}^{(k)})\boldsymbol{u} \approx \frac{F(\boldsymbol{x}^{(k)}+\epsilon\boldsymbol{u})-F(\boldsymbol{x}^{(k)})}{\epsilon}$, where we choose $\epsilon = \frac{1}{n\|\boldsymbol{u}\|_2} \sum_{i=1}^{n} \sqrt{\varepsilon_p}(1 + |x_i^{(k)}|)$ and ε_p denotes the machine precision. The value of the finite difference interval ϵ is chosen so that the approximation of the Jacobian is sufficiently accurate and is not spoiled by the roundoff errors.

2 Matrix-free multigrid preconditioner

The multigrid method is one of the most efficient techniques for solving linear systems of equations stemming from the discretization of the PDEs. In the case of geometric multigrid methods, we employ a hierarchy of nested meshes $\{\mathcal{T}_\ell\}_{\ell=0}^{L}$, which encapsulate the computational domain Ω. Through the following, we use the subscript $\ell = 0, \dots, L$ to denote a level, where L denotes the finest level and 0 denotes the coarsest level. We denote the number of unknowns on a given level by $\{n_\ell\}_{\ell=0}^{L}$.

The multigrid method relies on three main ingredients. Firstly, a set of transfer operators is required to pass the information between the subsequent levels of the multilevel hierarchy. Secondly, suitable smoothers are needed to damp the high-frequency components of the error associated with a given level ℓ. Finally, an appropriate coarse level solver is required to eliminate the low-frequency components of the error. As the JFNK methods are inherently matrix-free, these ingredients have to be adapted, such that they give rise to a matrix-free multigrid preconditioner.

Transfer Operators: In the standard multigrid method, the interpolation $\boldsymbol{I}_{\ell-1}^{\ell} : \mathbb{R}^{n_{\ell-1}} \rightarrow \mathbb{R}^{n_\ell}$ and restriction $\boldsymbol{R}_{\ell}^{\ell-1} : \mathbb{R}^{n_\ell} \rightarrow \mathbb{R}^{n_{\ell-1}}$ operators are employed to prolongate the correction to a finer level and restrict the residual to a coarser level, respectively. The presented multigrid method requires an evaluation of the action of a Jacobian on a vector on all levels of the multilevel hierarchy. Therefore, the current Newton iterate also has to be transferred to the coarser levels. With this aim, we employ a projection operator $\boldsymbol{P}_{\ell}^{\ell-1} : \mathbb{R}^{n_\ell} \rightarrow \mathbb{R}^{n_{\ell-1}}$. In our numerical experiments, we use $\boldsymbol{R}_{\ell}^{\ell-1} := (\boldsymbol{I}_{\ell-1}^{\ell})^\top$ and $\boldsymbol{P}_{\ell}^{\ell-1} = 2^{-d}(\boldsymbol{I}_{\ell-1}^{\ell})^\top$, where d denotes the spatial dimension in which the problem is defined. The scaling factor 2^{-d} in the definition of the projection operator $\boldsymbol{P}_{\ell}^{\ell-1}$ is added to ensure that the constant functions are preserved when projecting them from a fine space to a coarse space.

Smoothers: We utilize the three-level Chebyshev semi-iterative method [2], as its implementation does not require explicit matrix representation. This method is convergent if all eigenvalues of the Jacobian lie within a bounded interval. Our aim here is to reduce only the high-frequency components of the error associated with a given level ℓ. Therefore, we focus on the interval $[0.06\lambda_\ell, 1.2\lambda_\ell]$, where λ_ℓ is an estimated largest eigenvalue of the Jacobian on the level ℓ. We estimate the eigenvalue λ_ℓ at the beginning of each Newton iteration. More precisely, we employ the Power method, which we terminate within 30 iterations or when the difference between the subsequent estimates is lower than 10^{-2}. As an initial guess for the Power method, a random vector is provided at the first Newton step. While for the subsequent Newton steps, we utilize the eigenvector associated with the largest eigenvalue, obtained during the previous eigenvalue estimation process, as an initial guess.

The coarse level solver: In the traditional multigrid method, a direct solver is used to eliminate the remaining low-frequency components of the error on the coarsest level. In the Jacobian-free framework, we replace the direct solver with a Krylov-subspace method, e.g., the CG method. However, to obtain an accurate solution, a large number of iterations may be required. To reduce the amount of work, we employ a preconditioner based on the limited memory BFGS (L-BFGS) quasi-Newton method [8]. The L-BFGS preconditioner is created during the very first call to the CG method by storing a few secant pairs. Following [8], we collect the secant pairs using the uniform sampling method, which allows us to capture the whole spectrum of the Jacobian.

By design, the CG method is suitable for solving the symmetric positive definite systems. When solving the non-convex problems, the arising linear systems might be indefinite, which can render the CG method ineffective. To ensure the usability of

Algorithm 1: Jacobian-free Multigrid - $V(\nu_1, \nu_2)$-cycle

Function: $\boldsymbol{s}_\ell \leftarrowtail \mathrm{MG}(\boldsymbol{x}_\ell^{(k)}, F(\boldsymbol{x}_\ell^{(k)}), \boldsymbol{b}_\ell, \ell)$
$\boldsymbol{s}_\ell \leftarrowtail \boldsymbol{0}$; ▷ Initialize correction
if $\ell \neq 0$ **then**
  $\boldsymbol{s}_\ell \leftarrowtail \mathrm{Smoother}(\boldsymbol{s}_\ell, \boldsymbol{x}_\ell^{(k)}, F(\boldsymbol{x}_\ell^{(k)}), \boldsymbol{b}_\ell, \nu_1)$; ▷ Pre-smoothing
  $\boldsymbol{r}_{\ell-1} \leftarrowtail \boldsymbol{R}_\ell^{\ell-1}(\boldsymbol{b}_\ell - J(\boldsymbol{x}_\ell^{(k)})\boldsymbol{s}_\ell)$; ▷ Restrict the residual
  $\boldsymbol{x}_{\ell-1}^{(k)} \leftarrowtail \boldsymbol{P}_\ell^{\ell-1}\boldsymbol{x}_\ell^{(k)}$; ▷ Restrict Newton iterate
  $\boldsymbol{c}_{\ell-1} \leftarrowtail \mathrm{MG}(\boldsymbol{x}_{\ell-1}^{(k)}, F(\boldsymbol{x}_{\ell-1}^{(k)}), \boldsymbol{r}_{\ell-1}, \ell-1)$; ▷ Recursion
  $\boldsymbol{s}_\ell \leftarrowtail \boldsymbol{s}_\ell + \boldsymbol{I}_{\ell-1}^{\ell}\boldsymbol{c}_{\ell-1}$; ▷ Update the correction
  $\boldsymbol{s}_\ell \leftarrowtail \mathrm{Smoother}(\boldsymbol{s}_\ell, \boldsymbol{x}_\ell^{(k)}, F(\boldsymbol{x}_\ell^{(k)}), \boldsymbol{b}_\ell, \nu_2)$; ▷ Post-smoothing
else
  $\lambda_{c+} \leftarrowtail 0$; ▷ Initialize shifting parameter
  $\boldsymbol{s}_0, \lambda_c \leftarrowtail \mathrm{CG}(\boldsymbol{s}_0, \boldsymbol{x}_0^{(k)}, F(\boldsymbol{x}_0^{(k)}), \boldsymbol{r}_0, \lambda_{c+}, \nu_*)$; ▷ Coarse level solver
  while $\lambda_c < 0$ **do**
    $\lambda_{c+} \leftarrowtail \gamma \min(\lambda_c, \lambda_{c+})$; ▷ Update shifting parameter
    $\boldsymbol{s}_0, \lambda_c \leftarrowtail \mathrm{CG}(\boldsymbol{s}_0, \boldsymbol{x}_0^{(k)}, F(\boldsymbol{x}_0^{(k)}), \boldsymbol{r}_0, \lambda_{c+}, \nu_*)$; ▷ Shifted CG solver

the CG method, we propose a few modifications. Firstly, we terminate the iteration process, as soon as the negative curvature is encountered [9]. At this point, we also compute the Rayleigh quotient, given as $\lambda_c = \left(\frac{\boldsymbol{p}^\top \boldsymbol{A}\boldsymbol{p}}{\boldsymbol{p}^\top \boldsymbol{p}}\right)$, which gives an estimate of the eigenvalue encountered at the current iterate (which will also be negative). Secondly, we shift the whole spectrum of the Jacobian by adding a multiple of the identity, given as $\boldsymbol{A}_s = \boldsymbol{A} + (-\lambda_c)\boldsymbol{I}$, where $\boldsymbol{I}$ denotes the identity matrix. The shifting strategy is applied recursively, until the modified $\boldsymbol{A}_s$ becomes positive definite. Please note, the application of the $\boldsymbol{A}_s$ to a vector can be evaluated trivially in the Jacobian-free framework. The shifting parameter γ has to be chosen to be large enough that we do not require many shifting iterations and it has to be small enough that $\lambda_{\min}(\boldsymbol{A}_s) \approx -\lambda_{\min}(\boldsymbol{A})$.

The multigrid algorithm equipped with the shifting strategy is described in Algorithm 1.

3 Numerical experiments

We investigate the performance of the proposed MG preconditioner through three examples. We note that, for these examples, the analytical form of the Jacobian can be computed, but following the JFNK methods, we restrict ourselves from using this information or assembling the Jacobian on the coarsest level. We use discretize then optimize approach, where the discretization is done with the first order FE method.

Bratu: Let us consider a domain $\Omega := (0, 1)^2$. The solution of the Bratu problem is obtained by solving the following energy minimization problem:

$$\begin{aligned} \min_{u \in H^1(\Omega)} \quad & \Psi_B(u) = \int_\Omega \frac{1}{2}\|\nabla u\|^2 - \lambda \exp(u)\, \mathrm{d}\boldsymbol{x}, \\ & \text{such that } \; u = 0 \text{ on } \Gamma, \end{aligned} \tag{1}$$

where we choose $\lambda = 5$ and $\Gamma = \partial\Omega$ denotes the boundary. In our experiments, the mesh $\mathcal{T}_0$ is triangular and consists of 25 elements in each direction.

Minimal Surface: We consider again a domain $\Omega := (0, 1)^2$. This experiment aims to find the surface of minimal area described by the function u by solving the following convex minimization problem:

$$\begin{aligned} \min_{u \in H^1(\Omega)} \quad & \Psi_M(u) = \int_\Omega \sqrt{(1 + \|\nabla u\|^2)}\, \mathrm{d}\boldsymbol{x}, \\ & \text{such that } \; u = 0 \qquad \text{on } \Gamma_{D_1}, \\ & \qquad\qquad\; u = x(1-x) \text{ on } \Gamma_{D_2}, \end{aligned} \tag{2}$$

where, $\Gamma_{D_1} = \{[0, y) \cup [1, y)\}$ and $\Gamma_{D_2} = \{(x, 0] \cup (x, 1]\}$. We consider mesh $\mathcal{T}_0$ as in the previous example.

Hyperelasticity: Lastly, we investigate a finite strain deformation of a beam, $\Omega = (0, 10) \times (0, 1) \times (0, 1)$, with the rotational deformation applied on the boundaries $\Gamma_{D_1} = \{0\} \times [0, 1] \times [0, 1]$, and $\Gamma_{D_2} = \{10\} \times [0, 1] \times [0, 1]$. We consider a neo-Hookean material model, and seek for the displacement field $\boldsymbol{u}$ by solving the following non-convex minimization problem:

$$\begin{aligned} \min_{\boldsymbol{u} \in [H^1(\Omega)]^3} \quad & \Psi_N(\boldsymbol{u}) = \int_\Omega \frac{\mu}{2}(I_C - 3) - \mu(\ln(J)) + \frac{\lambda}{2}(\ln(J))^2\, \mathrm{d}\boldsymbol{x}, \\ & \text{such that } \; \boldsymbol{u} = \boldsymbol{0} \quad \text{on } \Gamma_{D_1}, \\ & \qquad\qquad\; \boldsymbol{u} = \boldsymbol{u}_2 \quad \text{on } \Gamma_{D_2}, \end{aligned} \tag{3}$$

where $\boldsymbol{u}_2 = (0, 0.5(0.5 + (y - 0.5)\cos(\pi/6) - (z - 0.5)\sin(\pi/6) - y), 0.5(0.5 + (y - 0.5)\sin(\pi/6) + (z - 0.5)\cos(\pi/6) - z))$. Here, $J := \det(\boldsymbol{F})$ denotes the determinant of the deformation gradient $\boldsymbol{F} := \boldsymbol{I} + \nabla\boldsymbol{u}$. The first invariant of the right Cauchy–Green tensor is computed as $I_C := \mathrm{trace}(\boldsymbol{C})$, where $\boldsymbol{C} = \boldsymbol{F}^\top\boldsymbol{F}$. For our experiment, the Lamé parameters $\mu = \frac{E}{2(1+\nu)}$ and $\lambda = \frac{E\nu}{(1+\nu)(1-2\nu)}$ are obtained by setting the value of Young's modulus $E = 10$ and Poisson's ratio $\nu = 0.3$. On the coarse level, the domain is discretized using hexahedral mesh, denoted by $\mathcal{T}_0$, with 10 elements in the x-direction and 1 element in the y and z directions.

Setup for the solution strategy: We solve the proposed numerical examples using the inexact JFNK (IN) method with a cubic backtracking line-search algorithm [3]. At each IN iteration, the search direction is required to satisfy

$\|J(\boldsymbol{x}^{(k)})\delta\boldsymbol{x}^{(k)}+F(\boldsymbol{x}^{(k)})\| \leq \eta^{(k)}\|F(\boldsymbol{x}^{(k)})\|$, where $\eta^{(k)} = \min(0.5, \|F(\boldsymbol{x}^{(k)})\|)$. The algorithm terminates if $\|F(\boldsymbol{x}^{(k)})\| < 10^{-6}$. We solve $J(\boldsymbol{x}^{(k)})\delta\boldsymbol{x}^{(k)} = -F(\boldsymbol{x}^{(k)})$, using three different solution strategies: the CG method without any preconditioner (CG), the CG method with L-BFGS preconditioner (CG-QN), and the CG method with the multigrid preconditioner (CG-MG). The L-BFGS preconditioner is constructed during the first inexact Newton iteration by storing 20 secant pairs. The V-cycle MG preconditioner performs 5 pre-smoothing and 5 post-smoothing steps. On the coarse level, we use the CG-QN method with the spectral shift, which is activated only if a negative curvature is encountered. We employ a shifting parameter $\gamma = 5$, in Algorithm 1. The coarse level solver terminates if $\|\boldsymbol{r}_0\| \leq 10^{-12}$, or if the maximum number of iterations, given by the number of unknowns, is reached.

The performance of all solution strategies is evaluated for increasing problem size on successively finer refinement levels. The refinement levels are denoted by $L0, L1 \ldots, L5$, where $L0$ denotes the coarse level, equipped with mesh $\mathcal{T}_0$. The number of levels in the multilevel hierarchy is increased with the refinement level, e.g., MG employs 2 levels for the $L1$ refinement level and 6 levels for the $L5$ refinement level. We assess the performance of the methods by measuring the number of required gradient evaluations (GE). In multilevel settings, the number of effective gradient evaluations is computed as $\mathrm{GE} = \sum_{\ell=0}^{L} 2^{-d(L-\ell)}\mathrm{GE}_\ell$, where GE_ℓ denotes the number of gradient calls on a given level ℓ.

We note that the discretization of the minimization problem is performed using the finite element framework libMesh [4], while the presented solution strategies are implemented as a part of the open-source library UTOPIA [10].

Influence of different preconditioners on the performance of the JFNK method: Table 1 and 2 illustrate the performance of the IN method with different linear solvers. As we can see, for the smaller problems ($L1, L2$), the IN method with the CG and the CG-QN outperforms the IN method with the CG-MG method. However, as the problem size increases, the IN method with CG-MG is significantly more efficient than with CG or CG-QN. For instance, for the Bratu example and $L5$ refinement level, the CG-MG method outperforms the other methods by an order of magnitude.

The nonlinearity of the Bratu problem is not affected by the problem size and therefore the number of IN iterations remains constant for all refinement levels. We can also observe that the behavior of the CG-MG method is level-independent. The number of required gradient evaluations is therefore bounded after a few refinements, as the cost of the coarse level solver becomes negligible. The same behavior cannot be observed for the minimal surface problem, as this problem is strongly nonlinear and the nonlinearity of the problem grows with increasing problem size. For this reason, the number of IN iterations and the total gradient evaluations also increases for the minimal surface problem. However, we note that increase is more prevalent for the IN method equipped with CG or CG-QN methods than with CG-MG.

For the hyperelasticity example, the stored energy functional is non-convex hence a negative curvature is quite often encountered on the coarse level. We notice that with increasing problem size, a negative curvature is encountered fewer times. As a consequence, a huge amount of coarse level gradient evaluations is required to shift

Levels	Bratu			Minimal surface			Hyperelasticity		
	CG	CG-QN	CG-MG	CG	CG-QN	CG-MG	CG	CG-QN	CG-MG
$L1$	176	107	264	360	229	596	467	546	868
$L2$	367	233	253	835	501	567	626	655	372
$L3$	767	476	244	2009	1170	662	1349	1464	426
$L4$	1582	1097	239	3544	2201	782	1971	1954	733
$L5$	3377	2345	238	6154	4316	931	–	–	–

Table 1: The number of total gradient evaluations required in inexact JFNK method.

Levels	Bratu			Minimal surface			Hyperelasticity		
	# IN	# CG-MG	# AGE	# IN	# CG-MG	# AGE	#IN	# CG-MG	# AGE
$L1$	3	7	39.32	6	13	45.85	9	28	34.66
$L2$	3	9	28.25	7	18	31.51	5	15	25.12
$L3$	3	9	27.10	8	25	26.50	5	20	20.95
$L4$	3	9	26.61	9	32	24.45	5	39	18.76
$L5$	3	9	26.53	9	41	22.79	–	–	–

Table 2: The total number of inexact JFNK iterations (# IN), the total number of CG-MG iterations (# CG-MG), and the average number of gradient evaluations per total linear iteration (# AGE).

Levels	CG			CG-QN			Shifted CG-QN		
	# IN	# CG-MG	#GE	# IN	# CG-MG	#GE	# IN	# CG-MG	#GE
$L1$	9	3010	51094	9	130	2057	9	28	868
$L2$	5	16	44866	5	1017	14814	5	15	372
$L3$	5	21	1265	5	26	512	5	20	426
$L4$	6	39	733	6	39	733	5	39	733

Table 3: The total number of inexact JFNK iterations (# IN), the total number of CG-MG iterations (# CG-MG), and the total number of gradient evaluations (# GE) with CG, CG-QN, and shifted CG-QN methods. The experiment was performed for the hyperelasticity example.

the spectrum of the Jacobian for smaller problems. Therefore, the average number of gradient evaluations per CG-MG decreases as the problem size increases, as we can observe in Table 2. Nevertheless, the IN method equipped with CG-MG outperforms the CG and the CG-QN methods, see Table 1. Interestingly, the use of the L-BFGS preconditioner is less effective, as in the first IN iteration, the CG method terminates before the whole spectrum of the Jacobian can be captured.

Effect of the coarse level solver on the performance of the multigrid: Due to the non-convexity of the stored energy function, for the hyperelasticity problem, it becomes essential to shift the spectrum of the Jacobian on the coarse level to retain the performance of the multigrid preconditioner. If only the CG or CG-QN method is used, the total number of effective gradient evaluations blows up, as we can see in Table 3. This is due to the fact that the coarse level solver (CG/CG-QN method) terminates as soon as a negative curvature is encountered. Therefore, the low-frequency components of the error are not eliminated and the multigrid preconditioner becomes unstable. In contrast, if we employ the shifting strategy, the multigrid preconditioner becomes stable and the total number of the gradient evaluations grows in proportion with the number of required linear iterations.

In conclusion, the performed experiments demonstrate that the proposed Jacobian-free multigrid is a robust and stable preconditioner when applied to problems of various types. Additionally, we observe level-independence behavior, if the nonlinearity or non-convexity of the problem is not influenced by the discretization parameter.

Acknowledgements The authors would like to thank the Swiss National Science Foundation for their support through the project and the Deutsche Forschungsgemeinschaft (DFG) for their support in the SPP 1962 "Stress-Based Methods for Variational Inequalities in Solid Mechanics: Finite Element Discretization and Solution by Hierarchical Optimization [186407]". Additionally, we would also like to gratefully acknowledge the support of Platform for Advanced Scientific Computing (PASC) through projects FASTER: Forecasting and Assessing Seismicity and Thermal Evolution in geothermal Reservoirs.

References

1. P. Bastian, E.H. Müller, S. Müthing and M. Piatkowski. Matrix-free multigrid block-preconditioners for higher order discontinuous Galerkin discretisations. *Journal of Computational Physics* **394**, 417–439 (2019).
2. D. Davydov, J.P. Pelteret, D. Arndt, M. Kronbichler and P. Steinmann. A matrix-free approach for finite-strain hyperelastic problems using geometric multigrid. *International Journal for Numerical Methods in Engineering* **121**(13), 2874–2895 (2020).
3. J.E. Dennis and R.B. Schnabel. *Numerical methods for nonlinear equations and unconstrained optimization*. Classics in Applied Math **16**. SIAM (1983).
4. B.S. Kirk, J.W. Peterson, R.H. Stogner and G.F. Carey. libmesh: a C++ library for parallel adaptive mesh refinement/coarsening simulations. *Engineering with Computers* **22**(3-4), 237–254 (2006).
5. D.A. Knoll and D.E. Keyes. Jacobian-free Newton–Krylov methods: a survey of approaches and applications. *Journal of Computational Physics* **193**(2), 357–397 (2004).
6. D.J. Mavriplis. An Assessment of Linear Versus Nonlinear Multigrid Methods for Unstructured Mesh Solvers. *Journal of Computational Physics* **175**(1), 302–325 (2002).
7. D.A. May, J. Brown and L. Le Pourhiet. A scalable, matrix-free multigrid preconditioner for finite element discretizations of heterogeneous Stokes flow. *Computer Methods in Applied Mechanics and Engineering* **290**, 496–523 (2015).
8. J.L. Morales and J. Nocedal. Automatic Preconditioning by Limited Memory Quasi-Newton Updating. *SIAM Journal on Optimization* **10**(4), 1079–1096 (2000).
9. J. Nocedal and S. Wright. *Numerical Optimization*. Springer (2000).
10. P. Zulian, A. Kopaničáková, M.C.G. Nestola, A. Fink, N. Fadel, A. Rigazzi, V. Magri, T. Schneider, E. Botter, J. Mankau and R. Krause. Utopia: A C++ embedded domain specific language for scientific computing. Git repository. https://bitbucket.org/zulianp/utopia (2016).

Overlapping DDFV Schwarz Algorithms on Non-Matching Grids

Martin J. Gander, Laurence Halpern, Florence Hubert, and Stella Krell

1 Introduction

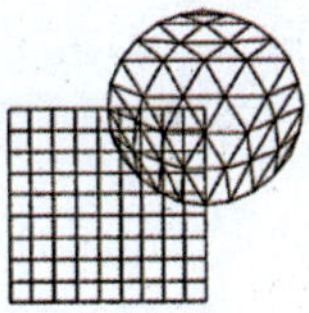

Ever since the publication of the first book on domain decomposition methods by Smith, Bjørstad, and Gropp [8], where non-matching grids were used for overlapping Schwarz methods (see on the right), and the methods worked very well, a theoretical understanding of their convergence remained open.

We are interested in a better understanding of such Schwarz methods for Discrete Duality Finite Volume (DDFV) discretizations for anisotropic diffusion,

$$\mathcal{L}(u) := -\mathrm{div}(A\nabla u) = f \text{ in } \Omega, \quad u = 0 \text{ on } \partial\Omega, \quad A(x,y) := \begin{pmatrix} A_{xx} & A_{xy} \\ A_{xy} & A_{yy} \end{pmatrix}, \tag{1}$$

where Ω is an open bounded domain of $\mathbb{R}^2$, and A is a uniformly symmetric positive definite matrix. DDFV optimized Schwarz methods have been developed for (1) in [4, 5], because these techniques are especially well suited for anisotropic diffusion [1, 3, 6]. We study here for the first time a new overlapping DDFV Schwarz algorithm with classical Dirichlet transmission conditions that can handle non-matching grids, due to carefully chosen additional unknowns in the DDFV scheme. We prove

Martin J. Gander
Section de Mathématiques, Université de Genève, e-mail: martin.gander@unige.ch

Laurence Halpern
LAGA, Université Sorbonne Paris-Nord, 93430 Villetaneuse,
e-mail: halpern@math.univ-paris13.fr

Florence Hubert
Aix-Marseille Université, CNRS, Centrale Marseille, I2M UMR 7373, 39 rue F. Joliot Curie, 13453 Marseille, Cedex 13, FRANCE, e-mail: florence.hubert@univ-amu.fr

Stella Krell
Université Côte d'Azur, Inria, CNRS, LJAD, France,
e-mail: stella.krell@univ-cotedazur.fr

S. C. Brenner et al. (eds.), *Domain Decomposition Methods in Science and Engineering XXVI*, Lecture Notes in Computational Science and Engineering 145,
https://doi.org/10.1007/978-3-030-95025-5_39

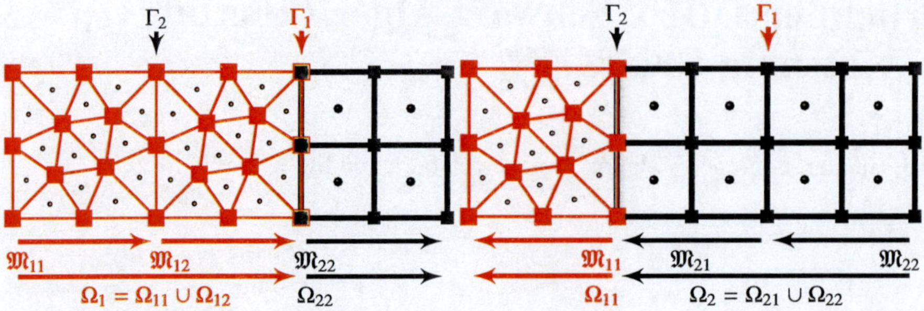

Fig. 1: Primal non-matching meshes associated to the decomposition $\Omega = \Omega_1 \cup \Omega_2$. Left: primal mesh $\mathfrak{M}_1 = \mathfrak{M}_{11} \cup \mathfrak{M}_{12}$ for Ω_1 in red. Right: primal mesh $\mathfrak{M}_2 = \mathfrak{M}_{21} \cup \mathfrak{M}_{22}$ for Ω_2 in black. Both meshes $\mathfrak{M}_j$ are completed to the entire domain to investigate the limit of the method.

convergence of the DDFV Schwarz algorithm in the case of matching grids, and show numerically that for some non-matching grids convergence is still achieved to monodomain DDFV solutions. Finally, under mesh refinement, the Schwarz limit always converges to the underlying continuous monodomain solution.

2 Overlapping DDFV Schwarz algorithm

The continuous parallel Schwarz method for (1) and two subdomains Ω_1 and Ω_2, $\bar{\Omega} = \bar{\Omega}_1 \cup \bar{\Omega}_2$ reads

$$-\operatorname{div}(A\nabla u_j^{l+1}) = f \text{ in } \Omega_j,\ u_j^{l+1} = 0 \text{ on } \partial\Omega_{j,\mathrm{D}},\ u_j^{l+1} = u_i^l \text{ on } \Gamma_j,\ j = 1,2, \tag{2}$$

where $i = j+1 \mod 2$ and $\partial\Omega_j = \partial\Omega_{j,\mathrm{D}} \cup \Gamma_j$ with $\Gamma_j \cap \partial\Omega = \emptyset$. Each subdomain $\bar{\Omega}_j$ can be partitioned into $\bar{\Omega}_{jj} \cup \bar{\Omega}_{ji}$ with $\Omega_{ji} = \Omega_{ij} = \Omega_j \cap \Omega_i$. We now introduce the technical description of DDFV, see [1] for more details.

The meshes. Consider for $j = 1,2$ a DDFV mesh $\mathcal{T}_j = (\mathfrak{M}_j, \mathfrak{M}_j^*, \partial\mathfrak{M}_j, \partial\mathfrak{M}_j^*)$ of the domain Ω_j defined as follows: the primal mesh $\mathfrak{M}_j = \mathfrak{M}_{jj} \cup \mathfrak{M}_{ji}$ is a set of disjoint open polygonal control volumes $\kappa \subset \Omega_j$ such that $\cup\overline{\kappa} = \overline{\Omega_j}$. Here $\mathfrak{M}_{jj}$ (resp. $\mathfrak{M}_{ji}$) stands for the control volumes in Ω_{jj} (resp. in Ω_{ji}). In particular, this implies that no primal control volume of $\mathfrak{M}_j$ is crossed by Γ_i. Note also that in general the meshes in the overlap need not be the same, $\mathfrak{M}_{ji} \neq \mathfrak{M}_{ij}$, as shown in Fig. 1. We call the special case when $\mathfrak{M}_{ji} = \mathfrak{M}_{ij}$ the conforming case, and otherwise the non-conforming case. We denote by $\partial\mathfrak{M}_j$ (resp. $\partial\mathfrak{M}_{j,\mathrm{D}}$, $\partial\mathfrak{M}_{\Gamma_j}$) the set of edges of the control volumes in $\mathfrak{M}_j$ included in $\partial\Omega_j$ (resp. $\partial\Omega_{j,\mathrm{D}}$, Γ_j) with $\partial\mathfrak{M}_j = \partial\mathfrak{M}_{j,\mathrm{D}} \cup \partial\mathfrak{M}_{\Gamma_j}$. To each primal cell κ, we associate a center x_κ. To each vertex x_{κ^*} of the primal mesh, we associate a dual cell as shown in Fig. 2, by joining the surrounding centers. We use analogous notation for the dual mesh, $\mathfrak{M}_j^*$, $\partial\mathfrak{M}_j^*$, $\partial\mathfrak{M}_{j,\mathrm{D}}^*$ and $\partial\mathfrak{M}_{\Gamma_j}^*$. The set of dual cells can be portioned into $\mathfrak{M}_j^* = \mathfrak{M}_{jj}^* \cup \mathfrak{M}_{ji}^* \cup \mathfrak{M}_{j,\Gamma_i}^*$ corresponding to cells included

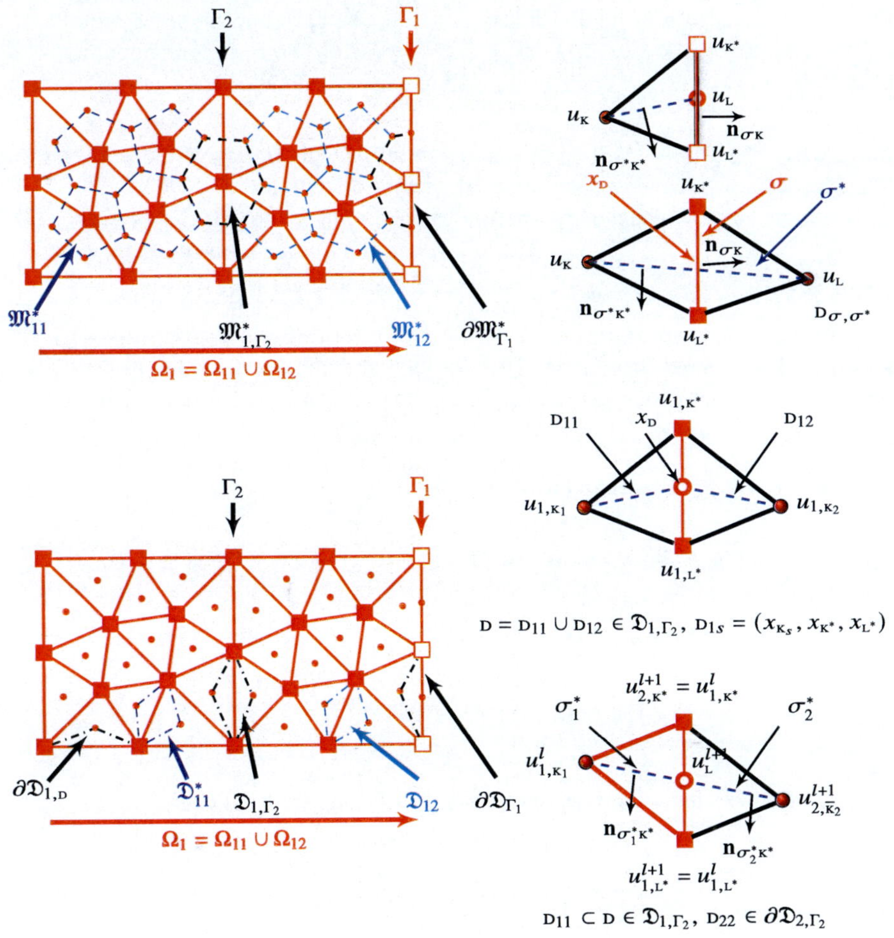

Fig. 2: Different dual cell sets (top left) and diamond cell sets (bottom left). Notations in the diamond cell (top right). Diamond cell in $\mathfrak{D}_{j,\Gamma_i}$ and $\partial\mathfrak{D}_{j,\Gamma_j}$ (bottom right).

in Ω_{jj}, Ω_{ji} or crossing Γ_i as shown in Fig. 2. For both meshes, the intersection of two control volumes that is not empty or reduced to a vertex is called an edge. We define the diamond cells $\mathrm{D}_{\sigma,\sigma^*}$ as the quadrangles whose diagonals are a primal edge $\sigma = \mathrm{K}|\mathrm{L} = (x_{\mathrm{K}^*}, x_{\mathrm{L}^*})$ and a corresponding dual edge $\sigma^* = \mathrm{K}^*|\mathrm{L}^* = (x_{\mathrm{K}}, x_{\mathrm{L}})$. The set of diamond cells is called the diamond mesh, denoted by $\mathfrak{D}_j$.

For any C in $\mathcal{T}_j$, we denote by m_{C} its Lebesgue measure, by $\mathcal{E}_{\mathrm{C}}$ the set of its edges, and $\mathfrak{D}_{\mathrm{C}} := \{\mathrm{D}_{\sigma,\sigma^*} \in \mathfrak{D}_j,\ \sigma \in \mathcal{E}_{\mathrm{C}}\}$. For $\mathrm{D} = \mathrm{D}_{\sigma,\sigma^*}$ with vertices $(x_{\mathrm{K}}, x_{\mathrm{K}^*}, x_{\mathrm{L}}, x_{\mathrm{L}^*})$, we denote by x_{D} the center of D, that is the intersection of the primal edge σ and the dual edge σ^*, by m_{D} its measure, by m_σ the length of σ, by m_{σ^*} the length of σ^*, by $m_{\sigma_{\mathrm{K}^*}}$ the length of $\partial\mathrm{K}^* \cap \Omega_j$, by $m_{\sigma_{\mathrm{L}}}$ the length of $\mathrm{D} \cap \partial\Omega_j$, and by $m_{\sigma_{\mathrm{K}}}$ the length of $[x_{\mathrm{K}}, x_{\mathrm{D}}]$. $\mathbf{n}_{\sigma\mathrm{K}}$ is the unit vector normal to σ oriented from x_{K} to x_{L}, and $\mathbf{n}_{\sigma^*\mathrm{K}^*}$ is the unit vector normal to σ^* oriented from x_{K^*} to x_{L^*}. We can split the set $\mathfrak{D}_j$ into $\mathfrak{D}_j^{int} \cup \mathfrak{D}_j^{ext}$ with $\mathfrak{D}_j^{int} = \mathfrak{D}_{jj} \cup \mathfrak{D}_{ji} \cup \mathfrak{D}_{j,\Gamma_i}$, $\mathfrak{D}_j^{ext} = \partial\mathfrak{D}_{j,D} \cup \partial\mathfrak{D}_{\Gamma_j}$ corresponding to cells included in Ω_{jj}, Ω_{ji} or crossing Γ_i or boundary diamond cells as shown in Fig. 2.

The unknowns. The DDFV method associates to all primal control volumes $K \in \mathfrak{M}_j \cup \partial\mathfrak{M}_j$ an unknown value $u_{j,K}$, and to all dual control volumes $K^* \in \mathfrak{M}_j^* \cup \partial\mathfrak{M}_j^*$ an unknown value u_{j,K^*}. We denote the approximate solution on the mesh $\mathcal{T}_j$ by $u_{\mathcal{T}_j} = \left((u_{j,K})_{K\in(\mathfrak{M}_j\cup\partial\mathfrak{M}_j)}, (u_{j,K^*})_{K^*\in(\mathfrak{M}_j^*\cup\partial\mathfrak{M}_j^*)}\right) \in \mathbb{R}^{\mathcal{T}_j}$. When f is a continuous function, we define $f_{\mathcal{T}_j} = \mathbb{P}_c^{\mathcal{T}_j} f$ the evaluation of f on the mesh $\mathcal{T}_j$ defined for all control volumes $C \in \mathcal{T}_j$ by $f_C := f(x_C)$.

Operators. DDFV schemes can be described by two operators: a discrete gradient $\nabla^{\mathfrak{D}_j}$ and a discrete divergence $\operatorname{div}^{\mathcal{T}_j}$, which are dual to each other, see [1]. Let $\nabla^{\mathfrak{D}_j} : u_{\mathcal{T}_j} \in \mathbb{R}^{\mathcal{T}_j} \mapsto \left(\nabla^D u_{\mathcal{T}_j}\right)_{D\in\mathfrak{D}_j} \in (\mathbb{R}^2)^{\mathfrak{D}_j}$ and $\operatorname{div}^{\mathcal{T}_j} : \xi_{\mathfrak{D}_j} = (\xi_{j,D})_{D\in\mathfrak{D}_j} \in (\mathbb{R}^2)^{\mathfrak{D}_j} \mapsto \operatorname{div}^{\mathcal{T}_j}\xi_{\mathfrak{D}_j} \in \mathbb{R}^{\mathcal{T}_j}$ be defined as

$$\nabla^D u_{\mathcal{T}_j} := \frac{1}{2m_D}\left((u_{j,L} - u_{j,K})m_\sigma \mathbf{n}_{\sigma K} + (u_{j,L^*} - u_{j,K^*})m_{\sigma^*}\mathbf{n}_{\sigma^* K^*}\right), \quad \forall D \in \mathfrak{D}_j,$$
$$\operatorname{div}^K \xi_{\mathfrak{D}_j} := \frac{1}{m_K}\sum_{D\in\mathfrak{D}_K} m_\sigma(\xi_{j,D}, \mathbf{n}_{\sigma K}), \ \forall K \in \mathfrak{M}_j,$$
$$\operatorname{div}^{K^*} \xi_{\mathfrak{D}_j} := \frac{1}{m_{K^*}}\sum_{D\in\mathfrak{D}_{K^*}} m_{\sigma^*}(\xi_{j,D}, \mathbf{n}_{\sigma^* K^*}), \ \forall K^* \in \mathfrak{M}_j^*.$$

DDFV scheme on Ω_j for Dirichlet boundary conditions on Γ_j. For $u_{\mathcal{T}_j} \in \mathbb{R}^{\mathcal{T}_j}$, $f_{\mathcal{T}_j} \in \mathbb{R}^{\mathcal{T}_j}$ and $h_{\mathcal{T}_j} \in \mathbb{R}^{\partial\mathfrak{M}_{\Gamma_j}\cup\partial\mathfrak{M}^*_{\Gamma_j}}$, the linear system denoted by $\mathcal{L}_{\Omega_j}^{\mathcal{T}_j}(u_{\mathcal{T}_j}, f_{\mathcal{T}_j}, h_{\mathcal{T}_j}) = 0$ refers to

$$-\operatorname{div}^K\left(A_{\mathfrak{D}_j}\nabla^{\mathfrak{D}_j}u_{\mathcal{T}_j}\right) = f_K, \quad \forall\, K \in \mathfrak{M}_j, \tag{3}$$
$$-\operatorname{div}^{K^*}\left(A_{\mathfrak{D}_j}\nabla^{\mathfrak{D}_j}u_{\mathcal{T}_j}\right) = f_{K^*}, \quad \forall\, K^* \in \mathfrak{M}_j^*, \tag{4}$$
$$u_{j,K} = 0, \ \forall\, K \in \partial\mathfrak{M}_{j,D}, \qquad u_{j,K^*} = 0, \ \forall\, K^* \in \partial\mathfrak{M}^*_{j,D}, \tag{5}$$
$$u_{j,L} - \frac{m_{D_i}}{m_D}u_{j,K_j} = h_{j,L}, \ \forall\, L \in \partial\mathfrak{M}_{\Gamma_j}, \quad u_{j,K^*} = h_{j,K^*}, \ \forall\, K^* \in \partial\mathfrak{M}^*_{\Gamma_j}, \tag{6}$$

where for all $L \in \partial\mathfrak{M}_{\Gamma_j}$, we note that the edge associated to L belongs both to a diamond cell $D \in \mathfrak{D}_{i,\Gamma_j}$ whose vertices are denoted by $x_{K_1}, x_{K_2}, x_{K^*}, x_{L^*}$ with $x_{K_s} \in \Omega_{is}$ and to a boundary diamond cell $D_{jj} \in \partial\mathfrak{D}_{i,\Gamma_j}$ whose vertices are denoted by $x_{\overline{K}_j}, x_{K^*}, x_{L^*}$. We denote by the half-diamond D_{ii} the triangle whose vertices are $x_{K_i}, x_{K^*}, x_{L^*}$ and by the half-diamond D_{jj} the triangle whose vertices are $x_{\overline{K}_j}, x_{K^*}, x_{L^*}$ (See Fig. 2 bottom right). It is classical to see that this discrete formulation is well posed, see [1].

DDFV Schwarz method. The overlapping DDFV Schwarz method performs for an arbitrary initial guess $h^0_{\mathcal{T}_j} \in \mathbb{R}^{\partial\mathfrak{M}_{\Gamma_j} \cup \partial\mathfrak{M}^*_{\Gamma_j}}$, and $l = 1, 2, \ldots$ the following steps (below either $(j,i) = (1,2)$ or $(j,i) = (2,1)$):

- Compute the solutions $u^{l+1}_{\mathcal{T}_j} \in \mathbb{R}^{\mathcal{T}_j}$ of $\mathcal{L}^{\mathcal{T}_j}_{\Omega_j}(u^{l+1}_{\mathcal{T}_j}, f_{\mathcal{T}_j}, h^l_{\mathcal{T}_j}) = 0$.
- Set $h^{l+1}_{j,\kappa^*} = u^{l+1}_{i,\kappa^*}$ for all $\kappa^* \in \partial\mathfrak{M}^*_{\Gamma_j}$, noting that $\kappa^* \in \mathfrak{M}^*_{i,\Gamma_j}$.
- Compute $h^{l+1}_{j,\mathrm{L}}$: there exists a unique value u^{l+1}_{L} such that

$$\left(A_{\mathrm{D}}\nabla^{\mathrm{D}_{ii}} u^{l+1}_{\mathcal{T}_i}, \mathbf{n}_{\sigma\kappa_i}\right) = \left(A_{\mathrm{D}}\nabla^{\mathrm{D}_{jj}} u^{l+2}_{\mathcal{T}_j}, \mathbf{n}_{\sigma\kappa_i}\right)$$

defined by

$$u^{l+1}_{\mathrm{L}} = \frac{m_{\mathrm{D}_i}}{m_{\mathrm{D}}} u^{l+2}_{j,\kappa_j} + \frac{m_{\mathrm{D}_j}}{m_{\mathrm{D}}} u^{l+1}_{i,\kappa_i} + \lambda_{\mathrm{D}}\left(u^{l+1}_{i,\mathrm{L}^*} - u^{l+1}_{i,\kappa^*}\right),$$

with $\lambda_{\mathrm{D}} = \frac{A_{\mathrm{D}}\mathbf{n}_{\sigma\kappa_j}\cdot\left(m_{\mathrm{D}_i} m_{\sigma^*_j}\mathbf{n}^j_{\sigma^*\kappa^*} - m_{\mathrm{D}_j} m_{\sigma^*_i}\mathbf{n}^i_{\sigma^*\kappa^*}\right)}{m_{\mathrm{D}} m_{\sigma} A_{\mathrm{D}}\mathbf{n}_{\sigma\kappa_1}\cdot\mathbf{n}_{\sigma\kappa_1}}$ which equals zero in the case of classical DDFV meshes, i.e. $x_{\mathrm{D}} = (x_{\kappa_j}, x_{\kappa_i}) \cap (x_{\kappa^*}, x_{\mathrm{L}^*})$, see Fig 2. We then obtain

$$h^{l+1}_{j,\mathrm{L}} := \frac{m_{\mathrm{D}_j}}{m_{\mathrm{D}}} u^{l+1}_{i,\kappa_i} + \lambda_{\mathrm{D}}\left(u^{l+1}_{i,\mathrm{L}^*} - u^{l+1}_{i,\kappa^*}\right).$$

3 Convergence of overlapping DDFV Schwarz

The main difficulty to prove convergence of a Schwarz algorithm on non-matching grids is to identify its limit. In the conforming case, we will show that the limit is the solution of a classical DDFV scheme on the entire domain, referred to as the monodomain solution. In the non-matching case, we will define two classical DDFV schemes on the entire domain, one associated to each subdomain, and then study numerically if convergence of the subdomain sequences occurs to their corresponding monodomain solution. To construct the monodomain solutions, consider $\bar{\mathcal{T}}_j$, the DDFV discretization of Ω associated to the primal mesh $\bar{\mathfrak{M}}_j = \mathfrak{M}_{jj} \cup \mathfrak{M}_{ji} \cup \mathfrak{M}_{ii}$. Note that in the conforming case, $\mathfrak{M}_{ji} = \mathfrak{M}_{ij}$, the extended meshes $\bar{\mathcal{T}}_1$ and $\bar{\mathcal{T}}_2$ coincide, and we denote them by $\bar{\mathcal{T}}$. The solution $\bar{u}^{\mathrm{DDFV}}_j$ of the classical monodomain DDFV scheme for homogeneous Dirichlet conditions is the solution of the variational formulation (see e.g. [3])

$$a_j(\bar{u}^{\mathrm{DDFV}}_j, \bar{v}_{\bar{\mathcal{T}}_j}) := \sum_{\mathrm{D}\in\bar{\mathfrak{D}}_j} m_{\mathrm{D}} A_{\mathrm{D}} \nabla^{\mathrm{D}} \bar{u}_{\bar{\mathcal{T}}_j} \cdot \nabla^{\mathrm{D}} \bar{v}_{\bar{\mathcal{T}}_j} = \frac{1}{2}\sum_{\kappa\in\bar{\mathfrak{M}}_j} m_{\kappa} f_{\kappa} \bar{v}_{\kappa} + \frac{1}{2}\sum_{\kappa^*\in\bar{\mathfrak{M}}^*_j} m_{\kappa^*} f_{\kappa^*} \bar{v}_{\kappa^*}.$$

In each subdomain, we solve $\mathcal{L}^{\mathcal{T}_j}_{\Omega_j}(u^{l+1}_{\mathcal{T}_j}, f_{\mathcal{T}_j}, h^l_{\mathcal{T}_j}) = 0$, and extend the solution $u^{l+1}_{\mathcal{T}_j}$ to $\mathbb{R}^{\bar{\mathcal{T}}_j}$ using the previous iterate on the neighboring domain,

$$\bar{u}_{\bar{\mathcal{T}}_j}^{l+1} = \begin{cases} u_{\mathcal{T}_j}^{l+1} \text{ on } \mathbb{R}^{\mathfrak{M}_j \cup \mathfrak{M}_j^*}, \\ u_{\mathcal{T}_i}^{l} \text{ on } \mathbb{R}^{\mathfrak{M}_{i,\Gamma_j}^* \cup \mathfrak{M}_{ii} \cup \mathfrak{M}_{ii}^*}. \end{cases} \tag{7}$$

Introducing $V_j = \{\bar{v}_{\bar{\mathcal{T}}_j} \in \mathbb{R}^{\bar{\mathcal{T}}_j}$ such that $\bar{v}^{\mathfrak{M}_{ii} \cup \mathfrak{M}_{ii}^* \cup \mathfrak{M}_{i,\Gamma_j}^*} = 0\}$, by construction of the extension, we have $\bar{u}_{\bar{\mathcal{T}}_j}^{l+1} - \bar{u}_{\bar{\mathcal{T}}_i}^{l} \in V_j$ and for all $\bar{v}_{\bar{\mathcal{T}}_j} \in V_j$ we have $a_j(\bar{u}_{\bar{\mathcal{T}}_j}^{l+1} - \bar{u}_j^{\text{DDFV}}, \bar{v}_{\bar{\mathcal{T}}_j}) = 0$ since there exists $(\psi_{\mathrm{K}^*}^{l+1})_{\mathrm{K}^* \in \mathfrak{M}_{i,\Gamma_j}^*}$ and $(\psi_{\mathrm{K}_i}^{l+1})_{\mathrm{L} \in \partial\mathfrak{M}_{\Gamma_j}}$ such that

$$a_j(\bar{u}_{\bar{\mathcal{T}}_j}^{l+1}, \bar{v}_{\bar{\mathcal{T}}_j}) = \frac{1}{2} \sum_{\mathrm{K} \in \bar{\mathfrak{M}}_j} m_{\mathrm{K}} f_{\mathrm{K}} \bar{v}_{\mathrm{K}} + \frac{1}{2} \sum_{\mathrm{K}^* \in \bar{\mathfrak{M}}_j^*} m_{\mathrm{K}^*} f_{\mathrm{K}^*} \bar{v}_{\mathrm{K}^*} + \sum_{\mathrm{K}^* \in \mathfrak{M}_{i,\Gamma_j}^*} \bar{v}_{\mathrm{K}^*} \psi_{\mathrm{K}^*}^{l+1} + \sum_{\mathrm{L} \in \partial\mathfrak{M}_{\Gamma_j}} \bar{v}_{\mathrm{K}_i} \psi_{\mathrm{K}_i}^{l+1}.$$

Theorem 1 *If the meshes are conforming, $\mathcal{M}_{ij} = \mathcal{M}_{ji}$, then the DDFV Schwarz algorithm converges in the discrete DDFV H^1 semi-norm*

$$\|\bar{u}_{\bar{\mathcal{T}}_j}\|_{H^1} := \Big(\sum_{D \in \bar{\mathfrak{D}}_j} m_D \|\nabla^D \bar{u}_{\bar{\mathcal{T}}_j}\|^2 \Big)^{\frac{1}{2}}. \tag{8}$$

Proof If $\mathcal{M}_{ij} = \mathcal{M}_{ji}$, then $a_j = a_i := a$, and we obtain that $a(\bar{u}_{\bar{\mathcal{T}}_j}^{l+1} - \bar{u}_{\bar{\mathcal{T}}_i}^{l}, \bar{v}_{\bar{\mathcal{T}}_j}) = 0$ for all $\bar{v}_{\bar{\mathcal{T}}_j} \in V_j$ and thus $\bar{u}_{\bar{\mathcal{T}}_j}^{l+1} - \bar{u}_{\bar{\mathcal{T}}_i}^{l}$ is the orthogonal projection of $\bar{u}_j^{\text{DDFV}} - \bar{u}_{\bar{\mathcal{T}}_i}^{l}$ onto V_j with respect to the scalar product induced by a. Now because $\mathbb{R}^{\bar{\mathcal{T}}} = V_1 + V_2$, we can apply [7, Lemma 2.12 and Theorem 2.15] (see also [2, Fig. 2.4]) to conclude that the proposed overlapping DDFV Schwarz method converges geometrically to the monodomain DDFV solution in the norm induced by a or equivalently for the discrete DDFV H^1 semi-norm (8). □

If the meshes are non-conforming, $\mathcal{M}_{ij} \neq \mathcal{M}_{ji}$, we have two monodomain solutions, one from extending each subdomain mesh to the overall domain, and neither convergence nor the limit of the DDFV Schwarz algorithm is known. We thus now study its convergence numerically, for both the conforming and non-conforming cases. We use a strong anisotropy $A = \begin{pmatrix} 1.5 & 0.5 \\ 0.5 & 15 \end{pmatrix}$ and a manufactured solution $u_e(x, y) = \sin(\pi x)\sin(\pi y)\sin(\pi(x + y))$ putting the corresponding source term f and non-homogeneous boundary conditions on $(-0.75, 0.75) \times (0, 1)$. The overlap is $(-0.25, 0.25) \times (0, 1)$. The meshes are built using refinements of the meshes shown in Fig. 1. For both families, $\mathfrak{M}_{11}$ is the triangle mesh and $\mathfrak{M}_{22}$ is the square mesh, and in the conforming case $\mathfrak{M}_{12}$ and $\mathfrak{M}_{21}$ are both the square mesh, while in the nonconforming case $\mathfrak{M}_{12}$ is the triangle mesh and $\mathfrak{M}_{21}$ is the square mesh. Note that the dual meshes exhibit a large variety of polygonal cells. Tables 1 and 2 show a detailed error analysis of the results we obtain, stopping the algorithm as soon as $\|u^l - u^{l-1}\|_{L^2} \leq 1e - 13$ with

$$\|\bar{u}_{\bar{\mathcal{T}}_j}\|_{L^2} := \sqrt{\sum_{\mathrm{K} \in \bar{\mathfrak{M}}_j} m_{\mathrm{K}} \bar{u}_{\mathrm{K}}^2} + \sqrt{\sum_{\mathrm{K}^* \in \bar{\mathfrak{M}}_j{}^* \cup \partial\mathfrak{M}_j^*} m_{\mathrm{K}^*} \bar{u}_{\mathrm{K}^*}^2}.$$

#cells	cellsize	$\|u^{l+1}-u^l\|_{H^1}$	$\|u^l-\bar{u}^{\mathrm{DDFV}}\|_{H^1}$	$\|u^l-\mathbb{P}_c^{\mathcal{T}}u_e\|_{H^1}$	order	$\|u^l-\mathbb{P}_c^{\mathcal{T}}u_e\|_{L^2}$	order
140	3.5E-01	2.080E-15	4.062E-16	1.355E-01	—	1.086E-01	—
458	1.8E-01	1.360E-15	1.022E-15	5.945E-02	1.19	2.897E-02	1.91
1634	8.8E-02	3.662E-14	1.911E-15	3.104E-02	0.94	7.998E-03	1.86
6146	4.4E-02	2.537E-14	8.703E-15	1.563E-02	0.99	2.071E-03	1.95
23810	2.2E-02	2.694E-14	1.945E-14	7.737E-03	1.01	5.228E-04	1.99
93698	1.1E-02	3.561E-14	3.139E-14	3.832E-03	1.01	1.311E-04	2.00

Table 1: Conforming overlap: convergence of the Schwarz algorithm $\|u^{l+1}-u^l\|_{H^1} \to 0$ and convergence to the monodomain solution $\bar{u}^{\mathrm{DDFV}}$ for all mesh sizes; convergence under mesh refinement of the limit of the Schwarz algorithm to the exact solution of order 1 in H^1 and order 2 in L^2.

#cells	cellsize	$\|u^{l+1}-u^l\|_{H^1}$	$\|u^l-\bar{u}^{\mathrm{DDFV}}\|_{H^1}$	order	$\|u^l-\mathbb{P}_c^{\mathcal{T}}u_e\|_{H^1}$	order	$\|u^l-\mathbb{P}_c^{\mathcal{T}}u_e\|_{L^2}$	order
166	3.5E-01	3.497E-15	3.875E-02	—	1.469E-01	—	1.207E-01	—
562	1.8E-01	1.079E-14	1.833E-02	1.08	5.987E-02	1.29	2.930E-02	2.04
2050	8.8E-02	8.451E-14	6.271E-03	1.55	3.233E-02	0.89	8.178E-03	1.84
7810	4.4E-02	6.264E-14	1.830E-03	1.78	1.677E-02	0.95	2.153E-03	1.93
30466	2.2E-02	4.514E-14	5.152E-04	1.83	8.448E-03	0.99	5.498E-04	1.97

Table 2: Non-conforming overlap: as for Table 1, but only convergence under mesh refinement to the monodomain solution $\bar{u}^{\mathrm{DDFV}}$.

#cells	cellsize	$\|u^l-u^{l-1}\|_{H^1}$	$\|u^l-\bar{u}^{\mathrm{DDFV}}\|_{H^1}$	$\|u^l-\mathbb{P}_c^{\mathcal{T}}u_e\|_{H^1}$	order	$\|u^l-\mathbb{P}_c^{\mathcal{T}}u_e\|_{L^2}$	order
166	3.54E-01	1.183E-13	2.412E-14	7.521E-03	—	1.366E-03	—
562	1.77E-01	7.843E-14	1.569E-14	3.769E-03	1.00	3.315E-04	2.04
2050	8.84E-02	7.024E-14	1.448E-14	1.886E-03	1.00	8.154E-05	2.02
7810	4.42E-02	7.071E-14	2.165E-14	9.434E-04	1.00	2.022E-05	2.01
30466	2.21E-02	7.668E-14	1.193E-13	4.718E-04	1.00	5.034E-06	2.00

Table 3: Case $u_e(x, y) = xy$ and $A = \mathrm{Id}$ and convergence of $\|u^l-\bar{u}^{\mathrm{DDFV}}\|_{H^1}$ as in the conforming case of Table 1, even thought the mesh is non-conforming!

In the third column we see that the algorithm converges in all cases in the relative discrete H^1-norm (8) defined for $u_{\mathcal{T}} - v_{\mathcal{T}}$ by $\|u_{\mathcal{T}} - v_{\mathcal{T}}\| := \frac{\|u_{\mathcal{T}}-v_{\mathcal{T}}\|_{H^1}}{\|v_{\mathcal{T}}\|_{H^1}}$. The fourth column in Table 1 shows convergence to the monodomain solution for conforming meshes as proved in Theorem 1, but only convergence under mesh refinement in the non-conforming case in Table 2. The remaining columns show that the limits of the Schwarz algorithm always converge under mesh refinement to the evaluation $\mathbb{P}_c^{\mathcal{T}}u_e$ of the exact solution u_e on the meshes, of order 1 in H^1 and order 2 in L^2. For an illustration of the converged solution, see Fig. 3.

We observe however also several cases where $\bar{u}^{\mathrm{DDFV}}$ corresponds to the limit of u^l even in the nonconforming case, e.g. for $u_e = 0$ or $u_e(x, y) = xy$ with $A = \mathrm{Id}$ as shown in Table 3. The complete understanding of convergence to the mono-domain solution in the non-conforming case thus requires a deeper study of the limiting equations of the overlapping Schwarz process when discretized by nonconforming DDFV.

Fig. 3: u_1^l (left) and u_2^l (right) after $l = 21$ iterations on the primal non-conforming meshes with refinement 2, corresponding to 562 unknowns.

References

1. B. Andreianov, F. Boyer and F. Hubert. Discrete duality finite volume schemes for Leray–Lions type problems on general 2D meshes. *Numerical Methods for PDE* **23**(1), 145–195 (2007).
2. G. Ciaramella and M.J. Gander. Analysis of the parallel Schwarz method for growing chains of fixed-sized subdomains: Part III. *ETNA* **49**, 201–243 (2018).
3. K. Domelevo and P. Omnes. Afinite volume method for the Laplace equation on almost arbitrary two-dimensional grids. *M2AN Math. Model. Numer. Anal.* **39**(6), 1203–1249 (2005).
4. M.J. Gander, L. Halpern, F. Hubert and S. Krell. Optimized overlapping DDFV Schwarz algorithms. In: R. Klöfkorn, E. Keilegavlen, F. Radu and J. Fuhrmann, editors, *Finite Volumes for Complex Applications IX – Methods, Theoretical Aspects, Examples. FVCA 2020*, volume 323 of Proceedings in Mathematics & Statistics, pp. 365–373, Springer (2020).
5. M.J. Gander, L. Halpern, F. Hubert and S. Krell. Optimized Schwarz Methods for Anisotropic Diffusion with Discrete Duality Finite Volume Discretizations. *Moroccan Journal of Pure and Applied Analysis* **7**(2), 182–213, July (2021).
6. F. Hermeline. Approximation of diffusion operators with discontinuous tensor coefficients on distorted meshes. *Comput. Methods Appl. Mech. Engrg.* **192**(16-18), 1939–1959 (2003).
7. P.-L. Lions. On the Schwarz alternating method. I. In: *First international symposium on domain decomposition methods for partial differential equations*, volume 1, p. 42, Paris, France (1988).
8. B. Smith, P. Bjørstad and W. Gropp. *Domain Decomposition: Parallel Multilevel Methods for Elliptic Partial Differential Equations*. Cambridge University Press (1996).

On the Nonlinear Dirichlet–Neumann Method and Preconditioner for Newton's Method

F. Chaouqui, M. J. Gander, P. M. Kumbhar, and T. Vanzan

1 Introduction

We consider a nonlinear Partial Differential Equation (PDE)

$$\mathcal{L}(u) = f \quad \text{in} \quad \Omega, \quad u = g \quad \text{on } \partial\Omega, \tag{1}$$

where $\Omega \subset \mathbb{R}^d$ for $d \in \{1,2,3\}$ is an open bounded domain with a polygonal boundary $\partial\Omega$, and $f, g \in L^2(\Omega)$. We suppose that (1) admits a unique weak solution in some Hilbert space $u \in \mathcal{X}$ (e.g. $H^1(\Omega)$). For instance, for a quasilinear operator $\mathcal{L}$ in divergence form, explicit assumptions can be found in [1] and references therein, see also [7, Chapter 8–9] and [5, Chapter 9]. Let us divide Ω into two nonoverlapping subdomains Ω_1 and Ω_2 and define $\Gamma_j = \partial\Omega_j \setminus \partial\Omega$, $j = 1, 2$. Let u_j be the restriction of u to Ω_j. The nonlinear Dirichlet–Neumann (DN) method starts from an initial guess λ^0 and computes for $n \geq 1$ until convergence

$$\begin{array}{llll} \mathcal{L}(u_1^n) = f_1, & \text{in} \quad \Omega_1, & \mathcal{L}(u_2^n) = f_2, & \text{in} \quad \Omega_2 \\ u_1^n \quad = g_1, & \text{on} \quad \partial\Omega_1 \setminus \Gamma, & u_2^n \quad = g_2, & \text{on} \quad \partial\Omega_2 \setminus \Gamma \\ u_1^n \quad = \lambda^n & \text{on} \quad \Gamma, & \mathcal{N}_2 u_2^n \ = -\mathcal{N}_1 u_1^n & \text{on} \quad \Gamma, \end{array} \tag{2}$$

where $\lambda^n = (1-\theta)\lambda^{n-1} + \theta u_{2|\Gamma}^{n-1}$, with $\theta \in (0,1)$, $f_j := f|_{\Omega_j}$ and $g_j := g|_{\partial\Omega_j \setminus \Gamma}$ for $j = 1, 2$. The operators $\mathcal{N}_j$ represent the outward nonlinear Neumann conditions that must

F. Chaouqui
Temple University, Philadelphia, USA, e-mail: faycal.chaouqui@temple.edu,

M. J. Gander
Université de Genève, Switzerland, e-mail: martin.gander@unige.ch,

P. M. Kumbhar
Karlsruher Institut für Technologie, Germany, e-mail: pratik.kumbhar@kit.edu,

T. Vanzan
École Polytecnique Fédérale de Lausanne, Switzerland, e-mail: tommaso.vanzan@epfl.ch

S. C. Brenner et al. (eds.), *Domain Decomposition Methods in Science and Engineering XXVI*, Lecture Notes in Computational Science and Engineering 145,
https://doi.org/10.1007/978-3-030-95025-5_40

be imposed on the interface Γ and are usually found through integration by parts of the variational formulation of the PDE. For instance, if $\mathcal{L}(u) = -\partial_x((1+\alpha u^2)\partial_x u)$, then $\mathcal{N}_j u = (-1)^{j+1}(1+\alpha u^2_{|\Gamma})\partial_x u_{|\Gamma}$. For the well-posedness of the Dirichlet–Neumann method, we further assume that $\mathcal{N}_j u$ defines a bounded linear functional over $\mathcal{X}$.

System (2) can be formulated as an iteration over the substructured variable λ as

$$\lambda^n = G(\lambda^{n-1}) := (1-\theta)\lambda^{n-1} + \theta \mathrm{NtD}_2\left(-\mathrm{DtN}_1\left(\lambda^{n-1},\psi_1\right),\psi_2\right), \tag{3}$$

where $\psi_j = (f_j, g_j)$, $j = 1, 2$, represent the force term and boundary conditions, while the nonlinear Dirichlet-to-Neumann (DtN_j) and Neumann-to-Dirichlet operators (NtD_j) are defined as $\mathrm{DtN}_j(\lambda,\psi_j) := \mathcal{N}_j u_j$, and $\mathrm{NtD}_j(\varphi,\psi_j) := v_j|_\Gamma$, with

$$\begin{aligned} \mathcal{L}(u_j) &= f_j \text{ in } \Omega_j, & \mathcal{L}(v_j) &= f_j \text{ in } \Omega_j,\\ u_j &= g_j \text{ on } \partial\Omega_j \setminus \Gamma, & v_j &= g_j \text{ on } \partial\Omega_j \setminus \Gamma,\\ u_j &= \lambda \text{ on } \Gamma & \mathcal{N}_j v_j &= \varphi \text{ on } \Gamma. \end{aligned} \tag{4}$$

If $u_{\mathrm{ex}} \in H^1(\Omega)$ is the solution of (1), then it must have continuous Dirichlet trace and Neumann flux along the interface Γ. Defining $u_\Gamma := u_{\mathrm{ex}}|_\Gamma$, $\varphi := \mathcal{N}_1 u_{\mathrm{ex}}|_\Gamma$ and using the operators DtN_j and NtD_j, these necessary properties are equivalent to

$$\mathrm{DtN}_1(u_\Gamma,\psi_1) = -\mathrm{DtN}_2(u_\Gamma,\psi_2), \quad \text{and} \quad \mathrm{NtD}_1(\varphi,\psi_1) = \mathrm{NtD}_2(-\varphi,\psi_2). \tag{5}$$

2 Nilpotent property and quadratic convergence

It is well known, see e.g. [4, 9], that if $\mathcal{L}$ is linear and the subdomain decomposition is symmetric, then the DN method converges in one iteration for $\theta = 1/2$. Indeed, if $\mathcal{L}$ is linear, one can work on the error equation, i.e. $\psi_j = 0$, and the symmetry of the decomposition is sufficient to guarantee $\mathrm{DtN}_1(\cdot,0) \equiv \mathrm{DtN}_2(\cdot,0)$, so that

$$\begin{aligned} \lambda^1 &= \frac{1}{2}\left(\lambda^0 + \mathrm{NtD}_2\left(-\mathrm{DtN}_1(\lambda^0,0)\right),0\right) = \frac{1}{2}\left(\lambda^0 + \mathrm{NtD}_2\left(-\mathrm{DtN}_2(\lambda^0,0)\right),0\right)\\ &= \frac{1}{2}(\lambda^0 - \mathrm{NtD}_2(\mathrm{DtN}_2(\lambda^0,0),0) = 0, \end{aligned} \tag{6}$$

where in the third equality we used linearity, and in the last $\mathrm{NtD}_2(\mathrm{DtN}_2(\lambda,\psi),\psi) = \lambda$. Can the nonlinear DN method also converge in one iteration?

On the one hand, the relation $\mathrm{NtD}_j(\mathrm{DtN}_j(\lambda,\psi),\psi) = \lambda$ holds even in the nonlinear case, simply because the nonlinear DtN_j operator is the inverse of the nonlinear NtD_j operator. On the other hand, due to the nonlinearity of $\mathcal{L}$, one cannot rely on the error equation, nor can one state that $\mathrm{NtD}_2(-\varphi) = -\mathrm{NtD}_2(\varphi)$, and the symmetry of the decomposition is not sufficient to guarantee $\mathrm{DtN}_1(\lambda,\psi_1) \equiv \mathrm{DtN}_2(\lambda,\psi_2)$, because of the boundary conditions and the force term.

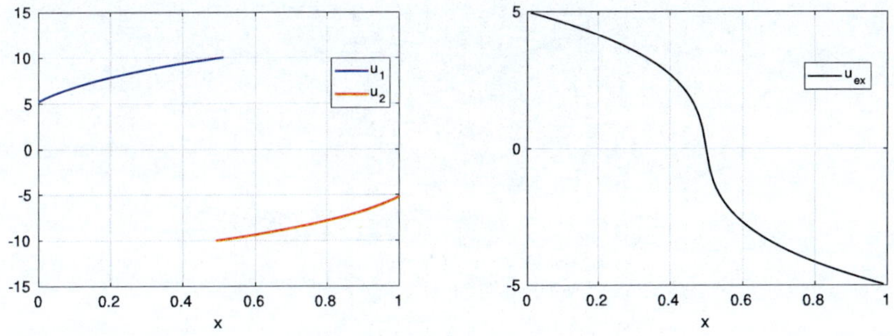

Fig. 1: Subdomain solutions of the nonlinear DN method after one iteration (left), and exact solution (right). The parameters are $g = 5$ and $k = 2$.

A straightforward observation is that if the nonlinear DN method converges in one iteration, then $G(\lambda) = \lambda_{\text{ex}}$, $\forall\lambda$, that is $G(\cdot)$ is a constant. A necessary and sufficient condition for the nonlinear DN method to converge in one iteration is then

$$0=G'(\lambda)=\frac{1}{2}+\frac{1}{2}(\text{NtD}_2\left(-\text{DtN}_1(\lambda,\psi_1),\psi_2\right))' \implies (\text{NtD}_2(-\text{DtN}_1(\lambda,\psi_1),\psi_2))'=-1. \tag{7}$$

Clearly, (7) is satisfied if $\text{NtD}_2(-\text{DtN}_1(\lambda,\psi_1),\psi_2) = -\lambda$. We consider a toy example in which this condition is satisfied. Let $\mathcal{L} = -\partial_x\left((1+u^2)\partial_x u\right)$, $u(0) = g \in \mathbb{R}^+$, $u(1) = -g$ and $f(x) = \sin((2k)\pi x)$. In the left plot of Fig. 1, we show the subdomain solutions u_1 and u_2 obtained from (2) after the first iteration. The two contributions sum to zero, which is the value of λ_{ex}. Thus, after one iteration we obtain the exact solution shown in the right panel.

Even though the nilpotent property does not hold in general, we show in the following Theorem that the nonlinear DN method can exhibit quadratic convergence.

Theorem 1 (Quadratic convergence of nonlinear DN) *For any one-dimensional nonlinear problem $\mathcal{L}(u) = f$ such that $DtN_1'(\lambda_{ex},\psi_1)\cdot DtN_2'(\lambda_{ex},\psi_2) > 0$ with $\lambda_{ex} := u_{ex}|_\Gamma$, there exists a $\theta \in (0,1)$ such that the nonlinear Dirichlet–Neumann method converges quadratically.*

Proof A sufficient condition for quadratic convergence is that the Jacobian of $G(\cdot)$, defined in (3), is zero at $\lambda_{\text{ex}} := u_{\text{ex}}|_\Gamma$, that is $G'(\lambda_{\text{ex}}) = 0$. A direct calculation shows

$$G'(\lambda) = (1-\theta) + \theta\text{NtD}_2'\left(-\text{DtN}_1(\lambda,\psi_1),\psi_2\right)\cdot\left(-\text{DtN}_1'(\lambda,\psi_1)\right). \tag{8}$$

Setting $\lambda = \lambda_{\text{ex}}$ and using the optimality condition $\text{DtN}_1(\lambda_{\text{ex}},\psi_1) = -\text{DtN}_2(\lambda_{\text{ex}},\psi_2)$ of (5), the above equation changes to

$$G'(\lambda_{\text{ex}}) = (1-\theta) + \theta\text{NtD}_2'\left(\text{DtN}_2(\lambda_{\text{ex}},\psi_2),\psi_2\right)\cdot\left(-\text{DtN}_1'(\lambda_{\text{ex}},\psi_1)\right). \tag{9}$$

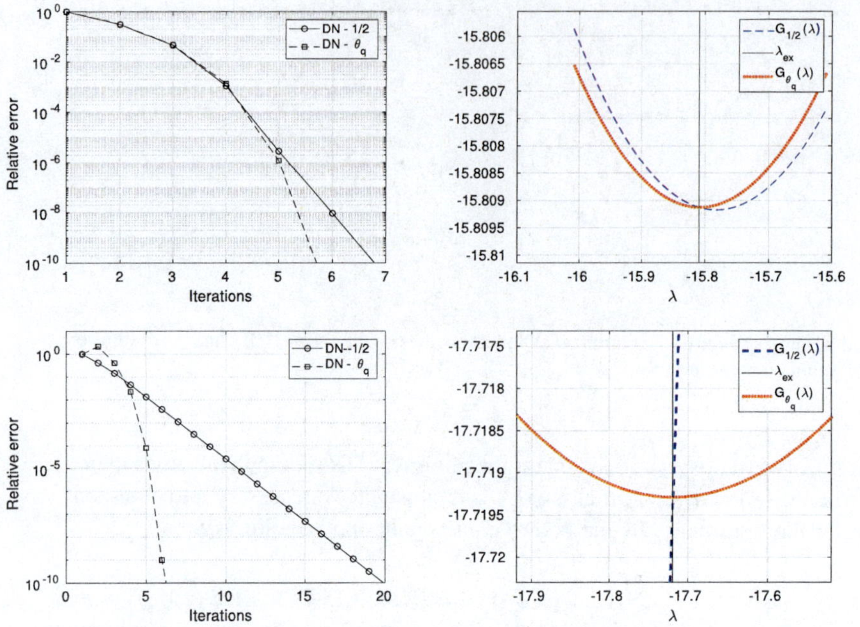

Fig. 2: In the left panels, we show the convergence curves, and in the right panels we plot $G_\theta(\lambda)$. The top-row refers to a symmetric decomposition, and the bottom-row to an asymmetric one.

If $\mathrm{DtN}_1'(\lambda_{\mathrm{ex}}, \psi_1) = \mathrm{DtN}_2'(\lambda_{\mathrm{ex}}, \psi_2)$ held true, then using the identity

$$\mathrm{NtD}_2'\left(\mathrm{DtN}_2(\lambda, \psi_2), \psi_2\right) \cdot \left(\mathrm{DtN}_2'(\lambda, \psi_2)\right) = 1,$$

obtained by differentiating $\mathrm{NtD}_j(\mathrm{DtN}_j(\lambda, \psi_j), \psi_j) = \lambda$, we would easily get that $\theta = 1/2$ leads to $G'(\lambda_{\mathrm{ex}}) = 0$. Nevertheless, variational calculus shows that to calculate $\mathrm{DtN}_j'(\lambda, \psi_j)$, one has to solve a linear PDE which does not depend on ψ_j anymore, but whose coefficients still depend on the subdomain solutions $u_{\mathrm{ex}}|_{\Omega_1}$ and $u_{\mathrm{ex}}|_{\Omega_2}$. In general then, $\mathrm{DtN}_1'(\lambda_{\mathrm{ex}}, \psi_1) \neq \mathrm{DtN}_2'(\lambda_{\mathrm{ex}}, \psi_2)$. However, DtN_j being one-dimensional functions, we have $\mathrm{DtN}_1'(\lambda_{\mathrm{ex}}, \psi_1) = \delta \mathrm{DtN}_2'(\lambda_{\mathrm{ex}}, \psi_2)$, for some $\delta \in \mathbb{R}^+$ if $\mathrm{DtN}_1'(\lambda_{\mathrm{ex}}, \psi_1) \cdot \mathrm{DtN}_2'(\lambda_{\mathrm{ex}}, \psi_2) > 0$. Inserting this into (9), we obtain $G'(\lambda_{\mathrm{ex}}) = 0$ if $\theta = \frac{1}{1+\delta} \in (0, 1)$. □

To illustrate Theorem 1 numerically, we consider $\mathcal{L}(u) = -\partial_x((1 + \alpha u^2)\partial_x u)$, $\Omega = (0, 1)$, $f(x) = 100x$, $u(0) = 0$ and $u(1) = -20$. In the top-row of Fig. 2, we set the interface Γ to $x = 1/2$. In the left panel, we plot the convergence curves for $\theta = 1/2$ and for $\theta_{\mathrm{q}} := \frac{1}{1+\delta}$. In this setting, $\delta = 1.006$ and $\theta_{\mathrm{q}} = 0.498$, so due to the symmetry of the decomposition, θ_{q} is still very close to $1/2$. In the right panel, we plot $G_\theta(\lambda)$ and see that as θ changes, the minimum of $G_\theta(\lambda)$ moves, such that it is attained at $\lambda = \lambda_{\mathrm{ex}}$ for $\theta = \theta_{\mathrm{q}}$.

Next, in the bottom row of Fig 2, we consider the same equation and boundary conditions, but Γ is now at $x = 0.3$. The decomposition is asymmetric, with $\delta = 0.43$ and $\theta_q = 0.699$. The left panel shows clearly that for $\theta = 1/2$ the convergence is linear, while for $\theta = \theta_q$, the DN method converges quadratically. In the right panel, we observe that $G_{1/2}(\lambda)$ does not have a local extremum at $\lambda = \lambda_{ex}$, while $G_{\theta_q}(\lambda)$ does. Theorem 1 does not easily generalize to higher dimensions, since DtN'_j are then matrices, and the relaxation parameter would have to be an operator. Numerically we observed for symmetric decompositions fast convergence for $\theta = 0.5$, while for asymmetric decompositions, θ needs to be tuned for good performance.

3 Mesh independent convergence

One of the attractive features of the DN method for linear problems is that it achieves mesh independent convergence. Does this also hold for the nonlinear DN method (2)? We first define the nonlinear DN method for multiple subdomains. Motivated by the definition of the DN method for the linear case in [2], we divide the domain $\Omega := (0, L) \times (0, L)$ into N nonoverlapping subdomains $\Omega_j = (\Gamma_{j-1}, \Gamma_j) \times (0, L)$, with $\Gamma_0 = 0$ and $\Gamma_N = L$. The nonlinear DN method for multiple subdomains is then defined for the interior subdomains by

$$
\begin{aligned}
\mathcal{L}(u_j^n) &= f_j && \text{in } \Omega_j,\\
\mathcal{N}_j u_j^n(\Gamma_{j-1}, \cdot) &= -\mathcal{N}_{j-1} u_{j-1}^n(\Gamma_{j-1}, \cdot) && \text{on } \Gamma_{j-1},\\
u_j^n(\Gamma_j) &= (1-\theta) u_j^{n-1}(\Gamma_j, \cdot) + \theta u_{j+1}^{n-1}(\Gamma_j, \cdot) && \text{on } \Gamma_j,
\end{aligned}
$$

where $\theta \in (0, 1)$, and for the left- and rightmost subdomains by

$$
\begin{aligned}
\mathcal{L}(u_1^n) &= f_1, \quad \text{in } \Omega_1, & \mathcal{L}(u_N^n) &= f_N, \quad \text{in } \Omega_N,\\
u_1^n(\Gamma, \cdot) &= g(0), & \mathcal{N}_N u_N^n(\Gamma_{N-1}, \cdot) &= -\mathcal{N}_{N-1} u_{N-1}^n(\Gamma_{N-1}, \cdot),\\
u_1^n(\Gamma_1, \cdot) &= (1-\theta) u_1^{n-1}(\Gamma_1, \cdot) + \theta u_2^{n-1}(\Gamma_1, \cdot), & u_N^n(L, \cdot) &= g(L).
\end{aligned}
$$

We perform two experiments, one in 1D and one in 2D. For the 1D case, we consider the nonlinear diffusion equation $-\partial_x\left((1+u^2)\partial_x u\right) = 0$, with $u(0) = 0$ and $u(1) = 20$. We divide the domain $\Omega = (0, 1)$ into ten equal subdomains. We then plot the relative error of the nonlinear DN for four different mesh sizes h = 1e-2, h = 2e-3, h = 1e-3, and h = 1e-4. The left plot in Fig. 3 shows that the convergence rate of the nonlinear DN is independent of mesh size, while it is quadratic for Newton's Method. We repeat a similar experiment in 2D, but now the domain $\Omega = (0, 1) \times (0, 1)$ is divided into four equal subdomains. Even in 2D, we observe the mesh independent convergence of the nonlinear DN method, see the right plot of Fig. 3.

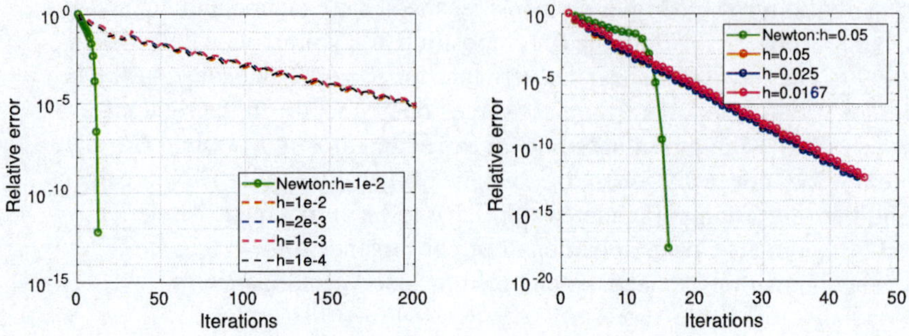

Fig. 3: Convergence behavior of nonlinear DN for different mesh sizes in 1D (left) and 2D (right).

4 Dirichlet–Neumann Preconditioned Exact Newton (DNPEN)

In Section 2, we observed that under some special conditions on the exact solution of the nonlinear problem and θ, the nonlinear DN method (2) can be nilpotent. Moreover, the nonlinear DN method can also converge quadratically. But to achieve this, we need to tune the parameter θ according to some a priori knowledge of the exact solution of the nonlinear problem. Thus in general, the nonlinear DN method converges linearly (as shown in Fig 3).

Iterative methods can be used as preconditioners to achieve faster convergence, see [4] for the linear case, and [8] for a historical introduction including also the nonlinear case. It was proposed in [3, 6] to use the nonlinear Restricted Additive Schwarz (RAS) and nonlinear Substructured RAS (SRAS) methods as preconditioner for Newton's method. We use the same idea here and apply Newton's method to the fixed point equation of the nonlinear DN method (3), which represents a systematic way of constructing non-linear preconditioners [8]. The fixed point version of (3) can be written as

$$\mathcal{F}(\lambda) := \lambda - G(\lambda) = \theta\lambda - \theta \mathrm{NtD}_2\left(-\mathrm{DtN}_1\left(\lambda, \psi_1\right), \psi_2\right). \tag{10}$$

Applying Newton to (10) we obtain a new method called the Dirichlet–Neumann Preconditioned Exact Newton (DNPEN) method.

We saw in Section 2 that the DN method can be nilpotent in certain cases. Can DNPEN still be nilpotent? Let λ_{ex} denote the fixed point of the iteration (3). Let us assume that the Dirichlet–Neumann method converges in one iteration. This means that G defined in (3) satisfies $\lambda_{\mathrm{ex}} = G(\lambda^0)$ for any initial guess λ^0. This shows that the map G is constant, and hence $\mathcal{F}'(\lambda)$ reduces to the identity matrix. Moreover, one step of Newton's method applied to (3) can then be written as

$$\lambda^1 = \lambda^0 - (\mathcal{F}'(\lambda^0))^{-1}\mathcal{F}(\lambda^0) = \lambda^0 - \mathcal{F}(\lambda^0) = G(\lambda^0) = \lambda_{\mathrm{ex}},$$

and hence DNPEN will also be nilpotent in that case. We further have the following result.

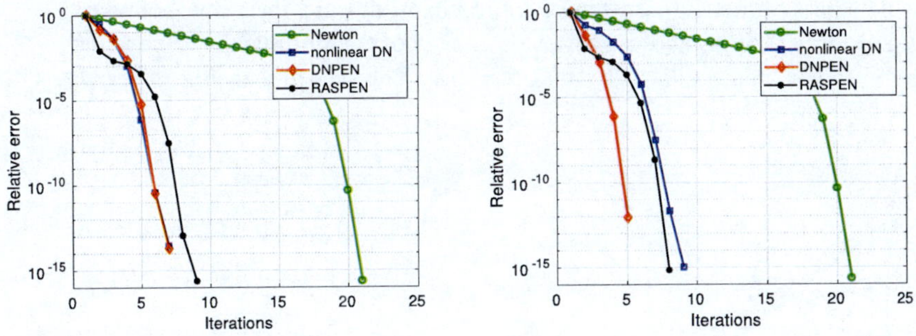

Fig. 4: Comparison of DNPEN (with optimal θ) with unpreconditioned Newton, nonlinear DN (with optimal θ) and RASPEN for a symmetric partition (left) and an asymmetric partition (right).

Theorem 2 *The convergence of DNPEN does not depend on the relaxation parameter θ in the DN preconditioner.*

Proof The function $\mathcal{F}$ from (10) corresponding to DNPEN can we rewritten as $\mathcal{F}(\lambda) = \theta\mathcal{K}(\lambda, \psi_1, \psi_2)$, where $\mathcal{K}(\lambda, \psi_1, \psi_2) := \lambda - \mathrm{NtD}_2\left(-\mathrm{DtN}_1\left(\lambda, \psi_1\right), \psi_2\right)$. Thus, Newton's iteration reads

$$\lambda^{k+1} = \lambda^k - \left(J\mathcal{F}(\lambda^k)\right)^{-1}\mathcal{F}(\lambda^k) = \lambda^k - \left(\theta J\mathcal{K}(\lambda^k)\right)^{-1}\theta\mathcal{K}(\lambda^k) = \lambda^k - \left(J\mathcal{K}(\lambda^k)\right)^{-1}\mathcal{K}(\lambda^k),$$

which shows that the Newton correction does not depend on the relaxation parameter θ. The iterates of Newton's method will thus only depend on $\mathcal{K}$, and DNPEN has θ independent convergence. □

The above theorem shows that when using DNPEN, one does not need to search for an optimal choice of θ, in contrast to the nonlinear DN method (2).

We now compare the convergence of DNPEN, the unpreconditioned Newton method, the nonlinear DN method (2) and RASPEN [6]. We consider the nonlinear diffusion problem $-\partial_x\left(\left(1+u^2\right)\partial_x u\right) = f$ on $\Omega = (0,1)$ decomposed into two equally sized subdomains, with $u(0) = 0$, $u(1) = 10$ and $f(x) = \sin(10\pi x)$. For both DN and DNPEN, we choose the optimal relaxation parameter provided in Theorem 1. The left plot in Fig. 4 shows that the iterative DN converges quadratically using the optimal parameter and is very similar to DNPEN with no significant gain in the number of iterations. The convergence curves also show that the unpreconditioned Newton method is slower than all preconditioned ones, and DNPEN has a slight advantage over RASPEN.

We repeat the same experiment but now using an asymmetric partition of the domain Ω. The right plot in Fig. 4 shows that for this configuration, DNPEN is the fastest while again unpreconditioned Newton is the slowest among the methods considered. Moreover, DNPEN is significantly faster than the nonlinear DN method.

Finally, we illustrate numerically that the convergence of DNPEN does not depend on θ. We know that in general, the nonlinear DN method converges linearly, and it is not always possible to find an optimal θ such that it converge quadratically. We again consider the symmetric partition of the domain and use the same boundary conditions and force term as above. However, instead of the optimal θ, we consider two non-optimal θ's, namely $\theta = 0.1$ and $\theta = 0.9$. The left plot in Fig. 5 shows

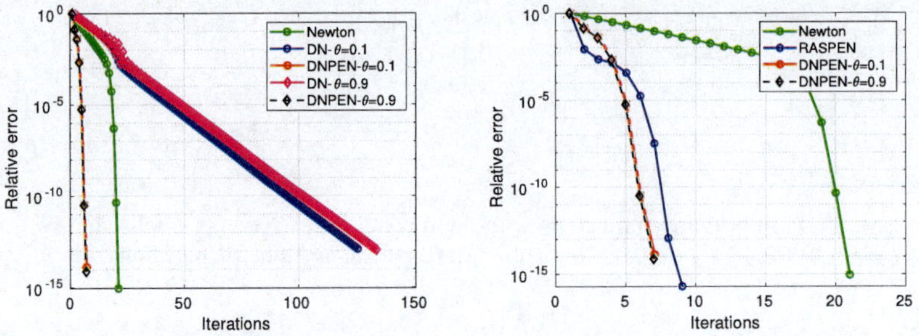

Fig. 5: Comparison of DNPEN with the unpreconditioned Newton method and nonlinear DN (left) and with RASPEN (right) for two different non optimal θ's.

the linear convergence of nonlinear DN for both $\theta = 0.1$, and $\theta = 0.9$, and both are slower than the unpreconditioned Newton method. However, DNPEN converges much faster than Newton's method and in the same number of iterations for the two different values $\theta = 0.1$ and $\theta = 0.9$. The right plot in Fig. 5 shows that DNPEN is still faster than RASPEN for both values θ considered.

5 Conclusion

While iterative DN methods are known to converge linearly, we proved that one can obtain quadratic converge for some one-dimensional nonlinear problems and for a well chosen relaxation parameter θ. Under specific conditions, the nonlinear DN method can also become a direct solver, like in the linear case. We then extended DN to multiple subdomains and numerically showed that its convergence is mesh independent. We finally introduced the nonlinear preconditioner DNPEN, proved that the convergence of DNPEN does not depend on the relaxation parameter θ, and observed numerically that DNPEN is faster than unpreconditioned Newton, nonlinear DN and RASPEN in all our examples.

Acknowledgements The third author gratefully acknowledges financial support from the Deutsche Forschungsgemeinschaft (DFG, German Research Foundation) – Project-ID 258734477 – SFB 1173.

References

1. X-C. Cai and M. Dryja. Domain decomposition methods for monotone nonlinear elliptic problems. *Domain Decomposition Methods in Scientific and Engineering* **180**, 21–27 (1994).
2. F. Chaouqui, G. Ciaramella, M.J. Gander and T. Vanzan. On the scalability of classical one-level domain-decomposition methods. *Vietnam J. Math* **46**(6), 1053–1088 (2018).
3. F. Chaouqui, M.J. Gander, P.M. Kumbhar and T. Vanzan. Linear and Nonlinear Substructured Restricted Additive Schwarz Iterations and Preconditioning. *arXiv preprint* arXiv:2103.16999 (2021).
4. G. Ciaramella and M.J. Gander. *Iterative Methods and Preconditioners for Systems of Linear Equations*. Accepted for publication, SIAM (2022).
5. P.G. Ciarlet. *Linear and Nonlinear Functional Analysis with Applications*. Applied mathematics. SIAM, Philadelphia, PA (2013).
6. V. Dolean, M.J. Gander, W. Kheriji, F. Kwok and R. Masson. Nonlinear preconditioning: How to use a nonlinear Schwarz method to precondition Newton's method. *SIAM Journal on Scientific Computing* **38**(6), A3357–A3380 (2016).
7. L.C. Evans. *Partial Differential Equations*. Graduate studies in mathematics. American Mathematical Society (2010).
8. M.J. Gander. On the origins of linear and non-linear preconditioning. In: *Domain Decomposition Methods in Science and Engineering XXIII*, pp. 153–161, Springer (2017).
9. A. Quarteroni and A. Valli. *Domain decomposition methods for partial differential equations*. Oxford University Press (1999).

A Nonlinear Optimized Schwarz Preconditioner for Elliptic Optimal Control Problems

Gabriele Ciaramella, Felix Kwok, and Georg Müller

1 Introduction

Consider the nonlinear optimal control problem

$$\begin{aligned} &\min_{y,u} J(y,u) := \frac{1}{2}\|y - y_d\|_{L^2}^2 + \frac{\nu}{2}\|u\|_{L^2}^2 + \beta\|u\|_{L^1}, \\ &\text{s.t.} \; -\Delta y + cy + b\varphi(y) = f + u \text{ in } \Omega, \; y = 0 \text{ on } \partial\Omega, \\ &\quad u \in U_{\text{ad}} := \{v \in L^2(\Omega) : |v| \leq \bar{u} \text{ in } \Omega\}, \end{aligned} \tag{1}$$

where $\|\cdot\|_{L^r}$ denotes the usual norm for $L^r(\Omega)$ with $1 \leq r \leq \infty$, the functions $y_d, f \in L^2(\Omega)$ are given, and the scalar parameters $b, c, \beta \geq 0$ and $\nu, \beta \geq 0$ are known. Our model includes problems such as the simplified Ginzburg–Landau superconductivity equation as well as inverse problems where L^1-regularization is used to enhance sparsity of the control function u. For simplicity, the domain $\Omega \subset \mathbb{R}^2$ is assumed to be a rectangle $(0, \widetilde{L}) \times (0, \widehat{L})$. The function $\varphi : \mathbb{R} \to \mathbb{R}$ is assumed to be of class C^2, with locally bounded and locally Lipschitz second derivative and such that $\partial_y \varphi(y) \geq 0$. These assumptions guarantee that the Nemytskii operator $y(\cdot) \mapsto \varphi(y(\cdot))$ is twice continuously Fréchet differentiable in $L^\infty(\Omega)$. In this setting, the optimal control problem (1) is well posed in the sense that there exists a minimizer $(y, u) \in X \times L^2(\Omega)$, with $X := H_0^1(\Omega) \cap L^\infty(\Omega)$, cf. [1, 7]. Our goal is to derive efficient nonlinear preconditioners for solving (1) using domain decomposition techniques.

G. Ciaramella
Politecnico di Milano, e-mail: gabriele.ciaramella@polimi.it

F. Kwok
Université Laval, e-mail: felix.kwok@mat.ulaval.ca

G. Müller
Universität Konstanz, e-mail: georg.mueller@uni-konstanz.de

S. C. Brenner et al. (eds.), *Domain Decomposition Methods in Science and Engineering XXVI*, Lecture Notes in Computational Science and Engineering 145,
https://doi.org/10.1007/978-3-030-95025-5_41

Let $(y,u) \in X \times L^2(\Omega)$ be a solution to (1). Then there exists an adjoint variable $p \in X$ such that (y,u,p) satisfies the system [6, Theorem 2.3]

$$
\begin{aligned}
-\Delta y + cy + b\varphi(y) &= f + u && \text{in } \Omega \text{ with } y = 0 \text{ on } \partial\Omega,\\
-\Delta p + cp + b\varphi'(y)p &= y - y_d && \text{in } \Omega \text{ with } p = 0 \text{ on } \partial\Omega,\\
u &= \mu(p),
\end{aligned}
$$

where $\mu : L^\infty(\Omega) \to L^2(\Omega)$ is

$$
\begin{aligned}
\mu(p) =& \max(0, (-\beta - p)/\nu) + \min(0, (\beta - p)/\nu)\\
&- \max(0, -\bar{u} + (-p - \beta)/\nu) - \min(0, \bar{u} + (-p + \beta)/\nu).
\end{aligned}
\tag{2}
$$

We remark that for $\beta = 0$, the previous formula becomes $\mu(p) = \mathbb{P}_{U_{\text{ad}}}(-p/\nu)$, which is the usual projection formula that leads to the optimality condition $u = \mathbb{P}_{U_{\text{ad}}}(-p/\nu)$; see [7]. Moreover, if $\beta = 0$ with $\bar{u} = \infty$, one obtains that $\mu(p) = -p/\nu$, which implies the usual optimality condition $\nu u + p = 0$, where $\nu u + p$ is the gradient of the reduced cost functional $\widehat{J}(u) = J(y(u), u)$ [7].

Eliminating the control using $\mu(p)$, the first-order optimality system becomes

$$
\begin{aligned}
-\Delta y + cy + b\varphi(y) &= f + \mu(p) && \text{in } \Omega \text{ with } y = 0 \text{ on } \partial\Omega,\\
-\Delta p + cp + b\varphi'(y)(p) &= y - y_d && \text{in } \Omega \text{ with } p = 0 \text{ on } \partial\Omega.
\end{aligned}
\tag{3}
$$

This nonlinear and nonsmooth system admits a solution $(y,p) \in X^2$ [1, 7].

2 An optimized Schwarz method and preconditioner

In this section, we introduce an optimized Schwarz method (OSM) for solving the optimality system (3). We consider the non-overlapping decomposition of Ω shown

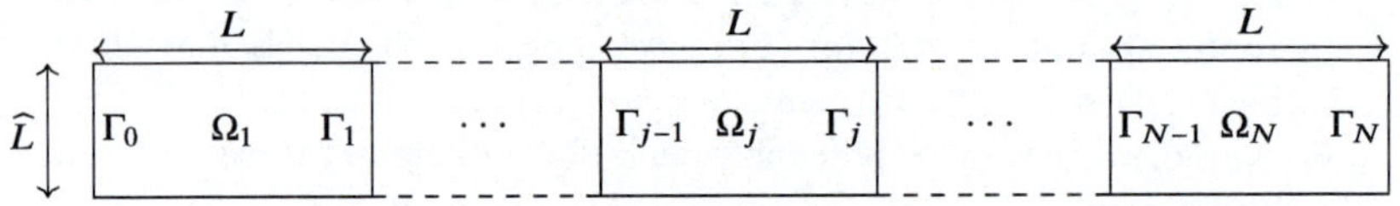

Fig. 1: Non-overlapping domain decomposition.

in Fig. 1 and given by disjoint subdomains Ω_j, $j = 1, \dots, N$ such that $\overline{\Omega} = \cup_{j=1}^N \overline{\Omega}_j$. The sets $\Gamma_j := \overline{\Omega}_j \cap \overline{\Omega}_{j+1}$, $j = 1, \dots, N-1$ are the interfaces. Moreover, we define $\Gamma_j^{\text{ext}} := \partial\Omega_j \cap \partial\Omega$, $j = 1, \dots, N$, which represent the external boundaries of the subdomains. The optimality system (3) can be written as a coupled system of N subproblems defined on the subdomains Ω_j, $j = 1, \dots, N$, of the form

$$
\begin{aligned}
-\Delta y_j + c y_j + b\varphi(y_j) &= f_j + \mu(p_j) && \text{in } \Omega_j, && \text{(4a)}\\
-\Delta p_j + c p_j + b\varphi'(y_j)(p_j) &= y_j - y_{d,j} && \text{in } \Omega_j, && \text{(4b)}\\
y_j = 0,\ p_j &= 0 && \text{on } \Gamma_j^{\text{ext}}, && \text{(4c)}\\
q\, y_j + \partial_x y_j &= q\, y_{j+1} + \partial_x y_{j+1} && \text{on } \Gamma_j, && \text{(4d)}\\
q\, p_j + \partial_x p_j &= q\, p_{j+1} + \partial_x p_{j+1} && \text{on } \Gamma_j, && \text{(4e)}\\
q\, y_j - \partial_x y_j &= q\, y_{j-1} - \partial_x y_{j-1} && \text{on } \Gamma_{j-1}, && \text{(4f)}\\
q\, p_j - \partial_x p_j &= q\, p_{j-1} - \partial_x p_{j-1} && \text{on } \Gamma_{j-1}, && \text{(4g)}
\end{aligned}
$$

for $j = 1, \dots, N$, where for $j \in \{1, N\}$ the boundary conditions at Γ_0 and Γ_N, respectively, must be replaced with homogeneous Dirichlet conditions. Here, $q > 0$ is a parameter that can be optimized to improve the convergence of the OSM; see, e.g, [2, 5]. The system (4) leads to the OSM, which, for a given $(y_j^0, p_j^0)_{j=1}^N$, consists of solving the subdomain problems below for $\mathbf{y}_j^k := (y_j^k, p_j^k)$, $k = 1, 2, 3, \dots$:

$$
\begin{aligned}
-\Delta y_j^k + c y_j^k + b\varphi(y_j^k) &= f_j + \mu(p_j^k) && \text{in } \Omega_j, && \text{(5a)}\\
-\Delta p_j^k + c p_j^k + b\varphi'(y_j^k)(p_j^k) &= y_j^k - y_{d,j} && \text{in } \Omega_j, && \text{(5b)}\\
\mathbf{y}_j^k &= 0, && \text{on } \Gamma_j^{\text{ext}}, && \text{(5c)}\\
q\, \mathbf{y}_j^k + \partial_x \mathbf{y}_j^k &= q\, \mathbf{y}_{j+1}^{k-1} + \partial_x \mathbf{y}_{j+1}^{k-1} && \text{on } \Gamma_j, && \text{(5d)}\\
q\, \mathbf{y}_j^k - \partial_x \mathbf{y}_j^k &= q\, \mathbf{y}_{j-1}^{k-1} - \partial_x \mathbf{y}_{j-1}^{k-1} && \text{on } \Gamma_{j-1}. && \text{(5e)}
\end{aligned}
$$

Now, we use the OSM to introduce a nonlinear preconditioner by setting $\mathbf{y}_j := (y_j, p_j)$, $j = 1, \dots, N$, and defining the solution maps S_j as

$$
\begin{aligned}
S_1(\mathbf{y}_2) &= \mathbf{y}_1 && \text{solution to (4) with } j = 1 \text{ and } \mathbf{y}_2 \text{ given},\\
S_j(\mathbf{y}_{j-1}, \mathbf{y}_{j+1}) &= \mathbf{y}_j && \text{solution to (4) with } 2 \le j \le N-1 \text{ and } \mathbf{y}_{j\pm1} \text{ given},\\
S_N(\mathbf{y}_{N-1}) &= \mathbf{y}_N && \text{solution to (4) with } j = N \text{ and } \mathbf{y}_{N-1} \text{ given}.
\end{aligned}
$$

Hence, using the variable $\mathbf{y} = (\mathbf{y}_1, \dots, \mathbf{y}_N)$, we can rewrite (4) as

$$
\mathcal{F}_{\mathrm{P}}(\mathbf{y}) = 0, \quad \text{where} \quad \mathcal{F}_{\mathrm{P}}(\mathbf{y}) := \begin{bmatrix} \mathbf{y}_1 - S_1(\mathbf{y}_2) \\ \mathbf{y}_2 - S_2(\mathbf{y}_1, \mathbf{y}_3) \\ \vdots \\ \mathbf{y}_{N-1} - S_{N-1}(\mathbf{y}_{N-2}, \mathbf{y}_N) \\ \mathbf{y}_N - S_N(\mathbf{y}_{N-1}) \end{bmatrix}. \tag{6}
$$

This is the nonlinearly preconditioned form of (3) induced by the OSM (4)–(5), to which we can apply a generalized Newton method. For a given initialization $\mathbf{y}^0$, a Newton method generates a sequence $(\mathbf{y}^k)_{k\in\mathbb{N}}$ defined by

$$
\text{solve} \quad D\mathcal{F}_{\mathrm{P}}(\mathbf{y}^k)(\mathbf{d}^k) = -\mathcal{F}_{\mathrm{P}}(\mathbf{y}^k) \quad \text{and update} \quad \mathbf{y}^{k+1} = \mathbf{y}^k + \mathbf{d}^k. \tag{7}
$$

Notice that at each iteration of (7) one needs to evaluate the residual function $\mathcal{F}_{\mathrm{P}}(\mathbf{y}^k)$, which requires the (parallel) solution of the N subproblems (4). The computational cost is therefore equivalent to one iteration of the OSM (5). As an inner solver for the subproblems, which involve the (mildly) non-differentiable function μ, a semismooth Newton can be employed.

We now discuss the problem of solving the Jacobian linear system in (7). Let $\mathbf{d} = (\mathbf{d}_1, \dots, \mathbf{d}_N)$, where $\mathbf{d}_j = (d_{y,j}, d_{p,j})$, $j = 1, \dots, N$. Then a direct calculation (omitted for brevity) shows that the action of the operator $D\mathcal{F}_{\mathrm{P}}(\mathbf{y})$ on the vector $\mathbf{d}$ is given by $D\mathcal{F}_{\mathrm{P}}(\mathbf{y})(\mathbf{d}) = \mathbf{d} - \widetilde{\mathbf{y}}(\mathbf{d})$, where $\widetilde{\mathbf{y}} := (\widetilde{\mathbf{y}}_1, \dots, \widetilde{\mathbf{y}}_N)$, and each $\widetilde{\mathbf{y}}_j = (\widetilde{y}_j, \widetilde{p}_j)$ satisfies the *linearized* subdomain problems

$$
\begin{aligned}
-\Delta \widetilde{y}_j + c\widetilde{y}_j + b\varphi'(y_j)\widetilde{y}_j &= D\mu(p_j)(\widetilde{p}_j) && \text{in } \Omega_j, && \text{(8a)}\\
-\Delta \widetilde{p}_j + c\widetilde{p}_j + b\varphi''(y_j)[p_j, \widetilde{y}_j] &= \widetilde{y}_j && \text{in } \Omega_j, && \text{(8b)}\\
\widetilde{\mathbf{y}}_j &= 0, && \text{on } \Gamma_j^{\mathrm{ext}}, && \text{(8c)}\\
q\,\widetilde{\mathbf{y}}_j + \partial_x \widetilde{\mathbf{y}}_j &= q\,\mathbf{d}_{j+1} + \partial_x \mathbf{d}_{j+1} && \text{on } \Gamma_j, && \text{(8d)}\\
q\,\widetilde{\mathbf{y}}_j - \partial_x \widetilde{\mathbf{y}}_j &= q\,\mathbf{d}_{j-1} - \partial_x \mathbf{d}_{j-1} && \text{on } \Gamma_{j-1}, && \text{(8e)}
\end{aligned}
$$

where

$$
\begin{aligned}
D\mu(p)(\widetilde{p}) = &\frac{1}{\nu}\Big[-\mathcal{G}_{\max}(-\beta - p) - \mathcal{G}_{\min}(\beta - p)\\
&+ \mathcal{G}_{\max}(-p - \beta - \nu\bar{u}) + \mathcal{G}_{\min}(-p + \beta + \nu\bar{u})\Big]\widetilde{p},
\end{aligned}
$$

with $\mathcal{G}_{\max}(v)(x) = \begin{cases} 1 & \text{if } v(x) > 0,\\ 0 & \text{if } v(x) \le 0,\end{cases}$ and $\mathcal{G}_{\min}(v)(x) = \begin{cases} 1 & \text{if } v(x) \le 0,\\ 0 & \text{if } v(x) > 0,\end{cases}$

and where the boundary values for $j \in \{1, N\}$ have to be modified as in (4). Note that this is the same linearized problem that must be solved repeatedly within the inner iterations of semismooth Newton, so its solution cost is only a fraction of the cost required to calculate $\mathcal{F}_{\mathrm{P}}(\mathbf{y})$. Our matrix-free preconditioned semismooth Newton algorithm that corresponds to the Newton procedure (7) is summarized in Algorithm 1.

3 Numerical experiments

We begin with a two subdomain case for $\Omega = (0, 1)^2$,

$$
\begin{aligned}
y_d(x, y) &= 10\sin(4\pi x)\sin(3\pi y),\\
f = 0, &\quad c = 1 \quad \text{and}\\
\varphi(y) &= y + \exp(y).
\end{aligned}
$$

The domain Ω is discretized with a uniform mesh of 51 interior points on each edge of the unit square. The discrete optimality system is obtained by the finite difference

Algorithm 1 Matrix-free preconditioned generalized Newton method

Require: Initial guess $\mathbf{y}^0$, tolerance ϵ, maximum number of iterations $k_{\max}$.
Compute $S_1(\mathbf{y}_2^0)$, $S_j(\mathbf{y}_{j-1}^0, \mathbf{y}_{j+1}^0)$, $j = 2, \dots, N-1$, and $S_N(\mathbf{y}_{N-1}^0)$.
Set $k = 0$ and assemble $\mathcal{F}_P(\mathbf{y}^0)$ using (6).
while $\|\mathcal{F}_P(\mathbf{y}^k)\| \geq \epsilon$ and $k \leq k_{\max}$ **do**
  Compute $\mathbf{d}^k$ by solving $D\mathcal{F}_P(\mathbf{y}^k)(\mathbf{d}^k) = -\mathcal{F}_P(\mathbf{y}^k)$ using a matrix-free Krylov method, e.g., GMRES (together with a routine for solving (8) to compute the action of $D\mathcal{F}_P(\mathbf{y}^k)$ on a vector $\mathbf{d}$).
  Update $\mathbf{y}^{k+1} = \mathbf{y}^k + \mathbf{d}^k$.
  Set $k = k + 1$.
  Compute $S_1(\mathbf{y}_2^k)$, $S_j(\mathbf{y}_{j-1}^k, \mathbf{y}_{j+1}^k)$, $j = 2, \dots, N-1$, and $S_N(\mathbf{y}_{N-1}^k)$.
  Assemble $\mathcal{F}_P(\mathbf{y}^k)$ using (6).
end while
Output: $\mathbf{y}^k$.

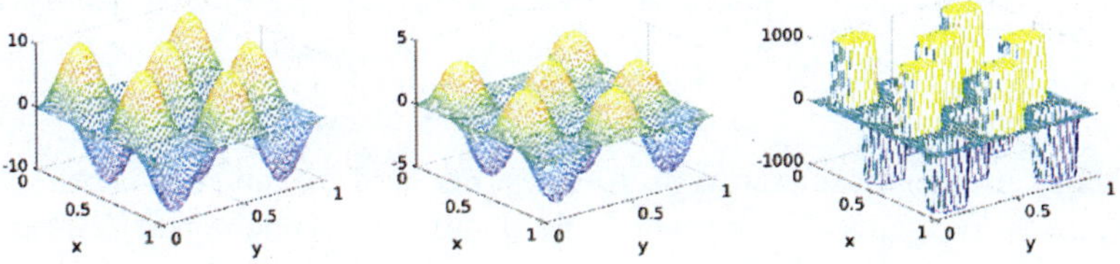

Fig. 2: Target y_d (left), optimal state y (middle), and optimal control u (right) computed for $b = 10$, $\nu = 10^{-7}$ and $\beta = 10^{-2}$.

method. Fig. 2 shows an example of the solution computed for $b = 10$, $\nu = 10^{-7}$, $\bar{u} = 10^3$ and $\beta = 10^{-2}$. Here, we can observe how the computed optimal state (middle) has the same shape as the target y_d (left), but the control constraints and the L^1-penalization prevent the control function from making the state equal to the desired target.

To study the efficiency and the robustness of the proposed numerical framework, we test the nonlinearly preconditioned Newton for several values of parameters ν, β, $\bar{u}$, b and q, and compare the obtained number of iterations with the ones performed by a (damped) semismooth Newton applied directly to (3). Moreover, to improve the robustness of our preconditioned Newton method, we implemented the following continuation procedure with respect to the regularization parameter ν: for $k = 1$, we set $\nu_1 = 10^{-1}$ and solve the Jacobian system (7) *once* to obtain $\mathbf{y}^2$. Next, we decrease ν by a factor of 4 ($\nu_2 = \nu_1/4$), do another solve and update step (7), and so on. When we reach the true ν prescribed by the problem, we set $\nu_k = \nu$ and repeat (7) until convergence; see [3] for convergence results for similar continuation procedures. We apply the same continuation procedure on semismooth Newton applied directly to (3) for comparison. Note that because only one Jacobian solve is performed before ν is updated, there are cases where semismooth Newton with continuation diverges, *even when its counterpart without continuation converges*, see Tab. 1. We initialize the four methods by randomly chosen vectors. The number of iterations performed

			$\bar{u} = 10^3$			$\bar{u} = \infty$		
	q	b	$\nu = 10^{-3}$	$\nu = 10^{-5}$	$\nu = 10^{-7}$	$\nu = 10^{-3}$	$\nu = 10^{-5}$	$\nu = 10^{-7}$
$\beta = 0$	1	0	4-5-2-5	6-9-11-11	4-11-41-12	3-5-2-5	3-9-2-9	3-12-3-12
	10	0	4-5-2-5	6-9-11-11	8-11-41-12	3-5-2-5	3-8-2-9	3-11-3-12
	100	0	3-5-2-5	6-9-11-11	×-11-41-12	3-5-2-5	3-9-2-9	3-11-3-12
	1	10	6-6-4-7	×-10-12-12	×-12-38-16	6-6-4-7	×-10-22-23	×-15-×-15
	10	10	5-6-4-7	7-10-12-12	×-12-38-16	5-6-4-7	×-10-22-23	×-14-×-15
	100	10	4-6-4-7	6-10-12-12	×-13-38-16	4-6-4-7	6-10-22-23	×-13-×-15
$\beta = 10^{-2}$	1	0	5-5-3-6	6-9-8-11	×-12-43-13	4-5-3-6	5-9-6-10	×-12-8-15
	10	0	4-5-3-6	6-9-8-11	×-11-43-13	4-5-3-6	4-9-6-10	×-12-8-15
	100	0	4-5-3-6	6-10-8-11	11-12-43-13	4-5-3-6	5-9-6-10	7-12-8-15
	1	10	6-6-4-6	×-11-10-12	×-12-×-17	6-6-4-6	×-10-18-×	×-13-×-15
	10	10	5-6-4-6	×-11-10-12	×-13-×-17	5-6-4-6	×-10-18-×	×-14-×-15
	100	10	4-6-4-6	6-11-10-12	9-13-×-17	4-6-4-6	6-10-18-×	×-13-×-15

Table 1: Two subdomains: outer iterations of preconditioned Newton (left value), preconditioned Newton with continuation (middle-left value), semismooth Newton applied to the original problem (middle-right value), and semismooth Newton with continuation applied to the original problem (right value).

by both methods to reach a tolerance of 10^{-8} are reported in Tab. 1, where the symbol × indicates divergence. These results show that if the preconditioned Newton converges, then it outperforms the semismooth Newton applied directly to the full system (3). However, the preconditioned Newton does not always converge due to the lack of damping. With continuation, however, our method always converges, with an iteration count comparable (for moderate values of ν) or much lower (for small ν) than for the semismooth Newton method.

To better gauge the cost of the continuation strategy, we show the total number of inner iterations required by 'pure' preconditioned Newton versus the one with continuation in Tab. 2. The reported numbers are computed as $\sum_k \max_{j=1,2} \mathrm{it}_{j,k}$, where k is the iteration count and $\mathrm{it}_{j,k}$, $j = 1, 2$, are the number of inner iterations required by the two subdomain solves performed at the kth outer iteration. (The max accounts for the fact that the two subdomain problems are supposed to be solved in parallel.) The results show that the continuation procedure actually *reduces* the total number of inner iterations for the most part, except for some very easy cases, such as $\beta = b = 0$, $\bar{u} = \infty$ (where the problem is in fact linear).

Finally, Tab. 3 shows the total number of GMRES iterations required for solving (7) (with or without continuation), together with the GMRES iteration count for semismooth Newton (with or without continuation); the latter is preconditioned by block Jacobi, using $-\Delta + cI$ as diagonal blocks. We see that for the "easy" case of $\nu = 10^{-3}$, semismooth Newton requires fewer GMRES steps than preconditioned Newton, but the situation reverses for smaller ν. In fact, for a well-chosen Robin parameter such as $q = 10$, the advantage of preconditioned Newton with continuation can be quite significant in these harder cases. All these numerical observations show clearly the efficiency of the proposed computational framework.

We now consider a multiple subdomain case. This time, the mesh is refined to have 101 interior points on each edge of Ω. We then fix $q = 100$ and repeat the

			$\bar{u} = 10^3$			$\bar{u} = \infty$		
	q	b	$\nu = 10^{-3}$	$\nu = 10^{-5}$	$\nu = 10^{-7}$	$\nu = 10^{-3}$	$\nu = 10^{-5}$	$\nu = 10^{-7}$
$\beta = 0$	1	0	6 - 5	31 - 12	× - 18	2 - 5	3 - 8	3 - 11
	10	0	5 - 5	26 - 11	96 - 19	2 - 5	3 - 8	3 - 11
	100	0	2 - 5	18 - 13	× - 19	2 - 5	2 - 8	3 - 11
	1	10	× - 17	× - 35	× - 47	27 - 17	× - 34	× - 60
	10	10	21 - 14	× - 31	103 - 43	21 - 14	× - 32	× - 53
	100	10	8 - 14	26 - 32	× - 43	8 - 14	45 - 30	× - 47
$\beta = 10^{-2}$	1	0	13 - 8	32 - 16	84 - 25	8 - 8	10 - 14	× - 25
	10	0	10 - 8	22 - 17	33 - 23	7 - 8	11 - 15	× - 24
	100	0	7 - 6	15 - 15	104 - 20	7 - 6	12 - 13	× - 22
	1	10	× - 17	× - 33	× - 45	28 - 17	× - 32	× - 47
	10	10	20 - 14	× - 33	× - 48	20 - 14	× - 30	× - 46
	100	10	10 - 14	23 - 30	125 - 44	10 - 14	40 - 26	× - 44

Table 2: Two subdomains: total number of inner iterations of preconditioned Newton (left value) and preconditioned Newton with continuation (right value).

			$\bar{u} = 10^3$			$\bar{u} = \infty$		
	q	b	$\nu = 10^{-3}$	$\nu = 10^{-5}$	$\nu = 10^{-7}$	$\nu = 10^{-3}$	$\nu = 10^{-5}$	$\nu = 10^{-7}$
$\beta = 0$	1	0	143 - 128 - 26 - 17	279-163-303-123	139-175-1255-155	100 - 128 - 26 - 17	170-140- 69 - 46	× -149-371- 66
	10	0	90 - 75 - 26 - 17	179-101-303-123	254-108-1255-155	64 - 75 - 26 - 17	114- 88 - 69 - 46	× - 92 -371- 66
	100	0	73 - 90 - 26 - 17	177-114-303-123	× -125-1255-155	73 - 90 - 26 - 17	60 - 88 - 69 - 46	54 - 93 -371- 66
	1	10	266 - 204 - 49 - 66	× -251-397-255	× -268-1479-457	256 - 204 - 49 - 66	× -240-2000-1172	× -379- × -928
	10	10	124 - 88 - 49 - 66	226-129-397-255	× -168-1479-457	117 - 88 - 49 - 66	× -152-2000-1172	× -190- × -928
	100	10	122 - 155 - 49 - 66	139-239-397-255	× -274-1479-457	122 - 155 - 49 - 66	161-234-2000-1172	× -250- × -928
$\beta = 10^{-2}$	1	0	226 - 164 - 31 - 42	290-198-187-123	× -223-1065-168	188 - 164 - 31 - 42	246-168- 183 - 109	× -218-522-380
	10	0	111 - 95 - 31 - 42	178-121-187-123	× -130-1065-168	115 - 95 - 31 - 42	143-124- 183 - 109	× -145-522-380
	100	0	135 - 118 - 31 - 42	179-158-187-123	333-175-1065-168	135 - 118 - 31 - 42	145-147- 183 - 109	165-173-522-380
	1	10	273 - 235 - 49 - 54	× -238-299-233	× -228- × -416	261 - 235 - 49 - 54	× -254-1362- ×	× -311- × -752
	10	10	139 - 124 - 49 - 54	× -158-299-233	× -164- × -416	138 - 124 - 49 - 54	× -161-1362- ×	× -179- × -752
	100	10	122 - 162 - 49 - 54	141-251-299-233	215-300- × -416	122 - 162 - 49 - 54	167-219-1362- ×	× -303- × -752

Table 3: Two subdomains: GMRES iterations of preconditioned Newton (left value), preconditioned Newton with continuation (middle-left value), semismooth Newton applied to the original problem (middle-right value), and semismooth Newton with continuation applied to the original problem (right value).

			$\bar{u} = 10^3$			$\bar{u} = \infty$		
	N	b	$\nu = 10^{-3}$	$\nu = 10^{-5}$	$\nu = 10^{-7}$	$\nu = 10^{-3}$	$\nu = 10^{-5}$	$\nu = 10^{-7}$
$\beta = 0$	4	0	94 - 116 - 131 - 157	× -185-541-329	× - 196 -1974- 436	92 - 112 - 131 - 157	64 -115- 151 -246	58 -128-513- 528
	8	0	178 - 157 - 176 - 217	× -244-503-365	× - 259 -2261- 455	121 - 155 - 176 - 217	83 -180- 140 -322	59 -204-438- 537
	16	0	228 - 229 - 217 - 301	× -383-747-509	× - 384 -2915- 554	159 - 238 - 217 - 301	130-281- 195 -444	× -314-201- 569
	4	10	145 - 184 - 202 - 234	218-287-447-517	× - 382 -1207- 910	143 - 186 - 202 - 234	171-276- 633 -506	× -296- × -1238
	8	10	192 - 247 - 232 - 314	286-381-453-557	× - 498 -1475- 979	196 - 247 - 232 - 314	214-368- 621 -552	× -368- × -1217
	16	10	272 - 346 - 364 - 425	× -625-648-777	× - 744 -1361-1180	280 - 345 - 364 - 425	327-549- 819 -765	× -510- × -1200
$\beta = 10^{-2}$	4	0	177 - 139 - 169 - 204	221-220-389-301	399- 255 -1745- 439	179 - 135 - 169 - 204	175-196- 340 -315	243-273-992- 649
	8	0	231 - 191 - 266 - 277	× -291-448-383	× - 340 -1594- 487	231 - 199 - 266 - 277	275-264- 350 -385	300-360-983- 658
	16	0	319 - 241 - 314 - 372	552-395-667-528	× - 452 -2244- 630	227 - 234 - 314 - 372	438-365- 517 -529	520-449-899-1714
	4	10	145 - 201 - 209 - 248	230-312-369-500	380- 408 -1536- 851	147 - 201 - 209 - 248	193-312- 982 -492	× -392- × -1320
	8	10	196 - 262 - 267 - 323	356-408-402-574	× - 641 -1625-1020	193 - 261 - 267 - 323	247-406-1041-568	× -488- × -1396
	16	10	268 - 365 - 368 - 444	538-660-577-794	× -1017-2000-1312	260 - 365 - 368 - 444	432-599-1425-790	× -702- × -1686

Table 4: Multiple subdomains: GMRES iterations of preconditioned Newton (left value), preconditioned Newton with continuation (middle-left value), semismooth Newton applied to the original problem (middle-right value), and semismooth Newton with continuation applied to the original problem (right value).

experiments above for $N = 4, 8, 16$ subdomains. In Tab. 4, we compare the GMRES iteration counts for preconditioned Newton (with and without continuation) to those of semismooth Newton applied to (3). We see that preconditioned Newton with continuation works well in all cases, and for smaller ν values, the iteration count is much lower than for semismooth Newton. The outer iteration counts are omitted for brevity, but we observed a behaviour similar to the two-subdomain case, and one which is robust for the mesh sizes $h = \frac{1}{26}, \frac{1}{51}, \frac{1}{101}$; see [2, 4] for related scalability discussions.

4 Further discussion and conclusion

This short manuscript represents a proof of concept for using domain decomposition-based nonlinear preconditioning to efficiently solve nonlinear, nonsmooth optimal control problems governed by elliptic equations. However, several theoretical and numerical issues must be addressed as part of a complete development of these techniques. From a theoretical point of view, to establish concrete convergence results based on classical semismooth Newton theory, it is crucial to study the (semismoothness) properties of the subdomain solution maps $\mathcal{S}_j$, which are implicit functions of semismooth maps. Another crucial point is the proof of well-posedness of the (preconditioned) Newton linear system. From a domain decomposition perspective, more general decompositions (including cross points) must be considered. Finally, a detailed analysis of the scalability of the GMRES iterations is necessary.

Acknowledgements G. Ciaramella is a member of INdAM GNCS. F. Kwok gratefully acknowledges support from the National Science and Engineering Research Council of Canada (RGPIN-2021-02595). The work described in this paper is partially supported by a grant from the ANR/RGC joint research scheme sponsored by the Research Grants Council of the Hong Kong Special Administrative Region, China and the French National Research Agency (Project no. A-HKBU203/19).

References

1. E. Casas, R. Herzog and G. Wachsmuth. Optimality conditions and error analysis of semilinear elliptic control problems with L1 cost functional. *SIAM J. Optim.* **22**(3), 795–820 (2012).
2. F. Chaouqui, G. Ciaramella, M.J. Gander and T. Vanzan. On the scalability of classical one-level domain-decomposition methods. *Vietnam Journal of Mathematics* **46**(4), 1053–1088 (2018).
3. G. Ciaramella, A. Borzi, G. Dirr and D. Wachsmuth. Newton methods for the optimal control of closed quantum spin systems. *SIAM J. Sci. Comp.* **37**(1), A319–A346 (2015).
4. G. Ciaramella and M.J. Gander. Analysis of the parallel Schwarz method for growing chains of fixed-sized subdomains: Part I. *SIAM J. Numer. Anal.* **55**(3), 1330–1356 (2017).
5. M.J. Gander. Optimized Schwarz methods. *SIAM J. Numer. Anal.* **44**(2), 699–731 (2006).
6. G. Stadler. Elliptic optimal control problems with L1-control cost and applications for the placement of control devices. *Comput. Optim. Appl.* **44**(2), 159–181 (2009).
7. F. Tröltzsch. *Optimal Control of Partial Differential Equations: Theory, Methods, and Applications*. American Mathematical Society (2010).

SParse Approximate Inverse (SPAI) Based Transmission Conditions for Optimized Algebraic Schwarz Methods

Martin J. Gander, Lahcen Laayouni, and Daniel B. Szyld

1 Introduction

There have been various studies on algebraic domain decomposition methods, see e.g. [1], [2], [6], [7], [8] and references therein. Algebraic Optimized Schwarz Methods (AOSMs) were introduced in [4] to solve block banded linear systems arising from the discretization of PDEs on irregular domains. AOSMs mimic Optimized Schwarz Methods (OSMs) [5] algebraically by optimizing transmission blocks between subdomains. We propose here a new approach for obtaining transmission blocks using SParse Approximate Inverse (SPAI) techniques [9]. SPAI permits the approximation of the required parts of an inverse needed in the optimal transmission blocks, without knowing the entire inverse, which would be infeasible in practice, and is naturally parallel, like the domain decomposition iteration itself. Using SPAI with different numbers of diagonals in a predefined sparsity pattern gives rise to approximations in the transmission blocks which can be interpreted as differential transmission operators at the continuous level of various degrees, and this can be used to compute a theoretical convergence factor of the resulting AOSM. We can therefore compare the performance of the SPAI AOSM theoretically, and show that a direct SPAI application without taking into account the entire non-linear structure of the convergence estimate of AOSM leads to suboptimal performance. We thus propose a modified SPAI-like technique that minimizes the entire convergence estimate and restores the expected performance.

Martin J. Gander
Section de Mathématiques, University of Geneva, Switzerland,
e-mail: Martin.Gander@unige.ch

Lahcen Laayouni
School of Science and Engineering, Al Akhawayn University, Avenue Hassan II, 53000 P.O. Box 1630, Ifrane, Morocco, e-mail: L.Laayouni@aui.ma

Daniel B. Szyld
Department of Mathematics, Temple University (038-16), 1805 N. Broad Street, Philadelphia, Pennsylvania 19122-6094, USA, e-mail: szyld@temple.edu

S. C. Brenner et al. (eds.), *Domain Decomposition Methods in Science and Engineering XXVI*, Lecture Notes in Computational Science and Engineering 145,
https://doi.org/10.1007/978-3-030-95025-5_42

2 Algebraic Optimized Schwarz methods

We are interested in solving linear systems of the form

$$Au = f,$$

where the $n \times n$ matrix A arises from a finite element or finite difference discretization of a partial differential equation, and has a block banded structure of the form

$$A = \begin{bmatrix} A_{11} & A_{12} & & \\ A_{21} & A_{22} & A_{23} & \\ & A_{32} & A_{33} & A_{34} \\ & & A_{43} & A_{44} \end{bmatrix}, \tag{1}$$

where A_{ij} are blocks of size $n_i \times n_j$, $i, j = 1, \ldots, 4$, and $n = \sum_i n_i$. We suppose that $n_1 \gg n_2$ and $n_4 \gg n_3$, representing two large subdomains; for generalizations to more subdomains, see [4, Section 6]. We consider Algebraic Optimized Schwarz methods of additive and multiplicative type, whose iteration operators are based on the following modifications inspired by OSM,

$$T_{ORAS} = I - \sum_{i=1}^{2} \tilde{R}_i^T \tilde{A}_i^{-1} R_i A, \text{ and } T_{ORMS} = \prod_{i=2}^{1} (I - \tilde{R}_i^T \tilde{A}_i^{-1} R_i A), \tag{2}$$

where

$$\tilde{A}_1 = \begin{bmatrix} A_{11} & A_{12} & \\ A_{21} & A_{22} & A_{23} \\ & A_{32} & S_1 \end{bmatrix}, \qquad \tilde{A}_2 = \begin{bmatrix} S_2 & A_{23} & \\ A_{32} & A_{33} & A_{34} \\ & A_{43} & A_{44} \end{bmatrix}, \tag{3}$$

with $S_1 = A_{33} + D_1$ and $S_2 = A_{22} + D_2$. Here D_1 and D_2 are transmission matrices to be chosen for fast convergence. The asymptotic convergence factor of AOSM depends on the product of the following two norms (see [4, Theorem 3.2]),

$$\| (I + D_1 B_{33})^{-1} [D_1 B_{12} - A_{34} B_{13}] \|, \; \| (I + D_2 B_{11})^{-1} [D_2 B_{32} - A_{21} B_{31}] \|, \tag{4}$$

where the B matrices involve certain columns of inverses of submatrices of A, namely

$$\begin{bmatrix} B_{31} \\ B_{32} \\ B_{33} \end{bmatrix} := \begin{bmatrix} A_{11} & A_{12} & \\ A_{21} & A_{22} & A_{23} \\ & A_{32} & A_{33} \end{bmatrix}^{-1} \begin{bmatrix} 0 \\ 0 \\ I \end{bmatrix}, \quad \begin{bmatrix} B_{11} \\ B_{12} \\ B_{13} \end{bmatrix} := \begin{bmatrix} A_{22} & A_{23} & \\ A_{32} & A_{33} & A_{34} \\ & A_{43} & A_{44} \end{bmatrix}^{-1} \begin{bmatrix} I \\ 0 \\ 0 \end{bmatrix}. \tag{5}$$

We can easily derive the optimal choice for the transmission matrices, see [4],

$$D_{1,\text{opt}} = -A_{34} A_{44}^{-1} A_{43} \qquad \text{and} \qquad D_{2,\text{opt}} = -A_{21} A_{11}^{-1} A_{12}, \tag{6}$$

which make (4) zero. The corresponding AOSM then converges in two iterations for ORAS, so one cannot do better than this. Computing these optimal blocks $D_{1,\text{opt}}$ and $D_{2,\text{opt}}$ is however equivalent to computing the Schur complements

$$S_{1,\text{opt}} = A_{33} - A_{34}A_{44}^{-1}A_{43} \quad \text{and} \quad S_{2,\text{opt}} = A_{22} - A_{21}A_{11}^{-1}A_{12} \tag{7}$$

corresponding to the submatrices

$$\begin{bmatrix} A_{33} & A_{34} \\ A_{43} & A_{44} \end{bmatrix} \quad \text{and} \quad \begin{bmatrix} A_{11} & A_{12} \\ A_{21} & A_{22} \end{bmatrix}, \tag{8}$$

and is thus very expensive, due to the large inverses A_{44}^{-1} and A_{11}^{-1}. In the next section we propose sparse approximations of the optimal transmission blocks using predefined sparsity patterns.

3 Sparse approximations of optimal transmission blocks

The new idea to determine approximations $D_{1,\text{app}}$ and $D_{2,\text{app}}$ that make the norms in (4) small and are cheap to compute is to use a SParse Approximate Inverse (SPAI) technique to make the differences

$$||D_1 B_{12} - A_{34}B_{13}|| \quad \text{and} \quad ||D_2 B_{32} - A_{21}B_{31}|| \tag{9}$$

small by approximating the inverse blocks A_{11}^{-1} and A_{44}^{-1} in (6). Due to the sparsity of A_{34}, A_{43}, A_{21}, and A_{12}, we need only to approximate small subblocks of A_{11}^{-1} and A_{44}^{-1} using SPAI.

To gain insight into the quality and performance of such SPAI approximations of $D_{1,\text{opt}}$ and $D_{2,\text{opt}}$, we consider the model problem $\Delta u = f$ in $\Omega = (0,1)^2$, discretized by a standard five point finite difference stencil, which leads to a system matrix of the form (1) with, e.g.,

$$A_{11} = \frac{1}{h^2}\begin{bmatrix} T & I & & \\ I & \ddots & \ddots & \\ & \ddots & \ddots & I \\ & & I & T \end{bmatrix}, \; A_{12} = \frac{1}{h^2}\begin{bmatrix} 0 & 0 & 0 \\ \vdots & & \vdots \\ 0 & 0 & 0 \\ I & 0 & 0 \end{bmatrix}, \; A_{21} = \frac{1}{h^2}\begin{bmatrix} 0 & \dots & 0 & I \\ 0 & \dots & 0 & 0 \\ 0 & \dots & 0 & 0 \end{bmatrix}, \tag{10}$$

where $T = \text{spdiag}([1, -4, 1])$. To approximate the block inverse A_{11}^{-1} in $D_{2,\text{opt}} = -A_{21}A_{11}^{-1}A_{12}$ with matrices from (10) using SPAI naively, we would solve for a matrix M of the same large size as A_{11} the minimization problem $||A_{11}M - I||_F \longrightarrow \min$, which requires solving a least squares problem for each column, and where one specifies a sparsity pattern for M. Because of the sparsity structure of A_{12} and A_{21} however in (10), we see that we need the SPAI approximation only of the last diagonal block (bottom right) of M, which we denote by M^{br}. Thus, it is not necessary to compute the entire SPAI approximation M, it is sufficient to just solve the least squares problems corresponding to the last few columns in M which contain M^{br}, and furthermore these least squares problems are also small due to the sparsity of A_{11}. Doing this for our model problem using a diagonal sparsity pattern for M leads to

$$D^{br}_{2,\text{app}} := M^{br} = -h^2 \begin{bmatrix} 0.2222 & & & & \\ & 0.2015 & & & \\ & & \ddots & & \\ & & & 0.2015 & \\ & & & & 0.2222 \end{bmatrix}. \tag{11}$$

In order to understand to what type of transmission conditions this approximation leads, it is best to look at the corresponding Schur complement approximation $S_{2,\text{app}}$ of $S_{2,\text{opt}}$ from (7), see also [3, Section 4.1], which is also modified only at the bottom right,

$$S^{br}_{2,\text{app}} = A^{br}_{22} - [A_{21} M A_{12}]^{br} = \frac{1}{h^2} T - \frac{1}{h^2} D^{br}_{2,\text{app}} \frac{1}{h^2}. \tag{12}$$

Rearranging this expression into

$$S^{br}_{2,\text{app}} = \frac{1}{h^2}\begin{bmatrix} -2 & 1/2 & & \\ 1/2 & -2 & 1/2 & \\ & \ddots & \ddots & \ddots \\ & & 1/2 & -2 & 1/2 \\ & & & 1/2 & -2 \end{bmatrix} + \frac{1}{2h^2}\begin{bmatrix} -2 & 1 & & \\ 1 & -2 & 1 & \\ & \ddots & \ddots & \ddots \\ & & 1 & -2 & 1 \\ & & & 1 & -2 \end{bmatrix} - \frac{1}{h^2}\begin{bmatrix} 0.7778 & & & & \\ & 0.7895 & & & \\ & & \ddots & & \\ & & & 0.7895 & \\ & & & & 0.7778 \end{bmatrix} \tag{13}$$

and neglecting the fact that the first and last entry from the diagonal SPAI approximation are slightly different from the others, we can interpret this as a second order transmission operator at the continuous level, see [3, Section 4.1],

$$\mathcal{B}_1 = -\frac{\partial u}{\partial n} + \frac{h}{2}\frac{\partial^2 u}{\partial y^2} - \frac{1}{h} 0.7895 u. \tag{14}$$

With the analogous result approximating the Schur complement $S_{1,\text{opt}}$ by SPAI, the corresponding OSM at the continuous level with overlap of one mesh size h would then have in Fourier space the convergence factor (see [3, Section 4.1])

$$\rho_1(k,h) = \left| \frac{|k| - 0.7895\frac{1}{h} - \frac{h}{2}k^2}{|k| + 0.7895\frac{1}{h} + \frac{h}{2}k^2} \right| e^{-kh}, \tag{15}$$

where $k > 0$ corresponds to the frequency in Fourier space, which allows us to asses the quality of this approximation theoretically for our model Poisson equation.

Using a tridiagonal SPAI approximation of the term A_{11}^{-1} leads to

$$D^{br}_{2,\text{app}} = -h^2 \begin{bmatrix} 0.2446 & 0.0504 & & & \\ 0.0552 & 0.2557 & 0.0521 & & \\ & 0.0516 & 0.2540 & 0.0516 & \\ & & 0.0516 & 0.2540 & 0.0516 \\ & & & \ddots & \ddots & \ddots \end{bmatrix}. \tag{16}$$

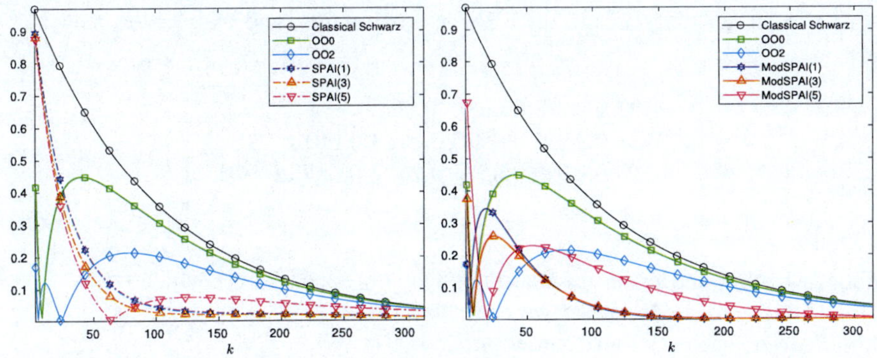

Fig. 1: Comparison of the convergence factors as functions of the Fourier frequency k for the classical Schwarz method, algebraic SPAI transmission conditions (left) and the modified SPAI transmission conditions (right).

As observed for the Schur complement in (12), the changes occur only in the bottom right block, which we can rewrite in the form (where we did not specify the slightly different boundary terms for simplicity in the last matrix)

$$S^{br}_{2,\mathrm{app}} = \frac{1}{h^2}\begin{bmatrix} -2 & 1/2 & & & \\ 1/2 & -2 & 1/2 & & \\ & \ddots & \ddots & \ddots & \\ & & 1/2 & -2 & 1/2 \\ & & & 1/2 & -2 \end{bmatrix} + \frac{0.0516}{h^2}\begin{bmatrix} -2 & 1 & & & \\ 1 & -2 & 1 & & \\ & \ddots & \ddots & \ddots & \\ & & 1 & -2 & 1 \\ & & & 1 & -2 \end{bmatrix} - \frac{1}{h^2}\begin{bmatrix} \ddots & & & \\ & \ddots & & \\ & & 0.6428 & \\ & & & \ddots \\ & & & & \ddots \end{bmatrix}. \tag{17}$$

This can again be interpreted as a second order transmission operator, namely

$$\mathcal{B}_3 = -\frac{\partial u}{\partial n} + 0.0516\, h\frac{\partial^2 u}{\partial y^2} - \frac{1}{h}0.6428u, \tag{18}$$

and the corresponding convergence factor in Fourier space with overlap h is

$$\rho_3(k,h) = \left|\frac{|k| - \frac{1}{h}0.6428 - 0.0516hk^2}{|k| + \frac{1}{h}0.6428 + 0.0516hk^2}\right| \mathrm{e}^{-kh}. \tag{19}$$

The two convergence factors ρ_1 from the diagonal SPAI approximation and ρ_3 from the tridiagonal SPAI approximation are very similar, there is no apparent benefit one would expect when going from a diagonal to a tridiagonal approximation, like when going from a zeroth order optimized (OO0) to a second order optimized (OO2) transmission condition [5, Theorem 4.5 and 4.8]. This is also clearly visible in Figure 1 on the left: SPAI(1) and SPAI(3) have a comparable and much larger low frequency (k small) contraction factor than OO0 and OO2.

We thus add further diagonals in the SPAI approximation, and obtain with five diagonals

$$D^{br}_{2,\text{app}} = -h^2 \begin{bmatrix} 0.2302 & 0.0478 & 0.0084 & 0.001 & & \\ 0.0521 & 0.2559 & 0.0570 & 0.0113 & 0.0017 & \\ & 0.0106 & 0.0573 & 0.2577 & 0.0573 & 0.0106 \\ & \ddots & \ddots & \ddots & \ddots & \ddots \end{bmatrix}. \tag{20}$$

Proceeding as before, and using the matrix spdiag($[1, -4, 6, -4, 1]$) which corresponds to a fourth-order derivative, we can show that the resulting transmission operator in Fourier space is a fourth-order operator given by

$$\mathcal{B}_5 = -\frac{\partial u}{\partial n} + q\, h\frac{\partial^2 u}{\partial y^2} - \frac{p}{h}u + h^3 0.0106\frac{\partial^4 u}{\partial y^4}, \tag{21}$$

where $q = 0.0573 + 4 \times 0.0106$, $r = 0.2577 - 6 \times 0.0106$, and $p = 1 - r - 2q$. The corresponding convergence factor in Fourier space is

$$\rho_5(k, h) = \left| \frac{|k| - \frac{1}{h}p - qhk^2 - h^3 0.0106k^4}{|k| - \frac{1}{h}p - qhk^2 - h^3 0.0106k^4} \right| e^{-kh}. \tag{22}$$

We see in Figure 1 on the left that this approximation now manages to put a zero into the convergence factor, like the OO0 does already with the diagonal approximation, but still the low frequency behavior of the SPAI transmission conditions is much worse than the low frequency behavior of the OO0 and OO2 transmission conditions. It seems that it is not sufficient to just minimize the norms (9) using SPAI approximations to obtain a transmission condition similar in the quality of the OO0 and OO2 transmission conditions.

We therefore now minimize instead the entire norms in (4) using a generic optimization algorithm, namely Nelder Mead, which leads to algebraic transmission conditions and associated AOSMs we call ModSPAI(1), ModSPAI(3), and ModSPAI(5), see Figure 1, right. More specifically, ModSPAI(1) is obtained by minimizing the norms in (4) with respect to the vectors d_0^i where $D_i = -\text{spdiags}(d_0^i, 0)$, $i = 1, 2$. ModSPAI(3) is obtained by minimizing the corresponding norms w.r.t to the vectors d_{-1}^i, d_0^i, and d_1^i such that $D_i = -\text{spdiags}([d_{-1}^i, d_0^i, d_1^i], -1 : 1)$, $i = 1, 2$. Similarly, ModSPAI(5) depends on the vectors d_{-2}^i, d_{-1}^i, d_0^i, d_1^i, and d_2^i where $D_i = -\text{spdiags}([d_{-2}^i, d_{-1}^i, d_0^i, d_1^i, d_2^i], -2 : 2)$, $i = 1, 2$. By introducing these quantities we expect to decrease significantly the corresponding convergence factors. We clearly see that the minimization of the entire product in (4) is essential to obtain AOSMs which have similar performance as OO0 and OO2. It is therefore important to develop an adapted nonlinear SPAI technique to make the norms (4) small, since the generic optimization we used here is too costly in practice, requiring the knowledge of the entire Schur complements to be performed.

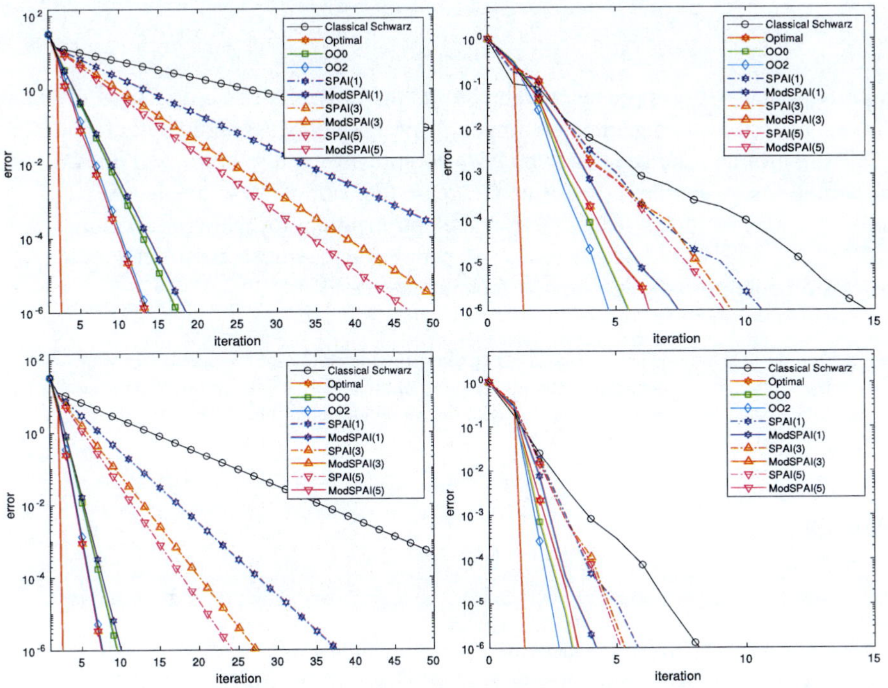

Fig. 2: Convergence history of SPAI-based AOSMs compared to the optimal choice of transmission blocks and OO0 and OO2. Left: iterative methods; Right: GMRES. Top: additive; Bottom: multiplicative.

4 Numerical experiments

To illustrate the performance of the new SPAI AOSMs we consider the advection-reaction-diffusion equation, $\eta u - \nabla \cdot (a\nabla u) + b \cdot \nabla u = f$, where $a = a(x, y) > 0$, $b = [b_1(x, y), b_2(x, y)]^T$, $\eta = \eta(x, y) \geq 0$, with $b_1 = y - \frac{1}{2}$, $b_2 = -x + \frac{1}{2}$, $\eta = x^2 \cos(x + y)^2$, and $a = 1 + (x + y)^2 e^{x-y}$. We perform the experiments on the unit square domain $\Omega = (0, 1) \times (0, 1)$, which we decompose into two subdomains $\Omega_1 = (0, \beta) \times (0, 1)$ and $\Omega_2 = (\alpha, 1) \times (0, 1)$, where $0 < \alpha \leq \beta < 1$. After discretization with a finite difference method, the corresponding matrix A is of size 1024×1024, with a decomposition into two subdomains where the blocks A_{11}, A_{12}, A_{21}, and A_{22} are of size 480×480, 480×32, 32×480, and 32×32 respectively. The parameter of OO0 is evaluated numerically and is given by $p = 51.72$. Similarly, the parameters of OO2 are given numerically by $p = 7.9515$ and $q = 0.3786$. In Figure 2, we present the evolution of the 2-norm of the error as a function of the number of iterations for our methods used as iterative solvers (left) and as preconditioners (right). Since our SPAI and ModSPAI AOSMs are purely algebraic, they can be applied to many different types of equations and discretizations.

Concluding remarks

We proposed a new SPAI approach which permits the inexpensive computation of transmission conditions in algebraic optimized Schwarz methods. Our analysis for a model Poisson problem showed that in order to completely capture optimized transmission conditions, it is either necessary to increase the bandwidth in the new SPAI approach, or to also include a second term in the optimization, for which a new nonlinear SPAI technique would need to be developed. For data-sparse approximations of transmission operators using H-matrix techniques, see [10].

Acknowledgements The second author would like to thank the hospitality of the Section de Mathématiques at the University of Geneva for the invitation in October 2019. The authors appreciate the referees' questions and comments, which helped improve the presentation.

References

1. M. Benzi, A. Frommer, R. Nabben and D.B. Szyld. Algebraic theory of multiplicative Schwarz methods. *Numerische Mathematik* **89**, 605–639 (2001).
2. A. Frommer and D.B. Szyld. Weighted max norms, splittings, and overlapping additive Schwarz iterations. *Numerische Mathematik* **83**, 259–278 (1999).
3. A. St-Cyr, M.J. Gander and S.J. Thomas. Optimized multiplicative, additive and restricted additive Schwarz preconditioning. *SIAM Journal on Scientific Computing* **29**, 2402–2425 (2007).
4. M.J. Gander, S. Loisel and D.B. Szyld. An optimal block iterative method and preconditioner for banded matrices with applications to PDEs on irregular domains. *SIAM Journal on Matrix Analysis and Applications* **33**, 653–680 (2012).
5. M.J. Gander. Optimized Schwarz Methods. *SIAM Journal on Numerical Analysis* **44**, 699–731 (2006).
6. M. Griebel and P. Oswald. On the abstract theory of additive and multiplicative Schwarz algorithms. *Numerische Mathematik* **70**, 163–180 (1995).
7. R. Nabben and S.B. Szyld. Schwarz iterations for symmetric positive semidefinite problems. *SIAM Journal on Matrix Analysis and Applications* **29**, 98–116 (2006).
8. W.P. Tang. Generalized Schwarz splittings. *SIAM Journal on Scientific and Statistical Computing* **13**, 573–595 (1992).
9. M.J. Grote and T. Huckle. Parallel preconditioning with sparse approximate inverses. *SIAM Journal on Scientific Computing* **18**, 838–853 (1997).
10. M.J. Gander and M. Outrata. Optimized Schwarz methods with data-sparse transmission conditions. *Preprint* (2021).

A Parareal Architecture for Very Deep Convolutional Neural Networks

Chang-Ock Lee, Youngkyu Lee, and Jongho Park

1 Introduction

Due to the large number of layers in deep neural networks (DNNs) [11, 12], DNN training is time-consuming and there are demands to reduce training time these days. Recently, multi-GPU parallel computing has become an important topic for accelerating DNN training [2, 6]. In particular, Günther et al. [6] considered the layer structure of ResNet [8] as the forward Euler discretization of a specific ODE and applied a nonlinear in-time multigrid method [3] by regarding the learning process of the network as an optimal control problem.

In this work, we propose a novel paradigm of multi-GPU parallel computing for DNNs, called a *parareal neural network*. In general, DNN has a feed-forward architecture. That is, the output of DNN is obtained from the input by sequential compositions of functions representing layers. We observe that sequential computations can be interpreted as time steps of a time-dependent problem. In the field of numerical analysis, after the pioneering work of Lions et al. [14], there have been numerous researches on parallel-in-time algorithms to solve time-dependent problems in parallel; see, e.g., [5, 15, 16]. Motivated by these works, we present a methodology to transform a given feed-forward neural network to another neural network called the parareal neural network which naturally adopts parallel computing. The parareal neural network consists of fine structures which can be processed in parallel and

Chang-Ock Lee
Department of Mathematical Sciences, KAIST, Daejeon 34141, Korea,
e-mail: colee@kaist.edu

Youngkyu Lee
Department of Mathematical Sciences, KAIST, Daejeon 34141, Korea,
e-mail: lyk92@kaist.ac.kr

Jongho Park
Natural Science Research Institute, KAIST, Daejeon 34141, Korea,
e-mail: jongho.park@kaist.ac.kr

S. C. Brenner et al. (eds.), *Domain Decomposition Methods in Science and Engineering XXVI*, Lecture Notes in Computational Science and Engineering 145,
https://doi.org/10.1007/978-3-030-95025-5_43

a coarse structure which approximates the fine structures by emulating one of the parallel-in-time algorithms called the parareal algorithm [14].

Note that both the proposed parareal neural network and the work of Günther et al. [6] seem to be very closely related in that they parallelize and accelerate the training of neural networks using a parallel time integration approach. However, unlike the work of Günther et al., the proposed network has the advantage of being more general by focusing on layer propagation in an arbitrary feed-forward network.

The parareal neural network can significantly reduce the time for inter-GPU communication because the fine structures do not communicate with each other but communicate only with the coarse structure. Therefore, the proposed methodology is effective in reducing the elapsed time for dealing with very deep neural networks. Numerical results confirm that the parareal neural network gives similar or better performance to the original network even with less training time.

2 The parareal algorithm

The parareal algorithm proposed by Lions et al. [14] is a parallel-in-time algorithm to solve time-dependent differential equations. For the purpose of description, the following system of ordinary differential equations is considered:

$$\dot{\mathbf{u}}(t) = A\mathbf{u}(t) \text{ in } [0,T], \quad \mathbf{u}(0) = \mathbf{u}_0, \tag{1}$$

where $A\colon \mathbb{R}^m \to \mathbb{R}^m$ is an operator, $T > 0$, and $\mathbf{u}_0 \in \mathbb{R}^m$. The time interval $[0,T]$ is decomposed into N subintervals $0 = T_0 < T_1 < \cdots < T_N = T$. First, an approximated solution $\{U_j^1\}_{j=0}^N$ on the coarse grid $\{T_j\}_{j=0}^N$ is obtained by the backward Euler method. The key step of the parareal algorithm is to correct residuals $\{S_j^k\}_{j=1}^{N-1}$ occurring in each interface. It is well-known that the algorithm converges to the exact solution uniformly [1, 4]. We briefly summarize the parareal algorithm in Algorithm 1.

3 Parareal neural networks

In this section, we propose a methodology to design a *parareal neural network* by emulating the parareal algorithm introduced in Section 2 from a given feed-forward neural network. The resulting parareal neural network has an intrinsic parallel structure and is suitable for parallel computation using multiple GPUs with distributed memory simultaneously.

Let $f_\theta\colon X \to Y$ be a feed-forward neural network, where X and Y are the spaces of inputs and outputs, respectively, and θ is a vector consisting of parameters. Since many modern neural networks such as [9, 10, 17] have block-repetitive substructures, we may assume that f_θ can be written as the composition of three functions

Algorithm 1: The parareal algorithm for (1)

Let $\Delta T_j = T_{j+1} - T_j$ and $0 = T_0 < T_1 < \cdots < T_N = T$.
for $j \leftarrow 0$ **to** $N-1$ **do**
  Solve $\dfrac{\mathbf{U}^1_{j+1} - \mathbf{U}^1_j}{\Delta T_j} = A\mathbf{U}^1_{j+1}, \ \mathbf{U}^1_0 = \mathbf{u}_0$
end
for $k \leftarrow 1, 2, \ldots$ **do**
  for $j \leftarrow 0$ **to** $N-1$ ***in parallel*** **do**
    Solve $\dot{\mathbf{u}}^k_j(t) = A\mathbf{u}^k_j(t)$ in $[T_j, T_{j+1}]$, $\mathbf{u}^k_j(T_j) = \mathbf{U}^k_j$.
  end
  for $j \leftarrow 0$ **to** $N-1$ **do**
    $\mathbf{S}^k_{j+1} = \mathbf{u}^k_j(T_{j+1}) - \mathbf{U}^k_{j+1}, \ \mathbf{S}^k_0 = 0.$
    Solve $\dfrac{\boldsymbol{\delta}^k_{j+1} - \boldsymbol{\delta}^k_j}{\Delta T_j} = A\boldsymbol{\delta}^k_{j+1} + \mathbf{S}^k_j, \ \boldsymbol{\delta}^k_0 = 0.$
    $\mathbf{U}^{k+1}_{j+1} = \mathbf{U}^k_{j+1} + \boldsymbol{\delta}^k_{j+1}.$
  end
end

$C_\delta \colon X \longrightarrow W_0$, $g_\varphi \colon W_0 \longrightarrow W_1$, and $h_\varepsilon \colon W_1 \longrightarrow Y$, i.e.,

$$f_\theta = h_\varepsilon \circ g_\varphi \circ C_\delta, \quad \theta = \delta \oplus \varphi \oplus \varepsilon,$$

where W_0 and W_1 are vector spaces, g_φ is a block-repetitive substructure of f_θ with parameters φ, C_δ is a *preprocessing operator* with parameters δ, and h_ε is a *postprocessing operator* with parameters ε. Note that $\oplus$ represents a concatenation.

For appropriate vector spaces $X_0, X_1, \ldots, X_N$, we further assume that g_φ can be partitioned into N subnetworks $\{g^j_{\varphi_j} \colon X_{j-1} \longrightarrow X_j\}_{j=1}^N$ which satisfy the following:

- $X_0 = W_0$ and $X_N = W_1$,
- $\varphi = \bigoplus_{j=1}^N \varphi_j$,
- $g_\varphi = g^N_{\varphi_N} \circ g^{N-1}_{\varphi_{N-1}} \circ \cdots \circ g^1_{\varphi_1}$.

In forward and backward propagations through g_φ, propagations are done sequentially through the subnetworks $\{g^j_{\varphi_j}\}_{j=1}^N$. Regarding the subnetworks as subintervals of a time-dependent problem and adopting the idea of the parareal algorithm introduced in Section 2, we construct a new neural network $\bar{f}_{\bar{\theta}} \colon X \longrightarrow Y$ which contains $\{g^j_{\varphi_j}\}_{j=1}^N$ as parallel subnetworks; the precise definition for parameters $\bar{\theta}$ will be given in (4).

Since the dimensions of the spaces $\{X_j\}_{j=0}^{N-1}$ are different for each j in general, we introduce preprocessing operators $C^j_{\delta_j} \colon X \longrightarrow X_{j-1}$ such that $C^1_{\delta_1} = C_\delta$ and $C^j_{\delta_j}$ for $j = 2, \ldots, N$ play similar roles to C_δ; particular examples will be given in Section 4. We write $\mathbf{x}_j \in X_{j-1}$ and $\mathbf{y}_j \in X_j$ as follows:

$$\mathbf{x}_j = C^j_{\delta_j}(\mathbf{x}) \text{ for } \mathbf{x} \in X, \quad \mathbf{y}_j = g^j_{\varphi_j}(\mathbf{x}_j). \tag{2}$$

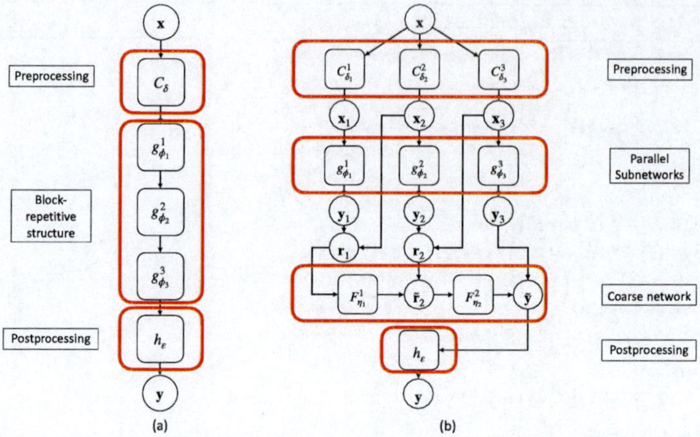

Fig. 1: A feed-forward neural network and its corresponding parareal neural network: **(a)** Feed-forward neural network f_θ, **(b)** Parareal neural network $\bar{f}_{\bar{\theta}}$ with N parallel subnetworks ($N = 3$).

Then, we consider neural networks $F^j_{\eta_j}\colon X_j \rightarrow X_{j+1}$ with parameters η_j for $j \geq 1$ such that it approximates $g^{j+1}_{\varphi_{j+1}}$ well while it has a cheaper computational cost than $g^{j+1}_{\varphi_{j+1}}$, i.e., $F^j_{\eta_j} \approx g^{j+1}_{\varphi_{j+1}}$ and $\dim(\eta_j) \ll \dim(\varphi_{j+1})$. Emulating the coarse grid correction of the parareal algorithm, we assemble a network called the *coarse network* with building blocks $F^j_{\eta_j}$. With inputs $\mathbf{x}_{j+1}$, $\mathbf{y}_j$, and an output $\mathbf{y} \in Y$, the coarse network is described as follows:

$$\mathbf{r}_N = \mathbf{0}, \quad \mathbf{r}_j = \mathbf{y}_j - \mathbf{x}_{j+1} \quad \text{for } j = 1, \dots, N-1, \tag{3a}$$

$$\tilde{\mathbf{r}}_1 = \mathbf{r}_1, \quad \tilde{\mathbf{r}}_{j+1} = \mathbf{r}_{j+1} + F^j_{\eta_j}(\tilde{\mathbf{r}}_j) \quad \text{for } j = 1, \dots, N-1, \tag{3b}$$

$$\tilde{\mathbf{y}} = \mathbf{y}_N + \tilde{\mathbf{r}}_N. \tag{3c}$$

That is, in the coarse network, the residual $\mathbf{r}_j$ at the interface between layers $g^j_{\varphi_j}$ and $g^{j+1}_{\varphi_{j+1}}$ propagates through shallow neural networks $F^1_{\eta_1}, \dots, F^{N-1}_{\eta_{N-1}}$. Then the propagated residual is added to the output.

Finally, the parareal neural network $\bar{f}_{\bar{\theta}}$ corresponding to the original network f_θ is defined as

$$\bar{f}_{\bar{\theta}}(\mathbf{x}) = h_\varepsilon(\tilde{\mathbf{y}}), \quad \bar{\theta} = \left(\bigoplus_{j=1}^{N} (\delta_j \oplus \varphi_j) \right) \oplus \left(\bigoplus_{j=1}^{N-1} \eta_j \right) \oplus \varepsilon. \tag{4}$$

That is, $\bar{f}_{\bar{\theta}}$ is composed of the preprocessing operators $\{C^j_{\delta_j}\}$, parallel subnetworks $\{g^j_{\varphi_j}\}$, the coarse network $\{F^j_{\eta_j}\}$, and the postprocessing operator h_ε. Figure 1(b) illustrates $\bar{f}_{\bar{\theta}}$.

Since each $g^j_{\varphi_j} \circ C^j_{\delta_j}$ lies in parallel, all computations related to $g^j_{\varphi_j} \circ C^j_{\delta_j}$ can be done independently. Therefore, multiple GPUs can be utilized to process $\{g^j_{\varphi_j} \circ C^j_{\delta_j}\}$ simultaneously for each j. In this case, one may expect significant decrease of the elapsed time for training $\bar{f}_{\bar{\theta}}$ compared to the original network f_θ. On the other hand, the coarse network cannot be parallelized since $\{F^j_{\eta_j}\}$ is computed in the sequential manner. One should choose $F^j_{\eta_j}$ whose computational cost is as cheap as possible in order to reduce the bottleneck effect of the coarse network.

Now, we want show that the proposed parareal neural network $\bar{f}_{\bar{\theta}}$ is consistently constructed in the sense that it recovers the original neural network f_θ in the setting where nonlinearity is removed. By collecting all the residuals in each interface and dealing with them sequentially, the following proposition is obtained.

Proposition 1 (Consistency) *Assume that the original network f_θ is linear and $F^j_{\eta_j} = g^{j+1}_{\varphi_{j+1}}$ for $j = 1, \ldots, N-1$. Then we have $\bar{f}_{\bar{\theta}}(\mathbf{x}) = f_\theta(\mathbf{x})$ for all $\mathbf{x} \in X$.*

Proposition 1 presents a guideline on how to design the coarse network of $\bar{f}_{\bar{\theta}}$. Under the assumption that f_θ is linear, a sufficient condition to ensure that $\bar{f}_{\bar{\theta}} = f_\theta$ is $F^j_{\eta_j} = g^{j+1}_{\varphi_{j+1}}$ for all j. Therefore, we can say that it is essential to design the coarse network with $F^j_{\eta_j} \approx g^{j+1}_{\varphi_{j+1}}$ to ensure that the performance of $\bar{f}_{\bar{\theta}}$ is as good as that of f_θ. Detailed examples will be given in Section 4.

4 Application to ResNet-1001

The proposed parareal neural network can be applied to a general feed-forward neural network. However, since most of the current very deep neural networks have residual structures, we applied it to ResNet-1001 [9], which is one of the typical very deep convolutional neural networks for classification problems. First, we describe the structure of ResNet-1001 with the terminology introduced in Section 3. Inputs for ResNet-1001 are 3-channel images with 32×32 pixels, i.e., $X = \mathbb{R}^{3\times 32\times 32}$. The output space Y is given by $Y = \mathbb{R}^m$, where m is the number of classes of images. ResNet-1001 has a block-repetitive substructure consisting of 333 residual units (RUs), so that we may set $g_\varphi \colon W_0 \to W_1$ as the composition of those RUs with $W_0 = \mathbb{R}^{16\times 32\times 32}$ and $W_1 = \mathbb{R}^{256\times 8\times 8}$. Then the preprocessing operator $C_\delta \colon X \to W_0$ is a single 3×3 convolution layer and the postprocessing operator $h_\varepsilon \colon W_1 \to Y$ consists of global average pooling and fully connected layers.

The design of a parareal neural network with N parallel subnetworks for ResNet-1001, denoted as *Parareal ResNet-N*, can be completed by specifying the structures $g^j_{\varphi_j}$, $C^j_{\delta_j}$, and $F^j_{\eta_j}$. For convenience, the original neural network ResNet-1001 is called Parareal ResNet-1. We assume that $N = 3N_0$ for some positive integer N_0. We note that g_φ can be decomposed as

$$g_\varphi = g^N_{\varphi_N} \circ \cdots \circ g^{2N_0+1}_{\varphi_{2N_0+1}} \circ g^{2N_0}_{\varphi_{2N_0}} \circ \cdots \circ g^{N_0+1}_{\varphi_{N_0+1}} \circ g^{N_0}_{\varphi_{N_0}} \circ \cdots \circ g^1_{\varphi_1},$$

where each $g^j_{\varphi_j}\colon X_{j-1} \to X_j$ consists of $\lceil 333/N \rceil$ RUs with

$$X_j = \begin{cases} \mathbb{R}^{64\times32\times32} & \text{for } j = 1,\dots,N_0, \\ \mathbb{R}^{128\times16\times16} & \text{for } j = N_0+1,\dots,2N_0, \\ \mathbb{R}^{256\times8\times8} & \text{for } j = 2N_0+1,\dots,N, \end{cases} \qquad \varphi = \bigoplus_{j=1}^{N} \varphi_j.$$

The main role of the preprocessing operator $C^j_{\delta_j}\colon X \to X_{j-1}$ is to transform an input $\mathbf{x} \in X$ to fit in the space X_{j-1}. In this perspective, we simply set $C^1_{\delta_1} = C_\delta$ and $C^j_{\delta_j}$ for $j > 1$ consists of a 1×1 convolution to match the number of channels after an appropriate number of 3×3 max pooling layers with stride 2 to match the image size. For the coarse network, we first define a coarse RU consisting of two 3×3 convolutions and skip-connection. If the downsampling is needed, then the stride of first convolution in coarse RU is set to 2. We want to define $F^j_{\eta_j}\colon X_j \to X_{j+1}$ having a smaller number of (coarse) RUs than $g^{j+1}_{\varphi_{j+1}}$ but a similar coverage to $g^{j+1}_{\varphi_{j+1}}$. Let N_c be the number of coarse RUs in $F^j_{\eta_j}$ of the coarse network. Note that the receptive field of $g^j_{\varphi_j}$ covers the input size 32×32. In the case of $N = 3$, even if we construct $F^j_{\eta_j}$ with $N_c = 4$ coarse RUs, it can cover 31×31 pixels which is a similar coverage to the parallel subnetwork $g^j_{\varphi_j}$. Generally, if we use N parallel subnetworks ($N \geq 3$), each $333/N$ RUs in $g^j_{\varphi_j}$ can be approximated by the N_c RUs in $F^j_{\eta_j}$ whenever we select $N_c = \lceil 12/N \rceil$.

5 Numerical results

In this section, we present numerical results of the Parareal ResNet-N with various N. First, we present details on the datasets we used. The CIFAR-m ($m = 10, 100$) dataset consists of 32×32 colored natural images and includes 50,000 training and 10,000 test samples with m classes. The SVHN dataset is composed of 32×32 colored digit images; there are 73,257 and 26,032 samples for training and testing, respectively, with an additional 531,131 training samples. However, we did not use the additional samples for training. MNIST is a classical dataset which contains handwritten digits encoded in 28×28 grayscale images. It includes 55,000 training, 5,000 validation, and 10,000 test samples. In our experiments, the training and validation samples are used as training data and the test samples as test data. We adopted a data augmentation technique in [13] for CIFAR datasets; four pixels are padded on each side of the images, and 32×32 crops are randomly sampled from the padded images and their horizontal flips.

All neural networks in this section were trained using the stochastic gradient descent with the batch size 128, weight decay 0.0005, momentum 0.9, and weights initialized as in [7]. The initial learning rate is set to 0.1, and is reduced by a factor of 10 in the 80th and 120th epochs. All networks were implemented in Python with

Table 1: Error rates (%) on the CIFAR-10, CIFAR-100, MNIST, and SVHN datasets of Parareal ResNet-N ($N = 1, 3, 6, 12, 18$) with $N_c = \lceil 12/N \rceil$.

N	Parameters per subnetwork	Parameters of coarse network	Total Parameters	CIFAR-10	CIFAR-100	MNIST	SVHN
1	-	-	10.3M	4.96	21.13	0.34	3.17
3	3.4M	5.6M	15.9M	4.61	21.14	0.31	3.11
6	1.7M	5.7M	16.1M	4.20	20.87	0.31	3.21
12	0.9M	5.8M	16.2M	4.37	20.42	0.28	3.25
18	0.6M	8.9M	19.4M	4.02	20.40	0.33	3.29

Table 2: Forward/backward computation time for Parareal ResNet-N ($N = 1, 3, 6, 12, 18$). The time is measured in one iteration for CIFAR-100 dataset input $\mathbf{x} \in \mathbb{R}^{3\times32\times32}$ with batch size 128.

	Virtual wall-clock time (ms)				
N	Preprocessing	Parallel subnetworks	Coarse network	Postprocessing	Total
1	0.25/6.46	443.81/1387.62	-	0.06/3.18	444.12/1397.26
3	0.25/6.45	131.92/458.87	10.01/97.60	0.06/3.71	142.24/566.63
6	0.27/6.42	67.59/219.72	14.68/137.08	0.06/3.33	82.60/366.55
12	0.28/6.59	48.47/113.33	17.97/149.52	0.06/3.63	66.78/273.07
18	0.29/6.17	30.40/77.84	27.87/163.25	0.06/3.64	58.62/250.90
24	0.29/6.58	22.71/58.04	41.03/242.87	0.06/3.54	64.09/311.03

PyTorch and all computations were performed on a cluster equipped with Intel Xeon Gold 5515 (2.4GHz, 20C), NVIDIA Titan RTX and the operating system Ubuntu 18.04 64bit.

With fixed $N_c = \lceil 12/N \rceil$, we report the classification results of Parareal ResNet with respect to various N on datasets CIFAR-10, CIFAR-100, SVHN, and MNIST. Table 1 shows that the error rates of Parareal ResNet-N are usually smaller than ResNet-1001.

Next, we investigate the elapsed time for forward and backward propagation of parareal neural networks. Table 2 shows the virtual wall-clock time for forward and backward computation of Parareal ResNet-N with various N for the input $\mathbf{x} \in \mathbb{R}^{3\times32\times32}$. As shown in Table 2, the larger N, the shorter the computing time of the parallel subnetworks $g^j_{\varphi_j}$, while the longer the computing time of the coarse network. This is because as N increases, the depth of each parallel subnetwork $g^j_{\varphi_j}$ becomes shallower while the number of $F^j_{\eta_j}$ in the coarse network increases. On the other hand, each preprocessing operator $C^j_{\delta_j}$ is designed to be the same as or similar to the preprocessing operator C_δ of the original neural network and the postprocessing operator h_ε is the same as the original one. Therefore, the computation time for the pre- and postprocessing operators does not increase even as N increases.

Table 3: The wall-clock time and relative speedup on the CIFAR-100 dataset. The wall-clock time is the total time taken to train a given network by 200 epochs.

Network	Parameters	Wall-clock time (h:m:s)	Relative speedup (%)
ResNet-1001	10.3M	22:44:53	0.0
Parareal ResNet-3	15.9M	16:28:38	27.6
Parareal ResNet-6	16.1M	11:48:13	48.1

Finally, we measure the wall-clock time of the Parareal ResNet with the CIFAR-100 dataset. Table 3 shows that Parareal ResNet's wall clock time is reduced by about half as N increases to 6.

In conclusion, despite the large number of layers, the parareal neural network can accelerate the training of the very deep CNN using multiple-GPU. To the best of our knowledge, the proposed methodology is a new kind of multi-GPU parallelism in the field of deep learning.

References

1. G. Bal. On the convergence and the stability of the parareal algorithm to solve partial differential equations. In: *Domain Decomposition Methods in Science and Engineering*, pp. 425–432, Springer, Berlin (2005).
2. C.C. Chen, C.L. Yang and H.Y. Cheng. Efficient and robust parallel DNN training through model parallelism on multi-GPU platform. *arXiv preprint* arXiv:1809.02839 (2018).
3. R.D. Falgout, S. Friedhoff, T.V. Kolev, S.P. MacLachlan and J.B. Schroder. Parallel time integration with multigrid. *SIAM Journal on Scientific Computing* **36**(6), C635–C661 (2014).
4. M.J. Gander and E. Hairer. Nonlinear convergence analysis for the parareal algorithm. In: *Domain Decomposition Methods in Science and Engineering XVII*, pp. 45–56, Springer, Berlin (2008).
5. M.J. Gander and S. Vandewalle. Analysis of the parareal time-parallel time-integration method. *SIAM Journal on Scientific Computing* **29**(2), 556–578 (2007).
6. S. Günther, L. Ruthotto, J.B. Schroder, E.C. Cyr and N.R. Gauger. Layer-parallel training of deep residual neural networks. *SIAM Journal on Mathematics of Data Science* **2**(1), 1–23 (2020).
7. K. He, X. Zhang, S. Ren and J. Sun. Delving deep into rectifiers: Surpassing human-level performance on ImageNet classification. In: *Proceedings of the IEEE International Conference on Computer Vision*, pp. 1026–1034 (2015).
8. K. He, X. Zhang, S. Ren and J. Sun. Deep residual learning for image recognition. In: *Proceedings of the IEEE Conference on Computer Vision and Pattern Recognition*, pp. 770–778 (2016).
9. K. He, X. Zhang, S. Ren and J. Sun. Identity mappings in deep residual networks. In: *European Conference on Computer Vision*, pp. 630–645, Springer, Cham (2016).
10. G. Huang, Z. Liu, L. Van Der Maaten and K.Q. Weinberger. Densely connected convolutional networks. In: *Proceedings of the IEEE Conference on Computer Vision and Pattern Recognition*, pp. 4700–4708 (2017).
11. A. Krizhevsky, I. Sutskever and G.E. Hinton. ImageNet classification with deep convolutional neural networks. In: *Advances in Neural Information Processing Systems*, pp. 1097–1105 (2012).

12. Y. LeCun, B. Boser, J.S. Denker, D. Henderson, R.E. Howard, W. Hubbard and L.D. Jackel. Backpropagation applied to handwritten zip code recognition. *Neural Computation* **1**(4), 541–551 (1989).
13. C.Y. Lee, S. Xie, P. Gallagher, Z. Zhang and Z. Tu. Deeply-supervised nets. In: *Artificial Intelligence and Statistics*, pp. 562–570 (2015).
14. J.L. Lions, Y. Maday and G. Turinici. Résolution d'EDP par un schéma en temps «pararél». *Comptes Rendus de l'Académie des Sciences-Series I-Mathematics* **332**(7), 661–668 (2001).
15. Y. Maday and G. Turinici. A parareal in time procedure for the control of partial differential equations. *Comptes Rendus Mathematique* **335**(4), 387–392 (2002).
16. M. Minion. A hybrid parareal spectral deferred corrections method. *Communications in Applied Mathematics and Computational Science* **5**(2), 265–301 (2011).
17. S. Xie, R. Girshick, P. Dollár, Z. Tu and K. He. Aggregated residual transformations for deep neural networks. In: *Proceedings of the IEEE Conference on Computer Vision and Pattern Recognition*, pp. 1492–1500 (2017).

Construction of 4D Simplex Space-Time Meshes for Local Bisection Schemes

David Lenz

1 Introduction

Space-time finite element methods (FEMs) approximate the solution to a PDE 'all-at-once' in the sense that a solution is produced at all times of interest simultaneously. This is achieved by treating time as just another variable and discretizing the entire space-time domain with finite elements. Naturally, discretizing the entire space-time domain creates a linear system with many more degrees of freedom (DOFs) than discretizing just the spatial domain. Adaptive mesh refinement can produce space-time discretizations that yield accurate solutions with relatively few degrees of freedom. For example, Langer & Schafelner [3] compared space-time FEMs using uniformly and adaptively refined meshes; they found that obtaining the same approximation error for their tests required more degrees of freedom on the uniform meshes by one to two orders of magnitude.

A technical challenge in the implementation of adaptive mesh refinement is, of course, the mesh refinement scheme. In order to refine a geometric element while introducing as few new DOFs as possible, algorithms typically employ a "red-green" approach (uniform refinement with closure) or an element bisection approach. Here we focus on bisection schemes in four dimensions but note that work on arbitrary-dimensional red-green schemes was recently undertaken by Grande [2]. Stevenson [7] has studied a bisection algorithm for arbitrary-dimensional simplicial meshes, which is a good candidate for four-dimensional space-time meshes. However, the algorithm relies on a mesh precondition that is difficult to satisfy.

In this article, we weaken this strict precondition for certain space-time meshes. We prove that the precondition on four-dimensional simplex meshes can be reduced to a precondition on an underlying three-dimensional mesh. This means that the condition need only be checked on a much smaller mesh. In addition, if one cannot

David Lenz
Argonne National Laboratory, Lemont, IL, USA, e-mail: dlenz@anl.gov

S. C. Brenner et al. (eds.), *Domain Decomposition Methods in Science and Engineering XXVI*, Lecture Notes in Computational Science and Engineering 145,
https://doi.org/10.1007/978-3-030-95025-5_44

immediately verify the precondition, refinements to this smaller three-dimensional mesh can be made to automatically satisfy the precondition.

In section 2, we describe a particular method for creating four-dimensional space-time meshes. Meshes of this type have a structure which will be exploited in section 3, where we summarize key concepts from Stevenson's bisection algorithm and then prove our main result. Finally, we conclude with some remarks on how this relaxed precondition can be used in practice.

2 Four-dimensional space-time mesh construction

When applying space-time FEMs to solve a PDE, it is generally necessary to create a space-time mesh that corresponds to a given spatial domain. A convenient method for doing so is to repeatedly extrude spatial elements (typically triangles or tetrahedra) into higher-dimensional space-time prisms and then subdivide these prisms into simplicial space-time elements (tetrahedra or pentatopes). We refer to mesh generation methods of this type as *extrusion-subdivision* schemes.

The method of extrusion-subdivision has appeared in several places in recent years. This idea was applied to moving meshes by Karabelas & Neumüller [6] and is discussed in a report by Voronin [9], where it is described in the context of an extension to the MFEM library [5]. For stationary (non-moving) domains, the extrusion step is straightforward, but subdividing the space-time prisms can be done in several ways. Behr [1] describes a method for subdividing space-time prisms using Delaunay triangulations, while subdivision based on vertex orderings is used in [6], [9].

In this paper, we consider space-time meshes produced by extrusion-subdivision where prism subdivision is defined in terms of vertex labels. In particular, we will assume that a k-coloring has been imposed on the mesh; that is, each vertex in the mesh has one of k labels (colors) attached to it and no two vertices connected by an edge share the same label.

Before we describe our particular prism subdivision method, we need to establish some notation. Let d be the spatial dimensionality of the problem and suppose $a = (a_0, a_1, \ldots, a_{d-1}) \in \mathbb{R}^d$. For any $r \in \mathbb{R}$, we define the map $\psi_r : \mathbb{R}^d \to \mathbb{R}^{d+1}$ by

$$\psi_r(a) = (a_0, a_1, \ldots, a_{d-1}, r). \tag{1}$$

For a simplex $T = \mathrm{conv}(a, b, c, d)$ (here $a, b, c, d \in \mathbb{R}^d$), we define

$$\psi_r(T) = \mathrm{conv}(\psi_r(a), \psi_r(b), \psi_r(c), \psi_r(d)); \tag{2}$$

that is, $\psi_r(T)$ is the embedding of T into $\mathbb{R}^{d+1}$ space-time within the plane $x_{d+1} = r$.

Next, the extrusion operator $\Phi_{r,s}$ produces the set of all points between $\psi_r(T)$ and $\psi_s(T)$. This is the convex hull of points in $\psi_s(T)$ and $\psi_r(T)$, which is a right tetrahedral prism:

$$\Phi_{r,s}(T) = \mathrm{conv}(\psi_r(T), \psi_s(T)). \tag{3}$$

We make one final notational definition to declutter the following exposition. Given a series of real values $\mathcal{S} = \{s_0, \ldots, s_M\}$, we define

$$\Phi_i^{\mathcal{S}}(T) = \Phi_{s_i, s_{i+1}}(T) \quad \text{where } s_i \in \mathcal{S}, \text{ for } 0 \le i \le M - 1. \tag{4}$$

We refer to $\mathcal{S}$ as a collection of "time-slices", which determine how spatial elements are extruded into space-time prisms. The value s_i is thus the "i^{th} time-slice". For most problems, the set of initial time-slices is fixed ahead of time, so it is often convenient to omit the superscript $\mathcal{S}$. We adopt this shorthand for the remainder of this paper. For an illustration of the operators ψ and Φ in two spatial dimensions, see figure 1.

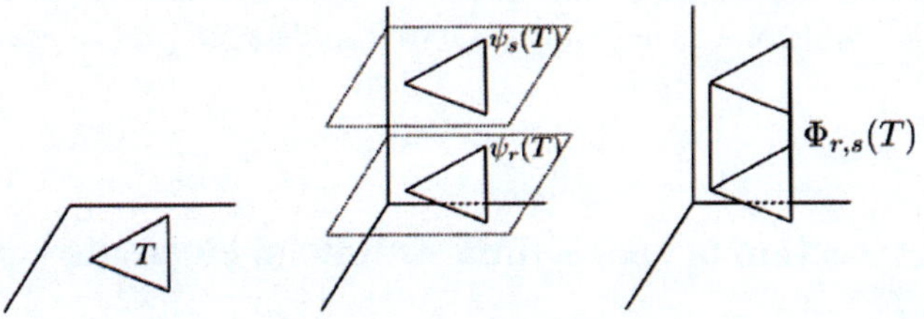

Fig. 1: Extrusion of a 2D simplex into a 3D simplex prism. Left: The spatial element. Center: Copies of the spatial element embedded in space-time. Right: The space-time prism element.

For the remainder of this article, we will focus on the case $d = 3$; that is, problems in four-dimensional space-time.[1] Let $\mathcal{T}$ be a conforming tetrahedral mesh with a 4-coloring,[2] and let the symbols A, B, C, D denote the four labels to be associated with each vertex of $\mathcal{T}$. Let $\mathcal{S}$ be a set of time-slices which define the extrusion of spatial elements into space-time prisms (cf. equation (4)). We will use the following rule to subdivide the tetrahedral prisms formed by extruding elements of $\mathcal{T}$.

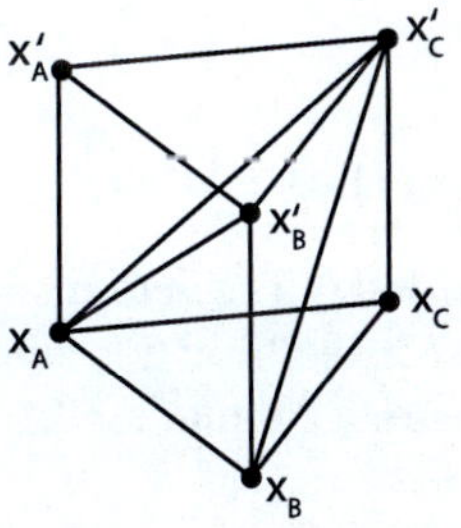

Fig. 2 Space-time triangular prism subdivision based on a coloring of the underlying spatial mesh.

[1] Due to the inherent difficulty in visualizing four-dimensional objects on a two-dimensional page, we will continue to illustrate figures in three-dimensional space-time. These figures are meant only as a guide for the reader to develop some geometric intuition.

[2] Not every tetrahedral mesh admits a 4-coloring, although all are 5-colorable. We discuss how to handle meshes which are not 4-colorable at the end of section 3.

Definition 1 (Subdivision Rule) Let $T \in \mathcal{T}$ be a tetrahedron with vertices v_A, v_B, v_C, v_D, where the subscript of each vertex denotes its color label. For a given time slice $s_i \in \mathcal{S}$, let $x_A = \psi_{s_i}(v_A)$ and $x'_A = \psi_{s_{i+1}}(v_A)$ (and likewise for B, C, D). The rule for subdividing $\Phi_i(T)$ is to create the pentatopes:

$$\begin{aligned}\tau_1 &= \operatorname{conv}(x_A, x_B, x_C, x_D, x'_D),\\ \tau_2 &= \operatorname{conv}(x_A, x_B, x_C, x'_C, x'_D),\\ \tau_3 &= \operatorname{conv}(x_A, x_B, x'_B, x'_C, x'_D),\\ \tau_4 &= \operatorname{conv}(x_A, x'_A, x'_B, x'_C, x'_D).\end{aligned} \tag{5}$$

An illustration of this subdivision rule is given in figure 2. Because the labeling of each vertex is shared by all elements, this subdivision scheme always produces a conforming mesh of pentatopes. However, we omit the detailed proof because of space constraints.

3 Conforming bisection of space-time simplicial elements

The aim of this section is to outline Stevenson's bisection algorithm [7],[3] with particular attention on the mesh precondition for its validity. Then, we show that space-time meshes produced by extrusion-subdivision (using definition 1) will always meet this precondition if the underlying spatial mesh is 4-colorable. The upshot of this is that all of the work to make an admissible mesh can be done in three dimensions instead of four. This is especially useful in light of the fact that there are many more meshing utilities for three-dimensional domains than four-dimensional domains.

The following definitions are due to Stevenson [7].

Definition 2 A *tagged pentatope* t is an ordering of the vertices of some pentatope $\tau = \operatorname{conv}(x_0, x_1, x_2, x_3, x_4)$, together with an integer $0 \le \gamma \le 3$ called the *type*. We write

$$t = (x_0, x_1, x_2, x_3, x_4)_\gamma \tag{6}$$

to denote this ordering-type pair.

Definition 3 The *reflection* of a tagged pentatope t is another tagged pentatope t_R such that the bisection rule produces the same child pentatopes for t and t_R. The unique reflection of $t = (x_0, x_1, x_2, x_3, x_4)_\gamma$ is

$$t_R = \begin{cases} (x_4, x_3, x_2, x_1, x_0)_\gamma & \text{if } \gamma = 0\\ (x_4, x_1, x_3, x_2, x_0)_\gamma & \text{if } \gamma = 1\\ (x_4, x_1, x_2, x_3, x_0)_\gamma & \text{if } \gamma = 2, 3.\end{cases} \tag{7}$$

[3] The bisection rule studied by Stevenson has also been studied by Maubach [4] and Traxler [8].

Definition 4 Two tagged pentatopes t and t' are *reflected neighbors* if they share a common hyperface, have the same integer type, and the vertex order of t' matches the vertex order of t or t_R in all but one position.

The notion of reflected neighbors is critical to our proof of the main result, so it is worth illustrating the concept with some examples.

Example 1 Let $t = (x_0, x_1, x_2, x_3, y)_1$ and $t' = (z, x_0, x_1, x_2, x_3)_1$. Then t and t' are NOT reflected neighbors. Although the relative ordering of their shared vertices is consistent, the taggings differ in every position.

Example 2 Let $t = (x_0, z, x_1, x_2, x_3)_1$ and $t' = (x_3, y, x_2, x_1, x_0)_1$. Then t and t' are reflected neighbors. To see this, we note that $t_R = (x_3, z, x_2, x_1, x_0)_1$; thus t_R and t' differ only in the second position. Likewise, we could have shown that t'_R and t differ in at most one position.

Definition 5 The tagging of two pentatopes $t = (x_0, x_1, x_2, x_3, x_4)_\gamma$ and $t' = (x'_0, x'_1, x'_2, x'_3, x'_4)_\gamma$ which share a hyperface is said to be *consistent* if the following condition is met:

1. If $\overline{x_0, x_4}$ or $\overline{x'_0, x'_4}$ is contained in the shared hyperface, then t and t' are reflected neighbors (N.B. these are the bisection edges for each element; see definition 6).
2. Otherwise, the two children of t and t' which share the common hyperface are reflected neighbors.

A *consistent tagging of a mesh* is a tag for each element in the mesh such that any two neighboring elements are consistently tagged.

In essence, definition 5 states that in a consistently tagged mesh, any pair of neighboring elements are either reflected neighbors or they do not share a common refinement edge. Furthermore, when two elements do not share a common refinement edge, their adjacent children will be reflected neighbors after one round of bisection.

Definition 6 (Bisection Rule) Given a tagged pentatope $t = (x_0, x_1, x_2, x_3, x_4)_\gamma$, applying the bisection rule produces the children:

$$t_1 = (x_0, x', x_1, x_2, x_3)_{\gamma'}, \qquad t_2 = \begin{cases} (x_4, x', x_3, x_2, x_1)_{\gamma'} & \text{if } \gamma = 0 \\ (x_4, x', x_1, x_3, x_2)_{\gamma'} & \text{if } \gamma = 1 \\ (x_4, x', x_1, x_2, x_3)_{\gamma'} & \text{if } \gamma = 2, 3 \end{cases}, \tag{8}$$

where $x' = (x_0 + x_4)/2$ and $\gamma' = (\gamma + 1) \mod 4$. We say that edge $\overline{x_0 x_4}$ is the *refinement edge*.

We can now state the main result of this section.

Proposition 1 *Let $\mathcal{T} \subset \mathbb{R}^3$ be a 4-colorable tetrahedral mesh and $\mathcal{T}' \subset \mathbb{R}^4$ be the pentatope mesh produced by extrusion-subdivision according to definition 1. Then $\mathcal{T}'$ admits a consistent tagging.*

Proof We will prove the result by constructing a consistent tagging of $\mathcal{T}'$ directly from a 4-coloring of $\mathcal{T}$. Every simplex is tagged according to its position within its extruded space-time prism from definition 1. For the above four pentatopes in definition 1, we make the following tagging (in each case t_i is a tagging of τ_i):

$$\begin{aligned} t_1 &= (x_D, x_C, x_B, x_A, x'_D)_0, & t_2 &= (x_C, x_B, x_A, x'_D, x'_C)_0, \\ t_3 &= (x_B, x_A, x'_D, x'_C, x'_B)_0, & t_4 &= (x_A, x'_D, x'_C, x'_B, x'_A)_0. \end{aligned} \tag{9}$$

To show that a tagging of $\mathcal{T}'$ is consistent, it suffices to consider an arbitrary element of $\mathcal{T}'$ and show that each of its neighbors satisfy the conditions in definition 5. Let $\tau \in \mathcal{T}'$ be an arbitrary element. Since $\mathcal{T}'$ is created by extrusion-subdivision, there is some time-slice s_i and $T \in \mathcal{T}$ such that $\tau \subset \Phi_i(T)$.

We will show that each neighbor τ' satisfies the consistency condition in definition 5. There are three cases, illustrated in figure 3:

1. τ and τ' are both pentatopes within the same space-time prism.
2. τ and τ' belong to different space-time prisms extruded from the same spatial element; for instance, $\tau \subset \Phi_i(T)$ and $\tau' \subset \Phi_{i+1}(T)$.
3. τ and τ' belong to different space-time prisms within the same space-time slab; for instance, $\tau \subset \Phi_i(T)$ and $\tau' \subset \Phi_i(T')$.

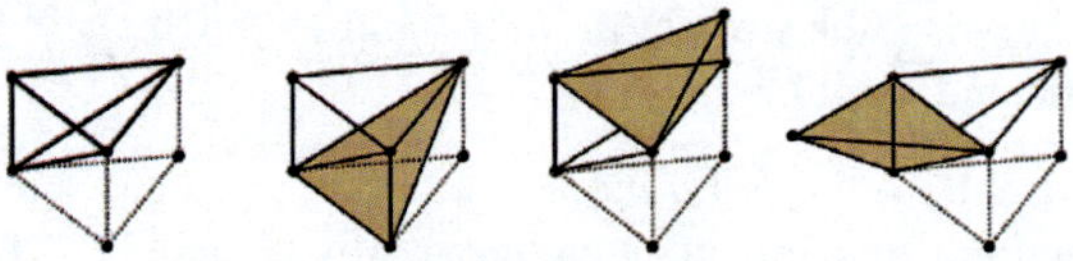

Fig. 3 Different types of neighboring elements. The three shaded elements are the three types of neighbors for the bold simplex given at left.

We verify consistency of the tagging scheme with repeated applications of definition 1 and few geometric arguments, which ensures that all cases are covered.

Consider Case (1). Since τ and τ' are neighbors, they must be a pair τ_i, τ_{i+1} ($i = 1, 2, 3$) from definition 1, since only consecutive pairs in our list are neighbors. In this case, the adjacent children of each pentatope are reflected neighbors. To make this point explicit, consider the neighboring tagged elements t_1, t_2. The children formed by bisecting these pentatopes are:

$$t_1 \to \begin{cases} (x_D, z_1, x_C, x_B, x_A)_1 \\ (x'_D, z_1, x_A, x_B, x_C)_1 \end{cases}, \qquad t_2 \to \begin{cases} (x_C, z_2, x_B, x_A, x'_D)_1 \\ (x'_C, z_2, x'_D, x_A, x_B)_1 \end{cases}, \tag{10}$$

where z_i are the new midpoints of the bisected edges. From here we note that the second child of t_1 is the reflected neighbor of t_2. The same exercise shows that the pairs t_2, t_3 and t_3, t_4 also share this property, and thus all are consistently tagged.

In Case (2), τ and τ' are neighbors in different space-time slabs; without loss of generality, $\tau \subset \Phi_i(T)$ and $\tau' \subset \Phi_{i+1}(T)$ for some $T \in \mathcal{T}$. With two space-time prisms, we have three sets of vertices at different time-slices. Denote vertices in the highest (latest) time hyperplane with double primes (”), the middle hyperplane with single primes ('), and the lowest (earliest) hyperplane with no primes; see figure 4.

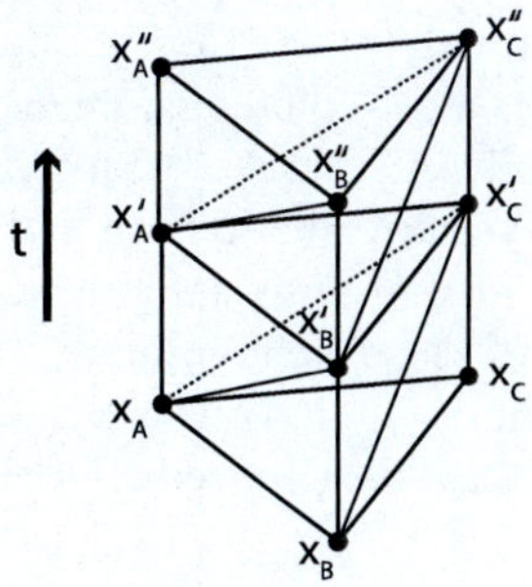

Fig. 4 Illustration of vertex labeling when considering consecutive space-time time prisms in two spatial dimensions.

Since τ and τ' belong to consecutive timeslices, the shared hyperface between the two must be $\text{conv}(x'_A, x'_B, x'_C, x'_D)$. Thus $\tau = \text{conv}(x_A, x'_A, x'_B, x'_C, x'_D)$ and $\tau' = \text{conv}(x'_A, x'_B, x'_C, x'_D, x''_D)$. According to the tagging scheme described above, the tags on these two pentatopes are

$$t = (x_A, x'_D, x'_C, x'_B, x'_A)_0, \quad t' = (x'_D, x'_C, x'_B, x'_A, x''_D)_0, \tag{11}$$

and thus their child elements are

$$t \to \begin{cases} (x_A, z, x'_D, x'_C, x'_B)_1 \\ (x'_A, z, x'_B, x'_C, x'_D)_1 \end{cases}, \qquad t' \to \begin{cases} (x'_D, z', x'_C, x'_B, x'_A)_1 \\ (x''_D, z', x'_A, x'_B, x'_C)_1 \end{cases}, \tag{12}$$

where z, z' are new vertices created by bisecting τ and τ'. This is indeed a consistent tagging, as the second child of t and the first child of t' are reflected neighbors.

Finally, consider Case (3). Since $\mathcal{T}'$ is conforming, when $\tau \subset \Phi_i(T)$ and $\tau' \subset \Phi_i(T')$ they must share a vertical edge like $\overline{x_A x'_A}$, which is always a bisection edge. Since vertex labels are "global" labels, both τ and τ' agree on the order in which the labeled vertices appear. Furthermore, τ and τ' share all but one vertex in common. Since all but one vertex is shared, both pentatopes have the same labels for the same vertices, and vertex order is uniquely determined by vertex label, the vertex orders agree in all but one position. Hence the tagged pentatopes are reflected neighbors.

Thus all three cases result in consistent taggings. Therefore, the tagging defined by equation (9) is consistent. □

The critical piece of this proof is the tagging scheme defined in equation (9). Furthermore, since the vertex orders are determined by the 4-coloring on $\mathcal{T}$, a consistent tagging of the space-time mesh can be constructed in linear time.

Corollary 1 *Let $\mathcal{T}$ and $\mathcal{T}'$ be as in proposition 1. Given a 4-coloring on $\mathcal{T}$, the space-time mesh $\mathcal{T}'$ can be consistently tagged in $O(N)$ time, where N is the number of vertices in $\mathcal{T}'$.*

Not every tetrahedral mesh is 4-colorable, but this can be worked around. First, we note that regular tetrahedral meshes are 4-colorable, so for rectilinear domains one can start with a coarse regular mesh and bisect until a desired resolution is met.

In addition, Traxler [8] has shown that tetrahedral meshes over simply connected domains are 4-colorable iff every edge is incident to an even number of tetrahedra.

Finally, any tetrahedral mesh can be made 4-colorable by dividing each element via barycentric subdivision and then choosing the following colors: each vertex of the original mesh is colored A, the new center of each edge is colored B, the new center of each face is colored C, and the new center of each tetrahedron is colored D. Barycentric subdivision creates new elements with one of each kind of point, which means that this is indeed a 4-coloring of the subdivided mesh.

4 Conclusions

We described a method for creating four-dimensional simplex space-time meshes from a given spatial mesh which has a 4-coloring. This procedure was based on the general extrusion-subdivision framework, with a new subdivision rule which is defined in terms of vertex labels (colors). We then proved that meshes of this form always satisfy the strict precondition of Stevenson's bisection algorithm, which can be used to adaptively refine space-time meshes. Finally, we showed that even when a tetrahedral mesh is not 4-colorable, the barycentric subdivision of the mesh will be.

Acknowledgements This work is supported by the U.S. Department of Energy, Office of Science, Advanced Scientific Computing Research under Contract DE-AC02-06CH11357, and the Exascale Computing Project (Contract No. 17-SC-20-SC), a collaborative effort of the U.S. Department of Energy Office of Science and the National Nuclear Security Administration.

References

1. M. Behr. Simplex space-time meshes in finite element simulations. *Internat. J. Numer. Methods Fluids* **57**(9), 1421–1434 (2008). DOI:10.1002/fld.1796
2. J. Grande. Red-green refinement of simplicial meshes in d dimensions. *Math. Comp.* **88**(316), 751–782 (2019). DOI:10.1090/mcom/3383
3. U. Langer and A. Schafelner. Adaptive space-time finite element methods for non-autonomous parabolic problems with distributional sources. *Comput. Methods Appl. Math.* **20**(4), 677–693 (2020). DOI:10.1515/cmam-2020-0042
4. J.M. Maubach. Local bisection refinement for n-simplicial grids generated by reflection. *SIAM J. Sci. Comput.* **16**(1), 210–227 (1995). DOI:10.1137/0916014
5. MFEM: Modular finite element methods library. mfem.org. DOI:10.11578/dc.20171025.1248
6. M. Neumüller and E. Karabelas. Generating admissible space-time meshes for moving domains in $(d + 1)$ dimensions. In: U. Langer, O. Steinbach (eds.) *Space-Time Methods*, pp. 185–206, De Gruyter, Berlin, Boston (2019). DOI:10.1515/9783110548488-006
7. R. Stevenson. The completion of locally refined simplicial partitions created by bisection. *Math. Comp.* **77**(261), 227–241 (2008). DOI:10.1090/S0025-5718-07-01959-X
8. C.T. Traxler. An algorithm for adaptive mesh refinement in n dimensions. *Computing* **59**(2), 115–137 (1997). DOI 10.1007/BF02684475
9. K. Voronin. A parallel mesh generator in 3d/4d. Tech. Rep. 11, Portland Institute for Computational Science Publications (2018).

Coefficient-Robust A Posteriori Error Estimation for H(curl)-elliptic Problems

Yuwen Li

1 Introduction

Adaptive mesh refinement (AMR) is a popular tool in numerical simulations as it is able to resolve singularities from nonsmooth data and irregular space domains. A building block of AMR is a posteriori error estimation, see, e.g., [10] for a classical introduction. On the other hand, preconditioners are discrete operators used to accelerate Krylov subspace methods for solving sparse linear systems (cf. [11]). Recently, [6, 7] introduced a novel framework linking a posteriori error estimation and *preconditioning* in Hilbert space. Such an approach yields many old and new error estimators for boundary value problems posed on de Rham complexes.

In particular, for the positive-definite H(curl) problem, [6] presents a new residual estimator robust w.r.t. high-contrast constant coefficients. In this paper, we extend the idea in [6] to the H(curl) interface problem and derive new a posteriori error estimates robust w.r.t. *both* extreme coefficient magnitudes as well as large coefficient jumps. The analysis avoids regularity assumptions used in existing works. We numerically compare the performance of the estimator in [6] with the one analyzed in [9].

1.1 H(curl)-elliptic problems

Let $\Omega \subset \mathbb{R}^d$ with $d \in \{2,3\}$ be a bounded Lipschitz domain, and n be a unit vector normal to $\partial\Omega$. Let $\nabla\times$ be the usual curl in $\mathbb{R}^3$, $\nabla\times = (\partial_{x_2}, -\partial_{x_1})\cdot$ in $\mathbb{R}^2$. We define

$$V = \left\{ v \in [L^2(\Omega)]^d : \nabla\times v \in [L^2(\Omega)]^{\frac{d(d-1)}{2}},\ v \wedge n = 0 \text{ on } \partial\Omega \right\},$$

Yuwen Li
Department of Mathematics, The Pennsylvania State University, University Park, PA 16802, USA, e-mail: yuwenli925@gmail.com

S. C. Brenner et al. (eds.), *Domain Decomposition Methods in Science and Engineering XXVI*, Lecture Notes in Computational Science and Engineering 145,
https://doi.org/10.1007/978-3-030-95025-5_45

where $v \wedge n = v \times n$ in $\mathbb{R}^3$, $v \wedge n = v \cdot n^\perp$ in $\mathbb{R}^2$ with $n^\perp$ the counter-clockwise rotation of n by $\frac{\pi}{2}$, and $[X]^d$ the Cartesian product of d copies of X. Let $(\cdot,\cdot)_{\Omega_0}$ denote the $L^2(\Omega_0)$ inner product and $(\cdot,\cdot) = (\cdot,\cdot)_\Omega$. Given $f \in L^2(\Omega)$ and positive $\varepsilon, \kappa \in L^\infty(\Omega)$, the H(curl)-elliptic boundary value problem seeks $u \in V$ s.t.

$$(\varepsilon\nabla \times u, \nabla \times v) + (\kappa u, v) = (f, v), \quad \forall v \in V. \tag{1}$$

The space V is equipped with the V-norm and energy inner product based on

$$(v, w)_V = (\varepsilon\nabla \times v, \nabla \times w) + (\kappa v, w), \quad \forall v, w \in V.$$

Let $\mathcal{T}_h$ be a conforming tetrahedral or hexahedral partition of Ω. Problem (1) is often discretized using the Nédélec edge element space $V_h \subset V$. The discrete problem is to find $u_h \in V_h$ s.t.

$$(\varepsilon\nabla \times u_h, \nabla \times v) + (\kappa u_h, v) = (f, v), \quad \forall v \in V_h. \tag{2}$$

The semi-discrete Maxwell equation is an important example of (1). In this case, ε is the reciprocal of the magnetic permeability and κ is proportional to $1/\tau^2$, where τ is the time stepsize. Therefore, we are interested in ε with large jump and potentially huge κ. In particular, we assume $\kappa > 0$ is a constant, $\Omega_1 \subset \Omega$, $\Omega_2 \subset \Omega$ are non-overlapping and simply-connected polyhedrons aligned with $\mathcal{T}_h$, $\bar{\Omega} = \bar{\Omega}_1 \cup \bar{\Omega}_2$, and

$$\varepsilon|_{\Omega_1} = \varepsilon_1, \quad \varepsilon|_{\Omega_2} = \varepsilon_2, \tag{3}$$

where $\varepsilon_1 \geq \varepsilon_2 > 0$ are constants. The interface is $\Gamma := \bar{\Omega}_1 \cap \bar{\Omega}_2$. A posteriori error analysis for more general ε, κ is beyond the scope of this work but is possible by making monotonicity-type assumptions on distributions of ε and κ, cf. [2, 3].

Throughout the rest of this paper, we say $\alpha \preccurlyeq \beta$ provided $\alpha \leq C\beta$, where C is an absolute constant depending solely on Ω, the aspect ratio of elements in $\mathcal{T}_h$, and the polynomial degree used in V_h. We say $\alpha \simeq \beta$ if $\alpha \preccurlyeq \beta$ and $\beta \preccurlyeq \alpha$. Given a Lipschitz manifold $\Sigma \subset \Omega$, we denote the $L^2(\Sigma)$ norm by $\|\cdot\|_\Sigma$.

2 Nodal auxiliary space preconditioning

The key idea in [6] is *nodal auxiliary space preconditioning*, originally proposed in [4] for solving discrete H(curl) and H(div) problems. The auxiliary H^1 space here is

$$W = \left\{w \in L^2(\Omega) : \nabla w \in [L^2(\Omega)]^d,\ w|_{\partial\Omega} = 0\right\},$$

endowed with the inner product

$$(w_1, w_2)_W = (\varepsilon\nabla w_1, \nabla w_2) + (\kappa w_1, w_2)$$

and the induced W-norm. The next regular decomposition (with *mixed* boundary condition, cf. [4, 6]) is widely used in the analysis of H(curl) problems.

Theorem 1 *Given $v \in V|_{\Omega_1}$, there exist $\varphi \in W|_{\Omega_1}$, $z \in [W|_{\Omega_1}]^d$, s.t. $v = \nabla\varphi + z$,*

$$\|z\|_{H^1(\Omega_1)} \leq C_0\|\nabla \times v\|,$$
$$\|\varphi\|_{H^1(\Omega_1)} \leq C_0(\|v\| + \|\nabla \times v\|),$$

where C_0 is a constant depending only on Ω_1.

To derive a posteriori error bounds for (2) uniform w.r.t. constant $\varepsilon \ll \kappa$, the work [6] utilizes the following modified regular decomposition.

Theorem 2 *Given $v \in V|_{\Omega_1}$, there exist $\varphi \in W|_{\Omega_1}$, $z \in [W|_{\Omega_1}]^d$, s.t. $v = \nabla\varphi + z$ and*

$$\|\varphi\|_{H^1(\Omega_1)} + \|z\| \leq C_1\|v\|,$$
$$|z|_{H^1(\Omega_1)} \leq C_1(\|v\| + \|\nabla \times v\|),$$

where C_1 is a constant depending only on Ω_1.

In the following, we give a new regular decomposition robust w.r.t. constant κ and piecewise constant ε. See also [5] for a weighted Helmholtz decomposition.

Theorem 3 *Given $v \in V$, there exist $\varphi \in W$ and $z \in [W]^d$, s.t. $v = \nabla\varphi + z$ and*

$$\|\kappa^{\frac{1}{2}}\varphi\|_{H^1(\Omega)} + \|z\|_W \leq C_2\|v\|_V,$$

where C_2 is a constant depending solely on Ω, Ω_1, Ω_2.

Proof The proof is divided into two cases. When $\varepsilon_1 \geq \kappa$, we use Theorem 1 on Ω_1 to obtain $\varphi_1 \in H^1(\Omega_1)$, $z_1 \in [H^1(\Omega_1)]^d$ both vanishing on $\partial\Omega_1\backslash\Gamma$ s.t.

$$\begin{aligned} v|_{\Omega_1} &= \nabla\varphi_1 + z_1, \\ \|z_1\|_{H^1(\Omega_1)} &\preccurlyeq \|\nabla \times v\|_{\Omega_1}, \\ \|\varphi_1\|_{H^1(\Omega_1)} &\preccurlyeq \|v\|_{\Omega_1} + \|\nabla \times v\|_{\Omega_1}. \end{aligned} \tag{4}$$

When $\varepsilon_1 < \kappa$, applying Theorem 2 to $v|_{\Omega_1}$ yields $\varphi_1 \in H^1(\Omega_1)$, $z_1 \subset [H^1(\Omega_1)]^d$ s.t.

$$\begin{aligned} &v|_{\Omega_1} = \nabla\varphi_1 + z_1, \quad \varphi_1|_{\partial\Omega_1\backslash\Gamma} = 0,\ z_1|_{\partial\Omega_1\backslash\Gamma} = 0, \\ &\|\varphi_1\|_{H^1(\Omega_1)} + \|z_1\|_{\Omega_1} \preccurlyeq \|v\|_{\Omega_1}, \\ &|z_1|_{H^1(\Omega_1)} \preccurlyeq \|v\|_{\Omega_1} + \|\nabla \times v\|_{\Omega_1}. \end{aligned} \tag{5}$$

In either case, it holds that

$$\|\kappa^{\frac{1}{2}}\varphi_1\|_{H^1(\Omega_1)} + \|z_1\|_{W|_{\Omega_1}} \preccurlyeq \|v\|_{V|_{\Omega_1}}. \tag{6}$$

First let $\hat{\varphi}_1 \in H^1(\mathbb{R}^d\backslash\Omega_2)$ and $\hat{z}_1 \in [H^1(\mathbb{R}^d\backslash\Omega_2)]^d$ be zero extensions of φ_1 and z_1 to $\mathbb{R}^d\backslash\Omega_2$, respectively. Then we take $\tilde{\varphi}_1 \in H^1(\Omega)$, $\tilde{z}_1 \in H^1(\Omega)$ to be the Stein universal extensions of $\hat{\varphi}_1$, $\hat{z}_1$ to $\mathbb{R}^d$ satisfying

$$\|\tilde{\varphi}_1\|_{\Omega_2} \preccurlyeq \|\varphi_1\|_{\Omega_1}, \quad \|\tilde{\varphi}_1\|_{H^1(\Omega_2)} \preccurlyeq \|\varphi_1\|_{H^1(\Omega_1)},$$
$$\|\tilde{z}_1\|_{\Omega_2} \preccurlyeq \|z_1\|_{\Omega_1}, \quad \|\tilde{z}_1\|_{H^1(\Omega_2)} \preccurlyeq \|z_1\|_{H^1(\Omega_1)}. \tag{7}$$

On Ω_2, applying Theorem 1 (if $\varepsilon_2 \geq \kappa$) or Theorem 2 (if $\varepsilon_2 < \kappa$) to $w = v|_{\Omega_2} - \nabla\tilde{\varphi}_1|_{\Omega_2} - \tilde{z}_1|_{\Omega_2}$ ($w \wedge n = 0$ on $\partial\Omega_2$), we have $\varphi_2 \in H_0^1(\Omega_2)$, $z_2 \in [H_0^1(\Omega_2)]^d$ s.t.

$$v|_{\Omega_2} - \nabla\tilde{\varphi}_1|_{\Omega_2} - \tilde{z}_1|_{\Omega_2} = \nabla\varphi_2 + z_2, \tag{8a}$$

$$\|\kappa^{\frac{1}{2}}\varphi_2\|_{H^1(\Omega_2)} + \|z_2\|_{W|_{\Omega_2}} \preccurlyeq \|v\|_{V|_{\Omega_2}} + \|\kappa^{\frac{1}{2}}\nabla\tilde{\varphi}_1\|_{\Omega_2} + \|\tilde{z}_1\|_{V|_{\Omega_2}}. \tag{8b}$$

Here (8b) follows from similar reasons for (6). Define $\varphi \in H_0^1(\Omega)$, $z \in [H_0^1(\Omega)]^d$ as

$$\varphi := \begin{cases} \varphi_1 & \text{on } \Omega_1 \\ \tilde{\varphi}_1 + \varphi_2 & \text{on } \Omega_2 \end{cases}, \quad z := \begin{cases} z_1 & \text{on } \Omega_1 \\ \tilde{z}_1 + z_2 & \text{on } \Omega_2 \end{cases},$$

and obtain $v = \nabla\varphi + z$ on Ω. If $\varepsilon_1 \geq \kappa$, it follows from (8b), (7), (4), $\varepsilon_2 \leq \varepsilon_1$ that

$$\begin{aligned} &\|\kappa^{\frac{1}{2}}\varphi\|_{H^1(\Omega_2)} + \|z\|_{W|_{\Omega_2}} \\ &\preccurlyeq \|v\|_{V|_{\Omega_2}} + \kappa^{\frac{1}{2}}\|\varphi_1\|_{H^1(\Omega_1)} + (\kappa^{\frac{1}{2}} + \varepsilon_2^{\frac{1}{2}})\|z_1\|_{\Omega_1} + \varepsilon_2^{\frac{1}{2}}|z_1|_{H^1(\Omega_1)} \\ &\preccurlyeq \|v\|_{V|_{\Omega_2}} + \kappa^{\frac{1}{2}}\|v\| + \varepsilon_1^{\frac{1}{2}}\|\nabla\times v\|_{\Omega_1}. \end{aligned} \tag{9}$$

Similarly when $\varepsilon_1 < \kappa$, it follows from (8b), (7), (5), $\varepsilon_2 \leq \varepsilon_1 < \kappa$ that

$$\|\kappa^{\frac{1}{2}}\varphi\|_{H^1(\Omega_2)} + \|z\|_{W|_{\Omega_2}} \preccurlyeq \|v\|_{V|_{\Omega_2}} + \kappa^{\frac{1}{2}}\|v\|_{\Omega_1}. \tag{10}$$

Combining (6), (9), (10) completes the proof. □

Remark 1 The work [12] gives a robust regular decomposition for the H(curl) interface problem with $\kappa = s\varepsilon$, $s \in (0, 1]$. In contrast, Theorem 3 is able to deal with large jump of ε as well as large $\kappa \gg \varepsilon$.

Given a Hilbert space X, let X' denote its dual space, and $\langle\cdot,\cdot\rangle$ the action of X' on X. We introduce bounded linear operators $A : V \to V'$, $A_\Delta : H_0^1(\Omega) \to H^{-1}(\Omega)$, $A_W : W^d \to ([W]^d)'$ as

$$\begin{aligned} \langle Av, w\rangle &= (\varepsilon\nabla\times v, \nabla\times w) + (\kappa v, w), \quad v, w \in V, \\ \langle A_\Delta v, w\rangle &= (\nabla v, \nabla w) + (v, w), \quad v, w \in H_0^1(\Omega), \\ \langle A_W v, w\rangle &= (\varepsilon\nabla v, \nabla w) + (\kappa v, w), \quad v, w \in [W]^d. \end{aligned}$$

Let $r \in V'$ be the residual given by

$$\langle r, v\rangle = (f, v) - (\varepsilon\nabla\times u_h, \nabla\times v) - (\kappa u_h, v), \quad v \in V. \tag{11}$$

Clearly the inclusion $I : [W]^d \hookrightarrow V$ and the gradient operator $\nabla : W \to V$ are uniformly bounded w.r.t. ε and κ. Then using such boundedness, Theorem 3, and the *fictitious space lemma* (cf. [4, 8] and Corollary 5.1 in [6]), we obtain the uniform spectral equivalence of two continuous operators

$$A^{-1} \simeq B := \nabla(\kappa A_\Delta)^{-1}\nabla' + IA_W^{-1}I', \tag{12}$$

where $I' : V' \to ([W]^d)'$ and $\nabla' : V' \to W'$ are adjoint operators. By $A^{-1} \simeq B$ from V' to V in (12) we mean $\langle R, A^{-1}R\rangle \simeq \langle R, BR\rangle$, $\forall R \in V'$. Note that $A(u - u_h) = r \in V'$. Therefore a direct consequence of (12) is

$$\begin{aligned} \|u - u_h\|_V^2 &= \langle A(u - u_h), u - u_h\rangle = \langle r, A^{-1}r\rangle \simeq \langle r, Br\rangle \\ &= \langle \nabla' r, (\kappa A_\Delta)^{-1}\nabla' r\rangle + \langle I'r, A_W^{-1}I'r\rangle = \kappa^{-1}\|\nabla' r\|_{H^{-1}(\Omega)}^2 + \|I'r\|_{([W]^d)'}^2. \end{aligned} \tag{13}$$

3 A posteriori error estimates

The goal of this paper is to derive a robust two-sided bound $\|u - u_h\|_V \simeq \eta_h$. The quantity η_h is computed from u_h and split into element-wise error indicators for AMR. Such local error indicators are used to predict element errors in the current grid and mark those tetrahedra/hexahedra with large errors for subdivision.

When deriving the error estimator, we assume that the source f is piecewise H^1-regular w.r.t. $\mathcal{T}_h$. By $\mathcal{S}_h$ we denote the collection of $(d-1)$-simplexes in $\mathcal{T}_h$ that are not contained in $\partial\Omega$. Each $S \in \mathcal{S}_h$ shared by $T_S^+, T_S^- \in \mathcal{T}_h$ is assigned with a unit normal n_S pointing from T_S^+ to T_S^-. Let h, h_s be the mesh size functions s.t. $h|_T = h_T := \mathrm{diam}(T)$ $\forall T \in \mathcal{T}_h$, $h_s|_S = h_S := \mathrm{diam}(S)$ $\forall S \in \mathcal{S}_h$. The weighted mesh size functions are

$$\bar{h} := \min\left\{\frac{h}{\sqrt{\varepsilon}}, \frac{1}{\sqrt{\kappa}}\right\}, \quad \bar{h}_s := \min\left\{\frac{h_s}{\sqrt{\varepsilon_s}}, \frac{1}{\sqrt{\kappa}}\right\},$$

where $\varepsilon_s|_S = \max\{\varepsilon_{T_S^+}, \varepsilon_{T_S^-}\}$ $\forall S \in \mathcal{S}_h$. For each $T \in \mathcal{T}_h$, $S \in \mathcal{S}_h$, let Ω_T denote the union of elements in $\mathcal{T}_h$ sharing an edge with T, and $\Omega_S = \cup_{S \in \mathcal{S}_h, S \subset \partial T}\Omega_T$. For each $S \in \mathcal{S}_h$, let $[\![\omega]\!]_S = \omega|_{T_S^+} - \omega|_{T_S^-}$ be the jump of ω across S. We define

$$\begin{aligned} R_1|_T &= -\nabla\cdot(f - \kappa u_h)|_T, \quad J_1|_S = [\![f - \kappa u_h]\!]_S \cdot n_S, \\ R_2|_T &= (f - (\nabla\times)^*(\varepsilon\nabla\times u_h) - \kappa u_h)|_T, \quad J_2|_S = -[\![\varepsilon\nabla\times u_h]\!]_S \wedge n_S, \end{aligned}$$

where $(\nabla\times)^* = \nabla\times$ in $\mathbb{R}^3$ and $(\nabla\times)^* = (-\partial_{x_2}, \partial_{x_1})$ in $\mathbb{R}^2$. By the element-wise Stokes' (in $\mathbb{R}^3$) or Green's (in $\mathbb{R}^2$) formula, we have

$$\langle \nabla' r, \psi\rangle = \langle r, \nabla\psi\rangle = \sum_{T\in\mathcal{T}_h}(R_1, \psi)_T + \sum_{S\in\mathcal{S}_h}(J_1, \psi)_S, \quad \psi \in H_0^1(\Omega), \tag{14}$$

$$\langle I'r, \varphi\rangle = \langle r, \varphi\rangle = \sum_{T\in\mathcal{T}_h}(R_2, \varphi)_T + \sum_{S\in\mathcal{S}_h}(J_2, \varphi)_S, \quad \varphi \in [W]^d. \tag{15}$$

In view of (13), it remains to estimate $\|\nabla' r\|_{H^{-1}(\Omega)}$ and $\|I'r\|_{([W]^d)'}$. Let $(\cdot,\cdot)_{\mathcal{S}_h}$ denote the inner product $\sum_{S\in\mathcal{S}_h}(\cdot,\cdot)_S$ and $\|\cdot\|_{\mathcal{S}_h}$ the corresponding norm. Let Q_h (resp. Q_h^s) be the L^2 projection onto the space of discontinuous and piecewise polynomials of fixed degrees on $\mathcal{T}_h$ (resp. $\mathcal{S}_h$). The estimation of $\|\nabla' r\|_{H^{-1}(\Omega)}$ is standard (cf. [6]) and given as

$$\|hR_1\| + \|h_s^{\frac{1}{2}} J_1\|_{\mathcal{S}_h} - \mathrm{osc}_{h,1} \preccurlyeq \|\nabla' r\|_{H^{-1}(\Omega)} \preccurlyeq \|hR_1\| + \|h_s^{\frac{1}{2}} J_1\|_{\mathcal{S}_h}, \tag{16}$$

where $\mathrm{osc}_{h,1} := \|h(R_1 - Q_h R_1)\| + \|h_s^{\frac{1}{2}}(J_1 - Q_h^s J_1)\|_{\mathcal{S}_h}$ is the data oscillation. We also need the second data oscillation $\mathrm{osc}_{h,2} := \|\bar{h}(R_2 - Q_h R_2)\| + \|\bar{h}_s^{\frac{1}{2}}(J_2 - Q_h^s J_2)\|_{\mathcal{S}_h}$. In the next lemma, we derive two-sided bounds for $\|I'r\|_{([W]^d)'}$.

Lemma 1 *It holds that*

$$\|\bar{h}R_2\| + \|\varepsilon^{-\frac{1}{4}}\bar{h}_s^{\frac{1}{2}} J_2\|_{\mathcal{S}_h} - \mathrm{osc}_{h,2} \preccurlyeq \|I'r\|_{([W]^d)'} \preccurlyeq \|\bar{h}R_2\| + \|\varepsilon^{-\frac{1}{4}}\bar{h}_s^{\frac{1}{2}} J_2\|_{\mathcal{S}_h}.$$

Proof The proof is similar to Lemma 4.4 of [6] except the use of the modified Clément-type interpolation $\widetilde{\Pi}_h : [L^2(\Omega)]^d \to V_h^0$ proposed in [3] for dealing with huge jump of ε. Here $V_h^0 \subseteq V_h$ is the lowest order edge element space. For any $v \in [W]^d$ and $T \in \mathcal{T}_h$, the analysis in Theorem 4.6 of [3] implies that

$$\|v - \widetilde{\Pi}_h v\|_T \preccurlyeq h_T \varepsilon|_T^{-\frac{1}{2}} \|\varepsilon^{\frac{1}{2}} \nabla v\|_{\Omega_T} \leq h_T \varepsilon|_T^{-\frac{1}{2}} \|v\|_{W|\Omega_T}, \tag{17}$$

$$\|\nabla(v - \widetilde{\Pi}_h v)\|_T \preccurlyeq \varepsilon|_T^{-\frac{1}{2}} \|\varepsilon^{\frac{1}{2}} \nabla v\|_{\Omega_T} \leq \varepsilon|_T^{-\frac{1}{2}} \|v\|_{W|\Omega_T}. \tag{18}$$

The L^2-boundedness of $\widetilde{\Pi}_h$ implies that

$$\|v - \widetilde{\Pi}_h v\|_T \preccurlyeq \|v\|_{\Omega_T} \leq \kappa^{-\frac{1}{2}} \|v\|_{W|\Omega_T}. \tag{19}$$

A direct consequence of (17) and (19) is

$$\|v - \widetilde{\Pi}_h v\|_T \preccurlyeq \bar{h}_T \|v\|_{W|\Omega_T}. \tag{20}$$

Given a face/edge $S \in \mathcal{S}_h$, let T be the element containing S over which ε is maximal. Using a trace inequality, (20), $h_S^{-1} \leq \bar{h}_S^{-1}\varepsilon_S^{-\frac{1}{2}}$, (18), $\bar{h}_S \simeq \bar{h}_T$, we have

$$\begin{aligned} \|v - \widetilde{\Pi}_h v\|_S^2 &\preccurlyeq h_S^{-1}\|v - \widetilde{\Pi}_h v\|_T^2 + \|v - \widetilde{\Pi}_h v\|_T \|\nabla(v - \widetilde{\Pi}_h v)\|_T \\ &\preccurlyeq h_S^{-1}\bar{h}_T^2 \|v\|_{W|\Omega_T}^2 + \bar{h}_T \varepsilon|_T^{-\frac{1}{2}} \|v\|_{W|\Omega_T}^2 \preccurlyeq \varepsilon|_T^{-\frac{1}{2}} \bar{h}_S \|v\|_{W|\Omega_T}^2. \end{aligned} \tag{21}$$

It follows from $r|_{V_h} = 0$, (15), and the Cauchy–Schwarz inequality that

$$\|I'r\|_{([W]^d)'} = \sup_{v\in[W]^d,\|v\|_W=1} \langle r, v\rangle = \sup_{v\in[W]^d,\|v\|_W=1} \langle r, v - \widetilde{\Pi}_h v\rangle$$
$$\leqslant \left(\|\bar{h}R_2\| + \|\varepsilon_s^{-\frac{1}{4}}\bar{h}_s^{\frac{1}{2}}J_2\|_{\mathcal{S}_h}\right) \sup_{\substack{v\in[W]^d\\ \|v\|_W=1}} \left(\|\bar{h}^{-1}(v - \widetilde{\Pi}_h v)\| + \|\varepsilon_s^{\frac{1}{4}}\bar{h}_s^{-\frac{1}{2}}(v - \widetilde{\Pi}_h v)\|_{\mathcal{S}_h}\right).$$

Then the upper bound of $\|I'r\|_{([W]^d)'}$ is a consequence of the above inequality and (20), (21). The uniform lower bound of $\|I'r\|_{([W]^d)'}$ w.r.t. ε, κ follows from the bubble function technique explained in [10] and extremal definitions of $\bar{h}$, $\bar{h}_s$, ε_s. □

For each $T \in \mathcal{T}_h$, we define the error indicator

$$\eta_h(T) = \kappa^{-1}h_T^2\|R_1\|_T^2 + \bar{h}|_T^2\|R_2\|_T^2 + \sum_{S\in\mathcal{S}_h, S\subset\partial T} \left\{\kappa^{-1}h_S\|J_1\|_S^2 + \bar{h}_s|_S\|\varepsilon^{-\frac{1}{4}}J_2\|_S^2\right\}.$$

Combining (13), (16) and Lemma 1 leads to the robust a posteriori error estimate

$$\sum_{T\in\mathcal{T}_h} \eta_h(T) - \mathrm{osc}_{h,1} - \mathrm{osc}_{h,2} \preccurlyeq \|u - u_h\|_V^2 \preccurlyeq \sum_{T\in\mathcal{T}_h} \eta_h(T). \tag{22}$$

Remark 2 Our analysis for (22) is based on regular decomposition and minimal regularity while the theoretical analysis of *recovery* estimators in [3] hinges on Helmholtz decomposition and full elliptic regularity of the underlying domain. Our estimator $\eta_h(T)$ is robust w.r.t. both large jumps of ε and extreme magnitudes of ε, κ.

4 Numerical demonstration of robustness

Lastly, we focus on (1) with *constant* and *positive* ε and κ, which is a special case of the interface problem considered before. In this case, the error indicator $\eta_h(T)$ reduces to the one derived in [6]. For constant ε and κ, the classical a posteriori error estimator for (2) (cf. [1, 9]) reads

$$\tilde{\eta}_h(T) = \kappa^{-1}h_T^2\|R_1\|_T^2 + \varepsilon^{-1}h_T^2\|R_2\|_T^2 + \sum_{S\in\mathcal{S}_h, S\subset\partial T} \left\{\kappa^{-1}h_S\|J_1\|_S^2 + \varepsilon^{-1}h_S\|J_2\|_S^2\right\}.$$

Although weighted with ε, κ, this estimator is not fully robust w.r.t. ε and κ. In fact, the ratio $\|u - u_h\|_V / (\sum_{T\in\mathcal{T}_h} \tilde{\eta}_h(T))^{\frac{1}{2}}$ may tend to zero as $\varepsilon \ll \kappa$, i.e., the constant $\underline{C}$ in the lower bound $\underline{C}(\sum_{T\in\mathcal{T}_h} \tilde{\eta}_h(T))^{\frac{1}{2}} \leq \|u - u_h\|_V$ + h.o.t. is not uniform.

To validate the result, we test $\eta_h(T)$ and $\tilde{\eta}_h(T)$ by the lowest order edge element discretization of (1) defined on $\Omega = [0, 1]^2$ with the exact solution $u(x_1, x_2) = \big(\cos(\pi x_1)\sin(\pi x_2), \sin(\pi x_1)\cos(\pi x_2)\big)$. The initial partition of Ω is a 4×4 uniform triangular mesh. A sequence of nested grids is computed by uniform quad-refinement. Let $e = \|u-u_h\|_V$, $\eta = (\sum_{T\in\mathcal{T}_h}\eta_h(T))^{\frac{1}{2}}$ and $\tilde{\eta} = (\sum_{T\in\mathcal{T}_h}\tilde{\eta}_h(T))^{\frac{1}{2}}$. Numerical results are shown in Table 1. In its last row, we compute the effectivity index "eff" of η (resp. $\tilde{\eta}$), which is the algorithmic mean of e/η (resp. $e/\tilde{\eta}$) over all grid levels. It is observed that the performance of η is uniformly effective for all ε, κ, while the efficiency of $\tilde{\eta}$ deteriorates for small ε and large κ.

Table 1: Convergence history of the lowest order edge element and error estimators

number of elements	e $\varepsilon=0.1$ $\kappa=10$	η $\varepsilon=0.1$ $\kappa=10$	$\tilde{\eta}$ $\varepsilon=0.1$ $\kappa=10$	e $\varepsilon=10^{-3}$ $\kappa=10^3$	η $\varepsilon=10^{-3}$ $\kappa=10^3$	$\tilde{\eta}$ $\varepsilon=10^{-3}$ $\kappa=10^3$	e $\varepsilon=10^{-5}$ $\kappa=10^5$	η $\varepsilon=10^{-5}$ $\kappa=10^5$	$\tilde{\eta}$ $\varepsilon=10^{-5}$ $\kappa=10^5$
32	8.42e-1	3.72	3.94	8.24	3.72e+1	1.46e+3	8.24e+1	3.72e+2	1.46e+6
128	4.35e-1	2.04	2.04	4.30	2.04e+1	3.80e+2	4.30e+1	2.04e+2	3.80e+5
512	2.19e-1	1.04	1.04	2.18	1.06e+1	9.70e+1	2.18e+1	1.06e+2	9.64e+4
2048	1.10e-1	5.26e-1	5.26e-1	1.10	5.36	2.48e+1	1.10e+1	5.36e+1	2.42e+4
8192	5.49e-2	2.64e-1	2.64e-1	5.49e-1	2.69	6.61	5.49	2.69e+1	6.06e+3
eff	N/A	2.13e-1	2.11e-1	N/A	2.09e-1	3.33e-2	N/A	2.09e-1	3.51e-4

References

1. R. Beck, R. Hiptmair, R.H.W. Hoppe and B. Wohlmuth. Residual based a posteriori error estimators for eddy current computation. *M2AN Math. Model. Numer. Anal.* **34**(1), 159–182 (2000). DOI:10.1051/m2an:2000136
2. C. Bernardi and R. Verfürth. Adaptive finite element methods for elliptic equations with nonsmooth coefficients. *Numer. Math.* **85**(4), 579–608 (2000). DOI:10.1007/PL00005393
3. Z. Cai and S. Cao. A recovery-based a posteriori error estimator for H(curl) interface problems. *Comput. Methods Appl. Mech. Engrg.* **296**, 169–195 (2015). DOI:10.1016/j.cma.2015.08.002
4. R. Hiptmair and J. Xu. Nodal auxiliary space preconditioning in H(curl) and H(div) spaces. *SIAM J. Numer. Anal.* **45**(6), 2483–2509 (2007). DOI:10.1137/060660588
5. Q. Hu, S. Shu and J. Zou. A discrete weighted Helmholtz decomposition and its application. *Numer. Math.* **125**(1), 153–189 (2013). DOI:10.1007/s00211-013-0536-6
6. Y. Li and L. Zikatanov. Nodal auxiliary a posteriori error estimates. *arXiv preprint* arXiv:2010.06774 (2020).
7. Y. Li and L. Zikatanov. A posteriori error estimates of finite element methods by preconditioning. *Comput. Math. Appl.* **91**, 192–201 (2021). DOI:10.1016/j.camwa.2020.08.001
8. S.V. Nepomnyaschikh. Decomposition and fictitious domains methods for elliptic boundary value problems. In: *Fifth International Symposium on Domain Decomposition Methods for Partial Differential Equations (Norfolk, VA, 1991)*, pp. 62–72. SIAM, Philadelphia, PA (1992).
9. J. Schöberl. A posteriori error estimates for Maxwell equations. *Math. Comp.* **77**(262), 633–649 (2008).
10. R. Verfürth. *A posteriori error estimation techniques for finite element methods. Numerical Mathematics and Scientific Computation*. Oxford University Press, Oxford (2013). DOI:10.1093/acprof:oso/9780199679423.001.0001

11. J. Xu. Iterative methods by space decomposition and subspace correction. *SIAM Rev.* **34**(4), 581–613 (1992). DOI:10.1137/1034116
12. J. Xu and Y. Zhu. Robust preconditioner for H(curl) interface problems. In: *Domain decomposition methods in science and engineering XIX*, Lect. Notes Comput. Sci. Eng., vol. 78, pp. 173–180, Springer, Heidelberg (2011). DOI:10.1007/978-3-642-11304-8_18

Convergence of PARAREAL for a Vibrating String with Viscoelastic Damping

Martin J. Gander, Thibaut Lunet, and Aušra Pogoželskytė

1 Model equation for a vibrating string

We consider an elastic string of length L, attached at its two end points, vibrating in a plane due to an initial deformation corresponding to a pinch in the middle, see Figure 1. Its deformation is represented by a scalar function $u(x, t)$, with $x \in [0, L]$, $t \in [0, T]$, and $u(0, t) = u(L, t) = 0$. This simple configuration is the basis for more complex problems that can model guitar and piano strings (see, e.g., [2, 6]), or similar musical instruments.

Generally, such a problem is modeled using the wave equation, possibly adding a first order damping term to produce the so called *Telegrapher's equation*,

$$\partial_{tt}u(x,t) = c^2\,\partial_{xx}u(x,t) - R\,\partial_t u(x,t)\ , \tag{1}$$

with c the wave velocity and R a damping parameter; setting $R = 0$ leads back to the wave equation. The vibration period of the string is defined as

$$T_W := 2L/c\ . \tag{2}$$

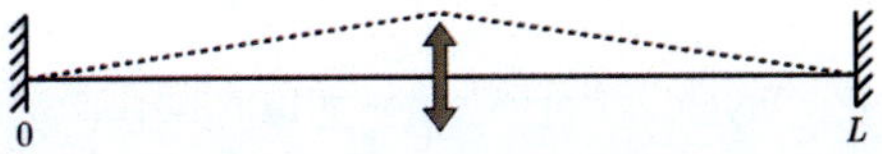

Fig. 1: Vibrating string attached at its two end points, with initial deformation after being plucked.

Martin J. Gander
University of Geneva, e-mail: martin.gander@unige.ch

Thibaut Lunet
University of Geneva, e-mail: thibaut.lunet@unige.ch

Aušra Pogoželskytė
University of Geneva, e-mail: ausra.pogozelskyte@unige.ch

S. C. Brenner et al. (eds.), *Domain Decomposition Methods in Science and Engineering XXVI*, Lecture Notes in Computational Science and Engineering 145,
https://doi.org/10.1007/978-3-030-95025-5_46

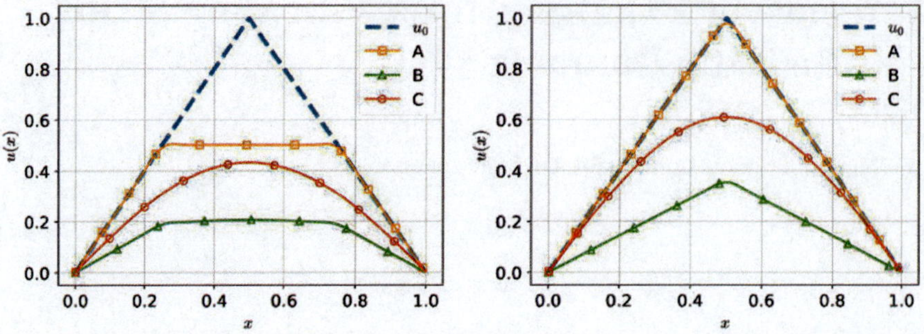

Fig. 2: Solution of the wave equation (A), the Telegrapher's equation with $R = 1$ (B), and the wave equation with viscoelastic damping $\gamma = 0.03$ (C). The dashed line is the initial condition, u_0, of all the equations for reference. Left: $t = \frac{9}{8} T_W$. Right: $t = T_W$.

Equation (1) does not, however, model the physical behavior of the string in Figure 1 accurately. To illustrate this, we show the numerical solution of (1) in Figure 2, at time $t = \frac{9}{8} T_W$ (left), and $t = T_W$ (right). Curve A represents the wave equation solution ($R = 0$): the string forms a central, unphysical plateau during the oscillation and comes back to the same plucked initial position after one oscillation ($t = T_W$). Adding the damping term ($R = 1$, curve B) reduces the amplitude of the vibration, but the shape still corresponds to the unphysical shape produced by the wave equation.

To correct this, we replace the damping term in (1) by a modified one,

$$\partial_{tt} u(x,t) = c^2 \, \partial_{xx} u(x,t) + \gamma \, \partial_{txx} u(x,t) \,, \tag{3}$$

where γ is a different damping parameter. We call this the *wave equation with viscoelastic damping*. A numerical solution of (3) is shown in Figure 2, curve C, which now looks closer to what we would expect from physics. The viscoelastic damping term in (3) is of prime importance when modeling string vibration. As shown in [5], $u(x,t)$ is a linear combination of string *modes*,[1]

$$\xi_n(x) := \sin\left(\frac{\kappa \pi x}{L}\right), \quad \kappa \in \mathbb{N}^* \,, \tag{4}$$

with κ the mode number. In the physical world, when vibrating, each mode is damped at a different rate: high frequency modes (large κ) are damped quickly while low frequency modes (small κ) persist longer. The viscoelastic term can model this behaviour while the damping term in (1), also called the *fluid term*,[2] introduces the same damping for all modes.

[1] Those modes are also the eigenfunctions of the one-dimensional Laplacian with Dirichlet boundary conditions.

[2] Actually, a more accurate model for a vibrating guitar string in [6] considers both fluid and viscoelastic terms. But for simplicity, we will consider here only the viscoelastic term.

To do Fourier analysis, we take the initial condition to be a string mode (4). This leads to a closed form solution of (3),

$$u(x,t) = \mathrm{e}^{-\mu_\kappa t}\left[\cos(\tilde{\omega}_\kappa t) + \frac{\omega_\kappa^2}{2}\sin(\tilde{\omega}_\kappa t)\right]\sin\left(\frac{\kappa\pi x}{L}\right), \tag{5}$$

with $\mu_\kappa := \gamma\frac{\kappa^2\pi^2}{2L^2}$, $\tilde{\omega}_\kappa := \omega_\kappa\sqrt{1-\left(\frac{\kappa\pi}{2}\frac{\gamma}{cL}\right)^2}$ and $\omega_\kappa := c\frac{\kappa\pi}{L}$, provided the mode number κ is low enough for the mode to still be oscillating. This is equivalent to the discriminant when solving (3) being negative, which means

$$\omega_\kappa^2 < \frac{4c^4}{\gamma^2} \iff \kappa < \frac{2}{\pi}\frac{cL}{\gamma} = \frac{2}{\pi}\mathcal{W}, \tag{6}$$

where we introduced $\mathcal{W} := \frac{cL}{\gamma}$, which is the equivalent of a Reynolds (or Peclet) number for advection. Indeed, equation (3) is purely hyperbolic when $\mathcal{W} \to +\infty$ (vibration with no damping) and purely parabolic when $\mathcal{W} \to 0$ (no vibration). Furthermore, the number of vibrating modes is limited by the value of $\mathcal{W}$ (see (6)). For one given mode, $\mathcal{W}$ also defines, together with T_W (see (2)), the lifespan of the vibration

$$\tau_\kappa := \frac{1}{\mu_\kappa} = \frac{\mathcal{W}}{\kappa^2\pi^2}T_W, \tag{7}$$

which represents the time for the mode amplitude to be reduced to $36.8\% = \mathrm{e}^{-1}$ of its initial value. As $\mathcal{W}$ gets larger, τ_κ increases, hence a numerical simulation may require more time steps to keep a good accuracy (for a more complete investigation, see Section 3). This can greatly increase computation time, so we now investigate the possibility of using time parallelization to speedup computations.

2 The PARAREAL algorithm

Time parallel time integration has received sustained attention over the last few decades. For a review, see [8]. Renewed interest in this area was sparked by the invention of the PARAREAL algorithm [13] for solving initial value problems of the form

$$\frac{\mathrm{d}\boldsymbol{u}}{\mathrm{d}t} = \mathcal{L}(\boldsymbol{u}(t),t), \quad \boldsymbol{u}(0) = \boldsymbol{u}_0, \quad t \in [0,T], \tag{8}$$

with $\mathcal{L}\colon \mathbb{R}^p \times \mathbb{R}^+ \to \mathbb{R}^p$, $\boldsymbol{u}(t) \in \mathbb{R}^p$, $\boldsymbol{u}_0 \in \mathbb{R}^p$, p being the total number of degrees of freedom, and T a positive real value. Problem (8) often arises from the spatial discretization of a (non-)linear system of partial differential equations (PDEs) through the method-of-lines.

For PARAREAL, one decomposes the global time interval $[0,T]$ into N time subintervals $[T_{n-1},T_n]$ of size ΔT, $n = 1,\ldots,N$, where N is the number of processes to be considered for the time parallelization. In the following, we denote by $\boldsymbol{U}_n$ the approximation of $\boldsymbol{u}$ at time T_n, i.e., $\boldsymbol{U}_n \approx \boldsymbol{u}(T_n)$. Let $\mathcal{F}^{n_F}_{T_{n-1}\to T_n}(\boldsymbol{U}_{n-1})$ denote the result of approximately integrating (8) on the time subinterval $[T_{n-1}, T_n]$ from a given starting value U_{n-1} using a fine propagator $\mathcal{F}$ and n_F time steps (with time step $\Delta t_F := (T_{n-1} - T_n)/n_F$). Similarly, PARAREAL also needs a coarse propagator $\mathcal{G}$ (using for example n_G time steps), which has to be much cheaper than $\mathcal{F}$ resulting in less accuracy (i.e. $n_F \gg n_G$).

The PARAREAL algorithm consists of a prediction step and a correction iteration. In the prediction step, PARAREAL computes an initial guess for the starting values U_n^0 at the beginning of each time subinterval using the coarse propagator,

$$U_0^0 = u_0, \quad U_n^0 = \mathcal{G}_{T_{n-1}\to T_n}^{n_G}(U_{n-1}^0), \; n = 1, \dots, N . \tag{9}$$

A correction iteration is then applied in PARAREAL, using the fine propagator $\mathcal{F}$ on each time subinterval concurrently,

$$U_n^k = \mathcal{F}_{T_{n-1}\to T_n}^{n_F}(U_{n-1}^{k-1}) + \mathcal{G}_{T_{n-1}\to T_n}^{n_G}(U_{n-1}^k) - \mathcal{G}_{T_{n-1}\to T_n}^{n_G}(U_{n-1}^{k-1}) , \tag{10}$$

where U_n^k denotes the approximation of u at time T_n at the k-th iteration of PARAREAL ($k = 1, \dots, K$, $n = 1, \dots, N$). While the application of $\mathcal{F}$ can be performed independently for each time subinterval, PARAREAL remains limited by the sequential nature of the coarse integration performed by $\mathcal{G}_{T_{n-1}\to T_n}^{n_G}$ in (10). PARAREAL will thus reduce the total computational time compared to a direct time-serial integration only if the application of $\mathcal{G}$ is cheap enough and if the total number of PARAREAL iterations K is small. A more complete description of parallel speedup for PARAREAL can be found in [1].

While this algorithm works well for parabolic problems, it is known to struggle when the problem of interest is close to hyperbolic (see, e.g., [7, 15, 16]). In our case, this happens when $\mathcal{W}$ becomes large in (3), similarly to what one obtains for the Navier–Stokes equations, as $\mathcal{W}$ plays the same role as the Reynolds number. In the latter case, keeping the same accuracy for the fine solver when the Reynolds number increases is very important (cf. [14, Sec. 3.5]), as the use of the fine solver with incorrect mesh resolution can lead to a misinterpretation of PARAREAL convergence results (see, e.g., [10, Sec. 4]). Hence in the next section, we investigate the accuracy of our space time discretization and its link to $\mathcal{W}$ and other parameters.

3 Space mesh requirement for fixed error tolerance

We solve (3) numerically using a uniform spatial mesh with n_x points ($x_j := j\,\Delta x$, with $\Delta x := \frac{L}{n_x+1}$). Denoting by $u_j(t) \approx u(x_j, t)$ and $\dot{u}_j(t) \approx \partial_t u(x_j, t)$, we define

$$\mathbf{v}(t) := [u_1(t), ..., u_{n_x}(t), \dot{u}_1(t), ..., \dot{u}_{n_x}(t)]^\top . \tag{11}$$

We use second-order centered finite differences, which leads to a tridiagonal square matrix A of size n_x. Applying the method-of-lines to (3) yields

$\mathcal{W}$	$\kappa = 1$			$\kappa = 2$			$\kappa = 4$		
	n_x	n_t	ϵ	n_x	n_t	ϵ	n_x	n_t	ϵ
100	305	619	1.008%	432	219	1.007%	610	78	1.009%
1000	965	19555	1.012%	1365	6916	1.012%	1929	2444	1.014%
10000	3050	618060	1.015%	4314	218550	1.014%	6100	77258	1.014%

Table 1: Mesh resolution needed for one percent relative error when varying n_x according to $\mathcal{W}$ and κ, with $\sigma = 10$. The number of time steps n_t is set to simulate $[0, T]$ with $T \approx \tau_\kappa$.

$$\frac{d\mathbf{v}}{dt} = \begin{pmatrix} 0 & I \\ c^2 A & \gamma A \end{pmatrix} \mathbf{v} = L\mathbf{v} , \tag{12}$$

with I the identity matrix of size n_x. For the time integration, we use a second order SDIRK2 scheme, integrating up to $T = \tau_\kappa$ with n_t time steps ($t_i := i\Delta t$ with $\Delta t := T/n_t$). We keep a constant CFL number,

$$\sigma := \frac{c\,\Delta t}{\Delta x} = 10 . \tag{13}$$

We define the relative numerical error as

$$\epsilon := \max_{t \in \{0, t_1, \ldots, t_{n_t}\}} \frac{\left\| \mathbf{u}(t) - \mathbf{u}_{\text{theory}}(t) \right\|_2}{\|\mathbf{u}(0)\|_2} , \tag{14}$$

with $\mathbf{u}(t)$ the part of $\mathbf{v}(t)$ containing only u values, and $\mathbf{u}_{\text{theory}}(t)$ the analytic solution from (5) evaluated at the grid points x_j.

Looking at similar problems (e.g. advection-diffusion [9]), one can expect that when we keep a fixed mesh resolution in space (and in time), the error increases with κ and $\mathcal{W}$. Hence, we assume that the minimal value of n_x for which the error ϵ is lower than a given tolerance follows a law of the form

$$n_{x,\min} = C\kappa^{\alpha}\mathcal{W}^{\beta} . \tag{15}$$

We compute $n_{x,\min}$ for different values of κ and $\mathcal{W}$ using a trial and error procedure, and then the parameters C, α and β are determined by least square regression. Setting $\epsilon \leq 0.01$ (i.e., less than 1% error) with $\sigma = 10$, we find for our space time discretization

$$n_{x,\min} \approx 30.5\sqrt{\kappa\mathcal{W}} . \tag{16}$$

In Table 1, we use (16) to give the values (n_x, ϵ) for several combinations of κ and $\mathcal{W}$, which confirms well our empirical law (16). Furthermore, we also indicate the number of time steps n_t required to compute the whole time interval $[0, \tau_\kappa]$ (i.e., the time period during which the mode vibrates, with τ_κ defined in (7)). This shows an important increase in the problem size with $\mathcal{W}$, since generally $n_t \gg n_x$, which motivates time parallelization for such problems.

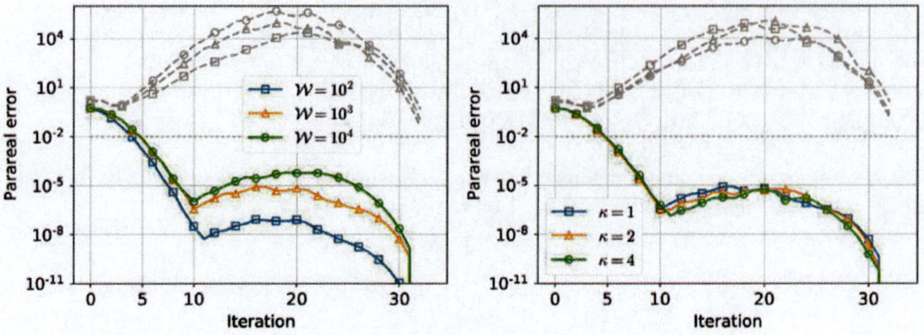

Fig. 3: Convergence of PARAREAL for the wave equation with viscoelastic damping, using $N = 32$ and $m = 8$. Left: $\kappa = 1$, varying $\mathcal{W}$. Right: $\mathcal{W} = 1000$, varying κ.

4 Numerical experiment with PARAREAL

We apply PARAREAL to the wave equation with viscoelastic damping (3), using N processors. We use the same space-time discretization as in Section 3, such that n_x and the time step Δt_F of the fine solver are fully determined by $\mathcal{W}$ and the mode number κ of the initial condition (4) (see Table 1). We denote by m the ratio between the coarse and the fine time step, i.e. $m = \Delta t_G / \Delta t_F$, and we set the number of time-steps per time-interval for the coarse and fine solver (n_F and n_G) such that $n_G \geq 1$ and that the final time of simulation T is close to τ_κ. Finally we compute the error of PARAREAL for each iteration as

$$E^k := \max_{n \in \{1,\ldots,N\}} \frac{\left\|\mathbf{U}_n^k - \mathbf{U}_n^F\right\|}{\|\mathbf{U}_0\|}, \tag{17}$$

where $\mathbf{U}_n^F$ is the solution at the end of each time sub-interval obtained by the fine solver run sequentially.

In Figure 3, we plot the PARAREAL error at each iteration for different values of the parameters κ and $\mathcal{W}$. We observe two convergence dynamics: for the first few iterations, the error decreases super-linearly and rapidly goes below the fine solver accuracy of 1‰, in around five iterations. Then, after about 10 iterations, divergence sets in and a bump forms until the last iteration when PARAREAL must converge to the fine solution after $k = N$ iterations [11]. This bump is due to the amplification of higher frequency modes in the PARAREAL iteration, and if PARAREAL is initialized with a random initial guess, we get the grey dashed convergence (or more divergence) curves in Figure 3, which shows how important the initialization here is and that PARAREAL struggles to generically solve such close to being hyperbolic problems. While this bump is not so much influenced by low initial modes κ, it does increase with $\mathcal{W}$. Especially when N gets large, this turns out to be problematic for larger values of $\mathcal{W}$, i.e., when the problem becomes more hyperbolic, even with a good coarse initial approximation, see Figure 4.

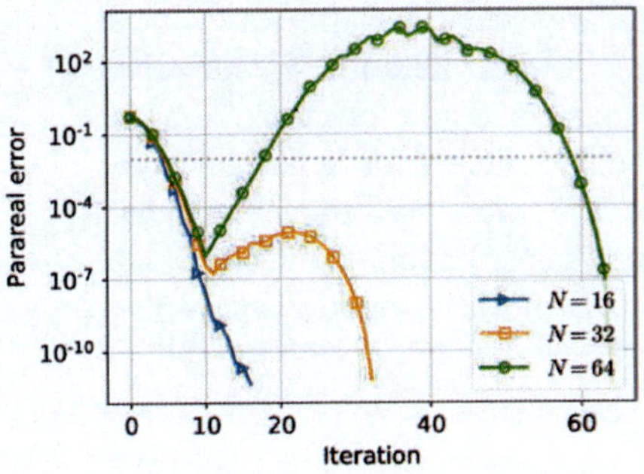

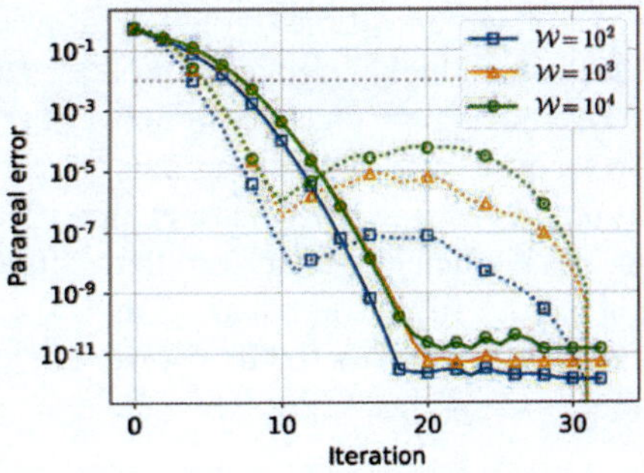

Fig. 4: Influence of N and the coarse solver's phase error on Parareal convergence. Left: $\mathcal{W} = 1000$, $m = 8$, varying N. Right: $N = 32$, $n_F = n_G$, and larger damping parameter for coarse solver, results from Figure 3 (left) in dotted lines and $\mathcal{W}$ is indicated for the fine solver.

A similar bump has been observed in [10] for the advection equation, and it is due to the amplification of high frequency error components in the Parareal iteration [10], because of the hyperbolic nature of advection: the Parareal correction step amplifies these high frequencies, which are present even in a smooth low frequency initial guess due to round-off error. The more processors one uses, the more these high frequency components are amplified, and the higher the bump becomes, even with a smooth low frequency initial guess. A theoretical way to avoid this problem is to impose very high regularity [3]. A more practical way is to reduce the number of processors N in order to limit the amplification induced by the Parareal iterations. For our problem, we show in Figure 4 (left) the impact of reducing N, and as expected, the bump is reduced and even disappears for low values of N. However, this limits the number of processors that can be used, and is thus also less useful in practice for parallel computations.

It has been shown for advection in [15] that removing the phase error of the coarse solver greatly improves convergence of Parareal. In order to simulate a coarse solver almost free of phase error, we consider using the same space-time discretization for both the coarse and fine solver, but with a larger damping parameter for the coarse solver. This is not useful in practice either since it makes the coarse solver as expensive as the fine solver (see [12], which can make using the same grids practical), but gives us further theoretical insight. We set the ratio between the coarse and fine damping parameter (around 5) such that the error between the two is equal to the one obtained when $\Delta t_G = m\,\Delta t_F$ with $m = 8$ (results in Figure 3 (left)). We plot the convergence of Parareal in Figure 4 (right), and see that now the convergence for the first iterations is slightly slower than in Figure 3, but the bump is no longer present in the later iterations.

To conclude, we have shown that under the condition that the fine solver has a sufficient mesh accuracy for the problem considered (determined by κ and $\mathcal{W}$), Parareal with a smooth low frequency initial guess obtained from the coarse solver converges for the first few iterations when applied to low frequency modes, which have the longest vibration time (see (7)). However divergence occurs afterward, due to the amplification of higher frequency modes by the PARAREAL iteration. We have then shown that removing the phase error between the coarse and fine solver can improve PARAREAL convergence. Designing inexpensive coarse solvers for hyperbolic-type problems that do not produce phase error with the fine solver may allow PARAREAL to become more efficient for such problems: for direct constructions using dispersion correction for the advection equation in 1D, see [4], and for a rapid coarse solver based on the same mesh as the fine solver but solved by diagonalization for general hyperbolic problems, see [12].

References

1. E. Aubanel. Scheduling of tasks in the Parareal algorithm. *Parallel Computing* **37**(3), 172–182 (2011).
2. J. Chabassier, A. Chaigne and P. Joly. Modeling and simulation of a grand piano. *The Journal of the Acoustical Society of America* **134**(1), 648–665 (2013).
3. X. Dai and Y. Maday. Stable Parareal in time method for first- and second-order hyperbolic systems. *SIAM Journal on Scientific Computing* **35**(1), A52–A78 (2013).
4. H. De Sterck, R.D. Falgout, S. Friedhoff, O.A. Krzysik and S.P. MacLachlan. Optimizing MGRIT and Parareal coarse-grid operators for linear advection. *arXiv preprint* arXiv:1910.03726 (2019).
5. G. Derveaux. Modélisation numérique de la guitare acoustique. Mathématiques. Ecole Polytechnique X (2002).
6. G. Derveaux, A. Chaigne, P. Joly and E. Bécache. Time-domain simulation of a guitar: Model and method. *The Journal of the Acoustical Society of America* **114**(6), 3368–3383 (2003).
7. M.J. Gander. Analysis of the Parareal algorithm applied to hyperbolic problems using characteristics. *Bol. Soc. Esp. Mat. Apl.* **42**, 21–35 (2008).
8. M.J. Gander. 50 years of time parallel time integration. In: T. Carraro, M. Geiger, S. Körkel, R. Rannacher (eds.) *Multiple Shooting and Time Domain Decomposition Methods*, pp. 69–114, Springer (2015).
9. M.J. Gander and T. Lunet. A Reynolds number dependent convergence estimate for the Parareal algorithm. In: *International Conference on Domain Decomposition Methods*, pp. 277–284, Springer (2018).
10. M.J. Gander and T. Lunet. Toward error estimates for general space-time discretizations of the advection equation. *Computing and Visualization in Science* **23**(1), 1–14 (2020).
11. M.J. Gander and S. Vandewalle. Analysis of the Parareal time-parallel time-integration method. *SIAM J. Sci. Comput.* **29**(2), 556–578 (2007).
12. M.J. Gander and S.L. Wu. A diagonalization-based parareal algorithm for dissipative and wave propagation problems. *SIAM Journal on Numerical Analysis* **58**(5), 2981–3009 (2020).
13. J.L. Lions, Y. Maday and G. Turinici. A "Parareal" in time discretization of PDE's. *C. R. Math. Acad. Sci. Paris* **332**(7), 661–668 (2001).
14. T. Lunet, J. Bodart, S. Gratton and X. Vasseur. Time-parallel simulation of the decay of homogeneous turbulence using parareal with spatial coarsening. *Computing and Visualization in Science* **19**(1), 31–44 (2018).
15. D. Ruprecht. Wave propagation characteristics of Parareal. *Computing and Visualization in Science* **19**(1-2), 1–17 (2018).
16. G.A. Staff and E.M. Rønquist. Stability of the Parareal algorithm. In: R. Kornhuber, et al (eds.) *Domain Decomposition Methods in Science and Engineering*, Lecture Notes in Computational Science and Engineering, vol. 40, pp. 449–456, Springer (2005).

Consistent and Asymptotic-Preserving Finite-Volume Robin Transmission Conditions for Singularly Perturbed Elliptic Equations

Martin J. Gander, Stephan B. Lunowa, and Christian Rohde

1 Introduction

Adaptive Dirichlet–Neumann and Robin–Neumann algorithms for singularly-perturbed advection-diffusion equations were introduced in [2], accounting for transport along characteristics, see also [6] for the discrete setting and damped versions using a modified quadrature rule to recover the hyperbolic limit. Non-overlapping Schwarz DDMs with Robin transmission conditions (TCs) applied to advection-diffusion equations were analyzed in [1, 10] and a stabilized finite-element method for singularly perturbed problems was discussed in [9], see also [3, 4] and references therein for heterogeneous couplings. However, the behavior of these DDMs in the limit of vanishing diffusion has not been addressed.

Our goal is to develop finite volume Robin TCs such that the associated non-overlapping DDM is consistent *and* asymptotic-preserving (AP). Consistent here means that, for fixed mesh size, the discrete DDM iterates converge to the discrete solution on the entire domain, and AP means that the singular limit in the DDM yields a convergent limit DDM (for more on AP, see e.g. [7]). We first show that the continuous DDM is only AP under a strict condition on the Robin transmission parameter, see Theorem 1. In contrast, our new discrete DDM is AP without restriction on this parameter, see Theorem 3, and fast convergence is automatically recovered in the hyperbolic limit. While our analysis is in 1D, we show numerical experiments also in 2D; for the nonlinear space-time case with triangular meshes, see [5].

Martin J. Gander
Université de Genève, Section de Mathématiques, Rue du Conseil-Général 7-9, CP 64, 1205 Genève, Suisse, e-mail: martin.gander@unige.ch

Stephan B. Lunowa
UHasselt – Hasselt University, Computational Mathematics, Agoralaan D, 3590 Diepenbeek, Belgium, e-mail: stephan.lunowa@uhasselt.be, https://orcid.org/0000-0002-5214-7245

Christian Rohde
University of Stuttgart, Institute for Applied Analysis and Numerical Simulation, Pfaffenwaldring 57, 70569 Stuttgart, Germany, e-mail: christian.rohde@mathematik.uni-stuttgart.de

S. C. Brenner et al. (eds.), *Domain Decomposition Methods in Science and Engineering XXVI*, Lecture Notes in Computational Science and Engineering 145,
https://doi.org/10.1007/978-3-030-95025-5_47

2 The continuous problem and non-overlapping DDM

We consider for $\nu \geq 0$, $a > 0$ and $f \in L^2(-1,1)$ the stationary advection-diffusion equation with homogeneous Dirichlet boundary conditions, i.e.,

$$\mathcal{L}(u) := \nu\partial_{xx}u - a\partial_x u = f \text{ in } \Omega := (-1,1), \quad u(-1) = 0\,, \quad \nu u(1) = 0\,. \tag{1}$$

In the singular limit $\nu = 0$, the PDE in (1) becomes (trivially) advective, and the boundary condition collapses into the inflow condition $u(-1) = 0$ only. It is easy to see that there exists a unique weak solution $u \in H^1(-1,1)$ of (1) for $\nu \geq 0$.

We apply a non-overlapping DDM with two sub-domains $\Omega_1 = (-1,0)$ and $\Omega_2 = (0,1)$ to (1). The problem (1) is then rewritten using at $x = 0$ the Robin TCs

$$\mathcal{B}_1(u) = \nu\partial_x u - au + \lambda u\,, \quad \mathcal{B}_2(u) = -\nu\partial_x u + au + \lambda u\,, \quad \lambda > 0\,. \tag{2}$$

Definition 1 (Continuous DDM) Let $u_2^0 \in H^1(\Omega_2)$. For $n \in \mathbb{N}$, the n-th (continuous) DDM-iterate $(u_1^n, u_2^n) \in H^1(\Omega_1) \times H^1(\Omega_2)$ is given as solution of

$$\nu\partial_{xx}u_j^n - a\partial_x u_j^n = f \qquad \text{in} \quad \Omega_j,\ j = 1,2\,, \tag{3}$$

$$u_1^n(-1) = 0\,, \qquad \nu u_2^n(1) = 0\,, \tag{4}$$

$$\nu\mathcal{B}_1(u_1^n) = \nu\mathcal{B}_1(u_2^{n-1})\,, \quad \mathcal{B}_2(u_2^n) = \mathcal{B}_2(u_1^n) \qquad \text{at} \quad x = 0\,. \tag{5}$$

Note that (3)–(5) is equivalent to (1) in the limit $n \to \infty$. In the limit when $\nu \to 0$, we get the stationary advection equation, and the two Robin TCs (5) degenerate into one Dirichlet TC. Note that the multiplication of $\mathcal{B}_1$ by ν is necessary to remove the TC in the limit $\nu \to 0$. The errors $e_j^n := u|_{\Omega_j} - u_j^n$ satisfy (3)–(5) with $f \equiv 0$ due to linearity. Therefore, we have by direct solution

$$\begin{aligned} e_1^n(x) &= A_1^n(\mathrm{e}^{ax/\nu} - \mathrm{e}^{-a/\nu})\,, & e_2^n(x) &= A_2^n(1 - \mathrm{e}^{a(x-1)/\nu}) & &\text{if } \nu > 0\,,\\ e_1^n &\equiv 0\,, & e_2^n &\equiv 0 & &\text{if } \nu = 0\,, \end{aligned}$$

where $A_1^n, A_2^n \in \mathbb{R}$ satisfy the recurrence relations

$$A_1^n = \frac{-a+\lambda(1-\mathrm{e}^{-a/\nu})}{a\mathrm{e}^{-a/\nu}+\lambda(1-\mathrm{e}^{-a/\nu})} A_2^{n-1}\,, \qquad A_2^n = \frac{-a\mathrm{e}^{-a/\nu}+\lambda(1-\mathrm{e}^{-a/\nu})}{a+\lambda(1-\mathrm{e}^{-a/\nu})} A_1^n\,.$$

This yields the following convergence result.

Theorem 1 (Convergence and AP property of the continuous DDM) *The sequence of continuous DDM-iterates $\{(u_1^n, u_2^n)\}_{n\in\mathbb{N}}$ converges pointwise to $(u|_{\Omega_1}, u|_{\Omega_2})$. For $\nu > 0$, the convergence is linear with convergence factor*

$$\rho = \left|\frac{(a-\lambda) + \lambda\mathrm{e}^{-a/\nu}}{(a+\lambda) - \lambda\mathrm{e}^{-a/\nu}}\right| \left|\frac{\lambda - (a+\lambda)\mathrm{e}^{-a/\nu}}{\lambda + (a-\lambda)\mathrm{e}^{-a/\nu}}\right| < 1\,. \tag{6}$$

Convergence in one iteration is achieved iff $\lambda = \frac{a}{1-e^{-a/\nu}}$ *or in the case* $\nu = 0$. *The continuous DDM* (3)–(5) *is AP if* $\lambda = \lambda(\nu)$ *satisfies* $|\lambda - a| = o(1)$ *as* $\nu \to 0$.

3 Cell-centered finite volume discretization

We discretize (1) and (3)–(5) by a cell-centered finite volume method. For given $I \in \mathbb{N}$, let the step-width be $h := 1/I$ and the volumes $V_i := [ih, (i+1)h]$ for $-I \leq i < I$ be given. Furthermore, define $f_i := \int_{V_i} f(x)\,\mathrm{d}x$. We denote the constant, cell-centered approximation of u in V_i by u_i, and encapsulate these for all V_i in the vector $\boldsymbol{u} := (u_i)_{i=-I}^{I-1} \in \mathbb{R}^{2I}$. Using centered differences for the diffusion and upwind fluxes for the advection, the discrete version of problem (1) reads

$$\tfrac{\nu}{h}(u_{i-1} - 2u_i + u_{i+1}) + a(u_{i-1} - u_i) = f_i \quad \text{for } -I < i < I-1, \tag{7}$$

$$\tfrac{\nu}{h}(-3u_{-I} + u_{-I+1}) - 2au_{-I} = f_{-I}\,, \tag{8}$$

$$\tfrac{\nu}{h}(u_{I-2} - 3u_{I-1}) + a(u_{I-2} - u_{I-1}) = f_{I-1}\,. \tag{9}$$

Here, we eliminated the ghost values u_{-I-1} and u_I using a linear interpolation of the boundary conditions. Analogously, one obtains the discrete version of (3) and (4), while (5) becomes

$$B_1(\boldsymbol{u}_1^n) = B_1(\boldsymbol{u}_2^{n-1})\,, \qquad B_2(\boldsymbol{u}_2^n) = B_2(\boldsymbol{u}_1^n)\,. \tag{10}$$

It remains to discretize the TC (2) to obtain B_1, B_2, and then to eliminate the ghost values $u_{1,0}$ and $u_{2,-1}$. For this, we use centered differences for the diffusion and linear combinations of the values in V_{-1} and V_0 for the other terms to obtain

$$B_1(\boldsymbol{u}) = \tfrac{\nu}{h}(u_0 - u_{-1}) - a((1-\alpha_1)u_{-1} + \alpha_1 u_0) + \lambda((1-\beta_1)u_{-1} + \beta_1 u_0)\,, \tag{11}$$

$$B_2(\boldsymbol{u}) = -\tfrac{\nu}{h}(u_0 - u_{-1}) + a((1-\alpha_2)u_{-1} + \alpha_2 u_0) + \lambda((1-\beta_2)u_{-1} + \beta_2 u_0)\,, \tag{12}$$

for some $\alpha_1, \alpha_2, \beta_1, \beta_2 \in [0,1]$. Note that $\alpha_j = \beta_j = 0$, $j = 1,2$, is an upwind discretization, while the centered choice $\alpha_j = \beta_j = 1/2$, $j = 1,2$, is typically used in the diffusion-dominated case $\nu \gg a$ to obtain second-order convergence in h.

To eliminate the ghost values $u_{1,0}$ and $u_{2,-1}$ in (7), we solve (11) for u_0 and (12) for u_{-1}. To eliminate $u_{2,-1}$ in (11) and $u_{1,0}$ in (12), we solve (7) for $u_{1,0}$ and $u_{2,-1}$. Inserting the resulting expressions and using (10), we obtain the following discrete DDM iteration.

Definition 2 (Discrete DDM) For given $\boldsymbol{u}_2^0 \in \mathbb{R}^I$, let $\tilde{B}_1(\boldsymbol{u}_2^0) := \frac{\nu B_1(\boldsymbol{u}_2^0)}{\nu - ah\alpha_1 + \lambda h\beta_1}$. For $n \in \mathbb{N}$, the n-th discrete DDM-iterate $(\boldsymbol{u}_1^n, \boldsymbol{u}_2^n) \in (\mathbb{R}^I)^2$ satisfies

$$\tfrac{\nu}{h}(u_{j,i-1}^n - 2u_{j,i}^n + u_{j,i+1}^n) + a(u_{j,i-1}^n - u_{j,i}^n) = f_i\,, \tag{13}$$

for $j = 1$, $-I < i < -1$ and for $j = 2$, $0 < i < I-1$,

$$\tfrac{\nu}{h}(-3u^n_{1,-I} + u^n_{1,-I+1}) - 2au^n_{1,-I} = f_{-I}\,, \tag{14}$$

$$\tfrac{\nu}{h}(u^n_{2,I-2} - 3u^n_{2,I-1}) + a(u^n_{2,I-2} - u^n_{2,I-1}) = f_{I-1}\,, \tag{15}$$

$$\tfrac{\nu}{h}(u^n_{1,-2} - 2u^n_{1,-1}) + a(u^n_{1,-2} - u^n_{1,-1}) + \tfrac{\nu}{h}c_1u^n_{1,-1} = f_{-1} - \tilde{B}_1(\boldsymbol{u}^{n-1}_2)\,, \tag{16}$$

$$\tfrac{\nu}{h}(-2u^n_{2,0} + u^n_{2,1}) - au^n_{2,0} + (\tfrac{\nu}{h} + a)c_2u^n_{2,0} = f_0 - \tilde{B}_2(\boldsymbol{u}^n_1)\,, \tag{17}$$

where

$$\tilde{B}_1(\boldsymbol{u}^n_2) = \tfrac{\nu}{h}u^n_{2,0} - \tfrac{\nu}{\nu+ah}c_1\big(f_0 - \tfrac{\nu}{h}(-2u^n_{2,0} + u^n_{2,1}) + au^n_{2,0}\big)\,, \tag{18}$$

$$\tilde{B}_2(\boldsymbol{u}^n_1) = (\tfrac{\nu}{h}+a)u^n_{1,-1} - \tfrac{\nu+ah}{\nu}c_2\big(f_{-1} - \tfrac{\nu}{h}(u^n_{1,-2}-2u^n_{1,-1}) - a(u^n_{1,-2}-u^n_{1,-1})\big)\,, \tag{19}$$

$$c_1 = \frac{\frac{\nu}{h}+a(1-\alpha_1)-\lambda(1-\beta_1)}{\frac{\nu}{h}-a\alpha_1+\lambda\beta_1}\,, \qquad c_2 = \frac{\frac{\nu}{h}-a\alpha_2-\lambda\beta_2}{\frac{\nu}{h}+a(1-\alpha_2)+\lambda(1-\beta_2)}\,. \tag{20}$$

Note that (13)–(19) is uniquely solvable for all $\nu \geq 0$ iff $c_1 = O(1/\nu)$ and $c_2 = O(\nu)$ as $\nu \longrightarrow 0$. The resulting system matrix for $\boldsymbol{u}^n_2$ is weakly chained diagonally dominant, and thus non-singular. The same holds for $\boldsymbol{u}^n_1$ if $c_1 \leq 1$. Further note that $\tilde{B}_1$ and $\tilde{B}_2$ in (16)–(19) are discrete Robin-to-Dirichlet operators, so that $c_1 = c_2 = 0$ corresponds to Dirichlet TCs, which do not lead to convergence without overlap.

We next investigate how the coefficients α_j, β_j, $j = 1, 2$, must be chosen to obtain a discrete DDM that is consistent with (7)–(9). Since the discretization (13)–(15) is the same as (7)–(9), consistency follows iff the solution to (16)–(19) in the limit when $n \longrightarrow \infty$ satisfies (7) and vice versa. The solution $\boldsymbol{u}$ of (7)–(9) solves (16)–(19), as can be directly seen when inserting it into (16)–(19) using (7) for $i = -1, 0$. This only requires that νc_1 and c_2/ν are well-defined for all $\nu \geq 0$ and all $\lambda > 0$. On the other hand, combining (16) and (18) as well as (17) and (19) yields

$$\begin{aligned}
&\tfrac{\nu}{h}(u_{1,-2} - 2u_{1,-1} + u_{2,0}) + a(u_{1,-2} - u_{1,-1}) \\
&\qquad = f_{-1} + \tfrac{\nu}{\nu+ah}c_1\big(f_0 - \tfrac{\nu}{h}(u_{1,-1} - 2u_{2,0} + u_{2,1}) - a(u_{1,-1} - u_{2,0})\big)\,, \\
&\tfrac{\nu}{h}(u_{1,-1} - 2u_{2,0} + u_{2,1}) + a(u_{1,-1} - u_{2,0}) \\
&\qquad = f_0 + \tfrac{\nu+ah}{\nu}c_2\big(f_{-1} - \tfrac{\nu}{h}(u_{1,-2} - 2u_{1,-1} + u_{2,0}) - a(u_{1,-2} - u_{1,-1})\big)\,.
\end{aligned}$$

We obtain equivalence with (7) iff $1 \neq c_1c_2$. Hence, we have proved the following theorem which provides choices for the TC parameters $\alpha_1, \alpha_2, \beta_1, \beta_2$ that ensure consistency for all $\lambda > 0$ and $\nu \geq 0$.

Theorem 2 (Consistency of the discrete DDM) *The limit of the discrete DDM iterates (13)–(19) as $n \longrightarrow \infty$ is equal to the solution of (7)–(9) for all $\lambda > 0$ if the following conditions hold:*

(A1) $\alpha_1 < \frac{\nu}{ah}$ *(or equal if $\beta_1 > 0$), and*
(A2) $\nu c_1 = O(1)$ *as $\nu \longrightarrow 0$, i.e. by (A1), $\nu = O(\nu - ah\alpha_1 + \lambda h\beta_1)$, and*
(A3) $c_2 = O(\nu)$ *as $\nu \longrightarrow 0$, i.e., $\alpha_2 + \beta_2 = O(\nu)$, and*
(A4) $c_1c_2 \neq 1$, *i.e.,*

$$0 \neq a^2(\alpha_2 - \alpha_1) + \lambda\big(\tfrac{2\nu}{h} + a(\beta_1 + \beta_2 - \alpha_1 - \alpha_2)\big) + \lambda^2(\beta_1 - \beta_2)\,.$$

Remark 1 Note that the simplest choice of the coefficients which satisfies Theorem 2 is $\alpha_1 = \alpha_2 = \beta_2 = 0$ and $\beta_1 = 1/2$. As shown below, this also yields convergence for any positive discrete Peclet number $\mathrm{Pe} := ah/\nu > 0$. Furthermore, this choice ensures that the discrete DDM is AP as $\nu \to 0$ for any $\lambda > 0$, as we show next.

We split the convergence analysis of the discrete DDM into two regimes due to the different types of solutions: the elliptic case $\nu > 0$ and the singular limit $\nu = 0$. For this, let $\boldsymbol{e}^n := \boldsymbol{u} - (\boldsymbol{u}_1^n, \boldsymbol{u}_2^n)$ be the error of the discrete DDM at iteration n. By linearity, $\boldsymbol{e}^n$ satisfies the discrete DDM (13)–(19) with $\boldsymbol{f} = \boldsymbol{0}$.

The elliptic case $\nu > 0$ **:** Then, (13)–(15) for $\boldsymbol{e}^n$ yield the solution

$$\boldsymbol{e}^n = \left(A_1^n \big(\xi^{(i+1)h} - (1 + \tfrac{\mathrm{Pe}}{2})\xi^{-1}\big)_{i=-I}^{-1},\ A_2^n \big(1 + \tfrac{\mathrm{Pe}}{2} - \xi^{(i+1)h-1}\big)_{i=0}^{I-1} \right),$$

where we defined $\xi := (1 + \mathrm{Pe})^I$. The constants $A_1^n, A_2^n \in \mathbb{R}$ are determined by (16)–(19), which yield the recurrence relations

$$A_1^n = -\frac{\lambda - a + \big(a\alpha_1 - \lambda(\mathrm{Pe}^{-1}+\beta_1)\big)\frac{2\mathrm{Pe}}{2+\mathrm{Pe}}\xi^{-1}}{\big(a\alpha_1 - \lambda(\mathrm{Pe}^{-1}+\beta_1)\big)\frac{2\mathrm{Pe}}{2+\mathrm{Pe}} + (\lambda - a)\xi^{-1}} A_2^{n-1}, \quad A_2^n = \frac{a\alpha_2 + \lambda(\mathrm{Pe}^{-1}+\beta_2) - (\lambda+a)\frac{2+\mathrm{Pe}}{2\mathrm{Pe}}\xi^{-1}}{(\lambda+a)\frac{2+\mathrm{Pe}}{2\mathrm{Pe}} - \big(a\alpha_2 + \lambda(\mathrm{Pe}^{-1}+\beta_2)\big)\xi^{-1}} A_1^n .$$

Therefore, the iteration is linearly convergent iff

$$\rho = \left| \frac{\lambda - a + \big(a\alpha_1 - \lambda(\mathrm{Pe}^{-1}+\beta_1)\big)\frac{2\mathrm{Pe}}{2+\mathrm{Pe}}\xi^{-1}}{\lambda + a - \big(a\alpha_2 + \lambda(\mathrm{Pe}^{-1}+\beta_2)\big)\frac{2\mathrm{Pe}}{2+\mathrm{Pe}}\xi^{-1}} \right| \left| \frac{a\alpha_2 + \lambda(\mathrm{Pe}^{-1}+\beta_2) - (\lambda+a)\frac{2+\mathrm{Pe}}{2\mathrm{Pe}}\xi^{-1}}{a\alpha_1 - \lambda(\mathrm{Pe}^{-1}+\beta_1) + (\lambda-a)\frac{2+\mathrm{Pe}}{2\mathrm{Pe}}\xi^{-1}} \right| < 1 . \quad (21)$$

Note that convergence in one iteration is possible for the choice

$$\lambda = \lambda_{\mathrm{opt}} := \frac{2\nu + ah - 2\alpha_1 ah\xi^{-1}}{2\nu + ah - 2(\nu + \beta_1 ah)\xi^{-1}} a \quad \overset{h\to 0}{\longrightarrow} \quad \frac{a}{1 - \mathrm{e}^{-a/\nu}}, \quad (22)$$

which is almost mesh independent when $\alpha_1 = 0$ and $\beta_1 = 1/2$. This is consistent with the continuous DDM and also yields $\lambda_{\mathrm{opt}} \to a$ as $\nu \to 0$.

Furthermore, note that (21) for $\alpha_1 = \alpha_2 = 0$ and $\beta_1 = \beta_2 = 1/2$ is satisfied for all $\lambda > 0$. But $\beta_2 = 1/2$ does not satisfy (A3) of Theorem 2, so that $\tilde{B}_2$ (and thus ρ) degenerate when $\nu \to 0$. However, choosing $\alpha_1 = \alpha_2 = \beta_2 = 0$ and $\beta_1 = 1/2$, Theorem 2 is satisfied for all $\nu > 0$, and (21) is satisfied for all $\lambda > 0$ due to $\mathrm{Pe} > 0$.

The singular limit $\nu = 0$ **:** Then, (13)–(15) for $\boldsymbol{e}^n$ yields

$$\boldsymbol{e}^n = \left((0)_{i=-I}^{-2}, A_1^n, (A_2^n)_{i=0}^{I-1} \right),$$

with $A_1^n, A_2^n \in \mathbb{R}$ determined by (16)–(19). To obtain $A_1^1 = 0$, i.e., the correct solution in Ω_1, this requires by (16)

$$0 = A_1^1 = \frac{-\tilde{B}_1(\boldsymbol{e}^0)}{\frac{\nu}{h}c_1 - a}, \qquad \tilde{B}_1(\boldsymbol{e}^0) = \frac{\nu B_1(\boldsymbol{e}^0)}{\nu - ah\alpha_1 + \lambda h\beta_1} .$$

Since $\nu c_1 = O(1)$ as $\nu \to 0$ by (A2), this holds iff $\lim_{\nu\to 0} \nu c_1 \neq ah$ and $\lim_{\nu\to 0} \nu/(\nu - ah\alpha_1 + \lambda h\beta_1) = 0$. Using (A1) of Theorem 2, this simplifies to $\nu/\beta_1 = o(1)$ as $\nu \to 0$ and implies $c_1 = o(1)$. For A_2^1, we then obtain by (17)–(19) and (A3) that $A_2^1 = 0$, i.e., convergence in one iteration. Then, $A_1^n = A_2^n = 0$ for all $n > 2$ follows by induction using (16)–(19).

Summarizing the above analysis, we obtain the following result.

Theorem 3 (Convergence and AP property of the discrete DDM) *Let (A1)-(A4) from Theorem 2 be satisfied. The sequence of discrete DDM iterates $\{(\boldsymbol{u}_1^n, \boldsymbol{u}_2^n)\}_{n\in\mathbb{N}}$ from (13)–(19) converges linearly to the solution of (7)–(9) for $\nu > 0$ iff (21) is satisfied.*
Convergence in one iteration is achieved if λ satisfies (22) or for $\nu = 0$ if the limit of the discrete DDM for $\nu/\beta_1 = o(1)$ as $\nu \to 0$ is used.
The discrete DDM (13)–(19) is AP if $|\lambda - a| = o(1)$ or $\nu/\beta_1 = o(1)$ as $\nu \to 0$.

Note that as shown above, the choice $\alpha_1 = \alpha_2 = 0$ and $\beta_1 = \beta_2 = 1/2$ yields linear convergence for $\nu > 0$, but the convergence rate degenerates for $\nu \to 0$. The choice $\alpha_1 = \alpha_2 = \beta_2 = 0$ and $\beta_1 = 1/2$ leads to linear convergence for $\nu > 0$ uniformly in ν with 1-step convergence for $\nu = 0$, and thus is AP.

Remark 2 (Convergence order and mass conservation) As the iterates of the discrete DDM converge to the solution of (7)–(9), which is a first-order convergent finite volume method (uniform in ν and a), the same holds for the discrete DDM at convergence (and before as soon as $\boldsymbol{e}^n = O(h)$). Furthermore, the finite volume method is locally mass conservative, so that mass conservation holds in each subdomain of the discrete DDM. At the interface between the subdomains, mass conservation is ensured at convergence, since the discrete DDM recovers the (implicit) mono-domain finite volume formulation. In contrast, methods based on an explicit splitting at the interface (see e.g. [8, 11]) directly ensure mass conservation, but require the usual time-step restriction of CFL-type when the diffusion vanishes ($\nu \to 0$).

4 Numerical examples

We now study numerically the convergence properties of the discrete DDM as $\nu \to 0$ for various choices of the parameters in the discrete Robin TCs. Since $\alpha_j = O(\nu)$, $j = 1, 2$, is required for convergence, we restrict our study to $\alpha_1 = \alpha_2 = 0$ and vary only β_1, β_2 and λ. We consider (1) for $f(x) = -\nu(k\pi)^2 \sin(k\pi x) - ak\pi\cos(k\pi x)$, which leads to the exact solution $u(x) = \sin(k\pi x)$. We fix $a = 1$, $k = 3$, $B_1(u_2^0) = 1$ and $I = 100$, and study the number of iterations required to reach an error of $\|\boldsymbol{e}^n\|_\infty < 10^{-12}$, see Fig. 1, both for experiments in 1D and 2D. As discussed above, the choice $\beta_1 = \beta_2 = 1/2$ leads to a degeneration as $\nu \to 0$, while the choice $\beta_1 = \beta_2 = \min(1/2, \nu/(ah))$ yields linear convergence, but is only AP for $\lambda \to a$. As predicted by Theorem 3, the convergence improves for all choices such that $\nu/\beta_1 = o(1)$ and $\beta_2 = O(\nu)$ as $\nu \to 0$. In particular, the number of iterations decreases

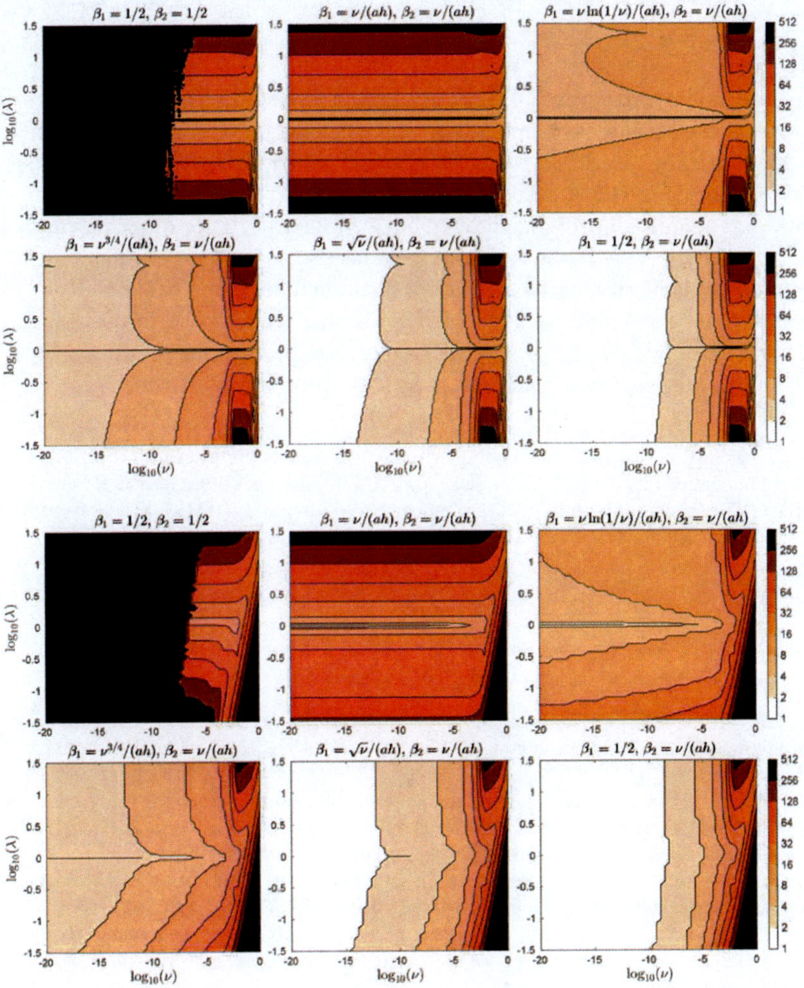

Fig. 1: Number of iterations for various β_1 and β_2 in 1D (top 6 panels) and 2D (bottom 6 panels).

faster when β_1 is large, which illustrates well the convergence factor ρ in (21), which satisfies $\rho = \frac{|\lambda-a|}{\lambda+a} O\left(\frac{\nu}{\nu+\beta_1}\right) + O\left(\nu^{I-1}\right)$. Note that the finite volume method permits a straightforward extension of the discrete DDM to higher dimensions. For our 2D example with equidistant rectangular mesh, the two-point fluxes across the edges on the interface between the subdomains can be constructed exactly as in 1D based on the TCs and ghost values. This leads to the 2D results in Fig. 1 for $\nu\Delta u - \nabla \cdot u = f$ in $(-1, 1) \times (0, 1)$, $u(-1, y) = u(x, 0) = 0$, $\nu u(1, y) = \nu u(x, 1) = 0$ for f chosen such that the exact solution is $u(x, y) = \sin(3\pi x)\sin(3\pi y)$. The technique developed here also works for non-linear time dependent advection-diffusion problems on triangular meshes, see [5].

5 Conclusion

The continuous non-overlapping DDM with Robin TCs applied to singularly-perturbed advection-diffusion problems is AP only when the transmission parameter λ tends to the advection speed as $\nu \to 0$. We showed that a much better result can be obtained for a discrete DDM based on a cell-centered finite volume method: in contrast to the continuous algorithm, a proper, but asymmetric choice of the discrete parameters $(\alpha_j, \beta_j, j = 1, 2)$ in the Robin TCs yields the AP property without any restriction on the transmission parameter λ. We illustrated the theoretical results by numerical examples in one and two spatial dimensions, see also the forthcoming work [5] where we show how the present techniques can be used for robust DDMs for nonlinear advection-diffusion equations in space-time on triangular meshes.

Acknowledgements S.B.L. thanks for the funding by Hasselt University (project BOF17NI01) and by the Research Foundation Flanders (FWO, project G051418N). C.R. thanks the German Research Foundation (DFG) for funding this work (project number 327154368 – SFB 1313). M.J.G. thanks the Swiss National Science Foundation for funding this work (project 200020_192064).

References

1. D. Bennequin, M.J. Gander, L. Gouarin and L. Halpern. Optimized Schwarz waveform relaxation for advection reaction diffusion equations in two dimensions. *Numerische Mathematik* **134**, 513–567 (2016).
2. C. Carlenzoli and A. Quarteroni. Adaptive domain decomposition methods for advection-diffusion problems. In: *Modeling, Mesh Generation, and Adaptive Numerical Methods for Partial Differential Equations*, IMA Vol. Math. Appl., vol. 75, Springer (1995).
3. M.J. Gander, L. Halpern and V. Martin. A new algorithm based on factorization for heterogeneous domain decomposition. *Numerical Algorithms* **73**, 167–195 (2016).
4. M.J. Gander, L. Halpern and V. Martin. Multiscale analysis of heterogeneous domain decomposition methods for time-dependent advection reaction diffusion problems. *J. Comp. Appl. Math.* **344**, 904–924 (2018).
5. M.J. Gander, S.B. Lunowa and C. Rohde. Non-overlapping Schwarz waveform-relaxation for nonlinear advection-diffusion equations (2021). Preprint available at `uhasselt.be/Documents/CMAT/Preprints/2021/UP2103.pdf`.
6. F. Gastaldi, L. Gastaldi and A. Quarteroni. ADN and ARN domain decomposition methods for advection-diffusion equations. In: *Ninth international conference on domain decomposition methods*, pp. 334–341 (1998).
7. S. Jin. Asymptotic preserving (AP) schemes for multiscale kinetic and hyperbolic equations: a review. *Riv. Math. Univ. Parma (N.S.)* **3**, 177–216 (2012).
8. D. Liang and Z. Zhou. The conservative splitting domain decomposition method for multicomponent contamination flows in porous media. *J. Comput. Phys.* **400**, 1–27 (2020).
9. G. Lube, L. Müller and F.C. Otto. A non-overlapping domain decomposition method for the advection-diffusion problem. *Computing* **64**, 49–68 (2000).
10. F. Nataf and F. Rogier. Factorization of the convection-diffusion operator and the Schwarz algorithm. *M3AS* **5**, 67–93 (1995).
11. Z. Zhou and D. Liang. The mass-preserving and modified-upwind splitting DDM scheme for time-dependent convection-diffusion equations. *J. Comput. Appl. Math.* **317**, 247–273 (2017).

Adaptive Schwarz Method for Crouzeix–Raviart Multiscale Problems in 2D

Leszek Marcinkowski*, Talal Rahman, and Ali Khademi

1 Introduction

In modeling real physical phenomena, we quite often see a heterogeneity of coefficients, e.g., in some ground flow problems in heterogeneous media. After applying a discretization method to the differential equations which model our physical phenomenon, e.g., a finite element method, we obtain a discrete problem which is usually very hard to solve by the standard preconditioned iterative methods, like, e.g., preconditioned CG (PCG) or preconditioned GMRES methods. A popular way of constructing parallel preconditioners is to use the Domain Decomposition Methods (DDMs) approach, in particular Schwarz methods, cf. e.g., [14]. In DDMs, it is very important to carefully construct coarse spaces. The overlapping and non-overlapping Schwarz methods were proposed over thirty years ago, and are extensively developed and analyzed, cf. [14] for overviews. The average Schwarz method was proposed in [2], cf. also [1, 6, 10, 12]. It is a non-overlapping Schwarz method with a very simple coarse space. This class of DDMs, along with other 'classical' DDMs constructed in the 1990s and 2000s, are well suited for the problems with coefficients that are constant or slightly varying in subdomains. However, when the coefficients may be highly varying and discontinuous almost everywhere, those 'classical' methods are not efficient. That is why many researchers started to look for new adaptive coarse spaces which are independent or robust for the jumps of the coefficients, i.e., the

Leszek Marcinkowski
Faculty of Mathematics, University of Warsaw, Banacha 2, 02-097 Warszawa, Poland,
e-mail: Leszek.Marcinkowski@mimuw.edu.pl

Talal Rahman and Ali Khademi
Faculty of Engineering and Science, Western Norway University of Applied Sciences, Inndalsveien 28, 5063 Bergen, Norway,
e-mail: Talal.Rahman@hvl.no, Ali.Khademi@hvl.no (akhademi.math@gmail.com)

* This work was partially supported by Polish Scientific Grant: National Science Center: 2016/21/B/ST1/00350.

S. C. Brenner et al. (eds.), *Domain Decomposition Methods in Science and Engineering XXVI*, Lecture Notes in Computational Science and Engineering 145,
https://doi.org/10.1007/978-3-030-95025-5_48

convergence of the constructed DDM is independent of the distribution and the magnitude of the coefficients of the original problem. We refer to [8], [13] and the references therein for similar earlier works on domain decomposition methods that used adaptivity in the construction of the coarse spaces. In recent years there have been many novel works in this direction cf. e.g., [4, 5, 7–9, 11] among others.

In our paper, we consider the nonconforming Crouzeix–Raviart element discretization, also called the nonconforming P_1 element discretization, and then construct an average Schwarz method with an adaptive coarse space. We extend the results from [10] when the conforming P_1 element is considered to the case of the average Schwarz method for CR non-conforming discretization applied to highly heterogeneous coefficients.

2 Discrete problem

Let us consider the following elliptic second order boundary value problem in 2D: Find $u^* \in H_0^1(\Omega)$

$$\int_\Omega \alpha(x)\nabla u^*\nabla v \,\mathrm{d}x = \int_\Omega fv \,\mathrm{d}x, \qquad \forall v \in H_0^1(\Omega), \tag{1}$$

where Ω is a polygonal domain in $\mathbb{R}^2$, $\alpha(x) \geq \alpha_0 > 0$ is a coefficient, α_0 is a positive constant, and $f \in L^2(\Omega)$.

We introduce $\mathcal{T}_h = \{K\}$ as the quasi-uniform triangulation of Ω consisting of open triangles such that $\bar{\Omega} = \bigcup_{K\in\mathcal{T}_h} \overline{K}$. Further, h_K denotes the diameter of K, and $h = \max_{K\in\mathcal{T}_h} h_K$ is the mesh parameter for the triangulation.

Let us consider a coarse non-overlapping partitioning of Ω into open, connected Lipschitz polygonal subdomains Ω_i, called substructures or subdomains, such that $\overline{\Omega} = \bigcup_{i=1}^N \overline{\Omega}_i$.

We also assume that those substructures are aligned with the fine triangulation, i.e., any fine triangle K of $\mathcal{T}_h$ is contained in one substructure. Thus each substructure

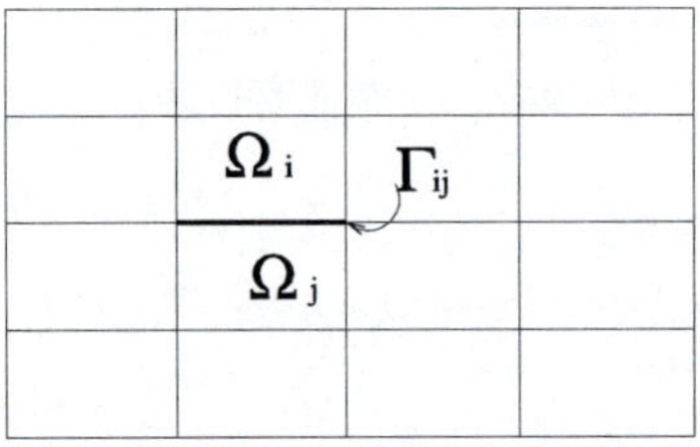

Fig. 1: An example of a coarse partition of Ω, where Γ_{ij} is an interface.

Ω_j has its local triangulation $T_h(\Omega_j)$ of triangles from T_h which are contained in $\overline{\Omega}_j$. For simplicity of presentation, we further assume that these substructures form a coarse triangulation of the domain which is shape regular in the sense of [3] and let $H = \max_j \operatorname{diam}(\Omega_j)$ be its coarse parameter.

We denote by Ω_h^{CR}, $\partial\Omega_h^{CR}$, $\Omega_{i,h}^{CR}$, $\partial\Omega_{i,h}^{CR}$, and $\Gamma_{ij,h}^{CR}$ the sets of midpoints of fine edges of the elements of $\mathcal{T}_h$, contained in Ω, $\partial\Omega$, Ω_i, $\partial\Omega_i$, and Γ_{ij} (the interface between Ω_i and Ω_j, see e.g., Figure 1), respectively. We call those sets the CR (Crouzeix–Raviart) nodal points of the respective sets.

Further, let us define the discrete space $S_h = S_h(\Omega)$ as the standard non-conforming Crouzeix–Raviart linear finite element space defined on the triangulation $\mathcal{T}_h$,

$$S_h(\Omega) := \{u \in L^2(\Omega) : u_{|K} \in P_1,\ K \in \mathcal{T}_h, u - \text{continuous}$$
$$\text{at CR nodal points and } u(x) = 0,\ x \in \partial\Omega_h^{CR}\}.$$

The degrees of freedom of a CR function on a fine triangle K are the values at the midpoints of its edges, cf. Figure 2.

Note that a function in S_h is multivalued on the boundaries of all fine triangles of $\mathcal{T}_h$ except the midpoints of the edges (CR nodal points). Thus $S_h \not\subset H_0^1(\Omega)$ as a space of discontinuous functions. S_h is only a subspace of $L^2(\Omega)$.

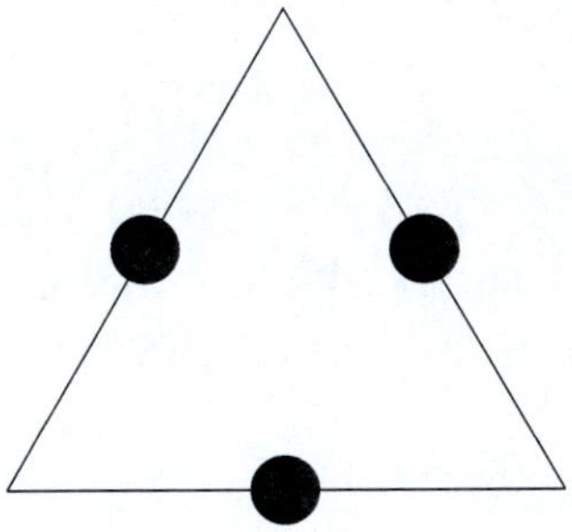

Fig. 2: The CR nodal points, i.e., the degrees of freedom of the Crouzeix–Raviart finite element space on a fine triangle.

We also introduce the local discrete space S_i as the subspace of S_h formed by all functions of S_h which are zeros at all CR nodal points which are NOT in Ω_i^{CR}, or equivalently, formed by functions which are restricted to Ω_i, are zero on $\partial\Omega_{i,h}^{CR}$, and extended by zero elsewhere. Naturally, formally S_i is a subspace of S_h but in practice, it is a local space of functions defined by the values at Ω_i^{CR}.

We consider the following Crouzeix–Raviart discrete problems: We want to find $u_h^* \in S_h$:

$$a_h(u_h^*.v) = f(v) \qquad \forall v \in S_h, \tag{2}$$

where $a_h(u, v) = \sum_{K \in T_h} \int_K \alpha_{|K}(x) \nabla u \nabla v \, dx$ is the so-called broken bilinear form. Note that ∇u_h for $u_h \in S_h$ is a piecewise constant over the fine triangles of T_h. We further assume that α is piecewise constant function over the elements of $\mathcal{T}_h$ since

$\int_K \alpha \nabla u \nabla v \, dx = (\nabla u)_{|K} (\nabla v)_{|K} \int_K \alpha(x) \, dx$. Since the broken form is S_h-elliptic, the discrete problem has a unique solution.

3 Additive Schwarz Method

In this section, we present our non-overlapping average Schwarz method for solving (2). Our method is based on the abstract Additive Schwarz Method framework, cf. e.g., [14].

The space S_h is decomposed into local sub-spaces and a global average Schwarz "spectrally enriched" coarse space. For the local spaces, we take $\{S_i\}_i$. We have that $S_h = \sum_{i=1}^N S_i$.

Coarse space

We introduce our spectrally enriched coarse space in this section.

First, we define the classical average Schwarz coarse space, see e.g. [2]. Let $I_{AS} : S_h \to S_h$ be the linear interpolating operator defined as follows:

$$I_{AS} u(x) = \begin{cases} u(x) & x \in \bigcup_{k=1}^N \partial\Omega_{i,h}^{CR}, \\ \overline{u}_i & x \in \Omega_{i,h}^{CR} \quad i = 1, \ldots, N, \end{cases} \tag{3}$$

where $\overline{u}_i = \frac{1}{M_i} \sum_{x \in \partial\Omega_{i,h}^{CR}} u(x)$ with $M_i = \#\partial\Omega_{i,h}^{CR}$, i.e., $\overline{u}_i$ is the CR discrete average of u over $\partial\Omega_i$. The standard coarse space of the average Schwarz method is the image of this interpolating operator:

$$V_{AS} = I_{AS} S_h. \tag{4}$$

We introduce two types of the local generalized eigenvalue problem, which is to find the eigenvalue and its associated eigenfunction: $(\lambda_i^j, \psi_i^j) \in \mathbb{R}_+ \times S_j$ such that

$$a_h(\psi_i^j, v) = \lambda_i^j b_j^{type}(\psi_i^j, v), \qquad \forall v \in S_j, \quad type \in \{\mathrm{I}, \mathrm{II}\}, \tag{5}$$

where

$$b_j^{type}(u, v) = \begin{cases} \sum_{K \in T_h(\Omega_j)} \int_K \underline{\alpha}_j \nabla u \nabla v \, dx & type = \mathrm{I} \\ \sum_{K \subset \Omega_j^\delta} \int_K \underline{\alpha}_j \nabla u \nabla v \, dx & \\ \quad + \sum_{K \subset \Omega_j \setminus \Omega_j^\delta} \int_K \alpha_j \nabla u \nabla v \, dx & type = \mathrm{II} \end{cases}$$

where $\underline{\alpha}_j := \inf_{x \in \overline{\Omega}_j} \alpha(x)$ and Ω_j^δ is the discrete boundary layer in Ω_j comprising those fine triangles of the local triangulation of Ω_j which have a fine edge on $\partial\Omega_j$.

Naturally, ψ_i^j should be denoted $\psi_i^{j,type}$ as it depends on the type of the RHS form, but, keeping in mind this dependence, we have tried to simplify the notation as much as possible.

Note that it follows from the definition that $a_h(u,u) \geq b_j^{type}(u,u)$ for any $u \in S_j$, thus all eigenvalues $\lambda_i^j \geq 1$ for both types of the form $b_j(\cdot,\cdot)$.

We arrange the eigenvalues in decreasing order as follows

$$\lambda_1^j \geq \lambda_2^j \geq \cdots \geq \lambda_{M_j}^j \geq 1$$

for $M_j = \dim(S_j)$. Next we introduce the local spectral component of the coarse space for all Ω_j and further the enriched coarse space V_0:

$$S_j^{eig} = \operatorname{Span}(\psi_i^j)_{i=1}^{n_j}, \tag{6}$$

where $0 \leq n_j \leq M_j$ is the number of eigenfunctions ψ_i^j selected by a user, e.g. in such a way that the eigenvalue $\lambda_{n_j}^j \geq \lambda$, where $\lambda \geq 1$ is a pre-selected threshold. Finally, the coarse space S_0 is introduced as:

$$S_0 = V_{AS} + \sum_{j=1}^{N} S_j^{eig}. \tag{7}$$

There are two types of this coarse space but the difference is not significant, and below S_0 means one of the described coarse spaces.

Average Schwarz operator T

Next we define the projection operators $T_i : S_h \to S_i$ as

$$a_h(T_i u, v) = a_h(u,v), \quad \forall v \in S_i, \qquad i = 0, \ldots, N. \tag{8}$$

Note that to compute $T_i u$, $i = 1, \ldots, N$, we have to solve N independent local problems.

Let $T := \sum_{i=0}^N T_i$ be the average Schwarz operator. We further replace (2) by the following equivalent problem: Find $u_h^* \in S_h$ such that

$$T u_h^* = g, \tag{9}$$

where $g = \sum_{i=0}^N g_i$ and $g_i = T_i u_h^*$. The functions g_i may be computed without knowing the solution u_h^* of (2), cf. e.g., [14].

The following theoretical estimate of the condition number can be obtained:

Theorem 1 *For all $u \in S_h$, the following holds,*

$$c\left(1+\max_j \lambda^j_{n_j+1}\right)^{-1} \frac{h}{H} a_h(u,u) \le a_h(Tu,u) \le C\, a_h(u,u),$$

where C and c are positive constants independent of the coefficient α, the mesh parameter h and the subdomain size H, and $\lambda^j_{n_j+1}$ is defined in (5) for both types of the coarse space.

The proof is based on the standard abstract ASM Method framework, cf. e.g. [14]. We have to prove three key assumptions, the most technical is the stable splitting assumption, namely we can show that for any $u \in S_h$ there exists $u_j \in S_j$, $j = 0, \dots, N$, such that $\sum_{j=0}^N a_h(u_j,u_j) \le c^{-1}\left(1+\max_j \lambda^j_{n_j+1}\right) a(u,u)$. The other two assumptions are easy to verify. Namely, the stability constant is equal to one since the broken form is used as the local forms. The third assumption is that the bound of the spectral radius of the matrix of constants of the strengthened Cauchy–Schwarz inequalities is also equal to one, since the local subspaces are a_h orthogonal subspaces to each other.

4 Numerical tests

Fig. 3: The location of all jumps in $\alpha(x)$, where $\Omega = [0,1] \times [0,1]$ is partitioned into 5×5 subdomains. The values of jumps on the white and green triangles are 1 and 1.0e4, respectively. To get numerical results, we use these green channels as the periodic patterns for different numbers of subdomains.

In this section, we consider the right-hand side function

$$f(x, y) = 2\pi^2 \sin(\pi x) \sin(\pi y),$$

where $(x, y) \in \Omega = [0, 1] \times [0, 1]$. To confirm the validity of the theoretical result numerically, we also divide all jumps in $\alpha(x)$ into $\alpha_b = 1$ and $\alpha_i = 1.0e4$ corresponding to the coefficients defined on the background and green channels, respectively, cf. Figure 3.

h	$H = 1/3$	$H = 1/6$	$H = 1/9$
1/18	7.2601e6	7.5289e6	1.3895e7
1/36	3.0394e7	2.8114e7	2.8563e7
1/54	7.4566e7	6.2314e7	5.8596e7

Table 1: The condition numbers of the non-preconditioned system for different values of H and h.

h	$H = 1/3$	$H = 1/6$	$H = 1/9$
1/18	57.2051 (47)	33.9312 (42)	20.7272 (38)
1/36	120.7130 (67)	54.8475 (62)	40.6837 (55)
1/54	177.2259 (85)	83.2981 (77)	56.8240 (65)

Table 2: The condition numbers of the additive average Schwarz preconditioner *type* = I, and the number of iterations of preconditioned CG method (in parentheses). Further, the given threshold to construct the enrichment coarse space is 100.

h	$H = 1/3$	$H = 1/6$	$H = 1/9$
1/18	57.2051 (44)	33.9312 (42)	20.7272 (38)
1/36	120.7130 (70)	54.8474 (60)	40.6747 (52)
1/54	177.2259 (93)	83.2881 (74)	56.8249 (67)

Table 3: The condition numbers of the additive average Schwarz preconditioner *type* = II, and the number of iterations of preconditioned CG method (in parentheses). Further, the given threshold to construct the enrichment coarse space is 100.

Table 1 presents the condition number of the non-preconditioned system. To see the efficiency of the enriched additive average Schwarz preconditioners for both types I and II, we refer to Tables 2 and 3. Those tables also present the numbers of iteration of the preconditioned CG method with the tolerance $1e-6$. For different values of H and h, the first observation is that there is a slight difference between the two types of enrichment in terms of the condition numbers and iteration numbers. The second observation is that the ratio of the condition numbers is proportional to the ratio of H/h, for instance, the condition numbers represented by purple color are very close together, where the ratio of H/h is identical. This means that

	type = I			*type* = II		
h	$H = 1/3$	$H = 1/6$	$H = 1/9$	$H = 1/3$	$H = 1/6$	$H = 1/9$
1/18	53	148	165	11	92	147
1/36	188	442	609	19	93	255
1/54	440	947	1317	26	122	298

Table 4: The number of eigenfunctions associated with the eigenvalues greater than 100 used in the construction of the enrichment part of the coarse space, where *type* $\in \{\mathrm{I}, \mathrm{II}\}$, $H \in \{1/3, 1/6, 1/9\}$ and $h \in \{1/18, 1/36, 1/54\}$.

the validity of Theorem 1 is confirmed numerically. Finally, Table 4 includes the number of eigenfunctions used in the construction of the enriched coarse space and shows that the second type of enrichment has a good performance throughout the implementation in comparison to the first type.

References

1. P.E. Bjørstad, M. Dryja and . Rahman. Additive Schwarz methods for elliptic mortar finite element problems. *Numer. Math.* **95**(3), 427–457 (2003).
2. P.E. Bjørstad, M. Dryja and E. Vainikko. Additive Schwarz methods without subdomain overlap and with new coarse spaces. In: *Domain decomposition methods in sciences and engineering (Beijing, 1995)*, pp. 141–157, Wiley, Chichester (1997).
3. S.C. Brenner and L.-Y. Sung. Balancing domain decomposition for nonconforming plate elements. *Numer. Math.* **83**(1), 25–52 (1999).
4. J.G. Calvo and O.B. Widlund. An adaptive choice of primal constraints for BDDC domain decomposition algorithms. *Electron. Trans. Numer. Anal.* **45**, 524–544 (2016).
5. T. Chartier, R.D. Falgout, V.E. Henson, J. Jones, T. Manteuffel, S. McCormick, J. Ruge and P.S. Vassilevski. Spectral AMGe (ρAMGe). *SIAM J. Sci. Comput.* **25**(1), 1–26 (2003).
6. M. Dryja and M. Sarkis. Additive average Schwarz methods for discretization of elliptic problems with highly discontinuous coefficients. *Comput. Methods Appl. Math.* **10**(2), 164–176 (2010).
7. E. Eikeland, L. Marcinkowski and T. Rahman. Overlapping Schwarz methods with adaptive coarse spaces for multiscale problems in 3D. *Numerische Mathematik*, **142**(1), 103–128, April (2019).
8. J. Galvis and Y. Efendiev. Domain decomposition preconditioners for multiscale flows in high-contrast media. *Multiscale Model. Simul.* **8**(4), 1461–1483 (2010).
9. A. Klawonn, P. Radtke and O. Rheinbach. A comparison of adaptive coarse spaces for iterative substructuring in two dimensions. *Electronic Transactions on Numerical Analysis* **45**, 75–106 (2016).
10. L. Marcinkowski and T. Rahman. Additive average Schwarz with adaptive coarse spaces: scalable algorithms for multiscale problems. *Electron. Trans. Numer. Anal.* **49**, 28–40 (2018).
11. F. Nataf, H. Xiang and V. Dolean. A two level domain decomposition preconditioner based on local Dirichlet-to-Neumann maps. *C. r. mathématique* **348**(21-22), 1163–1167 (2010).
12. T. Rahman, X. Xu and R. Hoppe. Additive Schwarz methods for the Crouzeix–Raviart mortar finite element for elliptic problems with discontinuous coefficients. *Numer. Math.* **101**(3), 551–572 (2005).

13. N. Spillane, V. Dolean, P. Hauret, F. Nataf, C. Pechstein and R. Scheichl. Abstract robust coarse spaces for systems of PDEs via generalized eigenproblems in the overlaps. *Numer. Math.* **126**, 741–770 (2014).
14. A. Toselli and O. Widlund. *Domain decomposition methods – algorithms and theory.* Volume 34 of Springer Series in Computational Mathematics, Springer-Verlag, Berlin (2005).

An Overlapping Waveform Relaxation Preconditioner for Economic Optimal Control Problems With State Constraints

Gabriele Ciaramella and Luca Mechelli

1 Introduction

This work is concerned with the numerical solution of so-called economic optimal control problems of the parabolic type. Let $\Omega = (-1,1)$, $T > 0$ and $\mathcal{U} := L^2(0,T;L^2(\Omega))$ endowed with its norm $\|\cdot\|_{\mathcal{U}}$. We want to solve

$$\min_{\mathcal{U}\times\mathcal{U}} \mathcal{J}(u,w) := \frac{1}{2}\|u\|_{\mathcal{U}}^2 + \frac{1}{2}\|w\|_{\mathcal{U}}^2, \tag{1a}$$

subject to the PDE-constraint

$$\begin{aligned} y_t(t,x) - \Delta y(t,x) &= f(t,x) + u(t,x), && \text{in } (0,T)\times\Omega,\\ y(t,-1) = y(t,1) &= 0, && \text{in } (0,T),\\ y(0,x) &= y_\circ(x), && \text{in } \Omega, \end{aligned} \tag{1b}$$

with $y_\circ \in L^2(\Omega)$ and $f \in \mathcal{U}$, and to mixed control-state constraints

$$|u(t,x)| < c_u, \quad |y(t,x) + \varepsilon w(t,x)| \le c_y(t), \quad \text{in } (0,T)\times\Omega, \tag{1c}$$

where $c_u, \varepsilon > 0$ and $c_y \in L^2(0,T)$ with $c_y(t) > 0$ for $t \in (0,T)$. Problem (1) is related to the virtual control approach [6, 8, 9], which is a regularization technique for pointwise state-constrained problems. Under further assumptions on w, in fact, one can show that, as $\varepsilon \to 0$, the solution to (1) converges to the one of the same optimal control problem with (1c) replaced by $|u(t,x)| \le c_u$ and $|y(t,x)| \le c_y(t)$ in $(0,T)\times\Omega$; see, e.g., [8]. Note that there are no weights in front of the control norms in (1a). This is because of the regularization parameter ε, which is also used

Gabriele Ciaramella
Politecnico di Milano, e-mail: gabriele.ciaramella@polimi.it

Luca Mechelli
Universität Konstanz, e-mail: luca.mechelli@uni-konstanz.de

S. C. Brenner et al. (eds.), *Domain Decomposition Methods in Science and Engineering XXVI*, Lecture Notes in Computational Science and Engineering 145,
https://doi.org/10.1007/978-3-030-95025-5_49

to tune the magnitude of the controls u and w. For example, the smaller is ε, the larger is $\|w\|_{\mathcal{U}}$. In contrast to classical optimal control problems, where the goal is to reach a precise target configuration, the focus of (1) is to find minimum-energy feasible controls such that the state solution to (1b) satisfies the bounds (1c). This difference is particularly evident in the cost functional $\mathcal{J}$ in (1a), where only the norm squared of the controls are considered, instead of typical tracking-type terms. For these reasons, problems of the type (1) are called economic optimal control problems. A typical example is the optimal heating and cooling of residual buildings [8]. Note that, for any given $u \in \mathcal{U}$, the state equation (1b) admits a unique (weak) solution $y = y(u) \in W(0,T) := \{\varphi \in L^2(0,T;H^1(\Omega)) | \varphi_t \in L^2(0,T;H^{-1}(\Omega))\}$; see, e.g., [9, 10]. We assume that the admissible set $\mathcal{U}^{\varepsilon}_{\text{ad}}$ has non-empty interior, where $\mathcal{U}^{\varepsilon}_{\text{ad}} := \{(u,w) \in \mathcal{U} \times \mathcal{U} | u \text{ and } y(u) + \varepsilon w \text{ satisfies (1c)}\} \subset \mathcal{U} \times \mathcal{U}$. This guarantees that (1) admits a unique solution $(\bar{u}, \bar{w}) \in \mathcal{U}^{\varepsilon}_{\text{ad}}$ [10]. The first-order necessary and sufficient optimality system [9, 10] of problem (1) is

$$\begin{aligned}
y_t(t,x) - \Delta y(t,x) &= \mathcal{P}(q(t,x)) + f(t,x), && \text{in } (0,T) \times \Omega,\\
y(t,-1) = y(t,1) &= 0, && \text{in } (0,T),\\
y(0,x) &= y_\circ(x), && \text{in } \Omega,\\
q_t(t,x) + \Delta q(t,x) &= Q^{\varepsilon}(y(t,x)), && \text{in } (0,T) \times \Omega,\\
q(t,-1) = q(t,1) &= 0, && \text{in } (0,T),\\
q(T,x) &= 0, && \text{in } \Omega,
\end{aligned} \tag{2}$$

where $Q^{\varepsilon}(y(t,x)) := \frac{1}{\varepsilon^2}(\max\{y(t,x) - c_y(t), 0\} + \min\{y(t,x) + c_y(t), 0\})$ and $\mathcal{P}(q(t,x)) := \max\{-c_u, \min\{c_u, q(t,x)\}\}$, for all $(t,x) \in (0,T) \times \Omega$, with q the so-called adjoint variable. The pair $(\bar{y}, \bar{q})$ is the solution to (2) if and only if $(\bar{u}(t,x), \bar{w}(t,x)) = (\mathcal{P}(\bar{q}(t,x)), -\varepsilon Q^{\varepsilon}(\bar{y}(t,x)))$, for $(t,x) \in (0,T) \times \Omega$, is the optimal solution to (1). System (2) can be rewritten in the form

$$\mathcal{F}(y,q) = 0 \tag{3}$$

and thus solved by using a semismooth Newton method; see, e.g., [5, 9].

As shown in [8], the semismooth Newton method lacks of convergence if the parameter ε is not sufficiently large. This is, however, in contrast with typical applications, where a sufficiently small ε is required [6, 8]. The goal of this paper is to tackle this problem by using a nonlinear preconditioning technique based on an overlapping optimized waveform-relaxation method (WRM) characterized by Robin transmission conditions [2, 3]. To the best of our knowledge, nonlinear preconditioning techniques have never been used for economic control problems. Therefore, this work aims to provide a first concrete study in order to show the applicability of WRM-based nonlinear preconditioners for this class of optimization problems. In particular, our goal is to assess the convergence behavior of the WRM nonlinear preconditioned Newton and its robustness against the regularization parameter ε. Our studies show that appropriate choices of the overlap L and of the Robin parameter p lead to a preconditioned Newton method with a robust convergence with respect to ε. Let us also mention that for elliptic optimal control problems, it is possible to consider different transmission conditions; see, e.g., [1, 4].

The paper is organized as follows. In Section 2, we introduce the WRM and present the algorithm for the proposed preconditioned generalized Newton. In Section 3, we report two numerical experiments that show the convergence behavior of the proposed computational framework in relation of the parameters characterizing problem (1) and the optimized WRM.

2 The waveform-relation and the preconditioned generalized Newton methods

Let Ω be decomposed into two overlapping subdomains $\Omega_1 = (-1, L)$ and $\Omega_2 = (-L, 1)$, where $2L \in (0,1)$ is the size of the overlap. Moreover, let $p > 0$ and consider the operator $\mathcal{R}_j$ defined as $\mathcal{R}_j(y) := y_x + (-1)^{3-j} p y$ for $j = 1, 2$. The WRM consists in iteratively solving, for $n \in \mathbb{N}$, $n \geq 1$, the system

$$\begin{aligned}
y_t^{j,n}(t,x) - \Delta y^{j,n}(t,x) &= \mathcal{P}(q^{j,n}(t,x)) + f(t,x), && \text{in } (0,T)\times\Omega_j, && \text{(4a)}\\
y^{j,n}(t,(-1)^j) &= 0, && \text{in } (0,T), && \text{(4b)}\\
\mathcal{R}_j(y^{j,n})(t,(-1)^{3-j}L) &= \mathcal{R}_j(y^{3-j,n-1})(t,(-1)^{3-j}L), && \text{in } (0,T), && \text{(4c)}\\
y^{j,n}(0,x) &= y_\circ(x), && \text{in } \Omega_j, && \text{(4d)}\\
q_t^{j,n}(t,x) + \Delta q^{j,n}(t,x) &= \mathcal{Q}^\varepsilon(y^{j,n}(t,x)), && \text{in } (0,T)\times\Omega_j, && \text{(4e)}\\
q^{j,n}(t,(-1)^j) &= 0, && \text{in } (0,T), && \text{(4f)}\\
\mathcal{R}_j(q^{j,n})(t,(-1)^{3-j}L) &= \mathcal{R}_j(q^{3-j,n-1})(t,(-1)^{3-j}L), && \text{in } (0,T), && \text{(4g)}\\
q^{j,n}(T,x) &= 0, && \text{in } \Omega_j, && \text{(4h)}
\end{aligned}$$

for $j = 1, 2$. We show first the well-posedness of the method.

Theorem 1 *Let $g_y^1, g_y^2, g_q^1, g_q^2 \in H^{1/4}(0,T)$ be initialization functions for the WRM, i.e., $\mathcal{R}_j(y^{j,1})(t,(-1)^{3-j}L) = g_y^j(t)$ and $\mathcal{R}_j(q^{j,1})(t,(-1)^{3-j}L) = g_q^j(t)$ for $t \in (0,T)$, with compatibility conditions $g_y^j(0) = \mathcal{R}_j(y_\circ)(t,(-1)^{3-j}L)$ and $g_q^j(0) = 0$ for $j = 1, 2$. Then the WRM (4) is well-posed.*

Proof For $j = 1, 2$, we define $H_j^{2,1} := L^2(0,T;H^2(\Omega_j)) \times H^1(0,T;L^2(\Omega_j))$ and $\mathcal{U}_j = L^2(0,T;L^2(\Omega_j))$. For given $g_y^j, g_q^j \in H^{1/4}(0,T)$, system (4) is the optimality system of an optimal control problem, which seeks to minimize $\mathcal{J}_{\text{aux}}(u^j, w^j) = \frac{1}{2}\|u^j\|_{\mathcal{U}_j}^2 + \frac{1}{2}\|w^j\|_{\mathcal{U}_j}^2 + \int_0^T g_q^j(t) y^j(t,(-1)^{3-j}L)\mathrm{d}t$, subject to the state equation (4a)–(4d). These auxiliary optimal control problems admit a unique optimal solution $(\bar{u}^j, \bar{w}^j) \in \mathcal{U}_j \times \mathcal{U}_j$ for $j = 1, 2$ and their optimality systems are uniquely solvable by $(\bar{y}^j, \bar{q}^j) \in H_j^{2,1} \times H_j^{2,1}$ such that

$$(\bar{u}^j(t,x), \bar{w}^j(t,x)) = (\mathcal{P}(\bar{q}^j(t,x)), -\varepsilon\mathcal{Q}^\varepsilon(\bar{y}^j(t,x))), \quad \text{in } (0,T)\times\Omega_j.$$

For more details see [3, 7, 10]. This proves well-posedness of the WRM for $n = 1$ and $j = 1, 2$. By iteratively applying the previous arguments is then easy to show that the WRM is well-posed for $n > 1$, because $y^{j,1}((-1)^j L)$, $y_x^{j,1}((-1)^j L)$, $q^{j,1}((-1)^j L)$, $q_x^{j,1}((-1)^j L) \in L^2(0,T)$. □

Theorem 1 implies that (4) admits a unique solution $(y^{j,n}, p^{j,n}) \in H_j^{2,1} \times H_j^{2,1}$ for $j = 1, 2$ and $n \geq 1$. Note that, at each iteration of the WRM, the solution at iteration n depends on the one at iteration $n-1$. Therefore, we can define the solution mappings $\mathcal{S}_j : H_{3-j}^{2,1} \times H_{3-j}^{2,1} \to H_j^{2,1} \times H_j^{2,1}$ for $j = 1, 2$ as

$$
\begin{aligned}
&(y^1, q^1) = \mathcal{S}_1(y^2, q^2) \text{ solves (4) for } j = 1,\, y^{2,n-1} = y^2 \text{ and } q^{2,n-1} = q^2,\\
&(y^2, q^2) = \mathcal{S}_2(y^1, q^1) \text{ solves (4) for } j = 2,\, y^{1,n-1} = y^1 \text{ and } q^{1,n-1} = q^1,
\end{aligned}
\tag{5}
$$

and the preconditioned form of (3) as

$$
\mathcal{F}_P(y^1, q^1, y^2, q^2) = (\mathcal{F}_1(y^1, q^1, y^2, q^2), \mathcal{F}_2(y^1, q^1, y^2, q^2)) = 0, \tag{6}
$$

where $\mathcal{F}_j(y^1, q^1, y^2, q^2) = (y^j, q^j) - \mathcal{S}_j(y^{3-j}, q^{3-j})$, for $j = 1, 2$. To solve (6), we apply a generalized Newton method. To do so, we assume that the maps $\mathcal{S}_j$, $j = 1, 2$, admit derivative[1] $D\mathcal{S}_j$. This allows us to characterize the derivative $D\mathcal{F}_P$ and its application to a direction $\mathbf{d}^{3-j} = (d_y^{3-j}, d_q^{3-j}) \in H_{3-j}^{2,1} \times H_{3-j}^{2,1}$, which is needed for the generalized Newton method. Let $z^j := (y^j, q^j) \in H_j^{2,1} \times H_j^{2,1}$ for $j = 1, 2$. Thus, we have that $z^j = \mathcal{S}_j(z^{3-j})$, according to the definition of the mapping $\mathcal{S}_j$ in (5). Moreover, we have that $\mathcal{F}_j(\mathcal{S}_j(z^{3-j}), z^{3-j}) = 0$. From this we formally obtain

$$
D_1\mathcal{F}_j(\mathcal{S}_j(z^{3-j}), z^{3-j}) D\mathcal{S}_j(z^{3-j})(\mathbf{d}^{3-j}) + D_2\mathcal{F}_j(\mathcal{S}_j(z^{3-j}), z^{3-j})(\mathbf{d}^{3-j}) = 0,
$$

which leads to $D\mathcal{S}_j(y^{3-j}, q^{3-j})(\mathbf{d}^{3-j}) = (\widetilde{y}^j, \widetilde{q}^j)$ where $(\widetilde{y}^j, \widetilde{q}^j)$ solves

$$
\begin{aligned}
\widetilde{y}_t^j(t,x) - \Delta \widetilde{y}^j(t,x) &= \widetilde{q}^j(t,x)\chi_{\mathcal{I}(q^j)}(t,x), && \text{in } (0,T) \times \Omega_j,\\
\widetilde{y}^j(t, (-1)^j) &= 0, && \text{in } (0,T),\\
\mathcal{R}_j(\widetilde{y}^j)(t, (-1)^{3-j}L) &= \mathcal{R}_j(d_y^{3-j})(t, (-1)^{3-j}L), && \text{in } (0,T),\\
\widetilde{y}^j(0,x) &= 0, && \text{in } \Omega_j,
\end{aligned}
\tag{7a}
$$

$$
\begin{aligned}
\widetilde{q}_t^j(t,x) + \Delta \widetilde{q}^j(t,x) &= \frac{\widetilde{y}^j(t,x)}{\varepsilon^2}\chi_{\mathcal{A}(y^j)}(t,x), && \text{in } (0,T) \times \Omega_j,\\
\widetilde{q}^j(t, (-1)^j) &= 0, && \text{in } (0,T),\\
\mathcal{R}_j(\widetilde{q}^j)(t, (-1)^{3-j}L) &= \mathcal{R}_j(d_q^{3-j})(t, (-1)^{3-j}L), && \text{in } (0,T),\\
\widetilde{q}^{j,n}(T,x) &= 0, && \text{in } \Omega_j,
\end{aligned}
\tag{7b}
$$

[1] Since the functions $\mathcal{S}_j$ are implicit functions of semismooth functions, one cannot directly invoke the implicit function theorem to obtain the desired regularity. Hence, investigating the existence and regularity of $D\mathcal{S}_j$ requires a detailed theoretical analysis, which is beyond the scope of this short manuscript.

for $j = 1, 2$, with $\chi_{\mathcal{I}(q^j)}$ and $\chi_{\mathcal{A}(y^j)}$ the characteristic functions of the sets

$$\mathcal{I}(q^j) := \{(t,x) \in (0,T) \times \Omega_j \big| \left|q^j(t,x)\right| \leq c_u\},$$
$$\mathcal{A}(y^j) := \{(t,x) \in (0,T) \times \Omega_j \big| \left|y^j(t,x)\right| > c_y(t)\}.$$

Note that (7) is a linearization of the WRM subproblems (4). Now, we can resume our preconditioned generalized Newton method in Algorithm 1.

Algorithm 1 WRM-preconditioned generalized Newton method

1: **Data:** Initial guess $y^{j,0}$ and $q^{j,0}$ for $j = 1, 2$, tolerance τ.
2: Perform one WRM step to compute $\mathcal{S}_j(y^{3-j,0}, q^{3-j,0})$;
3: Assemble $\mathcal{F}_P(y^{1,0}, q^{1,0}, y^{2,0}, q^{2,0})$ and set $k = 0$;
4: **while** $\|\mathcal{F}_P(y^{1,k}, q^{1,k}, y^{2,k}, q^{2,k})\| \geq \tau$ **do**
5: Compute $\mathbf{d}^1, \mathbf{d}^2$ solving $D\mathcal{F}_P(y^1, q^1, y^2, q^2)(\mathbf{d}^1, \mathbf{d}^2) = -\mathcal{F}_P(y^1, q^1, y^2, q^2)$ by using a matrix-free Krylov method, e.g., GMRES, and considering that $D\mathcal{F}_P(y^1, q^1, y^2, q^2)(\mathbf{d}^1, \mathbf{d}^2) = (\mathbf{d}^1 - (\widetilde{y}^1, \widetilde{q}^1), \mathbf{d}^2 - (\widetilde{y}^2, \widetilde{q}^2))$, with $(\widetilde{y}^j, \widetilde{q}^j)$ solution to the linearized subproblems (7) for $j = 1, 2$;
6: Update $(y^{j,k+1}, q^{j,k+1}) = (y^{j,k}, q^{j,k}) + \mathbf{d}^j$ and set $k = k + 1$;
7: Perform one WRM step to compute $\mathcal{S}_j(y^{3-j,k}, q^{3-j,k})$;
8: Assemble $\mathcal{F}_P(y^{1,k}, q^{1,k}, y^{2,k}, q^{2,k})$;
9: **end while**

3 Numerical experiments

In this section, we study the behavior of the preconditioned generalized Newton method (Algorithm 1) and its robustness against the Robin parameter p, the regularization ε and the overlap L. It is well known that the convergence of the semismooth Newton method applied to (3) deteriorates fast for decreasing values of ε, since in the case of a pure pointwise state-constrained problem, the adjoint variable q lacks L^2-regularity; cf. [8, 10]. The focus is on understanding if the WRM can be a valid (nonlinear) preconditioner and in which cases. We will perform two numerical experiments. In both tests we discretize the domain Ω with $n_x = 161$ points and we apply a centered finite-difference scheme. Furthermore, we consider $n_t = 21$ time discretization points and apply the implicit Euler method. The initial guesses $y^{j,0}$ and $q^{j,0}$ are chosen randomly but feasible, i.e. such that $\left(\mathcal{P}(q^{j,0}(t,x)), -\varepsilon Q^{\varepsilon}(y^{j,0}(t,x))\right) \in \mathcal{U}^{\varepsilon}_{\text{ad}}$, since we noticed that choosing feasible initial guesses improves the convergence of the method. We set the stopping tolerance $\tau = 10^{-10}$ for the norm of the Newton residual (see Algorithm 1) and the maximum number of outer (inner) iterations to 200 (500). For the first test we choose $T = 1$, $y_\circ(x) = 5\sin(\pi x)$, $f(t,x) = 20$, $c_u = 30$ and $c_y(t) = 10(1-t) + 3$ for all $(t,x) \in (0,1) \times \Omega$. As one can see from Table 1, for a decreasing ε the number of iterations of the semismooth Newton method applied to (3) increases and its convergence deteriorates fast. In contrast, the number of iterations of Algorithm 1 is almost constant as ε varies (when it converges). Choosing

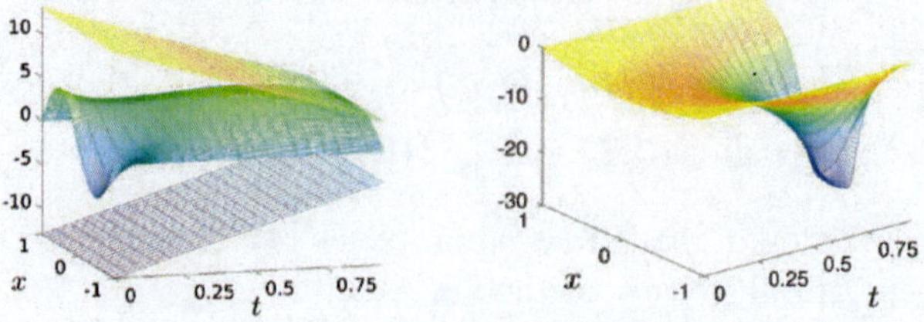

Fig. 1: Test 1: Optimal state with bound c_y (left) and control (right) for $\varepsilon = 5 \times 10^{-4}$.

L	p \ ε	10^{-1}	5×10^{-2}	10^{-2}	5×10^{-3}	10^{-3}	5×10^{-4}
Δx	10^{-6}	4(5–2)	4(6–2)	5(12–2)	6(13–2)	7(35–2)	8(45–2)
Δx	10^{-4}	4(5–2)	4(6–2)	5(13–2)	6(13–2)	7(34–2)	8(45–2)
Δx	10^{-2}	4(6–2)	4(6–2)	5(11–2)	6(13–2)	7(30–2)	8(43–2)
Δx	10^{0}	5(4–2)	5(5–2)	5(9–2)	6(12–2)	max(112–2)	max(123–3)
Δx	10^{2}	6(4–2)	6(5–2)	8(8–2)	9(9–2)	6(22–2)	9(37–2)
Δx	10^{4}	6(5–2)	6(5–2)	9(7–2)	9(10–2)	8(23–2)	max(65–4)
Δx	10^{6}	6(5–2)	6(5–2)	9(7–2)	9(10–2)	max(33–2)	max(92–3)
$2\Delta x$	10^{-6}	4(5–2)	4(7–2)	5(11–2)	6(13–2)	7(39–2)	6(51–2)
$2\Delta x$	10^{-4}	4(5–2)	4(7–2)	5(11–2)	6(13–2)	7(41–2)	6(48–2)
$2\Delta x$	10^{-2}	4(6–2)	4(7–2)	5(12–2)	5(13–2)	7(23–2)	6(54–2)
$2\Delta x$	10^{0}	5(4–2)	5(6–2)	5(9–2)	6(11–2)	7(27–2)	max(107–3)
$2\Delta x$	10^{2}	6(4–2)	6(5–2)	8(8–2)	8(10–2)	8(26–2)	9(37–2)
$2\Delta x$	10^{4}	6(5–2)	6(5–2)	8(8–2)	9(10–2)	8(19–2)	9(41–2)
$2\Delta x$	10^{6}	6(5–2)	6(5–2)	8(8–2)	9(9–2)	8(19–2)	9(41–2)
$4\Delta x$	10^{-6}	4(5–2)	4(7–2)	5(11–2)	6(13–2)	6(30–2)	max(126–6)
$4\Delta x$	10^{-4}	4(5–2)	4(7–2)	5(11–2)	6(13–2)	6(30–2)	max(98–4)
$4\Delta x$	10^{-2}	4(5–2)	4(7–2)	5(12–2)	6(13–2)	6(30–2)	11(124–2)
$4\Delta x$	10^{0}	4(5–2)	4(6–2)	5(9–2)	6(11–2)	6(27–2)	max(152 – 5)
$4\Delta x$	10^{2}	6(4–2)	6(5–2)	8(8–2)	8(10–2)	10(23–2)	15(40 – 2)
$4\Delta x$	10^{4}	6(4–2)	6(5–2)	8(8–2)	8(10–2)	9(26–2)	max(183–3)
$4\Delta x$	10^{6}	6(4–2)	6(5–2)	8(8–2)	8(10–2)	9(26–2)	max(45–2)
Sem. New.		4	5	10	13	30	44

Table 1: Test 1: Number of outer iterations (maximum number - minimum number of inner iterations) for preconditioned generalized Newton varying L, p and ε and number of iterations for the semismooth Newton applied to (3) (last row).

$p = 10^2$ guarantees that the method is convergent for any choice of ε and L. In particular, for small ε, such as 10^{-3} and 5×10^{-4}, the speed-up in terms of number of iterations is also significant. According to Table 1, there are some combinations for which Algorithm 1 reaches a maximum number of iterations (indicated in the tables with max). This issue can be related to the fact that $y^{j,k}$ and $q^{j,k}$ might become unfeasible during Algorithm 1 and when traced to the interface of the other subdomain

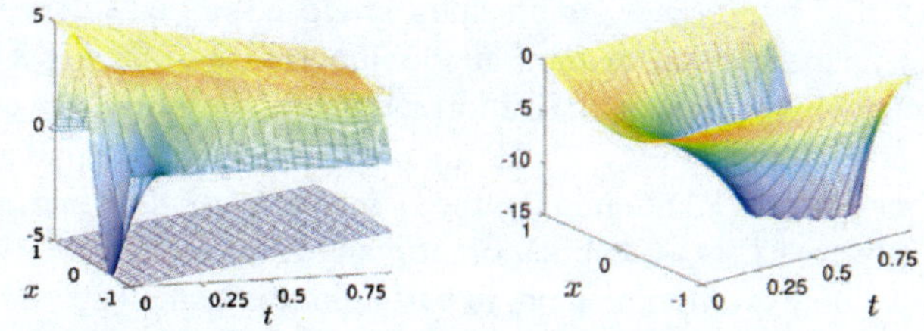

Fig. 2: Test 2: Optimal state with bound c_y (left) and control (right) for $\varepsilon = 5 \times 10^{-4}$.

L	ε / p	10^{-1}	5×10^{-2}	10^{-2}	5×10^{-3}	10^{-3}	5×10^{-4}
Δx	10^{-6}	5(5–2)	6(7–2)	10(10–2)	max(61–2)	max(102–2)	max(297–4)
Δx	10^{-4}	5(5–2)	6(7–2)	10(10–2)	max(32–2)	max(246–2)	max(145–2)
Δx	10^{-2}	5(5–2)	6(7–2)	8(10–2)	max(25–2)	max(max–2)	max(max–4)
Δx	10^{0}	5(5–2)	6(6–2)	6(10–2)	9(11–2)	max(122–4)	max(193–2)
Δx	10^{2}	6(4–2)	7(5–2)	9(8–2)	9(10–2)	9(20–2)	10(25–2)
Δx	10^{4}	6(4–2)	7(5–2)	9(8–2)	9(11–2)	11(20–2)	max(32–2)
Δx	10^{6}	6(4–2)	7(5–2)	9(8–2)	9(11–2)	11(20–2)	max(67–4)
$2\Delta x$	10^{-6}	5(6–2)	6(7–2)	12(11–2)	max(29–2)	max(123–2)	max(206–3)
$2\Delta x$	10^{-4}	5(6–2)	6(7–2)	12(11–2)	max(28–2)	max(91–2)	max(196–3)
$2\Delta x$	10^{-2}	5(6–2)	6(7–2)	11(11–2)	max(25–2)	max(max–4)	max(max–4)
$2\Delta x$	10^{0}	5(5–2)	6(6–2)	6(9–2)	7(10–2)	max(166–5)	max(183–2)
$2\Delta x$	10^{2}	6(4–2)	7(5–2)	8(8–2)	9(11–2)	9(20–2)	10(29–2)
$2\Delta x$	10^{4}	6(4–2)	7(5–2)	9(7–2)	9(11–2)	10(20–2)	9(26–2)
$2\Delta x$	10^{6}	6(4–2)	7(5–2)	9(7–2)	9(11–2)	10(19–2)	10(26–2)
$4\Delta x$	10^{-6}	5(5–2)	6(7–2)	10(11–2)	max(32–2)	max(313–4)	max(187–4)
$4\Delta x$	10^{-4}	5(5–2)	6(7–2)	10(11–2)	max(27–2)	max(145–4)	max(148–4)
$4\Delta x$	10^{-2}	6(5–2)	6(7–2)	9(11–2)	max(35–3)	max(296 – 4)	max(max – 4)
$4\Delta x$	10^{0}	5(5–2)	5(6–2)	6(8–2)	8(11–2)	max(136 – 3)	max(max – 3)
$4\Delta x$	10^{2}	6(4–2)	7(5–2)	6(8–2)	8(11–2)	11(20–2)	14(44–2)
$4\Delta x$	10^{4}	6(4–2)	7(5–2)	8(8–2)	8(11–2)	10(20–2)	12(26–2)
$4\Delta x$	10^{6}	6(4–2)	7(5–2)	8(8–2)	8(11–2)	10(20–2)	13(25–2)
Sem. New.		4	6	10	12	23	30

Table 2: Test 2: Number of outer iterations (maximum number – minimum number of inner iterations) for preconditioned generalized Newton varying L, p and ε and number of iterations for the semismooth Newton applied to (3) (last row).

might cause oscillations. For the second test we choose $T = 1$, $y_\circ(x) = 5\sin(\pi x)$, $f(t,x) = 18$, $c_u = 15$ and $c_y(t) = 2(1-t) + 3$ for $(t,x) \in (0,1) \times \Omega$. In this case, there are more points in the space-time domain for which both bounds become active (cf. Figures 1–2). This makes the problem even more difficult to solve by the WRM, since its nonlinearities are more strongly activated. In Table 2, in fact, the number of cases for which Algorithm 1 does not converge increases with respect to the first numerical experiment, particularly for ε small. We observe that transmission conditions of Dirichlet type and large-enough overlap L guarantee that the number of unfeasible points at the interface is significantly reduced, so that Algorithm 1 converges. This confirms the previous remark on the importance of having feasible iterations. As a rule of thumb, if the regularization ε is small, we suggest to choose a sufficiently large parameter p (e.g., $p \geq 10^2$) so that the Dirichlet part of the transmission conditions of the WRM dominates the Neumann part. Note that, also in the second test, there always exists a combination of p and L for which Algorithm 1 is faster than the semismooth Newton method, in particular for a small ε.

In conclusion, the WRM is a valid preconditioner for solving (3), although there are combinations of p and L for which the method may not converge. As observed, a crucial point for the convergence is to keep the iteration feasible. Preserving such a feasibility, together with other important aspects (e.g., multiple subdomains decomposition and the study of an optimal parameter p) will be the focus of a future work.

References

1. J.-D. Benamou. A domain decomposition method with coupled transmission conditions for the optimal control of systems governed by elliptic partial differential equations. *SIAM J. Numer. Anal.* **33**(6), 2401–2416 (1996).
2. V. Dolean, M.J. Gander, W. Kheriji, F. Kwok and R. Masson. Nonlinear preconditioning: How to use a nonlinear Schwarz method to precondition Newton's method. *SIAM J. Sci. Comput.* **38**(6), A3357–A3380 (2016).
3. M.J. Gander and L. Halpern. Optimized Schwarz waveform relaxation methods for advection reaction diffusion problems. *SIAM J. Numer. Anal.* **45**(2), 666–697 (2007).
4. M. Heinkenschloss and H. Nguyen. Neumann–Neumann domain decomposition preconditioners for linear-quadratic elliptic optimal control problems. *SIAM J. Sci. Comput.* **28**(3), 1001–1028 (2006).
5. M. Hintermüller, K. Ito and K. Kunisch. The primal-dual active set strategy as a semismooth Newton method. *SIAM J. Optim.* **13**(3), 865–888 (2002).
6. K. Krumbiegel and A. Rösch. A virtual control concept for state constrained optimal control problems. *Comput. Optim. Appl.* **43**, 213–233 (2009).
7. J.L. Lions and E. Magenes. *Non-homogeneous boundary value problems and applications (Vol II).* Die Grundlehren der mathematischen Wissenschaften. Springer-Verlag, Berlin, Heidelberg (1972).
8. L. Mechelli. *POD-based state-constrained economic Model Predictive Control of convection-diffusion phenomena.* PhD thesis, University of Konstanz (2019).

9. L. Mechelli and S. Volkwein. POD-based economic optimal control of heat-convection phenomena. In: M. Falcone, R. Ferretti, L. Grüne and W.M. McEneaney, editors, *Numerical Methods for Optimal Control Problems*, pp. 63–87, Springer International Publishing, Cham (2018).
10. F. Tröltzsch. *Optimal Control of Partial Differential Equations: Theory, Methods and Applications*. American Mathematical Society (2010).

Optimized Schwarz Methods With Data-Sparse Transmission Conditions

Martin J. Gander and Michal Outrata

1 Introduction and model problem

Optimized Schwarz methods (OSMs) use optimized transmission operators between subdomains adapted to the equation to be solved to maximize the convergence rate. OSMs have been studied in detail for *localized* transmission operators, see [6] and references therein, which after discretization become structurally sparse, banded matrices. We consider here *non-localized* transmission operators that become after discretization *data-sparse* matrices – a complementary approach to structural sparsity. Our main focus is on how to optimize OSMs within the classes of data-sparse approximations of the Schur complement.

As a model problem we consider the Poisson equation on $\Omega = (-a, a) \times (0, 1) \subset \mathbb{R}^2$,

$$-\Delta u = f \quad \text{in } \Omega, \quad \text{and} \quad u = g \quad \text{on } \partial\Omega, \quad f \text{ and } g \text{ given.} \tag{1}$$

We decompose Ω into two overlapping subdomains $\Omega_1 = (-a, L) \times (0, 1)$ and $\Omega_2 = (-L, a) \times (0, 1)$ with interfaces Γ_1 and Γ_2, overlap O and complements $\Theta_2 := \Omega_1^C$ and $\Theta_1 := \Omega_2^C$, see Figure 1. Creating an equidistant mesh on Ω with mesh-size h, we denote by N_r the number of grid rows and N_c the number of grid columns, and we discretize (1) with a finite difference scheme, obtaining the block tridiagonal system matrix

$$A = \begin{bmatrix} A_{\Theta_1} & A_{\Theta_1,\Gamma_2} & & & \\ A_{\Gamma_2,\Theta_1} & A_{\Gamma_2} & A_{\Gamma_2,O} & & \\ & A_{O,\Gamma_2} & A_O & A_{O,\Gamma_1} & \\ & & A_{\Gamma_1,O} & A_{\Gamma_1} & A_{\Gamma_1,\Theta_2} \\ & & & A_{\Theta_2,\Gamma_1} & A_{\Theta_2} \end{bmatrix}. \tag{2}$$

Martin J. Gander
University of Geneva, e-mail: martin.gander@unige.ch

Michal Outrata
University of Geneva, e-mail: michal.outrata@unige.ch

S. C. Brenner et al. (eds.), *Domain Decomposition Methods in Science and Engineering XXVI*, Lecture Notes in Computational Science and Engineering 145,
https://doi.org/10.1007/978-3-030-95025-5_50

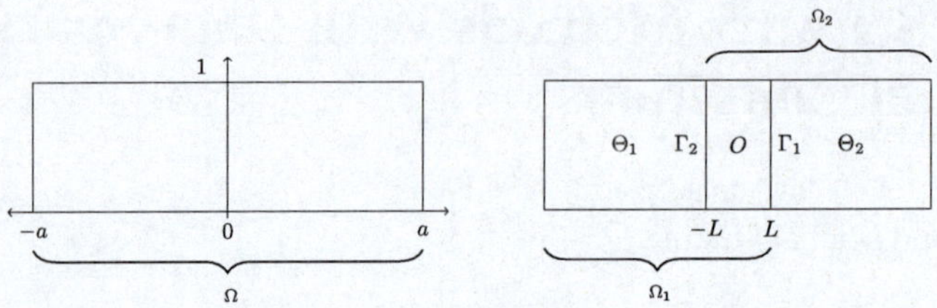

Fig. 1: The physical domain on the left and its subdomains on the right.

2 Parallel optimized Schwarz method

To solve the discretized problem with the *parallel optimized Schwarz method* (POSM), see [6, Section 6.1], and also the equivalent optimized Restricted Additive Schwarz formulations in [5, 10], we form the augmented system matrix (see [10, Section 3.4])

$$A_{\text{aug}} := \begin{bmatrix} A_{\Omega_1} & A_{\Omega_1,\Omega_2} \\ A_{\Omega_2,\Omega_1} & A_{\Omega_2} \end{bmatrix} := \begin{bmatrix} A_{\Theta_1} & A_{\Theta_1,\Gamma_2} & & & & & & \\ A_{\Gamma_2,\Theta_1} & A_{\Gamma_2} & A_{\Gamma_2,O} & & & & & \\ & A_{O,\Gamma_2} & A_O & A_{O,\Gamma_1} & & & & \\ & & A_{\Gamma_1,O} & \tilde{A}_{\Gamma_1} & & & \tilde{A}_{\Gamma_1,\Gamma_1} & A_{\Gamma_1,\Theta_2} \\ A_{\Gamma_2,\Theta_1} & \tilde{A}_{\Gamma_2,\Gamma_2} & & & \tilde{A}_{\Gamma_2} & A_{\Gamma_2,O} & & \\ & & & & A_{O,\Gamma_2} & A_O & A_{O,\Gamma_1} & \\ & & & & & A_{\Gamma_1,O} & A_{\Gamma_1} & A_{\Gamma_1,\Theta_2} \\ & & & & & & A_{\Theta_2,\Gamma_1} & A_{\Theta_2} \end{bmatrix},$$

where the transmission conditions are in the last block row of $[A_{\Omega_1}\ A_{\Omega_1,\Omega_2}]$ and first block row of $[A_{\Omega_2,\Omega_1}\ A_{\Omega_2}]$, which contain

$$\tilde{A}_{\Gamma_1} := A_{\Gamma_1} + S_1,\ \tilde{A}_{\Gamma_1,\Gamma_1} := -S_1 \quad \text{and} \quad \tilde{A}_{\Gamma_2} := A_{\Gamma_2} + S_2,\ \tilde{A}_{\Gamma_2,\Gamma_2} := -S_2.$$

Here S_1 and S_2 are transmission matrices that can be chosen to get fast convergence (classical parallel Schwarz would use $S_1 = S_2 = 0$). POSM for the augmented system has as iteration matrix T the non-overlapping block Jacobi iteration matrix for A_{aug},

$$T = I - \sum_{i=1}^{2} R_{\Omega_i}^T A_{\Omega_i}^{-1} R_{\Omega_i} A_{\text{aug}} \quad \text{with } R_{\Omega_1} = [I\ 0],\ R_{\Omega_2} = [0\ I], \tag{3}$$

where R_{Ω_i} is the discrete restriction operator to the subdomain Ω_i. Setting

$$E_{\Gamma_2}^{\Omega_1} := \left[0_{\Theta_1} I_{\Gamma_2} 0_O 0_{\Gamma_1}\right]^T,\ E_{\Gamma_1}^{\Omega_1} := \left[0_{\Theta_1} 0_{\Gamma_2} 0_O I_{\Gamma_1}\right]^T,\ E_{\Theta_1}^{\Omega_1} := \left[A_{\Gamma_2,\Theta_1} 0_{\Gamma_2} 0_O 0_{\Gamma_1}\right]^T,$$
$$E_{\Gamma_2}^{\Omega_2} := \left[I_{\Gamma_2} 0_O 0_{\Gamma_1} 0_{\Theta_2}\right]^T,\ E_{\Gamma_1}^{\Omega_2} := \left[0_{\Gamma_2} 0_O I_{\Gamma_1} 0_{\Theta_2}\right]^T,\ E_{\Theta_2}^{\Omega_2} := \left[0_{\Gamma_2} 0_O 0_{\Gamma_1} A_{\Gamma_1,\Theta_2}\right]^T,$$

we formulate a convergence result for POSM, analogous to [5, Theorem 3.2].

Theorem 1 [5, Section 3, Lemma 3.1, Theorem 3.2] *The POSM iteration matrix T in (3) has the structure*[1]

$$T = \begin{bmatrix} 0 & K \\ L & 0 \end{bmatrix}, \quad \begin{aligned} K &:= A_{\Omega_1}^{-1} E_{\Gamma_1}^{\Omega_1} \left[I + S_1 (A_{\Omega_1}^{-1})_{\Gamma_1,\Gamma_1} \right]^{-1} \left(S_1 (E_{\Gamma_1}^{\Omega_2})^T - (E_{\Theta_2}^{\Omega_2})^T \right), \\ L &:= A_{\Omega_2}^{-1} E_{\Gamma_2}^{\Omega_2} \left[I + S_2 (A_{\Omega_2}^{-1})_{\Gamma_2,\Gamma_2} \right]^{-1} \left(S_2 (E_{\Gamma_2}^{\Omega_1})^T - (E_{\Theta_1}^{\Omega_1})^T \right). \end{aligned} \tag{4}$$

Moreover, the asymptotic convergence factor $\rho\,(T)$ of POSM satisfies the bound

$$\begin{aligned} \rho\,(T) &\le \sqrt{\|M_1 B_1\|_2 \cdot \|M_2 B_2\|_2}, \\ M_1 &:= \left[I + S_1 (A_{\Omega_1}^{-1})_{\Gamma_1,\Gamma_1} \right]^{-1} \left(S_1 + A_{\Gamma_1,\Theta_2} A_{\Theta_2}^{-1} A_{\Theta_2,\Gamma_1} \right), \; B_1 := (A_{\Omega_2}^{-1})_{\Gamma_1,\Gamma_2}, \\ M_2 &:= \left[I + S_2 (A_{\Omega_2}^{-1})_{\Gamma_2,\Gamma_2} \right]^{-1} \left(S_2 + A_{\Gamma_2,\Theta_1} A_{\Theta_1}^{-1} A_{\Theta_1,\Gamma_2} \right), \; B_2 := (A_{\Omega_1}^{-1})_{\Gamma_2,\Gamma_1}. \end{aligned} \tag{5}$$

Remark 1 Because of the symmetry of the subdomains and the problem we have $S_1 = S_2 =: S$, implying[2] $M_1 = M_2 =: M$ and $B_1 = B_2 =: B$. Notice that both the spectral radius and the norm of T are minimized, becoming identically zero by taking for S the *exact* Schur complement transmission matrix, i.e., $S = S^\star := -A_{\Gamma_1,\Theta_2} A_{\Theta_2}^{-1} A_{\Theta_2,\Gamma_1}$.

3 Data-sparse transmission conditions

The term *data-sparse matrix* refers to a low-rank matrix or a matrix with a low-rank structure in some of its blocks. Taking $\mathcal{D}$ as the set of data-sparse matrices of a particular type, e.g., low-rank, we focus on the minimization problem

$$\min_{S\in\mathcal{D}} \|MB\|_2 = \min_{S\in\mathcal{D}} \left\| \left[I + S (A_{\Omega_1}^{-1})_{\Gamma_1,\Gamma_1} \right]^{-1} \left(S + A_{\Gamma_1,\Theta_2} A_{\Theta_2}^{-1} A_{\Theta_2,\Gamma_1} \right) B \right\|_2 \tag{6}$$

$$\le \min_{S\in\mathcal{D}} \left\| \left[I + S (A_{\Omega_1}^{-1})_{\Gamma_1,\Gamma_1} \right]^{-1} \left(S + A_{\Gamma_1,\Theta_2} A_{\Theta_2}^{-1} A_{\Theta_2,\Gamma_1} \right) B \right\|_F. \tag{7}$$

The Schur complement $S^\star$ makes the second term in the norms, the *numerator* part of M, zero but in general does not lie in $\mathcal{D}$. Minimizing only this term over $\mathcal{D}$ might not suffice, since the *denominator* part of M can also play an important role, as shown for structurally sparse transmission conditions in [5, Lemma 5.1, 5.3], [11, Section 2.5, pp. 80]. We call *NumOpt* minimizing the *numerator* part of MB in norm, and *FracOpt* minimizing the *entire expression*. The *NumOpt* solution is in general given by the truncated SVD of $S^\star$ or its blocks. For (7) this solution is unique (as the singular values of $S^\star$ are distinct), while for (6) we in general don't have uniqueness.

---

[1] The notation $(A_{\Omega_1}^{-1})_{\Gamma_1,\Gamma_1}$ is an abbreviation for $(E_{\Gamma_1}^{\Omega_1})^T A_{\Omega_1}^{-1} E_{\Gamma_1}^{\Omega_1}$. By analogy we also define $(A_{\Omega_1}^{-1})_{\Gamma_1,\Gamma_2}$ and the counterparts for $A_{\Omega_2}^{-1}$.

[2] We also use that both A_{Ω_1} and A_{Ω_2} are symmetric Toeplitz matrices and thus their inverses are symmetric and also *persymmetric*, see [7, Section 4.7].

On the other hand, (6) is a sharper bound on the convergence factor. We choose to work with (7) and comment on the differences where appropriate.

A direct computation shows that all the matrices defining M and B in (5), except possibly S_1 and S_2, can be diagonalized with the *1D discrete Fourier sine basis*

$$W = \left[\mathbf{w}_1, \ldots, \mathbf{w}_{N_r-2}\right] \text{ with } \mathbf{w}_k = \left[\sin\left(\tfrac{k\pi}{N_r-1} j\right)\right]_{j=1}^{N_r-2} \in \mathbb{R}^{N_r-2}. \tag{8}$$

We first take S to be a symmetric, rank r matrix of the form

$$S = \sum_{k=1}^{r} \gamma_k \mathbf{v}_k \mathbf{v}_k^T, \tag{9}$$

where $\gamma \in \mathbb{R}\backslash\{0\}$ and $\mathbf{v}_k \in \mathbb{R}^{N_r-2}$, linearly independent with $\|\mathbf{v}_k\| = 1$. Note that (9) *cannot* capture the diagonal singularity of $S^\star$ well with r small.[3] Taking

$$\mathbf{v}_k = \mathbf{w}_k \quad \text{for } k = 1, \ldots, r, \tag{10}$$

the matrices M, B can be diagonalized by W, and denoting the spectra by

$$\{\alpha_k\}_1^{N_r-2} := \operatorname{sp}\left((A_{\Omega_1}^{-1})_{\Gamma_1,\Gamma_1}\right), \ \{\mu_k\}_1^{N_r-2} := \operatorname{sp}\left(S^\star\right) \quad \text{and} \quad \{\beta_k\}_1^{N_r-2} := \operatorname{sp}\left(B\right),$$

we obtain for the eigenvalues[4] λ_k of MB the formula

$$\lambda_k = \frac{\gamma_k + \mu_k}{1 + \gamma_k \alpha_k} \beta_k. \tag{11}$$

Hence, in this special case, *NumOpt* and *FracOpt* give identical results, namely $\gamma_k = -\mu_k$! This is *atypical* for OSMs, see [5, 11], but is due to the fact that there is no interaction between the choices of γ_k for different k due to orthogonality. We show this in Figure 2.

We note that in this case (7) is overestimating but its minimizer solves (6) as well, i.e., *the method itself is optimized, just the bound is not sharp.*

For arbitrary vectors $\mathbf{v}_1, \ldots, \mathbf{v}_r \in \mathbb{R}^{N_r-2}$ in (9), even a rank one approximation can now interact with all of the eigenmodes, and $\|MB\|$ cannot be easily evaluated: we obtain the formula

$$W^T MBW = \begin{bmatrix} \alpha_1 & & \\ & \ddots & \\ & & \alpha_{N_r-2} \end{bmatrix}^{-1} \left[\begin{bmatrix} \alpha_1 & & \\ & \ddots & \\ & & \alpha_{N_r-2} \end{bmatrix} + \hat{S}\right]^{-1} \left(\hat{S} + \begin{bmatrix} \mu_1 & & \\ & \ddots & \\ & & \mu_{N_r-2} \end{bmatrix}\right) \begin{bmatrix} \beta_1 & & \\ & \ddots & \\ & & \beta_{N_r-2} \end{bmatrix}$$

---

[3] The Schur complement converges to a Green's function when the mesh is refined, and the Green's function for Laplace's equation has a diagonal singularity, see [6, Section 5.3, Remark 16 and below].

[4] Since MB is diagonalized by W, it is symmetric and thus the eigenvalues correspond to the singular values up to a sign.

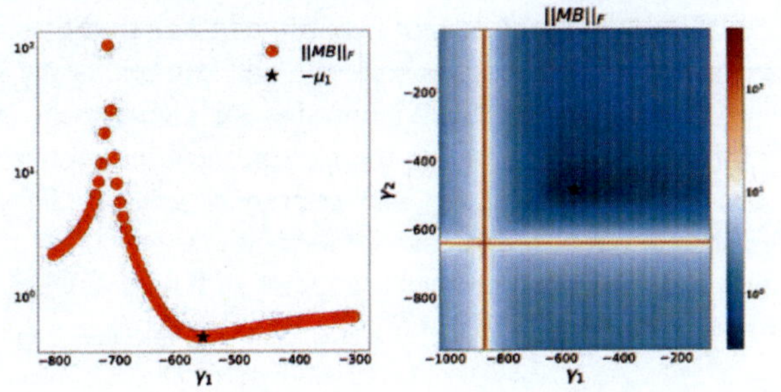

Fig. 2: $\|MB\|_F$ for the parameters γ_i close to the eigenvalues of $S^\star$ for $r = 1$ (left) and for $r = 2$ (right), with *NumOpt* $\gamma_i = -\mu_i$ highlighted by $\star$.

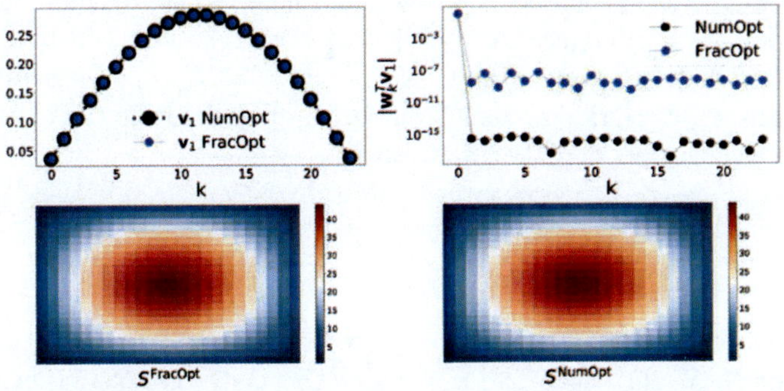

Fig. 3: *NumOpt* and *FracOpt* for $N_r = 26$, $L = h$ and $r = 1$. Top: coordinates of the resulting normalized vector $\mathbf{v}$ in the basis W. Bottom: corresponding transmission matrices S^{NumOpt}, S^{FracOpt}.

with $\hat{S} = W^T SW$, and the denominator and numerator are given as two different diagonal matrices with the same rank-r modification. Recalling [7, Theorem 8.5.3] for $r = 1$, a lengthy but direct calculation gives that the matrices can be diagonalized with the same transformation if and only if (10) holds. Using numerical optimization,[5] extensive experiments showed that *NumOpt* and *FracOpt* lead to the same optimal value *numerically*. For an example, see Figure 3. Again for (6) *the minimizer is not unique* but offers a sharper estimate on the convergence factor, and, in spite of having a worse bound in (7), the actual minimizer also solves the 2-norm problem in (6) and thus only the bound is affected, not the method. These observations remained consistent, changing any meaningful parameters of both the problem and the various optimization routines.

We next investigate hierarchical matrices, which are well suited to approximate the singularity appearing on the diagonal of the Schur complement for elliptic problems [1], and which were proposed for transmission for Helmholtz problems in [3]. Hierarchical matrices were developed to approximate directly in norm, corre-

[5] We used the routine `scipy.optimize()` with the option `shgo` (see [2]) for global optimization and options `Nelder-Mead` and `BFGS` for local optimization.

sponding to *NumOpt*, and in practice this might be sufficient due to the astounding accuracy *and* efficiency of the hierarchical formats, see our example at the end. We study whether *NumOpt* and *FracOpt* are equivalent also for hierarchical matrices. The eigenvalue theory for hierarchical matrices focuses on localization of eigenvalues through iterative processes, see, e.g., [9] and references therein. The explicit computation of $||MB||$ is hence out of reach, and we focus on numerical exploration.

We consider the simplest setting – a one-level hierarchy with the HODLR format[6] – and we assume $N_r = 2n$ for some $n \in \mathbb{N}$. Taking a 2-by-2 blocking of S,

$$S = \begin{bmatrix} S^1 & S_{\text{off-diag}} \\ S^T_{\text{off-diag}} & S^2 \end{bmatrix}, \quad \text{with} \quad S^1, S^2, S_{\text{off-diag}} \in \mathbb{R}^{n\times n},$$

the minimization problem (7) is posed over S with S^1 and S^2 equal to their counterparts in $S^\star$ and with $S_{\text{off-diag}}$ of rank r. As $S^\star$ is persymmetric, so are its off-diagonal blocks, and taking J as the exchange matrix,[7] we observe that $S^\star_{\text{off-diag}}J$ is symmetric and thus permits a symmetric low-rank approximation of the form (9).[8] Letting $\mathbf{q}_1, \ldots, \mathbf{q}_n$ be the eigenvectors of $S^\star$, we first consider

$$S_{\text{off-diag}}J = \sum_{i=1}^{r} \gamma_i \mathbf{q}_i \mathbf{q}_i^T, \tag{12}$$

where $\gamma_i \in \mathbb{R}\backslash\{0\}$. In our numerical experiments now *FracOpt* outperformed *NumOpt* slightly. For $N_r = 26$, $L = h$ and $r = 1$ *FracOpt* converges 6% faster than *NumOpt*.[9] This drops even lower for $r > 1$ (as the off-diagonal blocks are low-rank, e.g., $r = 2$ gives 2%) and seems to be quite stable under mesh refinement ($N_r = 52$ gives 10%, $N_r = 258$ gives 13%). For (6) instead of (7), we observe better bounds but with the same tendencies when changing r or N_r. We show the results for $r = 1, 2$ in Figure 4. Comparing to the global low-rank case, the situation *qualitatively* changed, as there is an improvement going from *NumOpt* to *FracOpt*.

For a more general set of vectors,

$$S_{\text{off-diag}}J = \sum_{i=1}^{r} \gamma_i \mathbf{v}_i \mathbf{v}_i^T, \tag{13}$$

with $\mathbf{v}_1, \ldots, \mathbf{v}_r \in \mathbb{R}^n$ normalized and linearly independent, the *FracOpt* approach gives again a minimizer $S^{\text{FracOpt}}_{\text{off-diag}}$ that is suboptimal in terms of approximating $S^\star_{\text{off-diag}}$, but minimizes $||MB||_F$ better than $S^{\text{NumOpt}}_{\text{off-diag}}$. For $r = 1$ we show a rep-

[6] Standing for *hierarchical off-diagonal low-rank*.

[7] The matrix with ones on the anti-diagonal and zeros elsewhere.

[8] The low-rank approximation of $S_{\text{off-diag}}$ can then be directly reconstructed from the one of $S_{\text{off-diag}}J$ by observing that $J = J^{-1}$.

[9] This refers to the improvement of the bound (5), e.g., *FracOpt* converging 6% faster than *NumOpt* means $\left(||M^{\text{NumOpt}}B||_F\right)^{1.06} = ||M^{\text{FracOpt}}B||_F$.

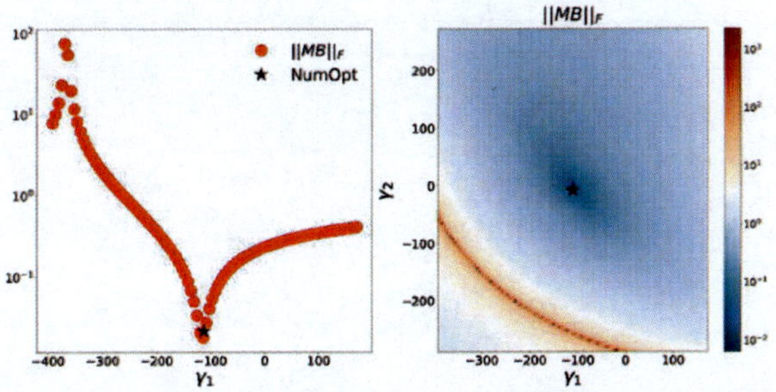

Fig. 4: $\|MB\|$ for the parameters γ_i of the off-diagonal block $S_{\text{off-diag}}J$ close to the eigenvalues of that block for $r = 1$ (left) and for $r = 2$ (right), with the *NumOpt* result for γ_i highlighted by $\star$.

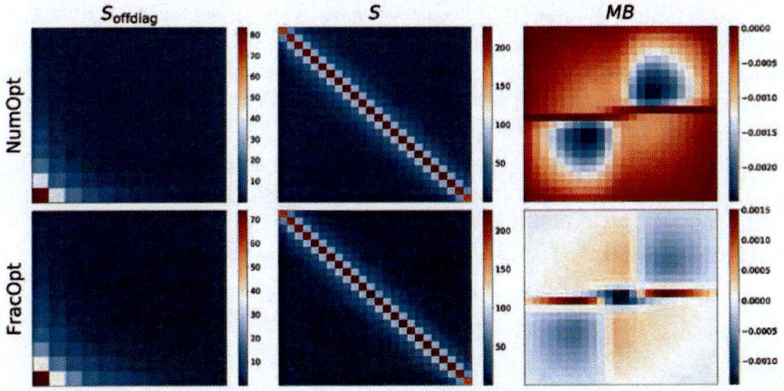

Fig. 5: Comparison of *NumOpt* (top) and *FracOpt* (bottom) for $N_r = 26$, $L = h$ and $r = 1$. We show $S_{\text{off-diag}}$ (left), then S (middle) and then MB (right). Although the difference in the second column seems to be almost negligible, its effect on MB is clearly visible.

resentative example[10] in Figure 5. For $N_r = 26$, $L = h$ and $r = 1$, *FracOpt* converges approximately 25% faster than *NumOpt* – in terms of the bound. This observation is in alignment with the performance, see Figure 6 later on. Taking $r > 1$ again diminishes this improvement (with $r = 2$ we get 13%) but refining h increases the improvement (with $N_r = 52$ we get 43%) – in contrast to the previous setting. We also observed that (6) and (7) now give different minimizers, which are comparable in both the bound and the method performance ((7) is slightly worse). In the context of OSMs, the optimization gains are quite small, see, e.g., [4, Section 3,4]. Thus, we observe that *FracOpt* and *NumOpt* are no longer equivalent for hierarchical formats but they seem to perform comparably for our model problem.

Finally, we show a numerical comparison of the iterative solver performance, including a full hierarchical approximation of $S^\star$ in the formats HODLR and $\mathcal{H}_2$ in Figure 6 (the full formats correspond to *NumOpt*; for more details see [8, Figure 2.1 and 2.3] and references therein). We see that simple low rank approximations

[10] In the sense that mesh refinement only refines these results but does not change their "shape".

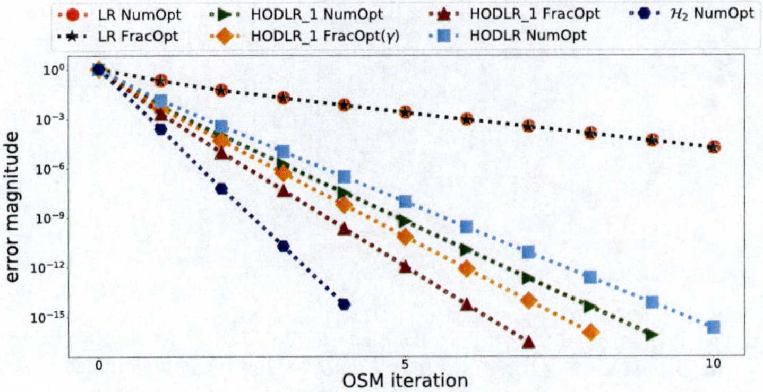

Fig. 6: Convergence of POSM for different choices of S. LR denotes the global low-rank, HODLR_1 the one-level HODLR format and FracOpt(γ), FracOpt denote the two variants (12) and (13) (we omit this for the global low-rank as there is numerically no difference in these). HODLR, $\mathcal{H}_2$ are $\mathcal{H}$-matrix formats corresponding to a binary partitioning with weak and standard admissibility conditions, see [8, Figure 2.1 and 2.3]. We take $N_r = 26$, $L = h$ and $r = 1$ wherever applicable.

of the entire Schur complement cannot perform very well as they miss the diagonal singularity. Hierarchical formats perform well, and follow our theoretical results.

References

1. M. Bebendorf and W. Hackbusch. Existence of H-matrix approximants to the inverse FE-matrix of elliptic operators with L^∞-coefficients. *Numerische Mathematik* **95**(1), 1–28 (2003).
2. S.C. Endres, W.W. Focke and C. Sandrock. A simplicial homology algorithm for Lipschitz optimisation. *Journal of Global Optimization* **72**(2), 181–217 (2018).
3. B. Engquist and L. Ying. Sweeping preconditioner for the Helmholtz equation: Hierarchical matrix representation. *Communications on Pure and Appl. Math.* **64**, 697–735 (2011).
4. M.J. Gander. Optimized Schwarz methods. *SIAM J. on Numer. Anal.* **44**(2), 699–731 (2006).
5. M.J. Gander, S. Loisel and D.B. Szyld. An optimal block iterative method and preconditioner for banded matrices with applications to PDEs on irregular domains. *SIAM Journal on Matrix Analysis and Applications* **33**(2), 653–680 (2012).
6. M.J. Gander and H. Zhang. A class of iterative solvers for the Helmholtz equation: factorizations, sweeping preconditioners, source transfer, single layer potentials, polarized traces, and optimized Schwarz methods. *SIAM Review* **61**(1), 3–76 (2019).
7. G.H. Golub and C.F. van Loan. *Matrix Computations*, Third edn. Johns Hopkins University Press, Baltimore (1996).
8. W. Hackbusch, B.N. Khoromskij and R. Kriemann. Hierarchical matrices based on a weak admissibility criterion. *Computing* **73**(3), 207–243 (2004).
9. T. Mach. *Eigenvalue algorithms for symmetric hierarchical matrices*. Ph.D. thesis, Chemnitz University of Technology (2012).
10. A. St-Cyr, M.J. Gander and S.J. Thomas. Optimized multiplicative, additive, and restricted additive Schwarz preconditioning. *SIAM Journal on Scientific Computing* **29**(6), 2402–2425 (2007).
11. T. Vanzan. *Domain decomposition methods for multiphysics problems*. Ph.D. thesis, University of Geneva (2020).

Space-Time Finite Element Tearing and Interconnecting Domain Decomposition Methods

Douglas R.Q. Pacheco and Olaf Steinbach

1 Introduction

Finite element tearing and interconnecting (FETI) domain decomposition methods [4] are well-established techniques for the parallel solution of elliptic problems. This is mainly due to their simple implementation and the availability of efficient and robust preconditioning strategies. Among other variants to deal with floating subdomains, total FETI [2] or all-floating FETI [8] methods handle all subdomains as *floating*, incorporating also Dirichlet boundary conditions via Lagrange multipliers. This can simplify the implementation, in particular when considering systems of partial differential equations. While the original derivation of the FETI method was based on a constrained minimization problem, related methods can be formulated for the Helmholtz [12] and Maxwell [13] equations as well, using tearing and interconnecting on the discrete level only. Nonetheless, domain decomposition and FETI methods have been so far mainly restricted to elliptic problems, or to time-dependent problems which are discretized through tensor-product ansatz spaces. Parallelization in time is in most cases based on the parareal algorithm [7] combining coarse and fine temporal grids.

In recent years, space-time discretization methods have become very popular, see, e.g., the review article [14] and the references given therein. These methods consider time as just another spatial coordinate, using a finite element discretization in the whole space-time domain [10]. As this allows an adaptive resolution in space and time simultaneously, solving the resulting algebraic system requires efficient solution strategies in parallel. Domain decomposition methods are a natural choice to provide efficient, robust preconditioning and allow parallelization when considering one subdomain per processor.

Douglas R.Q. Pacheco and Olaf Steinbach
Institut für Angewandte Mathematik, TU Graz, Steyrergasse 30, 8010 Graz, Austria,
e-mail: pacheco@math.tugraz.at, o.steinbach@tugraz.at

S. C. Brenner et al. (eds.), *Domain Decomposition Methods in Science and Engineering XXVI*, Lecture Notes in Computational Science and Engineering 145,
https://doi.org/10.1007/978-3-030-95025-5_51

While the work presented in Ref. [11] considers standard domain decomposition methods [1, 5] for the heat equation, the focus of the present contribution is on FETI methods applied to the Stokes system and the heat equation. In Section 2 we describe the space-time finite element discretization of the related model problems. For the solution of the resulting linear systems we present in Section 3 a FETI method, including a discussion on floating subdomains. When considering all subdomains as floating, we end up with an all-floating FETI method. First numerical results in Section 4 indicate the great potential of space-time FETI domain decomposition methods, including parallel-in-time algorithms.

2 Space-time finite element methods

We start with the homogeneous Dirichlet problem for the transient heat equation:

$$\begin{aligned} \partial_t u - \Delta_x u &= f \quad \text{in } Q, \\ u &= 0 \quad \text{on } \Sigma \cup \Sigma_0, \end{aligned} \tag{1}$$

where for a bounded domain $\Omega \subset \mathbb{R}^d$, $d = 1, 2$ or 3, and a finite time horizon T we have the space-time domain $Q := \Omega \times (0,T) \subset \mathbb{R}^{d+1}$ with lateral and bottom boundaries $\Sigma := \partial\Omega \times (0,T)$ and $\Sigma_0 := \Omega \times \{0\}$, respectively. For simplicity, we only consider homogeneous boundary and initial conditions, but inhomogeneous data and other types of boundary conditions can be handled as well. The space-time variational formulation of (1) reads to find $u \in X := L^2(0,T; H^1_0(\Omega)) \cap H^1_{0,}(0,T; H^{-1}(\Omega))$ such that

$$\int_0^T \int_\Omega \Big[\partial_t u v + \nabla_x u \cdot \nabla_x v\Big] \, dx \, dt = \int_0^T \int_\Omega f \, v \, dx \, dt \tag{2}$$

is satisfied for all $v \in Y := L^2(0,T; H^1_0(\Omega))$. Note that the ansatz space X covers zero boundary and initial conditions. For a space-time finite element discretization of (2), we introduce conforming finite element spaces $X_h \subset X$ and $Y_h \subset Y$, assuming $X_h \subset Y_h$. In particular, we use the finite element spaces $X_h = Y_h$ of continuous, piecewise linear basis functions, defined with respect to some admissible decomposition of the space-time domain Q into shape-regular simplicial finite elements. Detailed stability and error analysis of this space-time finite element method can be found in Refs. [10, 11]. The space-time finite element discretization of (2) results in a large linear system of algebraic equations which we shall solve using an appropriate tearing and interconnecting domain decomposition method.

As a second model problem, we consider the time-dependent Stokes system

$$\begin{aligned} \partial_t u - \mu \Delta_x u + \nabla_x p &= f \quad \text{in } Q, \\ \nabla_x \cdot u &= 0 \quad \text{in } Q, \\ u &= 0 \quad \text{on } \Sigma \cup \Sigma_0, \end{aligned} \tag{3}$$

once again assuming homogeneous boundary and initial conditions, for simplicity. The variational formulation of (3) seeks $u \in X^d$ and $p \in L^2(Q)$ such that

$$\int_0^T \int_\Omega \Big[\partial_t u \cdot v + \mu \nabla_x u : \nabla_x v - p \nabla_x \cdot v\Big] \mathrm{d}x\,\mathrm{d}t = \int_0^T \int_\Omega f \cdot v \,\mathrm{d}x\,\mathrm{d}t, \tag{4}$$

$$\int_0^T \int_\Omega q \nabla_x \cdot u \,\mathrm{d}x\,\mathrm{d}t + \int_0^T \left(\int_\Omega p \,\mathrm{d}x \int_\Omega q \,\mathrm{d}x \right) \mathrm{d}t = 0 \tag{5}$$

is satisfied for all $v \in Y^d$ and $q \in L^2(Q)$. Note that the additional term in (5) ensures the scaling condition $p \in L^2_0(\Omega)$ for all $t \in (0, T)$. The space-time variational formulation (4)–(5) can be analyzed similarly to what was done in Ref. [10] in the case of the heat equation, extending to the space-time setting the spatial inf-sup stability condition for the divergence. Note that inhomogeneous essential boundary and initial conditions g and u_0 can be handled through homogenization by using suitable extensions of such data into the space-time domain. For the space-time finite element discretization of (4) and (5) we use inf-sup stable pairs to approximate u_h and p_h. In particular, we extend the well established Taylor–Hood elements to the space-time setting using simplicial finite elements. As an alternative we may also use prismatic space-time Taylor–Hood elements, see Ref. [9] for first numerical results. A more detailed stability and error analysis will be published elsewhere.

3 Tearing and interconnecting domain decomposition methods

The space-time finite element discretization of the heat equation (1) and of the Stokes system (3) results in very large systems of algebraic equations which must be solved in parallel and, if possible, simultaneously in space and time. One possible approach is to use space-time finite element tearing and interconnecting methods, which are well established for elliptic problems. Here we generalize this approach to parabolic time-dependent problems. The space-time domain $Q = \Omega \times (0, T)$ is decomposed into s non-overlapping space-time subdomains Q_i which can be rather general, see Fig. 1 for a selection of possible simple decompositions. With respect to this space-time domain decomposition we consider the localized problems, where the continuity of the primal unknowns along the interface is enforced via discrete Lagrange multipliers. This results in the global linear system

$$\begin{pmatrix} K_1 & & & B_1^\top \\ & \ddots & & \vdots \\ & & K_s & B_s^\top \\ B_1 & \cdots & B_s & \end{pmatrix} \begin{pmatrix} \underline{u}_1 \\ \vdots \\ \underline{u}_s \\ \underline{\lambda} \end{pmatrix} = \begin{pmatrix} \underline{f}_1 \\ \vdots \\ \underline{f}_s \\ \underline{0} \end{pmatrix}, \tag{6}$$

where the K_i are the local space-time finite element stiffness matrices and the B_i are Boolean matrices. While (6) corresponds directly to the heat equation (1), it formally also includes the Stokes problem (3) with all quantities defined accordingly. Although we have chosen to enforce the interface continuity of the pressure field, this is in principle not necessary since the variational problem allows $p \in L^2(Q)$.

At this time, we assume that all local matrices K_i are invertible, so that when using direct solvers locally we end up with the Schur complement system

$$\sum_{i=1}^{s} B_i K_i^{-1} B_i^\top \underline{\lambda} = \sum_{i=1}^{s} B_i K_i^{-1} \underline{f}_i \,. \tag{7}$$

The heat equation can be seen as a diffusion equation with convection in the temporal direction. Since there is no difference between the spatial and temporal mesh size h, we conclude a spectral condition number of $O(h^{-2})$ for (6), and of $O(h^{-1})$ for the Schur complement system (7). The global linear system (7) is solved here by a GMRES method, either without preconditioning or with a simple diagonal preconditioner. More advanced preconditioning strategies also including some coarse grid contributions seem to be mandatory for more complex problems, being a topic of further research.

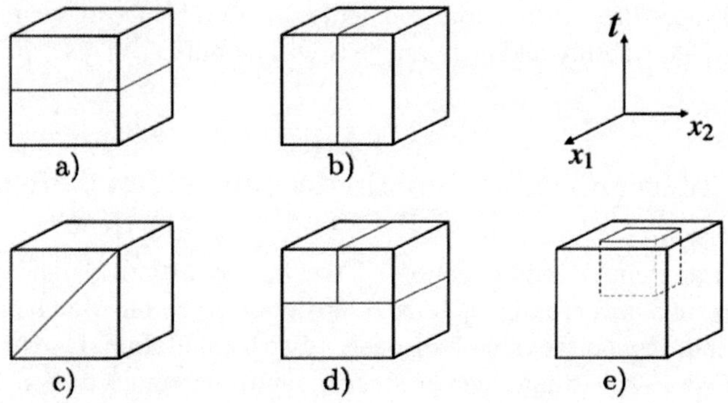

Fig. 1: Different decompositions for the space-time domain $Q = \Omega \times (0, T) \subset \mathbb{R}^3$.

In what follows, we discuss the more general situation in which a local matrix K_i is not invertible, i.e., when the subdomain Q_i is *floating*. Using a pseudo-inverse K_i^+ of K_i, we can describe the solutions of the local subproblems as

$$\underline{u}_i = K_i^+ (\underline{f}_i - B_i^\top \underline{\lambda}) + R_i \underline{\alpha}_i, \tag{8}$$

where the local matrices R_i describe the kernels $\mathcal{N}(K_i)$ of K_i, and $\underline{\alpha}_i$ are coefficients to be determined. The application of the pseudo-inverse K_i^+ also requires the solvability condition $\underline{f}_i - B_i^\top \underline{\lambda} \in \mathcal{R}(K_i)$, which is equivalent to

$$\widetilde{R}_i^\top (\underline{f}_i - B_i^\top \underline{\lambda}) = \underline{0},$$

where the local matrices $\widetilde{R}_i$ describe the kernels $\mathcal{N}(K_i^\top)$. In the case of floating subdomains we therefore end up with the Schur complement system

$$\begin{pmatrix} S & -G \\ \widetilde{G}^\top & 0 \end{pmatrix} \begin{pmatrix} \underline{\lambda} \\ \underline{\alpha} \end{pmatrix} = \begin{pmatrix} \underline{d} \\ \underline{e} \end{pmatrix}, \tag{9}$$

where

$$S = \sum_{i=1}^{s} B_i K_i^+ B_i^\top, \quad G = \Big(B_1 R_1, \cdots, B_s R_s\Big), \quad \widetilde{G} = \Big(B_1 \widetilde{R}_1, \cdots, B_s \widetilde{R}_s\Big),$$

$$\underline{d} = \sum_{i=1}^{s} B_i K_i^+ \underline{f}_i, \quad \underline{e} = \begin{pmatrix} \widetilde{R}_1^\top \underline{f}_1 \\ \vdots \\ \widetilde{R}_s^\top \underline{f}_s \end{pmatrix}.$$

Similarly as in FETI methods for elliptic problems, we introduce a projection

$$P := I - G(\widetilde{G}^\top G)^{-1} \widetilde{G}^\top,$$

and it remains to solve the constrained linear system

$$PS\underline{\lambda} = P\underline{d}, \quad \widetilde{G}^\top \underline{\lambda} = \underline{e}, \tag{10}$$

which can be done via a GMRES method [6]. Afterwards we can compute

$$\underline{\alpha} = (\widetilde{G}^\top G)^{-1} \widetilde{G}^\top (S\underline{\lambda} - \underline{d}).$$

Notice that the square matrix $\widetilde{G}^\top G$ is small, since it does not depend on the finite element mesh but only on the number s of subdomains. In fact, its dimension is simply s for the heat equation, or sd for the Stokes problem. Therefore, the inverse $(\widetilde{G}^\top G)^{-1}$ can be computed directly and works as a coarse-grid solver.

It remains to characterize the kernels $\mathcal{N}(K_i)$ and $\mathcal{N}(K_i^\top)$ of the local stiffness matrices K_i and their transposed matrices, respectively. For this we consider the heat equation in $Q_i = \Omega_i \times (t_{i-1}, t_i)$, where K_i corresponds to the space time discretization with zero Neumann boundary conditions and *without* initial or terminal conditions at t_{i-1} or t_i, respectively. In the continuous case, the solution in Q_i is given by

$$u_i(x,t) = \sum_{k=0}^{\infty} u_{i,k} e^{-\lambda_{i,k} t} v_{i,k}(x) \quad \text{for } (x,t) \in Q_i, \tag{11}$$

where $v_{i,k}$ are the eigenfunctions of the Neumann eigenvalue problem for the spatial Laplacian in Ω_i, with eigenvalues $\lambda_{i,k} \geq 0$. For the space-time finite element discretization we use continuous, piecewise linear basis functions as a partition of unity in Q_i, i.e., $v_{i,0} \in X_{h|Q_i}$ for $\lambda_{i,0} = 0$. Due to the exponential decay in the solution (11) for $k \geq 1$, no more eigenfunctions are represented in the local finite element space $X_{h|Q_i}$, and hence we conclude $\mathcal{N}(K_i) = \{\underline{1}\}$ in the case of the heat equation

(1). Similarly, for the Stokes problem (3) we have d constant eigenfunctions for the velocity, and additionally null pressure [15]. In both cases, the constant eigenfunctions remain true for general space-time subdomains Q_i. While the kernel $\mathcal{N}(K_i)$ is trivially constructed, the basis for $\mathcal{N}(K_i^\top)$ is in general mesh-dependent. Such bases are however easily obtained as subproducts of numerical techniques for computing pseudo-inverses K_i^+, see Ref. [3].

To simplify the implementation and to include all subdomains in the coarse-grid matrix $\widetilde{G}^\top G$, we may consider all subdomains as floating, incorporating Dirichlet boundary conditions via Lagrange multipliers as well. This results in the all-floating [8] or total [2] FETI approach.

4 Numerical results

As a first numerical example we consider the Stokes system (3) in the spatial domain $\Omega = (0,1)^2$ for $T = 1$, i.e., $Q = (0,1)^3$. To check the expected order of convergence we consider for $\mu = 1$ the manufactured solution

$$\begin{aligned} u_1(x,t) &= 2(1-\mathrm{e}^{-t})(x_2 - 3x_2^2 + 2x_2^3)[x_1(1-x_1)]^2, \\ u_2(x,t) &= 2(1-\mathrm{e}^{-t})(3x_1^2 - x_1 - 2x_1^3)[x_2(1-x_2)]^2, \\ p(x,t) &= (1 + x_1 - \mathrm{e}^{-x_1 x_2 t})t^2, \end{aligned}$$

with the right-hand side f computed accordingly. In this first example we consider decompositions of the space-time domain Q into only a few subdomains, see Fig. 1. Our particular interest is in the effect of the interface orientation on the number of required GMRES iterations to reach a given relative accuracy of $\varepsilon = 10^{-6}$, see also the discussion in Ref. [11] in the case of a standard domain decomposition approach for the heat equation. We solve the global Schur complement system without any preconditioning (I), or with a simple diagonal preconditioner (D). In all cases we observe a significant reduction in the number of iterations, with the best results appearing when considering a decomposition in time (a) or space (b) only, and for the diagonal decomposition (c). The results are not as good when considering the decomposition (d) and the inclusion (e). In general, some coarse-grid preconditioner should be used to further reduce the number of iterations.

In the second example we have the heat equation (1) in the spatially one-dimensional domain $\Omega = (0,1)$ and with the final time $T = 1$, i.e., $Q = (0,1)^2$. As a solution we have chosen $u(x,t) = \sin\frac{1}{2}\pi t \, \sin\pi x$. Here we consider a decomposition of the space-time domain Q into up to 64 time slabs, applying both the space-time FETI approach and the all-floating (AF) formulation. The results are given in Table 2, where we observe a reasonable number of iterations in all cases. Note that the number of degrees of freedom is significantly larger when using the all-floating approach instead of the standard FETI method. Although the latter requires fewer iterations in most examples, this is not always the case (cf. Table 2). Based on previous experiences [2, 8], we expect that this behaviour can be further

Table 1: Space-time FETI domain decomposition method for the time-dependent Stokes system in $Q = (0,1)^3$. Number of GMRES iterations for the Schur complement system without (I) and with diagonal (D) preconditioning, for different numbers N_e of elements.

					Domain decomposition									
					a)		b)		c)		d)		e)	
N_e	$\|\nabla_x(u-u_h)\|_{L^2(Q)}$		$\|p-p_h\|_{L^2(Q)}$		I	D	I	D	I	D	I	D	I	D
192	6.86e-3		2.63e-2		15	11	26	13	31	15	36	19		
1536	2.19e-3	1.64	6.53e-3	2.01	25	13	54	17	57	20	79	29	72	28
12288	5.82e-4	1.92	1.57e-3	2.05	36	17	94	22	105	27	165	44	181	50
98304	1.47e-4	1.98	3.81e-4	2.04	55	22	180	34	206	39	374	66	325	83

improved by using appropriate preconditioners for the all-floating scheme. Also note that this approach is strongly related to the parareal algorithm [7] where the coarse grid corresponds to the time slabs of the domain decomposition, see also the results in Ref. [11].

Table 2: Classical and all-floating (AF) space-time FETI methods for the heat equation. Number of GMRES iterations for a sequence of time slabs and meshes.

	$s=2$		$s=4$		$s=8$		$s=16$		$s=32$		$s=64$	
N_e	FETI	AF	FETI	AF	FETI	AF	FETI	AF	FETI	AF	FETI	AF
128	5	12	7	12	9	12						
512	7	12	8	14	12	18	17	17				
2048	8	13	10	15	14	21	23	29	34	27		
8192	9	15	11	18	16	24	26	36	40	53	69	49
32768	9	18	12	23	17	29	28	44	47	68	79	104

5 Conclusions

In this contribution, we have presented and described first results for space-time finite element tearing and interconnecting domain decomposition methods, including also the all-floating approach. Model problems include the heat equation and the Stokes system, but more complex partial differential equations can be considered as well. The space-time finite element discretization and the tearing and interconnecting approach follow the lines of the FETI method for elliptic problems, considering time as just an additional spatial coordinate. The main distinction here stems from the asymmetry of the space-time stiffness matrix, which requires a modified projection operator and also a numerical procedure to construct local kernels. First numerical results show the potential of the proposed method, in particular when using state-of-the-art parallel computing facilities for time-dependent problems. It is clear that a more detailed numerical analysis, in particular with respect to suitable preconditioning strategies for general space-time domain decompositions, is required. Related results will be investigated and published elsewhere.

Acknowledgements The authors acknowledge Graz University of Technology for the financial support of the Lead-project: Mechanics, Modeling and Simulation of Aortic Dissection.

References

1. J.H. Bramble, J.E. Pasciak and A.H. Schatz. The construction of preconditioners for elliptic problems by substructuring. I. *Math. Comp.* **47**, 103–134 (1986).
2. Z. Dostál, D. Horák and R. Kučera. Total FETI – an easier implementable variant of the FETI method for numerical solution of elliptic PDE. *Comm. Numer. Methods Engrg.* **22**, 1155–1162 (2006).
3. C. Farhat and M. Géradin. On the general solution by a direct method of a large-scale singular system of linear equations: application to the analysis of floating structures. *Int. J. Numer. Meth. Engrg.* **41**, 675–696 (1998).
4. C. Farhat and F.-X. Roux. A method of finite element tearing and interconnecting and its parallel solution algorithm. *Int. J. Numer. Meth. Engrg.* **32**, 1205–1227 (1991).
5. G. Haase, U. Langer and A. Meyer. The approximate Dirichlet domain decomposition method. Part I: An algebraic approach. *Computing* **47**, 137–151 (1991).
6. R. Kučera, T. Kozubek, A. Markopoulus, J. Haslinger and L. Mocek. Projected Krylov methods for solving non-symmetric two-by-two block linear systems arising from fictitious domain formulations. *Adv. Electr. Electron. Eng.* **12**, 131–143 (2014).
7. J.-L. Lions, Y. Maday and G. Turinici. Résolution d'EDP par un schéma en temps "pararéel". *C. R. Acad. Sci. Paris Sér. I Math.* **332**, 661–668 (2001).
8. G. Of and O. Steinbach. The all-floating boundary element tearing and interconnecting method. *J. Numer. Math.* **17**, 277–298 (2009).
9. D.R.Q. Pacheco and O. Steinbach. Space-time Taylor–Hood elements for incompressible flows. *Computer Meth. Material Sci.* **19**, 64–69 (2019).
10. O. Steinbach. Space-time finite element methods for parabolic problems. *Comput. Meth. Appl. Math.* **15**, 551–566 (2015).
11. O. Steinbach and P. Gaulhofer. On space-time finite element domain decomposition methods for the heat equation. In: S. Brenner, E. Chung, A. Klawonn, F. Kwok, J. Xu and J. Zou (eds.) *Domain Decomposition Methods in Science and Engineering XXVI*, Lecture Notes in Computational Science and Engineering Vol. 145, Springer, Cham, pp. 547–554 (2022).
12. O. Steinbach and M. Windisch. Stable boundary element domain decomposition methods for the Helmholtz equation. *Numer. Math.* **118**, 171–195 (2011).
13. O. Steinbach and M. Windisch. Stable BETI methods in electromagnetics. In: R. Bank, M. Holst, O. Widlund and J. Xu (eds.) *Domain Decomposition Methods in Science and Engineering XX*, Lecture Notes in Computational Science and Engineering, vol. 91, pp. 223–230, Springer, Heidelberg (2013).
14. O. Steinbach and H. Yang. Space-time finite element methods for parabolic evolution equations: Discretization, a posteriori error estimation, adaptivity and solution. In: U. Langer and O. Steinbach (eds.) *Space-Time Methods. Applications to Partial Differential Equations*, Radon Series on Computational and Applied Mathematics, vol. 25, pp. 207–248, de Gruyter, Berlin (2019).
15. B. Vereecke, H. Bavestrello and D. Dureisseix. An extension of the FETI domain decomposition method for incompressible and nearly incompressible problems. *Comput. Methods Appl. Mech. Engrg.* **192**, 3409–3429 (2003).

Localized Reduced Basis Additive Schwarz Methods

Martin J. Gander and Stephan Rave

1 Introduction

Reduced basis (RB) methods [6, 9] are a family of model order reduction schemes for parameterized PDEs, which can speed up the repeated solution of such equations by orders of magnitude. In the so-called *offline* phase, RB methods construct a problem-adapted low-dimensional approximation space by computing solutions of the PDE for selected *snapshot* parameters using a given high-fidelity discretization of the PDE. In the following *online* phase, the PDE is solved for arbitrary new parameters by computing the (Petrov–)Galerkin projection of its solution onto the precomputed reduced approximation space. While RB methods have been proven successful in various applications, for very large problems the computation of the solution snapshots in the offline phase may still be prohibitively expensive. To mitigate this issue, localized RB methods [4, 7] have been developed which construct the global approximation space from spatially localized less expensive problems. These local problems largely fall into two classes:

Training procedures construct local approximation spaces without knowledge of the global problem by, e.g., solving the equation on an enlarged subdomain with arbitrary boundary values and then restricting the solution to the domain of interest, or by solving related eigenvalue problems. As such, these training approaches have a strong connecting with numerical multiscale methods and the construction of spectral coarse spaces in domain decomposition methods.

In this contribution, however, we will focus on the construction of local RB spaces via *online enrichment*, where these spaces are iteratively built by solving localized

Martin J. Gander
Section de Mathématiques, Université de Genève, 2-4 rue du Lièvre, CP 64, CH-1211, Genève, Suisse, e-mail: Martin.Gander@unige.ch

Stephan Rave
Mathematics Münster, University of Münster, Einsteinstrasse 62, 48149 Münster, Germany, e-mail: Stephan.Rave@uni-muenster.de

S. C. Brenner et al. (eds.), *Domain Decomposition Methods in Science and Engineering XXVI*, Lecture Notes in Computational Science and Engineering 145,
https://doi.org/10.1007/978-3-030-95025-5_52

corrector problems for the residual of the current reduced solution. In particular we mention the use of online enrichment in context of the LRBMS [8], GMsFEM [5] and ArbiLoMod [3] methods. These enrichment schemes share strong similarities with Schwarz methods, and it is the main goal of this contribution to shed some light on the connections between these methods. We will do so by introducing a simple localized RB additive Schwarz (LRBAS) method which is phrased in the language of the abstract Schwarz framework but incorporates the central ingredients of online adaptive localized RB methods. In particular, we hope that LRBAS will help the analysis of localized RB methods from the perspective of Schwarz methods. Following [3], we will consider arbitrary but localized changes of the problem instead of parametric variations. In Section 2.1 we will see that LRBAS can indeed be interpreted as a locally adaptive version of a multi-preconditioned CG method.

Compared to Schwarz methods, a distinctive feature of LRBAS is that updates are only computed in high-residual regions, which can lead to a significant reduction of the number of local updates and a concentration of the updates to a few regions affected by the localized changes (cf. Section 3). This property might be exploited for the reduction of the overall power consumption and to balance the computational load among a smaller amount of compute nodes, in particular in cloud environments, where additional computational resources can be easily allocated and deallocated again.

2 A localized reduced basis additive Schwarz method

Our goal is to efficiently solve a sequence, indexed by k, of linear systems

$$A^{(k)}x^{(k)} = f \tag{1}$$

with $A^{(k)} \in \mathbb{R}^{n\times n}$ symmetric, positive definite and $x^{(k)}, f \in \mathbb{R}^n$, up to some fixed error tolerance ε. To this end, let $n \times n_i$ matrices R_i^T of rank n_i be given for $1 \le i \le I$ and $n \times n_0^{(k)}$ matrices $R_0^{(k)T}$ of rank $n_0^{(k)}$. Typically, $R_1, \dots, R_I$ will be the restriction matrices corresponding to a finite element basis associated with an overlapping domain decomposition Ω_i of the computational domain Ω, and the columns of $R_0^{(k)T}$ contain a basis of a suitable coarse space for $A^{(k)}$. In particular, we assume that each R_i is non-orthogonal to only a few neighboring spaces, i.e., there are a small constant C and index sets $O_i \subset \{1, \dots, I\}$ with $\#O_i \le C \cdot I$ such that

$$R_j \cdot R_i^T = 0_{n_j \times n_i} \qquad \text{whenever } j \notin O_i. \tag{2}$$

As usual, we define the local matrices

$$A_0^{(k)} := R_0^{(k)} A^{(k)} R_0^{(k)T} \qquad \text{and} \qquad A_i^{(k)} := R_i A^{(k)} R_i^T.$$

We are interested in the case where $A^{(k+1)}$ is obtained from $A^{(k)}$ by an arbitrary but local modification in the sense that

$$A_i^{(k+1)} = A_i^{(k)} \qquad \text{for } i \notin C^{(k+1)}, \tag{3}$$

where the sets $C^{(k)}$ contain the indices of the spaces affected by the change, generally assuming that $\#C^{(k)} \ll I$.

Over the course of the computation of the solutions $x^{(k)}$ we will build local low-dimensional reduced bases $\tilde{R}_i^{(k,l)T} \in \mathbb{R}^{n_i \times N_i^{(k,l)}}$ for $i \geq 1$ such that there are local coefficients $\tilde{x}_i^{(k,l)} \in \mathbb{R}^{N_i^{(k,l)}}$ and $\tilde{x}_0^{(k,l)} \in \mathbb{R}^{n_0^{(k)}}$ such that

$$\tilde{x}^{(k,l)} := R_0^{(k)T} \tilde{x}_0^{(k,l)} + \sum_{i=1}^{I} R_i^T \tilde{R}_i^{(k,l)T} \tilde{x}_i^{(k,l)} \tag{4}$$

is a good approximation of x^k for sufficiently large l. We obtain such an approximation via Galerkin projection onto the global reduced basis space spanned by the images of $R_0^{(k)T}$ and all $R_i^T \tilde{R}_i^{(k,l)T}$, i.e., $\tilde{x}^{(k,l)}$ is determined by the $(n_0^{(k)} + \sum_{i=1}^{I} N_i^{(k,l)})$-dimensional linear system

$$\begin{aligned} R_0^{(k)} A^{(k)} \tilde{x}^{(k,l)} &= R_0^{(k)} f, \\ \tilde{R}_i^{(k,l)} R_i A^{(k)} \tilde{x}^{(k,l)} &= \tilde{R}_i^{(k,l)} R_i f, \qquad 1 \leq i \leq I. \end{aligned} \tag{5}$$

Thanks to the locality (2) of the space decomposition, the matrix of the system (5) has a block structure allowing us to efficiently assemble and solve it.

To build the local reduced bases $\tilde{R}_i^{(k,l)T}$ we use an iterative enrichment procedure where the basis is extended with local Schwarz corrections $y_i^{(k,l)} \in \mathbb{R}^{n_i}$ for the current residual,

$$A_i^{(k)} y_i^{(k,l)} = r_i^{(k,l)} := R_i(f - A^{(k)} \tilde{x}^{(k)}). \tag{6}$$

In view of (3), the corrections are only computed in subdomains i with large residual norm $\|r_i^{(k,l)}\|$. In particular, for finite-element discretizations of elliptic PDEs without high-conductivity channels, we expect that with increasing k the number of enriched bases will be of the same order as the cardinality of C^{k+1}. The exact definition of the enrichment scheme is given in Algorithm 1. There are various possibilities to choose the criterion for the localized enrichment in line 9 of Algorithm 1. In this work we simply select those reduced spaces for enrichment for which the quotient between the norm of the local residual and the norm of the global residual is larger than a fixed constant that scales with the number of the subdomains.

Note that an important property of localized enrichment is that after an enrichment step only those blocks (i, j) of the matrix corresponding to (5) have to be updated for which either $\tilde{R}_i^{(k,l)}$ or $\tilde{R}_j^{(k,l)}$ have been enriched. Using reduced basis techniques [2] it is further possible to evaluate the residual norms $\|R_i r^{(k,l)}\|$ and $\|r^{(k,l)}\|$ using only reduced quantities, which again only have to be updated for local bases $\tilde{R}_i^{(k,l)T}$ affected by the enrichment. Thus, in a distributed computing environment only the main compute node solving (5) and those nodes associated with the enriched bases have to perform any operations, while the other compute nodes lay at rest.

Algorithm 1 Localized Reduced Basis Additive Schwarz method (LRBAS)

procedure LRBAS($A^{(k)}, f, R_0^T, R_i^T, \varepsilon, \varepsilon_{\text{loc}}$)
  $\tilde{R}_i^{(1,1)T} \leftarrow 0_{n_i \times 0}, \quad 1 \le i \le I$ ▹ initialize local bases
  for $k \leftarrow 1, \dots, \infty$ **do**
    $\tilde{x}^{(k,1)}, \tilde{x}_i^{(k,1)} \leftarrow$ solutions of (4), (5) ▹ initial solution
    $r^{(k,1)} \leftarrow f - A^{(k)} \tilde{x}^{(k,1)}$ ▹ initial residual
    $l \leftarrow 1$
    while $\|r^{(k,l)}\| / \|f\| > \varepsilon$ **do** ▹ loop until converged
      for $i \leftarrow 1, \dots I$ **do** ▹ enrichment procedure
        if $\|R_i r^{(k,l)}\|^2 > \varepsilon_{\text{loc}} \cdot I^{-1} \cdot \|r^{(k,l)}\|^2$ **then**
          $y_i^{(k,l)} \leftarrow$ solution of (6)
          $\tilde{R}_i^{(k,l+1)T} \leftarrow [\tilde{R}_i^{(k,l)T} \; y_i^{(k,l)}]$
        else
          $\tilde{R}_i^{(k,l+1)} \leftarrow \tilde{R}_i^{(k,l)}$
        end if
      end for
      $\tilde{x}^{(k,l+1)}, \tilde{x}_i^{(k,l+1)} \leftarrow$ solutions of (4), (5) ▹ update solution
      $r^{(k,l+1)} \leftarrow f - A^{(k)} \tilde{x}^{(k,l+1)}$ ▹ update residual
      $l \leftarrow l + 1$
    end while
    for $i \leftarrow 1, \dots I$ **do** ▹ update bases for next problem
      if $\tilde{R}_i^{(k,l)T} \neq \tilde{R}_i^{(k,1)T}$ **then** ▹ basis enriched at least once?
        $\tilde{R}_i^{(k+1,1)T} \leftarrow [\tilde{R}_i^{(k,1)T} \; \tilde{R}_i^{(k,l)T} \tilde{x}_i^{(k,l)}]$ ▹ only keep local solution in basis
      else
        $\tilde{R}_i^{(k+1,1)T} \leftarrow \tilde{R}_i^{(k,1)T}$
      end if
    end for
  end for
end procedure

We remark that several extensions to the LRBAS method are possible. In particular, we assumed for simplicity that all matrices $A^{(k)}$ are of the same dimension. This, for instance, is the case when coefficient functions of the PDE underlying (1) are modified, but the computational mesh remains unchanged. However, local geometry changes that lead to remeshing can also be handled by resetting all local bases that are supported on the changed geometry. In this context we note that, as another simplification, in the definition of LRBAS we have chosen to keep all basis vectors when transitioning from $A^{(k)}$ to $A^{(k+1)}$, including bases $\tilde{R}_i^{(k,l)T}$ affected by the change, even though these retained bases will generally not contribute to the convergence of the scheme. Finally, in many applications, a local or global parametric variation of $A^{(k)}$, e.g. the change of some material parameters, in addition to the considered non-parametric modifications may be of interest. In such cases, parametric model order reduction techniques such as greedy basis generation algorithms or offline/online decomposition of the reduced order system (5) can be incorporated into the scheme. In particular we refer to [3] where both additional parameterization of $A^{(k)}$ as well as the reinitialization of the local bases after non-parametric changes from $A^{(k)}$ to $A^{(k+1)}$ are discussed.

2.1 LRBAS as an additive-Schwarz multi-preconditioned CG method

Consider the solution of the systems (1) with the preconditioned conjugate gradient (PCG) algorithm, where we choose as preconditioner the additive Schwarz operator $(M^{(k)})^{-1} := R_0^{(k)T}(A_0^{(k)})^{-1}R_0^{(k)} + \sum_{i=1}^{I} R_i^T(A_i^{(k)})^{-1}R_i$. Let $x_{\text{pcg}}^{(k,l)}$ denote the l-th iterate of the PCG algorithm, starting with $x^{(k,0)} = 0$ as the initial guess. Then it is well known that $x_{\text{pcg}}^{(k,l)}$ lies in the search space $\mathcal{S}_{\text{pcg}}^{(k,l)}$ given by the Krylov space $\mathcal{K}^l\left((M^{(k)})^{-1}A^{(k)}, (M^{(k)})^{-1}f\right)$ and that the error $x^{(k)} - x_{\text{pcg}}^{(k,l)}$ is $A^{(k)}$-orthogonal to this space. Denoting by $r_{\text{pcg}}^{(k,l)} := f - A^{(k)}x_{\text{pcg}}^{(k,l)}$ the l-th residual, one readily checks that $\mathcal{S}_{\text{pcg}}^{(k,l)}$ is equivalently given by

$$\mathcal{S}_{\text{pcg}}^{(k,l)} := \operatorname{span}\left\{(M^{(k)})^{-1}r_{\text{pcg}}^{(k,0)}, \ldots, (M^{(k)})^{-1}r_{\text{pcg}}^{(k,l-1)}\right\},$$

i.e., in each iteration the search space is extended by the vector obtained from the application of the preconditioner to the current residual. The idea of multi-preconditioning [1] is to enlarge this search space by including each local preconditioner $(A_i^{(k)})^{-1}$ application into the search space individually, leading to

$$\begin{aligned}\mathcal{S}_{\text{mpcg}}^{(k,l)} := \operatorname{span}\Big(&\left\{R_0^{(k)T}(A_0^{(k)})^{-1}R_0^{(k)}r_{\text{mpcg}}^{(k,t)} \,\middle|\, 0 \le t \le l-1\right\}\\ &\cup\left\{R_i^T(A_i^{(k)})^{-1}R_i r_{\text{mpcg}}^{(k,t)} \,\middle|\, 1 \le i \le I,\, 0 \le t \le l-1\right\}\Big),\end{aligned}$$

with $r_{\text{mpcg}}^{(k,l)}$ denoting the multi-preconditioned CG residuals. Conversely, we easily see from (5) and (6) that for $\varepsilon_{\text{loc}} = 0$ the LRBAS iterates $\tilde{x}^{(k,l)}$ lie within the search space

$$\begin{aligned}\mathcal{S}_{\text{lrbas},0}^{(k,l)} := \operatorname{Im}&\left(\left[R_0^{(k)T}\ \ R_1^T\tilde{R}_1^{(k,1)}\ \ \ldots\ \ R_I^T\tilde{R}_I^{(k,1)}\right]\right)\\ &+ \operatorname{span}\left\{R_i^T(A_i^{(k)})^{-1}R_i r^{(k,t)} \,\middle|\, 1 \le i \le I,\, 1 \le t \le l-1\right\},\end{aligned}$$

and that the error $x^{(k)} - \tilde{x}^{(k,l)}$ is $A^{(k)}$-orthogonal to this space. Hence, LRBAS with $\varepsilon_{\text{loc}} = 0$ can be seen as a projected multi-preconditioned CG method for solving (5), where the projection space is given by the span of the coarse space and the initial local reduced bases and where the new solution iterate $\tilde{x}^{(k,l)}$ is obtained by direct solution of the reduced system (5) instead of an incremental update in order to preserve the locality of the reduced bases.

For $\varepsilon_{\text{loc}} > 0$ we arrive at an adaptive version of multi-preconditioning similar to [10]. However, in contrast to [10] where either all local search directions or their global sum are added to the search space, LRBAS is locally adaptive in the sense that only those local search directions are computed and included where a large local residual has to be corrected.

3 Numerical experiment

We consider the test case from [3] and solve a sequence of five elliptic problems

$$
\begin{aligned}
\nabla \cdot \big(-\sigma^{(k)}(x,y)\nabla u^{(k)}(x,y)\big) &= 0, && x, y \in (0,1),\\
u^{(k)}(0,y) &= 1, && y \in (0,1),\\
u^{(k)}(1,y) &= -1, && y \in (0,1),\\
-\sigma^{(k)}(x,y)\nabla u^{(k)}(x,y)\cdot \mathbf{n}(x,y) &= 0, && x \in (0,1),\ y \in \{0,1\},
\end{aligned}
\tag{7}
$$

where the coefficient $\sigma^{(k)}(x)$ is given as in Fig. 1. The problem is discretized using bilinear finite elements over a uniform 200×200 mesh. The resulting solutions are visualized in Fig. 2. We decompose the computational domain uniformly into 10×10 subdomains with an overlap of 4 mesh elements. For $R_0^{(k)}$ we choose GenEO [11] basis functions with eigenvalues below 0.5, yielding between two and five functions per subdomain. When connecting or disconnecting the high-conductivity channels, we expect enrichment to be required along the subdomains adjacent to the channels, whereas the other subdomains should be largely unaffected by the local change.

In Table 1 we compare the total number of iterations for all five problems and the total number of Schwarz corrections (6) required to reach a relative error tolerance of $\varepsilon = 10^{-6}$ for the following solution strategies: 1. the additive Schwarz preconditioned CG method with zero initial guess or with a localized RB solution as initial guess, where the localized basis is obtained from the linear span of previous solutions $x^{(k)}$ decomposed using the GenEO partition of unity; 2. LRBAS with and without local adaptivity ($\varepsilon_{\text{loc}} = 0.25$ or 0); 3. a version of LRBAS where the entire bases $\tilde{R}_i^{(k,l)T}$ are preserved when transitioning to $k+1$ instead of only the final solution $\tilde{R}_i^{(k,l)T}\tilde{x}_i^{(k,l)}$. As we see, LRBAS with locally adaptive enrichment significantly outperforms the PCG method with or without initial guess, both regarding the number of required iterations as well as the number of Schwarz corrections. Compared to non-adaptive

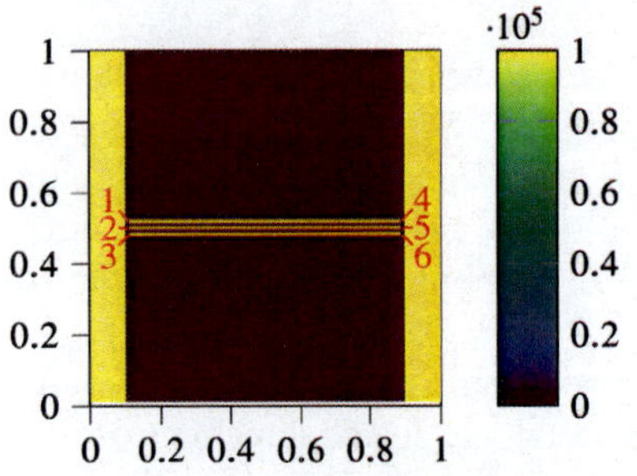

$k =$	1	2	3	4	5
ch. 1				✓	✓
ch. 2	✓				
ch. 3					
ch. 4					
ch. 5	✓	✓			✓
ch. 6					

Fig. 1: Definition of the coefficient functions $\sigma^{(k)}$ for the numerical test case (7); left: function $\sigma^{(0)}$, taking the values $10^5 + 1$ inside the high-conductivity regions and 1 elsewhere; right: $\sigma^{(k)}$ is obtained from $\sigma^{(0)}$ by connecting the three channels to the boundary regions at the marked locations.

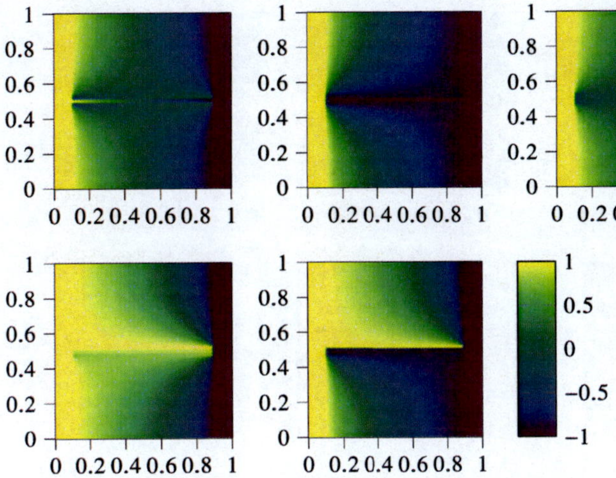

Fig. 2: Solutions of the test problem (7) for $k = 1, 2, 3$ (top row) and $k = 4, 5$ (bottom row).

	iterations	local enrichments (6)
PCG	107	10700
PCG + LRB solution as initial value	63	6300
LRBAS ($\varepsilon_{\text{loc}} = 0$)	33	3300
LRBAS ($\varepsilon_{\text{loc}} = 0.25$)	39	1386
LRBAS ($\varepsilon_{\text{loc}} = 0$, $\tilde{R}_i^{(k+1,1)} := \tilde{R}_i^{(k,l)}$)	28	2800
LRBAS ($\varepsilon_{\text{loc}} = 0.25$, $\tilde{R}_i^{(k+1,1)} := \tilde{R}_i^{(k,l)}$)	34	1335

Table 1: Total number of iterations and local Schwarz corrections (6) required to reach a relative error tolerance $\varepsilon = 10^{-6}$ for the test problem (7).

multi-preconditioning, i.e. LRBAS with $\varepsilon_{\text{loc}} = 0$, the number of local corrections is more than halved at the expense of a slightly increased number of iterations. Keeping all of $\tilde{R}_i^{(k,l)}$ improves the convergence of the method only slightly. Finally, in Fig. 3 we depict the number of required Schwarz corrections per subdomain for each k. We observe a good localization of the computational work among the subdomains most affected by the local changes.

Acknowledgements Funded by the Deutsche Forschungsgemeinschaft (DFG, German Research Foundation) under Germany's Excellence Strategy EXC 2044 –390685587, Mathematics Münster: Dynamics–Geometry–Structure.

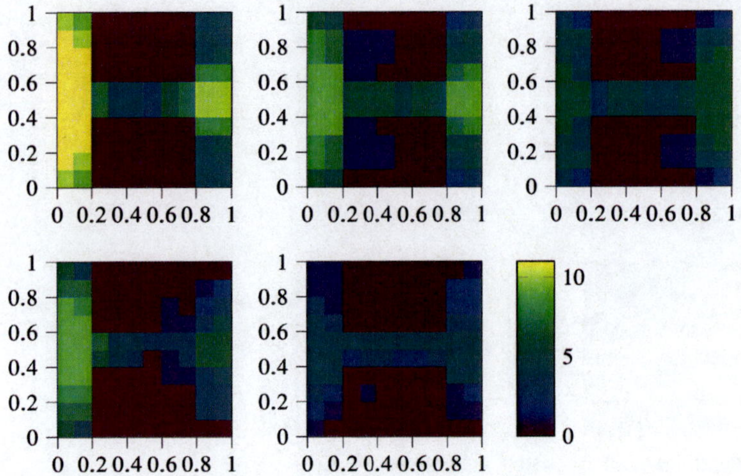

Fig. 3: Number of local Schwarz corrections (6) required by the LRBAS method with $\varepsilon_{\text{loc}} = 0.25$ to solve the five test problems (7) up to a relative error tolerance of $\varepsilon = 10^{-6}$.

References

1. R. Bridson and C. Greif. A multipreconditioned conjugate gradient algorithm. *SIAM Journal on Matrix Analysis and Applications* **27**(4), 1056–1068, Jan. (2006).
2. A. Buhr, C. Engwer, M. Ohlberger and S. Rave. A Numerically Stable A Posteriori Error Estimator for Reduced Basis Approximations of Elliptic Equations. In: E. Oñate, X. Oliver, and A. Huerta, editors, *11th World Congress on Computational Mechanics*, pp. 4094–4102, International Center for Numerical Methods in Engineering, Barcelona (2014).
3. A. Buhr, C. Engwer, M. Ohlberger and S. Rave. ArbiLoMod, a simulation technique designed for arbitrary local modifications. *SIAM J. Sci. Comput.* **39**(4), A1435–A1465 (2017).
4. A. Buhr, L. Iapichino, M. Ohlberger, S. Rave, F. Schindler and K. Smetana. Localized model reduction for parameterized problems. In: P. Benner, S. Grivet-Talocia, A. Quarteroni, G. Rozza, W. Schilders and L.M. Silveira, editors, *Model Order Reduction (Volume 2)*, De Gruyter, Berlin, Boston (2021).
5. E.T. Chung, Y. Efendiev and W.T. Leung. Residual-driven online generalized multiscale finite element methods. *Journal of Computational Physics* **302**, 176–190 (2015).
6. J.S. Hesthaven, G. Rozza and B. Stamm. *Certified Reduced Basis Methods for Parametrized Partial Differential Equations*. SpringerBriefs in Mathematics. Springer International Publishing, 1st edition (2016).
7. Y. Maday and E.M. Rønquist. A reduced-basis element method. *Comptes Rendus Mathematique* **335**(2), 195–200 (2002).
8. M. Ohlberger and F. Schindler. Error control for the localized reduced basis multiscale method with adaptive on-line enrichment. *SIAM J. Sci. Comput.* **37**(6), A2865–A2895 (2015).
9. A. Quarteroni, A. Manzoni and F. Negri. *Reduced Basis Methods for Partial Differential Equations*. Volume 92 of La Matematica per il 3+2, Springer International Publishing (2016).
10. N. Spillane. An adaptive MultiPreconditioned conjugate gradient algorithm. *SIAM Journal on Scientific Computing* **38**(3), A1896–A1918, Jan (2016).
11. N. Spillane, V. Dolean, P. Hauret, F. Nataf, C. Pechstein and R. Scheichl. Abstract robust coarse spaces for systems of PDEs via generalized eigenproblems in the overlaps. *Numerische Mathematik* **126**(4), 741–770, Aug (2013).

Micromechanics Simulations Coupling the deal.II Software Library With a Parallel FETI-DP Solver

S. Köhler, O. Rheinbach, and S. Sandfeld

1 Introduction

We consider adaptive finite elements, using the open source finite element library deal.II [1], and an implementation [11] of the FETI-DP (Finite Element Tearing and Interconnecting Dual–Primal) method based on PETSc, for the solution of problems from dislocation micromechanics. The library deal.II is well known for its adaptive finite element approach based on hanging node constraints. The parallel data structures in deal.II are meant to be used with global parallel matrices, which are assembled across the interface. However, in FETI-DP [6] or BDDC [4] methods, access to the Neumann matrices for each subdomain is needed. Here, we show that the deal.II infrastructure can still be used to efficiently construct the FETI-DP preconditioner. We have reported on first computational results of our approach in [9]; different improvements, including the construction of the coarse space, have been made since. A related implementation of a BDDC method, using adaptive mesh refinement not based on deal.II, has obtained good scalability to up to 2048 cores in [10].

Stephan Köhler and Oliver Rheinbach
Institut für Numerische Mathematik und Optimierung, Fakultät für Mathematik und Informatik, Technische Universität Bergakademie Freiberg, Akademiestr. 6, 09596 Freiberg,
e-mail: oliver.rheinbach@math.tu-freiberg.de,
stephan.koehler@math.tu-freiberg.de,
url: http://www.mathe.tu-freiberg.de/nmo/mitarbeiter/oliver-rheinbach

Stefan Sandfeld
Forschungszentrum Jülich, Institute for Advanced Simulation – Materials Data Science and Informatics (IAS-9), Wilhelm-Johnen-Straße, 52428 Jülich and RWTH Aachen University, Chair of Materials Data Science and Materials Informatics, Faculty 5, 52072 Aachen

S. C. Brenner et al. (eds.), *Domain Decomposition Methods in Science and Engineering XXVI*, Lecture Notes in Computational Science and Engineering 145,
https://doi.org/10.1007/978-3-030-95025-5_53

2 Micromechanical model problem

To compute the stresses associated with dislocations within a specimen for the characterization of the microstructure [12, 13], we start by considering a linear elastic model described by

$$\operatorname{div}\sigma = 0, \quad \sigma = \sigma^T, \quad \sigma = C : \varepsilon^{\mathrm{el}}, \quad \text{and} \quad \varepsilon^{\mathrm{el}} = \frac{1}{2}(\nabla u + (\nabla u)^T)$$

to be solved for the displacements u. Here, σ is the stress tensor, $\varepsilon^{\mathrm{el}}$ the elastic strain tensor, and C the stiffness tensor. Dislocations are one-dimensional defects present in crystalline materials. They are the boundary of a planar area over which two subdomains of a crystal have been displaced relative to each other with the directions given by the Burgers vector $\boldsymbol{b}$.

In the linear elastic context, dislocations may be modeled using an eigenstrain approach [5] by expressing the total strain by $\varepsilon^{\mathrm{tot}} = \varepsilon^{\mathrm{el}} + \varepsilon^{\mathrm{eig}}$, where $\varepsilon^{\mathrm{eig}}$ is the eigenstrain contribution caused by the dislocation microstructure. The area enclosed by a dislocation is described by an orthogonal vector $\boldsymbol{A}$. The eigenstrain contributions $\mathrm{d}\varepsilon^{\mathrm{eig}} = \frac{1}{2}(\boldsymbol{b} \otimes \mathrm{d}\boldsymbol{A} + \mathrm{d}\boldsymbol{A} \otimes \boldsymbol{b})$, where $\otimes$ denotes the outer product, are regularized using the non-singular formulation proposed in [3], similarly to [7]. The eigenstrain of a dislocation is a contribution to the body force term occurring in the elasticity problem.

As a benchmark problem, we chose an artificial dislocation structure which, however, reflects already many details of realistic microstructures that can be found in dislocation simulations. First of all, the considered sample is a cubic box with edge lengths of 1 μm; see also section 5. Single crystalline copper was used as a material, which has the anisotropic elastic constants $C_{11} = 168.4\,\mathrm{GPa}$, $C_{12} = 121.4\,\mathrm{GPa}$, and $C_{44} = 75.4\,\mathrm{GPa}$. Copper is a material that has a "face centered cubic" crystallographic structure with 12 possible slip systems on which dislocations can nucleate and move. In this artificial dislocation microstructure, all 28 dislocations are considered to be closed, circular loops; their center points and radii have been chosen randomly.

3 Parallel mesh handling in deal.II

For simplicity, let us first consider a domain $\Omega \subset \mathbb{R}^2$ decomposed into two subdomains $\Omega_1 \subset \Omega$ and $\Omega_2 \subset \Omega$; see Figure 1. In deal.II, each cell is owned by exactly one MPI rank, the *locally owned cells*. Each MPI rank has information about its locally owned cells and one additional layer of *ghost cells* of the neighboring subdomains; see Figure 2.

The degrees of freedom (dofs) have a global numbering. Each dof belongs to exactly one MPI rank; all dofs belonging to an MPI rank form the *locally owned dofs* of this rank. Each locally owned dof belongs to a locally owned cell, but some dofs of a locally owned cell may belong to the locally owned dofs of a different rank;

see Figure 3. The union of all dofs of all locally owned cells is called the *locally active dofs*. The union of the locally active dofs and the degrees of freedom of the cells of the ghost layer is called the *locally relevant dofs*; for details, see, e.g., [2] and Figure 3.

4 Subdomain Neumann matrices in deal.II

In deal.II, the global stiffness matrix K can be assembled by an instance of the class `AffineConstraints`, which also handles the hanging node constraints and the Dirichlet boundary values.

For nonoverlapping domain decomposition methods, such as FETI-DP and BDDC methods, we have to assemble the local subdomain stiffness matrices $K^{(i)}$ for each subdomain Ω_i, $i = 1, \ldots, N$. These local subdomain matrices are not assembled across the interface. There is currently no built-in support in the deal.II library for this operation. We have therefore added a layer on top of deal.II to implement the necessary functionality.

Computing the local subdomain Neumann matrices To assemble a local stiffness matrix, $K^{(i)}$, we need a local sparsity pattern, the local constraints and a local numbering $1, \ldots, n_i$ of all locally relevant dofs of this subdomain. We construct a local sparsity pattern and the local constraints from the global ones by copying the entries and the values with respect to the local numbering. The Dirichlet boundary needs some special care; see section 4.

Computing the interface, and faces, edges, and vertices In FETI-DP and related methods, the interface and its decomposition into faces, edges, and vertices are needed. The dofs of the interface can be computed as the intersection of the locally active dofs and the locally relevant dofs. But, due to the hanging node constraints, such an index set is not always appropriate for FETI-DP methods, where we need to introduce Lagrange multipliers on the interface. For hanging nodes, we therefore replace the hanging node dofs by those non hanging node dofs which constrain them; see, e.g., Figure 4.

Fig. 1: Left: Domain Ω. **Right:** Decomposition into Ω_1 and Ω_2; | Interface.

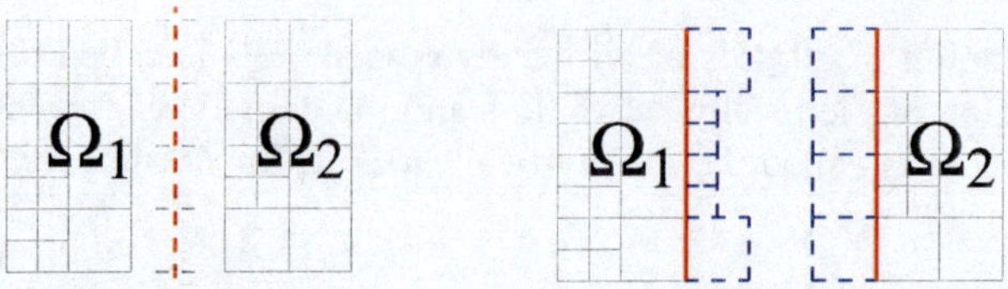

Fig. 2: Left: Locally owned cells of Ω_1, Ω_2. **Right:** Locally owned cells; - - ghost cells.

Fig. 3: An example of the classification of the degrees of freedom with Q_1 elements. **Left:** Subdomain Ω_1: • locally owned dofs; ∘ locally active dofs. Subdomain Ω_2: • locally owned dofs; ∘ locally active dofs. **Right:** Subdomain Ω_1: • locally relevant dofs; Subdomain Ω_2: ∘ locally relevant dofs.

Fig. 4: Partition into locally inner and locally active interface dofs. **Left:** Subdomain Ω_1: • locally inner dofs. Subdomain Ω_2: • locally inner dofs. **Right:** Subdomain Ω_1: • locally active interface dofs. Subdomain Ω_2: • locally active interface dofs.

We denote the interface dofs, as outlined above, of Ω_i by Γ_i and name them as *locally active interface* dofs. Let us remark that not all dofs on the geometric interface belong to Γ_i and, vice versa, see, e.g., Figure 5.

For $\Omega \subset \mathbb{R}^3$, we compute vertices, edges, and faces as follows: The basic idea is to compute the faces of all subdomains. After that, we build edges as intersections of faces and vertices as intersections of edges. Let us remark that, although the computation of faces is completely local to all MPI ranks, and, therefore, also the computation of edges and vertices, we need to communicate all computed edges and vertices to the neighboring subdomains, see, Figure 5; here, "neighboring" means subdomains which have a non-empty intersection of the locally relevant dofs; due to hanging node constraints, the result can be counterintuitive; see Figure 5 for the case of two dimensions.

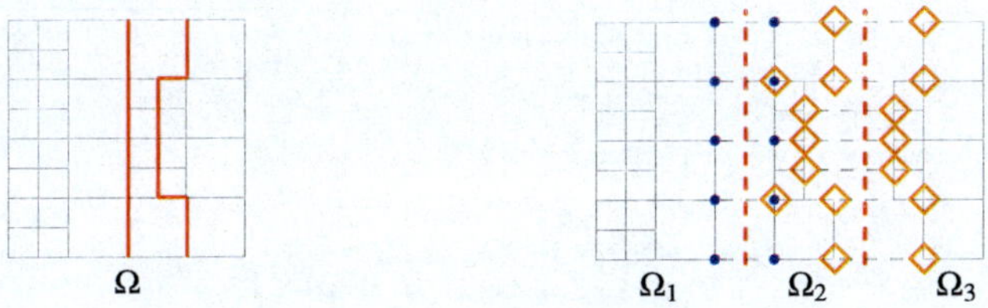

Fig. 5: Domain Ω partitioned into Ω_1, Ω_2 and Ω_3 with edge dofs between the subdomains. Only subdomain Ω_2 computes the vertices as intersection of the edges. These vertices are not seen by Ω_1 and Ω_3. **Left:** Domain Ω | interface. **Right:** • edge dofs between Ω_1 and Ω_2. ◇ edge dofs between Ω_2 and Ω_3.

The use of p4est (based on space filling curves), which is standard in deal.II's parallel distributed mesh class, does not guarantee that a subdomain is connected, and it may only be connected through vertices or edges of cells. This can be dealt with but it will typically increase the coarse problem size.

Furthermore, a subdomain may not have enough vertices or edges to ensure the invertibility of certain subdomain matrices in FETI-DP methods. Here, we sometimes need to introduce additional primal constraints by subdivision of faces or edges to constrain the low energy modes of all components of a subdomain. Let us remark that our method can still lead to faces or edges that are not connected.

As mentioned in section 4, we have to take care of the Dirichlet boundary condition. These are also handled by the `AffineConstraints` class, as the hanging node constraints. Therefore, for the computation of the interface, we need to extract information about the hanging nodes dofs from an instance where the Dirichlet boundary condition has not been set.

Construction of the FETI-DP jump operator A crucial element of FETI-DP methods is the jump operator B which imposes the continuity of the solution. This operator has a row for each Lagrange multiplier and each row consists of exact two entries, a $+1$ and a -1.

The Lagrange multipliers are related to the locally active interface dofs. Hence, we partition them, and manage the computation of the local parts of B, by the locally owned part of the interface, see Figure 6.

5 Numerical results

We use Q1 finite elements. The deal.II library uses the p4est library to compute the domain decomposition, as in [10]. Our FETI-DP implementation is based on [8, 11]. Our coarse space uses vertices, edges, and certain additional point constraints on faces. We perform 5 mesh refinement steps, using the Kelly error estimator. We use GMRES.

Denote the locally active interface of Ω_i by Γ_i and the locally owned dofs by I_i^o.

1. Decompose Γ_i into $\Gamma_i^o = \Gamma_i \cap I_i^o$ and $\Gamma_i^{o,C}$
2. Determine the number of rows m of B_i:
   (a) Compute the number $s_\lambda^{(i)}$ of all multipliers related to Γ_i^o.
   (b) Compute $m = \sum_i^N s_\lambda^{(i)}$.
3. Send the index set $\Gamma_i^{o,C}$ to all neighboring subdomains.
4. For each received set $\Gamma_j^{o,C}$ from a neighboring subdomain Ω_j compute $\Gamma_{i,j}^o = \Gamma_i^o \cap \Gamma_j^{o,C}$. Send to subdomain Ω_j the information for which multipliers, associated with the dofs in $\Gamma_{i,j}^o$, a +1 or −1 has to be inserted into B_j.

Fig. 6: Construction of the local jump operators B_i.

In Figure 7 (left) we show the eigenstrain distributions that are non-zero inside the loops and zero outside. This quick transition of the eigenstrain value is responsible for very high (for non-regularized problems: diverging) stresses that require a sufficiently fine mesh for obtaining an accurate solution.

In Tables 1 and 2, we report on the global problem size ("Global"), the size of the coarse problem ("Coarse"), the number of Krylov iterations ("it."), the solver and assembly time ("solve" and "ass."). We also report timings to build the interface Γ, to build faces, edges, and vertices (denoted "f/e/v"), and to build the FETI-DP jump operator B.

First, we observe that the deal.II infrastructure can provide, within a fraction of a second (for the smaller problems) to a few seconds (for the larger problems), the necessary connectivity information to construct the FETI-DP preconditioner, i.e., the interface, the face, edges, and vertices, and the information to build the B-matrix.

We also observe that the number of iterations increases slightly when refining the mesh. Note that the problem is anisotropic (see section 2), which results in higher iteration counts compared to standard benchmark problems.

For the same refinement cycle, the problem sizes for 512 cores are larger by a factor between 7 and 8 compared with 64 cores. Since the number of cores is larger by a factor of 8, we can roughly compare the timings for 512 cores and 64 cores in the sense of weak scaling. In this sense, when comparing refinement step 5, we observe acceptable parallel scalability for the solver time (29.8s vs. 18.8s) and the total time (203s vs. 174s). This is also the case when summing the total time over all refinement steps, i.e., we have 305s (512 cores) and 237s (64 cores). Since this is not weak scalability in the strict sense, we refrain from providing parallel efficiency numbers. Note that the assembly does not scale perfectly since a certain load imbalance is introduced by the additional computations involved with the dislocations.

Performing the same computations using a larger number of cores, i.e., 216 and 1728 cores, we see that the solver time starts to be dominated by the coarse solver, since the coarse problem is quite large, i.e., > 70 000 dof for the last two refinement cycles. This is also a result of our attempts to create a robust coarse space. As a result of the deteriorating solver scalability, the total time to solution, summed over all refinement steps, is 510s (1728 cores) to be compared with 277s (216 cores). This indicates that we need to reduce the coarse problem size by modifying our coarse space. Alternatively, we can move to a three- or multi-level method as in [10].

Fig. 7: Left: eigenstrain resulting from 28 dislocation loops (red color denotes a non-zero eigenstrain). **Middle:** Solution. **Right:** Solution and adaptive mesh for the fifth refinement cycle; problem size: 10.4 million dofs using 512 MPI ranks.

Table 1: Results for 5 refinement cycles on 64 cores and 512 cores.

		d.o.f.			Time in s					
#Cores	Refinem.	Global	Coarse	it.	solve	ass.	build Γ	f/e/v	build B	total time
64	1	14 739	405	23	0.16	9.41	< 0.01	0.01	< 0.01	10.0
	2	49 173	1 569	39	0.54	12.6	0.06	0.03	0.02	13.3
	3	153 420	1 680	46	1.23	20.3	0.24	0.03	0.01	22.1
	4	476 502	1 785	58	4.19	47.6	1.06	0.06	0.03	53.6
	5	1 475 034	1 896	55	18.8	150	3.80	0.19	0.06	174
	sum				24.9	240	5.17	0.32	0.13	273
512	1	107 811	4 557	22	1.39	5.07	0.02	0.01	0.02	6.94
	2	349 404	16 842	49	4.84	9.25	0.10	0.04	0.03	14.6
	3	1 087 689	19 275	51	6.57	15.3	0.23	0.05	0.06	22.8
	4	3 353 052	20 604	56	10.6	45.2	0.94	0.11	0.12	58.0
	5	10 358 751	22 254	53	29.8	165	4.59	0.22	0.23	203
	sum				53.2	240	5.88	0.86	0.46	305

Acknowledgements The authors acknowledge computing time on the Compute Cluster of the Fakultät für Mathematik und Informatik of Technische Universität Freiberg (DFG project number 397252409), operated by the university computing center (URZ). The first and second author would like to thank Guido Kanschat and Daniel Arndt for the fruitful discussions on the deal.II data structures. The third author acknowledges financial support from the European Research Council through the ERC Grant Agreement No. 759419 MuDiLingo ("A MultiscaleDislocation Language for Data-Driven Materials Science").

Table 2: Results for 5 refinement cycles on 216 cores and 1728 cores.

		d.o.f.			Time in s					
#Cores	Refinem.	Global	Coarse	it.	solve	ass.	build Γ	f/e/v	build B	total time
216	1	46 875	1 725	23	0.38	7.53	0.01	0.01	0.01	8.12
	2	153 801	6 492	43	3.07	10.1	0.07	0.03	0.02	13.5
	3	479 475	7 701	44	3.63	16.9	0.27	0.05	0.03	21.3
	4	1 477 617	7 839	52	7.29	41.6	0.91	0.08	0.06	50.7
	5	4 570 413	8 139	54	23.9	152	4.90	0.17	0.11	183
	sum				46	223	6.16	0.33	0.22	277
1728	1	352 947	17 061	22	6.56	4.39	0.02	0.02	0.07	11.8
	2	1 133 949	61 266	50	32.1	8.05	0.10	0.05	0.11	41.6
	3	3 509 349	67 716	49	35.9	14.1	0.25	0.07	0.18	52.0
	4	10 820 382	71 811	63	47.1	61.4	0.79	0.11	0.34	112
	5	33 427 005	76 125	60	76.9	207	3.78	0.22	0.68	292
	sum				199	295	4.94	0.47	1.34	510

References

1. D. Arndt, W. Bangerth, B. Blais, T.C. Clevenger, M. Fehling, A.V. Grayver, T. Heister, L. Heltai, M. Kronbichler, M. Maier, P. Munch, J.P. Pelteret, R. Rastak, I. Thomas, B. Turcksin, Z. Wang and D. Wells. The deal.II library, version 9.2. *J. Numer. Math.* **28**(3), 131–146 (2020). DOI:10.1515/jnma-2020-0043
2. W. Bangerth, C. Burstedde, T. Heister and M. Kronbichler. Algorithms and data structures for massively parallel generic adaptive finite element codes. *ACM Transactions on Mathematical Software (TOMS)* **38**(2), 14 (2011).
3. W.Cai, A.Arsenlis, C.R.Weinberger and V.V.Bulatov. A non-singular continuum theory of dislocations. *J. Mech. Phys. Solids* **54**(3), 561–587 (2006). DOI:10.1016/j.jmps.2005.09.005
4. C.R. Dohrmann. A preconditioner for substructuring based on constrained energy minimization. *SIAM J. Sci. Comput.* **25**(1), 246–258 (2003). DOI:10.1137/S1064827502412887
5. J.D. Eshelby and R.E. Peierls. The determination of the elastic field of an ellipsoidal inclusion, and related problems. *Pro. of the Royal Society of London. Series A. Math. and Phy. Sciences* **241**(1226), 376–396 (1957). DOI:10.1098/rspa.1957.0133
6. C. Farhat, M. Lesoinne, P. LeTallec, K. Pierson and D. Rixen. FETI-DP:ADual–Primal Unified FETI Method, part I: A faster alternative to the two-level FETI method. *Internat. J. Numer. Methods Engrg.* **50**(7), 1523–1544 (2001). DOI:10.1002/nme.76
7. O. Jamond, R. Gatti, A. Roos and B. Devincre. Consistent formulation for the discrete-continuous model: Improving complex dislocation dynamics simulations. *International Journal of Plasticity* **80**, 19–37 (2016). DOI:10.1016/j.ijplas.2015.12.011
8. A. Klawonn and O. Rheinbach. A parallel implementation of dual-primal FETI methods for three-dimensional linear elasticity using a transformation of basis. *SIAM J. Sci. Comput.* **28**(5), 1886–1906 (2006). DOI:10.1137/050624364
9. S. Köhler, O. Rheinbach, S. Sandfeld and D. Steinberger. FETI-DP Solvers and Deal.II for Problems in Dislocation Mechanics. *PAMM* **19**(1), e201900292 (2019). DOI:10.1002/pamm.201900292
10. P. Kůs and J. Šístek. Coupling parallel adaptive mesh refinement with a nonoverlapping domain decomposition solver. *Advances in Engineering Software* **110**, 34–54 (2017). DOI:10.1016/j.advengsoft.2017.03.012
11. O. Rheinbach. Parallel iterative substructuring in structural mechanics. *Arch. Comput. Methods Eng.* **16**(4), 425–463 (2009). DOI:10.1007/s11831-009-9035-4

12. S. Sandfeld, M. Monavari and M. Zaiser. From systems of discrete dislocations to a continuous field description: stresses and averaging aspects. *Modelling and Simulation in Materials Science and Engineering* **21**(8), 085006 (2013). DOI:10.1088/0965-0393/21/8/085006
13. D. Steinberger, R. Gatti and S. Sandfeld. A Universal Approach Towards Computational Characterization of Dislocation Microstructure. *JOM* **68**(8), 2065–2072 (2016). DOI:10.1007/s11837-016-1967-1

A Three-Level Extension for Fast and Robust Overlapping Schwarz (FROSch) Preconditioners with Reduced Dimensional Coarse Space

Alexander Heinlein, Axel Klawonn, Oliver Rheinbach, and Friederike Röver

1 Fast and Robust Overlapping Schwarz preconditioners

The Fast and Robust Overlapping Schwarz framework [7, 8], which is part of the Trilinos Software library [18], contains a parallel implementation of the *generalized Dryja–Smith–Widlund* (GDSW) preconditioner. The GDSW preconditioner is a two-level overlapping Schwarz domain decomposition preconditioner [17] with an energy minimizing coarse space [3, 4]. It is constructed based on a domain decomposition of the computational domain Ω into N nonoverlapping subdomains $\{\Omega_i\}_{i=1,\dots,N}$. These are then extended by k layers of elements, resulting in a corresponding overlapping domain decomposition $\{\Omega_i'\}_{i=1,\dots,N}$. The two-level GDSW preconditioner can then be written as

$$M_{\mathrm{GDSW}}^{-1} = \underbrace{\Phi K_0^{-1} \Phi^T}_{\text{coarse level}} + \underbrace{\sum\nolimits_{i=1}^{N} R_i^T K_i^{-1} R_i}_{\text{first level}}, \tag{1}$$

where Φ contains the coarse basis functions. Contrary to the classical approach, where the coarse basis functions are chosen as nodal finite element functions on

Alexander Heinlein
Delft University of Technology, Faculty of Electrical Engineering Mathematics & Computer Science, Delft Institute of Applied Mathematics, Mekelweg 4, 72628 CD Delft, Netherlands, e-mail: a.heinlein@tudelft.nl

Axel Klawonn
Department Mathematik/Informatik, Universität zu Köln, Weyertal 86-90, 50923 Köln, Germany, e-mail: axel.klawonn@uni-koeln.de
and
Center for Data and Simulation Science, University of Cologne.
https://www.cds.uni-koeln.de

Oliver Rheinbach and Friederike Röver
Institut für Numerische Mathematik und Optimierung, Technische Universität Bergakademie Freiberg, Akademiestr. 6, 09599 Freiberg, Germany,
e-mail: {oliver.rheinbach, friederike.roever}@math.tu-freiberg.de.

S. C. Brenner et al. (eds.), *Domain Decomposition Methods in Science and Engineering XXVI*, Lecture Notes in Computational Science and Engineering 145,
https://doi.org/10.1007/978-3-030-95025-5_54

a coarse triangulation, for the GDSW preconditioner, these are chosen as discrete harmonic extensions of certain interface functions Φ_Γ to the interior of each subdomain. In particular, the functions Φ_Γ are restrictions of the null space of the global Neumann matrix to the vertices, edges, and faces, which form a nonoverlapping decomposition of the domain decomposition interface. The matrix $K_0 = \Phi^T K \Phi$ is the coarse matrix and the matrices $K_i = R_i K R_i^T$, $i = 1, \ldots, N$, correspond to the overlapping subdomain problems on the first level. The local subspaces corresponding to the overlapping subdomains are denoted as $V_1, \ldots, V_N$, and the GDSW coarse space is denoted by V_0. For scalar elliptic problems, the condition number is bounded by

$$\kappa(M_{\text{GDSW}}^{-1} K) \leq C \left(1 + \frac{H}{\delta}\right) \left(1 + \log\left(\frac{H}{h}\right)\right)^2, \tag{2}$$

where C is a constant independent of the finite element size h, the size H of the nonoverlapping subdomains, and the width of the overlap $\delta = kh$; see [3]. The GDSW coarse space can be constructed in an algebraic fashion, i.e., without geometric information. For a further reduction of the coarse space, the FROSch framework provides an implementation of a reduced dimensional coarse space (RGDSW) [11]. For the reduced dimensional GDSW coarse space, the basis functions are constructed from nodal interface functions. Two options are currently available in FROSch: a fully algebraic version *(Option 1)* [5, 11], where the interface values are defined through the number of adjacent vertices, or the less algebraic version *(Option 2.2)* [5, 11], where the interface values are defined through the distance to the adjacent vertices; cf. [5, 11]. In general, the two options result in different partitions of unity. The interior values of each subdomain are determined as in the classical GDSW approach.

2 Three-level extension

For a large number of subdomains, the coarse problem of the two-level (R)GDSW preconditioners may become too large to be solved by a sparse direct solver. As in the three-level BDDC methods [19], we can resolve this by applying the GDSW preconditioner recursively to the coarse problem [9, 10]. This technique can be extended to a multi-level version, as in multi-level BDDC [1, 16] (which compete with inexact FETI-DP methods [13]), multilevel Schwarz methods [14, 15], or multigrid methods. We only discuss the three-level extension in this paper.

To apply the (R)GDSW preconditioner to the coarse problem, we need to define an additional layer of decomposition. We therefore decompose the domain into non-overlapping subregions Ω_{i0} of diameter H_c, whereas each subregion is a union of subdomains. To obtain overlapping subregions Ω'_{i0}, we extend each subregion by recursively adding layers of subdomains, as we do with finite elements on the subdomain level; see Figure 1. We denote the subregion overlap by Δ. The notation on the subdomain level is kept consistent with the two-level method.

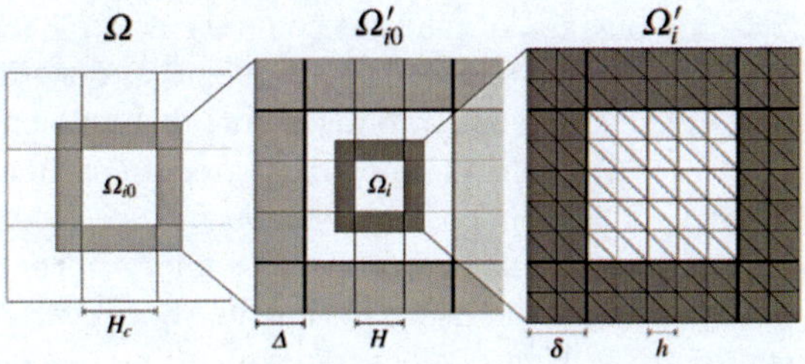

Fig. 1: Structured decomposition of an exemplary two-dimensional computational domain Ω into nonoverlapping subregions Ω_{i0}(left), a zoom into one overlapping subregion Ω'_{i0} consisting of subdomains Ω_i (middle), and a zoom into one overlapping subdomain Ω'_i (right). Each level of zoom corresponds to one level of the preconditioner; image from [9].

We define the three-level GDSW preconditioner [9, 10] by

$$M^{-1}_{\mathrm{GDSW-3L}} = \Phi\Big(\overbrace{\Phi_0 K_{00}^{-1}\Phi_0^T}^{\text{third level}} + \overbrace{\sum\nolimits_{i=1}^{N_0} R_{i0}^T K_{i0}^{-1} R_{i0}}^{\text{second level}}\Big)\Phi^T + \overbrace{\sum\nolimits_{j=1}^{N} R_j^T K_j^{-1} R_j}^{\text{first level}}, \qquad (3)$$

where the first level and the matrices Φ are defined as in the two-level method and where $K_{00} = \Phi_0^T K_0 \Phi_0$ and $K_{i0} = R_{i0} K_0 R_{i0}^T$. The restriction operators, restricting to the overlapping subregions Ω'_{i0}, are defined as $R_{i0} : V^0 \to V_i^0 := V^0(\Omega'_{i0})$ for $i = 1, ..., N_0$. The respective coarse space is denoted as V_{00} and spanned by the coarse basis functions Φ_0.

3 Implementation

The Fast and Robust Overlapping Schwarz (FROSch) framework [7, 8] is part of the package ShyLU from the Trilinos software library [18]. It contains parallel implementations of the GDSW and RGDSW preconditioners based on the Trilinos linear algebra interface Xpetra; it enables the use of both Trilinos linear packages Epetra and Tpetra. To test the three-level extension to the FROSch implementation, we considered a linear elasticity model problem on the unit cube $[0, 1]^3$ with homogenous Dirichlet boundary condition on $\partial\Omega$. We use piecewise trilinear finite elements and a structured decomposition of the computational domain. To assemble the stiffness matrix we apply the Trilinos package Galeri. Here, each process owns the same number of rows of stiffness matrix resulting in different subdomain sizes. We use a generic right-hand side vector in which each entry is set to one. If the coarse space is constructed as described in Section 1, the columns of the matrix Φ will be a generating set of the coarse space. However, for our model problem, the columns will not be linearly independent and, hence, not form a basis of the coarse space. This is because the restriction of the six-dimensional null space, consisting of translations and linearized rotations, to an interface component may yield linearly dependent vectors. For instance, the restriction of the null space to a single vertex yields only a three dimensional space. In order to make sure that the coarse matrix K_0 is invertible, we have to deal with this in our implementation. In particular, before building K_0, we replace linearly dependent coarse functions by null vectors until all other basis functions are linearly independent; in order to identify linear dependencies, we perform local orthonormalization using LAPACK's SGEQRF routine for computing a QR factorization using Householder transformations. This procedure yields zero rows and columns in K_0. Therefore, in order to make K_0 invertible, we finally replace those rows and columns by the corresponding unit vectors, leaving a one on the diagonal and zeros otherwise. This also has the nice side effect that the size of the coarse matrix is always the number of interface components times the dimension of the null space. The coarse level is decomposed into subregions in an unstructured way using the Parallel Hypergraph and Graph Partitioning (PGH) from the Trilinos package Zoltan2 [20]; see also [12]. As a Krylov iteration method, we apply the preconditioned conjugate gradient method (PCG) provided by the Trilinos package Belos (BelosPseudoBlockCG). The implementation offers a condition number estimate using the tridiagonal matrix constructed in the Lanczos process. We use the relative stopping criterion $\|r^k\|_2/\|r^0\|_2 \leq 10^{-6}$, where r^k is the residual in the k-th iteration step and r^0 is the initial residual. For all tests, we chose $20^3 * 3$ rows of the stiffness matrix for each process and approximately 8^3 subdomains per nonoverlapping subregion. The overlap is obtained by extending each subdomain by one layer of elements and by extending each subregion by one layer of subdomains. We performed all numerical tests on the GCS supercomputer SuperMUC-NG. The INTEL 19.0 compiler is used. The sparse linear subproblems arising in the preconditioner are solved using the sparse direct linear solver PardisoMKL [2].

4 Weak parallel scalability results for the three-Level extension

In this section, we focus on weak parallel scalability results for the three-level GDSW preconditioner with a reduced dimensional coarse space. We always use *Option 1* to construct the coarse basis functions. In Trilinos the data is distributed among the processes via the map object. We use a repeatedly decomposed map to determine the interface Γ. This map can be passed as an input to the FROSch framework.

For our weak parallel scalability tests, we consider three different setups to determine the interface Γ, which result in different sizes and sparsity patterns for the coarse problem; see Figure 2. We either use the *Geometric Map*, which is constructed from the structured non-overlapping domain decomposition on the first level, or the *Algebraic Map* [6], which is built algebraically from the uniquely decomposed row map of the input matrix. In particular, the interfaces and hence the vertices, edges, and faces may differ slightly for the two different maps; this effect may be more pronounced for unstructured domain decompositions.

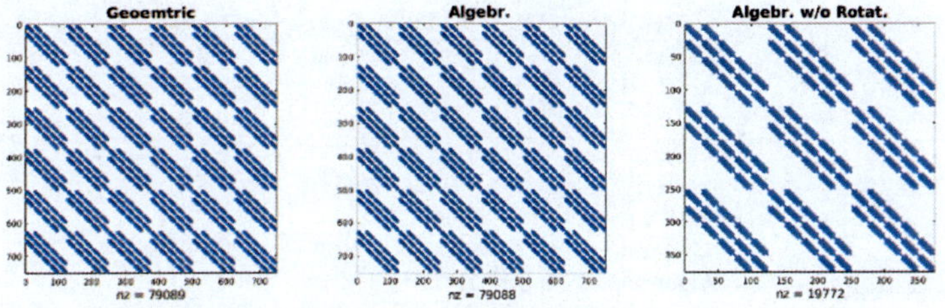

Fig. 2: Sparsity of the coarse matrix K_0 for our linear elasticity model problem in three dimensions with 216 subdomains; using the *Geometric Map* (left) and the *Algebraic Map*, with rotations (middle) and without linearized rotations (right). The subdomain size is chosen such that each process of the uniquely decomposed map owns 30^3 nodes.

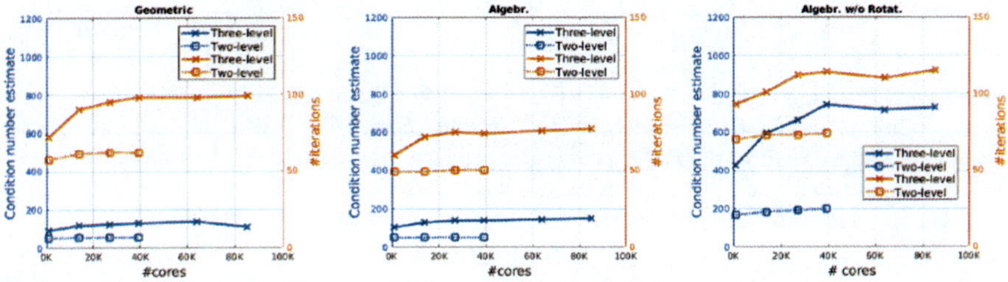

Fig. 3: Weak numerical scalability for the three- and two-level method with a reduced dimensional coarse space; see Table 1 for the data; using the *Geometric Map* and the *Algebraic Map* with and without rotations.

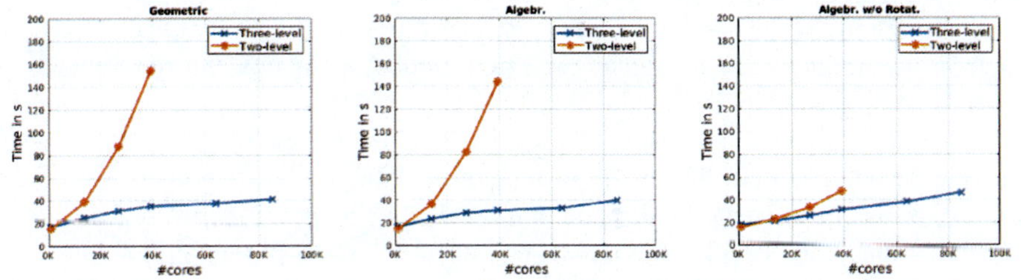

Fig. 4: Weak parallel scalability for the three- and two-level method with a reduced dimensional coarse space; see Table 1 for the data.

When using the *Algebraic Map*, we also consider the case where the rotations are neglected (*Algebr. w/o Rotat.*). In Figure 2, we only see minor differences in the sparsity pattern of K_0 using the *Geometric* and the *Algebraic Map*. For higher numbers of subdomains, the differences between these two approaches will be more visible: for our largest test case with 85 184 subdomains, we have 539 460 as a maximum nonzero entries per core in K_0 for the *Geometric Map*; this compares to 578 340 maximum nonzeros per core for the *Algebraic Map*. For all input maps, the

# Subd.	# Subr.	# Dofs	Map	Two-level			Three-level		
				$\kappa(M^{-1}K)$	Iter	Solver Time	$\kappa(M^{-1}K)$	iter	Solver Time
1 000	4	$2.4 \cdot 10^7$	Geom.	51.45	57	15.11s	90.46	72	16.99s
			Algebr.	50.73	49	14.54s	103.02	60	16.12s
			Algebr. w/o Rotat.	166.68	70	15.48s	429.05	93	17.91s
13 824	27	$3.3 \cdot 10^8$	Geom.	53.61	61	38.40s	116.19	90	24.89s
			Algebr.	51.08	49	36.39s	127.91	72	23.38s
			Algebr. w/o Rotat.	182.46	73	22.75s	594.97	101	21.58s
27 000	64	$6.5 \cdot 10^8$	Geom.	53.77	62	87.28s	122.18	95	30.87s
			Algebr.	51.12	50	82.01s	137.42	75	28.46s
			Algebr. w/o Rotat.	191.12	73	33.17s	663.44	112	26.02s
39 304	125	$9.4 \cdot 10^8$	Geom.	53.82	62	153.88s	128.39	98	35.12s
			Algebr.	51.12	50	144.01s	137.96	74	30.63s
			Algebr. w/o Rotat.	198.05	74	47.48s	745.26	114	31.40s
64 000	216	$1.5 \cdot 10^9$	Geom.	-	-	-	135.58	98	37.29s
			Algebr.	-	-	-	143.87	76	32.81s
			Algebr. w/o Rotat.	-	-	-	717.06	110	38.24s
85 184	275	$2.0 \cdot 10^9$	Geom.	-	-	-	108.49	99	40.80s
			Algebr.	-	-	-	150.37	77	39.87s
			Algebr. w/o Rotat.	-	-	-	729.14	115	46.45s

Table 1: Data corresponding to Figure 3 and 4. By *Iter*, we denote the number of PCG iterations, and κ is the condition number of the preconditioned operator. *Solver Time* is the time to build the preconditioner and to perform the Krylov iterations; see also Figure 3 and 4. The subdomain size is chosen such that each process of the uniquely decomposed map owns 20^3 nodes. We have $H_c/H \approx 8$. One layer of finite elements respectively one layer of subdomains is chosen as the overlap for each level.

two- and the three-level method are numerically scalable, whereas the *Algebraic Map without Rotations* yields the highest iteration counts and condition number estimates; cf. Figure 3 and Table 1. Replacing the direct solver for the coarse problem (used in the two-level method) by the application of the RGDSW preconditioner for the three-level method generally results in higher condition number estimates and iteration counts.

However, the three-level extension of the FROSch framework shows a better parallel weak scalability than the two-level method; cf Figure 4 and Table 1. The *Solver Time* is the time to build the preconditioner and to perform the Krylov iterations. The time includes the factorization and forward backward substitution for the sparse direct solvers. For the three-level method the time for the unstructured decomposition of the coarse problem is also included. For all test settings, the three-level method is faster for 13 824 and more cores. Moreover, at 39 304 cores the three-level method is faster by more than a factor of four: Using the three-level method, we obtain a *Solver Time* of 35.12 s using the *Geometric Map* and 30.63 s using *Algebraic Map*. This compares to a *Solver Time* of 153.88 s for the *Geometric Map* and 144.01 s for the *Algebraic Map* in the two-level method. Using *Algebraic Map without Rotation* results in a smaller coarse problem, making the two-level methods more competitive. Here, the three-level method (*Solver Time* 31.40 s) is still faster by a factor of 1.5 than the two-level method (*Solver Time* 47.48 s). As the results are clear, we did not perform tests beyond the 39 304 cores for the two-level method. To illustrate the strong influence of the size of the coarse problem on the preconditioner time, we consider the test case of 39 304 cores in Table 2. For this test case, the solution of coarse problem K_0 in the *Geometric Map* setup takes 78% (120.19 s) of the total *Solver Time* (153.88 s).

				Two-level method		Three-level method	
# Subd.	# Subr.	# Dofs	Map	Size K_0	K_0 Solve Time	Size K_{00}	K_{00} Solve Time
1 000	4	$2.4 \cdot 10^7$	Geom.	4 374	0.24s	6	<1e-5s
			Algebr.	4 374	0.22s	6	<1e-5s
			Algebr. w/o Rotat.	2 184	0.08s	3	<1e-5s
13 824	27	$3.3 \cdot 10^8$	Geom.	73 002	12.03s	366	0.01s
			Algebr.	73 002	12.04s	366	0.01s
			Algebr. w/o Rotat.	36 501	2.02s	174	0.003s
27 000	64	$6.5 \cdot 10^8$	Geom.	146 334	59.38s	1 056	0.08s
			Algebr.	146 334	45.75s	1 116	0.08s
			Algebr. w/o Rotat.	73 167	11.69s	546	0.04s
39 304	125	$9.4 \cdot 10^8$	Geom.	215 622	120.19s	2 508	0.29s
			Algebr.	215 622	114.06s	2 556	0.25s
			Algebr. w/o Rotat.	107 811	22.14s	1 290	0.11s
64 000	216	$1.5 \cdot 10^9$	Geom.	355 914	-	4 980	0.81s
			Algebr.	355 914	-	4 938	0.63s
			Algebr. w/o Rotat.	177 957	-	2 319	0.21s
85 184	275	$2.0 \cdot 10^9$	Geom.	477 042	-	6 432	0.63s
			Algebr.	477 042	-	6 660	0.72s
			Algebr. w/o Rotat.	238 521	-	3 222	0.16s

Table 2: Cost for solving the problem on the coarsest level. *Solve Coarse Problem Time* includes the time of the factorization of the problem as well as the forward and backward substitution in the Krylov iterations.

For this test case, the solution of coarse problem K_0 in the *Geometric Map* setup takes 78% (120.19 s) of the total *Solver Time* (153.88 s). This time compares to less than a second (0.29 s) to solve the coarse problem corresponding to K_{00} in the three-level method. Similar results are obtained for the *Algebraic Map* where 114.06 s for the two-level method compare with 0.25 s for the three-level method. For the *Algebraic Map without Rotations* the size of the coarse problem is reduced by a factor of two; cf. Table 2. Therefore, the cost of the coarse problem reduces to 22.14 s for the two-level method, which compares to 0.11 s for the three-level method. Although the *Algebraic Map* has the largest coarse problem size (see Table 2) this is consistently the fastest setup of the three-level method. The stronger connectivity given by this coarse problem improves the iteration count and therefore decreases the *Solver Time*. Resulting in the highest number of iterations (cf. Table 1) the *Algebraic Map without rotations* is the slowest test case.

Acknowledgements The authors gratefully acknowledge the Gauss Centre for Supercomputing e.V. (`www.gauss-centre.eu`) for providing computing time on the GCS Supercomputer SuperMUC-NG at Leibniz Supercomputing Centre (`www.lrz.de`).

The author acknowledge computing time on the Compute Cluster of the Fakultät für Mathematik und Informatik of Technische Universität Freiberg DFG project number 397252409), operated by the university computing center (URZ).

The third and fourth author would like to acknowledge funding by the Deutsche Forschungsgemeinschaft (DFG) under the DFG project number 441509557 within the DFG SPP 2256.

References

1. S. Badia, A.F. Martín and J. Principe. Multilevel balancing domain decomposition at extreme scales. *SIAM J. Sci. Comput.* **38**(1), C22–C52 (2016).
2. M. Bollhöfer, O. Schenk, R. Janalik, S. Hamm and K. Gullapalli. State-of-the-art sparse direct solvers. In: A. Grama and A.H. Sameh, editors, *Parallel Algorithms in Computational Science and Engineering*, pp. 3–33, Springer, Cham (2020).
3. C.R. Dohrmann, A. Klawonn and O.B. Widlund. Domain decomposition for less regular subdomains: overlapping Schwarz in two dimensions. *SIAM J. Numer. Anal.* **46**(4), 2153–2168 (2008).
4. C.R. Dohrmann, A. Klawonn and O.B. Widlund. A family of energy minimizing coarse spaces for overlapping Schwarz preconditioners. In: *Domain decomposition methods in science and engineering XVII*, volume 60 of LNCSE, pp. 247–254, Springer, Berlin (2008).
5. C.R. Dohrmann and O.B. Widlund. On the design of small coarse spaces for domain decomposition algorithms. *SIAM J. Sci. Comput.* **39**(4), A1466–A1488 (2017).
6. A. Heinlein, C. Hochmuth and A. Klawonn. Fully algebraic two-level overlapping Schwarz preconditioners for elasticity problems. In: F.J. Vermolen and C. Vuik, editors, *Numerical Mathematics and Advanced Applications ENUMATH 2019*, pp. 531–539, Springer, Cham (2021).
7. A. Heinlein, A. Klawonn, S. Rajamanickam and O. Rheinbach. FROSch: A fast and robust overlapping Schwarz domain decomposition preconditioner based on Xpetra in Trilinos. In: *Domain Decomposition Methods in Science and Engineering XXV*, pp. 176–184, Springer, Cham (2020).
8. A. Heinlein, A. Klawonn and O. Rheinbach. A parallel implementation of a two-level overlapping Schwarz method with energy-minimizing coarse space based on Trilinos. *SIAM J. Sci. Comput.* **38**(6), C713–C747 (2016).
9. A. Heinlein, A. Klawonn, O. Rheinbach and F. Röver. A three-level extension of the GDSW overlapping Schwarz preconditioner in two dimensions. In: *Advanced Finite Element Methods with Applications: Selected Papers from the 30th Chemnitz Finite Element Symposium 2017*, pp. 187–204, Springer, Cham (2019).
10. A. Heinlein, A. Klawonn, O. Rheinbach and F. Röver. A three-level extension of the GDSW overlapping Schwarz preconditioner in three dimensions. In: *Domain Decomposition Methods in Science and Engineering XXV*, pp. 185–192, Springer, Cham (2020).
11. A. Heinlein, A. Klawonn, O. Rheinbach and O.B. Widlund. Improving the parallel performance of overlapping Schwarz methods by using a smaller energy minimizing coarse space. In: *Domain Decomposition Methods in Science and Engineering XXIV*, pp. 383–392, Springer, Cham (2018).
12. A. Heinlein, O. Rheinbach and F. Röver. Choosing the subregions in three-level FROSch preconditioners. In: *14th WCCM-ECCOMAS Congress 2020*, volume 700 (2021).
13. A. Klawonn and O. Rheinbach. Inexact FETI-DP methods. *IJNME* **69**(2), 284–307 (2007).
14. F. Kong and X.-C. Cai. A highly scalable multilevel Schwarz method with boundary geometry preserving coarse spaces for 3D elasticity problems on domains with complex geometry. *SIAM J. Sci. Comput.* **38**(2), C73–C95 (2016).
15. S. Scacchi. A hybrid multilevel Schwarz method for the bidomain model. *Comput. Methods Appl. Mech. Engrg.* **197**(45-48), 4051–4061 (2008).

16. J. Šístek, B. Sousedík, J. Mandel and P. Burda. Parallel implementation of multilevel BDDC. In: A. Cangiani, R. Davidchack, E. Georgoulis, A. Gorban, J. Levesley and M. Tretyakov, editors, *Numerical mathematics and advanced applications 2011*, pp. 681–689, Springer, Berlin, Heidelberg (2013).
17. A. Toselli and O. Widlund. *Domain decomposition methods – algorithms and theory*. Volume 34 of Springer Series in Computational Mathematics. Springer-Verlag, Berlin (2005).
18. Trilinos public git repository. Web, 2018. https://github.com/trilinos/trilinos.
19. X. Tu. Three-level BDDC in two dimensions. *IJNME* **69**, 33–59 (2007).
20. The Zoltan2 Project Team. https://trilinos.github.io/zoltan2.html.

Space-Time Hexahedral Finite Element Methods for Parabolic Evolution Problems

Ulrich Langer and Andreas Schafelner

1 Introduction

We consider the parabolic initial-boundary value problem (IBVP), find u such that

$$\partial_t u - \operatorname{div}_x(\alpha \nabla_x u) = f \text{ in } Q, \;\; u = u_D := 0 \text{ on } \Sigma, \;\; u = u_0 := 0 \text{ on } \Sigma_0, \qquad (1)$$

as a model problem typically arising in heat conduction and diffusion, where $Q = \Omega\times(0,T)$, $\Sigma = \partial\Omega\times(0,T)$, and $\Sigma_0 = \Omega\times\{0\}$. The spatial domain $\Omega \subset \mathbb{R}^d$, $d = 1, 2$, is assumed to be bounded and Lipschitz, $T > 0$ is the terminal time, $f \in L_2(Q)$ denotes a given source, and $\alpha \in L_\infty(Q)$ is a given uniformly positive (almost everywhere) coefficient that may discontinuously depend on the spatial variable $x = (x_1, \ldots, x_d)$ and the time variable t, but $\alpha(x,t)$ should be of bounded variation in t for almost all $x \in \Omega$. Then there is a unique weak solution $u \in V_0 := \{v \in L_2(0,T;H_0^1(\Omega)) : \partial_t u \in L_2(0,T;H^{-1}(\Omega)), v = 0 \text{ on } \Sigma_0\}$ of the IBVP (1); see, e.g., [4, 14]. Moreover, $\partial_t u$ and $Lu := -\operatorname{div}_x(\alpha \nabla_x u)$ belong to $L_2(Q)$; see [3]. The latter property is called maximal parabolic regularity. In this case, the parabolic partial differential equation $\partial_t u - \operatorname{div}_x(\alpha \nabla_x u) = f$ holds in $L_2(Q)$. This remains valid even for inhomogeneous initial conditions $u_0 \in H_0^1(\Omega)$.

Time-stepping methods in combination with some spatial discretization method like the finite element method (FEM) are still the standard approach to the numerical solution of IBVPs like (1); see, e.g., [16]. This time-stepping approach as well as the more recent discontinuous Galerkin, or discontinuous Petrov–Galerkin methods based on time slices or slabs are in principle sequential. The sequential nature of these methods hampers the full space-time adaptivity and parallelization; but

Ulrich Langer
Institute for Computational Mathematics, Johannes Kepler University, Altenbergerstr. 69, 4040 Linz, Austria, e-mail: ulanger@numa.uni-linz.ac.at

Andreas Schafelner
Doctoral Program "Computational Mathematics", Johannes Kepler University, Altenbergerstr. 69, 4040 Linz, Austria, e-mail: andreas.schafelner@dk-compmath.jku.at

S. C. Brenner et al. (eds.), *Domain Decomposition Methods in Science and Engineering XXVI*, Lecture Notes in Computational Science and Engineering 145,
https://doi.org/10.1007/978-3-030-95025-5_55

see the overview paper [5] for parallel-in-time methods. Space-time finite element methods on fully unstructured decomposition of the space-time cylinder Q avoid these bottlenecks; see [15] for an overview of such space-time methods.

In this paper, we follow our preceding papers [9–11], and construct locally stabilized, conforming space-time finite element schemes for solving the IBVP (1), but on hexahedral meshes that are more suited for anisotropic refinement than the simplicial meshes used in [9–11]. We mention that SUPG/SD and Galerkin/least-squares stabilizations of time-slice finite element schemes for solving transient problems were already used in early papers; see, e.g., [8] and [6]. Section 2 recalls the construction of locally stabilized space-time finite element schemes, the properties of the corresponding discrete bilinear form, and the a priori discretization error estimates from [9–11]. In Section 3, we derive new anisotropic a priori discretization estimates for hexahedral tensor-product meshes, and we provide anisotropic adaptive mesh refinement strategies that are based on a posteriori error estimates, anisotropy indicators, and anisotropic adaptive mesh refinement using hanging nodes. In Section 4, we present and discuss numerical results for an example where a singularity occurs in the spatial gradient of the solution. The large-scale system of space-time finite element equations is always solved by means of the Flexible Generalized Minimal Residual (FGMRES) method preconditioned by space-time algebraic multigrid.

2 Space-time finite element methods

In this section, we will briefly describe the space-time finite element method based on localized time-upwind stabilizations; for details of the construction and analysis, we refer to our previous work [9–11]. Let $\mathcal{T}_h$ be a shape regular decomposition of the space-time cylinder Q, i.e., $\overline{Q} = \bigcup_{K \in \mathcal{T}_h} \overline{K}$, and $K \cap K' = \emptyset$ for all K and K' from $\mathcal{T}_h$ with $K \neq K'$; see, e.g., [2] for more details. Furthermore, we assume that α is piecewise smooth, and possible discontinuities are aligned with the triangulation as usual. On the basis of the triangulation $\mathcal{T}_h$, we define the space-time finite element space

$$V_{0h} = \{v \in C(\overline{Q}) : v(x_K(\cdot)) \in \mathcal{P}_p(\hat{K}),\ \forall K \in \mathcal{T}_h,\ v = 0 \text{ on } \overline{\Sigma} \cup \overline{\Sigma}_0\},$$

where $x_K(\cdot)$ denotes the map from the reference element $\hat{K}$ to the finite element $K \in \mathcal{T}_h$, and $\mathcal{P}_p(\hat{K})$ is either the space of polynomials of at most degree p on the reference element $\hat{K}$, or the space of polynomials of degree p in each variable on $\hat{K}$, for simplicial or tensor-product decompositions, respectively. Since we are in the maximal parabolic regularity setting, the parabolic Partial Differential Equation (PDE) is valid in $L_2(Q)$. Multiplying the PDE (1), restricted to $K \in \mathcal{T}_h$, by a locally scaled upwind test function $v_{h,K}(x,t) := v_h(x,t) + \theta_K h_K \partial_t v_h(x,t)$, $v_h \in V_{0h}$, integrating over K, summing over all elements, applying integration by parts, and incorporating the Dirichlet boundary conditions, we obtain the variational consistency identity

$$a_h(u, v_h) = \ell_h(v_h), \quad \forall v_h \in V_{0h}, \tag{2}$$

with the mesh-dependent bilinear form

$$a_h(u, v_h) = \sum_{K \in \mathcal{T}_h} \int_K \big[\partial_t u v_h + \theta_K h_K \partial_t u \partial_t v_h + \alpha \nabla_x u \cdot \nabla_x v_h - \theta_K h_K \mathrm{div}_x(\alpha \nabla_x u) \partial_t v_h\big] dK, \tag{3}$$

and the mesh-dependent linear form

$$\ell_h(v_h) = \sum_{K \in \mathcal{T}_h} \int_K \big[f v_h + \theta_K h_K f \partial_t v_h\big] dK.$$

Now we apply the Galerkin principle, i.e., we look for a finite element approximation $u_h \in V_{0h}$ to u such that

$$a_h(u_h, v_h) = \ell_h(v_h), \quad \forall v_h \in V_{0h}. \tag{4}$$

Using Galerkin orthogonality (subtracting (4) from (2)), and coercivity and extended boundedness of the bilinear form (3), we can show the following Céa-like best approximation estimate; see [9–11] for the proofs.

Theorem 1 *Let $u \in H_0^{L,1}(Q) := \{v \in V_0 \cap H^1(Q) : Lv := -\mathrm{div}_x(\alpha \nabla_x v) \in L_2(Q)\}$ and $u_h \in V_{0h}$ be the solutions of the parabolic IBVP* (1) *and the space-time finite element scheme* (4)*, respectively. Then the discretization error estimate*

$$\|u - u_h\|_h \le \inf_{v_h \in V_{0h}} \left(\|u - v_h\|_h + \frac{\mu_b}{\mu_c} \|u - v_h\|_{h,*} \right) \tag{5}$$

is valid provided that $\theta_K = O(h_K)$ is sufficiently small, where

$$\|v\|_h^2 = \frac{1}{2} \|v(\cdot, T)\|_{L_2(\Omega)}^2 + \sum_{K \in \mathcal{T}_h} \Big[\theta_K h_K \|\partial_t v\|_{L_2(K)}^2 + \|\alpha^{1/2} \nabla_x v\|_{L_2(K)}^2\Big],$$

$$\|v\|_{h,*}^2 = \|v\|_h^2 + \sum_{K \in \mathcal{T}_h} \Big[(\theta_K h_K)^{-1} \|v\|_{L_2(K)}^2 + \theta_K h_K \|\mathrm{div}_x(\alpha \nabla_x v)\|_{L_2(K)}^2\Big],$$

and μ_c and μ_b are the coercivity and extended boundedness constants of the bilinear form (3)*, respectively.*

The best-approximation error estimate (5) now leads to convergence rate estimates under additional regularity assumptions. If the solution u of (1) belongs to $H_0^{L,1}(Q) \cap H^l(Q)$, $l > 1$, then $\|u - u_h\|_h \le c(u) h^{s-1}$, where $s = \min\{l, p+1\}$, $h = \min_{K \in \mathcal{T}_h} h_K$, and $c(u)$ depends on the regularity of u. We refer the reader to [9, Theorem 13.3] and [10, Theorem 3] for the proof of more detailed estimates in terms of the local mesh-sizes h_K and the local regularity of the solution u.

3 Anisotropic a priori and a posteriori error estimates

The convergence rate estimates presented at the end of the previous section consider only isotropic finite elements, but in many applications the solution u evolves differently with respect to time and space directions. So, we should permit anisotropic finite elements with different mesh sizes in different directions. This raises the question whether we can obtain (localized) a priori estimates that are explicit in spatial and temporal mesh sizes as well as in spatial and temporal regularity assumptions imposed on the solution u. We refer to [1] for a comprehensive summary of anisotropic finite elements. For the remainder of this section, we will now assume that K is a *brick element* (hexahedral element for the case $d = 2$), i.e., the edges of K are parallel to the coordinate axes. Moreover, we assume that $u \in H_0^{L,1}(Q) \cap H^m(Q) \cap H^l(\mathcal{T}_h)$, $m, l \in \mathbb{N}$ with $m > (d+1)/2$ and $l > d/2+2$. Let $h_{K,i} = \max\{|x_i - x_i'| : x, x' \in \overline{K}\}$, and let $h_{K,x} = \max_{i=1,\dots,d} h_{K,i} \geq h_{K,i} \geq c h_{K,x}$, and $h_{K,t} = h_{K,d+1}$. Now we replace h_K by $h_{K,x}$ at all places in Sect. 2. Furthermore, let $e_h = u - I_h u$ and $s = \min\{l, p+1\}$, where I_h is the Lagrange interpolation operator, and p is the polynomial degree of the finite element shape functions in every coordinate direction. Using the anisotropic interpolation error estimates from [1], we get

$$\|e_h\|^2_{L_2(K)} \leq c\left(\sum_{j=1}^{d} h^{2s}_{K,j}\|\partial^s_{x_j} u\|^2_{L_2(K)} + h^{2s}_{K,t}\|\partial^s_t u\|^2_{L_2(K)}\right),$$

$$\|\partial_{x_i}(e_h)\|^2_{L_2(K)} \leq c\left(\sum_{j=1}^{d} h^{2(s-1)}_{K,j}\|\partial_{x_i}\partial^{(s-1)}_{x_j} u\|^2_{L_2(K)} + h^{2(s-1)}_{K,t}\|\partial_{x_i}\partial^{(s-1)}_t u\|^2_{L_2(K)}\right),$$

$$\|\partial^2_{x_i}(e_h)\|^2_{L_2(K)} \leq c\left(\sum_{j=1}^{d} h^{2(s-2)}_{K,j}\|\partial^2_{x_i}\partial^{(s-2)}_{x_j} u\|^2_{L_2(K)} + h^{2(s-2)}_{K,t}\|\partial^2_{x_i}\partial^{(s-2)}_t u\|^2_{L_2(K)}\right),$$

for $i = 1, \dots, d+1$, where c denotes generic positive constants. In particular, we use [1, Thm. 2.7] for $d = 1$, and [1, Thm. 2.10] for $d = 2$. These estimates of the interpolation error and its derivatives immediately lead to the corresponding interpolation error estimates with respect to the norms $\|\cdot\|_h$ and $\|\cdot\|_{h,*}$. Now let $\mathcal{T}_h$ be a decomposition of Q into brick elements. Then we can derive the anisotropic interpolation error estimates

$$\|u - I_h u\|_h \leq \left(\sum_{K\in\mathcal{T}_h} h^{2(s-1)}_{K,x}\mathfrak{c}_1(u,K) + h^{2(s-1)}_{K,t}\mathfrak{c}_2(u,K)\right)^{1/2}, \tag{6}$$

$$\|u - I_h u\|_{h,*} \leq \left(\sum_{K\in\mathcal{T}_h} h^{2(s-1)}_{K,x}\mathfrak{c}_{1,*}(u,K) + h^{2(s-1)}_{K,t}\mathfrak{c}_{2,*}(u,K)\right)^{1/2}, \tag{7}$$

where $s = \min\{l, p+1\}$, and $\mathfrak{c}_1(u,K)$, $\mathfrak{c}_2(u,K)$, $\mathfrak{c}_{1,*}(u,K)$ and $\mathfrak{c}_{2,*}(u,K)$ can easily be computed from the interpolation error estimates given above. Here, $\mathfrak{c}_1(u,K)$,

$\mathfrak{c}_2(u,K)$, and $\mathfrak{c}_{1,*}(u,K)$ do not depend on the aspect ratio of the spatial and temporal mesh-sizes, whereas $\mathfrak{c}_{2,*}(u,K)$ depends on $h_{K,x}/h_{K,t}$ and $h_{K,t}/h_{K,x}$ quadratically. Combining the interpolation error estimates (6) and (7) with the best approximation estimate (5), where h_K must be replaced by $h_{K,x}$ in the definition of the norms, we can immediately derive an a priori discretization error estimate.

Theorem 2 *Let the best approximation estimate* (5) *hold, and let the anisotropic interpolation error estimates* (6) *and* (7) *be fulfilled. Then the anisotropic a priori discretization error estimate*

$$\|u-u_h\|_h \leq \left(\sum_{K\in\mathcal{T}_h} h_{K,x}^{2(s-1)}\mathfrak{C}_1(u,K)+h_{K,t}^{2(s-1)}\mathfrak{C}_2(u,K)\right)^{1/2}$$

is valid, where $s=\min\{l,p+1\}$*, and* $\mathfrak{C}_1(u,K)$ *and* $\mathfrak{C}_2(u,K)$ *can be computed from* (6) *and* (7).

In computational practice, we would like to replace the uniform mesh refinement by an adaptive space-time mesh refinement that takes care of possible anisotropic features of the solution in space and time. Here brick finite elements with hanging nodes, as implemented in MFEM (see next section), are especially suited. To drive anisotropic adaptive mesh refinement, we need a localizable a posteriori error estimator providing local error indicators, and an anisotropy indicator defining the refinement directions in each brick element $K\in\mathcal{T}_h$.

We use the functional a posteriori error estimators introduced by Repin; see his monograph [13, Sect. 9.3]. Repin proposed two error majorants $\mathfrak{M}_1$ and $\mathfrak{M}_2$ from which the local error indicators

$$\eta_{1,K}^2(u_h)=\frac{1}{\delta}\int_K(1+\beta)\big[|\mathbf{y}-\alpha\nabla_x u_h|^2+\frac{1}{\beta}c_{F\Omega}^2|f-\partial_t u_h+\mathrm{div}_x\mathbf{y}|^2\big]\mathrm{d}K \text{ and}$$

$$\begin{aligned}\eta_{2,K}^2(u_h)=&\frac{1}{\delta}\int_K(1+\beta)\big[|\mathbf{y}-\alpha\nabla_x u_h+\nabla_x\vartheta|^2+\frac{c_{F\Omega}^2}{\beta}|f-\partial_t u_h-\partial_t\vartheta+\mathrm{div}_x\mathbf{y}|^2\big]\mathrm{d}K\\&+\gamma\|\vartheta(\cdot,T)\|_\Omega^2+2\int_K\big[\nabla_x u_h\cdot\nabla_x\vartheta+(\partial_t u_h-f)\vartheta\big]\mathrm{d}K\end{aligned}$$

can be derived for each element $K\in\mathcal{T}_h$, where $\mathbf{y}\in H(\mathrm{div}_x,Q)$ is an arbitrary approximation to the flux, $\vartheta\in H^1(Q)$ is also an arbitrary function, $\delta\in(0,2]$, $\beta>\mu$, $\mu\in(0,1)$, and $\gamma>1$. The positive constant $c_{F\Omega}$ denotes the constant in the inequality $\|v\|_{L_2(Q)}\leq c_{F\Omega}\|\sqrt{\alpha}\nabla_x v\|_{L_2(Q)}$ for all $v\in L_2(0,T;H_0^1(\Omega))$, which is nothing but the Friedrichs constant for the spatial domain Ω in the case $\alpha=1$. Both majorants provide a guaranteed upper bound for the errors

$$|\!|\!|u-u_h|\!|\!|_{(1,2-\delta)}^2\leq\sum_{K\in\mathcal{T}_h}\eta_{1,K}^2(u_h)\quad\text{and}\quad|\!|\!|u-u_h|\!|\!|_{(1-\frac{1}{\gamma},2-\delta)}^2\leq\sum_{K\in\mathcal{T}_h}\eta_{2,K}^2(u_h),$$

where $|\!|\!|v|\!|\!|_{(\epsilon,\kappa)}^2:=\kappa\|\sqrt{\alpha}\nabla_x v\|_{L_2(Q)}^2+\epsilon\|v(\cdot,T)\|_{L_2(\Omega)}^2$. Once we have computed the local error indicators $\eta_K(u_h)$ for all elements $K\in\mathcal{T}_h$, we use *Dörfler marking* to determine a set $\mathcal{M} \subseteq \mathcal{T}_h$ of elements that will be marked for refinement. The set $\mathcal{M}$ is of (almost) minimal cardinality such that

$$\sigma \sum_{K \in \mathcal{T}_h} \eta_K(u_h)^2 \leq \sum_{K \in \mathcal{M}} \eta_K(u_h)^2,$$

where $\sigma \in (0, 1)$ is a bulk parameter. Let $\mathbf{E}_K \in \mathbb{R}^{d+1}$ with entries $E_i^{(K)}$, $i = 1, \ldots, d + 1$, and $\chi \in (0, 1)$. In order to determine how to subdivide a marked element, we use the following heuristic: for each $K \in \mathcal{M}$, subdivide K in direction x_i iff $E_i^{(K)} > \chi |\mathbf{E}_K|$. In particular, we choose

$$\left(E_i^{(K)}\right)^2 := \begin{cases} \int_K (y_i - \alpha\, \partial_{x_i} u_h)^2 \,\mathrm{d}K, & i \leq d, \\ \int_K (\mathbf{d}_{th} - \partial_t u_h)^2 \,\mathrm{d}K, & i = d + 1, \end{cases}$$

as our local anisotropy vector $\mathbf{E}_K$, where $\mathbf{y}_h = (y_i)_{i=1}^d$, $\mathbf{d}_{th} = R_h(\partial_t u_h)$, and R_h is some nodal averaging operator like in a Zienkiewicz–Zhu approach.

4 Numerical results

Now let $\{p^{(j)} : j = 1, \ldots, N_h\}$ be the finite element nodal basis of V_{0h}, i.e., $V_{0h} = \mathrm{span}\{p^{(1)}, \ldots, p^{(N_h)}\}$, where N_h is the number of all space-time unknowns (dofs). Then we can express the approximate solution u_h in terms of this basis, i.e., $u_h(x, t) = \sum_{j=1}^{N_h} u_j\, p^{(j)}(x, t)$. Inserting this representation into (4), and testing with $p^{(i)}$, we get the linear system $K_h \underline{u}_h = \underline{f}_h$ for determining the unknown coefficient vector $\underline{u}_h = (u_j)_{j=1,\ldots,N_h} \in \mathbb{R}^{N_h}$, where $K_h = (a_h(p^{(j)}, p^{(i)}))_{i,j=1,\ldots,N_h}$ and $\underline{f}_h = (\ell_h(p^{(i)}))_{i=1,\ldots,N_h}$. The system matrix K_h is non-symmetric, but positive definite due to the coercivity of the bilinear form $a_h(\cdot, \cdot)$. Thus, in order to obtain a numerical solution to the IBVP (1), we just need to solve one linear system of algebraic equations. This is always solved by means of the FGMRES method preconditioned by space-time algebraic multigrid (AMG). We use the finite element library MFEM [12] to implement our space-time finite element solver. The AMG preconditioner is realized via *BoomerAMG*, provided by the linear solver library *hypre* [7]. We start the linear solver with initial guess $\mathbf{0}$, and stop once the initial residual has been reduced by a factor of 10^{-8}. In order to accelerate the solver in case of adaptive refinements, we also employ *Nested Iterations*. Here, we interpolate the finite element approximation from the previous mesh to the current mesh, and use that as an initial guess for FGMRES. Moreover, we stop the linear solver earlier, e.g. once the residual is reduced by a factor of 10^{-2}. Furthermore, we will use the notation $h = N_h^{-1/(d+1)}$ to indicate the corresponding convergence rates.

Let $Q = \Omega \times (0, 1)$, where $\Omega = (0, 1)^2 \setminus \{(x_1, 0) \in \mathbb{R}^2 : 0 \leq x_1 < 1\}$ is a "slit domain" that is not Lipschitz. Moreover, we choose the constant diffusion coefficient $\alpha \equiv 1$, and the manufactured solution $u(r, \varphi, t) = t\, r^\lambda \sin(\lambda \varphi)$, where (r, φ) are polar coordinates with respect to (x_1, x_2), and $\lambda = 0.5$[1]. The singularity leads to a reduced convergence rate $O(h^{0.5})$ for the polynomial degrees $p = 1, 2$; see the upper plots of Fig. 1.

In order to properly realize the adaptive refinement strategies, we need to choose appropriate $\mathbf{y}$ and ϑ. For the first majorant, we reconstruct an improved flux $\mathbf{y}_h^{(0)} = R_h(\nabla_x u_h)$, where R_h is a nodal averaging operator. We then improve this flux by applying a few CG steps to the minimization problem $\min_{\mathbf{y}} \mathfrak{M}_1$, obtaining the final flux $\mathbf{y}_h^{(1)}$ that is then used in the estimator. For the second majorant, we follow the same procedure, but right before postprocessing the flux, we first apply some CG iteration to another minimization problem $\min_{\vartheta} \mathfrak{M}_2$.

For linear finite elements, we observe at least optimal convergence rates for both error estimators. Anisotropic refinements, with the anisotropy parameter $\chi = 0.1$, manage to obtain a better constant than isotropic refinements; see Fig. 1 (upper left). For quadratic finite elements, anisotropic adaptive refinements, with $\chi = 0.15$, manage to recover the optimal rate of $O(h^2)$, while isotropic adaptive refinements result in a reduced rate of $O(h^{1.25})$; see Fig. 1 (upper right). The efficiency indices are rather stable for isotropic refinements, while some oscillations can be observed for anisotropic refinements; see Fig. 1 (lower right).

Acknowledgements This work was supported by the Austrian Science Fund (FWF) under grant W1214, project DK4.

References

1. T. Apel. *Anisotropic finite elements: Local estimates and applications.* Teubner, Stuttgart (1999).
2. P.G. Ciarlet. *The Finite Element Method for Elliptic Problems.* North-Holland Publishing Co., Amsterdam-New York-Oxford (1978).
3. D. Dier. Non-autonomous maximal regularity for forms of bounded variation. *J. Math. Anal. Appl.* **425**(1), 33–54 (2015).
4. L.C. Evans. *Partial differential equations.* Volume 19 of Graduate Studies in Mathematics. American Mathematical Society, Providence, RI, second edition (2010).
5. M.J. Gander. 50 years of time parallel time integration. In: *Multiple shooting and time domain decomposition methods*, volume 9 of Contrib. Math. Comput. Sci., pp. 69–113, Springer, Cham (2015).
6. T.J.R. Hughes, L.P. Franca and G.M. Hulbert. A new finite element formulation for computational fluid dynamics. VIII. The Galerkin/least-squares method for advective-diffusive equations. *Comput. Methods Appl. Mech. Engrg.* **73**(2), 173–189 (1989).
7. hypre: High performance preconditioners. http://www.llnl.gov/casc/hypre.
8. C. Johnson and J. Saranen. Streamline diffusion methods for the incompressible Euler and Navier–Stokes equations. *Math. Comp.* **47**(175), 1–18 (1986).
9. U. Langer, M. Neumüller and A. Schafelner. Space-time finite element methods for parabolic evolution problems with variable coefficients. In: *Advanced finite element methods with applications. Selected papers from the 30th Chemnitz finite element symposium, St. Wolfgang/Strobl, Austria, September 25–27, 2017*, pp. 247–275, Springer, Cham (2019).

[1] https://math.nist.gov/amr-benchmark/index.html

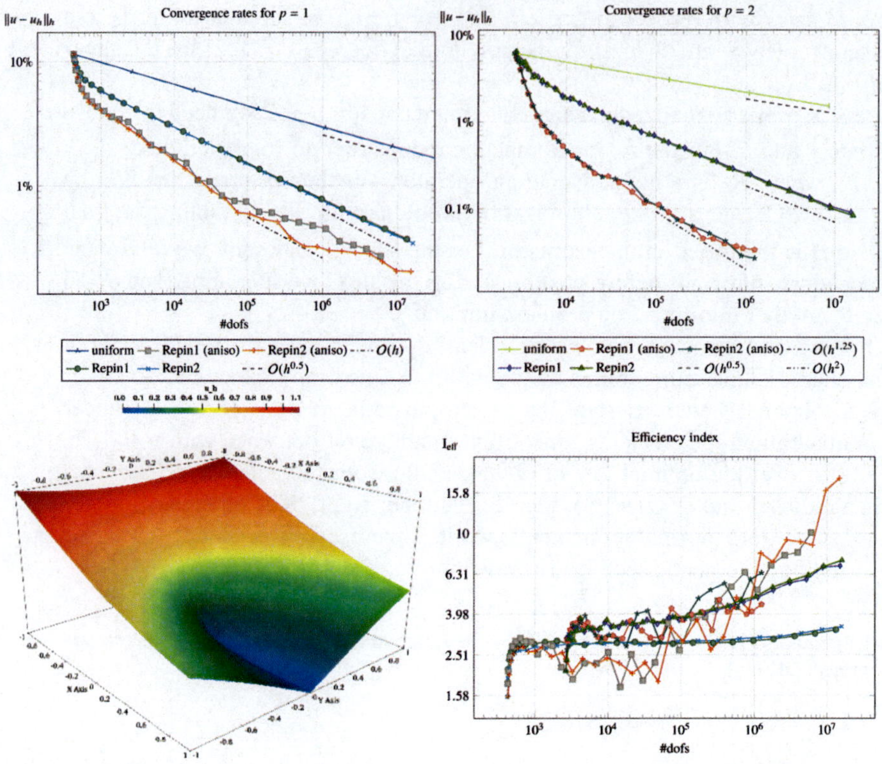

Fig. 1: Convergence rates for $p = 1$ (upper left); convergence rates for $p = 2$ (upper right); plot of $u(\cdot, \cdot, 1)$ (lower left); efficiency indices for $p = 1, 2$, with the respective styles from the upper plots (lower right).

10. U. Langer and A. Schafelner. Adaptive space-time finite element methods for nonautonomous parabolic problems with distributional sources. *Comput. Methods Appl. Math.* **20**(4), 677–693 (2020).
11. U. Langer and A. Schafelner. Space-time finite element methods for parabolic initial-boundary value problems with non-smooth solutions. In: *Large-scale scientific computing. 12th international conference, LSSC 2019, Sozopol, Bulgaria, June 10–14, 2019. Revised selected papers*, pp. 593–600, Springer, Cham (2020).
12. MFEM: Modular finite element methods library. `mfem.org`.
13. S. Repin. *A posteriori estimates for partial differential equations*. Volume 4 of Radon Series on Computational and Applied Mathematics, Walter de Gruyter GmbH & Co, KG, Berlin (2008).
14. O. Steinbach. Space-time finite element methods for parabolic problems. *Comput. Methods Appl. Math.* **15**(4), 551–566 (2015).
15. O. Steinbach and H. Yang. Space-time finite element methods for parabolic evolution equations: discretization, a posteriori error estimation, adaptivity and solution. In: *Space-time methods. Applications to partial differential equations*, pp. 207–248, Berlin: De Gruyter (2019).
16. V. Thomée. *Galerkin finite element methods for parabolic problems*. Volume 25 of Springer Series in Computational Mathematics, Springer-Verlag, Berlin, second edition (2006).

Towards a IETI-DP Solver on Non-Matching Multi-Patch Domains

Rainer Schneckenleitner and Stefan Takacs

1 Introduction

Isogeometric Analysis (IgA), see [7], is a method for discretizing partial differential equations (PDEs). The goal of its development has been to enhance the interface between computer-aided design (CAD) and simulation. Current state-of-the-art CAD tools use B-splines and NURBS for the representation of the computational domain. In IgA, the same kind of bases are also utilized to discretize the PDEs. Complex domains for real-world applications are usually the union of many patches, parametrized with individual geometry functions (multi-patch IgA). We focus on non-overlapping patches.

If the grids are not conforming and/or the interfaces between the patches do not consist of whole edges then discontinuous Galerkin (dG) methods are the discretization techniques of choice. A well-studied representative is the symmetric interior discontinuous Galerkin (SIPG) method, cf. [1]. It has already been adapted and analyzed in IgA, cf. [9, 10, 15] and others. An obvious choice to solve discretized PDEs on domains with many non-overlapping patches is to use tearing and interconnecting methods. The variant we are interested in is the dual-primal approach, see [4] for FETI-DP and [5, 6, 8] for its extension to IgA, which is called accordingly the dual-primal isogeometric tearing and interconnecting method (IETI-DP). This method is similar to Balancing Domain Decomposition with Constraints (BDDC) methods, that have also been adapted to IgA, see [2, 17] and references therein. In [14, 15], the authors have presented a p- and h-robust convergence analysis for IETI-DP. The authors have assumed that the interfaces consist of whole edges. If the vertices are

Rainer Schneckenleitner (corresponding author)
Institute of Computational Mathematics, Johannes Kepler University Linz, Altenberger Straße 69, 4040 Linz, Austria, e-mail: schneckenleitner@numa.uni-linz.ac.at,

Stefan Takacs
RICAM, Austrian Academy of Sciences, Altenberger Straße 69, 4040 Linz, Austria, e-mail: stefan.takacs@ricam.oeaw.ac.at

S. C. Brenner et al. (eds.), *Domain Decomposition Methods in Science and Engineering XXVI*, Lecture Notes in Computational Science and Engineering 145,
https://doi.org/10.1007/978-3-030-95025-5_56

chosen as primal degrees of freedom, it was shown that the condition number of the preconditioned Schur complement system is, under proper assumptions, bounded by

$$C\,p\,\left(1+\log p+\max_{k=1,\ldots,K}\log\frac{H_k}{h_k}\right)^2, \tag{1}$$

where p is the spline degree, h_k is the grid size on patch $\Omega^{(k)}$ and H_k is the diameter of $\Omega^{(k)}$ and $C > 0$ is a constant independent of these quantities. In this paper, we construct a new IETI-DP method that can deal with interfaces that do not consist of whole edges. This means that the patches can meet in T-junctions, which increases the flexibility of the geometric model significantly. In this IETI-DP variant, the construction of the coarse space is based on the idea of "fat vertices": We consider every basis function that is supported on a vertex or T-junction as primal degree of freedom. The numerical experiments indicate that a similar condition number bound to (1) might hold.

The remainder of this paper is organized as follows. In Section 2 we describe the model problem. In Section 3 we introduce the IETI-DP solver and we end this paper with numerical experiments in Section 4.

2 The problem setting

Let $\Omega \subset \mathbb{R}^2$ be open, simply connected and bounded with Lipschitz boundary $\partial\Omega$. $L_2(\Omega)$ and $H^1(\Omega)$ are the common Lebesgue and Sobolev spaces. As usual, $H^1_0(\Omega) \subset H^1(\Omega)$ denotes the subspace of functions that vanish on $\partial\Omega$.

We consider the following model problem: Find $u \in H^1_0(\Omega)$ such that

$$\int_\Omega \nabla u \cdot \nabla v \,\mathrm{d}x = \int_\Omega f v \,\mathrm{d}x \qquad \text{for all} \qquad v \in H^1_0(\Omega) \tag{2}$$

with a given source function $f \in L_2(\Omega)$. We assume that Ω is a composition of K non-overlapping patches $\Omega^{(k)}$, where every patch $\Omega^{(k)}$ is parametrized by a geometry function

$$G_k : \widehat{\Omega} := (0,1)^2 \to \Omega^{(k)} := G_k(\widehat{\Omega}) \subset \mathbb{R}^2,$$

that has a continuous extension to the closure of $\widehat{\Omega}$ and such that $\nabla G_k \in L_\infty(\widehat{\Omega})$ and $(\nabla G_k)^{-1} \in L_\infty(\widehat{\Omega})$.

We consider the case where the pre-images of the (Dirichlet) boundary consist of whole edges. The indices of neighboring patches $\Omega^{(\ell)}$ of $\Omega^{(k)}$, that share at least a part of their boundaries, is collected in the set

$$\mathcal{N}_\Gamma(k) := \{\ell \neq k \;:\; \text{meas}\,(\partial\Omega^{(k)} \cap \partial\Omega^{(\ell)}) > 0\},$$

where meas T is the measure of T. For any $\ell \in \mathcal{N}_\Gamma(k)$, we write $\Gamma^{(k,\ell)} = \partial\Omega^{(k)} \cap \partial\Omega^{(\ell)}$. The endpoints of $\partial\Omega^{(k)} \cap \partial\Omega^{(\ell)}$ that are not located on the (Dirichlet) boundary of Ω are referred to as junctions. A junction could be a common vertex or a T-junction.

For the IgA discretization spaces, we first construct a B-spline space $\widehat{V}^{(k)}$ on the parameter domain $\widehat{\Omega}$ by tensorization of two univariate B-spline spaces. The function spaces on the physical domain are then defined by the pull-back principle: $V^{(k)} := \widehat{V}^{(k)} \circ G_k^{-1}$.

The product of the local spaces gives the global approximation space $V := V^{(1)} \times \cdots \times V^{(K)}$. On this discretization space, we introduce the SIPG formulation, cf. [1, 15]. Since we are interested in a domain decomposition approach, we need patch-local formulations of SIPG.

3 The dG IETI-DP solver

For our patch-local formulations, we adapt the ideas of [3, 5, 6] and others. We choose local function spaces $V_e^{(k)}$ to be the product space of $V^{(k)}$ and the neighboring trace spaces $V^{(k,\ell)}$, which are the restrictions of $V^{(\ell)}$ to $\Gamma^{(k,\ell)}$. A function $v_e^{(k)} \in V_e^{(k)}$ is represented as a tuple $v_e^{(k)} = \big(v^{(k)}, (v^{(k,\ell)})_{\ell\in\mathcal{N}_\Gamma(k)}\big)$, where $v^{(k)} \in V^{(k)}$ and $v^{(k,\ell)} \in V^{(k,\ell)}$. Note that the traces of the basis functions for $V^{(\ell)}$ restricted to $\Gamma^{(k,\ell)}$ form a basis of $V^{(k,\ell)}$. The basis for $V_e^{(k)}$ consists of the basis functions of $V^{(k)}$ and the basis functions for $V^{(k,\ell)}$. The basis functions on $V^{(k,\ell)}$ are usually visualized as living on artificial interfaces.

On each patch, we consider the local problem: Find $u_e^{(k)} \in V_e^{(k)}$ such that

$$a_e^{(k)}(u_e^{(k)}, v_e^{(k)}) = \langle f_e^{(k)}, v_e^{(k)} \rangle \quad \text{for all} \quad v_e^{(k)} \in V_e^{(k)}, \text{where}$$

$$\begin{aligned}
a_e^{(k)}(u_e^{(k)}, v_e^{(k)}) &:= a^{(k)}(u_e^{(k)}, v_e^{(k)}) + m^{(k)}(u_e^{(k)}, v_e^{(k)}) + r^{(k)}(u_e^{(k)}, v_e^{(k)}),\\
\langle f_e^{(k)}, v_e^{(k)} \rangle &:= \int_{\Omega^{(k)}} f v^{(k)} \mathrm{d}x,\\
a^{(k)}(u_e^{(k)}, v_e^{(k)}) &:= \int_{\Omega^{(k)}} \nabla u^{(k)} \cdot \nabla v^{(k)} \, \mathrm{d}x,\\
m^{(k)}(u_e^{(k)}, v_e^{(k)}) &:= \sum_{\ell\in\mathcal{N}_\Gamma(k)} \int_{\Gamma^{(k,\ell)}} \frac{\partial u^{(k)}}{\partial n_k} (v^{(k,\ell)} - v^{(k)}) \, \mathrm{d}s,\\
&\quad + \sum_{\ell\in\mathcal{N}_\Gamma(k)} \int_{\Gamma^{(k,\ell)}} \frac{\partial v^{(k)}}{\partial n_k} (u^{(k,\ell)} - u^{(k)}) \, \mathrm{d}s,\\
r^{(k)}(u_e^{(k)}, v_e^{(k)}) &:= \sum_{\ell\in\mathcal{N}_\Gamma(k)} \int_{\Gamma^{(k,\ell)}} \frac{\delta p^2}{h_{k\ell}} (u^{(k,\ell)} - u^{(k)})(v^{(k,\ell)} - v^{(k)}) \, \mathrm{d}s
\end{aligned}$$

and n_k denotes the outward unit normal vector and δ is the dG penalty parameter, which has to be chosen large enough in order to guarantee that the bilinear form $a_e^{(k)}(\cdot,\cdot)$ is coercive. In [16], it was shown that δ can be chosen independently of p.

The discretization of $a_e^{(k)}(\cdot,\cdot)$ and $\langle f_e^{(k)},\cdot\rangle$ gives a local system, which we write as

$$\begin{pmatrix} A_{\mathrm{II}}^{(k)} & A_{\mathrm{I}\Gamma}^{(k)} \\ A_{\Gamma\mathrm{I}}^{(k)} & A_{\Gamma\Gamma}^{(k)} \end{pmatrix} \begin{pmatrix} \underline{u}_{\mathrm{I}}^{(k)} \\ \underline{u}_{\Gamma}^{(k)} \end{pmatrix} = \begin{pmatrix} \underline{f}_{\mathrm{I}}^{(k)} \\ \underline{f}_{\Gamma}^{(k)} \end{pmatrix}, \tag{3}$$

where the index I refers to the basis functions that are only supported in the interior of $\Omega^{(k)}$ and the index Γ refers to the remaining basis functions, i.e., those living on the patch boundary and on the artificial interfaces. We eliminate the interior degrees of freedom in (3) for every $k = 1,\ldots,K$ to get the block diagonal Schur complement system

$$S\underline{w} = \underline{g},$$

where the individual blocks of S are given by $S^{(k)} = A_{\Gamma\Gamma}^{(k)} - A_{\Gamma\mathrm{I}}^{(k)}\left(A_{\mathrm{II}}^{(k)}\right)^{-1}A_{\mathrm{I}\Gamma}^{(k)}$.

The IETI-DP method requires carefully selected primal degrees of freedom to be solvable. We choose the degrees of freedom associated to the basis functions which are non-zero on a junction to be primal. For every standard corner, we only have one primal degree of freedom per patch, as in [15]. On a T-junction however, the number of non-zero basis functions grows linearly with p. Since we take all of them, we refer to “fat vertices” in this context.

$C = \operatorname{diag}(C^{(1)},\ldots,C^{(K)})$ is the constraint matrix, i.e., it is defined such that $C\underline{w} = 0$ if and only if the associated function w vanishes at the primal degrees of freedom. The matrix Ψ represents the energy minimizing basis functions for the space of primal degrees of freedom.

Furthermore, we introduce the jump matrix B, which models the jumps of the functions between the patch boundaries and the associated artificial interfaces. Each row corresponds to one degree of freedom (coefficient for a basis function) on the patch boundary and one artificial interface; as usual, each row has only two non-zero coefficients that are -1 and 1. Primal degrees of freedom are excluded. For a visualization, see Fig. 1, where the primal degrees of freedom are marked with solid lines and the dotted arrows show the action of the jump matrix B. The basis functions on the artificial interfaces are labeled with the same symbols from the original spaces.

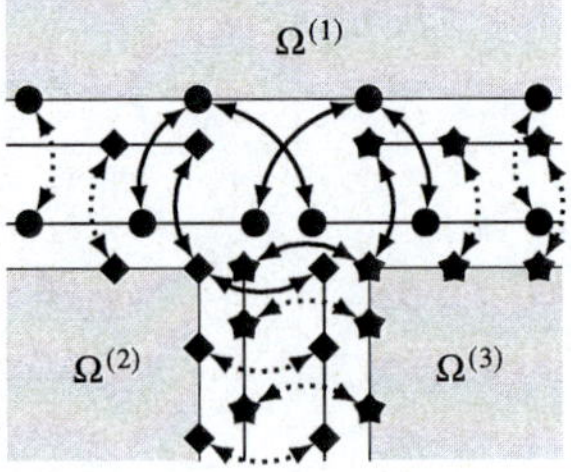

Fig. 1: Action of matrix B (dotted lines) and primal degrees of freedom (solid lines).

The following problem is equivalent to the SIPG discretization of (2), cf. [11]: Find $(\underline{w}_\Delta, \underline{\mu}, \underline{w}_\Pi, \underline{\lambda})$ such that

$$\begin{pmatrix} S & C^\top & & B^\top \\ C & & & \\ & & \Psi^\top S \Psi & (B\Psi)^\top \\ B & & B\Psi & \end{pmatrix} \begin{pmatrix} \underline{w}_\Delta \\ \underline{\mu} \\ \underline{w}_\Pi \\ \underline{\lambda} \end{pmatrix} = \begin{pmatrix} \underline{g} \\ 0 \\ \Psi^\top \underline{g} \\ 0 \end{pmatrix}.$$

We obtain the solution of the original problem by $\underline{w} = \underline{w}_\Delta + \Psi \underline{w}_\Pi$. We build a Schur complement of this system to get the linear problem

$$F\,\underline{\lambda} = \underline{d}. \tag{4}$$

We solve (4) with a preconditioned conjugate gradient (PCG) solver with the scaled Dirichlet preconditioner $M_{\text{sD}} := BD^{-1}SD^{-1}B^\top$,where D is a diagonal matrix defined based on the principle of multiplicity scaling, cf. [13, 14].

4 Numerical results

We consider the model problem

$$\begin{aligned} -\Delta u(x, y) &= 2\pi^2 \sin(\pi x)\sin(\pi y) && \text{for} \quad (x, y) \in \Omega \\ u &= 0 && \text{on} \quad \partial\Omega, \end{aligned}$$

on the geometries depicted in Fig. 2. Both represent the same computational domain with an inner radius of 1 and an outer radius of 2. The ring in Fig. 2a consists of 20 patches each of which has a width of 0.2. For the ring in Fig. 2b, we consider again 4 patches per layer, where the thin layer has a width of 0.02 and the other layers have a correspondingly larger width. We use NURBS of degree 2 to parametrize all patches. In the coarsest setting, i.e., $r = 0$, the discretization spaces on all patches consist of global polynomials only. The discretization spaces for $r = 1, 2, 3, \ldots$ are obtained by uniform refinement steps. We use a PCG solver to solve system (4) with the preconditioner M_{sD} and to estimate the condition number $\kappa(M_{\text{sD}}F)$, where we use the zero vector as initial guess. All experiments are carried out in the C++ library G+Smo, cf. [12] and are executed on the Radon1[1] cluster in Linz.

In Table 1, we report on the iteration counts (it) and the condition numbers (κ) for various refinement levels r and various spline degrees p, where we chose C^s-smoothness with $s = p - 1$ within the patches. The tables show the expected behavior with respect to h. The condition number decreases when we increase the spline degree p, which is better than one would expect from the theory in [15]. Although the width of the thin patches in Fig. 2b is one tenth of the width of the

[1] https://www.ricam.oeaw.ac.at/hpc/

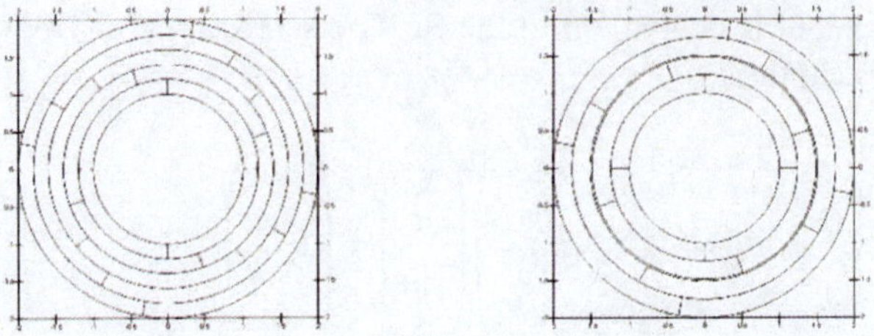

(a) Ring (b) Ring with thin gap

Fig. 2: Computational domains and the decomposition into patches.

patches in Fig. 2a, the condition number grows only by a factor between 5 and 6. Also the iteration counts grow only mildly.

	Fig. 2a								Fig. 2b							
	$p = 2$		$p = 3$		$p = 6$		$p = 7$		$p = 2$		$p = 3$		$p = 6$		$p = 7$	
r	it	κ	it	κ	it	κ	it	κ	it	κ	it	κ	it	κ	it	κ
4	9	3.7	9	3.5	8	2.4	8	2.1	12	18.0	13	18.0	13	12.9	11	11.6
5	10	4.6	10	4.5	9	3.8	9	3.5	20	24.1	19	23.4	19	19.3	18	18.0
6	10	5.8	10	5.5	10	4.9	10	4.8	22	31.7	22	29.9	21	25.9	20	24.8
7	11	6.3	11	6.2	10	5.6	10	5.5	24	37.2	24	36.3	22	31.3	22	30.1
8	11	6.7	11	6.7	11	6.3	10	5.6	24	43.2	24	42.3	24	36.5	24	31.6

Table 1: Iterations it and condition numbers κ; degree p; refinement level r.

Table 2 presents the parallel solving times for n processors. We only consider the domain in Fig. 2a again with $s = p - 1$. We see that the speedup rate with respect to n is a bit smaller than the expected rate of 2. This is probably caused by the rather small number of patches in the computational domain. In Table 3 we report on the iteration counts and the condition numbers for the decomposition in Fig. 2a when we change the smoothness s of the B-splines within the patches. The numbers in the table show the behavior for $r = 5$. We see that for a fixed smoothness s the condition number grows slightly with respect to the spline degree p. For a fixed degree p, we observe a decline in the condition number when we increase the smoothness s.

Acknowledgements The first author was supported by the Austrian Science Fund (FWF): S117 and W1214-04. The second author has also received support from the Austrian Science Fund (FWF): P31048.

r	$p=3$ $n=1$	$n=2$	$n=4$	$n=8$	$n=16$	$p=7$ $n=1$	$n=2$	$n=4$	$n=8$	$n=16$
6	3.8	2.8	2.4	1.2	0.8	10.0	6.5	5.0	2.5	1.75
7	24.0	16.1	13.6	6.4	4.1	47.0	31.7	26.1	12.2	9.3
8	107.0	81.4	66.8	29.5	19.5	220.0	158.7	129.4	56.7	45.4

Table 2: Solving times (sec.); degree p; refinement level r; n processors; Fig. 2a.

s	$p=2$ it	κ	$p=3$ it	κ	$p=4$ it	κ	$p=5$ it	κ	$p=6$ it	κ	$p=7$ it	κ
0	10	5.0	10	5.3	10	5.4	10	5.5	10	5.6	10	5.6
1	10	4.6	10	5.2	10	5.3	10	5.4	10	5.5	10	5.5
2			10	4.5	10	5.0	10	5.3	10	5.4	10	5.5
3					9	4.2	10	4.9	10	5.1	10	5.3
4							9	4.0	10	4.7	10	5.0
5									9	3.8	9	4.5
6											9	3.5

Table 3: Iterations it and condition number κ; refinement level $r = 5$; degree p; smoothness s; Fig. 2a.

References

1. D. Arnold. An interior penalty finite element method with discontinuous elements. *SIAM J. Numer. Anal.* **19**(4), 742–760 (1982).
2. L.B. da Veiga, L.F. Pavarino, S. Scacchi, O.B. Widlund and S. Zampini. Isogeometric BDDC Preconditioners with Deluxe Scaling. *SIAM J. Sci. Comput.* **36**(3), A1118–A1139 (2014).
3. M. Dryja, J. Galvis and M. Sarkis. A FETI-DP Preconditioner for a Composite Finite Element and Discontinuous Galerkin Method. *SIAM J. Numer. Anal.* **51**(1), 400–422 (2013).
4. C. Farhat, M. Lesoinne, P.L. Tallec, K. Pierson and D. Rixen. FETI-DP: A dual-primal unified FETI method I:A faster alternative to the two-level FETI method. *Int. J. Numer. Methods Eng.* **50**, 1523–1544 (2001).
5. C. Hofer. Analysis of discontinuous Galerkin dual-primal isogeometric tearing and interconnecting methods. *Math. Models Methods Appl. Sci.* **28**(1), 131–158 (2018).
6. C. Hofer and U. Langer. Dual-primal isogeometric tearing and interconnecting solvers for multipatch dG-IgA equations. *Comput. Methods Appl. Mech. Eng.* **316**, 2–21 (2017).
7. T.J.R. Hughes, J.A. Cottrell and Y. Bazilevs. Isogeometric Analysis: CAD, Finite Elements, NURBS, Exact Geometry and Mesh Refinement. *Comput. Methods Appl. Mech. Eng.* **194**(39-41), 4135–4195 (2005).
8. S. Kleiss, C. Pechstein, B. Jüttler and S. Tomar. IETI-Isogeometric Tearing and Interconnecting. *Comput. Methods Appl. Mech. Eng.* **247-248**, 201–215 (2012).
9. U. Langer, A. Mantzaflaris, S.E. Moore and I. Toulopoulos. Multipatch Discontinuous Galerkin Isogeometric Analysis. In: B. Jüttler and B. Simeon, editors, *Isogeometric Analysis and Applications 2014*, pp. 1–32, Springer International Publishing (2015).
10. U. Langer and I. Toulopoulos. Analysis of multipatch discontinuous Galerkin IgA approximations to elliptic boundary value problems. *Comp. Vis. Sci.* **17**(5), 217–233 (2015).

11. J. Mandel, C.R. Dohrmann and R. Tezaur. An algebraic theory for primal and dual substructuring methods by constraints. *Appl. Numer. Math.* **54**(2), 167–193 (2005).
12. A. Mantzaflaris, R. Schneckenleitner, S. Takacs and others (see website). G+Smo (Geometry plus Simulation modules). `http://github.com/gismo` (2020).
13. C. Pechstein. *Finite and Boundary Element Tearing and Interconnecting Solvers for Multiscale Problems*. Springer, Heidelberg (2013).
14. R. Schneckenleitner and S. Takacs. Condition number bounds for IETI-DP methods that are explicit in h and p. *Math. Models Methods Appl. Sci.* **30**(11), 2067–2103 (2020).
15. R. Schneckenleitner and S. Takacs. Convergence Theory for IETI-DP Solvers for Discontinuous Galerkin Isogeometric Analysis that is Explicit in h and p. *Comput. Methods Appl. Math.* (2021). Online first.
16. S. Takacs. Aquasi-robust discretization error estimate for discontinuous Galerkin Isogeometric Analysis (2019). Submitted. `https://arxiv.org/pdf/1901.03263.pdf`.
17. O.B. Widlund, S. Zampini, S. Scacchi and L.F. Pavarino. Block FETI–DP/BDDC preconditioners for mixed isogeometric discretizations of three-dimensional almost incompressible elasticity. *Math. Comp.* (2021). Has appeared electronically.

The Parallel Full Approximation Scheme in Space and Time for a Parabolic Finite Element Problem

Oliver Sander, Ruth Schöbel, and Robert Speck

1 Introduction

The parallel full approximation scheme in space and time (PFASST, [4]) can integrate multiple time-steps simultaneously by using inner iterations of spectral deferred corrections (SDC, [3]) on a space-time hierarchy. It mimics a full approximation scheme (FAS, [12]) for a sequence of coupled collocation problems. For the simulation of space–time dependent problems, PFASST has been used in combination with finite differences, e.g., in [8, 10], but also in connection with particle simulations [11] and spectral methods [5]. In this work we combine PFASST with a finite element discretization in space. Using a simple, nonlinear reaction-diffusion equation, we will derive the discretized "composite collocation problem" PFASST aims to solve in parallel and show the correct handling of the mass matrix. There exist two different ways to write down the composite collocation problem with a non-trivial mass matrix and we will demonstrate that we can avoid inversion of the mass matrix with the added benefit of a better order of accuracy in time per PFASST iteration. The choice of restriction and prolongation in space plays a major role and we will show the correct formulation and placement of those. Both mass matrix handling and choice of transfer operators mark key differences to using standard finite differences in space, both in terms of theoretical formulation and simulation results. Using a concrete example, we numerically test the order of accuracy per iteration of PFASST and compare it with SDC.

Oliver Sander
Faculty of Mathematics
TU Dresden, Germany, e-mail: oliver.sander@tu-dresden.de

Ruth Schöbel
Jülich Supercomputing Centre
Forschungszentrum Jülich GmbH, Germany, e-mail: r.schoebel@fz-juelich.de

Robert Speck
Jülich Supercomputing Centre
Forschungszentrum Jülich GmbH, Germany, e-mail: r.speck@fz-juelich.de

S. C. Brenner et al. (eds.), *Domain Decomposition Methods in Science and Engineering XXVI*, Lecture Notes in Computational Science and Engineering 145,
https://doi.org/10.1007/978-3-030-95025-5_57

2 PFASST and finite elements in space

We consider the reaction-diffusion equation

$$\begin{aligned} v_t(x,t) &= \Delta v(x,t) + g(v(x,t)), && x \in \Omega, t \in [t_0, T], \\ v(x,t) &= 0, && x \in \partial\Omega, \end{aligned} \tag{1}$$

with suitable initial conditions for $t = t_0$ and $g : \mathbb{R} \to \mathbb{R}$ continuously differentiable. Here $\Omega \subset \mathbb{R}$ is a polyhedral domain with boundary $\partial\Omega$, and Δ denotes the Laplace operator.

2.1 Finite element discretization in space

We define test functions φ^h in a finite-dimensional space $\mathbf{V}^h \subset \mathbf{H}_0^1(\Omega)$, multiply (1) by these test functions, and integrate by parts. Thus, $v^h(\cdot, t) \in \mathbf{V}^h$ is given by

$$\int_\Omega \varphi^h v_t^h \,\mathrm{d}x = -\int_\Omega \nabla\varphi^h \nabla v^h \,\mathrm{d}x + \int_\Omega \varphi^h \mathrm{g}(v^h) \,\mathrm{d}x \quad \forall \varphi^h \in \mathbf{V}^h. \tag{2}$$

We choose a basis $\varphi_1, \ldots, \varphi_N$ of $\mathbf{V}^h$ and approximate $\mathrm{g}(v^h)$ by an element of $\mathbf{V}^h$ and express v^h and $\mathrm{g}(v^h)$ as

$$v^h(x,t) = \sum_{i=1}^N v_i(t)\varphi_i(x), \qquad \mathrm{g}(v^h)(x,t) \approx \sum_{i=1}^N \mathrm{g}(v_i(t))\varphi_i(x), \tag{3}$$

where the coefficients $v_i(t)$, $i = 1, \ldots, N$, are time-dependent functions. Inserting (3) into equation (2) yields

$$\mathbf{M}u_t = -\mathbf{A}u + \mathbf{M}g(u) =: f(u). \tag{4}$$

Here, $u := (v_1, \ldots, v_N)$ is a vector holding the coefficients v_i, and $g : \mathbb{R}^N \to \mathbb{R}^N$, $g := (\mathrm{g}(v_1), \ldots, \mathrm{g}(v_N))^T$. The matrix $\mathbf{M} \in \mathbb{R}^{N\times N}$ is the mass matrix and $\mathbf{A} \in \mathbb{R}^{N\times N}$ the stiffness matrix

$$\mathbf{M}_{ij} := \int_{\Omega_i} \varphi_i \varphi_j \,\mathrm{d}x, \qquad \mathbf{A}_{ij} := \int_{\Omega_i} \nabla\varphi_i \nabla\varphi_j \,\mathrm{d}x.$$

2.2 The collocation problem and SDC

For the temporal discretization, we decompose the interval $[t_0, T]$ into time-steps $t_0 < t_1 < \cdots < t_L = T$, $L \in \mathbb{N}$. For one time-step $[t_l, t_{l+1}]$, the Picard formulation of (4) is

$$\mathbf{M}u(t) = \mathbf{M}u_{l,0} + \int_{t_l}^{t} f(u(s))\,\mathrm{d}s, \qquad t \in [t_l, t_{l+1}], \tag{5}$$

where $u_{l,0} := u(t_l)$. To approximate the integral we use a spectral quadrature rule on $[t_l, t_{l+1}]$ with M quadrature nodes $\tau_{l,1}, \ldots, \tau_{l,M}$ such that $t_l < \tau_{l,1} < \ldots < \tau_{l,M} = t_{l+1}$.

For each of the M nodes we introduce a set of M quadrature weights $q_{m,j} := \int_{t_l}^{\tau_{l,m}} L_j(s)\,\mathrm{d}s$, $m, j = 1, \ldots, M$, where $L_1, \ldots, L_M$ are the Lagrange polynomials for the nodes $\tau_{l,1}, \ldots, \tau_{l,M}$. We can then approximate the integral in (5) from t_l to $\tau_{l,m}$ by

$$\Delta t \sum_{j=1}^{M} q_{m,j} f(u_{l,j}) \approx \int_{t_l}^{\tau_{l,m}} f(u(s))\,\mathrm{d}s, \quad m = 1, \ldots, M,$$

where $\Delta t := t_{l+1} - t_l$ denotes the time-step size. Using this in Equation (5) the unknown values $u(\tau_{l,1}), \ldots, u(\tau_{l,M})$ can be approximated by a solution $u_{l,1}, \ldots, u_{l,M} \in \mathbb{R}^N$ of the nonlinear system of equations

$$\mathbf{M}u_{l,m} = \mathbf{M}u_{l,0} + \Delta t \sum_{j=1}^{M} q_{m,j} f(u_{l,j}) \quad \text{for } m = 1, \ldots, M.$$

This is the so-called "collocation problem", which we can rewrite as

$$\mathbf{C}_{\boldsymbol{f}}^{\mathrm{coll}}(\boldsymbol{u}_l) := (\mathbf{I}_M \otimes \mathbf{M} - \Delta t(\mathbf{Q} \otimes \mathbf{I}_N)\boldsymbol{f})\,(\boldsymbol{u}_l) = (\mathbf{I}_M \otimes \mathbf{M})\boldsymbol{u}_{l,0}, \tag{6}$$

where $\mathbf{I}_X \in \mathbb{R}^{X \times X}$, $X \in \mathbb{N}$ is the identity matrix, $\otimes$ denotes the Kronecker product, $\boldsymbol{u}_l := (u_{l,1}, \ldots, u_{l,M})^T \in \mathbb{R}^{MN}$, $\boldsymbol{u}_{l,0} := (u_{l,0}, \ldots, u_{l,0})^T \in \mathbb{R}^{MN}$, $\mathbf{Q} := (q_{ij}) \in \mathbb{R}^{M \times M}$, and the vector function $\boldsymbol{f} : \mathbb{R}^{MN} \to \mathbb{R}^{MN}$ is given by $\boldsymbol{f}(\boldsymbol{u}_l) := (f(u_{l,1}), \ldots, f(u_{l,M}))^T$.

With this matrix notation, spectral deferred corrections can simply be seen as a preconditioned Picard iteration [6, 9]. More precisely, for a lower triangular matrix $\mathbf{Q}_\Delta \in \mathbb{R}^{M \times M}$ we define the preconditioner

$$\mathbf{P}_{\boldsymbol{f}}^{\mathrm{sdc}}(\boldsymbol{u}_l) := (\mathbf{I}_M \otimes \mathbf{M} - \Delta t(\mathbf{Q}_\Delta \otimes \mathbf{I}_N)\boldsymbol{f})\,(\boldsymbol{u}_l).$$

Then the preconditioned iteration reads

$$\mathbf{P}_{\boldsymbol{f}}^{\mathrm{sdc}}(\boldsymbol{u}_l^{k+1}) = (\mathbf{P}_{\boldsymbol{f}}^{\mathrm{sdc}} - \mathbf{C}_{\boldsymbol{f}}^{\mathrm{coll}})(\boldsymbol{u}_l^{k}) + (\mathbf{I}_M \otimes \mathbf{M})\boldsymbol{u}_{l,0}, \quad k = 1, \ldots, K. \tag{7}$$

The properties of $\mathbf{P}_{\boldsymbol{f}}^{\mathrm{sdc}}$ depend first and foremost on the choice of the matrix $\mathbf{Q}_\Delta$. For this work, we use the backward Euler approach. We refer to [6, 9, 13] for more details on the notation and its relationship to the original description of SDC as in [3]. The key difference is the appearance of the mass matrix $\mathbf{M}$, which for finite differences is just the identity matrix.

2.3 The composite collocation problem and PFASST

For L time-steps, the composite collocation problem is

$$\begin{pmatrix} \mathbf{C}_f^{\text{coll}} & & & \\ -\mathbf{H} & \mathbf{C}_f^{\text{coll}} & & \\ & \ddots & \ddots & \\ & & -\mathbf{H} & \mathbf{C}_f^{\text{coll}} \end{pmatrix} \begin{pmatrix} \boldsymbol{u}_1 \\ \boldsymbol{u}_2 \\ \vdots \\ \boldsymbol{u}_L \end{pmatrix} = \begin{pmatrix} (\mathbf{I}_M \otimes \mathbf{M})\boldsymbol{u}_{0,0} \\ \mathbf{0} \\ \vdots \\ \mathbf{0} \end{pmatrix}, \tag{8}$$

where in the simplest case $\mathbf{N} \in \mathbb{R}^{M\times M}$ just holds ones in the last column and zeros elsewhere. Then, $\mathbf{H} := \mathbf{N} \otimes \mathbf{M}$ provides the value at the last quadrature node $\tau_{l,M}$ of a time-step $[t_l, t_{l+1}]$ as initial value for the following time-step. Defining the global state vector $\boldsymbol{u} := (\boldsymbol{u}_1, ..., \boldsymbol{u}_L)^T \in \mathbb{R}^{LMN}$, the vector $\boldsymbol{b} := ((\mathbf{I}_M \otimes \mathbf{M})\boldsymbol{u}_{0,0}, \mathbf{0}, ..., \mathbf{0})^T \in \mathbb{R}^{LMN}$, and $\boldsymbol{F} : \mathbb{R}^{LMN} \to \mathbb{R}^{LMN}$ with $\boldsymbol{F}(\boldsymbol{u}) := (\boldsymbol{f}(\boldsymbol{u}_1), \dots, \boldsymbol{f}(\boldsymbol{u}_L))^T$, we can write this in the more compact form as $\mathbf{C}_{\boldsymbol{F}}(\boldsymbol{u}) = \boldsymbol{b}$, where $\mathbf{C}_{\boldsymbol{F}}$ is the lower block-bidiagonal, nonlinear operator on the left of (8). Using the definition (6) of $\mathbf{C}_f^{\text{coll}}$ we write (8) as

$$(\mathbf{I}_{LM} \otimes \mathbf{M} - \Delta t(\mathbf{I}_L \otimes \mathbf{Q} \otimes \mathbf{I}_N)\boldsymbol{F} - \mathbf{E} \otimes \mathbf{H})(\boldsymbol{u}) = \boldsymbol{b} \tag{9}$$

where the matrix $\mathbf{E} \in \mathbb{R}^{L\times L}$ has ones on the lower off-diagonal and zeros elsewhere, accounting for the transfer of the solution from one step to the next.

There are two fundamentally different ways to solve this system iteratively with SDC. We can choose either (note the $\mathbf{Q}_\Delta$ instead of the $\mathbf{Q}$)

$$\mathbf{P}_{\boldsymbol{F}}^{\text{par}}(\boldsymbol{u}) := (\mathbf{I}_{LM} \otimes \mathbf{M} - \Delta t(\mathbf{I}_L \otimes \mathbf{Q}_\Delta \otimes \mathbf{I}_N)\boldsymbol{F})(\boldsymbol{u})$$

or

$$\mathbf{P}_{\boldsymbol{F}}^{\text{seq}}(\boldsymbol{u}) := (\mathbf{I}_{LM} \otimes \mathbf{M} - \Delta t(\mathbf{I}_L \otimes \mathbf{Q}_\Delta \otimes \mathbf{I}_N)\boldsymbol{F} - \mathbf{E} \otimes \mathbf{H})(\boldsymbol{u}),$$

where the latter still has the $\mathbf{H}$ matrices in the lower off-diagonal. $\mathbf{P}_{\boldsymbol{F}}^{\text{par}}$ is a parallel preconditioner, which performs SDC iterations on each step simultaneously, while $\mathbf{P}_{\boldsymbol{F}}^{\text{seq}}$ propagates a single SDC iteration sequentially forward in time.

The idea of PFASST now is to couple both preconditioners in a two-level space-time full approximations scheme: the parallel $\mathbf{P}_{\boldsymbol{F}}^{\text{par}}$ is used on the original problem in space and time (the "fine" level), while the sequential $\mathbf{P}_{\boldsymbol{F}}^{\text{seq}}$ with better convergence properties is used on a coarser, cheaper level with reduced accuracy in space and/or time to reduce the impact of its sequential nature. To create the coarse level, we reduce the number of degrees of freedom in space and choose a finite element subspace $\tilde{\mathbf{V}}^h \subset \mathbf{V}^h$. Three different transfer operations are then needed for PFASST:

1. Restriction of a coefficient vector $\boldsymbol{u}_{l,m}$, representing an object in $\mathbf{V}^h$, to the representation of an object in $\tilde{\mathbf{V}}^h$.
2. Restriction of the residual $\mathbf{C}_{\boldsymbol{F}}(\boldsymbol{u}) - \boldsymbol{b}$.

3. Prolongation of a coefficient vector $\tilde{\boldsymbol{u}}_{l,m}$, representing an object in $\tilde{\mathbf{V}}_h$, to the representation of an object in $\mathbf{V}_h$.

Using Lagrange polynomials, operations 2 and 3 can be done using the canonical injection $\mathbf{T}^N \in \mathbb{R}^{N\times\tilde{N}}$ for prolongation of the coefficient vector and its transpose $(\mathbf{T}^N)^T \in \mathbb{R}^{\tilde{N}\times N}$ for restriction of the residual. For Operation 1, we use the matrix $\mathbf{R}^N \in \mathbb{R}^{\tilde{N}\times N}$ that represents the Lagrange interpolation of functions from $\mathbf{V}^h$ in $\tilde{\mathbf{V}}^h$. By $\mathbf{T} := \mathbf{I}_{LM} \otimes \mathbf{T}^N$ and $\mathbf{R} := \mathbf{I}_{LM} \otimes \mathbf{R}^N$ we define global transfer operators. Using the tilde symbols to indicate entities on the coarse level, one iteration of PFASST reads:

1. Restrict current iterate to the coarse level: $\tilde{\boldsymbol{u}}^k = \mathbf{R}\boldsymbol{u}^k$.
2. Compute FAS correction: $\boldsymbol{\tau} = \tilde{\mathbf{C}}_{\boldsymbol{F}}(\tilde{\boldsymbol{u}}^k) - \mathbf{T}^T\mathbf{C}_{\boldsymbol{F}}(\boldsymbol{u}^k)$.
3. Compute $\tilde{\boldsymbol{u}}^{k+1}$ by solving: $\tilde{\mathbf{P}}_{\boldsymbol{F}}^{\text{seq}}(\tilde{\boldsymbol{u}}^{k+1}) = (\tilde{\mathbf{P}}_{\boldsymbol{F}}^{\text{seq}} - \tilde{\mathbf{C}}_{\boldsymbol{F}})(\tilde{\boldsymbol{u}}^k) + \tilde{\boldsymbol{b}} + \boldsymbol{\tau}$.
4. Apply coarse grid correction: $\boldsymbol{u}^{k+\frac{1}{2}} = \boldsymbol{u}^k + \mathbf{T}(\tilde{\boldsymbol{u}}^{k+1} - \mathbf{R}\boldsymbol{u}^k)$.
5. Compute $\boldsymbol{u}^{k+1}$ by solving: $\mathbf{P}_{\boldsymbol{F}}^{\text{par}}(\boldsymbol{u}^{k+1}) = (\mathbf{P}_{\boldsymbol{F}}^{\text{par}} - \mathbf{C}_{\boldsymbol{F}})(\boldsymbol{u}^{k+\frac{1}{2}}) + \boldsymbol{b}$.

In contrast to the description in [1, 2], the mass matrices are now included in $\mathbf{P}_{\boldsymbol{F}}^{\text{par}}(\boldsymbol{u})$, $\tilde{\mathbf{P}}_{\boldsymbol{F}}^{\text{seq}}(\tilde{\boldsymbol{u}})$ as well as in $\mathbf{C}_{\boldsymbol{F}}$ and $\tilde{\mathbf{C}}_{\boldsymbol{F}}$. This approach is preferable to others, including the naive one where the collocation problem (6) is multiplied by $\mathbf{M}^{-1}$ from the left. The collocation problem (6) then reads

$$\bar{\mathbf{C}}_{\bar{\boldsymbol{f}}}^{\text{coll}}(\boldsymbol{u}_l) := (\mathbf{I}_M \otimes \mathbf{I}_N - \Delta t(\mathbf{Q} \otimes \mathbf{I}_N)\bar{\boldsymbol{f}})(\boldsymbol{u}_l) = \boldsymbol{u}_{l,0}, \tag{11}$$

for $\bar{\boldsymbol{f}}(\boldsymbol{u}_l) := (\bar{f}(u_{l,1}), ..., \bar{f}(u_{l,M}))^T$ and $\bar{f}(u_{l,m}) = \mathbf{M}^{-1}f(u_{l,m})$. Similarly, the composite collocation problem then is

$$\bar{\mathbf{C}}_{\bar{\boldsymbol{F}}}(\boldsymbol{u}) := (\mathbf{I}_{LMN} - \Delta t(\mathbf{I}_L \otimes \mathbf{Q} \otimes \mathbf{I}_N)\bar{\boldsymbol{F}} - \mathbf{E} \otimes \mathbf{N} \otimes \mathbf{I}_N)(\boldsymbol{u}) = \bar{\boldsymbol{b}}, \tag{12}$$

where $\bar{\boldsymbol{b}} := (\boldsymbol{u}_{0,0}, \mathbf{0}, ..., \mathbf{0})^T$ and $\bar{\boldsymbol{F}} := \left(\bar{\boldsymbol{f}}(\boldsymbol{u}_1), \ldots, \bar{\boldsymbol{f}}(\boldsymbol{u}_L)\right)^T$. SDC and PFASST can then be derived precisely as in the literature, using the modified right-hand side $\bar{f}$. Note that the inversion of the mass matrix is only necessary, if the actual residual of (11) or (12) needs to be computed, which, e.g., is necessary for the FAS correction. There, $\bar{\boldsymbol{F}}$ has to be evaluated on the fine and the coarse level, both containing the inverse of the respective mass matrix. While seemingly attractive in terms of writing a generic code, inversion of the mass matrix can be costly and, as we will see later, convergence of PFASST is much worse in this case. Note that the components of the residual of (11) and (12) in contrast to (6) and (9) are not elements of the dual space of $\mathbf{V}^h$ and therefore cannot be restricted exactly. In this case, the obvious choice is to use $\mathbf{R}^N$ to transfer both residual and coefficient vectors to the coarse level.

3 Numerical results

We now investigate numerically the convergence behavior of PFASST with finite elements in space. In [7] it was shown that for a discretization with finite differences, the single-step version of PFASST (i.e. multilevel SDC) can gain two orders of accuracy per iteration, provided very high-order transfer operators in space are used [1, 7]. We will now show numerically that with finite elements in space, this is no longer necessary.

We use the following nonlinear differential equation

$$u_t = \Delta u + u^2(1-u) \quad \text{on } [0,2] \times [-20,20]. \tag{13}$$

In all simulations, we use 4 Gauss–Raudau nodes to discretize a single time-step. In the following we use SDC for serial time-step calculations and PFASST to calculate 4 time-steps simultaneously. The spatial domain $[-20, 20]$ is discretized using Lagrange finite elements of either order 1 or order 3. We use the initial value $u(x,0) = (1 + (\sqrt{2}-1)e^{-\sqrt{6}/6x})^{-2}$ as the initial guess for the iteration.

For the first test case we use a third-order Lagrange basis to approximate the solution. We coarsen the problem in space by restricting to a second-order Lagrange basis. Figure 1 shows the results for SDC and PFASST. They show the absolute error of the method in the infinity norm for different time-step sizes, in relation to a reference solution calculated with a much smaller Δt and SDC. While SDC gains one order per iteration as expected, PFASST can gain up to two orders per iteration, at least after some initial iterations have been performed. This "burn-in" phase causes a loss of parallel efficiency when actual speedup is measured. After this phase, however, PFASST shows ideal convergence behavior gaining two orders of accuracy per iteration. There is not yet a theoretical explanation for the "burn-in" or the "ideal" phase.

For a second test case we use a first-order Lagrange basis and coarsen the problem in space by doubling the element size. Figure 2 shows the results for SDC and PFASST. In the same way as in the high-order example before, SDC gains one order or accuracy per iteration, while PFASST can gain up to two orders after a few initial iterations. Note that in the case of a finite difference discretization, the order of the interpolation is crucial to obtain two orders per iteration [1, 7]. The usage of $\mathbf{T}^N$ as exact interpolation for nested finite element spaces removes this, so far, persistent and irritating limitation.

Finally, Figure 3 shows SDC and PFASST applied to the (composite) collocation problem (12) with inverted mass matrix and a first-order Lagrange basis. SDC behaves exactly as before, while PFASST fails to show any reasonable convergence. In particular, increasing the number of iterations does not increase the order of accuracy beyond 1.

The advantage of using finite elements together with PFASST in the way we demonstrated here is not yet analyzed analytically. We intend to address this in a follow-up work. Also, the important fact that two orders of accuracy per iteration

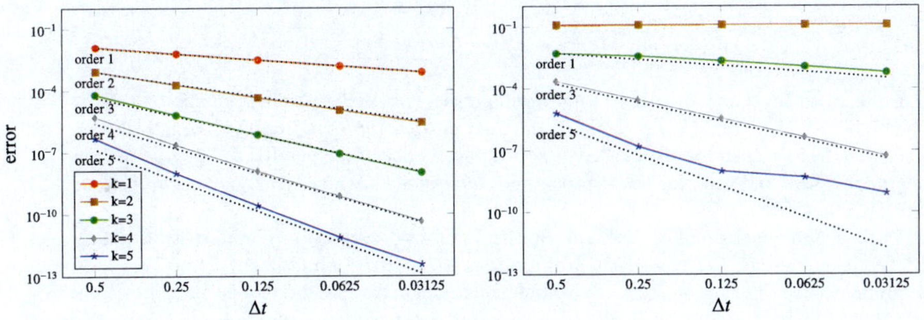

Fig. 1: SDC (left) and PFASST (right) errors for different Δt and number of iterations k, 128 spatial elements, order 3. The dashed lines indicate the expected order of accuracy in time.

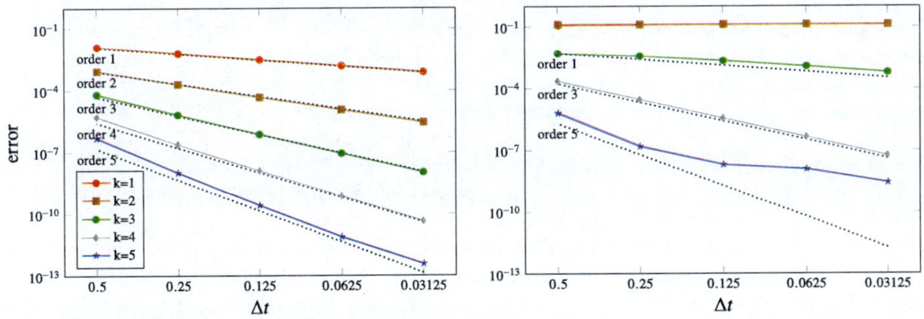

Fig. 2: SDC (left) and PFASST (right) errors for different Δt and number of iterations k, 512 spatial elements, order 1. The dashed lines indicate the expected order of accuracy in time.

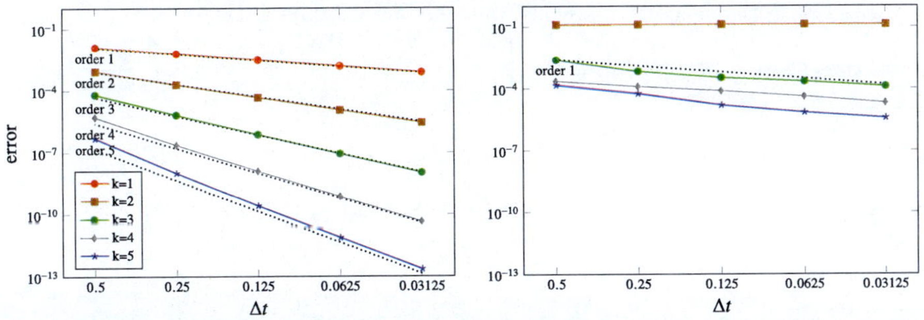

Fig. 3: Naive approach with inverted mass matrix: SDC (left) and PFASST (right) errors for different Δt and number of iterations k, 512 spatial elements, order 1.

is possible even with a low-order spatial interpolation does not have a theoretical explanation. A corresponding analysis is work in progress.

References

1. M. Bolten, D. Moser and R. Speck. A multigrid perspective on the parallel full approximation scheme in space and time. *Numerical Linear Algebra with Applications* **24**(6), e2110 (2017).
2. M. Bolten, D. Moser and R. Speck. Asymptotic convergence of the parallel full approximation scheme in space and time for linear problems. *Numerical Linear Algebra with Applications* **25**(6), e2208 (2018).
3. A. Dutt, L. Greengard and V. Rokhlin. Spectral deferred correction methods for ordinary differential equations. *BIT Numerical Mathematics* **40**(2), 241–266 (2000).
4. M. Emmett and M.L. Minion. Toward an efficient parallel in time method for partial differential equations. *Communications in Applied Mathematics and Computational Science* **7**, 105–132 (2012).
5. S. Götschel and M.L. Minion. Parallel-in-time for parabolic optimal control problems using PFASST. In: *International Conference on Domain Decomposition Methods*, pp. 363–371, Springer (2017).
6. J. Huang, J. Jia and M. Minion. Accelerating the convergence of spectral deferred correction methods. *Journal of Computational Physics* **214**(2), 633–656 (2006).
7. G. Kremling and R. Speck. Convergence analysis of multi-level spectral deferred corrections. *Communications in Applied Mathematics and Computational Science* **16**(2), 227–265 (2021).
8. M.L. Minion, R. Speck, M. Bolten, M. Emmett and D. Ruprecht. Interweaving PFASST and parallel multigrid. *SIAM Journal on Scientific Computing* **37**(5), S244–S263 (2015).
9. D. Ruprecht and R. Speck. Spectral deferred corrections with fast-wave slow-wave splitting. *SIAM Journal on Scientific Computing* **38**(4), A2535–A2557 (2016).
10. R. Schöbel and R. Speck. PFASST-ER: Combining the parallel full approximation scheme in space and time with parallelization across the method. *Computing and Visualization in Science* **23**(1), 1–12 (2020).
11. R. Speck, D. Ruprecht, R. Krause, M. Emmett, M. Minion, M. Winkel and P. Gibbon. A massively space-time parallel N-body solver. In: *SC'12: Proceedings of the International Conference on High Performance Computing, Networking, Storage and Analysis*, pp. 1–11, IEEE (2012).
12. U. Trottenberg, C. Oosterlee and A. Schuller. *Multigrid.* Academic Press (2000).
13. M. Weiser. Faster SDC convergence on non-equidistant grids by DIRK sweeps. *BIT Numerical Mathematics* **55**(4), 1219–1241 (2014).

A New Coarse Space for a Space-Time Schwarz Waveform Relaxation Method

Martin J. Gander, Yao-Lin Jiang and Bo Song

1 Introduction and model problem

Coarse spaces are in general needed to achieve scalability in domain decomposition methods, see [16] and references therein. There are however exceptions, where one level domain decomposition methods are scalable, which can be due to geometry and/or the operator, see [2] and references therein. In particular for space-time problems this can happen when solving parabolic problems on short time intervals, see [11] for a continuous analysis, [1] for Additive Schwarz applied to each time step, and [4] for hyperbolic problems.

We are interested here in space-time parallel solvers for parabolic problems over longer time intervals, where a coarse correction is needed for scalability. While for elliptic problems there are new coarse spaces constructed by directly improving general condition number estimates, like GenEO [15] and GDSW [12], there are so far no such estimates for evolution problems. We thus base our new coarse space construction for space-time problems on approximating an optimal coarse space, optimal in the sense that the resulting method converges after one coarse correction, see [8, 9] and references therein for elliptic problems.

With the invention of the parareal algorithm [13], research activity increased again tremendously to develop space-time parallel solvers, see the review [3] and references therein. While the parareal algorithm can be combined with Schwarz waveform relaxation [10] to obtain a general space-time parallel solver [5, 14],

Martin J. Gander
Université de Genève, Section de Mathématiques, Geneva, Switzerland,
e-mail: martin.gander@unige.ch

Yao-Lin Jiang
School of Mathematics and Statistics, Xi'an Jiaotong University, Xi'an 710049, China,
e-mail: yljiang@mail.xjtu.edu.cn

Bo Song
Corresponding author. School of Mathematics and Statistics, Northwestern Polytechnical University, Xi'an 710072, China, e-mail: bosong@nwpu.edu.cn

S. C. Brenner et al. (eds.), *Domain Decomposition Methods in Science and Engineering XXVI*, Lecture Notes in Computational Science and Engineering 145,
https://doi.org/10.1007/978-3-030-95025-5_58

whose convergence was analyzed in [6], we design here a new space-time two level Schwarz waveform relaxation method for evolution problems. For simplicity, we consider the one-dimensional heat equation

$$\mathcal{L}u := \partial_t u - \partial_{xx} u = f, \quad \text{in } \Omega \times (0, T), \tag{1}$$

where $\Omega = (a, b), a < b$, with initial condition $u(x, 0) = u_0(x), x \in \Omega$, and boundary conditions $u(a, t) = g_1(t)$ and $u(b, t) = g_2(t), t \in [0, T]$.

2 New two level Schwarz waveform relaxation

We divide the spatial domain (a, b) into I overlapping subdomains $\Omega_i := (a_i, b_i)$, $i = 1, 2, \ldots, I$, with $a_1 := a$, $b_I := b$, and decompose the time interval $(0, T)$ into N time subintervals, $0 =: T_0 \leq \cdots \leq T_n := n\Delta T \leq \cdots \leq T_N := T$, $\Delta T := T/N$. This defines the space-time subdomains $\Omega_{i,n} := \Omega_i \times (T_n, T_{n+1})$, $i = 1, 2, \ldots, I$, $n = 0, \ldots, N-1$. In [5, 6], the initial conditions in the space-time subdomains were updated using a parareal mechanism, while the boundary conditions were updated using Schwarz waveform relaxation techniques. In contrast, our new two level space-time Schwarz waveform relaxation algorithm consists of iterating two steps: a solve on each space-time subdomain, and a new coarse grid correction. The solver on each space-time subdomain $\Omega_{i,n}$ solves for given initial value $u_{i,n,0}$ and boundary value $\mathcal{B}_{i,n}\bar{u}$

$$\begin{aligned} \mathcal{L}u_{i,n} &= f, && \text{in } \Omega_{i,n}, \\ u_{i,n}(x, T_n) &= u_{i,n,0}, && x \in \Omega_i, \\ \mathcal{B}_{i,n}u_{i,n} &= \mathcal{B}_{i,n}\bar{u}, && \text{on } \partial\Omega_i \times (T_n, T_{n+1}). \end{aligned} \tag{2}$$

Here the operators $\mathcal{B}_{i,n}$ are transmission operators, which can be of Dirichlet, Robin or higher order type. We discretize (1) by a centered finite difference scheme in space and backward Euler in time, to get the linear space-time system $L^h \boldsymbol{u} = \boldsymbol{f}$. We denote by Ω^h, $\Omega^h_{i,n}$ the discretized spaces corresponding to Ω and $\Omega_{i,n}$, $i = 1, 2, \ldots, I$, $n = 0, 1, \ldots, N-1$. Also denoting by $\Gamma_{ij,n} := \partial\Omega_{i,n} \cap \Omega_{j,n}$ the interfaces, and $\Gamma_{i,n}$ the initial line for the space-time subdomain $\Omega_{i,n}$, $\Gamma^h_{ij,n}$ and $\Gamma^h_{i,n}$ are the corresponding discretized spaces. Furthermore, we let $N_{\Gamma^h_{ij,n}}$ and $N_{\Gamma^h_{i,n}}$ be the number of degrees of freedom (DOFs) on the interface $\Gamma^h_{ij,n}$ and the initial line $\Gamma^h_{i,n}$ for the space-time subdomain $\Omega^h_{i,n}$.

Then for any initial guess of the initial values $\boldsymbol{u}^0_{i,n,0}$ on the initial line $\Gamma^h_{i,n}$ and the interface values $\mathcal{B}^h_{i,n}\boldsymbol{u}^0$ for the space-time subdomain $\Omega^h_{i,n}$, our new two level Schwarz waveform relaxation method computes iteratively for $k = 0, 1, \ldots$, and for all subdomain indices $i = 1, 2, \ldots, I$, $n = 0, 1, \ldots, N-1$:

Step I. Solve the subdomain problems on each space-time subdomain $\Omega^h_{i,n}$,

$$\begin{aligned} L^h \boldsymbol{u}^{k+1/2}_{i,n} &= \boldsymbol{f}, && \text{in } \Omega^h_{i,n},\\ \boldsymbol{u}^{k+1/2}_{i,n}(x, T_n) &= \boldsymbol{u}^0_{i,n,0}, && x \in \Gamma^h_{i,n},\\ \mathcal{B}_{i,n}\boldsymbol{u}^{k+1/2}_{i,n} &= \mathcal{B}_{i,n}\boldsymbol{u}^k, && \text{on } \Gamma^h_{ij,n}. \end{aligned} \tag{3}$$

Step II. Denoting by $\bar{\boldsymbol{u}}^{k+1/2}$ a composed approximate solution from the subdomain solutions $\boldsymbol{u}^{k+1/2}_{i,n}$ using a partition of unity, the coarse correction step reads

$$\boldsymbol{u}^{k+1} = \bar{\boldsymbol{u}}^{k+1/2} + R_c^T L_c^{-1} R_c(\boldsymbol{f} - L^h \bar{\boldsymbol{u}}^{k+1/2}), \tag{4}$$

where R_c is a restriction matrix to a coarse space, and $L_c := R_c L^h R_c^T$. Finally we set $\boldsymbol{u}^{k+1}_{i,n,0} = \boldsymbol{u}^{k+1}$ on the initial lines $\Gamma^h_{i,n}$.

Definition 1 (Complete coarse space) A complete coarse space for the two level space-time Schwarz waveform relaxation method (3)–(4) for the model problem (1) is given by R_c such that (3)–(4) converges after one iteration for an arbitrary initial guess $\boldsymbol{u}^0_{i,n,0}$ and $\mathcal{B}_{i,n}\boldsymbol{u}^0$, i.e. the method becomes a direct solver.

To give an example of such a complete coarse space for the two level space-time Schwarz waveform relaxation method (3)–(4) for the model problem (1), we define $\boldsymbol{\varphi}^l_{ij,n,\text{cs}}$ for each DOF $l = 1, \ldots, N_{\Gamma^h_{i,n}}$ on the interface $\Gamma^h_{ij,n}$ to be the extension

$$\begin{aligned} L^h \boldsymbol{\varphi}^l_{ij,n,\text{cs}} &= 0 && \text{in } \Omega^h_{i,n},\\ \boldsymbol{\varphi}^l_{ij,n,\text{cs}} &= 1 && \text{at DOF } l \text{ of } \Gamma^h_{ij,n},\\ \boldsymbol{\varphi}^l_{ij,n,\text{cs}} &= 0 && \text{on } \Gamma^h_{i,n} \text{ and the rest of } \Gamma^h_{ij,n} \text{ and } \Omega^h. \end{aligned} \tag{5}$$

Similarly, we define $\boldsymbol{\varphi}^l_{i,n,\text{cs}}$ for each DOF $l = 1, \ldots, N_{\Gamma^h_{i,n}}$ on $\Gamma^h_{i,n}$ to be the extension

$$\begin{aligned} L^h \boldsymbol{\varphi}^l_{i,n,\text{cs}} &= 0 && \text{in } \Omega^h_{i,n},\\ \boldsymbol{\varphi}^l_{i,n,\text{cs}} &= 1 && \text{at DOF } l \text{ of } \Gamma^h_{i,n},\\ \boldsymbol{\varphi}^l_{i,n,\text{cs}} &= 0 && \text{on } \Gamma^h_{ij,n} \text{ and the rest of } \Gamma^h_{i,n} \text{ and } \Omega^h. \end{aligned} \tag{6}$$

We then define our complete coarse space by

$$V_{0,\text{cs}} := \text{span}\{\{\boldsymbol{\varphi}^l_{ij,n,\text{cs}}\}_{l=1}^{N_{\Gamma^h_{ij,n}}}\}_{i=1,n=1}^{i=I,n=N-1} \cup \{\{\boldsymbol{\varphi}^l_{i,n,\text{cs}}\}_{l=1}^{N_{\Gamma^h_{i,n}}}\}_{i=1,n=1}^{i=I,n=N-1}. \tag{7}$$

Theorem 1 *A complete coarse space for the two level space-time Schwarz waveform relaxation method* (3)–(4) *for the model problem* (1) *is given by* R_c *containing in its columns the vectors of* $V_{0,cs}$ *from* (7).

Proof The proof is technical [7]. For an illustration see Section 3. □

The dimension of the complete coarse space (7) corresponds only to the size of the interfaces and initial lines, but can still become prohibitively large, when the size of the problem increases, and we need to consider approximations of (7), which we call optimized coarse spaces, formed by extensions of linear and spectral functions along the interfaces $\Gamma^h_{ij,n}$ and initial lines $\Gamma^h_{i,n}$. The linear functions on the interfaces are ψ^{-1}_{ij} with $\psi^{-1}_{ij}(T_n) = 0$, $\psi^{-1}_{ij}(T_{n+1}) = 1$, and ψ^0_{ij} with $\psi^0_{ij}(T_n) = 1$, $\psi^0_{ij}(T_{n+1}) = 0$, and the spectral functions are $\psi^l_{ij} = \sin(\frac{l\pi(t-T_n)}{T_{n+1}-T_n})$, $t \in [T_n, T_{n+1}]$. Let $\varphi^l_{ij,n,\text{app}}$ be defined by the extension

$$\begin{aligned} L^h\varphi^l_{ij,n,\text{app}} &= 0 && \text{in } \Omega^h_{i,n}, \\ \varphi^l_{ij,n,\text{app}} &= \psi^l_{ij} && \text{on } \Gamma^h_{ij,n},\ l = -1, 0, 1, \dots, \ell_t, \\ \varphi^l_{ij,n,\text{app}} &= 0 && \text{on } \Gamma^h_{i,n} \text{ and the rest of } \Gamma^h_{ij,n} \text{ and } \Omega^h. \end{aligned} \tag{8}$$

Similarly the linear functions along the initial lines $\Gamma^h_{i,n}$ are ψ^{-1}_i with $\psi^{-1}_i(a_i) = 0$, $\psi^{-1}_i(b_i) = 1$, and ψ^0_i with $\psi^0_i(a_i) = 1$, $\psi^0_i(b_i) = 0$, and the spectral functions are $\psi^l_i = \sin(\frac{l\pi(x-a_i)}{b_i-a_i})$, $x \in [a_i, b_i]$. Let $\varphi^l_{i,n,\text{app}}$ be defined by the extension

$$\begin{aligned} L^h\varphi^l_{i,n,\text{app}} &= 0 && \text{in } \Omega^h_{i,n}, \\ \varphi^l_{i,n,\text{app}} &= \psi^l_{i} && \text{on } \Gamma^h_{i,n},\ l = -1, 0, 1, \dots, \ell_x, \\ \varphi^l_{i,n,\text{app}} &= 0 && \text{on } \Gamma^h_{ij,n} \text{ and the rest of } \Omega^h. \end{aligned} \tag{9}$$

Our optimized coarse space is then given by

$$V_{0,\text{cs-1}} := \text{span}\{\{\varphi^l_{ij,n,\text{app}}\}_{l=-1}^{\ell_t}\}_{i=1,n=1}^{i=I,n=N-1} \cup \{\{\varphi^l_{i,n,\text{app}}\}_{l=-1}^{\ell_x}\}_{i=1,n=1}^{i=I,n=N-1}. \tag{10}$$

3 Numerical experiments

We solve the model problem on $\Omega \times (0, T) := (0, 1) \times (0, 1)$ with the source term $f \equiv 0$, zero boundary conditions, and the initial value $u_0 = \exp(-3(0.5 - x)^2)$, discretized by centered finite differences in space using an overlap $4h$ with $h = 1/40$ being the mesh parameter and backward Euler in time with time step $\Delta t = 1/40$. The initial guesses along the interfaces and the initial lines of the space-time subdomains are all random. We first decompose the domain Ω into two overlapping subdomains and the time interval $(0, T)$ also into two time subintervals. Figure 1 shows two examples of basis functions from the complete coarse space: one coming from the interface (left) and one from the initial line (right). In Figure 2 we show on the left the residual after the first **Step I** of the new space-time Schwarz waveform relaxation algorithm, which shows that the residual is only non-zero along the interfaces and the initial line of the space-time subdomains. On the right we show the effect of the following coarse correction **Step II** using the complete coarse space, which reduces the residual to machine precision: the method becomes a direct solver.

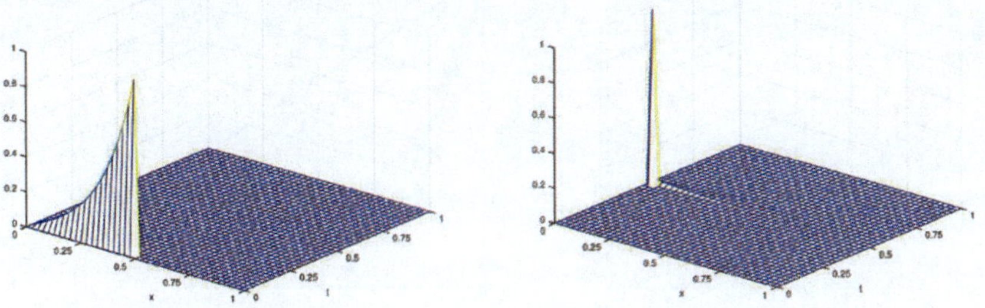

Fig. 1: First basis function of the complete coarse space from the interface $\Gamma^h_{12,1}$ of $\Omega^h_{1,1}$ (left) and from the initial line of $\Omega^h_{1,2}$ (right).

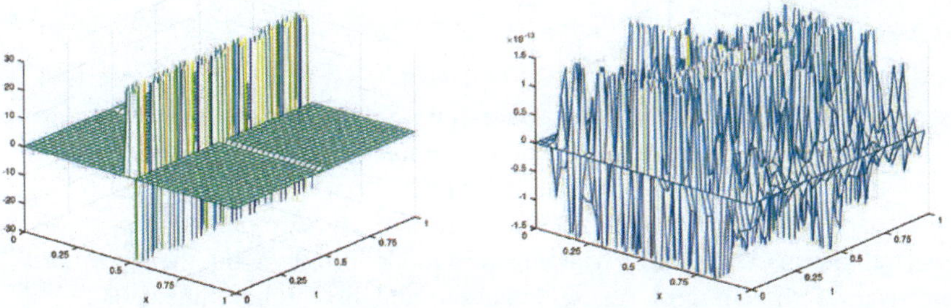

Fig. 2: Residual after the first execution of **Step I** of our new space-time Schwarz waveform relaxation algorithm (left) and after the following coarse correction **Step II** with the complete coarse space (right, note the different scale!).

We next show basis functions of our optimized coarse space: in Figure 4 basis functions from the interface of the space-time subdomains, and in Figure 5 basis functions from the initial line of the space-time subdomain. We show in Figure 3 the influence on the convergence of the optimized coarse space for both 2 spatial subdomains with 2 time subintervals (left) and 4 spatial subdomains with 4 time subintervals (right). The size of the coarse problem is $12, 18, 24, 30$ corresponding to $\ell = 0, 1, 2, 3$ for the first case, and $72, 108, 144, 180$ for the second case. Here the size of the fine problem is 1560 for the all-at-once discretization. We see that the coarse space indeed makes the new two level space-time Schwarz waveform relaxation method scalable, and increasing the number of spectral functions ℓ in the enrichment improves convergence.

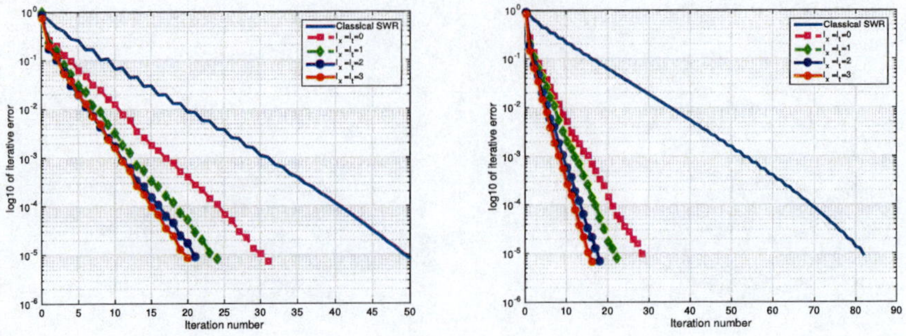

Fig. 3: Comparison of the numerically measured convergence rates using the new optimized coarse space with ℓ spectral function enrichment for 2 spatial subdomains with 2 time subintervals (left) and 4 spatial subdomains with 4 time subintervals (right).

4 Conclusions

We presented a new two level parallel space-time Schwarz waveform relaxation method. The method alternates between solving subproblems in space-time subdomains in parallel, and a new coarse correction which is a spectral approximation of a complete coarse space in space-time. We tested both the complete coarse space and its spectral approximation for a heat equation model problem, but the algorithm definition is valid for much more general equations and also higher dimensions.

Acknowledgements This work was supported by the Natural Science Foundation of China (NSFC) under grant 11801449, 11871393, the International Science and Technology Cooperation Program of Shaanxi Key Research & Development Plan under grant 2019KWZ-08, the Natural Science Basic Research Plan in Shaanxi Province of China (Program No. 2019JQ-617), China Postdoctoral Science Foundation under grant 2020M673465, and the Fundamental Research Funds for the Central Universities under grant G2018KY0306.

References

1. X.C. Cai. Additive Schwarz algorithms for parabolic convection-diffusion equations. *Numerische Mathematik* **60**(1), 41–61 (1991).
2. F. Chaouqui, G. Ciaramella, M.J. Gander and T. Vanzan. On the scalability of classical one-level domain-decomposition methods. *Vietnam Journal of Mathematics* **46**(4), 1053–1088 (2018).
3. M.J. Gander. 50 years of time parallel time integration. In: *Multiple Shooting and Time Domain Decomposition Methods*, pp. 69–113, Springer (2015).
4. M.J. Gander, L. Halpern and F. Nataf. Optimal Schwarz waveform relaxation for the one dimensional wave equation. *SIAM Journal on Numerical Analysis* **41**(5), 1643–1681 (2003).

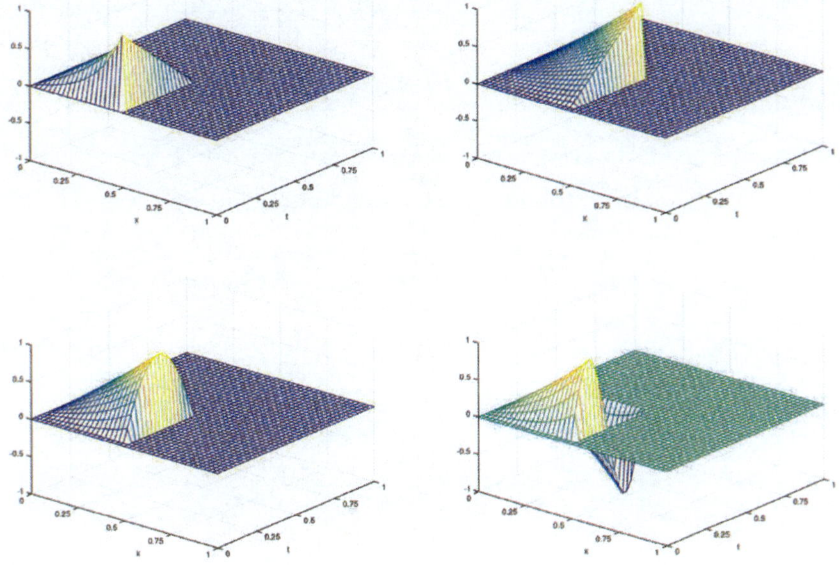

Fig. 4: First two linear basis functions extended to $\Omega^h_{1,1}$ from the interface (top), and first two spectral basis functions for the same subdomain (bottom).

5. M.J. Gander, Y.L. Jiang and R.J. Li. Parareal Schwarz waveform relaxation methods. In: *Domain Decomposition Methods in Science and Engineering XX*, pp. 451–458, Springer (2013).
6. M.J. Gander, Y.L. Jiang and B. Song. A superlinear convergence estimate for the parareal Schwarz waveform relaxation algorithm. *SIAM Journal on Scientific Computing* **41**(2), A1148–A1169 (2019).
7. M.J. Gander, Y.L. Jiang and B. Song. Space-time coarse corrections for the heat equation. In preparation (2021).
8. M.J. Gander and A. Loneland. SHEM: An optimal coarse space for RAS and its multiscale approximation. In: *Domain Decomposition Methods in Science and Engineering XXIII*, Springer (2016).
9. M.J. Gander, A. Loneland and T. Rahman. Analysis of a new harmonically enriched multiscale coarse space for domain decomposition methods. *arXiv preprint* arXiv:1512.05285 (2015).
10. M.J. Gander and A.M. Stuart. Space-time continuous analysis of waveform relaxation for the heat equation. *SIAM Journal on Scientific Computing* **19**(6), 2014–2031 (1998).
11. M.J. Gander and H. Zhao. Overlapping Schwarz waveform relaxation for the heat equation in *n* dimensions. *BIT Numerical Mathematics* **42**(4), 779–795 (2002).
12. A. Heinlein, A. Klawonn, J. Knepper and O. Rheinbach. Adaptive GDSW coarse spaces for overlapping Schwarz methods in three dimensions. *SIAM Journal on Scientific Computing* **41**(5), A3045–A3072 (2019).
13. J.L. Lions, Y. Maday and G. Turinici. A "parareal" in time discretization of PDE's. *Comptes Rendus de l'Académie des Sciences-Series I-Mathematics* **332**(7), 661–668 (2001).

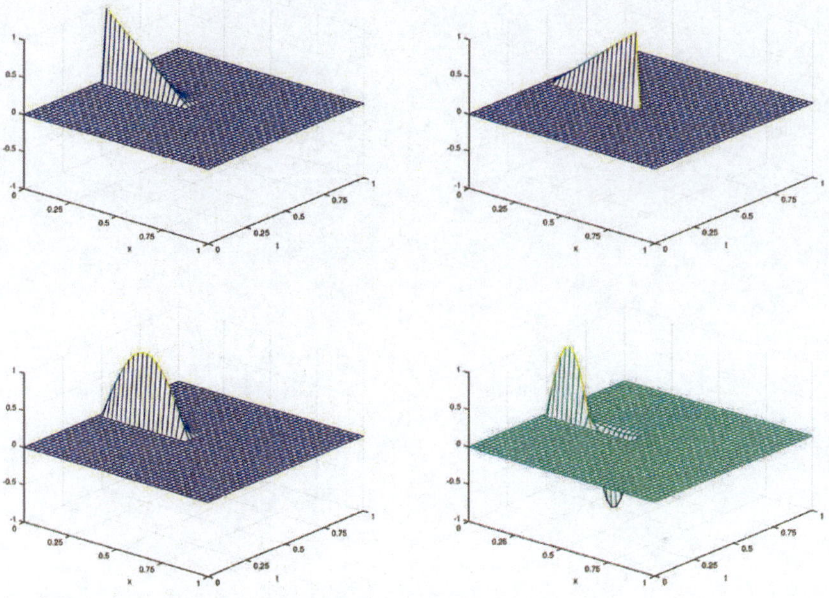

Fig. 5: First two linear basis functions extended to $\Omega_{1,2}^h$ from the initial line (top), and first two spectral basis functions for the same subdomain (bottom).

14. Y. Maday and G. Turinici. The parareal in time iterative solver: a further direction to parallel implementation. *Lecture Notes in Computational Science and Engineering* **40**, 441–448 (2005).
15. N. Spillane, V. Dolean, P. Hauret, F. Nataf, C. Pechstein and R. Scheichl. Abstract robust coarse spaces for systems of PDEs via generalized eigenproblems in the overlaps. *Numerische Mathematik* **126**(4), 741–770 (2014).
16. A. Toselli and O.B. Widlund. *Domain decomposition methods – algorithms and theory.* Volume 34 of Springer Series in Computational Mathematics, Springer-Verlag, Berlin (2005).

On Space-Time Finite Element Domain Decomposition Methods for the Heat Equation

Olaf Steinbach and Philipp Gaulhofer

1 Introduction

Space-time discretisation methods have become very popular in recent years, see, for example, the review article [12], and the references given therein. Applications in mind involve not only the direct simulation of time-dependent partial differential equations in fixed or moving domains, but also problems from optimisation, optimal control, and inverse problems. The solution of the latter applications can be characterised by a coupled problem of a primal forward problem, and an adjoint backward problem, which motivates the use of space-time methods for the solution of the global problem in the space-time domain. As an example, we mention a distributed control problem for the heat equation as considered in [6]. Space-time discretisation methods also allow the use of general and unstructured finite elements, and therefore an adaptive resolution in space and time simultaneously. But the solution of the overall global system in space and time requires the use of appropriate iterative solution strategies in parallel. Besides a pure parallelisation strategy using distributed memory and matrix vector products in parallel, domain decomposition methods can be used for both the parallelisation and the construction of suitable preconditioners. When doing a domain decomposition in space only, we may use the possibility to parallelise in time, where the latter can be done by using the parareal algorithm [8].

Following the well established approaches for domain decomposition methods for elliptic problems, e.g., [1, 5], we first consider the global space-time finite element discretisation of the heat equation, using, e.g., lowest order piecewise linear continuous basis functions. Using a non-overlapping domain decomposition of the space-time domain, and reordering the global stiffness matrix accordingly, we end up with a block system of linear equations, where we can eliminate all local degrees of freedom, e.g., using direct solution methods locally. The resulting Schur complement

Olaf Steinbach and Philipp Gaulhofer
Institut für Angewandte Mathematik, TU Graz, Steyrergasse 30, 8010 Graz, Austria,
e-mail: o.steinbach@tugraz.at, philipp.gaulhofer@student.tugraz.at

S. C. Brenner et al. (eds.), *Domain Decomposition Methods in Science and Engineering XXVI*, Lecture Notes in Computational Science and Engineering 145,
https://doi.org/10.1007/978-3-030-95025-5_59

system is then solved by a global GMRES iteration. In the case of a one-dimensional spatial domain, we will consider different space-time domain decomposition methods, e.g., domain decompositions in space, in time, in space and time, and interfaces which are oblique in space and time. Although we will not consider preconditioning strategies in detail at this time, we will discuss possible preconditioners for the situations mentioned above. In the particular case of a domain decomposition into time slabs, our approach is strongly related to the parareal algorithm. In any case, the numerical results as presented in this contribution indicate the great potential of space-time domain decomposition methods.

2 Space-time finite element methods

As a model problem, we consider the Dirichlet boundary value problem for the heat equation,

$$\begin{aligned} \partial_t u(x,t) - \Delta_x u(x,t) &= f(x,t) && \text{for } (x,t) \in Q := \Omega \times (0,T), \\ u(x,t) &= 0 && \text{for } (x,t) \in \Sigma := \partial\Omega \times (0,T), \\ u(x,0) &= 0 && \text{for } x \in \Omega, \end{aligned} \tag{1}$$

where $\Omega \subset \mathbb{R}^n$, $n = 1, 2, 3$, is some bounded Lipschitz domain, $T > 0$ is a finite time horizon, and f is some given source. For simplicity, we only consider homogeneous boundary and initial conditions, but inhomogeneous data as well as other types of boundary conditions can be handled as well.

The variational formulation of (1) is to find $u \in X$ such that

$$a(u,v) := \int_0^T \int_\Omega \Big[\partial_t u\, v + \nabla_x u \cdot \nabla_x v \Big]\, dx\, dt = \int_0^T \int_\Omega f\, v\, dx\, dt = \langle f, v\rangle_Q \tag{2}$$

is satisfied for all $v \in Y$. Here we use the standard Bochner spaces

$$X := \Big\{ u \in Y : \ \partial_t u \in Y^*, \ u(x,0) = 0, \ x \in \Omega \Big\}, \quad Y := L^2(0,T; H_0^1(\Omega)),$$

including zero boundary and initial conditions, with the norms

$$\|v\|_Y := \|\nabla_x v\|_{L^2(Q)}, \quad \|u\|_X := \sqrt{\|\partial_t u\|_{Y^*}^2 + \|u\|_Y^2} = \sqrt{\|w\|_Y^2 + \|u\|_Y^2},$$

where $w \in Y$ is the unique solution of the variational problem

$$\int_0^T \int_\Omega \nabla_x w \cdot \nabla_x v\, dx\, dt = \int_0^T \int_\Omega \partial_t u\, v\, dx\, dt \quad \text{for all } v \in Y. \tag{3}$$

Unique solvability of the variational problem (2) is based on the inf-sup stability condition [10, 11]

$$\frac{1}{\sqrt{2}} \|u\|_X \leq \sup_{0 \neq v \in Y} \frac{a(u,v)}{\|v\|_Y} \quad \text{for all } u \in X. \tag{4}$$

For the discretisation of the variational formulation (2), we introduce conforming space-time finite element spaces X_h and Y_h, where we assume $X_h \subset Y_h$, i.e., we use the finite element spaces $X_h = Y_h$ of piecewise linear and continuous basis functions, defined with respect to some admissible decomposition of the space-time domain Q into shape regular simplicial finite elements. Then, the Galerkin formulation of (2) is to find $u_h \in X_h$ such that

$$a(u_h, v_h) = \langle f, v_h \rangle_Q \quad \text{for all } v_h \in Y_h, \tag{5}$$

and unique solvability of (5) follows from the discrete inf-sup stability condition

$$\frac{1}{\sqrt{2}} \|u_h\|_{X_h} \leq \sup_{0 \neq v_h \in Y_h} \frac{a(u_h, v_h)}{\|v_h\|_Y} \quad \text{for all } u_h \in X_h. \tag{6}$$

Note that in (6), we use the discrete norm

$$\|u\|_{X_h} := \sqrt{\|w_h\|_Y^2 + \|u\|_Y^2} \quad \text{for } u \in X,$$

where $w_h \in Y_h$ is the unique solution of the Galerkin variational formulation

$$\int_0^T \int_\Omega \nabla_x w_h \cdot \nabla_x v_h \, dx \, dt = \int_0^T \int_\Omega \partial_t u \, v_h \, dx \, dt \quad \text{for all } v_h \in Y_h \tag{7}$$

of (3). From (6), we then conclude the quasi-optimal a priori error estimate

$$\|u - u_h\|_{X_h} \leq 3 \inf_{v_h \in X_h} \|u - v_h\|_X.$$

Assuming $u \in H^2(Q)$, we finally obtain, when using piecewise linear continuous basis functions,

$$\|\nabla_x (u - u_h)\|_{L^2(Q)} \leq c \, h \, |u|_{H^2(Q)}. \tag{8}$$

Once the finite element basis is chosen, i.e., $X_h = \text{span}\{\varphi_k\}_{k=1}^M$, the Galerkin variational formulation (5) can be rewritten as a linear system of algebraic equations, $K_h \underline{u} = \underline{f}$, where the stiffness matrix K_h and the load vector $\underline{f}$ are given as, for $k, \ell = 1, \ldots, M$,

$$K_h[\ell, k] = \int_Q \Big[\partial_t \varphi_k(x,t) \, \varphi_\ell(x,t) + \nabla_x \varphi_k(x,t) \cdot \nabla_x \varphi_\ell(x,t)\Big] dx \, dt, \tag{9}$$

$$f_\ell = \int_Q f(x,t) \, \varphi_\ell(x,t) \, dx \, dt \, .$$

For a more detailed numerical analysis of this space-time finite element method, we refer to [11], and the review article [12].

3 A space-time domain decomposition method

The finite element stiffness matrix K_h as defined in (9) is invertible. It is non-symmetric, but positive definite. Hence, we will use the GMRES method as iterative solver. Since we are discretising the problem in the whole space-time domain Q, preconditioning and parallelisation both in space and time are mandatory.

One possible approach is to use a non-overlapping domain decomposition method as originally designed for elliptic problems, see, e.g., [1, 5]. For the space-time domain $Q := \Omega \times (0,T) \subset \mathbb{R}^{n+1}$ we consider a domain decomposition into p non-overlapping subdomains,

$$\overline{Q} = \bigcup_{i=1}^{p} \overline{Q}_i, \quad Q_i \cap Q_j = \emptyset \quad \text{for } i \neq j.$$

We do not necessarily assume any tensor-product structure of the subdomains Q_i as shown in Fig. 1 a), b), d), f), we may also consider subdomains arbitrary in space and time, see, e.g., Fig. 1 c), e). We assume that the underlying space-time finite element mesh is resolved by the domain decomposition. In this case, we can rearrange all global degrees of freedom $\underline{u}$ into local interior ones per subdomain, i.e.,

$$\underline{u}_I = \left(\underline{u}_{I,1}^\top, \ldots, \underline{u}_{I,p}^\top\right)^\top,$$

and the remaining global degrees of freedom $\underline{u}_C$ on the coupling boundaries. Hence, we can rewrite the linear system $K_h \underline{u} = \underline{f}$ as

$$\begin{pmatrix} K_{II} & K_{CI} \\ K_{IC} & K_{CC} \end{pmatrix} \begin{pmatrix} \underline{u}_I \\ \underline{u}_C \end{pmatrix} = \begin{pmatrix} \underline{f}_I \\ \underline{f}_C \end{pmatrix}, \tag{10}$$

with the block-diagonal matrix

$$K_{II} = \operatorname{diag}\left(K_{11}, \ldots, K_{pp}\right),$$

where the block matrices K_{ii} correspond to the interior degrees of freedom in the subdomain Q_i. Instead of (10), we now consider the Schur complement system

$$S_C \underline{u}_C := (K_{CC} - K_{IC} K_{II}^{-1} K_{CI}) \underline{u}_C = \underline{f}_C - K_{IC} K_{II}^{-1} \underline{f}_I =: \underline{f}, \tag{11}$$

which will be solved by using some global iterative solver such as GMRES. At this time we will not focus on preconditioning the Schur complement system (11), but we will consider different cases of possible space-time domain decompositions as given in Fig. 1 and the influence of the resulting interface in the space-time domain.

Fig. 1: Domain decomposition of the space-time domain $Q = \Omega \times (0, T) \subset \mathbb{R}^2$.

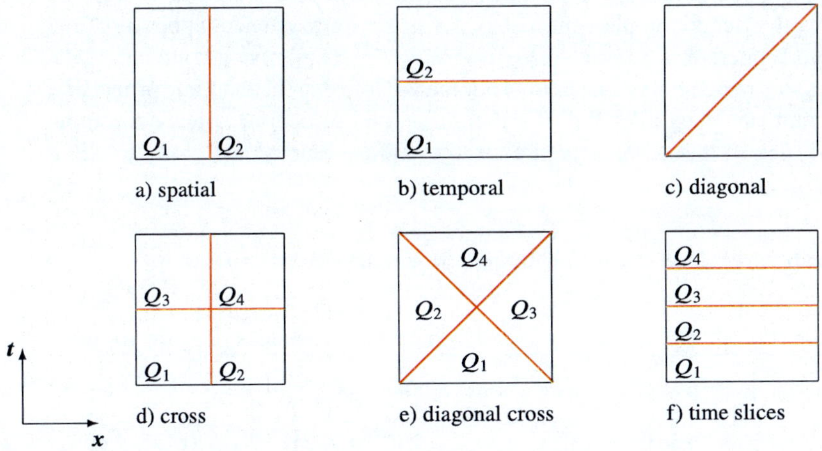

4 Numerical results

In this section, we present first numerical results for the space-time finite element domain decomposition method in the case of a one-dimensional spatial domain $\Omega = (0, 1)$ and the time horizon $T = 1$, i.e., $Q = (0, 1)^2$. In all examples, we consider the smooth function $u(x, t) = \cos \pi t \, \sin \pi x$ to ensure optimal linear convergence in $L^2(0, T; H_0^1(\Omega))$ as expected from the a priori error estimate (8). In Table 1, we also present the error in $L^2(Q)$ where we observe second order convergence in h. Note that DoF denotes the global number of degrees of freedom, Iter is the number of GMRES iterations without preconditioning to reach a relative accuracy of $\varepsilon = 10^{-7}$.

In Table 1, we first present the results for the case without domain decomposition (no), and for the domain decompositions a)–d) as depicted in Fig. 1. We observe that the spatial domain decomposition a) and the diagonal decomposition c) give rather good and comparable results, while the results for the other two cases show a more significant dependence on the mesh size h. When considering the Schur complement system (11) in the case of the spatial decomposition a), we note that the Schur complement matrix S_C is the finite element approximation of a continuous operator $S : H^{1/4}(\Gamma_C) \to H^{-1/4}(\Gamma_C)$, representing the interface conditions along the coupling boundary Γ_C in time only. Since S behaves like the heat potential hypersingular boundary integral operator, e.g., [3], in particular it is an operator of order $\frac{1}{2}$, the spectral condition number of S_C behaves like $h^{-1/2}$, and hence, the number of iterations to reach a given accuracy grows as $h^{-1/4}$, which corresponds to a factor of 1.19 in the case of a uniform mesh refinement. This behaviour is clearly seen in the last three refinement steps. To bound the number of required iterations independent of the mesh level, we can use a suitable preconditioning strategy. One possible option is the use of operator preconditioning, i.e., the Galerkin discretisation of the single layer heat potential as described in [4]. The situation is similar in the case c) of a diagonal domain decomposition. Here, the interface Γ_C is the diagonal $t = x$, and so the Schur complement matrix S_C is the finite element approximation of a continuous operator $S : H^{1/2,1/4}(\Gamma_C) \to H^{-1/2,-1/4}(\Gamma_C)$, using boundary trace spaces of anisotropic Sobolev spaces in the domain. Note that the mapping properties of related boundary integral operators remain true, and so operator preconditioning can be used also in this case, as well as in the higher dimensional case $\Omega \subset \mathbb{R}^n$, $n = 2, 3$.

Table 1: Numerical results for different space-time domain decompositions.

					GMRES iterations				
DoF	$\|u - u_h\|_{L^2(Q)}$		$\|\nabla_x(u - u_h)\|_{L^2(Q)}$		no	a) spatial	b) temporal	c) diagonal	d) cross
12	7.072 –2		5.969 –1		12	4	3	3	6
56	1.912 –2	1.89	3.057 –1	0.97	44	8	7	7	14
240	4.863 –3	1.98	1.538 –1	0.99	135	16	15	15	30
992	1.219 –3	2.00	7.705 –2	1.00	363	26	31	23	56
4.032	3.048 –4	2.00	3.855 –2	1.00	928	33	63	29	95
16.256	7.618 –5	2.00	1.928 –2	1.00	2414	39	127	38	161
65.280	1.907 –5	2.00	9.638 –3	1.00	6536	46	201	48	285

As the uniform finite element meshes used for the domain decompositions a)–d) can be described within time-slabs, the proposed space-time domain decomposition method can be applied also in the case of general space-time finite element meshes. In the case e) of a diagonal cross domain decomposition we apply a recursive red refinement within the subdomains, as depicted in Table 2. We observe similar results as in the case d) of a cross decomposition. Again, we may use a suitable preconditioning strategy which has to take care of the coarse grid involved. A possible approach is the combination of opposite operator preconditioning locally, with global preconditioning using BDDC [7].

Table 2: Space-time finite element mesh and numerical results for case e) of a diagonal cross domain decomposition with 4 subdomains.

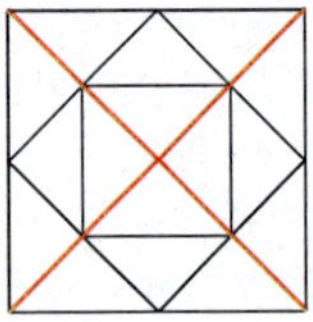

DoF	$\|u - u_h\|_{L^2(Q)}$		$\|\nabla_x(u - u_h)\|_{L^2(Q)}$		Iter
27	2.57 –2		3.52 –1		6
119	6.93 –3	1.89	1.77 –1	0.99	14
495	1.82 –3	1.93	8.89 –2	1.00	28
2.015	4.64 –4	1.97	4.45 –2	1.00	51
8.127	1.17 –4	1.99	2.23 –2	1.00	85
32.639	2.92 –5	2.00	1.11 –2	0.99	138

The last example covers case f) of a domain decomposition into time slabs, where we consider up to $p = 16$ temporal subdomains, see Table 3. Even without preconditioning of the global Schur complement system (11), we observe a rather good behaviour in the number of required iterations. It is obvious that this approach is strongly related to the parareal algorithm [8] where the coarse grid corresponds to the time slabs of the domain decomposition.

Table 3: Numerical results for a domain decomposition into time slabs.

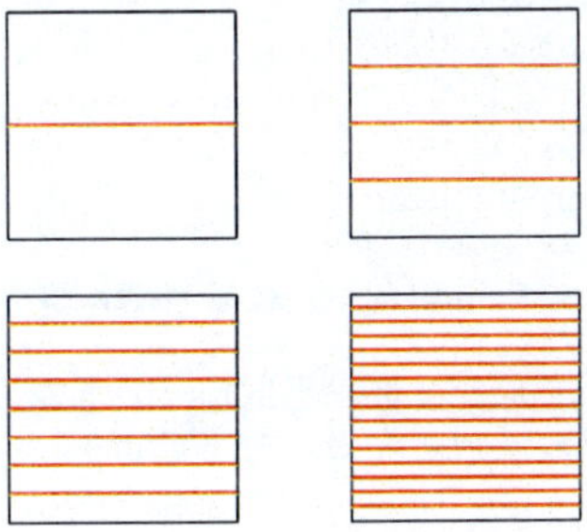

DoF	$p = 2$	$p = 4$	$p = 8$	$p = 16$
12	3			
56	7	17		
240	15	29	49	
992	31	36	65	101
4.032	63	65	91	132
16.256	127	128	133	198
65.280	201	226	255	280
261.632	281	311	383	497

5 Conclusions

In this note we have presented first numerical results for the numerical solution of the heat equation by using standard domain decomposition methods. This approach is based on a space-time finite element discretisation, where the resulting global stiffness matrix is, as in standard domain decomposition methods for elliptic problems, reordered with respect to some non-overlapping domain decomposition of the space-time domain. Eliminating the local degrees of freedom, we finally solve the resulting Schur complement system by a GMRES method without preconditioning. In the case of rather simple domain decompositions of the space-time domain for a one-dimensional spatial domain, we discuss the influence of the choice of the interface in the space-time setting. Since the single layer heat potential boundary integral operator can be defined for any manifold in the space-time domain, it can be used for operator preconditioning of the global Schur complement system, in combination with some coarse grid preconditioning as in BDDC [7], or in space-time FETI methods [9]. On the other hand, the global Schur complement matrix is spectrally equivalent to the global Galerkin matrix of the hypersingular heat potential boundary integral operator, which then allows the use of multigrid methods for an iterative solution, see [2] for a related discussion in the case of boundary element domain decomposition methods for elliptic problems. In the case of a space-time domain decomposition into time slabs, the proposed approach is obviously related to the parareal algorithm [8].

This contribution only gives first numerical results for space-time finite element domain decomposition methods for parabolic problems, and there are many open problems to be resolved in future work. In addition to different preconditioning strategies as already discussed, this covers the parallel implementation in the case of two- or three-dimensional spatial domains, and the application to more complex parabolic equations including problems from fluid mechanics. Some of these topics are already ongoing work, and related results will be published elsewhere.

References

1. J.H. Bramble, J.E. Pasciak and A.H. Schatz. The construction of preconditioners for elliptic problems by substructuring. I. *Math. Comp.* **47**(175), 103–134 (1986).
2. C. Carstensen, M. Kuhn and U. Langer. Fast parallel solvers for symmetric boundary element domain decomposition equations. *Numer. Math.* **79**(3), 321–347 (1998).
3. S. Dohr, K. Niino and O. Steinbach. Space-time boundary element methods for the heat equation. In: *Space-time methods. Applications to partial differential equations*. Volume 25 of Radon Series on Computational and Applied Mathematics, pp. 1–60, de Gruyter, Berlin (2019).
4. S. Dohr and O. Steinbach. Preconditioned space-time boundary element methods for the onedimensional heat equation. In: *Domain decomposition methods in science and engineering XXIV*, volume 125 of Lect. Notes Comput. Sci. Eng., pp. 243–251, Springer, Cham (2018).
5. G. Haase, U. Langer and A. Meyer. The approximate Dirichlet domain decomposition method. I. An algebraic approach. *Computing* **47**(2), 137–151 (1991).
6. U. Langer, O. Steinbach, F. Tröltzsch and H. Yang. Unstructured space-time finite element methods for optimal control of parabolic equations. *SIAM J. Sci. Comput.* **43**(2), A744–A771 (2021).
7. U. Langer and H. Yang. BDDC preconditioners for a space-time finite element discretization of parabolic problems. In: *Domain decomposition methods in science and engineering XXV*, volume 138 of Lect. Notes Comput. Sci. Eng., pp. 367–374, Springer, Cham (2020).
8. J.-L. Lions, Y. Maday and G. Turinici. Résolution d'EDP par un schéma en temps "pararéel". *C. R. Acad. Sci. Paris Sér. I Math.* **332**(7), 661–668 (2001).
9. D.R.Q. Pacheco and O. Steinbach. Space-time finite element tearing and interconnecting domain decomposition methods. In: *Domain decomposition methods in science and engineering XXVI*, Lect. Notes Comput. Sci. Eng. Vol. 145, pp. 479–486, Springer, Cham (2022).
10. C. Schwab and R. Stevenson. Space-time adaptive wavelet methods for parabolic evolution problems. *Math. Comp.* **78**(267), 1293–1318 (2009).
11. O. Steinbach. Space-time finite element methods for parabolic problems. *Comput. Methods Appl. Math.* **15**(4), 551–566 (2015).
12. O. Steinbach and H. Yang. Space-time finite element methods for parabolic evolution equations: Discretization, a posteriori error estimation, adaptivity and solution. In: *Space-time methods. Applications to partial differential equations*, volume 25 of Radon Series on Computational and Applied Mathematics, pp. 207–248, de Gruyter, Berlin (2019).

IETI-DP for Conforming Multi-Patch Isogeometric Analysis in Three Dimensions

Rainer Schneckenleitner and Stefan Takacs

1 Introduction

We are interested in fast domain decomposition solvers for multi-patch Isogeometric Analysis (IgA; [4]). We focus on variants of FETI-DP solvers, see [2, 10] and references therein. Such methods have been adapted to IgA in [5], where the individual patches of the multi-patch discretization are used as subdomains for the solver. This method is sometimes referred to as the dual-primal isogeometric tearing and interconnecting (IETI-DP) method. These methods are similar to Balancing Domain Decomposition by Constraints (BDDC) methods, which have also been adapted for IgA, see [1, 11] and references therein. That the spectra of FETI-DP and BDDC are almost identical is established in [6].

Much progress for the IETI-DP methods has been made in the PhD-thesis by C. Hofer, including the extension to various discontinuous Galerkin formulations, see [3]. Recently, the authors of this paper have extended the condition number bounds for the preconditioned Schur complement system to be explicit not only in the grid size but also in the spline degree, see [8] for the conforming case and [9] for an extension to the discontinuous Galerkin case. The analysis follows the framework from [6]. One key ingredient for the analysis in [8] has been the construction of a bounded harmonic extension operator for splines, which follows the ideas of [7]. The analysis in [8] treats the two-dimensional case. As usual for FETI-like methods, the extension of the analysis to three dimensions is not effortless. The goal of this paper is to demonstrate that the proposed method also performs well for higher spline degrees in three dimensions.

Rainer Schneckenleitner
Institute of Computational Mathematics, Johannes Kepler University Linz, Altenberger Straße 69, 4040 Linz, Austria, e-mail: schneckenleitner@numa.uni-linz.ac.at

Stefan Takacs (corresponding author)
RICAM, Austrian Academy of Sciences, Altenberger Straße 69, 4040 Linz, Austria, e-mail: stefan.takacs@ricam.oeaw.ac.at

S. C. Brenner et al. (eds.), *Domain Decomposition Methods in Science and Engineering XXVI*, Lecture Notes in Computational Science and Engineering 145,
https://doi.org/10.1007/978-3-030-95025-5_60

The remainder of this paper is organized as follows. In Section 2, we introduce the model problem, discuss its discretization and the proposed IETI-DP algorithm. In Section 3, numerical experiments for a three-dimensional example are presented.

2 Model problem and its solution

We consider a standard Poisson model problem. Let $\Omega \subset \mathbb{R}^d$ be a bounded Lipschitz domain. For given $f \in L_2(\Omega)$, we are interested in solving for $u \in H^1(\Omega)$ such that

$$-\Delta u = f \quad \text{in} \quad \Omega \qquad \text{and} \qquad u = 0 \quad \text{on} \quad \partial\Omega$$

holds in a weak sense. We assume that the closure of the computational domain Ω is the union of the closure of K non-overlapping patches $\Omega^{(k)}$ that are parametrized with geometry functions

$$G_k : \widehat{\Omega} := (0,1)^d \to \Omega^{(k)} := G_k(\widehat{\Omega})$$

such that for any $k \neq \ell$, the intersection $\overline{\Omega^{(k)}} \cap \overline{\Omega^{(\ell)}}$ is empty, a common vertex, a common edge, or (in three dimensions) a common face (cf. [8, Ass. 2]). We assume that both ∇G_k and $(\nabla G_k)^{-1}$ are in $L_\infty(\widehat{\Omega})$ for all patches. For the analysis, we need a uniform bound on the L_∞-norm and a uniform bound on the number of neighbors of each patch, cf. [8, Ass. 1 and 3].

For each of the patches, we introduce a tensor B-spline discretization on the parameter domain $\widehat{\Omega}$. The discretization is then mapped to the physical patch $\Omega^{(k)}$ using the pull-back principle. We use a standard basis as obtained by the Cox–de Boor formula. We need a fully matching discretization, which means that for each basis function that has a non-vanishing trace on one of the interfaces, there is exactly one basis function on each of the patches sharing this interface with the traces of the basis functions agreeing (cf. [8, Ass. 5]). This is a standard assumption for any multi-patch setting that is not treated using discontinuous Galerkin methods. For the analysis, we assume quasi-uniformity of grids within each patch, cf. [8, Ass. 4].

In the following, we explain how the IETI-DP solver is set up. Here, we loosely follow the notation used in the IETI-DP solution framework that was recently included in the public part of the G+Smo library. We choose the patches as IETI subdomains. We obtain patch-local stiffness matrices $A^{(k)}$ by evaluating the bilinear forms $a^{(k)}(u,v) = \int_{\Omega^{(k)}} \nabla^\top u(x) \nabla v(x) \mathrm{d}x$ using the basis functions of the corresponding patch. We set up matrices $C^{(k)}$ such that their null spaces are the coefficient vectors of the patch-local functions that vanish at the primal degrees of freedom. In [8], we have considered corner values, edge averages, and the combination of both. In the three-dimensional case, we can choose corner values, edge averages, face averages, and any combination thereof. We set up fully redundant jump matrices $B^{(k)}$. We omit the corner values if and only if the corners are chosen as primal degrees of freedom. We set up the primal problem in the usual way, i.e., we first, for $k = 1, \ldots, K$, compute a basis by

$$\Psi^{(k)} := (I\ 0)\,(\widetilde{A}^{(k)})^{-1}\begin{pmatrix}0\\ R_c^{(k)}\end{pmatrix}, \quad \text{where} \quad \widetilde{A}^{(k)} := \begin{pmatrix}A^{(k)} & (C^{(k)})^\top\\ C^{(k)} & \end{pmatrix}$$

and $R_c^{(k)}$ is a binary matrix that relates the primal constraints (with their patch-local indices) to the degrees of freedom of the primal problem (with their global indices) and set then

$$\widetilde{A}^{(K+1)} := \sum_{k=1}^{K} (\Psi^{(k)})^\top A^{(k)} \Psi^{(k)}, \quad \text{and} \quad \widetilde{B}^{(K+1)} := \sum_{k=1}^{K} B^{(k)} \Psi^{(k)}.$$

We consider the Schur complement problem $F\underline{\lambda} = \underline{g}$, where

$$F := \sum_{k=1}^{K+1} \widetilde{B}^{(k)} (\widetilde{A}^{(k)})^{-1} (\widetilde{B}^{(k)})^\top \quad \text{and} \quad \widetilde{B}^{(k)} := \begin{pmatrix}B^{(k)} & 0\end{pmatrix} \text{ for } k = 1, \dots, K.$$

The derivation of $\underline{g}$ is a patch-local preprocessing step. We solve the Schur complement problem using a preconditioned conjugate gradient (PCG) solver with the scaled Dirichlet preconditioner

$$M_{\mathrm{sD}} := \sum_{k=1}^{K} B_\Gamma D_k^{-1} \left(A_{\Gamma\Gamma}^{(k)} - A_{\Gamma I}^{(k)} (A_{II}^{(k)})^{-1} A_{I\Gamma}^{(k)}\right) D_k^{-1} (B_\Gamma)^\top,$$

where the index Γ refers to the rows/columns of $A^{(k)}$ and the columns of $B^{(k)}$ that refer to basis functions with non-vanishing trace, the index I refers to the remaining rows/columns, and the matrix D_k is a diagonal matrix defined based on the principle of multiplicity scaling. For the analysis, it is important that its coefficients are constant within each interface. The solution u itself is obtained from $\underline{\lambda}$ using the usual patch-local steps, cf. [8].

Under the presented assumptions, the condition number of the preconditioned Schur complement system is in the two-dimensional case bounded by

$$C\, p \left(1 + \log p + \max_{k=1,\dots,K} \log \frac{H_k}{h_k}\right)^2, \tag{1}$$

where p is the spline degree, H_k is the patch size, and h_k the grid size, see [8]. The constant C is independent of these quantities, the number of patches, and the smoothness of the splines within the patches. The authors conjecture that such a condition number estimate also holds for the three-dimensional case, except when only the vertex values are chosen as primal degrees of freedom.

3 Numerical results

In the following, we present numerical results for a three-dimensional domain and refer to the original paper [8] for the two-dimensional case. The computational domain Ω is a twisted version of a Fichera corner, see Fig. 1. The original geometry consists of 7 patches. We subdivide each patch uniformly into $4 \times 4 \times 4$ patches to obtain a decomposition into 448 patches.

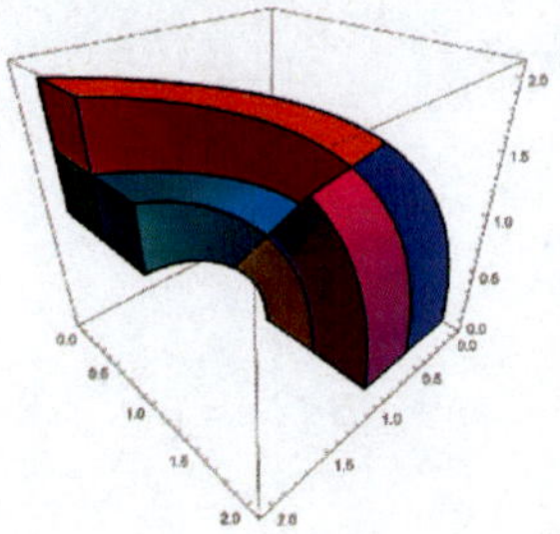

Fig. 1: Computational domain

We solve the model problem $-\Delta u(x, y, z) = 3\pi^2 \sin(\pi x) \sin(\pi y) \sin(\pi z)$ for $(x, y, z) \in \Omega$ with homogeneous Dirichlet boundary conditions on $\partial\Omega$ by means of the IETI-DP solver outlined in the previous sections. Within the patches, we consider tensor-product B-spline discretizations of degree p and maximum smoothness C^{p-1}. We consider several grid sizes, the refinement level $r = 0$ corresponds to a discretization of each patch with polynomials. The next refinement levels $r = 1, 2, \ldots$ are obtained by uniform refinement. All experiments have been carried out in the C++ library G+Smo[1] and have been executed on the Radon1 cluster[2] in Linz. All computations have been performed with a single core.

Concerning the choice of the primal degrees of freedom, we consider all possibilities. For the two-dimensional case, the common choices are the corner values, the edge averages, and a combination of both. We have seen in [8] that all approaches work, typically the corner values are better than when using the edge averages. As expected, the combination of both yields the best results. For the three-dimensional case, we have more possibilities. We report on these approaches in the Tables 1 (vertex values = V), 2 (edge averages = E) and 3 (face averages = F). The combinations V+E, V+F, E+F and V+E+F are only included in the diagrams. In any case, we report on the number of iterations (it) required by the PCG solver to reduce the residual with a random starting vector by a factor of 10^{-6} compared to the right-hand side. Moreover, we report on the condition numbers (κ) of the preconditioned system as estimated by the PCG solver.

[1] `https://github.com/gismo/gismo`, example file `examples/ieti_example.cpp`.

[2] `https://www.ricam.oeaw.ac.at/hpc/`

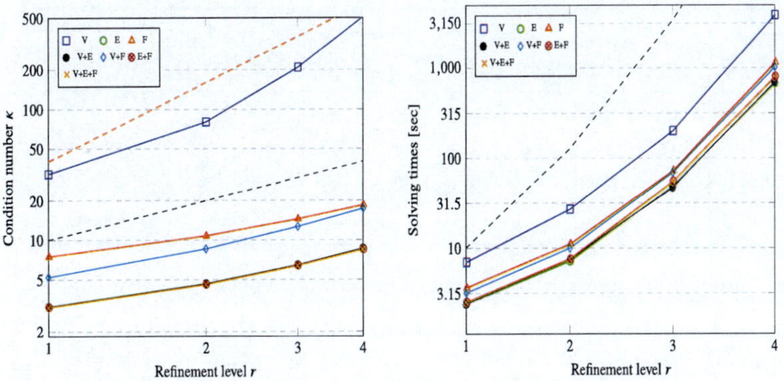

Fig. 2: Condition numbers and solving times for $p = 3$

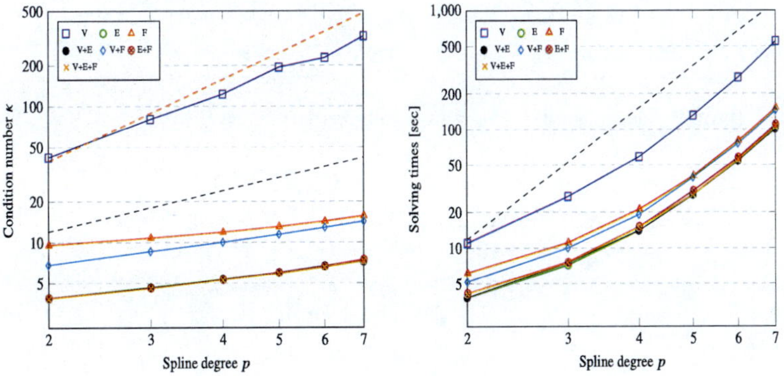

Fig. 3: Condition numbers and solving times for $r = 2$

	$p = 2$		$p = 3$		$p = 4$		$p = 5$		$p = 6$		$p = 7$	
r	it	κ	it	κ	it	κ	it	κ	it	κ	it	κ
1	33	14	51	32	64	45	89	84	108	109	136	178
2	57	42	79	80	98	122	124	193	148	227	176	326
3	94	116	123	208	149	315	175	439	199	566	OoM	
4	146	275	176	509	OoM		OoM		OoM		OoM	

Table 1: Iterations (it) and condition number (κ); Vertex (V)

In Figure 2, the dependence on the refinement level is depicted. Here, we have chosen the spline degree $p = 3$ and have considered all of the possibilities for primal degrees of freedom. Here, we have 44 965 ($r = 1$), 133 629 ($r = 2$), 549 037 ($r = 3$), and 2 934 285 ($r = 4$) degrees of freedom (dofs). We observe that choosing only vertex values as primal degrees of freedom leads to the largest condition numbers.

	$p=2$		$p=3$		$p=4$		$p=5$		$p=6$		$p=7$	
r	it	κ	it	κ	it	κ	it	κ	it	κ	it	κ
1	14	2.5	17	3.1	20	3.8	23	4.4	27	5.1	29	5.5
2	18	3.9	21	4.6	23	5.3	26	6.0	29	6.7	32	7.3
3	23	5.6	25	6.4	28	7.3	30	8.0	33	8.8	OoM	
4	27	7.5	30	8.6	OoM		OoM		OoM		OoM	

Table 2: Iterations (it) and condition number (κ); Edges (E)

	$p=2$		$p=3$		$p=4$		$p=5$		$p=6$		$p=7$	
r	it	κ	it	κ	it	κ	it	κ	it	κ	it	κ
1	22	6.1	26	7.4	29	8.3	33	9.5	37	10.4	41	11.5
2	29	9.5	31	10.7	34	11.8	37	12.9	42	14.2	46	15.5
3	35	13.1	38	14.4	41	15.9	43	17.0	47	18.3	OoM	
4	41	17.1	44	18.4	OoM		OoM		OoM		OoM	

Table 3: Iterations (it) and condition number (κ); Faces (F)

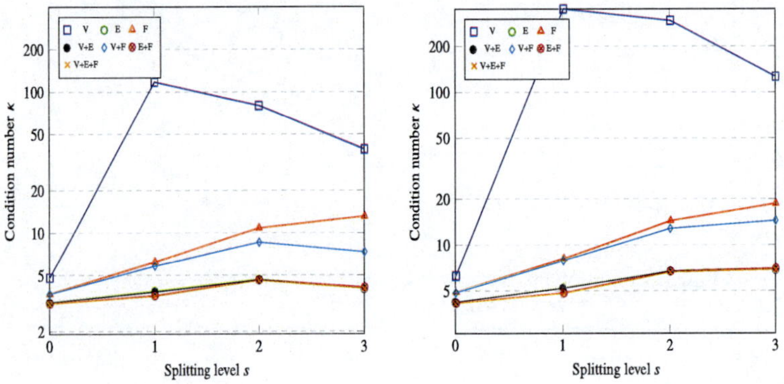

Fig. 4: Condition numbers for $p = 3$ (left) and $p = 6$ (right)

We observe that in this case the condition number grows like r^2 (the dashed red line indicates the slope of such a growth). This corresponds to a growth like $(1+\log H/h)^2$, as predicted by the theory for the two-dimensional case. All other options yield significantly better results, particularly those that include edge averages. In these cases, the growth seems to be less than linear in $r \approx \log H/h$ (the dashed black line shows such a slope). In the right diagram, we can see that choosing a strategy with smaller condition numbers also yields a faster method. Since the dimensions and the bandwidths of the local stiffness matrices grow like $(H_k/h_k)^3$ and $(H_k/h_k)^2$, respectively, the complexity of the LU decompositions grows like $\sum_{k=1}^{K}(H_k/h_k)^7$.

The complexity analysis indicates that they are the dominant factor. The dashed black line indicates such a growth.

In Figure 3, the dependence on the spline degree is presented, where we have chosen $r = 2$. Here, the number of dofs ranges from 66 989 ($p = 2$) to 549 037 ($p = 7$). Also in this picture, we see that the vertex values perform worst and the edge averages best. Again, we obtain a different asymptotic behavior for the corner values. For those primal degrees of freedom, the condition number grows like p^2 (the dashed red line indicates the corresponding slope). All the other primal degrees of freedom seem to lead to a growth that is smaller than linear in p (the dashed black line indicates the slope of a linear growth). Note that for the two-dimensional case, the theory predicts a growth like $p(1 + \log p)^2$. In the right diagram, we can see that the solving times grow like p^4 (the dashed line shows the corresponding slope). This seems to be realistic since the number of non-zero entries of the stiffness matrix grows like Np^d, where N is the number of unknowns. For $d = 3$, this yields in combination with the condition number bound the observed rates.

In Figure 4, we present results for the case that the number of patches is increased. We split each of the 7 patches, depicted in Fig. 1, uniformly into 8^s sub-patches. Within each patch, we consider a grid obtained by $r = 4 - s$ uniform refinement steps. The condition numbers grow slightly in s, where the curve seems to flatten for large values of s. If only the vertex values are primal degrees of freedom, the condition numbers seem to decline for $s > 1$.

Concluding, in this paper we have seen that the IETI method as described in [8] can indeed be extended to the three-dimensional case. If not only the vertex values are chosen as primal degrees of freedom, the condition number of the preconditioned system seems to obey the bound (1).

The extension of the proposed solver to more general elliptic differential equations, like problems with heterogeneous diffusion coefficients, is possible. Robustness in such coefficients is a possible future research direction.

Acknowledgements The first author was supported by the Austrian Science Fund (FWF): S117-03 and W1214-04. The second author has also received support from the Austrian Science Fund (FWF): P31048.

References

1. L. Beirão da Veiga, D. Cho, L. Pavarino and S. Scacchi. BDDC preconditioners for isogeometric analysis. *Math. Models Methods Appl. Sci.* **23**(6), 1099–1142 (2013).
2. C. Farhat, M. Lesoinne, P.L. Tallec, K. Pierson and D. Rixen. FETI-DP: A dual-primal unified FETI method I: A faster alternative to the two-level FETI method. *Int. J. Numer. Methods Eng.* **50**, 1523–1544 (2001).
3. C. Hofer. Analysis of discontinuous Galerkin dual-primal isogeometric tearing and interconnecting methods. *Math. Models Methods Appl. Sci.* **28**(1), 131–158 (2018).
4. T.J.R. Hughes, J.A. Cottrell and Y. Bazilevs. Isogeometric analysis: CAD, finite elements, NURBS, exact geometry and mesh refinement. *Comput. Methods Appl. Mech. Eng.* **194**(39-41), 4135–4195 (2005).

5. S. Kleiss, C. Pechstein, B. Jüttler and S. Tomar. IETI-Isogeometric Tearing and Interconnecting. *Comput. Methods Appl. Mech. Eng.* **247-248**, 201–215 (2012).
6. J. Mandel, C.R. Dohrmann and R. Tezaur. An algebraic theory for primal and dual substructuring methods by constraints. *Appl. Numer. Math.* **54**(2), 167–193 (2005).
7. S.V. Nepomnyaschikh. Optimal multilevel extension operators. https://www.tu-chemnitz.de/sfb393/Files/PDF/spc95-3.pdf (1995).
8. R. Schneckenleitner and S. Takacs. Condition number bounds for IETI-DP methods that are explicit in h and p. *Math. Models Methods Appl. Sci.* **30**(11), 2067–2103 (2020).
9. R. Schneckenleitner and S. Takacs. Convergence theory for IETI-DP solvers for discontinuous Galerkin Isogeometric Analysis that is explicit in h and p. *arXiv preprint* arXiv.2005.09546 (2020).
10. A. Toselli and O.B. Widlund. Domain Decomposition Methods – Algorithms and Theory. Springer, Berlin (2005).
11. O.B. Widlund, S. Zampini, S. Scacchi and L.F. Pavarino. Block FETI–DP/BDDC preconditioners for mixed isogeometric discretizations of three-dimensional almost incompressible elasticity. *Math. Comp.* (2021). Has appeared electronically.

Coupling of Navier–Stokes Equations and Their Hydrostatic Versions and Simulation of Riverbend Flow

Wenbin Dong, Hansong Tang, and Yingjie Liu

1 Introduction

It has become necessary to develop modeling capabilities to simulate multiscale, multiphysics ocean flows directly. An example of such flows is the 2010 Gulf of Mexico oil spill, in which the spill started as a high-speed plume at the seabed and then evolved into drifting oil patches on the ocean surface [2]. Since the plume and the patches are phenomena distinct in physics at temporal and spatial scales and better described by different sets of partial differential equations (PDEs), they are referred to as multiscale and multiphysics flows [3, 15].

Coupling the Navier–Stokes (NS) equations, which describe complex small-scale local flows, and hydrostatic versions of the Navier–Stokes (HNS) equations, which depict the large-scale background ocean flows, is a natural approach to realize simulations of multiscale and multiphysics ocean flows. Example topics of efforts on such coupling include optimized interface conditions for convergence speedup of Schwarz iteration between the NS and HNS equations [1], appropriate interface conditions [7], simulation of multiscale, multiphysics ocean flows [5, 6, 10, 13, 15], etc. Due to the complexity of the coupling, these earlier efforts are mostly simple and crude in methods (e.g., one-way coupling), and they are sporadic in both theoretical analysis and desired computation implementation (particularly two-way coupling). Actually, problems such as non-physical solutions have been reported in simulations

Wenbin Dong
Civil Engineering Department, City College, City University of New York, NY 10031, USA,
e-mail: wdong000@citymail.cuny.edu

Hansong Tang*, corresponding author
Civil Engineering Department, City College, City University of New York, NY 10031, USA,
e-mail: htang@ccny.cuny.edu

Yingjie Liu
School of Mathematics, Georgia Institute of Technology, Atlanta, GA 30332, USA,
e-mail: yingjie@math.gatech.edu

S. C. Brenner et al. (eds.), *Domain Decomposition Methods in Science and Engineering XXVI*, Lecture Notes in Computational Science and Engineering 145,
https://doi.org/10.1007/978-3-030-95025-5_61

based on such coupling [14]. A more detailed review can be found in [10, 13]. It is fair to say that we are still at an exploration stage of such coupling.

This work presents a discussion on the coupling of the NS and HNS equations to capture physical phenomena correctly. A numerical example is provided to illustrate the necessity for coupling the NS and HNS equations and the influence of different transmission conditions.

2 Governing equations

A flow domain, Ω_0, is divided into a near field assigned with the NS equations, Ω_{NS}, and a far field applied with the NHS equations, Ω_{HNS}, see Fig. 1. The near field wraps a small-scale, complex, local flow, and the far field covers its large-scale background flow. For a near field, the governing equations consist of the continuity equation and the NS equations [12, 13]:

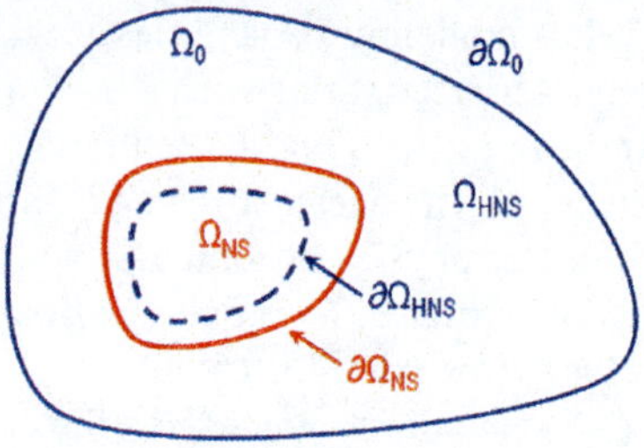

Fig. 1: Division of flow domain, Ω_0 $(= \Omega_{NS} \cup \Omega_{HNS})$. $\partial\Omega_0$ is its boundary, and $\partial\Omega_{NS}$ and $\partial\Omega_{HNS}$ are the interfaces for the NS and HNS equations, respectively.

$$\begin{aligned} &\nabla \cdot \mathbf{u} = 0, \\ &\mathbf{u}_t + \nabla \cdot \mathbf{uu} = \nabla \cdot ((\nu + \nu_t)\, \nabla\, \mathbf{u}) - \nabla p_d/\rho - g\, \nabla_H\, \eta. \end{aligned} \tag{1}$$

Here, $\mathbf{u} = (u, v, w)$, the velocity vector, with u and v as its components in x and y direction, respectively, on the horizontal plane, and w as its component in z direction, or, the vertical direction. η is the water surface elevation, and $p_d = p - \rho g(\eta - z)$, with p being pressure. ν is the viscosity, ν_t the turbulence viscosity, ρ the density, and g the gravity. ∇_H is the gradient in the horizontal plane. When $\rho = const$, p_d becomes the dynamic pressure, and it is introduced to facilitate the coupling of the NS to the HNS equations [13].

In a far field, the governing equations are the continuity equation and the HNS equations (the latter is simplified from the NS equations according to the hydrostatic assumption, i.e., $p = \rho g(\eta - z)$), and they read as

$$\begin{aligned} &\nabla \cdot \mathbf{u} = 0, \\ &\mathbf{v}_t + \nabla \cdot \mathbf{uv} = \nabla \cdot ((\nu + \nu_t)\, \nabla\, \mathbf{v}) - g\, \nabla_H\, \eta, \end{aligned} \tag{2}$$

where $\mathbf{v} = (u, v)$, the velocity vector in the horizontal plane.

As a practical approach, the following interface conditions are adopted in computation:

$$\begin{aligned} &\mathbf{u}|_{NS} = \mathbf{u}|_{HNS},\ \partial p_d/\partial n = 0, &&\text{on } \partial\Omega_{NS}, \\ &\mathbf{u}|_{HNS} = \mathbf{u}|_{NS}, &&\text{on } \partial\Omega_{HNS}, \end{aligned} \tag{3}$$

which requires the continuity of $\mathbf{u}$ across an interface. Another approach is

$$\begin{aligned} &\left(\mathbf{u}_n\mathbf{v} + p_{d_n}/\rho - (\nu + \nu_t)\partial\mathbf{v}/\partial n\right)|_{NS}, \\ &\quad = (\mathbf{u}_n\mathbf{v} - (\nu + \nu_t)\partial\mathbf{v}/\partial n)\,|_{HNS}, &&\text{on } \partial\Omega_{NS}, \\ &\mathbf{u}_\tau|_{NS} = \mathbf{u}_\tau|_{HNS},\ \partial p_d/\partial n = 0, &&\text{on } \partial\Omega_{NS}, \\ &\mathbf{u}|_{HNS} = \mathbf{u}|_{NS}, &&\text{on } \partial\Omega_{HNS}, \end{aligned} \tag{4}$$

here the subscript τ indicates the tangential direction. (4) is the same as (3) except that, on an interface of the NS equations and in its normal direction, the continuity of velocity is replaced by continuity of momentum flux. Condition (4) is adopted/modified from previous investigations [1, 12].

3 Computational methods

The NS equations (1) are computed using the Solver of Incompressible Flow on Overset Meshes (SIFOM) developed by us (e.g., [8, 11]). The solver discretizes the equations in curvilinear coordinates using a second-order accurate backward difference in time and central difference on non-staggered, composite structured grids [11]. The HNS equations are solved by utilizing the Finite Volume Method Coastal Ocean Model (FVCOM), which is an operational model in the ocean science community [4]. In this model, the HNS equations (2) are transformed and solved in the following form:

$$\begin{aligned} &\eta_t + \nabla_H \cdot (\mathbf{v}D) + \omega_\sigma = 0, \\ &(\mathbf{v}D)_t + \nabla_H \cdot (\mathbf{v}\mathbf{v}D) + (\mathbf{v}\omega)_\sigma = -gD\,\nabla_H\,\eta + \nabla_H \cdot (\kappa\mathbf{e}) \\ &\qquad\qquad\qquad\qquad\qquad\qquad\qquad + (\lambda\mathbf{v}_\sigma)_\sigma/D + \mathbf{I}, \end{aligned} \tag{5}$$

in which σ is a vertical coordinate, ω the vertical velocity in this coordinate, $\mathbf{e}$ the strain rate, subscript σ the derivative over σ, and $\mathbf{I}$ the other terms. κ and λ are coefficients. Actually, in FVCOM, equations (5) are solved together with another set of equations, which essentially result from integrating the NS equations in the vertical direction [4]. The model adopts the second-order accurate Runge–Kutta method in time and a second-order accurate finite volume method on a triangular grid in the horizontal plane and a σ-grid in the vertical direction. The grids of SIFOM and FVCOM overlap arbitrarily with each other (i.e., Chimera grids), and their solutions are exchanged at interfaces between the two grids via interpolation [13].

Let the discretized NS and HNS equations be respectively expressed as

$$\mathbf{F}(\mathbf{f}) = \mathbf{0}, \quad \mathbf{H}(\mathbf{h}) = \mathbf{0} \tag{6}$$

in which $\mathbf{f} = (\mathbf{u}, p_d)$, and $\mathbf{h} = (\mathbf{v}, \omega, \eta)$, being the solution for the NS and HNS equations, respectively. Then, the computation of their coupling when moving from time level n to $n+1$ is formulated as

$$\bar{\mathbf{f}}^0 = \mathbf{f}^n, \ \bar{\mathbf{h}}^0 = \mathbf{h}^n$$

$Do\ 1\ m = 1, M$

$$\begin{cases} \mathbf{F}(\bar{\mathbf{f}}^m) = \mathbf{0}, & \mathbf{x} \in \Omega_{NS} \\ \bar{\mathbf{f}}^m = \mathbf{p}(\mathbf{h}^{m-1}), & \mathbf{x} \in \partial\Omega_{NS} \end{cases} \qquad \begin{cases} \mathbf{H}(\bar{\mathbf{h}}^m) = \mathbf{0}, & \mathbf{x} \in \Omega_{HNS} \\ \bar{\mathbf{h}}^m = \mathbf{q}(\mathbf{f}^{m-1}), & \mathbf{x} \in \partial\Omega_{HNS} \end{cases} \tag{7}$$

$1\ End\ Do$

$$\mathbf{f}^{n+1} = \bar{\mathbf{f}}^M, \ \bar{\mathbf{h}}^{n+1} = \bar{\mathbf{h}}^M$$

in which m is the Schwarz iteration index, and M is a prescribed integer. $\mathbf{p}$ and $\mathbf{q}$ are operators for solution exchange between the NS and HNS equations.

4 Numerical simulation

Numerical experiments are made on a transient water flow in a riverbend, see Fig. 2. The water body is stationary initially, and it flows as velocity is imposed at its entrance, with the following initial and boundary conditions:

$$\begin{aligned} &\mathbf{u} = 0, \ p = \gamma(\eta - z), \ \ t = 0 \\ &u = 0.25(1 - \mathrm{e}^{-0.01t}), \ \text{ at entrance; } \eta = 15, \ \text{ at exit} \end{aligned} \tag{8}$$

Here, length is in m, and time is in s, and velocity is in m/s.

The mesh for the HNS equations covers the whole channel, with 10,000 triangular cells in the horizontal plane and 20 layers in the vertical direction, and the grid for the NS equations occupies the bend section, with a grid of $111\times13\times13$ in the streamwise, lateral, and vertical direction, respectively. The time step is set as $\Delta t = 0.01$, and $M = 1$ is used in (7).

The simulated instantaneous water surface elevation by the HNS equations and those by the coupled HNS and NS equations are depicted in Fig. 3. At $t = 60$, in the transient stage, the simulated elevations by both approaches are essentially identical. When $t = 1200$, at which the flow is about steady, the elevation obtained with the HNS/NS equations differs from that with the HNS equations in patterns. Additionally, the elevations computed by the coupled equations with interface conditions (3) and (4) are similar in main patterns, but they do have a little difference in magnitude.

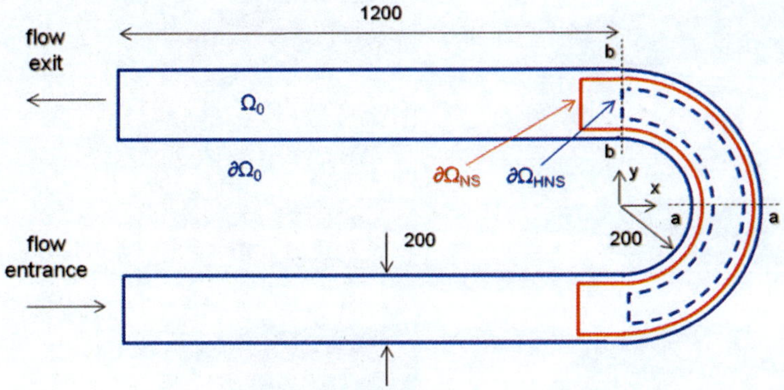

Fig. 2: Riverbend configuration and subdomains.

Fig. 4 illustrates the velocity field at cross-section a-a in the curved section of the channel (Fig. 2). The figure shows that the HNS simulation presents a vortex covering the whole cross-section. Whereas the HNS/NS simulations lead to two counter-rotation vortices in the middle of the cross-section, plus one at the left lower corner, one at the right lower corner, and one at the upper right corner. Such multiple vortices have been reported in previous investigations based on the NS equations [9], and the difference indicates the necessity of the NS/HNS coupling. Additionally, it is seen that, with the coupled equations, although both interface conditions (3) and (4) lead to three vorticities on the cross-section, their positions have changed in some degree (Fig. 4). This indicates that interface conditions (3) and (4) could produce a difference in solutions.

To further examine the difference, the vertically averaged velocities are plotted in Fig. 5. It is seen that the coupled HNS/NS equations provide solutions for both streamwise and lateral velocity that are distinct from those obtained with the HNS equations. Note the streamwise velocity is y-velocity and x-velocity at cross-section a-a and b-b, respectively. Moreover, the two interface conditions lead to similar streamwise velocities but distinct lateral velocities. To illustrate this in more detail, we present a quantification of the difference in the solutions obtained with the two interface conditions in Table 1. The numbers in the table show that the difference in the lateral velocities is more pronounced, indicating that the two interface conditions lead to a big difference in the secondary flows in the bend's cross-section.

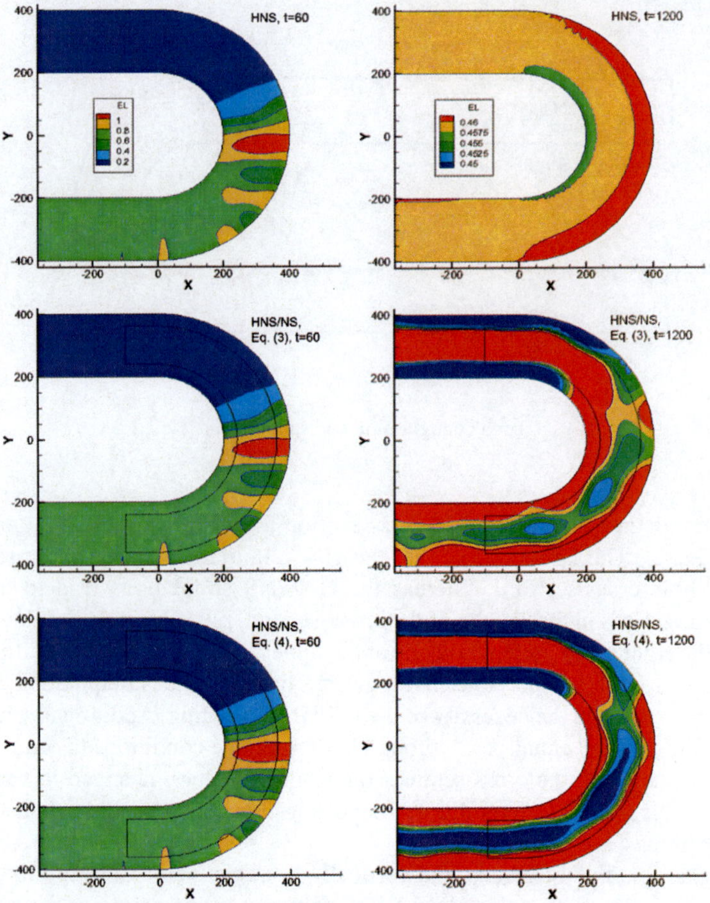

Fig. 3: Simulated instantaneous water surface elevation by the HNS equations and coupled NHS/NS equations with interface condition (3) and (4).

Table 1: Difference of the HNS/NS solutions obtained with different interface conditions. $\bar{u}$ and $\bar{v}$ are vertically integrated x- and y-velocity, respectively, and subscripts 3 and 4 depict interface condition (3) and (4), respectively.

cross section	$\|\|\bar{u}_4 - \bar{u}_3\|\|_2/\|\bar{u}_3\|_{max}$	$\|\|\bar{v}_4 - \bar{v}_3\|\|_2/\|\bar{v}_3\|_{max}$
a-a	1.809	0.385
b-b	0.268	2.196

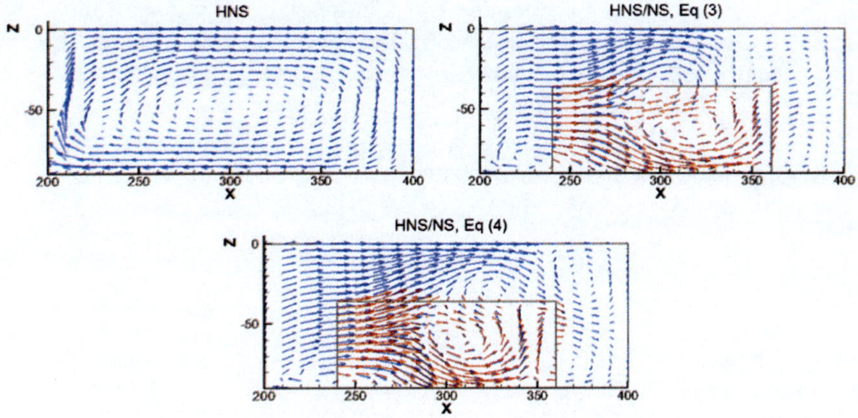

Fig. 4: Simulated cross-section flow field, at cross section a-a in Fig. 2, $t = 1200$.

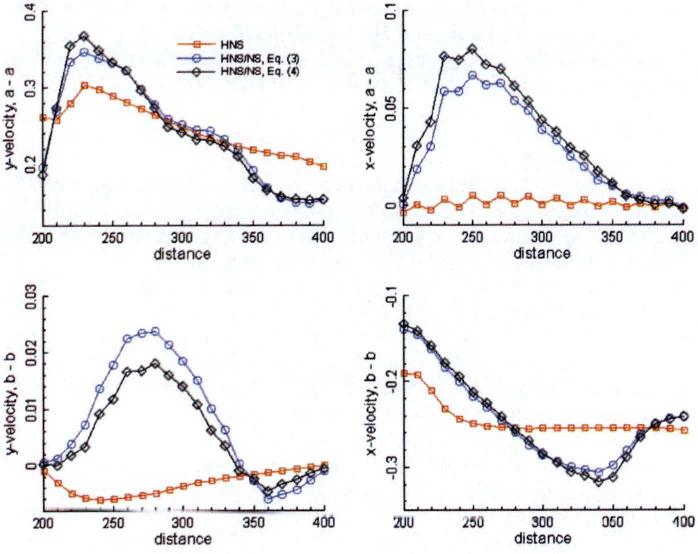

Fig. 5: Vertically averaged velocity at cross sections a-a and b-b in Fig. 2, $t = 1200$.

5 Concluding remarks

We present a discussion on the coupling of the NHS and NS equations. The numerical experiment on a riverbend flow shows that the coupling can capture complex flow phenomena that the HNS equations cannot resolve. It also indicates that different interface conditions may lead to different solutions, especially those for the secondary flows in cross-sections of the bend.

Further investigation is necessary for the two transmission conditions in this paper. Particularly, examining their performance against benchmarks plus theoretical analysis is expected to be the next step, followed by domain decomposition techniques to achieve desired computational efficiency.

Acknowledgements This work is supported by the NSF (DMS-1622453, DMS-1622459).

References

1. E. Blayo and A. Rousseau. About interface conditions for coupling hydrostatic and nonhydrostatic Navier–Stokes flows. *Discrete and Continuous Dynamical Systems - S* **9**, 1565–1574 (2016).
2. R. Camilli, C.M. Reddy, D.R. Yoerger, B.A.S. Van Mooy, M.V. Jakuba, J.C. Kinsey, C.P. McIntyre, S.P. Sylva and J.V. Maloney. Tracking hydrocarbon plume transport and biodegradation at Deepwater Horizon. *Science* **330**, 208–211 (2010).
3. A.S. Candy. An implicit wetting and drying approach for non-hydrostatic baroclinic flows in high aspect ratio domains. *Adv. Water Resour.* **102**, 188–205 (2017).
4. C. Chen, H. Liu and R.C. Beardsley. An unstructured, finite-volume, three-dimensional, primitive equation ocean model: application to coastal ocean and estuaries. *J. Atm. & Oceanic Tech.* **20**, 159–186 (2003).
5. O.B. Fringer, J.C. McWilliams and R.L. Street. A new hybrid model for coastal simulations. *Oceanography* **19**, 64–77 (2006).
6. K. Fujima, K. Masamura and C. Goto. Development of the 2D/3D hybrid model for tsunami numerical simulation. *Coastal Eng J.* **44**, 373–397 (2002).
7. P.C. Gallacher, D.A. Hebert and M.R. Schaferkotter. Nesting a nonhydrostatic model in a hydrostatic model: The boundary interface. *Ocean Modelling* **40**, 190–198 (2011).
8. L. Ge and F. Sotiropoulos. 3D unsteady RANS modeling of complex hydraulic engineering flows. I: numerical model. *J. Hydraul. Eng.* **131**, 800–808 (2005).
9. B. Li, X. Zhang,H.S. Tang and R. Tsubaki. Influence of deflection angles on flow behaviours in open channel bends. *J. Mountain Science* **15**, 2292–2306 (2018).
10. K. Qu, H.S. Tang and A. Agrawal, Integration of fully 3D fluid dynamics and geophysical fluid dynamics models for multiphysics coastal ocean flows: Simulation of local complex free-surface phenomena. *Ocean Modelling* **135**, 14–30 (2019).
11. H.S. Tang, S.C. Jones and F. Sotiropoulos. An overset-grid method for 3D unsteady incompressible flows. *J. Comput. Phys.* **191**, 567–600 (2003).
12. H.S. Tang and Y.J. Liu. Coupling of Navier–Stokes equations and their hydrostatic versions for ocean flows: A discussion on algorithm and implementation. In: R. Haynes, S. MacLachlan, X.-C. Cai, L. Halpern, H.H. Kim, A. Klawonn and O. Widlund (eds), *Domain Decomposition Methods in Science and Engineering XXV*, Lecture Notes in Computational Science and Engineering **138**, pp. 326–333, Springer (2020).
13. H.S. Tang, K. Qu and X.G. Wu. An overset grid method for integration of fully 3D fluid dynamics and geophysical fluid dynamics models to simulate multiphysics coastal ocean flows. *J. Comput. Phys.* **273**, 548–571 (2014).
14. H.S. Tang, K. Qu, X.G. Wu and Z.K. Zhang. Domain decomposition for a hybrid fully 3D fluid dynamics and geophysical fluid dynamics modeling system: A numerical experiment on a transient sill flow. In: T. Dickopf, M.J. Gander, L. Halpern, R. Krause and L.F. Pavarino (eds), *Domain Decomposition Methods in Science and Engineering XXII*, Lecture Notes in Computational Science and Engineering **104**, pp. 407-414, Springer (2016).
15. H.S. Tang and X.G. Wu. Multi-scale coastal flow simulation using coupled CFD and GFD models. In: D.A. Swayne, W. Yang, A.A. Voinov, A. Rizzoli and T. Filatova, (Eds.), *Modelling for Environment's Sake, 5th Biennial Meeting*. Ottawa (2010).

On the Links Between Observed and Theoretical Convergence Rates for Schwarz Waveform Relaxation Algorithm for the Time-Dependent Problems

Sophie Thery

1 Context

We study an application of the Schwarz waveform relaxation algorithm for time-dependent problems to a linear multiphysics problem on two non-overlapping physical domains Ω_1 and Ω_2:

$$\begin{cases} \partial_t u_j(x,t) - \mathcal{A}_j u_j(x,t) = F_j(x,t) & \text{on } \Omega_j \times]0,T[\\ \mathcal{B}_j u_j(x,t) = G_j(x,t) & \text{on } \partial\Omega_j^{\text{ext}} \times]0,T[\\ u_j(x,0) = u_{j,0}(x) & \text{in } \Omega_j \end{cases} \tag{1a}$$

$$\begin{cases} C_{1,1} u_1|_\Gamma(t) = C_{1,2} u_2|_\Gamma(t) & \text{on } [0,T[\\ C_{2,2} u_2|_\Gamma(t) = C_{2,1} u_1|_\Gamma(t) & \text{on } [0,T[\end{cases} \tag{1b}$$

where T can be a finite or infinite time. The Schwarz waveform relaxation algorithm is applied on problem (1a) with interface conditions (1b). For a given first guess $u_j^0|_\Gamma(t)$ on the interface Γ, the state of the algorithm is given at each iteration $n \in \mathbb{N}$ by (2). We suppose here the well-posedness of the initial problem (1) and of the algorithm (2). This means there exists a unique solution to (1) in $\mathcal{L}^2(0,T;\mathcal{L}(\Omega_j))$ denoted $\bar{u}$ and there exists a unique $u_j^n \subset \mathcal{L}^2(0,T;\mathcal{L}^2(\Omega_j))$ for all iterations n.[1] Some results on the well-posedness of such problems can be found in [1, 2] (for problems on a finite time window) and in a more general framework in [3] (for problems on a finite or infinite time window).

S. Thery
Univ. Grenoble-Alpes, Inria, CNRS, Grenoble INP, LJK, Grenoble, France
Tel.: +33 4 76 63 12 63
Fax: +33 4 76 61 52 52
e-mail: sophie.thery@univ-grenoble-alpes.fr

[1] For example, for parabolic problems we need to have $F \in \mathcal{L}^2(0,T;\mathcal{L}^2(\Omega_j))$ and $u_{j,0} \in \mathcal{L}^2(\Omega_j)$.

S. C. Brenner et al. (eds.), *Domain Decomposition Methods in Science and Engineering XXVI*, Lecture Notes in Computational Science and Engineering 145,
https://doi.org/10.1007/978-3-030-95025-5_62

$$\begin{cases} \partial_t u_1^n(z,t) - \mathcal{A}_1 u_1^n(x,t) = F_1(x,t) & \text{on } \Omega_1 \times [0,T[\\ \mathcal{B}_1 u_1^n(x,t) = G_1(x,t) & \text{on } \partial\Omega_1^{\text{ext}} \times [0,T[\\ u_1^n(x,0) = u_{1,0}(x) & \text{in } \Omega_1 \\ C_{1,1} u_1^n(x,t) = C_{1,2} u_2^{n-1}(x,t) & \text{on } \Gamma \times [0,T[\end{cases} \tag{2a}$$

$$\begin{cases} \partial_t u_2^n(x,t) - \mathcal{A}_2 u_2^n(x,t) = F_2(x,t) & \text{on } \Omega_2 \times [0,T[\\ \mathcal{B}_2 u_2^n(x,t) = G_2(x,t) & \text{on } \partial\Omega_2^{\text{ext}} \times [0,T[\\ u_2^n(x,0) = u_{2,0}(x) & \text{in } \Omega_2 \\ C_{2,2} u_2^n(x,t) = C_{2,1} u_1^n(x,t) & \text{on } \Gamma \times [0,T[\end{cases} \tag{2b}$$

From now on we also suppose $u_j^n(x) \in \mathcal{L}^2(]0,T[)$ for all $x \in \Omega_j$.[2] To quantify and possibly optimize the convergence of algorithm (2), it is appropriate to calculate the convergence rate as $\rho^{\text{obs}}_{\mathcal{A}_{\{1,2\}},\mathcal{B}_{\{1,2\}},C_{\{\{1,2\},\{1,2\}\}},j,n} = \left\|e_j^n\right\| / \left\|e_j^{n-1}\right\|$, where $e_j^n = u_j^n - \widetilde{u}|_{\Omega_j}$ is the error at each iteration n. In the rest of the paper, indices $_{\mathcal{A},\mathcal{B},C}$ are neglected to simplify the notation.

Remark 1 From now on we assume that Ω_j are one-dimensional domains. Since all convergence factors are calculated in Fourier space, all results explained here can be extended to higher space dimensions parallel to the interface.[3] We consider here Schwarz algorithms applied to multiphysics problems (for nonoverlapping domains) but the following results are also valid in the presence of an overlap.

2 Convergence for problems on an infinite time window

We first consider that the simulation is made on an infinite time window, i.e. $T = +\infty$.

Convergence factor in Fourier space: For time-dependent problems, the observed convergence factor cannot be calculated analytically. Thus the usual approach consists in applying a time Fourier transform to the error system. In the case where $T = +\infty$ and considering that the error is equal to zero for negative times, the convergence is determined in the Fourier space by solving the following system:

$$\begin{cases} i\omega \widehat{e}_1^n(x,\omega) - \mathcal{A}_1 \widehat{e}_1^n(z,\omega) = 0 & \text{on } \Omega_1 \times \mathbb{R} \\ \mathcal{B}_1 \widehat{e}_1^n(x,\omega) = 0 & \text{on } \partial\Omega_1^{ext} \times \mathbb{R} \\ C_{1,1} \widehat{e}_1^n(x,\omega) = C_{1,2} \widehat{e}_2^{n-1}(x,\omega) & \text{on } \Gamma \times \mathbb{R} \end{cases} \tag{3a}$$

$$\begin{cases} i\omega \widehat{e}_2^n(x,\omega) - \mathcal{A}_2 \widehat{e}_2^n(x,\omega) = 0 & \text{on } \Omega_2 \times \mathbb{R} \\ \mathcal{B}_2 \widehat{e}_2^n(x,\omega) = 0 & \text{on } \partial\Omega_2^{ext} \times \mathbb{R} \\ C_{2,2} \widehat{e}_2^n(x,\omega) = C_{2,1} \widehat{e}_1^n(x,\omega) & \text{on } \Gamma \times \mathbb{R} \end{cases} \tag{3b}$$

[2] For example, for parabolic problems we need to have $u_j^n \in \mathcal{L}^2(0,T;\mathcal{H}^1(\Omega_j))$, which is satisfied if G_j and first guess is regular enough (see [4]).

[3] This involves applying Fourier transforms in all directions parallel to the interface. Fourier transforms on spatial dimensions do not give rise to the problem that we expose here, which is specific to the temporal dimension.

Suppose that (3) can be solved for any $\omega \in \mathbb{R}$, then the convergence factor ϱ and convergence rate ρ in the Fourier space can be calculated as:

$$\varrho(\omega) := \frac{\widehat{e}_j^n|_\Gamma(\omega)}{\widehat{e}_j^{n-1}|_\Gamma(\omega)} \qquad \rho(\omega) := |\varrho(\omega)|. \tag{4}$$

It can be shown that ρ is independent of the space variable x and of the domain j. Without lack of generality, from now on suppose ρ is calculated from the errors at the interface Γ. General methods to study the convergence of Schwarz algorithms can be found in [5–7].

Observed convergence factor: From the well-posedness properties of the algorithm, we assume that $e_j^n(x,\cdot) \in \mathcal{L}^2(\mathbb{R})$ for all $x \in \Omega_j$. Then $\widehat{e}_j^n(x,\cdot) \in \mathcal{L}^2(\mathbb{R})$ and the following inequality is obviously satisfied:

$$\inf_{\omega\in\mathbb{R}} \rho(\omega) \left\|\widehat{e}_j^{n-1}(x,\cdot)\right\|_2 \leq \left\|\widehat{e}_j^n(x,\cdot)\right\|_2 \leq \sup_{\omega\in\mathbb{R}} \rho(\omega) \left\|\widehat{e}_j^{n-1}(x,\cdot)\right\|_2. \tag{5}$$

Thanks to Parseval's theorem, we can thus provide the following bounds to the observed convergence factor:

$$\inf_{\omega\in\mathbb{R}} \rho(\omega) \leq \frac{\left\|e_j^n(x,\cdot)\right\|_2}{\left\|e_j^{n-1}(x,\cdot)\right\|_2} =: \rho_{j,n}^{\text{obs}}(x) \leq \sup_{\omega\in\mathbb{R}} \rho(\omega). \tag{6}$$

Thus, if $\sup_{\omega\in\mathbb{R}} \rho(\omega) < 1$ then algorithm (2) converges in the $\mathcal{L}^2(]0,+\infty[)$ norm $\|e_j^n(x,\cdot)\|_2 \underset{n\to\infty}{\longrightarrow} 0$. Moreover, because ρ given by (4) is the same for all $x \in \Omega_j$, the previous bounds (5) and (6) are also valid for

$$\rho_{j,n}^{\text{obs}} := \left\|e_j^n\right\|_{\mathcal{L}^2(]0,\infty[,\mathcal{L}^2(\Omega))} \Big/ \left\|e_j^{n-1}\right\|_{\mathcal{L}^2(]0,\infty[,\mathcal{L}^2(\Omega))}.$$

Finally, the theoretical convergence rate ρ provides bounds for the observed convergence in the $\mathcal{L}^2(]0,\infty[,\mathcal{L}^2(\Omega_j))$ norms.

Discrete in time problems: Let us first assume that we simulate the semi-discrete problem over an infinite time window with a similar time step δt in both subdomains. The observed numerical error is denoted $E_j^{n,m}(x)$ with $m \in \mathbb{N}$, and can be seen as the result of a Dirac comb on the continuous error $e_j^n(x,\cdot)$:

$$E_j^{n,m}(x) = U_j^{n,m}(x) - u(x,t^m) \qquad \text{and} \qquad E_j^{n,\cdot}(x) = \Delta_{\delta t} e_j^n(x,\cdot)$$

with U the solution of the discrete problem, $\Delta_{\delta t}$ the Dirac comb of period δt and $e_j^n(x,t)$ the error on the continuous problem.[4] Frequencies higher than $\pi/\delta t$ are not generated by the temporal grid [8]. Applying Shannon's theorem leads us to restrict to the study of the errors in Fourier space $\widehat{E_j^{n,\cdot}}$ on an interval $I_\omega := \left[-\frac{\pi}{\delta t}; \frac{\pi}{\delta t}\right]$ and $\widehat{E_j^{n,\cdot}}(x,\omega) = \widehat{e_j^n}(x,\omega)$ for all $\omega \in I_\omega$. Details of the process can be found in [9]. Thus the $\mathcal{L}^2$ norm of $\widehat{E_j^{n,\cdot}}$ can be calculated using the result of the continuous case. Parseval's theorem can be used to obtain bounds on the observed convergence rate in the $\mathcal{L}^2$ norm:

$$\min_{|\omega|\leq\pi/\delta t} \rho(\omega) \leq \rho_{j,n}^{\mathrm{obs}}(x) := \frac{\left\|E_j^{n,\cdot}(x)\right\|_2}{\left\|E_j^{n-1,\cdot}(x)\right\|_2} \leq \max_{|\omega|\leq\pi/\delta t} \rho(\omega). \tag{7}$$

Consequently, the same bounds apply to the convergence rate $\rho_{j,n}^{\mathrm{obs}}$ in the $\mathcal{L}^2(]0,\infty[,\mathcal{L}^2(\Omega_j))$ norm.

3 Convergence for problems on a finite time window

Bold notation is used to describe the solution $\mathbf{u}_j^n$ and the error $\mathbf{e}_j^n$ over a finite window of time $[0,T]$ with $0 < T < +\infty$. We will consider $\omega \in I_\omega$ with $I_\omega = \mathbb{R}$ if we consider a continuous simulation, and I_ω defined in section 2 if we consider a discrete problem.

Difficulties expressing error over a finite time window: Applying the Fourier transform to the windowed signal would lead to a search for the solution of an equation of the type:

$$i\omega\widehat{\mathbf{e}_j^n}(x,\omega) + \mathcal{A}(\widehat{\mathbf{e}_j^n}(x,\omega)) = -\mathbf{e}_j^n(x,T)\exp(-i\omega T). \tag{8}$$

Without more knowledge about the error at time T, one cannot solve the differential equation (8), therefore cannot express the error $\widehat{\mathbf{e}_j^n}(z,\omega)$ according only to the parameters of the equation. Nevertheless, the error can be expressed by $\widehat{\mathbf{e}} = \widehat{e} * \widehat{P}_{[0,T]}(\omega)$ where P is the rectangular function on $[0,T]$ and $\widehat{P}_{[0,T]} = T\exp(-i\omega T/2)\,\mathrm{sinc}(\omega T/2)$ The convergence rate for the error $\mathbf{e}$ for a given frequency ω thus reads:

$$\rho(\omega) = \left|\frac{\widehat{e^{n+1}}|_\Gamma * \widehat{P_{[0,T]}}(\omega)}{\widehat{e^n}|_\Gamma * \widehat{P_{[0,T]}}(\omega)}\right| = \left|\frac{\int \varrho(\theta)\,\widehat{e^n}|_\Gamma(\theta)\widehat{P_{[0,T]}}(\omega-\theta)\,\mathrm{d}\theta}{\int \widehat{e^n}|_\Gamma(\theta)\widehat{P_{[0,T]}}(\omega-\theta)\,\mathrm{d}\theta}\right|, \tag{9}$$

[4] Here the discrete solution U (resp. the discrete error E) is obtained by discretizing the continuous solution u (resp. the continuous error e). The discrete signal obtained with a numerical simulation is an approximation of U depending on the numerical scheme.

which clearly shows that ρ and $\boldsymbol{\rho}$ are different functions, except in the exceptional case where $\varrho(\omega)$ is a constant. Also, definition (9) supports that the function ρ cannot be seen as the convergence factor at a given frequency: $\rho(\omega) \neq |\widehat{\mathbf{e}_j^n}(\omega)|/|\widehat{\mathbf{e}_j^{n-1}}(\omega)|$.

Bound on observed convergence: The bound on the convergence factor given by (9) is complicated to determine. However it is possible to directly bound the error:

Theorem 1 (Bound on the $\mathcal{L}^2$ norm)

$$\left\|\mathbf{e}_j^n(x,\cdot)\right\|_2 \leq \left(\sup_{\omega\in I_\omega} \rho(\omega)\right)^n \left\|\mathbf{e}_j^0(x,\cdot)\right\|_2, \tag{10}$$

which implies a bound for the n-product on the observed convergence rate:

$$\prod_{k=1}^{n} \boldsymbol{\rho}_{j,k}^{obs}(x) \leq \left(\sup_{\omega\in I_\omega} \rho(\omega)\right)^n \qquad \forall x \in \Omega_j. \tag{11}$$

This ensures the convergence of the error for the windowed algorithm as long as $\sup_{\omega\in I_\omega} \rho(\omega) < 1$. *This bounds also works for* $\left\|\mathbf{e}_j^n\right\|_{L(]0,T[,\mathcal{L}^2(\Omega_j))}$.

Proof It is possible to link the convergence of the windowed problem to a corresponding infinite-in-time problem. We can bound the $\mathcal{L}^2$ norm of error on the windowed problem by the corresponding error of an infinite in time problem: $\left\|\mathbf{e}_j^n(x,\cdot)\right\|_2 \leq \left\|e_j^n(x,\cdot)\right\|_2 \leq \left(\sup_{\omega\in I_\omega} \rho(\omega)\right)^n \left\|e_j^0(x,\cdot)\right\|_2$, where e_j^0 is any first guess extended to infinite time. Using the particular extension $e_j^0|_{[0,T]} = \mathbf{e}_j^0$ and $e_j^0|_{]T,\infty[} = 0$ leads to (10). Then combining $\|\mathbf{e}_j^n(x,\cdot)\|_2/\|\mathbf{e}_j^0(x,\cdot)\|_2 = \prod_{k=1}^{n} \boldsymbol{\rho}_{j,k}^{obs}(x)$ and (10) leads to (11). □

Remark 2 A bound on the convergence factor given by (9) was already calculated in [10]. This bound is difficult to calculate and then hardly usable. This paper makes a global remark on the possible influence of the time windowing. It is explained why the method used in [1, 11] needs special conditions and cannot be applied in a general context.[5]

Range of influencing frequencies: For a given problem discretized in time δt over a time window $[0.T]$, we estimate that in the general framework:

$$\min_{\pi/T\leq|\omega|\leq\pi/\delta t} \rho(\omega) \leq \boldsymbol{\rho}_{j,n}^{\text{obs}}(x) \leq \max_{\pi/T\leq|\omega|\leq\pi/\delta t} \rho(\omega) \tag{12}$$

and the interval $|\omega| \in \left[\frac{\pi}{T}, \frac{\pi}{\delta t}\right]$ is called the *influencing frequencies*. This interval of frequencies is usually considered for optimizing the convergence rate. As discussed

[5] It requires the determination of the inverse Fourier transform $\left\|\mathcal{F}^{-1}\rho\right\|_1$, which may not exist or can be hard to calculate.

in section 2, it is justified to consider that $\pi/\delta t$ is the maximum frequency. However the choice of the minimum frequency is justified only for time-independent problems, whereas for time-dependent problems it is an empirical estimate. We can still find justifications for this choice by considering the definition (9). First, it shows that convergence is influenced by $\int \rho(\theta)\mathrm{d}\theta$ more than by its value at a given frequency ω. Moreover, thanks to the property of $\widehat{P}(\omega-\theta)$, frequencies such that $|\omega| \ll \pi/T$ have a low impact on the convergence.[6] That said, the relevance of the minimum influencing frequency π/T is to be proved.

Remark 3 (minimal frequency for time-independent problems) The frequency $\omega_{\min}$ is justified in some cases of time-independent problems. If the conditions on the border of the dimension parallel to the interface (the one where the Fourier transform is applied) are satisfied, then the corresponding error system is periodic and a Discrete Fourier Transform (DFT) can be applied (for example see [12]). In our case, we can apply a Fourier transform on our discretised signal but, for the reasons evoked in section 3, we cannot guarantee that $|DFT(\mathbf{e}_j^{n+1})(\omega_i)|/|DFT(\mathbf{e}_j^n)(\omega_i)|$ is equal to $\rho(\omega_i)$.

Remark 4 (optimization) Usually the optimisation of the convergence speed is made by choosing interface conditions $C_{\{\{1,2\},\{1,2\}\}}$ under such conditions C, such that $\inf_{C_{\{\{1,2\},\{1,2\}\}}\in C} \max_{\pi/T\leq|\omega|\leq\pi/\delta t} \rho_{C_{\{\{1,2\},\{1,2\}\}}}(\omega)$. From (12) this guarantee a minimal upper bound to the observed convergence rate and consequently a fast convergence of $\left\|e_j^n\right\|_{\mathcal{L}^2(0,T,\mathcal{L}^2(\Omega))}$ to zero.

4 Numerical illustration

We propose to illustrate the previous properties on the coupling of two diffusion equations, with Dirichlet–Neumann interface conditions with a non-overlapping interface in $x = 0$.

$$\begin{cases} \partial_t u_j(x,t) - \nu_j \partial_x^2 u_j(x,t) = 0 & \text{on } \Omega_j \times]0,T[\\ u_j(x,t) = 0 & \text{on } \partial\Omega_j^{\text{ext}} \times]0,T[\\ u_j(x,t=0) = 0 & \text{on } \Omega_j \end{cases} \tag{13a}$$

$$\begin{cases} u_1(0,t) = u_2(0,t) & \text{on } [0,T[\\ \nu_2 \partial_x u_2(0,t) = \nu_1 \partial_x u_1(0,t) & \text{on } [0,T[\end{cases} \tag{13b}$$

with $\Omega_1 = [h_1, 0]$ and $\Omega_2 = [0, h_2]$. We simulate the problem via an implicit finite difference scheme. Schwarz's algorithm on this problem is performed on 20 iterations, with a time step $\delta t = 1000$s and parameters $h_1 = -50$m, $h_2 = 300$m, $\nu_1 = 0.12\mathrm{m}^2\mathrm{s}^{-1}$ and $\nu_2 = 0.6\mathrm{m}^2\mathrm{s}^{-1}$. In figures 1 and 2 we compare the theoretical convergence rate given in the Fourier domain $\rho(\omega)$ with the observed convergence

[6] For a given ω, frequencies such that $|\omega-\theta| \ll \pi/T$ drown in the integration in (9).

rate $\rho^{obs}_{n,j}$ and the convergence rate measured on the DFT of the error at the interface $|DFT(\mathbf{e}_j^{n+1})(0,\omega_i)|/|DFT(\mathbf{e}_j^{n})(0,\omega_i)|$, which can be seen as an approximation of $\rho(\omega_i)$. First guesses are initialised by a random signal which generates a large frequency spectrum. In these two figures, we find that bounds of the observed convergence verify the estimate (12) and thus also verify theorem 1. The evolution of the $\mathcal{L}^2$ norm of the error is not explicitly given here but it can be deduced from $\rho^{obs}_{j,n}$ (middle panel in 1 and 2). As expected the convergence observed on a given frequency ω_i does not correspond to the theoretical convergence $\rho(\omega_i)$ and conversely we tend towards equality for a window of assumed infinite size. Other examples on such problems were made in [13] and corroborate the estimate.

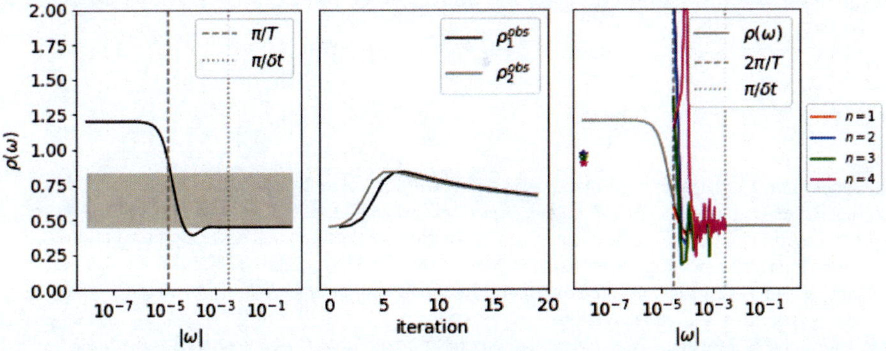

Fig. 1: For a finite time window with $T = 200\delta t$. Left panel: theoretical convergence rate $\rho(\omega)$, influencing frequencies are given by vertical lines and grey zones give the reached values of $\rho^{\text{obs}}_{j,n}$. The observed convergence factor $\rho^{\text{obs}}_{j,n}$ is given in the middle panel as a function of the iteration number n for the two domains. Right panel: the observed rate $|DFT(\mathbf{e}_1^n)(0,\omega_i)|/|DFT(\mathbf{e}_1^{n-1})(0,\omega_i)|$ is compared to the theoretical convergence rate ρ for the first four iterations.

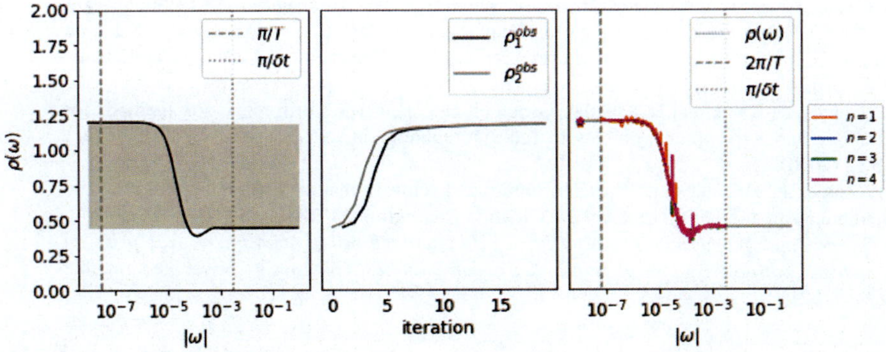

Fig. 2: Same as Figure 1 with $T = 10^5$. It is considered to be close to an infinite time window.

5 Conclusion

In the context of a time-dependent problem, the convergence rate ρ calculated in Fourier space can only be taken as such on problems considering an infinite time window. Thanks to Parceval's theorem, information on the algorithm in physical space can be obtained on the $\mathcal{L}^2$ norm of the error. It is therefore possible to bound the observed convergence rate ρ^{obs} with the bounds of the theoretical convergence ρ. For a finite time window, we can no longer consider ρ as a convergence rate for a given frequency. Yet, bounds on the observed convergence rate are still relevant and we can make these bounds precise by estimating an interval of influencing frequencies. In future work, it may be relevant to determine how to choose optimized interface conditions using the results on the observed convergence rate.

References

1. M. Gander and L. Halpern. Optimized Schwarz waveform relaxation methods for advection reaction diffusion problems. *SIAM J. Numerical Analysis* **45**(1), 666–697 (2007).
2. V. Martin. An optimized Schwarz waveform relaxation method for the unsteady convection diffusion equation in two dimensions. *Computers & Fluids* **33**(6), 829–837 (2004).
3. R. Dautray and J.-L. Lions. *Mathematical Analysis and Numerical Methods for Science and Technology*, volume 5. Springer (2000).
4. J.-L. Lions and E. Maganes. *Non-Homogeneous Boundary Value Problems and Applications*. Springer (1972).
5. C. Carlenzoli and A. Quarteroni. Adaptive domain decomposition methods for advection-diffusion problems. In: *Modeling, Mesh Generation, and Adaptive Numerical Methods for Partial Differential Equations*, pp. 165–186, Springer, New York (1995).
6. M. Gander, L. Halpern and F. Nataf. Optimized Schwarz methods. In: *12th International Conference on Domain Decomposition Methods*, pp. 15–27, Chiba, (Japan) (2000).
7. M.J. Gander. Schwarz methods over the course of time. *Electron. Trans. Numer. Anal.* **31**, 228–255 (2008).
8. O. Dubois, M. Ganger, S. Loisel, A. St-Cyr and B. Daniel. The optimized Schwarz method with a coarse grid correction. *SIAM J. Sci. Comput.* **34**, A421–A458 (2012).
9. C. Gasquet and P. Witomski. *Fourier Analysis and Applications*. Sciences sup. Dunod (2000).
10. S. Thery, C. Pelletier, F. Lemarié and E. Blayo. Analysis of Schwarz Waveform Relaxation for the Coupled Ekman Boundary Layer Problem with Continuously Variable Coefficients. *Numerical Algorithms*, July (2021).
11. M. Gander, F. Kwok and B. Mandal. Dirichlet–Neumann waveform relaxation methods for parabolic and hyperbolic problems in multiple subdomains. *BIT Numerical Mathematics* **61**(8), (2020).
12. M. Gander, F. Magoulès and F. Nataf. Optimized schwarz methods without overlap for the Helmholtz equation. *SIAM Journal on Scientific Computing* **24**, 38–60 (2002).
13. S. Thery. *Etude numérique des algorithmes de couplage océan-atmosphère avec prise en compte des paramétrisations physiques des couches limites*. PhD thesis, Université Grenoble Alpes (2021).

Construction of Grid Operators for Multilevel Solvers: a Neural Network Approach

Claudio Tomasi and Rolf Krause

1 Introduction

Multigrid (MG) methods are among the most successful strategies for solving linear systems arising from discretized elliptic equations. The main idea is to combine different levels of approximation in a multilevel hierarchy to compute the solution: it is possible to show that this algorithm is effective on the entire spectrum, thus leading to an optimal convergence property [2, 3]. Common to all these strategies is the need for the transfer of data or information between the different grids, or meshes. Therefore, a crucial point for reaching fast convergence is the definition of transfer operators, but they are generally problem-dependent. Except for the case of nested meshes, the computation of these operators is very expensive, and domain knowledge is always required.

The ever-increasing application of Machine Learning (ML) as support for methods in scientific computing makes it a natural solution to be employed in the definition of transfer operators, reducing the costs of their construction. In [9], the learning of a mapping between PDEs and operators has been proposed. Another approach is presented in [10], where restriction and prolongation matrices are optimized while minimizing the spectral radius of the iteration matrix. As an alternative, the method proposed in [13] uses Graph Neural Networks, for learning AMG prolongation operators, having classes of sparse matrices as input.

In this paper, we propose a methodology based on Deep Neural Networks to define transfer operators based on the concept of L^2-projection. We take information from the domain to create several examples and to make our model learn from experience. Therefore, our focus is the construction of a suitable training set and a correct loss function definition to create a model that can be employed in MG solvers. The actual state of the method presents some limitations related to the mesh

Claudio Tomasi and Rolf Krause
Università della Svizzera Italiana,Via Buffi 13, CH-6904 Lugano,
e-mail: claudio.tomasi@usi.ch, rolf.krause@usi.ch

S. C. Brenner et al. (eds.), *Domain Decomposition Methods in Science and Engineering XXVI*, Lecture Notes in Computational Science and Engineering 145,
https://doi.org/10.1007/978-3-030-95025-5_63

structure. An extension to a wider range of scenarios should be considered in future works.

2 Problem definition

Let $\Omega \subset \mathbb{R}^n$ be a domain with Lipschitz boundary and let $H_0^1(\Omega)$ be the Sobolev space of one-time weakly differentiable functions on Ω, with weak derivatives in $L^2(\Omega)$. We consider a multigrid method for the solution of the following problem:

$$\text{find } u \in V : a(u, v) = f(v) \;\; \forall v \in V, \tag{1}$$

where $V \subset H_0^1(\Omega)$, $a : V \times V \longrightarrow \mathbb{R}$ is a continuous symmetric elliptic bilinear form and $f : V \longrightarrow \mathbb{R}$ is a continuous linear functional. Let $V_h \subset V$ be the associated finite elements space, where $\dim(V_h) = n_h$ and $h > 0$, and consider a conforming shape-regular triangulation $\mathcal{T}_h$. For a more rigorous explanation, see e.g. [14]. Furthermore, let $I_H^h : \mathbb{R}^{n_H} \longrightarrow \mathbb{R}^{n_h}$ be a transfer operator which transfers information between $\mathcal{T}_h$ and a coarser grid $\mathcal{T}_H$, with $H > h$ and $n_H < n_h$. We denote by $A_h u_h = b_h$ the linear system arising from the finite element discretization of (1).

Let us consider a 2-grid correction scheme for solving the linear system. The extension to a general multigrid scenario is straightforward. To restrict or prolong information between coarse and fine grids, we apply I_H^h. Moreover, we define the coarse problem using the expression $A_H = (I_H^h)^T A_h \, I_H^h$. Hence, the definition of the transfer operator plays a central role in obtaining a fast convergence of the method. In [11], a general definition of transfer operators between meshes is discussed. Here, we focus on the L^2-projection as transfer operator. Let us call it Q:

$$Q = M_h^{-1} B_h,$$

where M_h is the mass matrix related to the fine level (grid), and B_h is a rectangular coupling operator matrix. The latter relates the two meshes, and it is computed through their intersection. Since the inverse of M_h is a dense matrix, the computation of Q might become expensive. Therefore, we use the pseudo-L^2-projection, where we invert the lumped mass matrix instead of M_h. For further reading, refer to [4–7].

2.1 Neural networks

ML algorithms are able to learn from data [8]; we refer to a single data object, calling it an example. An example is a collection of features together with a corresponding target. A feature is a property that has been measured from some object or event. The target is the correct response to the features, that the system should be able to reproduce. We represent an example as a couple $(\mathbf{x}, \mathbf{y})$, where $\mathbf{x} \in \mathbb{R}^n$ is the feature set and $\mathbf{y} \in \mathbb{R}^m$ is the target. ML can solve different tasks, such as classification, transcription, and so on. Our focus is on regression: we ask the model to predict numerical values given some inputs. In order to solve this task, the model is asked to output a function $f : \mathbb{R}^n \rightarrow \mathbb{R}^m$. To evaluate the ML algorithm abilities, we define a measure of its performance, called the loss function: for regression, we select the Mean Squared Error (MSE) indicator.

Neural Networks (NNs) belong to the class of supervised ML algorithms. They consist of layers of neurons, connected by weighted synapses. An NN defines a mapping $\mathbf{y} = f(\mathbf{x}; \boldsymbol{\theta})$ and learns the values of the parameters $\boldsymbol{\theta}$, providing the best function approximation. More details can be found in [1, 12].

2.2 Training transfer operators

We aim to define an NN model to learn and then predict the transfer operator Q. Specifically, we do not learn directly Q, but the coupling operator B_h. Once the model is optimized, we employ it as a black box for solving linear systems of equations in an MG fashion. We proceed by coarsening: we take M_h on the fine level, and we extract the features to input into the NN. More details on the data extraction from M_h are given in Section 3.1. The model produces parts of B_h that combined give rise to the full operator. We then retrieve the transfer operator Q, and we employ it in the MG algorithm. Furthermore, we use the predicted transfer operators to define the coarser mass matrix M_H using the so-called *Galerkin operator*, i.e., $M_H = Q^\top M_h Q$. We recursively apply this procedure to define coarser problems, giving rise to a multilevel hierarchy.

3 Training set

In order to allow the NN to learn, we provide a large number of examples (or records). We need several distinct examples to be sure of avoiding overfitting, occurring when the model predictions correspond too closely or exactly to a particular set of data. Thus, we define classes of examples, and we choose a fixed amount of records for each class. This allows us to create an unbiased training set without preferring some classes over others. The definition of class is related to the mesh from which we extrapolate the records. We set a number of elements N: all the records coming from meshes with N elements belong to class C_N.

3.1 Records

Given a subset of pre-identified coarse nodes, we extract a record for each of them. Let j be a coarse node. An example contains information related to *patch(*j*)*; here, *patch(*j*)* is the set consisting of node j together with its neighbors. For each node $k \in$ *patch(*j*)*, we define features and target as the non-zero entries of the kth rows of M_h and B_h, respectively.

We consider different examples of 2-grid scenarios, where we associate to each fine mesh one coarse mesh, in order to approximate an actual function. Since we consider several examples for the same class C_N, we need a strategy to avoid duplication inside the dataset. For this purpose, in each example, we consider the fine mesh and we move the nodes along the edges by a random quantity, proportional to the step size h. Therefore, we create different elements and consequently different records. We generate the examples in C_N, and we proceed to the next class by increasing N. Since NN models allow only fixed input and output dimensions, we define distinct models for 1D and 2D scenarios.

3.2 One-dimensional model

The records related to one-dimensional meshes are extracted from scenarios obtained by coarsening: starting from a randomly generated fine mesh, we decide which nodes to keep for defining the coarse grid. Here, *patch(*j*)* consists only of j itself, together with its left and right neighbors. For each coarse node, we take the information on *patch(*j*)* from M_h and B_h following the strategy explained in Section 3.1, to define each example.

3.3 Two-dimensional model

Let us call *patch-size* the number of nodes in a specific patch: given a node j, its patch-size is the cardinality of the set *patch(*j*)*. In 2D, even in the same mesh, we can have nodes with different patch-sizes. Hence, we start by considering only a fixed triangulation, such that the nodes have the same neighborhood pattern. We will focus on dealing with different patch-sizes later in the paper, referring to their treatment in the context of NNs.

A crucial point is to find a correct distribution of data in the training set, in terms of magnitude of the values. Since the NN should not prefer some examples – thus, some classes – over others, we need to define a correct and even filling of the training set. As a first approach, we relate the concept of class to the procedure of refinement. Mesh refinement is a strategy to increase the accuracy of the solution of a discretized problem. It works as an iterative procedure applied to the single elements of a mesh. Here we consider two different strategies: bisection, which halves each element, and mid-point refinement, which takes the mid-points of each edge and joins them to create new elements. When we refine, we deal with a new class of examples. Applying a training algorithm on these data results in a poor ability of approximation and a large prediction error, making an NN model unfit to work in an MG setting. The refinement procedure makes the number of elements scale by a factor of 2 (bisection) or 4 (mid-point). In terms of domains of training examples, this means that the initial classes of records, i.e., C_N with N small, are close to each other, while their distance grows when N increases. This turns out to produce an uneven training set, without a good balance in terms of data distribution. For this reason, we need a linear increase in the number of elements. If the classes of examples are evenly spaced in terms of domains, the network does not prefer some classes over others. Therefore, a second approach changes the definition of class, independent of the concept of refinement: we start from a number N, and we create a mesh having exactly N elements. Once we extract enough records, we proceed to the next class, increasing N by a constant K, and create a new mesh with $N + K$ elements. For each class, we extract the same number of examples. Following this simple procedure, the resulting training set is effectively unbiased and with a good distribution of the examples. A learning algorithm applied to these data produces the expected good approximation. Therefore, a model trained on this dataset can be applied inside an MG scenario.

4 Model training

A NN optimizes its parameters in order to reduce the prediction error. Employing MSE as the loss function results in good predictions, but the model does not gain a good generalization property, i.e., the ability to perform well on previously unobserved inputs.

Regularization helps us overcome this issue: it reduces the hypothesis space, allowing the NN to have a higher probability of choosing the most correct function. We introduce in the loss function some penalty terms related to the domain knowledge. These terms force constraints during the training phase in order to respect properties that the transfer operator must satisfy.

4.1 Regularization

During the construction of the training set, for each coarse node j, we extract patches of M_h and parts of B_h. We use this information to ask the model to force some rules on the rows of the predicted coupling operator.

We define the jth predicted and actual rows of the operator Q as

$$Q_j^{\text{pred}} = \frac{1}{\sum M_j} B_j^{\text{pred}}, \qquad Q_j^{\text{true}} = \frac{1}{\sum M_j} B_j^{\text{true}}, \tag{2}$$

where Φ_j denotes the jth row of the operator $\Phi = \{M_h, B_h\}$.

We know that the predicted transfer operator should preserve constants (more details can be found in [4, Section 3.2]). Hence, we consider the following penalty terms to specialize our loss function:

$$\|Q_j^{\text{pred}} \cdot \mathbb{1}_H - \mathbb{1}_h\|_2, \qquad \|Q_j^{\text{pred}} - Q_j^{\text{true}}\|_2, \tag{3}$$

where $\|\cdot\|_2$ denotes the Euclidean norm, and $\mathbb{1}_H$ and $\mathbb{1}_h$ the all-ones vectors of dimensions n_H and n_h, respectively. We then define, for all the nodes $k \in patch(j)$

$$p_k = \frac{1}{\alpha}\|Q_k^{\text{pred}} \cdot \mathbb{1}_H - \mathbb{1}_h\|_2 + \frac{1}{\beta}\|Q_k^{\text{pred}} - Q_k^{\text{true}}\|_2, \tag{4}$$

where $0 < \alpha, \beta < 1$.

Therefore, we define the loss function as

$$\mathcal{L}(y_{\text{true}}, y_{\text{pred}}) = MSE(y_{\text{true}}, y_{\text{pred}}) + \sum_k p_k. \tag{5}$$

Adopting the latter during the training phase, in addition to minimizing the simple distance between target and prediction, we aim to respect the above properties related to the transfer operator.

4.2 Model details

We use a classic splitting for our dataset: 20% for test and 80% for training, where the latter is divided again into 20% for the validation set and the remaining for the training phase. For a preliminary examination of the method, we used around 500,000 examples. For both one- and two-dimensional models, we adopt Adam as optimizer. Regarding the architecture, we report here only the structure for the 2D scenario used as initial test: we need at least 20 hidden layers, where for each of them we use 800 neurons, for a total of 12 million parameters. Through further investigations and several tests, the NN complexity can be improved, giving rise to less expensive computations. We initialize our weights using a normal distribution, using the methods provided by Tensorflow. In the context of the NN definition, more extensive work is needed to study the sensitivity of the predictions while changing the NN parameters.

5 Numerical results

We test our NNs for both prediction accuracy and their application in an MG setting. We compare our method with the Semi-Geometric multigrid (SGMG) method (see [4, Chapter 3]), which adopts the L^2-projection computed through intersections between meshes. In addition to the convergence, we consider the difference in the time spent to assemble the transfer operator. For our method, we take into account patches extraction, predictions, assembly of B and the computation of the operator. Regarding the computation of the actual L^2-projection we consider the time spent for intersecting fine and coarse mesh, triangulation for each intersecting polygon and numerical integration. We test the method on one-dimensional examples, and the results are good as expected, considering two or more levels. Our method converges with the same number of iterations as the SGMG method. Comparing the CPU time spent in creating the transfer operators, we see that the predicted one is assembled faster than the other since it depends only on the problem dimensions.

During the test on two-dimensional settings we need to deal with different patch-sizes, as described in Section 3.3. Even if we consider a simple regular mesh, the nodes near the boundaries have fewer neighbors than the internal nodes. A preliminary solution requires the mesh to be extended, to make all the nodes have the same patch-size. Virtually, we add neighbors to those nodes having a smaller patch in the given mesh. Using this expedient, the method works, and we can test the convergence against the SGMG method. Extending the mesh is useful for an initial application of MG, but it is very expensive in terms of computations. Increasing the degrees of freedom (dofs), we would have more and more virtual nodes to add and heavier computations to carry out. Therefore, we consider different NNs, each of them defined and optimized for a specific patch-size.

Fig. 1 shows the performance of the method on a two-dimensional scenario: in the left picture, we compare the convergence of our method against the convergence of Semi-Geometric MG; in the right picture we compare the CPU time spent in both methods.

6 Conclusion

This work presented the study and definition of a methodology to construct NNs to predict transfer operators for MG solvers. Starting from a one-dimensional case, we built an unbiased training set allowing the optimization of a model, which brought very good results in an MG context. Reproducing the same methodology, we approached the two-dimensional setting, which gave us the chance to better define a training set for this kind of method. Furthermore, we could test our method using different input-sized neural networks, resulting in fast convergence and bringing a great speedup in the computation of the transfer operator. The same procedure can be employed for constructing models to deal with a general N-dimensional scenario. Given the limitations of this method in its current state, further investigations should

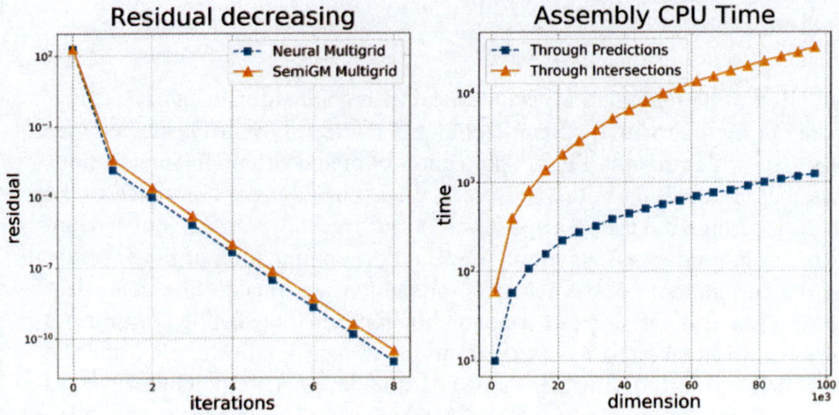

Fig. 1: Convergence of 2D Neural MG against SGMG on an example of 100,000 dofs (left). Comparison of CPU time between the two methods, increasing the dofs (right).

be devoted to overcome the necessity of having multiple NNs modeled on different patch-sizes, in order to define a general strategy for solving arbitrary problems. Future work should extend the method to deal with a wider class of triangulations, and for applications in other Multilevel scenarios.

References

1. C.M. Bishop. *Pattern recognition and machine learning*. Information Science and Statistics. Springer, New York (2006).
2. D. Braess. *Finite elements: Theory, fast solvers, and applications in solid mechanics*. Cambridge University Press (2007).
3. W.L. Briggs, V.E. Henson and S.F. McCormick. *A multigrid tutorial*. Society for Industrial and Applied Mathematics (SIAM), Philadelphia, PA, second edition (2000).
4. T. Dickopf. *On multilevel methods based on non-nested meshes*. PhD Thesis. Universitäts-und Landesbibliothek Bonn (2010).
5. T. Dickopf and R. Krause. A pseudo-L^2-projection for multilevel methods based on non-nested meshes. *INS Preprint* **908** (2009).
6. T. Dickopf and R. Krause. A study of prolongation operators between non-nested meshes. In: *Domain decomposition methods in science and engineering XIX*, volume 78 of Lect. Notes Comput. Sci. Eng., pp. 343–350, Springer, Heidelberg (2011).
7. T. Dickopf and R. Krause. Evaluating local approximations of the L^2-orthogonal projection between non-nested finite element spaces. *Numer. Math. Theory Methods Appl.* **7**(3), 288–316 (2014).
8. I. Goodfellow, Y. Bengio and A. Courville. Deep learning. Adaptive Computation and Machine Learning. MIT Press, Cambridge, MA, 2016.
9. D. Greenfeld, M. Galun, R. Basri, et al. Learning to optimize multigrid pde solvers. In: *International Conference on Machine Learning*, pp. 2415–2423, PMLR (2019).
10. A. Katrutsa, T. Daulbaev and I. Oseledets. Black-box learning of multigrid parameters. *J. Comput. Appl. Math.* **368**(12), 112524 (2020).

11. R. Krause and P. Zulian. A parallel approach to the variational transfer of discrete fields between arbitrarily distributed unstructured finite element meshes. *SIAM J. Sci. Comput.* **38**(3), C307–C333 (2016).
12. H. Larochelle, Y. Bengio, J. Louradour and P. Lamblin. Exploring strategies for training deep neural networks. *Journal of machine learning research* **10**(1) (2009).
13. I. Luz, M. Galun, H. Maron, et al. Learning algebraic multigrid using graph neural networks. In: *International Conference on Machine Learning*, pp. 6489–6499, PMLR (2020).
14. A. Quarteroni. *Numerical models for differential problems*, volume 2. Springer (2009).

Coarse Corrections for Schwarz methods for Symmetric and Non-symmetric Problems

Martin J. Gander and Serge Van Criekingen

1 Introduction

As is well known, domain decomposition methods applied to elliptic problems require in most cases a coarse correction to be scalable (for exceptions, see [5, 6]), the choice of the coarse space being critical to achieve good performance. We present here four new coarse spaces for the Restricted Additive Schwarz (RAS) method of Cai and Sarkis [4], both for symmetric and non-symmetric problems, and implement them in the PETSc library [1–3]. We compare them to a coarse space named `Q1` here from [10], originating from [7] and [9], and more classical coarse spaces. In particular, we introduce the new *adapted* coarse spaces `Q1_adapt` and `Q1_inner_adapt` using basis functions that locally solve the problem considered also with advection and turn out to be more robust for strong advection. We also introduce the `Half_Q1` coarse space that halves the coarse space dimension compared to `Q1` by using a selected combination of its basis functions and turns out to be the fastest, and the new enriched coarse space `Enriched_Q1` which leads to the lowest iteration counts. We further present results of the optimized method ORAS obtained by introducing optimized transmission conditions at subdomain interfaces [7, 8, 16].

Throughout the paper, our model problem for the symmetric case is the Laplace problem, while for the non-symmetric case we consider

$$-\Delta u + \mathbf{a} \cdot \nabla u = 0 \tag{1}$$

with an upwind scheme on the unit interval (in 1-D) or unit square (in 2-D) using the 5-point finite difference discretization and homogeneous boundary conditions.

Martin J. Gander
University of Geneva, e-mail: martin.gander@unige.ch

Serge Van Criekingen
CNRS/IDRIS and Maison de la Simulation, e-mail: serge.van.criekingen@idris.fr

S. C. Brenner et al. (eds.), *Domain Decomposition Methods in Science and Engineering XXVI*, Lecture Notes in Computational Science and Engineering 145,
https://doi.org/10.1007/978-3-030-95025-5_64

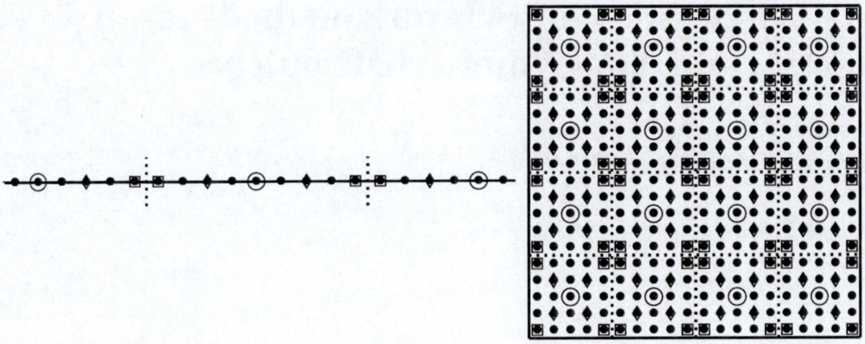

Fig. 1: Coarse grid point choice in 1-D and 2-D for the `Q1` (squares), `Q1_fair` (diamonds) and `Middle` (circles) options.

2 Two-level RAS with classical and new coarse spaces

We consider the solution of $A\mathbf{x} = \mathbf{b}$ on a domain Ω decomposed into a set of possibly overlapping subdomains Ω_j and introduce a restriction operator R_j onto each Ω_j. We also introduce a partition of Ω into non-overlapping subdomains $\tilde{\Omega}_j$ as well as the corresponding restriction operators $\tilde{R}_j$ for RAS. Obtaining a two-level method through coarse correction requires a restriction operator R_c to a coarse space, such that the resulting coarse system matrix reads $A_c = R_c\, A\, R_c^T$. The two-level coarse corrected RAS method with multiplicative coarse correction (denoted RAS2 in what follows) can then be written as

$$\mathbf{x}^{n+1/2} = \mathbf{x}^n + \sum_{j=1}^{J} \tilde{R}_j^T\, A_j^{-1}\, R_j\, (\mathbf{b} - A\mathbf{x}^n), \tag{2}$$

$$\mathbf{x}^{n+1} = \mathbf{x}^{n+1/2} + R_c^T\, A_c^{-1}\, R_c\, (\mathbf{b} - A\mathbf{x}^{n+1/2}), \tag{3}$$

where the first half iteration is the RAS method as defined by Cai and Sarkis [4].

The definition of the coarse space is critical to obtain an efficient two-level method. We consider here the following classical and new coarse spaces:

"`MidBasic`": The classical `MidBasic` coarse space, also called the Nicolaides coarse space, defined by using a constant coarse basis function in each subdomain.

"`Middle`": The classical `Middle` coarse space taking the fine mesh points in the middle of each subdomain as coarse grid points, along with linear (bilinear in 2-D) basis functions centered on these points. This is illustrated in Fig. 1 in 1- and 2-D.

"`Q1`": The `Q1` coarse space [7, 9] based on linear basis functions with coarse grid points chosen as illustrated in Fig. 1, namely placed on each side of the subdomain interfaces (in 1-D) or around each cross point (in 2-D) of the non-overlapping decomposition. It was shown in [9] that, for the Laplace equation, the `Q1` coarse correction yields convergence in two iterations in 1-D (or at iteration 1 in PETSc, where iteration count starts at 0).

"`Q1_fair`": This coarse space uses linear basis functions and the same number of coarse mesh points as `Q1`, but equally distributed as illustrated in Fig. 1. It is introduced for a fair comparison with `Q1` in terms of coarse space dimensions.

"`Q1_adapt`": The new `Q1_adapt` coarse space using the same coarse points as `Q1`, but *computed* ("adapted") basis functions that solve the homogeneous equation considered in each subdomain. In the Laplace case, the `Q1_adapt` basis functions are thus the same as the `Q1` (i.e., linear) functions, while with advection, the basis functions are different. In 1-D, `Q1_adapt` gives convergence of the two-level method at iteration 1 in PETSc, even in the non-symmetric case when advection is present, like `Q1` for the Laplace problem in [9].

In 2-D, the `Q1_adapt` basis functions are computed in two steps, first on the edges with a 1-D stencil obtained by lumping (i.e., summing up) the system matrix coefficients in the perpendicular direction, then inside each subdomain using the computed edge functions as boundary conditions, a bit like in MsFEM.

"`Q1_inner_adapt`": Defined in 2-D only, this new coarse space differs from `Q1_adapt` in that the coarse basis functions are "adapted" only inside each subdomain: the first of the two steps in `Q1_adapt` is skipped, and linear edge functions are used as boundary conditions to compute the basis functions within each subdomain.

"`Half_Q1`": The new `Half_Q1` coarse space is motivated by the eigenmodes of the RAS iteration matrix corresponding to its eigenvalues closest to 1 in modulus. In Fig. 2a, we computed them with SLEPc [13] (`https://slepc.upv.es`), for the Laplace test case and a 2×2 subdomain decomposition using minimal overlap (no algebraic overlap, i.e., block Jacobi). If q_1, q_2, q_3, q_4 are the `Q1` basis functions at a cross point, it can be observed that these modes appear to be $q_1 + q_2 + q_3 + q_4$ and $q_1 - q_2 + q_3 - q_4$, respectively. The `Half_Q1` coarse space is therefore obtained by taking these 2 combinations as basis functions, thus with 2 basis functions per cross point instead of 4 in the `Q1` case. (To add to Fig. 2a which gives only the first two eigenvalues, note that the next eigenvalues are .975571 (double), −.975571 (double), −.969651 and .969651).

With minimal overlap, we observed that the property of having the two largest eigenmodes in modulus corresponding to one continuous and one discontinuous mode remains verified when increasing the number of subdomains. We also observed this property when introducing various types of advection. This is illustrated in Fig. 2b for the case with 25 subdomains on our model problem (1) with rotating fluid advection $a_x = -10y$ and $a_y = 10x$ (for this case, the next eigenvalues are complex: $-0.989110 \pm 0.001513i$, $0.989110 \pm 0.001513i$ and $0.983268 \pm 0.002304i$).

With more than minimal overlap (i.e., non-zero algebraic overlap), even if we observed exceptions (typically when using more than 4 subdomains and a relatively low fine mesh resolution), the largest two modes tend to remain one continuous and one discontinuous, but the corresponding eigenvalues are then different in modulus,

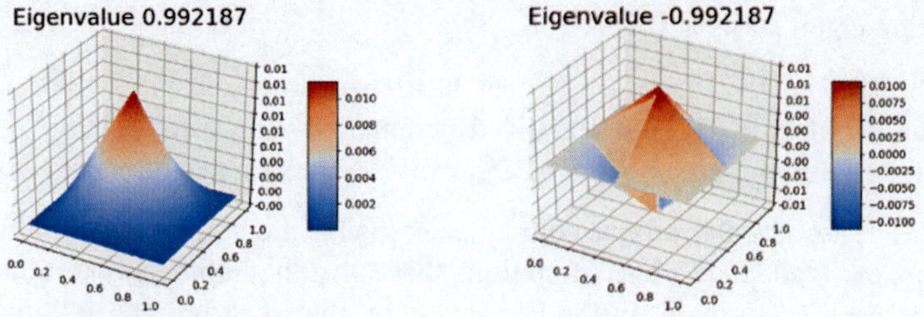

(a) 2×2 subdomains, no advection, minimal overlap.

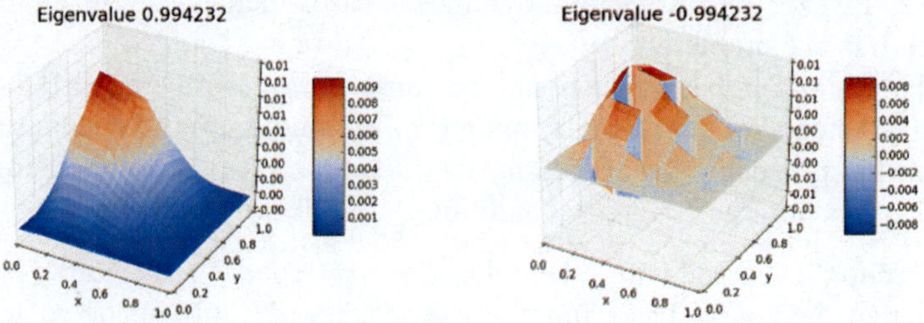

(b) 5×5 subdomains, advection $a_x = -10y$ and $a_y = 10x$, minimal overlap.

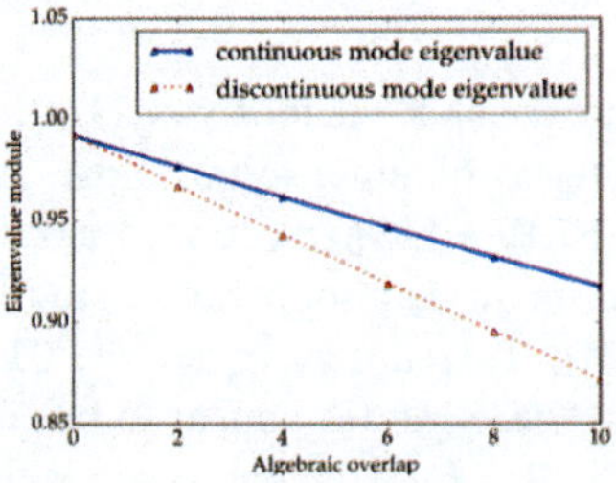

(c) 2×2 subdomains, no advection: evolution of the two largest eigenvalues in modulus with the overlap.

Fig. 2: In (a) and (b), eigenmodes of the RAS iteration operator corresponding to the two largest eigenvalues in modulus using a 256×256 fine mesh resolution; continuous modes on the left and discontinuous modes on the right. In (c), evolution of the two largest eigenvalues in modulus.

with a difference that increases when increasing the overlap. We illustrate this for the 2×2 subdomain decomposition by displaying in Fig. 2c the evolution of the two largest eigenvalues in modulus when increasing the overlap.

"`Enriched_Q1`": This new coarse space is obtained by adding extra linear basis functions to the Q1 coarse space, namely (in 2-D) with one extra coarse point placed in the middle of each edge and corresponding extra linear basis function. The goal is to come a step closer to the 2D grid representing a complete coarse space, leading

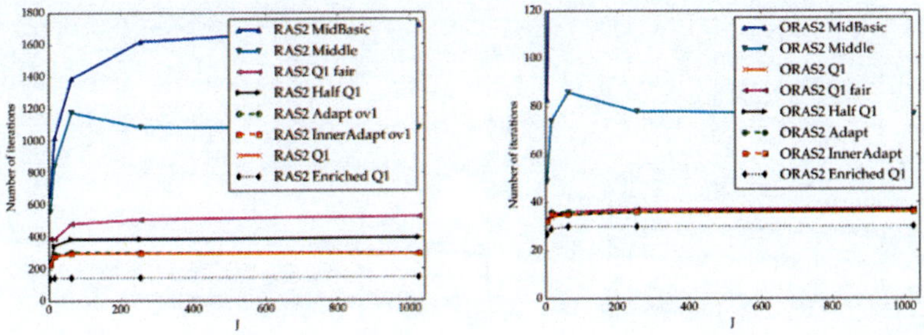

(a) RAS2 (b) ORAS2

Fig. 3: Results with advection $a_x = -10y$ and $a_y = 10x$.

to convergence in two iterations [9, Fig. 8]. This coarse space is thus twice as big as Q1.

3 Numerical results

Fig. 3 shows the iteration count for a weak scalability analysis on our non-symmetric model problem (1) with rotating fluid advection $a_x = -10y$ and $a_y = 10x$. This analysis consists in increasing the size of the problem while maintaining constant the workload per subdomain. The subdomain decomposition ranges from 2×2 to 32×32, each subdomain having a 256×256 fine mesh and being handled by one CPU core. The number of cores J ranges thus from 4 to 1024 here, and the coarse space dimension is J for `MidBasic` and `Middle`, $4J$ for `Q1`, `Q1_fair`, `Q1_adapt` and `Q1_inner_adapt`, $2J$ for `Half_Q1` and $8J$ for `Enriched_Q1` . An algebraic overlap of 2 is considered, which means one extra mesh layer for both subdomains at an interface and corresponds to an overlap of 1 in the PETSc sense. The corresponding Laplace results are very similar, we thus only show them in Table 1 for comparison.

While Fig. 3a displays the result for the RAS2 method, Fig. 3b displays the results for the optimized ORAS2 method obtained by modifying the local A_j matrices in the RAS2 iterations (2)–(3) to express Robin interface conditions [16], with a first-order accurate discretization of the normal derivative and two-level optimized coefficients (determined for the symmetric case) as defined in [7].

We observe that, except for the larger `Enriched_Q1` coarse space, the `Q1` coarse space gives the lowest iteration count when used with (non-optimized) RAS. Using adapted basis functions (i.e., `Q1_adapt` or `Q1_inner_adapt`) does not reduce the iteration count in the present case. However, these adapted coarse spaces appear more robust than `Q1` when increasing the advection strength, as can be seen in Table 1 with a five times larger advection: some of the stationary iterations appear to diverge using the `Q1` and/or `Half_Q1` coarse spaces with a rotating fluid advection of magnitude 50, while this is not the case with magnitude 10 (Fig. 3).

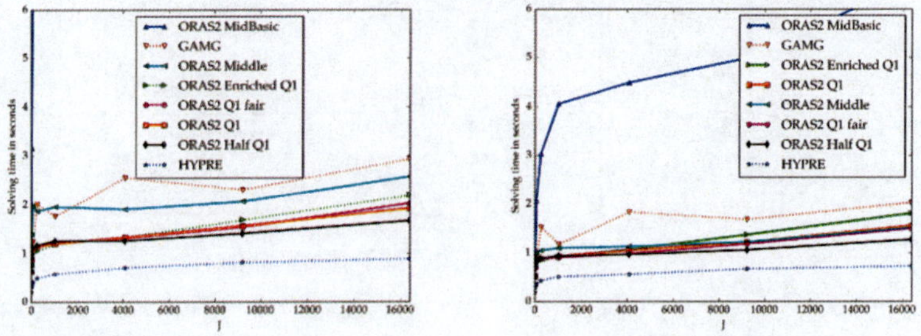

(a) Without GMRES acceleration

(b) With GMRES acceleration

Fig. 4: Computation times (s.) for the weak scaling experiment for the non-symmetric model problem with $a_x = -10y$ and $a_y = 10x$.

We also observe from Fig. 3 that Q1_fair and Half_Q1 take more advantage of the application of the optimized ORAS method than Q1, since their iteration counts then all become quasi-identical.

Timing results are presented in Fig. 4 for our weak scalability analysis, this time using up to $128 \times 128 = 16,384$ CPU cores (one per subdomain) of the CPU partition of the Jean Zay supercomputer at the Institute for Development and Resources in Intensive Scientific Computing (CNRS/IDRIS). A relative tolerance of 1.e-8 is used as convergence criteria. Note that PETSc's native direct solver is used for the local serial subdomain solves, while the coarse solve is performed in parallel with the MUMPS direct solver, after agglomeration of the coarse unknowns on a subset of the processors (here maximum 64) using PETSc's "Telescope" tool [14]. Beside the results obtained with the various coarse corrections introduced above, timings obtained with two algebraic multigrid options available through PETSc are also presented, namely HYPRE/BoomerAMG [12] (with tuning form [17]) and PETSc's native algebraic multigrid preconditioner GAMG (with smoothed aggregation and CG eigenvalue estimator [2]).

As already observed in [10] for the symmetric case, we see here that results with the ORAS2 method can be competitive with the multigrid options also in the non-symmetric case when using one of the Q1, Q1_fair, Half_Q1 or Enriched_Q1 coarse spaces (or even Middle with GMRES acceleration). Among

J	4	16	64	256	1024	4	16	64	256	1024
	RAS2					ORAS2				
Q1_fair	179(420)	357(424)	481(479)	506(509)	521(522)	80(34)	39(35)	37(36)	37(37)	37(37)
Q1	Div(255)	Div(295)	259(303)	281(298)	288(291)	68(35)	45(35)	35(35)	35(36)	36(36)
Half_Q1	Div(257)	Div(348)	Div(380)	494(385)	409(391)	Div(32)	Div(34)	172(35)	41(36)	37(37)
Q1_adapt	145	237	291	310	313	66	35	37	37	36
Q1_inner_ad.	145	213	261	282	289	70	36	35	35	36

Table 1: Number of RAS2 and ORAS2 stationary iterations with advection $a_x = -50y$, $a_y = 50x$, where "Div" means that the iterations are diverging. Laplace results are in parentheses, with "adapted" results then the same as Q1.

the various coarse spaces considered, `Half_Q1` exhibits the fastest computational times, most presumably thanks to its lower dimensionality, which does not significantly impact the iteration count (as observed in Fig. 3 up to 1024 cores and as can be verified up to 16,384 cores). This remains true when plotting not only the solving times as in Fig. 4, but the total timings including the setup/assembly phase.

4 Conclusions

We considered several coarse space options for the two-level RAS method applicable to non-symmetric problems and implemented them in the PETSc library. The `Q1` option, which enables a solution in two iterations on a 1-D Laplace test case, shows good performance on our 2-D non-symmetric model problem as well (using coarse points placed around the cross points), in that it has a better iteration count than the `Q1_fair` option (which uses as many but equally distributed coarse points). The new `Q1_adapt` and `Q1_inner_adapt` coarse spaces enable a solution in two iterations for a non-symmetric 1-D advection-diffusion test case, as in the Laplace case in [10]. Despite this promising feature, iteration counts on our 2-D model problem did not show improvements compared to the `Q1` option for moderate advection, but increased robustness was observed for strong advection. The `Enriched_Q1` coarse space, with its higher dimensionality, yields lower iteration counts but appears not to improve the overall computation time. Finally, the new `Half_Q1` coarse space shows promising performance in that the increase in iteration count due to its lower dimensionality appears very moderate and virtually disappears if optimized transmission conditions are introduced (ORAS method). In turn, this option provided the best computational time results in our weak scaling analysis, of the same order of magnitude as multigrid options. Other harmonic coarse spaces like GenEO [15] and GDSW/RGDSW [11] that target improving condition number estimates of Additive Schwarz, in contrast to accelerating low frequency continuous and discontinuous modes of RAS like our new coarse spaces, are also intrinsically based on MsFEM techniques. A more extensive comparison of all these coarse spaces will appear elsewhere.

Acknowledgements This work was performed using HPC resources from GENCI-IDRIS.

References

1. S. Balay, S. Abhyankar, M.F. Adams, J. Brown, P. Brune, K. Buschelman, L. Dalcin, A. Dener, V. Eijkhout, W.D. Gropp, , D. Karpeyev, D. Kaushik, M.G. Knepley, D.A. May, L. Curfman McInnes, R. Tran Mills, T. Munson, K. Rupp, P. Sanan, B.F. Smith, S. Zampini, H. Zhang and H. Zhang. PETSc Web page. http://www.mcs.anl.gov/petsc (2019).
2. S. Balay, S. Abhyankar, M.F. Adams, J. Brown, P. Brune, K. Buschelman, L. Dalcin, A. Dener, V. Eijkhout, W.D. Gropp, , D. Karpeyev, D. Kaushik, M.G. Knepley, D.A. May, L. Curfman

McInnes, R. Tran Mills, T. Munson, K. Rupp, P. Sanan, B.F. Smith, S. Zampini, H. Zhang and H. Zhang. PETSc users manual. Technical Report ANL-95/11 - Revision 3.14, Argonne National Laboratory (2020).
3. S. Balay, W.D. Gropp, L. Curfman McInnes and B.F. Smith. Efficient management of parallelism in object oriented numerical software libraries. In: E. Arge, A.M. Bruaset and H.P. Langtangen, editors, *Modern Software Tools in Scientific Computing*, pp. 163–202, Birkhäuser Press (1997).
4. X.-C. Cai and M. Sarkis. A restricted additive Schwarz preconditioner for general sparse linear systems. *SIAM J. Sci. Comp.* **21**(2), 239–247 (1999).
5. G. Ciaramella and M.J. Gander. Analysis of the parallel Schwarz method for growing chains of fixed-sized subdomains: Part I. *SIAM Journal on Numerical Analysis* **55**(3), 1330–1356 (2017).
6. G. Ciaramella and M.J. Gander. Analysis of the parallel Schwarz method for growing chains of fixed-sized subdomains: Part II. *SIAM Journal on Numerical Analysis* **56**(3), 1498–1524 (2018).
7. O. Dubois, M.J. Gander, S. Loisel, A. St-Cyr and D.B. Szyld. The optimized Schwarz methods with a coarse grid correction. *SIAM J. Sci. Comp.* **34**(1), A421–A458 (2012).
8. M.J. Gander. Optimized Schwarz methods. *SIAM J. Numer. Anal.* **44**(2), 669–731 (2006).
9. M.J. Gander, L. Halpern and K. Santugini. A new coarse grid correction for RAS/AS. In: *Domain Decomposition Methods in Science and Engineering XXI*, Lecture Notes in Computational Science and Engineering, pp. 275–284, Springer-Verlag (2014).
10. M.J. Gander and S. Van Criekingen. New coarse corrections for restricted additive Schwarz using PETSc. In: *Domain Decomposition Methods in Science and Engineering XXV*, Lecture Notes in Computational Science and Engineering, pp. 483–490, Springer-Verlag (2019).
11. A. Heinlein, C. Hochmuth and A. Klawonn. Reduced dimension GDSW coarse spaces for monolithic Schwarz domain decomposition methods for incompressible fluid flow problems. *International Journal for Numerical Methods in Engineering* **121**(6), 1101–1119 (2020).
12. V.E. Henson and U.M. Yang. Boomeramg: a parallel algebraic multigrid solver and preconditioner. *Applied Numerical Mathematics* **41**, 155–177 (2002).
13. V. Hernandez, J.E. Roman, and V. Vidal. SLEPc: A scalable and flexible toolkit for the solution of eigenvalue problems. *ACM Trans. Math. Software* **31**(3), 351–362 (2005).
14. A. May, P. Sanan, K. Rupp, M.G. Knepley and B.F. Smith. Extreme-scale multigrid components within PETSc. In: *Proceedings of the Platform for Advanced Scientific Computing Conference* (2016).
15. N. Spillane, V. Dolean, P. Hauret, F. Nataf, C. Pechstein and R. Scheichl. Abstract robust coarse spaces for systems of pdes via generalized eigenproblems in the overlaps. *Numer. Math.* **126**(4), 741–770 (2014).
16. A. St-Cyr, M.J. Gander and S.J. Thomas. Optimized multiplicative, additive, and restricted additive Schwarz preconditioning. *SIAM J. Sci. Comp.* **29**(6), 2402–2425 (2007).
17. P.S. Vassilevski and U.M. Yang. Reducing communication in algebraic multigrid using additive variants. *Numer. Linear Algebra Appl.* **21**(2), 275–296 (2014).

A Numerical Algorithm Based on Probing to Find Optimized Transmission Conditions

Martin J. Gander, Roland Masson, and Tommaso Vanzan

1 Motivation

Optimized Schwarz Methods (OSMs) are very versatile: they can be used with or without overlap, converge faster compared to other domain decomposition methods [5], are among the fastest solvers for wave problems [10], and can be robust for heterogeneous problems [7]. This is due to their general transmission conditions, optimized for the problem at hand. Over the last two decades such conditions have been derived for many Partial Differential Equations (PDEs), see [7] for a review.

Optimized transmission conditions can be obtained by diagonalizing the OSM iteration using a Fourier transform for two subdomains with a straight interface. This works surprisingly well, but there are important cases where the Fourier approach fails: geometries with curved interfaces (there are studies for specific geometries, e.g. [8, 9, 11]), and heterogeneous couplings when the two coupled problems are quite different in terms of eigenvectors of the local Steklov–Poincaré operators [6]. There is therefore a great need for numerical routines which allow one to get cheaply optimized transmission conditions, which furthermore could then lead to OSM black-box solvers. Our goal is to present one such procedure.

Let us consider the simple case of a two nonoverlapping subdomain decomposition, that is $\Omega = \Omega_1 \cup \Omega_2$, $\Omega_1 \cap \Omega_2 = \emptyset$, $\Gamma := \overline{\Omega_1} \cap \overline{\Omega_2}$, and a generic second order linear PDE

$$\mathcal{L}(u) = f, \text{ in } \Omega, \quad u = 0 \text{ on } \partial\Omega. \tag{1}$$

Martin J. Gander
Section de mathématiques, Université de Genève, e-mail: martin.gander@unige.ch

Roland Masson
Université Côte d'Azur, CNRS, Inria, LJAD, e-mail: roland.masson@unice.fr

Tommaso Vanzan
CSQI Chair, Institute de mathématiques, Ecole Polytechnique Fédérale de Lausanne, e-mail: tommaso.vanzan@epfl.ch

S. C. Brenner et al. (eds.), *Domain Decomposition Methods in Science and Engineering XXVI*, Lecture Notes in Computational Science and Engineering 145,
https://doi.org/10.1007/978-3-030-95025-5_65

The operator $\mathcal{L}$ could represent a homogeneous problem, i.e. the same PDE over the whole domain, or it could have discontinuous coefficients along Γ, or even represent a heterogeneous coupling. Starting from two initial guesses u_1^0, u_2^0, the OSM with double sided zeroth-order transmission conditions computes at iteration n

$$\begin{aligned} \mathcal{L}(u_1^n) &= 0 \text{ on } \Omega_1, \quad (\partial_{n_1} + s_1)u_1^n = (\partial_{n_1} + s_1)u_2^{n-1} \text{ on } \Gamma, \\ \mathcal{L}(u_2^n) &= 0 \text{ on } \Omega_2, \quad (\partial_{n_2} + s_2)u_2^n = (\partial_{n_2} + s_2)u_1^{n-1} \text{ on } \Gamma, \end{aligned} \tag{2}$$

where $s_1, s_2 \in \mathbb{R}$ are the parameters to optimize.

At the discrete level, the original PDE (1) is equivalent to the linear system

$$\begin{pmatrix} A_{II}^1 & 0 & A_{I\Gamma}^1 \\ 0 & A_{II}^2 & A_{I\Gamma}^2 \\ A_{\Gamma I}^1 & A_{\Gamma I}^2 & A_{\Gamma\Gamma} \end{pmatrix} \begin{pmatrix} \mathbf{u}_1 \\ \mathbf{u}_2 \\ \mathbf{u}_\Gamma \end{pmatrix} = \begin{pmatrix} \mathbf{f}_1 \\ \mathbf{f}_2 \\ \mathbf{f}_\Gamma \end{pmatrix},$$

where the unknowns are split into those interior to domain Ω_i, that is $\mathbf{u}_i$, $i = 1, 2$, and those lying on the interface Γ, i.e. $\mathbf{u}_\Gamma$. It is well known that the Dirichlet–Neumann and Neumann–Neumann methods can be seen as Richardson type methods to solve the discrete Steklov–Poincaré equation

$$\Sigma \mathbf{u}_\Gamma = \mu,$$

where $\Sigma := \Sigma_1 + \Sigma_2$, $\Sigma_i := A_{\Gamma\Gamma}^i - A_{\Gamma I}^i (A_{II}^i)^{-1} A_{I\Gamma}^i$, $\mu := \mu_1 + \mu_2$, $\mu_i := \mathbf{f}_\Gamma^i - A_{\Gamma I}^i (A_{II}^i)^{-1} \mathbf{f}_i$, $i = 1, 2$. It is probably less known that the OSM (2) can be interpreted as an Alternating Direction Implicit scheme (ADI, see e.g. [2]), for the solution of the continuous Steklov–Poincaré equation. This interesting point of view has been discussed in [1, 3]. At the discrete level, it results in the equivalence between a discretization of (2) and the ADI scheme

$$(s_1 E + \Sigma_1)\lambda^{n+\frac{1}{2}} = (s_1 E - \Sigma_2)\lambda^n + \mu, \quad (s_2 E + \Sigma_2)\lambda^{n+1} = (s_2 E - \Sigma_1)\lambda^{n+\frac{1}{2}} + \mu,$$

where E is either the mass matrix on Γ using a Finite Element discretization, or simply an identity matrix using a Finite Difference stencil. From now on, we will replace E with the identity I without loss of generality. Working on the error equation, the iteration operator of the ADI scheme is

$$T(s_1, s_2) := (s_2 I + \Sigma_2)^{-1}(s_2 I - \Sigma_1)(s_1 I + \Sigma_1)^{-1}(s_1 I - \Sigma_2), \tag{3}$$

and one would like to minimize the spectral radius, $\min_{s_1,s_2} \rho(T(s_1, s_2))$. It would be natural to use the wide literature available on ADI methods to find the optimized parameters s_1, s_2 for OSMs. Unfortunately, the ADI literature contains useful results only in the case where Σ_1 and Σ_2 commute, which is quite a strong assumption. In our context, the commutativity holds for instance if $\Omega_1 = \Omega_2$ and $\mathcal{L}$ represents a homogeneous PDE. Under these hypotheses, Fourier analysis already provides good estimates of the optimized parameters. Indeed it can be shown quite generally that the Fourier analysis and ADI theory lead to the same estimates. Without the commutativity assumption, the ADI theory relies on rough upper bounds which do not lead to precise estimates of the optimized parameters. For more details on the links between ADI methods and OSMs we refer to [13, Section 2.5].

Let us observe that if one used more general transmission conditions represented by matrices $\widetilde{\Sigma}_1$ and $\widetilde{\Sigma}_2$, (3) becomes

$$T(\widetilde{\Sigma}_1, \widetilde{\Sigma}_2) = (\widetilde{\Sigma}_2 + \Sigma_2)^{-1}(\widetilde{\Sigma}_2 - \Sigma_1)(\widetilde{\Sigma}_1 + \Sigma_1)^{-1}(\widetilde{\Sigma}_1 - \Sigma_2).$$

Choosing either $\widetilde{\Sigma}_1 = \Sigma_2$ or $\widetilde{\Sigma}_2 = \Sigma_1$ leads to $T = 0$, and thus one obtains that the local Steklov-Poincaré operators are optimal transmission operators [12].

2 An algorithm based on probing

Our algorithm to find numerically optimized transmission conditions has deep roots in the ADI interpretation of the OSMs and it is based on the probing technique. By probing, we mean the numerical procedure through which we estimate a generic matrix G by testing it over a set of vectors. In mathematical terms, given a set of vectors $\mathbf{x}_k$ and $\mathbf{y}_k := G\mathbf{x}_k$, $k \in \mathcal{K}$, we consider the problem

$$\text{Find } \widetilde{G} \quad \text{such that} \quad \widetilde{G}\mathbf{x}_i = \mathbf{y}_i, \forall i \in \mathcal{I}. \tag{4}$$

As we look for matrices with some nice properties (diagonal, tridiagonal, sparse, ...), problem (4) does not always have a solution. Calling D the set of admissible matrices, we prefer to consider the problem

$$\min_{\widetilde{G} \in D} \max_{k \in \mathcal{K}} \|\mathbf{y}_k - \widetilde{G}\mathbf{x}_k\|. \tag{5}$$

Having remarked that the local Steklov–Poincaré operators represent optimal transmission conditions, it would be natural to approximate them using probing. Unfortunately, this idea turns out to be very inefficient. To see this, let us carry out a continuous analysis on an infinite strip, $\Omega_1 = (-\infty, 0) \times (0, 1)$ and $\Omega_2 = (0, \infty) \times (0, 1)$. We consider the Laplace equation and, denoting by $\mathcal{S}_i$ the continuous Steklov–Poincaré operators, due to symmetry we have $\mathcal{S}_1 = \mathcal{S}_2 =: \mathcal{S}_e$. In this simple geometry, the eigenvectors of $\mathcal{S}_e$ are $v_k = \sin(k\pi y)$, $k \in \mathbb{N}^+$, with eigenvalues $\mu_k = k\pi$ so that $\mathcal{S}_e v_k = \mu_k v_k =: y_k$, see [5]. We look for an operator $S = sI$, $s \in \mathbb{R}^+$, which corresponds to a Robin transmission condition with parameter s. As probing functions, we choose the normalized functions v_k, $k = 1, ..., N_h$, where N_h is the number of degrees of freedom on the interface. Then (5) becomes

$$\begin{aligned} \min_{S=sI,\, s\in\mathbb{R}^+} \max_{k\in[1,N_h]} \|y_k - Sv_k\| &= \min_{s\in\mathbb{R}^+} \max_{k\in[1,N_h]} \|\mu_k v_k - s v_k\| \\ &= \min_{s\in\mathbb{R}^+} \max_{k\in[1,N_h]} |k\pi - s|. \end{aligned} \tag{6}$$

The solution of (6) is $s^* = \frac{N_h \pi}{2}$ while, according to a Fourier analysis and numerical evidence [5], the optimal parameter is $s^{\text{opt}} = \sqrt{N_h \pi}$. This discrepancy is due to the fact that problem (6) aims to make the parenthesis $(s_i I - \Sigma_{3-i})$, $i = 1, 2$, as small as possible, but it completely neglects the other terms $(s_i I + \Sigma_i)$.

This observation suggests to consider the minimization problem

$$\min_{\widetilde{\Sigma}_1, \widetilde{\Sigma}_2 \in D} \max_{k \in \mathcal{K}} \frac{\|\Sigma_2 \mathbf{x}_k - \widetilde{\Sigma}_1 \mathbf{x}_k\|}{\|\Sigma_1 \mathbf{x}_k + \widetilde{\Sigma}_1 \mathbf{x}_k\|} \frac{\|\Sigma_1 \mathbf{x}_k - \widetilde{\Sigma}_2 \mathbf{x}_k\|}{\|\Sigma_2 \mathbf{x}_k + \widetilde{\Sigma}_2 \mathbf{x}_k\|}. \tag{7}$$

We say that this problem is consistent in the sense that, assuming Σ_1, Σ_2 share a common eigenbasis $\{\mathbf{v}_k\}_k$ with eigenvalues $\{\mu_k^i\}$, $\widetilde{\Sigma}_i = s_i I$, $i = 1, 2$, $k = 1, \ldots, N_h$, then choosing $\mathbf{x}_k = \mathbf{v}_k$, we have

$$\min_{\widetilde{\Sigma}_1, \widetilde{\Sigma}_2 \in D} \max_{k \in \mathcal{K}} \frac{\|\Sigma_2 \mathbf{x}_k - \widetilde{\Sigma}_1 \mathbf{x}_k\|}{\|\Sigma_1 \mathbf{x}_k + \widetilde{\Sigma}_1 \mathbf{x}_k\|} \frac{\|\Sigma_1 \mathbf{x}_k - \widetilde{\Sigma}_2 \mathbf{x}_k\|}{\|\Sigma_2 \mathbf{x}_k + \widetilde{\Sigma}_2 \mathbf{x}_k\|} = \min_{s_1, s_2} \max_{k \in \mathcal{K}} \left| \frac{s_1 - \mu_k^2}{s_1 + \mu_k^1} \frac{s_2 - \mu_k^1}{s_2 + \mu_k^2} \right| = \min_{s_1, s_2 \in \mathbb{R}^+} \rho(T(s_1, s_2)),$$

that is, (7) is equivalent to minimizing the spectral radius of the iteration matrix.

We thus propose our numerical procedure to find optimized transmission conditions, summarized in Steps 2–4 of Algorithm 1.

Algorithm 1

Require: A set of vector $\mathbf{x}_k$, $k \in \mathcal{K}$, a characterization of $\widetilde{\Sigma}_1, \widetilde{\Sigma}_2$.

1: [Optional] For $i = 1, 2$, perform N iterations of the power method to get approximations of selected eigenvectors $\mathbf{x}_k^i$, $i = 1, 2$, $k \in \mathcal{K}$. Map $\mathbf{x}_j^i$ into $\mathbf{x}_k$, for $i = 1, 2$, $j \in \mathcal{K}$ and $k = 1, \ldots, 2|\mathcal{K}|$. Redefine $\mathcal{K} := \{1, \ldots, 2|\mathcal{K}|\}$.
2: Compute $y_k^i = \Sigma_i \mathbf{x}_k$, $k \in \mathcal{K}$.
3: Call an optimization routine to solve (7).
4: Return the matrices $\widetilde{\Sigma}_j$, $j = 1, 2$.

It requires as input a set of probing vectors and a characterization for the transmission matrices $\widetilde{\Sigma}_i$, that is if the matrices are identity times a real parameter, diagonal, or tridiagonal, sparse etc. We then precompute the action of the local Schur complement on the probing vectors. We finally solve (7) using an optimization routine such as fminsearch in MATLAB, which is based on the Nelder–Mead algorithm.

The application of Σ_i to a vector $\mathbf{x}_k$ requires a subdomain solve, thus Step 2 requires $2|\mathcal{K}|$ subdomain solves which are embarrassingly parallel. Step 3 does not require any subdomain solves, and thus is not expensive.

As discussed in Section 3, the choice of probing vectors plays a key role to obtain good estimates. Due to the extensive theoretical literature available, the probing vectors should be heuristically related to the eigenvectors associated to the minimum and maximum eigenvalues of Σ_i. It is possible to set the probing vectors $\mathbf{x}_k$ equal to lowest and highest Fourier modes. This approach is efficient when the Fourier analysis itself would provide relatively good approximations of the parameters. However there are instances, e.g. curved interfaces or heterogeneous problems, where it is preferable to have problem-dependent probing vectors. We thus include an additional optional step (Step 1), in which, starting from a given set of probing vectors, e.g. Fourier modes, we perform N iterations of the power method, which essentially correspond

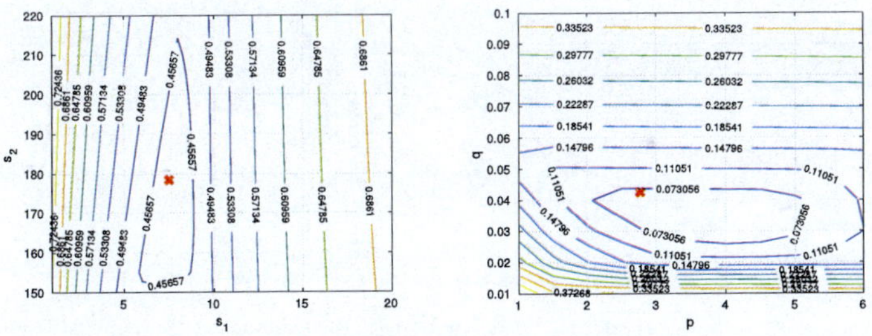

Fig. 1: Contour plot of the spectral radius of the iteration matrix $T(\widetilde{\Sigma}_1, \widetilde{\Sigma}_2)$ with $\widetilde{\Sigma}_i = s_i I$ (left) and of $T(\widehat{\Sigma}_1, \widehat{\Sigma}_2)$ with $\widehat{\Sigma}_i = pI + qH$ (right). The red crosses are the parameters obtained through Alg. 1.

to N iterations of the OSM, to get more suitable problem-dependent probing vectors. To compute the eigenvector associated to the minimum eigenvalue of Σ_i, we rely on the inverse power method which requires to solve a Neumann boundary value problem. Including Step 1, Algorithm 1 requires in total $2|\mathcal{K}|(N+2)$ subdomain solves, where $|\mathcal{K}|$ is the number of probing vectors in the input.

3 Numerical experiments

We start with a sanity check considering a Laplace equation on a rectangle Ω, with $\Omega_1 = (-1, 0)\times(0, 1)$, $\Omega_2 = (0, 1)\times(0, 1)$ and $\Gamma = \{0\}\times(0, 1)$. Given a discretization of the interface Γ with N_h points, we choose as probing vectors the discretization of

$$x_1 = \sin(\pi y), \quad x_2 = \sin(\sqrt{N_h}\pi y), \quad x_3 = \sin(N_h \pi y), \tag{8}$$

motivated by the theoretical analysis in [5], which shows that the optimized parameters s_i satisfy equioscillation between the minimum, the maximum and a medium frequency which scales as $\sqrt{N_h}$. We first look for matrices $\widetilde{\Sigma}_i = s_i I$ representing zeroth order double sided optimized transmission conditions. Then, we look for matrices $\widehat{\Sigma}_i = pI + qH$, where H is a tridiagonal matrix $H := \text{diag}(\frac{2}{h^2}) - \text{diag}(\frac{1}{h^2}, -1) - \text{diag}(\frac{1}{h^2}, +1)$, where h is the mesh size. At the continuous level, $\widehat{\Sigma}_i$ represent second order transmission conditions. Fig. 1 shows that Alg. 1 permits us to obtain excellent estimates in both cases with just three probing vectors. We emphasize that Alg. 1 requires 6 subdomain solves, which can be done in parallel, and leads to a convergence factor of order ≈ 0.07 for second order transmission conditions. It is clear that, depending on the problem at hand, this addition of 6 subdomain solves is negligible, considering the advantage of having such a small convergence factor.

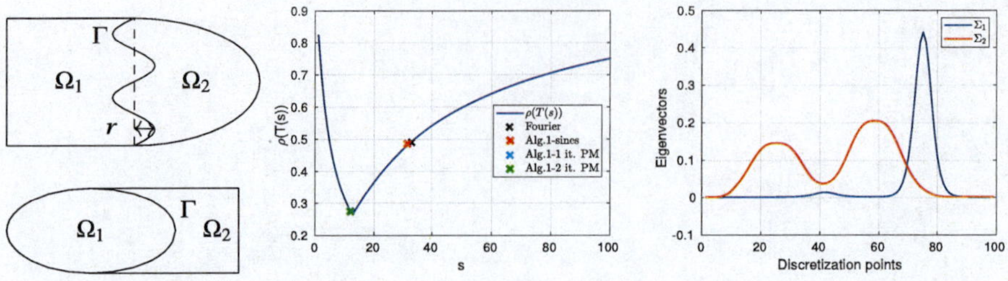

Fig. 2: Left: Ω decomposed into Ω_1 and Ω_2. Middle: optimized parameters obtained using Fourier analysis or Algorithm 1 with different sets of probing vectors. Right: eigenvectors associated to the smallest eigenvalues of Σ_j, $j = 1, 2$.

We now look at a more challenging problem. We solve a second order PDE

$$-\nabla \cdot \nu(\mathbf{x})\nabla u + \mathbf{a}(\mathbf{x})^\top \cdot \nabla u + \eta(\mathbf{x})u = f \quad \text{in } \Omega, \tag{9}$$

where Ω is represented in Fig. 2 on the top-left.

The interface Γ is the parametric curve $\gamma(t) : [0,1] \to (r\sin(\widehat{k}\pi t), t)$, with $r \in \mathbb{R}^+$. The coefficients are set to $\nu(\mathbf{x}) = 1$, $\mathbf{a}(\mathbf{x}) = (10(y+x^2), 0)^\top$, $\eta(\mathbf{x}) = 0.1(x^2+y^2)$ in Ω_1, $\nu(\mathbf{x}) = 100$, $\mathbf{a}(\mathbf{x}) = (10(1-x), x)^\top$, $\eta(\mathbf{x}) = 0$ in Ω_2, $f(\mathbf{x}) = x^2 + y^2$ in Ω. The geometric parameters are $r = 0.4$, $\widehat{k} = 6$ and the interface is discretized with $N_h = 100$ points. Driven by the theoretical analysis [7], we rescale the transmission conditions according to the physical parameters, setting $S_i := f_i(s)I$, where $f_i := \nu_i(s^2 + \frac{a_{i1}^2}{4\nu_i^2} + \frac{a_{i2}^2}{4\nu_i^2} + \frac{\eta_i}{\nu_i})^{1/2} - \frac{a_{i1}}{2}$. The center panel of Fig. 2 shows a comparison of the optimized parameters obtained by a Fourier analysis to the one obtained by Alg. 1 using as probing vectors the sine frequencies (8). It is evident that both do not deliver efficient estimates. The failure of Alg. 1 is due to the fact that, in contrast to the Laplace case, the sine frequencies do not contain information about the slowest modes. On the right panel of Fig 2, we plot the lowest eigenvectors of Σ_i, which clearly differ significantly from the simple lowest sine frequency. We therefore consider Alg. 1 with the optional Step 1 and as starting probing vectors we only use the lowest and highest sine frequencies. The center panel of Fig. 2 shows that Alg. 1 delivers efficient estimates with just one iteration of the power method. Let us now study the computational cost. To solve (9) up to a tolerance of 10^{-8} on the error, an OSM using the Fourier estimated parameters (black cross in Fig 2) requires 21 iterations, while only 12 are needed by Algorithm 1 with only one iteration of the power method. In the offline phase of Algorithm 1, we need to solve 4 subdomain problems in parallel in Step 1, and further 8 subdomain problems again in parallel in Step 2. Therefore the cost of the offline phase is equivalent to two iterations of the OSM in a parallel implementation, and consequently Alg. 1 is computationally attractive even in a single-query context.

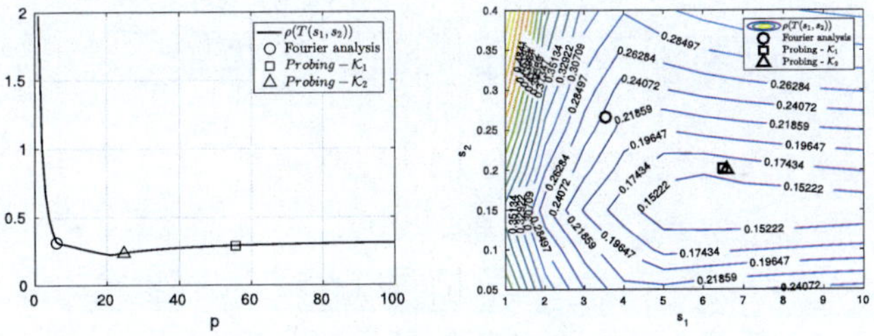

Fig. 3: Comparison between the optimized parameters obtained through Fourier analysis and Alg. 1 for single sided Robin boundary conditions (left) and double sided Robin boundary conditions (right).

Fourier estimates depend on the choice of $k_{\min}$ and $k_{\max}$ and in Fig. 2, we set $k_{\min} = \pi$ and $k_{\max} = \pi/h$. Inspired by [8] and a reviewer's comment, we optimized with $k_{\min} = \pi/|\Gamma| \approx \pi/4.96$ obtaining $s = 14.41$, which is very close to the optimal s^*. However, rescaling $k_{\min}$ with $|\Gamma|$ is not generally a valid approach. Considering Ω_1 as the ellipse of boundary $(\cos(2\pi t), 0.5\sin(2\pi t))$, $t \in (0,1)$, and $\Omega_2 = [0,2] \times [0,1] \setminus \Omega_1$, see Fig. 2 bottom-left, then $s^* = 40$, while $s_{k_{\min}=\pi} = 31.5$ and $s_{k_{\min}=\pi/|\Gamma|} = 20.44$. Thus, rescaling $k_{\min}$ worsens the Fourier estimate.

Next, we consider the Stokes–Darcy system in Ω, with $\Omega_1 = (-1,0) \times (0,1)$, $\Omega_2 = (0,1) \times (0,1)$ and $\Gamma = \{0\} \times (0,1)$ with homogeneous Dirichlet boundary conditions along $\partial\Omega$. Refs. [6, 13] show that the Fourier analysis fails to provide optimized parameters since the two subproblems do not share a common separation of variable expansion in bounded domains, unless periodic boundary conditions are enforced, see also [7][Section 3.3]. Thus, the sine functions do not diagonalize the OSM iteration, even in the simplified domain Ω with straight interface. Nevertheless, we apply Alg. 1 using two different sets of sines as probing vectors, corresponding to frequencies $\mathcal{K}_1 = \{1, \sqrt{N_h}, N_h\}$ and $\mathcal{K}_2 = \{1, 2, \sqrt{N_h}, N_h\}$. In $\mathcal{K}_2$ the first even frequency is included because in Ref. [6] it was observed that the first odd Fourier frequency converges extremely fast.

Fig. 3 shows the estimated parameters for single and double sided zeroth order transmission conditions obtained through a Fourier analysis [4] and using Alg. 1. The left panel confirms the intuition of [6], that is, the first even frequency plays a key role in the convergence. The right panel shows that Alg. 1, either with $\mathcal{K}_1$ or $\mathcal{K}_2$ provides better optimized parameters than the Fourier approach.

Next, we consider the stationary heat transfer model coupling the diffusion equation $\nabla\cdot(-\lambda\nabla u_1(\mathbf{x})) = 0$ in the porous medium domain $\Omega_1 = (0,L)\times(5,15)$ with the convection diffusion equation $\nabla\cdot(u_2(\mathbf{x})\mathbf{V}_t(y) - \lambda_t(y)\nabla u_2(\mathbf{x})) = 0$ in the free flow domain $\Omega_2 = (0,L) \times (0,5)$. Both the turbulent velocity $\mathbf{V}_t = (V_t(y), 0)^T$ and the thermal conductivity $\lambda_t(y)$ exhibit a boundary layer at the interface $\Gamma = (0,L)\times\{5\}$ and are computed from the Dittus–Boelter turbulent model. Dirichlet boundary conditions are prescribed at the top of Ω_1 and on the left of Ω_2, homogeneous Neumann boundary conditions are set on the left and right of Ω_1 and at the bottom of Ω_2, and a zero Fourier flux is imposed on the right of Ω_2. Flux and temperature continuity

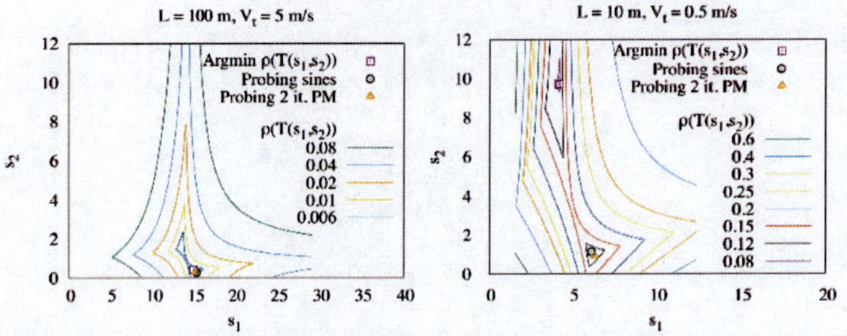

Fig. 4: For $L = 100$ m, $\overline{V}_t = 5$ m/s (left) and $L = 10$ m, $\overline{V}_t = 0.5$ m/s (right), comparison of the double sided Robin parameters s_1 and s_2 obtained from the probing algorithm using either the 3 sine vectors or the 6 vectors obtained from the 3 sines vectors by 2 PM iterations on both sides. It is compared with the minimizer of the spectral radius $\rho(T(s_1, s_2))$.

is imposed at the interface Γ. The model is discretized by a Finite Volume scheme on a Cartesian mesh of size 50×143 refined on both sides of the interface. Figure 4 shows that the probing algorithm provides a very good approximation of the optimal solution for the case $L = 100$ m, $\overline{V}_t = 5$ m/s (mean velocity) both with the 3 sine vectors (8) and with the 6 vectors obtained from the power method starting from the sine vectors. In the case $L = 10$ m, $\overline{V}_t = 0.5$ m/s, the spectral radius has a narrow valley with two minima. In that case the probing algorithm fails to find the best local minimum but still provides a very efficient approximation.

References

1. V. Agoshkov and V. Lebedev. Generalized Schwarz algorithm with variable parameters. *Russian J. Num. Anal. and Math. Model.* **5**(1), 1–26 (1990).
2. O. Axelsson. *Iterative Solution Methods*. Cambridge University Press (1994).
3. M. Discacciati. *Domain decomposition methods for the coupling of surface and groundwater flows*. Ph.D. thesis, Ecole Polytecnique Fédérale de Lausanne (2004).
4. M. Discacciati and L. Gerardo-Giorda. Optimized Schwarz methods for the Stokes–Darcy coupling. *IMA J. Num. Anal.* **38**(4), 1959–1983 (2018).
5. M.J. Gander. Optimized Schwarz methods. *SIAM J. Num. Anal.* **44**(2), 699–731 (2006).
6. M.J. Gander and T. Vanzan. On the derivation of optimized transmission conditions for the Stokes–Darcy coupling. In: *International Conference on Domain Decomposition Methods XXV*, pp. 491–498, Springer (2018).
7. M.J. Gander and T. Vanzan. Heterogeneous optimized Schwarz methods for second order elliptic PDEs. *SIAM J. Sci. Comput.* **41**(4), A2329–A2354 (2019).
8. M.J. Gander and Y. Xu. Optimized Schwarz methods for circular domain decompositions with overlap. *SIAM J. Numer. Anal.* **52**, 1981–2004 (2014).
9. M.J. Gander and Y. Xu. Optimized Schwarz methods for domain decompositions with parabolic interfaces. In: *Domain Decomposition Methods in Science and Engineering XXIII*, pp. 323–331, Springer (2017).
10. M.J. Gander and H. Zhang. A class of iterative solvers for the Helmholtz equation: Factorizations, sweeping preconditioners, source transfer, single layer potentials, polarized traces, and optimized Schwarz methods. *SIAM Review* **61**(1), 3–76 (2019).

11. G. Gigante, G. Sambataro and C. Vergara. Optimized Schwarz methods for spherical interfaces with application to fluid-structure interaction. *SIAM J. Sci. Comput.* **42**(2), A751–A770 (2020).
12. F. Nataf, F. Rogier and E. De Sturler. Optimal interface conditions for domain decomposition methods. Tech. rep., École Polytechnique de Paris (1994).
13. T. Vanzan. *Domain decomposition methods for multiphysics problems*. Ph.D. thesis, Université de Genève (2020).

Additive Schwarz Preconditioners for C^0 Interior Penalty Methods for a State Constrained Elliptic Distributed Optimal Control Problem

Susanne C. Brenner, Li-Yeng Sung, and Kening Wang

1 Introduction

Let Ω be a bounded convex polygon in $\mathbb{R}^2$, $f \in L_2(\Omega)$, and $\beta > 0$ be a constant. We consider the following elliptic optimal control problem: Find $(y, u) \in H_0^1(\Omega)\times L_2(\Omega)$ that minimize the functional

$$J(y,u) = \frac{1}{2}\int_\Omega (y-f)^2 + \frac{\beta}{2}\int_\Omega u^2\,\mathrm{d}x$$

subject to

$$-\Delta y = u \ \text{ in } \ \Omega, \qquad y = 0 \ \text{ on } \ \partial\Omega,$$

and $y \le \psi$ in Ω, where $\psi \in W^{3,p}(\Omega)$ for $p > 2$, and $\psi > 0$ on $\partial\Omega$.

By elliptic regularity (cf. [6]), we can reformulate the model problem as follows: Find $y \in K$ such that

$$y = \underset{v\in K}{\operatorname{argmin}}\left[\frac{1}{2}a(v,v) - (f,v)\right], \tag{1}$$

where $K = \{v \in H^2(\Omega)\cap H_0^1(\Omega) : v \le \psi \ \text{ in } \ \Omega\}$,

$$a(v,w) = \beta\int_\Omega \nabla^2 v : \nabla^2 w\,\mathrm{d}x + \int_\Omega vw\,\mathrm{d}x \quad \text{and} \quad (f,v) = \int_\Omega fv\,\mathrm{d}x.$$

Here $\nabla^2 v : \nabla^2 w = \sum_{i,j=1}^2 \frac{\partial^2 v}{\partial x_i \partial x_j}\frac{\partial^2 w}{\partial x_i \partial x_j}$ is the inner product of the Hessian matrices of v and w. Once y is calculated, then u can be determined by $u = -\Delta y$.

Susanne C. Brenner and Li-Yeng Sung
Department of Mathematics and Center for Computation and Technology, Louisiana State University, Baton Rouge, LA 70803, USA, e-mail: brenner@math.lsu.edu, sung@math.lsu.edu

Kening Wang
Department of Mathematics and Statistics, University of North Florida, Jacksonville, FL 32224, USA, e-mail: kening.wang@unf.edu

S. C. Brenner et al. (eds.), *Domain Decomposition Methods in Science and Engineering XXVI*, Lecture Notes in Computational Science and Engineering 145,
https://doi.org/10.1007/978-3-030-95025-5_66

A quadratic C^0 interior penalty method for the minimization problem (1) was analyzed in [4]. The goal of this paper is to apply the ideas in [3] for an obstacle problem of clamped Kirchhoff plates to develop and analyze additive Schwarz preconditioners for the discrete problem in [4].

2 The C^0 interior penalty method

Let $\mathcal{T}_h$ be a quasi-uniform triangulation of Ω consisting of convex quadrilaterals, and let $V_h \subset H_0^1(\Omega)$ be the standard Q_k finite element space (the space of polynomials of degree $\leq k$ in each variable) associated with $\mathcal{T}_h$.

The discrete problem of the optimal control problem (1) resulting from the C^0 interior penalty method is to find

$$y_h = \operatorname*{argmin}_{v \in K_h} \left[\frac{1}{2} a_h(v, v) - (f, v) \right], \tag{2}$$

where

$$K_h = \{v \in V_h : v(p) \leq \psi(p), \quad \forall\, p \in \mathcal{N}_h\},$$

$$a_h(v,w) = \beta \Big[\sum_{D \in \mathcal{T}_h} \int_D \nabla^2 v : \nabla^2 w \, dx + \sum_{e \in \mathcal{E}_h^i} \frac{\eta}{|e|} \int_e \left[\!\left[\frac{\partial v}{\partial n} \right]\!\right] \left[\!\left[\frac{\partial w}{\partial n} \right]\!\right] ds$$

$$+ \sum_{e \in \mathcal{E}_h^i} \int_e \left(\left\{\!\!\left\{ \frac{\partial^2 v}{\partial n^2} \right\}\!\!\right\} \left[\!\left[\frac{\partial w}{\partial n} \right]\!\right] + \left\{\!\!\left\{ \frac{\partial^2 w}{\partial n^2} \right\}\!\!\right\} \left[\!\left[\frac{\partial v}{\partial n} \right]\!\right] \right) ds \Big] + \sum_{D \in \mathcal{T}_h} \int_D vw \, dx,$$

$\mathcal{N}_h$ is the set of nodes in Ω associated with V_h, $\mathcal{E}_h^i$ is the set of edges in $\mathcal{T}_h$ that are interior to Ω, $\eta > 0$ is a sufficiently large penalty parameter, and the jump $[\![\cdot]\!]$ and the average $\{\!\{\cdot\}\!\}$ are defined as follows. Let e be an interior edge shared by two elements, D_- and D_+, and n_e be the unit normal vector pointing from D_- to D_+, we define

$$\left[\!\left[\frac{\partial v}{\partial n} \right]\!\right] = \frac{\partial v_+}{\partial n_e} - \frac{\partial v_-}{\partial n_e} \quad \text{and} \quad \left\{\!\!\left\{ \frac{\partial^2 v}{\partial n^2} \right\}\!\!\right\} = \frac{1}{2} \left(\frac{\partial^2 v_+}{\partial n_e^2} + \frac{\partial^2 v_-}{\partial n_e^2} \right).$$

Note that $a_h(\cdot,\cdot)$ is a consistent bilinear form for the biharmonic equation with the boundary conditions of simply supported plates (cf. [4]).

It follows from the standard theory that the discrete problem (2) has a unique solution $y_h \in K_h$ characterized by the discrete variational inequality

$$a_h(y_h, v_h - y_h) \geq (f, v_h - y_h) \qquad \forall\, v_h \in K_h. \tag{3}$$

Moreover, there exists a positive constant C independent of h such that (cf. [4])

$$\|y - y_h\|_h \leq C h^{\alpha},$$

where $\|\cdot\|_h$ is the mesh-dependent energy norm defined by

$$\|v\|_h^2 = \beta\Big(\sum_{D\in\mathcal{T}_h} |v|_{H^2(D)}^2 + \sum_{e\in\mathcal{E}_h^i} \frac{1}{|e|}\|\,[[\partial v/\partial n]]\,\|_{L_2(e)}^2\Big) + \|v\|_{L_2(\Omega)}^2,$$

h is the mesh size of the triangulation, and $\alpha \in (0, 1]$ is the index of elliptic regularity that is determined by the interior angles of Ω.

3 The primal-dual active set algorithm

By introducing a Lagrange multiplier $\lambda_h : \mathcal{N}_h \to \mathbb{R}$, the discrete variational inequality (3) is equivalent to

$$a_h(y_h, v) - (f, v) = -\sum_{p\in\mathcal{N}_h} \lambda_h(p)v(p) \qquad \forall v \in V_h, \tag{4}$$

$$y_h(p) - \psi(p) \geq 0,\ \ \lambda_h(p) \geq 0 \text{ and } (y_h(p) - \psi(p))\lambda_h(p) = 0\ \forall p \in \mathcal{N}_h. \tag{5}$$

Moreover, the optimality conditions (5) can be written concisely as

$$\lambda_h(p) = \max(0, \lambda_h(p) + c(y_h(p) - \psi(p))) \qquad \forall p \in \mathcal{N}_h, \tag{6}$$

where c is a large positive number. The system (4) and (6) can then be solved by a primal-dual active set (PDAS) algorithm (cf. [7, 8]).

Given the k-th approximation (y_k, λ_k), the $(k + 1)$-st iteration of the PDAS algorithm is to find (y_{k+1}, λ_{k+1}) such that

$$a_h(y_{k+1}, v) - (f, v) = -\sum_{p\in\mathcal{N}_h} \lambda_{k+1}(p)v(p) \qquad \forall v \in V_h, \tag{7a}$$

$$y_{k+1}(p) = \psi(p) \qquad \forall p \in \mathcal{A}_k, \tag{7b}$$

$$\lambda_{k+1}(p) = 0 \qquad \forall p \in \mathcal{I}_k, \tag{7c}$$

where $\mathcal{A}_k = \{p \in \mathcal{N}_h : \lambda_k(p) + c(y_k(p) - \psi(p)) > 0\}$ is the active set determined by (y_k, λ_k), and $\mathcal{I}_k = \mathcal{N}_h\backslash\mathcal{A}_k$ is the inactive set. The iteration terminates when $\mathcal{A}_{k+1} = \mathcal{A}_k$. Given a sufficiently accurate initial guess, the PDAS algorithm converges superlinearly to the unique solution of (3) (cf. [7]).

From (7b) and (7c), we can reduce (7a) to an auxiliary system that only involves the unknowns of $y_{k+1}(p)$ for $p \in \mathcal{I}_k$. But even so, for small h, the reduced auxiliary system is still large, sparse, and ill-conditioned. To solve such systems more efficiently, we can apply the preconditioned conjugate gradient method.

Let $\widetilde{\mathcal{N}}_h$ be a subset of $\mathcal{N}_h$. We define $\widetilde{T}_h : V_h \to V_h$, the truncation operator, by

$$(\widetilde{T}_h v)(p) = \begin{cases} v(p) & \text{if } p \in \widetilde{\mathcal{N}}_h, \\ 0 & \text{if } p \in \mathcal{N}_h\backslash\widetilde{\mathcal{N}}_h. \end{cases}$$

Then $\widetilde{T}_h$ is a projection from V_h onto $\widetilde{V}_h = \widetilde{T}_h V_h$. Moreover, let $\widetilde{A}_h : \widetilde{V}_h \to \widetilde{V}_h'$ be defined by

$$\langle \widetilde{A}_h v, w\rangle = a_h(v,w) \qquad \forall\, v, w \in \widetilde{V}_h,$$

where $\langle\cdot,\cdot\rangle$ is the canonical bilinear form on $\widetilde{V}_h' \times \widetilde{V}_h$.

In the context of solving (3), the set $\widetilde{\mathcal{N}}_h$ represents the inactive set that appears in an iteration of the PDAS algorithm and $\widetilde{A}_h$ represents the stiffness matrix for the corresponding auxiliary system. Our goal is to develop preconditioners for $\widetilde{A}_h$ whose performance is independent of $\widetilde{\mathcal{N}}_h$.

4 A one-level additive Schwarz preconditioner

Let $\Omega_j, 1 \le j \le J$, be overlapping subdomains of Ω such that $\Omega = \bigcup_{j=1}^J \Omega_j$, diam $\Omega_j \approx H$, and the boundaries of Ω_j are aligned with $\mathcal{T}_h$. We assume that there exist nonnegative $\theta_j \in C^\infty(\bar{\Omega})$ for $1 \le j \le J$ such that

$$\theta_j = 0 \qquad \text{on } \Omega \setminus \Omega_j,$$

$$\sum_{j=1}^J \theta_j = 1 \qquad \text{on } \bar{\Omega},$$

$$\|\nabla\theta_j\|_{L_\infty(\Omega)} \le \frac{C_\dagger}{\delta}, \qquad \|\nabla^2\theta_j\|_{L_\infty(\Omega)} \le \frac{C_\dagger}{\delta^2},$$

where $\nabla^2\theta_j$ is the Hessian of θ_j, $\delta > 0$ measures the overlap among subdomains, and $C_\dagger$ is a positive constant independent of h, H, and J. Moreover, we assume that

$$\text{any point in } \Omega \text{ can belong to at most } N_c \text{ many subdomains,}$$

where the positive integer N_c is independent of h, H, J and δ.

Let $\widetilde{V}_j$ be the subspace of $\widetilde{V}_h$ whose members vanish at all nodes outside Ω_j, and let $\widetilde{A}_j : \widetilde{V}_j \to \widetilde{V}_j'$ be defined by

$$\langle \widetilde{A}_j v, w\rangle = a_{h,j}(v,w) \qquad \forall\, v, w \in \widetilde{V}_j,$$

where

$$\begin{aligned} a_{h,j}(v,w) = \beta\Bigg[&\sum_{D\in\mathcal{T}_h}\int_D \nabla^2 v : \nabla^2 w \, dx + \sum_{\substack{e\in\mathcal{E}_h^i \\ e\subset\overline{\Omega}_j\setminus\partial\Omega}} \frac{\eta}{|e|}\int_e \left[\!\!\left[\frac{\partial v}{\partial n}\right]\!\!\right]\left[\!\!\left[\frac{\partial w}{\partial n}\right]\!\!\right] ds \\ &+ \sum_{\substack{e\in\mathcal{E}_h^i \\ e\subset\overline{\Omega}_j\setminus\partial\Omega}} \int_e \left(\left\{\!\!\left\{\frac{\partial^2 v}{\partial n^2}\right\}\!\!\right\}\left[\!\!\left[\frac{\partial w}{\partial n}\right]\!\!\right] + \left\{\!\!\left\{\frac{\partial^2 w}{\partial n^2}\right\}\!\!\right\}\left[\!\!\left[\frac{\partial v}{\partial n}\right]\!\!\right]\right) ds\Bigg] + \sum_{D\in\mathcal{T}_h}\int_D vw\, dx. \end{aligned}$$

The one-level additive Schwarz preconditioner $B_{OL} : \widetilde{V}_h' \to \widetilde{V}_h$ is then defined by

$$B_{OL} = \sum_{j=1}^{J} \widetilde{I}_j \widetilde{A}_j^{-1} \widetilde{I}_j^t,$$

where $\widetilde{I}_j : \widetilde{V}_j \to \widetilde{V}_h$ $(1 \leq j \leq J)$ is the natural injection operator, and $\widetilde{I}_j^t : \widetilde{V}_h' \to \widetilde{V}_j'$ is the transpose of $\widetilde{I}_j$.

With similar arguments as in [3], we can obtain the following result.

Theorem 1 *It holds that*

$$\kappa(B_{OL}\widetilde{A}_h) = \frac{\lambda_{\max}(B_{OL}\widetilde{A}_h)}{\lambda_{\min}(B_{OL}\widetilde{A}_h)} \leq C_1 \delta^{-4},$$

where the positive constant C_1 is independent of H, h, j, δ and $\widetilde{N}_h$.

Remark 1 The condition number estimate given in Theorem 1 is identical to the one for the plate bending problem without obstacles, which indicates that the obstacle is invisible to the one-level additive Schwarz preconditioner.

5 A two-level additive Schwarz preconditioner

A two-level additive Schwarz preconditioner contains not only subdomain solves, but also a coarse grid solve. Let $\mathcal{T}_H$ be a coarse quasi-uniform triangulation for Ω whose mesh size is comparable to the diameters of the subdomains $\Omega_j, 1 \leq j \leq J$, and $V_H \subset H_0^1(\Omega)$ be the Q_k finite element space associated with $\mathcal{T}_H$.

Since the Q_{k+2} Bogner–Fox–Schmit (BFS) tensor product element is a C^1 relative of the Q_k tensor product element (cf. [2]), we define $W_H \subset H^2(\Omega) \cap H_0^1(\Omega)$ to be the Q_{k+2} BFS finite element space associated with $\mathcal{T}_H$. The two spaces V_H and W_H can be connected by an enriching operator E_H which is constructed by the averaging technique (cf. [2, 3]).

Now we define $I_0 : V_H \to V_h$ by

$$I_0 = \Pi_h \circ E_H$$

where $\Pi_h : C^0(\bar{\Omega}) \to V_h$ is the nodal interpolation operator.

Let $\widetilde{V}_0 \subset \widetilde{V}_h$ be defined by

$$\widetilde{V}_0 = \widetilde{T}_h I_0 V_H,$$

and let the operator $\widetilde{A}_0 : \widetilde{V}_0 \to \widetilde{V}_0'$ be defined by

$$\langle \widetilde{A}_0 v, w \rangle = a_h(v, w) \qquad \forall\, v, w \in \widetilde{V}_0.$$

Then the two-level additive Schwarz preconditioner $B_{TL} : \widetilde{V}_h' \to \widetilde{V}_h$ is given by

$$B_{TL} = \sum_{j=0}^{J} \widetilde{I}_j \widetilde{A}_j^{-1} \widetilde{I}_j^t,$$

where $\widetilde{I}_j : \widetilde{V}_j \to \widetilde{V}_h$ $(0 \le j \le J)$ is the natural injection operator, and $\widetilde{I}_j^t$ is the transpose of $\widetilde{I}_j$.

Following the arguments in [3], we can obtain an estimate on the condition number of $B_{TL}\widetilde{A}_h$.

Theorem 2 *It holds that*

$$\kappa(B_{TL}\widetilde{A}_h) \le C_2 \min\left((H/h)^4, \delta^{-4}\right), \tag{8}$$

where C_2 is a positive constant independent of H, h, j, δ and $\widetilde{N}_h$.

Remark 2 When the obstacle is present, it is necessary to include the truncation operator in the construction of $\widetilde{V}_0$. Therefore, the condition number estimate (8) for the two-level additive Schwarz preconditioner is different from the one for the plate bending problem without obstacles (cf. [5]) which takes the form

$$\kappa(B_{TL}A_h) \le C_* \left(1 + (H/\delta)^4\right).$$

6 Numerical results

We consider the obstacle problem (cf. [1]) with $\Omega = (-0.5, 0.5)^2, \beta = 0.1, \psi = 0.01$, and $f = 10(\sin(2\pi(x_1 + 0.5)) + (x_2 + 0.5))$. We discretize the model problem by the C^0 interior penalty method that is based on a rectangular mesh, and choose V_h to be the standard Q_2 finite element space with the mesh size $h = 2^{-\ell}$, where ℓ is the refinement level. The resulting discrete variational inequalities are solved by the PDAS algorithm, in which we choose the constant c to be 10^8. The initial guess for the PDAS algorithm is taken to be the solution at the previous level or zero when $\ell = 1$.

The graphs of the numerical solution y_h and the discrete active set $\mathcal{A}_k$ at refinement level 7 are given in Figure 1.

For comparison, we first calculate the condition number of the un-preconditioned auxiliary system $\widetilde{A}_h$ in each iteration of the PDAS algorithm and then take the average. The average condition numbers and numbers of iterations of the PDAS algorithm for various levels are presented in Table 1.

We apply the one-level and two-level additive Schwarz preconditioners to the auxiliary system in each iteration of the PDAS algorithm. The average condition numbers of both preconditioned auxiliary systems for 4, 16, 64, and 256 subdomains with small overlap, $\delta = h$, are reported in Table 2 and Table 3 respectively. Comparing

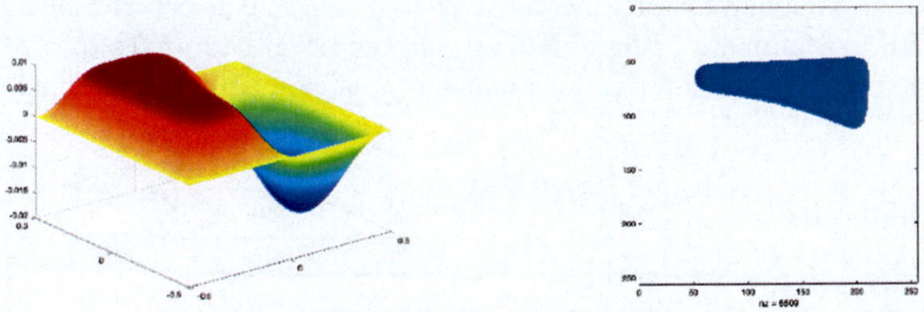

Fig. 1: The numerical solution y_h (left) and the discrete active set $\mathcal{A}_k$ (right) at refinement level 7.

	$\kappa(\widetilde{A}_h)$	PDAS Iterations
$\ell = 1$	1.7604×10^1	5
$\ell = 2$	2.2085×10^2	10
$\ell = 3$	4.3057×10^3	5
$\ell = 4$	6.7740×10^4	8
$\ell = 5$	1.0849×10^6	12
$\ell = 6$	1.8038×10^7	15

Table 1: Average condition number of $\widetilde{A}_h$, and number of iterations of the PDAS algorithm.

with the condition numbers of the unpreconditioned auxiliary systems in Table 1, both one-level and two-level algorithms show dramatic improvements.

	$J = 4$	$J = 16$	$J = 64$	$J = 256$
$\ell = 2$	5.8672×10^0	—	—	—
$\ell = 3$	1.9350×10^1	5.1410×10^1	—	—
$\ell = 4$	9.9423×10^1	2.4134×10^2	6.6698×10^2	—
$\ell = 5$	6.9235×10^2	1.7965×10^3	3.4752×10^3	1.0282×10^4
$\ell = 6$	5.6185×10^3	1.4676×10^4	2.8898×10^4	5.6312×10^4

Table 2: Average condition number of $B_{OL}\widetilde{A}_h$ with small overlap.

	$J = 4$	$J = 16$	$J = 64$	$J = 256$
$\ell = 2$	5.4489×10^0	—	—	—
$\ell = 3$	8.1290×10^0	1.2913×10^1	—	—
$\ell = 4$	3.6660×10^1	1.8647×10^1	3.4614×10^1	—
$\ell = 5$	2.1670×10^2	4.0108×10^1	4.6832×10^1	7.9579×10^1
$\ell = 6$	1.5552×10^3	2.4043×10^2	5.5854×10^1	1.0981×10^2

Table 3: Average condition number of $B_{TL}\widetilde{A}_h$ with small overlap.

Moreover, similar simulations for generous overlap $\delta = H$ are also performed. The average condition numbers of the one-level and two level additive Schwarz preconditioned auxiliary systems for various number of subdomains are presented in Tables 4 and 5 .

	$J = 4$	$J = 16$	$J = 64$	$J = 256$
$\ell = 2$	1.0000×10^0	—	—	—
$\ell = 3$	1.0000×10^0	1.1796×10^1	—	—
$\ell = 4$	1.0000×10^0	1.2828×10^1	1.1154×10^2	—
$\ell = 5$	1.0000×10^0	1.3457×10^1	1.1315×10^2	1.5925×10^3
$\ell = 6$	1.0000×10^0	1.4041×10^1	1.1760×10^2	1.6453×10^3

Table 4: Average condition number of $B_{OL}\widetilde{A}_h$ with generous overlap.

	$J = 4$	$J = 16$	$J = 64$	$J = 256$
$\ell = 2$	1.2500×10^0	—	—	—
$\ell = 3$	1.2500×10^0	7.8441×10^0	—	—
$\ell = 4$	1.2500×10^0	9.1917×10^0	2.4105×10^1	—
$\ell = 5$	1.2500×10^0	9.9897×10^0	2.5678×10^1	5.8649×10^1
$\ell = 6$	1.2500×10^0	1.0569×10^1	2.6729×10^1	6.3733×10^1

Table 5: Average condition number of $B_{TL}\widetilde{A}_h$ with generous overlap.

7 Conclusion

We present additive Schwarz preconditioners for the auxiliary systems that appear in a primal-dual active set algorithm for solving a state constrained elliptic distributed optimal control problem discretized by a C^0 interior penalty method. Both the one-level and two-level preconditioners improve the condition numbers of the auxiliary systems significantly.

Acknowledgements The work of the first two authors was supported in part by the National Science Foundation under Grant No. DMS-19-13035.

References

1. S.C. Brenner, C.B. Davis and L.-Y. Sung. *Additive Schwarz preconditioners for a state constrained elliptic distributed optimal control problem discretized by a partition of unity method.* Lecture Notes in Computational Science and Engineering, **138**, 100–107 (2020).
2. S.C. Brenner and L.-Y. Sung. *C^0 interior penalty methods for fourth order elliptic boundary value problems on polygonal domains. J. Sci. Comput.* **22/23**, 83–118 (2005).
3. S.C. Brenner, L.-Y. Sung and K. Wang. *Additive Schwarz Preconditioners for C^0 Interior Penalty Methods for the Obstacle Problem of Clamped Kirchhoff Plates.* Accepted by *Numer. Methods Partial Differential Equations.*
4. S.C. Brenner, L.-Y. Sung and Y. Zhang, *A Quadratic C^0 Interior Penalty Method for an Elliptic Optimal Control Problem with State Constraints. IMA Vol. Math. Appl.* **157**, 97–132 (2013).
5. S.C. Brenner and K. Wang. *Two-level additive Schwarz preconditioners for C^0 interior penalty methods. Numer. Math.* **102**, 231–255 (2005).
6. P. Grisvard. *Elliptic Problems in Non Smooth Domains.* Pitman, Boston (1985).
7. M. Hintermüller, K. Ito and K. Kunisch. *The primal-dual active set strategy as a semismooth Newton method. SIAM J. Optim.* **13**, 865–888 (2002).
8. K. Ito and K. Kunisch. *Lagrange Multiplier Approach to Variational Problems and Applications.* SIAM, Philadelphia (2008).

Space-Time Finite Element Methods for Initial Temperature Reconstruction

Ulrich Langer, Olaf Steinbach, Fredi Tröltzsch, and Huidong Yang

1 Introduction

In this work, we investigate the applicability of unstructured space-time methods to the numerical solution of inverse problems, considering as our model problem the classical inverse problem of the reconstruction of the initial temperature in the heat equation from an observation of the temperature $u_T^\delta \in L^2(\Omega)$ at a finite time horizon: Find the initial temperature $u_0^\delta(\cdot) := u(\cdot, 0) \in L^2(\Omega)$ on Σ_0 of the solution u of the backward heat equation

$$\partial_t u - \Delta_x u = 0 \quad \text{in } Q, \quad u = 0 \quad \text{on } \Sigma, \quad u = u_T^\delta \quad \text{on } \Sigma_T, \tag{1}$$

where $Q := \Omega \times (0, T)$ denotes the space-time cylinder with boundary $\partial Q = \overline{\Sigma} \cup \overline{\Sigma}_0 \cup \overline{\Sigma}_T$, $\Sigma := \partial\Omega \times (0, T)$, $\Sigma_0 := \Omega \times \{0\}$, $\Sigma_T := \Omega \times \{T\}$, for a bounded Lipschitz domain $\Omega \subset \mathbb{R}^d$, $d \in \{1, 2, 3\}$, and a finite time horizon $T > 0$. The observed terminal temperature u_T^δ may contain some noise characterized by the noise level $\delta \geq 0$,

$$\|u_T^\delta - u_T\|_{L^2(\Omega)} \leq \delta, \tag{2}$$

Ulrich Langer
Institute of Computational Mathematics, Johannes Kepler University Linz, Altenberger Straße 69, 4040 Linz, Austria, e-mail: ulanger@numa.uni-linz.ac.at

Olaf Steinbach
Institut für Angewandte Mathematik, Technische Universität Graz, Steyrergasse 30, 8010 Graz, Austria, e-mail: o.steinbach@tugraz.at

Fredi Tröltzsch
Institut für Mathematik, Technische Universität Berlin, Straße des 17. Juni 136, 10623 Berlin, Germany, e-mail: troeltzsch@math.tu-berlin.de

Huidong Yang
Johann Radon Institute for Computational and Applied Mathematics, Altenberger Straße 69, 4040 Linz, Austria, e-mail: huidong.yang@oeaw.ac.at

S. C. Brenner et al. (eds.), *Domain Decomposition Methods in Science and Engineering XXVI*, Lecture Notes in Computational Science and Engineering 145,
https://doi.org/10.1007/978-3-030-95025-5_67

where $u_T = u(\cdot, T) \in L^2(\Omega)$ represents the unpolluted exact data.

In contrast to the forward heat equation with known initial data, the backward heat equation (1) is severely ill-posed; see [2, Example 2.9]. In fact, the solution of (1) does not continuously depend on the data u_T^δ even when the solution exists. Following the notation in [2], the problem (1) may be reformulated as an abstract operator equation in a more general setting: Find $u_0 \in \mathcal{X}$ such that

$$Su_0 = u_T, \tag{3}$$

where $S : \mathcal{X} \to \mathcal{Y}$ denotes a bounded linear operator between two Hilbert spaces $\mathcal{X}$ and $\mathcal{Y}$. It is clear that there does not exist a continuous inverse operator $S^{-1} : \mathcal{Y} \to \mathcal{X}$ in general. Therefore, we consider a regularized solution, depending on the choice of Tikhonov's regularization parameter $\varrho := \varrho(\delta)$,

$$u_0^{\delta,\varrho} := (S^*S + \varrho I)^{-1}\, S^* u_T^\delta,$$

as the unique minimizer of the Tikhonov functional [9]

$$\mathcal{J}_\varrho(z) := \frac{1}{2}\, \|Sz - u_T^\delta\|_{\mathcal{Y}}^2 + \frac{\varrho}{2}\, \|z\|_{\mathcal{X}}^2. \tag{4}$$

It is well known that we have the convergence

$$\lim_{\delta\to 0} u_0^{\delta,\varrho} = u_0^\dagger \text{ in } \mathcal{X}, \text{ if the conditions } \lim_{\delta\to 0} \varrho(\delta) = 0 \text{ and } \lim_{\delta\to 0} \frac{\delta^2}{\varrho(\delta)} = 0$$

are satisfied. Here, $u_0^\dagger$ denotes the best-approximated solution to the operator equation (3); see [2, Theorem 5.2] for a more detailed discussion, and also [1, 7].

The main focus of this work is to describe a space-time finite element method (FEM) on fully unstructured simplicial meshes to solve the minimization problem (4) subject to the solution of the backward heat equation (1). Such a space-time method has been studied for the forward heat equation in [8], and for other parabolic optimal control problems in [5, 6].

The remainder of this paper is structured as follows: In Section 2, we discuss the related optimal control problem. Its solution is obtained by the optimality system consisting of the (forward) heat equation, the adjoint heat equation, and the gradient equation. Based on the Banach–Nečas–Babuška theory [3], we establish unique solvability of the resulting coupled system, when eliminating the unknown initial datum. In Section 3, for the numerical solution of the inverse problem (1), we first consider the discrete optimal control problem, which is based on the space-time discretization of the forward problem. The solution is characterized by a discrete gradient equation, which turns out to be the Schur complement system of the discretized coupled variational formulation. First numerical results are reported in Section 4. These results show the potential of the space-time approach proposed. Finally, some conclusions are drawn in Section 5.

2 The related optimal control problem

In our case, the Hilbert spaces $\mathcal{X}$ and $\mathcal{Y}$ are specified as $\mathcal{X} = \mathcal{Y} = L^2(\Omega)$, and the image Sz of the operator $S : L^2(\Omega) \to L^2(\Omega)$ in the Tikhonov functional (4) is defined by the solution $u \in X := L^2(0,T;H_0^1(\Omega)) \cap H^1(0,T;H^{-1}(\Omega))$ of the forward heat conduction problem

$$\partial_t u - \Delta_x u = 0 \quad \text{in } Q, \quad u = 0 \quad \text{on } \Sigma, \quad u = z \quad \text{on } \Sigma_0, \tag{5}$$

and its evaluation on Σ_T, i.e., $(Sz)(x) = u(x,T)$, $x \in \Omega$. Here, the control $z \in L^2(\Omega)$ represents the initial data in (5). Rewriting the minimization of the functional (4) in terms of z, we obtain the *optimal control problem*

$$\mathcal{J}_\varrho(z) := \frac{1}{2}\,\|u(x,T) - u_T^\delta\|^2_{L^2(\Omega)} + \frac{\varrho}{2}\,\|z\|^2_{L^2(\Omega)} \to \min_{z \in L^2(\Omega)}, \tag{6}$$

where the *state* $u \in X$ is associated to the *control* z subject to (5).

To set up the necessary and sufficient optimality conditions for the optimal control z with associated state u, we introduce the adjoint equation

$$-\partial_t p - \Delta_x p = 0 \quad \text{in } Q, \quad p = 0 \quad \text{on } \Sigma, \quad p = u - u_T^\delta \quad \text{on } \Sigma_T. \tag{7}$$

It has a unique solution $p \in X$, the *adjoint state*. The adjoint equation can be derived by a formal Lagrangian technique as in [10]. If z is the optimal control with associated state $u \in X$, then a unique adjoint state $p \in X$ solving (7) exists such that the *gradient equation*

$$p + \varrho\, z = 0 \quad \text{on } \Sigma_0 \tag{8}$$

is satisfied. Using this equation, we can eliminate the unknown initial datum z in the state equation (5) to conclude

$$\partial_t u - \Delta_x u = 0 \quad \text{in } Q, \quad u = 0 \quad \text{on } \Sigma, \quad u = -\frac{1}{\varrho}\, p \quad \text{on } \Sigma_0 \tag{9}$$

for the optimal state u. The *reduced optimality system* (7),(9) is necessary and sufficient for optimality of u with associated adjoint state p. In what follows, we will describe a space-time finite element approximation of this system.

The space-time variational formulation of the heat equation in (9) (without initial condition) is to find $u \in X$ such that

$$b(u,v) := \int_0^T \int_\Omega \Big[\partial_t u(x,t) v(x,t) + \nabla_x u(x,t) \cdot \nabla_x v(x,t)\Big]\, dx\, dt = 0 \tag{10}$$

is satisfied for all $v \in Y := L^2(0,T;H_0^1(\Omega))$. The spaces X and Y are equipped with the norms

$$\|v\|_Y = \|\nabla_x v\|_{L^2(Q)} \quad \text{and} \quad \|u\|_X = \sqrt{\|\partial_t u\|^2_{Y^*} + \|u\|^2_Y} = \sqrt{\|w_u\|^2_Y + \|u\|^2_Y},$$

with $w_u \in Y$ being the unique solution of the variational problem

$$\int_0^T \int_\Omega \nabla_x w_u(x,t) \cdot \nabla_x v(x,t)\,\mathrm{d}x\,\mathrm{d}t = \int_0^T \int_\Omega \partial_t u(x,t)\, v(x,t)\,\mathrm{d}x\,\mathrm{d}t \quad \forall\, v \in Y.$$

We multiply the adjoint heat equation (7) by a test function $q \in X$, integrate over Q, and apply integration by parts both in space and time. Then we insert the terminal data $u(T) - u_T^\delta$ of p in the arising term $p(T)$, and substitute the term $p(0)$ by $-\rho z = -\rho u(0)$ in view of (8). In this way, we arrive at the weak form of the adjoint problem (7)

$$\begin{aligned} 0 &= \int_0^T \int_\Omega \Big[-\partial_t p(x,t)\, q(x,t) - \Delta_x p(x,t)\, q(x,t) \Big] \mathrm{d}x\,\mathrm{d}t \\ &= -\int_\Omega [u(x,T) - u_T^\delta(x)]\, q(x,T)\,\mathrm{d}x - \varrho \int_\Omega u(x,0)\, q(x,0)\,\mathrm{d}x \\ &\quad + \int_0^T \int_\Omega \Big[p(x,t)\, \partial_t q(x,t) + \nabla_x p(x,t) \cdot \nabla_x q(x,t) \Big] \mathrm{d}x\,\mathrm{d}t\,. \end{aligned}$$

We end up with the variational problem to find $(u,p) \in X \times Y$ such that

$$\mathcal{B}(u,p;v,q) = \langle u_T^\delta, q(T) \rangle_{L^2(\Omega)} \quad \forall\, (v,q) \in Y \times X, \tag{11}$$

where the bilinear form $\mathcal{B}(\cdot,\cdot;\cdot,\cdot)$ is given as

$$\mathcal{B}(u,p;v,q) := b(u,v) - b(q,p) + \langle u(T), q(T) \rangle_{L^2(\Omega)} + \varrho\, \langle u(0), q(0) \rangle_{L^2(\Omega)}\,.$$

We note that the bilinear form $b(\cdot,\cdot)$, as defined by (10), is bounded:

$$|b(u,v)| \le \sqrt{2}\, \|u\|_X \|v\|_Y \quad \forall\, u \in X, v \in Y.$$

Since X is continuously embedded in $C([0,T];L^2(\Omega))$, there is a positive constant μ such that $\|u(0)\|_{L^2(\Omega)} \le \mu\, \|u\|_X$ and $\|u(T)\|_{L^2(\Omega)} \le \mu\, \|u\|_X$,

$$\mu = \left(1 + \frac{1}{2} \left[\frac{c_F}{T} \right]^2 + \sqrt{ \frac{1}{4} \left[\frac{c_F}{T} \right]^4 + \left[\frac{c_F}{T} \right]^2 } \right)^{1/2},$$

where c_F is the constant in Friedrichs' inequality in $H_0^1(\Omega)$. With these ingredients, we are in the position to prove that the bilinear form $\mathcal{B}(\cdot,\cdot;\cdot,\cdot)$ is bounded, i.e., for all $(u,p), (q,v) \in X \times Y$, there holds

$$|\mathcal{B}(u,p;v,q)| \le 2\,(1+\varrho)\, \mu^2 \sqrt{\|u\|_X^2 + \|p\|_Y^2}\, \sqrt{\|q\|_X^2 + \|v\|_Y^2}\,.$$

Moreover, we can establish the following inf-sup stability condition, which can be proved similarly to [5, Lemma 3.2].

Lemma 1 *For simplicity, let us assume $\varrho \in (0,1]$. Then there holds the inf-sup stability condition*

$$\frac{3}{10}\varrho\sqrt{\|u\|_X^2+\|p\|_Y^2} \le \sup_{0\neq(v,q)\in Y\times X}\frac{\mathcal{B}(u,p;v,q)}{\sqrt{\|q\|_X^2+\|v\|_Y^2}} \quad \forall (u,p)\in X\times Y.$$

Moreover, for any $(0,0)\neq(v,q)\in Y\times X$, there exist $(\overline{u},\overline{p})\in X\times Y$ satisfying

$$\mathcal{B}(\overline{u},\overline{p};v,q)>0.$$

Now, using the Banach–Nečas–Babuška theorem (see, e.g., [3]), we can ensure well-posedness of the variational optimality problem (11) for any fixed positive regularization parameter ϱ.

3 Space-time finite element methods

For the space-time finite element discretization of the variational formulation (11), we first introduce conforming finite element spaces $X_h \subset X$ and $Y_h \subset Y$. In particular, we consider $X_h = Y_h$ spanned by piecewise linear continuous basis functions which are defined with respect to some admissible decomposition of the space-time domain Q into shape regular simplicial finite elements. In addition, we will use the subspace $Y_{0,h} \subset Y_h$ of basis functions with zero initial values. Moreover, $Z_h \subset L^2(\Omega)$ is a finite element space to discretize the control z. The space-time finite element discretization of the forward problem (5) reads to find $u_h \in X_h$ such that

$$b(u_h,v_h)=0 \quad \forall v_h\in Y_{0,h}, \quad \langle u_h - z_h, v_h\rangle_{L^2(\Sigma_0)} = 0 \quad \forall v_h \in Y_h\backslash Y_{0,h}. \tag{12}$$

When denoting the degrees of freedom of u_h at Σ_0, at Σ_T, and in Q by $\underline{u}_0$, $\underline{u}_T$, and $\underline{u}_I$, respectively, the variational formulation (12) is equivalent to the linear system

$$\begin{pmatrix} M_{00} & & \\ K_{0I} & K_{II} & K_{TI} \\ & K_{IT} & K_{TT}\end{pmatrix}\begin{pmatrix}\underline{u}_0\\ \underline{u}_I\\ \underline{u}_T\end{pmatrix} = \begin{pmatrix} M_h^\top \underline{z}\\ \underline{0}\\ \underline{0}\end{pmatrix},$$

where the block entries of the stiffness matrix K_h and the mass matrices M_{00} and M_h are defined accordingly. After eliminating $\underline{u}_0$, the resulting system corresponds to the space-time finite element approach as considered in [8]. In particular, we can compute $\underline{u}_T = A_h\underline{z}$ to determine $u_h(T)$ in dependency on the initial datum z_h, where

$$A_h = \left(K_{TT} - K_{IT}K_{II}^{-1}K_{TI}\right)^{-1} K_{IT}K_{II}^{-1}K_{0I}M_{00}^{-1}M_h^\top = \widetilde{A}_h M_h^\top .$$

Instead of the cost functional (6), we now consider the discrete cost functional

$$\begin{aligned}\mathcal{J}_{\varrho,h}(z_h) &= \frac{1}{2}\,\|u_h(x,T) - u_T^\delta\|^2_{L^2(\Omega)} + \frac{\varrho}{2}\,\|z_h\|^2_{L^2(\Omega)} \\ &= \frac{1}{2}\,(A_h^\top M_{TT} A_h \underline{z}, \underline{z}) - (A_h^\top \underline{f}, \underline{z}) + \frac{1}{2}\,\|u_T^\delta\|^2_{L^2(\Omega)} + \frac{\varrho}{2}\,(\overline{M}_h \underline{z}, \underline{z}),\end{aligned}$$

whose minimizer is given as the solution of the linear system

$$A_h^\top (M_{TT} A_h \underline{z} - \underline{f}) + \varrho\, \overline{M}_h \underline{z} = \underline{0}. \tag{13}$$

Note that M_{TT} is the mass matrix formed by the basis functions of X_h at Σ_T, $\overline{M}_h$ is the mass matrix related to the control space Z_h, and $\underline{f}$ is the load vector of the target u_T^δ tested with basis functions from X_h at Σ_T. When inserting $\underline{u}_T = A_h \underline{z}$ and introducing $\underline{p}_0 := \widetilde{A}_h^\top (M_{TT} \underline{u}_T - \underline{f})$, $\underline{p}_T := (K_{TT} - K_{IT} K_{II}^{-1} K_{TI})^{-\top} (M_{TT} \underline{u}_T - \underline{f})$, $\underline{p}_I := -K_{II}^{-\top} K_{IT}^\top \underline{p}_T$, this finally results in the linear system to be solved:

$$\begin{pmatrix} & & & & -M_{00} & -K_{0I}^\top & \\ & & & & & -K_{II}^\top & -K_{IT}^\top \\ & & M_{TT} & & & -K_{TI}^\top & -K_{TT}^\top \\ & & & \varrho \overline{M}_h & M_h & & \\ M_{00} & & & -M_h^\top & & & \\ K_{0I} & K_{II} & K_{TI} & & & & \\ & K_{IT} & K_{TT} & & & & \end{pmatrix} \begin{pmatrix} \underline{u}_0 \\ \underline{u}_I \\ \underline{u}_T \\ \underline{z} \\ \underline{p}_0 \\ \underline{p}_I \\ \underline{p}_T \end{pmatrix} = \begin{pmatrix} \underline{0} \\ \underline{0} \\ \underline{f} \\ \underline{0} \\ \underline{0} \\ \underline{0} \\ \underline{0} \end{pmatrix}. \tag{14}$$

In the particular case, when $Z_h = Y_{h|\Sigma_0} \subset H_0^1(\Omega)$ is the space of piecewise linear basis functions as well, the mass matrices $M_{00} = \overline{M}_h = M_h$ coincide, and therefore we can eliminate $\underline{z} = \underline{u}_0$ and $\underline{p}_0 = -\varrho \underline{z} = -\varrho \underline{u}_0$ to obtain

$$\begin{pmatrix} \varrho M_{00} & & & -K_{0I}^\top & \\ & & & -K_{II}^\top & -K_{IT}^\top \\ & & M_{TT} & -K_{TI}^\top & -K_{TT}^\top \\ K_{0I} & K_{II} & K_{TI} & & \\ & K_{IT} & K_{TT} & & \end{pmatrix} \begin{pmatrix} \underline{u}_0 \\ \underline{u}_I \\ \underline{u}_T \\ \underline{p}_I \\ \underline{p}_T \end{pmatrix} = \begin{pmatrix} \underline{0} \\ \underline{0} \\ \underline{f} \\ \underline{0} \\ \underline{0} \end{pmatrix}. \tag{15}$$

Note that (15) is nothing but the Galerkin discretization of the variational formulation (11) when using $X_h \subset X$ and $Y_{0,h} \subset Y$ as finite element ansatz and test spaces. Obviously, the linear system (13) and, therefore, (15) are uniquely solvable.

In practice, the noise level $\delta \geq 0$ is usually given by the measurement environment, and one has to choose suitable discretization and regularization parameters h and ρ. This is well investigated for linear inverse problems; see, e.g., the classical book by Tikhonov and Arsenin [9] and the more recent publications [2, 4]. In our numerical experiments presented in the next section, we only play with the parameters δ and h for a fixed small ϱ.

4 Numerical results

We take $\Omega = (0,1)$ and $T = 1$, i.e., $Q = (0,1)^2$, and consider the manufactured observation data $u_T^\delta(x) := e^{-\pi^2}\sin(\pi x) + \delta\sin(10\pi x)$ with some noise represented by the second term; see exact and noisy data with $\delta \in \{0, 10^{-5}, 5\cdot 10^{-6}, 2.5\cdot 10^{-6}\}$ in Fig. 1. To study the convergence of the space-time finite element solution to

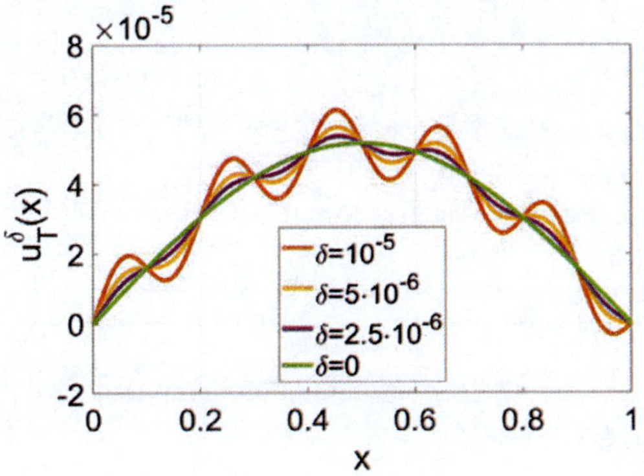

Fig. 1: Comparison of the exact ($\delta = 0$) and noisy ($\delta > 0$) observation data.

the exact initial datum $\sin(\pi x)$, we use the target $u_T(x) = e^{-\pi^2}\sin(\pi x)$ without any noise. The reconstructed initial data with respect to a varying mesh size are illustrated in the left plot of Fig. 2, where $\varrho = 10^{-14}$. We clearly see the convergence of the approximations to the exact initial datum with respect to the mesh refinement. The right plot of Fig. 2 shows the reconstructed initial approximation with different noise levels δ. For a decreasing δ, we observe an improved reconstruction.

5 Conclusions

We have applied the space-time FEM from [8] to the numerical solution of the classical inverse heat conduction problem to determine the initial datum from measured observation data at some time horizon T. The numerical results show the potential of this approach for more interesting inverse problems. The space-time FEM is very much suited for designing smart adaptive algorithms along the line proposed in [4] determining the optimal choice of ϱ and h for a given noise level δ in a multilevel (nested iteration) setting.

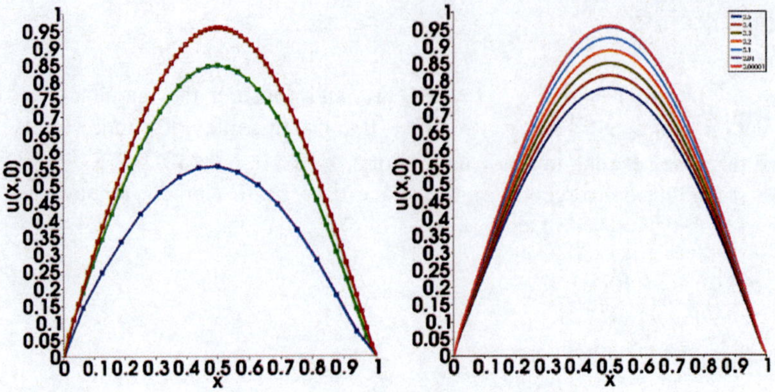

Fig. 2: Convergence of the reconstructed initial data with respect to the mesh refinement $h \in \{1/16, 1/32, 1/64\}$, $\delta = 0$, $\varrho = 10^{-14}$ (left), and convergence with respect to the noise level $\delta \in \{0.5, 0.4, 0.3, 0.2, 10^{-1}, 10^{-3}, 10^{-5}\}$, $h = 1/64$, $\varrho = 10^{-14}$ (right).

References

1. D.-H. Chen, B. Hofmann and J. Zou. Regularization and convergence for ill-posed backward evolution equations in Banach spaces. *J. Differential Equations* **265**, 3533–3566 (2018).
2. H.W. Engl, M. Hanke and G. Neubauer. *Regularization of Inverse Problems*. Springer, Netherlands (1996).
3. A. Ern and J.-L. Guermond. *Theory and Practice of Finite Elements*. Springer, New York (2004).
4. A. Griesbaum, B. Kaltenbacher and B. Vexler. Efficient computation of the Tikhonov regularization parameter by goal-oriented adaptive discretization. *Inverse Problems* **24**, 025025 (20pp.) (2008).
5. U. Langer, O. Steinbach, F. Tröltzsch and H. Yang. Space-time finite element discretization of parabolic optimal control problems with energy regularization. *SIAM J. Numer. Anal.* **59**, 675–695 (2021).
6. U. Langer, O. Steinbach, F. Tröltzsch and H. Yang. Unstructured space-time finite element methods for optimal control of parabolic equations. *SIAM J. Sci. Comput.* **43**, A744–A771 (2021).
7. D. Leykekhman, B. Vexler and D. Walter. Numerical analysis of sparse initial data identification for parabolic problems. *ESAIM Math. Model. Numer. Anal.* **54**, 1139–1180 (2020).
8. O. Steinbach. Space-time finite element methods for parabolic problems. *Comput. Methods Appl. Math.* **15**(4), 551–566 (2015).
9. A.N. Tikhonov and V.Y. Arsenin. Solution of Ill-posed Problems. *Winston & Sons* (1977).
10. F. Tröltzsch. *Optimal control of partial differential equations: Theory, methods and applications*. Volume 112 of Graduate Studies in Mathematics. American Mathematical Society, Providence, Rhode Island (2010).

Numerical Results for an Unconditionally Stable Space-Time Finite Element Method for the Wave Equation

Richard Löscher, Olaf Steinbach, and Marco Zank

1 Introduction

As a model problem, we consider the Dirichlet boundary value problem for the wave equation,

$$\left.\begin{aligned} \partial_{tt}u(x,t) - \Delta_x u(x,t) &= f(x,t) && \text{for } (x,t) \in Q := \Omega \times (0,T), \\ u(x,t) &= 0 && \text{for } (x,t) \in \Sigma := \partial\Omega \times [0,T], \\ u(x,0) = \partial_t u(x,t)_{|t=0} &= 0 && \text{for } x \in \Omega, \end{aligned}\right\} \tag{1}$$

where $\Omega \subset \mathbb{R}^d$, $d = 1, 2, 3$, is some bounded Lipschitz domain, $T > 0$ is a finite time horizon, and f is some given source. For simplicity, we only consider homogeneous boundary and initial conditions, but inhomogeneous data or other types of boundary conditions can be handled as well. To compute an approximate solution of the wave equation (1), different numerical methods are available. Classical approaches are time-stepping schemes together with finite element methods in space, see [1] for an overview. An alternative is to discretize the time-dependent problem without separating the temporal and spatial variables. However, on the one hand, most space-time approaches are based on discontinuous Galerkin methods, see, e.g., [3, 6]. On the other hand, conforming tensor-product space-time discretizations with piecewise polynomial, continuous ansatz and test functions are of Petrov–Galerkin type, see,

Richard Löscher
Fachbereich Mathematik, TU Darmstadt, Dolivostraße 15, 64293 Darmstadt, Germany,
e-mail: loescher@mathematik.tu-darmstadt.de

Olaf Steinbach
Institut für Angewandte Mathematik, TU Graz, Steyrergasse 30, 8010 Graz, Austria,
e-mail: o.steinbach@tugraz.at

Marco Zank
Fakultät für Mathematik, Universität Wien, Oskar–Morgenstern–Platz 1, 1090 Wien, Austria,
e-mail: marco.zank@univie.ac.at

S. C. Brenner et al. (eds.), *Domain Decomposition Methods in Science and Engineering XXVI*, Lecture Notes in Computational Science and Engineering 145,
https://doi.org/10.1007/978-3-030-95025-5_68

e.g., [7, 8, 12], where a stabilization is needed to avoid a CFL condition, i.e., a relation between the time mesh size and the spatial mesh size.

In this work, we use a modified Hilbert transformation to introduce a new space-time variational formulation of the wave equation (1), where ansatz and test spaces are equal. Conforming discretizations of this new variational setting, using polynomial, globally continuous ansatz and test functions, lead to space-time Galerkin–Bubnov finite element methods, which are unconditionally stable and provide optimal convergence rates in $\|\cdot\|_{L^2(Q)}$ and $|\cdot|_{H^1(Q)}$, respectively. The rest of the paper is organized as follows: In Section 2, a modified Hilbert transformation and its main properties are given. Section 3 states the space-time variational setting for the wave equation and introduces the new space-time Galerkin–Bubnov finite element method. Numerical examples for a one- and a two-dimensional spatial domain are presented in Section 4. Finally, we draw some conclusions in Section 5.

2 A modified Hilbert transformation

In this section, we summarize the definition and some of the most important properties of the modified Hilbert transformation $\mathcal{H}_T$ as introduced in [8], see also [9, 11]. Since the modified Hilbert transformation covers the dependency in time only, in this section, we consider functions $u(t)$ for $t \in (0,T)$, where a generalization to functions in (x,t) is straightforward.

For $u \in L^2(0,T)$, we consider the Fourier series expansion

$$u(t) = \sum_{k=0}^{\infty} u_k \sin\left(\left(\frac{\pi}{2} + k\pi\right)\frac{t}{T}\right), \quad u_k := \frac{2}{T}\int_0^T u(t)\,\sin\left(\left(\frac{\pi}{2} + k\pi\right)\frac{t}{T}\right)\,\mathrm{d}t,$$

and we define the modified Hilbert transformation $\mathcal{H}_T$ as

$$(\mathcal{H}_T u)(t) = \sum_{k=0}^{\infty} u_k \cos\left(\left(\frac{\pi}{2} + k\pi\right)\frac{t}{T}\right), \quad t \in (0,T). \tag{2}$$

By interpolation, we introduce $H^s_{0,}(0,T) := [H^1_{0,}(0,T), L^2(0,T)]_s$ for $s \in [0,1]$, where the space $H^1_{0,}(0,T)$ covers the initial condition $u(0) = 0$ for $u \in H^1(0,T)$. Analogously, we define $H^s_{,0}(0,T)$ for $s \in [0,1]$. With these notations, the mapping $\mathcal{H}_T \colon H^s_{0,}(0,T) \to H^s_{,0}(0,T)$ is an isomorphism for $s \in [0,1]$, where the inverse is the $L^2(0,T)$ adjoint, i.e., $\langle \mathcal{H}_T u, w\rangle_{L^2(0,T)} = \langle u, \mathcal{H}_T^{-1} w\rangle_{L^2(0,T)}$ for all $u, w \in L^2(0,T)$. In addition, the relations

$$\begin{aligned} \langle v, \mathcal{H}_T v\rangle_{L^2(0,T)} &> 0 && \text{for } 0 \neq v \in H^s_{0,}(0,T), 0 < s \leq 1, \\ \langle \partial_t \mathcal{H}_T u, v\rangle_{L^2(0,T)} &= -\langle \mathcal{H}_T^{-1} \partial_t u, v\rangle_{L^2(0,T)} && \text{for } u \in H^1_{0,}(0,T), v \in L^2(0,T) \end{aligned}$$

hold true. For the proofs of these aforementioned properties, we refer to [8, 9, 11]. Furthermore, the modified Hilbert transformation (2) allows a closed representation [8, Lemma 2.8] as Cauchy principal value integral, i.e., for $u \in L^2(0,T)$,

$$(\mathcal{H}_T u)(t) = \text{v.p.} \int_0^T \frac{1}{2T} \left(\frac{1}{\sin \frac{\pi(s+t)}{2T}} + \frac{1}{\sin \frac{\pi(s-t)}{2T}} \right) u(s)\, \mathrm{d}s, \quad t \in (0,T).$$

This representation can be used for an efficient realization, also using low-rank approximations of related discrete matrix representations, see [9] for a more detailed discussion.

3 Space-time variational formulations

A possible space-time variational formulation for the Dirichlet boundary value problem (1) is to find $u \in H^{1,1}_{0;0,}(Q) := L^2(0,T;H^1_0(\Omega)) \cap H^1_{0,}(0,T;L^2(\Omega))$ such that

$$-\langle \partial_t u, \partial_t v\rangle_{L^2(Q)} + \langle \nabla_x u, \nabla_x v\rangle_{L^2(Q)} = \langle f, v\rangle_{L^2(Q)} \tag{3}$$

is satisfied for all $v \in H^{1,1}_{0;,0}(Q) := L^2(0,T;H^1_0(\Omega)) \cap H^1_{,0}(0,T;L^2(\Omega))$. Note that the space $H^1_{0,}(0,T;L^2(\Omega))$ covers zero initial conditions, while the space $H^1_{,0}(0,T;L^2(\Omega))$ involves zero terminal conditions at $t = T$. For $f \in L^2(Q)$, there exists a unique solution u of (3), satisfying the stability estimate

$$\|u\|_{H^{1,1}_{0;0,}(Q)} := |u|_{H^1(Q)} := \sqrt{\|\partial_t u\|^2_{L^2(Q)} + \|\nabla_x u\|^2_{L^2(Q)}} \leq \frac{1}{\sqrt{2}} T \,\|f\|_{L^2(Q)},$$

see [4, 8, 12]. Note that the solution operator $\mathcal{L}\colon L^2(Q) \to H^{1,1}_{0;0,}(Q)$, $\mathcal{L}f := u$, is not an isomorphism, i.e., $\mathcal{L}$ is not surjective, see [10] for more details.

A direct numerical discretization of the variational formulation (3) would result in a Galerkin–Petrov scheme with different ansatz and test spaces, being zero at the initial and the terminal time, respectively. Hence, introducing some bijective operator $A\colon H^{1,1}_{0;0,}(Q) \to H^{1,1}_{0;,0}(Q)$, we can express the test function v in (3) as $v = Aw$ for $w \in H^{1,1}_{0;0,}(Q)$ to end up with a Galerkin–Bubnov scheme. While the time reversal map $\kappa_T w(x,t) := w(x, T-t)$ as used, e.g., in [2], is rather of theoretical interest, in the case of a tensor-product space-time finite element discretization, one may use the transformation $Aw_h(x,t) := w_h(x,T) - w_h(x,t)$, see [8]. However, the resulting numerical scheme is only stable when a CFL condition is satisfied, e.g., $h_t < h_x/\sqrt{d}$ when using piecewise linear basis functions and a tensor-product structure also in space. Although it is possible to derive an unconditionally stable scheme by using some stabilization approach, see [7, 12], our particular interest is in using an appropriate transformation A to conclude an unconditionally stable scheme without any further stabilization. A possible choice is the use of the modified Hilbert transformation $\mathcal{H}_T$ as introduced in Section 2. So, with the properties of $\mathcal{H}_T$, given in Section 2, we conclude that

$$-\langle \partial_t u, \partial_t \mathcal{H}_T w\rangle_{L^2(Q)} = \langle \partial_t u, \mathcal{H}_T^{-1}\partial_t w\rangle_{L^2(Q)} = \langle \mathcal{H}_T \partial_t u, \partial_t w\rangle_{L^2(Q)}$$

for all $u, w \in H^{1,1}_{0;0,}(Q)$, which leads to the variational formulation to find $u \in H^{1,1}_{0;0,}(Q)$ such that

$$\langle \mathcal{H}_T \partial_t u, \partial_t w\rangle_{L^2(Q)} + \langle \nabla_x u, \nabla_x \mathcal{H}_T w\rangle_{L^2(Q)} = \langle f, \mathcal{H}_T w\rangle_{L^2(Q)} \tag{4}$$

is satisfied for all $w \in H^{1,1}_{0;0,}(Q)$. Since the mapping $\mathcal{H}_T\colon H^{1,1}_{0;0,}(Q) \to H^{1,1}_{0;,0}(Q)$ is an isomorphism, unique solvability of the new variational formulation (4) follows from the unique solvability of the variational formulation (3).

Let $V_h = \operatorname{span}\{\phi_i\}_{i=1}^M \subset H^{1,1}_{0;0,}(Q)$ be some conforming space-time finite element space. The Galerkin–Bubnov formulation of the variational formulation (4) is to find $u_h \in V_h$ such that

$$\langle \mathcal{H}_T \partial_t u_h, \partial_t w_h\rangle_{L^2(Q)} + \langle \nabla_x u_h, \nabla_x \mathcal{H}_T w_h\rangle_{L^2(Q)} = \langle f, \mathcal{H}_T w_h\rangle_{L^2(Q)} \tag{5}$$

is satisfied for all $w_h \in V_h$. Note that for any conforming space-time finite element space $V_h \subset H^{1,1}_{0;0,}(Q)$, the related bilinear form in (5) is positive definite, since both summands are discretizations of second-order differential operators, which lead, together with the properties of $\mathcal{H}_T$, to two positive definite bilinear forms. Further details on the numerical analysis of this new Galerkin–Bubnov variational formulation (5) are far beyond the scope of this contribution, we refer to [5]. The discrete variational formulation (5) corresponds to the linear system $K_h \underline{u} = \underline{f}$ with the stiffness matrix $K_h = A_h + B_h$, and

$$A_h[i,j] = \int_0^T \int_\Omega \mathcal{H}_T \partial_t \phi_j(x,t)\, \partial_t \phi_i(x,t)\,\mathrm{d}x\,\mathrm{d}t,$$

$$B_h[i,j] = \int_0^T \int_\Omega \nabla_x \phi_j(x,t) \cdot \nabla_x \mathcal{H}_T \phi_i(x,t)\,\mathrm{d}x\,\mathrm{d}t$$

for $i, j = 1, \ldots, M$. Since the realization of the modified Hilbert transformation $\mathcal{H}_T$ is much easier for solely time-dependent functions, see [9, 11], here we choose as a special case a tensor-product ansatz. For this purpose, let the bounded Lipschitz domain $\Omega \subset \mathbb{R}^d$ be an interval $\Omega = (0, L)$ for $d = 1$, polygonal for $d = 2$, or polyhedral for $d = 3$. We consider admissible decompositions

$$\overline{Q} = \overline{\Omega} \times [0,T] = \bigcup_{i=1}^{N_x} \overline{\omega_i} \times \bigcup_{\ell=1}^{N_t} [t_{\ell-1}, t_\ell]$$

with $N := N_x \cdot N_t$ space-time elements, where the time intervals $(t_{\ell-1}, t_\ell)$ with mesh sizes $h_{t,\ell} = t_\ell - t_{\ell-1}$ are defined via the decomposition

$$0 = t_0 < t_1 < t_2 < \cdots < t_{N_t-1} < t_{N_t} = T$$

of the time interval $(0, T)$. The maximal and the minimal time mesh sizes are denoted by $h_t := h_{t,\max} := \max_\ell h_{t,\ell}$, and $h_{t,\min} := \min_\ell h_{t,\ell}$, respectively. For the spatial domain Ω, we consider a shape-regular sequence $(\mathcal{T}_\eta)_{\eta\in\mathbb{N}}$ of admissible decompositions $\mathcal{T}_\eta := \{\omega_i \subset \mathbb{R}^d : i = 1, \ldots, N_x\}$ of Ω into finite elements $\omega_i \subset \mathbb{R}^d$ with mesh sizes $h_{x,i}$ and the maximal mesh size $h_x := \max_i h_{x,i}$. The spatial elements ω_i are intervals for $d = 1$, triangles for $d = 2$, and tetrahedra for $d = 3$. Next, we introduce the finite element space $Q^1_{h,0}(Q) := S^1_{h_x,0}(\Omega) \otimes S^1_{h_t,0,}(0, T)$ of piecewise multilinear, continuous functions, i.e.,

$$\begin{aligned} S^1_{h_x,0}(\Omega) &:= S^1_{h_x}(\Omega) \cap H^1_0(\Omega) = \operatorname{span}\{\psi^1_j\}_{j=1}^{M_x}, \\ S^1_{h_t,0,}(0, T) &:= S^1_{h_t}(0, T) \cap H^1_{0,}(0, T) = \operatorname{span}\{\varphi^1_\ell\}_{\ell=1}^{N_t}, \end{aligned}$$

where ψ^1_j, $j = 1, \ldots, M_x$, are the spatial nodal basis functions, and φ^1_ℓ, $\ell = 1, \ldots, N_t$, are the temporal nodal basis functions. In fact, $S^1_{h_t}(0, T)$ is the space of piecewise linear, continuous functions on intervals, and $S^1_{h_x}(\Omega)$ is the space of piecewise linear, continuous functions on intervals ($d = 1$), triangles ($d = 2$), and tetrahedra ($d = 3$).

Choosing $V_h = Q^1_{h,0}(Q)$ in (5) leads to the space-time Galerkin–Bubnov variational formulation to find $u_h \in Q^1_{h,0}(Q)$ such that

$$\langle \mathcal{H}_T \partial_t u_h, \partial_t w_h \rangle_{L^2(Q)} + \langle \nabla_x u_h, \nabla_x \mathcal{H}_T w_h \rangle_{L^2(Q)} = \langle Q^0_h f, \mathcal{H}_T w_h \rangle_{L^2(Q)} \tag{6}$$

for all $w_h \in Q^1_{h,0}(Q)$. Here, for an easier implementation, we approximate the right-hand side $f \in L^2(Q)$ by

$$f \approx Q^0_h f \in S^0_{h_x}(\Omega) \otimes S^0_{h_t}(0, T), \tag{7}$$

where $Q^0_h \colon L^2(Q) \to S^0_{h_x}(\Omega) \otimes S^0_{h_t}(0, T)$ is the $L^2(Q)$ projection on the space $S^0_{h_x}(\Omega) \otimes S^0_{h_t}(0, T)$ of piecewise constant functions. The discrete variational formulation (6) is equivalent to the global linear system

$$K_h \underline{u} = \underline{\widetilde{f}} \tag{8}$$

with the system matrix

$$K_h = A^{\mathcal{H}_T}_{h_t} \otimes M_{h_x} + M^{\mathcal{H}_T}_{h_t} \otimes A_{h_x} \in \mathbb{R}^{N_t \cdot M_x \times N_t \cdot M_x},$$

where $M_{h_x} \in \mathbb{R}^{M_x \times M_x}$ and $A_{h_x} \in \mathbb{R}^{M_x \times M_x}$ denote spatial mass and stiffness matrices given by

$$M_{h_x}[i, j] = \langle \psi^1_j, \psi^1_i \rangle_{L^2(\Omega)}, \quad A_{h_x}[i, j] = \langle \nabla_x \psi^1_j, \nabla_x \psi^1_i \rangle_{L^2(\Omega)}, \quad i, j = 1, \ldots, M_x,$$

and $M^{\mathcal{H}_T}_{h_t} \in \mathbb{R}^{N_t \times N_t}$ and $A^{\mathcal{H}_T}_{h_t} \in \mathbb{R}^{N_t \times N_t}$ are defined by

$$M^{\mathcal{H}_T}_{h_t}[\ell, k] := \langle \varphi^1_k, \mathcal{H}_T \varphi^1_\ell \rangle_{L^2(0,T)}, \quad A^{\mathcal{H}_T}_{h_t}[\ell, k] := \langle \mathcal{H}_T \partial_t \varphi^1_k, \partial_t \varphi^1_\ell \rangle_{L^2(0,T)}$$

for $\ell, k = 1, \ldots, N_t$. The matrices $M_{h_t}^{\mathcal{H}_T}$, $A_{h_t}^{\mathcal{H}_T}$ are nonsymmetric, but positive definite, which follows from the properties of $\mathcal{H}_T$, given in Section 2. Additionally, the matrices M_{h_x}, A_{h_x} are positive definite. Thus, standard properties of the Kronecker product yield that the system matrix K_h is also positive definite. Hence, the global linear system (8) is uniquely solvable.

4 Numerical results

In this section, numerical examples for the Galerkin–Bubnov finite element method (6) for a one- and a two-dimensional spatial domain are given. For both cases, the number of degrees of freedom is given by dof $= N_t \cdot M_x$. The assembling of the matrices $A_{h_t}^{\mathcal{H}_T}$, $M_{h_t}^{\mathcal{H}_T}$ is done as proposed in [11, Subsection 2.2]. Further, to accelerate the computations, data-sparse approximations as known from boundary element methods, e.g., hierarchical matrices, can be used, see [9]. The integrals for computing the projection $Q_h^0 f$ in (7) are calculated by using high-order quadrature rules. The global linear system (8) is solved by a direct solver.

For the first numerical example, we consider the one-dimensional spatial domain $\Omega := (0, 1)$ with the terminal time $T = 10$, i.e., the rectangular space-time domain

$$Q := \Omega \times (0, T) := (0, 1) \times (0, 10). \tag{9}$$

As an exact solution, we choose

$$u_1(x, t) = t^2 \sin(10\pi x) \sin(t\, x), \quad (x, t) \in Q. \tag{10}$$

The spatial domain $\Omega = (0, 1)$ is decomposed into nonuniform elements with the vertices

$$x_0 = 0, \quad x_1 = 1/4, \quad x_2 = 1, \tag{11}$$

whereas the temporal domain $(0, T) = (0, 10)$ is decomposed into nonuniform elements with the vertices

$$t_0 = 0, \quad t_1 = 5/4, \quad t_2 = 5/2, \quad t_3 = 10 = T, \tag{12}$$

see Fig. 1 for the resulting space-time mesh. We apply a uniform refinement strategy for the meshes (11), (12). The numerical results for the smooth solution u_1 in (10) are given in Table 1, where we observe unconditional stability, quadratic convergence in $\|\cdot\|_{L^2(Q)}$, and linear convergence in $|\cdot|_{H^1(Q)}$.

For the second numerical example, the two-dimensional spatial Γ-shaped domain

$$\Omega := (-1, 1)^2 \setminus ([0, 1] \times [-1, 0]) \subset \mathbb{R}^2 \tag{13}$$

and the terminal time $T = 2$ are considered for the solution

$$u_2(x_1, x_2, t) = \sin(\pi x_1) \sin(\pi x_2) (\sin(t x_1 x_2))^2, \quad (x_1, x_2, t) \in Q = \Omega \times (0, T). \tag{14}$$

Table 1: Numerical results of the Galerkin–Bubnov finite element discretization (6) for the space-time cylinder (9) for the function u_1 in (10) for a uniform refinement strategy.

dof	$h_{x,\max}$	$h_{x,\min}$	$h_{t,\max}$	$h_{t,\min}$	$\|u_1 - u_{1,h}\|_{L^2(Q)}$	eoc	$\|u_1 - u_{1,h}\|_{H^1(Q)}$	eoc
3	0.7500	0.2500	7.5000	1.2500	5.0e+02	-	3.2e+03	-
18	0.3750	0.1250	3.7500	0.6250	4.2e+02	0.3	2.7e+03	0.2
84	0.1875	0.0625	1.8750	0.3125	3.2e+02	0.4	2.5e+03	0.1
360	0.0938	0.0312	0.9375	0.1562	8.4e+01	1.9	2.1e+03	0.2
1488	0.0469	0.0156	0.4688	0.0781	2.6e+01	1.7	1.0e+03	1.0
6048	0.0234	0.0078	0.2344	0.0391	7.2e+00	1.9	5.0e+02	1.1
24384	0.0117	0.0039	0.1172	0.0195	1.8e+00	2.0	2.5e+02	1.0
97920	0.0059	0.0020	0.0586	0.0098	4.7e-01	2.0	1.2e+02	1.0
392448	0.0029	0.0010	0.0293	0.0049	1.2e-01	2.0	6.2e+01	1.0
1571328	0.0015	0.0005	0.0146	0.0024	2.9e-02	2.0	3.1e+01	1.0

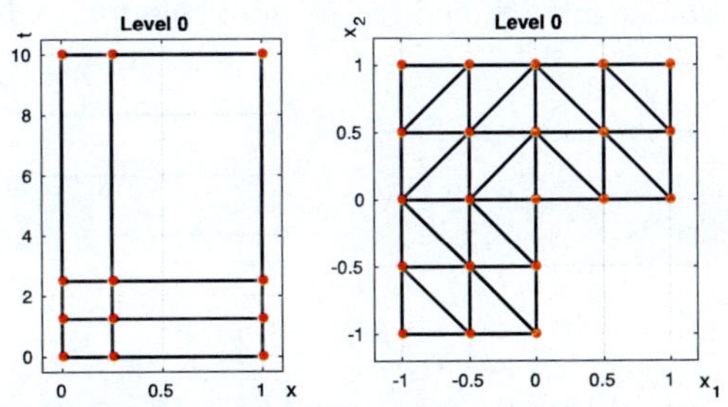

Fig. 1 Starting meshes for the one-dimensional spatial domain (left) and the two-dimensional spatial domain (right).

Table 2: Numerical results of the Galerkin–Bubnov finite element discretization (6) for the Γ-shape (13) and $T = 2$ for the function u_2 in (14) for a uniform refinement strategy.

dof	h_x	$h_{t,\max}$	$h_{t,\min}$	$\|u_2 - u_{2,h}\|_{L^2(Q)}$	eoc	$\|u_2 - u_{2,h}\|_{H^1(Q)}$	eoc
20	0.3536	1.5000	0.1250	1.756e-01	-	1.331e+00	-
264	0.1768	0.7500	0.0625	6.370e-02	1.5	6.882e-01	1.0
2576	0.0884	0.3750	0.0312	1.903e-02	1.7	3.439e-01	1.0
22560	0.0442	0.1875	0.0156	5.206e-03	1.9	1.730e-01	1.0
188480	0.0221	0.0938	0.0078	1.306e-03	2.0	8.555e-02	1.0
1540224	0.0110	0.0469	0.0039	3.284e-04	2.0	4.268e-02	1.0

The spatial domain Ω is decomposed into uniform triangles with uniform mesh size h_x as given in Fig. 1 for the first level. The temporal domain $(0, 2) = (0, T)$ is decomposed into nonuniform elements with the vertices

$$t_0 = 0, \quad t_1 = 1/8, \quad t_2 = 1/4, \quad t_3 = 1/2, \quad t_4 = 2 = T. \tag{15}$$

When a uniform refinement strategy is applied for the temporal mesh (15) and for the spatial mesh, the numerical results for the smooth solution u_2 are given in Table 2, where unconditional stability is observed and the convergence rates in $\|\cdot\|_{L^2(Q)}$ and $|\cdot|_{H^1(Q)}$ are optimal.

5 Conclusions

In this work, we introduced new conforming space-time Galerkin–Bubnov methods for the wave equation. These methods are based on a space-time variational formulation, where ansatz and test spaces are equal, using also integration by parts with respect to the time variable and the modified Hilbert transformation $\mathcal{H}_T$. As discretizations of this variational setting, we considered a conforming tensor-product approach with piecewise multilinear, continuous basis functions. However, a generalization to piecewise polynomials of higher-order degree is straightforward. We gave numerical examples, where the unconditional stability, i.e., no CFL condition is required, and optimal convergence rates in space-time norms were illustrated. For a more detailed stability and error analysis, we refer to our ongoing work [5]. Other topics include the realization for arbitrary space-time meshes, a posteriori error estimates and adaptivity, and the parallel solution including domain decomposition methods.

References

1. W. Bangerth, M. Geiger and R. Rannacher. Adaptive Galerkin finite element methods for the wave equation. *Comput. Methods Appl. Math.* **10**(1), 3–48 (2010).
2. M. Costabel. Boundary integral operators for the heat equation. *Integral Equations Operator Theory* **13**(4), 498–552 (1990).
3. W. Dörfler, S. Findeisen and C. Wieners. Space-time discontinuous Galerkin discretizations for linear first-order hyperbolic evolution systems. *Comput. Methods Appl. Math.* **16**(3), 409–428 (2016).
4. O.A. Ladyzhenskaya. *The boundary value problems of mathematical physics*. Applied Mathematical Sciences, vol. 49, Springer-Verlag, New York (1985).
5. R. Löscher, O. Steinbach and M. Zank. An unconditionally stable space-time finite element method for the wave equation (2022). In preparation.
6. A. Moiola and I. Perugia. A space-time Trefftz discontinuous Galerkin method for the acoustic wave equation in first-order formulation. *Numer. Math.* **138**(2), 389–435 (2018).
7. O. Steinbach and M. Zank. A stabilized space-time finite element method for the wave equation. In: *Advanced Finite Element Methods with Applications. Selected papers from the 30th Chemnitz FEM Symposium 2017*, Lect. Notes Comput. Sci. Eng., vol. 128, pp. 341–370, Springer, Cham (2019).
8. O. Steinbach and M. Zank. Coercive space-time finite element methods for initial boundary value problems. *Electron. Trans. Numer. Anal.* **52**, 154–194 (2020).
9. O. Steinbach and M. Zank. A note on the efficient evaluation of a modified Hilbert transformation. *J. Numer. Math.* **29**(1), 47–61 (2021).
10. O. Steinbach and M. Zank. A generalized inf-sup stable variational formulation for the wave equation. *J. Math. Anal. Appl.* **505**(1), Paper No. 125457, 24 (2022).
11. M. Zank. An exact realization of a modified Hilbert transformation for space-time methods for parabolic evolution equations. *Comput. Methods Appl. Math.* **21**(2), 479–496 (2021).
12. A.A. Zlotnik. Convergence rate estimates of finite-element methods for second-order hyperbolic equations. In: *Numerical methods and applications*, pp. 155–220, CRC, Boca Raton, FL (1994).

Décomposition de Domaine et Problème de Helmholtz: Thirty Years After and Still Unique

Martin J. Gander and Hui Zhang

1 Introduction

In 1990, Bruno Després published a short note [5] in *Comptes rendus de l'Académie des sciences. Série 1, Mathématique*. "The aim of this work is, after construction of a domain decomposition method adapted to the Helmholtz problem, to show its convergence." The idea has been further developed in [12], [3], [9], [4] and [2] by employing radiation conditions with a special structure, subdomains without overlap and iterations in parallel or one-sweep. As of today, it seems the *unique* means by which Schwarz iterations (not Krylov–Schwarz) for the Helmholtz equation have been proved to converge in general geometry and variable media; otherwise, e.g., using PML ([1]) as boundary conditions requires the (sub)domain to be convex. This paper is to show that those algorithmic parameters are difficult to perturb even in a rectangle while maintaining convergent Schwarz iterations.

To this end, we consider the Helmholtz equation in $\Omega = (X_0^-, X_N^+) \times (0,1)$:

$$(\Delta + k^2)u = f \text{ in } \Omega,\ \mathcal{B}^{\mp}u = 0 \text{ at } \{X_0^-, X_N^+\} \times (0,1),\ \mathcal{C}u = 0 \text{ at } (X_0^-, X_N^+) \times \{0,1\}, \quad (1)$$

where $k > 0$, and $\mathcal{B}^{\mp}$, $\mathcal{C}$ are some trace operators. In the free space problem $\mathcal{C} = \partial_{\mathbf{n}} - ik$ and in the waveguide problem $\mathcal{C} = \partial_{\mathbf{n}}$ ($\mathbf{n}$ being the unit outer normal vector). Assume that $\Omega = \cup_{l=1}^N \Omega_l$ with $\Omega_l = (X_{l-1}^-, X_l^+) \times (0,1)$, $X_l^{\pm} := lH \pm \frac{L}{2}$, $H > 0$ and $L \geq 0$. The optimized Schwarz method iteratively solves (1) restricted to Ω_l for $u_l \approx u|_{\Omega_l}$ in parallel or in some order of $l = 1, \ldots, N$ with the transmission conditions $\mathcal{B}^- u_l = \mathcal{B}^- u_{l-1}$ at $\{X_{l-1}^-\} \times (0,1)$, $l > 1$ and $\mathcal{B}^+ u_l = \mathcal{B}^+ u_{l+1}$ at $\{X_l^+\} \times (0,1)$, $l < N$.

Martin J. Gander
University of Geneva, Section of Mathematics, Rue du Conseil-Général 7-9, CP 64, 1205 Geneva, e-mail: martin.gander@unige.ch

Hui Zhang (corresponding author)
Xi'an Jiaotong-Liverpool University, Department of Applied Mathematics & Laboratory for Intelligent Computing and Financial Technology (KSF-P-02, RDF-19-01-09), Suzhou 215123, China, e-mail: mike.hui.zhang@hotmail.com

S. C. Brenner et al. (eds.), *Domain Decomposition Methods in Science and Engineering XXVI*, Lecture Notes in Computational Science and Engineering 145,
https://doi.org/10.1007/978-3-030-95025-5_69

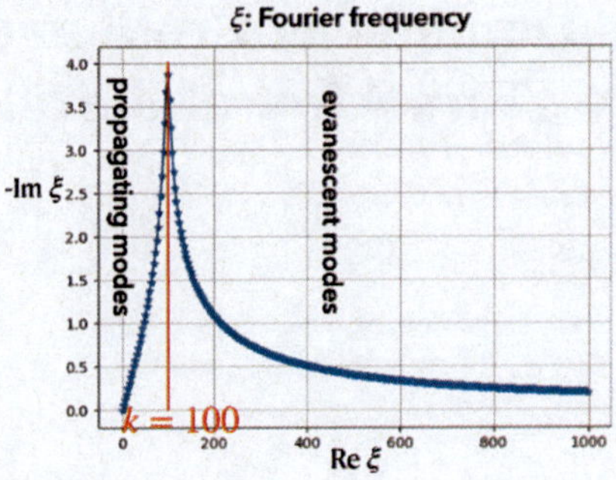

Fig. 1: Fourier frequencies from the Sturm–Liouville problem with $C = \partial_{\mathbf{n}} - ik$ for $k = 100$.

From the Sturm–Liouville problem $-\varphi'' = \xi^2\varphi$ in $(0, 1)$, $C\varphi = 0$ at $\{0, 1\}$, the expansion $u(x, y) = \sum_\xi \hat{u}(x, \xi)\varphi(y; \xi)$ transforms (1) to an ODE for $\hat{u}(x, \xi)$ for each ξ, and the iteration operator acting on $\{g_l^\mp := \mathcal{B}^\mp u_l$ at $\{X_{l-1}^-, X_l^+\} \times (0, 1)\}$ to a matrix for each ξ; see [8]. The spectral radius of the iteration matrix as a function of ξ or $\mathrm{Re}\,\xi/k$ is called the convergence factor ρ. The Schwarz iterations converge geometrically if and only if $\sup_\xi \rho < 1$. When $C = \partial_{\mathbf{n}}$, $\xi \in \{0, \pm\pi, \pm 2\pi, \pm 3\pi, \ldots\}$. When $C = \partial_{\mathbf{n}} - ik$, ξ's are complex roots of a nonlinear equation; see e.g. Figure 1.

2 What can we change from Després' original method?

In the original method, $\mathcal{B}^\mp = \partial_{\mathbf{n}} - ik$. From Figure 2 we see that all the curves are below $\rho = 1$, albeit $\rho \to 1$ as $\mathrm{Re}\,\xi \to \infty$, and the curves seemingly have a limit profile for a fixed N but increasing k, while they move higher up as N grows.

Can we add overlap? The short answer is 'no' for large overlap and 'yes' for small overlap; see Figure 3 and Figure 4. This is in contrast to the Laplace equation for which [10] claimed its proof "also applies ... with overlapping subdomains", though, an actual proof appeared only in [11] by rather different techniques.

Can we reorder subdomain iterations? Després' original method uses parallel iterations between subdomains. The sequential iterations from Ω_1 to Ω_N or one-sweep iterations through other orderings (e.g. red-black) of $\{\Omega_1, \ldots, \Omega_N\}$ behave similarly. In contrast, the double sweep iterations with the forward sweep from Ω_1 to Ω_N followed by a backward sweep from Ω_{N-1} to Ω_1 can diverge; see Figure 5. It is less divergent for larger kH, which suggests the next question.

What if we fix the subdomain size? In this case, the double sweeps converge very well and even better with large overlap; see Figure 6.

Can we add a real part to the Robin coefficient? Yes, if the real parts of the Robin coefficients for two adjacent subdomains are equal in absolute value but opposite in sign, as shown in [4, 9]; otherwise it may diverge. See Figure 7 and Figure 8.

Can we use second-order conditions? Yes, if the imaginary part of the Robin coefficient (i.e., tangential operator) is sign-definite on interfaces and outer boundaries,

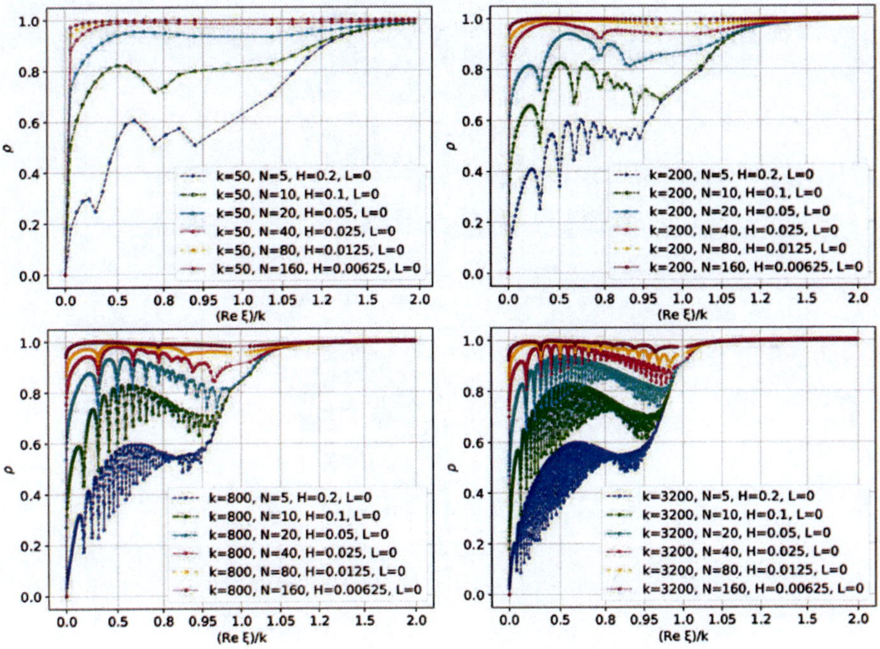

Fig. 2: ρ of Després' method for free space on $[0, 1]^2$.

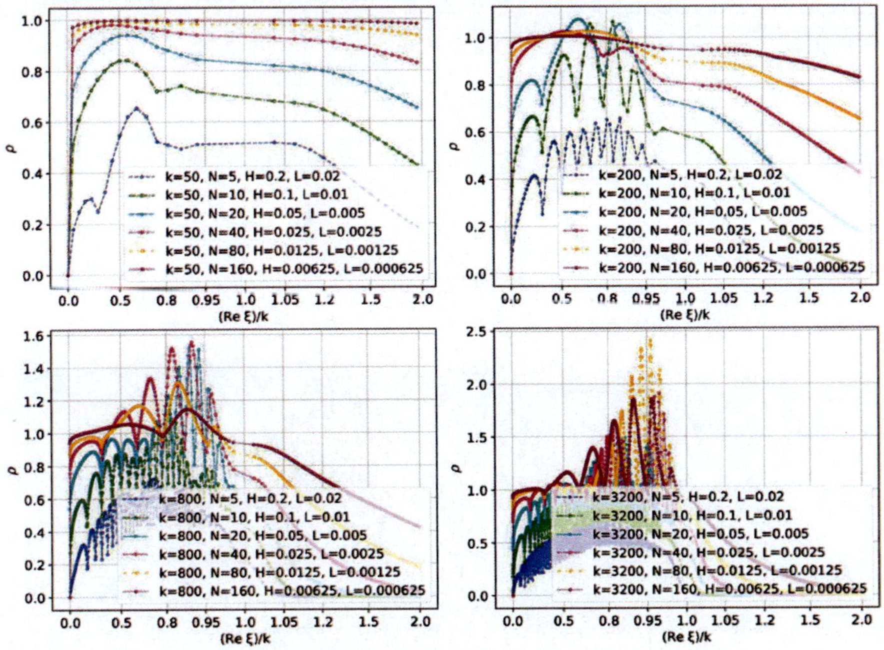

Fig. 3: ρ of Després' method plus $L = \frac{H}{10}$ overlap for free space on $[-\frac{L}{2}, 1 + \frac{L}{2}] \times [0, 1]$.

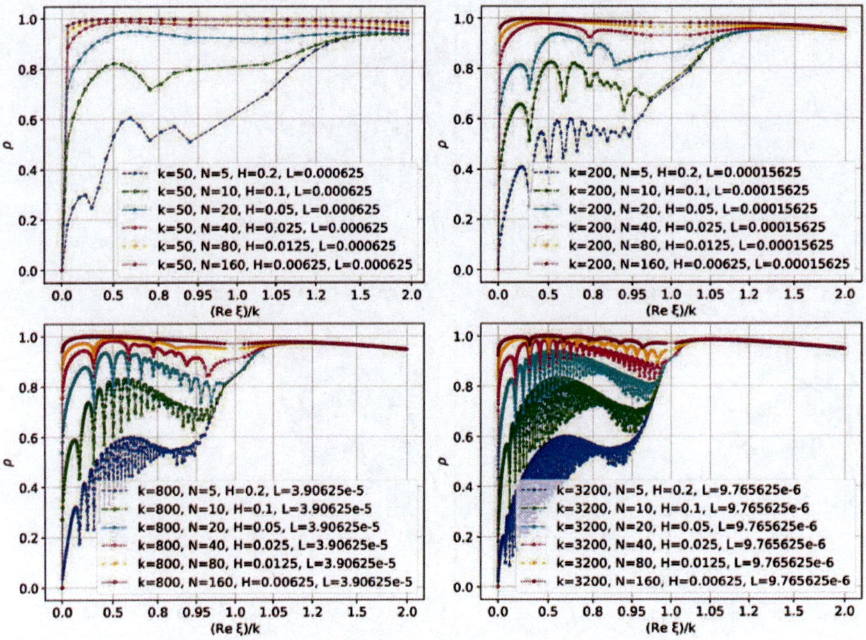

Fig. 4: ρ of Després' method plus $L = \frac{1}{32k}$ overlap for free space on $[-\frac{L}{2}, 1 + \frac{L}{2}] \times [0, 1]$.

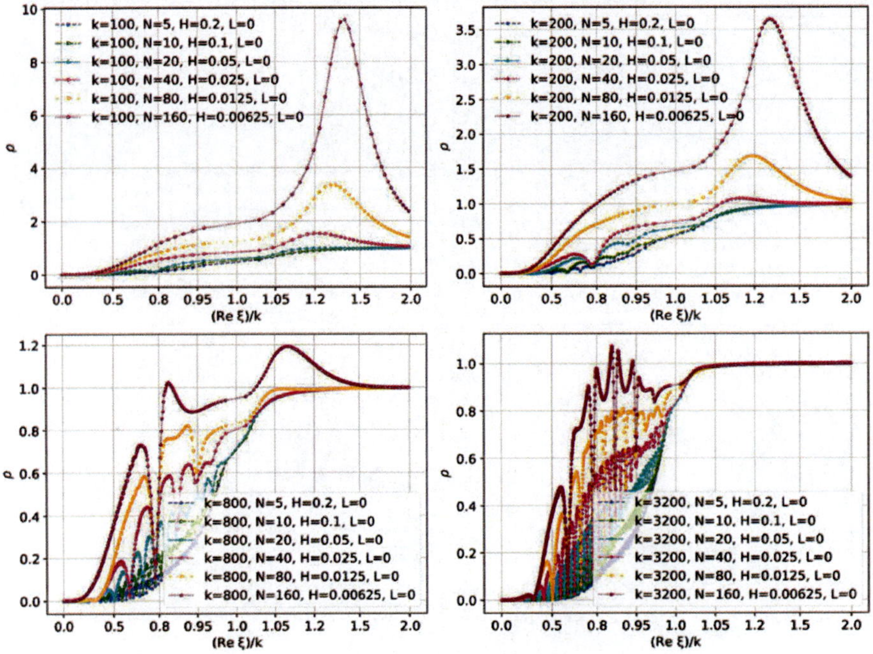

Fig. 5: ρ of Després' method in double sweep for free space on $[0, 1]^2$.

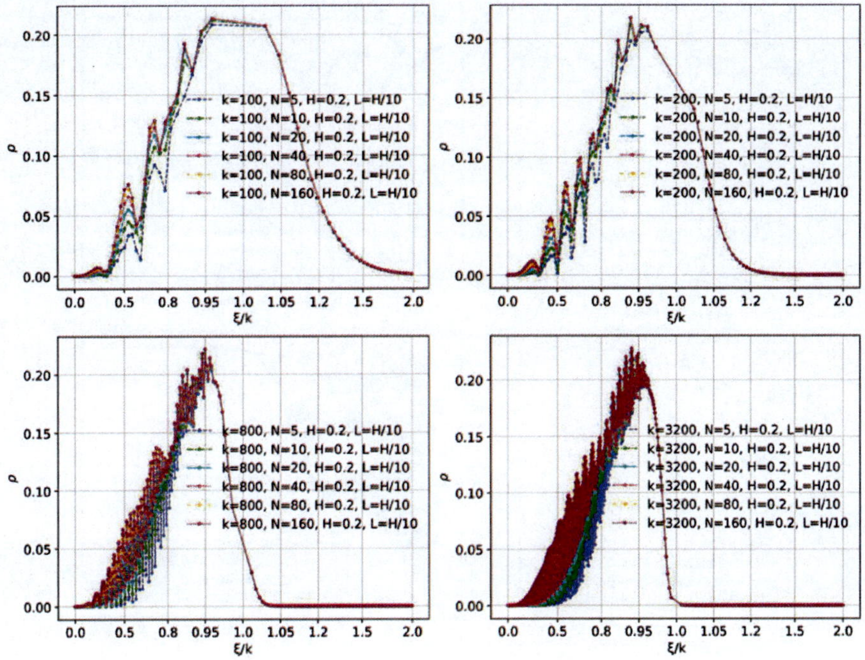

Fig. 6: ρ of Desprès' method plus $L = \frac{H}{10}$ overlap in double sweep for free space on $[-\frac{L}{2}, \frac{N}{5} + \frac{L}{2}] \times [0, 1]$.

as proved in [3, 12]. The 2nd-order Taylor approximation $\sqrt{k^2 - \xi^2} \approx k(1 - \frac{\xi^2}{2k^2})$ is sign-changing across $\xi^2 = 2k^2$ and thus $\mathcal{B}^{\mp} = \partial_{\mathbf{n}} - ik(1 + \frac{\Delta_S}{2k^2})$ (Δ_S is the Laplacian on interfaces) falls outside the theory. So, [12] proposed to reverse the sign in front of Δ_S, and [6] uses $\mathcal{B}^{\mp} = \partial_{\mathbf{n}} - ik(1 - \frac{\Delta_S}{2k^2})^{-1}$. See Figure 9 for comparison.

Can we change the outer boundary conditions? Yes, the proof in [5] and others all work as long as on part of the outer boundary a radiation condition is imposed with imaginary part of the tangential operator being sign-definite. It is thus interesting to check, e.g., with $C = \partial_{\mathbf{n}}$ and $\mathcal{B}^{\mp}$ the 2nd-order Taylor; see Figure 10.

Can we treat variable media? Yes, if the Robin coefficients for any two adjacent subdomains have equal imaginary parts, as proved in [3]. What if the symmetry is broken? For example, the optimal Schwarz method (see, e.g., the review [7]) uses the Dirichlet-to-Neumann maps from the two sides of an interface, which are generally not equal for propagating modes in variable media. In Desprès' method, to keep the symmetry one can use an average wavenumber on the interfaces, e.g., $k_{ij} = \sqrt{(k_i^2 + k_j^2)/2}$ for $\mathcal{B}^{\mp} = \partial_{\mathbf{n}} - ik_{ij}$ on $\partial\Omega_i \cap \partial\Omega_j$. To mimic the optimal Schwarz method, one can use the wavenumber from the other side, e.g., $\mathcal{B}^{\mp} = \partial_{\mathbf{n}} - ik_{i\mp 1}$ for u_i on $\partial\Omega_i \cap \partial\Omega_{i\mp 1}$. In our example, we split the first dimension into five equal layers on $[0, 1]^2$ and assume the wavenumber k in $\mathbb{R}^2 \backslash [0, 1]^2$ is a constant; see Figure 11.

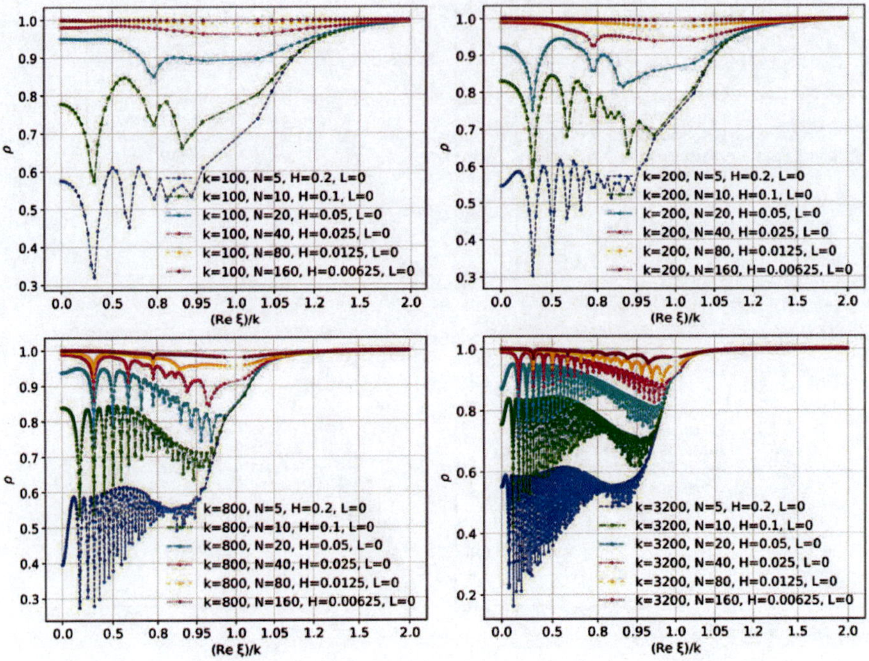

Fig. 7: ρ of Després' method with left/right Robin coeff. $-\mathrm{i}k(1 \pm 0.1\mathrm{i})$ for free space on $[0, 1]^2$.

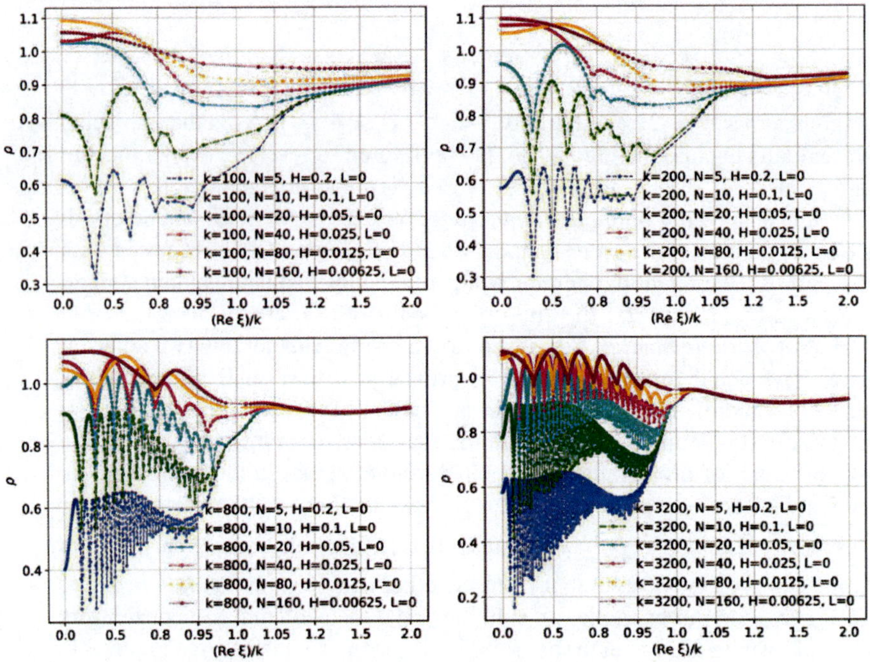

Fig. 8: ρ of Després' method with interface Robin coefficient $-\mathrm{i}k(1{+}0.1\mathrm{i})$ for free space on $[0, 1]^2$.

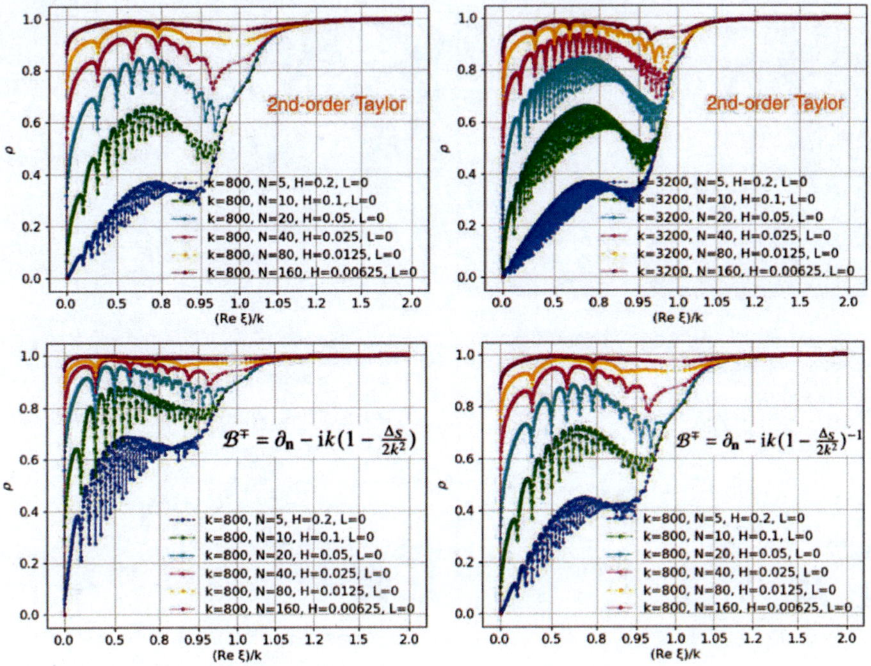

Fig. 9: ρ of Després' method plus a 2nd-order term for free space on $[0, 1]^2$.

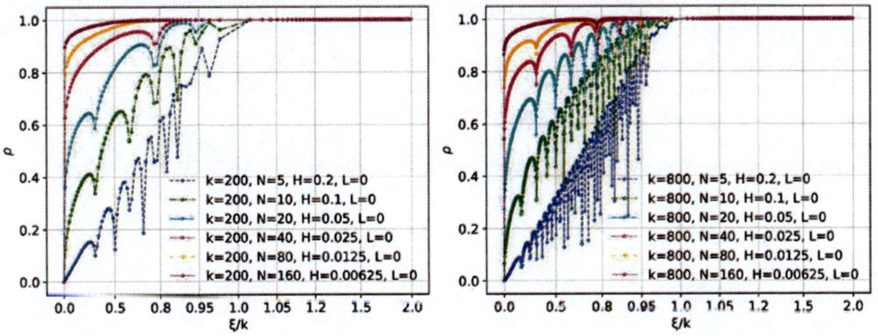

Fig. 10: ρ of Després' method with $\mathcal{B}^{\mp}$ the 2nd-order Taylor for waveguide on $[0, 1]^2$.

References

1. W.C. Chew, J.M. Jin and E. Michielssen. Complex coordinate stretching as a generalized absorbing boundary condition. *Microw. Opt. Technol. Lett.* **15**, 363–369 (1997).
2. X. Claeys, F. Collino, P. Joly and E. Parolin. A discrete domain decomposition method for acoustics with uniform exponential rate of convergence using non-local impedance operators. In: R. Haynes, S. MacLachlan, X.C. Cai, L. Halpern, H.H. Kim, A. Klawonn and O. Widlund (eds.) *Domain Decomposition Methods in Science and Engineering XXV*, pp. 310–317, Springer International Publishing, Cham (2020).

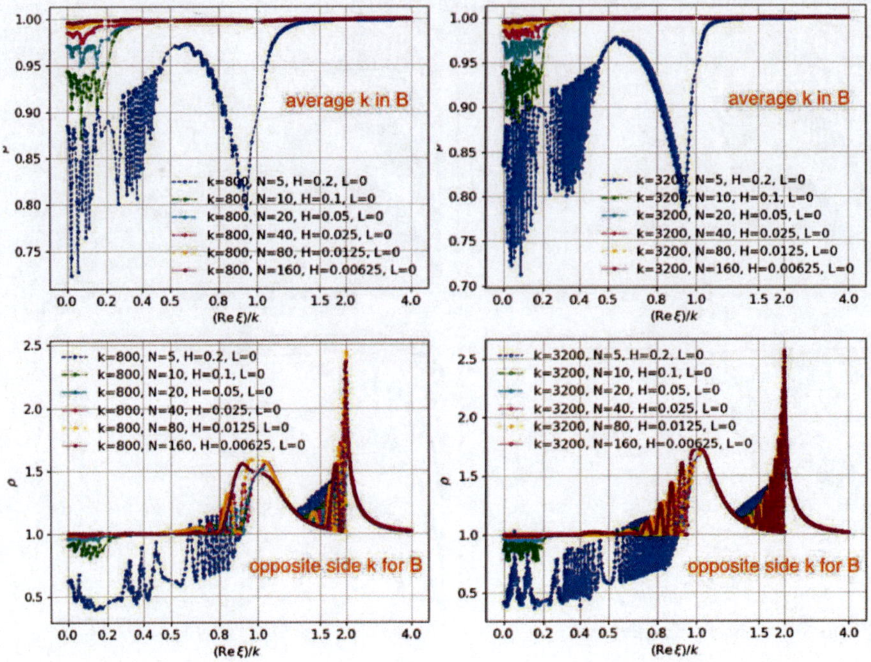

Fig. 11: ρ of Després' method on $[0, 1]^2$ with exterior wavenumber k & interior $k/[1, 5, 0.5, 2, 1]$.

3. F. Collino, S. Ghanemi and P. Joly. Domain decomposition method for harmonic wave propagation: a general presentation. *Comput. Methods Appl. Mech. Engrg.* **184**, 171–211 (2000).
4. F. Collino, P. Joly and M. Lecouvez. Exponentially convergent non overlapping domain decomposition methods for the Helmholtz equation. *ESAIM: M2AN* **54**, 775–810 (2020).
5. B. Després. Décomposition de domaine et problème de Helmholtz. *C. R. Acad. Sci. Paris t.* **311**, 313–316 (1990).
6. B. Després, A. Nicolopoulos and B. Thierry. Corners and stable optimized domain decomposition methods for the Helmholtz problem. Hal-02612368 (2020).
7. M.J. Gander and H. Zhang. A class of iterative solvers for the Helmholtz equation: Factorizations, sweeping preconditioners, source transfer, single layer potentials, polarized traces, and optimized Schwarz methods. *SIAM Review* **61**(1), 3–76 (2019).
8. M.J. Gander and H. Zhang. Analysis of double sweep optimized Schwarz methods: the positive definite case. In: R. Haynes, S. MacLachlan, X.C. Cai, L. Halpern, H.H. Kim, A. Klawonn, O. Widlund (eds.) *Domain Decomposition Methods in Science and Engineering XXV*, pp. 53–64, Springer International Publishing, Cham (2020).
9. M. Lecouvez, B. Stupfel, P. Joly and F. Collino. Quasi-local transmission conditions for nonoverlapping domain decomposition methods for the Helmholtz equation. *Comptes Rendus Physique* **15**, 403–414 (2014).
10. P.L. Lions. On the Schwarz alternating method III: A variant for nonoverlapping subdomains. In: T. Chan, R. Glowinski, J. Periaux, O.B. Widlund (eds.)/ *Third international symposium on domain decomposition methods for partial differential equations*, pp. 202–223, SIAM (1990).
11. S. Loisel and D.B. Szyld. On the geometric convergence of optimized Schwarz methods with applications to elliptic problems. *Numerische Mathematik* **114**(4), 697–728 (2010).
12. A. Piacentini and N. Rosa. An improved domain decomposition method for the 3D Helmholtz equation. *Comput. Methods Appl. Mech. Engrg.* **162**(1), 113–124 (1998).

Part III
Contributed Talks

Space-Time Parallel Methods for Evolutionary Reaction-Diffusion Problems

Andrés Arrarás, Francisco J. Gaspar, Laura Portero, and Carmen Rodrigo

1 Introduction and problem setting

In recent years, the gradual saturation of parallelization in space has been a strong motivation for the design and analysis of new parallel-in-time algorithms. Among these methods, the parareal algorithm, first introduced by Lions, Maday and Turinici [9], has received significant attention. This scheme has been formulated in the literature as a multiple shooting method [7], a predictor-corrector scheme [13], and a two-level multigrid method in time (see [3, 5] in the linear setting, and [7] for nonlinear problems using the full approximation storage (FAS) multigrid solver).

The key idea of the parareal method is to decompose the time interval into a certain number of subintervals, and solve the original problem concurrently over each one of them. In doing so, it defines two propagation operators which provide fine and coarse approximations to the exact solution. Since the coarse propagator usually considers large stepsizes, implicit time integrators are often used in this case to ensure stability. Choosing the implicit Euler method as the coarse propagator, different fine propagators have been analyzed in the literature: implicit Euler [7, 11, 18],

Andrés Arrarás
Institute for Advanced Materials and Mathematics (INAMAT^2), Department of Statistics, Computer Science and Mathematics, Public University of Navarre (UPNA), 31006 Pamplona, Spain, e-mail: andres.arraras@unavarra.es

Francisco J. Gaspar
Institute for Mathematics and Applications (IUMA), Department of Applied Mathematics, University of Zaragoza, 50009 Zaragoza, Spain, e-mail: fjgaspar@unizar.es

Laura Portero
Institute for Advanced Materials and Mathematics (INAMAT^2), Department of Statistics, Computer Science and Mathematics, Public University of Navarre (UPNA), 31006 Pamplona, Spain, e-mail: laura.portero@unavarra.es

Carmen Rodrigo
Institute for Mathematics and Applications (IUMA), Department of Applied Mathematics, University of Zaragoza, 50009 Zaragoza, Spain, e-mail: carmenr@unizar.es

S. C. Brenner et al. (eds.), *Domain Decomposition Methods in Science and Engineering XXVI*, Lecture Notes in Computational Science and Engineering 145,
https://doi.org/10.1007/978-3-030-95025-5_70

trapezoidal rule [7, 11, 18], Radau IIA [7], diagonally implicit Runge–Kutta (DIRK) [18] and Gauss Runge–Kutta [18], among others. Several combinations of A- and L-stable singly diagonally Runge–Kutta (SDIRK) fine and coarse propagators have been further studied in [6].

The main contribution of this work is to consider domain decomposition splitting time integrators as the fine and coarse propagators of the parareal algorithm. Since these methods are related to an overlapping decomposition of the spatial domain, spatial parallelization can also be exploited. Consequently, the resulting algorithms allow for parallelization in both time and space. This class of splitting methods was introduced in [15] in the context of regionally-additive schemes, and has been subsequently extended for solving linear parabolic problems [1, 2, 10, 12, 16] (see [4] for a recent work on nonlinear degenerate parabolic equations). The advantage of the new algorithms with respect to related existing methods (as parareal Schwarz waveform relaxation methods) is that they do not require any iteration to adjust the boundary conditions of the subdomains. As shown later, they are robust with respect to the discretization parameters, the number of disjoint components in each subdomain, the overlapping size and the coarsening factor under consideration.

In the rest of this section, we introduce the time-dependent reaction-diffusion problem to be solved, and derive the stiff system of ordinary differential equations resulting from the spatial discretization. More precisely, let us consider an initial-boundary value problem of the form

$$\begin{cases} u_t + Lu = f, & \text{in } \Omega \times (0,T], \\ u = g, & \text{on } \Gamma \times (0,T], \\ u = u_0, & \text{in } \overline{\Omega} \times \{0\}, \end{cases} \tag{1}$$

where $\Omega \subset \mathbb{R}^2$ is a bounded connected Lipschitz domain with boundary $\Gamma = \partial\Omega$, and $L = L(\mathbf{x})$ is an elliptic operator such that $Lu = -\nabla \cdot (K\nabla u) + cu$. Herein, $K = K(\mathbf{x}) \in \mathbb{R}^{2\times 2}$ is a symmetric tensor with coefficients $K_{i,j} \in L^\infty(\Omega)$, for $i, j \in \{1, 2\}$, that satisfies

$$\kappa_* \, \xi^T \xi \le \xi^T K \, \xi \le \kappa^* \, \xi^T \xi \qquad \forall \xi \in \mathbb{R}^2 \text{ and for almost all } \mathbf{x} \in \Omega,$$

for some $0 < \kappa_* \le \kappa^* < \infty$. In addition, the functions $u = u(\mathbf{x}, t)$, $f = f(\mathbf{x}, t)$, $g = g(\mathbf{x}, t)$, $u_0 = u_0(\mathbf{x})$ and $c = c(\mathbf{x})$, with $c \ge 0$, are assumed to be sufficiently smooth, and g and u_0 further satisfy suitable compatibility conditions, so that problem (1) admits a unique weak solution (see [14] for details).

Following the method of lines, we first define a suitable mesh Ω_h covering the spatial domain Ω, where h refers to the maximal grid spacing. Then, using a suitable discretization of the spatial variables (by means of finite difference, finite element or finite volume schemes), we obtain the initial value problem[1]

[1] If we consider a standard finite element method for discretizing (1), we initially obtain a system of ordinary differential equations of the form $M_h U_h'(t) + L_h U_h(t) = F_h(t)$, which is similar to the first equation in (2), but involves two symmetric and positive definite matrices, usually referred to as the mass (M_h) and stiffness (L_h) matrices. Now, considering the Cholesky decomposition

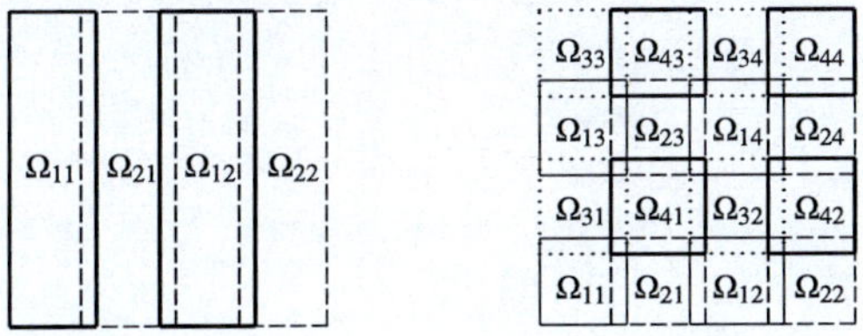

Fig. 1: Overlapping decompositions $\{\Omega_k\}_{k=1}^s$ of the unit square Ω into $s = 2$ (left) and $s = 4$ (right) subdomains. Each subdomain Ω_k is further decomposed into $\{\Omega_{kl}\}_{l=1}^{s_k}$ disjoint connected components, with $s_k = 2$ (left) and $s_k = 4$ (right).

$$\begin{cases} U_h'(t) + L_h U_h(t) = F_h(t), & t \in (0,T], \\ U_h(0) = \mathcal{R}_h u_0, \end{cases} \tag{2}$$

where $\mathcal{R}_h$ stands for an appropriate restriction or projection operator acting on the initial condition. If we denote by M the number of degrees of freedom in Ω_h for any $t \in [0,T]$, $U_h(t) \in \mathbb{R}^M$ and $L_h \in \mathbb{R}^{M\times M}$ denote the corresponding approximations to $u(\mathbf{x},t)$ and $L(\mathbf{x})$, respectively. Finally, $F_h(t) \in \mathbb{R}^M$ includes the approximation to $f(\mathbf{x},t)$ and the contribution of the boundary data $g(\mathbf{x},t)$.

2 Domain decomposition splitting methods

In this section, we describe how to construct a smooth partition of unity subordinate to an overlapping decomposition of the spatial domain Ω. In addition, we define suitable splittings for the discrete operator L_h and the right-hand side F_h, and further use them in a time integrator with a multiterm partitioning structure.

Let $\{\Omega_k\}_{k=1}^s$ be an overlapping decomposition of Ω into s subdomains, i.e., $\Omega = \bigcup_{k=1}^s \Omega_k$. Each subdomain $\Omega_k \subset \Omega$ is further defined as an open set involving s_k connected components $\Omega_k = \bigcup_{l=1}^{s_k} \Omega_{kl}$, for $k = 1, 2, \ldots, s$, that are considered to be pairwise disjoint ($\Omega_{ki} \cap \Omega_{kj} = \emptyset$, for $i \neq j$). The overlapping size is denoted by ε. Figure 1 shows two different decompositions of the unit square into $s = 2$ and $s = 4$ subdomains, each consisting of $s_k = 2$ and $s_k = 4$ disjoint connected components, respectively.

Subordinate to this decomposition, we define a smooth partition of unity consisting of a family of s non-negative and $C^\infty(\overline{\Omega})$ functions $\{\rho_k(\mathbf{x})\}_{k=1}^s$. Each function $\rho_k : \overline{\Omega} \to [0,1]$ is chosen to be

$M_h = N_h N_h^T$, where N_h is a lower triangular matrix with positive diagonal entries, we can define the new unknown $V_h(t) = N_h^T U_h(t)$. It is immediate to see that $V_h(t)$ satisfies a system like (2); in particular, $V_h'(t) + \hat{L}_h V_h(t) = \hat{F}_h(t)$, where $\hat{L}_h = N_h^{-1} L_h N_h^{-T}$ is symmetric and positive definite and $\hat{F}_h(t) = N_h^{-1} F_h(t)$ (cf. [8]).

$$\rho_k(\mathbf{x}) = \begin{cases} 0, & \text{if } \mathbf{x} \in \overline{\Omega} \setminus \overline{\Omega}_k, \\ h_k(\mathbf{x}), & \text{if } \mathbf{x} \in \bigcup_{l=1;\, l\neq k}^{s} (\overline{\Omega}_k \cap \overline{\Omega}_l), \\ 1, & \text{if } \mathbf{x} \in \overline{\Omega}_k \setminus \bigcup_{l=1;\, l\neq k}^{s} (\overline{\Omega}_k \cap \overline{\Omega}_l), \end{cases}$$

where $h_k(\mathbf{x})$ is $C^\infty(\Omega)$ and such that $0 \leq h_k(\mathbf{x}) \leq 1$ and $\sum_{k=1}^{s} h_k(\mathbf{x}) = 1$, for any $\mathbf{x} \in \bigcup_{l=1;\, l\neq k}^{s} (\overline{\Omega}_k \cap \overline{\Omega}_l)$. By construction, the family of functions $\{\rho_k(\mathbf{x})\}_{k=1}^{s}$ satisfies

$$\operatorname{supp}(\rho_k(\mathbf{x})) \subset \overline{\Omega}_k, \qquad 0 \leq \rho_k(\mathbf{x}) \leq 1, \qquad \textstyle\sum_{k=1}^{s} \rho_k(\mathbf{x}) = 1, \tag{3}$$

for any $\mathbf{x} \in \overline{\Omega}$. In practice, $h_k(\mathbf{x})$ may not necessarily be $C^\infty(\Omega)$, but only a continuous and piecewise smooth function [10].

In this framework, given the parabolic problem (1), we can define a domain decomposition operator splitting $L = L_1 + L_2 + \cdots + L_s$ and $f = f_1 + f_2 + \cdots + f_s$ such that each split term is given by

$$L_k u = -\nabla \cdot (\rho_k K \nabla u) + \rho_k c u, \qquad f_k = \rho_k f, \qquad \text{for } k = 1, 2, \ldots, s. \tag{4}$$

Accordingly, in the discrete setting (2), we may introduce the domain decomposition matrix splitting $L_h = L_{1h} + L_{2h} + \cdots + L_{sh}$ and $F_h = F_{1h} + F_{2h} + \cdots + F_{sh}$, where each term L_{kh} and F_{kh} is defined to be a suitable spatial discretization of its continuous counterpart (4), for $k = 1, 2, \ldots, s$. Typically, the discrete split terms L_{kh} have a simpler structure than L_h, but they do not commute pairwise. This lack of commutativity demands the use of suitable time integrators which preserve the unconditional stability even in the non-commuting case. The simplest of such methods is given by the so-called fractional implicit Euler scheme, first proposed by Yanenko in [19] and described in the sequel.

Let us divide the time interval $[0, T]$ into N_t subintervals $[t_n, t_{n+1}]$, with stepsize $\Delta t = t_{n+1} - t_n = T/N_t$, for $n = 0, 1, \ldots, N_t - 1$. We further define the fully discrete solution $U_h^n \approx U_h(t_n)$ at times $t_n = n\Delta t$, for $n = 0, 1, \ldots, N_t$. Then, given $U_h^0 = \mathcal{R}_h u_0$, the fractional implicit Euler method can be written recursively, for $n = 0, 1, \ldots, N_t - 1$, as

$$(I + \Delta t\, L_{kh})\, U_h^{n+k/s} = U_h^{n+(k-1)/s} + \Delta t\, F_{kh}(t_{n+1}), \qquad \text{for } k = 1, 2, \ldots, s. \tag{5}$$

Note that one integration step with (5) can be seen as s consecutive steps with the implicit Euler method, each with a different right-hand side function. In consequence, this time integrator is first-order convergent [17]. Eliminating the internal stages $U_h^{n+k/s}$, for $k = 1, 2, \ldots, s - 1$, (5) can be expressed as

$$U_h^{n+1} = \left(\prod_{k=1}^{s} (I + \Delta t\, L_{kh}) \right)^{-1} U_h^n + \sum_{j=1}^{s} \left(\prod_{k=j}^{s} (I + \Delta t\, L_{kh}) \right)^{-1} \Delta t\, F_{jh}(t_{n+1}). \tag{6}$$

For later use, we will denote the right-hand side of this expression by $\mathcal{S}_{\Delta t}(U_h^n)$. The linear system to be solved at the k-th internal stage of (5) involves just the split

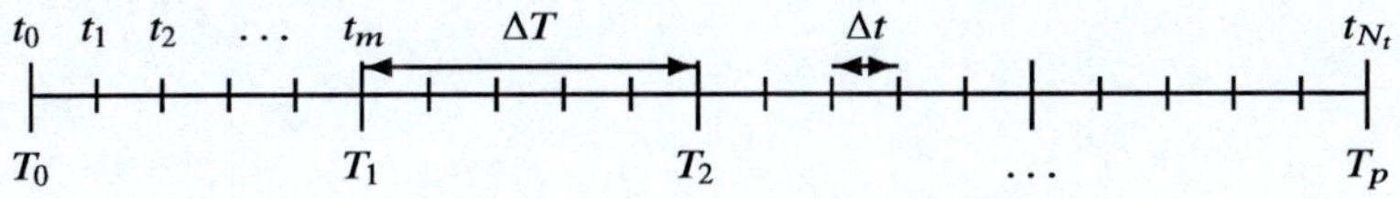

Fig. 2: Fine and coarse time grids considered in the parareal method.

term L_{kh} in the system matrix. As stated in (3), the function $\rho_k(\mathbf{x})$ has compact support on $\overline{\Omega}_k$. Hence, by construction, the entries of L_{kh} corresponding to the nodes that lie outside of this subdomain are zero. Moreover, since Ω_k involves s_k disjoint connected components Ω_{kl}, the previous linear system is indeed a collection of s_k uncoupled subsystems, which can be solved in parallel.

3 The parareal method

In this section, we briefly review the basis for the parareal algorithm, and further establish the connection with domain decomposition splitting schemes in order to derive our new proposal.

Let us first divide the time interval $[0, T]$ into p large time subintervals $[T_n, T_{n+1}]$, for $n = 0, 1, \ldots, p-1$, with stepsize $\Delta T = T_{n+1} - T_n = T/p$. Thus, $T_n = n\Delta T$, for $n = 0, 1, \ldots, p$. Subsequently, we further divide each $[T_n, T_{n+1}]$ into $m \geq 2$ smaller time subintervals $[t_n, t_{n+1}]$, for $n = 0, 1, \ldots, N_t - 1$, with stepsize $\Delta t = \Delta T/m = T/N_t$, where $N_t = pm$. In this case, $t_n = n\Delta t$, for $n = 0, 1, \ldots, N_t$. The parameter m is sometimes referred to as coarsening factor. A representation of these fine and coarse grids is shown in Figure 2.

In this setting, the parareal method makes use of two propagation operators which provide fine and coarse approximations to the solution of (2). We will denote by $\mathcal{F}_{\Delta t}$ the fine propagator, with stepsize Δt, and by $\mathcal{G}_{\Delta T}$ the coarse propagator, with stepsize ΔT. Essentially, the algorithm generates a sequence of iterations $U_h^{n,\ell}$, for $\ell = 0, 1, \ldots$, which converges to the solution of (2). To this end, we sequentially obtain an initial approximation to the numerical solution at the coarse time levels by using the coarse propagator $\mathcal{G}_{\Delta T}$ on the interval $[0, T]$: given $U_h^{0,0} = \mathcal{R}_h u_0$,

$$U_h^{n+1,0} = \mathcal{G}_{\Delta T}(U_h^{n,0}), \quad \text{for } n = 0, 1, \ldots, p-1. \tag{7}$$

Then, for $\ell = 0, 1, \ldots$, until convergence, we do:

1. On each subinterval $[T_n, T_{n+1}]$, we solve on the fine grid using the fine propagator $\mathcal{F}_{\Delta t}$: given $\tilde{U}_h^{nm} = U_h^{n,\ell}$, for $n = 0, 1, \ldots, p-1$,

$$\tilde{U}_h^{nm+j+1} = \mathcal{F}_{\Delta t}(\tilde{U}_h^{nm+j}), \quad \text{for } j = 0, 1, \ldots, m-1. \tag{8}$$

2. On the interval $[0, T]$, we solve on the coarse grid using the coarse propagator $\mathcal{G}_{\Delta T}$: given $U_h^{0,\ell+1} = \mathcal{R}_h u_0$,

$$U_h^{n+1,\ell+1} = \mathcal{G}_{\Delta T}(U_h^{n,\ell+1}) + \tilde{U}_h^{nm} - \mathcal{G}_{\Delta T}(U_h^{n,\ell}), \quad \text{for } n = 0, 1, \dots, p-1. \quad (9)$$

As suggested in [18], if we denote $\tilde{U}_h^{nm}$ by $\mathcal{F}_{\Delta t}^m(U_h^{n,\ell})$, indicating that we are taking m steps of the fine propagator with initial value $U_h^{n,\ell}$ and a stepsize Δt, the previous algorithm can be compactly written as

$$U_h^{n+1,\ell+1} = \mathcal{G}_{\Delta T}(U_h^{n,\ell+1}) + \mathcal{F}_{\Delta t}^m(U_h^{n,\ell}) - \mathcal{G}_{\Delta T}(U_h^{n,\ell}), \quad \text{for } n = 0, 1, \dots, p-1.$$

Based on this expression, the parareal method can be interpreted as a predictor-corrector scheme in which $\mathcal{G}_{\Delta T}(U_h^{n,\ell+1})$ plays the role of the predictor, while $\mathcal{F}_{\Delta t}^m(U_h^{n,\ell}) - \mathcal{G}_{\Delta T}(U_h^{n,\ell})$ is the correction term. Note that, at the $(\ell+1)$-th iteration, we can use p processors to compute both $\{\mathcal{F}_{\Delta t}^m(U_h^{n,\ell})\}_{n=1}^p$ and $\{\mathcal{G}_{\Delta T}(U_h^{n,\ell})\}_{n=1}^p$ in parallel.

Now, we are in position to introduce the new family of parareal domain decomposition splitting methods by suitably combining the fractional implicit Euler method (5) with the parareal algorithm (7)=-(9). More precisely, we propose using (5) for solving the fine- and coarse-grid problems in the parareal method. Recalling the definition of $\mathcal{S}_{\Delta t}(U_h^n)$ as the right-hand side of (6), we shall consider $\mathcal{F}_{\Delta t}(\cdot) = \mathcal{S}_{\Delta t}(\cdot)$ in (8), and $\mathcal{G}_{\Delta T}(\cdot) = \mathcal{S}_{\Delta T}(\cdot)$ in (7) and (9). In consequence, the resulting method allows for parallelization in both space and time. Remarkably, unlike related existing schemes (e.g., parareal Schwarz waveform relaxation methods), our proposal does not require Schwarz iterations, since the internal stages in (5) are solved sequentially (i.e., interface conditions need not be imposed on subdomains during the solution process). In the next section, we illustrate the performance of the new algorithm as compared to the classical parareal method using implicit Euler propagators $\mathcal{F}_{\Delta t}$ and $\mathcal{G}_{\Delta T}$.

4 Numerical experiments

Let us consider the two-dimensional heat equation with a simple reaction term (i.e., $K = I$ and $c = 1$) on $\overline{\Omega} \times [0, T] = [0, 1]^3$, with homogeneous initial and Dirichlet boundary conditions, whose right-hand side is chosen such that the exact solution is $u(x, y, t) = te^{-t}x(1-x)y(1-y)$. We consider a five-point finite difference spatial discretization with $N = N_x = N_y$ spatial nodes on each direction (so that the total number of degrees of freedom for the spatial discretization is $M = N^2$), and the parareal time integrator with p coarse intervals, each containing m fine subintervals. Both the fine and coarse propagators, $\mathcal{F}_{\Delta t}$ and $\mathcal{G}_{\Delta T}$, are chosen to be either the implicit Euler method or the fractional implicit Euler method. In the sequel, we will refer to these methods as Euler and DD-Euler, respectively. In the latter case, Ω is further

Table 1: Number of iterations, varying the number k of disjoint components (left) and the overlapping size ε (right), for a fixed value of $\Delta t = T/(pm)$ and increasing values of N.

Parameters: $p = 16$, $m = 20$, $\varepsilon = 2^{-6}$						
	N	10	20	40	80	160
Euler		8	8	8	8	8
DD-Euler	$k = 2$	11	13	14	14	14
	$k = 4$	12	13	15	14	15
	$k = 8$	9	14	15	14	15

Parameters: $p = 16$, $m = 20$, $k = 4$						
	N	10	20	40	80	160
Euler		8	8	8	8	8
DD-Euler	$\varepsilon = 2^{-4}$	12	12	13	14	13
	$\varepsilon = 2^{-5}$	12	13	12	14	15
	$\varepsilon = 2^{-6}$	12	13	15	14	15

Table 2: Number of iterations, varying the number k of disjoint components (left) and the overlapping size ε (right), for a fixed value of N and decreasing values of $\Delta t = T/(pm)$.

Parameters: $p = 16$, $N = 160$, $\varepsilon = 2^{-6}$						
	m	20	40	80	160	320
Euler		8	8	8	8	8
DD-Euler	$k = 2$	14	15	15	15	15
	$k = 4$	15	15	15	15	15
	$k = 8$	15	16	16	16	16

Parameters: $p = 16$, $N = 160$, $k = 4$						
	m	20	40	80	160	320
Euler		8	8	8	8	8
DD-Euler	$\varepsilon = 2^{-4}$	13	13	14	14	14
	$\varepsilon = 2^{-5}$	15	15	16	16	16
	$\varepsilon = 2^{-6}$	15	15	15	15	15

decomposed into $s = 2$ subdomains, each consisting of k disjoint components, with overlapping size ε. Figure 1 (left) illustrates the case $s = 2$ and $k = 2$.

Tables 1 and 2 show the asymptotic dependence of the two parareal algorithms on the parameters N and m. In addition, for the DD-Euler method, we also show the asymptotic dependence on the values k and ε. In all the cases, we stop the iteration process when the difference between the iterate and the target solution[2] is less than 10^{-8}. Notice that the number of iterations for the DD-Euler method does not increase when considering either a larger number k of disjoint connected components or a smaller overlapping size ε. Although not reported here, a similar number of iterations is obtained for larger values of p. In conclusion, the newly proposed algorithms are robust with respect to the discretization parameters, the number of disjoint components k, the overlapping size ε, and the coarsening factor m.

Finally, for the implicit Euler method, if we have a time grid with pm nodes, we need to solve sequentially pm linear systems with N^2 unknowns. If we perform it_E iterations of the parareal Euler method to satisfy the stopping criterion, we need to solve sequentially $it_E\ (m + p + 1)$ linear systems with N^2 unknowns. Thus, for large values of m with respect to p, the parallelization of computations make the effective cost of the parareal Euler method smaller than that of the classical Euler scheme. In turn, if we perform it_{DD} iterations of the parareal DD-Euler method, considering s

[2] The target solution is the solution obtained at the coarse time levels using the fine propagator $\mathcal{F}_{\Delta t}$ on the whole time interval in a sequential way.

subdomains and k disjoint connected components, and assuming $\varepsilon \approx 0$, we need to solve sequentially it_{DD} $(m+p+1)$ s linear systems with $N^2/(sk)$ unknowns. Thus, for large values of m with respect to p and large values of k, the effective cost of the parareal DD-Euler method will be even smaller than that of the parareal Euler method.

Acknowledgements. The work of Andrés Arrarás and Carmen Rodrigo was supported by the Spanish State Research Agency under project PGC2018-099536-A-I00 (MCIU/AEI/FEDER, UE). The work of Francisco J. Gaspar and Laura Portero was supported by the Spanish State Research Agency under project PID2019-105574GB-I00 (AEI/10.13039/501100011033).

References

1. A. Arrarás, K.J. in 't Hout, W. Hundsdorfer and L. Portero. Modified Douglas splitting methods for reaction-diffusion equations. *BIT* **57**(2), 261–285 (2017).
2. A. Arrarás and L. Portero. Improved accuracy for time-splitting methods for the numerical solution of parabolic equations. *Appl. Math. Comput.* **267**, 294–303 (2015).
3. V.A. Dobrev, T. Kolev, N.A. Petersson and J.B. Schroder. Two-level convergence theory for multigrid reduction in time (MGRIT). *SIAM J. Sci. Comput.* **39**(5), S501–S527 (2017).
4. M. Eisenmann and E. Hansen. Convergence analysis of domain decomposition based time integrators for degenerate parabolic equations. *Numer. Math.* **140**(4), 913–938 (2018).
5. R.D. Falgout, S. Friedhoff, T.V. Kolev, S.P. MacLachlan and J.B. Schroder. Parallel time integration with multigrid. *SIAM J. Sci. Comput.* **36**(6), C635–C661 (2014).
6. S. Friedhoff and B.S. Southworth. On "optimal" h-independent convergence of parareal and multigrid-reduction-in-time using Runge–Kutta time integration. *Numer. Linear Algebra Appl.* **28**(3), Paper No. e2301, 30pp. (2021).
7. M.J. Gander and S. Vandewalle. Analysis of the parareal time-parallel time-integration method. *SIAM J. Sci. Comput.* **29**(2), 556–578 (2007).
8. C. Johnson. *Numerical Solution of Partial Differential Equations by the Finite Element Method.* Cambridge University Press, Cambridge (1987).
9. J.L. Lions, Y. Maday and G. Turinici. Résolution d'EDP par un schéma en temps «pararéel». *C. R. Acad. Sci. Paris Sér. I Math.* **332**(7), 661–668 (2001).
10. T.P. Mathew, P.L. Polyakov, G. Russo and J. Wang. Domain decomposition operator splittings for the solution of parabolic equations. *SIAM J. Sci. Comput.* **19**(3), 912–932 (1998).
11. T.P. Mathew, M. Sarkis and C.E. Schaerer. Analysis of block parareal preconditioners for parabolic optimal control problems. *SIAM J. Sci. Comput.* **32**(3), 1180–1200 (2010).
12. L. Portero, A. Arrarás and J.C. Jorge. Contractivity of domain decomposition splitting methods for nonlinear parabolic problems. *J. Comput. Appl. Math.* **234**(4), 1078–1087 (2010).
13. G.A. Staff and E.M. Rønquist. Stability of the parareal algorithm. In: R. Kornhuber, R.H.W. Hoppe, J. Périaux, O. Pironneau, O.B. Widlund, J. Xu (eds.) *Domain Decomposition Methods in Science and Engineering*, Lect. Notes Comput. Sci. Eng., vol. 40, pp. 449–456. Springer, Berlin (2005).
14. V. Thomée. *Galerkin Finite Element Methods for Parabolic Problems.* Second edition, Springer Ser. Comput. Math., vol. 25. Springer-Verlag, Berlin (2006).
15. P.N. Vabishchevich. Difference schemes with domain decomposition for solving nonstationary problems. *U.S.S.R. Comput. Math. Math. Phys.* **29**(6), 155–160 (1989).
16. P.N. Vabishchevich. Domain decomposition methods with overlapping subdomains for the time-dependent problems of mathematical physics. *Comput. Methods Appl. Math.* **8**(4), 393—405 (2008).

17. J.G. Verwer. Contractivity of locally one-dimensional splitting methods. *Numer. Math.* **44**(2), 247–259 (1984).
18. S.L. Wu and T. Zhou. Convergence analysis for three parareal solvers. *SIAM J. Sci. Comput.* **37**(2), A970–A992 (2015).
19. N.N. Yanenko. *The Method of Fractional Steps. The Solution of Problems of Mathematical Physics in Several Variables*. Springer-Verlag, New York-Heidelberg (1971).

Parallel Domain Decomposition Solvers for the Time Harmonic Maxwell Equations

Sven Beuchler, Sebastian Kinnewig, and Thomas Wick

1 Introduction

The time harmonic Maxwell (THM) equations are of great interest in applied mathematics [5, 6, 11, 12, 14, 15] and current physics applications, e.g., the excellence cluster PhoenixD.[1] However, the numerical solution is challenging. This is specifically true for high wave numbers. Various solvers and preconditioners have been proposed, while the most promising are based on domain decomposition methods (DDM) [16]. In [5], a quasi-optimal domain decomposition (DD) algorithm was proposed, mathematically analyzed and demonstrated to perform well for several numerical examples.

The goal of this work is to employ the domain decomposition method from [5] and to re-implement the algorithm in the modern finite element library deal.II [2]. Therein, the construction of the subdomain interface conditions is a crucial aspect for which we use Impedance Boundary Conditions. Instead of handling the resulting linear system with a direct solver, which is typically done for the THM, we apply a well chosen block preconditioner to the linear system so we can solve it with an iterative solver like GMRES (generalized minimal residuals). Additionally high polynomial Nédélec elements are used in the implementation of the DDM, see [17].

This implementation is computationally compared to several other (classical) preconditioners such as incomplete LU, additive Schwarz, Schur complement. These comparisons are done for different wave numbers. Higher wave numbers are well-known to cause challenges for numerical solution.

Sven Beuchler, Sebastian Kinnewig, Thomas Wick
Leibniz University Hannover, Institute of Applied Mathematics, Welfengarten 1, 30167 Hannover, Germany, e-mail: {beuchler,kinnewig,wick}@ifam.uni-hannover.de

and
Cluster of Excellence PhoenixD (Photonics, Optics, and Engineering – Innovation Across Disciplines), Leibniz Universität Hannover, Germany

[1] https://www.phoenixd.uni-hannover.de/en/

S. C. Brenner et al. (eds.), *Domain Decomposition Methods in Science and Engineering XXVI*, Lecture Notes in Computational Science and Engineering 145,
https://doi.org/10.1007/978-3-030-95025-5_71

The outline of this work is as follows: In the section 2 we introduce some notation. In section 3 we introduce the domain decomposition method (DDM) for the THM, furthermore we introduce a block preconditioner which will allow us to solve the THM with iterative solves instead of direct solvers inside of DDM. In section 4 we will compare the results of the block preconditioner with the performance of different preconditioners. Moreover we will present some results of the combination of the preconditioner and the DDM for two benchmark problems.

2 Equations and finite element discretization

Let $\Omega \subset \mathbb{R}^d$, $d \in \{2,3\}$, be a bounded domain with sufficiently smooth boundary Γ. The latter is partitioned into $\Gamma = \Gamma^\infty \cup \Gamma^{\text{inc}}$. Furthermore, the time harmonic Maxwell equations are then defined as follows: Find the electric field $\boldsymbol{E} \in \mathbf{H}(\text{curl}, \Omega) := \{v \in \mathcal{L}^2(\Omega),\ \text{curl}\,(v) \in \mathcal{L}^2(\Omega)\}$ such that

$$\begin{cases} \text{curl}\left(\mu^{-1}\,\text{curl}\,\boldsymbol{E}\right) - \omega^2 \boldsymbol{E} & = \mathbf{0} & \text{in } \Omega \\ \mu^{-1}\gamma^t\left(\text{curl}(\boldsymbol{E})\right) - i\kappa\omega\gamma^T\left(\boldsymbol{E}\right) = \mathbf{0} & & \text{on } \Gamma^\infty \\ \gamma^T\left(\boldsymbol{E}\right) & = -\gamma^T\left(\boldsymbol{E}^{\text{inc}}\right) & \text{on } \Gamma^{\text{inc}} \end{cases}, \tag{1}$$

where $\boldsymbol{E}^{\text{inc}} : \mathbb{R}^d \to \mathbb{C}^d$, $d \in \{2,3\}$ is some given incident electric field, $\omega > 0$ is the wave number which is defined by $\omega := \frac{2\pi}{\lambda}$, where $\lambda > 0$ is the wave length, $\mu > 0$ is the relative permeability and $\kappa > 0$ is the relative permittivity. Let Ω be a domain with smooth interface. Following [9, 15], we define traces $\gamma^t : \mathbf{H}(\text{curl}, \Omega) \to \mathbf{H}_\times^{-1/2}(\text{div}, \Gamma)$ and $\gamma^T : \mathbf{H}(\text{curl}, \Omega) \to \mathbf{H}_\times^{-1/2}(\text{curl}, \Gamma)$ by

$$\gamma^t\left(\boldsymbol{v}\right) = \boldsymbol{n} \times \boldsymbol{v} \text{ and } \gamma^T\left(\boldsymbol{v}\right) = \boldsymbol{n} \times \left(\boldsymbol{v} \times \boldsymbol{n}\right),$$

where the vector $\boldsymbol{n}$ is the normal to Ω, $\mathbf{H}_\times^{-1/2}(\text{div}, \Gamma) := \{\boldsymbol{v} \in \mathbf{H}^{-1/2}(\Gamma) : \boldsymbol{v} \cdot \boldsymbol{n} = 0, \text{div}_\Gamma\, \boldsymbol{v} \in \mathbf{H}^{-1/2}(\Gamma)\}$ is the space of well-defined surface divergence fields, and $\mathbf{H}_\times^{-1/2}(\text{curl}, \Gamma) := \{\boldsymbol{v} \in \mathbf{H}^{-1/2} : \boldsymbol{v} \cdot \boldsymbol{n} = 0, \text{curl}_\Gamma\, \boldsymbol{v} \in \mathbf{H}^{-1/2}(\Gamma)\}$ is the space of well-defined surface curls.

System (1) is called time harmonic, because the time dependence can be expressed by $e^{i\omega\tau}$, where $\tau \geq 0$ denotes the time. For the implementation with the help of a Galerkin finite element method, we need the discrete weak form. Let $\mathcal{N}_h^p := \{v_h \in X : v_h|_K(x) = a_K(x) + \left(x \times b_K(x)\right),\ a_K, b_K \in [P^p(K)]^3\ \forall\, K \in \tau_h(\Omega)\}$ be the Nédélec space [15], where $X = \{\boldsymbol{v} \in \mathbf{H}(\text{curl}, \Omega) : \boldsymbol{v} \times \boldsymbol{n}|_{\Gamma^{\text{inc}}} = \boldsymbol{v} \times \boldsymbol{n}|_{\Gamma^\infty} \in L^2(\Gamma^\infty)\}$. Based on the de-Rham cohomology, basis functions can be developed, [17]. Find $\boldsymbol{E}_h \in \mathcal{N}_h^p(\Omega)$ such that

$$\int_\Omega \left(\mu^{-1}\,\text{curl}\left(\boldsymbol{E}_h\right)\text{curl}\left(\boldsymbol{\varphi}_h\right) - \omega^2 \boldsymbol{E}_h \boldsymbol{\varphi}_h\right) \mathsf{d}x$$
$$+ \int_{\Gamma^\infty} i\kappa\omega\gamma^T\left(\boldsymbol{E}_h\right)\gamma^T\left(\boldsymbol{\varphi}_h\right) \mathsf{d}s = \int_{\Gamma^{\text{inc}}} \gamma^T\left(\boldsymbol{E}_h^{\text{inc}}\right)\gamma^T\left(\boldsymbol{\varphi}_h\right) \mathsf{d}s \quad \forall \boldsymbol{\varphi}_h \in \mathcal{N}_h^p(\Omega). \tag{2}$$

In order to obtain a block system for the numerical solution process, we define the following elementary integrals

$$
\begin{aligned}
(A)_{u,v} &= \int_{\Omega} \mu^{-1} \operatorname{curl}(\varphi_u) \operatorname{curl}(\varphi_v), \quad (M)_{u,v} = \int_{\Omega} \varphi_u \varphi_v \\
(B)_{u,v} &= \int_{\Gamma^{\infty}} i\kappa\omega\gamma^T(\varphi_u)\,\gamma^T(\varphi_v), \quad (s)_u = \int_{\Gamma^{\text{inc}}} \gamma^T\left(\boldsymbol{E}^{\text{inc}}\right)\gamma^T(\varphi_u),
\end{aligned} \tag{3}
$$

where $\varphi_u, \varphi_v \in \mathcal{N}_h^p(\Omega)$. To this end, System (1) can be written in the form

$$
\begin{pmatrix} A - \omega^2 M & -B \\ B & A - \omega^2 M \end{pmatrix} \begin{pmatrix} \boldsymbol{E}_{RE} \\ \boldsymbol{E}_{IM} \end{pmatrix} = \begin{pmatrix} s_{RE} \\ s_{IM} \end{pmatrix}, \tag{4}
$$

where $\boldsymbol{E} = \boldsymbol{E}_{RE} + i\boldsymbol{E}_{IM}$ and $s = s_{RE} + is_{IM}$, where i denotes the imaginary number.

3 Numerical solution with domain decomposition and preconditioners

3.1 Domain decomposition

Due to the difficult structure of the time harmonic Maxwell equations, a successful approach to solve the THM is based on the DDM [16]. As the name suggests, the domain is divided into smaller subdomains. As these subdomains become small enough they can be handled by a direct solver. To this end, we divide the domain as follows: $\Omega = \bigcup_{i=0}^{N_{\text{dom}}} \Omega_i$ where N_{dom} is the number of domains, since we consider a non-overlapping DDM $\Omega_i \cap \Omega_j = \emptyset$, if $i \neq j\ \forall i, j \in \{1, \ldots, N_{\text{dom}}\}$ and we denote the interface from two neighbouring cells by $\partial\Omega_i \cap \partial\Omega_j = \Sigma_{ij} = \Sigma_{ji},\ \forall i, j \in \{1, \ldots, N_{\text{dom}}\}$.

The second step of the DD is an iterative method, indexed by k, to compute the overall electric field $\boldsymbol{E}$. Therefore we begin by solving System (1) on each subdomain Ω_i, we denote the solution of every subsystem by $\boldsymbol{E}_i^{k=0}$. From this we can compute the first interface condition by

$$
g_{ji}^{k=0} := -\mu^{-1}\gamma_i^t\left(\operatorname{curl}\left(\boldsymbol{E}_i^{k=0}\right)\right) - ikS\left(\gamma_i^T\left(\boldsymbol{E}_i^{k=0}\right)\right), \tag{5}
$$

where S describes some boundary operator, which we will discuss in more detail below. Afterward, we obtain the next iteration step $\boldsymbol{E}_i^{k+1}$ via:

$$\begin{cases} \operatorname{curl}\left(\mu^{-1} \operatorname{curl}\left(\boldsymbol{E}_i^{k+1}\right)\right) - \omega^2 \boldsymbol{E}_i^{k+1} & = \boldsymbol{0} & \text{in } \Omega_i \\ \mu^{-1} \gamma_i^t \left(\operatorname{curl}\left(\boldsymbol{E}_i^{k+1}\right)\right) - i\kappa\omega\gamma_i^T\left(\boldsymbol{E}_i^{k+1}\right) & = 0 & \text{on } \Gamma_i^{\infty} \\ \gamma_i^T\left(\boldsymbol{E}_i^{k+1}\right) & = -\gamma_i^T\left(\boldsymbol{E}_i^{\text{inc}}\right) & \text{on } \Gamma_i^{\text{inc}} \\ \mu^{-1} S\left(\gamma_i^t\left(\operatorname{curl}\left(\boldsymbol{E}_i^{k+1}\right)\right)\right) - i\kappa\omega\gamma_i^T\left(\boldsymbol{E}_i^{k+1}\right) & = g_{ji}^k & \text{on } \Sigma_{i,j}. \end{cases} \tag{6}$$

Once $\boldsymbol{E}_i^{k+1}$ is computed, the interface is updated by

$$g_{ji}^{k+1} = -\mu^{-1}\gamma_i^t\left(\operatorname{curl}\left(\boldsymbol{E}_i^{k+1}\right)\right) - ikS\left(\gamma_i^T\left(\boldsymbol{E}_i^{k+1}\right)\right) = -g_{ij}^k - 2ikS\left(\gamma_i^T\left(\boldsymbol{E}_i^{k+1}\right)\right), \tag{7}$$

where $\boldsymbol{E}_i^k \to \boldsymbol{E}|_{\Omega_i}$ as $k \to \infty$. This convergence depends strongly on the chosen surface operator S. For a convergence analysis when the IBC are considered, see [7]

This iteration above can be interpreted as one step of the Jacobi fixed point method for the linear system

$$(\mathbb{1} - \mathcal{A})\boldsymbol{g} = \boldsymbol{b} \tag{8}$$

where $\mathbb{1}$ is the identity operator, $\boldsymbol{b}$ is the vector of the incident electric field, $\mathcal{A}$ is defined by $\mathcal{A}\boldsymbol{g}^k = \boldsymbol{g}^{k+1}$ and Equations (6), (7). Convergence is achieved for $\|(1 - \mathcal{A})g^k - b\| < TOL$ with some small tolerance $TOL > 0$. Often, $TOL = 10^{-6}, \ldots, 10^{-8}$. Instead of a Jacobi fixed point method one can also use a GMRES method to solve (8) more efficiently.

The crucial point of the DD is the choice of the interface conditions between the subdomains. The easiest choice is a non-overlapping Schwarz decomposition, where Dirichlet like interface conditions are used. For large wave numbers, e.g. the parameter ω becomes large, the system is highly indefinite. Consequently, a convergence of this algorithm for the time harmonic Maxwell equations for all ω cannot be expected; see [8, 10]. An analysis for an overlapping additive Schwarz method is given in [4].

Rather, we need more sophisticated tools in which the easiest choice are Impedance Boundary Conditions (IBC), which can be classified as Robin like interface conditions

$$S = \mathbb{1}. \tag{9}$$

3.2 Preconditioner

As is clear, the DDM is an iterative method, where we have to solve system (6) on each subdomain in each iteration step k. Usually, this is done by a direct solver, but instead, we can use a GMRES solver, which is preconditioned by an approximation of the block system

$$\begin{pmatrix} A - \omega^2 M & 0 \\ 0 & A - \omega^2 M \end{pmatrix}^{-1}. \tag{10}$$

Therefore we need to compute an approximation of $(A - \omega^2 M)^{-1}$, and we obtain this approximation by applying the AMG preconditioner provided by MueLu [3], where for the level transitions a direct solver is used. The latter is necessary, since otherwise the AMG preconditioner does not perform well for the THM. On the one hand, this procedure is cost expensive. On the other hand, we can reuse the preconditioner each time we solve system (6).

Another possible choice is to use the AMG preconditioner to compute directly an approximation of

$$\begin{pmatrix} A - \omega^2 M & -B \\ -B & A - \omega^2 M \end{pmatrix}^{-1}.$$

With this preconditioner only a few GMRES iterations are needed to solve the system (6). Since we are computing an approximation of the complete inverse this comes with a much higher memory consumption than using (10) as a preconditioner. Actually the memory consumption while using an iterative solver with (10) as a preconditioner is even lower than the memory consumption from a direct solver, which we show numerically in the next section. Therefore the block diagonal preconditioner is used in the following.

4 Numerical tests

In this section, we compare the performance of different preconditioners for two numerical examples. We choose a simple wave guide as our benchmark problem, moreover we test the performance of our method on a Y beam splitter. Our implementation is based on the open-source finite element library `deal.II` [2] with Trilinos [13] and MueLu [3]. As a direct solver, MUMPS (Multifrontal Massively Parallel Sparse Direct Solver) [1] is used. We perform an additive domain decomposition and compute each step in parallel with MPI. For the computations an Intel Xeon Platinum 8268 CPU was used with up to 32 cores.

4.1 Example 1: Block benchmark

Before we test the domain decomposition method, we want to compare the performance of different preconditioners on a single domain. Therefore we consider a simple 2D squared domain decomposed of a material with a higher refractive index in the center a carrier material with a lower refractive index beside it, see Figure 1.

Table 1 displays the GMRES iterations with a relative accuracy of $\epsilon = 10^{-8}$ for different preconditioners:

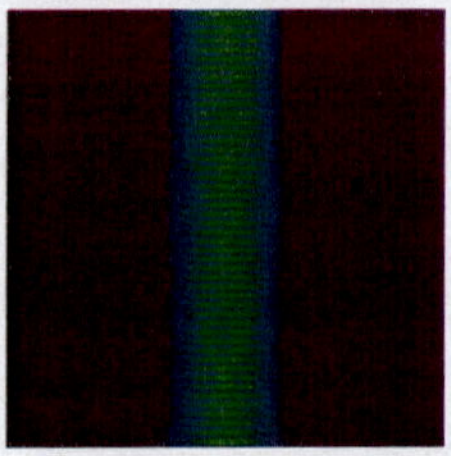

Fig. 1: As a benchmark problem, we consider a 1×1 square with different wave numbers (here $\lambda = 50$). In the center is a material with the refractive index $n_{\text{center}} = 1.516$ and as cladding the refractive index of air was used $n_{\text{air}} = 1.0$. For the discretization Nédélec elements with the polynomial degree $p = 1$ are used.

- ILU, the incomplete LU decomposition of (4),
- the implemented additive Schwarz preconditioner of [2, 13],[2]
- a Schur complement preconditioner based on [3],
- the block preconditioner (10).

Overall, the GMRES iteration numbers grow for large ω. In the case of the block preconditioner the GMRES iteration number first decreases and then increases for higher wave numbers.

Table 1: Example 1: GMRES iterations with different preconditioners.

wave number ω	GMRES iterations with the preconditioner			
	ILU	additive Schwarz	Schur complement	block preconditioner
5.0	165	515	156	75
10.0	349	750	161	52
20.0	833	>2000	172	26
40.0	>2000	-	diverged	25
60.0	-	-	diverged	38
80.0	-	-	diverged	49

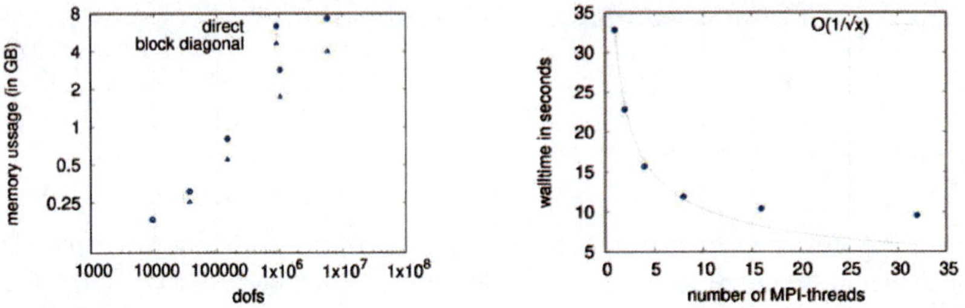

Fig. 2: On the left side: memory usage in dependence of the number of dofs. On the right side: walltime in dependence on the number of MPI-threads.

[2] https://www.dealii.org/current/doxygen/deal.II/classTrilinosWrappers_1_1PreconditionSSOR.html

[3] https://www.dealii.org/current/doxygen/deal.II/step_22.html

4.2 Example 2: Y beam splitter

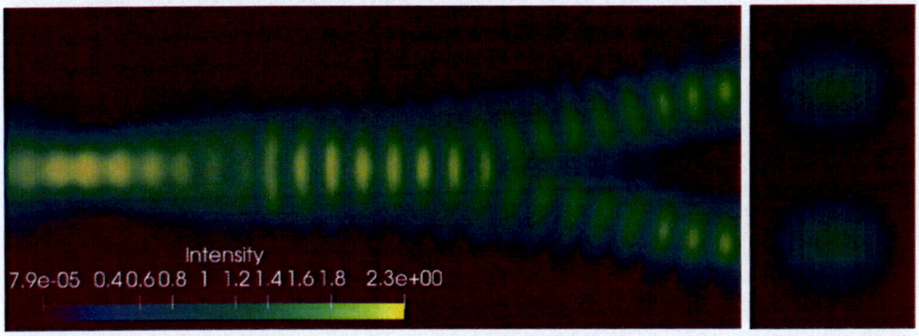

Fig. 3: Intensity plot of the y beam splitter, on the left side is the intensity on the x-y plane and on the right side is the intensity at the output.

Similar as in the simple wave guide, we consider for the Y beam splitter a material with a higher refractive index placed inside of an carrier material with a lower refractive index. Here we consider a 3D model of a Y beam splitter. The mesh was divided into 9 subdomains, and the average number of GMRES iterations to solve the subdomains are given in table 2, for the wave number $\lambda = 20$. For the discretization Nédélec elements with the polynomial degree $p = 3$ are used.

Table 2: Example 2: GMRES iterations on each domain for the block preconditioner

subdomain id	1	2	3	4	5	6	7	8	9
average number of GMRES iterations	34	40	41	31	35	39	37	33	32

5 Conclusion

In this contribution, we implemented a domain decomposition method with a block preconditioner for the time harmonic Maxwell equations. Therein, a crucial aspect is the construction of the subdomain interface conditions. Our algorithmic developments are demonstrated for two configurations of practical relevance, namely a block benchmark and a Y beam splitter.

Acknowledgements Funded by the Deutsche Forschungsgemeinschaft (DFG) under Germany's Excellence Strategy within the Cluster of Excellence PhoenixD (EXC 2122, Project ID 390833453).

References

1. P.R. Amestoy, A. Buttari, J.-Y. L'Excellent and T. Mary. Performance and scalability of the block low-rank multifrontal factorization on multicore architectures. *ACM Trans. Math. Software* **45**(1), Art. 2, 26 (2019).
2. D. Arndt, W. Bangerth, B. Blais, et al. The deal.II library, Version 9.2. *J. Numer. Math.* **28**(3), 131–146 (2020).
3. L. Berger-Vergiat, C.A. Glusa, J.J. Hu, M. Mayr, A. Prokopenko, C.M. Siefert, R.S. Tuminaro and T.A. Wiesner. MueLu multigrid framework. `http://trilinos.org/packages/muelu` (2019).
4. M. Bonazzoli, V. Dolean, I.G. Graham, E.A. Spence and P.-H. Tournier. Domain decomposition preconditioning for the high-frequency time-harmonic Maxwell equations with absorption. *Math. Comp.* **88**(320), 2559–2604 (2019).
5. M. El Bouajaj, B. Thierry, X. Antoine and C. Geuzaine. A quasi-optimal domain decomposition algorithm for the time-harmonic Maxwell's equations. *Journal of Computational Physics* **294**(1), 38–57 (2015).
6. M. Bürg. Convergence of an automatic hp-adaptive finite element strategy for Maxwell's equations. *Applied Numerical Mathematics* **72**, 188–204, 10 (2013).
7. V. Dolean, M.J. Gander and L. Gerardo-Giorda. Optimized Schwarz methods for Maxwell's equations. *SIAM J. Sci. Comput.* **31**(3), 2193–2213 (2009).
8. O.G. Ernst and M.J. Gander. Why it is difficult to solve Helmholtz problems with classical iterative methods. In: Numerical analysis of multiscale problems, Lect. Notes Comput. Sci. Eng. vol. 83, pp. 325–363, Springer, Heidelberg (2012).
9. V. Girault and P.-A. Raviart. *Finite element methods for Navier–Stokes equations. Theory and algorithms.* Volume 5 of Springer Series in Computational Mathematics. Springer-Verlag, Berlin (1986).
10. I.G. Graham, E.A. Spence and E. Vainikko. Domain decomposition preconditioning for high-frequency Helmholtz problems with absorption. *Math. Comp.* **86**(307), 2089–2127 (2017).
11. A. Grayver and T. Kolev. Large-scale 3D geoelectromagnetic modeling using parallel adaptive high-order finite element method. *GEOPHYSICS* **80**, E277–E291, 11 (2015).
12. G. Haase, M. Kuhn, U. Langer, S. Reitzinger and J. Schöberl. Parallel Maxwell Solvers. In: *Scientific Computing in Electrical Engineering*, pp. 71–78. Lecture Notes in Computational Science and Engineering, Vol. 18. Springer, Germany (2001).
13. M.A. Heroux, R.A. Bartlett, V.E. Howle, R.J. Hoekstra, J.J. Hu, T.G. Kolda, R.B. Lehoucq, K.R. Long, R.P. Pawlowski, E.T. Phipps, A.G. Salinger, H.K. Thornquist, R.S. Tuminaro, J.M. Willenbring, A. Williams and K.S. Stanley. An overview of the trilinos project. *ACM Trans. Math. Softw.* **31**(3), 397–423 (2005).
14. U. Langer, D. Pauly and S. Repin, editors. *Maxwell's Equations*. volume 24 of Radon Series on Computational and Applied Mathematics, de Gruyter, Berin (2019).
15. P. Monk. *Finite Element Methods for Maxwell's Equations*. Oxford Science Publications (2003).
16. A. Toselli and O. Widlund. *Domain decomposition methods – algorithms and theory*. Volume 34 of Springer Series in Computational Mathematics. Springer, Berlin, Heidelberg (2005).
17. S. Zaglmayr. *High Order Finite Element Methods for Electromagnetic Field Computation*. PhD thesis, Johannes Kepler University Linz (2006).

Adaptive Finite Element Thin-Plate Spline With Different Data Distributions

Lishan Fang and Linda Stals

1 Introduction

The thin-plate spline is a data fitting technique that possesses many favourable properties like insensitivity to noise [6]. One obstacle of its usage is the high computational cost and memory requirement for large data sets. The finite element thin-plate spline (TPSFEM) was proposed by Roberts, Hegland and Altas [11] to efficiently interpolate large data sets with similar smoothing properties as the thin-plate spline. It uses simple H^1 finite elements resulting in a sparser system of equations as opposed to ones with higher-order finite elements used in [5]. A detailed formulation of the TPSFEM is provided by Stals and Roberts [16] and a brief description is given below similar to the one shown by Fang [8].

Let $\{(\boldsymbol{x}_{(i)}, y_{(i)}) : i = 1, 2, \ldots, n\}$ be the observed data of size n and dimension d on a domain Ω, where $\boldsymbol{x}_{(i)} \in \mathbb{R}^d$ and $y_{(i)} \in \mathbb{R}$ are the i-th predictor value and response value, respectively. The TPSFEM s is defined as a combination of piecewise linear basis functions $\boldsymbol{b}$, where $s(\boldsymbol{x}) = \boldsymbol{b}(\boldsymbol{x})^T \boldsymbol{c}$ and $\boldsymbol{c}$ are coefficients of the basis functions. The TPSFEM s minimises the functional

$$J(\boldsymbol{c}, \boldsymbol{g}_1, \ldots, \boldsymbol{g}_d) = \boldsymbol{c}^T A \boldsymbol{c} - 2\boldsymbol{d}^T \boldsymbol{c} + \boldsymbol{y}^T \boldsymbol{y}/n + \alpha \sum_{k=1}^{d} \boldsymbol{g}_k^T L \boldsymbol{g}_k, \tag{1}$$

subject to the constraint $L\boldsymbol{c} = \sum_{k=1}^{d} G_k \boldsymbol{g}_k$, where $\boldsymbol{g}_k$ are coefficients of gradient approximations of s in dimension k, $A = \frac{1}{n} \sum_{i=1}^{n} \boldsymbol{b}(\boldsymbol{x}_{(i)}) \boldsymbol{b}(\boldsymbol{x}_{(i)})^T$, $\boldsymbol{d} = \frac{1}{n} \sum_{i=1}^{n} \boldsymbol{b}(\boldsymbol{x}_{(i)}) y_{(i)}$, $\boldsymbol{y} = [y_{(1)}, \ldots, y_{(n)}]^T$, L is a discrete approximation to the neg-

Lishan Fang
Mathematical Sciences Institute, Australian National University,
e-mail: Lishan.Fang@anu.edu.au

Linda Stals
Mathematical Sciences Institute, Australian National University,
e-mail: Linda.Stals@anu.edu.au

S. C. Brenner et al. (eds.), *Domain Decomposition Methods in Science and Engineering XXVI*, Lecture Notes in Computational Science and Engineering 145,
https://doi.org/10.1007/978-3-030-95025-5_72

ative Laplacian and G_k is a discrete approximation to the gradient operator in dimension k.

The smoothing parameter α balances the goodness of fit and smoothness of s. It is estimated iteratively using a stochastic estimator of the generalised cross-validation from Hutchinson [9]. More details are provided in [7]. It may also be calculated using alternate approaches discussed in [5]. Minimiser (1) is solved using Lagrange multipliers and the size of the resulting system of equations is proportional to the number of basis functions. This system is more efficiently solved than that of the thin-plate spline. A comparison using data from Section 4 between the TPSFEM and compactly supported basis functions (CSRBFs) from Wendland [17] with radius 0.5 is shown in Table 1. The TPSFEM achieves a similar root mean square error (RMSE) and maximum errors (MAX) as the CSRBFs using a system with significantly fewer nonzero entries. A comprehensive comparison is in progress and will be provided in [15].

Table 1: Computational cost

Technique	No. basis	Dimension	No. nonzero	RMSE	MAX
TPSFEM	900	3603	52,060	0.027	0.20
TPSFEM	1600	6403	93,538	0.014	0.091
CSRBFs	1024	1024	496,274	0.0098	0.157

The remainder of the article is organised as follows. In Section 2, we show adaptive refinement and error indicators of the TPSFEM. In Section 3, four two-dimensional data distribution patterns are displayed and compared regarding their influence on the maximum distance to any data. In Section 4, a numerical experiment is presented to examine the influence of these patterns. In Section 5, we summarise this article and the findings of the experiment.

2 Adaptive refinement

The accuracy of finite element approximations depends on the mesh size of the finite element grid [10]. The accuracy is improved by adaptive refinement that adapts the accuracy of the approximation within sensitive regions, like peaks, dynamically during an iterative process. An error indicator marks regions that require finer elements to achieve higher accuracy for refinement. Many error indicators have been developed for approximating partial differential equations but they may not be applicable for the TPSFEM.

The formulation of the TPSFEM is different from that of the traditional finite element method and it may not provide the information required by some error indicators. For example, the TPSFEM uses the observed data instead of given functions of partial differential equations and the data is often perturbed by noise or irregularly distributed. Fang [8] adapted the iterative adaptive refinement process and three error indicators of the finite element method for the TPSFEM. In this article, we will focus on the performance of these three error indicators, which are the auxiliary problem error indicator, recovery-based error indicator and norm-based error indicator.

The auxiliary problem error indicator evaluates approximation quality by solving a local approximation, which is the TPSFEM built on a union of elements [1, 8, 10]. It solves Minimiser (1) using a small subset of the observed data within those elements. The local approximations are locally more accurate than the global TPSFEM s and the approximation quality is measured by the difference between them. The recovery-based error indicator estimates errors by post-processing the gradient approximations of the TPSFEM [18]. It improves the discontinuous gradient approximations of s with piecewise linear basis functions and calculates the error as the difference between the two gradient approximations. The norm-based error indicator uses an error bound on the L_∞ norm of the TPSFEM to optimise the approximation [13]. It approximates second-order derivatives of s to identify regions that change rapidly and refines them to improve accuracy.

These three error indicators use different information of the TPSFEM to indicate regions with large errors. The recovery-based error indicator and norm-based error indicator only use $\boldsymbol{c}$ values in Minimiser (1) and they were adapted without major changes. In contrast, the original auxiliary problem error indicator solves boundary value problems and was modified to use data instead of a given function [2, 7]. Consequently, it is more susceptible to changes in the data like noise [8]. When the data distribution pattern changes, these error indicators may behave differently.

As opposed to the finite element method, the error of the TPSFEM may not converge with a smaller mesh size h. Roberts, Hegland and Altas [11] proved that the error convergence of the TPSFEM depends on the smoothing parameter α, maximum distance to data $d_{\boldsymbol{x}}$ and h. The new iterative adaptive refinement process updates the optimal α after the grid is refined and prevents it from dominating the error [8]. Besides, previous studies tested the performance of the error indicators using uniformly distributed data sets of size 1,000,000, which provide sufficiently small $d_{\boldsymbol{x}}$ [8]. When the data is irregularly distributed, we may not have sufficiently small $d_{\boldsymbol{x}}$ over the whole domain and the error convergence will be affected.

3 Data distribution

The observed data distributes differently depending on the application. For example, data is stored as maps of pixels for digital images or sampled randomly for surface reconstructions [4, 12]. Previous studies on the TPSFEM already deployed several data distribution patterns, including the uniform distribution [8], uniform distribution with missing data [14], random distribution [16] and random normal distribution [11]. We focus on these four data distribution patterns in this article.

The uniform distribution places data uniformly on the domain with fixed $d_{\boldsymbol{x}}$, which minimises its influence on the error convergence of the TPSFEM. However, other data distribution patterns may have varied $d_{\boldsymbol{x}}$ across the domain. Data points in certain regions may be missing and the TPSFEM needs to recover these surfaces.

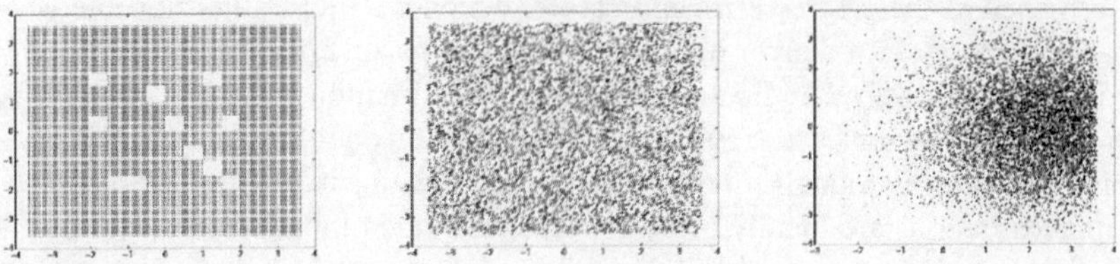

(a) (b) (c)

Fig. 1: Data distribution of 10,000 data points with (a) uniform distribution with data missing; (b) random distribution; and (c) random normal distribution

An example is shown in Figure 1(a), where data points in eleven square regions are missing and some of them neighbour each other. The error in these regions may not be improved by smaller h since $d_{\boldsymbol{x}}$ is large. Besides, the auxiliary problem error indicator uses data that is not available in those regions and the performance may be affected.

In many applications, the predictor values of data are sampled randomly with equal probabilities instead of a perfect uniform distribution, as shown in Figure 1(b). The random normal distribution places data points using a probability density function defined by a mean and a variance [3]. An example with variance 1.5; and mean 2.5 and 0 for predictor values $\boldsymbol{x}_1$ and $\boldsymbol{x}_2$ is shown in Figure 1(c). The density of the predictor values is higher at their mean than the rest of the domain. When a randomly distributed data set contains a large number of data points, the data density will be close across the domain. While it may not have a significant influence on the interpolant s due to sufficiently small $d_{\boldsymbol{x}}$, it may affect the error indicators as some elements may contain few data points. In comparison, a randomly normally distributed data set has different data densities across the domain with varied $d_{\boldsymbol{x}}$. While the error convergence behaviour may not be affected by randomly distributed data, the error indicators were not developed to handle data with various densities and their performance may be affected [5].

4 Numerical experiment

A numerical experiment was conducted to test the error indicators using these data distribution patterns. The data consists of 10,000 data points limited inside $[-3.6, 3.6]^2$ and is modeled by the peaks function f from MATLAB, where $f(\boldsymbol{x}) = 3(1-x_1)^2 \mathrm{e}^{-x_1^2-(x_2+1)^2} - 10(x_1/5 - x_1^3 - x_2^5)\mathrm{e}^{-x_1^2-x_2^2} - \frac{1}{3}\mathrm{e}^{-(x_1+1)^2-x_2^2}$. It has oscillatory surfaces at the center of the domain and flat surfaces near its boundaries [11]. Gaussian noise with mean 0 and standard deviation 0.01 is also included in some data sets to assess the performance in the presence of noise. The distribution patterns of irregularly distributed data sets are shown in Figure 1.

We focus on the efficiency of uniform and adaptive grids, which is measured by the error metric versus the number of nodes in the grid. A grid that achieves a low error metric with a smaller number of nodes is considered more efficient. We consider both the root mean square error (RMSE) and approximate error, which measures how closely s fits data and reproduces f, respectively [7, 8]. The approximate error e is defined as $e = \sqrt{\sum_{i=1}^{m} h_i^2 e_i^2}$, where e_i is the difference between s and f at the i-th node, h_i is the longest edge connected to the i-th node and m is the number of nodes in the grid. The efficiency of final grids are calculated as products of the error metric and the number of nodes and is provided in the legend of each convergence plot.

4.1 Results

The convergence of the RMSE for data sets with the four distribution patterns is shown in Figure 2. Adaptive refinement focuses on refining the oscillatory surfaces at the centre and error convergence rates of all three adaptive grids are higher than that of the uniform grid in Figure 2(a). When the data is uniformly distributed, the three error indicators have similar performance and produce adaptive grids more than twice as efficient as the uniform grid.

Figure 2(c) shows similar error convergence of uniform and adaptive grids with random distribution. The TPSFEM and error indicators are not affected as d_x remains sufficiently small within a large number of randomly distributed data points. In contrast, d_x is large in some regions where data points are missing or the data is randomly normally distributed shown in figures 1(a) and 1(c), respectively. While this does not markedly affect the TPSFEM, it slightly weakens the performance of the auxiliary problem error indicator, as shown in figures 2(b) and 2(d), respectively. Local approximations of the auxiliary problems are built with different numbers of data points and the accuracy deteriorates.

The convergence of the approximate error for data sets with the four distribution patterns is shown in Figure 3. The error convergence rates of the approximate error with the uniform or random distribution in figures 3(a) and 3(c) are similar to those of the RMSE in figures 2(a) and 2(c). The TPSFEM closely reproduces the original smooth function f when d_x is sufficiently small in these two distribution patterns. When the data is scarce in some regions, the TPSFEM interpolates smooth surfaces, which may not recover f especially when it is oscillatory. Consequently, the convergence of the approximate errors for the uniform distribution with missing data or random normal distribution slows down in the last few iterations as shown in figures 3(b) and 3(d), respectively. Similarly, the auxiliary problem error indicator underperforms compared to the other two error indicators.

In the presence of noise, the RMSE values of the TPSFEM stop decreasing at some point depending on the noise level of data as demonstrated by Fang [8]. Therefore, we only consider the approximate error here. The convergence of the approximate error for data sets with noise is shown in Figure 4. All error convergence rates are lower than those without noise since the TPSFEM may not reproduce f from noisy data.

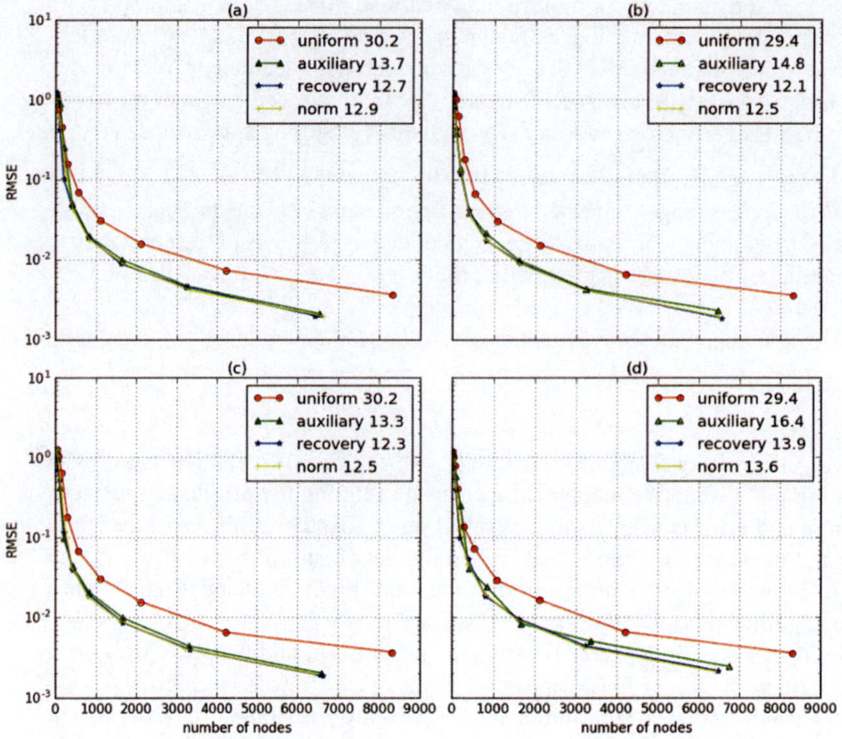

Fig. 2: RMSE for (a) uniform distribution; (b) uniform distribution with missing data; (c) random distribution; and (d) random normal distribution.

The error convergence rates with the uniform or random distribution are higher than the others in Figure 3. Elements in these two distribution patterns contain a similar number of data points and the effects of noise are cancelled out when data points are projected on the finite element grid. In comparison, the error with the uniform distribution with missing data and random normal distribution stops decreasing at the last two iterations. The error convergence rates with random normal distribution in Figure 3(d) are the lowest of four distribution patterns. Since some elements may contain few data points, it is more sensitive to noise in data, which leads to a marked difference to f.

The three error indicators perform differently in the experiment. The performance of the auxiliary problem error indicator worsens when data is perturbed by noise, especially for the random distribution in Figure 4(c). Since some elements may contain few data points, the accuracy of the local approximation is more susceptible to noise and may indicate large errors incorrectly. In contrast, the recovery-based error indicator and norm-based error indicator use $\boldsymbol{c}$ values to indicate large errors. Since the effects of the data distribution pattern and noise have been minimised by

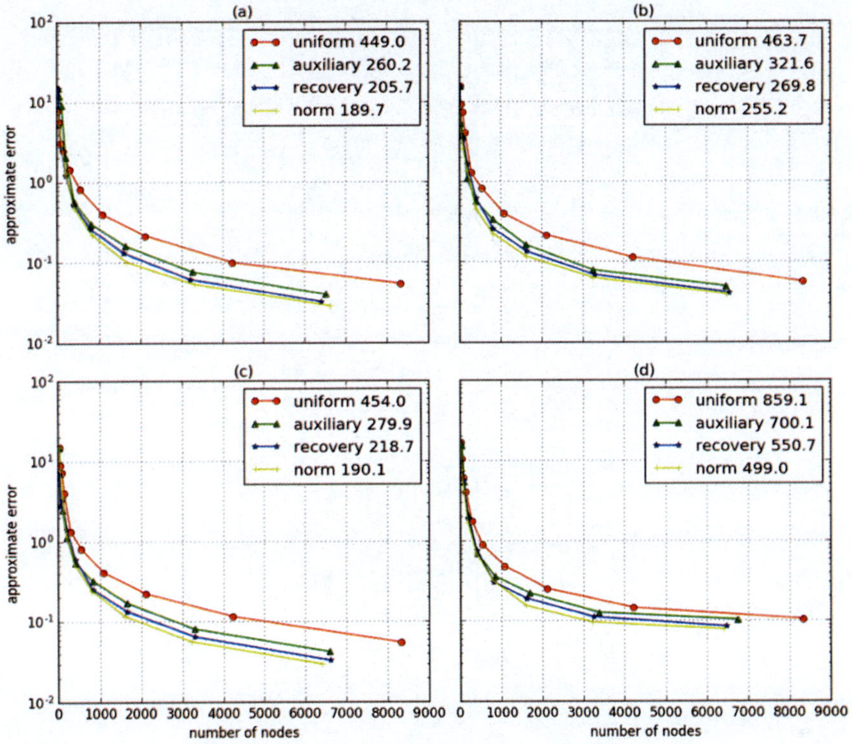

Fig. 3: Approximate error for (a) uniform distribution; (b) uniform distribution with missing data; (c) random distribution; and (d) random normal distribution.

the TPSFEM, these two error indicators produce efficient adaptive grids for data sets with noise in Figure 4.

5 Conclusion

In this article, we explore four data distribution patterns and investigate their effects on the efficiency of adaptive grids generated using three error indicators. The four data distribution patterns lead to different maximum distances to data and affect the performance of the TPSFEM and its error indicators. While the TPSFEM may not restore the original function in regions with scarce data, it recovers a smooth surface to closely interpolate the data. Besides, the uniform and random distributions have close data densities across the domain and thus have less influence on the TPSFEM than the uniform distribution with missing data and random normal distribution. We also find that all the error indicators significantly improve the efficiency of

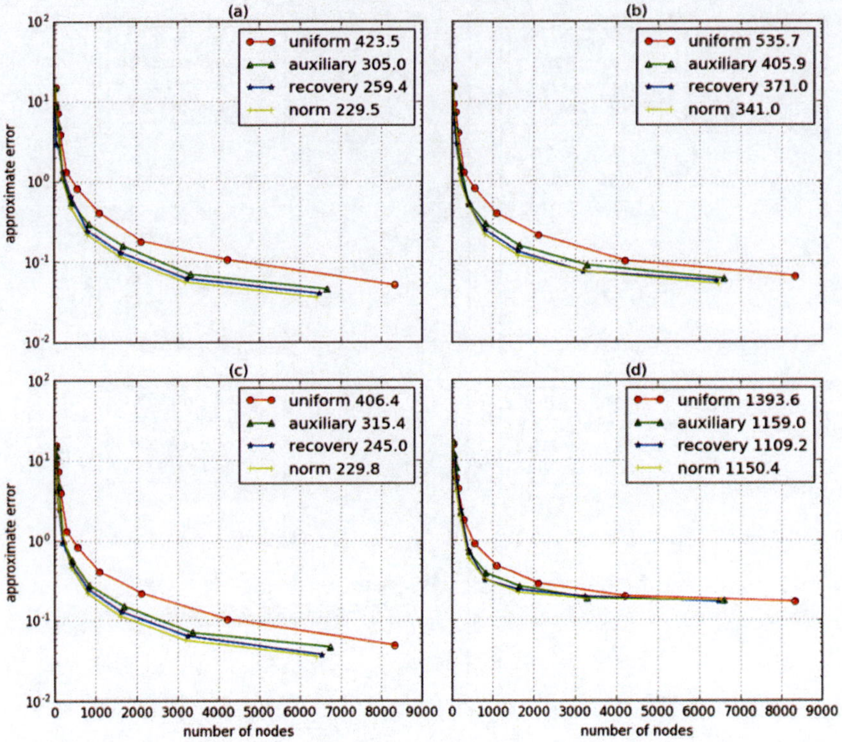

Fig. 4: Approximate error for data sets perturbed by noise with (a) uniform distribution; (b) uniform distribution with missing data; (c) random distribution; and (d) random normal distribution.

adaptive grids with all four data distribution patterns. The auxiliary problem error indicator uses data for local approximations and is more vulnerable to changes in the data distribution patterns and noise. In contrast, the recovery-based error indicator and norm-based error indicator only use the information of the TPSFEM and are insensitive to these two factors.

References

1. M. Ainsworth and J.T. Oden. Aposteriori error estimation in finite element analysis. *Computer Methods in Applied Mechanics and Engineering* **142**(1-2), 1–88 (1997).
2. I. Babuvška and W.C. Rheinboldt. Error estimates for adaptive finite element computations. *SIAM Journal on Numerical Analysis* **15**(4), 736–754 (1978).
3. J.S. Bendat and A. Piersol. *Random data: analysis and measurement procedures*. Wiley series in probability and statistics volume 729, Wiley (2011).

4. J.C. Carr, R.K. Beatson, J.B. Cherrie, T.J. Mitchell, W.R. Fright, B.C. McCallum and T.R. Evans. Reconstruction and representation of 3D objects with radial basis functions. In: *Proceedings of the 28th annual conference on Computer graphics and interactive techniques*, pp. 67–76, ACM (2001).
5. Z.M. Chen, R. Tuo and W.L. Zhang. Stochastic convergence of a nonconforming finite element method for the thin plate spline smoother for observational data. *SIAM Journal on Numerical Analysis* **56**(2), 635–659 (2018).
6. J. Duchon. Splines minimizing rotation-invariant semi-norms in sobolev spaces. In: *Constructive theory of functions of several variables*, pp. 85–100, Springer (1977).
7. L. Fang. *Error estimation and adaptive refinement of finite element thin plate spline*. PhD thesis, The Australian National University (2020).
8. L. Fang and L. Stals. Adaptive discrete thin plate spline smoother. *ANZIAM Journal* (Accepted).
9. M.F. Hutchinson. A stochastic estimator of the trace of the influence matrix for Laplacian smoothing splines. *Communications in Statistics – Simulation and Computation* **19**(2), 433–450 (1990).
10. W.F. Mitchell. A comparison of adaptive refinement techniques for elliptic problems. *ACM Transactions on Mathematical Software* **15**(4), 326–347 (1989).
11. S. Roberts, M. Hegland and I. Altas. Approximation of a thin plate spline smoother using continuous piecewise polynomial functions. *SIAM Journal on Numerical Analysis* **41**(1), 208–234 (2003).
12. R.J. Schalkoff. *Digital image processing and computer vision*. Volume 286. Wiley New York (1989).
13. E.G. Sewell. *Analysis of a finite element method: PDE/PROTRAN*. Springer Science & Business Media (2012).
14. L. Stals. Efficient solution techniques for a finite element thin plate spline formulation. *Journal of Scientific Computing* **63**(2), 374–409 (2015).
15. L. Stals. Radial basis functions: Comparison between compact basis and finite element basis. (In preparation).
16. L. Stals and S. Roberts. Smoothing large data sets using discrete thin plate splines. *Computing and Visualization in Science* **9**(3), 185–195 (2006).
17. H. Wendland. Piecewise polynomial, positive definite and compactly supported radial functions of minimal degree. *Advances in Computational Mathematics* **4**(1), 389–396 (1995).
18. O.C. Zienkiewicz and J.Z. Zhu. A simple error estimator and adaptive procedure for practical engineerng analysis. *International Journal for Numerical Methods in Engineering* **24**(2), 337–357 (1987).

A Multirate Accelerated Schwarz Waveform Relaxation Method

Ronald D. Haynes and Khaled Mohammad

1 Introduction

Schwarz Waveform relaxation (SWR) [1, 2, 6] is an iterative algorithm for solving time dependent partial differential equations (PDEs) in parallel. The domain of the PDE is partitioned into overlapping or non-overlapping subdomains, then the PDE is solved iteratively on each subdomain. The emphasis has focused on developing artificial transmission conditions which exchange information between neighboring subdomains and lead to fast convergence.

The initial guess at the subdomain boundaries is often chosen to be a constant (maybe a continuation of the initial condition for the PDE). We show here that, in some situations, we can dramatically reduce the number of SWR iterations to convergence by computing an improved initial guess at the subdomain boundaries using a multirate (MR) time integrator. The MR time integrator naturally produces a spatial splitting over time windows, while the SWR portion of the algorithm can fix a potential loss of accuracy in the MR approach. The efficacy of the resulting accelerated SWR (ASWR) algorithm is demonstrated for a test problem.

Ronald D. Haynes
Department of Mathematics & Statistics, Memorial University of Newfoundland, St. John's, Newfoundland, Canada, A1C 5S7, e-mail: rhaynes@mun.ca

Khaled Mohammad
Department of Mathematics & Statistics, Memorial University of Newfoundland, St. John's, Newfoundland, Canada, A1C 5S7, e-mail: km2605@mun.ca

S. C. Brenner et al. (eds.), *Domain Decomposition Methods in Science and Engineering XXVI*, Lecture Notes in Computational Science and Engineering 145,
https://doi.org/10.1007/978-3-030-95025-5_73

2 Background material

We assume the PDE has been semi-discretized in space using finite differences, leading to a system of ordinary differential equations (ODEs) of the form

$$\begin{aligned} y' &= f(t, y), \\ y(t_0) &= y_0, \; y \in R^N. \end{aligned} \tag{1}$$

We integrate (1) using an MR method largely due to Savenco et al. [5].

Consider the embedded Rosenbrock method given by

$$\begin{aligned} (I - \gamma \Delta t f_y(t_{n-1}, y_{n-1}))\mathcal{K}_1 &= \Delta t f(t_{n-1}, y_{n-1}) + \gamma \Delta t^2 f_t(t_{n-1}, y_{n-1}), \\ (I - \gamma \Delta t f_y(t_{n-1}, y_{n-1}))\mathcal{K}_2 &= \Delta t f(t_{n-1} + \Delta t, y_{n-1} + \mathcal{K}_1) \\ &\quad - \gamma \Delta t^2 f_t(t_{n-1}, y_{n-1}) - 2\mathcal{K}_1, \\ y_n &= y_{n-1} + \mathcal{K}_1, \\ \hat{y}_n &= y_{n-1} + \frac{3}{2}\mathcal{K}_1 + \frac{1}{2}\mathcal{K}_2, \end{aligned}$$

where $f_y(t_n, y_n)$ is the Jacobian matrix and $f_t(t_n, y_n)$ is the time derivative. In what follows, f_t is estimated using a forward difference. The first order approximation (ROS1), y_n, is used as the time integrator to obtain the numerical results presented in this paper, while the second order approximation (ROS2), $\hat{y}_n$, provides an estimate of the local error vector, E_n. In our tests we use $\gamma = 1/2$ which results in A-stable ROS1 and ROS2 methods [4]. The approximation is linearly implicit, requiring a linear solve at each time step. This can be efficient for non-linear problems.

ROS1 and ROS2 can be used together to produce an adaptive (single rate) time stepper based on local error control. The local error of the ith component for the ODEs (1) at time $t = t_n$, $E_{n,i}$, can be estimated as $E_{n,i} = |y_{n,i} - \hat{y}_{n,i}|$, for $i = 1, \ldots, N$. If $\|E_n\|_\infty$ (obtained with time step Δt) is less than the required tolerance, the integration proceeds with a (possibly larger) new time step, otherwise the step is repeated with a smaller step size. In either case the new time step is given by $\Delta t_{\text{new}} = \theta \Delta t (tol/\|E_n\|_\infty)^{1/2}$, where $\theta < 1$ is a safety factor and tol is the tolerance.

3 A multirate approach

The local error control mechanism can also be used as the basis for an MR approach, see [5]. Suppose a local error, E_n, is obtained with a time step Δt. We can estimate the time step required by each component of the ODE system, $\Delta t_{n,i}$ to achieve the tolerance tol as $\Delta t_{n,i} = \theta \Delta t (tol/E_{n,i})^{1/2}$, for $i = 1, 2, \ldots, N$. We denote the minimum time step required by any component by $\Delta t_{\min} = \min_{i=1,\ldots,N} \Delta t_{n,i}$. Figure 1 shows two scenarios for the size of the local error during the integration of parabolic PDEs of interest here.

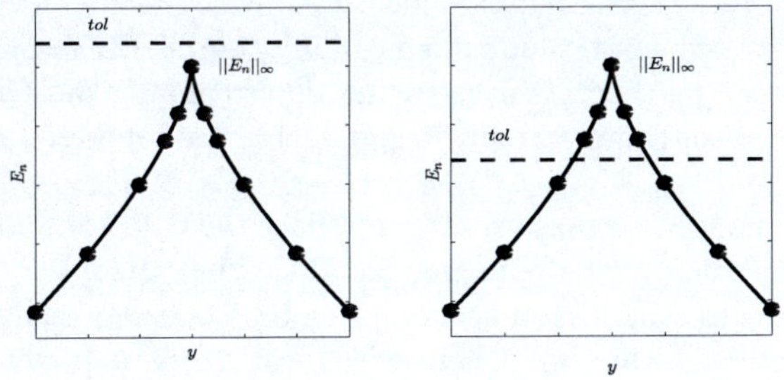

Fig. 1: Identifying *fast* components using the local error.

In the figure on the left all of the components of the local error are below the required tolerance. In this case the time step is accepted, and likely increased for the next step. The plot on the right shows a situation where some components of the local error are larger than the required tolerance. In the MR approach, **only** these (fast) components are recomputed (using the smaller time step, $\Delta t_{\min}$). The other (slow) components are accepted without further computation. Coupling between the fast and slow components is typically handled by interpolation or using dense output formulae. The single rate approach, in contrast, would recompute all components with a smaller time step if the norm of the local error is larger than the tolerance. The process is then repeated for the next global time step. In [5] the size of the global time step is chosen using a MR factor which is controlled by a heuristic based on the estimated computational savings.

In [5] uniform or recursive refinements are suggested for the fast components. An error analysis for linear systems and the θ-method with one level of refinement is given in [3]. For parabolic time-dependent PDEs which have groups of components evolving at different time scales, the MR method demonstrates a gain in efficiency. In our experience, however, the approach is quite sensitive to the choice of slow and fast components and the accuracy of the interpolation method.

To illustrate this we consider the traveling wave equation

$$u_t = \epsilon u_{xx} + \xi u^2(1-u), \tag{2}$$

for $0 < x < 5, 0 < t \leq T = 3$, with initial and boundary conditions $u(x,0) = (1 + e^{\lambda(x-1)})^{-1}$ and $u_x(0,t) = u_x(5,t) = 0$, where $\epsilon = 10^{-2}$, $\xi = 1/\epsilon$ and $\lambda = \sqrt{\xi/2\epsilon}$. In space, u is discretized with $N = 1000$ grid points and standard second order differences. For comparison purposes a single rate reference solution has been integrated in time using Matlab's $ode15s$ with tolerance 10^{-10}. The solution is a travelling wave solution with a sharp interface between $u = 1$ and $u = 0$ moving to the right.

In Tables 1 and 2, we use Savenco's code, see [5], for both the single rate and MR approaches. We modify the inputs to control the MR time step size, the number of points added to fast region identified by the local error test, and the interpolation used to generate the slow components needed during the refinement of the fast components. The errors at the final time are measured by subtracting the single or the MR solution from the reference solution in the infinity norm. The work estimates are based on the cost of the linear solves in the timestepping. The CPU times (in seconds) for both the single rate and MR approaches are reported for various tolerances.

Table 1 shows that the MR approach is able to reduce the CPU time, albeit with some decrease in the accuracy. The reduction in CPU time is more dramatic for smaller required tolerances. The loss in accuracy can be reduced by adding points to the fast regions identified by the component-wise local error test or by increasing the accuracy of the interpolation used at the interfaces of the regions, see Table 2.

	Single-rate			Multirate		
Tol	**Error**	**Work**	**CPU**	**Error**	**Work**	**CPU**
1.00e-03	3.204e-03	1639638	3.790	1.406e-02	131260	3.020
5.00e-04	1.924e-03	2256254	5.530	2.586e-03	167978	2.990
1.00e-04	4.835e-04	4862858	3.990	6.812e-03	319690	4.530
5.00e-05	2.541e-04	6816810	5.580	3.294e-03	442186	4.120
1.00e-05	5.427e-05	15057042	12.120	5.460e-04	971304	6.880

Table 1: Errors, Work and CPU time in seconds at $T = 3$ of Savcenco's MR approach with uniform refinement and using the dense output method.

Added Points	Error$_L$	Error$_Q$	Error$_D$
0	8.392e-03	3.407e-03	3.271e-03
5	2.061e-03	1.052e-03	1.028e-03
10	7.418e-04	5.623e-04	5.582e-04
15	5.062e-04	4.751e-04	4.744e-04
20	4.654e-04	4.600e-04	4.599e-04

Table 2: Errors obtained using linear and quadratic interpolation and dense output for (2) at $T = 3$ using a fixed MR time step $\Delta_m t = 2\Delta_s t$ with $tol = 10^{-4}$ while varying the number of points added to the fast region.

The number of added points which allows the MR algorithm to recover the single rate error for a given tolerance depends on the MR time step size, the final integration time, the PDE being solved, and the discretizations used. This is difficult to determine a priori.

4 An accelerated SWR approach

Consider our test problem discretized using 1000 uniformly spaced points on [0, 5]. We solve the global domain problem with MR time steps of $\Delta_m t = m\Delta_s t$ with a multirate factor m and $\Delta_s t = 0.01$ (a time step which keeps the local error below a tolerance of $tol = 5 \times 10^{-3}$ for the single rate (global) algorithm). In Figure 2, the horizontal lines show multirate time steps with $m = 20$. The local error estimate is used to identify the fast region (shown in red) and the slow regions, during each MR time step.

To implement an SWR iteration the domain is partitioned into ten equal subdomains, as shown in the left of Figure 2. We refer to this as a static partitioning. Overlapping subdomains are obtained by adding a small overlap (not shown) to the left and right of the interior interfaces. We generate initial guesses for the SWR iteration as follows. If an interface lies in a slow region then an interpolant in time, constructed using the solution obtained from the MR time step, is used. If an interface lies in a fast region then an improved initial guess is constructed by refining the fast region using a single rate method with a time step of $\Delta_s t$, as described in Section 3. A (classical) SWR iteration is used from these initial guesses, here the SWR iterates are also computed using $\Delta_s t$ (in practice one could use an adaptive time stepping for the subdomain solves). The process is then repeated over the next $\Delta_m t$, and so on. To demonstrate, in Figure 3 we plot the results of this experiment for ASWR with static partitioning (S-ASWR) on the second (left) and fourth (right) time windows. The vertical axis shows the error between the single rate and SWR solutions. The two norm of the error (in time) is calculated along all interfaces. SWR is accelerated if *any* of the subdomain boundaries lie in a fast region and hence is able to benefit from the refined solution. The reduction in the iteration count on each time window depends on the position of the interface in the fast region. For this example, we will see that with a good placement of the interface one SWR iteration is able to correct the loss of accuracy inherent in the MR algorithm.

Motivated by the improvement, should a subdomain boundary lie in a fast region, we can build an improved dynamic partitioning algorithm. After completing a global MR time step, assuming a sufficient number of processors we partition the whole domain by introducing an interface in each fast region, and partition the rest of the domain so that the subdomains are of (approximately) equal size. This is illustrated in the right plot in Figure 2. Placing the interface in the middle of the fast region attempts to minimize the coupling between the fast and slow components. With this dynamic partitioning D-ASWR accelerates convergence in an approximately uniform way over all time windows, see Figure 4 where the SWR errors are shown on the second time window for two different multirate time steps.

The difficulty in choosing the appropriate number of points to add to the fast region and the interpolation required in the MR method is pushed aside and instead the refined fast solution can be used to accelerate a correction using SWR. The computation of the global time step and the subsequent partitioning from the MR algorithm provides: information that can guide the SWR partitioning, improved initial guesses at the interfaces for the subsequent SWR correction, and information about the single rate or SWR time step required to globally achieve the local error tolerance.

A general algorithm would handle multiple fast regions during a multirate time step. Interfaces are introduced into each fast region and SWR initial guesses are obtained by refining the fast regions (in parallel). A global time step for the SWR iteration can be chosen to be the smallest time step used over all the fast regions. Again with a sufficient number of processors a well load-balanced splitting is possible while keeping interfaces in the fast regions.

The number of SWR iterations can be further minimized by introducing a non-overlapping splitting and an optimized SWR iteration.

5 A comparison

In Table 3, we provide a comparison of the single rate, MR, static and dynamic ASWR algorithms. Single rate results are given, then the local error estimate is used to identify and refine the fast region. MR results (using the algorithm in Section 3) with 0 and 20 points added to the identified fast region are provided. Finally, one classical ASWR iteration is used with static and dynamic partitioning with $S = 15$ subdomains for $\Delta_s t = 0.01$, $S = 26$ for $\Delta_s t = 0.005$, $S = 30$ for $\Delta_s t = 0.0025$, $S = 34$ for $\Delta_s t = 0.00125$ and only one point of overlap. A multirate factor of $m = 10$ is used for the MR and ASWR results.

	Single-rate		MR (0)		MR (20)		S-ASWR		D-ASWR	
$\Delta_s t$	Error	Work	Error	Work	Error	Work	Error	Work	Error	Work
0.01	0.0273	300000	0.0345	51910	0.0274	63930	0.0279	72198	0.0274	74505
0.005	0.0131	600000	0.2126	84760	0.0138	108600	0.0243	115085	0.0130	110360
0.025	0.0042	1200000	0.0950	162710	0.0043	210680	0.0107	215412	0.0037	207800
0.0125	0.0012	2400000	0.0391	317990	0.0012	413980	0.0309	423535	0.0002	400996

Table 3: Errors and work at $T = 3$ for the single rate method, MR with 0 and 20 added points to the fast region, and static and dynamic ASWR.

Table 3 shows that the MR method without points added to the fast region loses accuracy compared to the single rate method. The refined fast region allow us to accelerate the SWR convergence recovering the lost accuracy with a cost less than the cost of the single rate solution. Increasing the number of subdomains further makes the simulation more efficient. The S-ASWR method (with static partitioning) has a higher error than the D-ASWR approach after one SWR correction. This is due to the somewhat random placement of the interfaces in the S-ASWR approach. One iteration of D-ASWR is sufficient to achieve the required tolerance for this problem.

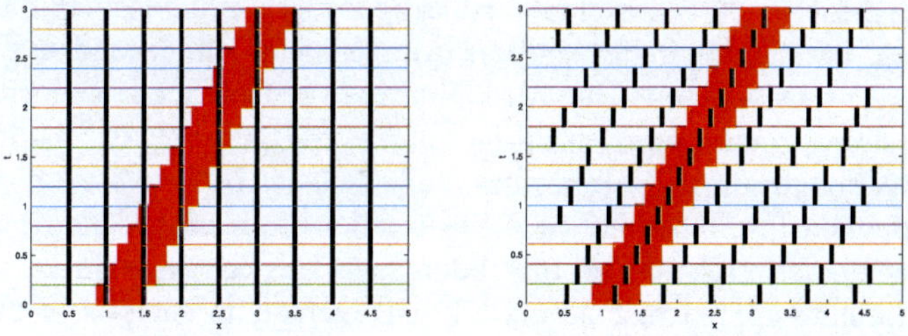

(a) Static Partitioning (b) Dynamic Partitioning

Fig. 2: Partitioning approaches for ASWR.

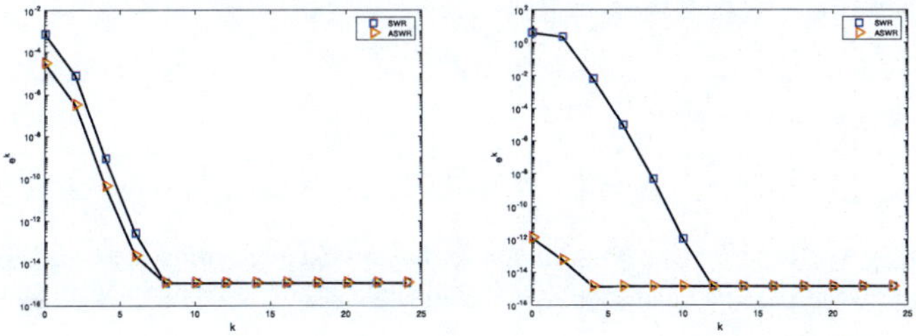

(a) $t \in [\Delta_m t, 2\Delta_m t]$ (b) $t \in [3\Delta_m t, 4\Delta_m t]$

Fig. 3: Convergence histories for classical S-ASWR with $S = 10$ and $m = 20$ on the second (left) and fourth (right) time window using a static partitioning. An overlap of 10 points is used during the SWR.

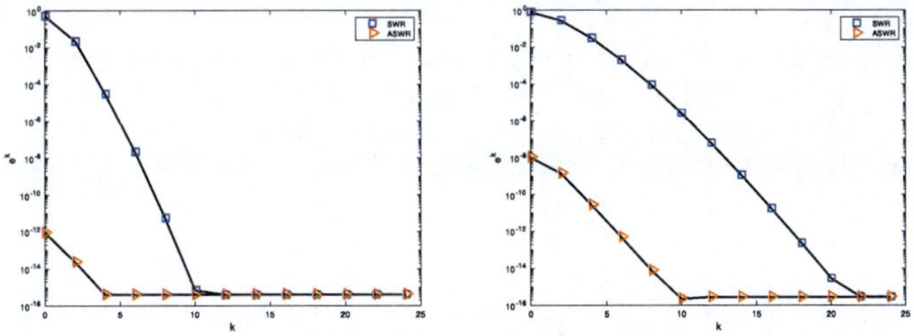

(a) $t \in [\Delta_m t, 2\Delta_m t]$ with $m = 20$ (b) $t \in [\Delta_m t, 2\Delta_m t]$ with $m = 10$

Fig. 4: Convergence histories for D-ASWR with $m = 20$, 10 points of overlap, on $S = 10$ subdomains (left) and $m = 10$, 5 points of overlap, on $S = 15$ subdomains (right) on the second time window using a dynamic partitioning.

6 Conclusions

The MR approach proposed in [5] provides an automatic way to identify the fast and slow components of a problem based on a local error estimate. The coupling between this fast-slow splitting leads to a loss in accuracy as compared to a single rate approach. The error can be reduced by increasing the size of the fast region (to reduce the coupling) but the required size of the overlap is problem dependent.

We propose algorithms which use the MR splitting to provide a decomposition of the space-time domain and improved initial guesses for the SWR (correction), resulting in an ASWR algorithm. The robustness and efficiency of the ASWR comes from the large reduction in the number of SWR iterations to reach the single rate accuracy and the increase in the number of subdomains. This can be achieved with the dynamic partitioning approach. Future work will include an analysis of these ASWR algorithms.

Acknowledgements The authors would like to thank E. Savcenco for providing his multirate code for experimental purposes.

References

1. V. Dolean, P. Jolivet and F. Nataf. *An introduction to domain decomposition methods: Algorithms, theory, and parallel implementation*. Society for Industrial and Applied Mathematics (SIAM), Philadelphia, PA (2015).
2. M.J. Gander and A.M. Stuart. Space-time continuous analysis of waveform relaxation for the heat equation. *SIAM Journal on Scientific Computing* **19**(6), 2014–2031, November (1998).
3. W. Hundsdorfer and V. Savcenco. Analysis of a multirate theta-method for stiff ODEs. *Applied numerical mathematics* **59**(3), 693–706 (2009).
4. W. Hundsdorfer and J.G. Verwer. *Numerical Solution of Time-Dependent Advection-Diffusion-Reaction Equations*. Volume 33 of Springer Series in Computational Mathematics. Springer, Berlin-Heidelberg (2003).
5. V. Savcenco, W. Hundsdorfer and J.G. Verwer. A multirate time stepping strategy for stiff ordinary differential equations. *BIT Numerical Mathematics* **47**(1), 137–155 (2007).
6. S.-L. Wu and M.D. Al-Khaleel. Optimized waveform relaxation methods for RC circuits: discrete case. *ESAIM: Mathematical Modelling and Numerical Analysis* **51**(1), 209–223, January (2017).

A Convergence Analysis of the Parallel Schwarz Solution of the Continuous Closest Point Method

Alireza Yazdani, Ronald D. Haynes, and Steven J. Ruuth

1 Introduction

Consider the surface intrinsic positive Helmholtz equation

$$(c - \Delta_S)u = f, \tag{1}$$

where Δ_S denotes the Laplace–Beltrami operator associated with the surface $S \subset \mathbb{R}^d$, and $c > 0$ is a constant. Discretization of this equation arises in many applications including the time-stepping of reaction-diffusion equations on surfaces [10], the comparison of shapes [15], and the solution of Laplace–Beltrami eigenvalue problems [9]. As a consequence, considerable recent work has taken place to develop efficient, high-speed solvers for this and other related PDEs on surfaces.

There are several methods to solve surface intrinsic differential equations (DEs). If a surface parameterization (a mapping from the surface to a parameter space) is known, then the equation can be solved in the parameter domain [4]. For triangulated surfaces, a finite element discretization can be created [5]. Alternatively, we can solve the DE in a neighborhood of the surface using standard PDE methods in the underlying embedding space [2, 3, 12, 16]. Here, we discretize via the *closest point method* (CPM), which is an embedding method suitable for the discretization of PDEs on surfaces. The closest point method leads to non-symmetric linear systems to solve. On complex geometries or when varying scales arise, iterative solvers can be slow despite the sparsity of the underlying systems. In order to develop an efficient iterative solver which is also capable of parallelism, Parallel Schwarz (PS) and Optimized Parallel Schwarz (OPS) algorithms have been applied to the CPM for

Alireza Yazdani, Steven J. Ruuth
Simon Fraser University, 8888 University Dr., Burnaby, BC V5A 1S6,
e-mail: alirezay@sfu.ca, sruuth@sfu.ca

Ronald D. Haynes
Memorial University of Newfoundland, 230 Elizabeth Ave, St. Johns, NL A1C 5S7,
e-mail: rhaynes@mun.ca

S. C. Brenner et al. (eds.), *Domain Decomposition Methods in Science and Engineering XXVI*, Lecture Notes in Computational Science and Engineering 145,
https://doi.org/10.1007/978-3-030-95025-5_74

(1) in [13]. Here, we study the convergence of the PS-CPM at the continuous level for smooth, closed 1-manifolds where periodicity is inherent in the geometry. As shown in Section 3, this problem, posed in $\mathbb{R}^d$, is equivalent to a one-dimensional periodic problem. This leads us to study the 1-dimensional periodic problem in detail.

While there has been substantial work carried out on Schwarz methods, they have not been widely used for solving surface DEs. The shallow-water equation is solved with a PS iteration on the cubed-sphere with a finite volume discretization in [17]. PS and OPS methods for the 2D positive definite Helmholtz problem are investigated on the unit sphere in [8]. In [8], the analysis is based on latitudinal subdomains that are periodic in longitude. Hence, the Fourier transform is a natural choice to solve the subproblems analytically and obtain the contraction factor. PS and OPS methods are also analyzed with an overset grid for the shallow-water equation in [14]. In that work, the discretization in 1D is reduced to the positive definite Helmholtz problem on the unit circle. The unit circle case is investigated with two equal-sized subdomains, and a convergence factor is derived for the configuration in terms of the overlap parameter. In addition, the 2D positive definite Helmholtz problem on the sphere is analyzed where the subdomains are derived from a Yin-Yang grid system. It is worth noting a key difference between our work and [14]. In our problem, domain subdivision is carried out in the underlying embedding space. As a consequence, the unequal-sized subdomain case is essential to our understanding of the problem.

The convergence of PS and OPS for general surfaces remains unknown. Section 2 reviews the CPM. Section 3 studies the PS-CPM combination for the surface intrinsic positive Helmholtz equation (1) by analyzing an equivalent one-dimensional periodic problem. This section proves convergence and derives convergence factors. Although (1) on 1-manifolds can be solved through parameterization, we only investigate the convergence of the PS-CPM for 1-manifolds in this paper with the hope of extending our work to higher dimensional manifolds in the future. Section 4 provides a numerical experiment in which the PS-CPM contraction factor converges to its PS counterpart by increasing the grid resolution. Finally, Section 5 gives conclusions.

2 The closest point method

The CPM was first introduced in [16] for explicitly solving evolutionary PDEs on surfaces. It is an embedding method and allows the use of standard Cartesian methods for the discretization of surface intrinsic differential operators. The surface representation and extension of quantities defined on the surface to the surrounding embedding space is done using the closest point mapping $\mathrm{cp}_{\mathcal{S}}(x) = \arg\min_{s\in\mathcal{S}} |x - s|$ for $x \in \mathbb{R}^d$. This mapping gives the closest point in Euclidean distance to the surface for any point x in the embedding space. It is smooth for any point in the embedding space within a distance R_0 of a smooth surface, where R_0 is a lower bound for the surface radii of curvature [3].

Suppose the closest point mapping of a manifold is smooth over a tubular neighborhood $\Omega \subset \mathbb{R}^d$ of the manifold. We introduce $\tilde{u} : \Omega \to \mathbb{R}$ as the solution to the embedding CPM problem. Two principles are fundamental to the CPM: *equivalence of gradients* and *equivalence of divergence* [16]. Assuming a smooth manifold $\mathcal{S}$, the equivalence of gradients principle gives us $\nabla\tilde{u}(\text{cp}_{\mathcal{S}}) = \nabla_{\mathcal{S}} u$ since $\tilde{u}(\text{cp}_{\mathcal{S}})$ is constant in the normal direction to the manifold. Further, applying the equivalence of divergence principle, $\nabla \cdot (\nabla\tilde{u}(\text{cp}_{\mathcal{S}})) = \nabla_{\mathcal{S}} \cdot (\nabla\tilde{u}(\text{cp}_{\mathcal{S}}))$ holds on the manifold. Therefore, on the manifold,

$$\Delta\tilde{u}(\text{cp}_{\mathcal{S}}) = \nabla \cdot (\nabla\tilde{u}(\text{cp}_{\mathcal{S}})) = \nabla_{\mathcal{S}} \cdot (\nabla\tilde{u}(\text{cp}_{\mathcal{S}})) = \nabla_{\mathcal{S}} \cdot (\nabla_{\mathcal{S}} u) = \Delta_{\mathcal{S}} u. \tag{2}$$

A modified version of (2) offers improved stability at the discrete level and is normally used in elliptic problems [7, 9, 11]. The regularized Laplace operator is

$$\Delta_h^{\#}\tilde{u} = \Delta\tilde{u}(\text{cp}_{\mathcal{S}}) - \frac{2d}{h^2}\left[\tilde{u} - \tilde{u}(\text{cp}_{\mathcal{S}})\right], \tag{3}$$

where $0 < h \ll 1$. As in [9, 11], we take the parameter h to be equal to the mesh spacing in the fully discrete setting.

Equation (3) gives our replacement for the Laplace–Beltrami operator. Applying it, and extending the function f off the manifold using the closest point mapping gives our embedding equation for (1):

$$(c - \Delta_h^{\#})\,\tilde{u} = f(\text{cp}_{\mathcal{S}}), \quad x \in \Omega. \tag{4}$$

Standard numerical methods in the embedding space may be applied to (4) to complete the discretization. In this paper, we apply standard second order finite differences on regular grids to approximate the derivative operators. Because discrete points do not necessarily lie on $\mathcal{S}$, an interpolation scheme is needed to recover surface values. Utilizing tensor product barycentric Lagrangian interpolation [1], an extension matrix $\mathbf{E}$ is defined to extend values off of the manifold. Note that the extension matrix may be viewed as a discretization of the closest point mapping.

Using a mesh spacing h and degree-p interpolation polynomials, it is sufficient to numerically approximate equation (4) in a narrow tube around $\mathcal{S}$ of radius $r = \sqrt{(d-1)(p+1)^2 + (p+3)^2}h/2$. A more thorough explanation of the CPM at the discrete level can be found in [11].

3 The PS-CPM convergence analysis

PS is an overlapping domain decomposition method which is designed to iteratively solve DEs over subdomains, distributing the computational costs. It is also capable of parallelism and can be combined with the CPM, a method whose underlying linear system is sparse. We assume $\mathcal{S}$ to be a smooth, closed 1-manifold in $\mathbb{R}^d$ with arclength L. We consider the case with two subdomains, but the discussion can be generalized to any finite number of subdomains [18]. We let the disjoint subdomains be $\tilde{\mathcal{S}}_1$ and $\tilde{\mathcal{S}}_2$. We parameterize the manifold by arclength s starting at a boundary of $\tilde{\mathcal{S}}_1$. Next, we let the overlapping subdomains be $\mathcal{S}_1 = [a_1, b_1]$ and $\mathcal{S}_2 = [a_2, b_2]$. Since overlapping subdomains are needed, we have $a_1 < 0$ and $b_2 > L$. Define $\ell_1 \equiv b_1 - a_1$ and $\ell_2 \equiv b_2 - a_2$ to be the subdomain lengths. Further, let $\delta_1 = b_1 - a_2$ and $\delta_2 = b_2 - (a_1 + L)$ denote the subdomain overlaps at $s = \ell_1$ and $s = \ell_2$, respectively. In addition, we assume $0 < \delta_1 + \delta_2 < \min\{\ell_1, \ell_2\}$. In the CPM, the overlapping subdomains Ω_1 and Ω_2, corresponding to $\mathcal{S}_1$ and $\mathcal{S}_2$, are constructed using a graph-based partitioning algorithm applied over the computational tube [13]. Then, the PS-CPM for equation (1) is: for $n = 0, 1, \ldots$ and for $j = 1, 2$ solve

$$\begin{cases} (c - \Delta_h^{\#})\, \tilde{u}_j^{n+1} = f(\mathrm{cp}_{\mathcal{S}}), & \text{in } \Omega_j, \\ \tilde{u}_j^{n+1} = \tilde{u}^n(\mathrm{cp}_{\mathcal{S}}), & \text{on } \Gamma_{jk}, k \neq j \end{cases} \tag{5}$$

where Γ_{jk} for $j, k = 1, 2$ are the boundaries of subdomains j and k.

To begin, an initial guess is needed over the subdomain boundaries. An iteration may then be completed by solving all subproblems. This gives new boundary values that can be used to initiate the next iteration, and so on, until convergence. In this form of the Schwarz algorithm, there is no concept of a global solution. In order to construct the global solution, a weighted average of subdomain solutions is utilized [6]. In this paper, at any time, the approximation of the global solution is given as the union of the disjoint subdomain solutions $u^n = u_1^n|_{\tilde{\mathcal{S}}_1} \cup u_2^n|_{\tilde{\mathcal{S}}_2}$. This is called restricted additive Schwarz (RAS), and we use the labels PS and RAS interchangeably. Our analysis examines the equivalent one dimensional periodic problem formulated below.

Theorem 1 *In the limit as $h \to 0$, and using two subdomains $\mathcal{S}_1 = [a_1, b_1]$ and $\mathcal{S}_2 = [a_2, b_2]$, the PS-CPM for the positive surface intrinsic Helmholtz equation* (5) *is equivalent to:*

$$\begin{cases} (c - \frac{\mathrm{d}^2}{\mathrm{d}s^2})u_1^{n+1} = f, & \text{in } \mathcal{S}_1, \\ u_1^{n+1}(a_1) = u_2^n(a_1 + L), \\ u_1^{n+1}(b_1) = u_2^n(b_1), \end{cases} , \quad \begin{cases} (c - \frac{\mathrm{d}^2}{\mathrm{d}s^2})u_2^{n+1} = f, & \text{in } \mathcal{S}_2, \\ u_2^{n+1}(a_2) = u_1^n(a_2), \\ u_2^{n+1}(b_2) = u_1^n(b_2 - L), \end{cases} \tag{6}$$

where L is the manifold length.

Proof For a smooth manifold $\mathcal{S}$, the regularized operator $\Delta_h^{\#}$ is consistent with the Laplace operator on the manifold [11]. Thus the CPM is consistent with the surface intrinsic PDE problems in the limit $h \to 0$ where h denotes the mesh size. Parameterizing a one-dimensional manifold $\mathcal{S}$ in $\mathbb{R}^d$ by arclength s, the differential operator $\Delta_{\mathcal{S}}$ becomes $\mathrm{d}^2/\mathrm{d}s^2$, yielding our result. □

In [14], the convergence of (6) is studied for an equal-sized partitioning. The partitioning arising from the PS-CPM problems in (5) is performed within the embedding space. As a consequence, our subdomains will be unequal. This motivates us to investigate the convergence of the method for an unequal-sized partitioning.

By defining the errors $\epsilon_j^n = u_j^n - u|_{S_j}$, $j = 1, 2$, and using the linearity of (1), iteration (6) is reduced to:

$$\begin{cases} (c - \frac{\mathrm{d}^2}{\mathrm{d}s^2})\epsilon_1^{n+1} = 0, & \text{in } \mathcal{S}_1, \\ \epsilon_1^{n+1}(a_1) = \epsilon_2^n(a_1 + L), \\ \epsilon_1^{n+1}(b_1) = \epsilon_2^n(b_1), \end{cases} \quad , \quad \begin{cases} (c - \frac{\mathrm{d}^2}{\mathrm{d}s^2})\epsilon_2^{n+1} = 0, & \text{in } \mathcal{S}_2, \\ \epsilon_2^{n+1}(a_2) = \epsilon_1^n(a_2), \\ \epsilon_2^{n+1}(b_2) = \epsilon_1^n(b_2 - L). \end{cases} \tag{7}$$

After solving the ODEs in (7), error values at the boundaries can be computed. At each iteration, these error values depend on the error values at the boundaries from the previous iteration. To state this concisely, we define an error vector at iteration n which is comprised of the error values at the boundaries:

$$\boldsymbol{\epsilon}^n := [\epsilon_1^n(b_2 - L), \epsilon_1^n(a_2), \epsilon_2^n(b_1), \epsilon_2^n(a_1 + L)]^T. \tag{8}$$

We obtain, in matrix form, $\boldsymbol{\epsilon}^{n+1} = \mathbf{M}_{\mathrm{PS}}\boldsymbol{\epsilon}^n$, where

$$\mathbf{M}_{\mathrm{PS}} = \begin{bmatrix} 0 & 0 & r_1 & p_1 \\ 0 & 0 & q_1 & s_1 \\ r_2 & p_2 & 0 & 0 \\ q_2 & s_2 & 0 & 0 \end{bmatrix} \tag{9}$$

is called the iteration matrix. It has entries

$$p_j = \frac{1 - \mathrm{e}^{2\sqrt{c}(\ell_j - \delta_{j-1})}}{1 - \mathrm{e}^{2\sqrt{c}\ell_j}} \mathrm{e}^{\sqrt{c}\delta_{j-1}}, \qquad r_j = \frac{1 - \mathrm{e}^{2\sqrt{c}\delta_{j-1}}}{1 - \mathrm{e}^{2\sqrt{c}\ell_j}} \mathrm{e}^{\sqrt{c}(\ell_j - \delta_{j-1})},$$

$$q_j = \frac{1 - \mathrm{e}^{2\sqrt{c}(\ell_j - \delta_j)}}{1 - \mathrm{e}^{2\sqrt{c}\ell_j}} \mathrm{e}^{\sqrt{c}\delta_j}, \qquad s_j = \frac{1 - \mathrm{e}^{2\sqrt{c}\delta_j}}{1 - \mathrm{e}^{2\sqrt{c}\ell_j}} \mathrm{e}^{\sqrt{c}(\ell_j - \delta_j)}, \tag{10}$$

for $j = 1, 2$ and $\delta_0 \equiv \delta_2$. The definitions of δ_j and ℓ_j may be found at the beginning of this section. The following lemma holds for the quantities in (10):

Lemma 1 **([18])** *Suppose* $0 < \delta_1 + \delta_2 < \min\{\ell_1, \ell_2\}$. *Then the scalars* p_j, q_j, r_j, s_j, $j = 1, 2$, *appearing in* (10) *satisfy* $0 < q_j + s_j < 1$ *and* $0 < p_j + r_j < 1$.

Now, we arrive at the most important result of this section.

Theorem 2 *Under the restrictions on the partitioning of the manifold $\mathcal{S}$ detailed in Lemma 1 above, the PS iteration* (6) *for the positive Helmholtz equation on any closed, smooth one-dimensional manifold converges globally.*

Proof We must show the spectral radius of the iteration matrix, $\rho(\mathbf{M}_{\mathrm{PS}})$, is less than 1. $\|\mathbf{M}_{\mathrm{PS}}\|_\infty$ bounds the spectral radius, $\rho(\mathbf{M}_{\mathrm{PS}}) \le \|\mathbf{M}_{\mathrm{PS}}\|_\infty = \max\{r_j + p_j, q_j + s_j\}$. In Lemma 1, we have shown that $0 < p_j + r_j < 1$ and $0 < q_j + s_j < 1$. Therefore, $\|\mathbf{M}_{\mathrm{PS}}\|_\infty < 1$, and consequently the algorithm converges. □

We define the convergence factor κ as the ratio of the ∞-norm of the error vector (8) at two steps $n+2$ and n, $\kappa = \|\epsilon^{n+2}\|_\infty / \|\epsilon^n\|_\infty$. Considering the inequality $\|\epsilon^{n+1}\|_\infty \leq \|\mathbf{M}_{\mathrm{PS}}\|_\infty \|\epsilon^n\|_\infty$, $\|\mathbf{M}_{\mathrm{PS}}\|_\infty^2$ is an upper bound for the convergence factor. That is, $\kappa \leq \|\mathbf{M}_{\mathrm{PS}}\|_\infty^2$. In the following corollary, we show that the our analysis for the equal-sized partitioning agrees with the one obtained in [14].

Corollary 1 *Assume an equal-sized partitioning for the PS iteration* (6). *That is,* $\mathcal{S}_1 = [-\delta, L/2+\delta]$, $\mathcal{S}_2 = [L/2-\delta, L+\delta]$. *Then, the convergence factor can be calculated as* $\kappa \leq (p+r)^2 = (\mathrm{e}^{\sqrt{c}L/2} + \mathrm{e}^{\sqrt{c}\delta})^2/(1+\mathrm{e}^{\sqrt{c}(L/2+\delta)})^2$.

Proof If we make the simplifying assumption that both subdomains are of equal size and have a common overlap size, then $q_1 = q_2 = p_1 = p_2 = p$ and $s_1 = s_2 = r_1 = r_2 = r$. The iteration matrix becomes a doubly stochastic matrix with row and column sums of $p+r$, and subsequently $\rho(\mathbf{M}_{\mathrm{PS}}) = p+r$. By a direct substitution for p and r, we obtain $\kappa = \rho(\mathbf{M}_{\mathrm{PS}})^2 = (\mathrm{e}^{\sqrt{c}L/2} + \mathrm{e}^{\sqrt{c}\delta})^2/(1+\mathrm{e}^{\sqrt{c}(L/2+\delta)})^2$. □

4 Numerical simulation

Here we numerically verify the results obtained in Section 3. Since numerical solutions of the PS-CPM and the PS algorithm will be compared, we use RAS as the domain decomposition method to build a global approximate solution. It is shown in [6] that RAS and PS are identical iterations and have the same convergence rate. Hence, we will use RAS-CPM instead of PS-CPM hereafter.

Theorem 1 shows that the CPM equipped with RAS as a solver is in the limit as $h \to 0$ equivalent to RAS applied to a 1D periodic problem. To verify this, we numerically solve (1) with $c = 1$ and $f(s) = \sin(2\pi s/L)$ using the RAS-CPM for the boundary of a Möbius strip with width 1, whose center circle has radius 1. The initial guess for the discrete solution is taken as $U^{(0)} = 0$. Two disjoint subdomains are created by splitting the length of the curve in a 1:2 ratio, and overlapping subdomains are formed using overlaps $\delta = \delta_1 = \delta_2 = 0.1L$. The solution using the RAS-CPM with grid spacing $h = 0.01$ and fourth degree barycentric Lagrangian interpolation applied in a dimension-by-dimension fashion is shown in Fig. 1 (left). Here, the disjoint subdomains are visualized as point clouds. Convergence histories for various grid spacings are depicted in Fig. 1 (right). Here, the RAS and the RAS-CPM contraction factors are compared with the theoretical result. The errors are defined as the max-norm of the difference of the DD solution and the single domain solution. As we observe in Fig. 1 (right), the RAS error has the same decay rate as that described in Theorem 1 (shown as the dashed line). In addition, the RAS-CPM error tends toward the RAS error as the mesh size is reduced.

As another experiment, (1) is solved with two equal-sized subdomains, assuming $\mathcal{S}$ is the unit circle. The disjoint subdomains are shown in Fig. 2 (left). Fig. 2 (right) shows the effect of the overlap parameter δ on RAS-CPM for three different grids ($h = 0.05, 0.01, 0.005$). For a given h and δ, the numerical convergence factor changes slightly as the iteration progresses, hence we present an average of the convergence

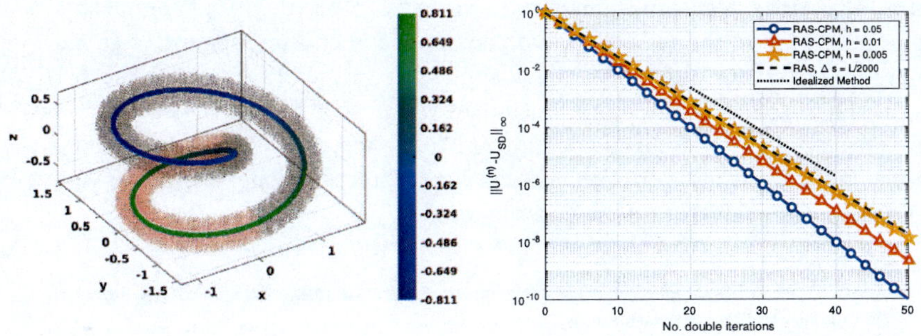

Fig. 1: Left: RAS-CPM solution of the surface intrinsic Helmholtz equation on edge of a Möbius strip. The disjoint subdomains are depicted. **Right:** Error versus the double iteration number.

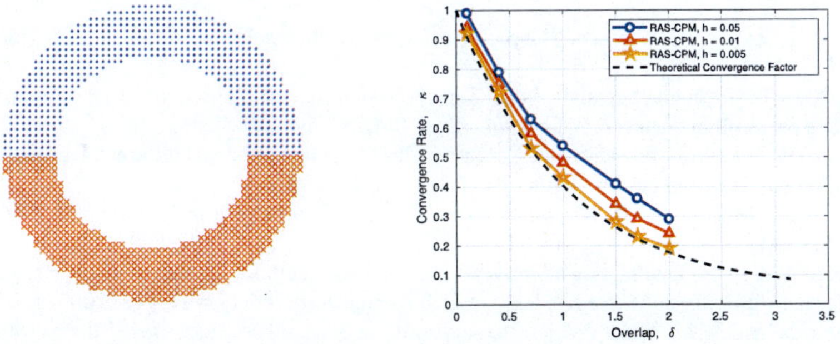

Fig. 2: Left: Equal-sized disjoint subdomains for the unit circle. **Right:** Comparison of the RAS-CPM convergence factor and theoretical convergence factor for different values of overlap parameter in an equal-sized subdomain configuration for the unit circle.

factor over all iterations. To compare with the result in Corollary 1, the theoretical convergence factor associated with a double iteration, $(e^{L/2} + e^{\delta})^2/(1 + e^{L/2+\delta})^2$, is shown in Fig. 2 (right) as a dashed line. The observed RAS-CPM contraction factor converges to the theoretical value as the grid quality improves. By increasing the overlap, κ is reduced and a better convergence factor is obtained.

5 Conclusion

Employing RAS as a solver for the CPM parallelizes the solution of PDEs on surfaces and enhances the performance for large scale problems. In this paper, convergence of the (continuous) CPM equipped with a restricted additive Schwarz solver was investigated for a one-dimensional manifold in $\mathbb{R}^d$. Convergence was shown for the two-subdomain case; extensions to any finite number of subdomains is under investigation [18]. Observed convergence rates agree with our theory as the mesh spacing is refined. Indeed, the results apply to any convergent discretization (e.g., a finite element discretization) of RAS solvers applied to surface PDEs as the mesh spacing approaches zero. Finally, note that other variants of Schwarz methods – sequential restricted additive Schwarz, optimized restricted additive Schwarz, and multiplicative methods – can be utilized as a solver or a preconditioner for the CPM. We plan to extend our analysis to these cases as well.

Acknowledgements The authors gratefully acknowledge the financial support of NSERC Canada (RGPIN 2016-04361 and RGPIN 2018-04881).

References

1. J.-P. Berrut and L.N. Trefethen. Barycentric Lagrange interpolation. *SIAM Review* **46**(3), 501–517 (2004).
2. M. Bertalmi, L.-T. Cheng, S. Osher and G. Sapiro. Variational problems and partial differential equations on implicit surfaces. *J. Comput. Phys.* **174**(2), 759–780 (2001).
3. J. Chu and R. Tsai. Volumetric variational principles for a class of partial differential equations defined on surfaces and curves. *Math. Sci.* **5**(2), 1–38 (2018).
4. P. Degener, J. Meseth and R. Klein. An adaptable surface parameterization method. *IMR* **3**, 201–213 (2003).
5. A. Demlow and G. Dziuk. An adaptive finite element method for the Laplace–Beltrami operator on implicitly defined surfaces. *SIAM J. Numer. Anal.* **45**(1), 421–442 (2007).
6. E. Efstathiou and M.J. Gander. Why restricted additive Schwarz converges faster than additive Schwarz. *BIT Numer. Math.* **43** (2003).
7. I. Glehn, T. März and C.B. Macdonald. An embedded method-of-lines approach to solving partial differential equations on surfaces. *arXiv preprint* arXiv:1307.5657 (2013).
8. S. Loisel, J. Côté, M.J. Gander, L. Laayouni and A. Qaddouri. Optimized domain decomposition methods for the spherical Laplacian. *SIAM J. Numer. Anal.* **48**(2), 524–551 (2010).
9. C.B. Macdonald, J. Brandman and S.J. Ruuth. Solving eigenvalue problems on curved surfaces using the closest point method. *J. Comput. Phys.* **230**(22), 7944–7956 (2011).
10. C.B. Macdonald, B. Merriman and S.J. Ruuth. Simple computation of reaction–diffusion processes on point clouds. *PNAS USA* **110**(23), 9209–9214 (2013).
11. C.B. Macdonald and S.J. Ruuth. The implicit closest point method for the numerical solution of partial differential equations on surfaces. *SIAM J. Sci. Comput.* **31**(6), 4330–4350 (2010).
12. L. Martin and Y. Tsai. Equivalent extensions of Hamilton–Jacobi–Bellman equations on hypersurfaces. *J. Sci. Comput.* **84**, 43 (2020).
13. I.C.T. May, R.D. Haynes and S.J. Ruuth. Schwarz solvers and preconditioners for the closest point method. *SIAM J. Sci. Comput.* **42**(6), A3584–A3609 (2020).
14. A. Qaddouri, L. Laayouni, S. Loisel, J. Côté and M.J. Gander. Optimized Schwarz methods with an overset grid for the shallow-water equations: preliminary results. Selected Papers from the Seventh IMACS International Symposium on Iterative Methods in Scientific Computing. *Appl Numer Math* **58**(4), 459–471 (2008).
15. M. Reuter, F.E. Wolter, M. Shenton and M. Niethammer. Laplace–Beltrami eigenvalues and topological features of eigenfunctions for statistical shape analysis. Selected Papers from the 2007 New Advances in Shape Analysis and Geometric Modeling Workshop. *CAD* **41**(10), 739–755 (2009).
16. S.J. Ruuth and B. Merriman. A simple embedding method for solving partial differential equations on surfaces. *J. Comput. Phys.* **227**(3), 1943–1961 (2008).

17. C. Yang, J. Cao and X.-C. Cai. A fully implicit domain decomposition algorithm for shallow water equations on the cubed-sphere. *SIAM J. Sci. Comput.* **32**(1), 418–438 (2010).
18. A. Yazdani. *Convergence study of the domain decomposition of the closest point method.* Master's thesis, Simon Fraser University. In progress.

Dual-Primal Preconditioners for Newton–Krylov Solvers for the Cardiac Bidomain Model

Ngoc Mai Monica Huynh, Luca F. Pavarino, and Simone Scacchi

1 Introduction

We present here an overview of Newton–Krylov solvers for implicit time discretizations of the cardiac Bidomain equations, preconditioned by Balancing Domain Decomposition with Constraints (BDDC) [5] or Dual-Primal Finite Element Tearing and Interconnecting (FETI-DP) [7] algorithms.

The Bidomain model describes the propagation of an electric signal in a cardiac tissue by means of two parabolic partial differential equations (PDEs) [3, 13]; it is coupled through the non-linear reaction term to a system of ordinary differential equations (ODEs), modeling the ionic currents through the cell membrane and the associated opening and closing process of ionic channel gates.

One of the main issues to face when computing these systems is the choice of an appropriate solver, which can combine computational efficiency and accuracy in representing the solution. As a matter of fact, the need to accurately represent phenomena both at macroscopic and at microscopic levels leads to time and space discretizations with millions of degrees of freedom (dofs) or more. The solution of the associated large discrete systems for increasing dimensions represent a challenging computation, requiring efficient parallel solvers [4, 14].

In this work we show some parallel numerical results obtained with two non-linear solvers, each of which derives from a different solution strategy: a monolithic (or *coupled*) solution approach and a staggered (or *decoupled*) approach. Both these approaches arise from an implicit time discretization of the Bidomain model, which

N.M.M. Huynh and L.F. Pavarino
Dipartimento di Matematica, Università degli Studi di Pavia, via Ferrata 5 - 27100 Pavia, Italy,
e-mail: ngocmaimonica.huynh01@universitadipavia.it, luca.pavarino@unipv.it

S. Scacchi
Dipartimento di Matematica, Università degli Studi di Milano, via Saldini 50 - 20133 Milano, Italy,
e-mail: simone.scacchi@unimi.it

S. C. Brenner et al. (eds.), *Domain Decomposition Methods in Science and Engineering XXVI*, Lecture Notes in Computational Science and Engineering 145,
https://doi.org/10.1007/978-3-030-95025-5_75

is solved coupled to or decoupled from the ionic equations, respectively, as in Refs. [11, 12].

In Sec. 2 a brief description of the model is provided, while in Sec. 3 we present our solution strategies. Parallel numerical experiments in Sec. 4 using the PETSc library [1] end this work.

2 The Bidomain cardiac electrical model

The propagation of the electrical impulse in the cardiac tissue is modeled by a system of two parabolic reaction-diffusion PDEs coupled through the non-linear reaction term to a system of ODEs describing the flow of ionic currents inward and outward the cell membrane:

$$\begin{cases} \chi C_m \dfrac{\partial v}{\partial t} - \operatorname{div}\left(D_i \cdot \nabla u_i\right) + I_{\text{ion}}(v, w) = 0 \\ -\chi C_m \dfrac{\partial v}{\partial t} - \operatorname{div}\left(D_e \cdot \nabla u_e\right) - I_{\text{ion}}(v, w) = -I^e_{\text{app}} \quad \text{in } \Omega \times (0, T), \\ \dfrac{\partial w}{\partial t} - R(v, w) = 0 \end{cases} \tag{1}$$

where u_i and u_e are the intra- and extracellular potentials, $v(x,t) = u_i(x,t) - u_e(x,t)$ is the transmembrane potential and w represents the opening and closing process of the ionic channel gates in the cell membrane. Here, C_m is the membrane capacitance, I_{ion} the ionic membrane current (both for unit area of the membrane surface), χ is the membrane surface to volume ratio and I_{app} is the applied external current. This system is known in the literature as the Bidomain model [3, 13].

In this work, we consider a phenomenological ionic model, named the Rogers-McCulloch ionic model [15]. More realistic and complex ionic models have been integrated in different numerical studies, see e.g. Refs. [4, 14].

3 Dual-primal Newton–Krylov methods

Space and time discretizations. The cardiac domain Ω is discretized in space with a structured quasi-uniform grid of hexahedral finite elements, leading to the semi-discrete system

$$\begin{cases} \chi C_m \mathcal{M} \dfrac{\partial}{\partial t} \begin{bmatrix} \mathbf{u}_i \\ \mathbf{u}_e \end{bmatrix} + \mathcal{A} \begin{bmatrix} \mathbf{u}_i \\ \mathbf{u}_e \end{bmatrix} + \begin{bmatrix} M\, \mathbf{I}_{\text{ion}}(\mathbf{v}, \mathbf{w}) \\ -M\, \mathbf{I}_{\text{ion}}(\mathbf{v}, \mathbf{w}) \end{bmatrix} = \begin{bmatrix} \mathbf{0} \\ -M\, \mathbf{I}^{\mathbf{e}}_{\text{app}} \end{bmatrix}, \\ \dfrac{\partial \mathbf{w}}{\partial t} = R\left(\mathbf{v}, \mathbf{w}\right), \end{cases} \tag{2}$$

with the stiffness and mass block-matrices $\mathcal{A} = \begin{bmatrix} A_i & 0 \\ 0 & A_e \end{bmatrix}$, $\mathcal{M} = \begin{bmatrix} M & -M \\ -M & M \end{bmatrix}$.

Regarding the time discretization, in the literature it is very common to adopt operator splitting strategies [2, 16] or implicit-explicit (IMEX) schemes [4, 17], in order to avoid the elevated computational costs related to the solution of the non-linear discrete problem. Here we propose two ways to solve the discretized system using the Backward Euler method: a monolithic, or *coupled*, solution strategy where at each time step we solve the non-linear system with the discrete Bidomain coupled with the ionic model, as in Refs. [8, 12], and a staggered, or *decoupled*, solution approach (as in Refs. [9, 11]). Both approaches rely on a preconditioned Krylov method nested within a Newton loop.

***Coupled* solution approach.** The monolithic strategy can be summarized in algorithmic steps as follows. At the n-th time step, solve the non-linear system $\mathbf{F}_{\text{coupled}}(\mathbf{s}^{n+1}) = \mathbf{0}$, with $\mathbf{s}^{n+1} = (\mathbf{u}_i^{n+1}, \mathbf{u}_e^{n+1}, \mathbf{w}^{n+1})$:

$$\mathbf{F}_{\text{coupled}}(\mathbf{s}^{n+1}) = \begin{cases} (\chi C_m \mathcal{M} + \tau\mathcal{A}) \begin{bmatrix} \mathbf{u}_i^{n+1} \\ \mathbf{u}_e^{n+1} \end{bmatrix} + \tau \begin{bmatrix} M\mathbf{I}_{\text{ion}}(\mathbf{v}^{n+1}, \mathbf{w}^{n+1}) \\ -M\mathbf{I}_{\text{ion}}(\mathbf{v}^{n+1}, \mathbf{w}^{n+1}) \end{bmatrix} - \mathbf{G} \\ \mathbf{w}^{n+1} - \tau R(\mathbf{v}^{n+1}, \mathbf{w}^{n+1}) - \mathbf{w}^n \end{cases}$$

with $\mathbf{G} = \chi C_m \mathcal{M} \begin{bmatrix} \mathbf{u}_i^n \\ \mathbf{u}_e^n \end{bmatrix} + \tau \begin{bmatrix} \mathbf{0} \\ -M\mathbf{I}_{\text{app}}^{\mathbf{e}} \end{bmatrix}$ and being $\tau = t_{n+1} - t_n$. This non-linear system is solved with a Newton method:

1. compute and solve the Jacobian linear system $\mathbf{DF}(\mathbf{s}^n)\, \mathbf{ds}^{n+1} = -\mathbf{F}(\mathbf{s}^n)$, where $\mathbf{ds}^{n+1} := (\mathbf{du}_i^{n+1}, \mathbf{du}_e^{n+1}, \mathbf{dw}^{n+1})$ is the increment at step $n+1$;
2. update $\mathbf{u}_i^{n+1} = \mathbf{u}_i^n + \mathbf{du}_i^{n+1}$, $\mathbf{u}_e^{n+1} = \mathbf{u}_e^n + \mathbf{du}_e^{n+1}$ and $\mathbf{w}^{n+1} = \mathbf{w}^n + \mathbf{dw}^{n+1}$.

Since the linear system in Step 1 is non-symmetric (due to the presence of the gating term), it is necessary to use the Generalized Minimal Residual method (GMRES) for its solution.

***Decoupled* solution approach.** As an alternative to the previous strategy, the staggered approach first requires the solution of the ionic model, then solving and updating the Bidomain equations. For each time step n,

a. given the intra- and extracellular potentials at the previous step, hence $\mathbf{v} := \mathbf{u}_i^n - \mathbf{u}_e^n$, compute the gating

$$\mathbf{w}^{n+1} - \tau R(\mathbf{v}, \mathbf{w}^{n+1}) = \mathbf{w}^n;$$

b. solve and update the Bidomain non-linear system. Given $\mathbf{u}_{i,e}^n$ at the previous time step and given $\mathbf{w}^{n+1}$, compute $\mathbf{u}^{n+1} = (\mathbf{u}_i^{n+1}, \mathbf{u}_e^{n+1})$ by solving the system $\mathbf{F}_{\text{decoupled}}(\mathbf{u}^{n+1}) = \mathbf{G}$

$$\mathbf{F}_{\text{decoupled}}(\mathbf{u}^{n+1}) = (\chi C_m \mathcal{M} + \tau\mathcal{A}) \begin{bmatrix} \mathbf{u}_i^{n+1} \\ \mathbf{u}_e^{n+1} \end{bmatrix} + \tau \begin{bmatrix} M\mathbf{I}_{\text{ion}}(\mathbf{v}^{n+1}, \mathbf{w}^{n+1}) \\ -M\mathbf{I}_{\text{ion}}(\mathbf{v}^{n+1}, \mathbf{w}^{n+1}) \end{bmatrix},$$

$$\mathbf{G} = \chi C_m \mathcal{M} \begin{bmatrix} \mathbf{u}_i^n \\ \mathbf{u}_e^n \end{bmatrix} + \tau \begin{bmatrix} \mathbf{0} \\ -M\mathbf{I}_{\text{app}}^{\mathbf{e}} \end{bmatrix}.$$

The Jacobian linear system associated to the non-linear problem in step (b) is symmetric, thus allowing us to use the Conjugate Gradient (CG) method within each Newton iteration.

Dual-primal preconditioners. In both approaches, a linear system has to be solved within each Newton loop, either by GMRES (for the *coupled* approach) or by CG method (for the *decoupled* case), preconditioned by a dual-primal substructuring algorithm.

In this work, we focus on the most common dual-primal iterative substructuring algorithms, the BDDC and FETI-DP methods. FETI-DP methods were first proposed in Ref. [7] and are based on the transposition of the linear system to a constrained minimization problem. Conversely, BDDC methods were introduced in Ref. [5] as an alternative to FETI-DP and provide a preconditioner for the discretized linear problem.

Convergence rate bound. In Ref. [10] these two algorithms are shown to be spectrally equivalent, thus allowing us to derive a convergence rate estimate for the preconditioned operator, which holds for both preconditioners if the same coarse space is chosen. In the *coupled* approach, the bound is related to the residual at the m-th iteration of GMRES, while in the *decoupled* strategy the bound is for the condition number. Details on the derivation of both bounds can be found in the works of the authors [8, 9].

4 Numerical experiments

Parallel numerical experiments are performed on an idealized left ventricular geometry, modeled as a portion of a half truncated ellipsoid, see Fig. 1.
Boundary and initial conditions represent an isolated tissue with resting potential. We simulate the initial excitation process on the time interval $[0, 2]$ ms following an extracellular stimulus.

Fig. 1: Computational domain.

Different supercomputers are taken into account: the Galileo cluster from the Cineca centre (a Linux Infiniband cluster equipped with 1084 nodes, each with 36 2.30 GHz Intel Xeon E5-2697 v4 cores and 128 GB/node, for a total of 39024 cores, www.hpc.cineca.it) for the tests related to the *coupled* solution approach and the weak scaling of the *decoupled* case; the Linux cluster Indaco at the University of Milan (a Linux Infiniband cluster with 16 nodes, each carrying 2 processors Intel

Xeon E5-2683 v4 2.1 GHz with 16 cores each, https://www.indaco.unimi.it/) for the strong scaling of the *decoupled* approach.

Our C code is based on the parallel library PETSc [1] from the Argonne National Laboratory. BDDC and FETI-DP preconditioners are built-in in the PETSc library, both applied with default parameters (coarse space made up of vertices and edge averages, direct subdomain solver with a LU factorization, etc), while *Boomer* Algebraic MultiGrid (bAMG) is from the Hypre library [6]. In our tests, we always assign one subdomain to each processor. In the strong scaling tests, part of the speedup comes from the superlinear computational complexity of the sparse subdomain solvers based on LU factorization.

We manually implement the Newton method for the coupled case, with an absolute residual stopping criterion with tolerance 10^{-4}, while for the decoupled case we use the default non-linear solver (SNES) from the PETSc library and we adopt the default SNES convergence test as the stopping criterion, based on the comparison of the L^2-norm of the non-linear function at the current iterate and at the current step (see the PETSc manual [1] for tolerance values and further details). The linear systems arising in Steps 1 and (b) of the two approaches are solved with the GMRES and CG methods respectively, both using PETSc default stopping criteria and default tolerances. We compare the following quantities: the average Newton iterations per time step *nit*, the average linear iterations per Newton iteration *lit* and the average CPU solution time per time step *time* in seconds.

Coupled tests. The linear system arising from the discretization of the Jacobian problem at each Newton step is solved with the GMRES method, preconditioned by BDDC preconditioners and bAMG.

Coupled weak scaling. We report here a weak scaling test. We fix the local mesh to $12 \cdot 12 \cdot 12$ elements and we increase the number of subdomains (and therefore the number of processors) from 32 to 256, yielding an ellipsoidal portion of increasing dimensions. It is clear from Table 1 that BDDC performs better than bAMG in terms of average number of linear iterations per non-linear step, as this parameter is lower for BDDC and does not increase with the number of processors. As a matter of fact, there is an increasing reduction rate up to 90% for the average linear iterations. In contrast, BDDC's average CPU time is higher than bAMG CPU time (we do not have a clear explanation of this fact), but we remark that BDDC timings do not increase significantly when the number of processors is increased from 32 to 256, while bAMG timings more than double.

Coupled strong scaling. We fix the global mesh to $128 \cdot 128 \cdot 24$ elements (resulting in more than 1 million global dofs) and we increase the number of processors from 32 to 256. As the number of processors increases, the local number of dofs decreases and BDDC's average number of linear iterations and CPU times decrease (see Table 2), while bAMG iterations increase and the CPU timings decrease less than expected, even if they are lower than BDDC timings. Moreover, in order to test the efficiency of the proposed solver on the parallel architecture, we compute the parallel speedup $\frac{T_1}{T_N}$, which is the ratio between the runtime T_1 needed by 1 (or N_1) processor and

Table 1: *Coupled* weak scaling test. Local mesh of $12 \cdot 12 \cdot 12$ elements. Comparison of Newton–Krylov solvers preconditioned by BDDC and bAMG. Cluster: Galileo.

procs.	global n. dofs	BDDC			bAMG		
		nit	lit	time	nit	lit	time
32	180,075	2	45	6.8	2	142	1.5
64	356,475	2	32	6.9	2	145	1.9
128	705,675	2	23	7.0	2	158	2.1
256	1,404,075	2	23	8.5	2	212	3.2

the average runtime T_N needed by N processors to solve the problem. Here, we set $N_1 = 32$. While bAMG is sub-optimal, BDDC outperforms the ideal linear speedup.

Table 2: *Coupled* strong scaling test. Global mesh of $128 \cdot 128 \cdot 24$ elements (1,248,075 dofs). Comparison of Newton–Krylov solvers preconditioned by BDDC and bAMG. Parallel speedup (S_p), with ideal speedup in brackets. Cluster: Galileo.

procs.	BDDC				bAMG			
	nit	lit	time	S_p	nit	lit	time	S_p
32	2	37	189.3	-	2	187	15.1	-
64	2	44	59.1	3.2 (2)	2	222	9.2	1.6 (2)
128	2	29	20.1	9.4 (4)	2	240	5.3	2.8 (4)
256	2	46	10.2	18.5 (8)	2	280	3.2	4.7 (8)

Decoupled tests. The outer Newton loop is solved with the non-linear solver SNES of the PETSc library, which implements a Newton method with cubic backtracking linesearch. The linear system arising from the discretization of the Jacobian problem at each Newton step is solved with the CG method, preconditioned by BDDC or FETI-DP preconditioners.

Decoupled weak scaling. We fix here the local mesh size to $16 \cdot 16 \cdot 16$ and we increase the number of processors from 32 to 2048. Also in this case, the good performance of the dual-primal algorithms is confirmed by the average number of linear iterations per Newton step, which is low and remains stable as the number of subdomains increases (see Table 3).

Decoupled strong scaling. We now compare the performance of the dual-primal preconditioners while varying the number of processors from 64 to 256 over a time interval of $[0, 100]$ ms, for a total of 2000 time steps. The global mesh is fixed to $192 \cdot 96 \cdot 24$ elements (936,050 dofs). We can observe an overall reduction of the CPU time while increasing the number of subdomains from 64 to 128. Concerning the FETI-DP behavior, the increase of average CPU time and average number of linear iterations between 128 and 256 processors is unexpected and needs further investigation (Figure 2).

Table 3: *Decoupled* weak scaling test. Local mesh of $16 \cdot 16 \cdot 16$ elements. Comparison of Newton–Krylov solvers preconditioned by BDDC and FETI-DP. Cluster: Galileo.

procs	dofs	BDDC nit	BDDC lit	BDDC time	FETI-DP nit	FETI-DP lit	FETI-DP time
32	278,850	1	30	5.4	1	20	4.7
64	549,250	1	37	6.2	1	20	6.5
128	1,090,050	1	26	7.5	1	19	6.6
256	2,171,650	1	25	8.7	1	17	10.7
512	4,309,890	1	27	10.5	1	18	11.4
1024	8,586,370	1	28	12.5	1	19	11.0
2048	17,139,330	1	28	26.6	1	19	21.4

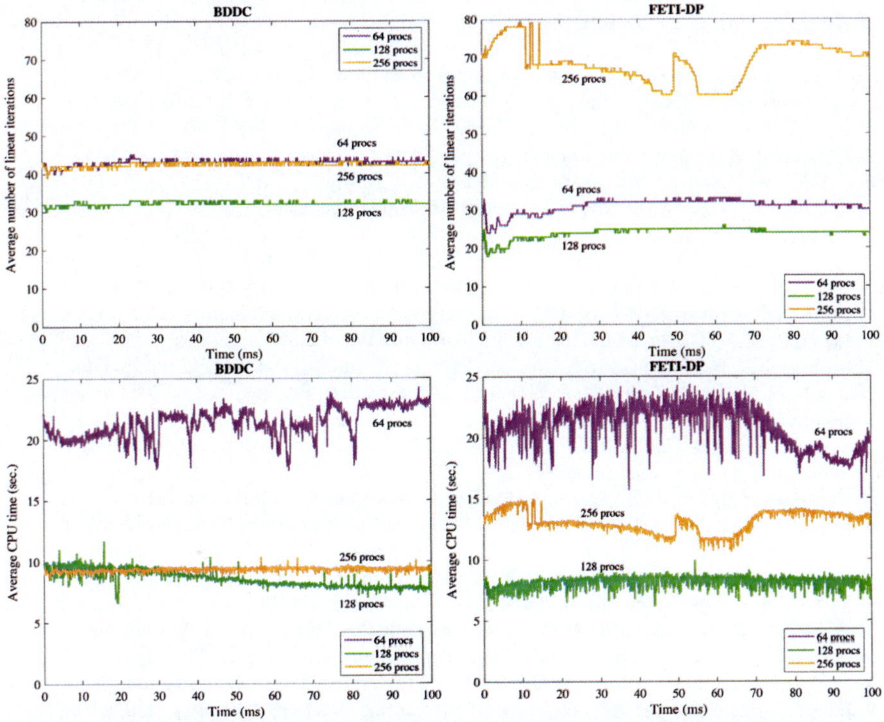

Fig. 2: *Decoupled* strong scaling. Global mesh of $192 \cdot 96 \cdot 24$ elements (936,050 dofs). Comparison between BDDC (left column) and FETI-DP (right column) preconditioners. Top: average number of linear iterations per time step; bottom: average CPU time in seconds of each SNES solver call. Cluster: Indaco.

5 Conclusion

We designed and numerically tested two different solution strategies for the solution of implicit time discretizations of the Bidomain model. Each of these solvers is preconditioned by a dual-primal substructuring algorithm, which perform better than the algebraic multigrid method in terms of number of iterations, scalability, and speedup, even if the computational times of algebraic multigrid are still better for these parameter settings. Future works should extend these solver to the solution of coupled cardiac electro-mechanical models and to more complex ionic models.

References

1. S. Balay et al. PETSc users manual. Technical Report ANL-95/11 – Revision 3.15, Argonne National Laboratory (2021).
2. H. Chen, X. Li and Y. Wang. A two-parameter modified splitting preconditioner for the Bidomain equations. *Calcolo* **56**(2), 1–24 (2019).
3. P. Colli Franzone, L.F. Pavarino and S. Scacchi. *Mathematical cardiac electrophysiology*. Volume 13 of MS&A. Modeling, Simulation and Applications. Springer, Cham (2014).
4. P. Colli Franzone, L.F. Pavarino and S. Scacchi. A numerical study of scalable cardiac electro-mechanical solvers on HPC architectures. *Frontiers in physiology* **9**, 268 (2018).
5. C.R. Dohrmann. A preconditioner for substructuring based on constrained energy minimization. *SIAM J. Sci. Comput.* **25**(1), 246–258 (2003).
6. R.D. Falgout and U.M. Yang. hypre: A library of high performance preconditioners. In: *International Conference on Computational Science*, pp. 632–641, Springer (2002).
7. C. Farhat et al. FETI-DP: a dual-primal unified FETI method. I. A faster alternative to the two-level FETI method. *Internat. J. Numer. Methods Engrg.* **50**(7), 1523–1544 (2001).
8. N.M.M. Huynh. Newton–Krylov-BDDC deluxe solvers for non-symmetric fully implicit time discretizations of the Bidomain model. *arXiv preprint* arXiv:2102.08736 (2021).
9. N.M.M. Huynh, L.F. Pavarino and S. Scacchi. Parallel Newton–Krylov-BDDC and FETIDP deluxe solvers for implicit time discretizations of the cardiac Bidomain equations. *arXiv preprint* arXiv:2101.02959 (2021).
10. J. Li and O.B. Widlund. FETI-DP, BDDC, and block Cholesky methods. *International journal for numerical methods in engineering* **66**(2), 250–271 (2006).
11. M. Munteanu and L.F. Pavarino. Decoupled Schwarz algorithms for implicit discretizations of nonlinear Monodomain and Bidomain systems. *Mathematical Models and Methods in Applied Sciences* **19**(07), 1065–1097 (2009).
12. M. Murillo and X.-C. Cai. A fully implicit parallel algorithm for simulating the non-linear electrical activity of the heart. *Numer. linear algebra with appl.* **11**(2-3), 261–277 (2004).
13. M. Pennacchio, G. Savaré and P. Colli Franzone. Multiscale modeling for the bioelectric activity of the heart. *SIAM Journal on Mathematical Analysis* **37**(4), 1333–1370 (2005).
14. A. Quarteroni, T. Lassila, S. Rossi and R. Ruiz-Baier. Integrated Heart—Coupling multiscale and multiphysics models for the simulation of the cardiac function. *Computer Methods in Applied Mechanics and Engineering* **314**, 345–407 (2017).
15. J.M. Rogers and A.D. McCulloch. A collocation-Galerkin finite element model of cardiac action potential propagation. *IEEE Trans. Biomed. Engrg.* **41**(8), 743–757 (1994).
16. J. Sundnes, G.T. Lines and A. Tveito. An operator splitting method for solving the bidomain equations coupled to a volume conductor model for the torso. *Mathematical biosciences* **194**(2), 233–248 (2005).
17. S. Zampini. Dual-primal methods for the cardiac bidomain model. *Mathematical Models and Methods in Applied Sciences* **24**(04), 667–696 (2014).

Domain Decomposition Algorithms for Physics-Informed Neural Networks

Hyea Hyun Kim and Hee Jun Yang

1 Introduction

Domain decomposition algorithms are widely used as fast solutions of algebraic equations arising from the discretization of partial differential equations. The original algebraic equations are partitioned and solved in each subdomain combined with an iterative procedure. The resulting solution for the original algebraic equations is then obtained from the iterative procedure. In such approaches, the convergence often gets slow as more subdomains are introduced. To accelerate the convergence, a global coarse problem is formed and combined in the iterative procedure. We refer to [8] for a general introduction to domain decomposition algorithms.

Recently, many successful approaches have been developed to solve partial differential equations using deep neural networks, see [4, 6, 7, 9]. The advantage of these new approaches is that they can be used for partial differential equations without much concern for the discretization methods suitable for the given problem. On the other hand, a suitable design of the neural network and a suitable choice of data sets for training the parameters are important for these new approaches. In general, both the network and the data set can be large The parameter training then becomes inefficient and even may encounter numerical instability.

The purpose of this study is to develop domain decomposition algorithms for solutions of partial differential equations using deep neural networks. The idea is similar to the classical domain decomposition methods. The problem is solved using independent smaller neural networks iteratively and the smaller neural networks are trained as solutions of local problems that are restrictions of the original problem to smaller subdomains. In previous pioneering studies by [2, 3], the same idea is

Hyea Hyun Kim
Department of Applied Mathematics and Institute of Natural Sciences, Kyung Hee University, Korea, e-mail: hhkim@khu.ac.kr

Hee Jun Yang
Department of Mathematics, Kyung Hee University, Korea, e-mail: yhjj109@khu.ac.kr

S. C. Brenner et al. (eds.), *Domain Decomposition Methods in Science and Engineering XXVI*, Lecture Notes in Computational Science and Engineering 145,
https://doi.org/10.1007/978-3-030-95025-5_76

used but there has been no study of accelerating the convergence of the iterative scheme. In this work, an additional global coarse network is introduced and it is trained as a solution of the global problem using a coarse data set. The global coarse network is then used to accelerate the convergence of the iterative solution obtained from the independent smaller neural networks. The smaller neural networks and global coarse network are trained in each iteration. Their parameter training can be done in parallel. Among several neural network approaches, we will consider the PINN (Physics Informed Neural Network) method of [6]. Our domain decomposition approach can also be applied to other methods of [4, 7, 9].

In this work we report the first successful result for parallel algorithms for PINN using both local networks and one global coarse network. The introduction of the global coarse network is noble and it accelerate the convergence of the iteration. Numerical results also show that the use of the global coarse network makes the parallel algorithm scalable, i.e., the number of iterations is robust under an increase in the number of subdomains.

This paper is organized as follows. In Section 2, we introduce the PINN method for solving partial differential equations and in Section 3 we propose a two-level overlapping domain decomposition algorithm for solving partial differential equations utilizing the PINN approach. In Section 4, numerical results are presented for a model elliptic problem in two dimensions and conclusions are given.

2 Physics informed neural networks (PINN)

We will introduce the physics-informed neural networks (PINN) which are trained to solve supervised learning tasks in order to satisfy any given laws of physics described by partial differential equations, see [6]. We consider a general differential operator with a boundary condition,

$$\begin{aligned} \mathcal{L}(u) &= f, \quad \text{in } \Omega, \\ \mathcal{B}(u) &= g, \quad \text{on } \partial\Omega, \end{aligned} \tag{1}$$

where $\mathcal{L}$ can be a differential operator defined for a function u and $\mathcal{B}$ describes a given boundary condition on u, and f, g are given functions. We assume that the model problem in (1) is well-posed and the solution u exists. We then approximate the solution u in (1) by a neural network, $U(x;\theta)$, that can be trained by minimizing the cost function $\mathcal{J}(\theta)$ consisting of the two terms

$$\mathcal{J}(\theta) = \mathcal{J}_{X_\Omega}(\theta) + \mathcal{J}_{X_{\partial\Omega}}(\theta),$$

where

$$\mathcal{J}_{X_\Omega}(\theta) := \frac{1}{|X_\Omega|} \sum_{x \in X_\Omega} |\mathcal{L}(U(x;\theta)) - f(x)|^2,$$

$$\mathcal{J}_{X_{\partial\Omega}}(\theta) := \frac{1}{|X_{\partial\Omega}|} \sum_{x \in X_{\partial\Omega}} |\mathcal{B}(U(x;\theta)) - g(x)|^2.$$

In the above, X_D denotes the collection of points chosen from the region D and $|X_D|$ denotes the number of points in the set X_D. The cost functions $\mathcal{J}_{X_\Omega}(\theta)$ and $\mathcal{J}_{X_{\partial\Omega}}(\theta)$ are designed so that the optimized neural network $U(x;\theta)$ satisfies the equations in (1) derived from physics laws.

3 A two-level overlapping algorithm for PINN

We consider the following model elliptic problem in a two-dimensional domain Ω,

$$\begin{aligned} -\Delta u &= f \text{ in } \Omega, \\ u &= g \text{ on } \Omega. \end{aligned} \tag{2}$$

We propose an iterative scheme to find its solution u by using an overlapping subdomain partition, $\{\Omega_i\}_i$, with an overlapping width δ. For a given $u^{(n)}$, we solve the following problem in each subdomain Ω_i to find $u_i^{(n+1)}$,

$$\begin{aligned} -\Delta u_i^{(n+1)} &= f \text{ in } \Omega_i, \\ u_i^{(n+1)} &= u^{(n)} \text{ on } \partial\Omega_i, \\ u_i^{(n+1)} &= u^{(n)} \text{ in } \overline{\Omega} \setminus \Omega_i. \end{aligned} \tag{3}$$

Using $u_i^{(n+1)}$, the next iterate is given by

$$u^{(n+1)} = (1 - N\tau)u^{(n)} + \tau \sum_{i=1}^{N} u_i^{(n+1)}, \tag{4}$$

where N denotes the number of subdomains and τ denotes the relaxation parameter. Let N_c be the maximum number of subdomains sharing the same geometric position in Ω. With $\tau \leq 1/N_c$, $u^{(n)}$ converges to the solution u of (2) under a suitably chosen space of functions, see [1, 5, 8, 10]. We can rewrite the above iteration formula as follows: for any x in Ω

$$u^{(n+1)}(x) = (1 - |s(x)|\tau)u^{(n)}(x) + \tau \sum_{i \in s(x)} u_i^{(n+1)}(x), \tag{5}$$

where $s(x)$ denotes the set of subdomain indices sharing x and $|s(x)|$ denotes the number of elements in the set $s(x)$. We introduce

$$\widehat{u}^{(n+1)}(x) := \frac{1}{|s(x)|} \sum_{i \in s(x)} u_i^{(n+1)}(x)$$

and rewrite the above iteration formula as

$$u^{(n+1)}(x) = (1 - |s(x)|\tau)u^{(n)}(x) + |s(x)|\tau\widehat{u}^{(n+1)}(x).$$

Using this formula, we can see that $\widehat{u}^{(n+1)}(x)$ also converges to $u(x)$.

For i in $s(x)$, the solution $u_i^{(n+1)}(x)$ is updated after solving the local problem in (3). We thus define $U_i(x;\theta_i^{(n+1)})$ as a neural network function to approximate $u_i^{(n+1)}(x)$ in each Ω_i. Using the PINN method, we can find the optimal parameters $\theta_i^{(n+1)}$. Using them, we define

$$\widehat{U}^{(n+1)}(x) := \frac{1}{|s(x)|}\sum_{i\in s(x)} U_i(x;\theta_i^{(n+1)}). \tag{6}$$

We now propose the following one-level method:

Algorithm 1: One-level method (input: $U^{(0)}$, output: $\widehat{U}^{(n+1)}$)
Step 0: Let $U^{(0)}(x)$ be given and $n = 0$.
Step 1: Find $\theta_i^{(n+1)}$ in $U_i(x;\theta_i^{(n+1)})$ for

$$-\triangle u = f \text{ in } \Omega_i,$$
$$u = U^{(n)} \text{ on } \partial\Omega_i.$$

Step 2: Update $U^{(n+1)}$ at each data set $X_{\partial\Omega_i}$ as, see (6),

$$U^{(n+1)}(x) = (1 - \tau|s(x)|)U^{(n)}(x) + \tau|s(x)|\widehat{U}^{(n+1)}.$$

Step 3: Go to **Step 1** with $n = n + 1$ or set the output as $\widehat{U}^{(n+1)}$ if the stopping condition is met.

Using sufficiently large neural network functions $U_i(x;\theta_i^{(n)})$, we can approximate $u_i^{(n+1)}(x)$ and $\widehat{U}^{(n+1)}(x)$ will thus approximate $\widehat{u}^{(n+1)}(x)$. Since $\widehat{u}^{(n+1)}(x)$ converges to $u(x)$, $\widehat{U}^{(n+1)}(x)$ will converge to $u(x)$. We note that if one wishes to take $U^{(n+1)}(x)$ as the final output then one needs to store all the parameters $\theta_i^{(m)}$ for all previous steps m. In addition, the evaluation of $U^{(n+1)}(x)$ at any given point x can be very expensive. We thus take $\widehat{U}^{(n+1)}(x)$ as the final solution in our algorithm. We will only need to store the parameters $\theta_i^{(n+1)}$ at the final step. Since the local problems in the above algorithm are solved by the PINN method, the function $\widehat{U}^{(n+1)}(x)$ needs to be evaluated at x in the data set $X_{\partial\Omega_i}$. In our algorithm, we only store these function values at each iteration and use them when we solve the local problems (3) using the PINN method.

As we can see in the numerical results provided in Section 4, the convergence of the one-level algorithm gets slower as more subdomains are introduced in the partition. We thus improve the one-level algorithm by enriching the boundary condition $U^{(n)}$ with a suitable coarse correction term. For a given $U^{(n)}$, we consider the following global problem:

$$
\begin{aligned}
-\Delta u_c^{(n)} &= f \text{ in } \Omega_\delta, \\
-\Delta u_c^{(n)} &= -\Delta U^{(n)} \text{ in } \Omega \setminus \Omega_\delta, \\
u_c^{(n)} &= g \text{ on } \partial\Omega,
\end{aligned} \tag{7}
$$

where Ω_δ denotes the overlapping region of the subdomain partition $\{\Omega_i\}_i$. For the solution $u_c^{(n)}$, we can obtain the following error equation,

$$
\begin{aligned}
-\Delta(u - u_c^{(n)}) &= 0 \text{ in } \Omega_\delta, \\
-\Delta(u - u_c^{(n)}) &= -\Delta(u - U^{(n)}) \text{ in } \Omega \setminus \Omega_\delta, \\
u - u_c^{(n)} &= 0 \text{ on } \partial\Omega.
\end{aligned} \tag{8}
$$

From the above error equation, we have $u - u_c^{(n)}$ with smaller errors than $u - U^{(n)}$. We will then find a coarse correction term $U_c^{(n)}(x; \theta_c^{(n)})$ that approximates $u_c^{(n)}$ with the parameters $\theta_c^{(n)}$ determined by the PINN method. Using the coarse correction term, for $\alpha > 0$ we set

$$\widetilde{U}^{(n)} = (1 - \alpha)U^{(n)} + \alpha U_c^{(n)}$$

and use it when we evaluate the boundary condition for the local problems in (3). We note that when we find $\theta_c^{(n)}$ using the PINN method we will only need to evaluate $-\Delta U^{(n)}$ at the data set $X_{\Omega\setminus\Omega_\delta}$ without the need to store the parameters $\theta_i^{(m)}$ for all previous steps m.

We now summarize the two-level method:

Algorithm 2: Two-level method (input: $U^{(0)}$, output: $\widehat{U}^{(n+1)}$)
Step 0: Let $U^{(0)}(x)$ be given and $n = 0$.
Step 1-1: Find $U_c^{(n)}(x; \theta_c^{(n)})$ for (7) and set

$$\widetilde{U}^{(n)}(x) = (1 - \alpha)U^{(n)}(x) + \alpha U_c^{(n)}(x; \theta_c^{(n)}).$$

Step 1-2: Find $\theta_i^{(n+1)}$ in $U_i(x; \theta_i^{(n+1)})$ for

$$
\begin{aligned}
-\Delta u - f &\text{ in } \Omega_i, \\
u = \widetilde{U}^{(n)} &\text{ on } \partial\Omega_i.
\end{aligned}
$$

Step 2: Update $U^{(n+1)}(x)$ at each data set $X_{\partial\Omega_i}$ as, see (6),

$$U^{(n+1)}(x) = (1 - \tau|s(x)|)U^{(n)}(x) + \tau|s(x)|\widehat{U}^{(n+1)}.$$

Step 3: Go to **Step 1-1** with $n = n + 1$ or set the output as $\widehat{U}^{(n+1)}$ if the stopping condition is met.

4 Numerical results

We perform numerical results of the proposed two algorithms for the model problem in (2) with f and g given according to the known exact solution $u(x, y)$ and with Ω as a unit rectangular domain. The domain Ω is partitioned into uniform rectangular subdomains with an overlapping width δ. For the iterates $U^{(n)}$, we stop the iteration when the relative l^2-error between the two successive iterates is less than 5×10^{-3}. When training parameters $\theta_i^{(n)}$ and $\theta_c^{(n)}$, we stop the iteration when the relative errors for cost function values between 100 steps is less than 10^{-4} or when the number of iterations is more than the maximum number of epochs, which is set as 5000. For local problems, we use neural network functions as a two block Resnet with each block consisting of 10 hidden layers and with Tanh as the activation function, which give 921 parameters $\theta_i^{(n)}$ for each local problem. To train the parameters, we use 200 data points for X_{Ω_i} and 40 data points for $X_{\partial\Omega_i}$. For the coarse problem, we use the same network and the same size of data sets.

In our method, we have two parameters τ and α. For τ, we can set τ as less than or equal to $1/N_c$ and α as a number between 0 and 1. When $\alpha = 0$, the two-level algorithm is identical to the one-level algorithm. When $\alpha > 0$, the method is enhanced with the coarse correction term.

In Table 1, we report the performance of the proposed method with various α and N for the exact solution $u(x, y) = \sin(\pi x)\sin(\pi y)$. The relative L^2-errors to the exact solution and the number of iterations are presented. We set τ as 1/4, and note $N_c = 4$. Without the coarse correction term, i.e., $\alpha = 0$, the one-level method shows that the number of iterations increases as N increases. For the other choices of $\alpha(> 0)$, the coarse correction term accelerates the convergence and the number of iterations seems robust under an increase in the number of subdomains.

Table 1: The performance with $\tau = 1/4$ depending on α and N (the subdomain partition): the numbers are relative L^2-errors to the exact solution and the numbers inside the parentheses are the number of iterations.

N	$\alpha = 0$	$\alpha = 0.25$	$\alpha = 0.5$	$\alpha = 1$
2×2	0.0098(22)	0.0073(14)	0.0117(19)	0.0082(21)
3×3	0.0282(34)	0.0243(17)	0.0260(14)	0.0310(14)
4×4	0.0402(52)	0.0070(18)	0.0237(10)	0.0431(8)
5×5	0.0769(67)	0.0255(21)	0.0298(13)	0.0392(10)

We compare our Algorithm 1 and that in the previous study by [3]. Under the same setting with Table 3 of [3], we apply our Algorithm 1 and obtain much fewer iterations and more accurate solutions, see Table 2.

To show the advantage of partitioning the problem, we consider a more difficult problem with the exact solution given by

$$u(x, y) = 100x(1 - x)y(1 - y)\sin((x - 0.5)(y - 0.5)/0.05). \tag{9}$$

Table 2: The performance of Algorithm 1 under the same setting for the model in [3, Table 3]: relative L^2-errors and the number of iterations (numbers inside the parentheses) depending on the number of layers (L), and the number of units (U).

N	L \ U	10	20	30	40	50	100
4	2	0.0040(2)	0.0037(2)	0.0030(2)	0.0043(2)	0.0023(2)	0.0029(2)
4	3	0.0042(2)	0.0034(2)	0.0046(2)	0.0030(2)	0.0037(2)	0.0061(2)
4	4	0.0075(2)	0.0046(2)	0.0038(2)	0.0045(2)	0.0060(2)	0.0047(2)

To approximate the highly oscillatory solution with high contrast, we use a single neural network with its number of parameters as 9109 and with 2000 interior points and 400 boundary points for training the parameters using 250000 epochs. With this, we solve the model problem in the whole domain $\Omega = (0\,1)^2$. For the same model problem, we partition the domain into 9 overlapping subdomains and employ a smaller neural network with 921 number of parameters. For the global coarse network, we use the same number of parameters. For training parameters in both local and coarse neural networks, 200 interior points and 40 boundary points are used. The computation time and the accuracy of trained solutions are compared in Table 3. We can observe the advantage of partitioning with much less computation time and less errors than the single domain case. When the local solutions are solved in parallel, the computation time can be further reduced. For the analysis of computational time, we let T_s be the training time for one local or coarse neural network, and T be the training time for the single neural network of the whole domain. Let *iter* be the number of iterations in our Algorithm 2. Assuming that the local networks are trained in parallel, the total computation time becomes $iter \times 2T_s$. With a proper size of local and coarse neural networks, the computation time T_s can become much smaller than T and the total computation time is thus expected to be much smaller than T.

Table 3: The performance of the proposed method for the model problem in (9): single domain and 9 subdomains with different α values, the numbers inside the parentheses are the number of iterations.

	single domain	$\alpha = 0$	$\alpha = 0.25$	$\alpha = 0.5$	$\alpha = 1.0$
L^2-error	0.0754	0.0820 (93)	0.0623 (56)	0.0705 (61)	0.0793 (61)
time(sec)	289840	74702	49980	54442	54442

In conclusion, a two-level algorithm suitable for deep neural network architecture is proposed and tested. By partitioning the large deep neural network, the computation time is greatly reduced with a more accurate solution in our test example. A more rigorous numerical study and convergence analysis will be done in a more complete paper.

Acknowledgements The authors are supported by NRF-2019R1A2C1010090.

References

1. M.J. Gander. Schwarz methods over the course of time. *Electron. Trans. Numer. Anal.* **31**, 228–255 (2008).
2. K. Li, K. Tang, T. Wu and Q. Liao. D3M: A deep domain decomposition method for partial differential equations. *IEEE Access* **8**, 5283–5294 (2019).
3. W. Li, X. Xiang and Y. Xu. Deep domain decomposition method: Elliptic problems. In: *Mathematical and Scientific Machine Learning*, pp. 269–286, PMLR (2020).
4. Z. Long, Y. Lu and B. Dong. PDE-Net 2.0: learning PDEs from data with a numeric-symbolic hybrid deep network. *J. Comput. Phys.* **399**(17), 108925 (2019).
5. J. Park. Additive Schwarz methods for convex optimization as gradient methods. *SIAM J. Numer. Anal.* **58**(3), 1495–1530 (2020).
6. M. Raissi, P. Perdikaris and G.E. Karniadakis. Physics-informed neural networks: a deep learning framework for solving forward and inverse problems involving nonlinear partial differential equations. *J. Comput. Phys.* **378**, 686–707 (2019).
7. J. Sirignano and K. Spiliopoulos. DGM: a deep learning algorithm for solving partial differential equations. *J. Comput. Phys.* **375**, 1339–1364 (2018).
8. A. Toselli and O. Widlund. *Domain decomposition methods – algorithms and theory*. Volume 34 of Springer Series in Computational Mathematics. Springer-Verlag, Berlin (2005).
9. E. Weinan, J. Han and A. Jentzen. Deep learning-based numerical methods for high-dimensional parabolic partial differential equations and backward stochastic differential equations. *Commun. Math. Stat.* **5**(4), 349–380 (2017).
10. J. Xu and L. Zikatanov. The method of alternating projections and the method of subspace corrections in Hilbert space. *J. Amer. Math. Soc.* **15**(3), 573–597 (2002).

Numerical Study of an Additive Schwarz Preconditioner for a Thin Membrane Diffusion Problem

Piotr Krzyżanowski

1 Introduction

In the biology of the cell one has to take into account the situation when two different materials – for example, the cytoplasm and the nucleus – are separated by a permeable membrane. Chemicals inside the cell diffuse not only inside both the nucleus and in the cytoplasm, but they also pass through the membrane as well. A mathematical model of such phenomenon, which gained some popularity (see e.g. [7, 14] and the literature therein) has been introduced by Kedem and Kachalsky, where a system of diffusive PDEs is coupled by specific boundary conditions on the inner interface. In this paper we will investigate a simplified problem, hoping our approach may be applicable to more complicated cases as well.

Let us denote by $\Omega \subset R^d$ $(d = 2, 3)$ the domain occupied by the cell. It naturally decomposes into disjoint open sets: the surrounding cytoplasm Ω_1 and $N-1$ organelles (the nucleus, mitochondria, etc.), denoted here $\Omega_2, \dots, \Omega_N$, so that $\bar{\Omega} = \bigcup_{i=1}^N \bar{\Omega}_i$ and $\Omega_i \cap \Omega_j = \emptyset$, cf. Figure 1. The interface between the i-th organelle and the outer cell will be denoted $\Gamma_i = \partial\Omega_1 \cap \partial\Omega_i = \partial\Omega_i$ and for simplicity of notation we set $\Gamma = \bigcup_{i=2}^N \bar{\Gamma}_i$. Our model problem reads:

$$-\operatorname{div}(\varrho_i \nabla u_i) + K_i u_i = F_i \text{ in } \Omega_i, \qquad i = 1, \dots, N, \tag{1}$$

with interface conditions

$$-\varrho_1 \nabla u_1 \cdot n_1 = G_i \cdot (u_1 - u_i) = \varrho_i \nabla u_i \cdot n_i \text{ on } \Gamma_i \tag{2}$$

for $i = 2, \dots, N$, where n_i denotes the unit outer normal vector to Ω_i. The system is completed with a non-permeability external boundary condition,

Piotr Krzyżanowski
Faculty of Mathematics, Informatics and Mechanics, University of Warsaw, Poland,
e-mail: p.krzyzanowski@mimuw.edu.pl

S. C. Brenner et al. (eds.), *Domain Decomposition Methods in Science and Engineering XXVI*, Lecture Notes in Computational Science and Engineering 145,
https://doi.org/10.1007/978-3-030-95025-5_77

$$-\varrho_1 \nabla u_1 \cdot n = 0 \text{ on } \partial\Omega. \tag{3}$$

Here, $\varrho_1, \ldots, \varrho_N$ and $K_1, \ldots, K_N$ are prescribed positive constants, which can be different between the subdomains. For the source terms we assume $F_i \in L^2(\Omega_i)$, $i = 1, \ldots, N$. The unknown functions u_i defined in $\bar{\Omega}_i$, $i = 1, \ldots, N$ may represent e.g. the hes1 mRNA concentration in the cell [14].

Positive constant parameters G_i model the thickness of the interface; roughly speaking, the permeability constant $G_i \sim 1/H_i$, where H_i is the thickness of the membrane between Ω_i and Ω_1; therefore for thin interfaces $G_i \gg 1$. In order to address the interface conditions (2), we incorporate them directly into the bilinear form, obtaining the following weak formulation of (1)–(3):

Problem 1 Find $(u_1, \ldots, u_N) \in V = H^1(\Omega_1) \times \cdots \times H^1(\Omega_N)$ such that

$$\sum_{i=1}^{N} \int_{\Omega_i} \varrho_i \nabla u_i \cdot \nabla \varphi_i + K_i u_i \varphi_i \,\mathrm{d}x + \sum_{i=2}^{N} \int_{\Gamma_i} G_i (u_i - u_1)(\varphi_i - \varphi_1) \,\mathrm{d}s = \sum_{i=1}^{N} \int_{\Omega_i} F_i \varphi_i \,\mathrm{d}x$$

for all $(\varphi_1, \ldots, \varphi_N) \in V$.

The bilinear form appearing in Problem 1 is symmetric and elliptic. Note that the interface integral term in Problem 1 results from the permeability condition (2) and it penalizes the jump of the solution across the interface Γ.

We discretize Problem 1 with a composite discontinuous Galerkin *h*-*p* finite element method [8]. Inside Ω_i, we use a continuous *h*-*p* method, while allowing for the discontinuity of the solution across Γ. In order not to complicate the exposition, we will assume from now on that each Ω_i is a polyhedron.

Let us define a simplicial, quasi-uniform, conforming triangulation $\mathcal{T}_h$ with mesh size h over Ω, whose elements are aligned with Ω_i, so that Γ crosses no element in $\mathcal{T}_h$. In this way each Ω_i, $i = 1, \ldots, N$, is supplied with its own triangulation $\mathcal{T}_h(\Omega_i)$. We define the corresponding local continuous finite element spaces as

$$V_h^p(\Omega_i) = \{v \in C(\Omega_i) : v_{|K} \in \mathcal{P}^p(K) \quad \forall K \in \mathcal{T}_h(\Omega_i)\},$$

where $\mathcal{P}^p$ is the space of polynomials of degree at most $p \geq 1$. The finite element approximation of Problem 1 then reads:

Problem 2 Find $u \in V_h^p = \{v \in L^2(\Omega) : v_{|\Omega_i} \in V_h^p(\Omega_i), \quad i = 1, \ldots, N\}$ such that $\mathcal{A}(u, v) = \sum_{i=1}^{N} \int_{\Omega_i} F_i \varphi_i \,\mathrm{d}x$ for all $\varphi \in V_h^p$, where

$$\mathcal{A}(u, v) = \sum_{i=1}^{N} \int_{\Omega_i} \varrho_i \nabla u_i \cdot \nabla \varphi_i + K_i u_i \varphi_i \,\mathrm{d}x + \sum_{i=2}^{N} \int_{\Gamma_i} G_i \cdot (u_i - u_1)(\varphi_i - \varphi_1) \,\mathrm{d}s.$$

Our goal in this paper is to describe and experimentally evaluate the performance of a preconditioner for Problem 2, based on the additive Schwarz method, see e.g. [15], in terms of the convergence rate of the preconditioned conjugate gradients iterative solver. The penalty constant G_i is an independent parameter of the original problem, in contrast to the analogous term in the interior penalty discontinuous Galerkin method. For the latter, a preconditioner for the Poisson equation with $\varrho_i = 1$ was developed and proved optimal with respect to discretization and the penalty constant in [5], where numerical evidence was provided that this method leads to a condition number which grows linearly with the contrast ratio in the diffusion coefficient. Another approach was considered in [9] and [12], where it was proved that the convergence rate is uniformly bounded with respect to diffusion coefficient jumps; however, the dependence on the penalizing constant was not investigated. Here, we provide extensive tests of the preconditioning properties of a method first introduced in [13], which is inspired by [5] and [12]. It turns out that the method considered here is robust with respect to both the problem's parameters and to discretization parameters as well.

The rest of paper is organized as follows. In Section 2, a preconditioner based on the additive Schwarz method for solving Problem 2 is presented. We report on its performance in a series of numerical experiments in Section 3. We conclude with final remarks in Section 4.

2 Additive Schwarz preconditioner

In this section we consider a preconditioner based on the nonoverlapping additive Schwarz method, first proposed, in a different setting, in [2] and later developed in many papers, including [3, 4, 9, 12]. The space V_h^p is decomposed as follows:

$$V_h^p = V_0 + \sum_{i=1}^{N} V_i,$$

where for $i = 1, \ldots, N$ the local spaces are

$$V_i = \{v \in V_h^p : v_{|\Omega_j} = 0 \text{ for all } j \neq i\},$$

so that V_i is a zero-extension of functions from $V_h^p(\Omega_i)$. Note that V_h^p is already a direct sum of these local spaces. In the setting of Problem 2, the main goal of the coarse space V_0 is to deal with the penalization term; we define V_0 as the finite element space of piecewise polynomial functions which are continuous in the entire Ω,

$$V_0 = \{v \in C(\Omega) : v_{|K} \in \mathcal{P}^p(K) \text{ for all } K \in \mathcal{T}_h\}.$$

The choice of the coarse space is inspired by the work by Antonietti et al. [5] for the standard Poisson problem and notably leads to a problem whose number of unknowns is smaller than the original only by a small fraction.

As mentioned above, $\mathcal{A}(\cdot, \cdot)$ is symmetric positive definite on $V_h^p \subset V$. We define operators $T_i : V_h^p \to V_i$, $i = 0, 1, \ldots, N$, by "inexact" solvers $A_i(T_i u, v) = \mathcal{A}(u, v)\, \forall v \in V_i$. We will assume that $A_i(\cdot, \cdot)$ are symmetric, positive definite, and they induce a linear operator which is spectrally equivalent to the operator induced by $\mathcal{A}(\cdot, \cdot)$ on V_i. The preconditioned operator is

$$T = T_0 + \sum_{i=1}^{N} T_i. \tag{4}$$

While all T_i, $i = 0, 1, \ldots, N$, can be applied in parallel, the performance of the preconditioner is affected by the specific choice of subspace solvers $A_i(\cdot, \cdot)$. In the experiments in the following section, we will choose the algebraic multigrid (AMG) solvers, see e.g. [16]. In particular, it is well known that AMG can be a robust preconditioner for discontinuous coefficient problems discretized with continuous finite elements, so a parallel AMG makes a reasonable choice for the inexact solver on V_0 (other choices, e.g. the additive average Schwarz method [10], are also possible).

From the definition of T_i it follows that virtually all degrees of freedom are solved twice when T is applied, so there is room for improvement of the complexity of the method. On the other hand, as it will be shown in the following section, the method converges independently of the size of the permeability coefficients.

3 Numerical experiments

Since the number of problem parameters is large we restrict ourselves to the case when $\varrho_1 = K_1 = 1$ and $\varrho_2 = \cdots = \varrho_N$, $K_2 = \cdots = K_N$ and $G_2 = \cdots = G_N = G$. Our goal in this section is to investigate the influence of various parameters of the problem: the diffusion coefficient contrast $\varrho = \varrho_2/\varrho_1$, the reaction coefficient contrast $K = K_2/K_1$, the value of the permeability coefficient G, the number of subdomains N, and discretization parameters: the mesh size and the polynomial degree, on the convergence rate of the preconditioned conjugate gradients (PCG) iteration and the condition number of T. Our implementation is based on the FEniCS software [1] with PETSc [6] as the linear algebra backend. For the inexact solvers on the subspaces we chose the algebraic multigrid method: BoomerAMG solver from the hypre library [11], with default parameters. We performed tests for Ω in 2D and 3D; example domains are depicted in Figure 1. The organelles were allowed to touch neither the boundary of the domain, nor other organelles.

The domain Ω was triangulated with an unstructured, quasi-uniform mesh with resolution parameter r, roughly proportional to h^{-1}. For the finite element polynomial degrees $1 \leq p \leq 3$ this resulted in discrete problem sizes summarized in Table 1.

In the tables below, we report the number of iterations required to reduce the initial residual norm by a factor of 10^8; in parentheses, we also provide the condition number estimate of T, with the mantissa rounded to the nearest integer. The initial guess was always equal to zero. If the convergence criterion was not reached in 100 iterations, we place a dash. Experiments which were not performed due to hardware limitations are marked with 'N/A'. For comparison, we also include results when the

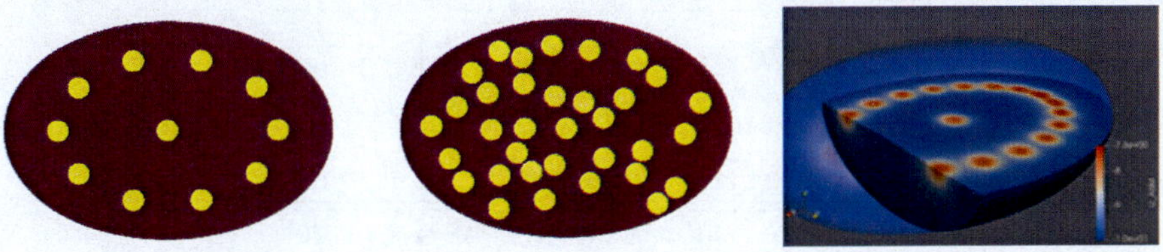

Fig. 1: Types of domains and subdomains. Left: elliptic shaped Ω with regularly placed circular $N = 11$ organelles. Center: elliptic shaped Ω with randomly placed nonoverlapping circular $N = 33$ organelles. Right: 3D ellipsoid with regularly placed organelles (visualized is a cross-section of the domain; colors reflect the value of the solution).

$\downarrow r \rightarrow p$	1	2	3
16	$4.4 \cdot 10^2$	$1.7 \cdot 10^3$	$3.7 \cdot 10^3$
32	$1.7 \cdot 10^3$	$6.5 \cdot 10^3$	$1.4 \cdot 10^4$
64	$6.2 \cdot 10^3$	$2.5 \cdot 10^4$	$5.5 \cdot 10^4$
128	$2.5 \cdot 10^4$	$9.7 \cdot 10^4$	$2.2 \cdot 10^5$

$\downarrow r \rightarrow p$	1	2	3
16	$1.3 \cdot 10^4$	$9.6 \cdot 10^4$	$3.2 \cdot 10^5$
24	$6.1 \cdot 10^4$	$4.7 \cdot 10^5$	N/A
32	$9.6 \cdot 10^4$	$7.4 \cdot 10^5$	N/A

Table 1: Approximate total number of degrees of freedom for various values of mesh resolution parameter r and polynomial degree p. Left: 2D case; right: 3D case.

problem was solved with the PCG, where the BoomerAMG was used to precondition the whole discrete system resulting from Problem 2.

While varying other parameters, if not specified otherwise, we assume default values $\varrho = K = 1$, $p = 2$, $N = 22$ and $r = 128$ in the 2D case or $r = 32$ in the 3D case. In Tables 2–3 we investigate the dependence of the convergence rate on r, p, ϱ, K for both moderate and very large value of G. It turns out that the performance of T is essentially uniform across the range (with some small degradation for certain extreme values of ϱ or K) regardless of G, while the AMG suffers for most combinations of parameters when G is large. Tables 6–7 confirm analogous behavior in 3D.

In Table 4 we repeat the first experiment with irregularly scattered organelles (cf. the middle picture in Figure 1) with no significant differences. From Table 5 it follows T performs well, independently of the number of inclusions, again, with some increase of the number of iterations for large ϱ.

Finally, in Table 8 we provide more detailed insight into the convergence rate of T for G in the range $10^0 \ldots 10^{12}$, while keeping other parameters fixed. It turns out that the number of iterations of T stays essentially constant.

$\downarrow r \rightarrow p$	1	2	3	1	2	3
16	10 ($2 \cdot 10^0$)	10 ($2 \cdot 10^0$)	10 ($2 \cdot 10^0$)	6 ($1 \cdot 10^0$)	6 ($1 \cdot 10^0$)	7 ($1 \cdot 10^0$)
32	10 ($2 \cdot 10^0$)	10 ($2 \cdot 10^0$)	10 ($2 \cdot 10^0$)	6 ($1 \cdot 10^0$)	6 ($1 \cdot 10^0$)	7 ($1 \cdot 10^0$)
64	10 ($2 \cdot 10^0$)	10 ($2 \cdot 10^0$)	10 ($2 \cdot 10^0$)	6 ($1 \cdot 10^0$)	6 ($1 \cdot 10^0$)	7 ($1 \cdot 10^0$)
128	10 ($2 \cdot 10^0$)	10 ($2 \cdot 10^0$)	11 ($2 \cdot 10^0$)	6 ($1 \cdot 10^0$)	7 ($1 \cdot 10^0$)	7 ($1 \cdot 10^0$)
16	7 ($2 \cdot 10^0$)	9 ($2 \cdot 10^0$)	11 ($3 \cdot 10^0$)	46 ($3 \cdot 10^5$)	96 ($3 \cdot 10^5$)	–
32	9 ($2 \cdot 10^0$)	9 ($2 \cdot 10^0$)	11 ($2 \cdot 10^0$)	69 ($3 \cdot 10^5$)	–	–
64	9 ($2 \cdot 10^0$)	10 ($2 \cdot 10^0$)	11 ($2 \cdot 10^0$)	92 ($3 \cdot 10^5$)	–	–
128	9 ($2 \cdot 10^0$)	10 ($2 \cdot 10^0$)	11 ($3 \cdot 10^0$)	–	–	–

Table 2: Iteration count (the condition number estimate in parentheses) for varying mesh resolution r and polynomial degree p: T (left) vs. AMG preconditioner (right). Top: $G = 10^0$; bottom: $G = 10^6$. 2D case, regularly placed $N = 22$ subdomains. $\varrho = K = 1$.

$\downarrow \varrho \rightarrow K$	10^{-6}	10^0	10^6	10^{-6}	10^0	10^6
10^{-6}	10 ($2 \cdot 10^0$)	10 ($2 \cdot 10^0$)	10 ($2 \cdot 10^0$)	13 ($3 \cdot 10^0$)	13 ($3 \cdot 10^0$)	7 ($1 \cdot 10^0$)
10^0	12 ($3 \cdot 10^0$)	10 ($2 \cdot 10^0$)	10 ($2 \cdot 10^0$)	9 ($2 \cdot 10^0$)	7 ($1 \cdot 10^0$)	7 ($1 \cdot 10^0$)
10^6	15 ($4 \cdot 10^0$)	13 ($3 \cdot 10^0$)	11 ($2 \cdot 10^0$)	11 ($2 \cdot 10^0$)	9 ($1 \cdot 10^0$)	7 ($1 \cdot 10^0$)
10^{-6}	11 ($2 \cdot 10^0$)	12 ($3 \cdot 10^0$)	11 ($2 \cdot 10^0$)	–	–	40 ($4 \cdot 10^1$)
10^0	10 ($2 \cdot 10^0$)	10 ($2 \cdot 10^0$)	11 ($2 \cdot 10^0$)	–	–	40 ($4 \cdot 10^1$)
10^6	14 ($3 \cdot 10^0$)	13 ($3 \cdot 10^0$)	10 ($2 \cdot 10^0$)	52 ($1 \cdot 10^6$)	44 ($7 \cdot 10^5$)	13 ($3 \cdot 10^0$)

Table 3: Iteration count (the condition number estimate in parentheses) for varying contrast ratios ϱ and K for T (left) vs. AMG preconditioner (right). Top: $G = 10^0$; bottom: $G = 10^6$. 2D case, regularly placed $N = 22$ subdomains. $r = 128$, $p = 2$.

$\downarrow r \rightarrow p$	1	2	3	1	2	3
16	9 ($2 \cdot 10^0$)	9 ($2 \cdot 10^0$)	10 ($2 \cdot 10^0$)	6 ($1 \cdot 10^0$)	6 ($1 \cdot 10^0$)	7 ($1 \cdot 10^0$)
32	10 ($2 \cdot 10^0$)	10 ($2 \cdot 10^0$)	10 ($2 \cdot 10^0$)	6 ($1 \cdot 10^0$)	6 ($1 \cdot 10^0$)	7 ($1 \cdot 10^0$)
64	10 ($2 \cdot 10^0$)	10 ($2 \cdot 10^0$)	10 ($2 \cdot 10^0$)	6 ($1 \cdot 10^0$)	6 ($1 \cdot 10^0$)	7 ($1 \cdot 10^0$)
128	10 ($2 \cdot 10^0$)	10 ($2 \cdot 10^0$)	10 ($2 \cdot 10^0$)	6 ($1 \cdot 10^0$)	7 ($1 \cdot 10^0$)	8 ($1 \cdot 10^0$)
16	7 ($2 \cdot 10^0$)	9 ($2 \cdot 10^0$)	10 ($2 \cdot 10^0$)	32 ($2 \cdot 10^5$)	64 ($3 \cdot 10^5$)	93 ($3 \cdot 10^5$)
32	8 ($2 \cdot 10^0$)	9 ($2 \cdot 10^0$)	11 ($2 \cdot 10^0$)	68 ($3 \cdot 10^5$)	–	–
64	9 ($2 \cdot 10^0$)	10 ($2 \cdot 10^0$)	11 ($2 \cdot 10^0$)	85 ($3 \cdot 10^5$)	–	–
128	10 ($2 \cdot 10^0$)	10 ($2 \cdot 10^0$)	12 ($3 \cdot 10^0$)	–	–	–

Table 4: Iteration count (the condition number estimate in parentheses) for varying mesh resolution r and polynomial degree p: T (left) vs. AMG preconditioner (right). Top: $G = 10^0$ bottom: $G = 10^6$. 2D case, irregularly placed $N = 19$ subdomains. $\varrho = K = 1$.

4 Conclusions

Numerical experiments indicate the preconditioner under consideration performs well in a broad range of problem parameters. The main advantage of the proposed preconditioner over the AMG preconditioner applied directly to the discrete prob-

$\rightarrow N$ $\downarrow \varrho$	19	35	51	19	35	51
10^{-6}	10 ($2 \cdot 10^0$)	10 ($2 \cdot 10^0$)	11 ($2 \cdot 10^0$)	12 ($3 \cdot 10^0$)	13 ($3 \cdot 10^0$)	16 ($4 \cdot 10^0$)
10^0	10 ($2 \cdot 10^0$)	10 ($2 \cdot 10^0$)	10 ($2 \cdot 10^0$)	7 ($1 \cdot 10^0$)	7 ($1 \cdot 10^0$)	7 ($1 \cdot 10^0$)
10^6	12 ($2 \cdot 10^0$)	13 ($3 \cdot 10^0$)	14 ($3 \cdot 10^0$)	8 ($1 \cdot 10^0$)	9 ($1 \cdot 10^0$)	10 ($1 \cdot 10^0$)
10^{-6}	11 ($2 \cdot 10^0$)	11 ($2 \cdot 10^0$)	11 ($2 \cdot 10^0$)	–	–	–
10^0	10 ($2 \cdot 10^0$)	10 ($2 \cdot 10^0$)	10 ($2 \cdot 10^0$)	–	–	–
10^6	13 ($3 \cdot 10^0$)	13 ($3 \cdot 10^0$)	13 ($3 \cdot 10^0$)	44 ($7 \cdot 10^5$)	61 ($7 \cdot 10^5$)	69 ($9 \cdot 10^5$)

Table 5: Iteration count (the condition number estimate in parentheses) for varying contrast ratios ϱ and number of subdomains N: T (left) vs. AMG preconditioner (right). Top: $G = 10^0$ bottom: $G = 10^6$. 2D case, irregularly placed subdomains. $r = 128$, $p = 2$, $K = 1$.

$\downarrow r \rightarrow p$	1	2	3	1	2	3
24	10 ($2 \cdot 10^0$)	10 ($2 \cdot 10^0$)	11 ($3 \cdot 10^0$)	5 ($1 \cdot 10^0$)	7 ($2 \cdot 10^0$)	9 ($2 \cdot 10^0$)
32	9 ($2 \cdot 10^0$)	9 ($2 \cdot 10^0$)	N/A	5 ($1 \cdot 10^0$)	6 ($1 \cdot 10^0$)	N/A
48	9 ($2 \cdot 10^0$)	10 ($2 \cdot 10^0$)	N/A	5 ($1 \cdot 10^0$)	6 ($1 \cdot 10^0$)	N/A
32	8 ($2 \cdot 10^0$)	10 ($2 \cdot 10^0$)	11 ($3 \cdot 10^0$)	86 ($1 \cdot 10^5$)	–	–
64	7 ($2 \cdot 10^0$)	9 ($2 \cdot 10^0$)	N/A	83 ($1 \cdot 10^5$)	–	N/A
128	8 ($2 \cdot 10^0$)	9 ($2 \cdot 10^0$)	N/A	92 ($1 \cdot 10^5$)	–	N/A

Table 6: Iteration count (the condition number estimate in parentheses) for varying mesh resolution r and polynomial degree p: T (left) vs. AMG preconditioner (right). Top: $G = 10^0$ bottom: $G = 10^6$. 3D case. $\varrho = K = 1$, $N = 22$.

$\downarrow \varrho \rightarrow K$	10^{-6}	10^0	10^6	10^{-6}	10^0	10^6
10^{-6}	9 ($2 \cdot 10^0$)	11 ($2 \cdot 10^0$)	12 ($2 \cdot 10^0$)	9 ($2 \cdot 10^0$)	10 ($2 \cdot 10^0$)	12 ($2 \cdot 10^0$)
10^0	10 ($2 \cdot 10^0$)	10 ($2 \cdot 10^0$)	12 ($2 \cdot 10^0$)	7 ($1 \cdot 10^0$)	6 ($1 \cdot 10^0$)	12 ($2 \cdot 10^0$)
10^6	12 ($3 \cdot 10^0$)	11 ($2 \cdot 10^0$)	9 ($2 \cdot 10^0$)	9 ($1 \cdot 10^0$)	8 ($1 \cdot 10^0$)	6 ($1 \cdot 10^0$)
10^{-6}	9 ($2 \cdot 10^0$)	11 ($2 \cdot 10^0$)	15 ($3 \cdot 10^0$)	–	–	42 ($3 \cdot 10^1$)
10^0	9 ($2 \cdot 10^0$)	9 ($2 \cdot 10^0$)	15 ($3 \cdot 10^0$)	–	–	42 ($3 \cdot 10^1$)
10^6	12 ($3 \cdot 10^0$)	12 ($3 \cdot 10^0$)	10 ($2 \cdot 10^0$)	34 ($6 \cdot 10^5$)	33 ($5 \cdot 10^5$)	14 ($4 \cdot 10^0$)

Table 7: Iteration count (the condition number estimate in parentheses) for varying contrast ratios ϱ and K. T (left) vs. AMG preconditioner (right). Top: $G = 10^0$; bottom: $G = 10^6$. 3D case, $r = 32$, $p = 2$, $N = 22$.

lem is the robustness of the former with respect to the permeability parameter G. Theoretical analysis of the preconditioner will be presented elsewhere.

Acknowledgements The author would like to thank two anonymous referees whose comments and remarks helped to improve the paper. This research was partially supported by the Polish National Science Centre grant 2016/21/B/ST1/00350.

G	10^0	10^2	10^4	10^6	10^8	10^{10}	10^{12}
T	10 ($2 \cdot 10^0$)	10 ($2 \cdot 10^0$)	10 ($2 \cdot 10^0$)	10 ($2 \cdot 10^0$)	10 ($2 \cdot 10^0$)	10 ($2 \cdot 10^0$)	10 ($2 \cdot 10^0$)
AMG	7 ($1 \cdot 10^0$)	39 ($3 \cdot 10^1$)	–	–	–	–	–
T	10 ($2 \cdot 10^0$)	10 ($2 \cdot 10^0$)	9 ($2 \cdot 10^0$)	9 ($2 \cdot 10^0$)	9 ($2 \cdot 10^0$)	9 ($2 \cdot 10^0$)	9 ($2 \cdot 10^0$)
AMG	6 ($1 \cdot 10^0$)	31 ($2 \cdot 10^1$)	–	–	–	–	–

Table 8: Iteration count (the condition number estimate in parentheses) for varying permeability coefficient G. Regularly placed $N = 18$ subdomains. Top: 2D case ($r = 128$), bottom: 3D case ($r = 32$). $\varrho = K = 1$, $p = 2$.

References

1. M.S. Alnæs, J. Blechta, J. Hake, A. Johansson, B. Kehlet, A. Logg, C. Richardson, J. Ring, M.E. Rognes and G.N. Wells. The FEniCS Project Version 1.5. *Archive of Numerical Software* **3**(100), 9–23 (2015).
2. P.F. Antonietti and P. Houston. A class of domain decomposition preconditioners for hp-discontinuous Galerkin finite element methods. *J. Sci. Comput.* **46**(1), 124–149 (2011).
3. P.F. Antonietti, P. Houston, G. Pennesi and E. Süli. An agglomeration-based massively parallel non-overlapping additive Schwarz preconditioner for high-order discontinuous Galerkin methods on polytopic grids. *Math. Comp.* **89**(325), 2047–2083 (2020).
4. P.F. Antonietti, P. Houston and I. Smears. A note on optimal spectral bounds for nonoverlapping domain decomposition preconditioners for hp-version Discontinuous Galerkin method. *Int. J. Numer. Anal. Model.* **13**(4), 513–524 (2016).
5. P.F. Antonietti, M. Sarti, M. Verani and L.T. Zikatanov. A uniform additive Schwarz preconditioner for high-order discontinuous Galerkin approximations of elliptic problems. *J. Sci. Comput.* **70**(2), 608–630 (2017).
6. S. Balay, S. Abhyankar, M.F. Adams, J. Brown, P. Brune, K. Buschelman, L. Dalcin, V. Eijkhout, W.D. Gropp, D. Kaushik, M.G. Knepley, D.A. May, L.C. McInnes, K. Rupp, P. Sanan, B.F. Smith, S. Zampini, H. Zhang and H. Zhang. PETSc users manual. Technical Report ANL-95/11 – Revision 3.8, Argonne National Laboratory (1995–).
7. M.A.J. Chaplain, C. Giverso, T. Lorenzi and L. Preziosi. Derivation and application of effective interface conditions for continuum mechanical models of cell invasion through thin membranes. *SIAM J. Appl. Math.* **79**, 2011–2031 (2019).
8. M. Dryja. On discontinuous Galerkin methods for elliptic problems with discontinuous coefficients. *Comput. Methods Appl. Math.* **3**(1), 76–85 (electronic) (2003).
9. M. Dryja and P. Krzyżanowski. A massively parallel nonoverlapping additive Schwarz method for discontinuous Galerkin discretization of elliptic problems. *Num. Math.* **132**(2), 347–367 (2015).
10. M. Dryja and M. Sarkis. Additive average Schwarz methods for discretization of elliptic problems with highly discontinuous coefficients. *Comput. Methods Appl. Math.* **10**(2), 164–176 (2010).
11. R.D. Falgout, J.E. Jones and U.M. Yang. The design and implementation of hypre, a library of parallel high performance preconditioners. In: *Numerical solution of Partial Differential Equations on Parallel Computers*, Lect. Notes Comput. Sci. Eng, pp. 267–294, Springer-Verlag (2006).
12. P. Krzyżanowski. On a nonoverlapping additive Schwarz method for h-p discontinuous Galerkin discretization of elliptic problems. *Num. Meth. PDEs* **32**(6), 1572–1590 (2016).
13. P. Krzyżanowski. Simple preconditioner for a thin membrane diffusion problem. In: R. Wyrzykowski, E. Deelman, J. Dongarra and K. Karczewski, editors, *Parallel Processing and Applied Mathematics*, volume 12044 LNCS, pp. 267–276, Springer International Publishing, Cham (2020).

14. M. Sturrock, A.J. Terry, D.P. Xirodimas, A.M. Thompson and M.A.J. Chaplain. Influence of the nuclear membrane, active transport, and cell shape on the Hes1 and p53–Mdm2 pathways: Insights from spatio-temporal modelling. *Bulletin of Mathematical Biology* **74**(7), 1531–1579, Jul (2012).
15. A. Toselli and O. Widlund. *Domain decomposition methods—algorithms and theory*. Volume 34 of Springer Series in Computational Mathematics. Springer-Verlag, Berlin (2005).
16. J. Xu and L. Zikatanov. Algebraic multigrid methods. *Acta Numerica* **26**, 591–721 (2017).

Additive Schwarz Methods for Convex Optimization – Convergence Theory and Acceleration

Jongho Park

1 Introduction

This paper is concerned with additive Schwarz methods for convex optimization problems of the form

$$\min_{u \in V} \{E(u) := F(u) + G(u)\}, \tag{1}$$

where V is a reflexive Banach space, $F: V \to \mathbb{R}$ is a Frechét differentiable convex function, and $G: V \to \overline{\mathbb{R}}$ is a proper, convex, lower semicontinuous function which is possibly nonsmooth. We further assume that E is coercive, so that (1) admits a solution $u^* \in V$. There are plenty of scientific problems of the form (1), e.g., nonlinear elliptic problems [13], variational inequalities [1, 12], and mathematical imaging problems [5, 10], and there has been much research on Schwarz methods corresponding to them.

In this paper, we present a unified view of some notable recent results [8, 9] on additive Schwarz methods for convex optimization (1). The starting point is the generalized additive Schwarz lemma presented in [9]. Based on the relevancy between additive Schwarz methods and gradient methods for (1) investigated in the generalized additive Schwarz lemma, two main results are considered: the abstract convergence theory [9] that generalizes some important existing results [1, 13, 15] and the momentum acceleration scheme [8] that greatly improves the convergence rate for additive Schwarz methods. In addition, we propose a novel backtracking strategy for additive Schwarz methods that further improves the convergence rate. We present numerical results for additive Schwarz methods equipped with the proposed backtracking strategy in order to highlight numerical efficiency.

Jongho Park
Natural Science Research Institute, KAIST, Daejeon 34141, Korea,
e-mail: jongho.park@kaist.ac.kr

S. C. Brenner et al. (eds.), *Domain Decomposition Methods in Science and Engineering XXVI*, Lecture Notes in Computational Science and Engineering 145,
https://doi.org/10.1007/978-3-030-95025-5_78

2 Additive Schwarz methods

In this section, we present an abstract additive Schwarz method for (1). In what follows, an index k runs from 1 to N. Let V_k be a reflexive Banach space and $R_k^*\colon V_k \to V$ be a bounded linear operator such that $V = \sum_{k=1}^N R_k^* V_k$ and its adjoint $R_k\colon V^* \to V_k^*$ is surjective. In order to describe local problems, we define $d_k\colon V_k \times V \to \overline{\mathbb{R}}$ and $G_k\colon V_k \times V \to \overline{\mathbb{R}}$ as functions which are proper, convex, and lower semicontinuous with respect to their first arguments. For positive constants τ and ω, an *additive Schwarz operator* $\mathrm{ASM}_{\tau,\omega}\colon V \to V$ is defined by

$$\mathrm{ASM}_{\tau,\omega}(v) = v + \tau \sum_{k=1}^N R_k^* \tilde{w}_k,$$

where

$$\tilde{w}_k \in \arg\min_{w_k \in V_k} \left\{ F(v) + \langle F'(v), R_k^* w_k \rangle + \omega d_k(w_k, v) + G_k(w_k, v) \right\}. \tag{2}$$

We note that (2) may admit nonunique minimizers; we take $\tilde{w}_k$ as any one among them in this case. If we set

$$d_k(w_k, v) = D_F(v + R_k^* w_k, v), \quad G_k(w_k, v) = G(v + R_k^* w_k), \quad \omega = 1 \tag{3a}$$

in (2), then the minimization problem is reduced to

$$\min_{w_k \in V_k} E(v + R_k^* w_k), \tag{3b}$$

which is the case of exact local problems. Here D_F denotes the Bregman distance

$$D_F(u, v) = F(u) - F(v) - \langle F'(v), u - v \rangle, \quad u, v \in V.$$

We note that other choices of d_k and G_k, i.e., cases of inexact local problems, include various numerical methods such as block coordinate descent methods and constraint decomposition methods [5, 12]; see [9, Sect. 6.4] for details.

The abstract additive Schwarz method for (1) is presented in Algorithm 1. Constants τ_0 and ω_0 in Algorithm 1 will be given in Section 3. Note that $\operatorname{dom} G$ denotes the effective domain of G, i.e., $\operatorname{dom} G = \{ v \in V : G(v) < \infty \}$.

Algorithm 1 Additive Schwarz method for (1)

Choose $u^{(0)} \in \operatorname{dom} G$, $\tau \in (0, \tau_0]$, and $\omega \geq \omega_0$.
for $n = 0, 1, 2, \ldots$

$$u^{(n+1)} = \mathrm{ASM}_{\tau,\omega}(u^{(n)})$$

end

An important observation made in [9, Lemma 4.5] is that Algorithm 1 can be interpreted as a kind of a gradient method equipped with a nonlinear distance function [14]. A rigorous statement is presented in the following.

Proposition 1 (generalized additive Schwarz lemma) *For $\tau, \omega > 0$, we have*

$$\mathrm{ASM}_{\tau,\omega}(v) = \operatorname*{arg\,min}_{u \in V} \left\{ F(v) + \langle F'(v), u - v \rangle + M_{\tau,\omega}(u, v) \right\}, \quad v \in V,$$

where the functional $M_{\tau,\omega} \colon V \times V \to \overline{\mathbb{R}}$ is given by

$$M_{\tau,\omega}(u, v) = \tau \inf \left\{ \sum_{k=1}^{N} (\omega d_k + G_k)(w_k, v) : u - v = \tau \sum_{k=1}^{N} R_k^* w_k,\ w_k \in V_k \right\} + (1 - \tau N)\, G(v), \quad u, v \in V.$$

In the field of mathematical optimization, there has been a great deal of research on gradient methods for solving convex optimization problems [4, 6, 14]. Therefore, invoking Proposition 1, we can adopt many valuable tools from the field of mathematical optimization in order to analyze and improve Schwarz methods. In particular, we present two fruitful results in the remainder of the paper: novel convergence theory [9] and acceleration [8] for additive Schwarz methods.

3 Convergence theory

This section is devoted to an abstract convergence theory of additive Schwarz methods for convex optimization. The convergence theory introduced in this section directly generalizes the classical theory for linear problems [15, Chapter 2] to convex optimization problems. Similar to [15, Chapter 2], the following three conditions are considered: stable decomposition, strengthened convexity, and local stability.

Assumption 1 (stable decomposition) There exists a constant $q > 1$ such that for any bounded and convex subset K of V, the following holds: for any $u, v \in K \cap \operatorname{dom} G$, there exists $w_k \in V_k$, $1 \leq k \leq N$, with $u - v = \sum_{k=1}^{N} R_k^* w_k$, such that

$$\sum_{k=1}^{N} d_k(w_k, v) \leq \frac{C_{0,K}^q}{q} \|u - v\|^q, \quad \sum_{k=1}^{N} G_k(w_k, v) \leq G(u) + (N-1) G(v),$$

where $C_{0,K}$ is a positive constant depending on K.

Assumption 2 (strengthened convexity) There exists a constant $\tau_0 \in (0, 1]$ which satisfies the following: for any $v \in V$, $w_k \in V_k$, $1 \le k \le N$, and $\tau \in (0, \tau_0]$, we have

$$(1 - \tau N) E(v) + \tau \sum_{k=1}^{N} E(v + R_k^* w_k) \ge E\left(v + \tau \sum_{k=1}^{N} R_k^* w_k\right).$$

Assumption 3 (local stability) There exists a constant $\omega_0 > 0$ which satisfies the following: for any $v \in \operatorname{dom} G$, and $w_k \in V_k$, $1 \le k \le N$, we have

$$D_F(v + R_k^* w_k, v) \le \omega_0 d_k(w_k, v), \quad G(v + R_k^* w_k) \le G_k(w_k, v).$$

Assumption 1 is compatible with various variants of stable decomposition presented in existing works [1, 13, 15]. Assumption 2 trivially holds with $\tau_0 = 1/N$ due to the convexity of E. However, a better value for τ_0 independent of N can be found by the usual coloring technique. In the same spirit as [15], Assumption 3 gives a one-sided measure of approximation properties of the local solvers. It was shown in [9, Sect. 4.1] that the above assumptions reduce to [15, Assumptions 2.2 to 2.4] if they are applied to linear elliptic problems. Under the above three assumptions, we have the following convergence theorem for Algorithm 1 [9, Theorem 4.7].

Theorem 1 *Suppose that Assumptions 1, 2, and 3 hold. In Algorithm 1, we have*

$$E(u^{(n)}) - E(u^*) = O\left(\frac{\kappa_{\mathrm{ASM}}}{n^{q-1}}\right),$$

where κ_{ASM} is the additive Schwarz condition number defined by $\kappa_{\mathrm{ASM}} = \omega C_0^q / \tau^{q-1}$.

Meanwhile, it is well-known that the Łojasiewicz inequality holds in many applications [11]; it says that the energy functional E of (1) is sharp around the minimizer u^*. We summarize this property in Assumption 4.

Assumption 4 (sharpness) There exists a constant $p > 1$ such that for any bounded and convex subset K of V satisfying $u^* \in K$, we have

$$\frac{\mu_K}{p} \|u - u^*\|^p \le E(u) - E(u^*), \quad u \in K,$$

for some $\mu_K > 0$.

We can obtain an improved convergence result for Algorithm 1 compared to Theorem 1 under an additional sharpness assumption on E [9, Theorem 4.8].

Theorem 2 *Suppose that Assumptions 1, 2, 3, and 4 hold. In Algorithm 1, we have*

$$E(u^{(n)}) - E(u^*) = \begin{cases} O\left(\left(1 - \left(1 - \frac{1}{q}\right) \min\left\{\tau, \left(\frac{\mu}{q\kappa_{\mathrm{ASM}}}\right)^{\frac{1}{q-1}}\right\}\right)^n\right), & \text{if } p = q, \\ O\left(\frac{(\kappa_{\mathrm{ASM}}^p/\mu^q)^{\frac{1}{p-q}}}{n^{\frac{p(q-1)}{p-q}}}\right), & \text{if } p > q, \end{cases}$$

where κ_{ASM} was defined in Theorem 1.

Theorems 1 and 2 are direct consequences of Proposition 1 in the sense that they can be easily deduced by invoking theories of gradient methods for convex optimization [9, Sect. 2].

4 Acceleration

An important observation on Schwarz methods for linear problems is that they can be interpreted as preconditioned Richardson iterations with appropriate preconditioners. Replacing Richardson iterations by conjugate gradient iterations with the same preconditioners, we can obtain improved algorithms that converge faster. Since Proposition 1 says that additive Schwarz methods for (1) are in fact gradient methods, in the same spirit, we may adopt some acceleration schemes for gradient methods (see, e.g., [4, 7]) in order to improve additive Schwarz methods. Motivated by the FISTA (Fast Iterative Shrinkage-Thresholding Algorithm) momentum [2] and the gradient adaptive restarting scheme [7], the following accelerated variant of Algorithm 1 was considered in [8].

Algorithm 2 Accelerated additive Schwarz method for (1)

Let $u^{(0)} = v^{(0)} \in \operatorname{dom} G$, $\tau > 0$, and $t_0 = 1$.
for $n = 0, 1, 2, \ldots$

$$u^{(n+1)} = \mathrm{ASM}_{\tau,\omega}(v^{(n)})$$

$$\begin{cases} t_{n+1} = 1, \ \beta_n = 0, & \text{if } \langle v^{(n)} - u^{(n+1)}, u^{(n+1)} - u^{(n)} \rangle > 0, \\ t_{n+1} = \frac{1+\sqrt{1+4t_n^2}}{2}, \ \beta_n = \frac{t_n - 1}{t_{n+1}}, & \text{otherwise.} \end{cases}$$

$$v^{(n+1)} = u^{(n+1)} + \beta_n (u^{(n+1)} - u^{(n)})$$

end

The major part of each iteration of Algorithm 2 is to compute the additive Schwarz operator $\mathrm{ASM}_{\tau,\omega}$; the computational cost for momentum parameters t_n and β_n is marginal. Therefore, the main computational cost of Algorithm 2 is the same as the one of Algorithm 1. Nevertheless, it was shown numerically in [8] that Algorithm 2 achieves much faster convergence to the energy minimum compared to Algorithm 1.

In the remainder of this section, we consider how to further improve Algorithm 2. More precisely, we present a backtracking strategy for additive Schwarz methods that allows for local optimization of the parameter τ. Mimicking [3, 6], at each iteration of additive Schwarz methods, we choose τ as large as possible satisfying

$$E(u^{(n+1)}) \leq F(u^{(n)}) + \langle F'(u^{(n)}), u^{(n+1)} - u^{(n)} \rangle + M_{\tau,\omega}(u^{(n+1)}, u^{(n)}).$$

An optimal τ can be found by a logarithmic grid search. Algorithm 2 accompanied with the backtracking strategy is presented in Algorithm 3. Note that the parameter $\rho \in (0, 1)$ in Algorithm 3 plays the role of an adjustment parameter for the grid search.

Algorithm 3 Accelerated additive Schwarz method for (1) with backtracking

---

Let $u^{(0)} = v^{(0)} \in \operatorname{dom} G$, $\tau > 0$, $t_0 = 1$, and $\rho \in (0, 1)$.
for $n = 0, 1, 2, \dots$

$$\tau \leftarrow \tau/\rho$$

repeat

$$u^{(n+1)} = \mathrm{ASM}_{\tau,\omega}(v^{(n)})$$

if $E(u^{(n+1)}) > F(u^{(n)}) + \langle F'(u^{(n)}), u^{(n+1)} - u^{(n)} \rangle + M_{\tau,\omega}(u^{(n+1)}, u^{(n)})$

$$\tau \leftarrow \rho\tau$$

end if
until $E(u^{(n+1)}) \leq F(u^{(n)}) + \langle F'(u^{(n)}), u^{(n+1)} - u^{(n)} \rangle + M_{\tau,\omega}(u^{(n+1)}, u^{(n)})$

$$\begin{cases} t_{n+1} = 1, \ \beta_n = 0, & \text{if } \langle v^{(n)} - u^{(n+1)}, u^{(n+1)} - u^{(n)} \rangle > 0, \\ t_{n+1} = \frac{1+\sqrt{1+4t_n^2}}{2}, \ \beta_n = \frac{t_n - 1}{t_{n+1}}, & \text{otherwise.} \end{cases}$$

$$v^{(n+1)} = u^{(n+1)} + \beta_n (u^{(n+1)} - u^{(n)})$$

end

Different from the existing works [3, 6], adopting the backtracking strategy for additive Schwarz methods has an own difficulty that evaluation of $M_{\tau,\omega}(u^{(n+1)}, u^{(n)})$ is not straightforward due to its complicated definition. The following proposition provides a way to evaluate $M_{\tau,\omega}(u^{(n+1)}, u^{(n)})$ without major computational cost.

Proposition 2 *If* $u = \mathrm{ASM}_{\tau,\omega}(v)$*, then*

$$M_{\tau,\omega}(u,v) = \tau \sum_{k=1}^{N} (\omega d_k + G_k)(\tilde{w}_k, v) + (1 - \tau N) G(v),$$

where $\tilde{w}_k$*,* $1 \leq k \leq N$*, were defined in (2). In particular, if the exact local problems (3) are used, then we have*

$$F(v) + \langle F'(v), u - v \rangle + M_{\tau,\omega}(u,v) = (1 - \tau N) E(v) + \tau \sum_{k=1}^{N} E(v + R_k^* \tilde{w}_k).$$

Proof See the proof of [9, Lemma 4.5]. □

Thanks to Proposition 2, one can compute $M_{\tau,\omega}(u^{(n+1)}, u^{(n)})$ in Algorithm 3 without solving the infimum in the definition of $M_{\tau,\omega}$. As discussed in [3], the backtracking strategy improves the convergence rate because it allows for adaptive adjustment of τ depending on the local flatness of the energy functional.

In order to show the computational efficiency of Algorithm 3, we present numerical results applied to a finite element s-Laplacian problem ($s \geq 1$). We set $\Omega = [0,1]^2 \subset \mathbb{R}^2$. We decompose the domain Ω into $\mathcal{N} = N \times N$ square subdomains $\{\Omega_k\}_{k=1}^{\mathcal{N}}$ in which each subdomain has the sidelength $H = 1/N$. Each subdomain Ω_k, $1 \leq k \leq \mathcal{N}$, is partitioned into $2 \times H/h \times H/h$ uniform triangles to form a global triangulation $\mathcal{T}_h$ of Ω. Similarly, we partition each Ω_k into two uniform triangles and let $\mathcal{T}_H$ be a coarse triangulation of Ω consisting of such triangles. Overlapping subdomains $\{\Omega_k'\}_{k=1}^{\mathcal{N}}$ are constructed so that Ω_k' is a union of Ω_k and its surrounding layers of fine elements in $\mathcal{T}_h$ with the width δ such that $0 < \delta < H/2$. The model finite element s-Laplacian problem is written as

$$\min_{u \in S_h(\Omega)} \left\{ \frac{1}{s} \int_{\Omega} |\nabla u|^s \, dx - \int_{\Omega} f u \, dx \right\}, \tag{4}$$

where $f \in (L^s(\Omega))^*$ and $V = S_h(\Omega)$ is the continuous piecewise linear finite element space on $\mathcal{T}_h$ with the homogeneous Dirichlet boundary condition. We set $V_k = S_h(\Omega_k')$, $1 \leq k \leq \mathcal{N}$, and take $R_k^* \colon V_k \to V$ as the natural extension operator, where $S_h(\Omega_k')$ is the continuous piecewise linear finite element space on the $\mathcal{T}_h$-elements in Ω_k' with the homogeneous Dirichlet boundary condition. As a coarse space, we set V_0 to be the continuous piecewise linear space $S_H(\Omega)$ on $\mathcal{T}_H$ and take $R_0^* \colon V_0 \to V$ as the natural interpolation operator.

For numerical experiments, we set $s = 4$, $f = 1$, and $u^{(0)} = 0$. Exact local and coarse solvers (3) were used; they were solved numerically by FISTA with gradient adaptive restarts [7]. The initial step size τ was chosen as $1/5$ (cf. [9, Sect. 5.1]).

Figure 1 plots the energy error $E(u^{(n)}) - E(u^*)$ of various additive Schwarz methods when $h = 1/2^6$, $H = 1/2^3$, and $\delta = 4h$. As shown in Figure 1(a), Algorithm 3 shows faster convergence to the energy minimum compared to Algorithm 2 for various values of ρ. Hence, we can say that the backtracking strategy proposed in this paper is effective for acceleration of convergence. Although Algorithm 3 shows

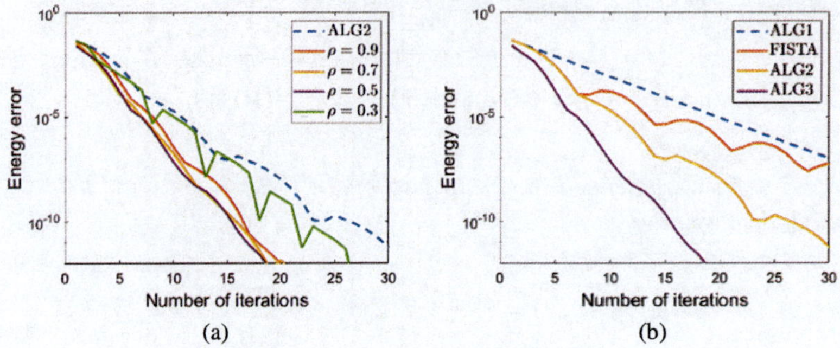

Fig. 1: Decay of the energy error $E(u^{(n)}) - E(u^*)$ in additive Schwarz methods ($\tau = 1/5$, $\omega = 1$) for the s-Laplacian problem (4) ($h = 1/2^6$, $H = 1/2^3$, $\delta = 4h$). (a) Algorithm 3 with various values of ρ. (b) Comparison of various additive Schwarz methods. FISTA denotes the FISTA momentum without restarts and ALG3 denotes Algorithm 3 with $\rho = 0.5$.

better performance than Algorithm 2 for all values of ρ, it remains as a future work to discover how to find an optimal ρ. Figure 1(b) presents a numerical comparison of Algorithm 1, Algorithm 1 equipped with the FISTA momentum, and Algorithms 2 and 3. We can observe that all of the FISTA momentum, adaptive restarting technique, and backtracking strategy provide positive effects on the convergence rate of additive Schwarz methods. Consequently, Algorithm 3, which assembles all of the aforementioned acceleration schemes, show the best convergence rate among all methods. Since the main computational costs of all algorithms are essentially the same, we conclude that Algorithm 3 numerically outperforms all the others.

References

1. L. Badea. One- and two-level additive methods for variational and quasi-variational inequalities of the second kind. Preprint, *Ser. Inst. Math. Rom. Acad.* (5) (2010).
2. A. Beck and M. Teboulle. A fast iterative shrinkage-thresholding algorithm for linear inverse problems. *SIAM J. Imaging Sci.* **2**(1), 183–202 (2009).
3. L. Calatroni and A. Chambolle. Backtracking strategies for accelerated descent methods with smooth composite objectives. *SIAM J. Optim.* **29**(3), 1772–1798 (2019).
4. A. Chambolle and T. Pock. An introduction to continuous optimization for imaging. *Acta Numer.* **25**, 161–319 (2016).
5. H. Chang, X.C. Tai, L.L. Wang and D. Yang. Convergence rate of overlapping domain decomposition methods for the Rudin–Osher–Fatemi model based on a dual formulation. *SIAM J. Imaging Sci.* **8**(1), 564–591 (2015).
6. Y. Nesterov. Gradient methods for minimizing composite functions. *Math. Program.* **140**(1), 125–161 (2013).
7. B. O'Donoghue and E. Candes. Adaptive restart for accelerated gradient schemes. *Found. Comput. Math.* **15**(3), 715–732 (2015).
8. J. Park. Accelerated additive Schwarz methods for convex optimization with adaptive restart. *arXiv preprint* arXiv:2011.02695 (2020).

9. J. Park. Additive Schwarz methods for convex optimization as gradient methods. *SIAM J. Numer. Anal.* **58**(3), 1495–1530 (2020).
10. J. Park. Pseudo-linear convergence of an additive Schwarz method for dual total variation minimization. *Electron. Trans. Numer. Anal.* **54**, 176–197 (2021).
11. V. Roulet and A. d'Aspremont. Sharpness, restart, and acceleration. *SIAM J. Optim.* **30**(1), 262–289 (2020).
12. X.C. Tai. Rate of convergence for some constraint decomposition methods for nonlinear variational inequalities. *Numer. Math.* **93**(4), 755–786 (2003).
13. X.C. Tai and J.Xu. Global and uniform convergence of subspace correction methods for some convex optimization problems. *Math. Comp.* **71**(237), 105–124 (2002).
14. M. Teboulle. A simplified view of first order methods for optimization. *Math. Program.* **170**(1), 67–96 (2018).
15. A. Toselli and O. Widlund. *Domain Decomposition Methods – Algorithms and Theory.* Springer, Berlin (2005).

Non-local Impedance Operator for Non-overlapping DDM for the Helmholtz Equation

Francis Collino, Patrick Joly and Emile Parolin

In the context of time harmonic wave equations, the pioneering work of B. Després [4] has shown that it is mandatory to use impedance type transmission conditions in the coupling of sub-domains in order to obtain convergence of non-overlapping domain decomposition methods (DDM). In later works [2, 3], it was observed that using non-local impedance operators leads to geometric convergence, a property which is unattainable with local operators. This result was recently extended to arbitrary geometric partitions, including configurations with cross-points, with provably uniform stability with respect to the discretization parameter [1].

We present a novel strategy to construct suitable non-local impedance operators that satisfy the theoretical requirements of [1] or [2, 3]. It is based on the solution of elliptic auxiliary problems posed in the vicinity of the transmission interfaces. The definition of the operators is generic, with simple adaptations to the acoustic or electromagnetic settings, even in the case of heterogeneous media. Besides, no complicated tuning of parameters is required to get efficiency. The implementation in practice is straightforward and applicable to sub-domains of arbitrary geometry, including ones with rough boundaries generated by automatic graph partitioners.

1 General approach for a two-domain decomposition

We consider the Helmholtz equation in a bounded Lipschitz domain $\Omega \subset \mathbb{R}^d$, $d \in \{1, 2, 3\}$, with a first order absorbing boundary condition imposed on the boundary Γ: Find $u \in H^1(\Omega)$ such that

$$(-\operatorname{div} \mathfrak{a}\nabla - \kappa^2 \mathfrak{n})u = f, \quad \text{in } \Omega, \qquad (\mathfrak{a}\partial_{\mathbf{n}} - \imath\kappa)u = g, \quad \text{on } \Gamma, \tag{1}$$

Francis Collino, Patrick Joly, Emile Parolin
POEMS, CNRS, INRIA, ENSTA Paris, Institut Polytechnique de Paris, 91120 Palaiseau, France,
e-mail: francis.collino@orange.fr; patrick.joly@inria.fr;
emile.parolin@inria.fr

S. C. Brenner et al. (eds.), *Domain Decomposition Methods in Science and Engineering XXVI*, Lecture Notes in Computational Science and Engineering 145,
https://doi.org/10.1007/978-3-030-95025-5_79

where $f \in L^2(\Omega)$ and $g \in L^2(\Gamma)$, κ denotes the wavenumber, $\mathfrak{a}$ and $\mathfrak{n}$ are two strictly positive and bounded functions (so that the medium is purely propagative) and $\mathbf{n}$ is the outward normal to Γ. The well-posedness of this problem is guaranteed by application of the Fredholm alternative and a unique continuation principle.

A geometrically convergent DD method. We consider a non-overlapping partition into two domains, excluding the presence of (boundary) cross-points, by introducing a *closed* Lipschitz interface Σ that splits the domain Ω into an interior domain Ω_1 and exterior domain Ω_2, see Figure 1 (left). The Domain Decomposition (DD) method consists in solving iteratively the Helmholtz equation in parallel in each sub-domain by imposing two transmission conditions. Introducing a boundary operator T on Σ we consider here impedance-like transmission conditions:

$$\begin{cases} (+\mathfrak{a}\partial_{\mathbf{n}_1} - \imath\kappa T)u_1 = (-\mathfrak{a}\partial_{\mathbf{n}_2} - \imath\kappa T)u_2, \\ (-\mathfrak{a}\partial_{\mathbf{n}_1} - \imath\kappa T)u_1 = (+\mathfrak{a}\partial_{\mathbf{n}_2} - \imath\kappa T)u_2, \end{cases} \quad \text{on } \Sigma,$$

where $\mathbf{n}_1$ (resp. $\mathbf{n}_2$) denotes the outward unit normal vector to Ω_1 (resp. Ω_2).

The DD method is best analysed in the form of an interface problem. Let us introduce $(w_1, w_2) \in H^1(\Omega_1) \times H^1(\Omega_2)$, a lifting of the source defined as follows

$$\begin{cases} (-\operatorname{div}\mathfrak{a}\nabla - \kappa^2\mathfrak{n})w_1 = f|_{\Omega_1}, & \text{in } \Omega_1, \\ (+\mathfrak{a}\partial_{\mathbf{n}_1} - \imath\kappa T)w_1 = 0, & \text{on } \Sigma, \end{cases} \quad \begin{cases} (-\operatorname{div}\mathfrak{a}\nabla - \kappa^2\mathfrak{n})w_2 = f|_{\Omega_2}, & \text{in } \Omega_2, \\ (+\mathfrak{a}\partial_{\mathbf{n}} - \imath\kappa)w_2 = g, & \text{on } \Gamma, \\ (+\mathfrak{a}\partial_{\mathbf{n}_2} - \imath\kappa T)w_2 = 0, & \text{on } \Sigma, \end{cases} \tag{2}$$

and we define for any $x_j \in H^{-1/2}(\Sigma)$, $L_j x_j := v_j \in H^1(\Omega_j)$, $j \in \{1, 2\}$, such that

$$\begin{cases} (-\operatorname{div}\mathfrak{a}\nabla - \kappa^2\mathfrak{n})v_1 = 0, & \text{in } \Omega_1, \\ (+\mathfrak{a}\partial_{\mathbf{n}_1} - \imath\kappa T)v_1 = x_1, & \text{on } \Sigma, \end{cases} \quad \begin{cases} (-\operatorname{div}\mathfrak{a}\nabla - \kappa^2\mathfrak{n})v_2 = 0, & \text{in } \Omega_2, \\ (+\mathfrak{a}\partial_{\mathbf{n}} - \imath\kappa)v_2 = 0, & \text{on } \Gamma, \\ (+\mathfrak{a}\partial_{\mathbf{n}_2} - \imath\kappa T)v_2 = x_2, & \text{on } \Sigma. \end{cases} \tag{3}$$

Assuming that the operator T is *self-adjoint positive definite*, one can prove that the local sub-problems appearing in (2) and (3) are well posed [3, Lem. 2.5]. Finally let us introduce for any $x \in H^{-1/2}(\Sigma)$ the so-called local *scattering* operators, $j \in \{1, 2\}$

$$S_j x := (-\mathfrak{a}\partial_{\mathbf{n}_j} - \imath\kappa T)L_j x,$$

and set
$$\mathbf{S} := \begin{bmatrix} S_1 & 0 \\ 0 & S_2 \end{bmatrix}, \qquad \mathbf{\Pi} := \begin{bmatrix} 0 & \mathrm{Id} \\ \mathrm{Id} & 0 \end{bmatrix}, \qquad \mathbf{b} := \begin{bmatrix} (-\mathfrak{a}\partial_{\mathbf{n}_2} - \imath\kappa T)w_2 \\ (-\mathfrak{a}\partial_{\mathbf{n}_1} - \imath\kappa T)w_1 \end{bmatrix}.$$

$\mathbf{S}$ is the *global* scattering operator and $\mathbf{\Pi}$ is referred to as the *exchange* operator since its action consists in swapping information between the two sub-domains. It can be shown [2, Th. 2] that if u satisfies the model problem (1) then the two (incoming) Robin traces $\mathbf{x} := ((+\mathfrak{a}\partial_{\mathbf{n}_1} - \imath\kappa T)u|_{\Omega_1}, (+\mathfrak{a}\partial_{\mathbf{n}_2} - \imath\kappa T)u|_{\Omega_2})$, satisfy the interface problem

$$(\mathrm{Id} - \mathbf{\Pi}\mathbf{S})\mathbf{x} = \mathbf{b}. \tag{4}$$

Reciprocally, if $\mathbf{x} = (x_1, x_2)$ satisfies the interface problem (4), then the concatenation of $(L_1 x_1 + w_1, L_2 x_2 + w_2)$ is a solution to the original problem (1).

One of the simplest iterative methods to solve (4) is the (relaxed) Jacobi algorithm. Let $\mathbf{x}^0$ and a relaxation parameter $0 < r < 1$ be given, then a sequence $(\mathbf{x}^n)_{n \in \mathbb{N}}$ is constructed using the (relaxed) Jacobi algorithm as follows

$$\mathbf{x}^{n+1} = [(1 - r)\mathrm{Id} + r\mathbf{\Pi S}]\, \mathbf{x}^n + r\mathbf{b}, \qquad n \in \mathbb{N}. \tag{5}$$

Theorem 1 [3, Th. 2.1] *If T is a positive self-adjoint* isomorphism *between the trace spaces $H^{1/2}(\Sigma)$ and $H^{-1/2}(\Sigma)$, then the above algorithm converges geometrically*

$$\exists\, 0 \le \tau < 1,\ C > 0,\ \|u_1 - (L_1 x_1^n + w_1)\|_{H^1} + \|u_2 - (L_2 x_2^n + w_2)\|_{H^1} \le C\tau^n.$$

Note that the isomorphism property is essential to ensure the geometric nature of the convergence, and, together with the positivity and self-adjointness properties, necessarily requires T to be *non-local*. Alternatively, a more efficient algorithm to use in practice is the GMRES algorithm. The convergence rate of the GMRES algorithm is necessarily better (i.e. the algorithm is always faster) than the convergence of the Jacobi algorithm, but much more delicate to analyse.

A suitable impedance operator. We propose to construct impedance operators that satisfy the above theoretical requirements of the convergence analysis from the elliptic (or dissipative) version of conventional Dirichlet-to-Neumann (DtN) maps. To do so, we introduce two strips $\mathcal{B}_1 \subset \Omega_1$ and $\mathcal{B}_2 \subset \Omega_2$ so that $\mathcal{B}_1$ (resp. $\mathcal{B}_2$) has two disconnected (and not intersecting) boundaries Σ and Σ_1 (resp. Σ_2), see Figure 1 (left). We do not exclude the case $\Sigma_1 = \emptyset$ for which we have $\mathcal{B}_1 = \Omega_1$. We denote by $\mathbf{n}_1$ (resp. $\mathbf{n}_2$) the outward unit normal vector to $\mathcal{B}_1$ (resp. $\mathcal{B}_2$).

We define two operators, for $j \in \{1, 2\}$ and any $x \in H^{1/2}(\Sigma)$,

$$T_j x := \kappa^{-1} \mathfrak{a} \partial_{\mathbf{n}_j} u_j, \qquad u_j \in H^1(\mathcal{B}_j), \begin{cases} (-\operatorname{div} \mathfrak{a}\nabla + \kappa^2 \mathfrak{n}) u_j = 0, & \text{in } \mathcal{B}_j, \\ \mathfrak{a}\partial_{\mathbf{n}_j} u_j + \kappa u_j = 0, & \text{on } \Gamma_j, \\ u_j = x, & \text{on } \Sigma. \end{cases} \tag{6}$$

The following result, which is a straightforward consequence of the surjectivity of the Dirichlet trace operator and the Lax–Milgram Lemma, guarantees that we fall within the situation of Theorem 1.

Proposition 1 *The impedance operator defined as $T = \frac{1}{2}(T_1 + T_2)$ is a self-adjoint positive isomorphism from $H^{1/2}(\Sigma)$ to $H^{-1/2}(\Sigma)$.*

2 Quantitative analysis for the wave-guide

The aim of this section is to derive convergence estimates to study in particular the influence of the width of the strip in the definition of the auxiliary problems, which has a direct influence on the computational cost of the proposed method. We consider the theoretical (because unbounded) configuration of an infinite wave guide of width L, so that $\Omega := \{(x, y) \in \mathbb{R}^2 | 0 < x < L\}$, see Figure 1 (right). The media is considered homogeneous ($\mathfrak{a} \equiv \mathfrak{n} \equiv 1$); we impose homogeneous Dirichlet boundary conditions on the sides $u(0, \cdot) = u(L, \cdot) = 0$ and require u to be outgoing [5].

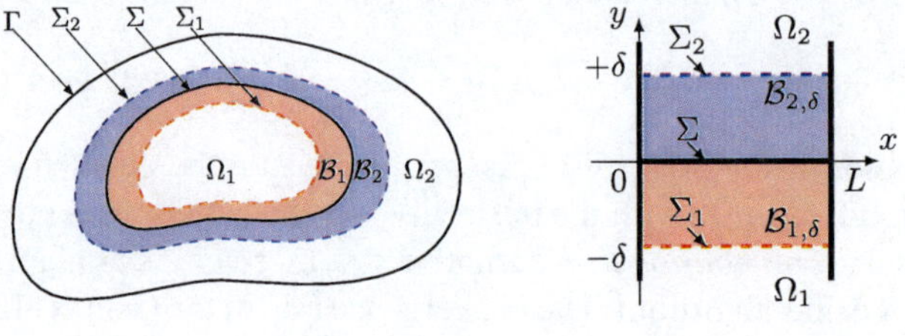

Fig. 1: Geometric configurations.

Remark 1 The above problem is well-posed except at cut-off frequencies $\kappa L \in \pi\mathbb{Z}$, configurations which are thus excluded in what follows.

The domain Ω is divided into its upper region $\Omega_2 := \{(x, y) \in \Omega | y > 0\}$ and lower region $\Omega_1 := \{(x, y) \in \Omega | y < 0\}$ and the interface is $\Sigma := (0, L) \times \{0\}$. Suppose that we have at hand a suitable impedance operator T (described below), in spite of the different geometry and the unboundedness, the same DD algorithm of Section 1 is formally applicable with minor adaptations. For completeness and because it will be important in the following, we simply provide the full definition of the local scattering operators, for $j \in \{1, 2\}$ and any $\mathrm{x} \in H^{-1/2}(\Sigma)$

$$S_j \mathrm{x} := (-\kappa^{-1}\partial_{\mathbf{n}_j} - \imath T) u_j|_{y=0}, \quad u_j \in H^1(\Omega_j), \quad \begin{cases} (-\Delta - \kappa^2) u_j = 0, & \text{in } \Omega_j, \\ u_j(0, \cdot) = u_j(L, \cdot) = 0, & \text{on } \partial\Omega_j \setminus \Sigma, \\ (\kappa^{-1}\partial_{\mathbf{n}_j} - \imath T) u_j = \mathrm{x}, & \text{on } \Sigma, \end{cases}$$

where u_j is assume to be outgoing.

A family of suitable impedance operators. We introduce now several possible impedance operators on the model of (6). The domain of the auxiliary problem that defines the impedance operator is bounded in the y-direction, for a positive parameter $\delta > 0$, let $\mathcal{B}_{j,\delta} := \{(x, y) \in \Omega_j | 0 \le |y| \le \delta\}$, $j \in \{1, 2\}$. We consider the operators, indexed by the width δ and the type of boundary condition $* \in \{D, N, R\}$ (for Dirichlet, Neumann and Robin), for $j \in \{1, 2\}$ and any $\mathrm{x} \in H^{1/2}(\Sigma)$

$$T^*_{j,\delta} \mathrm{x} := \kappa^{-1} \partial_{\mathbf{n}_j} v^*_j|_{y=0},$$

where v_j^* solves the (elliptic) problem,

$$\begin{cases}(-\Delta+\kappa^2)v_j^*=0, & \text{in } \mathcal{B}_{j,\delta},\\ v_j^*(0,y)=v_j^*(L,y)=0, & |y|\le\delta,\\ v_j^*(\cdot,0)=\mathrm{x}, & \text{on } \Sigma,\end{cases} \quad \text{and} \quad \begin{cases}v_j^D=0,\\ \partial_{\mathbf{n}_j}v_j^N=0, & \text{on } \Sigma_j.\\ (\partial_{\mathbf{n}_j}+\kappa)v_j^R=0,\end{cases}$$

The impedance operators are then $T_\delta^* := \frac{1}{2}(T_{1,\delta}^*+T_{2,\delta}^*)$. The aim of this section is to investigate the effect on the convergence of the type of boundary condition $* \in \{D,N,R\}$; as well as the shrinking of the width δ of the strips $\mathcal{B}_{1,\delta}$ and $\mathcal{B}_{2,\delta}$.

Modal analysis, convergence factor. Because of the separable geometry, we are able to conduct a quantitative study. The main tool for this is the Hilbert basis $\{\sin(k_m x)\}_{m\in\mathbb{N}}$ of $L^2(]0,L[)$, where we introduced the mode numbers $k_m := m\frac{\pi}{L}$, $m\in\mathbb{N}$. All the operators involved are diagonalized on this basis.

Symbol of the impedance operators. By symmetry, we need only to study the upper half-region. Standard computations show that the coefficients $(v_{m,2}^*)_{m\in\mathbb{N}}$ of v_2^* satisfy,

$$v_{m,2}^*(y)=\hat{x}_m\frac{\mathrm{e}^{-\mu_m y}+\alpha_{\delta,m}^*\mathrm{e}^{\mu_m y}}{1+\alpha_{\delta,m}^*},\quad 0\le y\le\delta \quad \text{where} \begin{cases}\alpha_{\delta,m}^D=-\mathrm{e}^{-2\mu_m\delta},\\ \alpha_{\delta,m}^N=\mathrm{e}^{-2\mu_m\delta},\\ \alpha_{\delta,m}^R=\frac{\kappa^{-1}\mu_m-1}{\kappa^{-1}\mu_m+1}\mathrm{e}^{-2\mu_m\delta},\end{cases}$$

where we set $\mu_m := \sqrt{k_m^2+\kappa^2}$, and introduce in addition the coefficients $(\hat{x}_m)_{m\in\mathbb{N}}$ of the decomposition of x on the same modal basis. The symbol of the transmission operator T_δ^* is then

$$\hat{t}_{\delta,m}^*=\kappa^{-1}\mu_m\frac{1-\alpha_{\delta,m}^*}{1+\alpha_{\delta,m}^*}>0,\qquad m\in\mathbb{N}.$$

Symbol of the scattering operators. Relying again on symmetry, we consider only $j=2$. From the definitions of the scattering operators S_j, we formally have

$$S_j=-\left(\Lambda_j+\imath T_\delta^*\right)\left(\Lambda_j-\imath T_\delta^*\right)^{-1},$$

where we introduce the (propagative) DtN operators, for any $\mathrm{x}\in H^{1/2}(\Sigma)$

$$\Lambda_j\mathrm{x} := \kappa^{-1}\partial_{\mathbf{n}_j}u_j|_{y=0},\quad u_j\in H^1(\Omega_j), \begin{cases}(-\Delta-\kappa^2)u_j=0, & \text{in } \Omega_j,\\ u_j(0,\cdot)=u_j(L,\cdot)=0, & \text{on } \partial\Omega_j\setminus\Sigma,\\ u_j(\cdot,0)=\mathrm{x}, & \text{on } \Sigma,\end{cases}$$

and u_j is assumed to be outgoing. The coefficients $(u_{2,m})_{m\in\mathbb{N}}$ of u_2 satisfy

$$u_{2,m}(y)=\hat{x}_m\mathrm{e}^{-\xi_m y},\qquad 0\le y,\qquad m\in\mathbb{N},$$

and we set
$$\xi_m := \begin{cases} -\imath\sqrt{\kappa^2 - k_m^2}, & \text{if } k_m \le \kappa, \\ \sqrt{k_m^2 - \kappa^2}, & \text{if } \kappa \le k_m, \end{cases} \qquad m \in \mathbb{N}.$$

The symbols of the operators Λ_j and the scattering operators S_j are then, for $m \in \mathbb{N}$,

$$\hat{\lambda}_{j,m} = \kappa^{-1}\xi_m, \quad \hat{s}^*_{\delta,j,m} = \frac{-\hat{\lambda}_{j,m} - \imath\hat{t}^*_{\delta,m}}{\hat{\lambda}_{j,m} - \imath\hat{t}^*_{\delta,m}} = -\frac{\hat{z}^*_{\delta,j,m} - \imath}{\hat{z}^*_{\delta,j,m} + \imath}, \quad \text{with} \quad \hat{z}^*_{\delta,j,m} = -\frac{\hat{\lambda}_{j,m}}{\hat{t}^*_{\delta,m}}.$$

Modal and global convergence factors. Finally, the modal and global convergence factors of the algorithm (5) can be estimated respectively by (we skip the technical details which can be found in [3, Th. 4.2])

$$\hat{\tau}^*_{\delta,m} := \max_{\pm} \left|(1-r) \pm r\sqrt{\hat{s}^*_{\delta,1,m}\hat{s}^*_{\delta,2,m}}\right|, \qquad \text{and} \qquad \hat{\tau}^*_{\delta} := \sup_{m\in\mathbb{N}} \hat{\tau}^*_{\delta,m}.$$

Study of the convergence factor $\hat{\tau}^*_{\delta}$. We stress that, ultimately, much of the analysis boils down to the properties of the Cayley transform $z \mapsto \frac{z-\imath}{z+\imath}$ in the complex plane, allowing us to get a rather deep understanding of the convergence [6, Lem. 6.5]. For instance, the positivity of T^*_{δ} implies that in the propagative regime ($k_m < \kappa$) the ratio $\hat{z}^*_{\delta,j,m} \in \imath\mathbb{R}^+ \setminus \{0\}$, whereas in the evanescent regime ($\kappa < k_m$) the ratio $\hat{z}^*_{\delta,j,m} \in \mathbb{R} \setminus \{0\}$. The properties of the Cayley transform imply, in turn, that the scattering operators S_j are contractions ($|\hat{s}^*_{\delta,j,m}| < 1$) [6, Cor. 6.6] so that all modal convergence factors satisfy $\hat{\tau}^*_{\delta,m} < 1$. To study the global convergence factor we will use the following technical result whose proof rests on simple Taylor expansions.

Lemma 1 *Let $z(\epsilon) \in \mathbb{C}$, $\epsilon > 0$. The asymptotic behavior of the modal convergence factor of the form*
$$\tau_z = \max_{\pm} \left|(1-r) \pm r\frac{z(\epsilon) - \imath}{z(\epsilon) + \imath}\right|,$$
as ϵ goes to 0 can be deduced from that of $z(\epsilon)$: we have

$$\begin{array}{c} z(\epsilon) \in \imath\mathbb{R}^+, \\ (\zeta \in \mathbb{R}^+) \end{array} \begin{cases} z(\epsilon) \sim \imath\zeta, \\ z(\epsilon) \sim \imath\zeta\epsilon, \\ z(\epsilon) \sim \imath\zeta\epsilon^{-1}, \end{cases} \Rightarrow \begin{cases} \tau_z = 1 - 2r(1+\zeta)^{-1}\min(1,\zeta) + O(\epsilon), \\ \tau_z = 1 - 2r\zeta\epsilon + O(\epsilon^2), \\ \tau_z = 1 - 2r\zeta^{-1}\epsilon + O(\epsilon^2), \end{cases}$$

$$\begin{array}{c} z(\epsilon) \in \mathbb{R}, \\ (\zeta \in \mathbb{R}) \end{array} \begin{cases} z(\epsilon) \sim \zeta, \\ z(\epsilon) \sim \zeta\epsilon, \\ z(\epsilon) \sim \zeta\epsilon^{-1}, \end{cases} \Rightarrow \begin{cases} \tau_z = 1 - 2r(1-r)(1+\zeta^2)^{-1}\min(1,\zeta^2) + O(\epsilon), \\ \tau_z = 1 - 2r(1-r)\zeta^2\epsilon^2 + O(\epsilon^3), \\ \tau_z = 1 - 2r(1-r)\zeta^{-2}\epsilon^2 + O(\epsilon^3). \end{cases}$$

Interest in using non-local operators (δ fixed). It is immediate to check that

$$\hat{z}^*_{\delta,j,m} \sim -1, \qquad \text{as } m \to \infty, \qquad \text{for } * \in \{D, N, R\}.$$

Lemma 1 (with $z(\epsilon) \equiv \hat{z}^*_{\delta,j,m}$, $\epsilon \equiv 1/m$), implies that $\lim_{m\to+\infty} \hat{\tau}^*_{\delta,m} = 1 - r(1-r) < 1$. Notice that the limit is independent of both δ and the type of boundary condition. This is not surprising as the highest modes "do not see", in some sense, the boundary condition. Since we have already established that $\hat{\tau}^*_{\delta,m} < 1$ for all m, it follows that,

$$\hat{\tau}^*_\delta < 1, \qquad \text{for } * \in \{D, N, R\}.$$

We see here a manifestation of the effect of choosing an operator with the "right" order that adequately deals with the highest frequency modes. For instance, if we were to use a multiple of the identity as proposed originally by Després [4], then in this case we would obtain $z(m^{-1}) \sim -\xi_m$ so that the asymptotic convergence factor would behave like $1 - O(m^{-2})$ and the global convergence rate would be 1.

Influence of the strip width δ. From the previous expressions, we obtain that all transmission operators become local in the limit $\delta \to 0$ and, for a fixed m,

$$\hat{z}^D_{\delta,j,m} \sim -\xi_m\delta, \qquad \hat{z}^N_{\delta,j,m} \sim -\xi_m\kappa^{-2}\delta^{-1}, \qquad \hat{z}^R_{\delta,j,m} \sim -\xi_m\kappa^{-1}, \qquad \text{as } \delta \to 0.$$

Lemma 1 (with $z(\epsilon) \equiv \hat{z}^*_{\delta,j,m}$, $\epsilon \equiv \delta$) implies that, in the cases $* \in \{D, N\}$, the modal convergence factor $\hat{\tau}^*_{\delta,m}$ converges to 1 as $O(\delta)$ in the propagative regime ($k_m < \kappa$) and as $O(\delta^2)$ in the evanescent regime ($\kappa < k_m$). In contrast, in the case $* = R$, the modal convergence factor $\hat{\tau}^R_{\delta,m}$ is bounded away from 1 in all regimes.

We wish to study now the global convergence factor $\hat{\tau}^R_\delta$. We report in Figure 2 (left) the mode number of the slowest converging mode with respect to δ/λ, for $\lambda := 2\pi/\kappa$ and $\kappa = 3\pi$. This reveals that, for $* \in \{D, N\}$, the maximum modal factor is attained for a fixed mode number m as $\delta \to 0$. Therefore, our theoretical and numerical analysis has demonstrated that

$$\hat{\tau}^*_\delta = 1 - O(\delta^2), \qquad \text{as } \delta \to 0, \qquad * \in \{D, N\}.$$

In contrast, in the case $* = R$, the maximum modal factor is attained for the mode number $m \propto \delta^{-1/2}$ as $\delta \to 0$. This motivates to study the case $\delta_m = k_m^{-2}$ in the limit $m \to +\infty$. We have

$$\hat{z}^R_{\delta_m,j,m} \sim \kappa(1+\kappa)^{-1}\,\delta_m^{-1/2}, \qquad \text{as } m \to +\infty \text{ with } \delta_m = k_m^{-2}.$$

Therefore, using Lemma 1 (with $z(\epsilon) \equiv \hat{z}^R_{\delta_m,j,m}$, $\epsilon \equiv \delta_m^{1/2}$), the above theoretical and numerical analysis shows that

$$\hat{\tau}^R_\delta = 1 - O(\delta), \qquad \text{as } \delta \to 0.$$

To conclude, we report in Figure 2 (right) the global convergence factor $\hat{\tau}^*_\delta$ with respect to δ/λ, for $\lambda := 2\pi/\kappa$ and $\kappa = 3\pi$. For δ large enough we observe that the convergence factor is constant and the same for all three cases. This can be explained by the dissipative nature of the auxiliary problems and the fact that the boundary condition $*$ is imposed far away from the source of the problem. For sufficiently small δ, the asymptotic regime is attained and corroborates our previous findings.

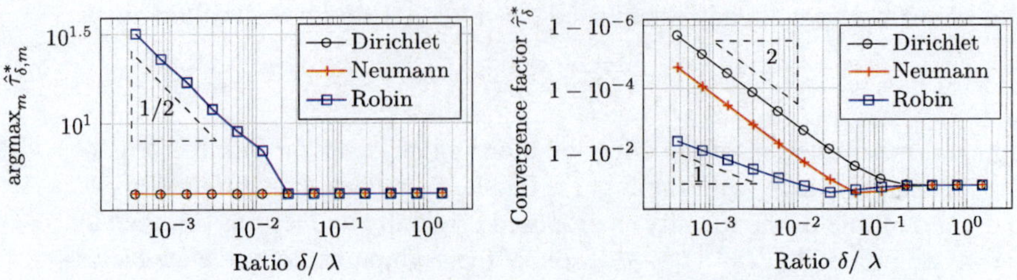

Fig. 2: Slowest-converging mode (left) and convergence factor $\hat{\tau}^*_\delta$ (right) for the wave-guide.

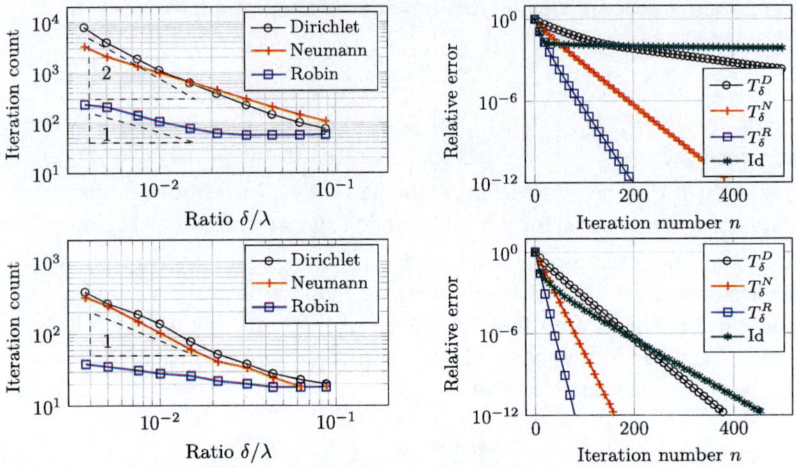

Fig. 3: Iteration count (left) and convergence history (right) in the circular configuration for the Jacobi algorithm (top row) and the GMRES algorithm (bottom row).

3 Finite element computations in a circular geometry

We provide the results of actual computations using $\mathbb{P}_1$-Lagrange finite elements with the relaxed Jacobi algorithm ($r = 1/2$) and the restarted GMRES algorithm (restart 20 iterations). The problem is (1) in a homogeneous ($\mathfrak{a} \equiv \mathfrak{n} \equiv 1$) disk of radius $R = 2$ with an interface at $R = 1$. We compute the relative error using the κ-weighted H^1-norm $\|u\|^2 := \|u\|^2_{L^2} + \kappa^{-2}\|\nabla u\|^2_{L^2}$. We report in the left column of Figure 3 the iteration count to reach a set tolerance of 10^{-8} with respect to δ/λ, with $\lambda := 2\pi/\kappa$ and $\kappa = 1$ and mesh size $h = \lambda/400$. We observe a quasi-quadratic growth for sufficiently smaller δ for the Dirichlet and Neumann conditions. In contrast, for the Robin condition, the growth is only linear and we still benefit of the non-local effect up to $\delta \approx \lambda/50$. We also report in the right column of Figure 3 the convergence history in the case $\kappa = 10$, mesh size $h = \lambda/40$ and $\delta = \lambda/20$ (i.e. strip width of two mesh cells). We added the results using the Després operator $T = \mathrm{Id}$

for comparison. The efficiency of the approach, using the Robin-type condition, is clearly demonstrated.

Acknowledgements This work was supported by the Research Grant ANR–15–CE23–0017–01.

References

1. X. Claeys and E. Parolin. Robust treatment of cross points in Optimized Schwarz Methods (2021). Accepted in *Numerische Mathematik*.
2. F. Collino, S. Ghanemi and P. Joly. Domain decomposition method for harmonic wave propagation: a general presentation. *CMAME* **184**(24), 171–211 (2000).
3. F. Collino, P. Joly and M. Lecouvez. Exponentially convergent non overlapping domain decomposition methods for the Helmholtz equation. *ESAIM: M2AN* **54**(3), 775–810 (2020).
4. B. Després. *Méthodes de décomposition de domaine pour la propagation d'ondes en régime harmonique*. PhD thesis, Université Paris IX Dauphine (1991).
5. I. Harari, I. Patlashenko and D. Givoli. Dirichlet-to-Neumann maps for unbounded wave guides. *J. Comput. Phys.* **143**(1), 200–223 (1998).
6. E. Parolin. *Non-overlapping domain decomposition methods with non-local transmission operators for harmonic wave propagation problems*. PhD thesis, Institut Polytech. de Paris (2020).

Asynchronous Multi-Subdomain Methods With Overlap for a Class of Parabolic Problems

Mohamed Laaraj and Karim Rhofir

1 Introduction

In previous work [3] and [5], we presented asynchronous iterations for solving second-order elliptical partial differential equations based on an overlapping domain decomposition. Asynchronous iterations are not only a family of algorithms suitable for asynchronous computations on multiprocessors, but also a general framework for formulating general iteration methods associated with a fixed point mapping on a product space, including the most standard ones such as the successive approximation method (Jacobi, Gauss–Seidel and their block versions). In this paper, we will associate with the alternate Schwarz method for second-order parabolic problems an affine fixed point map whose linear part is a contraction in the uniform norm. In this context we will develop a method of analyzing the multi-subdomain case, as well as asynchronous iterations for parabolic problems. We will give a new technical result to update the Hopf maximum principle and construct a new exponential weighted norm. This work is devoted to the framework of a class of second-order parabolic problems with Dirichlet condition. We associate with a method based on the resolution of sub-problems on subdomains with overlap a fixed point map defined by the restrictions to subdomains without overlap. We examine the contraction of this fixed point map with respect to a new norm. One important feature of the results presented here is the use of exponential weighted norms, which allows us to obtain a stronger convergence property than the usual uniform norm. We thus obtain a convergence result for asynchronous iterations with respect to a norm finer than the usual one, which includes the basic situation of the very traditional

M. Laaraj
Univ. Hassan II Mohammadia Casablanca ENSAM Avenue Nile 150, Casablanca, Morocco,
e-mail: mohamed.laaraj@gmail.com

K. Rhofir
Univ. Sultan Moulay Slimane LISERT-ENSA Bd. Beni Amir, BP. 77, Khouribga, Morocco,
e-mail: k.rhofir@usms.ma

S. C. Brenner et al. (eds.), *Domain Decomposition Methods in Science and Engineering XXVI*, Lecture Notes in Computational Science and Engineering 145,
https://doi.org/10.1007/978-3-030-95025-5_80

alternate method of Schwarz. At the level of subdomains having a common border portion with the boundary of the domain, this requires the implementation of Hopf's maximum principle. Compared to the previously described method, this formalism is particularly effective at examining the influence of the size of the overlaps and comparing the contraction constant of the fixed point map. After the introduction, in the second section we formulate the problem, introduce the notation used in the sequel and give our new technical result. In the third section we define the linear mapping $\mathcal{T}$ which defines the substructured solution process. Then we define the linear fixed point mapping T which is the composition of $\mathcal{T}$ with a suitable restriction operator R. We prove that T is a linear mapping in a suitable function space context. We also study the contraction property of T . We finally introduce an affine mapping whose linear part is T and whose fixed point is the solution of a parabolic partial differential equation. We state in the closing proposition the convergence of asynchronous iterations applied to the approximation of this affine fixed point mapping.

2 Notation and assumptions

Let Ω be an open bounded domain of $\mathbb{R}^n$ with boundary $\partial\Omega$ and m an integer such that $m \geq 2$. In order to formulate our algorithm, we need an overlapping decomposition of Ω with certain overlap properties. We build such a decomposition by decomposing Ω into m non-overlapping open subdomains $\widetilde{\Omega}_i$ as

$$\widetilde{\Omega}_i \cap \widetilde{\Omega}_j = \emptyset \text{ if } i \neq j \text{ and } \cup_{i=1}^m \overline{\widetilde{\Omega}}_i = \overline{\Omega} \tag{1}$$

and $\partial\Omega$ (resp. $\partial\widetilde{\Omega}_i$) the boundary of Ω (resp. $\widetilde{\Omega}_i$) and $\widetilde{\Gamma}_i = \partial\widetilde{\Omega}_i \cap \Omega\,;\widetilde{\Gamma}'_i = \partial\widetilde{\Omega}_i \cap \partial\Omega$ such that $\Omega = \cup_{i=1}^m \left(\widetilde{\Gamma}_i \cup \widetilde{\Omega}_i\right)$. From this non-overlapping decomposition of Ω we get the desired overlapping decomposition which will be used by our algorithm. To $\widetilde{\Omega}_i$, we associate Ω_i, the overlapping multi-subdomain decomposition: $\widetilde{\Omega}_i \subset \Omega_i \subset \Omega$, $\Omega = \cup_{i=1}^m \Omega_i$, and

$$\Gamma_i = \partial\Omega_i \cap \Omega\,;\Gamma'_i = \partial\Omega_i \cap \partial\Omega \tag{2}$$

such that

$$\overline{\widetilde{\Omega}}_i \cap \overline{\Gamma}_i = \emptyset,\ i = 1, \ldots, m. \tag{3}$$

For the exchange of information between subdomains, we will also employ the index notation

$$\Gamma_{i,j} = \Gamma_i \cap \widetilde{\Omega}_j,\ j \in J(i) \tag{4}$$

where the index set J is defined by

$$J(i) = \left\{j : \Gamma_i \cap \widetilde{\Omega}_j \neq \emptyset, j \neq i\right\}. \tag{5}$$

2.1 Technical result

Consider a bounded domain D of $\mathbb{R}^n$, the boundary $\partial D = \Gamma \cup \Gamma'$ and another domain $\widetilde{D} \subset D$ such that $\partial D \cap \partial \widetilde{D} \subset \Gamma'$ and $\overline{\Gamma} \cap \overline{\widetilde{D}} = \emptyset$

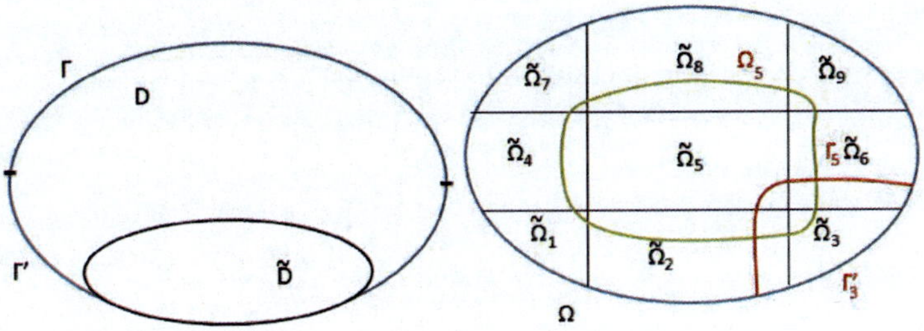

Fig. 1: Illustrative example of decomposition.

Lemma 1 *Consider the kernel $k(x, y)$ defined on $\overline{\widetilde{D}} \times \overline{\Gamma}$. Suppose that $k(x, y)$ is continuously differentiable ($\frac{\partial k}{\partial x_j}(x, y)$ exist and are continuous on $\overline{\widetilde{D}} \times \overline{\Gamma}$) then for all integrable functions $g : y \to g(y)$, $g \in L^1(\overline{\Gamma})$, the function $r(x) = \int_{\overline{\Gamma}} k(x, y) g(y) \mathrm{d}y$ admits continuous partial derivatives with respect to the components x_j of x on $\overline{\widetilde{D}}$, which can be expressed by*

$$\frac{\partial r}{\partial x_j}(x) = \int_{\overline{\Gamma}} \frac{\partial k}{\partial x_j}(x, y) g(y) \mathrm{d}y.$$

2.2 Problem statement

We introduce the time interval $[0; \bar{t}]$. Let us define

$$\begin{cases} Q = \Omega \times [0; \bar{t}] \; ; \widetilde{Q}_i = \widetilde{\Omega}_i \times [0; \bar{t}] \; ; Q_i = \Omega_i \times [0; \bar{t}] \\ \partial \widetilde{Q}_i = \partial \widetilde{\Omega}_i \times [0; \bar{t}] \; ; \partial Q_i = \partial \Omega_i \times [0; \bar{t}] \; ; \partial Q = \partial \Omega \times [0; \bar{t}] \end{cases} \tag{6}$$

and

$$\begin{cases} \Sigma_i = \Gamma_i \times [0; \bar{t}] \; ; \Sigma_i' = \Gamma_i' \times [0; \bar{t}] \\ \Sigma_{i,j} = \Gamma_{i,j} \times [0; \bar{t}] \,. \end{cases} \tag{7}$$

For $0 < t \leq \bar{t}$, we define

$$\begin{cases} \widetilde{Q}_i^a = \overline{\widetilde{\Omega}}_i \times [0; \bar{t}] \; ; Q_i^a = \overline{\Omega}_i \times [0; \bar{t}] \; ; Q^a = \overline{\Omega} \times [0; \bar{t}] \\ \Sigma_i^a = \overline{\Gamma}_i \times [0; \bar{t}] \; ; \Sigma_i'^a = \overline{\Gamma}_i' \times [0; \bar{t}] \; ; \Sigma_{i,j}^a = \overline{\Gamma}_{i,j} \times [0; \bar{t}] \end{cases} \tag{8}$$

and

$$\Gamma = \cup_{i=1}^{m} \cup_{j\in J(i)} \Gamma_{i,j} = \cup_{i=1}^{m}\Gamma_i,\ \Sigma = \cup_{i=1}^{m} \cup_{j\in J(i)} \Sigma_{i,j} = \cup_{i=1}^{m}\Sigma_i. \tag{9}$$

Suppose that

$$L^1(\Sigma) = \prod_{i=1}^{m} \prod_{j\in J(i)} L^1\left(\Sigma_{i,j}\right) \tag{10}$$

and

$$\left\{\begin{array}{l} A \text{ is a second-order elliptic operator with regular coefficients on } \Omega \text{ and} \\ \text{suppose that there exists an } e \in C(\overline{\Omega}),\ e > 0, \text{ such that: } Ae = \lambda e,\ \lambda \in R,\ \lambda > 0 \end{array}\right. \tag{11}$$

and for integers p, q, r

$$f \in L^p(Q)\ ;\ g \in L^q(\Sigma)\ ;\ u^0 \in L^r(\Omega),\ p, q > 1 \text{ and } r \geq 1. \tag{12}$$

We consider the linear parabolic problem with Cauchy conditions

$$\left\{\begin{array}{rl} \dfrac{\partial u}{\partial t} + Au = & f_{/Q} \\ u = & g_{/\partial Q} \\ u(x,0) = & u^0_{/\Omega}. \end{array}\right. \tag{13}$$

Assume that the problem (13) has a unique solution u^* in a suitable function space. On $\overline{Q} = \overline{\Omega} \times \left[0, \overline{t}\right]$, we define the weighted norm

$$|u|^{\overline{t}}_{e,\infty} = \max_{(x,t)\in\overline{Q}} \frac{|u(x,t)|}{e(x)}. \tag{14}$$

We can see that if the initial condition is $u^0_{/\Omega} = 0$, then $|u|^{\overline{t}}_{e,\infty} = \max_{(x,t)\in Q^a} \frac{|u(x,t)|}{e(x)}$.

3 Fixed point mappings

3.1 The linear mapping $\mathcal{T}$

Consider the function space $C^{\overline{t}} = \prod_{i=1}^{m} C\left(C(\overline{\overline{\Omega}}_i); \left]0, \overline{t}\right]\right)$ and define the linear mapping

$$\mathcal{T} : L^1(\Sigma) \to C^{\overline{t}}, \mathcal{T} : \widetilde{w} \to \widetilde{v}.$$

Note that $C^{\overline{t}} \neq C\left(C(\overline{\Omega}; \left]0, \overline{t}\right]\right)$. For each given function $\widetilde{w} \in L^1(\Sigma)$,

$$\widetilde{w} = \{\ldots, \widetilde{w}_i, \ldots\}_{i=1,\ldots,m},\ \widetilde{w}_i = \left\{\ldots, \widetilde{w}_{i,j}, \ldots\right\}_{j\in J(i)} \in L^1(\Sigma_i),$$

we compute using the solutions v_i of the subproblems

$$\frac{\partial v_i}{\partial t} + Av_i = 0 \text{ in } Q_i, v_{i/\Sigma_{i,j}} = \widetilde{w}_{i,j},\ j \in J(i), v_{i/\Sigma'_i} = 0, v_i(x,0) = 0 \text{ on } \Omega_i, \quad (15)$$

where we assume the following regularity of subdomain solutions

$$v_i \in \begin{cases} C^\infty(\widetilde{Q}_i^a), & \text{if } \Gamma'_i \neq \emptyset, \\ C^1(\widetilde{Q}_i^a), & \text{otherwise.} \end{cases} \quad (16)$$

Now we take the restriction $\widetilde{v}_i = v_{i/\overline{\overline{Q}}_i}$ and we define the linear operator $\mathcal{T}_i$ by $\mathcal{T}_i(\widetilde{w}) = \widetilde{v}_i$. Finally, we set

$$\widetilde{v} = \{\ldots, \widetilde{v}_i, \ldots\} = \{\ldots, \mathcal{T}_i(\widetilde{w}), \ldots\} = \mathcal{T}(\widetilde{w}).$$

Proposition 1 $\mathcal{T} \in \mathcal{L}(L^1(\Sigma); C^{\bar{t}})$ *is a linear isotone mapping with respect to the natural order.*

3.2 The linear mapping *T*

Let $C_e(\overline{\overline{Q}}_j)$ be the space formed by all elements of $C(\overline{\overline{Q}}_j)$ endowed with the norm $\left|w_j\right|^{\bar{t}}_{e,\infty,j} = \max_{(x,t)\in\overline{\overline{Q}}_j} \frac{|w_j(x,t)|}{e(x)}$, where $e(x)$ denotes the eigenfunction in (11). We define $C_e^{\bar{t}} = \prod_{j=1}^{m} C_e(\overline{\overline{Q}}_j)$ equipped with the norm $|w|^{\bar{t}}_{e,\infty} = \max_{j=1,\ldots,m} \left|w_j\right|^{\bar{t}}_{e,\infty,j}$, where $w = \{\ldots, w_i, \ldots\} \in C_e$.

Define R'_i, the restriction operator from $C^{\bar{t}}$ to $\prod_{j\in J(i)} C(\overline{\overline{Q}}_j)$, by $R'_i(w) = \overline{w}_i = \left\{\ldots, w_j, \ldots\right\}_{j\in J(i)}$ and R''_i the restriction operator from $\prod_{j\in J(i)} C(\overline{\overline{Q}}_j)$ to $\prod_{j\in J(i)} C(\Sigma^a_{i,j})$ which associates to each $\overline{w}_i$ the element $\left\{\ldots, \overline{w}_{i,j}, \ldots\right\}_{j\in J(i)}$, where the $\overline{w}_{i,j}$ are defined by

$$\overline{w}_{i,j} = w_{j/\Sigma^a_{i,j}}$$

and

$$R_i = R''_i \circ R'_i.$$

Then

$$R(w) = \{R_1(w), \ldots, R_m(w)\} \text{ and } R \in \mathcal{L}\Big(\prod_i C(\overline{\overline{Q}}_i); \prod_{i=1}^{m} \prod_{j\in J(i)} C(\Sigma^a_{i,j})\Big).$$

We define the mapping $T = \{\dots, T_i, \dots\}$ at each $i \in \{1, \dots, m\}$ by

$$\widetilde{v}_i = T_i(w) = \mathcal{T}_i \circ R_i(w) = \mathcal{T}_i \circ R(w),$$

then

$$\widetilde{v} = \{\dots, \widetilde{v}_i, \dots\}_{i=1,\dots,m} = T(w).$$

Proposition 2 *$T \in \mathcal{L}(C_e^{\bar{t}})$ is a linear isotone mapping with respect to the natural order.*

3.3 Contraction property of T

Using (11), we take the restriction $\widetilde{\Psi^{\bar{t}}_{e,i}} = \Psi^{\bar{t},e}_{i/\overline{\overline{Q}}_i}$, where $\Psi_i^{\bar{t},e}$ is the solution of

$$\begin{cases} \frac{\partial \Psi_i^{\bar{t},e}}{\partial t} + A\Psi_i^{\bar{t},e} = 0_{/Q_i} \\ \Psi^{\bar{t},e}_{i/\Sigma_i \cup \Sigma_i'} = e_{/\Sigma_i \cup \Sigma_i'} \\ \Psi_i^{\bar{t},e}(\cdot, 0) = 0_{/\Omega_i}. \end{cases} \tag{17}$$

Lemma 2 $\frac{\widetilde{\Psi^{\bar{t}}_{e,i}}(t,x)}{e(x)}$ *is well defined and continuous on $\overline{\overline{Q}}_i$, and*

$$\max_{(x,t)\in\overline{\overline{Q}}_i} \frac{\widetilde{\Psi^{\bar{t}}_{e,i}}(t,x)}{e(x)} \le \mu_i < 1. \tag{18}$$

Proposition 3 *$T \in \mathcal{L}(C_e^{\bar{t}})$ is a contraction with contraction constant*

$$\mu = \max_{i=1,\dots,m} \mu_i, \text{ where } \mu_i = \max_{(x,t)\in\overline{\overline{Q}}_i} \frac{\widetilde{\Psi^{\bar{t}}_{i,e}}(x,t)}{e(x)}. \tag{19}$$

First, we resolve the subproblems for $i = 1, \cdots, m$: $\frac{\partial u_i}{\partial t} + Au_i = f_{i/Q_i}$, $u_{i/\Sigma_i} = 0_{/\Sigma_i}$, $u_{i/\Sigma_i'} = g_{i/\Sigma_i'}$, $u_i(x,0) = u^0$ on Ω_i. Restricting u_i to $\overline{u}_i = u_{i/\overline{\overline{Q}}_i}$ and considering the new subproblems $\frac{\partial v_i}{\partial t} + Av_i = f_{i/Q_i}$, $v_{i/\Sigma_i} = w_{/\Sigma_i}$, $v_{i/\Sigma_i'} = g_{i/\Sigma_i'}$, $v_i(x,0) = u^0$, we get the restricted values $\widetilde{v}_i = v_{i/\overline{\overline{Q}}_i}$ so that the fixed point is given by

$$\widetilde{v}_i = \overline{u}_i + T_i(w). \tag{20}$$

Proposition 4 *The asynchronous iterations initialized by u^0, applied to the affine fixed point mapping $F(w) = T(w) + \overline{u}$, give rise to a sequence of iterates which converges, with respect to the uniform weighted norm $|\cdot|^{\overline{t}}_{e,\infty}$, towards u^*, the solution of problem (13).*

Proof The proofs of the above propositions will be given in the extended version of this paper. □

4 Constants of contraction comparison with respect to the weighted exponential norm

Let A_0 be an operator satisfying the previous conditions and $\alpha \in \mathbb{R}$. We define the operator A_α by $A_\alpha = A_0 + \alpha I$.

We consider two open subdomains Ω_i^k, $k = 1, 2$ such that $\Omega_i^1 \subset \Omega_i^2$, then $Q_i^1 = \Omega_i^1 \times \left]0; \overline{t}\right]$; $Q_i^2 = \Omega_i^2 \times \left]0; \overline{t}\right]$; $Q_i^1 \subset Q_i^2$. Define

$$\begin{cases} \Gamma_i^k = \partial\Omega_i^k \cap \Omega \,;\, \Gamma_i'^k = \partial\Omega_i^k \cap \partial\Omega \\ \Sigma_i^k = \Gamma_i^k \times \left]0; \overline{t}\right] \;;\, \Sigma_i'^k = \Gamma_i'^k \times \left]0; \overline{t}\right] \end{cases}$$

and assume that for $k = 1, 2$: Γ_i^k, $\Gamma_i'^k$ satisfy (3), and $\Psi_i^{\overline{t},e,k}$, $\Psi_{e,i}^{\overline{t},k}$ are obtained by (17) and (18) with respect to Ω_i^k. Let T_1 (resp. T_2) the fixed point mapping associated to Q^1 (resp. Q^2), and μ_1 (resp. μ_2) the contraction constant of T_1 (resp T_2) defined by (19).

Proposition 5 *Under the previous notations, $\widetilde{\Psi}_{e,i}^{\overline{t},2} < \widetilde{\Psi}_{e,i}^{\overline{t},1}$ and $\mu_2 < \mu_1 < 1$.*

Let us solve the following problem, with $A = A_\beta$ for $\beta \in \mathbb{R}$:

$$\begin{cases} \dfrac{\partial u}{\partial t} + Au = f_{/Q} \\ u = g_{/\partial Q} \\ u(x, 0) = u^0_{/\Omega}. \end{cases} \tag{21}$$

Let $\mathcal{D}$ be a bounded domain of $\mathbb{R}^n$, $\overline{t} \in \mathbb{R}_+$. Let $\alpha \in \mathbb{R}$ and $e_{\mathcal{D}}$ be a positive function on $\overline{\mathcal{D}}$.

We define on $\overline{\mathcal{D}} \times [0, \overline{t}]$ the weighted exponential norm $|\cdot|^{\overline{t}}_{e_{\mathcal{D}},\infty,\alpha}$ by:

$$| u|^{\overline{t}}_{e_{\mathcal{D}},\infty,\alpha} = \max_{(x,t)\in\overline{\mathcal{D}}\times[0,\overline{t}]} \left| \frac{\exp(-\alpha t)u(x,t)}{e_{\mathcal{D}}(x)} \right|. \tag{22}$$

Replacing $\overline{\mathcal{D}}$ (resp. $e_{\mathcal{D}}$) by $\overline{\Omega}_i$ (resp. $e_{\overline{\Omega}_i}$), we can define on $C_e(\overline{Q}_i)$ the norm $|\cdot|^{\overline{t}}_{e,\infty,\alpha,i}$ by $|u_i|^{\overline{t}}_{e,\infty,\alpha,i} = |u_i|^{\overline{t}}_{e_{\overline{\Omega}_i},\infty,\alpha} = \max_{(x,t)\in\overline{\Omega}_i\times[0,\overline{t}]} \left|\frac{\exp(-\alpha t)u(x,t)}{e(x)}\right|$.

Then, we define on $C_e^{\overline{t}}$ the norm $|\cdot|^{\overline{t}}_{e,\infty,\alpha}$ by $|u|^{\overline{t}}_{e,\infty,\alpha} = \max_{i=1,\ldots,m} |u_i|^{\overline{t}}_{e,\infty,\alpha,i}$.

The function $v = \exp(-\alpha t)u$ then satisfies

$$\begin{cases} \exp(\alpha t)\dfrac{\partial v}{\partial t} + \exp(\alpha t)Av + \exp(\alpha t)\alpha v = f_{/Q} \\ \exp(\alpha t)v = g_{/\partial Q} \\ v(x,0) = u^0_{/\Omega}. \end{cases} \tag{23}$$

If $\overline{A}_\alpha = A + \alpha I$, then the problem becomes

$$\begin{cases} \dfrac{\partial v}{\partial t} + \overline{A}_\alpha v = \exp(-\alpha t) f_{/Q} \\ v = \exp(-\alpha t) g_{/\partial Q} \\ v(x,0) = u^0_{/\Omega}, \end{cases} \tag{24}$$

where $\overline{A}_\alpha = A_0 + (\alpha + \beta) I = A_{\alpha+\beta}$ and if we choose $\alpha = -\beta$ then $\overline{A}_\alpha = A_0$.

Proposition 6 *Let $\alpha \geq 0$ and $w \in C_e^{\bar{t}}$. For the subproblems $\frac{\partial u_i}{\partial t} + Au_i + \alpha u_i = f_{i/Q_i}$, $u_{i/\Sigma_i} = w_{/\Sigma_i}$, $u_{i/\Sigma_i'} = 0_{/\Sigma_i'}$, $u_i(x,0) = 0_{/\Omega_i}$, we associate to T_α the affine fixed point map and to μ_α its constant contraction. Then μ_α is strictly decreasing as a function of α.*

Let $w \in C_e^{\bar{t}}$, and suppose that u_i is the solution of the subproblems

$$\begin{cases} \dfrac{\partial u_i}{\partial t} + Au_i = f_{/Q_i}, \\ u_{i/\Sigma_i'} = g_{i/\Sigma_i'}, \\ u_{i/\Sigma_i} = w_{/\Sigma_i'}, \\ u_i(x,0) = u^0_{/\Omega_i}. \end{cases} \tag{25}$$

We can define the affine fixed point map F by $u_i = F_i(w) = T_i(w) + \overline{u}$.

Proposition 7 *F is a fixed point mapping with respect to the weighted exponential norm with contraction constant μ, where μ is a contraction constant of the fixed point mapping $\mathbb{F}$ associated to A_0 with respect to the weighted norm $|\cdot|^{\bar{t}}_{e,\infty,\alpha}$ defined by (22).*

Proof The proof of the above propositions will be given in the extended version of this paper. □

Proposition 8 *The asynchronous iterations initialized by u^0, applied to the affine fixed point mapping $F(w) = T(w) + \overline{u}$, give rise to a sequence of iterates which converges, with respect to the uniform weighted norm $|\cdot|^{\bar{t}}_{e,\infty,\alpha}$, towards u^*, the solution of problem (25).*

The proof is based on the use of El Tarazi's theorem [2].

Remark 1

1. If $\beta > 0$, the norm $|\cdot|^{\bar{t}}_{e,\infty,\alpha}$, for $\alpha = -\beta$, is finer than $|\cdot|^{\bar{t}}_{e,\infty}$ (which means it converges faster).
2. If $\beta < 0$, we obtain the convergence for the norm $|\cdot|^{\bar{t}}_{e,\infty,\alpha}$ with $\alpha = -\beta$, which is less fine than $|\cdot|^{\bar{t}}_{e,\infty}$. For this last norm and any $\beta < 0$ there is no convergence in general.
3. Let λ the smallest positive eigenvalue of A_0. For $\beta \in]-\lambda, 0]$, $A = A_\beta$ satisfies the previous conditions but with the contraction constant satisfying $\mu \leq \mu_\beta < 1$ and when $\beta \searrow -\lambda$, $\mu_\beta \nearrow 1$.
4. If we make the change $v = s^{(-\alpha t)}u$, $s \in R^{+*}$, instead of $v = \exp(-\alpha t)u$, then all the results are still valid.

References

1. L. Badea. On the Schwarz alternating method with more than two subdomains for nonlinear monotone problems. *SIAM J. Numer. Anal.* **28**(1), 179–204 (1991).
2. M.N. El Tarazi. Some convergence results for asynchronous algorithms. *Numer. Math.* **39**, 325–340 (1982).
3. J.-C. Miellou, M. Laaraj and M.J. Gander. Overlapping multi-subdomain asynchronous fixed point methods for elliptic boundary value problems. In: *7th International Colloquium on Numerical Analysis and Computer Science with Applications, PlovDiv, Bulgaria 1998* (1998).
4. G.A. Meurant. A domain decomposition method for parabolic problem. *Applied Numerical Mathematics* **8**(4-5), 427–441 (1991).
5. K. Rhofir and M. Laaraj. Asynchronous overlapping weighted multi-subdomain decomposition for elliptic problem. *International Journal of Mathematical Analysis* **10**(20), 981–999 (2016).
6. L. Xiaoqin, J. Lilli and H. Thi-Thao-Phuong. Overlapping domain decomposition based exponential time differentiacing methods for semilinear parabolic equations. *BIT Numerical Mathematics* **63**, 1–36 (2020).

Toward a New Fully Algebraic Preconditioner for Symmetric Positive Definite Problems

Nicole Spillane

1 Introduction

We set out to solve the linear system $\mathbf{A}\mathbf{x}_* = \mathbf{b}$, for a given symmetric positive definite (spd) matrix $\mathbf{A} \in \mathbb{R}^{n \times n}$. There exist a variety of two-level methods for which fast convergence is guaranteed without making assumptions on the number of subdomains, their shape, or the distribution of the coefficients in the underlying PDE (see e.g. [3, 5, 6, 8, 11–13, 15, 16, 18, 19]). In all of these methods one selects vectors for the coarse space by computing low- or high-frequency eigenvectors of well-chosen generalized eigenvalue problems (of the form $\mathbf{M}_A\mathbf{y} = \lambda\mathbf{M}_B\mathbf{y}$) posed in the subdomains. To the best of the author's knowledge, none of these methods can be applied if the so-called local *Neumann* matrices are not known. Specifically, the definition of either $\mathbf{M}_A$ or $\mathbf{M}_B$ is based on a family of symmetric positive semi-definite (spsd) matrices $\mathbf{N}^s$ that satisfy

$$\exists C > 0, \text{ such that } \sum_{s=1}^{N} \mathbf{x}^\top \mathbf{R}^{s\top} \mathbf{N}^s \mathbf{R}^s \mathbf{x} \leq C\, \mathbf{x}^\top \mathbf{A}\mathbf{x};\ \forall\, \mathbf{x} \in \mathbf{R}^n, \quad (1)$$

where it has been assumed that there are N subdomains with restriction operators $\mathbf{R}^s$. The Neumann matrices are a natural choice for $\mathbf{N}^s$ and the above estimate then holds with constant C equal to the maximal multiplicity of a mesh element. This limitation is very well known (and stated clearly in e.g. [1, 2]).

In this work, it is proposed to relax the assumptions on the matrices $\mathbf{N}^s$ in (1) by allowing them to be symmetric (but not necessarily positive semi-definite). Such matrices $\mathbf{N}^s$, then denoted $\mathbf{B}^s$, can always be defined algebraically. Special treatment must be applied to the non-positive part of $\mathbf{B}^s$ and this will be reflected in the cost of setting up and applying the preconditioner. In Section 2, the new preconditioner

Nicole Spillane
CNRS, CMAP, Ecole Polytechnique, Institut Polytechnique de Paris, 91128 Palaiseau Cedex, France, e-mail: nicole.spillane@cmap.polytechnique.fr

S. C. Brenner et al. (eds.), *Domain Decomposition Methods in Science and Engineering XXVI*, Lecture Notes in Computational Science and Engineering 145,
https://doi.org/10.1007/978-3-030-95025-5_81

is defined and the result on the condition number is given. In Section 3, some preliminary numerical illustrations are provided. Finally, Section 4 offers up some conclusive remarks about the new preconditioner, as well as some of its current limitations that are addressed in the full length article [7].

2 Definition of the new preconditioner and theory

This section introduces the new preconditioner $\mathbf{H}(\tau)$ and proves the resulting bound for the condition number of $\mathbf{H}(\tau)\mathbf{A}$. The methodology is as follows. In Subsection 2.1, some elements of the abstract Schwarz setting are defined in their algebraic form. Then, in Subsection 2.2, a new matrix $\mathbf{A}_+$ is introduced for which an algebraic splitting into spsd matrices is available by construction (i.e., (1) is satisfied). The availability of this splitting makes it possible to apply the abstract GenEO theory [14] to choose a coarse space. Hence, in Subsection 2.3, a two-level preconditioner $\mathbf{H}_+(\tau)$, with a GenEO coarse space parametrized by a threshold τ, is defined for $\mathbf{A}_+$. The spectral bound for $\mathbf{H}_+(\tau)\mathbf{A}_+$ is given. Finally, in Subsection 2.4, the Woodbury matrix identity [17] is applied to find a formula for $\mathbf{A}^{-1} - \mathbf{A}_+^{-1}$ and this (provably low-rank) term is added to $\mathbf{H}_+(\tau)$ in order to form the new preconditioner $\mathbf{H}(\tau)$ for $\mathbf{A}$. A spectral bound for $\mathbf{H}(\tau)\mathbf{A}$ follows.

2.1 Algebraic domain decomposition

Let $\Omega = [\![1, n]\!]$ be the set of all indices in $\mathbb{R}^n$. In all that follows, it is assumed that Ω has been partitioned into a family of subdomains $(\Omega^s)_{s=1,\ldots,N}$ and that the partition has minimal overlap in the sense given by Definition 1. The usual restriction operators are also defined.

Definition 1 A set $(\Omega^s)_{s=1,\ldots,N}$ of $N \in \mathbb{N}$ subsets of $\Omega = [\![1, n]\!]$ is called a partition of Ω if $\Omega = \bigcup_{s=1}^{N} \Omega^s$. Each Ω^s is called a subdomain. The partition is said to have at least minimal overlap if: for any pair of indices $(i, j) \in [\![1, n]\!]^2$, denoting by A_{ij} the coefficient of $\mathbf{A}$ at the i-th line and j-th column,

$$A_{ij} \neq 0 \Rightarrow (\exists\, s \in [\![1, N]\!] \text{ such that } \{i, j\} \subset \Omega^s) .$$

Moreover, for each $s \in [\![1, N]\!]$, let n^s be the cardinality of Ω^s. Finally, let the restriction matrix $\mathbf{R}^s \in \mathbb{R}^{n^s \times n}$ be zero everywhere except for the block formed by the columns in Ω^s which is the $n^s \times n^s$ identity matrix.

2.2 Definition of $\mathbf{A}_+$ and related operators

The starting point for the algebraic preconditioner is to relax condition (1) by allowing symmetric, but possibly indefinite, matrices in the splitting of $\mathbf{A}$.

Definition 2 Let $\mathbf{B} \in \mathbb{R}^{n\times n}$ be the matrix whose (i, j)-th entry is

$$B_{ij} := \begin{cases} \frac{A_{ij}}{\#\{s; \{i,j\}\subset\Omega^s\}} & \text{if } A_{ij} \neq 0, \\ 0 & \text{otherwise.} \end{cases}$$

Then, for each $s = 1, \dots, N$, let $\mathbf{B}^s := \mathbf{R}^s\mathbf{B}\mathbf{R}^{s\top}$ $(\in \mathbb{R}^{n^s\times n^s})$.

Theorem 1 *Thanks to the minimal overlap assumption, the symmetric matrices* $\mathbf{B}^s$ *are well-defined and satisfy* $\mathbf{A} = \sum_{s=1}^N \mathbf{R}^{s\top}\mathbf{B}^s\mathbf{R}^s$.

The proof is given in [7, Theorem 3.2]. In particular, (1) holds with $\mathbf{N}^s = \mathbf{B}^s$ and $C = 1$. Next, each $\mathbf{B}^s$ is split into an spsd and a symmetric negative semi-definite part.

Definition 3 Let $s \in [\![1, N]\!]$. Since $\mathbf{B}^s$ is symmetric, there exist a diagonal matrix $\mathbf{\Lambda}^s$ and an orthogonal matrix $\mathbf{V}^s$ such that $\mathbf{B}^s = \mathbf{V}^s\mathbf{\Lambda}^s\mathbf{V}^{s\top}$. It can further be assumed that the diagonal entries of $\mathbf{\Lambda}^s$ (which are the eigenvalues of $\mathbf{B}^s$) are sorted in non-decreasing order and that

$$\mathbf{\Lambda}^s = \begin{pmatrix} \mathbf{\Lambda}^s_- & \mathbf{0} \\ \mathbf{0} & \mathbf{\Lambda}^s_+ \end{pmatrix}, \quad \mathbf{V}^s = \left[\mathbf{V}^s_- | \mathbf{V}^s_+\right], \quad \mathbf{\Lambda}^s_+ \text{ is spd}, \quad -\mathbf{\Lambda}^s_- \text{ is spsd.}$$

Finally, let

$$\mathbf{A}^s_+ := \mathbf{V}^s_+\mathbf{\Lambda}^s_+\mathbf{V}^{s\top}_+ \text{ and } \mathbf{A}^s_- := -\mathbf{V}^s_-\mathbf{\Lambda}^s_-\mathbf{V}^{s\top}_-.$$

In words, the positive (respectively, non-positive) eigenvalues of $\mathbf{B}^s$ are on the diagonal of $\mathbf{\Lambda}^s_+$ (respectively, $\mathbf{\Lambda}^s_-$) and the corresponding eigenvectors are in the columns of $\mathbf{V}^s_+$ (respectively, $\mathbf{V}^s_-$). It is also clear that

$$\mathbf{B}^s = \mathbf{A}^s_+ - \mathbf{A}^s_-, \quad \mathbf{A}^s_+ \text{ is spsd, and } \mathbf{A}^s_- \text{ is spsd.}$$

In the next definition, these new local matrices are assembled into global matrices and in particular the all important matrix $\mathbf{A}_+$ is defined.

Definition 4 Let $\mathbf{A}_+$ and $\mathbf{A}_-$ be the two matrices in $\mathbb{R}^{n\times n}$ defined by

$$\mathbf{A}_+ := \sum_{s=1}^N \mathbf{R}^{s\top}\mathbf{A}^s_+\mathbf{R}^s, \text{ and } \mathbf{A}_- := \sum_{s=1}^N \mathbf{R}^{s\top}\mathbf{A}^s_-\mathbf{R}^s.$$

It is clear that $\mathbf{A} = (\mathbf{A}_+ - \mathbf{A}_-)$ and $\mathbf{A}_-$ is spsd. As a result, $\mathbf{A}_+$ is spd.

2.3 Two-level preconditioner for $\mathbf{A}_+$ with a GenEO coarse space

Following [14], there are many possible choices for a two-level preconditioner for $\mathbf{A}_+$ with a GenEO coarse space. This is not the novelty here so only one is given with no further comment on other possibilities.

Theorem 2 *Let $\tau > 1$ be a threshold. Let $\mathbf{H}_+(\tau)$ be defined by*

$$\mathbf{H}_+(\tau) := \sum_{s=1}^{N} \mathbf{R}^{s\top}(\mathbf{R}^s\mathbf{A}_+\mathbf{R}^{s\top})^{-1}\mathbf{R}^s + \mathbf{R}^0(\tau)^\top(\mathbf{R}^0(\tau)\mathbf{A}_+\mathbf{R}^0(\tau)^\top)^{-1}\mathbf{R}^0(\tau),$$

where the lines of $\mathbf{R}^0(\tau)$ form a basis for the GenEO coarse space $V^0(\tau)$. The coarse space is in turn defined according to [14, Definition 5] by

$$V^0(\tau) := \sum_{s=1}^{N} \operatorname{span}\left\{\mathbf{R}^{s\top}\mathbf{y}^s;\, (\lambda^s, \mathbf{y}^s) \in \mathbb{R}^+ \times \mathbb{R}^{n^s} \text{ solution of (2) and } \lambda^s < \tau^{-1}\right\}.$$

where the generalized eigenvalue problem is

$$(\mathbf{D}^s)^{-1}\mathbf{A}_+^s(\mathbf{D}^s)^{-1}\mathbf{y}^s = \lambda^s\mathbf{R}^s\mathbf{A}_+\mathbf{R}^{s\top}\mathbf{y}^s;\ \textit{for } \mathbf{D}^s := \mathbf{R}^s\left(\sum_{t=1}^{N}\mathbf{R}^{t\top}\mathbf{R}^t\right)^{-1}\mathbf{R}^{s\top}. \quad (2)$$

If $\tau > 1$ and $\mathcal{N}_+$ is the minimal number of colors that are needed to color each subdomain in such a way that two subdomains with the same color are $\mathbf{A}_+$-orthogonal, then the eigenvalues of the preconditioned operator satisfy

$$\lambda(\mathbf{H}_+(\tau)\mathbf{A}_+) \in \left[((1+2\mathcal{N}_+)\tau)^{-1}, \mathcal{N}_+ + 1\right]. \quad (3)$$

Proof This is the result in [14, Remark 3, Corollary 4, Assumption 6]. □

2.4 New preconditioner for A

Definition 5 Let $n_- = \operatorname{rank}(\mathbf{A}_-)$. Let $\mathbf{\Lambda}_- \in \mathbb{R}^{n_- \times n_-}$ and $\mathbf{V}_- \in \mathbb{R}^{n \times n_-}$ be the diagonal matrix and the orthogonal matrix that are obtained by removing the null part of $\mathbf{A}_-$ from its diagonalization in such a way that $\mathbf{A}_- = \mathbf{V}_-\mathbf{\Lambda}_-\mathbf{V}_-^\top$ with $\mathbf{\Lambda}_-$ spd.

It now holds that $\mathbf{A} = \mathbf{A}_+ - \mathbf{V}_-\mathbf{\Lambda}_-\mathbf{V}_-^\top$ and the Woodbury matrix identity [17] applied to computing the inverse of $\mathbf{A}$, viewed as a modification of $\mathbf{A}_+$, gives

$$\mathbf{A}^{-1} = \mathbf{A}_+^{-1} + \mathbf{A}_+^{-1}\mathbf{V}_-\left(\mathbf{\Lambda}_-^{-1} - \mathbf{V}_-^\top\mathbf{A}_+^{-1}\mathbf{V}_-\right)^{-1}\mathbf{V}_-^\top\mathbf{A}_+^{-1}. \quad (4)$$

This leads to the main theorem in this article in which the new algebraic preconditioner for $\mathbf{A}$ is defined and the corresponding spectral bound is proved.

Theorem 3 *For $\tau > 1$, let the new preconditioner be defined as*

$$\mathbf{H}(\tau) := \mathbf{H}_{+}(\tau) + \mathbf{A}_{+}^{-1}\mathbf{V}_{-}\left(\mathbf{\Lambda}_{-}^{-1} - \mathbf{V}_{-}^{\top}\mathbf{A}_{+}^{-1}\mathbf{V}_{-}\right)^{-1}\mathbf{V}_{-}^{\top}\mathbf{A}_{+}^{-1}.$$

The eigenvalues of the preconditioned operator satisfy

$$\lambda(\mathbf{H}(\tau)\mathbf{A}) \in \left[((1 + 2\mathcal{N}_{+})\tau)^{-1}, \mathcal{N}_{+} + 1\right], \tag{5}$$

where, once more, $\mathcal{N}_{+}$ is the coloring constant with respect to the operator $\mathbf{A}_{+}$.

Proof The estimate for the eigenvalues of $\mathbf{H}_{+}(\tau)\mathbf{A}_{+}$ in (3) is equivalent to

$$((1 + 2\mathcal{N}_{+})\tau)^{-1}\langle \mathbf{x}, \mathbf{A}_{+}^{-1}\mathbf{x}\rangle \leq \langle \mathbf{x}, \mathbf{H}_{+}(\tau)\mathbf{x}\rangle \leq (\mathcal{N}_{+} + 1)\langle \mathbf{x}, \mathbf{A}_{+}^{-1}\mathbf{x}\rangle, \ \forall \mathbf{x} \in \mathbb{R}^{n}.$$

Adding $\langle \mathbf{x}, \mathbf{A}_{+}^{-1}\mathbf{V}_{-}\left(\mathbf{\Lambda}_{-}^{-1} - \mathbf{V}_{-}^{\top}\mathbf{A}_{+}^{-1}\mathbf{V}_{-}\right)^{-1}\mathbf{V}_{-}^{\top}\mathbf{A}_{+}^{-1}\mathbf{x}\rangle$ to each term, it holds that

$$((1 + 2\mathcal{N}_{+})\tau)^{-1}\langle \mathbf{x}, \mathbf{A}^{-1}\mathbf{x}\rangle \leq \langle \mathbf{x}, \mathbf{H}(\tau)\mathbf{x}\rangle \leq (\mathcal{N}_{+} + 1)\langle \mathbf{x}, \mathbf{A}^{-1}\mathbf{x}\rangle, \ \forall \mathbf{x} \in \mathbb{R}^{n},$$

where (4) was applied as well as $\mathcal{N}_{+} \geq 1$ and $\tau \geq 1$. This is equivalent to (5). □

Remark 1 (Cost of the new preconditioner) In order to apply the preconditioner, the matrix $\mathbf{A}_{+}^{-1}\mathbf{V}_{-}$ must be formed. This can be done by solving iteratively n_{-} linear systems preconditioned by $\mathbf{H}_{+}(\tau)$. It is likely that block Krylov methods would be advantageous. Note that unfortunately $\mathbf{A}_{+}^{-1}\mathbf{V}_{-}$ is dense, as is $\left(\mathbf{\Lambda}_{-}^{-1} - \mathbf{V}_{-}^{\top}\mathbf{A}_{+}^{-1}\mathbf{V}_{-}\right)$. Setting up and applying the second coarse problem $\mathbf{A}_{+}^{-1}\mathbf{V}_{-}\left(\mathbf{\Lambda}_{-}^{-1} - \mathbf{V}_{-}^{\top}\mathbf{A}_{+}^{-1}\mathbf{V}_{-}\right)^{-1}\mathbf{V}_{-}^{\top}\mathbf{A}_{+}^{-1}$ is the most costly part of the algorithm.

The good news is that the number n_{-} of columns in $\mathbf{V}_{-}$ (which equals the rank of $\mathbf{A}_{-}$) satisfies $n_{-} \leq \sum_{s=1}^{N} n^{s} - n$. Consequently, the rank of $\mathbf{A}_{-}$ is low compared to the rank n of $\mathbf{A}$ ($n_{-} \ll n$) as long as there is little overlap between subdomains. Note that n_{-} can be (and hopefully is) much smaller even than $\sum_{s=1}^{N} n^{s} - n$.

3 Numerical illustration

The results in this section are obtained using the software FreeFem++ [9], GNU Octave [4] and METIS [10]. The linear systems that are considered arise from discretizing with $\mathbb{P}_1$ finite elements some two-dimensional linear elasticity problems.

Fig. 1: Testcase 1 – partition ($N = 4$) and distribution of E (10^8 if white and 10^3 if dark).

The first test case is posed on the domain $\Omega = [4, 1]$ discretized by 112×28 elements. The problem size is $n = 6496$ degrees of freedom. The coefficients in the linear elasticity equation are $\nu = 0.3$ for Poisson's ratio and

$$E(x, y) = \begin{cases} 10^8 & \text{if } y \in [1/7, 2/7] \cup [3/7, 4/7] \cup [5/7, 6/7]; \\ 10^3 & \text{otherwise.} \end{cases}$$

The domain is partitioned into 4 subdomains with Metis. No overlap is added. Figure 1 shows both the partition into subdomains and the distribution of E. For this problem, the coloring constants with respect to $\mathbf{A}$ and $\mathbf{A}_+$ are $\mathcal{N} = 2$, and $\mathcal{N}_+ = 3$. The problem is solved with the one-level Additive Schwarz (AS), the two-level AS with the GenEO coarse space from [14, Section 5.2.2] and the new method. The value of the threshold τ for the last two methods is chosen to be $\tau = 10$. The theoretical bounds for GenEO and the new method is that the eigenvalues are in the interval $[1/50 = 0.02, 3]$ and $[1/70 \approx 0.014, 4]$, respectively. The $\mathbf{A}$-norm of the error at each iteration of the preconditioned conjugate gradient is represented in Figure 2. The quantities of interest are in Table 1. The one-level method is not efficient on this problem. This was to be expected. Both the GenEO solver and the new solver converge fast. With $\tau = 10$ in both methods, the coarse space for the new method is larger than with GenEO (58 *versus* 49 coarse vectors). For the new method there is also an additional problem of size 49. The results show that the new preconditioner converges a little bit faster than GenEO. A study with more values of all the parameters is needed to compare GenEO and the new solver as the parameter τ does not play exactly the same role in the setup of both preconditioners. Since there is a lot more information injected into GenEO (through the Neumann matrices), it is expected that GenEO will be more efficient. However the new method has the very significant advantage of being algebraic, and being almost as efficient as GenEO would be an achievement.

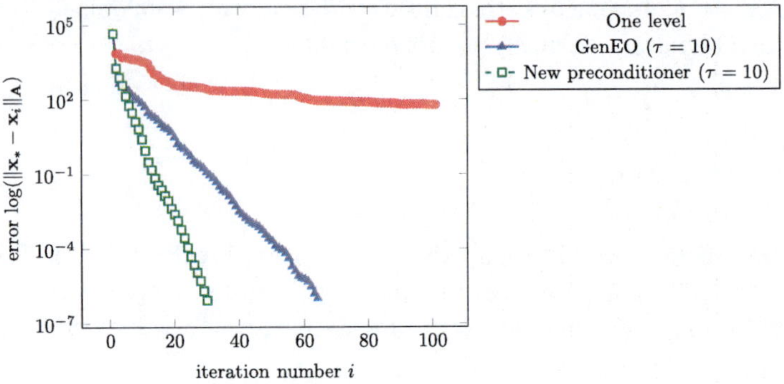

Fig. 2: Testcase 1 – Convergence history for the one-level method, the two-level GenEO method and the new method.

	$\lambda_{\min}$	$\lambda_{\max}$	κ	It	$\#V^0$	n_-
One-level AS	$2 \cdot 10^{-4}$	2.0	$1.0 \cdot 10^4$	>100	0	0
Two-level AS with GenEO	0.059	3.0	51	65	49	0
New method	0.24	2.93	12	30	58	49

Table 1: Testcase 1 – Extreme eigenvalues ($\lambda_{\min}$ and $\lambda_{\max}$), condition number (κ), iteration count (It), size of coarse space ($\#V^0$), and size of second coarse space in new method ($n_- = \text{rank}(\mathbf{A}_-)$).

It is very good news that the coarse space and the space V_- did not explode on the previous test case. The second test case is a rather easy problem posed on $\Omega = [1, 1]$ with a distribution of both coefficients that is homogeneous: $\nu = 0.3$ and $E = 10^8$. Two partitions are considered: one into $N = 16$ regular subdomains and the other into $N = 16$ subdomains with Metis. No overlap is added to the subdomains. The results are presented in Table 2. For the problem with regular subdomains, the new method selects a coarse space of size 44 (*versus* 40 for GenEO). This means, that even without the knowledge of the Neumann matrix, a coarse space is constructed that has almost the same number of vectors as the optimal coarse space for this problem which consists of $3 \times 12 = 36$ rigid body modes (there are 4 non-floating subdomains). Of course the second coarse space also adds to the cost.

	$N = 16$ **regular subdomains**						$N = 16$ **subdomains with Metis**					
	$\lambda_{\min}$	$\lambda_{\max}$	κ	It	$\#V^0$	n_-	$\lambda_{\min}$	$\lambda_{\max}$	κ	It	$\#V^0$	n_-
One-level AS	$2 \cdot 10^{-3}$	4.0	1996	97	0	0	$1.7 \cdot 10^{-3}$	3.0	1817	>100	0	0
Two-level AS with GenEO	0.07	4.0	60	61	40	0	0.095	3.4	36	54	74	0
New method	0.19	4.0	21	39	44	24	0.26	3.0	11.3	31	117	94

Table 2: Testcase 2 – Extreme eigenvalues ($\lambda_{\min}$ and $\lambda_{\max}$), condition number (κ), iteration count (It), size of coarse space ($\#V^0$), and size of second coarse space in new method ($n_- = \text{rank}(\mathbf{A}_-)$).

4 Conclusion

A new algebraic preconditioner was defined for the first time and bounds for the spectrum of the resulting preconditioned operator were proved. They are independent of the number of subdomains and any parameters in the problem. The new preconditioner has two coarse spaces. One of them is dense and a sparse approximation is under investigation. The full length article [7] proposes variants of the new preconditioner that have cheaper choices for $\mathbf{H}_+$ and less exotic coarse solves.

References

1. E. Agullo, L. Giraud and L. Poirel. Robust preconditioners via generalized eigenproblems for hybrid sparse linear solvers. *SIAM Journal on Matrix Analysis and Applications* **40**(2), 417–439 (2019).
2. H. Al Daas and L. Grigori. A class of efficient locally constructed preconditioners based on coarse spaces. *SIAM Journal on Matrix Analysis and Applications* **40**(2), hal-01963067 (2018).
3. V. Dolean, F. Nataf, R. Scheichl and N. Spillane. Analysis of a two-level Schwarz method with coarse spaces based on local Dirichlet-to-Neumann maps. *Comput. Methods Appl. Math.* **12**(4), 391–414 (2012).
4. J.W. Eaton, D. Bateman, S. Hauberg and R. Wehbring. *GNU Octave version 5.2.0 manual: a high-level interactive language for numerical computations* (2020).
5. Y. Efendiev, J. Galvis, R. Lazarov and J. Willems. Robust domain decomposition preconditioners for abstract symmetric positive definite bilinear forms. *ESAIM Math. Model. Numer. Anal.* **46**(5), 1175–1199 (2012).
6. M.J. Gander and A. Loneland. SHEM: An optimal coarse space for RAS and its multiscale approximation. In: *Domain Decomposition Methods in Science and Engineering XXIII*, pp. 313–321, Springer (2017).
7. L. Gouarin and N. Spillane. Fully algebraic domain decomposition preconditioners with adaptive spectral bounds. `https://hal.archives-ouvertes.fr/hal-03258644` (2021).
8. R. Haferssas, P. Jolivet and F. Nataf. An additive Schwarz method type theory for Lions's algorithm and a symmetrized optimized restricted additive Schwarz method. *SIAM Journal on Scientific Computing* **39**(4), A1345–A1365 (2017).
9. F. Hecht. New development in FreeFem++. *J. Numer. Math.* **20**(3-4), 251–265 (2012).
10. G. Karypis and V. Kumar. A fast and high quality multilevel scheme for partitioning irregular graphs. *SIAM J. Sci. Comput.* **20**(1), 359–392 (electronic) (1998).
11. A. Klawonn, M. Kuhn and O. Rheinbach. Adaptive coarse spaces for FETI-DP in three dimensions. *SIAM Journal on Scientific Computing* **38**(5), A2880–A2911 (2016).
12. J. Mandel and B. Sousedík. Adaptive selection of face coarse degrees of freedom in the BDDC and the FETI-DP iterative substructuring methods. *Comput. Methods Appl. Mech. Engrg.* **196**(8), 1389–1399 (2007).
13. C. Pechstein and C.R. Dohrmann. A unified framework for adaptive BDDC. *Electron. Trans. Numer. Anal.* **46**(3), 273–336 (2017).
14. N. Spillane. An abstract theory of domain decomposition methods with coarse spaces of the GenEO family. `https://hal.archives-ouvertes.fr/hal-03186276` (2021).
15. N. Spillane, V. Dolean, P. Hauret, F. Nataf, C. Pechstein and R. Scheichl. Abstract robust coarse spaces for systems of PDEs via generalized eigenproblems in the overlaps. *Numer. Math.* **126**(4), 741–770 (2014).
16. N. Spillane and D.J. Rixen. Automatic spectral coarse spaces for robust FETI and BDD algorithms. *Int. J. Numer. Meth. Engng.* **95**(11), 953–990 (2013).
17. M.A. Woodbury. *Inverting modified matrices*. Statistical Research Group (1950).
18. Y. Yu, M. Dryja and M. Sarkis. From Additive Average Schwarz Methods to Non-overlapping Spectral Additive Schwarz Methods. *arXiv preprint* arXiv:2012.13610 (2020).
19. S. Zampini. PCBDDC: a class of robust dual-primal methods in PETSc. *SIAM Journal on Scientific Computing* **38**(5), S282–S306 (2016).

Aitken–Schwarz Heterogeneous Domain Decomposition for EMT-TS Simulation

H. Shourick, D. Tromeur-Dervout, and L. Chedot

1 Introduction

The introduction of renewable energies into the power grid leads to the use of more components based on power electronics which have to be well dimensioned in order not to be damaged by electrical disturbances. These components imply faster dynamics, for power system safety simulations, which cannot be handled by traditional Transient Simulations (TS) with dynamic phasors. Nevertheless, for large power grids, it can be expected that the need of high level details requiring Electro-Magnetic Transient (EMT) modeling will be localized close to disturbances, as other parts of the network still use TS modeling. This paper deals with a proof of concept to develop heterogeneous Schwarz domain decomposition with different modeling (EMT-TS) between the sub-domains. Hybrid (Jacobi type) EMT-TS co-simulation has to face several locks [4]: EMT and TS do not use the same time step size, the transmission of values is also a problem as the solutions do not have the same representation and are subject to some information loss. Our approach doesn't use waveform relaxation [5], and the domain partitioning is not based on cutting the transmission lines [2, 6, 7] as we want to be able to define an overlap between the two representations. On the contrary, we want to use the traditional Schwarz DDM but also where the transmission conditions can lead to divergent DDM. The pure linear convergence/divergence of the linearized problems is then used to accelerate the convergence to the solution by Aitken's technique. In Section 2, we describe the EMT and TS modeling and perform homogeneous Schwarz DDM accelerated by Aitken's acceleration of the convergence technique. Section 3 gives behavior results

H. Shourick and D. Tromeur-Dervout
University of Lyon, UMR5208 U.Lyon1-CNRS, Institut Camille Jordan,
e-mail: damien.tromeur-dervout@univ-lyon1.fr

H. Shourick and L. Chedot
Supergrid-Institute, 14 rue Cyprien, 69200 Villeurbanne,
e-mail: helena.shourick, laurent.chedot@supergrid-institute.com

S. C. Brenner et al. (eds.), *Domain Decomposition Methods in Science and Engineering XXVI*, Lecture Notes in Computational Science and Engineering 145,
https://doi.org/10.1007/978-3-030-95025-5_82

obtained for each modeling. Section 4 describes the heterogeneous EMT-TS DDM and gives first results obtained before concluding in section 5

2 EMT and TS modeling

Simulation of the power grid consists in solving a system of differential algebraic equations (DAE) where the unknowns are currents and voltages. This system is built using the Modified Augmented Nodal Analysis [8], where each component of the grid contributes through relations between currents and voltages and Kirshoff's laws give the algebraical constraints. Let x (respectively y) be the differential (respectively algebraical) unknowns. For the EMT modeling, we have to solve the DAE:

$$F(t, x(t), \dot{x}(t), y(t)) = 0, \text{ with Initial Conditions.} \tag{1}$$

The linearized BDF time discretization of (1) (Backward Euler here) leads to solving the linear system (2) to integrate the state space representation of the DAE from time step t^n to time step t^{n+1} (the operator $\mathbb{I}$ represents the difference between two potentials or the identity for intensity variables, G represents the voltage/intensity sources)

$$\underbrace{\begin{pmatrix} \mathbb{I} - \Delta t A & B \\ C & D \end{pmatrix}}_{\mathbf{H}_{\Delta t}} \begin{pmatrix} x^{n+1} \\ y^{n+1} \end{pmatrix} = \begin{pmatrix} \mathbb{I} & 0 \\ 0 & 0 \end{pmatrix} \begin{pmatrix} x^n \\ y^n \end{pmatrix} + G^{n+1}. \tag{2}$$

For TS modeling the variables are assumed to oscillate with a specific angular frequency $\omega_0 = \dfrac{2\pi}{T}$ (where T is the period) and its selected harmonics taken from a subset $I = \{\ldots, -1, 0, 1, \ldots\}$

$$z(t) = \sum_{k \in I} z_k(t) e^{ik\omega_0 t}, \ z = \{x, y\}. \tag{3}$$

Introducing (3) into (1) leads after simplification (i.e. orthogonality of the functions $e^{ik\omega_0 t}$ with respect to the dot product $[f, g] = \frac{1}{T}\int_t^{t+T} f(z)g(z)dz$) to another DAE system that takes into account the differential property of the dynamic phasor. The resulting DAE system has smoother dynamics. The number of TS variables is then multiplied by the number of harmonics chosen, and the number of equations must be multiplied accordingly.

For example, on the right is the structure of the matrix H_{TS} by choosing two harmonics $k = a$ and $k = c$ and by solving the imaginary and real part separately and with $\mathbf{S}$ the matrix taking into account the differential property of the dynamic phasor modeling.

$$\mathbf{H}_{TS} = \left(\begin{array}{c|c} \begin{array}{c|c} \mathbf{H}_{\Delta T} & -a\,\omega_0\,\mathbf{S} \\ \hline a\,\omega_0\,\mathbf{S} & \mathbf{H}_{\Delta T} \end{array} & 0 \\ \hline 0 & \begin{array}{c|c} \mathbf{H}_{\Delta T} & -c\,\omega_0\,\mathbf{S} \\ \hline c\,\omega_0\,\mathbf{S} & \mathbf{H}_{\Delta T} \end{array} \end{array}\right).$$

Let x_T^{n+1} (respectively x_E^{n+1}) be the algebraic and differential unknowns of TS (respectively EMT) modeling associated to the linear system $H_{TS}x_T^{n+1} = b_T^n$ (respectively $H_E x_E^{n+1} = b_E^n$).

3 EMT and TS Schwarz homogeneous DDM

We consider the linear RLC circuit of Figure 1 to develop the proof of concept of the Schwarz DDM on TS and EMT models.

By adapting the notations of [1], we consider a non-singular matrix $H \in \mathbb{R}^{n \times n}$ having a non-zero pattern and the associated directed graph $G = (W, F)$, where the set of vertices $\Omega = \{1, , n\}$ represents the n unknowns and the set of edges $F = \left\{(i, j)|a_{i,j} \neq 0\right\}$ represents the pairs of vertices that are coupled by a non-zero element in H. Next, we assume that a graph partitioning was applied and resulted in N non-overlapping subsets Ω_i^0 whose union is Ω. Let Ω_i^p be the p-overlap partition of Ω, obtained by including all the vertices immediately neighboring the vertices of Ω_i^{p-1}. Let $R_i^p \in \mathbb{R}^{n_i \times n}$ be the operator which restricts $x \in \mathbb{R}^n$ to the components of x belonging to Ω_i^p. Let $\tilde{R}_i^0 \in \mathbb{R}^{n \times n}$ be the operator which restricts $x \in \mathbb{R}^n$ to the components of x belonging to Ω_i^0 and 0 otherwise. Let $\Omega_{i,e}^p = \Omega_i^{p+1} \backslash \Omega_i^p$ and $R_{i,e}^p \in \mathbb{R}^{n_{i,e} \times n}$ the restriction operator which restricts $x \in \mathbb{R}^n$ to the components of x belonging to $W_{i,e}^p$. By defining $H_i = R_i^p H R_i^{pT}$, $F_i = R_i^p H (R_{i,e}^p)^T$, $x_i = R_i^p x$ and $\mathrm{b}_i = R_i^p b$, $x_{i,e} = R_{i,e}^p x$, then the Restrictive Additive Schwarz (RAS) iteration $k + 1$ to solve $Hx^\infty = b \in \mathbb{R}^n$ is written locally for the Ω_i^p partition

$$x_i^{k+1} = H_i^{-1}(b_i - F_i x_{i,e}^k). \tag{4}$$

The previous paragraph presents the general way of proceeding and among other things to set up the overlap. However, in this work we have chosen another overlap for optimization reasons, because of the small size of our circuit.

The small linear system associated with the RLC circuit is partitioned into two subdomains using graph partitioning without overlap (Figure 2 top) and with an overlap of 1 (Figure 2 bottom). Each subdomain needs two values from the other to solve its equations.

The RAS applied to each time step has a pure linear convergence, i.e. the error operator P does not depend on the RAS iteration.

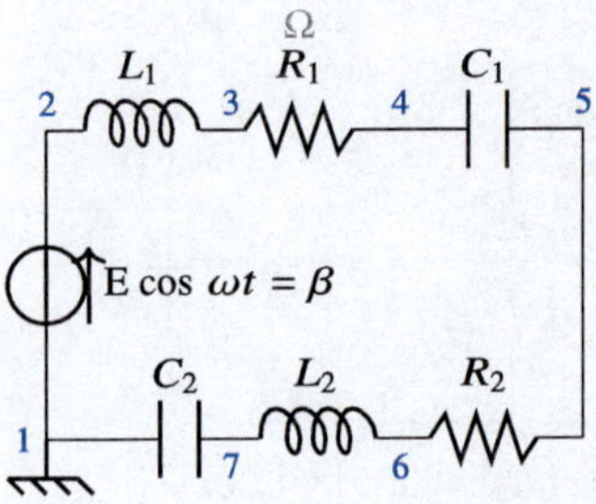

$$v_1 = 0, \quad (5)$$

$$v_2 - v_1 - E - Z_s i_{12} = 0, \quad (6)$$

$$v_3 - v_2 - L_1 \frac{\mathrm{d}i_{23}}{\mathrm{d}t} = 0, \quad (7)$$

$$v_4 - v_3 - R_1 i_{34} = 0, \quad (8)$$

$$C_1\left(\frac{\mathrm{d}v_5}{\mathrm{d}t} - \frac{\mathrm{d}v_4}{\mathrm{d}t}\right) - i_{45} = 0, \quad (9)$$

$$v_6 - v_5 - R_2 i_{56} = 0, \quad (10)$$

$$v_7 - v_6 - L_2 \frac{\mathrm{d}i_{67}}{\mathrm{d}t} = 0, \quad (11)$$

$$C_2\left(\frac{\mathrm{d}v_1}{\mathrm{d}t} - \frac{\mathrm{d}v_7}{\mathrm{d}t}\right) - i_{71} = 0, \quad (12)$$

$$i_{12} - i_{23} = 0, \quad (13)$$

$$i_{23} - i_{34} = 0, \quad (14)$$

$$i_{34} - i_{45} = 0, \quad (15)$$

$$i_{45} - i_{56} = 0, \quad (16)$$

$$i_{56} - i_{67} = 0, \quad (17)$$

$$i_{67} - i_{71} = 0. \quad (18)$$

Fig. 1: Linear RLC circuit and its associated EMT modeling DAE system with $x = \{v_1, i_{23}, v4, v5, i_{67}, v_7\}$ and $y = \{v_2, i_{12}, v_3, i_{34}, i_{45}, i_{56}, v_6, i_{71}\}$. $L_1 = L_2 = 0.7$, $C_1 = C_2 = 1.10^{-6}$, $R_1 = R_2 = 77$, $Z_s = 1.10^{-6}$, $\omega = 2\pi\, 50$, $E = 5$.

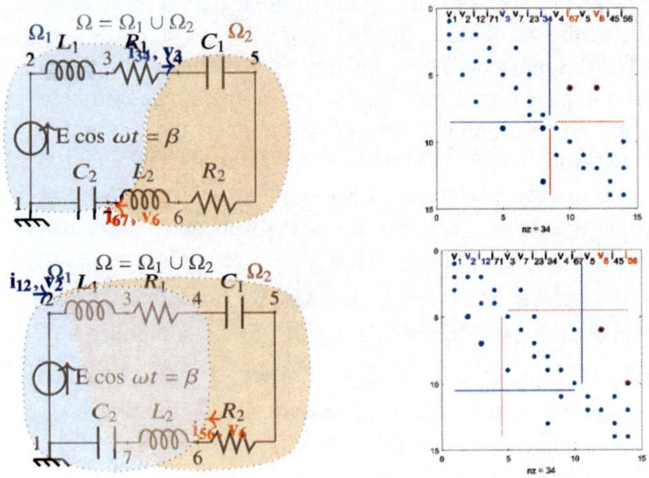

Fig. 2: Graph partitioning of the RLC circuit in two subdomains and the associated matrix partitioning without overlap (top) and with overlap of 1 (bottom).

$$x^{m+1,p+1} - x^{m+1,\infty} = P(x^{m+1,p} - x^{m+1,\infty}). \quad (19)$$

Thus, if it does not stagnate, it can be accelerated with Aitken's acceleration of the convergence, using (19), to obtain the true solution regardless of its convergence or divergence [3]:

$$x^{m+1,\infty} = (I_d - P)^{-1}(x^{m+1,1} - Px^{m+1,0}). \quad (20)$$

$\lambda(P)$	without overlap	with overlap	Schwarz	time step
EMT	$\pm$ 6.0638i	$\pm$ 6.0638i	RAS	2.10^{-4}
TS $_{k=1}$	$\pm0.3667 \pm 6.0635$i	$\pm0.3667 \pm 6.0635$i	RAS	2.10^{-4}
TS $_{k=0}$	±6.0638i	±6.0638i	RAS	2.10^{-4}
TS $_{k=1}$	$\pm$ 0.2114 ±1.1548i	$\pm$ 0.2114 ±1.1548i	RAS	2.10^{-3}
TS $_{k=0}$	±1.1946i	±1.1946i	RAS	2.10^{-3}

Table 1: Larger eigenvalue for P error operator for RAS and EMT modeling ($\Delta t = 2.10^{-4}$), and for RAS and TS $k = 0, 1$ ($\Delta T = 2.10^{-4}$, $\Delta T = 2.10^{-3}$) modeling.

For this small problem it can be directly computed working on the matrix partitioning.

$$P = -[(\tilde{R}_1)^t A_1^{-1} E_{1,e} R_{1,e} + (\tilde{R}_2)^t A_2^{-1} E_{2,e} R_{2,e}]. \tag{21}$$

Table 1 gives the larger eigenvalue in modulus for the P RAS (Restricted Additive Schwarz) error operator for the EMT modeling and for the P RAS error operator for the TS modeling harmonics $k = 0, 1$ applied to the RLC circuit. In both cases (EMT and TS modeling) the eigenvalue modulus is greater than one, so the method diverges. We observe that the overlap does not impact the divergence of the method. The time step increasing from $\Delta T = 2.10^{-4}$ to $\Delta T = 2.10^{-3}$ has a beneficial effect on the TS-TS DDM divergence. Nevertheless, the divergence is purely linear and Aitken's acceleration (20) can be performed after the first iteration if P is known (here by Eq(20)). If P is unknown, the pure linear convergence property also holds for the solution iterated at the global artificial interface $\Gamma = \left\{y \in \mathbb{R}^{n_\Gamma} | y = (x_{1,e}^T, x_{2,e}^T)^T\right\}$. Let R_Γ be the restriction operator from $\mathbb{R}^n$ to Γ, $y^{m+1,p} = R_\Gamma x^{m+1,p}$ be the RAS iterated solution restricted to Γ and $e^{p+1} = y^{m+1,p+1} - y^{m+1,p}$ be the error between two consecutive iterations. Then, from $e^{p+1} = P_\Gamma e^p$, one can build $P_\Gamma = [e^{n_\Gamma+1}, \dots, e^2][e^{n_\Gamma}, \dots, e^1]^{-1}$ with $n_\Gamma + 1$ RAS iterations and the true solution at interface $y^{m+1,\infty}$ is obtained with $y^{m+1,\infty} = (I_d - P_\Gamma)^{-1}(y^{m+1,n_\Gamma+1} - P_\Gamma y^{m+1,n_\Gamma})$. Then one local solve gives $x^{m+1,\infty}$.

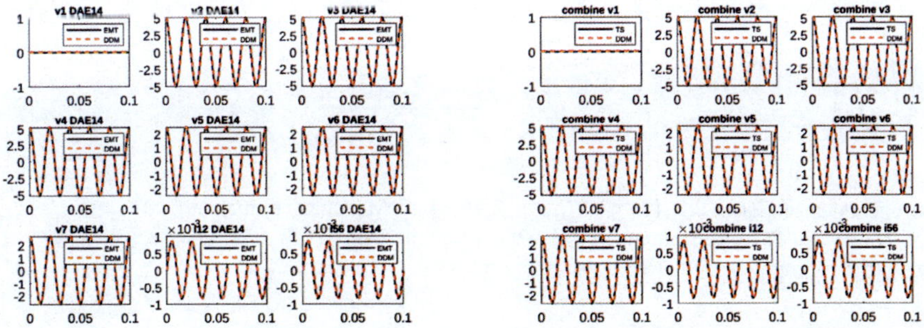

Fig. 3: Homogeneous DDM results comparison with DAE monodomain: (Left) RAS for EMT modeling with $\Delta t_E = 1.10^{-4}$ and (right) RAS for TS modeling with $\Delta t_T = 2.10^{-3}$.

4 Heterogeneous DDM EMT-TS

Our goal is to simulate the electrical network using heterogeneous RAS DDM, using TS modeling to model one part, which can use large time steps ΔT, and using EMT modeling for the other part, which requires smaller time steps Δt as the high oscillations remain.

These two representations TS and EMT of the solution imply having some operators E_{emt}^{TS} (respectively E_{TS}^{emt}) to transfer the solution from the subdomain EMT (respectively TS) to the other TS (respectively EMT). The E_{TS}^{emt} operator needs to compute the fundamental harmonic and other harmonics chosen from the history of the EMT solution. The history time length is one period. This is performed by the FFT of the solution over the time period and keeping the mode corresponding to the chosen harmonics.

The E_{emt}^{TS} operator is simpler as it consists in recombining the TS modes of the solution with the appropriate Fourier basis modes.

Let us consider a linear electrical network with the TS modeling. The time discretisation of the DAE to integrate from T^N to T^{N+1}, assuming that $\Delta T = m\Delta t$ can be written as

$$\underbrace{\begin{pmatrix} \mathbb{I} - \Delta T A_{TS} & B_{TS} \\ C_{TS} & D_{TS} \end{pmatrix}}_{H_{TS}} \underbrace{\begin{pmatrix} x_{TS}^{N+1} \\ y_{TS}^{N+1} \end{pmatrix}}_{w_{TS}^{N+1}} = \underbrace{\begin{pmatrix} \mathbb{I} & 0 \\ 0 & 0 \end{pmatrix}}_{\Theta_{TS}} \begin{pmatrix} x_{TS}^{N} \\ y_{TS}^{N} \end{pmatrix} + \underbrace{\begin{pmatrix} E_{TS}^{A} & E_{TS}^{B} \\ E_{TS}^{C} & E_{TS}^{D} \end{pmatrix}}_{E_{TS}^{emt}} \begin{pmatrix} x_{emt}^{m} \\ y_{emt}^{m} \end{pmatrix} + G_{TS}^{N+1}.$$

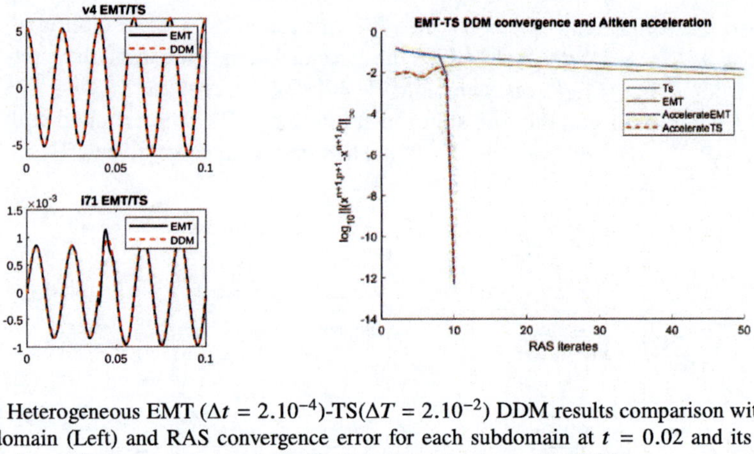

Fig. 4: Heterogeneous EMT ($\Delta t = 2.10^{-4}$)-TS($\Delta T = 2.10^{-2}$) DDM results comparison with DAE monodomain (Left) and RAS convergence error for each subdomain at $t = 0.02$ and its Aitken acceleration with P_Γ computed numerically from 9 iterates ($n_\Gamma = 8$) (right).

Similarly one time step for the EMT side to integrate from t^n to t^{n+1} can be written as

$$\underbrace{\begin{pmatrix} \mathbb{I}-\Delta t A_{emt} & B_{emt} \\ C_{emt} & D_{emt} \end{pmatrix}}_{H_{emt}} \underbrace{\begin{pmatrix} x^{n+1}_{emt} \\ y^{n+1}_{emt} \end{pmatrix}}_{w^{n+1}} = \underbrace{\begin{pmatrix} \mathbb{I} & 0 \\ 0 & 0 \end{pmatrix}}_{\Theta_{emt}} \begin{pmatrix} x^{n}_{emt} \\ y^{n}_{emt} \end{pmatrix} + \underbrace{\begin{pmatrix} E^A_{emt} & E^B_{emt} \\ E^C_{emt} & E^D_{emt} \end{pmatrix}}_{E^{TS}_{emt}} \underbrace{\begin{pmatrix} x^{N+1}_{TS}(t^{n+1}) \\ y^{N+1}_{TS}(t^{n+1}) \end{pmatrix}}_{W^{N+1}(t^{n+1})} + G^{n+1}_{emt}.$$

The m time steps can be gathered into one larger system considering $t^n = T^N$

$$\underbrace{\begin{pmatrix} I & & & \\ -\Theta_{emt} & H_{emt} & & \\ & \ddots & \ddots & \\ & & -\Theta_{emt} & H_{emt} \\ & & & -\Theta_{emt} \;\; H_{emt} \end{pmatrix}}_{\mathbb{H}_{emt}} \underbrace{\begin{pmatrix} w^n_{emt} \\ w^{n+1}_{emt} \\ \vdots \\ w^{n+m-1}_{emt} \\ w^{n+m}_{emt} \end{pmatrix}}_{\mathbb{W}_{emt}} =$$

$$\underbrace{\begin{pmatrix} I & & & \\ & E^{TS}_{emt} & & \\ & & \ddots & \\ & & & E^{TS}_{emt} \\ & & & & E^{TS}_{emt} \end{pmatrix}}_{\mathbb{E}^{TS}_{emt}} \underbrace{\begin{pmatrix} (x^n, y^n)^t \\ W^{N+1}(t^{n+1}) \\ \vdots \\ W^{N+1}(t^{n+m-1}) \\ W^{N+1}(t^{n+m}) \end{pmatrix}}_{\mathbb{W}^{N+1}_{TS}} + \underbrace{\begin{pmatrix} 0 \\ G^{n+1}_{emt} \\ \vdots \\ G^{n+m-1}_{emt} \\ G^{n+m}_{emt} \end{pmatrix}}_{\mathbb{G}^{N+1}_{emt}}. \tag{22}$$

This system needs the values that the TS solution connected to the EMT part has taken on the small time steps. The two domains are connected via the connected or flowing variables. Since these variables should be the solution at time T^{N+1}, we need the Schwarz iterative algorithm to obtain the exact values. We then iterate the iteration $p+1$ by taking the connected values, at the iteration p, from the other subdomain. We can use the multiplicative form or the additive form as follows:

$$\begin{cases} H_{TS}\, w^{N+1,\mathbf{p}+1}_{TS} = \Theta_{TS}\, w^{N}_{TS} + E^{emt}_{TS}\, w^{m,\mathbf{p}}_{emt} + G^{N+1}_{TS}, \\ \mathbb{H}_{emt}\, \mathbb{W}^{N+1,\mathbf{p}+1}_{emt} = \mathbb{E}^{TS}_{emt}\, \mathbb{W}^{N+1,\mathbf{p}}_{TS} + \mathbb{G}^{N+1}_{emt}. \end{cases} \tag{23}$$

Figure 4 (left) shows the solutions v_4 EMT and i_{71} TS of heterogeneous DDM EMT ($\Delta t = 2.10^{-4}$) -TS ($\Delta T = 2.10^{-2}$) compared with the DAE solution on the monodomain. We proceed to a jump in amplitude at $t = 0.04$ for the voltage source. Figure 4 (right) gives the $\log_{10}$ of the error between two consecutive RAS iterates at $t = 0.02$. It shows a linear convergence behavior and can therefore be accelerated by Aitken's acceleration of the convergence technique after the 9 iterates needed to numerically construct the error operator P_Γ.

5 Conclusion

A Schwarz heterogeneous DDM was used to co-simulate an RLC electrical circuit where a part of the domain is modeled with EMT modeling and the other part with TS modeling. We showed the convergence/divergence property of the homogeneous DDM EMT-EMT and TS-TS and of the heterogeneous DDM TS-EMT, with or without overlap, and we used the pure linear divergence/convergence of the method to accelerate it toward the true solution with Aitken's acceleration of the convergence technique. The domain partitioning is only based on connectivity considerations since we want, in the long term, for the electrical network, to take advantage of the two TS and EMT representations on the overlap in order to identify the loss of information between the two models. We would like then to use this knowledge to work on transmission conditions other that the Dirichlet condition to conserve some invariants such as electrical power.

References

1. X. Cai and M. Sarkis. A restricted additive schwarz preconditioner for general sparse linear systems. *SIAM J. Sci. Comput.* **21**, 792–797 (1999).
2. M.J. Gander, M. Al-Khaleel and A.E. Ruchli. Optimized Waveform Relaxation Methods for Longitudinal Partitioning of Transmission Lines. *IEEE Transactions on Circuits and Systems I: Regular Papers* **56**(8), 1732–1743 (2009).
3. M. Garbey and D. Tromeur-Dervout. On some Aitken-like acceleration of the Schwarz method. *International Journal for Numerical Methods in Fluids* **40**(12), 1493–1513 (2002).
4. V. Jalili-Marandi, V. Dinavahi, K. Strunz, J.A. Martinez and A. Ramirez. Interfacing Techniques for Transient Stability and Electromagnetic Transient Programs IEEE Task Force on Interfacing Techniques for Simulation Tools. *IEEE Transactions on Power Delivery* **24**(4), 2385–2395 (2009).
5. E. Lelarasmee, A. Ruehli and A. Vincentelli. The Waveform Relaxation Method for Time-Domain Analysis of Large Scale Integrated Circuits. *IEEE Transactions on Computer-Aided Design of Integrated Circuits and Systems* **1**, 131–145 (1982).
6. K. Mudunkotuwa and S. Filizadeh. Co-simulation of electrical networks by interfacing EMT and dynamic-phasor simulators. *Electric Power Systems Research* **163**, 423–429 (2018).

7. F. Plumier, P. Aristidou, Ch. Geuzaine and Th. Van Cutsem. Co-Simulation of Electromagnetic Transients and Phasor Models: A Relaxation Approach. *IEEE Transactions on Power Delivery* **31**(5), 2360–2369 (2016).
8. L. Wedepohl and L. Jackson. Modified nodal analysis: an essential addition to electrical circuit theory and analysis. *Engineering Science and Education Journal* **11**(3), 84–92 (2002).

Parareal Schwarz Waveform Relaxation Method for the Time-Periodic Parabolic Problem

Bo Song, Yao-Lin Jiang, and Kang-Li Xu

1 Introduction and model problem

Time-periodic problems appear typically in special physical situations, for example in eddy current simulations [1], or when periodic forcing is used, like for periodically forced reactors, see [14, 15]. The numerical simulation of time-periodic problems is a special area of research, since the time periodicity modifies the problem structure and solution methods significantly. When the scale of the problems increases, it is desirable to use parallel methods to solve such problems.

For the time-dependent problems, Schwarz waveform relaxation algorithms are parallel algorithms based on a spatial domain decomposition [10]. More recently, time-parallel methods were also considered to increase the parallelism in time [5], i.e., the parareal method proposed by Lions, Maday, and Turinici in the context of virtual control to solve evolution problems in parallel; see [12]. Two parareal algorithms for time-periodic problems were proposed in [9]: one with a periodic coarse problem (PP-PC), and one with a non-periodic coarse problem (PP-IC). Further, based on these two algorithms, new applications and parallel methods for time-periodic problems were also considered; see [2, 11].

The paper [13] was the first time that Schwarz waveform relaxation and the parareal algorithm were combined. Further, in [7], a new parallel algorithm named the parareal Schwarz waveform relaxation algorithm (PSWR) was introduced, where there is no order between the Schwarz waveform relaxation algorithm and the parareal

Bo Song
Corresponding author. School of Mathematics and Statistics, Northwestern Polytechnical University, Xi'an 710072, China, e-mail: bosong@nwpu.edu.cn

Yao-Lin Jiang
School of Mathematics and Statistics, Xi'an Jiaotong University, Xi'an 710049, China, e-mail: yljiang@mail.xjtu.edu.cn

Kang-Li Xu
School of Mathematics and Statistics, Xi'an Jiaotong University, Xi'an 710049, China, e-mail: klxuqd@163.com

S. C. Brenner et al. (eds.), *Domain Decomposition Methods in Science and Engineering XXVI*, Lecture Notes in Computational Science and Engineering 145,
https://doi.org/10.1007/978-3-030-95025-5_83

algorithm. A superlinear convergence estimate of this algorithm has been provided in [8]. Recently, a new space-time algorithm which uses the optimized Schwarz waveform relaxation algorithm as the inner iteration of the parareal algorithm was also provided [4].

In this work, we consider a new PSWR algorithm for the following time-periodic parabolic problem

$$\begin{aligned} \frac{\partial u}{\partial t} &= \mathcal{L}u + f && \text{in } \Omega \times (0,T), \\ u(x,0) &= u(x,T) && \text{in } \Omega, \\ u &= g && \text{on } \partial\Omega \times (0,T), \end{aligned} \tag{1}$$

where $\mathcal{L}$ is the Laplace operator, $f(x,0) = f(x,T)$, $g(x,0) = g(x,T)$, and $\Omega \subset \mathbb{R}^d$, $d = 1,2,3$.

2 PSWR for time-periodic parabolic problem

We first introduce a parareal algorithm for time-periodic problems [7]. We decompose the time interval $[0,T]$ into N subintervals $[T_n, T_{n+1}]$, $n = 0, 1, \ldots, N-1$, with $0 = T_0 < T_1 < \cdots < T_{N-1} < T_N = T$. We define the so-called coarse propagator $G(T_{n+1}, T_n, U_n, f, g)$, which provides a rough approximation in time of the solution $u_n(x, T_{n+1})$ of (2)

$$\begin{aligned} \frac{\mathrm{d}u_n}{\mathrm{d}t} &= \mathcal{L}u_n + f \text{ in } \Omega \times (T_n, T_{n+1}), \\ u_n(x,T_n) &= U_n(x) \text{ in } \Omega, && (2) \\ u_n &= g \text{ on } \partial\Omega \times (T_n, T_{n+1}). && (3) \end{aligned}$$

with a given initial condition $u_n(x, T_n) = U_n(x)$, right-hand side source term f and boundary condition g. We also define a fine propagator $F(T_{n+1}, T_n, U_n, f, g)$, which gives a more accurate approximation in time of the same solution of (2).

Then starting with an initial guess U_n^0 at the coarse time points $T_0, T_1, T_2, \ldots, T_{N-1}$, e.g., solving the model problem on the coarse time points, the periodic parareal algorithm with initial-value coarse problem (PP-IC) for the time-periodic problem (1) performs for $k = 0, 1, 2, \ldots$ the correction iteration

$$\begin{aligned} U_0^{k+1} &= U_N^k, \\ U_{n+1}^{k+1} &= F(T_{n+1}, T_n, U_n^k, f, g) + G(T_{n+1}, T_n, U_n^{k+1}, f, g) - G(T_{n+1}, T_n, U_n^k, f, g), \\ &\quad n = 0, 1, \ldots, N-1. \end{aligned} \tag{4}$$

Furthermore, we introduce the Schwarz waveform relaxation algorithm for the model problem (1) based on a spatial decomposition only, in the most general case into overlapping subdomains $\Omega = \cup_{i=1}^{I} \Omega_i$. The Schwarz waveform relaxation algorithm solves iteratively for $k = 0, 1, 2, \dots$ the space-time subdomain problems

$$\begin{aligned} \frac{\partial u_i^{k+1}}{\partial t} &= \mathcal{L} u_i^{k+1} + f && \text{in } \Omega_i \times (0, T), \\ u_i^{k+1}(x, 0) &= u_0 && \text{in } \Omega_i, \\ \mathcal{B}_i u_i^{k+1} &= \mathcal{B}_i \bar{u}^k && \text{on } \partial\Omega_i \times (0, T). \end{aligned}$$

Here $\bar{u}^k$ denotes a composed approximate solution from the previous subdomain solutions u_i^k using for example a partition of unity, and an initial guess $\bar{u}^0$ is needed to start the iteration. The operators $\mathcal{B}_i$ are transmission operators: in the case of the identity, it will be a Dirichlet transmission condition and we have the classical Schwarz waveform relaxation algorithm; for Robin or higher order transmission conditions, we obtain an optimized Schwarz waveform relaxation algorithm, if the parameters in the transmission conditions are chosen to optimize the convergence of the algorithm.

Finally, according to the reference [8], which designed the PSWR algorithm for parabolic problems, we construct here PSWR for the time-periodic parabolic problem (1). We decompose the spatial domain Ω into I overlapping subdomains $\Omega = \cup_{i=1}^{I} \Omega_i$, and the time interval $(0, T)$ is divided into N time subintervals (T_n, T_{n+1}) with $0 = T_0 < T_1 < \cdots < T_N = T$. Therefore we can get a sequence of space-time subdomains $\Omega_{i,n} = \Omega_i \times (T_n, T_{n+1})$, $i = 1, 2, \dots, I$, $n = 0, \dots, N-1$.

Like in the parareal algorithm, we introduce a fine subdomain solver $F_{i,n}(U_{i,n}^k, \mathcal{B}_i \bar{u}_n^k)$ and a coarse subdomain solver $G_{i,n}(U_{i,n}^k, \mathcal{B}_i \bar{u}_n^k)$, where, to avoid further increasing the complexity of the notation, we do not explicitly state the dependence of these solvers on the time interval, the right-hand side f and the original Dirichlet boundary condition g. There is also a further important notational difference with parareal: here the fine solver F returns the entire solution in space-time, not just at the final time, since this solution is also needed in the transmission conditions of the algorithm. Then for any initial guess of the initial values $U_{i,n}^0$ and the interface values $\mathcal{B}_i u_n^0$, a new PSWR algorithm (named PSWR-IC) for the time-periodic parabolic problem (1) computes for iteration index $k = 0, 1, 2, \dots$ and all spatial and time indices $i = 1, 2, \dots, I$, $n = 0, 1, \dots, N-1$:

Step I: Use the more accurate evolution operator to calculate

$$u_{i,n}^{k+1} = F_{i,n}(U_{i,n}^k, \mathcal{B}_i \bar{u}_n^k).$$

Step II: Update new initial conditions using a parareal step both in space and time for $n = 0, 1, \dots, N-1$

$$U_{i,n+1}^{k+1} = u_{i,n}^{k+1}(\cdot, T_{n+1}) + G_{i,n}(U_{i,n}^{k+1}, \mathcal{B}_i \bar{u}_n^{k+1}) - G_{i,n}(U_{i,n}^k, \mathcal{B}_i \bar{u}_n^k).$$

Step III: Update initial conditions at $t = 0$: $U_{i,0}^{k+1} = U_{i,N}^{k}$.

Here $\bar{u}_n^k$ is a composed approximate solution from the subdomain solutions $u_{i,n}^k$ using for example a partition of unity, e.g., $\bar{u}_n^k = u_{i,n}^k$ in $\Omega_{i,n} \cup_{j=1,j\neq i}^{I} \setminus (\Omega_{i,n} \cap \Omega_{j,n})$, and $\bar{u}_n^k$ is the average value of $u_{i,n}^k$ and $u_{j,n}^k$ in the overlap $\Omega_{i,n} \cap \Omega_{j,n}$, $j = 1, 2, \ldots, I$ and $j \neq i$. An initial guess $\bar{u}_n^0$ and $U_{i,n}^0$ is needed to start the iteration (the latter can for example be computed by a time-periodic problem on the coarse subdomain using the coarse propagator once the former is chosen). Note that the first step in the proposed PSWR-IC algorithm, which is the expensive step involving the fine propagator $F_{i,n}$, can be performed in parallel over all space-time subdomains $\Omega_{i,n}$, since both the initial and boundary data are available from the previous iteration. The cheap second step in the proposed PSWR-IC algorithm involving only the coarse propagator $G_{i,n}$ to compute a new initial condition for most space-time subdomains on $T_1, T_2, \ldots, T_{N-1}$, is still in parallel in space, but now sequential in time, like in the parareal algorithm. In step III, we use the idea of the PP-IC algorithm in [7] to update the initial condition at $t = 0$, which is a relaxation of $U_{i,0}^{k+1} = U_{i,N}^{k+1}$, avoiding solving a coupled system on the time coarse points T_i.

We have the following convergence result for the PSWR-IC algorithm.

Theorem 1 *If the fine propagator F is the exact solver, and the coarse propagator G is Backward Euler, then PSWR-IC with Dirichlet transmission conditions and overlap L in the two subdomain case for the* 1*-dimensional heat equation converges linearly on bounded time intervals* $(0, T)$.

The proof is technical (see [16]). For an illustration, see Section 3.

3 Numerical experiments

To investigate numerically how the convergence of the PSWR-IC algorithm for time-periodic problems depends on the various parameters in the space-time decomposition, we use the following time-periodic 1-dimensional model problem

$$\begin{aligned} \frac{\partial u(x,t)}{\partial t} &= \frac{\partial^2 u(x,t)}{\partial x^2} + f(x,t) && (x,t) \in \Omega \times (0,1), \\ u(x,t) &= 0 && (x,t) \in \partial\Omega \times (0,1), \\ u(x,0) &= u(x,T) && x \in \Omega, \end{aligned} \tag{5}$$

where the domain $\Omega = (0, 3)$, and the exact solution of the model problem is $u = x(x-3)\sin(2\pi t)$. The model problem (5) is discretized by a second-order centered finite difference scheme with mesh size $h = 3/128$ in space and by the Backward Euler method with $\Delta t = 1/100$ in time. The time interval is divided into N time subintervals, while the domain Ω is decomposed into I equal spatial subdomains with overlap L. We define the relative error of the infinity norm of

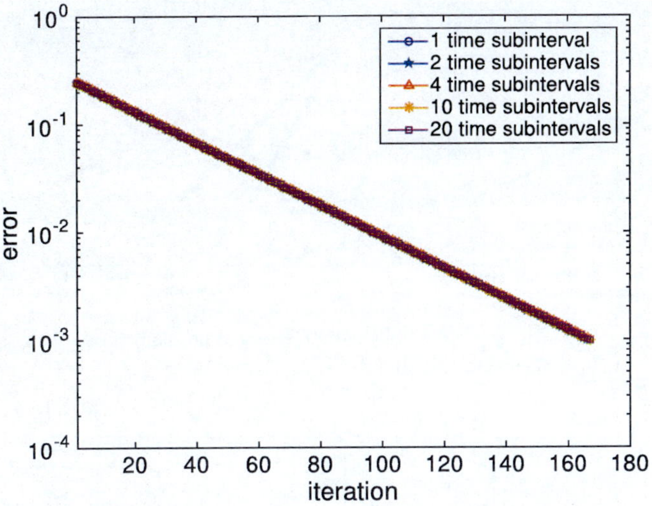

Fig. 1: Dependence of the PSWR-IC algorithm for the time-periodic problem (5) on the number of time subintervals.

the errors along the interface and initial time in the space-time subdomains as the iterative error of our new algorithm.

We decompose the domain Ω into two spatial subdomains with overlap $L = 2h$. The total time interval length is $T = 1$. We show in Figure 1 on the left the convergence of the PSWR-IC algorithm when the number of time subintervals equals 1 (classical Schwarz waveform relaxation for time-periodic problems), 2, 4, 10, and 20. This shows that the convergence of the PSWR-IC algorithm does indeed not depend on the number of time subintervals, which is the same as the PSWR algorithm for the initial value problem. Here we also observe that the PSWR-IC algorithm converges linearly, in contrast to the superlinear convergence of the PSWR algorithm for the initial value problem.

We next study the dependence on the overlap. We use $L = 2h$, $4h$, $8h$ and $16h$, and divide the time interval $(0, T)$ with $T = 1$ into 10 time subintervals, still using the same two subdomain decomposition of Ω as before. We see on the left in Figure 2 that increasing the overlap substantially improves the convergence speed of the algorithm. However, this also increases the cost of the method, since bigger subdomain problems need to be solved.

We then investigate numerically if the convergence result we derived for two subdomains also holds for the case of many subdomains. We decompose the domain Ω into 2, 4, and 8 spatial subdomains, keeping again the overlap $L = 2h$. For each case, we divide the time interval $(0, T)$ with $T = 1$ into 10 time subintervals. We see in Figure 2 on the right that using more spatial subdomains makes the algorithm converge more slowly, like the PSWR algorithm for the initial value problem.

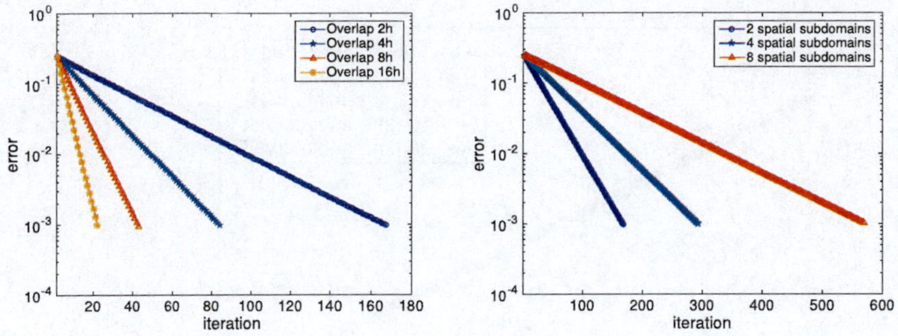

Fig. 2: Dependence of the PSWR-IC algorithm for the time-periodic problem (5) on the overlap (left), and on the number of spatial subdomains (right).

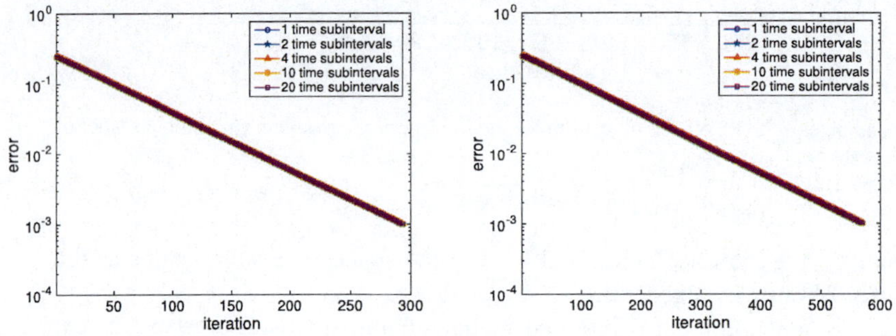

Fig. 3: Independence of the PSWR-IC algorithm on the number of time subintervals for four spatial subdomains (left), and eight spatial subdomains (right).

We further investigate whether the convergence of the algorithm still does not depend on the number of time subintervals for the case of many subdomains. We see in Figure 3 that the convergence behavior for four spatial subdomains (left), and eight spatial subdomains (right) is the same as the convergence behavior for two spatial subdomains.

Finally, we compare the convergence behavior of the PSWR-IC algorithm for the time-periodic problem (5) with Dirichlet and optimized transmission conditions. Using optimized transmission conditions leads to much faster, so-called optimized Schwarz waveform relaxation methods, see for example [3, 6]. We divide the time interval $(0, T)$ with $T = 1$ into 10 time subintervals, and the domain Ω is decomposed into 2, 4 and 8 spatial subdomains. We use first-order transmission conditions and choose for the parameters $p = 1$, $q = 1.75$ (for the terminology, see [3]), which is the same as optimized Schwarz waveform relaxation and optimized PSWR for the initial value problem. The corresponding convergence curves in Figure 4 show that using the optimized transmission conditions of these parameters may not even lead to convergence. Then we chose numerically optimized parameters $p = 10.5$, $q = 0$, which leads to a substantially better performance of the PSWR-IC algorithm, even

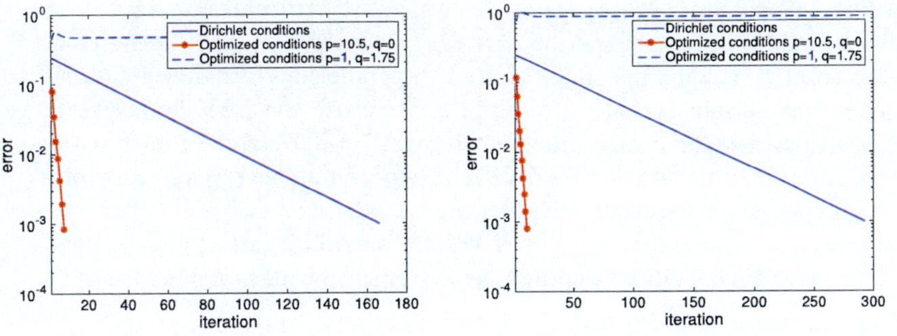

Fig. 4: Comparison of the PSWR-IC algorithm with Dirichlet and optimized transmission conditions for two spatial subdomains (left) and four spatial subdomains (right).

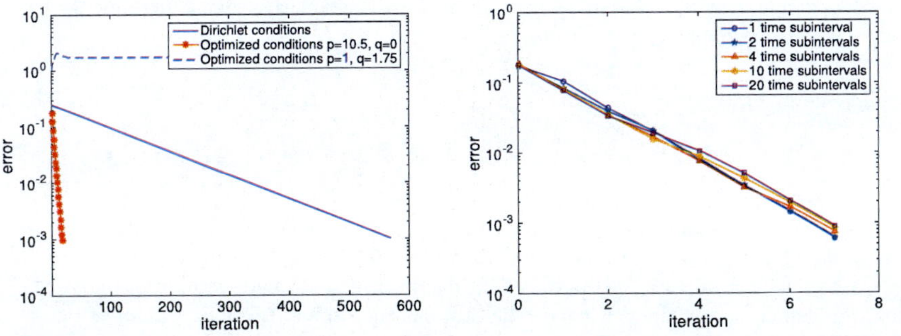

Fig. 5: Left: comparison of the PSWR-IC algorithm with Dirichlet and optimized transmission conditions for eight spatial subdomains. Right: dependence of the PSWR-IC algorithm on the number of time subintervals with optimized transmission conditions.

better than very generous overlap, and this at no additional cost. We also investigate the dependence on the number of time subintervals (on the right in Figure 5), where we choose the problem configuration as in the case of the Dirichlet transmission conditions in Figure 1. We observe that convergence is much faster with optimized transmission conditions (less than 10 iterations instead of over 100), and convergence is still linear, indicating that there is a different convergence mechanism dominating now, due to the optimized transmission conditions.

4 Conclusions

We designed a new parareal PSWR algorithm for time-periodic problems, i.e., the PSWR-IC algorithm. This algorithm is based on a domain decomposition of the entire space-time domain into smaller space-time subdomains, i.e., the decomposition is both in space and in time. The new algorithm iterates on these space-time subdomains using two different updating mechanisms: the Schwarz waveform relaxation approach for boundary condition updates, and the parareal mechanism for initial condition updates. All space-time subdomains are solved in parallel, both in space and in time. For the time-periodic problem, in particular, we use the periodic parareal algorithm with initial-value coarse problem to update initial condition at $t = 0$. The numerical results illustrate that the PSWR-IC algorithm converges linearly on bounded time intervals when using Dirichlet transmission conditions in space, in contrast to the superlinear convergence of PSWR for the initial value problem, and optimized transmission conditions improve the convergence behavior significantly.

Acknowledgements This work was supported by the Natural Science Foundation of China (NSFC) under grant 11801449, 11871393, the International Science and Technology Cooperation Program of Shaanxi Key Research & Development Plan under grant 2019KWZ-08, the Natural Science Basic Research Plan in Shaanxi Province of China (Program No. 2019JQ-617), China Postdoctoral Science Foundation under grant 2020M673465, and the Fundamental Research Funds for the Central Universities under grant G2018KY0306.

References

1. F. Bachinger, U. Langer and J. Schöberl. Efficient solvers for nonlinear time-periodic eddy current problems. *Computing and Visualization in Science* **9**(4), 197–207 (2006).
2. D. Bast, I. Kulchytska-Ruchka, S.Schöps and O.Rain. Accelerated steady-state torque computation for induction machines using parallel-in-time algorithms. *IEEE Transactions on Magnetics* **56**(2), 1–9 (2020).
3. D. Bennequin, M.J. Gander and L. Halpern. A homographic best approximation problem with application to optimized Schwarz waveform relaxation. *Mathematics of Computation* **78**(265), 185–223 (2009).
4. D.Q. Bui, C. Japhet, Y. Maday and P. Omnes. Coupling parareal with optimized Schwarz waveform relaxation for parabolic problems. `hal.inria.fr/hal-03167571` (2021).
5. M.J. Gander. 50 years of time parallel time integration. In: *Multiple Shooting and Time Domain Decomposition Methods*, pp. 69–113, Springer (2015).
6. M.J. Gander and L. Halpern. Optimized Schwarz waveform relaxation methods for advection reaction diffusion problems. *SIAM J. Numer. Anal.* **45**(2), 666–697 (2007).
7. M.J. Gander, Y.L. Jiang and R.J. Li. Parareal Schwarz waveform relaxation methods. In: *Domain Decomposition Methods in Science and Engineering XX*, pp. 451–458, Springer (2013).
8. M.J. Gander, Y.L. Jiang and B. Song. A superlinear convergence estimate for the parareal Schwarz waveform relaxation algorithm. *SIAM Journal on Scientific Computing* **41**(2), A1148–A1169 (2019).
9. M.J. Gander, Y.L. Jiang, B. Song and H. Zhang. Analysis of two parareal algorithms for time-periodic problems. *SIAM Journal on Scientific Computing* **35**(5), A2393–A2415 (2013).
10. M.J. Gander and A.M. Stuart. Space-time continuous analysis of waveform relaxation for the heat equation. *SIAM Journal on Scientific Computing* **19**(6), 2014–2031 (1998).
11. I. Kulchytska-Ruchka and S. Schöps. Efficient parallel-in-time solution of time-periodic problems using a multiharmonic coarse grid correction. *SIAM Journal on Scientific Computing* **43**(1), C61–C88 (2021).
12. J.L. Lions, Y. Maday and G. Turinici. A "parareal" in time discretization of PDE's. *Comptes Rendus de l'Académie des Sciences-Series I-Mathematics* **332**(7), 661–668 (2001).

13. Y. Maday and G. Turinici. The parareal in time iterative solver: a further direction to parallel implementation. *Lecture Notes in Computational Science and Engineering* **40**, 441–448 (2005).
14. T.L. van Noorden, S.V. Lunel and A. Bliek. The efficient computation of periodic states of cyclically operated chemical processes. *IMA Journal of Applied Mathematics* **68**(2), 149–166 (2003).
15. B.A. van de Rotten, S.M.V. Lunel and A. Bliek. Efficient simulation of periodically forced reactors with radial gradients. *Chemical engineering science* **61**(21), 6981–6994 (2006).
16. B. Song and Y.L. Jiang. Parareal Schwarz waveform relaxation method for the heat equation. In preparation (2021).

Acceleration of the Convergence of the Asynchronous RAS Method

D. Tromeur-Dervout

1 Introduction

Nowadays high performance computers have several thousand cores and more and more complex hierarchical communication networks. For these architectures, the use of a global reduction operation such as the dot product involved in the GMRES acceleration can be a bottleneck for the performance. In this context domain decomposition solvers with local communications are becoming particularly interesting. Nevertheless, the probability of temporarily failures/unavailabilty of a set of processors/clusters is non-zero, which leads to the need for fault tolerant algorithms such as asynchronous Schwarz-type methods. With the asynchronism the transmission conditions (TC) at artificial interfaces generated by the domain decomposition may not have been updated for some subdomains and for some iterations. The message passing interface MPI-3 standard provides a one-sided communication protocol where a process can directly write on the local memory of an another process without synchronizing. This can also occur in the OpenMP implementation. For asynchronous methods, it is very difficult to know if the update has been performed and most papers fail to give the level of asynchronism in their implementation results.

From the numerical point of view, this asynchronism affects the linear operator of the interface problem. In this context Aitken's acceleration of convergence should not be applicable as it is based on the pure linear convergence of the DDM [6, 10, 11], i.e. there exists a linear operator P independent of the iteration that connects the error at the artificial interfaces of two consecutive iterations. This paper focuses on Aitken's acceleration of the convergence of the asynchronous Restricted Additive Schwarz (RAS) iterations. We develop a mathematical model of the Asynchronous RAS allowing us to set the percentage of the number of randomly chosen local artificial interfaces where transmission conditions are not updated. Then we show

D. Tromeur-Dervout
University of Lyon, UMR5208 U.Lyon1-CNRS, Institut Camille Jordan,
e-mail: Damien.tromeur-dervout@univ-lyon1.fr

S. C. Brenner et al. (eds.), *Domain Decomposition Methods in Science and Engineering XXVI*, Lecture Notes in Computational Science and Engineering 145,
https://doi.org/10.1007/978-3-030-95025-5_84

how this ratio deteriorates the convergence of the Asynchronous RAS and how some regularization techniques on the traces of the iterative solutions at artificial interfaces allow us to accelerate the convergence to the true solution.

The plan of the paper is the following. Section 2 gives the notation and the principles of the Aitken–Schwarz method using some low-rank approximation of the interface error operator. Section 3 presents the modeling of the asynchronous RAS on a 2D Poisson problem allowing us to define the level of asynchronism. Section 4 presents the results of the acceleration with respect to the level of asynchronism and the enhancement of this acceleration with regularisation techniques before concluding in section 5.

2 Aitken–Schwarz method principles

By adapting the notations of [3], we consider a non-singular matrix $A \in \mathbb{R}^{n\times n}$ having a non-zero pattern and the associated graph $G = (W, F)$, where the set of vertices $W = \{1, \ldots, n\}$ represents the n unknowns and the edge set $F = \left\{(i, j)|a_{i,j} \neq 0\right\}$ represents the pairs of vertices that are coupled by a nonzero element in A. Then we assume that a graph partitioning has been applied and has resulted in N nonoverlapping subsets W_i^0 whose union is W. Let W_i^p be the p-overlap partition of W, obtained by including all the immediate neighboring vertices of the vertices from W_i^{p-1}. Let $W_{i,e}^p = W_i^{p+1} \backslash W_i^p$. Then let $R_i^p \in \mathbb{R}^{n_i \times n}$ ($R_{i,e}^p \in \mathbb{R}^{n_{i,e}\times n}$ and $\tilde{R}_i^0 \in \mathbb{R}^{n_i\times n}$ respectively) be the operator that restricts $x \in \mathbb{R}^n$ to the components of x belonging to W_i^p ($W_{i,e}^p$ and W_i^0 respectively, and the operator $\tilde{R}_i^0 \in \mathbb{R}^{n_i\times n}$ assigns 0 to those unknowns belonging to $W_i^p\backslash W_i^0$). We define the operators $A_i = R_i^p A R_i^{pT}$ and $E_i = R_i^p A R_{i,e}^{pT}$, the vectors $x_i = R_i^p x$, $b_i = R_i^p b$, and $x_{i,e} = R_{i,e}^p x$, then the RAS iteration $k + 1$ reads locally for the partition W_i^p:

$$x_i^{k+1} = A_i^{-1}(b_i - E_i x_{i,e}^k). \tag{1}$$

By defining $M_{RAS}^{-1} \overset{\text{def}}{=} \sum_{i=0}^{N-1} \tilde{R}_i^{0T} A_i^{-1} R_i^p$ and adding the contribution of each partition W_i^p, RAS can be viewed as a Richardson's process:

$$\sum_{i=0}^{N-1} \tilde{R}_i^{0T} R_i^p x^{k+1} = \sum_{i=0}^{N-1} \tilde{R}_i^{0T} A_i^{-1} R_i^p b - \sum_{i=0}^{N-1} \tilde{R}_i^{0T} A_i^{-1} R_i^p A R_{i,e}^{pT} x^k, \tag{2}$$

$$x^{k+1} = M_{RAS}^{-1} b - M_{RAS}^{-1} A x^k + x^k = x^k + M_{RAS}^{-1}(b - Ax^k). \tag{3}$$

The Richardson's process (3) is deduced from (2) (see [5, Theorem 3.7]) by using the property $R_i^p A = R_i^p A(R_i^{pT} R_i^p + R_{i,e}^{pT} R_{i,e}^p)$. It can be reduced to a problem with the unknowns on the interface (see [12, eq. (2.12) and (2.13)]).

The restriction of (3) to the interface $\Gamma = \left\{W^p_{0,e}, \ldots, W^p_{N-1,e}\right\}$ of size $n_\Gamma = \sum_{i=0}^{N-1} n_{i,e}$, by defining $R_\Gamma = (R^p_{0,e}, \ldots, R^p_{N-1,e})^T \in \mathbb{R}^{n_\Gamma \times n}$ and by using the property $R^{pT}_{i,e} R^p_{i,e} R^T_\Gamma R_\Gamma = R^{pT}_{i,e} R^p_{i,e}$, reads as

$$\underbrace{R_\Gamma x^{k+1}}_{y^{k+1}} = \underbrace{R_\Gamma \left(I - M^{-1}_{RAS} A\right) R^T_\Gamma}_{P} \underbrace{R_\Gamma x^k}_{y^k} + \underbrace{R_\Gamma M^{-1}_{RAS} b}_{c} . \tag{4}$$

The pure linear convergence of the RAS at the interface given by $y^k - y^\infty = P(y^{k-1} - y^\infty)$ (the error operator P does not depend on the iteration k) allows us to apply Aitken's acceleration of convergence technique to obtain the true solution y^∞ on the interface Γ: $y^\infty = (I - P)^{-1}(y^k - Py^{k-1})$, and thus after another local resolution, the true solution x^∞. Let us note that we can accelerate the convergence to the solution for a convergent or a divergent iterative method. The only need is that 1 is not one of the eigenvalues of P. Considering $e^k = y^k - y^{k-1}, k = 1, \ldots,$ the operator $P \in \mathbb{R}^{n_\Gamma \times n_\Gamma}$ can be computed algebraically after $n_\Gamma + 1$ iterations as $P = [e^{n_\Gamma+1}, \ldots, e^2][e^{n_\Gamma}, \ldots, e^1]^{-1}$. Nevertheless, for 2D or 3D problems, the value n_Γ may be too large to have an efficient method. So a low-rank approximation of P is computed using the iterated interface solutions and Aitken's acceleration is performed on the low-rank space of dimension $n_\gamma \ll n_\Gamma$. As we are searching for the converged interface solution y^∞, we build from the singular value decomposition [9] of the matrix $Y = [y^0, \ldots y^q] = U\Sigma V^T$ a low-rank space by selecting the n_γ singular vectors associated to the most significant singular values.

Algorithm 1 Approximated Aitken's acceleration

Require: x^0 an arbitrary initial condition , $\epsilon > 0$ a given tolerance, $y^0 = R_\Gamma x^0$,
1: **repeat**
2: **for** $k = 1 \ldots q$ **do**
3: $x^k = x^{k-1} + M^{-1}_{RAS}\left(b - Ax^{k-1}\right), y^k = R_\Gamma x^k$ // RAS iteration
4: **end for**
5: Compute SVD of $[y^0, y^1, \ldots, y^q] = U\Sigma V'$,
keep the n_γ singular vectors $U_{1:n_\gamma}$ such that $\sigma_{n_\gamma+1} < \epsilon$
6: Compute $[\hat{y}^{q-n_\gamma-2}, \ldots, \hat{y}^q] = U^T_{1:n_\gamma}[y^{q-n_\gamma-2}, \ldots, y^q]$, and $\hat{e}^k = \hat{y}^k - \hat{y}^{k-1}$
7: Compute $\widehat{P} = [\hat{e}^{q-n_\gamma} \ldots, \hat{e}^q][\hat{e}^{q-n_\gamma-1}, \ldots, \hat{e}^{q-1}]^{-1}$
8: $y^0 \leftarrow U_{1:n_\gamma}\left(I - \widehat{P}\right)^{-1}\left(\hat{y}^q - \widehat{P}\hat{y}^{q-1}\right)$
9: **until** convergence

This low-rank approximation of the acceleration has been very efficient at solving 3D Darcy flow with a highly heterogeneous and randomly generated permeability field [1]. Step 7 of the algorithm may be subject to bad conditioning and matrix inversion can be replaced by pseudo inversion. Other techniques developed in [1] avoid the matrix inversion. For 1D partitioning (i.e. $\forall i = \{0, \ldots N-1\}, W^p_{i,e} \cap W^0_j = \emptyset, \forall j \neq \{i-1, i+1\}$), we can use the sparsity of P to define a sparse Aitken acceleration, which is made numerically more efficient by using local SVD for each subdomain [2].

3 Modeling the asynchronous RAS

If the Schwarz DDM converges then the asynchronous Schwarz does the same [8, Theorem 5 with assumption 2], under the additional hypothesis that the TC have been generated before their use, no subdomain stops updating its components and no subdomain has a TC that is never updated.

We consider the 2D Poisson problem:

$$\begin{cases} -(\frac{\partial^2}{\partial z_1^2} + \frac{\partial^2}{\partial z_2^2})\, x(z_1, z_2) = b(z_1, z_2),\ (z_1, z_2) \in]0,1[\times]0,1[, \\ \text{with homogeneous Dirichlet B.C.} \end{cases} \tag{5}$$

We discretize (5) with second-order centered finite differences on a regular Cartesian mesh of $n^g_{z_1} \times n^g_{z_2} = n$ points.

Given a non-prime number $N \in \mathbb{N}$, we split the domain $[0,1]^2$ into $N = N_{z_1} \times N_{z_2}$ overlapping partitions W^p_i. For the sake of simplicity, we consider each partition W^p_i to have $n_i = n^l_{z_1} \times n^l_{z_2}$ discretizing points and we define $n^g_{z_1}$ and $n^g_{z_2}$ accordingly. Due to the Cartesian mesh discretizing, the set $W^p_{i,e}$, for each i, can be split into a maximum of four parts corresponding to the four local artificial interfaces generated by the partitioning. Two: $W^{O,p}_{i,e}$ and $W^{E,p}_{i,e}$ (respectively $W^{S,p}_{i,e}$ and $W^{N,p}_{i,e}$) are in the z_1 (respectively z_2) direction.

The asynchronous RAS algorithm does not wait until the updates of the transmission conditions (TC) (the term $E_i x^k$ in (1)) are done before starting the next iteration. Consequently, the TC of one partition might not have been totally or partially updated. As there is no control on the restraining of the communication network, it is difficult to evaluate the number of updates of the local TC that are missing.

In order to modelize the asynchronous RAS, we propose a model where each of the four TCs of each subdomain is either totally updated or not, following a random draw of four numbers $(l^O_i, l^W_i, l^S_i, l^N_i)$ per W^p_i. Only if the draw associated to a local TC is greater than a fixed limit l is this local TC updated. The value l gives the percentage of missing TC updates. The synchronous RAS algorithm is obtained setting $l = 0$ and we denote by l-RAS the asynchronous RAS with level of asynchronism l. The l-RAS iterates until $R_\Gamma x^k$ does not evolve anymore. Figure 1 (left) shows that the level of asynchronism deteriorates the convergence of the RAS. The error between two consecutive iterations oscillates quite strongly with l. These oscillations are smoother for the error with the true solution. Table 1 shows the $\log_{10}$ of the error with the true solution of the asynchronous l-RAS for 240 iterations and the associated Aitken's acceleration of convergence. The results for l-RAS, with respect to the asynchronism level l, have an increasing variance but the min, max and mean values of the error are close. The Aitken acceleration of

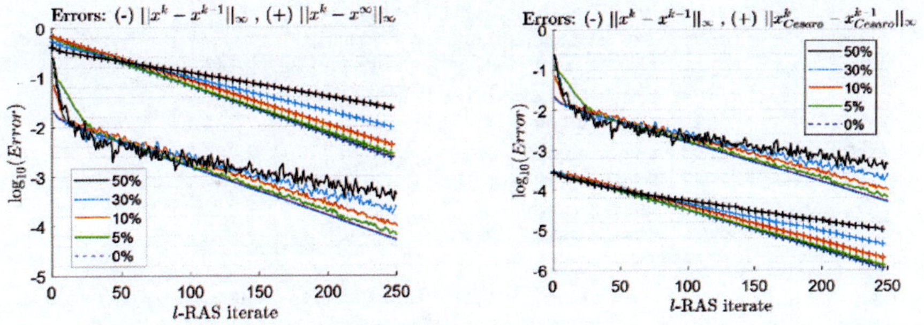

Fig. 1: l-RAS convergence with respect to the level of asynchronism l: for two consecutive iterations (continuous line) and (left) with the true solution (+), (right) two consecutive iterations after Césaro summation (+). ($n^l_{z_1} = n^l_{z_2} = 10$, $N_{z_1} = N_{z_2} = 5$, $n_\gamma = 40$.)

convergence, using the set of 240 l-RAS iterations, still accelerates even at a high level l of asynchronism, even though the acceleration deteriorates with increasing l. These results have a more stable variance and the mean value is closer to the max value than to the min value. We limited n_γ to be 40 for $l \neq 0$ and to be 20 for $l = 0$ due to the strong decreasing of the first singular values. Let us notice that for this test case $n_\Gamma = 544$ and the low-rank space is of size $n_\gamma = 40$.

	Aitken l-RAS				l-RAS				Update failures			
l	min	max	mean	σ	min	max	mean	σ	min	max	mean	σ
0.0‰	-11.12	-11.12	-11.12	2e-14	-2.543	-2.543	-2.543	3e-15	0	0	0	0
0.5‰	-3.666	-5.839	-4.969	4.0e-1	-2.527	-2.556	-2.533	4.8e-3	99	145	120.7	9.9
1.0‰	-2.814	-5.440	-4.751	4.7e-1	-2.513	-2.544	-2.524	7.1e-3	202	277	239.48	15.6
5.0‰	-2.521	-5.023	-4.284	4.2e-1	-2.415	-2.479	-2.443	1.4e-2	1121	1286	1197.3	34.3
10.‰	-1.729	-4.707	-3.956	5.3e-1	-2.303	-2.406	-2.347	2.1e-2	2267	2502	2397.9	43.6
30.‰	-1.037	-4.005	-3.280	4.6e-1	-1.868	-2.089	-1.974	4.7e-2	7044	7349	7203.3	66.5
50.‰	0.548	-3.613	-2.643	6.1e-1	-1.472	-1.961	-1.678	9.3e-2	11860	12199	12013	66.1

Table 1: Statistics (min, max, mean and variance σ), based on 100 runs, of $\log_{10}(||x^{240} - x^\infty||_\infty)$, with respect to l, for the asynchronous l-RAS and its Aitken acceleration of convergence (with the same data). ($n^l_{z_1} = n^l_{z_2} = 10$, $N_{z_1} = N_{z_2} = 5$, $n_\gamma = 40$.)

4 Regularization of the Aitken acceleration of convergence of the asynchronous RAS

At first glance, previous results on Aitken's acceleration of convergence of the l-RAS are surprising as the pure linear convergence of the RAS is destroyed with the asynchronism, i.e. the error operator depends on the iteration: $y^{k+1} - y^k = P_k(y^k - y^{k-1})$. The explanation comes from the low-rank space built with the SVD. Let $Y_l = [y^0_l, \ldots, y^q_l]$ be the matrix of the iterated l-RAS interface solutions. As the

	Aitken Césaro l-RAS				Upper	Aitken l-RAS	l-RAS
l	min	max	σ	mean	bound	mean	mean
0.0%	-12.42	-12.42	8.e-15	-12.42	-12.27	-11.111	-2.543
0.5%	-4.059	-6.968	4.5e-1	-6.284	-6.120	-4.969	-2.533
1.0%	-4.667	-6.856	3.9e-1	-6.096	-5.902	-4.751	-2.524
5.0%	-4.184	-6.383	4.9e-1	-5.546	-5.434	-4.284	-2.443
10.%	-3.844	-6.047	4.5e-1	-5.294	-5.106	-3.956	-2.347
30.%	-3.457	-5.261	3.9e-1	-4.500	-4.431	-3.280	-1.974
50.%	-2.505	-4.553	4.7e-1	-3.841	-3.794	-2.643	-1.678

Table 2: Statistics (min, max, mean and variance σ) for 100 runs of $\log_{10}$ of the error with the true solution of the Aitken acceleration of convergence of l-RAS with Cesaro's mean with respect to the asynchronism level l. ($n^l_{z_1} = n^l_{z_2} = 10$, $N_x = N_y = 5$, $n_\gamma = 40$, $m = 200$.)

asynchronous l-RAS converges, we can write $Y_l = Y_0 + E_l$, where E_l is a perturbation matrix with smaller and smaller entries with respect to the iterations. Then using the Fan inequality [4, Theorem 2, p. 764] of the SVD of a perturbation matrix, we have: $\sigma_{r+s+1}(Y_0 + E_l) \le \sigma_{r+1}(Y_0) + \sigma_{s+1}(E_l)$ with $r, s \ge 0$, $r + s + 1 \le q + 1$.

Setting $s = 0$, we have $|\sigma_{r+1}(Y_0 + E_l) - \sigma_{r+1}(Y_0)| \le \sigma_1(E_l) = ||E_l||_2, \forall r \le q$. By using Schmidt's Theorem [7, Theorem 2.5.3] on the SVD approximation, we can write:

$$\begin{aligned}
\min_{X,\text{rank}\,X=k}(||Y_l - X||_2) = \sigma_{k+1}(Y_l) &= \min_{X,\text{rank}\,X=k}(||Y_l - Y_0 + Y_0 - X||_2 \\
&\le ||Y_l - Y_0||_2 + \min_{X,\text{rank}\,X=k} ||Y_0 - X||_2 \\
&\le \sigma_1(E_l) + \sigma_{k+1}(Y_0). \qquad (6)
\end{aligned}$$

This result implies that:

- the low-rank space U_l built from Y_l is an approximation of U_0 with a small perturbation $||E_l||_2 = \sigma_1(E_l)$.
- As $\lim_{k\to\infty} y_l^k \to y^\infty$, the columns of the perturbation matrix E_l have a decreasing 2-norm. Thus, a better acceleration is obtained by considering the last q iterations to build U_l.

This last result suggests an improvement of Aitken's acceleration of convergence with the Césaro mean of the iterated interface solutions. We transform the sequence (y_l) into another sequence $(\tilde{y}_l)$ defined as $\tilde{y}_l^i = \frac{1}{m}\sum_{j=0}^{m-1} y_l^{i+j}$. The summation still preserves the pure linear convergence of the synchronous 0%-RAS: $\tilde{y}_0^{k+1} - y^\infty = P(\tilde{y}_0^k - y^\infty)$ and will smooth the perturbation E_l. Figure 1 (right) shows the $\log_{10}$ of the error with the true solution of the iterated interface solution with the Césaro mean with $m = 200$. This allows us to smooth the error oscillations on the convergence of l-RAS. The difference between two consecutive iterations of the sequence $(\tilde{y}_l)$ has a smaller amplitude than for the original sequence (y_l). This leads to a low-rank space U_l built from this $(\tilde{y}_l)$ more representative of the space where the true solution lives.

Table 2 gives the statistics for 100 runs of Aitken's acceleration of convergence for the l-RAS using the Césaro mean with respect to l. The acceleration of convergence is enhanced using $(\tilde{y}_l)$ instead of (y_l). The variance and the amplitude between the min and the max values of the results are smaller. Even the 0%-RAS is better accelerated. Moreover, it shows an upper bound for the mean acceleration of the l-RAS with the Césaro mean being $\frac{1}{\sqrt{m}}$ of the mean acceleration of the l-RAS. Figure 2 gives the singular values (σ_i) of the SVD of Y_l obtained with l-RAS with

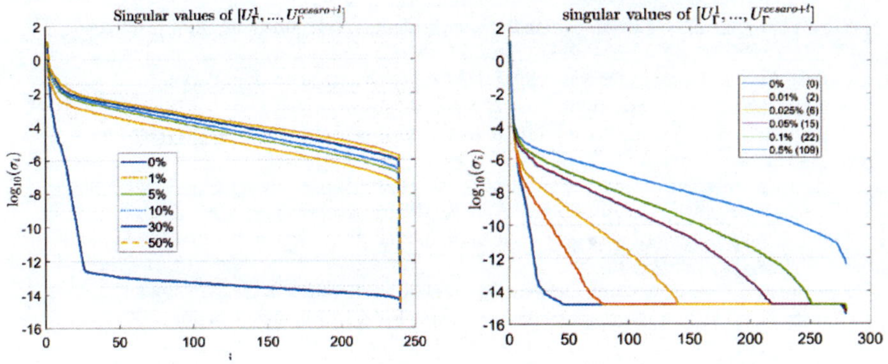

Fig. 2: Singular values of one sample of 250 l-RAS iterations for $l = \{0\%, 1\%, 5\%, 10\%, 30\%, 50\%\}$ (left) and for $l = \{0\%, 0.01\%, 0.025\%, 0.05\%, 0.1\%, 0.5\%\}$ and 300 iterations with the number of transmission condition update failures in brackets (right). ($N_x^l = N_y^l = 10, P_x = P_y = 5, p = 40.$)

respect to the level l of asynchronism. It shows that the fast decreasing of (σ_i) is lost with the asynchronism. It still exhibits some decreasing of (σ_i) that allows Aitken's acceleration of convergence. The right figure shows that the decreasing of σ_i is deteriorated even with a very small level l of asynchronism and with few TC update failures (the total number of updated for 300 0%-RAS iterations is $300 \times (4 \times 2 + 12 \times 3 + 9 \times 4) = 24000$).

5 Conclusion

We have successfully accelerated the asynchronous RAS with Aitken's acceleration of convergence technique based on the low-rank approximation of the error operator with the SVD of the matrix of interface iterated solutions. The SVD allows the asynchronous effect to be smoothed over the iterations. We proposed a model for setting the level of asynchronism. It can be used to estimate the asynchronism in real applications. Knowing the observed convergence rate of the real application, we can extrapolate the level of asynchronism of the implementation. The model proposed here considers a uniform probability for TC update failure (the worst case) but we can also consider that only certain parts of the domain decomposition may be temporarily at fault. Finally, we proposed a regularisation technique based on the Césaro mean of the l-RAS iterated interface solutions that improves Aitken's acceleration of convergence even on the synchronous RAS.

References

1. L. Berenguer, T. Dufaud and D. Tromeur-Dervout. Aitken's acceleration of the Schwarz process using singular value decomposition for heterogeneous 3D groundwater flow problems. *Computers & Fluids* **80**, 320–326 (2013).
2. L. Berenguer and D. Tromeur-Dervout. Sparse Aitken–Schwarz with application to Darcy flow. https://hal.archives-ouvertes.fr/hal-03090521 (2020).
3. X.-C. Cai and M. Sarkis. A restricted additive Schwarz preconditioner for general sparse linear systems. *SIAM J. Sci. Comput.* **21**(2), 792–797 (electronic) (1999).
4. K. Fan. Maximum properties and inequalities for the eigenvalues of completely continuous operators. *Proc. of the National Academy of Sciences* **37**(11), 760–766 (1951).
5. M.J. Gander. Schwarz methods over the course of time. *Electronic Transactions on Numerical Analysis*, pp. 228–255 (2008).
6. M. Garbey and D. Tromeur-Dervout. On some Aitken-like acceleration of the Schwarz method. LMS Workshop on Domain Decomposition Methods in Fluid Mechanics (London, 2001).*Internat. J. Numer. Methods Fluids* **40**(12), 1493–1513 (2002).
7. G.H. Golub and Ch.F. Van Loan. *Matrix Computations*. The Johns Hopkins University Press, third edition (1996).
8. F. Magoules, D.B. Szyld and C. Venet. Asynchronous optimized Schwarz methods with and without overlap. *Numerische Mathematik* **137**(1), 199–227, Sep (2017).
9. G.W. Stewart. On the early history of the singular value decomposition. *SIAM Rev.* **35**(4), 551–566 (1993).
10. D. Tromeur-Dervout. Meshfree Adaptative Aitken–Schwarz Domain Decomposition with application to Darcy Flow. In: B.H.V. Topping and P. Ivanyi, editors, *Parallel, Distributed and Grid Computing for Engineering*, volume 21 of CSET Series, pp. 217–250, Saxe-Coburg Publications (2009).
11. D. Tromeur-Dervout. Approximating the trace of iterative solutions at the interfaces with nonuniform Fourier transform and singular value decomposition for cost-effectively accelerating the convergence of Schwarz domain decomposition. *ESAIM: PROCEEDINGS* **42**, 34–60 (2013).
12. P. Wilders and E. Brakkee. Schwarz and Schur: An algebraical note on equivalence properties. *SIAM J. Sci. Comput.* **20**(6), 2297–2303, Jul 22 (1999).